Compound Semiconductors 1995

Other titles in the Series

The Institute of Physics Conference Series regularly features papers presented at important conferences and symposia highlighting new developments in physics and related fields. Recent titles include:

96: **Gallium Arsenide and Related Compounds 1988**
Papers presented at the 15th International Symposium, Atlanta, Georgia, USA
Edited by J Harris

106: **Gallium Arsenide and Related Compounds 1989**
Papers presented at the 16th International Symposium, Karuizawa, Japan
Edited by T Ikoma and H Watanabe

112: **Gallium Arsenide and Related Compounds 1990**
Papers presented at the 17th International Symposium, Jersey, Channel Islands
Edited by K E Singer

120: **Gallium Arsenide and Related Compounds 1991**
Papers presented at the 18th International Symposium, Seattle, Washington, USA
Edited by G B Stringfellow

129: **Gallium Arsenide and Related Compounds 1992**
Papers presented at the 19th International Symposium, Karuizawa, Japan
Edited by T Ikegami, F Hasegawa and Y Takeda

136: **Gallium Arsenide and Related Compounds 1993**
Papers presented at the 20th International Symposium, Freiburg im Braunschweig, Germany
Edited by H S Rupprecht and G Weimann

141: **Compound Semiconductors 1994**
Papers presented at the 21st International Symposium, San Diego, California, USA
Edited by H Goronkin and U Mishra

Compound Semiconductors 1995

Proceedings of the Twenty-Second International Symposium on Compound Semiconductors held in Cheju Island, Korea, 28 August to 2 September 1995

Edited by Jong-Chun Woo and Yoon Soo Park

Prior to the 1994 meeting, previous symposia were published as Gallium Arsenide and Related Compounds in the Institute of Physics Conference Series

CRC Press
Taylor & Francis Group
Boca Raton London New York

CRC Press is an imprint of the
Taylor & Francis Group, an **informa** business

Institute of Physics Conference Series Number 145

CRC Press
Taylor & Francis Group
6000 Broken Sound Parkway NW, Suite 300
Boca Raton, FL 33487-2742

First issued in paperback 2019

© 1996 by Taylor & Francis Group, LLC
CRC Press is an imprint of Taylor & Francis Group, an Informa business

No claim to original U.S. Government works

ISBN-13: 978-0-7503-0342-2 (hbk)
ISBN-13: 978-0-367-40135-1 (pbk)

CODEN IPHSAC 145 1–1312 (1996)

British Library Cataloguing in Publication Data

A catalogue record for this book is available from the British Library.

Library of Congress Cataloging-in-Publication Data are available

**Visit the Taylor & Francis Web site at
http://www.taylorandfrancis.com**

**and the CRC Press Web site at
http://www.crcpress.com**

International Symposium on Compound Semiconductors Award and Heinrich Welker Gold Medal

The International Symposium on Compound Semiconductors Award was initiated in 1976; the recipients are selected by the International Symposium on Compound Semiconductors Award Committee for outstanding research in the area of III–V compound semiconductors. The Award consists of $2500 and a plaque citing the recipient's contribution to the field. The Award is accompanied by the Heinrich Welker Gold Medal, established by Siemens AG, Munich, in honour of the foremost pioneer in III–V compound semiconductor development.

The winners of the Symposium Award and the Heinrich Welker Medal are:

1976	Nick Holonyak	for developing the first practical light-emitting diodes
1978	Cyril Hilsum	for contributions in the fields of transferred electron logic devices (TELDs) and GaAs MESFETs
1980	Hisayoshi Yanai	for his work on TELDs, GaAs MESFETs and ICs, and laser diode modulation with TELDs
1981	Gerald L Pearson	for research and teaching in compound semiconductors physics and device technology
1982	Herbert Kroemer	for his work on hot-electron effects, Gunn oscillators and III–V heterostructure devices
1984	Izuo Hayashi	for development and understanding of room temperature operation of DH lasers
1985	Heinz Beneking	for his contributions to III–V semiconductor technology and novel devices
1986	Alfred Y Cho	for pioneering work on molecular beam epitaxy and his contribution to III–V semiconductor research
1987	Zhores I Alferov	for outstanding contributions in theory, technology and devices, especially epitaxy and laser diodes
1988	Jerry Woodall	for introducing the III–V alloy AlGaAs and fundamental contributions to III–V physics
1989	Don Shaw	for pioneering work on epitaxial crystal growth by chemical vapour deposition

1990 George S Stillmann for the characterization of high-purity GaAs and developing avalanche photodetectors

1991 Lester F Eastman in recognition of his dedicated work in the field, especially on ballistic electron transport, δ-doping, buffer layer technique, and AlInAs/GaInAs heterostructures

1992 Harry C Gatos for his contribution to science and technology of GaAs and related compounds, particularly in relating growth parameters, composition and structure to electronic properties

1993 James A Turner for pioneering the development of GaAs MESFETs, MMICs, circuit fabrication and analytical techniques

1994 Federico Capasso for leading work on bandgap engineering of semiconductor devices and discovery of many new phenomena in artificially structured semiconductors

The 1995 Award Committee of the 22nd International Symposium on Compound Semiconductors has selected Professor Isamu Akasaki of Meijo University for the Symposium Award and the Heinrich Welker Gold Medal for his pioneering and outstanding contributions in the field of III–V nitride compound semiconductor research.

Professor Isamu Akasaki is a true pioneer and has made numerous major contributions in the field of GaN based III–V semiconductor research. He has been working on the vapour phase epitaxial growth of GaN, AlN, AlGaN and GaInN for over twenty years. His persistent efforts led to a number of important breakthroughs. These include successful p-doping using low energy electron beam irradiation, resulting in efficient blue p–n junction LEDs in cooperation with H Amano. He also significantly improved the quality of MOVPE GaN by inserting a thin AlN buffer layer. This approach has been widely adopted in nitride epitaxial growth. All these achievements have led to the commercial development of high performance blue LEDs and very low threshold UV stimulated emission, establishing III–V nitrides as promising material for optoelectronics applications in the blue–UV spectral region.

He has been the leading researcher on nitride systems for the past two decades and continues to do outstanding work. As a result of his pioneering work, researchers are now actively pursuing nitride-related research and great progress is being made which opens a new field in optoelectronics. Without Professor Akasaki's continued foresight, leadership and discoveries over 20 years, nitride research and development would not have been so stimulated today.

Professor Akasaki also contributed significantly to other compound semiconductor studies. It is also important to note his early successes in the areas of GaP single-crystal growth by LEC, the brightest GaP LED and his ultra-pure GaAs growth by VPE. These show that he has made truly exceptional general contributions of importance in the compound semiconductor area.

These achievements were recognized by the JACG Prize, the CHU-NICHI Cultural Award, the Optoelectronic Conference Special Award and the JACG Technological Contribution Award.

Since 1992 Dr Akasaki has been a Professor Emeritus at Nagoya University. He is currently a visiting professor in the Research Center for Interface Quantum Electronics at Hokkaido University.

Young Scientist Award

The International Advisory Committee of the International Symposium on Compound Semiconductors has established a Young Scientist Award to recognize technical achievements in the field of compound semiconductors by a scientist under the age of forty. The Award consists of a financial reward and a plaque citing the recipient's contribution. This year's Award was donated by Kopin Corporation.

The Young Scientist Award recipients are:

1986 Russel D Dupuis — for work in the development of organometallic vapour phase epitaxy of compound semiconductors

1987 Naoki Yokoyama — for contributions to self-aligned gate technology for GaAs MESFETs and ICs and the resonant tunnelling hot-electron transistor

1989 Russel Fischer — for demonstration of state of the art performance, at DC and microwave frequencies, of MESFETs, HEMTs and HBTs using (AlGa)As on Si

1990 Yasuhiko Arakawa — for pioneering work on low-dimensional semiconductor lasers, showing the superior performance of quantum wire and quantum box devices

1992 Umesh K Mishra — for pioneering and outstanding work on AlInAs–GaInAs HEMTs and HBTs

1993 Young-Kai Chen — for significant advancements in the fields of high-speed III–V electronic and optoelectronic devices

1994 Michael A Haase — for contributions on II–VI based blue LEDs and ZnSe-based electro-optic modulators

The 1995 Award Committee of the 22nd International Symposium on Compound Semiconductors has selected Dr John Ralston of SDL, Inc. for the Young Scientist Award for his pioneering and outstanding contributions in the field of high-speed high-power semiconductor lasers.

John D Ralston received BSc degrees in both Physics and Electrical Engineering from the Massachusetts Institute of Technology in 1982. After working in the Optical Fiber Systems Group at the Advanced Technology Laboratories of Bell Northern Research on the devolopment of active repeaters for optical fibre local-area networks, he began his graduate studies at Cornell University. He completed his PhD degree in Electrical Engineering in 1988 under Professor Lester Eastman's supervision where he focused on MBE growth and advanced fabrication techniques for MQW optoelectronic devices.

During his $6\frac{1}{2}$ years at the Fraunhofer-Institut für Angewandte Festkörperphysik in Freiburg, Germany, he collaborated on the design, epitaxial growth, fabrication, characterization and integration of III–V heterostructure and quantum-well devices for application in very-high-speed optoelectronic interconnects, high-power laser sources, microwave systems, and IR imaging. He established, managed, and was intimately involved in the research programme which led to the demonstration of the first semiconductor lasers to achieve direct modulation bandwidths of 30, and later 40, GHz. He also helped to develop the chemically-assisted ion-beam etching (CAIBE) system and etch processes used to fabricate the above ultra-high-speed MQW lasers and the first monolithic high-speed MQW-laser/MODFET-driver OEICs. He was awarded the 1994 Prize of the German Microelectronics Society for his achievements. He also established and directed projects focusing on the development of high-speed, low-chirp, directly-modulated 1.55 μm DFB lasers and high-power, filamentation-free GaAs-based tapered MQW lasers. He participated in the design and development of the first MBE system equipped with valved crackers for the all-solid-source epitaxial growth of III–V arsenides, antimonides, and phosphides.

John Ralston joined the Research Department at SDL, Inc., where he is currently involved in the commercialization of semiconductor diode lasers for both high-speed and high-power applications.

He has authored or co-authored more than 100 professional publications, and has been awarded 2 US and 1 German patents.

Preface

The 22nd International Symposium on Compound Semiconductors (ISCS-22), the successor to the International Symposium on GaAs and Related Compounds, was held in Cheju Island, Korea from 28 August to 2 September 1995. More than 460 scientists and engineers from 15 countries came to this remote volcanic island making the 22nd Symposium truly international. A total of 253 papers including four plenary session papers were selected from over 300 submissions for oral and poster presentations at the conference.

The theme of the symposium was 'Technologies for Future Electronics and Optoelectronics Industries', and the areas of interest covered almost the entire spectrum of modern device technologies and applications in compound semiconductors. Korea, where emerging technologies and future developments are emphasized, was a very fitting place for the symposium and for the theme. The country, currently the third largest producer of electronics, is embarked on the process called globalization, a slogan reflecting her efforts of transition from a technology follower to a leader. Both in the opening plenary speech of Mr Jin Ku Kang, President and CEO of Samsung Electronics, and the banquet address of the Minister of Science and Technology of Korea, Dr Kun-Mo Chung (read by the Vice-Minister, Dr Bohn-Young Koo), the importance of compound semiconductor technology was emphasized as one of the potential areas with explosive growth in the forthcoming information age.

The semiconductor industry is expected to be a US $200 billion industry worldwide by the year 2000, and the compound semiconductor industry revenues are projected to be as large as a few per cent of the overall semiconductor market. It may be fair to say that the compound semiconductor has not quite lived and may not live up to its high expectations in terms of the market share and of carving a larger niche at the expense of silicon microelectronics. Nevertheless, the importance of the compound semiconductor technology should not be evaluated in terms of the sales volume alone. Firstly, the compound semiconductor technology is indispensable in many important areas, such as satellite communications. And, secondly, new technologies may emerge that will dramatically enhance the importance of compound semiconductors. During the symposium, one such breakthrough is highlighted in the area of wide band gap semiconductors.

It was fitting that the 17th Compound Semiconductor Award and the Heinrich Welker Gold Medal were presented to Professor Akasaki for his pioneering work on GaN light emitting diodes. The Young Scientist Award was presented to Dr John Ralston of SDL, Inc., formerly of the Fraunhofer-Institut in Freiburg, Germany, for his outstanding work on the development of high-speed lasers.

The areas prominently featured at the symposium included blue and UV light emitters, high speed and surface emitting lasers, characterization, modelling, simulation, fabrication, and parameter extraction for devices. Plasma-wave electronic devices, which are expected to operate in the terahertz range, were one of the new devices presented here. Other topics of interest included quantum cascade lasers, novel quantum devices, quantum dot devices, laser–microwave interaction, optical interconnects, and new *in situ* characterization techniques.

It was the first international symposium of this scale in the field of semiconductors to be held in Korea. The list of the presenters seems to read as a Who is Who of Compound Semiconductors in the world. However, the symposium owes a great deal of its success to the devotion of the participants. On behalf of the organizers, we would like to express our sincere gratitude to all of the participants for their support and contribution shown throughout the symposium.

We would like to express our special thanks to the Secretariat, Professor Young Se Kwon, Treasurer, Professor Jin Koo Rhee, and the members of the technical program committee for their invaluable service and sacrifice. Our appreciation is extended to Dr Doyeol Ahn, Professor Keon-Ho Yoo, Professor Chi-Kyu Choi, Professor Namic Kwon, Dr Seuk-Joo Rhee, Mr Do-Hyun Kim, Ms Chu Mi Kang, Ms Hea Yong Kang, Ms Janie Lee, Ms Yong Ja Shin and many others for their help.

Jong-Chun Woo
Seoul National University, Korea
Program Chair

Yoon Soo Park
Office of Naval Research, USA
Conference Chair

The organizers of the symposium gratefully acknowledge the sponsors for their generous contribution

LG Electronics Research Center
LG Semicon Co. Ltd
Samsung Electronics
Hyundai Electronics Industries Co.
Kukje Corporation, Electronics Division
Sammi Technology and Industries Co. Ltd
Korean Air
Korea Research Foundation, Ministry of Education, Korea
Ministry of Science and Technology, Korea

Contents

Chapter 3: Characterization

Chapter 4: High Power, High Temperature Devices

Chapter 5: Heterojunction Transistors

Chapter 6: Quantum Effect

Chapter 7: Nanoelectronics and Nanophotonics

Chapter 8: Optoelectronic and Optical Devices

Inst. Phys. Conf. Ser. No 145: Chapter 1
Paper presented at 22nd Int. Symp. Compound Semiconductors, Cheju Island, Korea, 28 August–2 September 1995
© *1996 IOP Publishing Ltd*

Evolution of Compound Semiconductor Microelectronics for Nanoelectronics

Takuo Sugano

Toyo University, 5-28-20 Hakusan Bunkyo-ku

Tokyo 112 Japan

The Institute of Physical and Chemical Research

2-1 Hirosawa Wako, Saitama 351-01 Japan

Abstract. In microelectronics, where the feature size of devices is of the order of μm, conduction electrons and positive holes can be treated as if they were to be classical particles. On the other hand in nanoelectronics, where the feature size of devices is of the order of nm, the wave nature of electronics appears explicitly. Nanoelectronics has advantages in miniaturization of devices, low power consumption and realization of new circuit configurations over microelectronics, but nanoelectronics and microelectronics are complimentary and will co-prosper.

1. Introduction

Compound semiconductor integrated electronics can be divided conceptually into two categories, microelectronics and nanoelectronics, in terms of the feature size of integrated semiconductor devices, if "micro" stands for micrometer (10^{-6}m) and "nano" stands for nanometer (10^{-9}m).

The feature size of existing semiconductor devices, that is, the smallest planar pattern size such as the gate length of MESFET or the smallest active region size such as the base thickness of HBT, is conveniently represented with micrometer as length unit. In this sense it is appropriate to call the present integrated electronics "microelectronics".
The miniaturization of semiconductor devices is continuing and the feature size has reached to deep submicron or even to nanometers. This shrinkage of device size does not mean the simple scale-down of devices, but new physics is involved in performance of such miniaturized devices and new electronics to be called "nanoelectronics" is emerging now.

Essential differences between microelectronics and nanoelectronics besides devices size will be discussed here.

2. Microelectronics as semi-classical electronics.

Major compound semiconductor devices used in integrated circuits are bipolar transistors such as HBTs and field effect transistor such as MESFETs and HEMTs. In order to study the performance of those devices, the understanding of semiconductor physics such as the energy band structure of semiconducting materials and the transport of electrons and holes is required at least. The concept of the energy band and that of conduction electrons and positive holes are quantum-mechanical ones, but, once those concepts are established, classical approach is a good approximation. That is, conduction electrons and positive holes can be treated or classical particles with the effective mass of the conduction electron and a negative elementary charge, and those with the effective mass of the positive hole and a positive elementary charge, respectively. If an energy band is approximated by parabolic one in momentum space, the constant effective mass aproximation is valid, but, if not, the effective mass depends on the momentum of conduction electron or positive hole, respectively.

So based on the effective mass approximation for electrons in an allowed band, conduction electrons and positive holes can be trated as if they were to be classical particles.

Another basic equation to describe the performance of semiconductor devices is the current equation, in which the density of electron current and that of hole current are expressed as the sum of the drift current density and the diffusion current density of each component. Those current equations are based on the Boltzmann's transport equation, in which the phase of electron wave is included as an important variable. However in the process of deriving the macroscopic current equations, the random phase approximation is used and the information related to the phase of electron wave is completely lost.

Current components, drift current and diffusion current, are represented with the transport coefficients, drift mobility and diffusion coefficient, and this formulated form is the same as that for molecules in a dilute gas. In this sense the current equations are classical and here again conduction electrons and positive holes in semiconductors are treated as if they were to be classical particles. So the physics used to understand the performance of semiconductor devices is to be called semi-classical.

Compound semiconductor devices have been scaled down and the feature size has reached to deep submicron or even below that. However during the scale-down process the physics of devices has not changed in essence and remains semi-classical even though some physical phenomena which are not included in the conventional formulation of electron transport, such as the electron velocity overshoot have been observed.

Considering that physics involved in microelectronics devices is semi-classical, the present microelectronics can be categorized as semi-classical electronics.

3. Nanoelectronics as quantum-mechanical electronics

The reason why the semi-classical approach is valid and useful for the present microelectronics is that the feature size of devices is longer than the mean free path of conduction electrons or positive holes determined by phonon scatterings and, of course, much longer than the wavelength of conduction electron or positive hole, too.

Phonon scattering is inelastic and changes the energy of the conduction electron or the positive hole, to loose the phase coherency. At room temperature the mean free path was significantly short in comparison with the feature size of devices. But in the course of miniaturization of devices the feature size of devices is now approaching the mean free path and the ballistic transport of conduction electronics has become a realistic matter. Then the phase coherency of an electron wave is maintained in the transport. In the regime that the feature size of devices approaches to the de Broglie wavelength of electron, the wave nature of electrons becomes more explicit and the phase coherency is of course maintained.

Here the physics of compound semiconductor devices is purely quantum-mechanical, and the operation of devices depends on physical principles different from those of exiting microelectronics devices, and a new electronics comes out. The feature size of devices used in the new electronics is conveniently represented with nanometer as length unit and this justifies that the new electronics shall be called nanoelectronics.

Various physical principles have the possibility to be used for nanoelectronics devices, and in particular the phase coherency of an electron wave and the Coulomb blockade of tunneling are attractive physical phenomena. The electronics based on phase coherency is called "Quantum Phase Electronics" and "Single Electronics" is based the Coulomb blockade of tunneling.

4. Interference of electron waves in compound semiconductor devices.

A classical demonstration of the phase coherency of waves is the interference phenomenon. A schematic of the device structure used by Aihara et.al.[1] to demonstrate the interference between two electron waves transmitted for different distance.   The device was made on high mobility AlGaAs/GaAs heterostructure. A quantum wire structure with a stub was fabricated by the split gate technique and the effective stub length was controlled by the stub-tuner gate voltage. Periodic conductance oscillations were clearly observed in terms of the stub-tuner gate voltage.

A quantum-mechanical demonstration of the phase coherency of electron wave in the Bloch reflection of electron wave. Nonomiya et. al.[2] showed the conductance change due to the Bloch reflection of electron wave in a quantum wire. Here the wave length of electron is fixed to 84.96Å and the distance between two impurities which have δ-function like potential, is changed.

For inter-impurity distance corresponding to the multiple of a half wave length the conductance of a quantum wire, in which two impurities are lined up, shows a peak. This result indicates the possibility to obtain negative resistance in such quantum wire by controlling the wave length of electron with a gate voltage.

Another quantum-mechanical demonstration of the phase coherency of electron wave is the interference phenomenon by the Aharanov-Bohm effect. An experimental confirmation of the Aharanov-Bohm effect on two-dimensional electron gas formed at GaAs-AlGaAs heterojunction interface was reported by Ishibashi et. al.[3] An electron wave is split to two pathes with geometrically same length. However a magnetic field is applied perpendicularly to the ring plane and phase difference arises between the electron waves by the Aharanov-Bohm effect. When two electron waves are out of phase at the outlet, the resistance between the inlet and the outlet show peaks.

In these electron wave interference experiments, electrons are expected to travel for several wavelengths without phonon scattering, so that those electrons are weakly localized ones. Phonon scattering destroys the phase coherency, and in consequence the room temperature operation of electron wave interference devices will not be materialized unless nanometer fabrication technology is available.

5. Coulomb blockade of tunneling

Another physical principle, which can be used to build nano-electronics device, is the Coulomb blockade of tunneling. The Coulomb blockade of tunneling is a phenomenon inherent to ultra-small structure, nanostructure, due to its very small capacitance. Capacitance of nanostructure is so small that the voltage change due to one electronic charge change can be large enough to block the succeeding tunneling.

If the capacitance is 10^{-18} F (1 at F), the voltage change due to one electronic charge is 0.16V and large enough to be detected at room temperature. In order to assure ample operational margin, 10^{-19} F or 0.1 at F must be chosen as capacitance and the edge length of the capacitor electrode is about 10nm. To delineate planar pattern of 10nm times 10nm with accuracy sufficient for intergration is not easy at present, but will be possible in future by innovating lithography technology, for instance STM related technique.

The Coulomb blockade of tunneling generates a threshold voltage in current-voltage relation and makes possible single electron transfer. The latter is a basis of single electron circuit, a new type of quantum-mechanical digital circuits, similar to the single flux quantum Josephson circuits.

An advantage of single electron circuits is the reduces power dissipation. Low power integrated circuits are concerned with the viewpoint of elongation of battery life and heat transfer from a chip. Here discussion will be focused on the issue of heat transfer from a chip.

As well known, power consuemed in an electron circuits has two sources. One is bias power and the other is signal power. Assuming CMOS like circuit configration, bias power shall be neglected and only signal power is taken into account here.

In thermal equilibrium electrical energy consumption on a chip is balanced with heat transfer from the chip. The number of gate per cm^2, which can be integrated from the viewpoint of heat transter, as a fundtion of signal energy per bit was studied. Clock frequency and heat transfer rate were used as paramenters. There the duty of the gates is assumed to be unity, that is, all getes are assumed to consume the signal energy in each clock cycle. Assuming a conventianal short channel MOSFET, whose input circuit capacitance is 100 fF and charged up to 1V, 10^{-12} Joule is consuemed par cycle and 6×10^5 electrons are transfered to the next stage as a packet. For the clock frequency as high as 1GHz, it has been found that the power consumption will limit the density of gates on a chip. However in single electron circuit the signal energy is so small that it hardly limits the density of gates on a chip.

With respect to single electron circuits one of the most inportant issues is how to build single electron digital circuits. Difficulty may arise from small voltage gain of single electron tansistor. Realization of single electron memory is more plausible than that of single electron logic circuits. Recently Yano et. al. [4]reported a single electron memory devices using quantum dots formed in poly crystalline Si film.

The von Neumann type configuration of logic circuits is not appropriate for single electron circuits because of the small voltage gain of single electron transistors. Quantum automaton is one of the possibilities. Another configuration appropriate for single electron circuits is the binary decision diagram, where an electron is used as a messenger of signal and the function of devices is just to steer the signal to one of two signal paths. In consequence this is so-called steering logic. The possibility of implementing steering logic with single electron circuit is now being explored by Amemiya[5].

6. Quantum-mechanical phenomena in a quantum dot

Classical aspects of the Coulomb blockade phenomenon have been discussed. However quantum-mechanical phenomena are taking place in a quantum dot.

The edge length of a quantum dot is still longer than or of the same order of the wavelength of electron. In consequence the wave nature of electron explicitly appears in a quantum dot. For instance by applying magnetic field perpendicular to the quantum dot surface, the Aharonov-Bohm effect [6] and the magneto-Coulomb oscillation[7] were observed.

7. Concluding Remarks

Nanoelectronics, especially single electronics, is an evolution of microelectronics with respect to
(1) miniaturization of devices
(2) low power consumptions
and
(3) realization of new circuit configurations
However nanoelectronics is complimentary with microelectronics and is not to be understood as post-microelectronics. In other words microelectronics and nanoelectronics can coprosper and microelectronics will not be replaced by nanoelectonics.

6

References

[1] Aihara K, Yamamoto M and Mizutani T 1992 *IEEE IEDM* 491

[2] Nonomiya S, Nakamura A , Aoyagi Y, Sugano T and Okiji A 1992 *SSDM* 750

[3] Ishibashi K, Takagaki Y, Gamo K, Namba S, Ishida S, Murase K, AoyagiY and   Kawabe M
     1987 *Solid State Comm* **64**. 4. 573

[4] Yano K, Ishii T, Hasimoto T, Kobayashi T, Murai F and Seki K 1994 *Ext. Abst. SSDM* 325

[5] Amemiya Y 1995 *JSAP Symposium Mav* 29

[6] Bird J P. Ishibashi K, Stopa M, Aoyagi Y and Sugano T 1994 *Phys. Rev.* **B 50** 14983

[7] Bird J P, Ishibashi K, Stopa M, Taylor R P. Aoyagi Y and Sugano T 1994 *Phys. Rev.* **B 49** 11488

Inst. Phys. Conf. Ser. No 145: Chapter 1
Paper presented at 22nd Int. Symp. Compound Semiconductors, Cheju Island, Korea, 28 August–2 September 1995

Optimization of GaAs-based HEMTs for microwave and millimeter wave IC applications

T Grave

Siemens AG, Corporate Research and Development,
D-81730 Munich, Germany

Abstract. The family of pseudomorphic HEMTs based on GaAs substrate devides into two main classes of transistors: the pseudomorphic single-heterojunction HEMT (SH-PHEMT) and the pseudomorphic double-heterojunction HEMT (DH-PHEMT). In this paper, the respective advantages and drawbacks of these two different device concepts are discussed in view of the intended frequency range and with respect to the question if a small or large signal amplifier application is envisaged.

1. Introduction

Today, high-electron mobility transistors (HEMTs) grown on GaAs substrate are a standard technology for the fabrication of microwave and millimeter wave ICs. However, other technologies exist which are potentially better suited to at least part of the applications GaAs HEMTs are used for. The first are heterobipolar transistors (HBTs) often judged to be superior for power applications in the lower part of the microwave frequency spectrum. The second possibility are HEMTs grown on InP which offer highest cutoff frequencies and constitute the ultimate choice for small-signal applications close to 100 GHz or even beyond. However, both alternatives are generally less mature with respect to manufacturability compared to GaAs HEMTs.

Most suppliers of analog ICs try to extend the applicability of GaAs-based HEMTs as far as possible in order to avoid the introduction of additional technologies into their production lines (particularly if they are based on extremely brittle InP substrate). Thus, the challenge arises to make the very best out of the possibilities inherent to GaAs HEMTs, particularly by properly optimizing the epitaxial layer sequences in view of the specific application in mind.

2. Requirements for major circuit applications

An ideal device would be a transistor which meets the different requirements for all important types of circuits simultaneously. Of course, such a device is hypothetic. Before we discuss which combination of properties can indeed be realized within a single device, we first want to check the features that a good transistor must show for the most frequent circuit applications.

For a transistor used in receiver amplifiers, the main requirement is a low noise figure F in combination with a high associated gain G_a. The noise figure at frequency f is to a very first order approximated by Fukui's expression [1]

$$F = 1 + 2\pi K f C_{GS} ((R_S + R_G)/g_m)^{1/2} \tag{1}$$

where K is an empirical fitting factor representing the transport properties of the channel, C_{GS} is the gate-source capacitance, R_S and R_G are the source and gate resistances, and g_m is the intrinsic transconductance. The best combination of small C_{GS} with reasonable g_m is usually achieved at small drain currents $I_D = 50...80$ mA/mm not very far above pinchoff. The gain is essential since the total noise figure F_{tot} of a multistage receiver combined of stages 1, 2, 3 ... is given by [2]

$$F_{tot} = F_1 + (F_2 - 1)/G_{a1} + (F_3 - 1)/G_{a1}G_{a2} + ... \tag{2}$$

Only in the absence of large input signals, the modulation of the transistor is truly "small signal", i.e. it is operated at a dc bias of drain-source voltage V_{DS} and gate-source voltage V_{GS}, and the RF signal superimposed to the dc gate voltage is of comparatively small amplitude. Hence, the properties of the transistor are only relevant at voltages close to the dc bias point.

The number of circuits which are small-signal ICs in this strict sense is limited. Even in low-noise amplifiers, the input signal may vary considerably in power. In principle, general purpose amplifiers should be able to operate over a certain dynamic range. Thus, in many cases, V_{GS} is modulated with an RF amplitude which is a considerable fraction of the transistor's maximum V_{GS} sweep. These amplifiers are small-signal circuits only in so far as their output power is comparatively small but the character of bias modulation is large-signal. A common requirement for such MMICs is linearity, i.e. constant gain for a large range of input signal amplitude. To achieve this, variations of g_m and output conductance g_{ds} with V_{DS} and V_{GS} must be kept to a minimum (among other parameters that also have to be optimized). Thus, a high g_m maximum is of little use if $g_m(V_{GS})$ is strongly peaked. The noise figure of these circuits is often not very critical, for instance when they serve as preamplifiers to the final power stages of transmitter systems.

This leads us to the requirements imposed by power amplifiers which are large-signal circuits in the twofold sense that both V_{GS} modulation amplitude and output power are large. The maximum output power of a transistor in class "A" operation is given by

$$P_{out\,max} = (1/8)\,\Delta V_{DS}\,\Delta I_D \tag{3}$$

where ΔV_{DS} and ΔI_D are the maximum possible drain voltage and drain current sweeps, respectively. Evidently, a power HEMT should simultaneously be capable of high V_{DS} (which requires both large gate-drain diode and open channel breakdown voltages) and high I_D (which requires high electron concentration n_s in the channel). Additionally, for a *linear* power amplifier, as needed for instance in the radio links of wireless telecommunication systems, the gain should be constant over a large range of voltage and current, leading to the consequences for g_m and g_{ds} already mentioned above. The gain does not only determine the amplification of the circuit but also plays a role for its power losses. Low losses are synonymous with a high power-added efficiency *PAE* defined by

$$PAE = (P_{out} - P_{in})/P_{dc} = P_{out}(1 - G_a^{-1})/P_{dc} \tag{4}$$

where P_{dc} is the power of the dc bias, and P_{in} is the RF input power.

Having identified the desirable device properties for a number of circuit applications, we now want to discuss what features are specific to commonly used types of GaAs-based HEMTs.

3. Properties of common epitaxial layer schemes

The three most popular versions of GaAs HEMTs are the pseudomorphic single-heterojunction HEMT (SH-PHEMT) with homogeneously doped supply, the delta-doped SH-PHEMT, and the pseudomorphic double-heterojunction HEMT (DH-PHEMT), mostly also used with delta-doping. Their conduction band diagrams are shown in Fig. 1. SH-PHEMTs usually possess an $Al_xGa_{1-x}As$ barrier between gate and channel with $0.20 \leq x \leq 0.25$ and an $In_yGa_{1-y}As$ channel typically 12 nm thick with $0.15 \leq y \leq 0.25$ where the GaAs buffer is located directly below the channel. The DH-PHEMT is distinguished from them by a second (doped) $Al_xGa_{1-x}As$ barrier between buffer and channel. A comparison of important properties specific to the three epitaxial layer schemes is given in Tab. 1.

First of all, the three devices differ with respect to the electron sheet concentrations n_s that can be achieved. The delta-doped SH-PHEMT is superior to the homogeneously doped one due to the better transfer of carriers from the donors to the channel which originates in the smaller average donor-channel distance. The DH-PHEMT is even better because it is doped on both sides of the quantum well which is additionally deeper. As a consequence of the doping scheme, the distance d of the gate/semiconductor interface to the heterojunction between channel and upper barrier is typically larger for a homogeneously doped PHEMT compared to the two delta-doped devices. However, the effective separation of the two-dimensional electron gas (2DEG) from the gate is not given by d but by a quantity $d_{eff} = d + d'$ which takes into account that the maximum of the electron wavefunction is not located at the channel/barrier heterojunction but somewhere inside the quantum well. The distance d' ist a function of V_{GS} (and, hence, also d_{eff}). At V_{GS} slightly above pinchoff when I_D is small, d' is at its maximum. It becomes continuously smaller when V_{GS} is increased, i.e. the 2DEG moves towards the gate. Even when the quantum well thicknesses are the same ($d_{QW} = 12$ nm, for

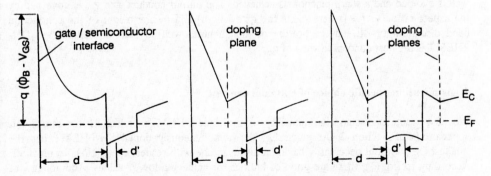

Fig. 1. Conduction band diagrams of GaAs-based PHEMTs.
E_F is the Fermi level, E_C the conduction band edge, and $q\Phi_B$ is the Schottky barrier height.

	homog. doped SH-PHEMT	delta-doped SH-PHEMT	delta-doped DH-PHEMT
n_s (cm^{-2})	$\leq 2.0 \times 10^{12}$	$\leq 2.5 \times 10^{12}$	$\leq 3.2 \times 10^{12}$
d (nm)	30...40	15...25	15...25
d' (nm) close to pinchoff	6	6	9
d' (nm) open channel	3	3	6
$g_{m\ max}$	fair	very high	high
g_m close to pinchoff	fair	fair	fair
C_{GS}	low	high	high
g_{ds}	high	low	very low

Tab. 1. Properties of different GaAs-based PHEMTs.
For the values of d', a quantum well thickness $d_{QW} = 12$ nm is assumed.

instance), d' is different for SH- and DH-PHEMTs: in an SH-PHEMT close to pinchoff, d' is approximately half of the channel thickness (d' = 6 nm in our example), and the electrons sit symmetrically in the channel. Under open-channel conditions ($V_{GS} > 0$), d' has decreased to about 3 nm. In a DH-PHEMT, the effective distance of the 2DEG to the upper barrier/quantum well interface is larger by the order of 3 nm. Therefore, at V_{GS} close to pinchoff, the maximum of the electron wavefunction is only 3 nm away from the lower heterojunction, whereas under open-channel conditions, the 2DEG is nearly centered in the quantum well [3].

Both C_{GS} and $g_{m\ max}$ are proportional to d_{eff}^{-1} which is reflected by the data given in Tab. 1. For small I_D, differences of g_m may not be very pronounced if the epitaxial layer sequence is properly designed, i.e. if parasitic conduction by electrons below the quantum well is avoided and a sharp pinchoff is achieved. The output conductance g_{ds} is governed by the aspect ratio L_G/d_{eff} (where L_G is the gate length) and by the height of the conduction band discontinuity ΔE_C at the bottom of the quantum well so that the delta-doped DH-PHEMT is the best concept in view of g_{ds}.

4. Application-specific choice of epitaxial concept

We now want to discuss the suitability of the three PHEMTs of Tab. 1 for the circuit applications of Section 2. At given L_G, the homogeneously doped SH-PHEMT has the smallest C_{GS} without necessarily having a disadvantage with respect to g_m close to pinchoff. According to Eq. (1), these are good conditions for achieving low F values. Additionally, at low I_D, the wave function is centered in the quantum well so that interface scattering is minimized and transport properties are favorable, i.e. the factor K in Eq. (1) is small. On the other hand, the large d_{eff} may lead to unsatisfactory aspect ratios $L_G/d_{eff} < 5$ for $L_G < 200$ nm, resulting in high g_{ds}. Hence, a most simple SH-PHEMT might be the best device for low-noise circuits at low frequencies (for instance, DBS receivers at $f = 12$ GHz), but it cannot be properly scaled for shorter gates and applications at higher frequencies.

The aspect ratio handicap is removed with the introduction of delta-doping, but the low d_{eff} responsible for the better g_{ds} simultaneously causes a significantly enhanced C_{GS}. This can only be compensated by a short L_G so that delta-doped SH-PHEMTs require short gates even when they are used for low-noise ICs at comparatively low frequencies. On the other hand, the small d_{eff} leads to the benefit of a high g_m under open channel conditions, and the aspect ratio can be properly scaled even for $L_G << 150$ nm [7]. This enables the delta-doped SH-PHEMT to be used for highest frequencies, for instance automotive radar applications at $f = 77$ GHz.

SH-PHEMTs offer only a small energy barrier between channel and buffer. This low ΔE_C causes poor electron confinement to the quantum well at high $V_{DS} > 5$ V when the carriers have gained high kinetic energies from the electric field. In the DH-PHEMT, the conduction band discontinuity beneath the channel is significantly enhanced due to the use of the lower AlGaAs barrier layer, and the number of electrons penetrating into the buffer is reduced. Additionally, the DH-PHEMT features the largest n_s of all GaAs-based HEMTs (see Tab. 1) and is, thus, capable of the highest I_D. These properties make the DH-PHEMT most suitable for power applications where both high V_{DS} and I_D are required [8].

However, DH-PHEMTs can be efficiently used for additional purposes. The better confinement due to the AlGaAs barrier below the channel causes the output conductance g_{ds} to be even smaller than in most delta-doped SH-PHEMTs. Therefore, the DH-PHEMT is particularly appropriate for short gate lengths, i.e. MMICs for very high frequencies. As often stated in the literature, high sheet concentration n_s is a property which enhances the modulation efficiency of a HEMT leading to the best possible current gain cutoff frequencies f_T and high transconductances g_m over a large range of currents [4,5]. This is illustrated by Fig. 2 which shows that among three PHEMTs with comparable f_T, the DH-PHEMT offers the highest maximum frequency of oscillation f_{max} and, hence, the highest gain over a large I_D interval. In Sec. 2, this was identified as a requirement for devices in linear amplifiers with

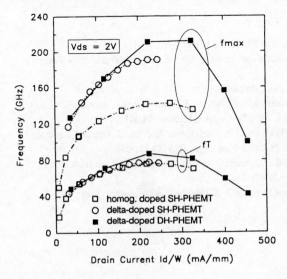

Fig. 2. Cutoff frequencies f_T and f_{max} of PHEMTs on GaAs substrate. All devices passivated, $L_G = 180...200$ nm.

large dynamic input range. Unfortunately, DH-PHEMTs also have some drawbacks. The peak g_m that can be achieved by minimizing d is not as large as in the delta-doped SH-PHEMT since d' is about 3 nm larger under open channel conditions which is a considerable effect if d is of the order of 15 nm. A second disadvantage is the inferior F as compared to SH-PHEMTs. This also originates in the larger d': under low-noise bias conditions, the 2DEG is close to the AlGaAs/InGaAs heterojunction at the bottom of the channel where its transport properties suffer from interface scattering. In our laboratory, passivated DH-PHEMTs have achieved $F = 0.63$ dB compared to $F = 0.55$ dB realized with SH-PHEMTs ($f = 12$ GHz). With increasing frequency, gain becomes more important (since it drops with a slope of 6 dB/octave) so that DH-PHEMTs represent an interesting general purpose device for K-band frequencies and above [6].

5. Conclusion

We have discussed possible applications for the three most common GaAs-based HEMT structures: the homogeneously doped SH-PHEMT, the delta-doped SH-PHEMT, and the delta-doped DH-PHEMT. The most simple homogeneously doped SH-PHEMT is still a suitable device for low-noise circuits at low frequencies whereas the delta-doped SH-PHEMT offers the highest peak g_m and may thus be used for the highest frequencies. The DH-PHEMT is not restricted to power applications but also very well suitable for general purposes in the intermediate frequency range, especially when linearity is required.

Acknowledgment

The author is indebted to H J Siweris for clarifying discussions concerning particularly Sec. 2 of this paper, and to A Mesquida Küsters and H Kniepkamp for critically reading the manuscript.

References

[1] Fukui H 1979 *IEEE Trans. Electron Devices* **ED-26** 1032-1037

[2] Vendelin G D 1982 *Design of Amplifiers and Oscillators by the S-Parameter Method* (New York: Wiley & Sons) 103

[3] Riechert H 1994 unpublished 1-dim. model calculations

[4] Foisy M C, Tasker P J, Hughes B and Eastman, L F 1988 *IEEE Trans. Electron Devices* **35** 871-878

[5] Nguyen L D, Larson L E and Mishra U K 1992 *Proc. of the IEEE* **80** 494-518

[6] Wu C S, Pao C K, Yau W, Kanber H, Hu M, Bar S X, Kurdoghlian A, Bardai Z, Bosch D, Seashore C and Gawronski M 1995 *IEEE Trans. Microw. Theo. Tech.* **43** 257-266

[7] Chao P C, Shur M S, Tiberio R C, Duh K H G, Smith P M, Ballingall J M, Ho P and Jabra A A 1989 *IEEE Trans. Electron Devices* **36** 461-473

[8] Smith P M, Chao P C, Ballingall J M and Swanson A W 1990 *Microw. Journal* **5** 71-86

Inst. Phys. Conf. Ser. No 145: Chapter 1
Paper presented at 22nd Int. Symp. Compound Semiconductors, Cheju Island, Korea, 28 August–2 September 1995

Photonics : An Information Age Technology

R. F. Leheny

ARPA/ETO, 3701 N. Fairfax Dr., Arlington, VA 22203, USA
rleheny@arpa.mil

Abstract. Information Age Technologies are those technologies associated with the generation, distribution, storage and display of information. Optoelectronic devices are enabling for many of these applications and are finding new applications as photonic device replace electronic devices. This paper reviews the status of compound semiconductor research as applied to information age photonic technologies .

Photonics is the science and technologies of using photons to perform the functions frequently associated with electronics in Information Age Systems; signal generation, modulation, manipulation, transmission, amplification and detection, Figure 1. The applications of photonics for fiber optic communication links and CD recorder memory access systems are so common that we have difficulty imagining these functions being performed in any other way, the advancement of photonic technology even further into the traditional realms of electronic functions for information systems continues to be a fruitful area of research.

Function	Important Device Properties
Photon Generation	Wavelength, Threshold, Quantum Efficiency, Temperature Dependence
Modulation	Direct or External; Digital (Bit-rate) or Analog (Linearity); Drive voltage
Manipulation	Active or Passive; Medium (Attenuation, Coupling Loss)
Switching	
Splitting	
Transmission	Fiber-Glass or Plastic (Transparency, Cost) Free Space
Amplification	Medium-Fiber, Semiconductor, Glass
Photon Detection	Speed, Sensitivity, Quantum Efficiency, Avalanche Gain

Figure 1 : Photonic Device Functions

An appreciation of these research challenges can be gained by considering the link length spanned by various applications illustrated in Figure 2. Photonics is unique in finding applications extending from the very small, on the order of mm's, to very long distances, on the order of 100's of kilometers. It would be truly remarkable if the same device technologies would find applications over the entire span of distances, however it is possible to divide this length scale into three distinct regions and recognize that within each region there is considerable commonality between the kinds of components required to meet information system needs. At the longest lengths, served by telecommunication systems, the technology choice is set by the low loss properties of glass fibers which have the best transmission characteristics at 1300 and 1550 nm. For these applications the requirements for single mode fiber and very high bit-rate transmission systems (2.5-10 Gb/s) drive the development of photonic components.

At intermediate distances, between about a meter and a kilometer, applications are much more cost sensitive since only a small number of users are served by the links. In this case multimode glass or plastic fibers are suitable and these transmission media can operate with low cost to manufacture shorter wavelength sources and in fact, for plastic fiber visible emitters (in the red-yellow spectrum) are favored. Bit-rates for even advanced systems are more modest and at the shorter distances parallel rather than serial interconnections can serve to provide very high information transfer rates even at relatively modest bit-rates per channel (100-1000 Mb/s). However, cost is a major issue for wide acceptance of photonic technology to meet these markets and only recently have developments, particularly in vertical cavity lasers and packaging, been encouraging for wide scale acceptance of photonic solutions to compete with electrical interconnections.

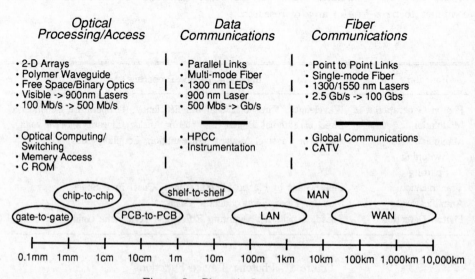

Figure 2 : Photonic Interconnection

For the shortest distance links present two opportunities for photonic technology. CD-ROM and laser printer applications where continued advances in optical disk memory requires research to realize even shorter wavelength lasers, pushing from red to blue emitters. The second application for photonics on this scale has been to develop parallel 2-D interconnections that can provide logical control of the optical interconnection to allow very fast memory access and optical signal processing functions. Full scale development of systems exploiting the advantages of photonics on this scale are still being researched, but again developments in device technologies, particularly vertical cavity surface emitting lasers are very encouraging.

This review discusses some of the compound semiconductor materials and materials processing technology developments that are driving photonic solutions for high performance information systems. While compound semiconductors play a pivotal role in photonics they are complimented by the electro-optical crystals, such a $LiNbO_3$, glass, including rare earth doped glasses, and polymer materials which all contribute to performing photonic functions required to realize a complete system. Cost and performance are the key elements in determining the best material choice for a given application.

Advances in compound semiconductor materials for photonics have been paced by advances in deposition technology, beginning with liquid phase epitaxy in the 1970's to today's more complex techniques based on gas source OMCVD and MBE. Advances in optical emitter technology reflected in reduction of threshold current requirements that has occurred over the last twenty five years is illustrated schematically in Figure 3. Starting with threshold currents of hundreds of milliamps for the early heterojunction devices, developments in materials and device designs have resulted in decreasing threshold to the order of a few milliamps in today's commercial devices. Key milestones to these improvements have been the development

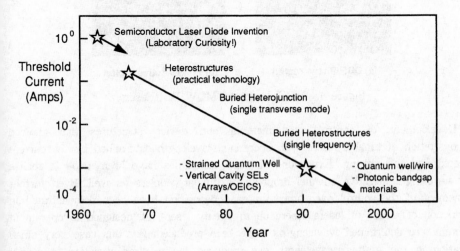

Figure 3 : Progress in semiconductor laser threshold reduction

of deposition technology to allow the preparation of extremely thin film Quantum Wells and the demonstration that the growth of strained material, once considered the death of an minority carrier device, can in fact be advantageous providing the film thickness is thin enough. Recent developments in oxide defined vertical cavity lasers have demonstrated thresholds below 10 microamps. These low threshold currents are critical to minimizing heating and can make possible applications such as operation with CMOS driver circuits and laser arrays that could only be dreamed with thresholds exceeding 100 milliamps.

Advances in growth technologies have gone hand-inland with advances in materials processing so that today it is possible to routinely produce structures such as the DFB laser structure illustrated in Figure 4. Not only has the technology for etching the grating with spacing on the order of 120 nm, but the regrowth of the waveguide material over the grating can be accomplished with minimum degradation. The critical materials issues in producing structures of this type are to control the material composition in the active region, and also in the waveguide itself since the effective optical spacing of the grating is determined by both the waveguide material dispersion and the waveguide dimensions.

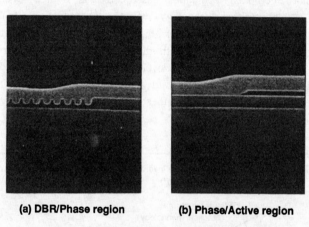

(a) DBR/Phase region   **(b) Phase/Active region**

Figure 4 : MOCVD-grown MQW DBR lasers

Development of control over material and device structures has allowed demonstration of tunable lasers with operating wavelength determined by the current injected into the grating. Extensions of these techniques have been made to realize the wavelength selective channel dropping directional couplers as well. Even further refinement of the control over processing and deposition have been demonstrated in production of arrays of lasers where up to twenty lasers, all designed to operate at different, pre-determined wavelengths, have been produced. For this case very small variation in the grating wavelength has produced the required variation in output wavelength. Such multiwavelength sources are finding applications in point-to-point links to increase the transmission capacity of fiber without the need for more expensive higher bandwidth electronics. Wavelength selective elements can route signals through an "all-optical" network without conversion of optical to electrical formats.

A critical issue for low cost semiconductor lasers operating in real-world environments remains the influence of temperature on device performance. The development of material structures designed to enhance the trapping of mobile charges against thermal excitation out of the active region are making possible high performance high temperature lasers. Significant differences in thermal variation of threshold and linearity of laser output are obtained for devices fabricated with strained multiquantum well AlGaInAs material compared with conventional unstrained (bulk) GaInAsP materials. The Al containing alloy provides a larger conduction band off-set to confine injected electrons even at higher temperatures. For the AlGaInAs lasers the operating characteristics at 100C are not significantly degraded from room temperature operation.

One new device structure made possible as a result of the advances in material control is the Vertical Cavity Surface Emitting Laser, or VCSEL. Because of their unique structure VCSELs can be tested on wafer during processing and can be easily mounted in standard diode packages and aligned with fiber, greatly decreasing the cost to manufacture. For VCSELs, not only is precise control of the active region and adjacent carrier confining layers critical to device performance, but also the deposition with subwavelength accuracy of large numbers of alternating materials to form high reflectivity mirrors represents a major challenge. While one of the mirrors can be made using dielectric materials, in the best designs both mirror stacks are semiconducting and current flows through these mirrors. Design for efficient optical and electrical characteristics is a challenge, but devices operating with threshold below a milliamp are possible.

VCSELs can be produced in arrays that can be inexpensively coupled to fiber ribbons. Moreover their low threshold and low thermal cross-talk (turning one device on doesn't influence the threshold for adjacent devices) allows them to be mounted close to drive circuits. These features make VCSELs ideal emitters for incorporation into optical interconnect systems as almost a replacement for lower speed, higher drive power LEDs. High performance workstation applications have been demonstrated with 32 bit wide parallel interconnects operating at 500Mb/s for a net data transfer rate of 16Gb/s. For future applications in optical signal processing, 2-D arrays of VCSELs will be combined with electronic logic circuits to realize so called "smart pixel" where turning on and off of optical signals will be determined by the logic states of the electronic components.

Two recent research results for VCSELs have achieved improvements which result directly from an advance in materials processing. First, the development of a stable oxide of Ga that can be used to form the mirror layers in a VCSEL while also controlling the current flow and guiding the laser mode. These devices have achieved the lowest thresholds for room temperature operation of lasers yet reported-under 10 microamps. Second, VCSEL lasers designs to date have been confined to the GaAs based materials and therefore short wavelength emission, however using wafer fusion technique that allows the bonding of films grown on different substrates. Using this technique room temperature operation of an InP based long wavelength VCSEL bonded to GaAs material containing efficient mirrors has been demonstrated.

In conclusion only a small fraction of the exciting research that is going on in compound semiconductor materials and devices for photonic applications has been touched on here. Perhaps one of the most exciting new topics is the race between GaN and the two-six materials to demonstrate a practical room temperature blue-to-green laser. Regardless of who wins this particular contest, what is particularly encouraging is the continued progress that has marked the last 30 years of research in this field. From the early work improving bulk material properties in the 1960's, to the pioneering work on epitaxy in the 1970's that lead to MBE and OMCVD with the spectacular development of quantum well structures and the application of strained materials to devices in the 1980's, to today's efforts to demonstrate what was only a dream as few years ago: to be able to deposit any material, on any substrate, limited only by the imagination of the device designer.

Inst. Phys. Conf. Ser. No 145: Chapter 1
Paper presented at 22nd Int. Symp. Compound Semiconductors, Cheju Island, Korea, 28 August–2 September 1995
© *1996 IOP Publishing Ltd*

Present and Future of Group III Nitride Semiconductors

Isamu Akasaki   and Hiroshi Amano

Department of Electrical and Electronic Engineering, Meijo University, 1-501 Shiogamaguchi, Tempaku-ku, Nagoya 468, Japan

Abstract. This paper reviews the recent progress of crystal growth and conductivity control of group III nitride semiconductors. Current status of nitride-based   short wavelength light emitters and future prospects of widegap nitrides will also be presented.

1.Introduction

The group III nitrides with the exception of BN, that is wurtzite polytypes of AlN,GaN, InN and their alloy AlGaInN, have the direct transition type bandstructure with   bandgap energy ranging 1.9eV for InN, 3.4eV for GaN and 6.2eV for AlN at room temperature (RT). Furthermore, these nitrides possess superior physical and chemical properties e.g. high thermal conductivity,   high electron saturation velocity, and physical and chemical stability. Therefore, they are one of the most promising materials for applications to short wavelength light emitters, such as light emitting diodes (LEDs) and laser diodes (LDs) in the green to ultraviolet (UV) regions, as well as high-temperature electronic devices. These short wavelength light emitters enable us to produce an all-solid-state, full-color flat-panel display system, a compact and high-density optical storage system, a high-speed printing system, a small medical apparatus and so on. The high-temperature electronic devices have many desirable electrical properties that are useful for the current high-performance engines and the future electronic components operated in a harsh environment.

   To realize such new devices, it is essential to grow high-quality epitaxial films and control their electrical conductivity. On the contrary to other III-V compounds such as GaAs and InP, however, it had been quite difficult to grow high-quality epitaxial nitride films on a flat surface free from cracks. This is mainly due to the lack of substrate materials with lattice constant and thermal expansion coefficient close to those of GaN and nitride alloys. Moreover, it has been well-known that because undoped nitrides were of strong n-type conductivity with high residual electron concentration, p-type nitrides had not been realized until recently. These problems had prevented from making the actual application of nitride devices for a long time.

2.Recent progress of crystal growth and conductivity control

During the last decades, significant advance in the heteroepitaxial growth of nitrides on highly-mismatched substrates (e.g. sapphire) [1],[2] has enabled us to grow high-quality GaN [1],[2],[3], AlGaN [4], GaInN [5][6] and their heterostructures [7],[8]. The conductivity control of both n-type [9] and p-type nitrides [10] has been also achieved. Furthermore, optical and carrier confinements have been achieved by using double heterostructure(DH) [11],[12], and the emitted wavelength has been controlled by the use of alloys [12].

These achievements as well as the earlier discovery of impurities, which form very efficient blue luminescence centers in nitrides [13], have led to the fabrication of high-brightness UV/blue [14], blue [12], and bluish green [15] LEDs with efficiencies in excess of 1%. UV stimulated emission from nitrides operating at RT by optical pumping [16] as well as nitride-based field effect transistor (FET)[17] have also been achieved.

3.Current status of nitride Research

At present, steady progress is made in the area of the metalorganic vapor phase epitxy (MOVPE) growth. Shown in Fig.1 is the photo-luminescence (PL) spectrum of undoped GaN (full line curve)grown on sapphire using the AlN buffer layer [18]. The PL is dominated by the intrinsic free exction (FE) with peaks at about 3.488, 3.495 and 3.506 eV, (labelled A,B and C), corresponding to the transition between valence band A,B and C, respectively. The intensity of main exciton peak FE(A) is about 2 orders of magnitude higher than those of impurity-bound excitons as they are donated ABE and DBE in the figure. The PL full width at a half maximum (FWHM) for the FE(A) is about 3 meV. We also observed the FE with a FWHM as narrow as 1.4meV for another undoped sample. Such a sharp and strong FE emission indicates the much improved crystalline quality.

Figure 2 (a) shows a typical example of a cross-sectional transmission electron micrograph (TEM) of a recent blue LED. The structure of p-GaN/p-AlGaN/ (GaN/GaInN) multi quantum well (MQW)/n-GaN/AlN buffer/ sapphire is shown in Fig 2 (b) [19]. The MQW structure with 5 GaInN wells is clearly seen. Fig.2(c) shows the In concentration profile by SIMS. In spite of the existence of high-density dislocations on the order of

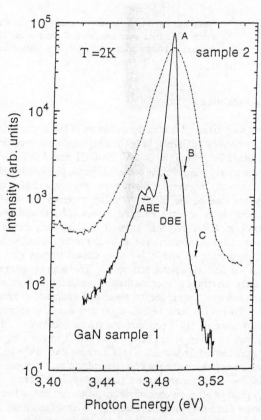

Fig.1 Photoluminescence spectrum of the undoped GaN sample 1 (full line curve) at 2K, showing the free exciton emissions (A,B and C), and the donor and acceptor bound excitons (denoted DBE and ABE, respectively) at lower photon energies. The dash-dotted curve is the corresponding spectrum for another slightly n-type sample 2, which shows a much broader edge spectrum.

10^9 cm^{-2} in the active layer, the degradation in output power of the blue emission peaking at about 450 nm has been small, and the LED was attained a brightness of about 3 cd at 20 mA. The forward voltage is as low as 3.3 V. This indicates that further progress can be made in blue LED performance.

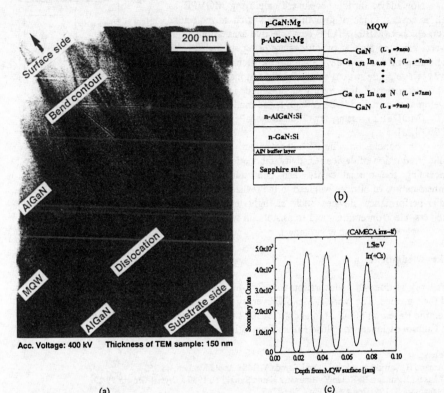

Acc. Voltage: 400 kV   Thickness of TEM sample: 150 nm

(a)   (c)

Fig.2(a) Crosssectional TEM photograph of blue LED, structure of which is schematically shown in (b). MQW structure with 5 GaInN wells is clearly seen. Interfaces between GaN and AlGaN in both n and p layers could not be distinguished in the TEM photograph. (c) shows the In concentration profile by SIMS.

4.Future prospects of group III nitrides

Until recently, hexagonal nitrides, grown by MOVPE on sapphire substrate, have been mostly used in devices mentioned above. In addition to hexagonal nitrides, cubic nitrides and so-called III-V nitrides, which include GaNAs, GaNP, AlNSiC and other materials containing nitrogen as one of the major constituents, are also attracting the attention by many researchers.

On the other hand, various kinds of substrates such as Si, GaAs, SiC, ZnO, MgAl$_2$O$_4$, MgO and NdGaO$_3$, which have less-mismatched and/or electrically conductive substrates, are being used for a variety of reasons. Some of them can be cleaved easily. This leads to the possibility of fabricating a good cavity mirror.

Figure 3 shows RT laser action with Fabry-Perot mode peaking at 377.20 nm by optical pumping from AlGaN/GaInN DH grown on 6H-SiC substrate [20]. The threshold power is about 27 KW·cm^{-2}, which is the lowest to date for this device. This is thought to be one of the most promising structures for the realization of LD by current injection.

Less-strained or strain-free nitride films are being prepared by the use of homoepitaxial growth on GaN substrate, which has been already successfully grown by hydride vapor phase epitaxy (HVPE)[21].

Nowadays, nitride people are employing MOVPE as well as several kinds of growth methods such as molecular beam epitaxy (MBE), HVPE etc. Selective area growth and several kinds of etching of nitrides are being studied for the fabrication of desirable device structures, such as the optical cavity, wave guide structure and low-dimensional structures.

As one of the results, it resulted in the recent development of a nitride-based high-speed modulation doped FET (MODFET) using two dimensional electron gas (2DEG)[22].

In conclusion, further improvement of crystalline quality, reduction of device resistivity, fabrication technology concerning the optical cavity, and understanding of the intrinsic nature of nitrides will lead to the realization of much higher-performance devices such as light emitters, light receivers, electron emitters and transistors, in which they are able to operate in a harsh environment.

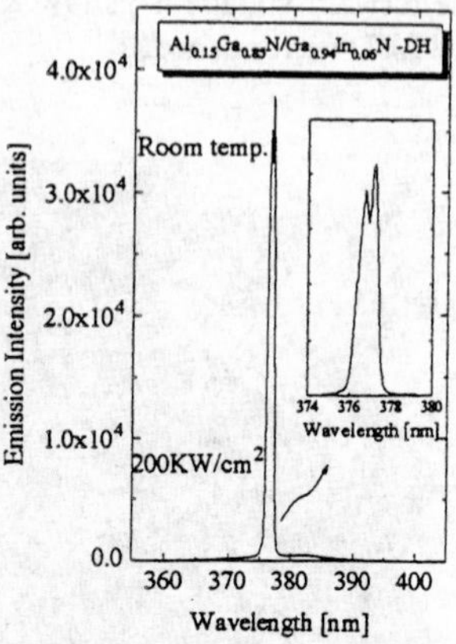

Fig.3 Edge emission mode spectrum from AlGaN/GaInN DH excited above the threshold input power density of about $27KW \cdot cm^{-2}$. The peak wavelength is 377.20 nm, and wavelength difference between modes is 0.03 nm.

Acknowledgments

This work is done in collaboration with Dr.M.Koike, Prof. L.F.Eastman, Prof. B.Monemar and their groups. The work at Meijo University was supported partly by the Grant-in-Aid for Scientific Research #06452114 and #07505012 both from the Ministry of Education, Science and Culture of Japan, and Hoso Bunka Foundation.

References
[1] Amano H, Sawaki N, Akasaki I and Toyoda Y, 1986 Appl.Phys.Lett. 48 353
[2] Akasaki I, Amano H, Koide Y, Hiramatsu K and Sawaki N, 1989 J.Crystal Growth 98 209
[3] Nakamura S, 1991 Jpn.J.Appl.Phys. 30 L1705
[4] Koide Y, Itoh H, Sawaki N and Akasaki I, 1986 J.Electrochem.Soc. 133 1956
[5] Yoshimoto N, Matsuoka T, Sasaki T and Katsui A, 1991 Appl.Phys.Lett. 59 2251
[6] Nakamura S and Mukai T, 1992 Jpn.J.Appl.Phys. 31 L1457
[7] Itoh K, Kawamoto T, Amano H, Hiramatsu K and Akasaki I, 1991 Jpn.J.Appl.Phys. 30 1924
[8] Nakamura S, Senoh M. Mukai T, 1993 Jpn.J.Appl.Phys. 32 L8
[9] Amano H and Akasaki I, 1990 Mat.Res.Soc. Ext.Abs. EA-21 165
[10] Amano H, Kito M, Hiramastu K and Akasaki I, 1989 Jpn.J.Appl.Phys. 28 L2112
[11] Akasaki I and Amano H, 1991 Mat.Res.Soc.Symp.Proc. 242 383
[12] Nakamura S, Mukai T and Senoh M, 1994 Appl.Phys.Lett. 64 1687
[13] Pankove J I, Miller E A, Richman D and Berkeyheiser J E, 1971, J.Lumin 4 63
[14] Akasaki I, Amano H, Itoh K, Koide N and Manabe K , 1992 Inst.Phys.Conf.Ser. 129 851
[15] Nakamura S, 1994 Ext.Abs.Int.Conf. on SSDM S-1-7-1 81
[16] Amano H, Asahi T and Akasaki I, 1990 Jpn.J.Appl.Phys. 29 L205
[17] Khan M A, Kuznia J N,Bhattarai A R and Olson D T, 1993 Appl.Phys.Lett. 62 1786
[18] Harris C I, Monemar B, Amano H and Akasaki I, 1995 accepted for publication in Appl.Phys.Lett.
[19] Koike M, Yamasaki S, Nagai S, Koide N and Asami S, Amano H and Akasaki I, to be submitted
[20] Amano H and Akasaki I, 1995 to be submitted to Fall Meeting, Mat.Res.Soc.Symp.AAA
[21] Detchprohm T, Amano H, Hiramatsu K and Akasaki I, 1992 Appl.Phys.Lett. 61 2688
[22] Burn J, Schaff W J, Eastman L F, Amano H and Akasaki I, 1995 to be submitted to ISCS

Inst. Phys. Conf. Ser. No 145: Chapter 1
Paper presented at 22nd Int. Symp. Compound Semiconductors, Cheju Island, Korea, 28 August–2 September 1995
© 1996 IOP Publishing Ltd

Gap between microelectronics and nanoelectronics

T. Ikoma*, **, T. Hiramoto and K. Hirakawa****

*Texas Instruments Tsukuba R&D Center, Miyukigaoka, Tsukuba 305, Japan
**Institute of Industrial Science, University of Tokyo, Roppongi, Minato-ku, Tokyo 106, Japan

Abstract. There is a gap between microelectronics and nanoelectronics. It should not be considered that nanoelectronics is existing on a simple extrapolation of the present micro-electronics. In order for nanoelectronics to move from research to more developmental, we have to clearly define what target nanoelectronics is trying to hit and all our effort should be directed towards that direction.

1. Introduction

Microelectronics, which includes Si LSI and its applications, is a major thrust to bring us into the networked-information society and itself has achieved a great progress. Microelectronic devices become smaller and smaller, the integration scale grows larger and larger, the circuit more and more complex and the investment cost higher and higher. Many people argue that all of these factors are now approaching to the limitation but in reality the progress is going on.

Nanoelectronics, which is expected to breakthrough the anticipated limitations of microelectronics was historically started with the proposal of man-made superlattices by Esaki and Tsu (1969) and with the formulation of the quantum transport of electrons in a very small structure by Landauer (1957). In addition, we can find another origin, which is a possibility to manipulate single electron proposed by Likharev (1987). Naturally we may think that the progress of microelectronics, in particular, the shrinkage of device size, shall bring us into the nanoelectronics era, where a new technology will give us a new paradigm. However, a question arises; Can microelectronics smoothly move to nanoelectronics? Is nanoelectronics really on the horizon of Si-microelectronics?

In this paper we address this issue and show there is a gap between microelectronics and nanoelectronics. A new approach is necessary to open up a new paradigm.

2. Scale-down

The shrinkage of device size in microelectronics is remarkable, still following Moore's law; the integration level increases by 4 times every 3 years, which leads to the minimum feature size shrinkage by 0.7 times per each generation. At present, in mass-production lines, 0.35 μm devices are common and in near future, 0.25 μm devices will be introduced in mass production. At a research stage, 0.05 to 0.04 μm MOS transistors were fabricated and tested (Ono et al. 1993, Hori et al. 1994), and we know these very small devices can be operated just in the conventional manner. For these devices, a gate insulator is as thin as 1.5 (Momose et al. 1994) to 3 nm, where the direct tunneling current is dominant.

In nanoelectronics, electrons are confined in a small structure and various quantum effects are observed. The major features are the discretized energy-states, the modification of state density, the (coherent) electron-wave nature and the granularity of electrons, which result in new

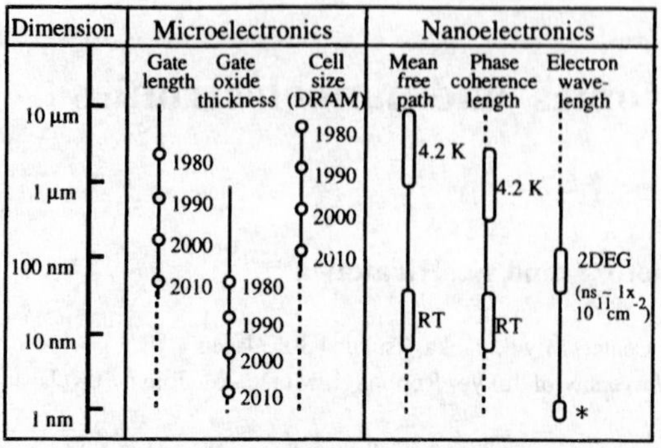

Dimension	Microelectronics			Nanoelectronics		
	Gate length	Gate oxide thickness	Cell size (DRAM)	Mean free path	Phase coherence length	Electron wave-length
10 μm			1980	4.2 K		
1 μm	1980 1990		1990 2000		4.2 K	
100 nm	2000 2010	1980	2010			2DEG (ns,~ 1x, 10^11 cm^-2)
10 nm		1990 2000		RT	RT	
1 nm		2010				0 *

Fig. 1. The characteristic sizes of microelectronics and nanoelectronics. *Localized electrons in Deep Traps.

Quantum Hall Effect	$R_H = \dfrac{h}{e^2}$
Quantized Conductance	$G = \dfrac{e^2}{h}$
Aharonov Bohm Effect	$\Phi = \dfrac{h}{e}$
Coulomb Blockade oscillation	$\Delta V = \dfrac{e}{C}$
Charging Energy	$E = \dfrac{e^2}{2C}$

Fig. 2. Material independent nature in nanoelectronics.

physical phenomena peculiar to nanoelectronics such as resonant tunneling, quantum Hall effect, electron-wave interference, quantized conductance, etc. Most of these phenomena are only observed at cryogenic temperature or below except for resonant tunneling. The characteristic sizes for nanoelectronics are Fermi wave-length , phase coherence length, mean free path, trapped electron wave-spreading, which are shown in Fig.1. If we compare these factors at low temperatures, these characteristic lengths are already in the same order of magnitudes as the minimum feature sizes in microelectronics. This means we are already in the nanoelectronics age. However, the gap between the two is rather large since in nanoelectronics, the measurement temperature is very low in the order of several K to mK and the current and voltage levels are extremely small. Many of the quantum effects are smeared out with increasing temperature and we have to use very often a lock-in amplifier to detect signals. From the view point of applications these are great drawback, although they are quite acceptable for a discovery of new physics.

In all of the conventional switching-devices we treat current as an electron flow just like a liquid, while in nanoelectronics the conductance is expressed by using the scattering matrix (Buttiker, 1988) similar to the S-matrix applied to the electro-magnetic waveguide circuit or electrons are treated as a granular particle with a discrete negative charge, where the charging energy is important to describe the current flow. Thus, microelectronics and nanoelectronics are standing on different foundations, classical and quantum. In single-electronics with the zeroth-order approximation, we can use the classical mechanics to describe its current-voltage characteristics, for instance, the charging energy and Coulomb blockade can be described with a simple classical expression. However, in small islands or quantum dots coupled with each other by tunneling, the many-body effect appears directly to the terminal characteristics and it is necessary to use the quantum mechanics.

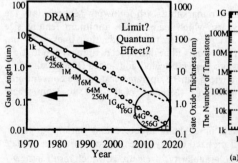

Fig. 3. Trend of gate length and gate oxide thickness of MOS transistors.

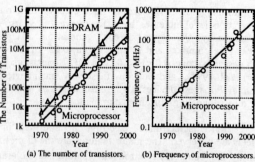

(a) The number of transistors.　(b) Frequency of microprocessors.

Fig. 4. Trend of the number of transistors on chip and frequency of microprocessors.

3. Material choice

III-V compound semiconductors are mainly used to study and realize nanoelectronics because the combination of different compound semiconductors can form narrow and sharp potential wells in which electrons are confined. Another reason to use compound semiconductors is that the electron effective mass is small and the mean free path is long in some of compound semiconductors and therefore, quantum effects are easily observed. The quest of higher mobility by confining electrons in one dimensional quantum wire has not been successful. It should be noted, however, that the concept of mobility no more applies to nanostructure devices because the conductance in such a quantum wire is expressed by Landauer formula in which the electron-wave scattering matrix is a key to determine the current flow and not the mobility. In nanostructure devices, averaging is not adequate to describe any physical quantities and current depends on the location of scatterers or impurities. Therefore, process technologies to fabricate nanostructure devices are very different from those for microelectronic devices. Furthermore, many of the new phenomena in nanoelectronics are material independent, i.e., quantum Hall resistance, quantized conductance, the period of Aharonov-Bohm oscillation, Coulomb blockade oscillation voltage and charging energy are expressed by the universal constants (plus capacitor in the latter two) as shown in Fig. 2. The phase coherence length is almost the same in GaAs and Si, since it is mainly determined by the electron-electron scattering at low temperatures.

On the other hand, Si is mainly used in microelectronics. There are many reasons for it, among which is that Si /SiO2 system is an ideal combination from the view point of process technologies as well as device and circuit operations. At present, the most advanced technologies are available on Si materials and processing. The interface is nearly perfect with a very sharp transition and a surface-state density as low as $1 \times 10^{10}/cm^2$. Since nanoelectronics requires a precise control of potential distribution and impurity location, the very-matured technology should be better used. Thus, Si nanoelectronics is worthy being investigated. For single-electronics, a key technology is the fabrication of very small capacitors coupled to each other by tunneling and hence a metal/insulator system may be the most suitable one. With transition from microelectronics to nanoelectronics we have more choices of material systems.

4. Trend in microelectronics (Si LSI)

SIA (Semiconductor Industrial Association) published the Semiconductor Technology Roadmap, where the technological trend and the issues to be solved are summarized up to 2010. It predicts at 2010 the minimum feature size is 0.07 μm and the integration scale of DRAM is 64 Gbits. This is based on the extrapolation from the past as shown in Fig3. There have been arguments on when we shall abandon the linear extrapolation on a semi-logarithmic scale. Some ten years ago, people thought that 0.2 μm would be a limit of optical lithography and the conventional operation of transistors. Now we see that at least down to 0.18 μm, we can use the conventional optical lithography with 193nm DUV and even for 0.13 μm we will be able to use the conventional DUV. We have already a proof that conventional CMOS inverters can be operated down to 0.07 μm (Takeuchi et al. 1995). In addition to the shrinkage of device feature size, the number of transistors and the chip frequency are increased as shown in Fig.4. By 2010 a microprocessor will have 90 M transistors/cm^2 and ASIC will have 40 M transistors/cm^2. The chip frequency will exceed 1.1 GHz by 2010.

To realize the predicted trend we have to address many complicated issues such as design and test, process integration, device and structure, lithography, interconnect, materials and bulk processes, assembly and packaging, factory integration, and environmental, health and safety. The most difficult problem is a huge investment cost, which leads to the alliance policy among semiconductor companies at world wide.

If we investigate more details of device development trend, we can see two different directions; high speed and low power. High speed is necessary to achieve high performance computers and low power is important for portable terminals in multi-media networks. The shrinkage of device size contributes to the both but to achieve low power, voltage and current

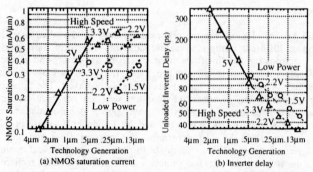

Fig. 5. Trend of NMOS saturation current and inverter delay.

should be lowered as much as possible keeping the frequency so low as to satisfy the specification. These trends are shown in Fig. 5, where drain current and delay time are plotted against the transistor nodes for various drain voltages (Hu 1994). We can see clearly the difference in low power and high speed applications. This trend is more clearly observed in Fig. 6 where delay time vs. power is plotted for CMOS inverters. For high speed application the curve moves downward and for the low power application it does horizontally.

In summary Si microelectronics will continue to progress; the minimum feature size will become less than 0.1 μm and the integration level will exceed 64 Gbits by 2010. To achieve this goal very complex problems must be solved both technically and economically. For this purpose any quantum effects are undesirable and should be avoided to appear. The most critical issue is a thin gate insulator, which will be less than 3 nm thick if SiO_2 is used. Direct tunneling current is dominant in such a thin insulating film. To avoid direct tunneling a material which has a higher dielectric constant must be used to keep the thickness thicker than the critical thickness for direct tunneling. This is an example that we have to suppress quantum effects in microelectronics.

5. Target of nanoelectronics

In nanoelectronics, many ideas to apply quantum effects to switch, logic, memory and cell-automaton have been proposed. Among them two cases are discussed here. One is an electron-wave switch which consists of two electron-waveguides coupled each other with distributed tunneling. Because of the quantum beat, an electron-wave oscillates between the two states when they are strongly coupled. When the gate region is equal to the transition length an electron-wave can be switched from one channel to the other. For this electron-wave switch to be nicely operated, the electron wave must preserve its phase (coherency), therefore the phase is a key quantity. Another example is a single electron transistor (Tucker 1992) or single electron memory (Nakazato et al. 1994), where the number of electrons is an important quantity. Since these two quantities are related each other with the uncertainty law, these are two extreme cases.

5.1. Electron-wave switch

The operational principle of the electron-wave switch is completely different from that of the conventional FET. In the latter, electrons are depleted in a channel by negative gate voltage to switch off the device. Electrons feel repulsive force when negative voltage is applied and attractive force when positive voltage is applied. Since there are scatterings, an electron motion induces power dissipation. In a C-MOS inverter, power dissipation takes place during the transition period, mainly due to the gate capacitance charging. If we can avoid the gate capacitance charging, power dissipation can be considerably reduced. In the electron-wave switch, electrons are transferred from one channel to the other by the quantum beat, which is a lossless process. The quantum beat takes place when the two waveguides have the same eigen energy. If the gate length is equal to the transition length the transfer rate is 1 in an ideal case. For a switching action, gate voltage is applied to align the two eigen energies, i.e. to modify the coupling efficiency. The gate current may flow due to change of the distance of electron location from the electrode, which is negligibly small. The switching speed is determined by the transition time, which is equal to the transit time of electron-wave across the gate length. Therefore, the switching speed is not so high as compared with the conventional transistor, since the gate length is approximately 0.1 μm in a realistic device. The advantage is low power.

The switching characteristic is very sensitive to impurity locations in the waveguides and the control of impurities and defects in a crystal is a key issue. At present the technology is not advanced enough to fabricate an ideal electron-wave switch and it is a great challenge to realize it.

5.2. Single-electronics

Single electronics is based on the Coulomb blockade; when an electron moves into an isolated region (a capacitor), the electrostatic energy at the region increases by the charging energy $e^2/2C$. At very low temperatures, where the thermal energy kT is much smaller than the charging energy, electron can not enter the island, that is, the capacitor can not be charged until the electron energy becomes higher than the charging energy. This means when the applied voltage is smaller than $e/2C$, current does not flow (Coulomb blockade) and does only at the voltage of $ne/2C$ (n;integer) (Coulomb staircase). With using this phenomenon, single electron memory and logic circuits have been proposed. To observe the Coulomb blockade, temperature should be very low or the capacitance should be very small.

The figure of merit of single-electronics is shown in Fig.6, where it is assumed that a small isolated electrode is placed between two tunneling junctions and the tunneling resistance is 100 kΩ (Mooij 1993). When C=1 aF, the power is 100 nW and the delay time is 100 fs. The voltage swing is 160 mV. When C=1 fF the power is as low as 0.1 pW and the delay time is 100 ps. The voltage swing is 0.16 mV. In the figure the two limitations are shown; the quantum limit caused by Heisenberg's uncertainty law and the thermal limit. To observe the Coulomb blockade, the temperature should be lower than 1 K for C= 1 fF, while it is observable at 300 K if C=1 aF.

From the figure we can find that the gap between CMOS inverters and single electronics is great .The difference is three to five orders of magnitude. Furthermore we can say that single-electronics is suitable for low power applications but not for high speed. To make the difference clearer, Fig 7 shows the number of electrons per bit/gate used in present circuits. The number of electrons is as high as 1×10^6 for DRAM and depends only on voltage. In future, it will decrease down to 1×10^5 at 1.0V. Nevertheless, a very large gap is present. In the present single electronics, one island accommodates about 100 electrons and single electron charging and discharging is detected. Perturbation and noise immunity become serious problems.

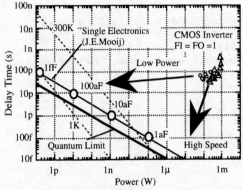

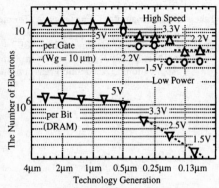

Fig. 6. Delay vs. power for CMOS and single electronics.   Fig. 7. The number of electrons per bit and per gate.

In summary, nanoelectronics is apparently directed toward low power applications but the discrepancy from the present microelectronic is very large. Therefore, it should not be considered that nanoelectronics is existing on a simple extrapolation of the present microelectronics. It is not a continuous evolution to replace microelectronics but a discontinuous revolution. It is needed to find out a new application field.

6. Summary

We have shown that nanoelectronics is still very far from microelectronics although the feature size is now similar to each other. The idea that the evolutional change from microelectronics should bring us into nanoelectronics may be wrong. In between the two, there will be a great possibility to develop new technologies which will be useful to empower the present Si technology. One of the challenges is resonant tunneling devices (RTDs). Various versions of RTDs and their circuits have been developed (Yokoyama et al. 1992, Seabaugh and Reed 1994). The advantage of RTD is room temperature operation and signal level compatible with the present Si LSI. The target of RTDs is neither high speed nor low power, but their functionalities. The RTDs are suitable, for instance, for multi-valued logic and a frequency multiplier, where the number of transistors can be reduced and the circuit can be simplified. This is one of the fields that nanoelectronics should be addressing. In order for nanoelectronics to move from research to more developmental, we have to clearly define what target nanoelectronics is trying to hit and all our effort should be directed towards that direction.

We would like to thank Dr. A. C. Seabaugh for his cooperation.

References

Buttiker 1988 *IBM J. Res. Develop.* 32 pp 317

Esaki L and Tsu R 1969 *IBM J. Res. Develop.* 14 pp 61

Hori A, Nakaoka H, Umimoto H, Yamashita K, Takase M, Shimizu N, Mizuno B and Odanaka S 1994 *IEDM Tech. Dig.* pp 485 - 488

Hu C 1994 *ISSCC Tech. Dig.* pp 86 - 87

Landauer 1957 *IBM J. Res. Develop.* 1 pp 223

Likharev K K 1987 *Mikroelektronikz* **16** pp 195

Momose H S, Ono M, Yoshitomi T, Ohguro T, Nakamura S, Saito M and Iwai H 1994 *IEDM Tech. Dig.* pp 593 - 596

Mooij J E 1993 *Extended Abstracts of International Conference on Solid State Devices and Materials* pp 339 - 340

Nakazato K, Blaikie R J and Ahmed H 1994 *J. Appl. Phys.* **75** pp 5123 - 5134

Ono M et al. 1993 *IEDM Tech. Dig.* pp.119 - 122

Seabaugh A C and Reed M 1994 "Heterostructures and Quantum Devices" Chapt. 11 (Academic Press) pp 351

Takeuchi K, Yamamoto T, Furukawa A, Tamura T and Yoshida K 1995 *VSLI Technology Symposium Tech. Dig.* pp 9 - 10

Tucker J R 1992 *J. Appl. Phys.* **72** pp 4399 - 4413

Yokoyama N et al. 1992 "Hot Carriers in Semiconductor Nanostructures; Physics and Applications" (Academic Press) pp 443

Inst. Phys. Conf. Ser. No 145: Chapter 1
Paper presented at 22nd Int. Symp. Compound Semiconductors, Cheju Island, Korea, 28 August–2 September 1995

One Million Transistor Circuits in GaAs

Louis R. Tomasetta

Vitesse Semiconductor Corporation
741 Calle Plano
Camarillo, CA 93012

Abstract. This paper describes the process innovations that have enabled GaAs digital ICs to achieve one million transistor circuits. While initial applications as ultra high speed digital processors have not been commercialized, the basic technology is now being widely used in telecommunications and data communications applications. Performance projections of GaAs are discussed and compared with CMOS, BiCMOS, GaAs Bipolar and Silicon ECL. The implications of these performance and cost projections are discussed.

1. Introduction

GaAs Digital IC technology has matured to the point that circuits with over one million transistors can be routinely fabricated. These circuits can include large blocks of high speed SRAM, analog functions (e.g. PLL, VCO, amplifiers, etc.), and of course various GaAs digital architectures that result in either extremely high speed (>5 Ghz) or extremely low power (<0.1 mW/gate at 750 Mhz) logic gates. Historical concerns about high cost, wafer breakage, manufacturability, and reliability have been overcome to the point where Vitesse digital GaAs ICs cost no more than comparable BiCMOS ICs and considerably less than Silicon ECL circuits, while providing substantial improvement in speed and/or power.

2. Process Technology

The Vitesse four layer aluminum metal, digital/analog IC process [1] is shown in Figure 1. The key innovations that we made to achieve the process control needed to reach these levels of complexity include the use of refractory metal self-aligned gates, stress-free planarization of all dielectric layers to reduce piezo-electric threshold voltage shifts, and high temperature ohmic contacts. Gold has been eliminated from the process.

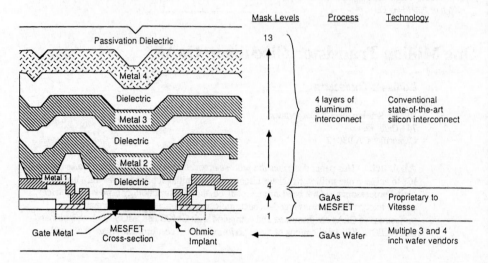

Figure 1

Figure 2 summarizes the innovations made at Vitesse compared to the more traditional RF/Microwave processes. In addition, a number of traditional silicon improvement programs, such as particle reduction programs, have been very aggressively pursued. Over the past 10 years, Vitesse has continued to improve both the performance of the transistors through traditional gate length reduction and channel depth scaling as well as overall circuit performance parameters (e.g. delay-power product) through interconnect scaling and capacitance reduction (e.g. polyimide). Figure 3 summarizes the different process generations developed by Vitesse and their performance.

## Vitesse	## Traditional GaAs
Tungsten gates, high temperature	Gold gates, low temperature
Self aligned	Not possible
Tight control on process	Large variations
Aluminimum interconnect	Gold interconnect
DCFL low power	High power, logic families

Figure 2

	Lg	f_T	Metal Layers	Delay-Power	Year
H-GaAs 1	1.2μm	15 GHz	2	20pJ	1987
H-GaAs 2	0.8μm	22 GHz	3	10pJ	1989
H-GaAs 3	0.6μm	28 GHz	4	4pJ	1991
H-GaAs 4	0.5μm	38 GHz	4	1-2pJ	1995

Figure 3

Moreover, modifications to the process are possible that enable the monolithic integration of optoelectronic devices through the use of selective epitaxial growth of MBE or MOCVD to fabricate optical emitters. Since the basic process is extremely temperature tolerant, epitaxial growth is typically done after all the electronic circuits are completed. Initial product performance of integrated GaAs MSM (Metal-Semiconductor-Metal) optical receivers will be presented.

3. Applications

One early and exciting application for GaAs VLSI technology was to try to fabricate a 500 Mhz GaAs microprocessor. This application was an early driver to improve the technology to implement circuits with greater than one million transistors. Results on integer and floating point test chips indicated that a factor of 2.5 to 3 improvement in clock speed could be achieved. Unfortunately, other architectural issues such as the size and speed of the cache memory created serious technical difficulties that could not be overcome. In addition, the cost associated with a custom processor design in GaAs is comparable to that of a CMOS processor and the economics dictated that CMOS designs would always provide a greater return on invested design effort. Examples of million transistor processor circuits and measured and simulated results are presented.

Other examples of current applications [2] that are in volume production include SONET/SDH (2.4 GB/s and 622 MB/s) transmission and switching circuits and fibre channel (1 GB/s) disk drive transceiver functions. Ten GB/s SONET transmission products under development with a 0.5 micron MESFET process (H-GaAs IV) are also discussed.

4. Technology Assessment

A comparative assessment of CMOS, BiCMOS, ECL , Ion Implanted, and Epitaxial GaAs technologies will be discussed both for current technologies (0.5-0.6 micron) and changes that will occur as gate lengths scale to 0.2 micron. The comparison will focus on the tradeoffs of performance, manufacturing cost and price /performance. Based upon these results, projections for future growth areas for GaAs will be presented.

References

[1]   Mikkelson J. M. *IEDM Technical Digest*, December 1991, p. 9.1.
[2]   Deyhimy I. .1995 *IEEE Spectrum*, **32** 33-40

Inst. Phys. Conf. Ser. No 145: Chapter 2
Paper presented at 22nd Int. Symp. Compound Semiconductors, Cheju Island, Korea, 28 August–2 September 1995
© *1996 IOP Publishing Ltd*

Cubic phase GaN: Correlation between growth kinetics and material quality

O Brandt[1], H Yang, M Ramsteiner, J Menniger and K H Ploog

Paul-Drude-Institut für Festkörperelektronik, Hausvogteiplatz 5–7, D-10117 Berlin, Germany

Abstract. We study the crystal quality of cubic GaN grown on GaAs by plasma-assisted molecular beam epitaxy and its dependence on the growth conditions. Even if epitaxial growth is established, the resulting layer may include significant volume fractions of the hexagonal phase. We find that the major parameter controlling the phase purity of cubic GaN films is the surface stoichiometry during growth. We present a method which allows the real-time monitoring of the surface stoichiometry and which thus enables us to achieve single-phase cubic GaN films.

1. Introduction

The growing interest in GaN and related compounds is based on their importance for opto-electronic devices emitting in the blue and UV spectral range as well as for high-temperature and high-power electronic devices (Strite and Morkoç 1992). The large lattice mismatch between group-III nitrides and the available substrate materials as well as their extreme hardness makes the growth challenging not only at the nucleation stage but also during the following growth. Compared to the hexagonal phase, the cubic modification of the nitrides has some potential advantages for device applications, such as easy cleavage with respect to the substrate, higher carrier mobility, and a substantially smaller bandgap. Molecular beam epitaxy (MBE) not only allows, in principle, the synthesis of the metastable cubic phase (Cheng *et al* 1995, Lei *et al* 1993, Liu *et al* 1993, Paisley *et al* 1989, Powell *et al* 1993, Strite *et al* 1991), but also the application of *in situ* diagnostic methods for investigating the growth in real time.

In this paper, we investigate the conditions required for the synthesis of epitaxial cubic GaN films on GaAs(001). We find that growth has to be carried out with a V/III ratio close to one to avoid the formation of hexagonal domains within the cubic matrix. Monitoring the surface reconstructions during growth provides a means for the real-time control of the surface stoichiometry and thus enables one to synthesize single-phase cubic GaN films.

---

[1] Corresponding author. E-mail: brandt@pdi.wias-berlin.de

2. Experimental

Growth of GaN films is carried out on (001)-oriented (offcut <0.1°), semi-insulating GaAs substrates in a custom-designed solid-source MBE chamber. Active N is generated by a high voltage ($\simeq 1.5$ keV) plasma glow-discharge. The plasma power is kept constant for all experiments at 30 W. The surface of the growing crystal is monitored *in situ* by reflection high-energy electron diffraction (RHEED), using an incident angle between 1° and 2° and an acceleration voltage of 15 kV. RHEED patterns are recorded by a CCD camera connected to an image processing system. The Ga flux is determined by RHEED oscillations during GaAs growth under N_2 background pressure, while the N flux is determined by evaluating the growth rate via the layer thickness as measured by scanning electron microscopy (SEM) for films grown under Ga rich conditions. X-ray measurements are performed using a double-crystal diffractometer equipped with a $Cu_{K\alpha 1}$ anode and a Ge(004) monochromator. The measurements are taken either with a wide open detector [x-ray rocking curve (XRC)] or with a 50 μm detector slit (ω scan). Depolarized Raman spectra are recorded in backscattering configuration with the samples cooled to 80 K using the 2.41 eV line of an Ar^+ ion laser for excitation. Cathodoluminescence (CL) spectra at 5 K are recorded by using a scanning electron microscope equipped with an Oxford mono-CL. Electron beam energy and current are set to 8 keV and 1 nA, respectively.

3. Results and Discussion

First, we briefly comment on the question under which conditions epitaxial growth of GaN on GaAs is actually achieved. We found (Yang *et al* 95) that epitaxy is obtained by using a high V/III ratio during the nucleation stage, whereas a low V/III ratio invariably results in polycrystalline columnar growth. In the following, we focus on results obtained on epitaxial GaN/GaAs films with thicknesses ranging between 100 and 1400 nm.

Figure 1 displays the linewidth (a) and integrated intensity (b) of the (002) x-ray reflection profile versus the thickness of epitaxial GaN films. The reflection linewidth in XRC includes contributions from mosaicity, lateral and vertical domain size, and inhomogeneous strain (Lei *et al* 1993). The two latter contributions are absent for measurements taken in the ω mode. Our best linewidth and reflectivity data are apparently well described by a power law with an exponent of about −0.5 for the former and 1 for the latter. In Fig. 1 (a), we have also included data from cubic GaN films fabricated by other groups. Interestingly, the narrowest linewidths obtained today are quite comparable to each other relative to the films' thickness, regardless of (i) the different plasma discharge mechanisms employed and (ii) of the different substrates whose lattice mismatch to GaN varies between 20% (GaAs) and -6.7% (MgO) with the closest match (-3.2%) provided by SiC. This finding suggests that the growth of GaN itself determines the film's structural quality. Some of our samples exhibit a significantly wider linewidth and, simultaneously, lower reflectivity than those represented by the solid lines which define the current state-of-the-art. In fact, all these samples were grown under N excess and, as revealed by x-ray reciprocal space maps (Brandt *et al* 1995), incorporate a significant volume fraction of hexagonal domains. The lower reflectivity may thus just arise from an effectively smaller volume of the cubic phase with respect to the nominal

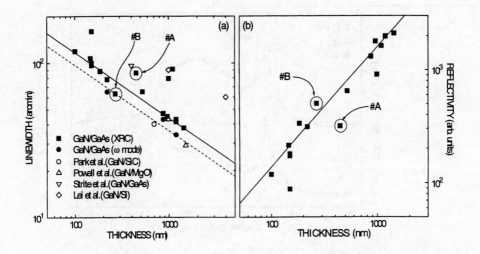

Figure 1. Linewidth and reflectivity of the (002) reflection profile of GaN films with different thickness. Solid symbols indicate our data and open symbols those obtained by other groups. The solid and dashed lines represent the trend of the best data with films' thickness. The two samples highlighted in the figure are discussed further in the text.

one derived from the overall thickness of the film. This effect might provide a convinient means for determining the phase purity of the GaN film. This conjecture is examined in the following by comparing two representative samples (#A and #B) highlighted in Fig. 1 by Raman and CL spectroscopy.

Figure 2 shows the Raman spectra of samples #A and #B. Both samples exhibit the LO and TO modes characteristic for cubic GaN. The LO mode is absent for polarized spectra as expected from the selection rules for the (001) oriented zincblende structure. The broad band inbetween these modes, which is clearly stronger for sample #A, arises from disorder-activated scattering. Most important, however, is the presence of an additional resonance in the case of sample #A. This resonance occurs at the spectral position of the the E_2 (high) mode of hexagonal GaN (Azuhata *et al* 1995), and its presencence is thus a proof of the existence of hexagonal domains in the cubic GaN matrix.

Figure 3 shows the CL spectra of #A and #B. The spectra of both samples are dominated by luminescence lines between 3.0 and 3.3 eV, i. .e., just beneath the bandedge of cubic GaN (Ramírez-Flores *et al* 1995). However, both samples also show luminescence lines above the cubic GaN bandedge. This contribution is significant in the case of sample #A, as expected from the results shown in Fig. 2. An actual surprise is the presence of luminescence associated to the hexagonal phase in the case of sample #B, for which hexagonal components were detected neither by x-ray nor by Raman measurements. This apparent inconsistency is resolved by SEM micrographs: sample #B, grown under slightly Ga rich conditions, exhibit large ($\simeq 1\mu m$) hexagonal crystallites grown underneath Ga droplets formed on the surface. Though these crystallites constitute a

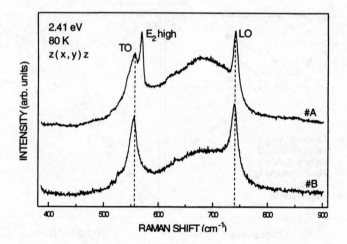

Figure 2. Depolarized Raman spectra of samples #A and #B.

negligible volume fraction compared to that of the cubic component, they can give rise to a strong luminescence signal because of their size which prohibits carrier relaxation towards the smaller gap cubic matrix.

It is clear from the above that the formation of hexagonal grains, be it within the bulk (sample #A) or on the surface (sample #B), is one of the prime problems we are confronted with when we attempt to fabricate high-quality cubic GaN. The highest volume fraction of hexagonal components is found for samples grown under N rich conditions. However, also Ga rich conditions ultimately lead to the growth of hexagonal GaN crystallites within Ga droplets on the surface. It thus seems that a careful control of the V/III ratio during growth is required.

The basis for such an achievement has been decribed in a previous work of us (Brandt *et al* 1995) where we have identified the major surface reconstructions of cubic GaN to be of (2×2) and c(2×2) symmetry with corresponding surface coverages of 0.5 and 1 ML Ga, respectively. Based on this understanding, the reconstruction transitions are used to probe the surface stoichiometry during GaN growth, i.e., with a permanently supplied flux of active N. Figure 4 shows RHEED intensity recordings of the half-order reconstruction streak along the [$\bar{1}$10] azimuth. Both upon opening and closing the Ga shutter, the RHEED intensity exhibits a peak-like behavior. Starting from an initially N terminated (1×1) surface, the maximum intensity for either peak is related to the formation of the (2×2) reconstruction at half-monolayer coverage. The intensity decrease after the first (Ga related) peak corresponds to an increase of the surface Ga coverage and the subsequent formation of the c(2×2) reconstruction. We point out that this decrease is sensitively dependent on the substrate temperature because of the comparatively low activation energy for thermal desorption of Ga above half monolayer coverage. The second (N related) peak arises from the consumption of Ga until a N terminated surface is established. Assuming an exponential decay of the Ga coverage θ upon supply of the effective N flux j_N, i.e., $d\theta/dt = -(j_N/N_S)\theta$, where N_S is the surface site density,

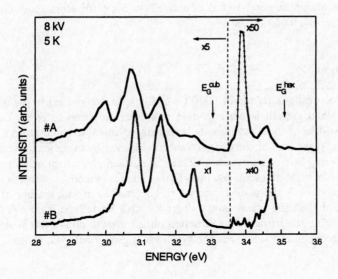

Figure 3. CL spectra of samples #A and #B.

we can thus determine j_N by simply measuring the time in which the RHEED intensity (assumed to be proportional to θ^2) drops to $1/e$ of its maximum value. In the present

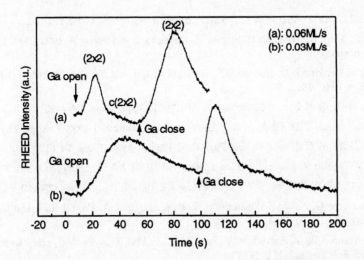

Figure 4. Intensity of the half-order reconstruction streak as a function of time. Traces (a) and (b) differ in the Ga flux used. Opening and closing of the Ga shutter is indicated by arrows.

case, we thus obtain an effective flux of active N of 5.4×10^{13} atoms/cm^2, corresponding to a growth rate of 0.055 ML/s.

4. Conclusion

The best structural quality of our GaN/GaAs films, as determined by x-ray diffraction, is obtained when growth is performed at a V/III ratio close to one. The reflection profiles' linewidths of these samples are essentially identical to the best data reported in the literature, regardless of the different substrates used by other groups corresponding to vastly differing lattice mismatches. Those of our samples which exhibit significantly larger linewidths exhibit volume fractions of the hexagonal modifications. It seems, thus, that phase mixture is the main factor which governs the crystal quality of cubic GaN films. We find that the phase purity degrades with deviations from a V/III ratio of unity. Monitoring the surface reconstruction during growth allows the *in situ* control of the surface stoichiometry, and thus the synthesis of purely cubic GaN films.

Acknowledgement

We acknowledge financial support of this work by the Bundesministerium für Bildung und Forschung der Bundesrepublik Deutschland.

References

Azuhata T, Sota T, Suzuki K and Nakamura S 1995 J. Phys. C **7** L129

Brandt O, Yang H, Jenichen B, Suzuki Y, Däweritz L and Ploog K 1995 Phys. Rev. B (to be published)

Cheng T S, Jenkins L C, Hooper S E, Foxon C T, Orton J W, and Lacklison D E 1995 Appl. Phys. Lett. **66**, 1509

Lei T, Ludwig Jr K F, and Moustakas T D 1993 J. Appl. Phys. **74** 4430

Lei T, Moustakas T D, Graham R J, He Y, and Berkowitz S J 1992 J. Appl. Phys. **71** 4933

Liu H, Vrenkel A C, Kim J G, and Park R M 1993 J. Appl. Phys. **74** 6124

Paisley M J, Sitar Z, Posthill J B, and Davis R F 1989 J. Vac. Sci. Technol. A **7** 701

Powell R C, Lee N-E, Kim Y-W, and Greene J E 1993 J. Appl. Phys. **73** 189

Ramírez-Flores G, Navarro-Contreras H, Lastras-Martínez A, Powell R C, and Greene J 1994 Phys. Rev B **50** 8433

Strite S, Ruan J, Li Z, Salvador A, Chen H, Smith D J, Choyke W J, and Morkoç H 1991 J. Vac. Sci. Technol. B **9** 1924.

Strite S, and Morkoç H 1992 J. Vac. Sci. Technol. B **10** 1237

Yang H, Brandt O and Ploog K H 1995 *5th Int. Conf. on the Formation of Semiconductor Interfaces, Princeton (USA)*

Inst. Phys. Conf. Ser. No 145: Chapter 2
Paper presented at 22nd Int. Symp. Compound Semiconductors, Cheju Island, Korea, 28 August–2 September 1995
© *1996 IOP Publishing Ltd*

Molecular Beam Epitaxy with Gaseous Sources: Growth and Applications of Mixed Group V Compounds and Selective-Area Growth

C.W. Tu, X.B. Mei, N.Y. Li, and H.K. Dong

Department of Electrical and Computer Engineering, University of California, San Diego, La Jolla, CA 92093-0407, U.S.A.

Abstract. Molecular beam epitaxy (MBE) with gaseous sources has the advantages of growing mixed arsenide-phosphide compounds and selective-area growth by laser irradiation or on patterned substrates. We shall first describe the growth of strained $InAs_xP_{1-x}/InP$ and strain-compensated $InAs_xP_{1-x}/Ga_yIn_{1-y}P$ multiple quantum wells (MQWs) using elemental group III sources and thermally cracked arsine and phosphine and their electro-absorption properties for light modulation at 1.3 μm or beyond. We then report on the use of all gaseous sources for selective-area growth, doping, and composition modification by argon ion laser irradiation and for selective-area regrowth of a carbon-doped AlGaAs/GaAs external base for improving the performance of heterojunction bipolar transistors (HBTs).

1. Introduction

Molecular beam epitaxy (MBE), with solid sources, has been proven to be a versatile thin-film growth technique for research, development, and production of semiconductor materials. Most of its applications involve various heterostructures of arsenides. Because of the high vapor pressure of phosphorus, it is difficult to grow heterostructures with arsenide in one layer and phosphide in the adjacent layer or with mixed group V compounds. Before the recent development of valved crackers for solid arsenic and phosphorus sources[1,2,3], mixed group V compounds have been grown with gas-source MBE, where arsenic and phosphorus dimers are derived from cracked arsine and phosphine, respectively. Gas-source MBE thus extends the capability of solid-source MBE to be as flexible as organometallic vapor phase epitaxy (OMVPE). When the group III sources are also in gaseous form, selective-area growth (SAG) by an external energy source, e.g., an electron beam or a laser beam, or SAG on insulator-patterned substrates, where growth occurs only in the openings, can be achieved. SAG can improve device performance or realize novel devices. We shall first describe the growth of strained $InAs_xP_{1-x}/InP$ and strain-compensated $InAs_xP_{1-x}/Ga_yIn_{1-y}P$ multiple quantum wells (MQWs) using elemental group III and doping sources and thermally cracked arsine and phosphine. These MQWs can be used for light modulation at 1.3 μm and even near 1.5 μm. When the group III sources are in gaseous forms, laser-assisted slective-area growth and doping, and issues of SAG on a patterned substrate, such as selectivity and sidewall facets, are discussed and SAG of an external base is applied to heterojunction bipolar transistors (HBTs).

2. A mixed group V compound: InAsP

Previously we have shown that $InAs_xP_{1-x}$ is a viable alternative quantum well material to the quaternary InGaAsP and InGaAlAs for light modulators operating at 1.3 μm [4] because of the independent control of thickness and composition [5] and reduced alloy scattering. Compressively strained InAsP/InGaAsP quantum wells have also become important for 1.3 μm laser applications [6,7]. Because As incorporates much more efficiently than P in InAsP, the As composition can be controlled easily by the As/In incorporation ratio when that ratio is less than about 0.5 [8].

In order to obtain a high contrast ratio in an $InAs_{0.4}P_{0.6}$/InP MQW modulator (wavelength near 1.3 μm), a relatively large number of periods is desirable. Since the $InAs_{0.4}P_{0.6}$ is in compression (strain of 1.3%), the whole MQW stack has a large net strain, which limits the number of periods without strain relaxation [9]. To solve this problem, $Ga_yIn_{1-y}P$, in tension with respect to InP, is employed as the barrier material, so the net strain of the MQW can be reduced and a stable structure with a large number of periods can be achieved. Moreover, the barrier strain introduces a new degree of freedom for structure design. Very sharp satellite peaks in the double-crystal X-ray rocking curves (XRC) and narrow peaks in low-temperature photoluminescence are obtained from $InAs_xP_{1-x}/Ga_yIn_{1-y}P$ MQWs of up to 50 periods. Fig. 1 shows (a) an XRC of an 11-period strain-uncompensated $InAs_{0.4}P_{0.6}$ (90 Å)/InP (130 Å) MQW structure, which clearly indicates lattice relaxation, and (b) an XRC of a 30-period strain-compensated $InAs_{0.4}P_{0.6}$ (86 Å)/$Ga_{0.17}In_{0.83}P$ (120 Å) MQW structure, which exhibits sharp superlattice satellite peaks, indicating good structural integrity and vertical periodicity.

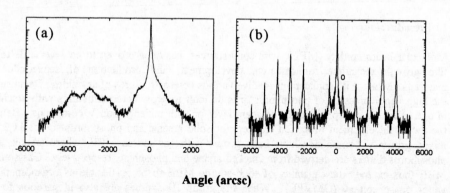

Angle (arcsc)

Fig. 1 X-ray rocking curves of (a) a strain-uncompensated $InAs_{0.4}P_{0.6}$ (90 Å)/InP (130 Å) MQW and (b) a strain-compensated $InAs_{0.4}P_{0.6}$ (86 Å)/$Ga_{0.17}In_{0.83}P$ (120 Å) MQW.

With strain compensation the strain in the quantum well layers can be even larger (>2%). Fig. 2 shows a cross-sectional transmission electron micrograph (XTEM) of a 20-period strain-compensated $InAs_{0.54}P_{0.46}$ (100 Å)/$Ga_{0.17}In_{0.83}P$ (140 Å) MQW structure. No dislocations are observed in the MQW active region. The XRC of this structure is of the same quality as that of Fig. 1(b).

Fig. 2 Cross-sectional TEM photograph of a strain-compensated 20-period
InAs$_{0.54}$P$_{0.46}$ (100 Å)/Ga$_{0.17}$In$_{0.83}$P (140 Å) MQW region.

P-i-n structures with the *i*-region consisting of strain-compensated MQWs can exhibit a large quantum-confined Stark effect near 1.3 μm [1] or 1.5 μm [10].

3. Selective-area CBE growth by laser irradiation

Using all gaseous sources in metal-organic MBE (MOMBE) or chemical beam epitaxy (CBE) has additional advantages of low thermal budget, replacing sources without opening the growth chamber, and SAG by argon ion laser irradiation or on oxide- or nitride-patterned substrates. We have also investigated the use of safer, alternative precursors to arsine and phosphine, in particular, tris-dimethylamino arsenic (TDMAAs) [11]. TDMAAs has no As-H nor As-C bonds, only As-N bonds. GaAs samples were grown with triethylgallium (TEGa) and TDMAAs, and p-type carbon doping is achieved with CBr$_4$ or Cl$_2$H$_2$ [12].

Fig. 3 shows the growth rate of GaAs as a function of substrate temperature, with and without irradiation from a multiline argon ion laser. Because of the long distance between the substrate and the viewport in a CBE/MBE growth chamber, the laser spot was focused to about 300 μm. The laser power density used was about 400 W/cm^2. The growth rate data are normalized to that at the high-temperature region (between 480 °C and 510 °C), where TEGa decomposition is complete. The substantial differences between laser-assisted growth using TDMAAs and AsH$_3$ [13]or As$_4$ [14] are that the threshold temperature for enhanced growth is about 100 °C lower and that the temperature range is about 100 °C wider. Furthermore, there are two different substrate-temperature regions for laser-enhanced growth, 340 to 440 °C and 265 to 340 °C [15]. The former is believed to be due to laser-induced heating of about 20-25 °C because the activation energy in this temperature range is similar to that without laser irradiation, whereas the latter indicates the effect of photochemical reaction at the surface.

Fig. 4 shows the hole concentration as a function of growth temperature, with and without laser irradiation. Carbon incorporation decreases as the substrate temperature increases because of the increased production of atomic hydrogen through the β-hydride elimination process of TDMAAs on a GaAs surface [16]. With laser irradiation, the hole

42

concentration increases, despite a laser-induced temperature rise, indicating photochemical effects. It is interesting to note that there can be two orders of magnitude difference in the hole concentration with and without laser irradiation, which may have potential applications.

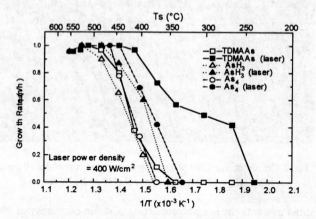

Fig. 3 The substrate temperature dependence of growth rates of boath laser-irradiated and nonirradiated areas. The solid lines denote the growth data obtained using TDMAAs, while the dashed and dotted lines denote the data from AsH₃ and As₄, respectively.

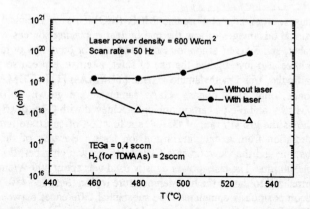

Fig. 4 Effect of Ar⁺ laser irradiation on the hole concentration vs. substrate temperature.

When laser is irradiated during growth of InGaAs in a pseudomorphic $In_xGa_{1-x}As$/GaAs quantum well structure at a normal substrate temperature (~500 °C), TEGa molecules apparently are desorbed, resulting in selective-area composition modification of the $In_xGa_{1-x}As$ quantum well layer. There is therefore a lateral variation of exciton emission wavelength [17]. Yamada et al. have used this kind of behavior to fabricate a dual-wavelength InP/InGaAsP laser diode [18].

4. Selective-area regrowth on patterned substrates

Many devices can be improved by selective-area regrowth on etched patterned substrates, e.g., as a current and light confining area in laser diodes or as an external base layer in HBTs. With TEGa and TDMAAs, true selective growth of GaAs:C on SiO_2-patterned GaAs, i.e., growth only on the openings, can be achieved above 470 °C, which is much lower than the 600 °C in the case of using TEGa and arsine [19]. With uncracked TDMAAs, the surface morphology is always textured and the carbon incorporation efficiency is low. Thermally cracking TDMAAs can produce a specular surface morphology and a much higher carbon incorporation efficiency. In Fig. 5 (a) the growth temperature is too low such that there is polycrystalline growth on the SiO_2 surface, and in Fig. 5 (b) true selectivity is achieved at 540 °C with good surface morphology.

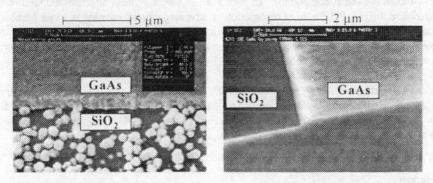

Fig. 5 Scanning electron micrographs of GaAs SAG. (a) Using uncracked TDMAAs at a substrate temperature of 440 °C; (b) Using cracked TDMAAs at 540 °C.

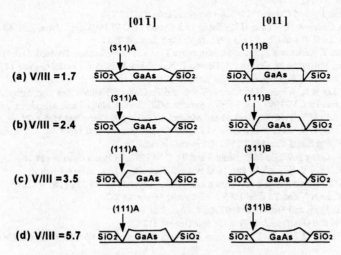

Fig. 6 Schematic cross section of GaAs grown at V/III ratios of (a) 1.7, (b) 2.4, (c) 3.5, (d) 5.7 along [01$\bar{1}$] and [011].

The regrown profile can be adjusted by changing the V/III ratio, or the surface migration length of adsorbed molecules, as shown in Fig. 6. To fill in the undercut area in our case, which is produced by wet etching after reactive-ion etching, so that the external base layer has a good contact with the intrinsic base region, a regrowth profile with (311) facets is used. Because the external base is thicker, the resulting HBTs have lower base sheet resistance [20]. Because the external base layer also passivates the sidewall of the internal base, the base recombination current is also reduced, resulting in higher current gain.

5. Summary

Using cracked arsine and phosphine as the As_2 and P_2 source, respectively, allows us to grow mixed As and P compounds. In particular, InAsP is a viable alternative quantum well material to InGaAsP. Strain compensation is advantageous in growing a large number of strained quantum wells. When the sources are all in gaseous forms, they bring out new possibilities for device design, e.g., lateral change in doping and composition, and selective-area regrowth on patterned substrates for HBT applications.

Acknowledgment: This work is partially supported by the ARPA Optoelectronic Techology Center and US Air Force Wright Laboratory. We wish to thank W.G. Bi, C.H. Yan, W.S. Wong, Y.M. Hsin, T. Nakamura, P.M. Asbeck, L. R. Chou, and K.C. Hsieh for their help and collaborations.

References

[1] Miller D L, Bose, S S and Sullivan G J 1990 *J. Vac. Sci. Technol. B* **8** 311
[2] Wicks G W, Koch M W, Varriano J A, Johnson F G, Wie C R, Kim H M and Colombo P 1991 *Appl. Phys. Lett.* **59** 342
[3] Baillargeon J N and Cho A Y 1995 *J. Vac. Sci. Technol. B* **13** 736
[4] Hou H Q, Cheng A N, Wieder H H, Chang W S C, and Tu C W 1993 *Appl. Phys. Lett.* **63** 1833
[5] Hou H Q, Tu C W and Chu S N G 1991 *Appl. Phys. Lett.* **58** 2954
[6] Fukushima T, Kasukawa A, Iwase M, Namegaya T, et al. *IEEE Photonic Technol. Lett.* 1993 **5** 117
[7] Sugiura H, Mitsuhara M, Oohashi A, Hirono T, Nakashima K 1995 *J. Crystal Growth* **147** 1
[8] Hou H Q and Tu C W 1992 *Appl. Phys. Lett.* **60** 1872
[9] Mei X B, Loi K K, Wieder H H, Chang W S C and Tu C W 1995 *Mater. Res. Soc. Symp.* **379**
[10] Mei X B and Tu C W 1995 *15th North American MBE Conf., College Park, Maryland, 18-20 Oct.*
[11] Dong H K, Li N Y, Tu C W, Geva M and Mitchel, W C 1995 *J. Electronic Mater.* **24** 69
[12] Li N Y, Dong H K, Tu C W and Geva M *J.* 1995 *Crystal Growth* **150** 245
[13] Sugiura H, Iga R and Yamada T 1992 *J. Crystal Growth* **120** 389
[14] Dong H K, Liang B W, Ho M C, Hung S and Tu C W 1992 *J. Crystal Growth* **124** 181
[15] Dong H K, Li N Y, Wong S W and Tu C W 1995 *Appl. Phys. Lett.*, to be published.
[16] Bohling D A, Abernathy C R and Jensen K F 1994 *J. Crystal Growth* **136** 118
[17] Dong H K, Li N Y and Tu C W 1995 *J. Electronic Mater.* **24** 827
[18] Yamada T, Iga R and Sugiura H 1992 *Appl. Phys. Lett.* **61** 2449
[19] Furuhata N and Okamoto A 1991 *J. Crystal Growth* **112** 1
[20] Li N Y, Hsin Y M, Dong H K, Nakamura T, Asbeck P M and Tu C W 1995 *J. Crystal Growth* **150** 562

Inst. Phys. Conf. Ser. No 145: Chapter 2
Paper presented at 22nd Int. Symp. Compound Semiconductors, Cheju Island, Korea, 28 August–2 September 1995
© *1996 IOP Publishing Ltd*

The dependence on growth temperatures of the electrical and structural properties of GaSb/InAs single quantum well structures grown by MBE

S.J. Chung, A.G. Norman*, W.T. Yuen, T.Malik and R.A. Stradling

Blackett Laboratory, Imperial College, Prince Consort Road, London SW7 2BZ
*Now at Materials Department, Oxford University, Parks Road, Oxford OX1 3PH,UK

Abstract. We have investigated a series of GaSb/InAs/GaSb single quantum well structures grown at several different temperatures with either 'InSb like' interfaces or 'GaAs-like' interfaces. During the growth, a minimum As overpressure was employed together with pause times of 30s. For all four samples grown at 400°C with 'GaAs-like' interfaces the 77K mobilities were as high as 95,000~115,000 cm²/Vs (4K mobilities ~200,000 cm²/Vs) indicating that As incorporation is not a problem at this growth temperature. These mobilities are even higher than those of samples with 'InSb-like' interfaces grown at the same temperature. The optimum growth temperature for 'InSb-like' interfaces appears to be 450°C and the 77K mobilities for these samples are no better than the mobilities found with samples grown at 400°C with 'GaAs-like' interfaces. We present a study of these samples using HRTEM with a view to understand the As-incorporation and their interfacial structures.

1. Introduction

The GaSb/InAs system has attracted attention in recent years because of possible infrared applications as well as the unique physical properties arising from the semimetallic type-II band alignment at the interface where the InAs conduction band minimum lies below the GaSb valence band edge.[1,2] The GaSb/InAs system, as well as the related AlSb/InAs system, is also of interest because the materials have neither common cations nor anions. As a result either 'InSb-like' interface bonding or 'GaAs-like' interface bonding can be promoted by careful control of the switching sequences. Without such control the interface bonding will be random. The bonding type at the interface can have a crucial effect on the physical properties of the system. In the case of AlSb/InAs system, it has been reported[3] that single quantum wells with 'InSb-like' interfaces can have much higher electron mobility and lower electron concentration than those with 'AlAs-like' interfaces.

In the case of both the GaSb/InAs and AlSb/InAs systems, there have been several reports of the structural properties[4,5,6] which indicate that roughening of the interface can be induced by the incorporation of excess As into the GaSb during the growth of a 'GaAs-like' interface. It was also claimed that the use of an As cracker cell could minimise the As incorporation into the GaSb layer and produce higher mobility samples.[7] The electrical properties of single quantum well structures with different types of interfaces should thus reflect the degree of interface roughness. However, the As-incorporation rate should depend both on growth temperature and the pause time employed during the growth of the interfaces.

2. Experiments

We have grown a series of single quantum well structures at several different temperatures with either 'InSb interfaces or 'GaAs-like' interfaces. All samples were grown on semi-insulating GaAs [001] substrates using a Vacuum Generators Semicon V80H MBE system. A GaAs buffer layer of 8000Å thickness was grown on the substrate at 580ºC, after desorbing the oxide layer. Substrate temperatures were measured with respect to the oxide desorbing temperature of the GaAs substrate which is known to be 590ºC. After growing the GaAs buffer layer, substrate temperature was reduced to the required growth temperature and 8000Å of GaSb bottom barrier , 200Å of InAs and 200Å of GaSb cap layer were grown in sequence. At each quantum well interface the appropriate shutter sequence to make 'InSb-like' interface bonding or 'GaAs-like' interface bonding was used. For all samples, the pause time under the group V elements (As for 'GaAs-like' interface and Sb for 'InSb-like' interface) at each interface was fixed at 30 seconds. During the growth of the InAs layer, the minimum As overpressure was maintained.

Low magnetic field Hall measurements in the temperature range 77K to 300K were performed to obtain electron mobilities and sheet carrier concentrations of the samples. Conventional Van der Pauw clover-leaf patterns were prepared. After this, four In contact points were made by annealing at 200ºC for 90 seconds in an atmosphere of hydrogen and nitrogen. 4K experiments were also performed to obtain low temperature data using a superconducting magnet in liquid He

<100> cross-sectional high resolution transmission electron microscopy (HRTEM) specimens were prepared using Ar$^+$ ion beam milling with the specimens cooled to liquid nitrogen temperature. The specimens were examined in a JEOL 4000EX operated at 300kV to minimise damage of the HRTEM specimens by high energy electrons. The Sherzer resolution was less than 0.2nm. Lattice images were obtained using an objective aperture which just included the {220}, {200} and (000) beams resulting in {200} and {220} fringes being present in the images.

3. Results and discussions

The bonding type of the interfaces has a crucial effect on the electron mobility. It has been claimed that structures with 'GaAs-like' interface[8] or 'AlAs-like' interface[3] in the case of InAs/AlSb quantum wells have substantially lower mobilities than those with InSb interfaces. During the growth of the first 'GaAs-like' interface the GaSb surface is exposed to an As flux and As incorporates into the GaSb layer. On the other hand the GaSb surface is exposed to an Sb-flux during the growth of 'InSb-like' interfaces. As a result of As incorporation into the GaSb layer, the interface becomes rougher than the 'InSb-like' interface which is not exposed to an As flux. However, As incorporation into the GaSb layer is dependent on the growth temperature[9] as well as the value of the As flux incident upon the GaSb surface. This means that the As incorporation into the GaSb layer can be controlled by varying the growth temperature and the amount of As flux. During the growth of quantum well structures, we have grown a series of samples at different growth temperatures but employing the minimum

Table-1. Electron mobilities and concentrations of GaSb/InAs/GaSb single quantum well samples

Samples	Interface type	Growth temperature(°C)	μ at 77K (cm^2/Vs)	n at 77K (x10^{12}cm^{-2})	n at 4K (x10^{12}cm^{-2})
IC442	GaAs	375	71100	1.10	-
IC405	GaAs	400	96900	1.19	0.82
IC419	GaAs	400	94100	1.03	0.83
IC469	GaAs	400	113400	0.96	0.72
IC446	GaAs	425	80400	0.96	0.68
IC440	GaAs	450	52900	1.08	-
IC462	InSb	400	68000	1.07	-
IC404	InSb	450	107000	1.50	0.99
IC454	InSb	450	97000	1.20	-
IC393	InSb	490	54000	1.08	0.80

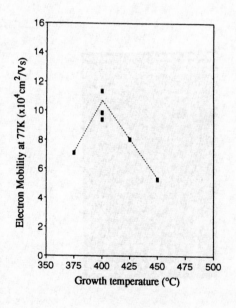

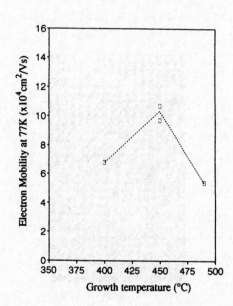

a) 'GaAs-like' interface samples   b) 'InSb-like' interface samples

Fig.1 Mobilities vs growth temperatures of two different types of interface samples

a) IC419 grown at 400 °C

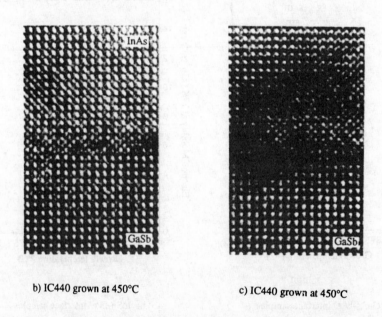

b) IC440 grown at 450°C

c) IC440 grown at 450°C

Fig.2 <100> cross-sectional HRTEM pictures of single quantum well structures
        with 'GaAs-like' interfaces

As flux which can maintain an As overpressure at the InAs growth surface. The electrical properties of those samples are given in Table-1. The 77K mobility of the 'GaAs-like' interface sample grown at 450°C is 52,900 cm^2/Vs while these of the 'InSb-like' interface samples grown at the same temperature are twice as high. We believe that this difference arises from interface roughening due to the As incorporation at the interface. On the other hand, for 'GaAs-like' interface samples grown at 400°C the 77K mobilities are as high as 94,100~113,400 cm^2/Vs while the 77K mobility of 'InSb-like' interface sample is only 68,000 cm^2/Vs. Fig.1 shows 77K electron mobilities of both types of interface samples as a function of the growth temperature. This implies that the effects of As incorporation into the GaSb is much reduced by decreasing the growth temperature. Even without the use of an As cracker cell, we obtained extremely high 4K mobilities from the 'GaAs-like' interface samples grown at 400°C and the 'InSb-like' interface samples grown at 450°C which can be as high as 200,000 cm^2/Vs for both types of interface samples. In Fig.2, (a) is shown a HRTEM picture of the lower interface of an InAs well with 'GaAs-like' interfaces grown at 400°C(IC419). The interface shows virtually no interfacial roughening and little evidence of any regions of strain contrast at the interface. In contrast, (b) and (c) of Fig.2 show HRTEM micrographs of the lower interface of an InAs well with 'GaAs-like' interfaces grown at 450°C(IC440). In this sample, some areas of the interface appear good with interfacial roughness of ~ 1 monolayer and relatively free from regions of strain contrast as shown in Fig.2 (b). Other areas, such as Fig.2 (c), show regions of strain contrast at the interface probably caused by incorporation of excess As into the GaSb layer during the 30s growth pause used at this interface. These areas of strain and slightly higher interfacial roughness observed for this sample probably account for the lower electron mobilities obtained in the 'GaAs-like' interface samples grown at higher temperature(450°C) in comparision to the 'GaAs-like' interface samples grown at 400°C. The upper interfaces of the wells were similar in quality in both samples and showed interfacial roughness and interdiffusion over approximately ~ 2 monolayers.

Considering the highest mobility samples of both types of interface, the low-temperature electron concentrations were consistently lower for the samples with 'GaAs-like' interfaces.(Table-1) This implies that the GaAs interface bonds reduce the overlap between the InAs conduction band and the GaSb valence band[10,11].

4. Conclusions

We have investigated a series of GaSb/InAs/GaSb single quantum well structures grown at several different temperatures with either 'InSb-like' interfaces or 'GaAs-like' interfaces. By reducing the growth temperature, we have achieved electron mobilities with samples having 'GaAs-like' interfaces as high as those of 'InSb-like' interface samples. HRTEM results also show that the 'GaAs-like' interface grown at lower temperature is less rough and less strained. The mobilities which we achieved without a cracker cell for As demonstrate that the As cracker cell is not necessary to obtain high mobility in this structure. From our results, we propose that the dominant factor which determines the electrical properties of this structure is not simply the bonding type at the interface but the amount of As incorporation into the GaSb layer which can be controlled by reducing the growth temperature.

50

Acknowledgement

The work was supplied by the Engineering and Physical Sciences Research Council(EPSRC).

References

[1] Chang L L and Esaki L 1980 Surface Science **98** 70 -89

[2] Chang L L, Kawai N, Sai-Halasz G A, Ludeke R and Esaki L 1979 Appl. Phys. Lett. **35** 939-942

[3] Tuttle G, Kroemer H and English J H 1990 J. Appl. Phys. **67** 3032-3037

[4] Spitzer J, Fuchs H D, Ethegoin P, Ilg M, Cardona M, Brar B and Kroemer H 1993 Appl. Phys. Lett. **62** 2274-2276

[5] Wagner J, Schmitz J, Maier M, Ralston J D and Koidl P 1993 Workbook of Sixth international conference on modulated semiconductor structures, Germany 687

[6] Twigg M E, Bennett B R, Shanabrook B V, Waterman J R, Davis J L and Wagner R J 1994 Appl. Phys. Lett. **64** 3476-3478

[7]. Schmitz J, Wagner J, Maier M, Obloch H, Hiesinger P, Koidl P and Ralston J D 1993 Inst. Phys. Conf. Ser. No 136:Chapter 6 367-372

[8]. Daly M S, Symons D M, Lakrimi M, Nicholas R J, Mason N J and Walker P J, unpublished

[9]. Yano M, Yokose H, Iwai Y and Inoue M J. 1991 J.Crystal Growth **111** 609-613

[10] Waterman J R, Shanabrook B V, Wagner R J, Yang M J, Davis J L and Omaggio J P 1993 Semicond. Sci. Technol. **8** S106-S111

[11] Hemstreet L A, Fong C Y and Nelson J S 1993 J. Vac. Sci. Technol. B **11** 1693-1696

Paper presented at 22nd Int. Symp. Compound Semiconductors, Cheju Island, Korea, 28 August–2 September 1995
© *1996 IOP Publishing Ltd*

Real-time study of dopant incorporation and segregation during MBE growth of GaAs(001):Si

L Däweritz[1,2], P Schützendübe[1], K Stahrenberg[3], M Maier[4] and K Ploog[1]

[1] Paul-Drude-Institut für Festkörperelektronik, 10117 Berlin, Germany
[2] Tokyo Institute of Technology, Research Center for Quantum Effect Electronics, Meguro-ku, Tokyo 152, Japan
[3] Institut für Festkörperphysik, Technische Universität Berlin, 10623 Berlin, Germany
[4] Fraunhofer-Institut für Angewandte Festkörperphysik, 79108 Freiburg, Germany

Abstract. The incorporation as well as segregation of dopant atoms during Si delta-doping of GaAs (001) has been studied in real-time by using reflection high-energy electron diffraction (RHEED) and reflectance anisotropy spectroscopy (RAS). Under conditions of enhanced Si adatom mobility distinct ordering processes are observed, which are promoted by ordered step arrays on the vicinal surface. The demonstrated real-time control of the complete delta-doping process is promising for a tailoring of doping structures.

1. Introduction

For the fabrication of novel devices by epitaxial growth it is of importance to control the position and concentration of dopant atoms. Motivated by the interest in quantum wire and edge quantum wire structures, the diffusion lengths of Ga and Al adatoms have been systematically studied by analysing scanning electron microscopy cross-sectional images of structures grown on patterned substrates [1]. For a detailed study of surface diffusion and segregation of dopant atoms, however, one has to search for other appropriate techniques.

In dependence on the surface conditions during the supply of Si atoms on GaAs(001) the distribution of the dopant atoms can vary between a random one, as expected in conventional delta-doping [2, 3], and an ordered one, as envisaged in wire-like doping by deposition of the dopant atoms on a regularly stepped vicinal surface under conditions favourable for adatom migration [4 - 6]. In this work, we investigate atomic processes during Si deposition on singular and vicinal GaAs(001) surfaces at enhanced adatom mobility. The corresponding changes in the long- and short-range order of the surface have been monitored by RHEED and RAS.

2. Experiment

The initial surfaces were prepared by GaAs growth on singular and $2°$ toward (111)Ga misoriented GaAs(001) substrates at conditions leading to a (2 x 4) reconstruction and to a step flow growth mode on the vicinal surface. Si was deposited with a flux of about 10^{11} atoms cm^{-2}s^{-1} in pulses of 60 s duration and 180 s interruption. The rate for GaAs overgrowth was 0.13 ML(monolayers)s^{-1}. The substrate temperature and the As$_4$ beam equivalent (BE) pressure were kept constant at 590 °C and 1 x 10^{-6} Torr, respectively, during the whole delta-doping process to facilitate the interpretation of the RHEED intensity recordings. Previous experiments on the dependence of ordering processes during Si deposition suggest that the Si surface migration is optimum at these conditions [6, 7].

Real-time RHEED measurements were performed with the electron beam parallel or perpendicular to the misorientation steps, i.e. in the [$\bar{1}$10] and [110] azimuth, respectively. A system consisting of a video camera, video recorder and image processing unit was used to analyse the time evolution of the specular and fractional order beams. RAS monitoring was done through a heatable strain free window mounted on the pyrometer viewport of the growth chamber. The monitored RAS signal consisted of the real part of the reflectance anisotropy

$$\Delta r / r = 2 (r_{[\bar{1}10]} - r_{[110]}) / (r_{[\bar{1}10]} + r_{[110]}).$$

At the 2.6 eV photon energy used it will be mostly affected by the As surface dimers [8]. RAS spectra in the range of 1.6 eV to 5.3 eV were also taken, but will be discussed elsewhere [9].

3. Results and discussion

3.1. Si deposition

Fig. 1 shows the behaviour of the specular beam RHEED intensity recorded in the [$\bar{1}$10] azimuth during deposition of 0.6 ML Si in 64 pulses on the vicinal surface and during GaAs overgrowth by 190 nm GaAs. It is evident that distinct ordering processes proceed during the Si flux interruption, even at high coverages. The intensity decrease during the pulse is explained by the build-up of a Si adatom concentration on the terraces and an increasing number of kinks formed during Si attachment at the step edges. The intensity recovery after the pulse is ascribed to the migration of Si adatoms to the step edges where they become incorporated on Ga sites in an ordered (3 x 2) structure.

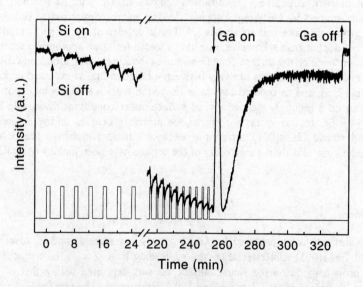

Fig. 1. RHEED intensity recorded in the [$\bar{1}$10] azimuth during Si deposition of 0.6 ML in 64 pulses on a vicinal GaAs(001) surface and during GaAs overgrowth. Deposition parameters: T, 590 °C, Si flux 5.7×10^{12} atoms cm^{-2} pulse^{-1}, pulse duration 60 s, interruption 180 s. GaAs overgrowth 0.13 MLs^{-1}

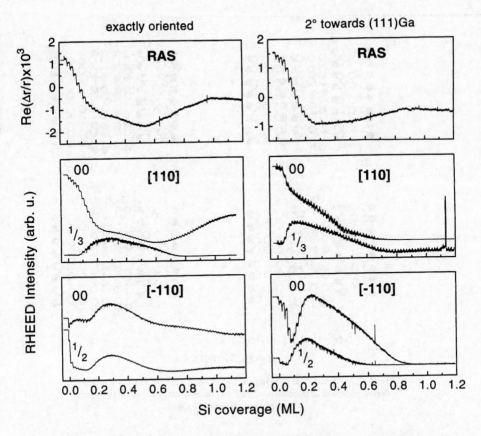

Fig. 2. RAS transients and recordings of the specular, 1/2-order and 1/3-order-beam RHEED intensities for pulsed Si deposition on the singular and vicinal GaAs(001) surface.Deposition parameters: T, 590 °C, $p_{BE}As_4$ 1x10⁻⁶ Torr, Si flux 2x10¹¹ atoms cm⁻²s⁻¹, pulse duration 60 s, interruption 180 s

For an analysis of the evolution of the (3 x 2) structure on the singular and on the vicinal surface we use the RAS transients and the RHEED intensity recordings for the 1/3- and 1/2-order beams due to the (3 x 2) (or (3 x 1)) reconstruction which are compiled in Fig. 2. The recordings of the fractional-order beams reveal that the (3 x 2) structure appears at lower Si coverages and evolves faster on the vicinal surface than on the singular surface, indicating the role of misorientation steps as preferred nucleation sites. The sudden change in reconstruction and strain can explain the rapid rise in the intensity of the specular RHEED beam in the [$\bar{1}$10] azimuth after passing a clear minimum, which is in clear contrast to the intensity behaviour for the singular surface.

The evolution of the (3 x 2) structure at higher Si coverages is discussed with the data for the singular surface. The fractional-order beams appear at a coverage of ~ 0.1 ML, reach the maximum intensity at a coverage of ~ 0.3 ML and disappear at ~ 0.65 ML. The 1/2-order spots appear at a slightly higher coverage than the 1/3-order spots. These observations are consistent with the model shown on Fig.3. The ordered attachment of Si dimers in a row perpendicular to the dimer axis at any preferred nucleation site leads to the (3 x 1) structure. Thereby the As atoms marked by a cross become undimerized and will reevaporate. A further

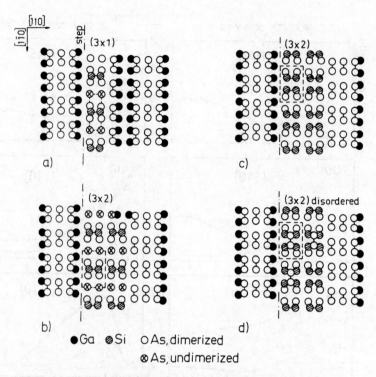

Fig. 3. Schemetic model showing the evolution of the (3 x 2) reconstruction
during pulsed Si deposition on the (2 x 4) reconstructed GaAs(001) surface
misoriented toward (111)Ga

attachment of Si dimers leads to a (3 x 2) symmetry with extended Si dimer rows (Fig. 3b). The reevaporation of undimerized As atoms exposes a row of Ga atoms that will dimerize. The reduction of the As dimer density during this process is evident from the decreasing RAS signal (Fig. 2). The pulsating behaviour of the signal reflects the ordering process with a destruction of dimers during the pulse and their partial reforming after flux interruption. In the ideal case, the structure considered in Fig. 3b would be completed at a coverage of 0.33 ML which is in good agreement with the coverages corresponding to the intensity maxima of the fractional-order beams. A (3 x 2) structure can be maintained if with increasing Si coverage Si dimers are incorporated between already existing dimer rows (Fig. 3c). In this growth stage a combined incorporation of Si and As atoms will take place. Once two neighbouring Si dimers are created, As atoms can condense on Si atoms as dimers (Fig. 3d). This leads to the observed degradation of the (3 x 2) structure. It can also explain characteristic changes in the slope of the RAS curve if we assume that the RAS signal at 2.6 eV is sensitive also to dimerized As atoms on top of Si. The increasing concentration of As dimers in the top layer explains the increase of the RAS signal above ~0.5 ML Si deposition. In the case of the vicinal surface this increase of the signal is observed at lower coverage. This is expected for preferential Si incorporation at step edges, which allows the formation of As dimers on top of Si atoms in an earlier growth stage. After completion of the first monolayer Si and beginning incorporation of Si atoms in a second layer one expects a 90° rotation of the As dimers in the top layer. In agreement with this the RAS curve changes the sign of the slope at a coverage of ~1 ML.

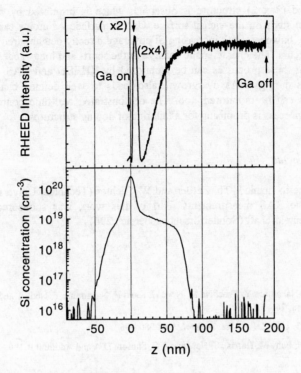

Fig. 4. Comparison of the RHEED intensity vs. GaAs layer thickness z plot
(cf. Fig. 1) with the SIMS profile of the Si concentration

3.2. GaAs overgrowth

To analyse the vertical spreading of the Si atoms during the long-time Si deposition and GaAs overgrowth, the RHEED intensity recording for GaAs overgrowth of 0.6 ML Si (right part in Fig. 1) is replotted in Fig. 4 as a function of the layer thickness and compared with the secondary ion mass spectrometry (SIMS) profile of the structure. It is evident that, besides some Si indiffusion, segregation effects become important when Si is incorporated in this extremely high concentration at high temperature. Under the initial rapid reduction in the Si concentration a surface smoothing process proceeds until a well-ordered (m x 2) structure is formed. The further Si depletion leads to a disordered (2 x 4) structure. From scanning tunneling microscopy studies of samples with different doping levels it is known that the incorporated Si atoms induce a concentration dependent number of kinks in the As dimer rows [10, 11]. Therefore in the present experiment the RHEED intensity increases until complete Si depletion. After growth interruption the intensity reaches the initial value for the clean and well-ordered surface.

4. Conclusions

In contrast to conventional delta-doping, distinct ordering processes occur when Si atoms are supplied to GaAs(001) at conditions leading to enhanced adatom mobility. The formation of a

56

highly ordered (3 x 2) structure is observed, which is promoted by an ordered array of misorientation steps on the vicinal surface. On this surface As atoms cannot be adsorbed as dimers. Once, however, with increasing Si coverage directly neighboured Si dimers exist, the adsorption occurs. The characteristic changes in the short- and long-range order of the surface accompanying these processes can be monitored by RHEED and RAS. The segregation of dopant atoms during GaAs overgrowth also leads to well defined changes in the surface structure and can be monitored, too. The demonstrated real-time control of the complete delta-doping process is promising for a tailoring of doping structures.

Acknowledgements

We would like to thank J.-Th. Zettler and W. Richter (Technical University Berlin) for their support of the RAS measurements. Part of this work was sponsored by the Deutsche Forschungsgemeinschaft (Sonderforschungsbereich 296).

References

[1] Koshiba S, Nakamura Y, Tsuchiya M, Noge H, Kano H, Nagamune Y, Noda T and Sakaki H 1994 *J. Appl. Phys.* **76** 4138 - 44

[2] Schubert E F 1990 *J. Vac. Sci. Technol.* A **8** 2980 - 96

[3] Ashwin M J, Fahy M, Harris J J, Newman R C, Sansom D A and Addinall R 1993 *J. Appl. Phys.* **73** 633 - 9

[4] Bauer G E W and van Gorkum A A 1990 in *Science and Engineering of One- and Zero-Dimensional Semiconductors*, edited by Beaumont S E and Sotomayor C M (New York: Plenum) p 133

[5] Wood C E C 1992 *J. Appl. Phys.* **71** 1760 - 3

[6] Däweritz L, Hagenstein K and Schützendübe P 1993 *J. Vac. Sci. Technol.* A **11** 1802 - 6

[7] Däweritz L and Kostial H 1994 *Appl. Phys. A* **58** 81 - 86

[8] Chang Y and Aspnes D E 1990 *J. Vac. Sci. Technol.* B **8** 896 -9

[9] Stahrenberg K, Zettler J-TH, Däweritz L, Schützendübe P and Richter W *to be published*

[10] Pashley M D and Haberern K W 1991 *Phys. Rev. Lett.* **67** 2697 - 2700

[11] Wassermeier M, Behrend J, Däweritz L and Ploog K *to be published*

Inst. Phys. Conf. Ser. No 145: Chapter 2
Paper presented at 22nd Int. Symp. Compound Semiconductors, Cheju Island, Korea, 28 August–2 September 1995
© *1996 IOP Publishing Ltd*

Lateral compositional change of InAlAs on non-planer substrates by molecular beam epitaxy

Takeyoshi SUGAYA, Tadashi NAKAGAWA and Yoshinobu SUGIYAMA

Electrotechnical Laboratory, 1-1-4, Umezono, Tsukuba 305 Japan

Abstract.   Lateral compositional change of InAlAs on non-planer substrates with truncated ridges is demonstrated.   In-rich InAlAs epitaxial layers are grown on top of ridges.   It is found that In molar fraction in the In-rich regions increases as the epitaxial growth proceeds.   This result indicates that self-formation of electron confining InAlAs wire structures can be fabricated.   Also, the wires lattice-matched to the substrate are obtained by the control of In content of the grown layers.

1. Introduction

The selective growth of III-V semiconductors on non-planer substrates has been studied as a promising method to fabricate nanostructure devices.   A large number of reports of the GaAs quantum wire structures fabricated by the selective growth on non-planer substrates have been demonstrated by metalorganic chemical vapor deposition (MOCVD) as well as molecular beam epitaxy (MBE).   The quantum wire structures with good optical properties have been formed at the bottom of V-grooves (Kapon et al. 1989, Tsukamoto et al. 1992) or on top of the mesas (Koshiba et al. 1994) by depositing an AlGaAs/GaAs/AlGaAs quantum well structure, where the GaAs well width at the bottom or top regions becomes thicker than elsewhere.   Also, the InGaAs quantum wires fabricated on top of the mesas by using the same method have been reported (Fujikura et al. 1995).   Recently, for the MBE of AlGaAs on patterned substrates, the spontaneous formation of wire structures due to the lateral change of Al content has been demonstrated(Kadoya et al. 1995).   This novel technique is attractive because the fabrication process of quantum wires is governed by self-ordering of adatoms.

The similar effect of the self-formation of the wire structures will also be expected for the MBE of InGaAs and InAlAs.   However, the phenomena for this material group arises a serious problem of the crystal deterioration.   In this paper, we demonstrate the lateral compositional change of InAlAs epitaxial layers on non-planer substrates. Furthermore, we report a method to keep the composition of InAlAs constant within wire structures.

2. Experimental

The non-planer structure with truncated ridge was prepared on (001) InP by the conventional photolithography followed by chemical etching in $HCl:CH_3COOH:H_2O_2$ (1:2:1 by volume) for 30 seconds. The depth of the etching was about 220nm. The various line and space patterns are formed along [1-10] direction. The substrates have flat (001) regions and (111)A slopes. After a thorough degreasing, the substrates were rinsed in deionized water and etched in the H_2SO_4-based etchant system to remove remaining surface damages and oxides. The substrates were then loaded into the MBE chamber and thermally cleaned by raising the temperature to 560°C in As_4 atmosphere. The growth rate of InAlAs was 0.9μm/h and the epitaxial layers of various thickness were grown at the temperature of 490°C with substrate rotation of 5rpm. During the growth, the flux intensities of As_4, In and Al were 1.3×10^{-3}Pa, 4.3×10^{-4}Pa and 1.7×10^{-4}Pa, respectively, as measured with an ion-gauge at the substrate position. The scanning electron microscopy (SEM) observations and the PL measurements at 12K were performed and the composition of the epitaxial layer was determined by the peak energy of the PL measurement.

3. Results and Discussions

Figure 1 shows the cross-sectional SEM photograph of the InAlAs layer grown on a truncated ridge together with the structure schematics. The In composition of 53% was expected when the substrate was entirely flat. The width of lines and spaces were 2μm and the truncated ridge width before the growth was 1.6μm. The thickness of the epitaxial layer was

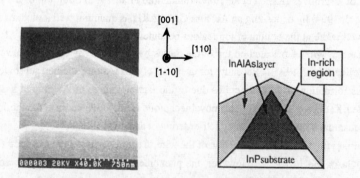

Fig.1 Cross-sectional view of InAlAs grown layer perpendicular to [1-10].
InAlAs layer is separated into two compositional different regions spontaneously.

1µm. In Fig.1, the InAlAs layer was separated into two compositional different regions spontaneously, one forming a triangular shape on the flat (001) ridges and the other on the (111)A slope of the ridge.

Figure 2 shows the PL spectra of the epitaxial layers on flat and patterned area with a schematic cross-sectional view of the structure. The PL measurements were performed at 12K using Ar^+ laser (λ=514.5nm) and GaAs photodetector. The width of lines and spaces were 4µm and the truncated ridge width before the growth was 3.5µm. In this sample, the ridge width is wide and the former triangular region (in Fig.1) becomes trapezoid as shown in Fig.2(b). The broken line in Fig.2(a) is the spectrum for the flat area, which correspond to $In_{0.53}Al_{0.47}As$ determined by the peak energy, and the solid line is that for the patterned area. Two peaks, one shorter and one longer in wavelength than the peak for the flat area, are observed in the latter spectrum. The peak at around 710nm and that at around 850nm originate from the (111)A slope and the other (001) region, respectively. It is because the former peak, in which the difference of In-content from the InAlAs on the flat area was larger than the latter peak, should originate from the smaller area in Fig.2(b). The InAlAs layers with the In content of 0.415 that is determined by the peak energy are grown on the (111)A slope and those of 0.542 are grown on the ridge and presumably bottom regions.

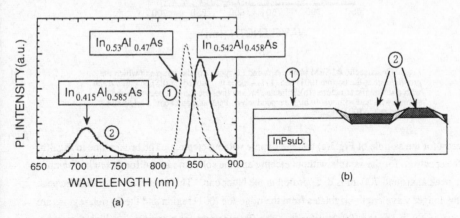

Fig.2 PL spectra of epitaxial layers on flat and patterned area with schematic structure.
Two peaks are observed in the spectrum for the patterned area.

In order to confirm the In-rich regions grown on the ridge-top (001) surfaces, the PL spectrum for a sample in which the epitaxial layer grown at the bottom (001) and (111)A slope was removed by the chemical etching was measured. Figure3(a) shows the cross-sectional SEM photograph of the etched sample with the schematic structure. This figure indicates that the epitaxial layer grown on the bottom and slope regions is removed by the chemical etching and a part of the grown layer on the ridge-top (001) is remained. Figure3(b) shows the PL

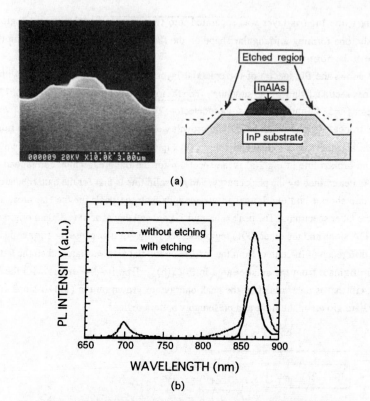

(a)

(b)

Fig.3   Cross-sectional SEM photograph and PL spectra of the sample in which the
grown layer at the bottom (001) and (111)A slope is removed. (a):SEM photograph
with a schematic structure. (b):Solid and broken lines are the PL spectra for the
sample with and without etching, respectively.   Peak at aroud 865nm originates
from ridge-top (001) regions.

spectra for the sample of Fig.3(a) and the sample without etching.   The broken line in Fig.3(b)
is the spectrum for the sample without etching and the solid line is that for the etched sample.
The peak at around 700nm is disappeared in the latter one.   This result indicates that the peak
of the longer wavelength originates from the ridge-top (001) region and the In-rich regions are
grown on the ridge-top (001) surfaces.   The PL spectrum for a sample in which the grown
layer on the bottom (001) was remained also has the peak of the longer wavelength and any
shorter peak.   Therefore, the In-rich regions are also grown at the bottom (001) surfaces and
the peak at around 700nm in Fig.2(a) and Fig.3(b) originates from the (111)A slope.

Figure 4 shows the PL spectra of the flat region and the In-rich ridge-top regions for
various layer thickness.   The observed   [001] ridge-top width decreases with increasing
epitaxial layer thickness.   The PL measurements were performed with Ge photodetector.   The
PL peaks at around 900nm originate from the InP substrates because these peaks appeares for
InP substrates without the InAlAs growth.   The broad peak which exists in the range of 900-

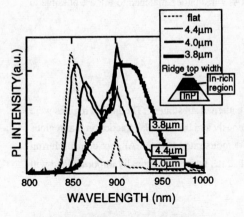

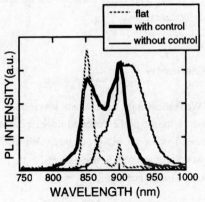

Fig.4 PL spectra of In-rich regions for various layer thicknesses. PL peak wavelength for InAlAs layers increases with decreasing ridge top width.

Fig.5 PL spectra of InAlAs grown with and without compositional control. InAlAs layers lattice-matched to InP are grown on the ridges by the compositional control.

970nm may be due to some impurities or defects in the InAlAs layer.   The peaks existed in 850-875nm correspond to the band edge emission from the InAlAs grown layers.   The In-content of the grown layer is changed from 0.54 to 0.56.   The PL peak wavelength for the InAlAs layer on the ridges increases with decreasing the ridge top width.   This result implies that the In content in the triangular shaped region (in Fig.1) increases gradually toward a top of the triangle and the electron confining InAlAs quantum wires can be fabricated by this method.

As regards to the growth mechanisms in this InAlAs wire fabrications, the difference in the diffusion length of In and Al is the predominant process.   The In and Al atoms arrived at the (111)A slope migrate to the (001) ridge top and bottom regions.   The composition of InAlAs on the flat (001) regions is decided by the migration of the In atoms because the diffusion length of the Al atoms, that is the order of 10nm (Koshiba et al. 1994), is much smaller than that of the In atoms($\sim\mu$m).   The large number of In atoms in comparison with the Al atoms migrate to the (001) regions and the In-rich regions grow on the (001) surfaces.   As the growth proceeds, the (111)A slope changes to the (311)A slope and the slopes become wider and the (001) regions become narrower.   The In content in the triangular shaped region (in Fig.1) increases gradually as the InAlAs growth proceeds, because the larger number of In atoms can be migrate into the narrower (001) regions.

The wire structure mentioned above, however, does not lattice-match with the substrate, and the compositional change may deteriorate the wire.   To alleviate this problem we tried the MBE growth where the In content of the ridge-top InAlAs was kept constant by the gradual increase of the Al flux during the growth.   The broken line in the Fig.5 is the spectrum of In$_{0.54}$Al$_{0.46}$As on the flat area.   The thick and thin solid lines are the spectra of the ridge-top InAlAs with and without compositional control, respectively.   The peak energies of the In$_{0.54}$Al$_{0.46}$As on the flat area and the ridge-top InAlAs grown with the flux control were

62

almost same. This figure indicates that the InAlAs wires lattice-matched to InP are possible to grow by this method.

4. Conclusions

We fabricated the InAlAs wire structures on patterned substrates. The In-rich regions were grown on the top of the truncated ridges spontaneously and the In molar fraction in that regions increased as the growth proceeded. We also demonstrated that the Al flux control during growth makes the wire lattice-matched to the substrate, which is practically important and will become a useful technique.

References

Fujikura H and Hasegawa H 1995 Workbook of 8th Int. Conf. MBE, Osaka, p.379.

Kadoya Y, Yoshida T, Nagamune Y, Noge H and Sakaki H. 1995 Workbook of 8th Int. Conf. MBE, Osaka, p.355.

Kapon E, Simhony S, Bhat R, and Hwang D M 1989 Appl. Phys. Lett. 55 1989.

Koshiba S, Nakamura Y, Tsuchiya M, Noge H, Kano H, Nagamune Y, Noda T and Sakaki H. 1994 Appl. Phys. Lett. 64 363.

Koshiba S, Noge H, Akiyama H, Inoshita T, Nakamura Y, Shimizu A, Nagamune Y, Tsuchiya M, Kano H, and Sakaki H. 1994 Appl. Phys. Lett. 64 363.

Tsukamoto S, Nagamune Y, Nishioka M and Arakawa Y. 1992 J. Appl. Phys. 71 533.

Inst. Phys. Conf. Ser. No 145: Chapter 2
Paper presented at 22nd Int. Symp. Compound Semiconductors, Cheju Island, Korea, 28 August–2 September 1995
© *1996 IOP Publishing Ltd*

Hydrogenation effect in nitrogen doped ZnSe films grown by molecular beam epitaxy

M. D. Kim, E. S. Oh, J. R. Kim, H. D. Jeong, B. J. Kim, H. S. Park, T. I. Kim, W. C. Choi[*]

Photonics Semiconductor Lab., Materials and Devices Research Center, Samsung Advanced Institute of Technology, P. O. Box 111, Suwon Korea
* *Department of Physics, Sung Kyun Kwan University, Suwon, 440-746, Korea*

Abstract. Photoluminescence and micro-Raman measurements were carried out in order to investigate the hydrogenation effect on the behavior of donor–acceptor band emission and bound exciton for nitrogen doped ZnSe:N(N_a-N_d=3~8 x 10^{17} cm^{-3}) grown by molecular beam epitaxy. The luminescence due to recombination of the donor bound-exciton(I_2) appeared, due to the donor–acceptor pair transition shift to higher energy, and the acceptor bound-exciton(I_1) line intensity remarkably decreased after hydrogenation. When the hydrogenated samples were annealed rapidly, the acceptor bound exciton peak was increased, indicating hydrogen atoms were released from the materials. After hydrogenation, the intensity of the unscreened GaAs LO phonon is significantly decreased compared with the plasmon mode(L1). The increase of I(L1)/I(LO) is decrease probably due to the defects, which is free from the plasma damage. These results indicate that hydrogen atoms passivated not only nitrogen acceptors but also nitrogen related compensating defect complexes.

1. Introduction

Due to the strong interest in fabricating blue laser diodes, there has been a vigorous research effort in the doping of wide-band-gap semiconductors. ZnSe, for example, can be easily doped as n-type but it is very difficult to achieve a high level of p-type doping. The origin of this behavior is currently a topic of controversy[1]. This doping problem has been partially overcome in ZnSe by the use of nitrogen as the dopant impurity. Progress in p-type doping of ZnSe by metalorganic vapor phase epitaxy(MOVPE) and gas source molecular beam epitaxy(GSMBE) has been hampered by unintentional hydrogen incorporation and that is believed to cause passivation of the nitrogen acceptor atoms[2]. Since the hydrogen atom can passivate the electrical behavior of dangling or defective bonds, the existence of these bonds in single-crystal and polycrystalline semiconductors is very interesting[3], and the improvements in the electrical and the optical properties of these materials have been observed[4-7]. However, most hydrogenation effects on semiconductors were studied on GaAs, Si[8,9], and even though Ho *et al*[10]. reported the incorporation of hydrogen in ZnSe:N and ZnSe:Cl layers grown by gas source molecular beam epitaxy(GSMBE) using elemental Zn and H_2Se as source material, very little work has been performed on II-VI compound semiconductors. The detailed mechanism of the improvement in the crystallinity of the II-VI semiconductors by hydrogenation has still to be investigated.

In this work, photoluminescence(PL) and micro-Raman measurements have been carried out in order to study hydrogenation effect on behavior of the donor-acceptor pair(DAP) band, the ZnSe/GaAs interfaces, and the neutralization of acceptors.

2. Experimental

The samples were grown by molecula beam epitaxy(MBE) in RIBER 32P MBE system using element Zn and Se source materials. The ZnSe epilayers were grown on semi-insulating GaAs(100) substrates using a Se:Zn flux ratio of 2:1 at a growth rate of 0.7 $\mu m/h$ and substrate temperature of 280℃. Nitrogen doping was carried out using an Oxford Applied Research plasma source. In order to maintain a stable nitrogen plasma throughout the film growth, nitrogen flow greater than 1.0×10^{-6} Torr and an rf power 150 W were used. The nitrogen doped ZnSe(ZnSe:N) samples used in the PL study have different N_a-N_d values being varied from 3×10^{17} cm^{-3} to 8×10^{17} cm^{-3} . N_a-N_d was determined by C-V profiling using Au as a Schottky barrier metal. As-grown ZnSe:N samples were exposed to the hydrogen plasma at a pressure of 0.9 Torr with a power density of 0.078 W/cm^2 using a capacitively coupled rf(13.56 MHz) plasma enhanced chemical vapor deposition system(PECVD). Hydrogenation was performed at 80℃ for 30 mins. Hydrogenated samples were rapid thermal annealing(RTA) using a halogen lamp at 250℃, 300℃, 350℃, and 400℃ in a N_2 atmosphere for 10 sec. The PL spectra were measured using the 3250 Å line from a He-Cd laser as an excitation source in a temperature range from 11 to 280 K. In the micro-Raman measurement, the beam was focused with the diameter of the laser spot being about 2 μm on the sample surface at room temperature. The excitation source was 5145 Å line of an Ar ion laser.

3. Results and Discussion

PL spectra of nitrogen doped ZnSe before and after hydrogenation are shown in Fig. 1. The as-grown spectrum shown in Fig. 1(a) is dominated by a strong donor-acceptor pair(DAP) emission, at 2.684 eV accompanied by its phonon replicas and the near bandedge emission is dominated by an deep acceptor-bound exciton peak(I_1^d) at 2.785 eV. The free exciton emission line(E_x) is also observed, but with significantly lower intensities. It is well known that the position of the DAP peak depends on the nitrogen acceptor concentrations. Also shown in Fig. 1(b) is the PL spectrum for the hydrogenated ZnSe:N epilayer. After hydrogenation, the intensity of the PL spectrum is increased: in particular, that of the exciton luminescence, and DAP band emission are changed remarkably as shown in Fig.1(b). The luminescence of the peak corresponding to I_1 disappeared, and a new peak of the donor-bound exciton at 2.798 eV(I_2) is observed. The observed DAP emission is shifted by a few meV to the higher energy(2.701 eV) than that for as-grown DAP peak position.

Figure 2 shows PL spectra obtained on a heavily doped sample, with a net acceptor concentration of about $N_a-N_d=8\times10^{17}$ cm^{-3}. As shown in Fig. 2(a), the as-grown spectrum is dominated by DAP with a phonon emission at 2.692 eV. No emission were observed in the excitonic region for these heavily doped layers. After hydrogenation, the intensity of the luminescence increased remarkably and the DAP peak energy is about 10 meV higher energy than that for as-grown DAP peak. In the excitonic energy region, the I_2 peak has appeared. Although the DAP line intensities of the PL spectra in Fig. 1(b) and Fig. 2(b) are very much larger than those of as-grown, the shapes of the DAP peaks do not change significantly. After hydrogenation we observe two main changes. Firstly, the luminescence intensities after hydrogenation in these samples are 10 times stronger than before. Secondly, the DAP band emission becomes narrower and move to the higher energy region(2.701 eV) regardless of the nitrgen concentrations. Even though the detailed origin of these results has to be further investigated, the increase in the emission intensity can be possibly explained by the following reason. When the hydrogen atom enters into a crystal, either it decomposes into an impurity or a carrier bound in a complex level inside the crystal or it passivates a nonradiative recombination

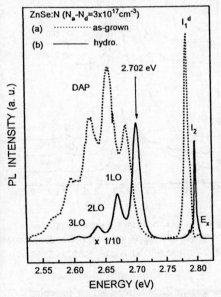

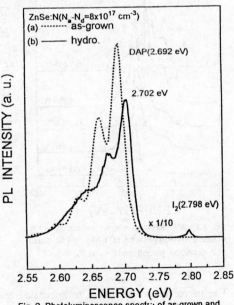

Fig. 1. Photoluminescence spectra of as-grown and hydrogenated ZnSe:N ($N_a-N_d=3\times10^{17}$ cm^{-3}) films measured at 11 K.

Fig. 2. Photoluminescence spectra of as-grown and hydrogenated ZnSe:N(Na-Nd=8x1017 cm-3) films measured at 11 K.

center at a deep level. The shrink and shift of the DAP band after hydrogenation are interpreted as arising from hydrogen neutralization the dopant impurities that contributes to the donor–acceptor pair transition. These results indicate that the nitrogen acceptors in ZnSe bonded hydrogen atoms, and hydrogen atoms are seen to act as shallow donor.

In Fig. 3, the bound exciton spectra of hydrogenated and RTA ZnSe:N layers are shown and Fig. 3(b)–4(e) correspond to a series of annealing temperature, (b) 250, (c) 300, (d) 350, and (e) 400°C under a N_2 atmosphere. The intensity of I_1 peak for hydrogenated ZnSe:N annealed at 300°C slightly increased and a new peak is observed at 2.785 eV, this peak is considered to be related to bound exciton due to the neutral acceptors. When the annealing temperature was 400°C, the intensity of I_1 peak is raised by as much as 2 times in comparison to the Fig. 3(a). This indicates that thermal energy apparently affects the rate of dissociaion of the hydrogen–nitrogen complex in ZnSe:N.

The results of the DAP band emission obtained from RTA hydrogenated ZnSe:N samples are shown in Fig. 4. As the annealing temperature increased to 400°C as shown Fig. 4(e), the intensity of the DAP peak slightly increased, but energy position of DAP peak did not change.

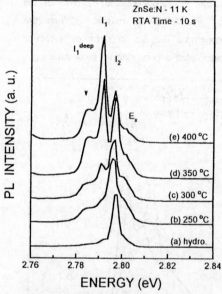

Fig. 3. Photoluminescence spectra obtained for different RTA temperature. (a) hydro., (b) annealed at 250, (c) 300, (d) 350, and (e) 400 °C.

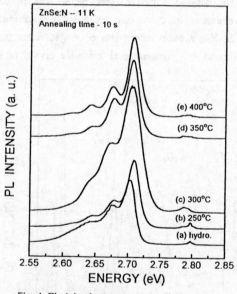

Fig. 4. Photoluminescence spectra obtained for different RTA temperature. (a) hydro., (b) annealed at 250, (c) 300, (d) 350, and (e) 400°C.

The micro-Raman spectra of hydrogenated and as-grown ZnSe:N revealed an intensity comparison of the ZnSe/GaAs interface between the two samples. as shown in Fig. 5. We observe the transverse and logitudinal optical phonons of ZnSe, plasmon-optical phonon coupling mode associated with high carrier densities in GaAs, and the GaAs logitudinal optical phonon, in the increasing order of frequency. The plasmon-phonon coupling mode has been extensively studied and it has been claimed that its intensity depends on the perfection of the interface, e.g., interfacial abruptness, defect density, interdiffusion, and surface treatment prior to ZnSe growth. Figure 5(b) shows that the incorporation of the H-atom significantly affects the relative intensity of the coupling mode with respect to that of the GaAs LO mode, clearly indicating that the H-atom had been diffused into the ZnSe/GaAs interface. It is appears likely that the H-atoms decrease some defects(e.g., point defects).

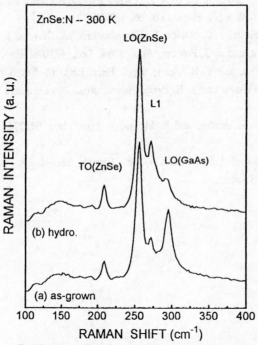

Fig. 5. Micro-Raman measurements of the as-grown and hydrogenated ZnSe:N.

4. Conclusions

Using PL and micro-Raman spectroscopy, we have discovered two new changes after hydrogenation. Firstly, a new peak of the donor-bound exciton at 2.798 eV(I_2) is observed. Secondly, the DAP band emission becomes narrower and move to the higher energy region(2.701 eV) regardless of the nitrgen concentrations. These results indicate that the nitrogen acceptors in ZnSe bonded hydrogen atoms and hydrogen atoms are seen to act as shallow donor.

68

References

[1]. D. B. Laks,n C. G. Van de Walle, G. F. Neumark, and S. T. Pantelide, Phys. Rev. Lett. 66, 648(1991).

[2]. J. A. Wolk, J. W. Ager, K. J. Duxstad, E. E. Haller, N. R. Taskar, D. R. Dorman, and D. J. Olego, Appl. Phys. Lett. 63, 2756(1993).

[3]. S. J. Pearton, J. W. Corbett, and T. S. Shi, Appl. Phys. A43, 153(1987).

[4]. R. Mostefaout, J. Chevallier, A. Jalil, J. C. Pesant, C. W. Tu, and R. F. Kopf, J. Appl. Phys. 64, 207(1988).

[5]. N. Pan, B. Lee, S. S. Bose, M. H. Kim, J. S. Hughes, G.E.Stillman, K. Arai, and Y. Nashimoto, Appl. Phys. Lett. 50, 1832(1987).

[6]. W. C. Dautremont-Smith, J. C. Nabity, V. Swaminathan, M. Stavola, J. Chevallier, C. W. Tu, and S. J. Pearton, Appl. Phys. Lett. 49,1098(1986).

[7]. Y. F. Chen, C. S. Tsai, and Y. H. Chang, Appl. Phys. Lett. 57, 70(1990).

[8]. J. I. Pankove, R. O. Wance and J. E. Berkeyheisen, Appl. Phys. Letters 45, 670(1984).

[9]. A. Jalil, J. Chevallier, R. Azulay, and R. Mircea, J. Appl. Phys. 59,3774 1986)

[10]. E. Ho, P. A. Fisher, J. L. House, G. S. Petrich, and L. A. Kolodziejski, J. Walker and N. M. Johnson.

Inst. Phys. Conf. Ser. No 145: Chapter 2
Paper presented at 22nd Int. Symp. Compound Semiconductors, Cheju Island, Korea, 28 August–2 September 1995

Real time in-situ thickness control of Fabry Perot cavities and thin quantum wells in MBE by ellipsometry

C. H. Kuo, M. D. Boonzaayer, and Sonu Daryanani

Department of Electrical Engineering, Center for Solid State Electronic Research
Arizona State University , Tempe, AZ 85287-6206

G. N. Maracas

Motorola Phoenix Corporate Research Labs., 2100 E. Elliot Road
Tempe, AZ 85284

Abstract. We demonstrate the use of closed loop-loop feedback control of MBE by in-situ spectroscopic ellipsometry to grow reproducible AlAs/GaAs Fabry Perot vertical cavity surface emitting laser cavities and 4 periods of thin (25Å) AlAs/GaAs quantum wells. Sample to sample reproducibility of the Fabry Perot cavity mode position was better than 0.2% among six samples grown at different times. Control of thin (25Å) quantum wells shows a variation of less than 0.5 ML in quantum well thickness. These results show the possibility of using bulk material optical constants in closed-loop feedback control with spectroscopic ellipsometry to increase the reproducibility of complex multilayer structures and thin quantum wells.

1. Introduction

Epitaxial growth techniques such as molecular beam epitaxy (MBE) and organometallic chemical vapor deposition (OMCVD) have been used to achieve heterostructure devices having complicated multilayer epitaxial structures or reproducible thin film growth. The strength of these epitaxial techniques is that alloy composition, thickness, and doping concentration can be achieved on thickness scales of monolayers. Application of these techniques can be used in the growth of quantum device such as vertical cavity surface emitting laser (VCSEL) and resonance tunneling diode (RTD).

Performance of such quantum devices are affected by the epitaxial layer thickness variation. To achieve reproducible epitaxial layer thickness control of complicate structure like VCSEL is a change in MBE growth. Several different techniques such as reflection mass spectrometry[1], desorption mass spectrometry[2], and laser induced fluorescence[3] have been used to monitor group III or group V desorption from substrate surface during the growth of epitaxial layers. The desorption flux measurement in real time can be used to monitor epitaxial layer growth rate and composition with proper calibration procedure.

Reflection high-energy electron diffraction(RHEED) is the standard equipment in most MBE chambers. The oscillation signal from RHEED signal is the most popular method to measure epitaxial layer growth rate and alloy composition. The disadvantage of the RHEED oscillation is that the oscillation signal was taken without substrate rotation which is different from the growth condition. The other disadvantage of RHEED is that it can not account for the flux fluctuation when shutter is opening or closing. The real time feedback control is not easy to implement by using all the techniques mentioned above.

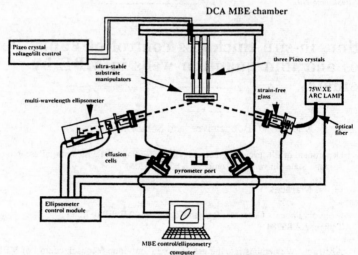

Fig 1 Implementation of a MWE on to a MBE system

Ellipsometry is becoming more popular ever since the earlier work by Aspen[4]. The ellipsometry technoque can be used not only for ex-situ to measure the thickness, alloy composition, and optimize the thin film growth condition from various growth techniques but also can be used in-situ to monitor and analyze data in real time for feedback control. The ellipsometric data can be taken with substrate rotation with normal rotation speed which is an advantage over RHEED. Since thickness information is measured in real time in MBE experiment, flux fluctuation does not affect the control of each epitaxial layer thickness. Ellipsometry is becoming one of the techniques to be used in the MBE experiment for in-situ control sensor.

2. Experiment and results

The detail theory of ellipsometry may be found elsewhere[5]. The hardware necessary for successful implemention of in-situ ellipsometer onto a MBE system is shown in Fig. 1. This growth chamber has two sets of optical ports whose axes coincide with the center of the wafer at an angle of 75° and 60°. Since the III-V epitaxial layer is used in the present study, 75° optical ports are used. The choice of 75°, which is close to the Brewster angle of III-V material, allows the maximum sensitivity in monitoring epitaxial layer thickness during MBE growth. Both optical ports are equipped with a strain-free window to prevent birefringence effect introduced in optical path. Both windows must be regularly, every other week, heated up to 300 °C to desorp arsenic coating.

During the growth of epitaxial layers, substrate rotation is necessary to ensure the uniformity of each epitaxial layer grown on the substrate. If the normal of the substrate surface is not parallel to the axis of the rotation, the reflected SE signal on the detector is not stable. The wobble of the substrate surface increases the uncertainty of the analyzed ellipsometric data. This in turn will increase the uncertainty of analyzed epitaxial layer thickness and make it more difficult to control layer thickness in real time with thickness variation of 0.5 % or less. To achieve the wobble of substrate during rotation, sections of the three stainless steel rods which support the manipulator were cut off and replaced with Pizeo crystals. By adjusting applied negative high voltage on the three Pizeo crystals, the length of the three Pizeo crystal can be adjusted and wobble of substrate surface can be reduced from a typical variation of ±0.2º to ±0.02º.

The multi-wavelength ellipsometer(MWE), 44 wavelength by J. A. Woollam Co. range from 4400Å to 7400Å, can obtain 44 wavelength ellipsometric data at the speed of 25 data points per second. MWE data was analyzed in real time to determine epitaxial layer thickness during epitaxial layer growth with the speed of 1 data point per second. The analyzed thickness information was passed to MBE growth control program. By analyzing epitaxial thickness in real time, the MBE growth control program closes and opens the shutter to grow the desired complicated quantum device structure.

2.1. Fabry Perot cavity

Growth of Fabry-Perot cavities used for vertical cavity surface emitting lasers (VCSEL) and electro-optic modulator was used to demonstrate in-situ real time MWE thickness control. Such structures consist of a cavity having an optical mode placed in between two dielectric mirrors. This particular structure was used to test MWE control of thick epitaxial layers because the position of the FP mode is very sensitive to the change of the cavity thickness. Distributed Bragg reflectors (DBR) consisting of alternating $\lambda/4$ high and low index of refraction layers produces mirrors that have a high reflectance wavelength band. High reflectivity indicates good thickness control of the DBR layers.

The FP cavity designed for this experiment consisted of 10 period of top and 10.5 periods of bottom DBRs having AlAs/GaAs thickness of 822.2Å and 679.4 Å respectively. The GaAs 1λ FP cavity was designed to have a mode at 970nm so the thickness of the GaAs was nominally 2717.5 Å.

To analyze ellipsometric data collected from MWE, the optical constants of AlAs and GaAs at growth temperature, 600 ºC, are needed. The bulk material file of GaAs and AlAs[6] from our previous studies was used in this experiment to obtain thickness information in real time. Another important aspect of real time control is to reduce the time required to obtain thickness information for a complicated structure like DBR. The virtual interface model was used to reduce the data analysis to a simple three phase model(vacuum/epitaxial/substrate).

Six samples of 1λ FP cavities were grown to demonstrate the reproducibility of the MWE thickness control. Sample 1 to sample 4 were grown under the same Al and Ga cell temperature conditions over the period of one month. The same Ga and Al cell temperatures were used for all four different samples. The variation of FP mode from four different samples were less than 2.1 nm. This indicates a difference of less than 0.2 % in thickness control of the 1λ GaAs cavity. The FP mode measured at the edge of the 2 inch wafer is a blue shift of 5 nm compared to the center of the 2 inch wafer as shown in Fig. 2. This indicates the layer thickness uniformity across the 2 inch wafer is about 0.5 %.

Sample 5 was grown with perturbed Al and Ga cell temperatures during the growth of AlAs and GaAs layers. Since the thickness control program is monitoring epitaxial layer thickness information in real time, we do not expect the flux fluctuation to affect epitaxial layer thickness control. As shown in Fig. 5, the FP mode from sample 5 stay at the center of variation from the early four samples. This indicates that constant calibration of growth rate for epitaxial layer is not necessary in MWE thickness control experiment.

Sample 6 was grown with Si doping, 1×10^{18}, in all the epitaxial layers. Since virtual interface model and bulk material optical constants are used in analyzing thickness information, a constant surface temperature is assumed durint the growth of epitaxial layers. The effect of the increasing surface temperature while Si shutter open and possible change of the epitaxial layer optical constants due to Si doping do not take into account in data analysis. However from sample 6 FP mode as shown in Fig. 2, the difference between sample 6 and the average value from sample 1 to 4 is less than 0.1 %. The effect

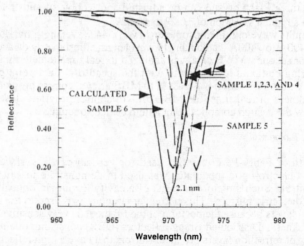

Fig2. Normal incidence reflectance of a 970 nm (design) FP cavity. Superimposed on the calculation is the measured curve showing FP mode from different samples.

of surface temperature changing and Si doping can be neglected in in-situ DBR thickness control experiment.

2.2   Thin quantum well

Another interesting aspect of real time in-situ control study is to grow multiple thin quantum wells (QW) or thin quantum barriers (QB). Are thin QW and QB optical constants the same as bulk material? Does the quantum confinement of QW affect the MWE thickness control in multiple QW structure? From the ex-situ two quantum wells study by Snyder[7] at room temperature, indicates the change of the optical constants due to thin film and quantum confinement. Are all these effects in optical constants at growth temperature preventing us from growing multiple thin QW by in-situ real time thickness control?

To answer the questions raised above, a simple four 25Å GaAs QW separated by 300 Å AlAs with MWE in-situ thickness control was grown. In order to obtain a better thickness control of 25Å GaAs thin film, the growth rate of 0.5Å/second was used for GaAs. 600 °C bulk material optical constants from our previous study were used in the MWE program to calculated GaAs layer thickness in real time. The shutter opening time during the growth of GaAs thin film ranges from 55 seconds to 65 seconds which indicates a possible change of GaAs layer thickness by more than 5 Å. From analyzing the ellipsometric data, the thickness difference for all four QW can not be more than 1 Å.

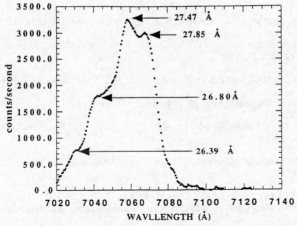

Fig. 3 77 K PL of 4 25 Å thin QW

77 K photoluminescence data from four 25Å GaAs/AlAs QW is shown in Fig. 3. Four excitonic peaks corresponding to the n=1 electron to heavy hole transitions can be clearly seen. Four PL peaks can be assigned to exciton transitions from 27.85Å, 27.47Å, 26.80Å, and 26.39Å QW respectively. The 2Å difference between assigned value from Fig. 3 and target MWE controlled value of 25Å could coming from either error in theoretical calculation for thin QW exciton transition or error in bulk material optical constants used in MBE growth control program. It is not clear which causes the error at the present time. Although we can not prove the absolute thin QW thickness control by MWE at the present, the maximum relative difference of 1.5 Å for four QW thickness does show a control of repeated thin QW thickness control. The problem of absolute QW thickness control can be resolved by a more detailed study at a later time.

3. Conclusion

Bulk material optical constants can be used by MWE in real time thickness control for thick epitaxial layers such as FP cavity structures. Sample to sample reproducibility of the FP mode position was better than 0.2% among four different samples grown at different times. Control of thin (25Å) QWs shows variation of less than 0.5 ML in QW thickness. These results show the possibility of using bulk material optical constants in closed loop feedback control with MWE to increase the reproducibility of complex multilayer structures.

Acknowledgments

This work was supported by ARPA/ULTRA under contract No. N000014-92-J-1931.

References

[1]   Tsao J. Y, Brennan T. M., Hammons B. E., J. Crystal Growth **111** 125 (1991)

[2]   Evans K. R., Kaspi R., Jones C. R., Sherriff R. E., Jogat V., and Reynolds D. C., J. crystal Growth, **127** 523 (1993)

[3]   Kuo C. H., Choi C., Maracas G. N., and Steimle T., J. Vac. Sci. Technol. **B 11** 833 (1993)

[4]   Aspnes D. E., Thin Solid Film, **89** 249 (1982)

[5]   Azzam R. M. A. and Bashara N. M., 1977, *Ellipsometry and Polarized Light* (North Holland Publishing Co.)

[6]   Kuo C. H., Anand S., Droopad R., Choi Y., and Maracas G. N., J Vac. Sci. Technol., **B 12** 1214 (1994)

[7]   Herzinger C. M., Snyder P. G., Celii F. G., Kao Y. C., Chow D., Johs B., and Woollam J. A.,Proceedings Int'l Confirence on Compound Semiconductors, San Diego, Ca(Sept, 1994)

Inst. Phys. Conf. Ser. No 145: Chapter 2
Paper presented at 22nd Int. Symp. Compound Semiconductors, Cheju Island, Korea, 28 August–2 September 1995

1.55 µm (GaIn)(AsP)-MQW-Laser Diodes Grown by Chemical Beam Epitaxy

G. Tränkle, R. Müller[‡], A. Nutsch, B. Torabi, and G. Weimann

Walter-Schottky-Institut, Technische Universität München, D-85478 Garching, Germany
[‡]now at : Siemens AG, HL OS TE 2, 93049 Regensburg, Germany

Abstract: (GaIn)(AsP)-heterostructures for 1.55 µm-MQW-laser diodes were grown by Chemical Beam Epitaxy. Laser properties compared favourably with state-of-the-art devices grown by MOCVD. Threshold current densities of 100 A/cm^2 per quantum well, internal quantum efficiencies of 0.7, internal losses of 15 - 20 cm^{-1} and characteristic temperatures of 65 K were achieved with lattice-matched active layers. Devices with strained active layers showed reduced threshold currents of 80 A/cm^2 per quantum well, increased internal quantum efficiencies of 0.8 - 0.9 and reduced internal losses of 5 - 15 cm^{-1}. Convincing T_0-values as high as 90 K were observed. Using strain symmetrization allowed increasing the number of QWs to 10.

1. Introduction

Future photonic circuits require the integration of optoelectronic components with very different layer structures, such as laser diodes and waveguides. Chemical beam epitaxy (CBE) has the inherent advantage of selective area growth on masked substrates with only little influence of aperture dimensions on layer thickness and composition [1]. The epitaxial growth of device quality of ternary and quaternary compounds is, however, limited by extremely narrow growth parameter windows, e.g. substrate temperature variations of ± 2 K [2]. In this paper we report on the CBE growth of (InGa)(AsP)-heterostructures for 1.55 µm multiple quantum well (MQW)-lasers and on the properties of broad-area and ridge wave guide devices with lattice-matched, strained and strain-symmetrized active layers.

2. CBE growth of (InGa)(AsP)-heterostructures

We have grown (GaIn)(AsP)-MQW-lasers with GaInAs-QWs and quaternary barriers and optical confinement layers using the alkyles trimethylindium (TMI) and triethylgallium (TEG) together with the pure hydrides AsH$_3$ and PH$_3$ on n-doped (100) InP-substates oriented 2 ° off to the nearest [110] direction. Fluxes of the alkyles were pressure-controlled to provide long-term stability. The hydrides were cracked in a low-pressure injector at 1050°C. With OEIC application as goal, full 2" wafers were used for growth, with In-free mounting and radiation heating. Elemental Si- and Be-sources were used for n- and p-doping.

To provide maxiumum photoluminescence yield and best material quality all (InGa)(AsP)-heterostructures were grown at high substrate temperatures $T_S \geq 535°C$, measured in the InP-buffer by optical pyrometer [2]. V/III-ratios had to be adjusted reproducibly and precisely in very narrow limits. For quarternary layers this is shown in Figure 1 at $T_S = 543°C$ for growth rates around 1 μm/h. Under these conditions perfect surface morphology was achieved with defect densities well below 100 cm^{-2}. Layer composition and thicknesses were very homogeneous over the full substrates.

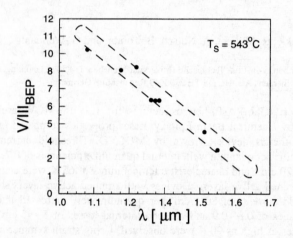

Figure 1.   Growth of quaternary (InGa)(AsP)-layers:
Optimized V/III-ratio (measured in beam equivalent pressures) at high substrate temperatures had to be controlled thoroughly in very narrow limits

3. Lattice-matched and strained (InGa)(AsP) laser diodes

Lattice matched MQW-laser structures contained 5 or 9 QWs of 7.5 nm thickness with quaternary barriers ($\lambda = 1.3$ μm, width 8 nm) centered in 200 nm wide quaternary optical confinement regions with the composition identical to the barriers. 1.5 μm thick n- and p-doped InP-claddings were used, with Be δ-doping near the surface to facilitate p-contacting. Device characteristics, measured on broad area and ridge waveguide lasers, included extrapolated threshold current densities $j_{th}(\infty)$ of 100 A/cm^2 per QW, internal quantum efficiencies η_i of 0.7, internal losses of 15 - 20 cm^{-1} and characteristic temperatures T_0 of 65 K. Single moded dfb-lasers with 9 QWs showed modulation frequencies of 12 GHz[1].

Incorporation of strain reduces effective hole masses, leading to increased symmetry in band structure and improved lasing characteristics. The beneficial role of strain was investigated in GaInAs-QWs with 1.2 % compressive and 0.6 % tensile strain by changing the In-content. Emission around 1.5 μm required well widths of 3.5 nm and 12.5 nm, respectively, while quaternary barrier compositions corresponding to $\lambda = 1.3$ μm and 1.1 μm were chosen. Lasers with 3 QWs, with either compressive or tensile strain, showed improved device performance with reduced $j_{th}(\infty) = 80 - 90$ A/cm^2 per QW, increased $\eta_i = 0.8 - 0.9$, reduced internal losses of 5 - 15 cm^{-1} and T_0-values between 80 and 90 K for both types.

---

1. dfb-lasers were processed in the research laboratories of the Deutsche Telekom AG, Darmstadt

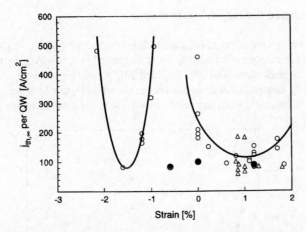

Figure 2. Extrapolated threshold current density $j_{th}(\infty)$ of CBE grown (InGa)(AsP)-laser diodes (full dots) compared to $j_{th}(\infty)$ of similar lasers grown by MOVCD (open symbols and lines; after Thijs [3])

In Figure 2 extrapolated threshold current density $j_{th}(\infty)$ of our CBE-grown (InGa)(AsP)-lasers are compared to $j_{th}(\infty)$ obtained in comparable MOCVD-grown lasers by several authors. For lattice-matched as well as for strained active layers the CBE-grown lasers show very low threshold current densities below 100 A/cm^2 per QW, among the lowest obtained in MOCVD-grown lasers.

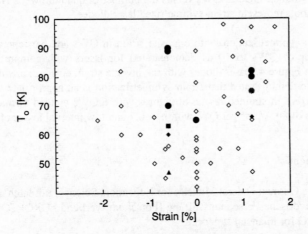

Figure 3. Characteristic temperatures T_0 of CBE grown (InGa)(AsP)-laser diodes (full symbols; large full dots: cavity length 1200 µm) compared to T_0 of similar lasers grown by MOVCD (open symbols, after Thijs [3])

In Figure 3 a similar comparison is made for the characteristic temperatures T_0. Especially for strained active layers T_0-values exceeded 80 K; these T_0-values again are higher than those generally obtained in MOCVD-grown lasers. Our CBE-grown laser diodes thus compared favourably with state-of-the-art devices.

78

4. Strain-symmetrized (InGa)(AsP) laser diodes

Gain saturation in QW-lasers necessitates a high number of QWs for high bit rate modulation. We found the number of strained QWs to be limited to 3, however, by the total critical thickness of the overall layer structure. Strained MQW-lasers with more than 3 QWs then revealed drastically increased threshold currents as shown in Figure 4 for compressively strained InGaAs-QWs. Internal losses increased as well due to the onset of lattice relaxation.

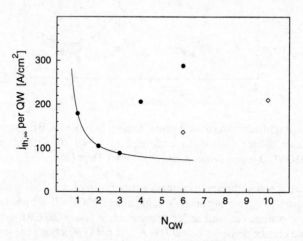

Figure 4.   Extrapolated threshold current density $j_{th}(\infty)$ of lasers with compressively strained InGaAs-QWs versus the number of quantum wells N_{QW}. Open symbols represent strain-symmetrized laser diodes

Using strain symmetrization, with opposite strain in QWs and barriers, allowed increasing the number of QWs to 10 as demonstrated for lasers with compressively strained InGaAs-QWs in Figure 4. Laser diodes with the inverse strain design, i.e. with tensile strain in the wells, also demonstrated that strain symmetrization is an appropriate way to increase the number of QWs in strained lasers. Structures with 6 QWs yielded unchanged threshold current densities of 80 A/cm^2 per QW, with $\eta_i = 0.7$ and low internal losses of 6 - 8 cm^{-1}.

Acknowledgement

We thank the German Federal Ministry for Education, Science, Research and Technology (BMBF), the Bayerische Forschungsstiftung (Forschungsverbund FOROPTO) and the Deutsche Telekom AG for financial support.

References

[1] Heinecke H 1993, *J . Crystal Growth* **127**, 126
[2] Rothfritz H, Müller R, Buchegger C, Tränkle G, Weimann G 1994, *J . Crystal Growth* **136**, 225
[3] Thijs P J A, Tiemeijer L F, Binsma J J M, van Dongen T 1994, *IEEE J. Quant. Electron.* **30**, 477

Inst. Phys. Conf. Ser. No 145: Chapter 2
Paper presented at 22nd Int. Symp. Compound Semiconductors, Cheju Island, Korea, 28 August–2 September 1995
© *1996 IOP Publishing Ltd*

Facet formation and selective area epitaxy of InGaAs by chemical beam epitaxy using unprecracked monoethylarsine

Sung-Bock Kim, Seong-Ju Park*, Jeong-Rae Ro, and El-Hang Lee

Research Department, Electronics and Telecommunications Research Institute, Yusong P. O. Box 106, Taejon, 305-600, Republic of Korea

Abstract. We have successfully grown various facets on the partially masked substrate by selective area epitaxy with chemical beam epitaxy (CBE) using unprecracked monoethylarsine. InGaAs selective growth mechanism was also examined on the GaAs(100) substrate patterned with various filling factors (the ratio of opening area to total area). The results show that the molecular reactants impinged on the mask are efficiently removed from the mask without gas phase diffusion or surface migration, thereby suppressing growth enhancement and compositional variation. We have also observed that the smooth (311), (377), and $(\overline{1}\,\overline{1}\,\overline{1})$ facets change only to (311) facet with increasing growth temperature in $[0\overline{1}1]$ direction of stripe mask. In case of [011] direction, however, (111) and $(0\overline{1}1)$ facets further developed into (111) facet with increasing growth temperature. Facet formation mechanisms are discussed using dangling bond model and migration properties.

1. Introduction

Epitaxial growth techniques, such as molecular beam epitaxy (MBE) and metalorganic chemical vapor deposition (MOCVD), have achieved atomic scale control in one dimension of the growth direction. This capability has resulted in a variety of new physical phenomena and device applications, producing one dimensional quantum confined structures such as quantum wells. Recently, there has been interest in fabricating two or three dimensional quantum confinement structure, so called quantum wires and dots, by simply restricting the lateral dimensions. These low dimensional quantum structures have received much attention because of their new physical phenomena and potential to greatly improve the device performance.

There have been several approaches to achieve lateral confinements using physical and chemical processing. Electron or ion beam lithography followed by dry etching(Kash *et al* 1989; Miller *et al* 1989), ion beam implantation for impurity-induced disordering(Cibert *et al* 1986), or selective diffusion(Zarem *et al* 1989) has been used to define these nanostructures. However, such techniques produce serious interface damages during the etching process or implantation. These damages are known to seriously reduce quantum efficiency. Thus, the selective area epitaxy without air-exposed and etching damaged interfaces is highly desirable. The control of facet formation using selective area epitaxy has been considered as one of the key issues for fabricating such quantum-confined structures with damage-free interfaces.

*permanent address : Department of Materials Science and Engineering, Kwangju Institute of Science and Technology, Kwangju, Republic of Korea

Since the first report of selective area epitaxy of GaAs(Duchemin *et al* 1978), selective area epitaxy of GaAs(Tokumitsu *et al* 1984; Davies *et al* 1992), InP(Heinecke *et al* 1992), and InGaAs(Chen *et al* 1992) using various growth techniques have been reported using SiO_2 or Si_3N_4 dielectric mask. Though various methods have demonstrated selective epitaxy, chemical beam epitaxy (CBE) was found to be an optimum technique showing excellent selectivity at normal growth temperature(Heinecke *et al* 1986).

In this study, we have investigated the facet formation of InGaAs and selective area epitaxy by CBE using unprecracked monoethylarsine (MEAs) under various growth conditions, such as pattern direction, growth temperature, and filling factor which is defined the ratio of the opening area to the total area. This study showed that InGaAs epilayers can be selectively grown using unprecracked MEAs at a wide range of temperatures from 540 $^{\circ}$C to 640 $^{\circ}$C and at the V/III ratio of 10. InGaAs was selectively grown at lower growth temperature than the conventional CBE using precracked hydride gas and the growth rate and indium composition corresponding to filling factor were not varied.

For facet formation, we observed that the smooth (311), (377) and ($\overline{1}\,\overline{1}\,\overline{1}$) InGaAs facets change to (311) facet with increasing growth temperature in [0$\overline{1}$1] direction of mask. In case of [011] direction, however, (111) and (0$\overline{1}$1) facets developed into (111) facet with increasing temperature. The facet growth mechanism could be explained by temperature-dependent migration length of group III atoms on sidewall and also by the number of dangling bonds of As atoms.

2. Experimentals

Si_3N_4 mask films for facet formation by selective area epitaxy were deposited on GaAs(100) substrate by plasma enhanced chemical vapor deposition (PECVD). These stripe masks consisted of two parts of constant opening area and constant mask area. In constant opening, 5 μm opening windows with various mask widths ranging from 1 to 30 μm were patterned into the Si_3N_4 film using standard photolithography and wet etching method. In constant mask, patterns consisted of 3 μm mask with opening widths ranging from 1 to 27 μm. These stripe masks were patterned in the [0$\overline{1}$1] or [011] direction to examine the dependence of facet formation on pattern direction.

Before growth, the patterned substrate were not given any treatment other than degreasing in an organic solution. They were sequentially heated at 150 $^{\circ}$C for 10 min in ultrahigh vacuum and at 600 $^{\circ}$C for 15 min in MEAs ambient of about 4×10^{-4} Torr in order to remove the moisture and the oxide layer from the surface, respectively. InGaAs epitaxial layers were grown at a pressure of 8×10^{-4} Torr from trimethylindium (TMIn), triethylgallium (TEGa) and unprecracked MEAs at temperatures ranging from 540 to 640 $^{\circ}$C. The source gases were introduced by automatic absolute pressure controlled leak valves without hydrogen carrier gas. The detailed configuration of system used in this work was described in the previous works(Park *et al* 1993; 1994).

The samples were inspected by optical Nomarski microscope and Auger electron spectroscopy (AES) for the measurement of selectivity and indium composition. The faceting morphologies and thickness of the selectively grown layers were also examined using a cross-sectional scanning electron microscope (SEM).

3. Results and discussion

Nomarski microscope and SEM photographs clearly show that InGaAs epilayers were all grown selectively on a GaAs substrate patterned with Si_3N_4 films at temperatures ranging from 540 to 640 °C and V/III ratio of 10.

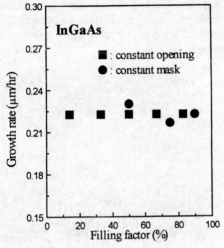

Fig.1 Effect of filling factors on InGaAs growth rates at a temperature of 555 °C and V/III ratio of 10

Fig. 2 Dependence of the relative indium composition on the filling factors

To investigate the area-dependent growth rate and composition variation in selective epitaxy by CBE using unprecracked hydride gas, we fabricated GaAs substrate patterned with stripe which has various filling factors. Fig. 1 shows that InGaAs growth rates corresponding to the filling factors have not been enhanced on these samples. Using a AES analysis, we also found that In composition is constant over a wide range of filling factors as shown in Fig. 2. These results can be explained by long collision length of source materials at low pressure. The desorbed materials from the mask do not contribute to surface diffusion and a subsequent growth of InGaAs in the opening GaAs, resulting in the constant growth rates and composition. Fig. 3 schematically illustrates the selective growth mechanism by adsorption-

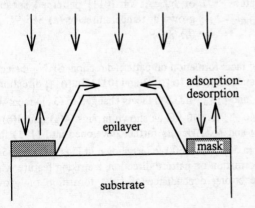

Fig. 3 A schematic diagram of selective growth mechanism by adsorption-desorption in CBE

desorption in CBE. The results show that the serious problems of variation in the film thickness and composition in epilayers, which occurs typically in MOCVD process, can be avoided using this method. Selective growth mechanism in MOCVD is mainly attributed to gas phase diffusion(Colas *et al* 1991) or surface migration on the mask(Hiruma *et al* 1990). In MOCVD, therefore, the difference of migration length and/or diffusion constant of the group III elements produce variation in the composition and thickness depending on the pattern shapes and sizes. However, in the CBE using unprecracked hydride reactants, the molecular species seem to have much lower sticking coefficient on the mask than an opened GaAs surface. The selective growth temperature is thus much lowered compared to the conventional CBE process where reactive atomic arsenics are employed.

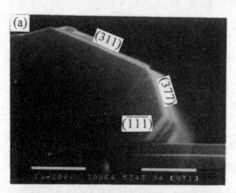

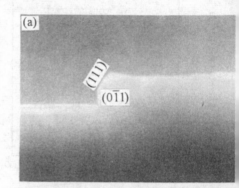

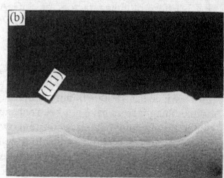

Fig. 4 Cross-sectional SEM photographs of InGaAs on [0$\bar{1}$1] patterned substrate grown at temperatures of (a) 560 °C and (b) 640 °C

Fig. 5 Cross-sectional SEM photographs of InGaAs on [011] patterned substrate grown at temperatures of (a) 555 °C and (b) 600 °C

To further study the dependence of facet formation on pattern direction, Si$_3$N$_4$ dielectric stripe patterns were prepared in the directions parallel to [0$\bar{1}$1] and [011]. In [0$\bar{1}$1] direction of Si$_3$N$_4$ stripe mask, smooth (311), (377), and ($\bar{1}\bar{1}$1) InGaAs facets changed to (311) facet with increasing growth temperatures from 540 °C to 640 °C as shown in Figs. 4(a) and 4(b). In case of [011] direction, however, (111) and (0$\bar{1}$1) facets further developed into (111) facet with increasing growth temperatures from 555 °C to 600 °C as shown in Figs. 5(a) and 5(b). In addition to the dependence of facet formation on pattern direction, a striking feature which is also noted in Figs. 4 and 5 is the strong dependence of facet formation on growth temperature. Below 560 ºC, deposition on the sidewalls and overgrowth on the mask seem to occur resulting in the formation of 2 or 3 facet planes as shown in Figs. 4(a) and 5(a). Above 600 ºC, however, InGaAs epliayers have only one stable facet. These results indicate that the surface migration length of reactants on the sidewalls strongly depends on the growth temperature. Deposition on the sidewall and the mask is attributed to the difference between the migration length of reactants on sidewalls and the length of these planes. At low growth temperature, the migration length of materials impinged on the sidewalls is relatively short and all of reactants can not transfer to the top surface, resulting in the sidewall growth and overgrowth. On the other hand, at high temperature, most of source materials migrate to (100) top surface due to the longer migration length, producing no growth on the sidewall and the mask.

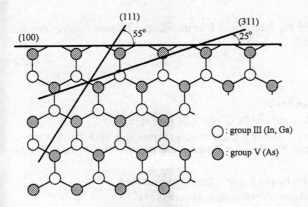

: group III (In, Ga)
: group V (As)

Fig. 6 A schematic diagram of atomic arrangement of InGaAs and their dangling bonds for each plane.

The (100) top surface has been always observed to be U-shaped in the stable facets as shown in Figs. 4(b) and 5(b). When two planes are adjacent during facet growth, the molecules tend to migrate from a plane with fewer dangling bonds to one with more dangling bonds. Therefore facet shapes indicate the migration profile of group III elements which come from the adjacent region. The facet growth mechanism may be explained by dangling bond model(Ando and Fukui 1989). Fig. 6 shows schematic diagram of atomic arrangement and dangling bonds for each plane. The relative dangling bond density of arsenic atoms per unit area is estimated to be 1 for (100), 0.6 for (311) and 0.58 for (111) plane. This means that when a (100) and (311) plane or (100) and (111) plane are adjacent, the adsorbates migrate from (311) or (111) to a (100) plane. Therefore, In and Ga atoms migrated from sidewalls are more easily adsorbed on (100) top surface than on the (311) or (111) facet producing U-shaped (100) top surface.

4. Summary

We have selectively grown InGaAs epilayers by CBE using unprecracked MEAs. The selective growth mechanism in CBE was investigated by measuring the growth rate and In compositional variation corresponding to the various filling factors. These results suggest that the reactants on the mask are removed from the mask region by reevaporation owing to long collision length at low pressure, resulting in no growth enhancement and invariation of composition. Also, InGaAs epilayers are selectively grown at the low temperature due to the unprecracked hydride which desorb more easily from mask than atomic arsenic.

Furthermore, we have investigated the facet formation of InGaAs corresponding to the pattern directions and growth temperatures. In $[0\bar{1}1]$ direction of stripe mask, (311), (377) and $(\bar{1}\bar{1}1)$ facets developed into (311) stable facet with increasing growth temperature. In cases of [011] direction, facets were finally reached to only one (111) facet with increasing temperature. These results indicate that the sidewall growth is closely related to the migration length of reactants on the facets, which is strongly dependent on the growth temperatures. Formation of the U-shaped (100) top surface could be explained by number of dangling bonds of As atoms and migration length of group III atoms.

Acknowledgment

This work was supported in part by the Ministry of Information and Telecommunications and Korea Telecom.

References

Ando S and Fukui T 1989 *J. Cryst. Growth* **98** 646

Chen Y, Zucker J E, Chiu T H, Marshall J L and Jones K L, 1992 *Appl. Phys. Lett.* **61** 10

Cibert J, Petroff P M, Dolan G J, Pearton S J, Gossard A C and Englich J H 1986 *Appl. Phys. Lett.* **49** 1275

Colas E, Shahar A, Soole J B D Tomlinson W J, Hayes J R, Caneau C and Bhat R 1991 *J. Cryst. Growth* **107** 226

Davies G J, Skevington P J, French C L and Foord J S 1992 *J. Cryst. Growth* **120** 369

Duchemin J P, Bonnet M, Koelsh F and Huyghe D 1978 *J. Cryst. Growth* **45** 181

Heinecke H, Baur B, Schimpe R, Matz R, Cremer C, Hö ger R and Miklis A 1992 *J. Cryst. Growth* **120** 376

Heinecke H, Brauers A, Grafahrrend F, Plass C, Pü tz N, Werner K, Weyers M, Lü th H and Balk P 1986 *J. Cryst. Growh* **77** 303

Hiruma K, Haga T and Miyazaki M 1990 *J. Cryst. Growth* **102** 717

Kash K, Scherer A, Worlock J M, Craighead H G and Tamargo M C 1986 *Appl. Phys. Lett.* **49** 1043

Miller B I, Shahar A, Koren U, and Corvini P J 1989 *Appl. Phys. Lett.* **54** 188

Park S J, Ro J R, Sim J K and Lee E H 1993 *Proc. Mater. Res. Soc.* **281** 37

Park S J, Ro J R, Sim J K and Lee E H 1994 *ETRI Journal* **16** 1

Tokumitsu E, Kudou Y, Konagai M and Takahashi K 1984 *J. Appl. Phys.* **55** 3136

Zarem H A, Sercel P C, Hoenk M E, Lebens J A and Vahala K J 1989 *Appl. Phys. Lett.* **54** 269

Inst. Phys. Conf. Ser. No 145: Chapter 2
Paper presented at 22nd Int. Symp. Compound Semiconductors, Cheju Island, Korea, 28 August–2 September 1995
© *1996 IOP Publishing Ltd*

Molecular beam epitaxy and mobility enhancement of $In_xGa_{1-x}As/In_{0.52}Al_{0.48}As/InP$ HEMT structure

Dong-Wan Roh, Hae-Gwon Lee, Jae-Jin Lee, Kwang-Eui Pyun, and Kee-Soo Nam

Research Department and Semiconductor Division, Electronics and Telecommunications Research Institute, Yusong P.O.Box 106, Taejon, 305-600, Korea

Abstract. The $In_xGa_{1-x}As/In_{0.52}Al_{0.48}As/InP$ HEMT structures for low noise application were grown by MBE onto InP substrates. The influence of growth temperature profile, growth interruption, and structural parameters on the electrical characteristics have been systematically studied based on Hall measurements. The growth of the whole layer with interruption results in increase of mobility due to a improvement of interface abruptness. For the lattice-matched conditions maximum mobilities with values amounting to as high as 11,400 cm^2/Vsec (300K), 50,300 cm^2/Vsec (77K) were obtained near $d_{spacer}=100$Å and $n_s=1.5 \times 10^{12}$ cm^{-2}. To improve the mobility characteristics, graded and pseudomorphic $In_xGa_{1-x}As$ were adopted as a channel layer. With the graded composition channel layer the mobilities of 11,800 cm^2/Vsec (300K), 50,900 cm^2/Vsec (77K) were obtained near $d_{spacer}=30$Å and $n_s=2.5 \times 10^{12}$ cm^{-2}.

1. Introduction

InGaAs/InAlAs heterostructures grown on InP have emerged as an important system for low noise application, thanks to the large band gap of InAlAs and the high n_s value due to high conduction band discontinuity (ΔE_c) at the interface. Moreover, high peak velocity due to large Γ-L valley separation, high mobility, a small effective mass of the electrons in the InGaAs layer, and reduced trapping and DX-related problem in InAlAs make this system very suitable for high speed devices e.g. HEMT and HBT[1-3].

InP based HEMT technology has evolved from research in a few labs into the mainstream of high speed microwave and millimeter wave circuits. InP HEMTs produce lower noise figures, higher gain, and higher cut-off frequencies in comparison to the best GaAs based FETs[4]. Generally, the electrical characteristics, especially electron mobility of InGaAs/InAlAs system can be described by the three components: i) material quality of individual InGaAs and InAlAs layer, ii) epitaxial structure including doping concentration, doping type, and thickness of spacer and channel, and iii) novel growth technique, for example growth interruption.

Advanced MBE-grown InGaAs/InAlAs HEMT layer structures were realized by carefully optimizing the growth conditions and the layer sequence. In order to clarify the effect of growth conditions on layer quality, the lattice-matched $In_{0.53}Ga_{0.47}As/In_{0.52}Al_{0.48}As$ HEMT structures were grown by MBE onto InP substrates. The influence of growth temperature profile, growth interruption between individual layers, and structural parameters on the electrical parameters has been systematically studied based on Hall measurements.

To improve the mobility characteristics, graded and pseudomorphic $In_xGa_{1-x}As$ were adopted as a channel layer. The channel layer for best 2DEG confinement consist of lattice-matched prelayer and graded composition layer.

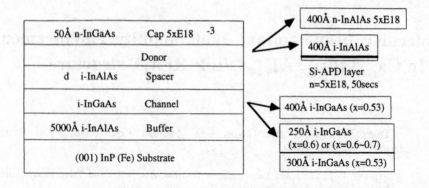

Fig.1 Epitaxial structures for modulation-doped HEMT devices.

2. Material growth and characterization

InGaAs/InAlAs modulation-doped structures were grown by MBE in a Riber MBE45 system on Fe-doped (100) InP substrates. The samples were three kinds of $In_xGa_{1-x}As/In_{0.52}Al_{0.48}As/InP$ HEMT structures consisting of i)lattice-matched (x=0.53), ii) pseudomorphic (x=0.6), and iii) graded (x=0.6~0.7) channel. Fig. 1 shows the generic epitaxial layer design studied in this work.

The growth temperature was measured by a thermocouple and referenced to the (2x4) to (4x2) surface reconstruction transition of the InP substrate. This transition was assumed to occur at 520 °C under an As_4 beam equivalent pressure of 1×10^{-5} torr. The native oxides on the substrate were removed prior to growth by heating to this temperature as discussed by

Gallet *et al.* [5], followed by growth of each layer at ~1μm/hr. The growth temperature of the structure with the exception of the strained channel was 540 °C. The V/III BEP ratio was 10 and the flux ratio between III elements of the lattice-matched layers, $F_{In}/(F_{In}+F_{Ga})$ and $F_{In}/(F_{In}+F_{Al})$ were 0.822 and 0.671, rspectively.

First growth interruption of 50 secs was introduce after the growth of the lattice-matched prechannel so that the Ga temperature could be adjusted for the new composition, and second one after the growth of the whole channel to improve the interface abruptness between channel and spacer layer.

Hall measurements at 300 and 77K were used to determine the 2DEG concentration and mobility and photoluminescence measurements were performed at 300, 77, and 10K. The compositions of $In_xGa_{1-x}As$ and $In_{0.52}Al_{0.48}As$ samples were verified by DXRD.

3. Results and discussion

3.1. Structural properties

PL measurements at 77K on these structures exhibit bandgap luminescence of InGaAs (0.816eV, FWHM=23meV) as shown in Fig. 2. Inset shows the photoluminescence spectrum of InAlAs at 10K (1.519eV, FWHM=18meV). These data indicate the high quality of the grown material. DXRD was also used to examine the structural quality and In composition of InGaAs. All the layers are lattice-matched to InP substrate except the 250Å $In_{0.6}Ga_{0.4}As$ channel as shown in Fig. 3.

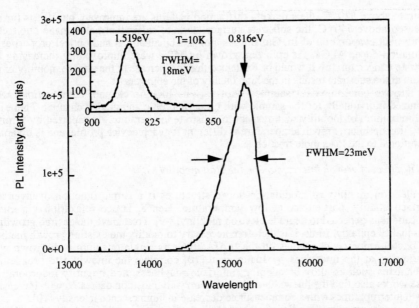

Fig. 2 Photoluminescence spectra for lattice-matched HEMT structure

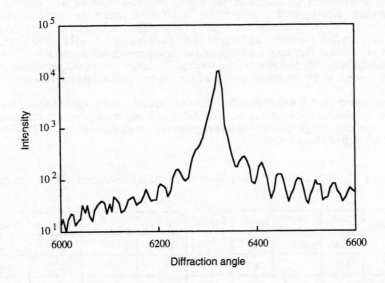

Fig. 3 Double crystal diffraction spectrum for lattice-matched HEMT structure

3.2 *The effect of the substrate temperature on the Hall mobility*

Table 1 shows the Hall mobility data and structural parameters for a number of modulation-doped structures in this work. The Hall mobility of Si δ-doped structure shows increase from 7850 cm^2/Vsec (300K) and 30000 cm^2/Vsec (77K) to 9600 cm^2/Vsec (300K) and 47500 cm^2/Vsec (77K) as the substrate temperature increase from 520 to 540 °C. Upon heating the substrate above 520 °C, the streaking diffraction patterns are improved, but if it is further increased above 540 °C, the substrate surface starts to suffer thermal damage due to the incongruent evaporation of In, Ga, and As atoms. The electrical and optical properties of modulation-doped $Al_xGa_{1-x}As$ epilayers grown by MBE were improved by increasing the substrate temperature[6]. It would be explained that the increase of the surface mobility of the adatom at the substrate results in the high quality of the epilayers.

The growth conditions for samples with various InGaAs channel compositions were optimized individually for the sample with pseudomorphic and graded channel. There is a local maximum of mobility with respect to substrate temperature. As reported by Streit *et al.*[7], this optimum growth temperature is different for each device profile, and is inversely proportional to the InAs mole fraction.

3.3 The effect of growth interruption on the Hall mobility

The electron mobility of modulation-doped structures is a strong function of layer and interface quality, particulary at low temperature. Hence, lattice relaxation via misfit dislocations is expected to degrade electron mobility[8-9]. To achieve the desired growth of high-quality epitaxial films, it is therefore necessary to modify the existing growth method. The photoluminescence studies on the InAlAs/InGaAs single quantum well grown with interruption at the interfaces by Juang *et al.*[10] correlate the linewidth to broadening mechanisms such as alloy broadening, interface roughness, and impurity incorporation. These works also find the linewidth of the quantum well transition decreases with increasing growth interrupt times with a commensurate decrease in luminescence intensity [10].

In this work, we report studies on the mobility enhancement of InGaAs/InAlAs HEMT structure grown with and without interruption as shown in Table 1. It has been known that the large difference in the In-As and Al-As bond strengths produces very different cation migration rates during growth, thereby making it difficult to control the interface quality. The difficulties may be overcome by novel growth approaches such as interrupting the growth before the interface formation and use of thin smoothening layers[11]. The interruption allows a greater time for cation migration and/or reevaporation and consequent smoothening of the surface before the formation of the interface. Therefore interrupted growth can result in an improvement in the structural quality of the heterointerface and enhancement of Hall mobility.

The interrupted growth which introduce between InGaAs donor and InAlAs spacer layer again has no effect on the enhancement of mobility. It is important to point out the abruptness at the interface of triangle potential well is key role to increase the mobility of 2DEG confined in potential well of channel layer.

Table 1. The Hall mobility characteristics of various lattice-matched HEMT structure.

sample	Sub. Temp (oC).	Growth Inter-ruption	Spacer Thick. (Å)	Doping Type	n_s ,1x10^{12} cm^{-2}		μ_H ,cm^2 /Vsec	
					300K	77K	300K	77K
A1	520	x	200	δ	2.95	2.0	7850	30000
A2	540	x	200	δ	2.3	1.8	9600	47500
B1	540	x	200	space	1.5	1.2	7060	22000
B2	540	o	200	space	1.5	1.2	9690	57400
A3	540	o	100	space	1.5	1.4	11400	43400
A4	540	o	30	space	2.64	1.86	9420	50300

It could be concluded that for the mobility enhancement of HEMT structure interface abruptness between channel and spacer layer is important to do not prevent the movement of 2DEG in the undoped InGaAs channel.

3.4 The effect of structural parameters on the Hall mobility

Table 1 also shows the Hall mobilities as a function of the thickness of the spacer layer. For the lattice-matched condition the maximum mobilities of 11,400 cm^2/Vsec (300K) and 50,300 cm^2/Vsec (77K) were obtained near d_{spacer}=100Å and n_s=1.5x10^{12} cm^{-2}, and the mobility decreased with a decrease in spacer thickness, possibly due to the large Coulombic interaction by ionized donors and ionized impurity scattering by high density dopants which were necessary to achieve a large n_s. Table 2 shows the 2DEG mobility for the two types of HEMT structure with pseudomorphic (x=0.6) and graded (x=0.6~0.7) channel layer as compared with lattice-matched one grown under same growth conditions. The InAlAs spacer thickness and the n-InAlAs doping density were fixed at 30Å and 5x10^{18} cm^{-3}, respectively. The channel consists of 300Å lattice-matched prechannel and 250Å second one.
The mobility at 300K increased with an increase in x from 9250 cm^2/Vsec at x=0.53 (lattice-matched conditions) to 11800 cm^2/Vsec at x=0.6~0.7 (graded In composition). The 77K mobility had the similar dependence on x and the maximum value was 50900 cm^2/Vsec for graded composition case. On the other hand, the 2DEG density, n_s remained almost constant at 2.5x10^{12} cm^{-2} at 300K. These results show that our new channel structure has a enhancement effect on the electron transport characteristics even if the 2DEG density is changed to fit the low noise application.

4. Conclusions

The optimization of the structural parameters and material growth technique of a lattice-matched InGaAs/InAlAs/InP HEMT is reported. The growth of the whole layer with interruption results in increase of mobility due to a improvement of interface abruptness. For the lattice-matched conditions maximum mobilities with values amounting to as high as 11,400 cm^2/Vsec (300K), 50,300 cm^2/Vsec (77K) were obtained near d_{spacer}=100Å and n_s=1.5x10^{12} cm^{-2}. The further improvement of mobility was obtained by using graded channel design. The mobility at 300K for the graded channel with d_{spacer}=30Å and n_s=2.5x10^{12} cm^{-2} increased with an increase in x from 9250 cm^2/Vsec at x=0.53 (lattice-matched conditions) to 11800 cm^2/Vsec at x=0.6~0.7 (graded In composition).

Table 2. The Hall mobility data for the strained HEMT structure. (Si δ-doped, d_{spacer}=30Å, and with growth interruption).

sample	Second channel type	n_s,1x10^{12} cm^{-2}		μ_H,cm^2/Vsec	
		300K	77K	300K	77K
A	lattice-matched (x=0.53)	2.5	2.0	9250	40400
B	pseudomorphic (x=0.6)	2.5	2.2	9800	43000
C	graded (x=0.6~0.7)	2.4	2.3	11800	50900

90

References

[1] Dickmann J, Riepe K, Daembkes H and Kunzel H 1993 *Proc. 5th Int. Conf. on InP and Related Materials* (April 19-22, 1993, Paris, France; by the IEEE Lasers and Electro-Optics Society and the IEEE Electron Devices Society, Piscataway, USA) 461-4

[2] Mishra U K and Shealy J B 1994 *Proc. 6th Int. Conf. on InP and Related Materials* (March 27-31, 1994, Santa Barbara, CA; by the IEEE Lasers and Electro-Optics Society and the IEEE Electron Devices Society, Piscataway, USA) 14-7

[3] Bhattacharya P ed. 1993 *Properties of Lattice-Matched and Strained Indium Gallium Arsenide* (INSPEC, Herts, UK) 290-305

[4] Nguyen L D, Brown A S, Thompson M A and Jelloian L M 1992 *IEEE Trans. on Electron. Dev.* **39** 2007-14

[5] Gallet D, Gendry M, Hollinger G, Overs A, Jacob G, Boudart B, Gauneau M, L'Haridon H and Lecrosnier D 1991 *J. Electron. Mat.* **20** 963-5

[6] Drummond T J, Fisher R, Morkoc H and Miller P 1982 *Appl. Phys. Lett.* **40** 430-2

[7] Streit D C, Block T R, Wojtowicz M, Pascua D, Lai R, Ng G I, Liu P H and Tan K L 1995 *J. Vac. Sci. Technol. B* **13** 774-6

[8] Tacano M, Sugiyama Y, Takeuchi Y and Ueno Y 1991 *J. Electon. Mat.* **20** 1081-5

[9] Chough K B, Chang T Y, Feuer M D and Lalevic B 1992 *Electron Lett.* **28** 329-31

[10] Juang F Y, Bhattacharya P K and Singh J 1986 *Appl. Phys. Lett.* **48** 290-2

[11] Goldstein L, Charasse M N, Jean-Louis A M, Leroux G, Allovon M and Marzin J Y 1985 *J. Vac. Sci. Technol. B* **3** 947-9

Inst. Phys. Conf. Ser. No 145: Chapter 2
Paper presented at 22nd Int. Symp. Compound Semiconductors, Cheju Island, Korea, 28 August–2 September 1995
© *1996 IOP Publishing Ltd*

Substrate orientation dependence of lateral composition modulation in $(GaP)_n(InP)_n$ strained short period superlattices grown by gas source MBE

S. J. Kim, H. Asahi, K. Asami, T. Ishibashi and S. Gonda

The Institute of Scientific and Industrial Research, Osaka University

8-1, Mihogaoka, Ibaraki, Osaka 567, Japan

Abstract. $(GaP)_1(InP)_1$ superlattices (SLs) were grown on GaAs (100), (311), (411) and (111)B substrates by gas source MBE. Atomic force microscopy and transmission electron microscopy observations show that the SLs grown on GaAs (100), (311)A and (411)A substrates have wire, wire-like and dot-like lateral composition modulations, respectively. 4.2K photoluminescence peak energies are greatly dependent on the substrate orientation and growth temperature, ranging from 1.722 eV to 1.979 eV. The lowest peak energy ever reported, 273 meV lower than that of the disordered $In_{0.5}Ga_{0.5}P$ alloy, is observed from the SL grown on GaAs (111)B substrate, where the strong CuPt-type ordering is confirmed.

1. Introduction

Low-dimensional confinements of carriers in quantum wire/dot structures are interesting for novel device applications and physics [1-3]. In laser diode using quantum structures increased differential gain, lower threshold current density and improved stability are predicted. However, the difficulty lies in the fabrication of these structures. Selective etching of planar quantum well structures and multi-layer growth on the vicinal/patterned substrates have been studied widely [4]. However, low dimensional structures fabricated by these methods are not satisfactory.

Recently, lateral periodic composition modulation was observed in the $(GaP)_n(InP)_n$ short period superlattices (SLs) grown on GaAs(100) substrates by gas source MBE and was applied to the fabrication of quantum wires [5]. The lateral composition modulation was found to occur along the [011] direction with a periodicity of about 200 Å and an average length of up to 3000 Å along the [01$\bar{1}$] direction, which forms the lateral quantum wires perpendicular to the growth direction. This lateral composition modulation was explained as being induced by strain present in the SLs [6]. However, the dimensions of the lateral quantum wire grown on the (100) just substrates are not completely uniform. One possible way to improve the uniformity may be the use of vicinal or high index surface substrates.

In this paper, we have studied the substrate orientation and substrate temperature dependence of the lateral composition modulation in $(GaP)_1(InP)_1$ SLs grown by gas source MBE.

2. Experimental

$(GaP)_1(InP)_1$ SLs were grown by gas source MBE. Elemental Ga, In and thermally cracked PH_3 were used as group III and group V sources, respectively. The substrates used were GaAs (100), (111)B 2° off toward $[\bar{1}\bar{1}0]$, (311)A, (311)B and (411)A. Reflection high energy electron diffraction (RHEED) intensity oscillations were used to calibrate GaP and InP growth rates. $(GaP)_1(InP)_1$ SLs of 530 periods (thickness: 0.3 μm) were grown at 400, 425, 460 and 490°C after growth of 100 nm GaAs buffer layer at 600°C. PH_3 flow rate during growth was 1.2 SCCM. Formed structures were investigated with atomic force microscopy (AFM), transmission electron microscopy (TEM) and 4.2 K photoluminescence (PL) measurements. The TEM observation was conducted with a 100 kV acceleration voltage. A 325 nm line of He-Cd laser was used as an excitation source in the PL measurement.

3. Results and discussion

Figure 1 shows the examples of the AFM images observed on the $(GaP)_1(InP)_1$ SLs grown on the GaAs (100) and (411)A substrates. Surface structures observed by AFM were dependent on the growth temperature and substrate orientation and showed that the surface structures on the $(GaP)_1(InP)_1$ SLs grown on the GaAs(100), (311)A and (411)A substrates

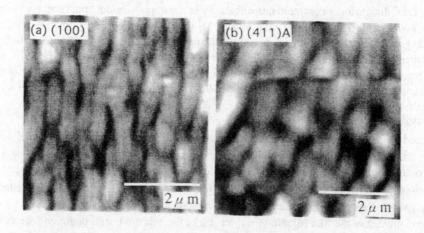

Fig.1   AFM images for the $(GaP)_1(InP)_1$ SLs on (a) GaAs (100), (b) GaAs (411)A. The images show wire and dot-like structures.

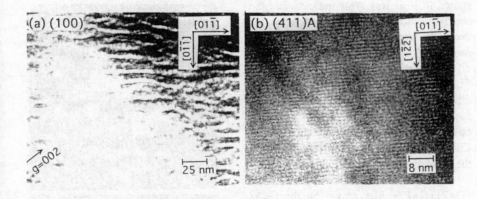

Fig. 2 (a) g=002 dark-field high-resolution plan-view TEM image for the $(GaP)_1(InP)_1$
SLs grown on GaAs (100) at 400°C and (b) bright-field high-resolution plan-view
TEM image for the $(GaP)_1(InP)_1$ SLs grown on GaAs (411)A at 460°C.

were wire, wire-like and dot-like, respectively. The size of the surface structures increased
with substrate temperature. The plan-view TEM images of the structures were similar to the
AFM surface structures. These results suggest that the growth of $(GaP)_n(InP)_n$ SLs on the
(n11) substrates probably makes it possible to form quantum dot structures as well as
quantum wire structures by gas source MBE, though further investigations are required. They
have a high density of structures which is very important to obtain high optical gain in the
optical device applications.

The relatively large scale lateral layer composition modulation observed was always along
the [011] direction. This composition modulation is probably due to the island formation
along the [01$\bar{1}$] direction during growth, which is primarily determined by the direction of the
group V dimer bonds related to the different surface energy and migration direction [7].

High resolution plan-view TEM image for the $(GaP)_1(InP)_1$ SL grown on GaAs (100)
substrate at 400°C shows the fine superstructures with a lateral periodicity of 120 Å along the
[011] direction, as shown in Fig. 2(a). Much more fine lateral superstructures were observed
in the SLs grown on GaAs (411)A and GaAs (311)A substrates, as shown in Fig.2(b). The
period of the lateral superstructures is about 13Å along the [$\bar{2}$33] direction and 11Å along the
[$\bar{1}$22] direction for (311)A and (411)A, respectively. Nötzel et al. [8] observed the lateral
superstructures of the periodicity of 13Å along the [$\bar{2}$33] direction in the GaAs/AlAs multilayer
structures grown on GaAs (311)A substrates with the high-resolution TEM. These periods
correspond to the terrace widths on these substrates.

The TEM diffraction pattern with an incident beam parallel to the [100] direction for the SL
grown on the GaAs(100) substrate clearly shows the sharp spots (0 1/2 1/2) and (0 $\bar{1/2}$ $\bar{1/2}$) as

shown in Fig. 3(a). This indicate that the superstructure with a period of twice the usual lattice constant in the [011] direction is formed [9]. On the other hand, in the TEM diffraction patterns for the $(GaP)_1(InP)_1$ SLs grown on GaAs (111)B with a [211] direction incident electron beam, the sharp superstructure spots {1/2 1/2 1/2}were observed (Fig. 3(b)). This indicates that the CuPt-type long-range ordering is formed in the [111] direction in these SLs.

Figure 4 shows 4.2K PL spectra for the $(GaP)_1(InP)_1$ SLs grown on the GaAs substrates having various orientations. The spectrum is greatly dependent on the substrate orientation. The SLs grown on the (311)A and (411)A surfaces show two sharp peaks, while other SLs exhibit a broad peak. The SL grown on the (411)A surface has the strongest PL intensity. The PL peak energies vary from 1.722 eV to 1.979 eV, depending on the substrate orientation and substrate temperature as shown in Fig. 5(a). The peak energy of the $(GaP)_1(InP)_1$ SLs grown on the (100) surface shows a slight shift toward lower energy with increasing substrate temperature as observed in Ref. [7]. This is attributed to the larger composition modulation in the lateral quantum well. The lower energy transition corresponds to the formation of larger In composition regions.

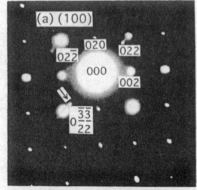

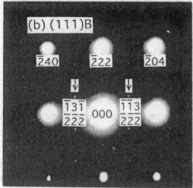

Fig.3   TEM diffraction patterns for the SLs grown on (a) GaAs (100) and (b) GaAs (111)B. Incident electron beam is parallel to (a) [100] and (b) [211] directions. Observations of the sharp spots (0 1/2 1/2) and (0 $\overline{1/2}$ $\overline{1/2}$) in Fig. (a) and {1/2 1/2 1/2} in Fig. (b) indicate the existence of the fine lateral superstructures along the [011] direction and the CuPt-type long-range ordering along the [111] direction, respectively.

The $(GaP)_1(InP)_1$ SLs grown on the (311)A and (411)A surfaces exhibit higher peak energy than those grown on other orientations and these PL peak energies increase with substrate temperature. These high peak energy are considered to be due to the fine lateral superstructures observed in Fig. 2(b). The lateral structure forms the lateral superlattices. The lateral confinement in the lateral superlattices shifts the PL peak toward higher energy.

The PL from the SL grown on the GaAs (111)B substrate shows the lowest peak energy of 1.722 eV. The CuPt-type spontaneous long-range ordering in the [111] direction was observed in the InGaP alloy layers grown on GaAs (100) substrates by MOVPE [10], and was reported to lower the band gap energy up to 50~160 meV compared with that for the disordered InGaP alloy [9, 11, 12]. The PL peak energy observed in the present work is about 273 meV lower than that (1.995 eV at 4.2K [12]) of the disordered InGaP alloy. The

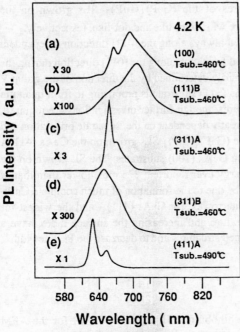

Fig.4 4.2K PL spectra for the $(GaP)_1(InP)_1$ SLs
grown on GaAs (a) (100), (b) (311)A, (c) (311)B,
(d) (311)B and (e) (111)B substrates.

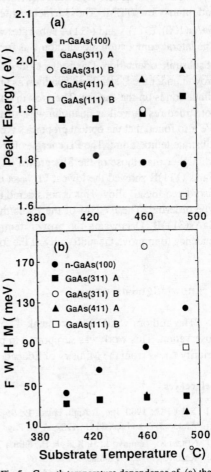

Fig.5 Growth temperature dependence of (a) the
4.2K PL peak energy and (b) the FWHM for the
$(GaP)_1(InP)_1$ SLs as a function of substrate
orientation.

formation of CuPt-type ordering was confirmed in Fig.3(b). The observed PL peak energy is the lowest value ever reported. [11, 12].

The full width at half maximum (FWHM) of the 4.2K PL peaks varies with substrate temperature and substrate orientation as shown in Fig. 5(b). The FWHM for the $(GaP)_1(InP)_1$ SL grown on the GaAs (100) substrate increases with substrate temperature due to the increased non-uniformity in the lateral ordering, which was confirmed in the plan-view TEM images. However, the FWHM for the GaAs (311)A showed only small substrate temperature variation. The widest FWHM was observed for the (311)B and the narrowest for the (411)A. In general, decreasing the growth temperature and increasing the surface index seem to improve the uniformity of the formed structures.

4. Summary

$(GaP)_1(InP)_1$ SLs were grown on GaAs(100) and (n11) substrates by gas source MBE.

The dependence of self-organized structures on the substrate orientation was studied by AFM and plan-view TEM. The surface structures of the $(GaP)_1(InP)_1$ SLs grown on the GaAs(100), (311)A and (411)A substrates were wire, wire-like and dot-like, respectively.
The lateral compositon modulation was formed always along the [011] direction independent of substrate orientation. This is due to the island formation along the [01$\bar{1}$] direction during the growth under the P-stabilized conditon as primarily determined by the direction of the group V dimer bonds on the surface. The growth on the (n11) substrate is promising to form quantum dot structures as well as quantum wire structures, though further investigations are required. We also found that the optical properties are greatly dependent on the substrate orientation and substrate temperature. The PL energies for the $(GaP)_1(InP)_1$ SLs grown on the GaAs (411)A are higher than those of the SLs grown on the GaAs (100) substrates. The SL grown on the GaAs (111)B showed the lowest PL peak energy ever reported: 273 meV lower than that of disordered InGaP alloy. This is considered to be due to the formation of much complete CuPt-type ordering. The FWHM of the PL is the narrowest for GaAs (411)A and the widest for GaAs(311)B. Decreasing the growth temperature and increasing the surface index have a tendency to improve the uniformity of the formed structures and to decrease the PL linewidth.

Acknowledgment

The authors would like to thank T. Yoshinobu and H. Iwasaki of ISIR for the AFM observation. This work was surpported in part by a Grand- in-Aid for Scientific Research on Priority Areas from the Ministry of Educations, Science and Culture (No. 07225210).

References

[1]   Sakaki H.; 1980, Jpn. J. Appl. Phys., **19**, 95

[2]   Arakawa Y. and Sakaki H.; 1982, Appl. Phys. Lett., **40**, 939

[3]   Nötzel R., Temmyo J., H. Kadada H., Furuta T. and Tamamura T.; 1994, Appl. Phys. Lett., **65**, 1

[4]   Kash K.; 1990, J. Luminescence, **46**, 69

[5]   Hsieh K. C., Baillargen J. N. and Cheng K. Y.; 1990, Appl. Phys. Lett., **57**, 2244

[6]   Pearah P. J., Chen A. C., Moy A. M., Hsieh K. C. and Cheng K. Y.; 1994, IEEE J. Quantum Electron., **QE-30**, 608

[7]   Chen A. C., Moy M., Chou L. J., Hsieh K. C. and Cheng K. Y.; 1995, Appl. Phys. Lett., **66**, 2694

[8]   Nötzel R., Ledentsov N. N., Däweritz L., Hohenstein M. and Ploog K.; 1991, Phys. Rev. Lett., **67**, 3812

[9]   Gomyo A., Suzuki T., Kobayashi K., Kawata S., Hino I. and Yuasa T.; 1987, Appl. Phys. Lett., **50**, 673

[10] Gomyo A., Susuki T., and Iijima S.; 1988, Phys. Rev. Lett., **60**, 2645

[11] McDermott B. T., Reid K. G., El-Masry N. A., Bedair S. M., Duncan W. M., Yin X. and Pollak H.; 1990, Appl. phys. Lett., **56**, 1171

[12] Su L. C., Ho I. H., Kobayashi N. and Stringfellow G. B.; 1994, J. Crystal. Growth., **145**, 140

Inst. Phys. Conf. Ser. No 145: Chapter 2
Paper presented at 22nd Int. Symp. Compound Semiconductors, Cheju Island, Korea, 28 August–2 September 1995
© *1996 IOP Publishing Ltd*

The influence of surface reconstructions on the GaAs/AlAs interface formation by MBE

N T Moshegov, L V Sokolov, A I Toropov, A K Bakarov, A K Kalagin, V V Tichomirov

Institute of Semiconductor Physics, Siberian Branch of Russian Academy
of Sciences, pr.Lavrentieva, 13, 630090 Novosibirsk, Russia

Abstract. Peculiarities of the RHEED specular beam oscillation behaviour during the formation of normal (AlAs on GaAs) and inverted (GaAs on AlAs) heterointerfaces were investigated. It was demonstrated that at intermediate As_4/Al flux ratios the stabilization of 3x2 structure on growing AlAs surface is controlled by segregation of Ga atoms into growing layer. The observed phase shift of oscillation curves in $[1\bar{1}0]$ azimuth, that accompanies the appearance of 3x2 reconstruction on growing AlAs surface, is induced by variation of the relationship between specular and Bragg components in measured signal. Contrary to the case of normal interface the process of oscillation phase variation during growth of GaAs on AlAs is completed over first 1-2 oscillations.

1. Introduction

One of the most important problems connected with the growth of quantum well heterostructures and short period superlattices is the structural disorder on the atomic scale at the normal (AlAs on GaAs) and inverse (GaAs on AlAs) heterointerfaces. As AlAs binding energy is higher and surface segregation of Ga and such impurities as C and O are enhanced the AlAs surface can be expected to be rougher than GaAs surface [1-3]. Usual method of *in situ* control during MBE is the precise monitoring of oscillation in the intensity of the specular reflected spot of the reflection high energy electron diffraction (RHEED). However, the main problem in the interpretation of dynamical RHEED pictures is the ability to separate diffraction-induced effects and changes in morphology of the growth front.

In this work the direct correlation between phase shift of oscillation curves and changes of the surface reconstruction was established. The observed phase shift of oscillation curves in $[1\bar{1}0]$ azimuth, that accompanies the appearance of 3x2 reconstruction on growing AlAs surface, is induced by variation of the relationship between specular and Bragg components in measured signal.

98

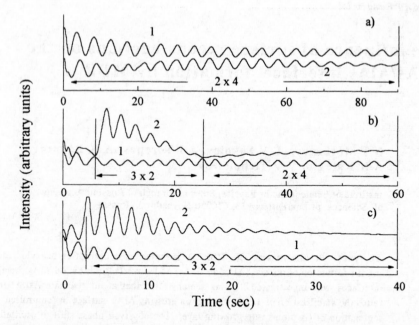

Fig.1. Dependence of RHEED intensity oscillation shapes on growth rate (AlAs on GaAs) for the [110]-(1) and [1$\bar{1}$0] -(2) azimuth of the incident electron beam. a) The growth rate of AlAs (V_{AlAs}) is 0.22 ML/s; b) V_{AlAs}=0.36 ML/s; c)V_{AlAs}=0.56 ML/s. T_s=620°C.

2. Experimental

The MBE growth was carried out on "Katun-C" machine on the singular GaAs (100) substrate with misorientation <0.1°. Before experiment buffer layer was grown to provide a clear (2x4)As reconstruction. The usual growth sequence was 15-20ML of GaAs and 15-20ML of AlAs separated by 60s recovery intervals. The fixed recovery times ensure a reproducible surface state at the beginning of growth. The growth rate for GaAs was 0.5ML/s and for AlAs ranged from 0.2 to 0.7ML/s. The substrate temperature (T_s) was varied from 420 to 620°C. We have compared the RHEED intensity oscillations recorded in the [110] and [1$\bar{1}$0] azimuths under the same growth condition. The data acquisition system we employ involves an optical photometer which output signal fed through interface into computer where it was analyzed by software.

3. Dependence of phase shift on As$_4$/Al flux ratio

In Fig.1 the dependencies of specular beam intensity (SBI) on time are shown for AlAs growth rates ranging from 0.2 to 0.6 ML/s under invariable As$_4$ specific pressure and substrate temperature of T_s=620°C. In the beginning the oscillations of SBI were recorded in the [110] azimuth. After this the substrate was rotated in azimutal direction by 90 degrees and recording was repeated with keeped growth conditions and without any tuning of electron beam.

At low Al fluxes the growth surface of AlAs is characterized by the same surface reconstruction (2x4)As as the initial GaAs. In the curves recorded in the [110] azimuth the

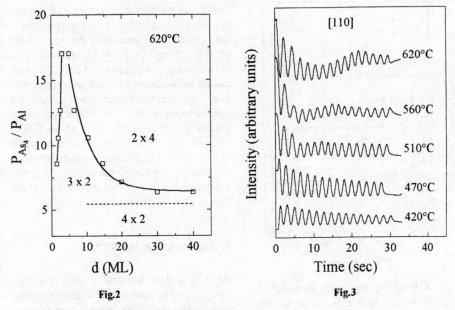

Fig.2. Kinetic phase diagram for growing surface of AlAs on GaAs.

Fig.3. Temperature dependence of the RHEED intensity oscillation shapes of growth AlAs on GaAs for the [110] azimuth.

beats are presented but as a whole the oscillations in curves 1 and 2 (Fig.1a) are in phase over a considerable time. With the increase of AlAs growth rate up to 0.5ML/s at SBI curve taken in the [1$\bar{1}$0] azimuth a "hump" evolves that is accompanied by, firstly, the variation of superstructure from (2x4)As to 3x2, and, secondly, shift of oscillation phase by π in reference to SBI curve in the orthogonal azimuth. After 10-15s AlAs growth SBI newly decreases and the curves recorded in the [110] and [1$\bar{1}$0] azimuths become synphased (Fig.1b).

Subsequent increase of Al flux (or decrease of As$_4$/Al flux ratio) results in that reconstruction 3x2 forms practically immediately after the beginning of AlAs growth and remains immutable with time (Fig.1c). In this case the phase of oscillations in the [1$\bar{1}$0] azimuth is shifted by π as to that in the [110] azimuth.

On the basis of RHEED data the kinetic phase diagram was constructed for growth surface of AlAs on GaAs at T_s=620°C (Fig.2). It is obvious that with the decrease of As$_4$/Al flux ratio the region of surface reconstruction 3x2 expands and captures 15-20 ML of AlAs. Moreover, with further decrease of As$_4$ pressure the structure is conserved over whole growth time.

4. Dependence of phase shift on substrate temperature

The manipulation with T_s keeping constant As$_4$ and Al fluxes makes possible the building up of AlAs in different parts of phase diagram. Low T_s=420°C corresponds to strong enriching both GaAs and growing AlAs surfaces with As. In this case any peculiarities were detected at the oscillation curves in the [110] and [1$\bar{1}$0] azimuths (Fig.3). Starting with

100

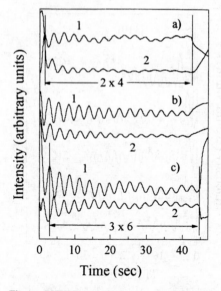

Fig.4. RHEED intensity curves for GaAs on GaAs growth. The surface do not change the reconstruction during growth-b); the growth is accompanied by the c(4x4)As→ (2x4)As- a), and (2x4)As → (3x6)-c) surface reconstruction transition. Azimuth [110]-1, [1Ī0]-2.

560°C the distinguished variation of the envelope of oscillations becomes noticeable that coincides in time with the appearance of SBI "hump" in the [1Ī0] azimuth. The development of characteristic dip is accompanied by variation of surface reconstruction, the fractional diffraction streaks 1/2 order disappear but 1/3 and 2/3 streaks arise gradually. An elevation of T_s above 630°C induces the reconstruction change from (2x4)As to 3x2 in accordance with the phase diagram of AlAs growth surface. It is significant that there is no additional phase shift in the [110] azimuth.

5. Segregation influence

As it was demonstrated in our previous study [4], the variation of oscillation phase may also take place during homoepitaxy growth when surface reconstruction changes. In the transition from the structure c(4x4)As highly enriched by As, that is a characteristic of initial GaAs surface, to growth structure (2x4)As the shift of the oscillations phase is observed in both azimuths (Fig.4a). So, the dependencies of SBI on time except first period are in antiphase to the curves recorded when the reconstruction change with growth starting is absent (Fig.4b). The increase of period and transformation of the shape of first oscillation allow to assume that two-dimensional islands do not form at the beginning of growth. According to [5] it may be considered that on the surface type c(4x4)As the additional As layer presents with the thickness of $a_0/4$. The arriving Ga interacts with the redundant As to form GaAs, in this process either normal or step-flow growth mechanism takes place. The transition of superstructure from c(4x4)As to (2x4)As has finished over one period and only after this the conditions appear for two dimensional nuclei formation. When the reconstruction changes from (2x4)As to 3x6 the phase shift of oscillations curves occurs only in the [1Ī0] azimuth as a result of variation of relationship between specular and Bragg components in measured signal.

Material change in the early stages of AlAs heteroepitaxy on GaAs naturally results in significant perturbations of surface microrelief, because the migration rate of Al adatoms is noticeably lower than that of Ga ones. As may be seen from the kinetic phase diagram of growth surface of AlAs on GaAs sublayer (Fig.2), in a definite field of T_s and As_4/Al flux ratio two sequential superstructure transitions occur in the early stages of the growth. The observed sequence of transformations (2x4)As→ 3x2→ (2x4)As with fixed T_s and other growth parameters may be attributed to variation of the composition of growth surface, particularly, presence of Ga in growing AlAs layer in a result of segregation processes. This is in a good agreement with the results of the study [6], where the different mechanisms of the appearance of additional phase shift in oscillation dependencies in the [1Ī0] azi-

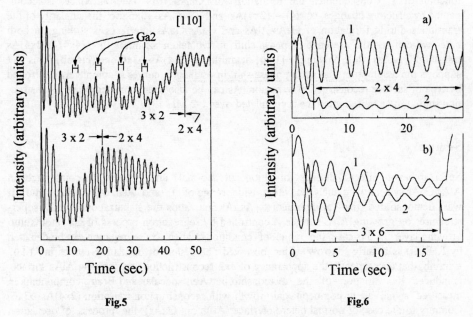

Fig.5

Fig.6

Fig.5. A changes in the envelope of the RHEED traces in [110] azimuth during AlAs on GaAs growth due to the additional Ga flux.

Fig.6. RHEED intensity oscillation shapes for inverse heterointerface GaAs on AlAs. The growth is accompanied by the c(4x4)As→ (2x4)As-a) and (2x4)As→ (3x6)-b) surface reconstruction transition. Azimuth [110]-1, [1 1̄ 0] -2.

muth during growth of homo-(AlAs/AlAs) and hetero-(AlAs/GaAs) interfaces were analyzed.

The following experiment would be considered as a strong evidence that just the presence of Ga atoms on the growth surface defines the stability of 3x2 structure over intermediate As4/Al flux ratios (Fig.5). During the growth of AlAs on GaAs in certain moments the Ga source was actuated for 1-2s to form a very weak flux of Ga atoms at the level of 0.1 ML/s. As a result, the time of observation of 3x2 structure can be modified over wide limits. The effect was clearly demonstrated in both azimuths. However, the range of either T_s or As4 effective pressure where the process of segregation can be detected by RHEED is fairly narrow. With the decrease of As4 pressure the growth surface of AlAs is characterized by 3x2 reconstruction irrespectively to Ga presence. The observed phase shift in the [1 1̄ 0] azimuth, that accompanies the appearance of 3x2 structure on AlAs growth surface, is associated from our opinion apparently with the redistribution of specular and Bragg components in measured signal as it was in the case of (2x4)As→3x6 transition during homoepitaxial growth.

6. Peculiarities of the formation of GaAs/AlAs heterointerface

In the majority of our experiments the conditions of GaAs growth were chosen near the boundary of phase transition (2x4)As→3x6. In this case the structure of AlAs surface after the pause for smoothing was characterized by c(4x4)As or (2x4)As structures in a function of T_s. Consequently, the formation of GaAs-AlAs heterointerface is accompanied by structure changes $c(4 \times 4) \rightarrow (2 \times 4)As$ and $(2 \times 4)As \rightarrow 3 \times 6$. So, the similarity of the transitions during formation of GaAs-AlAs and GaAs-GaAs interfaces is striking: in both cases we can see the presence of phase shift in two taken azimuths for $c(4 \times 4) \rightarrow (2 \times 4)As$ transition (Fig.6a) while only in the $[1\bar{1}0]$ azimuth for $(2 \times 4)As \rightarrow 3 \times 6$ one (Fig.6b). This fact supports the assumption that strong phase shift in oscillation curves is completely controlled by type of surface reconstruction transition. As can be also seen from Fig.6 the reconstruction shift at the inverted interface is completed over 1-2 ML.

Summary

Specular beam effects during formation of normal and inverse heterointerfaces in AlAs/GaAs system were analyzed within wide range of T_s and $As_4/Al(Ga)$ flux ratio. It was demonstrated that at intermediate As_4/Al flux ratios the stabilization of 3×2 superstructure on growing AlAs surface is controlled by segregation process of Ga atoms into growing layer. In this case the range of conditions when Ga segregation can be detected by RHEED is a fairly narrow. The observed phase shift of oscillation curves in $[1\bar{1}0]$ azimuth, that accompanies the appearance of 3×2 reconstruction on growing AlAs surface, is induced by variation of the relationship between specular and Bragg components in measured signal as in homoepitaxial growth with reconstruction transition $(2 \times 4)As \rightarrow 3 \times 6$. Contrary to the case of normal heterointerface (AlAs on GaAs) the process of oscillation phase variation during growth of GaAs on AlAs is completed over first 1-2 oscillations.

Acknowledgements. The authors would like to thank the International Science Foundation and Russian Fundamental Research Foundation (grant 94-02-05472-a) for financial support.

References

[1] Jusserand B, Mollot F 1992 *Appl. Phys. Lett.* **61** 423-5
[2] Moison J M, Guille C, Houzay F, et al 1989 *Phys.Rev.B* **40** 6149-6161
[3] Chand N 1993 *Thin Solid Films* **231** 143-157
[4] Moshegov N T, Sokolov L V, Toropov A I, to be published
[5] Wassermeier M, Kamiya I, Aspnes D E, et al 1991 *J.Vac. Sci. Technol.B* **9** 2263-7
[6] Brown W, Ploog K N 1994 *J.Appl.Phys.***75** 1993-2001

Inst. Phys. Conf. Ser. No 145: Chapter 2
Paper presented at 22nd Int. Symp. Compound Semiconductors, Cheju Island, Korea, 28 August–2 September 1995
© *1996 IOP Publishing Ltd*

Growth and characterization of diluted magnetic semiconductor Zn$_{1-x}$Mn$_x$Se/ZnSe strained-layer superlattice

C.X. Jin, Z. Ling, G.C. Yu , J. Wang and Xun Wang

Surface Physics Laboratory, Fudan University, Shanghai 200433, China

Abstract. Zn$_{1-x}$Mn$_x$Se/ZnSe superlattices are grown by molecular beam epitaxy on GaAs(100) substrates and characterized by x-ray diffraction and Raman scattering. Higher-order satellite peaks up to 5th order in x-ray diffraction spectra have been observed for the first time, which indicates the sharpness and abruptness of the superlattice interfaces. Well resolved LO phonon peaks of Raman scattering from Zn$_{1-x}$Mn$_x$Se barrier and ZnSe well layers are observed and the relatively small FWHM and ratio of Γ_a/Γ_b confirm the formation of high quality superlattice. The frequency shifts of LO phonons due to the misfit strain are calculated. The results are in good agreement with the experiments.

1. Introduction

The great interests in studying the Mn-based II-VI compound diluted magnetic semiconductors (DMS) are stimulated by their novel physical properties and possible applications[1-3]. The outstanding magneto-optical properties are originated from the sp-d exchange interaction between electron/hole band states and the Mn^{2+} 3d^5 electron states in the materials[4-6]. A novel concept of spin separation in DMS ZnMnSe/ZnSe quantum wells or superlattices was demonstrated[7-8], in which the electrons or holes with different spin polarizations are confined in different well regions depending on the applied magnetic field. The first achievement of growing ZnMnSe/ZnSe was reported by Kolodziejski and his coworkers[6]. The crystalline quality was characterized by using transmission electron microscopy (TEM), photoluminescence (PL), photoluminescence excitation spectroscopy and reflection spectroscopy.

In this paper, we report the growth of high quality Zn$_{1-x}$Mn$_x$Se/ZnSe superlattice(SL) with large thicknesses of well and barrier layers and the investigation of their characteristics by x-ray diffraction and Raman scattering. Up to 5th high order satellite peaks have been observed in the x-ray diffraction spectra,and well resolved LO phonon peaks from Zn$_{1-x}$Mn$_x$Se barrier and ZnSe well layers are observed in the Raman scattering spectra. It seems that the lattice mismatch between the well and barrier layers of the sample is totally accommodated by the elastic strain giving rise to high quality strained-layer superlattice.

2. Experimental

The $Zn_{1-x}Mn_xSe/ZnSe$ superlattices were grown by molecular beam epitaxy(MBE) on n type GaAs(100) substrates with carrier concentration of $2x10^{18}cm^{-3}$. The substrate wa ultrasonically cleaned in the solutions of carbon tetrachloride (CCl_4), acetone and alcohol sequentially, and then was chemically etched by a solution of $H_2SO_4+H_2O_2+H_2O(5:1:1$ with stirring at 60^oC for 3 min.. After the chemical etching, the substrate was introduced into the growth chamber, whose background pressure was lower than $5x10^{-9}$ Torr. Prior to the growth, the substrate was heated to 580^oC for 3 min. to remove the native oxides on the surface.

$Zn_{1-x}Mn_xSe$ and ZnSe layers were grown alternately by periodically openning and closing the shutter of the Mn source. Three effusion cells containing Zn(6N), Se(6N) and Mn(6N) were used as the beam sources of MBE. The temperatures of Zn and Se source were typically 300^oC and 150^oC, respectively, the Mn source temperature , however ranged from 850-950°C to provide various Mn mole fractions. During the growth, the substrate temperature was kept at 300^oC with the growth rate of 3-4nm/min., and the pressure in the chamber was kept below $5x10^{-8}$ Torr. A ZnSe buffer layer with the thickness of 200-300nm was deposited before growing the superlattice.The thicknesses of well and barrier layers were controlled to the same for each sample.

The x-ray diffraction measurements were performed on a computer-controlled D/MAX-YB x-ray diffractometer with $CuK_{\alpha1}$ and $CuK_{\alpha2}$ radiations.. The Raman spectra of the superlattice was obtained at room temperature in the backscattering geometry under the excitation of a 488nm line provided by an Ar^+ ion laser operated at 300mW. The scattered light was dispersed through a SPEX-1403 triple monochromator with the resolution of $2.4cm^{-1}$ in accordance with the strength of the.Raman signals and detected by a standard photon-counting system.

3. Results and discussion

Figure 1 shows the high angle diffraction patterns of two superlattice samples in the vicinity of the (400) reflection of GaAs substrate. The substrate peak around $2\theta=66.06^o$, which splits into $K_{\alpha1}$ and $K_{\alpha2}$ components, and the ZnSe buffer layer peak around $2\theta=65.86^o$, whose $K_{\alpha2}$ component overlaps with the $K_{\alpha1}$ component of GaAs, can be easily distinguished, The other peaks are considered to be SL's satellite peaks. Up to 5 orders of well resolved satellite peaks are assigned, showing successful growth of modulated superlattice structures with sharp and abrupt interfaces. Due to the mismatch between the average lattice of $Zn_{1-x}Mn_xSe/ZnSe$ superlattice and GaAs substrate , the 0th order peak does not overlap with the (400) diffraction peak from GaAs substrate. The greater intensities of the satellite peaks at lower angle side compared to that at higher angle side are due to the asymmetry distribution of the built-in strain caused by the lattice mismatch.

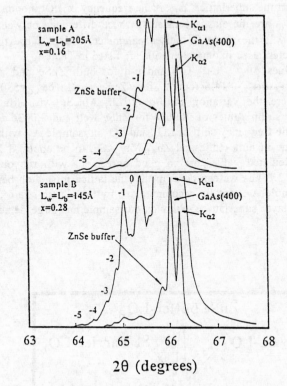

Fig.1 XRD spectra of $Zn_{1-x}Mn_xSe/ZnSe$ SLs

Figure 2 gives the Raman scattering spectrum of sample A. Three LO phonon modes around $250cm^{-1}(LO_w)$, $251.5cm^{-1}(LO_B)$ and $255cm^{-1}(LO_b)$ come from ZnSe well, ZnSe buffer and $Zn_{0.84}Mn_{0.16}Se$ barrier layers, respectively. In addition to the LO phonon of GaAs substrate at $291cm^{-1}$, the peaks around $206cm^{-1}$ and $268cm^{-1}$, which correspond to the ZnSe-TO phonon and GaAs-TO phonon, respectively, also appear. The LO_2 mode originating from Mn^{2+} impurity mode is not observed. It could be seen that the frequency of LO_b shifted by around 2 cm^{-1} from the value of 253 cm^{-1} for the bulk $Zn_{0.84}Mn_{0.16}Se[13]$. LO_B occurs at approximately the same frequency as that of bulk crystal, while LO_w appears to be shifted by 1.5 cm^{-1} to a lower frequency. It has been proved that the LO frequency shift is resulted from the misfit strains which are known to exist in the heterostructure and superlattice[9,10].

The shift of the LO phonon duo to the strain is calculated from the following equation:[11,12]

$$\delta\omega_{LO}=-2\gamma\omega_{LO}[(C_{11}-C_{12})/C_{11}]\varepsilon-[(p-q)/(3\omega_{LO})][(C_{11}+2C_{12})/C_{11}]\varepsilon \qquad (1)$$

where ε is the strain in the superlattice, ω_{LO} is the frequency of LO phonon for the sample free of strain. C_{11}, C_{12} are the elastic constants, whose difference between the well and barrier is neglected, γ is the mode Gr neisen parameter, and p, q are the deformation constants. The parameters used in our calculation are listed in Table I.

Taking the values of γ , C_{11}, C_{12} and p-q for bulk ZnSe and $Zn_{0.84}Mn_{0.16}Se$. repectively. we get $\delta\omega_{LO}=314.6\varepsilon(cm^{-1})$ for ZnSe and $\delta\omega_{LO}=530.3\varepsilon(cm^{-1})$ for $Zn_{0.84}Mn_{0.16}Se$. (where the variation of p-q of $Zn_{1-x}Mn_xSe$ with Mn fraction x is neglected). Using the strain values of -0.0038 for ZnSe well and 0.0036 for $Zn_{1-x}Mn_xSe$ barrier. we estimate the frequency shift of LO_w and LO_b in sample A , with respect to the frequency of LO mode in bulk ZnSe and $Zn_{0.84}Mn_{0.16}Se$, to be about -1.2 and $2.2cm^{-1}$. respectively. The calculated values are in good agreement with the observed Raman spectra shown in Fig.2. Thus we can confirm that the lattice mismatch between the well and the barrier of sample A is totally accommodated by the elastic strain , giving rise to high quality strained-layer superlattice. although our sample has large demensions of well and barrier layers.

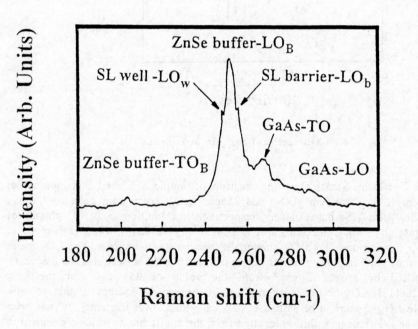

Fig.2 Raman spectra of $Zn_{0.84}Mn_{0.16}Se/ZnSe$ SL

Table I Parameters of ZnSe and $Zn_{0.84}Mn_{0.16}Se$ used in our calculation

Material	ω_{LO} (cm^{-1})	C_{11} (10^6kg/cm^2)	C_{12}	γ	$(p-q)/2\omega_{LO}^2$
ZnSe	252.5	0.826	0.498	0.79	0.42
$Zn_{0.84}Mn_{0.16}Se$	253	0.826	0.498	1.06	0.99

The crystalline quality of the epilayer could also be illustrated by the line shape symmetry and the full width at half maximum(FWHM) of the LO mode[13,14]. Figure 3 shows the expanded versions of LO peaks in the frequency range of 240-270cm^{-1} for (A) ZnSe film. (B) $Zn_{0.92}Mn_{0.08}Se$ film, (C) $Zn_{0.84}Mn_{0.16}Se$/ZnSe SL and (D) $Zn_{0.67}Mn_{0.33}Se$ SL grown by Suh et al [11]. For our superlattice, the LO phonon from ZnSe well and $Zn_{0.84}Mn_{0.16}Se$ barrier layers are masked by the tail of LO phonon from the ZnSe buffer

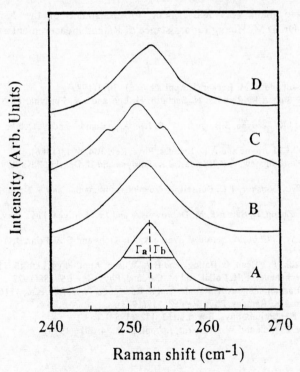

Fig.3. The expanded versions of the Raman spectra in LO
frequency range of 240-270cm^{-1}
A: ZnSe film;
B: $Zn_{0.92}Mn_{0.08}Se$ film;
C: $Zn_{0.84}Mn_{0.16}Se$/ZnSe SL;
D: $Zn_{0.67}Mn_{0.33}Se$/ZnSe SL grown by Suh. et al[11]

Table II   The values of FWHM and Γ_a/Γ_b of Raman spectras

sample	A	B	C	D
FWHM	5.6	8.2	8.7	8.8
Γ_a/Γ_b	1.1	1.15	1.14	1.5

layer. From Fig. 3, the values of the FWHM and the ratios of the left to right halves of the peak (Γ_a/Γ_b) can be determined and are listed in Table II. The FWHM of our superlattice i almost the same as that of $Zn_{0.92}Mn_{0.08}Se$ film and $Zn_{0.67}Mn_{0.33}Se/ZnSe$ SL , while the ratio of (Γ_a/Γ_b) of our superlattice is smaller than that of the other samples except the ZnSe film. This comparison seems quite encouraging, since the lattice mismatch of ou superlattice is larger than that of $Zn_{0.92}Mn_{0.08}Se$ film.

Acknowledgments

The authors wish to thank Mr. X.L. Sheng for assistance of the X-ray diffraction measurements, and Dr. D.M. Huang for assistance of Raman measurements.

References

[1] R.B. Bylsma, J. Kossut and W.M. Becker, J. Appl. Phys. 61, 3011(1987).
[2] R.G. Alonso, E.-K. Suh, A.K. Ramdas, N. Samarth, H. Luo and J.K. Furdyna, Phys. Rev. B40, 3720 (1989).
[3] N. Samarth, H. Luo, J.K. Furdyna, S.B. Qudri, Y. R. Lee, A.K. Ramdas and N. Otsuka, Appl. Phys. Lett. 54, 2680(1989).
[4] D.U. Bartholomew, J.K. Furdyna and A.K. Ramdas, Phys. Rev. B34, 6943(1986).
[5] Eunsoon Oh, D.U. Bartholomew, A.K.Ramdas , J.K. Furdyna and U. Debska, Phys.Rev. B38, 13183(1988).
[6] L.A. Kolodziejski, R.L. Gunshor, T.C. Bonsett, R. V6enkatasubramanian and S. Datta, Appl. Phys. Lett 47, 169(1985).
[7] N. Dai, H. Luo, F.C. Zhang, N. Samarth, M. Dobrowolska and J.K. Furdyna, Phys. Rev. Lett67, 3824(1991).
[8] B.T. Jonker, X. Liu, W.C. Chou, A. Petrou, J. Warnock, J.J. Krebs and G.A. Prinz, J. Appl. Phys 69, 6098(1991).
[9] F. Cerdeira, A. Pinczuk, J.C. Bean, B. Batlogy and B.A. Wilson, Appl. Phys. Lett 45, 1138(1984).
[10] F. Cerdeira, C.J. Bechenauer, F.H. Pollak and M. Cardona, Phys. Rev. B5, 580(1972).
[11] E. -K. Suh , D. U. Bartholomew, A.K. Ramdas and S. Rodriguez, Phys. Rev.B36, 4316(1987).
[12] S. Venugopalan and A.K. Ramdas, Phys. Rev B8, 717(1973).
[13] P. Parayanthal and F.H. Pollak, Phys. Rev. Lett52, 1822(1984).
[14] S. Dauer, H. Berger, P. Link and W. Gebhardt, J. Appl. Phys.74, 3916(1993)

Inst. Phys. Conf. Ser. No 145: Chapter 2
Paper presented at 22nd Int. Symp. Compound Semiconductors, Cheju Island, Korea, 28 August–2 September 1995
© *1996 IOP Publishing Ltd*

Selective Epitaxial Growth of $Si_{1-x}Ge_x$ on SiO_2-patterned Si Substrate Using Elemental Source Molecular Beam Epitaxy

Sun Jin Yun, Seung-Chang Lee, Jae-Jin Lee, and Kee-Soo Nam

Electronics and Telecommunications Research Institute, Yusong P. O. Box 106, Taejon, Korea

Abstract. The selective epitaxial growth (SEG) of $Si_{1-x}Ge_x$ ($0.0 \leq x \leq 1.0$) on SiO_2-patterned Si(100) substrate was investigated using the elemental source molecular beam epitaxy. The dependence of the behavior of incident Si and Ge atoms was studied on the substrate temperature and the flux. The fluxes of source beams and the substrate temperatures were in the range of 1.0×10^{13} - 2.9×10^{14} $cm^{-2}s^{-1}$ and 600 - 810°C, respectively. Under SEG condition, the Ge atoms were reevaporated from the oxide surface and were not solely effective in etching SiO_2. However, Ge contributed to etch the oxide layer with accompanying Si flux. The etch rate at a Si flux of 1.0×10^{13} $cm^{-2}s^{-1}$ was about 0.85Å/min, and then increased to 1.7Å/min with accompanying Ge flux of about 2.0×10^{13} $cm^{-2}s^{-1}$. It was also shown that the SEG of $Si_{1-x}Ge_x$ was achieved by the SiO_2 etch due to the concurrent atomic fluxes of Si and Ge, and in part by the reevaporation of Ge atoms from SiO_2 surface. The effect of source ions and electrons naturally generated from an electron beam evaporator was also investigated on the SEG characteristics of Ge. The result exhibited that the incident charged particles enhanced the nucleation of source atoms on SiO_2 surface to negatively affect the SEG of $Si_{1-x}Ge_x$.

1. Introduction

$Si_{1-x}Ge_x$ ($0.0 \leq x \leq 1.0$) alloy has been successfully applied to the fabrication of heterojunction devices [1]. Especially, Ge is a very promising material as a lattice-matched template of GaAs and the related compounds. In the epitaxial growth of $Si_{1-x}Ge_x$, the deposition can be restricted only on crystalline area and no material is deposited on masking area of SiO_2 or Si_3N_4 by achieving the selective epitaxial growth (SEG). The SEG of $Si_{1-x}Ge_x$ has advantages for preparing complex and self-aligned structures as well as simplifying post-growth process of semiconductor devices.

The SEG of $Si_{1-x}Ge_x$ has been demonstrated with growth techniques using chemical source beams (e.g., SiH_4, SiH_2Cl_2, GeH_4, etc.), such as the gas source molecular beam epitaxy (GSMBE) and chemical vapor deposition (CVD) [2,3]. In contrast to the outstanding progress in the SEG technology using GSMBE and CVD, the SEG of $Si_{1-x}Ge_x$ using the elemental solid-source molecular beam epitaxy (SSMBE) has been rarely reported[4]. However, in view of device fabrication, as the SSMBE of $Si_{1-x}Ge_x$ can yield a good controllability with a simple growth mechanism resulting in the composition and growth rate independent of growth temperature, it is advantageous over the chemical beam techniques.

In the present work, the interactions of elemental Si and Ge fluxes with SiO_2 film were studied as a preliminary work of $Si_{1-x}Ge_x$ ($0.0 \leq x \leq 1.0$) SEG using SSMBE. The dependence of the growth behavior was investigated on the substrate temperature and the fluxes of Si and Ge. The contribution of Ge flux to the oxide etch was also determined at a constant Si flux by varying the Ge flux.

The use of electron beam evaporator (EBE) is known to generate a small fraction of low energy source ions caused by ionizing interactions between source atoms and the primary electron beam [5]. The earlier work also presented that the bombardment of source ions greatly improved the surface morphology of $Si_{1-x}Ge_x$ film grown on Si [6]. Then, the effect of source ions on the SEG behavior was investigated as well as that of electrons scattered from EBE by applying dc-potentials ranging from -1.6 kV to 1.5 kV to the substrate.

2. Experiments

About 40 - 110Å thick oxide films on 5 inch p-type Si(100) substrate were prepared by thermal oxidation at 900°C in an oxygen ambient. The oxide thickness on each wafer was measured using an ellipsometer to determine the etch rate due to the source fluxes. To prepare substrates patterned with SiO_2, Si wafers were oxidized to 2500Å thickness at 925°C in an water vapor ambient. Then, the SiO_2 films were patterned by a photolithography and a magnetically enhanced reactive ion etching (MERIE). The MERIE process resulted in the vertically patterned oxide (the angle from the surface >85°).

The basic experiments for SEG of Si and SiGe were carried out using RIBER SIVA45 MBE system equipped with a SIMBA II Temescal electron beam evaporator controlled by a Sentinel III flux monitor. The fluxes of Si and Ge were in the range of 1.0×10^{13} - $2.9 \times 10^{14} cm^{-2}s^{-1}$. The substrate temperatures were ranging from 600°C to 810°C.

3. Results and Discussion

In principle, SEG of Si and $Si_{1-x}Ge_x$ films can proceed with the etching of SiO_2 masking material or the reevaporation of source vapors from SiO_2 surface along with the simultaneous epitaxial growth on open Si surface.

The oxide etching due to impinging Si has been explained by the reaction mechanism (1). An impinging Si atom interacts with a SiO_2 to produce two volatile SiO molecules as shown by the reaction (1).

$$Si + SiO_2 \text{-----------}> 2 SiO \uparrow \tag{1}$$

The reaction (1) competes with the deposition of Si nucleated by clustering of Si atoms on SiO_2 surface. The formation of nucleation islands tends to be promoted as increasing the flux of source atoms and the rate of reaction (1) decreases as decreasing the substrate temperature [4]. Therefore, a higher flux requires a higher growth temperature to achieve SEG.

In the SEG of the chemical beam technique using GeH_4, earlier studies have suggested for oxide etching that the molecules, GeH_4, decompose to Ge and $2H_2$ by the thermal decomposition or the electron bombardment in a plasma. Then the oxide can be etched by forming volatile GeO through the reactions (2) and (3) [7,8].

$$Ge \; + \; SiO_2 \; \text{-------------->} \; GeO_2 \; + \; Si \tag{2}$$

$$Ge \; + \; GeO_2 \; \text{--------------->} \; 2GeO \uparrow \tag{3}$$

Ge deposition using GeH_4 is also known to be naturally selective, i.e., Ge does not deposit on SiO_2 but removes SiO_2[7].

In contrast, Ge atomic flux is used as a source beam in the SSMBE of $Si_{1-x}Ge_x$. The interaction between Ge atoms and SiO_2 surface was investigated to determine if the oxide etching proceeded through the reactions (2) and (3). An elemental Ge flux of 2.9×10^{13} $cm^{-2}s^{-1}$ was impinged on the 110Å oxide film for 2000s at 810°C (sample A) and the 40Å oxide for 6000s at 710°C (sample B).

The cross sectional transmission electron microscopy (XTEM) image and the x-ray photoelectron spectroscopy (XPS) spectrum of the sample A are presented in Figs. 1(a) and (b), showing the microstructure and the chemical composition of the surface region. The XTEM micrograph of Fig. 1(a) illustrates that the thickness of oxide was not reduced and there was no Ge overlayer on SiO_2 surface. The surface-sensitive XPS spectrum in Fig. 1(b) also demonstrates that Ge was not deposited on the oxide surface. The XPS spectrum shows only Si 2p photoelectron peak from the oxidized Si at the binding energy of about 110 eV which was shifted about 7 eV from normal peak position, 103 eV. The energy shift of photoelectrons is caused by the surface charging due to electron emission from the insulating oxide surface. There is no trace of Ge photoelectron peaks in the XPS spectrum. In the XTEM and XPS analyses of sample B, the same results were obtained with sample A. These experimental data suggest that the elemental Ge was not solely effective in etching of the SiO_2 layer and did not deposit on the oxide surface at a temperature higher than 710°C with a flux of 2.9×10^{13} $cm^{-2}s^{-1}$. Of course, the temperature to achieve SEG strongly depends on the flux of source atoms. Ge islands were nucleated on SiO_2 with a flux of $6.2 \times 10^{13} cm^{-2}s^{-1}$ at 640°C.

However, although the Ge flux did not etch SiO_2 film solely, Ge atoms promoted the SiO_2 etching of Si flux. The contribution of Ge flux to the oxide etching was investigated quantitatively at a constant Si flux of 1.0×10^{13} $cm^{-2}s^{-1}$ at 810°C. In Fig. 2, the etch rate of SiO_2 was plotted against the Ge flux ranging from 0.0 to 5.3×10^{13} $cm^{-2}s^{-1}$.

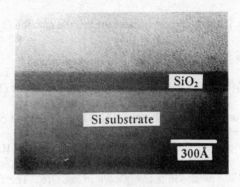

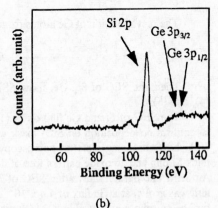

(a)   (b)

Figure 1. (a) The XTEM micrograph and (b) the XPS spectrum, showing the thickness and surface morphology of SiO_2 film, and the chemical composition of the surface region after impinging Ge atoms of flux rate, 2.9×10^{13} $cm^{-2}s^{-1}$ on 110Å SiO_2 for 2000s at 810°C.

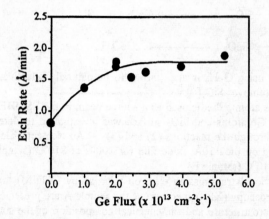

Figure 2. Plot of SiO₂ etch rate with respect to Ge flux at a constant Si flux of 1.0 x 10¹³ cm⁻²s⁻¹ and a substrate temperature of 810°C.

For these experiments, the substrates with about 80Å thick oxide layers were used and the endpoint of oxide etching was detected by observing *in situ*-reflection high energy electron diffraction (RHEED) patterns. As shown in Fig. 2, the etch rate was gradually enhanced with increasing Ge flux from 0.0 to about 2.0 x 10^{13} cm⁻²s⁻¹, and then the etch rate was nearly constant at Ge fluxes higher than 2.0 x 10^{13} cm⁻²s⁻¹. The etch rate with Si flux of 1.0 x 10^{13} cm⁻²s⁻¹ was about 0.85Å/min and then increased to 1.7Å/min by adding Ge flux of 2.0 x 10^{13} cm⁻²s⁻¹. Comparing to the result of the Si flux-only, the etch rate was increased about 100% by adding Ge flux of about 2.0 times Si flux. At a higher Ge flux, the excess Ge atoms did not contribute to the SiO₂ etching.

The results shown in Figs. 1 and 2 indicate that the SiO₂ etching by impinging Ge atoms would not proceed through the separate reactions described by the chemical reactions (2) and (3), but a reaction mechanism including the Si flux as suggested by the reaction (4).

$$Ge + Si + 2SiO_2 \text{-----------} > GeO \uparrow + 3SiO \uparrow \tag{4}$$

The reevaporation of Ge atoms from SiO₂ surface can be described as the reaction (5).

$$Ge + SiO_2 \text{------------} > Ge \uparrow + SiO_2 \tag{5}$$

Then, the SEG of Si$_{1-x}$Ge$_x$ using SSMBE can be explained to follow the reactions (1), (4), and (5).

The SEG of Si and Ge film on Si (100) substrate patterned with SiO₂ was carried out to confirm no deposition on SiO₂ mask and selective growth on the open Si area. Figure 3(a) shows scanning electron microscopy (SEM) micrographs of selectively grown 110Å thick Si film and an oxide pattern free of deposit at 810°C with a Si flux of 2.8 x 10^{13} cm⁻²s⁻¹. An SEM micrograph illustrating SEG of 300Å thick Ge is also shown in Fig. 3(b). The Ge film was grown at a Ge flux of 4.4 x 10^{13} cm⁻²s⁻¹ at 660°C. The Ge film has a typical rough surface morphology caused by 4.2 % lattice mismatch between Ge overlayer and Si substrate which is one of the important problems to be solved in lattice-mismatched heteroepitaxy.

The effect of charged particles on the growth characteristics of Ge was studied to clarify if it is possible to achieve SEG of Ge film with a smooth surface morphology by bombarding charged particles toward the substrate.

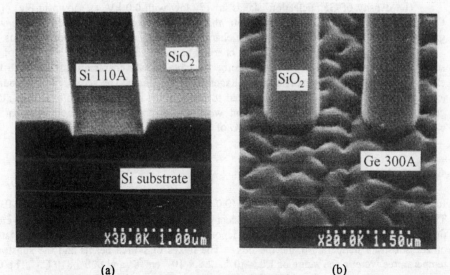

(a)   (b)

Figure 3. SEM photographs showing the oxide patterns free of deposit and the selective epitaxial growth of Si and Ge: (a) 110Å thick Si film grown with Si flux of 2.8×10^{13} cm^{-2}s^{-1} at 810°C and (b) 300Å thick Ge film grown with the flux of 4.4×10^{13} cm^{-2}s^{-1} at 660°C.

It is also very important to confirm the effect of charged particles on the SEG behavior because the presence of a small fraction of source ions and scattered electrons is inevitable in SSMBE using EBE. In this work, a dc-voltage in the range of -1 6 kV to 1.5 kV was applied to the substrate at a substrate temperature of 600°C.

SEM photographs in Fig. 4 show the effect of source ion bombardment and electron showering on the nucleation of Ge atoms on SiO$_2$ surface. The surface was observed after impinging Ge atoms of flux, 2.9×10^{14} cm^{-2}s^{-1}, on SiO$_2$ surface for 400s.

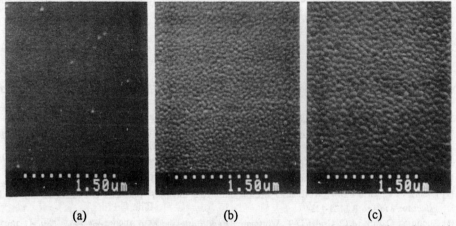

(a)   (b)   (c)

Figure 4. SEM photographs showing the effect of source ion bombardment and electron showering on the nucleation of Ge atoms on SiO$_2$ surface at a substrate temperature of 600°C as compared with that of 0.0 kV. The applied voltages were (a) 0.0 kV, (b) 1.5 kV, and (c) -1.6 kV.

114

The density of Ge nucleation islands is very low with 0.0 kV, as shown in Fig. 4(a). Contrarily, Figs. 4(b) and (c) demonstrate that the density of Ge nucleation islands was greatly increased with the bombardment of charged particles, i.e., source ions or electrons.

The result exhibits that the incident charged particles could enhance the nucleation of source atoms on SiO_2 surface to degrade the selective growth characteristics although the effect was exaggerated by applying dc-voltages to the substrate. The acceleration of source ions toward the substrate has been reported to improve the surface morphology of $Si_{1-x}Ge_x$ grown on Si [6]. However, the present work demonstrates that the bombardment of charged particles negatively affects the SEG of $Si_{1-x}Ge_x$ film.

4. Conclusion

In the present work, the SEG of $Si_{1-x}Ge_x$ ($0.0 \leq x \leq 1.0$) was investigated using SSMBE. The dependence of the behavior of the incident elemental Si and Ge fluxes, i.e., the SiO_2 etch, the reevaporation of source atoms, or the deposition of polycrystalline films was investigated on the substrate temperature and the flux. The fluxes of source beams and the substrate temperatures were in the range of 1.0×10^{13} - 2.5×10^{14} $cm^{-2}s^{-1}$ and 600 - 810°C. The Ge atoms were reevaporated from the oxide surface and were not solely effective in etching SiO_2 under SEG condition. However, Ge contributed to etch the oxide layer with accompanying Si flux. The etch rate with a Si flux of 1.0×10^{13} $cm^{-2}s^{-1}$ was about 0.85Å/min, and then increased to 1.7Å/min with accompanying Ge flux of about 2.0×10^{13} $cm^{-2}s^{-1}$.

The SEG of Si and Ge film on Si(100) substrate patterned with SiO_2 was also demonstrated to confirm no deposition on SiO_2 pattern and selective epitaxial growth on the open Si area. In the present work, it was also shown that the SEG of $Si_{1-x}Ge_x$ was achieved by the SiO_2 etch due to the concurrent atomic fluxes of Si and Ge, and in part by the reevaporation of Ge atoms from SiO_2 surface.

The effect of source ions and electrons naturally generated from EBE was also investigated on the SEG characteristics of Ge. The result exhibited that the incident charged particles enhanced the nucleation of source atoms on SiO_2 surface to degrade the selective growth characteristics. Conclusively, the bombardment of charged particles was shown to negatively affect the SEG of $Si_{1-x}Ge_x$.

References

[1] Iyer S, Patton G L, Stork J, Meyerson B S and Harame D 1989 *IEEE Trans. Electron Devices* **ED-36** 2043-2064

[2] Regolini J L, Bensahel D, Scheid E and Mercier J 1989 *Appl. Phys. Lett.* **54** 658-659

[3] Sedgwick T O, Grutzmacher D A, Zaslavsky A and Kesan V P 1992 *J. Vac. Sci. Technol.* **B11** 1124-1128

[4] Yun S J, Lee S-C, Kim B-W and Kang S-W 1994 *J. Vac. Sci. Technol.* **B12** 1167-1169

[5] Kubiak R A A, Leong W Y and Parker E H C 1985 *J. Electrochem. Soc.* **132** 2738-2742

[6] Yun S J, Lee S-C, Lee J-J and Park S-C 1995 *J. Vac. Sci. Technol.* **B13** 728-730

[7] Ozturk M C, Grider D T, Wortman J J, Littlejohn M A, Zhong Y, Batchelor D and Russel P 1990 *J. Electron. Mater.* **19** 1129-1134

[8] Zhong Y, Ozturk M C, Grider D T, Wortman J J and Littlejohn M A 1990 *Appl. Phys. Lett.* **57** 2092-2094

Inst. Phys. Conf. Ser. No 145: Chapter 2
Paper presented at 22nd Int. Symp. Compound Semiconductors, Cheju Island, Korea, 28 August–2 September 1995

A RHEED study of the growth of InAs on InAs(111)A

T. Nomura[a),*] , **I. Kamiya**[a),b),+], **M. R. Fahy**[a)], **J. H. Neave**[a)]
and B. A. Joyce[a)]

a) Interdisciplinary Research Centre for Semiconductor Materials
    The Blackett Laboratory, Imperial College of Science, Technology and Medicine
    Prince Consort Road, London SW7 2BZ, UK
b) Research Development Corporation of Japan (JRDC)
*   Permanent address:
    Research Institute of Electronics, Shizuoka University
    3-5-1 Johoku, Hamamatsu 432, Japan
+   Present Address:
    Quantum Transition Project
    4-7-6 Komaba, Meguro-ku, Tokyo 153, Japan.

Abstract.   Growth of InAs on InAs (111)A by molecular beam epitaxy has been studied using reflection high energy electron diffraction intensity oscillations. The period of the intensity corresponds to the flux of In for lower substrate temperatures and/or higher As fluxes and is independent of growth conditions. However, for higher substrate temperatures and/or lower As fluxes, the period is longer than the value expected from the In flux and strongly depends on the growth conditions. A critical growth condition, defined as the boundary between the two regions, is a strong function of the As flux and substrate temperature. The increase of the period is explained by the deficiency of As on the growing surface due to its low sticking coefficient.

1. Introduction

Recently the growth of compound semiconductors on novel index surfaces has been extensively studied. Growth on the {111} orientation is of particular interest because it is often exposed as a facet plane of patterned substrate structures [1-7] . An understanding of the migration of group III elements and formation of facet planes on (001)-{111} nonplanar substrates is required to control this type of growth. Strained superlattices grown on (111) surfaces also have the possibility of optoelectronic device application because of the modification of the band structure and optical properties by a built-in piezoelectric potential [8-11]. We have previously studied the growth process and Si doping behaviour of GaAs, AlAs and AlGaAs on (111)A surfaces by molecular beam epitaxy (MBE) [12-15]. The period of reflection high energy electron diffraction (RHEED) intensity oscillations is strongly dependent on the growth conditions and may be explained by a model which assumes a low sticking coefficient of As on the (111)A surface [15]. A detailed understanding of InAs growth will be essential for fabrication of InGaAs MQW structures and the clarification of migration effects and compositional variation on patterned substrates. However, there are few reports of the growth process of InAs on (111)A. In this paper, we report on the growth of InAs on InAs (111)A by MBE as studied by RHEED and in particular discuss the change in the intensity oscillation period as a function of growth conditions has been examined.

2. Experiments

The equipment used is a purpose-built solid source MBE system. A cracker cell, with a cracking efficency of virtually 100% (as measured by time-of-flight mass spectrometry), is used to produce an As_2 beam. Fluxes were calibrated by the periods of intensity oscillations of RHEED; the In flux by InAs growth on InAs (001) and the As flux by As induced oscillations on GaAs (001), respectively. The substrate temperature was monitored by optical pyrometer. Epi-ready GaAs(111)A wafers supplied by American Xtal Technology were used as substrates. A 100nm GaAs buffer layer was grown to make a well-defined surface and fozzowing confirmation of a GaAs (111)-2x2 reconstruction by RHEED, a 100nm InAs buffer layer was grown prior to the measurement of RHEED intensity oscillations. A two dimensional growth mode is maintained from the initial growth stage of InAs and the relaxation of lattice mismatch between the GaAs substrate and the InAs was confirmed by the spacing of diffraction spots.

RHEED intensity oscillations were measured as a function of substrate temperature and As flux. The diffraction conditions were as follows; incident angle 1°, incident azimuth [110], acceleration voltage 15kV. The sample was annealed at 520°C for 10 minutes after each measurement of RHEED intensity oscillations to recover a well-defined surface condition.

3. Results and discussions

For (001) substrates, the period of RHEED intensity oscillations is relatively insensitive to substrate temperature and As flux, and is mainly determined by the flux of group III material, but for growth on (111)A the period strongly depends on growth conditions. Figure 1 shows typical intensity oscillations of the specular spot measured during the growth of InAs for various substrate temperatures. The In flux is 3.6×10^{14} atoms cm^{-2} s^{-1} and the atomic flux of As_2 is 8×10^{14} atoms cm^{-2} s^{-1}: the nominal V/III ratio is 2.2. Arrows in Fig. 1 indicate closure of the In shutter. Steady intensity oscillations are observed for a substrate temperature of 360°C and the period of the oscillation is in good agreement with the In flux, but by 480°C the oscillations damp rapidly and their period increases. The average RHEED intensity also

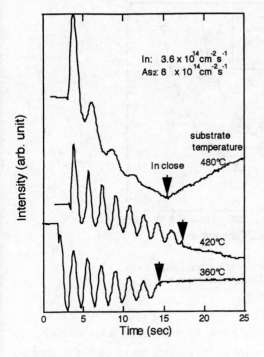

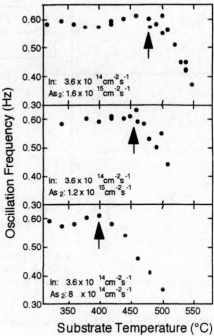

Fig. 1 RHEED intensity oscillations of InAs on (111)A for various temperatures. Arrows indicate closure of the In shutter.

Fig. 2 The variation of growth rate as a function of substrate temperature and As/In flux ratio. Arrows indicate the critical growth condition.

decreases during growth, followed by an increase after the In shutter is closed. The period of the oscillations, calculated by the time interval of adjacent peaks, is plotted as a function of substrate temperature and As flux in Fig. 2. For each V/III ratio in the lower substrate temperature regime, the oscillation period is approximately in agreement with the In flux, although there is a small but significant non-monotonic variation of the oscillation period as a function of substrate temperature, even though the period is expected to be determined only by the In flux. On the other hand, for the higher temperature regime, the period decreases substantially as the substrate temperature is increased. We note that this decrease of the oscillation period is commensurate with the decrease in average intensity. This coincidence suggests that a different process dominates the growth of InAs in each temperature regime and the decrease of the oscillation period is caused by a change of the growth dynamics. A critical growth condition is defined as the boundary of these two temperature regimes and is shown by an arrow in Fig. 2 for each As flux. The temperature of the critical growth condition increases as the As flux is increased for a constant In flux.

A plot of the critical growth condition as a function of substrate temperature and In flux is shown in Fig. 3. The experimental results for GaAs growth on GaAs (111)A [15] are also

118

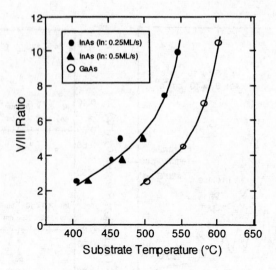

Fig. 3 The variation of V/III ratio at the critical growth condition as a function of substrate temperature.

plotted. In the lower substrate temperature or higher As flux regions below each line, the period of the intensity oscillation is determined by the In flux, but in the higher substrate temperature or lower As flux regions above each line, the period is strongly dependent on growth conditions. For equivalent flux ratios the critical temperature for InAs is 70°C lower than that of GaAs, which indicates that the critical growth condition depends on the cation species. Although the relationship between the crystal quality and the variation of the oscillation period has not yet been quantified, a higher As flux seems to be required for InAs than GaAs.

For growth on (001), the growth rate is mainly determined by the flux of Group III material and is insensitive to the other growth parameters under proper growth conditions: the growth rate is independent of the substrate temperature up to the start of evaporation of the Group III element, and of the As flux under excess As conditions. An alternative growth mechanism has to be established to explain the dependence of the period of the RHEED intensity oscillations on growth conditions for the case of InAs growth on InAs (111)A. The addition of a step flow mode to the two-dimensional nucleation growth mode is known to increase the period of RHEED intensity oscillations, since adatoms incorporated at existing steps do not contribute to the intensity oscillation [16] and the effect of this step-flow mode would be more significant for the longer migration lengths corresponding to higher substrate temperatures and lower As fluxes. This tendency is in good agreement with the dependence of the period as shown in fig. 2. For lower As fluxes, the substrate temperature of the critical growth condition decreases since the In migration is enhanced by reduction of the As/In flux ratio. The introduction of a step-flow mode cannot, however, explain the decrease of the average RHEED intensity which occurs concomitantly with the increase of period, so although the effect of a step-flow mode cannot be ruled out, it is not the only cause of the increase of the period.

A second possible mechanism is a low apparent sticking coefficient of As on a (111)A surface. In the case of an insufficient supply of As in growth on (001) substrates, the period of the RHEED intensity oscillations is determined by the incorporation rate of As and the average RHEED intensity decreases as droplets of Group III element are formed on the surface. Assuming the As sticking coefficient on (111)A surface is strongly dependent on substrate temperature, the dependence of the period of the intensity oscillation on the substrate temperature could be explained as follows. The actual As flux supplied to the surface is nominally high enough compared to the In flux, but the effective As flux, which is the product of the flux and its sticking coefficient, decreases as the substrate temperature is increased due to reduction of the sticking coefficient. The effective As flux becomes equal to the In flux at the critical growth condition. For higher substrate temperatures, the period increases as the sticking coefficient and effective As flux decrease with increasing substrate temperature. At the same time, the formation of In droplets due to the In stable growth decreases the average RHEED intensity. However, the change of the surface terminating chemical species cannot be observed by RHEED, since the surface reconstruction does not change with surface stoichiometry on (111)A [17]. An increased substrate temperature for the critical growth condition at higher As/Ga flux ratios is well explained by an increase of the effective As flux. We believe the deficiency of As caused by its low sticking coefficient on (111)A surfaces is the dominant effect causing the increase of the oscillation period.

For the case of GaAs, the sticking coefficient of Ga is considered to be unity, but some consideration should be given to the desorption of In during growth of InAs. The In desorption may cause an increase of the oscillation period for higher substrate temperatures but cannot explain the strong dependence on the As flux of the substrate temperature at which the period begin to increase. The steeper slope of the higher substrate temperature region for an As flux of $1.6 \times 10^{15} \text{cm}^{-2}\text{s}^{-1}$ compared with the other As fluxes observed in fig. 2, may, however reflect In desorption from InAs.

Finally, we should consider the small variations in the RHEED intensity oscillation period when it is effectively determined by the In flux and represents the growth rate. We do not have a definitive explanation, but suggest that it may relate to the very low reactivity between As species and In, so that even when As is apparently in excess, small fluctuations in incorporation of In can still occur.

4. Conclusion

The MBE growth process of InAs on InAs (111)A has been studied by RHEED. The oscillation period during growth strongly depends on growth conditions. The period is in agreement with In flux for lower substrate temperatures or higher As fluxes but increases for higher substrate temperature and lower As fluxes. The increase can be explained by the lower sticking coefficient of As on (111)A than on (100).

Acknowledgments

The financial support of EPSRC under grant no. GR/J 97540 and of the Ministry of Education, Science and Culture, Japan (for TN) is gratefully acknowledged.

References

[1] E. Kapon, M. C. Tamargo and D. M. Hwang, Appl. Phys. Lett. 50 (1987) 347.

[2] J. S. Winer, G. Danan, A. Pinczuk, J. Valladeres, L. N. Pfeiffer and K. West, Phys. Rev. Lett. 63 (1989) 1641.

[3] M. E. Hoenk, C. W. Nieh, H. Z. Chen and K. J. Vahala, Appl. Phys. Lett. 55 (1989) 53.

[4] R. Nötzel and K Ploog, J. Vac. Sci. Technol. A 10 (1992) 617.

[5] T. Isu, M. Hata, Y. Morishita, Y. Nomuraa, S. Goto and Y. Katayama, J. Crystal Growth 120 (1992) 45.

[6] X. Q. Chen, M. Tanaka, K. Wada, and T. Nishinaga, J. Crystal Growth 135 (1994) 85.

[7] X. Q. Chen and T. Nishinaga, J. Cryst. Growth 146 (1995) 374.

[8] B. K. Laurich, K. Elcess, C. G. Fonstad, J. G. Berry, C. Mailhiot, and D. L. Smith, Phys. Rev. Lett. 62, 649 (1989).

[9] E. A. Caridi, T. Y. Chang, K. W. Goosen, and L. F. Eastman, Appl. Phys. Lett. 56, 659 (1990).

[10] D. L. Smith and C. Mailhiot, Phys. Rev. Lett. 58, 1264 (1987)

[11] T. Hayakawa, K. Takahashi, M. Kondo, T. Sunayama, S. Yamaoto, and T. Hjikata, Phys. Rev. Lett. 60 (1988) 349.

[12] M. R. Fahy, J. H. Neave, M. J. Ashwin, R. Murray, R. C. Newman, B. A. Joyce, Y. Kadoya and H. Sakaki, J. Crystal Growth 127 (1993) 871.

[13] M. R. Fahy, K. Sato and B. A. Joyce, Appl. Phys. Lett. 64 (1994) 190.

[14] M. R. Fahy, K. Sato and B. A. Joyce, Appl. Surface Sci. 82/83 (1994) 14.

[15] K. Sato, M. R. Fahy and B. A. Joyce, Surface Sci. 315, 105 (1994).

[16] T. Shitara, J. Zhang, J. H. Neave and B. A. Joyce, J. Appl. Phys. 71 (1992) 4299.

[17] D. A. Woolf, D. I. Westwood and R. H. Williams, Appl. Phys. Lett. 62 (1993) 1370.

Inst. Phys. Conf. Ser. No 145: Chapter 2
Paper presented at 22nd Int. Symp. Compound Semiconductors, Cheju Island, Korea, 28 August–2 September 1995
© *1996 IOP Publishing Ltd*

Atomic Force Microscopy and Growth Modeling of GaN Nucleation Layers on (001) GaAs by Metalorganic Chemical Vapor Deposition

Kun Wang, Dimitris Pavlidis, and Jasprit Singh

Solid State Electronics Laboratory, Department of Electrical Engineering &
Computer Science, The University of Michigan, Ann Arbor, MI48109, U.S.A

Abstract: Nucleation images of GaN buffers grown by metalorganic chemical vapor deposition (MOCVD) on (001) GaAs substrates were obtained by atomic force microscopy (AFM) and were employed to investigate the growth temperature and time dependence of the nucleation mechanisms. The growth mode corresponds to 2D island nucleation at low temperatures, while 3D island growth is observed at high temperatures. Large grain sizes and good surface coverage was obtained for 3min growth. Optimum criteria were established for buffer growth and 0.5μm thick GaN films were grown on different buffers for comparison. Their morphology and phase dependence on nucleation layer characteristics are discussed. Theoretical modeling by Monte-Carlo techniques compliments these studies.

1 Introduction

GaN is a wide bandgap semiconductor with potential applications in optoelectronic devices from blue to ultraviolet wavelengths, as well as, high-power and high-temperature devices[1,2]. However, successful use of GaN for device applications has been hindered by the lack of lattice matched substrates. A possible solution to these difficulties is the introduction of a buffer layer between the substrate and the nitride layer of interest. This approach appears to greatly improve the material quality, as already demonstrated by various experimental results [3~5]. Good understanding of the growth parameters and growth mechanisms for the buffer layer growth are therefore of importance for achieving better quality of subsequently grown GaN films[6~8]. It is the purpose of this paper to provide such an understanding by investigating GaN nucleation layers by means of Atomic force microscopy (AFM) and Monte-carlo modeling.

2. Growth mode considerations

The issue of epitaxial growth is complex and depends on many factors such as rates of impingement of species, growth temperature, surface migration rates of reactants, the bond strengths and bond lengths of the substrate and overgrowth atoms, and the nucleation processes. Under equilibrium conditions, the growth mode adopted in a growth system depends on the surface energy of epilayer σ_e, the surface energy of the substrate σ_s, the interfacial free energy γ, and the strain energy σ_{st} due to lattice mismatch. Under non-equilibrium conditions, lattice mismatched growth as occurring in the case of nitrides grown in MOCVD is strongly influenced by the initial stages of growth and the resulting nucleation layers since the latter determine the growth mode. Nucleation and growth involves atoms in the vapor phase condensing on the substrate, and then forming clusters of adatoms. The formation of the clusters is controlled by the energy terms of the system under study. Due to its wide bandgap, GaN should display a very strong bonding strength. Should such a bonding strength be stronger than the bonding between the deposit and the substrate, then σ_e would be larger than γ. In this case, the contact angle of the critical nuclei would be expected to be large[9]. As a result, "wetting" of the substrate by the epitaxial layer is very little or equivalently speaking nucleation would be difficult. Nucleation is also governed by the growth temperature and the nucleation rate for a growth system is given by Eq. 1.[9]

$$c(i)=c(1)\exp(-\Delta G(i)/kT) \qquad (1)$$

where $c(i)$ is the concentration of the clusters of size i of adatoms, $\Delta G(i)$ is the formation energy for such a cluster, k is the Boltzmann constant, and T is the temperature. As one can see, the nucleation rate can be increased by decreasing the growth temperature. However, the migration rate of the adatoms would decrease as the temperature decreases and low migration rates could lead to a rough nucleation layer surface. Another factor to be considered is the reduction of reactant decomposition rates as the temperature decreases. These factors set a limit to the minimum temperature that can be used for growth. Based on the preceding discussions it is clear that the growth temperature plays a key role in achieving nucleation layers of good morphology. The studies undertaken in this work attempt for this reason to provide an understanding of layer nucleation as a function of temperature and examine growth at various stages by examining samples grown over different times.

The GaN nucleation layers investigated here were grown on (001) GaAs by MOCVD. Their morphology was examined using AFM which is a powerful technique for surface topography analysis. Vertical layer thickness variations in the nanometer range can easily be detected by this technique making it an ideal tool for studying the formation of the initial nuclei and their subsequent growth and coalescence. The roles of evaporation rate and surface migration were also studied by Monte Carlo simulation. The quality of films grown on different buffers is finally evaluated.

3. Growth of GaN and Techniques Employed for Film Characterization

The GaN films were grown in our low pressure (60 Torr) MOCVD system using H_2 carrier gas. Epi-ready (001) GaAs substrates were used. They were loaded in a horizontal reactor and annealed in H_2 and AsH_3 at 670°C for 5 minutes to remove the surface oxide. After cooling down to the desired growth temperature, NH_3 alone was supplied for 10 minutes so that surface nitridation could take place. Trimethylgallium (TMGa) and NH_3 were then supplied simultaneously and GaN films were grown. The TMGa flow rate was fixed at 13.4 μmol/min for all the experiments reported here. The NH_3 flow was 1000 sccm. The H_2 supply rate was 2 slm. The samples were grown at a fixed V/III ratio of 3300 and substrate temperatures ranging from 450°C to 600°C. The growth rate of the buffer layer was determined by growing a thick nitride film using the same layer growth condition as for the buffer and examining its cross section using scanning electron microscopy. The morphology of the nucleation layer was examined by using a Digital Instruments Nanoscope III AFM system with tapping mode. The structural properties of the subsequently grown GaN films were characterized by X-ray diffraction(XRD).

4. Characterization and Modeling Results

A first set of samples was first examined with GaN layers grown over a fixed time of 1min. The growth temperature of these samples varied from 450 °C to 600 °C. AFM analysis of the layers was performed and the results are depicted in Fig. 1. The morphology of the nucleation layers can here be examined as a function of the growth temperature (a, 450°C ; b, 500°C; c, 550°C; d, 600°C).

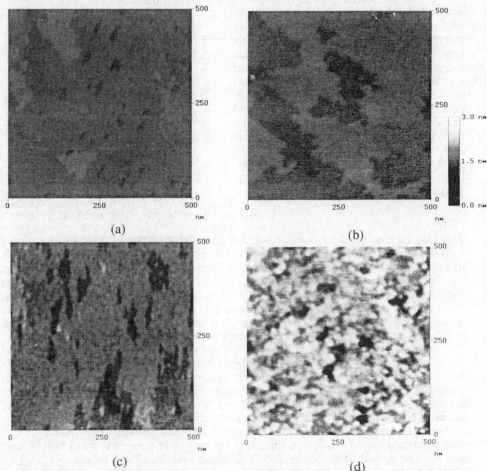

(a)   (b)

(c)   (d)

Fig.1 AFM images of GaN nucleation layers grown on (001) GaAs for 1 min. at a temperature of (a) 450°C; (b) 500°C; (c) 550°C; (d) 600°C.

At the growth temperature of 450°C, small regions of 2D nucleation islands (~50nm) could be observed in this case. The results show that the growth mode at low temperatures corresponds to two dimensional (2D) nucleation. Surface diffusion rates are apparently increased at growth temperatures of 500°C and the size of the 2D nucleation islands increased considerably (~300nm). To confirm the 2D features of this nucleation layer, a cross sectional view was obtained by AFM and the monolayer step was observed to be about 0.24nm.

When the temperature was increased further to 550°C, 2D nucleation remained still the dominant mechanism but three dimensional (3D) islands could also be observed. 3D island growth became finally the dominant mechanism at 600°C and the surface roughness increased significantly. These experiments suggest that growth temperature conditions would be optimum if a temperature of 500°C is chosen.

To complete the study layers were grown under the above mentioned optimum temperature conditions of 500°C but different growth times were employed. AFM analysis was again performed and Fig.2 shows the layer roughness estimated by AFM as a function of the growth time. Surface roughness is shown here to be minimum at a growth time of 3 minutes. The thickness of the film corresponding to this growth condition is around 200Å as estimated based on growth of thick layers at the same temperature. A 3-minute growth time also corresponds to large grain sizes and good surface coverage as found by AFM. Fig. 2 also shows that the roughness increases as the film becomes thicker.

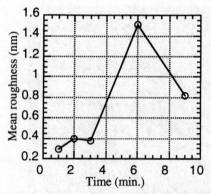

Fig.2 The AFM extracted roughness of GaN nucleation layers grown at 500°C vs growth time.

This may be explained by considering that once the substrate is completely covered with the nucleation layer, the subsequent GaN film is grown under homoepitaxial-like conditions. Thus the growth difficulties encountered at the initial growth stages are no longer faced and the deposited GaN film wets adequately the surface of the nucleation layer. The nucleation rate will in this case increase significantly. However, due to the relatively low growth temperature, the surface migration rates of the adatoms would be very small. Vertical growth would consequently take place faster than lateral growth and as a result, 3D growth will be dominant.

To understand the role of kinetics in GaN growth, a model was developed and recently reported by the authors[10]. A basic mechanism considered in this model is that due to the very high nitrogen evaporation, the nitrogen atom incorporation occurs only when Ga atoms impinge and cause nitrogen atoms to be incorporated on the substrate. The model is implemented in a Monte Carlo simulation approach and can be used to study the impact of substrate temperature, Ga flux and V/III (group V element to group III element) ratio on growth rate and growth front quality. Fig. 3 shows the morphology of layers simulated for growth at temperatures of 500 °C and 600 °C. The growth front of the T=500°C case corresponds to 6 monolayers (MLs), while the growth front of the T=600°C case corresponds to 3 MLs. As can be seen, the morphology is improved as the temperatures increases. This can be understood by considering that the migration rate increases as the temperature increases. Higher growth temperatures are consequently preferred for achieving surface thermal equilibrium and promoting lateral growth. These results suggest that once the substrate is covered by a thin nucleation layer, which acts as a buffer, the subsequent film should be grown at a higher temperature for improving quality. Growth should consequently take place in two steps. A thin buffer layer should first be grown at a low temperature, and then followed by a thick film grown at a higher temperature.

Using the above described two step approach, 0.5 μm thick GaN films were grown on buffers realized using different growth parameters. Fig. 4 shows the morphologies of 0.5 μm GaN films grown on buffers which were realized at (a) 530 °C for 5 minutes, (b) 500 °C for 3 minutes; the latter corresponds to the optimized growth condition for buffer layers. As can been seen, films grown on the optimized buffer layer show improved morphology. It should also be pointed out that the morphology of the thick GaN films depends on the roughness of the nucleation layers; smooth buffer layers resulted in good morphology of the subsequently grown thick layers. X-ray diffraction measurements were finally carried out on the two samples and the results are shown in Fig. 5. Enhancement of the GaN cubic phase is shown for films grown on optimized buffer layers as suggested by the improved full width at half maximum and intensity of such layers.

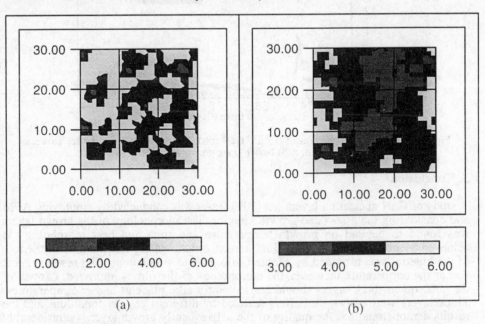

(a) (b)

Fig.3 Simulated growth front contours: (a) At growth temperature of 500°C ; (b) At growth temperature of 600°C . The legend on the bottom gives the number of surface monolayers.

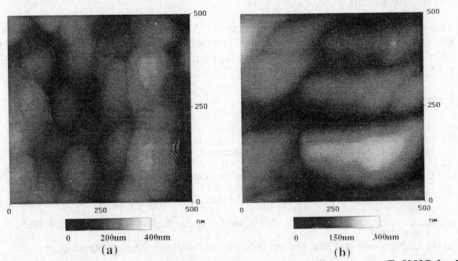

(a) (b)

Fig. 4 AFM images of 0.5 μm GaN thick layers grown on (a) buffer grown at T=530°C for 5 minutes ; (b) buffer layer grown at T=500°C for 3 minutes. The legends show the vertical dimension.

126

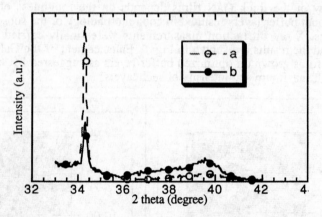

Fig. 5 X-ray diffraction patterns of 0.5 μm GaN thick layers grown on (a) buffer grown at T=530°C for 5 min. ; (b) buffer layer grown at T=500°C for 3 min.

5. Conclusion

A study of GaN nucleation layers on (001) GaAs was conducted by employing AFM and considering the associated growth kinetics. The morphology of the grown layers was found to depend on growth temperature and time and best morphology is achieved by growth at 500 °C for 3 minutes. Complimentary information on the role of the kinetics was provided by Monte Carlo growth simulations. The result suggest that as the temperature increases, the morphology of the film is improved. Growth of the subsequent thick layers should consequently take place at higher temperatures. Thick layers were grown on buffers realized by different growth conditions and the results demonstrate that the quality of the subsequently grown layer is considerably controlled by the morphology of the buffer layer.

Acknowledgments

The authors would like to thank Max Yoder of ONR for his support and encouragement, Dr. Nasserdine Draidia for helpful discussion, and Dr. J. Mansfield for help in AFM characterization. Work supported by ONR Contract No: N00014-92-J-1552

References

[1]   R. F. Davis, Proc. IEEE **79**, 702 (1991).
[2]   S. Strite and H. Morkoç, J. Vac. Sci. Technol. B **10** , 1237 (1992).
[3]. H. Amano, N. Sawaki, I. Akasaki, and Y. Toyoda,Appl. Phys. Lett. 48, 3531,1986
[4]. S. Nakamara, Jpn J. Appl. Phys., 30, L1705, 1991
[5]. C. H. Hong,, K. Wang, and D. Pavlidis, J. Electron Materials, 24(4), 213, 1995
[6]. J.N. Kuznia, M.A. Khan, D.T. Olson, R. Kaplan, and J. Freitas, . J. Appl. Phys. 73,4700 (1993)
[7]. Y. Koide, N. Itoh, K. Itoh, N. Sawaki, and I. Akasaki,. Jpn J. Appl. Phys. Vol. 27, No.7, 1156, (1988)
[8]. K. Doverspike, L.B. Rowland, D.K. Gaskill, and J.A. Freitas, Jr., J. of Elec. Mater. Vol. 24, No. 4, 269, (1995)
[9]   B. K. Chakraverty, in *Crystal Growth: An Introduction*, edited by P. Hartman, North holland, 1973.
[10]. K. Wang, D. Pavlidis, and J. Singh, J. Appl. Phys. 76 (6), 1, 1994

Inst. Phys. Conf. Ser. No 145: Chapter 2
Paper presented at 22nd Int. Symp. Compound Semiconductors, Cheju Island, Korea, 28 August–2 September 1995
© *1996 IOP Publishing Ltd*

Effects of the mask geometry on the selective-area growth by metalorganic chemical vapor deposition

T. Itagaki, M. Takemi, T. Takiguchi, T. Kimura, Y. Mihashi and S. Takamiya

Optoelectronic & Microwave Devices Laboratory, Mitsubishi Electric Corporation
4-1, Mizuhara, Itami, Hyogo 664, Japan

Abstract. We have investigated the growth-rate enhancement effect in the selective-area growth of InP by metalorganic chemical vapor deposition using SiO_2 masks. The experiments using isolated masks of rectangular shapes with various widths and lengths have proved that the growth rate as well as its distribution in the area adjacent to the mask is affected not only by the size of the mask but also the aspect ratio of it. Based on this result, the shape of the mask has been modified in order to obtain uniform layer thickness along the stripe direction between a pair of stripe masks. The results have shown that the uniformity in growth-rate distribution between the paired stripe masks is successfully improved by attaching small additional rectangles to the ends of the stripe masks.

1. Introduction

Selective-area growth by metalorganic chemical vapor deposition (MOCVD) using dielectric mask patterns is an attractive technique to change the thickness of the epitaxial layer around the masks from that in the area far from the masks [1][2][3]. It has been reported that, using a pair of stripe masks, the bandgap energy of the multiple quantum-well (MQW) structure formed between the paired masks can be successfully controlled by varying either of the mask width and opening width between the masks [2][3]. Using different mask patterns with various widths and the opening widths, the plural functional components (i.e., laser diodes, optical modulators, tapered-waveguide lenses, and so on) have been monolithically integrated by one-step growth [4][5][6]. In order to obtain the excellent characteristics of each component, it is important to realize uniform distribution in thickness along the optical axis (i.e., along the stripe direction) within the short device region (for instance, several hundred micro-meters).

In this paper, we report on the thickness distribution around isolated rectangular masks with various lengths and widths in detail. Especially, attention has been focused on the effects of the width-to-length ratio (aspect ratio) of the mask. Furthermore, it is reported that the modification of the mask shape beyond the simple rectangular shape is effective to improve the uniformity in the layer thickness in the case of paired stripe masks.

2. Growth-rate distribution around the isolated mask

2.1. Selective-area growth

InP layers of 200 nm in thickness were grown on (100) InP substrates using a conventional low-pressure MOCVD system with source materials of trimethylindium (TMIn) and phosphine (PH$_3$). The growth pressure, temperature, and V/III ratio were 150 Torr, 650°C, and 120, respectively. The SiO$_2$ mask patterns with rectangular shapes were formed so that their sides are parallel to [011] and [01$\bar{1}$] (see Fig. 1 (A)). Those masks were settled away from each other so that mutual interaction can be neglected. Hereafter, we denote the lengths of the rectangular mask along [011] and [01$\bar{1}$] as the mask length and width, respectively. The mask width (Wm) was varied by changing the mask width/length ratio from 0.1 to 1.25, for the different mask lengths (Lm) of 125, 250, and 500 μm. No deposition of InP polycrystal was observed on the masks in all cases.

Fig. 1 (B) shows the schematic illustration of the cross-sectional view of the selectively-grown layer. The layer thickness of the selectively-grown InP became maximum at the growth edge, which terminated in a (111)B facet. Along the mask side, the height of the growth edge became maximum at the midpoint of the mask side. We designate the growth rate at the growth edge at the midpoint of the mask side as Rg-max. Growth-rate enhancement factor was defined as Rg-max/Ro, where Ro denotes the growth rate in the area far from the mask. An X-Y stylus profiler was used for the layer-thickness measurements.

In the following, we describe the effects of the mask geometry on the growth-rate enhancement factor, Rg-max/Ro.

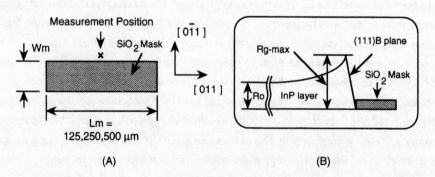

(A)

(B)

Fig. 1 : The measurement point against the mask geometry (A) and typical cross-sectional view of the selectively grown layer (B)

2.2. *Dependence of the growth-rate enhancement factor on the mask width*

Fig. 2 shows the growth-rate enhancement factor as a function of the mask width for three different mask lengths. It has been reported that the growth-rate enhancement factor is almost proportional to the mask width when the mask is narrow [2][4]. As can be seen in the Fig. 2, however, the growth-rate enhancement factor was proportional to the mask width only when Wm is small, and showed a saturation behavior when the mask width is increased. The saturation behavior depends on the mask length : For smaller value of Lm (i.e., for higher value of the Wm/Lm ratio), the saturation occurred earlier, i.e., at smaller value of Wm.

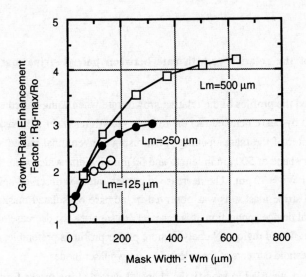

Fig. 2 : Growth-rate enhancement factor as a function of the mask width for three different mask lengths.

2.3. *Growth-rate profile along the mask edge*

In order to explain the saturation behavior of the growth-rate enhancement factor, we investigated the distribution of the growth rate along the mask edge. Fig. 3 shows the profiles of the relative growth rate measured at the top of the (111)B facet for masks with a length of Lm = 250 μm and various Wm/Lm ratios, in which the growth rate is normalized by Ro. As can be seen in Fig. 3, high uniformity in the relative growth rate was obtained at a Wm/Lm ratio of 0.1, while convex profiles appeared for higher Wm/Lm ratios.

    The decrease in the relative growth rate at the mask-end region is considered to be caused by the relative increase in the amount of the diffusing species along the lateral direction ([011] direction). The saturation behavior of the growth-rate enhancement factor shown in Fig. 2 should also be attributed to the increase in lateral diffusion with increasing the Wm (Wm/Lm ratio).

130

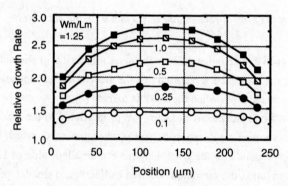

Fig. 3 : Relative growth-rate profiles along the mask edge for masks with a length of Lm = 250 μm and various Wm/Lm ratios.

3. Distribution of the relative growth rate between paired stripe masks

Next, we investigated the profiles of the relative growth rate when using paired stripe masks. In Fig. 4, we plotted by curve denoted (A) the relative growth rate near the mask-end region measured at the center of the mask opening, when using conventional paired stripe masks having a rectangular shape of 500 μm in length and 50 μm in width, which were separated by the opening width of Wo = 70 μm. The decrease in growth rate was observed near the end of conventional paired stripe masks, as was observed in the case of isolated masks. From the viewpoint of practical device application, it is necessary to suppress the decrease of the relative growth rate in the mask-end region. Namely, a more planar profile is preferable, for instance, to obtain the low threshold currents of the selectively-grown laser diodes.

For this purpose, we tried to modify the shape of paired stripe masks from the simple rectangle. The concept of the modification is as follows. It is known that the relative growth rate between paired stripe masks increases with decreasing the opening width [1][2]. Therefore, in order to flatten the convex profiles, the modification of the mask shape should be designed so as to narrow the opening width in the mask-end region. It is, therefore, considered that, by attaching a small additional rectangles to the mask-end region, as illustrated in Fig. 4, the relative growth rate in the mask-end region can be increased.

Based on this idea, firstly, we investigated the controllability of the relative growth rate, using the modified paired stripe masks. Fig. 4 shows the profiles of the relative growth rate for paired stripe masks measured at the center of the mask opening area for three types of modified paired masks (B, C, and D), comparing with the profile obtained when using the conventional simple rectangular paired masks. In these masks, Wm, Lm, and Wo were fixed at 50, 500, and 70 μm, respectively. The additional rectangles for the modification were 20 μm in width and 15, 25, and 50 μm in length, in modified mask B, C, and D, respectively.

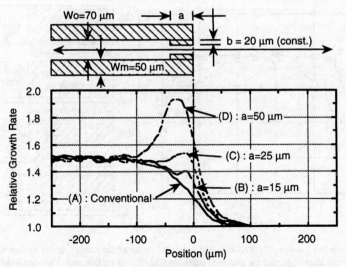

Fig. 4 : Profiles of relative growth rate measured at the center of the mask opening using conventional paired stripe masks (A) and three types of modified paired stripe masks [(B) : a= 15 μm, (C) : a= 25 μm, and (D) : a= 50 μm)]. Wm, Lm, and Wo were fixed at 50, 500, and 70 μm, respectively.

In the case of mask B, the relative growth rate in the mask-end region was still smaller than that in the mask-center region. On the other hand, in the case of mask C, the relative growth rate in the mask-end region was nearly equal to that in the mask-center region. In the case of mask D, the relative growth rate at the mask-end region was excessively larger than that in the mask-center region. From these results, it is shown that the relative growth rate in the mask-end region is increased with increasing the length of additional rectangles, and that the most uniform profile of the relative growth rate was obtained at 25 μm in length of additional rectangles.

For the application of selective-area growth to the tapered-waveguide-lens integrated laser diodes [6], the uniform growth rate is required for higher Wm/Lm ratios, since higher relative growth rates are required. Therefore, we next investigated the case of a high Wm/Lm ratio of 0.4. In Fig. 5, we plotted the profiles of the relative growth rate measured at the center of the mask-opening area for a pair of modified stripe masks (B), comparing with that obtained when using the conventional simple rectangular paired mask (A). Wm, Lm, and Wo were fixed at 200, 500, and 100 μm, respectively. In this case, the optimum dimensions of the additional rectangles were found to be 20 μm in width and 100 μm in length. As can be seen in Fig. 5, a uniform profile of the relative growth rate was obtained by the mask modification even for a higher Wm/Lm ratio of 0.4. It has thus been demonstrated that adequate modification of the mask shape by attaching small rectangles is effective to control the profiles of the relative growth rate in the mask-end regions of paired masks. This method must be very useful for the mask designing of practical devices, such as laser diodes, optical modulators, and so on.

132

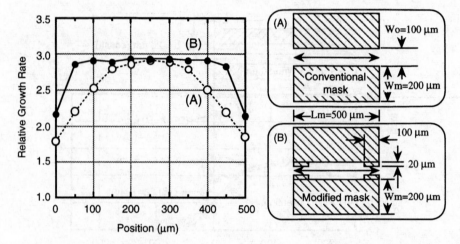

Fig. 5 : Relative growth-rate profiles measured at the center of the mask opening for conventional paired stripe masks (A) and for modified paired masks (B). Wm, Lm, and Wo were fixed at 200, 500, and 100 μm, respectively.

4.   Conclusions

We have investigated the growth-rate enhancement effect in the selective-area growth by MOCVD using isolated masks of rectangular shapes with various widths and lengths. It has been found that the growth-rate enhancement factor in the selective area growth is influenced not only by the mask width but also by the mask length, thus it depends on the Wm/Lm ratio. In addition, the uniformity in growth-rate distribution along the stripe direction between paired stripe masks is successfully improved by attaching small additional rectangles to the ends of the stripe masks. These findings must be very useful for realizing excellent characteristics of monolithically integrated photonic devices.

Acknowledgment

The authors would like to thank Mr. Ishimura, Mr. Takemoto, Mr. Goto, and Dr. Kajikawa for their fruitful discussions.

References

[1]   Kim M S 1992 J. Crystal Growth **123** 69-74
[2]   Suzuki M 1994 J. Crystal Growth **145** 249-255
[3]   Itagaki T 1994 J. Crystal Growth **145** 256-262
[4]   Aoki M 1993 IEEE J. Quantum Electronics **29** 2088-2096
[5]   Sasaki T 1994 J. Crystal Growth **145** 846-851
[6]   Kobayashi H 1994 IEEE Photonics Tech. Letters **6** 1080-1081

Inst. Phys. Conf. Ser. No 145: Chapter 2
Paper presented at 22nd Int. Symp. Compound Semiconductors, Cheju Island, Korea, 28 August–2 September 1995
© *1996 IOP Publishing Ltd*

Production of P-type GaN in a Multi-wafer-rotating-disc Reactor

**C Yuan, R Walker, T Salagaj, A Gurary, W Kroll, R A Stall,
M Schurman[1], C -Y Hwang[1], Y Li[1], Y Lu[1], W E Mayo[1],
S Krishnankutty[2], R M Kolbas[2] and S J Pearton[3]**

EMCORE Corporation, 35 Elizabeth Ave., Somerset, New Jersey 08873
1) Rutgers University, Piscataway, New Jersey 08855;
2) Department of Electrical & Computer Engineering, North Carolina State University, Raleigh, NC 27695;
3) Department of Materials Science and Engineering, University of Florida, Gainesville, FL 32611

Abstract. The production of high quality p-type GaN materials has been a potential problem for optoelectronic device applications. High quality GaN based material depositions have been achieved in small scale reactors but not on a mass production scale. For the first time we report here on the optical, electrical, and material properties of high quality p-GaN epitaxial deposition in a production scale MOCVD system which can produce 6 x 50 mm wafers per run.

1. Introduction

The direct transition type band structure, superior physical properties [1,2] and promising optical properties have put GaN based material system into one of the best choices for blue, green, violet and ultraviolet device applications, especially for light emitting diodes (LEDs) and laser diodes (LDs). Practically, high quality III-Nitride films have been achieved by several advanced epitaxial techniques, such as metalorganic chemical vapor deposition (MOCVD) and molecular beam epitaxy (MBE). For high volume production of GaN based material, MOCVD has proven to be the most promising technique. In general, p-type *in-situ* Mg doping in GaN has been a primary potential problem. By combining the post growth treatments [3-7], converting the highly resistive Mg-doped GaN into p-type GaN, high quality Mg-doped p-type GaN films have been achieved in small size single wafer reactors, but not yet on a mass production scale. In this paper, we investigate the *in-situ* Mg doping of GaN grown on c-sapphire in a production scale MOCVD system.

2. Experiment

The GaN growth system is a fully computer-controlled, multi-wafer, rotating disk reactor, capable of growing up to 6 x 50 mm wafers in a run. The system, shown in Fig. 1, is equipped with a double walled water-cooled stainless steel chamber, a two zone resistance heater, and a UHV loadlock. A high speed motor driven by a toothed-belt generates the spindle rotation. A susceptor on top of the spindle provides support and restraint for the removable wafer carriers. We demonstrated a wafer carrier (7"), temperature uniformity of ± 3.5°C at 1050°C and < ± 1.0°C at 550°C, which ensures the temperature requirements for mass production. In addition, the high speed rotating disk generates a sharp temperature gradient normal to the wafer which is desirable for the minimization of pre-reactions that can occur in the III-nitride system.

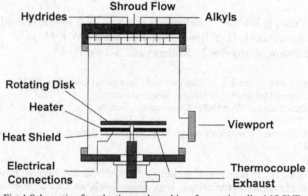

Fig. 1 Schematic of production scale multi-wafer-rotating-disc MOCVD.

In preparation for epitaxial p-doped GaN growth, trimethylgallium (TMG) and ammonia (NH$_3$, 100%) were utilized as the primary chemical precursors. The c-sapphire substrates first undergo *ex situ* cleaning by HCl, HNO$_3$ and H$_2$O. This is followed by *in situ* cleaning in a pure H$_2$ flow at 1070 °C. Next, a GaN buffer layer is grown in order to overcome the large lattice mismatch between the GaN and substrate. A NH$_3$ flow of 4-20 l/min and TMG flow of 10-30 sccm were used to produce a 200~300Å layer of amorphous or polycrystalline GaN at 530°C. This thin GaN layer is crystallized by a ramped thermal process. Finally, p-type GaN epitaxial films are grown by the reaction of NH$_3$ and TMG at 1040 °C with a flow rate of 4-20 l/min and 30-60 sccm, respectively. The pressure was 30-300 Torr, and the rotating speed was 500-1000 RPM during the growth process. The biscyclopentadienyl magnesium (Cp$_2$Mg) was used as the p-type doping precursor. Before the p-type doping, a low background carrier concentration of n ~ 5 x 10^{16} cm^{-3} was achieved. A growth rate of 2 ~ 3 μm/h was achieved during the growth cycle. Transparent and featureless 2" wafers with smooth surface morphology were obtained over a wide range of Cp$_2$Mg flow (0.1 - 20 μmole/min).

3. Results and discussions

For the photoluminescence experiments a Spectra Physics argon-ion laser (275-305 nm output wavelength, 750 mW maximum CW power) was used as the photo-excitation source. The 77 and 300K characteristic emission spectra of Mg doped p-type GaN after post annealing in N$_2$ ambient at ~700°C for an hour are shown in Fig. 2. The spectra are dominated by an intense, violet, emission band peaking around 427.5 nm (2.9 eV). The FWHM of these violet peaks are approximately 300 meV which is comparable to the narrowest linewidths from highly p-type Mg-doped GaN [5]. The peak intensity of emission spectra at 77K from the sample with $p=5 \times 10^{17}$ cm^{-3} is at least an order of magnitude larger than for the sample with 1×10^{17} cm^{-3}. However, even the lowest intensity emission is clearly visible to the naked eye at room temperature. This intense, narrow linewidth, violet emission at room temperature clearly indicates the high optical quality of our p-type, Mg-doped GaN films.

Transmission electron microscopy (TEM) is a powerful tool to study material structure, defects, and crystal quality. We prepared TEM specimens by conventional techniques [8,9]. Fig. 3 is the plan-view of the Mg-doped GaN. The black and white contrast near the end of the threading dislocations is due to the surface relaxation of the dislocations [10]. From plan-view TEM micrographs, we estimate a defect density of 4×10^{9} cm^{-2} from Mg-doped GaN compared to 1.5×10^{9} cm^{-2} from undoped high quality GaN. Both of these values are lower than most of the reported data. However, minimizing such defects should result in further improvement of the crystal quality of the epi-layer.

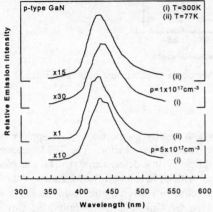

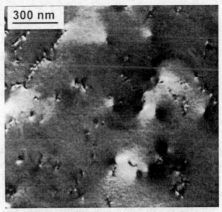

Fig. 2 PL spectra from two p-type GaN samples at 300 and 77K.

Fig. 3 Plan-view TEM: Mg-doped GaN with defect density of ~ 4×10^{9} cm^{-2}.

To obtain the average carrier concentration in the Mg doped GaN films, Hall measurements were performed at room temperature. Fig. 4 shows the hole concentration and mobility, calculated by assuming a single carrier type, as a function of corresponding Cp$_2$Mg mole fraction on Mg-doped GaN samples after post annealing. One can see that when the Cp$_2$Mg/TMG ratio is less than 1.1, the Mg-doped GaN are still n-type or highly compensated materials. When the Cp$_2$Mg mole fraction passes the threshold, the Mg-GaN samples were converted into p-type after annealing. For instance, for a 2.1 μm thick sample, we obtained a hole concentration of 5.2×10^{18} cm^{-3} with a mobility of 20 cm^2 V^{-1}s^{-1}. The resistivity of the same p-GaN sample was measured to be as low as 0.1 ohm-cm by both Hall effect and contactless resistivity measurement. The p-type nature of the GaN was also confirmed by a hot probe test. From the temperature dependence of the carrier density, we obtained an activation energy of ~ 155 meV, consistent with previously reported results on Mg-doped GaN [11]. For this ionization level, only ~ 3×10^{-3} of the Mg acceptors will be active at room

136

temperature, suggesting in our most highly doped films we are incorporating $\sim 10^{21}$ Mg atoms cm^{-3}. The carrier concentration drops for a higher Cp_2Mg mole fractions, which is also associated with the epi-layer cracking along the a-axis. Under these conditions the active Mg concentration decreases, and the mobilities also are degraded.

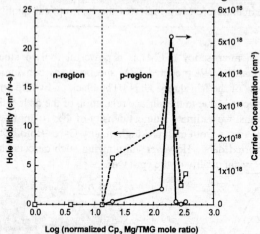

Fig. 4 Cp_2Mg mole flow rate effect on carrier concentrations and hole mobilities.

4. Conclusions

In summary, we have achieved for the first time high quality p-type GaN on c-sapphire by MOCVD from a mass production scale multi-wafer-rotating-disc reactor. The TEM structure study indicates high quality Mg-doped GaN with relatively low defect density. The p-doping concentrations range from 1 x 10^{17} to 5 x 10^{18} cm^{-3} with an hole mobility of 2-20 $cm^2V^{-1}s^{-1}$. The post thermal annealing process proved to be effective in converting the highly resistive as grown Mg-doped GaN into highly conductive p-type GaN material. In addition, the p-type epitaxial layers were single crystal, and exhibited excellent surface morphology, uniformity, and optical quality. The research at EMCORE was supported by BMDO-IST under contract N00014-93-C-0269 and managed by Mr. M. Yoder at ONR.

References

[1] Morkoç H, Strite S, Gao G B, Lin M E, Sverdlov B and Burns M 1994 *J. Appl. Phys.* **76** 3
[2] Strite S and Morkoç H 1992 *J. Vac. Sci. Technol.* **B10** 4
[3] Amano H, Sawaki N, Akasaki I and Toyoda Y 1986 *Appl. Phys. Lett.* **48** 353
[4] Amano H, Kito M, Hiramatsu K and Akasaki I 1989 *Jpn. J. Appl. Phys.* **28** L2112
[5] Nakamura S, Senoh M, Mukai T 1991 *Jpn. J. Appl. Phys.* **30** L1708
[6] Nakamura S, Mukai T, Senoh M and Iwasa N 1992 *Jpn. J. Appl. Phys.* **31** L139
[7] Nakamura S 1991 *Jpn. J. Appl. Phys.* **30** 1620
[8] Bravman J C and Sinclair R 1984 *J. Electron Microsc. Tech.* **1** 53
[9] Helmersson U and Sundgren J 1986 *ibid* **4** 361
[10] Tunstall W J, Hirsh P B and Steeds J 1964 *J. Phil. Mag.* **9** 99
[11] Tanaka T, Watanabe A, Amano H, Kobayashi Y, Akasaki I, Yamazaki S and Koike M 1994 *Appl. Phys. Lett.* **65** 593

Inst. Phys. Conf. Ser. No 145: Chapter 2
Paper presented at 22nd Int. Symp. Compound Semiconductors, Cheju Island, Korea, 28 August–2 September 1995
© *1996 IOP Publishing Ltd*

Lateral growth rate enhancement on patterned GaAs substrates with CCl4 by MOCVD

Seong-Il Kim, Yong Kim, Seong-Min Hwang, Moo-Sung Kim, and Suk-Ki Min

Semiconductor Materials Research Center, Korea Institute of Science and Technology, P.O. Box 131, Cheongryang, Seoul 130-650, Korea

Abstract. We have investigated the CCl4 doping effect during MOCVD on patterned GaAs substrate. It was shown that the CCl4 flow rate was a very important parameter which enhanced the lateral growth rate of GaAs/AlGaAs on patterned substrates. With supplying of CCl4, the increase of GaAs lateral growth rate was remarkable and the increment could be described as a linear function of CCl4 flow rate. The lateral growth rate increased up to 700 ℃, but it decreased for more elevated growth temperature. With increasing the V/III ratio, the lateral growth rate increased, but at higher V/III ratio the increment seems to be saturated. This novel characteristics can be utilized in fabricating very novel quantum wire like structure.

1. Introduction

Low dimensional structures such as quantum well, quantum wire, or quantum dot structures can be obtained by defining a confinement pattern by electron beam or ion beam lithography and by etching or impurity induced disordering. Such methods are accompanied with surface or interface damages. However, high quality interfaces are required in order to achieve efficient luminescence and allow investigation of their intrinsic optical properties. To reduce the nonradiative recombination due to the damages on the surface or in the interface, growth on patterned substrate is more preferable than any other methods. Recently, this technology of growth on patterned substrate has been intensively studied by many authors [1-4], because this technology provide an easy method to fabricate various nanometer scale structures which have high optical quality are easily obtained. [1]

With suppling of CCl4, which has been utilized as a p-type dopant source for carbon doped GaAs epilayers during MOCVD growth, very novel facetting behavior of GaAs/AlGaAs multilayers were observed on the patterned GaAs substrate by MOCVD [1-4]. Growth rates on various facets are different from each other and they depend strongly on the growth conditions. By controlling these facet properties, it is possible to fabricate quantum structures. In this work we have investigated the CCl4 doping effects on pattered

GaAs substrate. The CCl_4 flow rate are shown to be very important parameter which control the lateral growth rate during MOCVD growth of GaAs/AlGaAs on patterned GaAs substrates.

2. Experimental procedures

The exact (100) oriented n+ GaAs wafers were used as substrates. Parallel stripes are defined by standard photolithography along the $\langle 01\bar{1}\rangle$ direction and a photomask consisting of both alternating stripe arrays and single isolated stripes (~2 μm width) is imployed for pattern transfer. Each stripe array consists of 40 stripes of 2 μm width each and separated by 2 μm. After the pattern transfer, the mesa and groove regions are generated by wet chemical etching (1:8:40 of H_2SO_4:H_2O_2:H_2O by volume ratio) using photoresist (PR) as an etching mask. The mesa heights and groove depths are about 1 μm. The exposed side wall planes of as-etched mesa are (111)A. All the experiments were carried out by the atmospheric pressure MOCVD [3-5]. Trimethylgallium(TMG), trimethylaluminum (TMA), AsH_3, and 1000 ppm H_2 diluted CCl_4 gas were utilized as the source reagents and palladium cell diffused pure hydrogen (H_2) gas with a total flow rate of 5 standard liters per minute was used as the carrier gas. The mole fraction of TMG to total carrier gas flow is 5.4×10^{-5}, and the mole fraction of TMA varies with the desired aluminum content of the AlGaAs layer. The desired amount of CCl_4 is supplied through a mass flow controller, with a maximum allowed flow rate of 0.3 sccm (i.e. mole fraction of 6×10^{-5}).

3. Results and Discussion

Fig. 1 shows the scanning electron microscope (SEM) pictures of the cross sections of the GaAs/$Al_{0.5}Ga_{0.5}As$ alternating layers grown on mesas with various CCl_4 flowing rates. The alternating GaAs (bright appearance) and $Al_{0.5}Ga_{0.5}As$ (dark appearance) layers are clearly resolved when stain etched in KOH:$KFe(CN)_6$:H_2O solution. The growth rate of undoped GaAs on non-patterned region with a V/III ratio of 60 at 750 ℃ is 0.062 μm/min. Thin $Al_{0.5}Ga_{0.5}As$ layer (~200 A) has a role of marker. No growth interruption is made during growth of these structures. Fig. 1(a) is for the sample without CCl_4 and Fig. 1(b), (c) and (d) correspond to the samples with CCl_4 flow rate of 0.05 sccm, 0.1 sccm, and 0.2 sccm, respectively. Markers represent 10 μm. For undoped multilayer, the evolutions of (433)A and (511)A facet planes are clearly visible. On the other hands, in the presence of CCl_4, (433)A and (511)A planes are simultaneously evolved on mesa side wall (Fig. 1(b)). However, (433)A are gradually disappeared and (511)A facet remains with increasing CCl_4. Additionally a distinguished extension of mesa top can be observed. The extension is due to the increase of lateral thickness of CCl_4-doped layer. If this property is adequately used, it will be possible to control the shapes of the epitaxial layers, especially in the active region of the optical device.

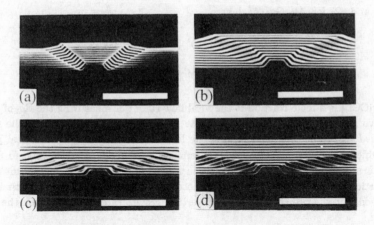

Fig. 1 SEM pictures of the cross sections of the GaAs/Al$_{0.5}$Ga$_{0.5}$As alternating layers grown on mesas. (a) is for the sample without CCl$_4$ and (b), (c), and (d) correspond to the CCl$_4$ flow rates of 0.05 sccm, 0.1 sccm, and 0.2 sccm, respectively.

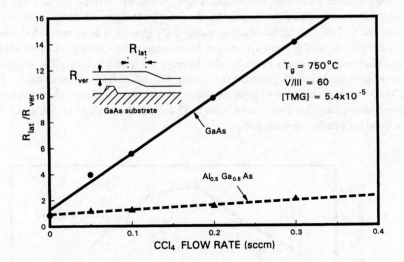

Fig. 2 The lateral (R$_{lat}$) / vertical (R$_{ver}$) growth rates of GaAs and AlGaAs as a function of CCl$_4$ flow rate. The inset illustrates the measurements of R$_{lat}$ and R$_{ver}$.

Fig. 2 shows the lateral (R$_{lat}$) / vertical (R$_{ver}$) growth rate ratios of GaAs and AlGaAs as a function of CCl$_4$ flow rate. The inset illustrates the measurements of R$_{lat}$ and R$_{ver}$. R$_{ver}$ decrease slightly as the CCl$_4$ flow rates increase, which is attributed to the etching effects. But due to the triple bonding of Ga on the side wall steps, it seems that the etching effect observed on (100) surface is not significant on the side wall. R$_{lat}$ increases in linear fashion as the CCl$_4$ flow rate increases. The relationship can be described as equation (1) for GaAs layers and (2) for $Al_{0.5}Ga_{0.5}As$ layers.

$$R_{lat}(\mu m/min) = 0.081 + 1.04 \times 10^4 \ [CCl_4], \quad \text{for GaAs} \tag{1}$$

$$R_{lat}(\mu m/min) = 0.099 + 0.20 \times 10^4 \ [CCl_4], \quad \text{for } Al_{0.5}Ga_{0.5}As \tag{2}$$

where $[CCl_4]$ is a mole fraction to the total flow rate. For measuring the R_{lat} of $Al_{0.5}Ga_{0.5}As$ we have grown another $Al_{0.5}Ga_{0.5}As/GaAs$ multilayers and in this case the GaAs layer was used as a marker. The ratio of the lateral growth rate to the vertical growth rate is significantly different between GaAs and AlGaAs layers. Particularly, in GaAs layer, the lateral growth rate is tremendously increased and an order of magnitude larger than the vertical growth rate. It should be noted that the GaAs lateral growth is remarkly increased, and the ratio of the lateral to vertical growth rates is about 14 for the CCl_4-doped sample.

Also in $Al_{0.5}Ga_{0.5}As$ layer, the lateral growth rate is increased larger than the vertical growth rate, though the increment is relatively small as compared with GaAs layer. This is due to the difference of mobilities between Ga and Al species. From these results, it is considered that the CCl_4 flow rate can be a very good parameter for the lateral growth control.

High surface mobilities (migration length) of column III species on (100) surface is suggested to be the cause of the increment of the lateral growth rates at increased V/III ratio of 60 at 750 ℃. In the case of using CCl_4 gas, it is also suggested that lateral gas phase diffusion of column III species (particularly Ga species) which is related with the complicated by-products involving the decomposed products from CCl_4 are the another cause of enhanced lateral growth rate during the MOCVD growth process. However, $Al_{0.5}Ga_{0.5}As$ layers which have the same colum III species with GaAs, show the remarkably different characteristics from GaAs layers. Thus our postulation is not clear yet. Thus more detailed studies are needed.

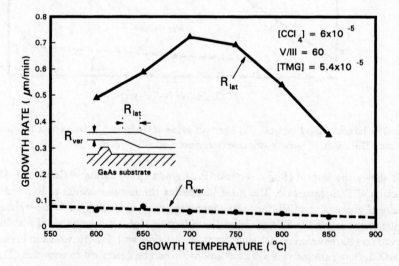

Fig. 3 The lateral growth rate of GaAs as a function of growth temperature.

Fig. 3 shows the lateral growth rate of GaAs as a function of growth temperature. The lateral growth rate increases up to 700 ℃, but it decreases for more elevated growth temperature. It is suggested that high surfaces mobilities (migration length) of column III species on (100) surface is responsible for the increment of the growth rates on inclined facets up to 700 ℃. But at growth temperature greater than 750 ℃, the gas phase depletion effects or increased desorption of reactant species seems to be the dominant process. Thus above 750 ℃, the lateral growth rate appears to be decreased. This phenomena is particularly outstanding in the side wall than on (100) surface, which is correspond to previous report [7].

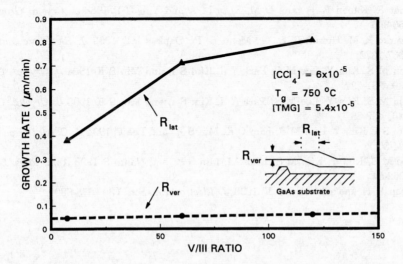

Fig. 4 The lateral growth rate of GaAs as a function of V/III ratio.

Fig. 4 shows the lateral growth rate of GaAs as a function of V/III ratio. With increasing the V/III ratio, the lateral growth rate increased, but at higher V/III ratio the increment seems to be saturated. Thus the lateral growth rate is represented as a sublinear function of V/III ratio. This phenomena seems to be similar to that of growth at low temperature regime, in other words kinetically controlled growth regime. At the given condition of constant supplyment of arsine, the increased lateral growth correspond to effective decrement of V/III ratio. Thus this phenomena seems to be closely related with decomposition of arsine. The characteristics of lateral growth enhancement can be applied to quantum wire fabrication to improve the thickness enhancement factor.

4. Conclusion

In this work we have investigated the CCl_4 doping effects on pattered GaAs substrate. With supplying of CCl_4, GaAs lateral growth is remarkably increased, and the maximum ratio of the lateral to vertical growth rates is about 14 for the CCl_4-doped sample. If this property is adequately used, it will be possible to control the shapes of the epitaxial layers, especially in the active region of the optical device. And the increase of GaAs lateral growth

142

rate was remarkable and the increment can be described as a linear function of CCl₄ flow rate. The lateral growth rate increases up to 700 ℃, but it decreases for more elevated growth temperature. With increasing the V/III ratio, the lateral growth rate increases, but at higher V/III ratio the increment seems to be saturated.

This work was supported in part by KIST-2000 program. The authors would like to thank Mr. Jae-Kyung Kim for SEM observation.

References

[1] Bhat R, Kapon E, Hwang D M, Koza M A and Yun C P 1988 *J. Crystal Growth* **93**, 850-856

[2] Dzurko K M, Hummell, S. G., Menu E. P., Dapkus P.D 1990 *J. Electronic Mater.* **19**, 1367-1372

[3] Kim M S, Kim Y, Lee M S, Park Y J, Kim S I, and Min S K 1994 *J. Crystal. Growth*, **139**, 231-237

[4] Kim M S, Kim Y, Lee M S, Park Y J, Kim S I, and Min S K 1995 *J. Crystal. Growth*, **146**, 482-88

[5] Kim S I, Kim Y, Kim M S, Kim C K, Min S K, and Lee C 1994 *J. Crystal Growth* **141**, 324-30

[6] Lee M S, Kim Y, Kim M S, Kim S I, Min S K, and Nahm S 1993 *Appl. Phys. Lett.* **63**, 3052-54

[7] Reep D H and Ghandhi S K 1983 *J. Electrochem. Soc.* **130**, 675-680

Inst. Phys. Conf. Ser. No 145: Chapter 2
Paper presented at 22nd Int. Symp. Compound Semiconductors, Cheju Island, Korea, 28 August–2 September 1995

The study of the interface between GaAs substrate and the regrowth GaAs layer formed by MBE and MOCVD

H Fujimoto, M Tanabe and A Tamura

Electronics Research Laboratory, Matsushita Electronics Corporation
3-1-1,Yagumo-nakamachi, Moriguchi, Osaka 570, JAPAN

Abstract. We have studied properties of interfaces between a GaAs substrate and a GaAs layer grown by MBE and MOCVD. DLTS measurement showed that hole trap levels exist in the interface by MBE, but no trap levels exist in the interface by MOCVD. SIMS measurement showed that carbon impurities of the growth interface by MOCVD are much lower than those by MBE. These results show that the MOCVD technique is suitable for obtaining good regrowth interface.

1. Introduction

In order to improve performances of field effect transistors (FETs) and heterojunction bipolar transistors (HBTs), many attempts to introduce regrowth techniques, including the selective regrowth technique, into the device fabrication process have been reported [1][2]. Combining the ion-implantation technique for a channel formation and the molecular beam epitaxy (MBE) growth technique for an undoped GaAs cap layer, we have developed a new power GaAs MESFET without a recessed gate structure[3]. Moreover, we have tried to introduce the metal-organic chemical vapor deposition (MOCVD) growth into our new FET fabrication process[4].

For these new fabrication processes, the interface characteristics between a substrate and an epi-layer is very important for electronic and optoelectronic devices. In this work, DLTS (Deep Level Transient Spectroscopy) [5] and SIMS (Secondary Ion Mass Spectroscopy) were used to investigate the interface characteristics between a GaAs substrate and a GaAs layer grown by MBE and MOCVD.

144

2. Experimental

We used undoped liquid encapsulated Czochralski (LEC) GaAs wafers with the (100) surface. N layers and n+ ohmic regions of experimental samples were formed by implanting Si ion directly into GaAs substrates. Three types of samples were studied. The fabrication processes are shown in Fig.1. For the first type (sample A), the SiO/WSiN capped furnace annealing [6] at 820°C for 15min was carried out to activate ion-implanted layers. Then an undoped GaAs layer was grown by MBE after normal preparation. Ohmic electrodes were formed by alloying AuGe/Ni/Au at 450°C after etching the undoped GaAs layer. Finally an Al/Ti Schottky electrode was formed on the undoped GaAs surface layer. For the second type (sample B), an undoped

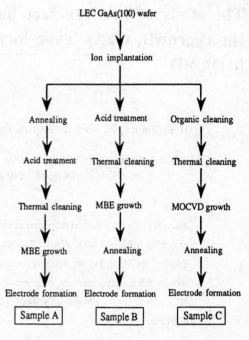

Fig.1 Fabrication process flow of experimental samples.

GaAs layer was grown by MBE after normal preparation. Then SiO/WSiN capped furnace annealing was done, and ohmic and Schottky electrodes were formed. For the third type (sample C), after organic cleaning and thermal cleaning in an MOCVD chamber were done, an undoped GaAs layer was grown by MOCVD. After annealing the sample, electrodes were formed.

By comparing the sample A with the sample B, we can investigate annealing effects on the growth interface. Comparing the sample B with the sample C, we can get the information of the difference between the MBE and MOCVD growth.

3. Results and Discussion

3.1 DLTS results of Sample A

Fig.2 shows typical DLTS spectra for the sample A under various measurement conditions. For curve (a), we used the condition: the reverse bias voltage (Vr) of 0.1V, the filling pulse amplitude (Vp) of 0.2V and the rate window time constant (τ) of 0.391msec. Under this condition, the DLTS measured region was set around the MBE growth interface. For curve (b), Vr, Vp and τ were -0.5V, 0.6V and 0.391msec, respectively. In this case, the DLTS measured region was set deeper than that of curve (a). The peak height of curve (a) is higher than that of curve (b). Therefore, it is found that hole trap levels exist near the growth interface. From Arrhenius plots of emission rate constants related to curves (a) and (b), the hole trap levels of curves (a) and (b) were 0.86eV and 0.83eV, respectively. Next, we tried to measure the deeper region than the measured region of curve (b). No trap levels could be observed. These results show that the hole trap density decreases toward the GaAs substrate and the hole traps exist only around the MBE growth interface.

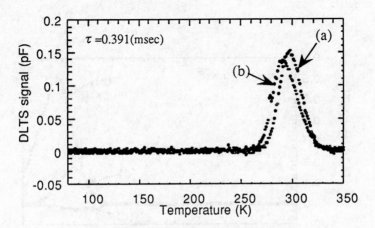

Fig.2 Typical DLTS spectra of sample A.The mesurement
conditions of curves (a) and (b) are as follows.
(a) Vr=0.1(V), Vp=0.2(V)
(b) Vr=-0.5(V), Vp=0.6(V)

3.2 DLTS results of Sample B and Sample C

Fig.3 shows typical DLTS spectra of the sample B under various measurement conditions. Measurement conditions of curves (c) and (d) were the same as curves (a) and (b), respectively. For curve (e), Vr, Vp and τ were -2.5V, 2.0V and 0.391msec, respectively. Curve (c) shows that an electron trap level exists. From Arrhenius plots of emission rate constants related to curve (c), the electron trap level was found to be 0.86eV. Curves (d) and (e) show that hole trap levels exist. From Arrhenius plots of the emission rate constants related to curves (d) and (e), the hole trap levels of curves (d) and (e) were found to be 0.86eV and 0.83eV, respectively. The hole trap level of curve (d) is equal to that of curve (a). The peak height of curve (e) is much higher than that of curve (d). These results show that the hole trap density increases toward the GaAs substrate, and the hole traps of the sample B exist in the deeper region than that of the sample A. The deeper the DLTS measurement region of the sample A becomes from the growth interface, the lower the signal peak height becomes. This shows that the origin of hole traps exist near the growth interface. On the other hand, the sample B shows the contrary results to the DLTS spectra of the sample A. The trap density of the sample B is much larger than that of the sample A. This result could be explained by the following hypothesis: At first, the main origin of traps exists on the GaAs substrate surface, and after MBE growth, this origin creates the hole traps of the sample A; In the case of the sample B, the origin of traps diffuses into the deeper region of the GaAs substrate during the annealing process at 820°C.

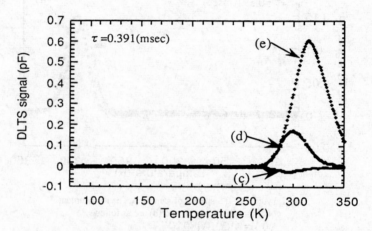

Fig.3 Typical DLTS spectra of sample B. The mesurement
conditions of curves (c), (d) and (e) are as follows.
(c) Vr=0.1(V), Vp=0.2(V)
(d) Vr=-0.5(V), Vp=0.6(V)
(e) Vr=-2.5(V), Vp=2.0(V)

For the sample C, no trap levels could be observed under various DLTS measurement conditions. This indicates that the MOCVD growth interface does not have the origin of traps.

3.3 SIMS results of Sample B and Sample C

In order to investigate the different DLTS results between MBE and MOCVD, we tried to analyze the distribution of atoms around the growth interface by using SIMS. Figs.4 and 5 show depth profiles of impurities (carbon and oxygen) of the sample B and the sample C, respectively. For the sample B, distinct pileup of carbon and oxygen atoms is observed around the MBE growth interface. On the other hand, no distinct pileup of carbon and oxygen atoms is observed for the sample C, although the sample C had the thinner surface undoped GaAs layer than that of the sample B. Moreover, carbon impurities of the growth interface by MOCVD is much lower than by MBE. Taking DLTS results into account, the hole trap levels of the sample B are considered to be caused by these impurities around the MBE growth interface.

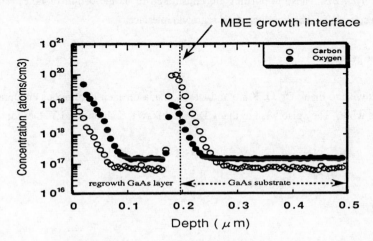

Fig.4 SIMS depth profile of carbon and oxygen atoms for the sample B.

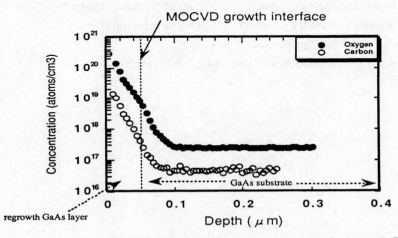

Fig.5 SIMS depth profile of carbon and oxygen atoms for the sample C.

4. Summary

We have studied characteristics of interfaces between a GaAs substrate and a GaAs layer formed by MBE and MOCVD, using DLTS and SIMS. DLTS results showed that hole trap levels exist in the MBE growth interface, but no trap levels exist in the MOCVD growth interface. SIMS measurements showed that distinct pileup of carbon and oxygen atoms exists in the MBE growth interface. Carbon impurities of the growth interface by MOCVD are much lower than by MBE. These impurities are considered to be the origin of hole trap levels. MOCVD is suitable for obtaining the good regrowth interface.

Acknowledgement

The authors wish to thank Dr. G. Kano and Dr. M. Inada for their continuous encouragement through this work. They also wish to thank Dr. O.Ishikawa, K. Inoue and Y. Ota for valuable discussion.

References

[1]Pao Y C, Yuen C, Madden C and Marsland R 1994 IEDM93, Technical Dig. pp217-220

[2]Nishibori K, Kitamura Y, Nagaoka M, Tanabe Y, Mihara M, Yoshimura M, Hirose M and Uchitomi N 1995 Technical Report of IEICE. ED94-113

[3]Fujimoto H, Tanabe M, Maeda M and Tamura A 1995 IEE Electronics Letters Vol.31 No.2 pp137-139

[4]Fujimoto H, Morimoto S, Masato H and Tamura A 1995 ISCS-22 to be published

[5]Lang D V 1974 J. Appl. Phys. 45 pp3023-3032

[6]Tamura A, Ikeda Y, Yokoyama T and Inoue K 1990 J. Appl. Phys. 67 pp6171-6174

Inst. Phys. Conf. Ser. No 145: Chapter 2
Paper presented at 22nd Int. Symp. Compound Semiconductors, Cheju Island, Korea, 28 August–2 September 1995
© *1996 IOP Publishing Ltd*

Erbium δ-doping to InP by OMVPE

Y. Fujiwara, N. Matsubara, J. Yuhara*, M. Tabuchi, K. Fujita, N. Yamada, Y. Nonogaki, Y. Takeda and K. Morita*

Department of Materials Science and Engineering, *Department of Crystalline Materials Science, School of Engineering, Nagoya University, Furo-cho, Chikusa-ku, Nagoya 464-01, Japan

Abstract. Erbium (Er) δ-doping to InP has been successfully performed *by OMVPE* for the first time and characterized systematically. Rutherford backscattering (RBS) measurements showed that the Er atoms located within the region of 6 nm which is the resolution limit of RBS, and that the Er sheet density was directly proportional to the Er source supply time. In 4.2 K PL measurements, a characteristic Er-related emission was observed in all the Er δ-doped specimens. X-ray crystal truncation rod (CTR) scattering measurements using synchrotron radiation revealed clearly the atomic-scale Er profiles in InP and the crystalline quality of InP overgrowth.

1. Introduction

There has been increasing interest in rare-earth (RE) doped III-V semiconductors due to their possible applications in optoelectronic devices. Feasibility of near-infrared light-emitting diodes (LED) and semiconductor lasers is based on the observation of the 4f-intrashell emission lines which are characteristic of the dopant, independent of the host crystal, and insensitive to temperature. Previous experimental results have exhibited that many kinds of RE-related luminescence centers are generally formed in one specimen [1], that the efficiency of luminescence is still very low ($\sim 10^{-6}$) [2], and that the luminescence intensity decreases steeply with increasing measurement temperature [3]. Much effort has been made to overcome these problems and to understand the characteristics of this class of materials.

We have been intensively investigating OMVPE growth of RE-doped III-V semiconductors by using less toxic alkyl sources such as TBP (tertiarybutylphosphine) and TBAs (tertiarybutyl-arsine) in stead of extremely toxic hydrides for group-V sources [4]. In the growth of InP doped uniformly with Er, Er concentration in the layer was well controlled by the change of the Er source temperature and the hydrogen flow rate through the source. The Er concentration as high as $\sim 10^{19}$ cm^{-3} was realized together with a specular surface. The low-temperature PL measurements revealed that the Er-related PL spectrum depends drastically on the Er concentration in the layer and the growth temperature.

In this study, δ-doping with Er to InP, which is an important technique necessary for future device applications, is performed by OMVPE and characterized systematically.

2. Experimental procedures of δ-doping

The low-pressure growth system with a vertical quartz reactor was utilized in this work.

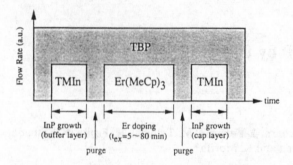

Fig. 1 Time sequence of source gases supply for the δ-doping with Er to InP.

Details of the growth system was described previously [5]. TMIn (trimethylindium) and TBP were used as source materials for InP growth. Er(MeCp)$_3$ (tris(methylcyclopentadienyl) -erbium) as the Er source was maintained at 100 °C and introduced in the reactor by H$_2$ flow of 125 ccm through the cylinder.

The sequence of the δ-doping with Er to InP, shown schematically in Fig. 1, was as follows: 1) The InP buffer layer of 100 nm in thickness was grown on a (100)-oriented Fe-doped InP substrate at 530 °C. 2) The supply of TMIn was stopped to suspend the growth. 3) The Er source was supplied at the growth temperature to form an atomic plane of the dopant on the surface. 4) After the supply of the Er source was stopped, the InP growth was restarted to cap the doped layer. Before and after the Er source supply, a purge period of 30 s was inserted to ensure the gas change. TBP was flowed continuously during a series of steps. The duration exposed to the Er source was varied from 5 to 80 min.

3. Results and Discussion

3.1. Surface morphology

The surface was mirror-like under Nomarski microscope observation in all the Er δ-doped InP specimens. Figure 2 shows Nomarski photomicrograph of the specimen prepared with the longest Er-exposure duration of 80 min.

Fig. 2 Nomarski photomicrograph of the δ-doped InP specimen surface prepared with the longest Er-exposure duration of 80 min. The marker represents 50 μm.

3.2. RBS measurements

The profile of Er atoms doped in InP was characterized by RBS technique.

The RBS measurement was performed with a 1.8 MeV He⁺ ion beam of 1 mm in diameter. The beam current was typically 4 nA and the angular divergence of the beam was about 0.01°. The probing beam induced damage to the specimen was reduced as low as possible by keeping a total ion influence less than 6×10^{15} ions/cm². Backscattered He⁺ ions were detected by a surface-barrier detector at a scattering angle of 160°, leading us to the depth resolution of 6 nm.

Figure 3 shows a random RBS spectrum of the specimen prepared with the Er-exposure duration of 80 min. The backscattering from Er is observed clearly at channel 415. The sharp spectral shape indicates that Er atoms are involved in the region of 6 nm which is the resolution limit.

Using the Rutherford cross section, the measured detector solid angle and the incident ion influence, the number of atoms per unit area visible to the incident ion beam was calculated with an estimated accuracy of ± 5 %. The number of Er atoms per unit area is plotted as a function of the Er-exposure duration in Fig. 4. The number of Er atoms per unit area increases linearly with the Er-exposure duration and exhibits no saturation even at 80 min.

Assuming that Er atoms substitute to In sites in InP, the sheet density for one monolayer (ML) is calculated to be 5.8×10^{14} cm⁻². The sheet density is the same in the case that Er atoms exist in the form of ErP with a NaCl crystal structure, if the layer is enough thin and distorted pseudomorphically to match to InP lattice. Therefore, as can be seen in Fig. 4, Er atoms, corresponding to 1 ML in number, are incorporated during the Er source supply of about 23 min.

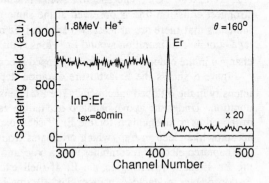

Fig. 3 Random RBS spectrum of the δ-doped InP specimen prepared with the Er-exposure duration of 80 min.

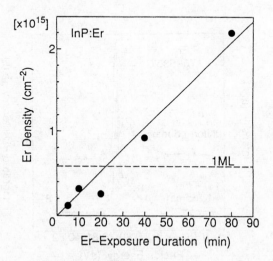

Fig. 4 Er-exposure duration dependence of the number of Er atoms per area in the δ-doped InP specimens.

3.3. PL measurements

The optical properties of the specimens were characterized by PL measurements.

PL measurements were carried out with the specimens directly immersed into liquid He at 4.2 K. The photoexcitation source was a cw mode Ar⁺ laser with a beam diameter of 1 mm and an incident power of 180 mW. The luminescence of the specimen was dispersed with a 1.25 m grating monochromator and detected by a Ge pin photodiode cooled by liquid N_2.

A characteristic Er-related emission was successfully observed in all the Er δ-doped specimens. In Fig. 5, the typical Er-related PL spectrum is shown and compared with that for the specimen doped uniformly at 530 °C under the same Er source supply conditions (125 ccm, 100 °C). The spectral shape is quite different from each other, though the strongest emission line is observed at the same photon energy of 0.8041 eV. This result indicates that there are at least two kinds of Er-related luminescence centers in which the local atomic configuration around Er^{3+} ions is different, and that their concentration ratio is changed in the δ-doped and uniformly doped specimens.

Figure 6 shows the Er-exposure duration dependence of the integrated Er-related PL intensity in the δ-doped specimens. The PL intensity decreased with increasing Er-exposure duration. Under the above-bandgap excitation, as used in this study, the energy transfer is proposed as a mechanism for the Er 4f-shell excitation [6]. According to the model, an acceptor-like electron trap, which originates from the doped Er ions, captures an electron. The captured electron recombines with a hole, and the recombination energy is transferred to the Er 4f-shell, resulting in the Er 4f-shell excitation. If InP overgrown on the thick Er-doped layer is degraded, photoexcited electron-hole pairs recombine nonradiatively, leading to low efficiency in the energy transfer for the Er 4f-shell excitation. The suppression of the Er-related PL intensity with increasing the Er-exposure duration might be due to degraded overgrowth of InP. Such degraded InP overgrowth was confirmed directly in the X-ray CTR scattering measurements described below.

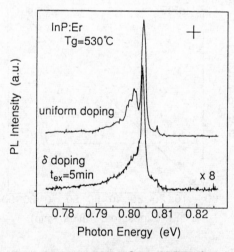

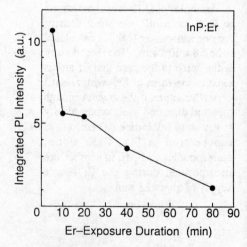

Fig. 5 PL spectrum for the δ-doped InP specimen prepared with the Er-exposure duration of 5 min, compared with that for the specimen doped uniformly with Er.

Fig. 6 Er-exposure duration dependence of the integrated Er-related PL intensity in the δ-doped InP specimens.

3.4. X-ray CTR scattering measurements

X-ray CTR is the diffraction spots extended normal to the surface for a finite crystal. Its shape is sensitively modified as the structure near the surface is changed [7, 8]. We have successfully applied the X-ray CTR scattering measurements to study the epitaxially grown semiconductor interfaces and have demonstrated that it is a very powerful technique to reveal the atomic-scale interface structures embedded in semiconductors [9, 10]. In this study, the atomic-level heterointerface structures of the Er δ-doped specimens were investigated by the X-ray CTR scattering measurements.

The X-ray CTR scattering measurements were conducted at the beam line BL6A$_2$ of the Photon Factory in the National Laboratory for High Energy Physics at Tsukuba using synchrotron radiation from the 2.5 GeV storage ring. The wavelength of the X-ray was tuned at 1.60 Å by a bent Si (111) monochromator. The diffraction pattern was recorded by a Weissenberg camera with IP (imaging plate) as a detector for 40 min for each specimen.

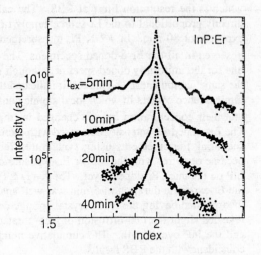

Fig. 7   Measured CTR spectra for the δ-doped InP specimens prepared with the Er-exposure duration of 5, 10, 20 and 40 min, respectively.

Figure 7 shows the measured CTR spectra around (002) Bragg point for the specimens with the Er-exposure duration of 5, 10, 20 and 40 min, respectively. Characteristic modulated structures become obscurer with the longer Er-exposure duration, indicating degraded overgrowth of InP on the thick Er-doped layer.

The X-ray CTR spectrum for the specimen prepared with the Er-exposure duration of 5 min was analyzed by comparing it with a theoretically calculated spectrum based on a model structure. Details of the procedures will be described elsewhere. On the assumption that Er atoms are incorporated substitutionally to In sites in InP crystal, the modulated structure on the left side of the main diffraction in the experimentally obtained spectrum could not be fitted theoretically. Using a model in which Er atoms are incorporated having the NaCl crystal arrangement with P atoms, on the other hand, we obtained successfully a calculated spectrum which fits completely to the measured one. This model is equivalent to the recently proposed model that almost all the Er atoms are located in tetrahedral interstitial sites in GaAs [11]. The resultant Er profiles is almost symmetric, implying slight thermal-diffusion of incorporated Er atoms at the stages of the Er-exposure and the InP overgrowth. In addition, the number of Er atoms at the peak of the profile corresponds to 0.04 ML and the full width at the half maxium is 4 ML (= 1.2 nm). The cumulative number of Er atoms in the specimen is equivalent to 0.2ML (= 1.2 x 10^{14} cm^{-2}), being well coincident with the RBS results described above.

154

4. Conclusions

We have successfully performed δ-doping with Er to InP by OMVPE for the first time and characterized systematically. No severe degradation in surface morphology was observed in all the specimens. In RBS measurements, the Er atoms located within the region of 6 nm which is the resolution limit of RBS. The calculated number of Er atoms per area was directly proportional to the Er source supply time and exhibited no saturation even for the exposure of 80 min. In 4.2 K PL measurements, a characteristic Er-related emission was observed in all the Er δ-doped specimens. The spectral shape was quite different from the one for the uniformly doped specimen, though the strongest emission line was observed at the same photon energy. This result indicated that there are at least two kinds of Er-related luminescence centers in which local atomic configuration around Er^{3+} ions is different, and that their concentration ratio is changed in the δ-doped and uniformly doped specimens. The Er-related PL intensity decreased with increasing Er-exposure duration. In X-ray CTR scattering measurements using synchrotron radiation, characteristic modulated structures became obscurer with the longer Er-exposure duration, implying degraded overgrowth of InP on the thick Er-doped layer. The X-ray CTR spectrum for the specimen prepared with the Er-exposure duration of 5 min was well analyzed using a model in which Er atoms are incorporated having the NaCl crystal arrangement with P atoms. The resultant Er profiles revealed slight thermal-diffusion of incorporated Er atoms at the stages of the Er-exposure and the InP overgrowth. The cumulative number of Er atoms in the specimen was well coincident with the RBS results.

Acknowledgments

This work was performed as a part of the project (Project No. 94G331) accepted by the Photon Factory Program Advisory Committee. The authors would like to thank Professors T. Ohyama and H. Nakata of Osaka University for PL measurements. The authors wish to acknowledge Tri Chemical Laboratory Inc. for the Er source, and Showa Denko K. K. for InP substrates. This work was supported in part by the Grant-in-Aid for Encouragement of Young Scientists No. 07750348, Scientific Research (B) No. 07455007 from the Ministry of Education, Science and Culture, and also in part by the Mazda Foundation's Research Grant and the Japan Securities Scholarship Foundation.

References

[1]   Pomrenke G S, Ennen H and Haydl W 1986 J. Appl. Phys. **59** 601-610.
[2]   Auzel F, Jean-Louis A M and Toudic Y 1989 J. Appl. Phys. **66** 3952-3955.
[3]   Takahei K, Whitney P S, Nakagome H and Uwai K 1989 J. Appl. Phys. **65** 1257-1260.
[4]   Fujiwara Y, Ito Y, Nonogaki Y, Matsubara N, Fujita K and Takeda Y 1995 to be published in *Proc. 18th Inter. Conf. Defects in Semiconductors, Mater. Sci. Forum* (Switzerland, Trans Tech Publications).
[5]   Fujiwara Y, Furuta S, Makita K, Ito Y, Nonogaki Y and Takeda Y 1995 J. Cryst. Growth **146** 544-548.
[6]   Taguchi A, Takahei K and Horikoshi Y 1994 J. Appl. Phys. **76** 7288-7295.
[7]   Kashihara Y, Kimura S and Harada J 1989 Surf. Sci. **214** 477-492.
[8]   Harada J, Shimura T, Tanaka M, Yakushiji K and Hori K 1990 J. Cryst. Growth **104** 773-776.
[9]   Takeda Y, Sakuraba Y, Fujibayashi K, Tabuchi M, Kumamoto T, Takahashi I, Harada J and Kamei H 1995 Appl. Phys. Lett. **66** 332-334.
[10]   Tabuchi M, Takeda Y, Sakuraba Y, Kumamoto T, Fujibayashi K, Takahashi I, Harada J and Kamei H 1995 J. Cryst. Growth **146** 148-151.
[11]   Nakata J, Jourdan N, Yamaguchi H, Takahei K, Yamamoto Y and Kido Y 1995 J. Appl. Phys. **77** 3095-3103.

Inst. Phys. Conf. Ser. No 145: Chapter 2
Paper presented at 22nd Int. Symp. Compound Semiconductors, Cheju Island, Korea, 28 August–2 September 1995
© *1996 IOP Publishing Ltd*

MOVPE Growth of Compressively Strained GaInP/AlGaInP Structure with MQB

M. Oh, H. Park, C. Lee, G. Lim, T. Kim, J. Kim and T. Kim

Materials & Devices Research Center, Samsung Advanced Institute of Technology,
P.O.Box 111, Suwon, Korea

Abstract. In this work, the strained multi-quantum wells(MQWs) and multi-quantum barrier(MQB) were applied to decrease the threshold current and improve the characteristic temperature(To). The threshold current was greatly reduced by the optimization of the compressive strain, well number and width for 650nm laser diode(+0.5%, four 4.5nm QWs). The MQB was grown with the thickness controllability of three monolayer. The high temperature performance was clearly improved by incorporating a MQB. This improvement was due to enhancement of the potential barrier height between the active layer and p-cladding layer.

1. Introduction

The GaInP/AlGaInP layers lattice-matched to GaAs substrate are known as attractive materials for 600nm band laser diode. The 650nm GaInP/AlGaInP laser diode has been very attractive as light sources for commercial applications such as digital video discs(DVD) and quad-speed CD-ROMs. The relatively small band offset in the conduction band between the active layer and cladding layer also hinders the realization of short wavelength laser with the good performance. This small energy bandgap difference causes a increase in the electron overflow and the maximum operation temperature is limited. It is also difficult to obtain high p-doping in the cladding layer. The misoriented substrate has been used to overcome such problems[1][2].

The MQB structure has been studied on the reduction of the electron overflow in 675nm AlGaInP lasers[3]. The incorporation of compressively strained multi-quantum wells(MQWs) has been used for decreasing the threshold current by reducing the hole density of states near the valence band edge[4][5]. The compressively strained active layer provides a larger effective barrier height than a lattice-matched one. A multiple QW active region has lower threshold at elevated temperature because bandfilling is reduced. However, the number of QWs should be optimized to minimize the pumped volume. The MQB has been proposed for improving the characteristic temperature(To) by enhancing the electron confinement between the active layer and cladding layer[6][7]. It is theoretically shown that interwell interference of electron levels can be obtained by a MQB, a series of the wells and barriers in the p-cladding layer, thereby raising the conduction level[8]. In this work, the effects of compressively strained MQW and a MQB are described on the reduction of threshold current and the improvement of temperature characteristics for 650nm GaInP/AlGaInP lasers.

2. Experiments

In the present work the epitaxial layers were grown by a low pressure metal-organic vapor phase epitaxy(MOVPE) on 7° off (001) Si-doped GaAs substrate toward [011] direction. The graphite susceptor rotating at 20rpm was heated inductively in a horizontal reactor. The process pressure, growth temperature and growth rate were 50torr, 740°C and 2μm/hr, respectively. The source materials used in this study were triethylgallium(TEGa), trimethylaluminum(TMAl), and trimethylindium(TMIn) for group III and 100% arsine(AsH_3) and phosphine(PH_3) for group V. Diethylzinc(DEZn) and silane(SiH_4, 0.1% in H_2) were used as p- and n-doping sources, respectively.

Thickness controllability strongly depended on the transition process in the MOVPE gas system. The accurate pressure balance was controlled within 0.1 torr between the vent line and process line to avoid the transition process. To examine the thickness controllability in our MOVPE system, the single quantum wells of GaInP with four different thickness of 1, 2, 4 and 8nm were grown on GaAs substrate. To evaluate the strain effect under the fixed lasing wavelength around 645nm, the MQWs structures for 0, +0.5 and +1.0% compressive strain were grown on exact and 7° off GaAs(001) substrate, respectively. The well number and thickness were adjusted to keep a constant optical confinement factor(Γ). They were composed of 3 wells and 7.5nm thickness at unstrained, 4 wells and 4.5nm thickness at +0.5% strained, and 5 wells and 3.0nm thickness at +1% strained MQW structures, respectively. The structure parameters of a MQB were determined from the results of the theoretical reflectivities of various combination based on reference 6. The MQB consisted of 10pairs of GaInP wells and $(Al_{0.7}Ga_{0.3})_{0.5}In_{0.5}P$ barriers and was grown with the thickness controllability of three monolayer. The thicknesses of GaInP wells and $(Al_{0.7}Ga_{0.3})_{0.5}In_{0.5}P$ barriers were 1.2 and 1.8nm, respectively. The thickness of first barrier was 22nm to suppress the electron tunneling from the active region to the MQB.

Interfacial coherency of epitaxial layers was checked by a double crystal x-ray diffractometer(DCD). The heterointerface abruptness and thickness were checked by low temperature photoluminesence(PL) and cross-sectional transmission electron microscopy(TEM). The carrier concentrations were measured by electrochemical C-V profiler.

3. Results and Discussion

The structure consisted of four GaInP Qws(1.0, 2.0, 4.0, 8.0nm), sandwitched by $(Al_{0.45}Ga_{0.55})_{0.5}In_{0.5}P$ barrier layers as shown from the cross-sectional TEM photograph in Fig1. Fig.2 shows low temperature PL spectra meaured at 10K. The PL peaks of four QWs were clearly observed. The fact that the PL peak of 1nm thick QW was clearly seen indicated the thickness controllability of less than three monolayers. It was shown that its thickness controllability was suitable for the growth of MQB structure, 1.2nm thick GaInP wells and 1.8nm thick $(Al_{0.7}Ga_{0.3})_{0.5}In_{0.5}P$ barriers.

It has been repoted that the lasing chracteristics was improved by employing the compressive strain in the GaInP active layer. However, the large compressive strain caused the shift of the lasing wavelength to become longer. As the compressive strain was varied from 0% to 1% to optimize a MQW structure, the well width was changed for a fixed lasing wavelength, around 645nm, which corresponded to around 630nm photoluminescence(PL) wavelength. The well number for each MQW structure was determined to maintain a constant optical confinement factor(Γ), approximately 0.1. The PL wavelength of MQW structure grown on the exact (001) GaAs substrate under the same growth conditions was about 645nm, 15nm longer than that grown on 7° off (001) GaAs substrate.

Fig.1 Crosectional TEM photograph of four GaInP Qws(1.0, 2.0, 4.0, 8.0nm) sandwitched by $(Al_{0.45}Ga_{0.55})_{0.5}In_{0.5}P$ barrier layers

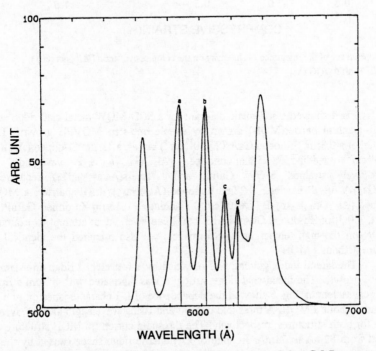

Fig.2 PL spectrum measured at 10K for (a) 1nm, (b) 2nm, (c) 4nm, (d) 8nm GaInP quantum well layers

158

This shift might be due to the GaInP ordering effect or the difference of indium incorporation rate on the different growth plane. As shown in Fig.3, the room temperature PL intensity significantly increased with the compressive strain(+0.5%). However, when the 3nm thick narrower wells at +1% compressive strain were used, it became smaller because a large number of electrons and holes tended to accumulate at the heterointerface between narrower wells and many carriers were recombined by nonradiative processes due to the defects at the interface. The integrated PL intensity at +0.5% compressive strain was about 1.4 times higher than that of lattice-matched MQWs.

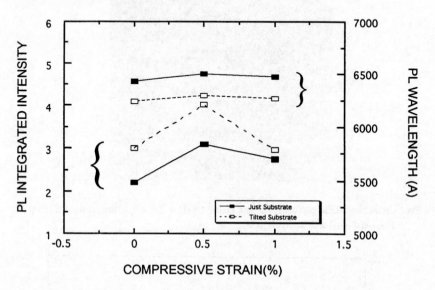

Fig.3 Dependence of the integrated PL intensity on the compressive strain [■: exact (001), □: 7° off (001)GaAs]

Fig.4 showed a schematic diagram for a SCH-MQW metal clad ridge structure with a multi-quantum barrier(MQB) grown by simple one-step MOVPE growth. The structure consisted of a 0.3µm Si-doped GaAs($2X10^{18}cm^{-3}$) buffer, a 1.1µm Si-doped $(Al_{0.7}Ga_{0.3})_{0.5}In_{0.5}P$ $(2X10^{18}cm^{-3})$ cladding, a 55nm undoped $(Al_{0.45}Ga_{0.55})_{0.5}In_{0.5}P$ waveguide, four 4.5nm compressively strained MQW GaInP active layers($\Delta a/a=+0.5\%$) separated by 7nm $(Al_{0.45}Ga_{0.55})_{0.5}In_{0.5}P$ barriers, a 22nm undoped-$(Al_{0.7}Ga_{0.3})_{0.5}In_{0.5}P$ barrier, a MQB, a 1.1µm Zn-doped $(Al_{0.7}Ga_{0.3})_{0.5}In_{0.5}P$ $(7X10^{17}cm^{-3})$ cladding, a 0.15µm Zn-doped GaInP($2X10^{18}cm^{-3}$) contact, a 0.3µm Zn-doped GaAs($1X10^{19}cm^{-3}$) cap layer. In an attempt to examine the effect of MQB on the high temperature performance, we also prepared the identical SCH-MQW structure without a MQB.

The lateral index guiding was obtained by a wet etched ridge after simple one step MOVPE growth. The metal clad ridge structure was fabricated with a 5µm stripe width and 400µm cavity length. Fig. 5 showed the dependence of L-I characteristics on two temperatues with and without a MQB. A threshold current and lasing wavelength at 25C° were 43mA and 648nm for both structures, respectively. The threshold current of MQB structure at 60°C was reduced from 82mA to 71mA. In spite of asymmetric ridge shape caused by the misoriented substrate, the far-field patterns were symmetric, and the beam divergencies parallel and normal to the junction planes were 8.2° and 29°, respectively.

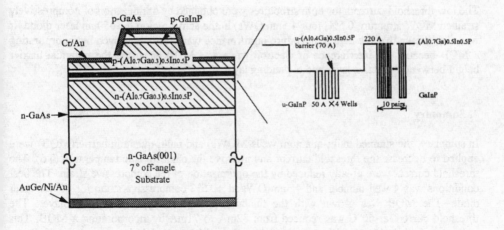

Fig.4 Schematic diagram for a SCH-MQW metal clad ridge structure wth a muli-quantum
barrier(MQB)

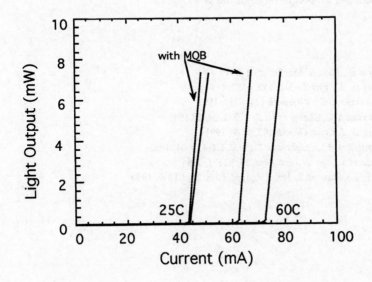

Fig.5 Dependence of I-L characteristics on temperature without a MQB

160

The low threshold currents for both structures were obtained by optimization of compressively strained MQW structure(0.5%, four 4.5nm QWs) in the active region for 650nm laser diode. It was also shown that the high temperature performance was clearly improved by incorporating a MQB because the interference of electron wave at a MQB enhanced the potential barrier height between the active layer and p-cladding layer.

4. Summary

In summary, the strained multi-quantum wells(MQWs) and multi-quantum barrier(MQB) were applied to decrease the threshold current and improve the characteristic temperature(To). The threshold current were greatly reduced by the optimization of the compressive strain. The best conditions was 4 well number and 4.5nm QWs at +0.5% compressive strain for 650nm laser diode. The MQB was grown with the thickness controllability of three monolayer. The threshold current at 60°C was reduced from 82mA to 71mA by incorporating a MQB. This improvement was due to enhancement of the potential barrier height between the active layer and p-cladding layer. The excellent performance of a 648nm laser diode was obtained with a metal clad ridge structure in this work. The further optimization of strained MQWs and MQB makes it possible to lower the threshold current and improve the characteristic temperature(To).

Acknowledgments

The authors would like to thank Managing Director, Dr. T.S. Park, for his continuous encouragement and support throughout this work.

References

[1] H. Hamada et al., *Electrn. Lett.* **28**, p501, 1992
[2] M. Susuki et al., *J. Cryst. Growth* **113**, p127, 1991
[3] K. Kadoiwa et al., *J. Cryst. Growth* **124**, p757, 1992
[4] G.D. Sandes and Y.C. Chang, *Phys. Rev.* **B32**, p4282,1985
[5] M. Kondo et al., *J. Cryst. Growth* **107**, p578, 1991
[6] E. Yablonovitch et al., *J. Lightwave Technol.* **LT-4**, p504, 1986
[7] S. Kamiyama et al., *Jpn. J. Appl. Phys.* **33**, p2571, 1994
[8] T. Takagi, F. Koyama and K. Iga, *J. J. Appl. Phys.* **29**, L1977, 1990

Inst. Phys. Conf. Ser. No 145: Chapter 2
Paper presented at 22nd Int. Symp. Compound Semiconductors, Cheju Island, Korea, 28 August–2 September 1995
© *1996 IOP Publishing Ltd*

Epitaxial growth and characterization of GaN films on 6H-SiC by MOVPE

Ok-Hyun Nam, Gyeungho Kim, Dalkeun Park*, Ji-Beom Yoo**, and Dong-Wha Kum

Division of Metals, Korea Institute of Science and Technology
Cheongryang P.O. Box 131, Seoul 136-650, Korea

* Division of Chemical Engineering, Korea Institute of Science and Technology
Cheongryang P.O. Box 131, Seoul 136-650, Korea

** Department of Materials Engineering, Sung Kyun Kwan University
300 Chunchun-Dong, Jangan-Ku, Suwon 440-330, Korea

Abstract Metalorganic vapor phase epitaxial growth and characterization of GaN layers on vicinal 6H-SiC(0001) substrates are reported. GaN films have been grown at 1020℃ and atmospheric pressure using trimethylgallium(TMG) and NH_3 in an IR lamp heated horizontal reactor. GaN films have optically flat surfaces free from cracks, and cross-sectional TEM has revealed that GaN films have relatively small number of defects such as dislocations and stacking faults near the GaN/SiC interface. The epitaxial relationship between GaN and SiC was (0001)GaN // (0001)SiC and the interface was highly coherent. Double crystal X-ray rocking curve(DCXRC) measurements on GaN films have revealed full width at half maximum(FWHM) values as low as 249 arcsec for GaN(0002) peak without correcting the instrumental broadening effects. Cross-sectional TEM and XRC results indicate that the crystal quality of GaN directly grown on SiC(0001) without a buffer layer is comparable to that of GaN with a buffer layer on sapphire(0001) substrates. Low temperature(T=11K) PL spectra of the GaN films grown to date were dominated by the near-band-gap emission band with the maximum at 358nm(3.46eV).

1. Introduction

Recently, wide band gap Ⅲ-Ⅴ nitrides and their alloys have attracted much attention, with potential application in blue and ultraviolet light emitting diodes(LEDs) and laser diodes(LDs), because wurtzite GaN has a band gap energy of 3.4eV at room temperature and forms continuous solid solution with InN(1.9eV) and AlN(6.2eV), which correspond to wavelengths from 650nm to 200nm[1-3].

Although advanced crystal growth techniques, especially MOVPE, have greatly improved the quality of GaN during the past few years, there remain some problems to be solved in the growth of III-V nitrides, such as the high n-type carrier concentration due to nitrogen vacancy[4], the difficulty of p-type doping[5,6] and the lack of a suitable substrate material[7].

Sapphire has been employed as a substrate for GaN growth despite its large thermal and lattice mismatch with GaN(about 25.4% and 16.1%, respectively). With the use of low temperature AlN or GaN thin buffer layers on sapphire, MOVPE technique has produced the high quality GaN films at the temperature of about 1000℃[8,9]. A new development on GaN growth is emerging as 6H-SiC substrate becomes commercially available. The small lattice mismatch(about -3.5%) between GaN and SiC makes 6H-SiC a good candidate on which to grow nitrides[10].

In this paper we report on the metalorganic vapor phase epitaxial growth and characterization of GaN films on 6H-SiC(0001) substrates using trimethylgallium and NH_3 in an IR lamp heated horizontal reactor. The crystal quality of GaN films was studied by double crystal X-ray diffraction, scanning electron microscopy and transmission electron microscopy. Low temperature photoluminescence and Hall measurements were carried out to investigate the optical and electrical properties of GaN films.

2. Experimental procedure

GaN films were grown directly on vicinal 6H-SiC(0001) substrates [3.5° off from (0001) toward $\langle 11\overline{2}0 \rangle$][11] using an IR lamp heated horizontal reactor at 1020℃ and atmospheric pressure. The SiC substrates were degreased and dipped into 10% HF solution for 10 minutes to remove protective oxide layer($\sim$750Å thick) and rinsed in deionized water. The source gases were trimethylgallium(TMG) and NH_3. N_2 was also used as a diluent and carrier gas of TMG. Flow rates of TMG, NH_3 and N_2 were 4.5-13.5 μmol/min, 1 slm and 1 slm, respectively. III and V source gases were separately introduced into the reactor. The typical growth rate and growth time of GaN films were 1.7-5.2 μm/h and 30-60min, respectively.

The surface morphology of GaN films was characterized using scanning electron microscopy. The crystal quality was evaluated by the full width at half maximum (FWHM) of (0002) reflection in the double crystal X-ray rocking curves(DCXRC). Cross-sectional transmission electron microscopy(Philips CM30, 200kV) was used to observe microstructural defects in GaN films. High resolution transmission electron microscopy(300kV, Scherzer defocus condition) was also used to analyze lattice image of 2H-GaN/6H-SiC interface.

The photoemission properties of GaN films were determined by low temperature (T=11K) PL using the 10 mW He-Cd laser(325nm) as an excitation source. GaN films were characterized by Hall effect measurement using the Van der Pauw method.

3. Results and discussion

The optical observation showed that the GaN films had flat surfaces free from cracks

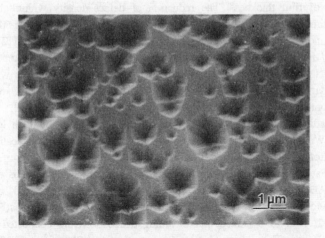

Fig. 1. Scanning electron micrograph of GaN film grown directly on 6H-SiC at 1020℃.

and were transparent in visible light. However, SEM observation revealed that many hexahedral pits smaller than about 1 μm in width were formed on the surface of the GaN films(Fig. 1). The formation of these hexahedral pits with (1$\bar{1}$01) facets is believed to be due to imperfect lateral growth of GaN islands.

Fig. 2 shows the cross-sectional TEM micrographs of the GaN film on SiC(0001). The GaN film has relatively small number of defects such as stacking faults on (0001) plane and dislocations along [0001] direction, and the defect density diminishes with

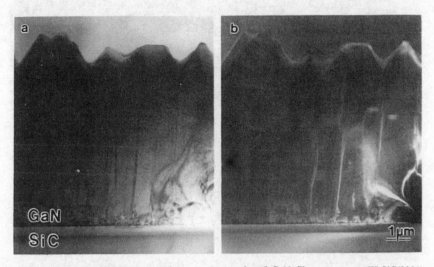

Fig. 2. Cross-sectional transmission electron micrographs of GaN film grown on 6H-SiC(0001). (a) bright field image and (b) dark field image under two-beam condition of [$\bar{2}$110] at zone axis of [01$\bar{1}$0]

the increase of film thickness. The reduction of defect density within about 500nm of the GaN films is likely due to the annihilation and coalescence through mutual interactions and the formation of half loops as reported by Qian et al[12]. In the study of GaN/AlN/α-Al$_2$O$_3$, Hiramatsu et al.[13] and Qian et al.[12] have reported that both AlN and GaN crystals at the interface region have columnar structures with low angle grain boundaries composed of threading dislocations. However, no appreciable diffraction contrast was observed in the GaN films as shown in Fig. 2. This suggests that the threading dislocations in GaN films on SiC(0001) do not form low angle grain boundaries. Weeks et al.[14] have proposed that the threading dislocations along the [0001] direction of the GaN films originated at steps on vicinal SiC surface as a result of the mismatch between the stacking sequence of 6H-SiC(ABCACB...) and that of 2H-GaN(ABAB...). High resolution TEM observation, however, did not show any supporting evidence that the origin of threading dislocations is the steps(Fig. 3).

In Fig. 2, it is to be noted that threading dislocations are connected to the bottom of hexahedral pits. These vertical threading dislocations, therefore, are believed to be formed during coalescence of GaN islands at the initial stage, based on the nucleation mechanism of GaN grown on α-Al$_2$O$_3$ with AlN buffer layer first proposed by Akasaki et al.[15]. However, unlike GaN layer on AlN/α-Al$_2$O$_3$, no diffraction contrast is observed among the regions separated by threading dislocations. This difference is considered to be due to the very small lattice mismatch between GaN and SiC. High resolution cross-sectional TEM micrograph of GaN film on vicinal SiC(0001), as shown in Fig. 3, confirms that the epitaxial relationship between GaN and SiC is (0001)GaN// (0001)SiC and the interface is highly coherent.

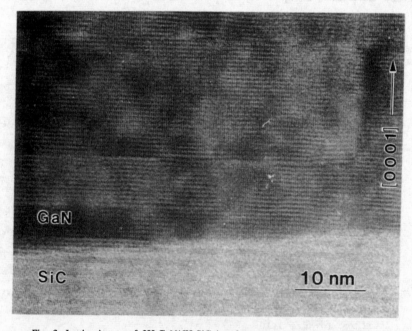

Fig. 3. Lattice image of 2H-GaN/6H-SiC interface at zone axis of [2$\bar{1}$$\bar{1}$0] .

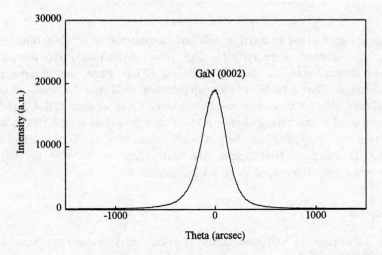

Fig. 4. Double crystal X-ray rocking curve of (0002) reflection of 1.7 μm thick GaN film on 6H-SiC.

Double crystal X-ray rocking curve for (0002) reflection of GaN film is shown in Fig. 4. A full width at half maximum was as low as 249 arcsec without correcting the instrumental broadening effects. Cross-sectional TEM and DCXRC results indicate that the crystal quality of GaN directly grown on SiC(0001) without a buffer layer is comparable to that of GaN with AlN or GaN buffer layer on sapphire(0001) substrate[8,9].

The PL spectra at T=11K from the majority of the GaN films were dominated by the near-band-gap emission band with the maximum at 358nm(3.46eV), attributed to the recombination of excitons bound to neutral donors(Fig. 5). GaN films showed n-type characteristics with carrier concentrations of about 10^{19}/cm^3 and electron mobilities of 50-60 cm^2/V $\cdot$ s at room temperature.

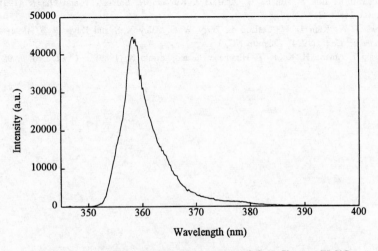

Fig. 5. Low temperature(T=11K) PL spectrum of GaN film on 6H-SiC.

4. Conclusions

Metalorganic vapor phase epitaxial growth and characterization of GaN films on vicinal 6H-SiC(0001) substrates were reported. GaN films on SiC(0001) had relatively small number of defects, such as stacking faults on (0001) plane and dislocations along [0001] direction. The epitaxial relationship between GaN and SiC was (0001)GaN // (0001)SiC and the interface was highly coherent. Cross-sectional TEM and DCXRC results indicate that the crystal quality of GaN directly grown on SiC(0001) without a buffer layer is comparable to that of GaN with AlN or GaN buffer layer on sapphire(0001) substrate. Nevertheless, The GaN films on SiC did not show good electrical properties, which need to be further studied.

Acknowledgement

This work was supported by KIST(Korea Institute of Science and Technology)-2000 Research Program. (Grant No. BSV00020-002-4)

References

[1] Strite S and Morkoc H (1992) *J. Vac. Sci. Technol.* **B10** 1237
[2] Akasaki I and Amano H (1994) *J. Electrochem. Soc.* **141** 2266
[3] Davis R F (1993) *Physica B* **185** 1
[4] Seifert W, Franzheld R, Butter E, Subotta H and Riede V (1983) *Cryst. Res. Technol.* **18** 383
[5] Amano H, Kito M, Hiramatsu K, Akasaki I (1989) *Jpn. J. Appl. Phys.* **28** L2112
[6] Nakamura S, Senoh M and Mukai T (1991) *Jpn. J. Appl. Phys.* **30** 1708
[7] Yoshida S, Misawa S and Gonda S (1983) *Appl. Phys. Lett.* **42** 427
[8] Amano H, Sawaki N, Akasaki I and Toyoda Y (1986) *Appl. Phys. Lett.* **48** 353
[9] Nakamura S (1991) *Jpn. J. Appl. Phys.* **30** L1705
[10] Lin M E, Sverdlov B, Zhou G L and Morkoc H (1993) *Appl. Phys. Lett.* **62** 3479
[11] *Cree Research Inc.* 2810 Meridia Pkwy. Duhram, NC 27713
[12] Qian W, Skowronski M and De Graef M (1995) *Appl. Phys. Lett.* **66** 1252
[13] Hiramatsu K, Itoh S, Amano H, Akasaki I, Kuwano N, Shiraish T and Oki K (1991) *J. Cryst. Growth* **115** 628
[14] Weeks T W, Kum D W, Carlson E, Perry W G, Ailey K S and Davis R F (1994) *High Temp. Electronic Conf.* (1994) Charotte NC, USA
[15] Akasaki I, Amano H, Koide Y, Hiramatsu K and Sawaki N (1989) *J. Cryst. Growth* **98** 209

Inst. Phys. Conf. Ser. No 145: Chapter 2
Paper presented at 22nd Int. Symp. Compound Semiconductors, Cheju Island, Korea, 28 August–2 September 1995
© *1996 IOP Publishing Ltd*

Effect of strain and growth temperature on In incorporation and properties of high power laser diodes in MOVPE grown (In,Ga)(As,P)/GaAs

F. Bugge, G. Erbert, S. Gramlich, I. Rechenberg, U. Zeimer, and M. Weyers

Ferdinand-Braun-Institut für Höchstfrequenztechnik Berlin, Rudower Chaussee 5, D-12489 Berlin, Fed. Rep. Germany

For adjustment laser emission wavelength the knowledge about the active zone is important. At high-power laser diodes emitting in the wavelength region between 800 nm and 1017 nm strained InGaAs(P) is usually employed as active zone. Calibration of composition by growing thick layers that allow a analysis is based on the assumption of a strain independent In incorporation. This assumption does not hold for MOVPE growth.

   The different incorporation behaviour of In for relaxed and strained layers is shown. By increasing growth temperature from 650°C to 750°C and higher TMIn/(TMIn+TMGa) ratio a increased In segregation is obtained, however in the range of detection limit not In desorption.

   By optimization of strain the threshold current density of broad area laser diodes with infinite cavity length could be reduced to 53 A/cm^2.

1. Introduction

High-power laser diodes emitting between 800 nm and 1017 nm with strained InGaAs(P) as active zone are attracting great interest, especially for pumping of solid state lasers or Er$^+$- or Pr$^+$-doped fibre amplifiers for optical communication. The characteristics and properties of laser diodes depend mainly on the structure of the active zone, a good electrical and optical confinement and the absence of misfit dislocations. By changing composition and thickness of strained quantum wells a tuning of wavelength and strain is possible. The strain affects the electrical confinement [1, 2] and also the degradation behaviour [3]. However, the composition in thin QWs is difficult to determine. The photoluminescence (PL) peak wavelength is a result of both composition and thickness of the QW. Calibration of the relationship between TMIn/(TMIn + TMGa) in the vapour phase (x_v) to In/(In + Ga) in the solid state (x_s) by growing thick layers that allow for a compositional analysis is based on the assumption of a strain-independent In-incorporation efficiency. This assumption does not hold for MOVPE growth.

   The different incorporation behaviour of In for relaxed and strained layers and the effect of growth temperature (T_g) on the transition from 2-dimensional to 3-dimensional in growth low pressure MOVPE was studied. Based on these studies the performance of laser diodes with a strained InGaAs QW in the active zone was optimized.

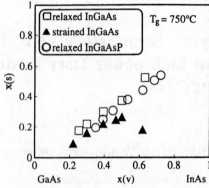
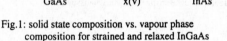
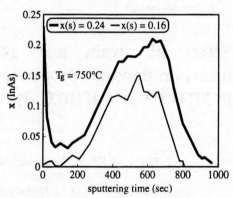

Fig.1: solid state composition vs. vapour phase
composition for strained and relaxed InGaAs

Fig.2: AES depth profiles of strained InGaAs
QWs with different In concentration

2. Experimental procedure

MOVPE growth of the layer structures was performed in a horizontal reactor at a total pressure of 70 hPa and at growth temperatures (T_g) between 600°C and 750°C. Starting materials were pure arsine and phosphine and the trimethyl compounds of aluminum, gallium and indium. Disilane diluted in hydrogen was used for n-type doping, dimethyl zinc for p-type doping. The laser structures is described in [4]. Additionally, test structures consisting only of the GaAs/InGaAs(P)/GaAs active region were grown on Si-doped (100) GaAs substrates (VGF-grown - "epiready" or Bridgman-grown after standard chemical cleaning, epd < 10^3 cm^{-2}).

Keeping all other growth parameters constant we have changed either the vapour phase ratio x_v or the thickness of the QW or the growth temperature. The different incorporation behaviour of In for relaxed and strained layers and the effect of different thicknesses and growth temperature (T_g) on the transition from 2-dimensional to 3-dimensional growth mode and on the In incorporation efficiency was studied by photoluminescence, high resolution x-ray diffractometry (HRXRD), Auger electron spectroscopy (AES) and scanning electron microscopy. This assessment of the layer properties was complemented by the characterization of broad area laser diodes (stripe width 80 μm and cavity length 300 μm to 1000 μm) under pulsed operation processed from the grown structures.

3. Results and discussions

For an optimization of laser diodes for a specific emission wavelength knowledge about strain in the active region is necessary. To determine the strain the In content and the thickness of the InGaAs QW has to be known. Fig. 1 show the different incorporation behaviour of In for relaxed and strained layers, where the composition of relaxed, about 500 nm thick InGaAs layers is estimated by electron microprobe and the composition of strained layers is measured and simulated by HRXRD. For strained layers a reduced In incorporation is found and the difference increases with increasing TMIn/(TMIn + TMGa) vapour phase ratio. The addition of phosphorus reduces the In incorporation in comparison to InGaAs.

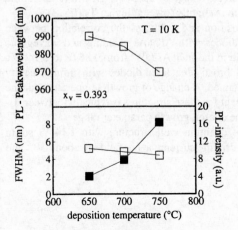

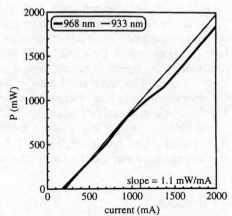

Fig. 3: PL wavelength, half width and
intensity for InGaAs-QW grown at
different temperatures

Fig. 4: output power vs. bias current
for unmounted laser diodes
with coated facettes

Probably this is an effect of prereactions between TMIn and PH_3. However, the slope of the incorporation curve for relaxed layers is not change. Since the As/P ratio also affects the strain, a direct comparison of strained and relaxed InGaAsP is difficult due to the wide range of different strain values in dependence of As/P and In/Ga ratio.

There are various possible reasons for the reduced In-uptake into compressively strained layers. It is known from studies on selective growth and orientation dependent growth that the decomposition of the metalorganic precursors can depend on the detailed structure of the surface. However, a suppressed release of In from the relatively unstable TMIn is not expected to play a major role at the growth conditions used in this study. While the amount of In supplied to the surface is expected to be the same, the incorporation of the big In into the strained lattice is reduced in comparison to relaxed layers having a bigger lattice constant. Earlier studies on the effect of growth interruptions indicate that unlike in the case of MBE [5], In desorption does not play a dominat role in MOVPE [4]. Instead, the excess In that is not incorporated into the lattice segregates on the surface. This strain induced In segregation is more pronounced at higher In content. The AES depth profile of an InGaAs-QW ($x_{InAs} \approx 0.25$) indicates up to 4 % In in the GaAs top layer (fig. 2). Additionally, the obtained x-ray rocking curves can only be successfully simulated when an In incorporation into the GaAs layer on the order of the one observed by AES is taken into account. At $x_{InAs} \approx 0.16$ in the QW the In content in the barrier is only slightly above the AES detection limit. This segregation also seems to put an upper limit to the In-uptake into strained layers which lies at about $x_s = 0.3$ at $T_g = 750°C$. An increase of x_v to above 0.5 does not result in a further increase of x_s. Cathodoluminescence images show the formation of "In-clusters" at higher x_v. Their density increases with increasing x_v. These clusters take up the excess In and potentially have deleterious effects on the laser performance.

By lowering the growth temperature from 750°C to 650°C the difference in the In incorporation between strained and unstrained layers is reduced due to a lower In segregation. Fig. 3 show the increase of the PL wavelength at lower T_g. From HRXRD and AES a reduction of In incorporation into the QW from $x_s = 0.24$ at 650°C to ≈ 0.21 at 750°C can be estimated. By lowering T_g also the critical thickness for the onset of relaxation processes is increased as revealed by cathodoluminescence images [6]. The transition from 2-dimensional to 3-dimensional growth is shifted from ≈ 15 nm at 750°C to ≈ 20 nm at

650°C for $x_s \approx 0.16$. While this appears to be an advantage of the lower growth temperature, the PL-intensity is higher and the PL peak width is slightly lower at higher T_g (fig. 3).

Based on the obtained data on composition and thickness the correlation between strain and threshold current density of laser diodes with AlGaAs cladding and waveguide layers has been studied. By increasing the strain in the InGaAs QW from 0.85 % to 1.14 % a reduction of the threshold current density of broad area laser diodes with infinite cavity length ($j_{th \infty}$) from 76 A/cm^2 to 53 A/cm^2 is obtained. A change of growth temperature for the QW does not affect j_{th}. The absorption coefficient for these structures is typically between 1 - 3 cm^{-1} and is not affected by strain and T_g in the chosen growth parameter range.

Ridge-waveguide lasers with a double quantum well structure with 1.14 % strain show pulsed output powers of 2 W at 2 A driving current, a far field of about 30° and degradation rates lower than 1×10^{-5} h^{-1} (fig. 4).

4. Conclusions

The In incorporation into MOVPE-grown strained InGaAs layers is reduced in comparison to relaxed layers. The reduction in In incorporation primarily is due to surface segregation. This segregation is more pronounced at high growth temperatures and high In content. At $T_g =$ 750°C the In incorporation into strained layers is limited to $x_s \approx 0.3$. Further increase of the In supply leads to 3-dimensional growth and the formation of "In-clusters".

For an understanding of the performance of strained active regions in laser diodes the above effects have to be taken into account. A measurement of the actual In content in the QW is indispensable. Extrapolation from results obtained on relaxed layers does not give valid results.

Acknowledgements

We thank T. Tessaro for excellent technical support and Dr. M. Procop (BAM, Berlin) for providing AES data. Dr. G. Beister and J. Maege are acknowledged for providing data on laser lifetime. The work was financially supported by the BMBF (13N6382).

References

[1]   Bour D B, Martinelli R U, Hawrylo F Z, Evans G A, Carlson N W and Gilbert D B 1990 *Appl. Phys. Lett.* **56** 318-321
[2]   Wang C A and Choi H K 1991 *J. Electronic Materials* **20** 929-934
[3]   Rechenberg I, Beister G, Bugge F, Erbert G, Gramlich S, Klein A, Maege J, Pilatzek M, Richter U, Ruvimov S S, Treptow H and Weyers M 1994 *Materials Science and Engineering* **B28** 310-313
[4]   Bugge F, Beister G, Erbert G, Gramlich S, Rechenberg I, Treptow H and Weyers M 1994 *J. Crystal Growth* **145** 907-910
[5]   Nagle J, Landesman J P, Larive M, Mottet C and Bois P 1993 *J. Crystal Growth* **127** 550-554
[6]   Zeimer U, Bugge F, Oster A and Weyers M *(to be published)*

Inst. Phys. Conf. Ser. No 145: Chapter 2
Paper presented at 22nd Int. Symp. Compound Semiconductors, Cheju Island, Korea, 28 August–2 September 1995
© *1996 IOP Publishing Ltd*

MOVPE growth of (Ga,In)(As,P)/GaAs for high-power laser diodes

A. Knauer, F. Bugge, G. Erbert, A. Oster and M. Weyers

Ferdinand-Braun-Institut für Höchstfrequenztechnik Berlin, Rudower Chaussee 5, D-12489 Berlin

$Ga_xIn_{1-x}As_yP_{1-y}$ lattice matched to GaAs has been grown by low pressure metal organic vapour phase epitaxy (LP-MOVPE) over the whole compositional range. The layer quality, the morphology and the p-type doping level of (In,Ga)(As,P) and (Ga,In)P show a strong dependence on substrate misorientation. The smoothest surface is obtained for GaInP on (100) off oriented to (11-1)B. For the quaternary the smoothest surfaces and best luminescence properties are obtained on (100) exact substrates. Growth of $Ga_xIn_{1-x}As_yP_{1-y}$ with small y on substrates misoriented to (11-1)B leads to ordering effects similar to the ternary InGaP. Preliminary tests on unmounted laser diodes ($\lambda = 808$ nm) indicate a high efficiency and very low internal losses ($\alpha \sim 2$ cm^{-1}).

1. Introduction

$Ga_xIn_{1-x}As_yP_{1-y}$ alloys lattice matched to GaAs are attracting increasing interest especially for the fabrication of high power laser diodes. Here the Al-free quaternary offers some advantages over the well established AlGaAs [1]. A lower tendency for oxidation facilitates the growth of buried laser structures and, together with a lower density of surface states should lead to laser diodes that are less prone to catastrophic optical degradation (COD) than AlGaAs/InGaAs/GaAs lasers. However, the growth of the quaternary requires tight compositional control to achieve lattice matching. Additionally, the existence of a miscibility gap makes the growth of certain compositions difficult [2]. While GaInAsP/GaAs lasers have already been successfully fabricated from layers grown by LPE [1, 3, 4], GSMBE [5] and MOVPE [2, 6-8], detailed studies on the growth of this material are scarce. For AlGaAs/InGaAs/GaAs lasers it is known that the best results are achieved on exact oriented (100) substrates or only slightly misoriented ones [9]. We have studied the growth of GaInAsP and GaInP on just oriented and differently misoriented substrates by LP-MOVPE at growth temperatures T_g between 580°C and 730°C. Ordering effects are not only observed in the ternary but also in the quaternary by x-ray diffraction and transmission electron diffraction.

Although the preferred orientation for the ternary GaInP is different from that of the quaternary, layer structures for laser diodes were grown successfully and processed into broad-area lasers showing high efficiency and low internal losses. For laser structures containing InGaP and InGaAsP with a composition near the miscibility gap the strong dependence of the layer quality on the different substrate orientations has to be taken into consideration.

2. Experimental

A horizontal MOVPE system with a rotating 2" susceptor was used at a reactor pressure of 70 hPa. Precursors were trimethylindium, trimethylgallium, dimethylzinc, disilane and pure arsine and phosphine in a hydrogen carrier gas flow. InGaAsP was grown at substrate temperatures of 650°C and 680°C, and InGaP between 580°C and 730°C. The V/III ratio was varied from 100 to 600. Substrate orientations of 0.5° to 2° off from the (100) to {111}A and B and 6° to (110) were studied using up to four different orientations in one growth run to allow for a valid comparison. The growth rate of the quaternary at a constant TMGa partial pressure of 0.37 Pa varies from 0.5 μm/h to 1.5 μm/h depending on the composition of the layer. The GaInP was grown with a growth rate of 2.4 μm/h. The lattice mismatch was smaller than 5×10^{-4}.

The surface morphology of the layers was inspected by optical microscopy with a Nomarski interference microscope. The surfaces and cleavage planes of the grown structures were examined by scanning electron microscopy in the backscattered secondary electron (BSE)-compo-mode. Additionally, the composition of the grown layers was determined by electron microprobe analysis (EPMA). The optical properties were assessed by photoluminescence (PL) at 10 K and room temperature (RT). Spatially resolved RT-PL maps yield information on the layer homogeneity. High resolution x-ray diffractometry (HRXRD) was performed in a four crystal diffractometer to check the lattice matching and to study ordering related diffraction peaks. Additionally, some samples were studied by transmission electron diffraction (TED). The electrical layer properties were assessed by Hall-measurements and electrochemical capacitance-voltage (ECV) profiling with a Polaron profiler. The laser properties were investigated on unmounted lasers with a stripe width of 100 μm and different resonator lengths under pulsed operation ($t_p = 500$ ns, $f_r = 5$ kHz).

3. Results and discussion

For InGaP layers the smoothest surface morphology and the best 10 K PL data are obtained on (100) GaAs with off orientation to (11-1)B at $T_g = 580°$ - 730°C. TED patterns show that in contrast to the just oriented one this substrate orientation leads to ordering only in one <11-1>B direction. This is confirmed by the observation of only one additional (115) reflex in HRXRD from an ordered superlattice [10, 11] and the lowest energy of the PL peak. The degree of ordering is related to T_g and has its maximum at $T_g \sim 650°C$.

$Ga_xIn_{1-x}As_yP_{1-y}$ lattice matched to GaAs has been grown over the whole compositional range. At $T_g = 650°C$ a low PL intensity together with a high FWHM is observed for $y \approx 0.2$ -

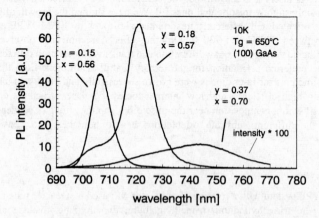

Fig. 1   10 K PL spectra of $Ga_xIn_{1-x}As_yP_{1-y}$ layers of different composition grown on exact
(100) GaAs ($T_g = 650°C$)

0.5 (fig. 1). These bad optical properties correlate with a considerable surface roughness. These effects are due to the predicted miscibility gap in this compositional region [12]. The negative effects of this miscibility gap are reduced by increasing the growth temperature. Fig. 2 shows that growth at 680°C leads to a tenfold increase of the PL intensity and a reduction of the peak half width of the layers with y = 0.5. Thus higher growth temperatures are favoured for laser structures containing layers of such composition.

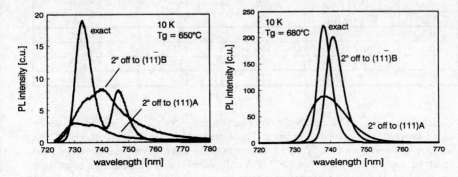

Fig. 2   10 K PL spectra of $Ga_xIn_{1-x}As_yP_{1-y}$ (y = 0.5) deposited at 650°C and 680°C on differently misoriented substrates

For quaternary layers with small arsenic content (y ≤ 0.2) the PL intensity is not so much affected by the growth temperature (fig. 3). However, in this compositional range close to InGaP the PL peak energy strongly depends on the substrate orientation at T_g = 650°C (fig. 3a). Exact orientation and a miscut to (11-1)B lead to similar peak positions of 1.774 eV at RT which is about 41 meV below that for the other two orientations. At higher T_g (680°C) this difference is reduced to 24 meV. The results of the 10 K PL analysis of layers grown at 680°C are summarized in table 1.

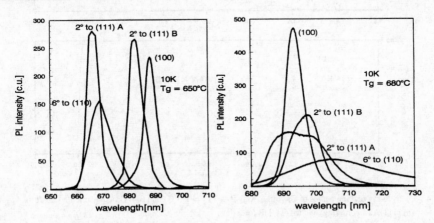

Fig. 3   10 K PL spectra of $Ga_xIn_{1-x}As_yP_{1-y}$ (y ≈ 0.15) deposited at 650°C and 680°C on differently misoriented substrates

Table 1   10 K PL wavelength, peak width and peak intensity (calibrated units) of $Ga_xIn_{1-x}As_yP_{1-y}$ layers grown on differently off-oriented GaAs substrates at 680°C, (exitation density 73 mW/cm^2)

y - value	off-orientation	λo [nm]	FWHM [nm]	I-PL [c.u.]
0.1	0°	691, (698)		1.6 E07
0.1	2° to {111}B	693	6.4	4.7 E07
0.1	2° to {111}A	687	10.7	2.0 E07
0.1	6° to (110)	704	28.8	7.9 E06
0.5	0°	738	4.4	2.2 E06
0.5	2° to {111}B	740	5.4	2.0 E06
0.5	2° to {111}A	736	11.8	8.9 E05
0.5	6° to (110)	745	14.9	2.3 E03

Like in the case of the ternary the observed differences of the band gap on differently oriented substrates are related to ordering. Fig. 4 shows the (115) rocking curves of two quaternary layers with y = 0.12. While for the miscut to (11-1)B a strong ordering related peak is observed, such a peak is absent on the (111)A misoriented sample. The presence of ordering again correlates to the occurence of superlattice spots in the TED patterns [13]. For small arsenic content the surface morphology is not significantly influenced by the substrate orientation.

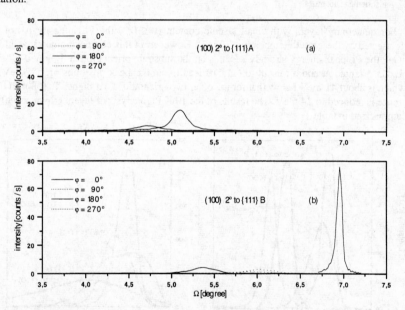

Fig. 4   (115) X-ray rocking curves of $Ga_{0.56}In_{0.44}As_{0.12}P_{0.88}$ (T_g = 680°C) on (100) GaAs misoriented 2°off to: (a) (111)A (no ordering) and (b) (11-1)B (ordering)

In contrast to that smooth surfaces of layers with y = 0.4 - 0.5 for T_g = 650°C could only be obtained on just oriented substrates. Compositional analysis by EPMA in conjunction with the very similar lattice mismatch suggests that the different PL energies observed for differently oriented layers with y = 0.5 (fig. 2) are not due to a different composition.

However, currently the existence of ordering is not confirmed by TED and HRXRD in this compositional region. Layers deposited at higher T_g (680°C) are of much higher quality and the PL data do not suggest the presence of ordering effects (fig. 2 and table 1).

The tendency for roughening of the surfaces of quaternary layers observed on misoriented substrates appears to be enhanced in the presence of strain. Similarly, p-type doping leads to enhanced roughness already for doping levels of a few 10^{17} cm^{-3}. Additionally, p-type doping is enhanced on off orientations to the (111)A plane and suppressed by a tilt to (11-1)B (fig. 5). The doping efficiency increases with the As content. No influence of n-type doping with disilane on the surface morphology was observed. The n-type doping level also does not show a significant dependence on the substrate orientation.

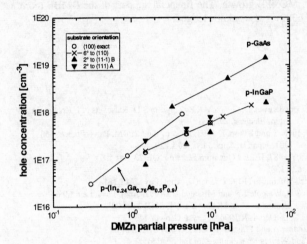

Fig. 5   Hole concentration versus DMZn partial pressure for GaAs, GaInP, and one quaternary composition (T_g = 650°C)

The different orientation dependence of the layer quality of InGaP (prefers off orientation) and especially of the quaternary layers with compositions near the miscibility gap and with p-type doping (prefer exact substrates) can restrict the design of a laser structure. Taking this into account (100)GaAs substrates with 2° off orientation to (11-1)B were chosen for the deposition of aluminum free broad area laser structures. The laser structures consist of GaInP-cladding layers, $Ga_{0.56}In_{0.44}As_{0.15}P_{0.85}$-waveguide layers and different layers as active region as listed in table 2.

Table 2 data of 1mm long, unmounted broad area laser in pulsed operation: t_p = 500ns, f_r = 5khz; $\Theta_\perp$: vertical far field angle; α_i: internal losses; η_i: internal efficiency; $j_{th\infty}$: extrapolated threshold current for L to ∞;

active region	x / y	$j_{th\infty}$ (A/cm^2)	α_i (cm^{-1})	η_i (%)	λ (nm)	$\Theta_\perp$ (grd)
InGa$_x$As-SQW	0.92 / 1	110	3	77	872	27
GaAs-SQW	1 / 1	98	1	68	853	26
InGa$_x$As$_y$P-SQW	0.87 / 0.8	115	1.5-3	82	818	26
InGa$_x$As$_y$P-SQW	0.58 / 0.77	230	1	56	807	29

First unmounted laser diodes indicate a high efficiency (1.7 W at 2 A driving current) and low losses ($\alpha \sim 2$ cm^{-1}). This demonstrates the high potential of this material for fabrication of high power laser diodes.

Acknowledgments

The authors greatfully acknowledge the support of U. Zeimer (EPMA), I. Rechenberg (TED), D. Gutsche, S.Gramlich (PL), E. Richter (ECV), K. Vogel (laser processing) and M. Gielow for her technical assistance in MOVPE-growth. The financial support of the BMBF (contract numbers 13 N 6375 and 13 N 6382) is also acknowledged.

References

[1] Garbuzov D Z, Antonishkis N Y, Bondarev A D, Gulakov A B, Zhigulin S N, Katsavets N I, Kochergin A V and Rafailov E V 1991, IEEE J. Quantum Electron. **27** 1531
[2] Iketani A, Ohkubo M, Namiki S, Ijichi T and Kikuta K 1993, Proc. of the 5.IPRM, Paris, France, 553
[3] Mukai S, Yajima H and Shimada J 1981, Japan. J. Appl. Phys. **20** L729
[4] Kishino K, Harada A and Kaneko Y 1987, IEEE J Quantum Electron. QE-**23** 180
[5] Zhang G 1993, Appl. Phys. Lett. **63** 1128
[6] Iwamoto T, Mori K, Mizuta M and Kukimoto H 1985, Japan. J. Appl. Phys. **24** L131
[7] Liau Z L, Palmateer S C, Groves S H, Walpole J N and Missagia L J 1992, Appl. Phys. Lett. **60** 6
[8] Razeghi M, He X G, Diaz J, Hoff J, Erdtmann M and Kolev E 1994, SPIE **2145** 23
[9] Hiramoto K, Tsuchiya T, Sagawa M and Uomi K 1994, J. Cryst. Growth **145** 133
[10] Liu Q, Lakner H, Scheffer F, Lindner A and Prost W 1993, J. Appl. Phys.**73** 2770
[11] Oster A, Knauer A, Gutsche D and Weyers M, submitted for publication
[12] Onabe K 1982, Japan. J. Appl. Phys. **21** 797
[13] Rechenberg I, Oster A, Knauer A and Weyers M, submitted for publication

Inst. Phys. Conf. Ser. No 145: Chapter 2
Paper presented at 22nd Int. Symp. Compound Semiconductors, Cheju Island, Korea, 28 August–2 September 1995
© *1996 IOP Publishing Ltd*

Carbon doped GaAs grown by MOVPE using CBr4

P. Kurpas, E. Richter, D. Gutsche, F. Brunner and M. Weyers

Ferdinand-Braun-Institut für Höchstfrequenztechnik Berlin, Rudower Chaussee 5, D-12489 Berlin, FR Germany

Abstract. Heavily carbon doped GaAs-layers with hole concentrations up to $1.6*10^{20}$ cm^{-3} were grown by low-pressure MOVPE using carbon tetrabromide as doping source. Specular layer morphologies and hole mobilities comparable with literature data were obtained in the temperature region 550°C - 670°C. At very high CBr_4 partial pressures the morphology deteriorates and the growth rate decreases due to increasing etching effects. Traces of bromine are detected by SIMS only in layers grown at low temperature ($6*10^{15}$ cm^{-3} at 550°C and $2*10^{14}$ cm^{-3} at 580°C). Carbon deactivation by hydrogen was observed in all layers. This passivation effect can be reduced by annealing in nitrogen for 15 minutes at 450°C. By photoluminescence mapping the excellent homogeneity of doping level (± 4% over entire 2inch wafer) was verified. These results demonstrate the suitability of CBr_4 for growth of device structures like GaInP/GaAs HBTs.

1. Introduction

Carbon has been established in the recent years as a standard dopant for p-GaAs due to its favourable properties (low diffusion coefficient, no memory effects, possibility of extremely high doping levels). GaAs:C layers can be fabricated by MOVPE using the precursors trimethylgallium (TMGa) and trimethylarsine (TMAs) as suppliers of CH_3-groups together with a small arsine (AsH_3) concentration. However, in this case the doping level is linked to the growth rate. More independent control of both parameters can be achieved by mixing of TMGa with a second gallium precursor [1]. Extrinsic carbon sources like CCl_4 allow for a better control of the doping process. With CCl_4 high quality GaAs:C layers can be obtained [2] but this halomethane has to be replaced due to enviromental restrictions.

The suitability of carbon tetrabromide CBr_4 for MOVPE growth was previously indicated [3]. CBr_4 has already replaced to some extent CCl_4 in gas source molecular beam epitaxy (GSMBE) [4, 5]. In this paper we present a systematic evaluation of CBr_4 as carbon doping source for low pressure MOVPE. The properties of the grown GaAs:C layers are characterized with respect to the application in GaInP/GaAs HBT layer structures. Especially, carbon deactivation by incorporated hydrogen and doping homogeneity over 2inch wafer are studied in detail.

2. Experimental

GaAs:C layers were grown in an AIX200 2inch single wafer MOVPE system with a horizontal reactor using a rotating graphite susceptor. The layers were grown on Cr-compensated and Si-doped GaAs(001) substrates tilted by 2° toward (111)B. The growth experiments were performed at low pressure (70 hPa) using hydrogen carrier gas. Trimethylgallium (TMGa) and arsine (AsH$_3$) were used as growth precursors. Constant TMGa partial pressure of 0.7 Pa which corresponds to a GaAs growth rate of 1.3 µm/h at 600°C was maintainted for all runs.

Electrochemical capacitance voltage measurements (ECV) calibrated by stylus depth measurements were used for the determination of the interface position between p-doped layer and n-doped substrate. The obtained results are well in agreement with those of secondary ion mass spectrometry (SIMS). Hole concentrations were determined by van der Pauw Hall effect measurements at room temperature and 77 K. Indium contacts were formed by spark discharge ensuring that the samples were not heated after the growth. Passivation effects of hydrogen were investigated by annealing the Hall samples in nitrogen atmosphere at 450°C three times for 5 minutes and subsequent Hall effect measurements. Further experiments verified that one annealing step for 15 minutes leads to the same amount of carbon activation while larger annealing times do not result in substantial additional activation. SIMS depth profiles of carbon, hydrogen and bromine were measured on selected as-grown and annealed samples. GaAs-layers as calibration standards for these elements were prepared by ion implantation. This allowed the quantitative determination of C, H and Br concentrations with an accuracy of better than 20 %. PL spectra were measured on as-grown and annealed samples at room temperature and 77 K using a commercial photoluminescence (PL) mapping equipment SPM 200 (Waterloo Scientific). The PL spectra were then used as a measure of the hole concentration.

3. Growth results

Fig. 1 shows the temperature dependence of carrier concentration and growth rate. The doping level increases with decreasing growth temperature. From the data in fig. 1 the activation energy E_a of the processes limiting carbon incorporation can be evaluated to be in the range of 36 - 47 kcal/mol. This activation energy is slightly higher than the one reported for CCl$_4$ (25 - 32 kcal/mol) [6] and in the range reported for CCl$_4$, CBr$_4$ and other halomethanes (30 - 60 kcal/mol) [3]. The positive slope of the curve indicates that not the pyrolysis of CBr$_4$ is the main limiting process but desorption of carbon containing species. The higher E_a in the case of CBr$_4$ indicates a stronger bond of the chemisorbed fragments (CBr$_x$, x = 1,2) to the surface than in the case of CCl$_4$. This probably is also responsible for the higher doping efficiency of CBr$_4$ observed in GSMBE [5].

The hole concentration depends sublinear ($p \sim p_{CBr4}^{0.7}$) on CBr$_4$ concentration as shown in fig. 2. An increase in carrier concentration after annealing up to 20 % was observed for samples with doping levels below $1*10^{20}$ cm^{-3}. At the highest doping levels carbon deactivation becomes even more pronounced.

The growth rate is not affected at CBr$_4$ partial pressures below 0.2 Pa at 600°C and 0.08 Pa at 670°C. At higher CBr$_4$ concentrations a decrease in the growth rate occurs (fig. 2) due to etching effects. At highest CBr$_4$ partial pressures the morphology of the layers starts to deteriorate from mirror-like to dull. All other samples show specular morphology in Nomarski interference microscopy.

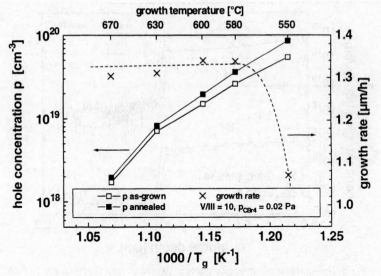

Fig. 1: Arrhenius-plot of hole concentration and growth rate dependence on temperature

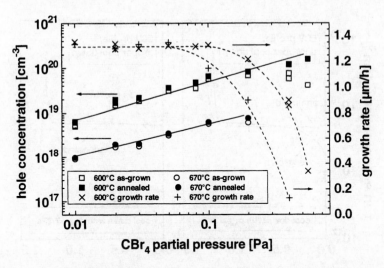

Fig. 2: Dependence of hole concentration and growth rate on CBr₄ partial pressure (V/III = 10)

Hole mobilities in GaAs:C layers grown with CBr₄ are as high as mobilities obtained in MOVPE with CCl₃ [7] or using TMGa and AsH₃ at V/III ratios near 1 [8]. The empirical formula given by Hilsum [9]

$$\mu = \mu_0 / [1 + (p / p_0)^m] \qquad \mu_0 = 430 \; cm^2/Vs, \; p_0 = 3*10^{17} \; cm^{-3}, \; m = 0.3$$

with the same parameters as suggested by Kim et al. [10] describes well the mobility dependence on hole concentration in the doping region above 10^{18} cm⁻³. This relationship is not affected by the used annealing procedure.

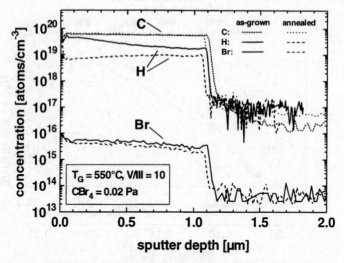

Fig. 3: Depth profiles of carbon (C), hydrogen (H) and bromine (Br) measured by SIMS
in a GaAs:C layer grown at 550°C

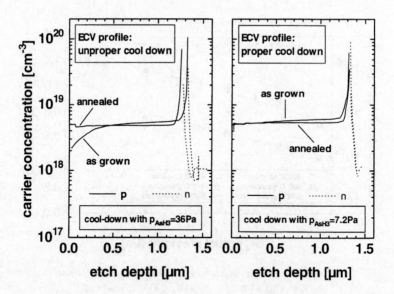

Fig. 4: Comparison of ECV profiles of samples stabilized under different arsine partial
pressures p_{AsH3} during the post-growth cooling-down period
(both layers grown at 600°C, p_{CBr4} = 0.01Pa, V/III = 10)

Fig. 3 shows depth profiles of carbon, hydrogen and bromine measured by SIMS on
the sample grown at the lowest temperature (550°C). Only in this GaAs:C layer and in the
sample grown at 580°C bromine was detected (detection limit below $1*10^{14}$ cm^{-3}). The
carbon and bromine profiles remain unchanged after annealing. In the as-grown sample a

high hydrogen concentration increasing from $1.8*10^{19}$ cm^{-3} in the bulk to $4.5*10^{19}$ cm^{-3} near the surface was found. Our annealing procedure decreases the hydrogen concentration by a factor of 6 to 10 below the carbon concentration (fig. 3). However, efficient hydrogen reduction from GaAs:C layers embedded in device heterostructures needs further investigations.

ECV measurements show that the hydrogen passivation adjacent to the surface depends on post-growth cooling-down conditions. The passivation becomes stronger if a higher AsH$_3$ partial pressure is used (fig. 4). This result proves that hydrogen atoms arising from arsine decomposition are responsible for carbon deactivation. Proper adjustment of the arsine concentration during and after GaAs:C growth is essential in order to minimize carbon deactivation in device structures.

4. Photoluminescence measurements

Characteristic energies in a PL spectrum from a GaAs:C layer can be correlated with the doping level due to significant band gap shrinkage in highly p-doped GaAs. According to the method given by Olego and Cardona [11] the band gap energy E_g can be extracted by linear extrapolation of the low energy side of the spectrum to the background. Fig. 5 shows the dependence of E_g on hole concentration measured at room temperature on as-grown and on annealed samples. Our results agree well with the data of Lu et.al. [12] represented by the solid line for E_g. Also the energy E_L corresponding to the half height on the low energy side of the spectra as visualized by the inset in fig. 5 can be used for analysis. Both E_g and E_L show the same dependence on hole concentration. The empirical fit E_L [eV] = 1.429 - $4.1*10^{-6} * p^{0.217}$ [12] describes well our data and gives an easy-to-use correlation between E_L and hole concentration.

E_L is more convenient than E_g for homogeneity assessment by PL mapping. The accuracy of E_L determination in our measurements is 0.3 meV. For a 2inch wafer an average E_L energy of 1356 meV with a standard deviation of 0.6 meV over the entire wafer area and 0.4 meV in its central part (75% of wafer area) was obtained. The derived doping level of $3.8*10^{19}$ cm^{-3} agrees well with the value of $3.6*10^{19}$ cm^{-3} obtained by Hall effect measurement. From the standard deviation of E_L the doping level homogeneity is assessed to be ± 3.9 % over the entire wafer and ± 2.6 % in the central part. These results prove that excellent homogeneity can be easily achieved with CBr$_4$.

5. Conclusions

The results presented here show that CBr$_4$ is a suitable source for homogeneous growth of GaAs:C layers. A stable growth rate at moderate CBr$_4$ partial pressures leading to technologically relevant doping levels in the middle 10^{19} cm^{-3} region allows for well controlled doping. Furthermore, a relatively high growth temperature of 600°C can be used making the GaAs:C growth conditions compatible with the use for GaInP/GaAs heterostructures of high quality.

The low bromine incorporation observed only at lower temperature (550°C and 580°C) does not affect the layer quality. However, carbon deactivation by incorporated hydrogen could decrease the performance of device structures. We have shown that this problem can be controlled by proper process conditions especially during the post-growth cooling phase. The application of CBr$_4$ for the doping of the base in GaInP/GaAs HBTs is currently being studied.

182

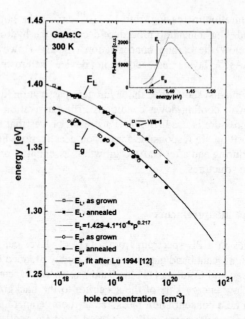

Fig. 5: Dependence of characteristic energies in PL spectra on hole concentration

Acknowledgements

The authors would like to thank O. Fink for help with MOVPE growth. We acknowledge RTG Mikroanalyse GmbH, Berlin for performing SIMS measurements. We are grateful to G. Schöne and H. Strusny for implantation of the calibration standards and F. Bugge and A. Knauer for valuable suggestions.

References

[1] Ungermanns C, Hardtdegen H, Raafat T, Hollfelder M 1995 *presented at EW-MOVPE VI, Gent, Belgium*

[2] Yang L W, Wright P D, Eu V, Lu Z H, Majerfeld A 1992 *J. Appl. Phys.* **72** 2063

[3] Buchan N I, Kuech T F, Scilla G, Cardone F 1991 *J. Cryst. Growth* **110** 405

[4] Houng Y-M, Lester S D, Mars D E, Miller J N 1993 *J. Vac. Sci. Technol B* **11** 915

[5] de Lyon T J, Buchan N I, Kirchner P D, Woodall J M, Scilla G J, Cardone F 1991 *Appl. Phys. Lett.* **58** 517

[6] Nittono T, Watanabe N, Ito H, Sugahara H, Nagata K, Nakajima O 1994 *Jpn. J. Appl. Phys.* **33** 6129

[7] Colomb C M, Stockman S A, Gardner N F, Curtis A P, Stillman G E, Low T S, Mars D E, Davito D B 1993 *J. Appl. Phys.* **73** 7471

[8] Richter E, Kurpas P, Gutsche D, Weyers M 1994 *presented at 7th OMVPE Workshop, Fort Myers, USA*

[9] Hilsum C 1974 *Electr. Lett.* **10** 259

[10] Kim S-I, Kim Y, Lee M S, Kim M-S, Min S, Lee C 1993 *Solid State Comm.* **88** 743

[11] Olego D, Cardona M 1980 *Phys. Rev. B* **22** 886

[12] Lu Z H, Hanna M C, Majerfeld A 1994 *Appl. Phys. Lett.* **64** 88

Inst. Phys. Conf. Ser. No 145: Chapter 2
Paper presented at 22nd Int. Symp. Compound Semiconductors, Cheju Island, Korea, 28 August–2 September 1995
© *1996 IOP Publishing Ltd*

MOVPE-Grown AlGaInP (411)A-like Micro-Facets on Nonplanar Substrates

C. Anayama and T. Tanahashi

Fujitsu Laboratories Ltd., 10-1 Morinosato-Wakamiya, Atsugi 243-01, Japan

Abstract. We studied the nonplanar growth-shape of AlGaInP by MOVPE, containing the (411)A-like inclined micro-facet on the stepped substrate. We found that the (411)A-like inclined micro-facet was stable in different growth conditions. We suggest that the micro-facets are an equilibrium form determined by factors such as surface energy, and that the growth shapes were determined by both the growth thickness distributions and the surface energy.

1. Introduction

Metal organic vapor phase epitaxy (MOVPE) on nonplanar substrates provides some particular device structures. It will be one of the key techniques for new device fabrication. The features and mechanisms of the nonplanar growth, however, are still unclear. Recently, much attention has been paid to (411)A-oriented substrates because of their good crystalline qualities and surface reconstruction stabilities.[1,2] The (411)A face has been known to appear as the inclined micro-facet of the nonplanar grown layer, and we also used this nonplanar growth for visible laser fabrication.[3,4] In this work, we studied the dependence of the nonplanar growth-shape on the growth condition. We then propose its growth mechanisms.

2. Experiments

We grew epitaxial layers in a low-pressure vertical MOVPE reactor with flow controlled multiple quartz gas injectors.[5] Using this system, we can obtain good thickness uniformity and high reproducibilities. The group-III sources used in this experiment were trimethylaluminium, triethylgallium, and trimetylindium for AlGaInP and GaInP, trimetylgallium for GaAs. The group-V sources were arsine and phosphin. The growth temperature was 670°C and V/III ratio was 50 for GaAs. We varied growth temperature and V/III ratio for AlGaInP around 700°C and 200, respectively. The growth rates for the AlGaInP and GaAs were about 2mm/h, and about 1μm/h for GaInP. The operating pressure was 50 Torr. We prepared the stepped form nonplanar substrate with chemical wet-etching. The stepped substrate has an inclined face whose orientation was determined by the relative etching rate between the etching depth and the lateral side-eching in the underlying layers on the etching-mask edge. We varied this orientation by controlling the etching solution composition.

3. Results and discussions

Figure 1 shows SEM images of cross-sectional and 30°-tilt cross-sectional view for the

184

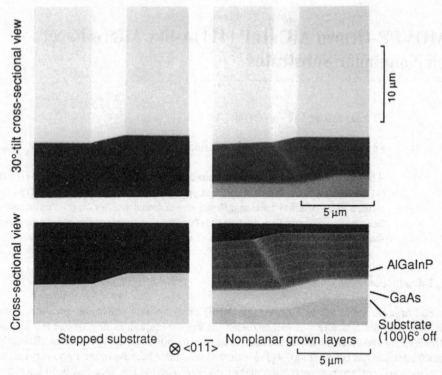

Figure 1   SEM images of substrate and grown layers

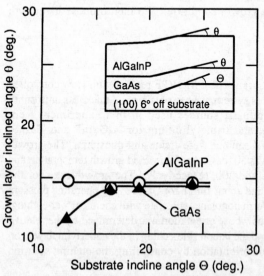

Figure 2   The dependence of the grown inclined face angle on substrate shape

(01$\bar{1}$)-cleaved nonplaner substrate and epitaxial layers on nonplanar stepped substrates. The grown layers are a GaAs buffer layer and AlGaInP layers with GaInP growth markers. A nearly-(411)A-oriented inclined face appeared on the stepped substrate. After epitaxial growth, the relatively rough surface morphology of the substrate's inclined surface was converted to a smooth surface. Another inclined face of nearly (100) 10°-offset face also appeared at the bottom of the (411)A-like inclined face. The nearly 10°-offset face gradually reduce the surface area of the (411)A-like inclined face as the growth proceeded. Figure 2 shows the grown inclined face orientation against the stepped substrate's inclined face orientation. We still

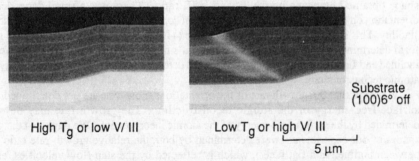

Substrate
(100)6° off

High T$_g$ or low V/ III   Low T$_g$ or high V/ III

|— 5 µm —|

Figure 3   Cross-sectional SEM images of growth shape variation

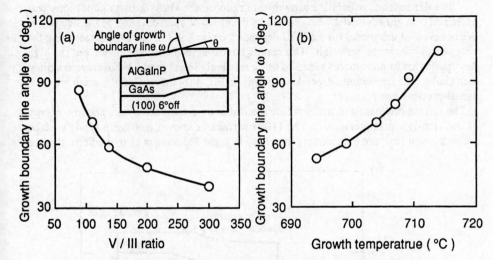

Figure 4 Growth shape dependence on growth conditions

found that the (411)A-like micro-facet appeared, given a sufficient growth thickness. The (411)A-like face orientation was not seriously affected by the substrate shapes.

We observed the growth rate on the inclined face strongly varied with the growth conditions. The cross-sectional images of growth shape variation are shown in the Fig. 3. Figures 4 (a) and (b) show the dependence of growth shape on the growth temperature (T$_g$) and V/III ratio. A high T$_g$ or a low V/III ratio yields a low growth rate on the inclined face, and a low T$_g$ or a high V/III ratio yields a high growth rate on the inclined face.   In the Figs.4, the growth rate on the inclined face was described by the intersection angle (ω) between the inclined face and the upper growth-boundary-line (GBL) which is the growth-boundary between the upper flat face and the inclined face. We use the intersection angle for characterizing the growth thickness ratio between the flat and inclined face because it sensitively reflected the growth rate on the inclined face. This relationship can be seen in the following simple equation.

$t_1 / t_2 = \sin \omega / \sin (\omega - \theta)$,

where t_1 is the thickness on inclined face, t_2 is the thickness on flat face, ω is the intersection angle between inclined face and GBL, and θ is the intersection angle between flat face and inclined face. Using the intersection angle, we can characterize the stability of the inclined face orientation against the growth shapes. Figure 5 shows its stability against the growth shape obtained by changing the T_g and V/III ratio. There was a small dependence of orientation on the growth conditions, but we can conclude that the (411)A-like face has high stability. This suggests that the formation of the (411)A-like micro-facet was not the growth form determined by such factors as the adatom's migration length difference between the inclined and flat faces, but the equilibrium form determined by the surface free-energy of the atomic reconstruction.

We think that the growth rate of the orientations, however, are not determined by the surface free-energy of the atomic reconstruction. The growth rate may be mainly determined by the surface free-energy of the atomic incorporation. We think that the growth shapes of our experiments were determined by both the relative growth-rate ratio (ratio between inclined and flat face), which is reflected by the step-flow velocities, and the surface orientations of the inclined face, which is reflected by the step-raw arrangements.

The dependence of relative growth-rate ratio on the GaAs growth conditions was explained by a growth model which considers the group-V atom coverage at the step site.6) We believe that this model for the V/III dependence also holds in our experiment using the phosphorus compound materials. Our results almost agreed with the model, but the high desorption rate of phosphorus seems to bring relatively lower group-V coverage conditon than GaAs. The temperature dependence shown in Fig. 4 (b) may reflect the small group-V partial-pressure case.

The surface orientation of inclined face seemed to be determined by surface free-energy of the step-law arrangements. The (100) surface is known to have a (2x4) surface reconstruction structure (the missing dimer model), and Shimomura et al. predicted that the

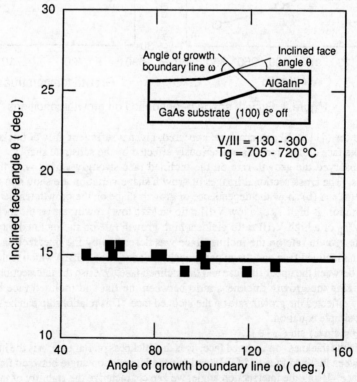

Fig. 5. Stability of (411)A-like inclined face orientation

(411)A face has a resemble structure to the (100) surface.[1] This suggests that the surface free-energy of the (411)A surface has a minimum value for the orientation. We believe that the high orientation stability of the (411)A-like inclined facet may be due to the local minimum surface free-energy for orientation.

However, the actual micro-facet orientations that appeared in our experiment were not exact (411)A but nearly-(411)A vicinal faces. They were not simply attributed to the free-energy of the exact (411)A flat-face. An inclined face has a free-energy of its own orientation per unit area. An orientation change of the inclined face involves the area change of both inclined and flat faces. The total free-energy of the nonplanar growth surface is determined by the area-integration of the surface free-energies. We presume the vicinal surface becomes a minimum free-energy surface when we consider this energy system as a region containing surrounding surfaces around the inclined face.

The dependence of inclined face orientation on substrate orientation is shown in Fig. 6. The (411)A-like face varied slowly with the substrate orientation. This substrate orientation dependence proves that the peripheral surface (flat surface) affects the determination of the inclined face orientation. In order to explain the dependnce of Fig. 6 qualitatively, we assumed free-energy function was a cosine-function which has a local minimum at the (411) face. Figure 7 shows the calculated relative free-energy with respect to the inclined face orientation. In the Fig. 7, the bold line is the assumed free-energy, the thinner lines are the calculated lines with the substrate orientation as parameters, and the dashed line is the variation of the local minimum of the free-energy. The free-energy's local minimum are not exact (411)A and varied slightly with the substrate orientation. We think these results qualitatively explain our experimental results in Fig. 6.

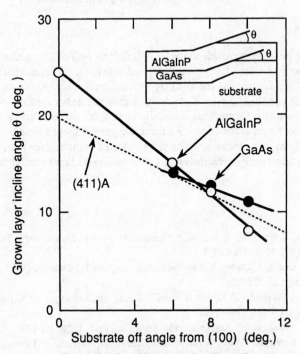

Fig. 6 Inclined face angle dependence
on substrate orientation

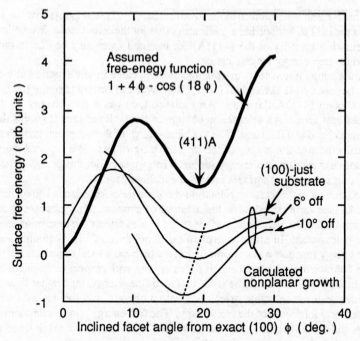

Fig. 7 Surface free-energy of the nonplanar grown
layers using assumed orientation energy

4. Conclusions

We showed the nonplanar growth-shape of AlGaInP by MOVPE, especially containing the (411)A-like inclined micro-facet on the stepped substrate. We then discussed its growth mechanisms. We found that the (411)A-like inclined micro-facet was very stable in different growth conditions, even if the relative growth rates at the facet were varied. We presume that the micro-facets were an equilibrium form determined by the surface free-energy of the atomic reconstruction. We think the growth shapes were determined by both the growth thickness distributions, which were reflected in the surface phosphorus coverage, and the surface free-energy, which contains surrounding surfaces around the inclined face.

References

[1] S. Shimomura, A. Wakejima, A. Adachi, Y. Okamoto, N. Sano, K. Murase, and S. Hiyamizu, 1993
    Jpn. J. Appl. Phys. 32 L1728-L1731.

[2] M. Kondo, N. Okada, K. Domen, K. Sugiura, C. Anayama, and T. Tanahashi, 1994
    J. Electron. Mater. 23 355-358

[3] C. Anayama, H. Sekiguchi, M. Kondo, H. Sudo, T. Fukushima, A. Furuya, and T. Tanahashi,
    1993 Appl. Phys. Lett. 63 1736-1738.

[4] A. Furuya, M. Kondo, and C. Anayama, 1994 Fujitsu Sci. Tech. J., 30 162-170

[5] M. Kondo, A. Kuramata, T. Fujii, C. Anayama, J. Okazaki, H. Sekiguchi, T. Tanahashi, S. Yamazaki,
    and K. Nakajima, 1992 J. Cryst. Growth 124 265-271

[6] H. Asai , 1987 J. Cryst. Growth 80 425-433.

Inst. Phys. Conf. Ser. No 145: Chapter 2
Paper presented at 22nd Int. Symp. Compound Semiconductors, Cheju Island, Korea, 28 August–2 September 1995
© *1996 IOP Publishing Ltd*

AFM observation of atomically flat heterointerface of GaInAs/InP grown by OMVPE

Michihiko SUHARA[1], Chuma Nagao, Yasuyuki Miyamoto and Kazuhito Furuya

Department of Electrical and Electronic Engineering, Tokyo Institute of Technology O-okayama, Meguro-ku, Tokyo 152, Japan

Abstract. Heterointerface roughness of GaInAs/InP grown by organometallic vapor phase epitaxy (OMVPE) is studied by atomic force microscopy (AFM). In order to achieve an atomically flat region on the initial InP substrates during the initial annealing process, a temperature around 600°C is found to be critical under a high flow rate of PH_3. Optimizing growth interruptions, monolayer steps and atomically flat terraces of 300 – 450 nm width were observed on the surface of a GaInAs quantum well layer on InP. Isotropic steps were oriented perpendicular to the $[0\bar{1}1]$ direction. Using the formation of atomically flat heterointerfaces, a reduction in the quantum well width fluctuation is expected to be useful for the estimation of the intrinsic the phase coherent length of hot electrons in semiconductors using RTDs.

1. Introduction

To realize quantum coherent devices[1], semiconductor heterointerfaces are required to be uniform on an atomic scale. We have estimated phase coherent length of hot electrons in semiconductors from the evaluation of the quantized energy level width using a resonant tunneling diode (RTD).[2] Since heterointerface roughness affects the variation of quantum well width, the apparent phase coherent length includes the effect of inhomogenious broadening of the energy level. For example, a two monolayer roughness in the heterointerface causes a well energy broadening of 20 meV in 4 nm width quantum well.[3]

Although it is known that the heterointerface roughness is related to the surface formation process during crystal growth, the surface structure during epitaxial growth has not been well understood on an atomic scale. For GaAs surfaces grown by organometallic vapor phase epitaxy (OMVPE), the mechanism of monolayer or multilayer step formation has been studied using atomic force microscopy (AFM)[4] and the mechanism of step bunching has been studied using scanning tunneling microscopy (STM).[5] For

[1] E-mail: suhara@pe.titech.ac.jp

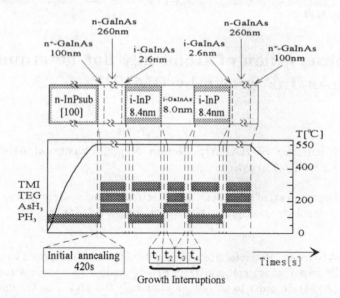

Figure 1. Device structure of RTD and OMVPE growth diagram. Initial annealing and growth interruptions were performed at the heterointerfaces.

GaAs, AlAs and AlGaAs growth surfaces, the formation of wide surface terraces has been observed by AFM.[6] However, for GaInAs/InP heterointerface structures, few investigations on the atomic scale have been reported[7, 8], although InP-based materials are expected to be useful for high speed quantum devices.

In this report we investigated OMVPE grown GaInAs/InP heterointerfaces using AFM and optimized growth conditions in order to form atomically flat interfaces. Atomically flat terraces with several hundred nm width and monolayer steps were obtained on an GaInAs surface on InP for the first time using long annealing intervals in the growth interruption at the heterointerface.

2. Experimental

$Ga_{0.47}In_{0.53}As$/InP RTD structures were fabricated by OMVPE using the growth sequence shown in Fig.1. The structure consists of 100 nm n^+- GaInAs, 260 nm n- GaInAs, 2.6 nm i- GaInAs for the spacer, 8.4 nm i- InP for the barrier, 8.0 nm i- GaInAs for quantum well, 8.4 nm InP for the barrier, 2.6 nm i- GaInAs for the spacer, 260 nm n- GaInAs, and 100 nm n^+- GaInAs for the top contact layer. From typical J-V characteristics of RTDs, a peak to valley current ratio of around 3 was obtained at 4K in samples having top Cr/Au electrode of 100 μm × 100 μm.

A Series of growths was performed on (100) oriented n-InP substrates whose misorientation magnitudes were within ± 0.2°. Previous to the growth, substrates were etched in H_2O:H_2SO_4:H_2O_2 (=1:3:1) for 90 sec and rinsed in deionized water.

The pressure and flow velocity in the OMVPE reactor were 76 Torr and 3 m/s, respectively. Triethylgallium (TEG) and Trimethylindium (TMI) were used for group

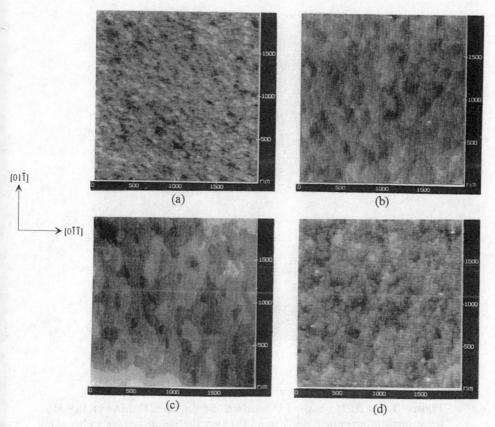

[01Ī]

[0ĪĪ]

(a)

(b)

(c)

(d)

Figure 2. AFM images of InP surfaces of (a) the pre-loaded substrate, annealed substrates in PH$_3$ flow rate of 4.46×10^{-3} mol/min, (b) 550°C, (c)600°C, (d)650°C, respectively.

III source materials and AsH$_3$ and PH$_3$ were used for group V materials. The growth rate and V/III ratio were 820 nm/h and 480 for GaInAs, 435 nm/h and 1976 for InP. Lattice mismatch of the GaInAs layer to an InP substrate was evaluated to be less than 0.05%.

Surface structures at heterointerfaces of samples were observed by AFM (Digital Instruments: Nanoscope II) in contact mode in air.

3. Annealed surface of InP substrates

The atomic topography of the initial surface of the substrate affects the finally grown surfaces[8]. To improve GaInAs/InP heterointerface roughness, initial annealing of InP substrate was investigated for various temperatures and PH$_3$ flow rate.

After the heating up process, the substrates are initially annealed for 7 min keeping the temperature constant. PH$_3$ was introduced into the reactor during both the

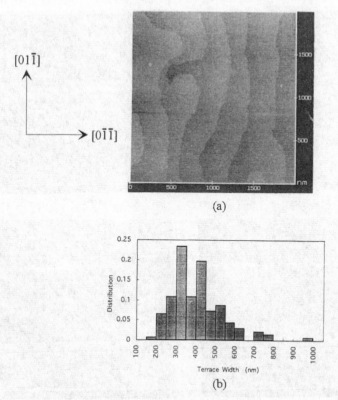

(a)

(b)

Figure 3. (a) AFM image of the surface of 8 nm thick GaInAs on InP after the growth interruption $t_3 = 7$ min. (b) Distribution of terrace width along [01$\bar{1}$] direction on the GaInAs surface. Data were examined in $5\mu m \times 5\mu m$ area

heating up and annealing periods in H_2 ambient. Figure 2. shows AFM images of 2 $\mu m \times 2$ μm area of InP surface of (a)pre-loaded substrate, just after the annealing at (b)550°C, (c)600 and (d)650°C under the constant PH_3 flow rate of 4.46×10^{-3} mol/min, respectively. Atomically flat regions were obtained on the annealed InP surface at 600°C without InP deposition as shown in Fig.2 (c). The step height was evaluated as 0.28 nm which corresponds to one monolayer thickness of an InP crystal. For other temperatures, the atomically flat region and steep steps were not observed as shown in Fig.2 (b), (d). Root mean square values of roughness in the scan area of $2\mu m \times 2\mu m$ were estimated as 0.31 nm and 0.33 nm for 550°C and 650°C, respectively. The temperature dependency of the formation of an InP atomic surface structure tends to be more critical than that of GaAs surface.[6]

For the low PH_3 flow rate (2.23×10^{-3} mol/min) at the critical temperature of 600°C, the surface became anisotropically rough and atomically flat regions and step structures were not observed.

4. Formation of atomically flat terraces on a GaInAs surface

During the growth of heterointerfaces in RTD, growth interruptions (GI) were performed as shown in Fig.1. Previously GI time intervals t_1=0.2 sec and t_2=0.2 sec were used for GaInAs to InP transitions, while t_3=5.0 sec and t_4=0.5 sec were used for InP to GaInAs transitions.[9] To improve interface roughness, initial annealing of the substrate and GI time were investigated.

At the first stage the InP substrate was annealed for 7 min at 600°C under high PH_3 flow of 4.46×10^{-3} mol/min to obtain monolayer steps described in the previous section. Both t_1 and t_3 GI times were increased for effective migration after the deposition of group III atoms to form the atomically flat region. To avoid total pressure fluctuations in the reactor when the exchange of AsH_3 and PH_3 was performed, the same flow rate for group V gases with ambient H_2 was carried out.

For AFM observations of the heterointerface, the growth was interrupted just after the interval t_3 and the sample was cooled down from the growth temperature to room temperature with AsH_3 atmosphere. Figure 3 (a) shows the AFM image of an 8.0 nm thick GaInAs surface with t_1=t_3=7 min. The growth temperature was 600°C. Atomically flat terraces with monolayer step structures were isotropically formed perpendicular to the [0$\bar{1}\bar{1}$] direction. Terrace width distribution is shown in Fig.3 (b). Terrace widths were examined along the [0$\bar{1}\bar{1}$] direction in 5μm ×5μm area. Wide terraces 300 – 450 nm width were obtained at GaInAs surface on InP. The misorientation angle of the substrate was estimated by AFM as 0.04° – 0.05° towards the [0$\bar{1}\bar{1}$] direction. This result is consistent the misorientation magnitude calculated from the peak value of terrace width distribution.

5. Conclusions

OMVPE grown GaInAs/InP heterointerface roughness is studied by AFM. In order to obtain atomically flat regions on initial InP substrates during the annealing process, a temperature of around 600°C is found to be critical under high flow rate of PH_3. Atomically flat terraces of 300 – 450 nm width were achieved on the surface of a GaInAs quantum well layer on InP with optimized GI times. Isotropic steps were oriented perpendicular to the [0$\bar{1}\bar{1}$] direction. Using the formation of atomically flat heterointerfaces, reduction of the quantum well width fluctuation is expected to be useful for the estimation of the intrinsic phase coherent length of hot electrons in semiconductors using RTDs.

Acknowledgments

The authors thank Professor S. Arai and Associate Professor M. Asada for fruitful discussions.

References

[1] K. Furuya, it J. Appl. Phys. **62** (1987) 1492.

194

[2] Y. C. Kang, K. Furuya, M. Suhara and Y. Miyamoto, *Jpn. J. Appl. Phys.*, (1994) 6491.

[3] Y. C. Kang, M. Suhara, K. Furuya and R. Koizumi, *Jpn. J. Appl. Phys.*, (1995) in print.

[4] T.Fukui and H.Saito: *Jpn.J.Appl.Phys.* 29 (1990) 483

[5] M. Kasu and N. Kobayashi, *J.Cryst.Growth*, 145 (1994) 120.

[6] M. Shinohara, M. Tanimoto, H. Yokoyama, and N. Inoue, *Appl.Phys.Lett.* **65**, (1994) 1418.

[7] T.Y. Wang, K.L. Fly, A. Persson, E.H. Reihlen, and G.B. Stringfellow, *J.Appl.Phys.* **63**, (1988) 2674.

[8] J. E. Epler, J. Söchtig, and H.C. Sigg, *Appl.Phys.Lett.* **65**, (1994) 1949.

[9] T. Sekiguchi, Y. Miyamoto and K. Furuya, *J.Cryst. Growth*, **124**, (1992) 807.

Inst. Phys. Conf. Ser. No 145: Chapter 2
Paper presented at 22nd Int. Symp. Compound Semiconductors, Cheju Island, Korea, 28 August–2 September 1995
© *1996 IOP Publishing Ltd*

Properties of GaInP-based Materials for UHB-LED Production grown in Multiwafer Planetary Reactors

R. Beccard, J. Knauf, G. Lengeling, D. Schmitz and H. Jürgensen

AIXTRON GmbH, Kackertstrasse 15-17, D-52072 Aachen, Germany

Abstract

Al(GaInP) is the basis material for the production of ultra high brightness LEDs. Since these devices are fabricated in mass production processes, it is necessary to perform the epitaxial growth in multiwafer machines. The requirements for such a machine are high throughput, high efficiency and a good uniformity of the grown structures. These requirements are met by AIXTRON Planetary Reactors®. In this study we discuss results obtained in an AIX 2400 machine used for the growth of GaInP based materials. In this reactor a total wafer area of more than 400 cm^2 can be grown in one run. GaInP has been grown both with phosphine and tertiary-butyl-phosphine as a group V precursor. The crystalline quality is demonstrated by DCXD half widths for GaInP layers similar to those of GaAs substrates. Room temperature PL mapping demonstrate the wavelength uniformity better 1 nm on a 2" GaInP wafer. Similar results were obtained for AlGaInP. GaInP grown using tertiarybutylphosphine show that V/III ratios as low as 26 can be used this way without affecting the quality of the GaInP films. Due to the high cracking efficiency for TBP growth could be performed at very low temperatures. Down to 615°C diffusion-controlled growth was found.

1. Introduction

One of the most important mass products made of III-V compounds is the light emitting diode (LED). Besides the standard LEDs fabricated in the GaAsP and AlGaAs materials systems since decades, there is a strong interest today in ultra high brightness LEDs (UHB-LEDs). These LEDs consist of rather sophisticated multilayer structures in the AlGaInP materials system. AlGaInP can be grown lattice matched to GaAs and has a direct energy gap between 1.9 eV and 2.3 eV at room temperature. The growth method which is employed for this purpose is metalorganic vapour phase epitaxy (MOVPE). Since these LEDs are real mass products, only state-of-the-art multiwafer MOVPE systems meet the requirements of the device manufacturers. These requirements are: high throughput, high growth efficiency and excellent uniformity of growth rate and composition.

2. Experimental

In this paper we describe the use of AIXTRON Planetary Reactors for the growth of (Al)GaInP. These reactors offer real multiwafer capability and a high degree of flexibility. Different types of reactors, all based on the same concept, are available. This way either 7x2" wafers (AIX 2000), 5x4" or 15x2"(AIX 2400) or 92x2" or 20x4" (AIX 3000) can be grown simultaneously in one run. All these reactors are based on Frijlink's concept [1]. Their main features are a double gas foil rotation, a fast vent/run switching manifold with zero dead volume and a precise PLC control of the entire process. The reactors make use of a controlled depletion of the gas phase allowing to combine outstanding uniformity and high growth efficiency (40-50% for the group III precursor). All experiments described in this paper were obtained on an AIX 2400 reactor using a 15x2" susceptor setup.

As group III precursors trimethylgallium, trimethylindium and trimethylaluminium were used. The group V precursors were undiluted AsH_3 and PH_3 and additionally tertiary-butyl-phosphine (TBP). Substrates were undoped and n-type GaAs wafers (2" and 4" diameter). Typical growth parameters were: a reactor pressure of 50 mbar, deposition temperature of 700 - 750 °C and a total flow rate of 17.2 l/min.

The thickness of the films was measured after stain etching using a microscope and in some cases additionally using DEKTAK. For DCXD measurements a Bede QC1 diffractometer was used. PL measurements were carried out at ISI, KfA Jülich and at Institut für Halbleitertechnik, RWTH Aachen.

3. Results

3.1 GaInP

The first part of the investigation was to find optimum growth temperatures. For this purpose we focused on GaInP since this is the base material for all the other compounds required for a LED. GaInP can be grown diffusion-controlled in a wide range of temperatures from 660 to 780 °C (Fig. 1). By using TBP instead of PH_3 even lower growth temperatures could be used together with reduced V/III ratios. The optimum growth temperatures for GaInP using PH_3 are in the range between 700 and 750 °C. In this temperature regime best morphologies and highest PL intensities were found. Using TBP, similar results were obtained at temperatures as low as 675 °C. Diffusion controlled growth was observed down to 620 °C. A more detailed comparison between PH3 and TBP-grown GaInP is given elsewhere [2]

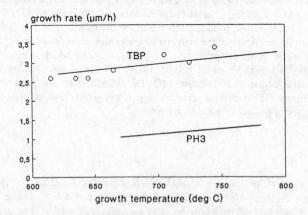

Fig. 1: Growth rate vs. growth temperature using PH_3 and TBP. Different growth rates are due to different parameter sets

The effect of the growth temperature on the GaInP composition is also rather predictable (Fig. 2). The lattice mismatch decreases monotonously by approx. 3000 ppm when increasing the growth temperature from 650 °C to 670 °C. This means that the indium content decreases slightly towards higher temperatures.

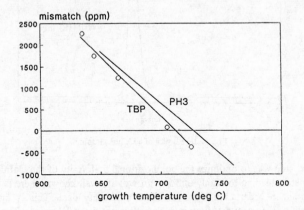

Fig. 2: Lattice mismatch of GaInP on GaAs vs. growth temperature using PH₃ and TBP

The high crystalline quality of the GaInP is demonstrated by DCXD (fig. 3). The half widths of the layer peaks are comparable with the GaAs substrate peaks (reference crystal was InP). The high quality is confirmed by low temperature PL. Fig. 4 shows a 2K PL plot of a GaInP sample revealing a FWHM of only 13 meV.

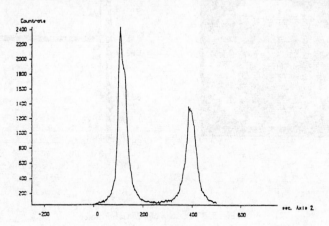

Fig. 3: DCXD plot of a GaInP sample

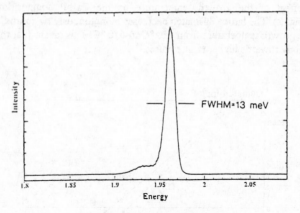

Fig. 4: 2K PL plot of a GaInP sample

The next step was to optimize the compositional uniformity of the GaInP layers. Since the basic parameters leading to highly uniform growth are well known, there is just some fine tuning to be done in order to achieve excellent uniformity data. This is mainly done by adjusting the group III to group V flow ratio (which is not the V/III ratio!). The result of such an optimization is demonstrated in fig. 5. The room temperature PL map reveals a mean wavelength of 679 nm and a standard deviation of only 0.3 nm.

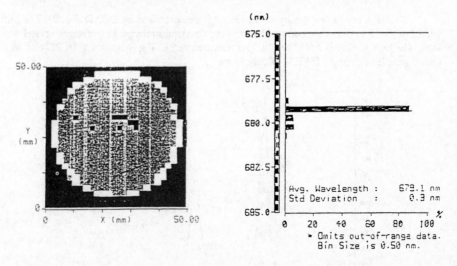

Fig. 5: RT PL map of a 2" GaInP wafer

3.2 AlGaInP and AlInP

For the growth of a complete UHB LED structure different $(Al_xGa_{1-x})InP$ alloys are required. For the active region AlGaInP with a low Al content x is needed. This material must be grown with a uniform composition in order to minimize fluctuations of the emission wavelength. In fig. 6 a room temperature PL map of an $(Al_{0.2}Ga_{0.8})InP$ layer is shown. It reveals a mean wavelength of 620 nm and a standard deviation of only 1.3 nm.

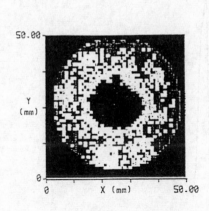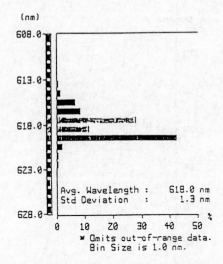

Fig. 6: RT PL map of a 2" $(Al_{0.2}Ga_{0.8})InP$ wafer

Furthermore, AlGaInP with a high Al content is needed. Above x=0.65 the material has an indirect bandgap. Thus PL cannot be used any more to evaluate the material quality. However, DCXD still gives some indication that the crystalline quality is not altered when the Al content is increased. In fig. 7 a DCXD plot of an $(Al_{0.7}Ga_{0.3})InP$ layer is shown revealing a sharp peak. To cover the entire range of AlGaInP alloys, pure AlInP was finally grown. For this material again a good crystalline quality was demonstrated by DCXD (Fig. 8).

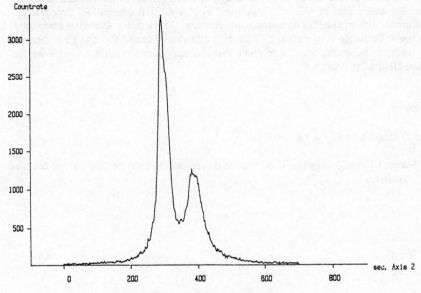

Fig. 7: DCXD plot of an $(Al_{0.7}Ga_{0.3})InP$ layer

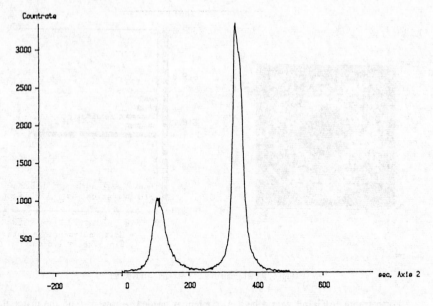

Fig. 8: DCXD plot of an AlInP layer

4. Conclusion

The results described in this paper demonstrate that MOVPE in Planetary Reactors is a powerful method for the fabrication of ultra high brightness LEDs in the AlGaInP materials system. The entire range of AlGaInP alloys from GaInP to AlInP is covered. All these material can be grown with an excellent compositional uniformity. Wavelength uniformities are in the 1 nm range or below. In combination with the high reproducibility and the high throughput of these reactors, the results once more prove that Planetary Reactors offer a unique way to fabricate UHB LED structures.

References

[1] F.M. Frijlink, *J. Crystal Growth* **93** (1988), 207

[2] R. Beccard, J. Knauf, G. Lengeling, D. Schmitz and H. Jürgensen, *Proc. 7th Int. Conf. InP and Rel. Mat.* (1995), 126

Inst. Phys. Conf. Ser. No 145: Chapter 2
Paper presented at 22nd Int. Symp. Compound Semiconductors, Cheju Island, Korea, 28 August–2 September 1995
© *1996 IOP Publishing Ltd*

Novel Empirical Expression and Kinetic Analysis for the Incorporation of As and P in InGaAsP Grown by Metal-Organic Vapor Phase Epitaxy

Seung Won Lee, Jeong Soo Kim, Hyung Mun Kim, Hong Man Kim, Kwang Eui Pyun and Hyung Moo Park

Optoelectronics Section , Electronics and Telecommunications Research Institute Yusong P.O. Box 106, Taejon, Korea

Abstract. We present the novel empirical expression for As and P incorporation properties for MOVPE grown InGaAsP. From the MOVPE grown InGaAsP and InAsP composition data measured by X-ray diffraction, photoluminescence, and Auger electron spectroscopy, we deduce the linear relation between the relative distribution coefficient and the solid composition of InGaAsP, and constant $k_{As/P}$ for InAsP. The relation between the relative distribution coefficient of two compete component As and P, i.e., $k_{As/P}$ and solid composition can give intuitive information about the incorporation behavior of solid component in InGaAsP. We also investigated the dependence of relative distribution coefficient on growth temperature.

1. Introduction

The quaternary InGaAsP grown by Metal-Organic Vapor Phase Epitaxy (MOVPE) technique is very promising material in long wavelength optoelectronic devices such as laser diode, photodiode, optical modulator and optical switch. For the growth of these epi-layers it is important to control the mixed group V component in $In_{1-x}Ga_xAs_yP_{1-y}$. There have been some studies in incorporation properties of InGaAsP grown by MOVPE with thermodynamic analysis [1] [2] kinetic analysis [3], and also grown by Gas Source Molecular Beam Epitaxy (GS-MBE) with kinetic analysis [4] [5]. However up to now, these studies show the nonlinear relation between vapor and solid compositions in group V component. So this relation is inconvenient to estimate the first trial growth conditions for unknown composition materials.

In this report we present the novel expression for As and P incorporation in InGaAsP. The relation between the relative distribution coefficient of two compete component As and P, i.e., $k_{As/P}$ and solid composition could give intuitive information.

The relative distribution coefficient $k_{As/P}$ is defined as $k_{As/P} = \dfrac{\left(x_{As}/f_{AsH_3}\right)}{\left(x_P/f_{PH_3}\right)}$, which has

derivative characteristics for source incorporation and therefore it could simplify the incorporation behavior of solid component in InGaAsP. This expression shows the linear relation between the relative distribution coefficient and the solid composition of group

V components As and P. This expression also could be useful for analyzing the kinetics of incorporation mechanism, especially multi III and V alloy. We also investigated the dependence of relative distribution coefficient on growth temperature.

2. Experiment

A LP-MOVPE system with rotating vertical reactor was used for the epitaxial growth. The growth were carried out at a reactor pressure of 60 torr, growth temperature of 600, 630 and 660℃, V/III ratio 475~700 and growth rate 0.6~1.2µm/hour. The conventional metal-organic and hydride source such as TMI, TEG, AsH₃ and PH₃ were used as precursors. The composition of InGaAsP and InAsP were determined by double crystal X-ray diffractometry, room temperature photoluminescence and Auger electron spectroscopy.

3. Results and Discussion

First we investigated the incorporation behavior of group III elements. The relation of TEG flow ratio to Ga solid composition is shown in fig. 1 (a) and (b).

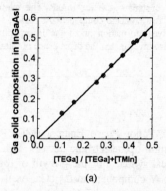
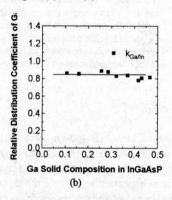

(a)   (b)

Fig.1. The comparison between the conventional expression (a) and the relative distribution coefficient expression (b) which shows the constant relative distribution coefficient $k_{Ga/In}$ and the value is nearly unity.

As shown in fig. 1 (a), the Ga composition is proportional to the TEGa flow ratio. Comparing to conventional expression, the relative distribution coefficient of Ga to In is constant and nearly unity for various Ga composition. The value of unity means that the incorporation probability ratio is one to one for Ga and In.

As already known for the incorporation of group V components, the relation of gas phase ratio to solid composition is nonlinear and the distribution coefficient of As increases with increasing As solid composition in both case of ternary GaAsP, InAsP or quaternary InGaAsP. It is also shown in our experimental data fig.2 (a).

The comparison between conventional expression and relative distribution coefficient expression are shown in fig.2. The conventional relation between group V source flow and As solid composition in InGaAsP and InAsP, which is compared to the relation between relative distribution coefficient of As $k_{As/P}$ and As solid composition. The conventional relation shows nonlinear relation and it is difficult to distinguish the different trend between InGaAsP and InAsP. However the relative distribution coefficient of As $k_{As/P}$ shows constant for ternary InAsP and linear relation for quaternary . It also indicates the group III dependence of As and P incorporation rate explicitly. The group III dependence of As incorporation ratio is expected qualitatively by G. Stringfellow [6] and we present the tool of quantitative analysis by using relative distribution coefficient relation.

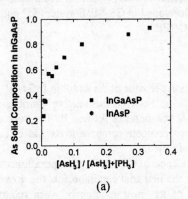

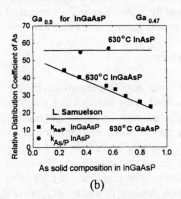

(a) (b)

Fig. 2. The conventional relation between group V source flow and As solid composition in InGaAsP and InAsP, which is compared to the relation between relative distribution coefficient of As $k_{As/P}$ and As solid composition. The conventional relation shows nonlinear relation and it is difficult to find out the different trend between InGaAsP and InAsP. However the relative distribution coefficient of As $k_{As/P}$ shows linear relation and explicitly indicate the group III dependence of As and P incorporation rate.

The temperature dependence of relative distribution coefficient is shown in fig. 3.

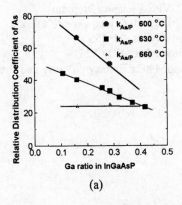

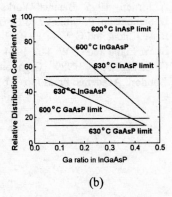

(a) (b)

Fig. 3. From the temperature dependence of $k_{As/P}$ on group III composition data, relative distribution coefficient expression gives more information about the mechanism of adsorption and desorption rate of As and P which is bonded with different group III material, In or Ga atom.

For low growth temperature the relative distribution coefficient changes largely with the change of Ga composition, on the other hand at high temperature the relative distribution coefficient is almost same like InAsP case. Therefore we assume that the tendency to temperature dependence of As incorporation properties could be shown as in fig 3 (b).

From the temperature dependence of $k_{As/P}$ on group III composition data, we suppose that the relative distribution coefficient expression gives more information about the mechanism of adsorption and desorption rate of As and P which is bonded with different group III material, In or Ga atom.

In addition we presume that our empirical expression coincides with B. Liang's kinetic model for mixed group V quaternary InGaAsP developed in GS-MBE system [5] and it proves the applicability of his kinetic model to MOVPE.

4. Summary

The novel empirical expression for the incorporation behavior of As and P in InGaAsP is presented. This expression shows the linear relation between the relative distribution coefficient and the solid composition of group V components As and P. The relation between the relative distribution coefficient of two compete component As and P, i.e., $k_{As/P}$ and solid composition could give intuitive information.

This expression also shows the group III dependence of As and P incorporation rate explicitly. The expression is convenient to obtain the first trial condition for the growth of new ternary or quaternary material because of its linear relation between relative distribution coefficient and solid composition.

From the temperature dependence of $k_{As/P}$ on group III composition data, it could be obtained more information about the adsorption and desorption mechanisms of As and P which are bonded with different group III material, In or Ga atom.

Reference

[1] A. Koukitu and H. Seki   1986   J. Crystal Growth **76** 233
[2] A. Jordan   1995   Biennial Workshop on MOVPE Technical Program
[3] L. Samuelson, P. Omling and H. Grimmeiss   1983   J. Crystal Growth **61** 425
[4] B. W. Liang and C. W. Tu   1993   J. Applied Physics **74** 255
[5] B. W. Liang and C. W. Tu   1994   J. Electronic Material **23** 1251
[6] G. Stringfellow   1989   Organometalic Vapor-Phase Epitaxy : Theory and Practice (London: Academic Press Inc.) p 324

Inst. Phys. Conf. Ser. No 145: Chapter 2
Paper presented at 22nd Int. Symp. Compound Semiconductors, Cheju Island, Korea, 28 August–2 September 1995
© *1996 IOP Publishing Ltd*

Study on HgCdTe/CdZnTe grown by Isothermal Vapor Phase Epitaxy

T S Lee†, J M Chang†, W S Song†, S U Kim†, M J Park†, Y T Jeoung‡, S M Park‡, H K Kim‡, J M Kim‡ and S H Kim §

†Korea University, Anamdong 5-1, Seongbugku, Seoul, Korea

‡Agency for Defense Development, Youseong P.O. Box 35, Taejeon 305-604, Korea

§Acquisition Policy Bureau, Ministry of National Defense, Seoul, Korea

Abstract. $Hg_{1-x}Cd_xTe$ films($x=0.3$) have been grown on CdZnTe substrates by isothermal vapor phase epitaxy. The size of HgCdTe epilayer is in the range of 50 ~500 mm^2 in area. The composition uniformity measured by infrared spectrometer is ±0.003 in x value across the entire surface. HgCdTe epilayers annealed under Hg pressure around 370°C are p-type with a carrier concentration of mid 10^{15} to low 10^{16} cm^{-3} at 77 K. The p-n junction diodes have been fabricated on p-HgCdTe epilayers. R_0A obtained from I-V curves of the diodes is in the range of 10^3~10^4 ohm-cm^2 at 77 K.

1. Introduction

$Hg_{1-x}Cd_xTe$ is a compound semiconductor with bandgap variable from 0.0 to 1.6 eV as x increases from 0.17 to 1.0. Especially, $Hg_{1-x}Cd_xTe$ alloys with $x=0.2$ and 0.3 that have bandgaps of 0.1 and 0.24 eV, respectively, are required to detect the wavelengths of 8-12 and 3-5 μm. During the past several decades, therefore, HgCdTe has found widespread applications as infrared detectors, which are very useful in a variety of military, space, medical and industrial systems.

Using this material photoconductive(PC) and photovoltaic(PV) infrared detectors have been developed, and for better resolution the number of array elements has been increased. The second generation of HgCdTe detectors is photovoltaic with focal plane arrays(FPA) up to 640x480 elements[1]. High density arrays such as FPA need large-area HgCdTe wafer with good uniformity in properties, especially in composition.

In this work we report on isothermal vapor phase epitaxial(ISOVPE) growth of large-area HgCdTe epilayers on lattice-matched CdZnTe substrates, electrical property control by

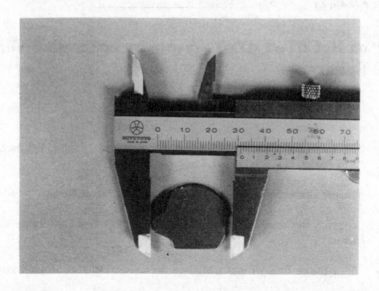

Figure 1. The photograph of HgCdTe wafer grown on (110) CdZnTe substrate by ISOVPE.

Hg vacancy annealing, and current-voltage characteristics after fabrication of p-n junction diodes.

2. Experimental

CdZnTe boules for substrate have been grown by vertical Bridgman method. The CdZnTe substrates with mirror-like surface are prepared by mechanical and chemical polishing. The $Hg_{0.368}Cd_{0.082}Te_{0.550}$ source for ISOVPE growth was prepared by evacuating and sealing the quartz ampoule containing the elements(6N purity) under 10^{-3} Torr. The source compound are synthesized at 450°C for 24 hrs and heated for complete mixing of the melt at 650°C for 10 hrs followed by quenching in water to avoid a segregation.

The vertical loading type ampoule is used for the ISOVPE growth to obtain better lateral compositional uniformity. HgCdTe layers are then grown on (110) or (111) CdZnTe substrates with various sizes ranging 50 to 700 mm^2. The growth temperature and time to obtain 40~50 μm thick epilayers are 550°C and 48 hrs, respectively. The largest HgCdTe wafer grown on (110) CdZnTe substrate is shown in figure 1.

3. Results and discussion

The surface and the cross-section of the as-grown HgCdTe epilayer have been observed with optical microscope for the examination of morphology and the epilayer thickness. The HgCdTe epilayer grown on (111) substrate has mirror-like surface and one grown on (110) substrate is, however, less shiny. The surface becomes more mirror-like as the deviation from [111] decreases and the best morphology is obtained on (111)A±0.5° substrate.

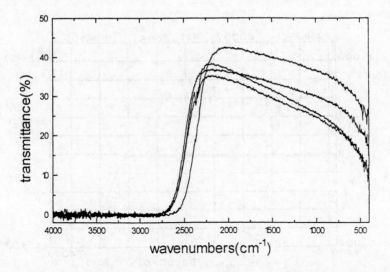

Figure 2, IR transmission curves over the entire surface of HgCdTe wafer after Hg vacancy annealing. The beam size is 1x1 mm^2 and the distance of each beam is 5 mm.

The depth profile of composition is measured by electron probe microanalysis(EPMA) in cross-section view. The typical composition depth profile through the thickness of HgCdTe epilayer is flat near the surface with a deviation of ±0.005 in x value within 10 μm of the surface. The lateral variation of composition over the whole epilayer has been measured by infrared spectrometer using an aperture of 1x1 mm^2 (figure 2). The lateral uniformity of composition is found to be very excellent for the HgCdTe epilayer obtained with vertical loading type; the deviation is ±0.003 in x value across the entire surface of 26 mm in diameter.

The HgCdTe epilayers are annealed under Hg pressure at various temperatures for carrier concentration control. The electrical properties of as-grown and annealed HgCdTe epilayers have been determined using van der Pauw-Hall measurements at 77 K. The as-grown HgCdTe epilayer is p-type and its carrier concentration and mobility are low 10^{17} cm^{-3} and 114 cm^2/Vs, respectively, at 77 K. The epilayers after Hg vacancy annealing around 370°C are p-type with a concentration of mid 10^{15} to low 10^{16} cm^{-3} and a mobility of 200 to 300 cm^2/Vs at 77 K (table 1).

Table 1. Electrical properties of ISOVPE grown Hg$_{0.7}$Cd$_{0.3}$Te epilayers.

Annealing Temperature(°C) / Time(hrs)	Conduction Type (at 77 K)	Concentration (cm^{-3}, at 77 K)	Mobility (cm^2/Vs, at 77 K)
as-grown	p	1.5×10^{17}	114
380 / 3	p	2.4×10^{16}	85
370 / 5	p	7.9×10^{15}	253
360 / 6	n	1.3×10^{16}	579
350 / 8	n	2.2×10^{15}	11161
310 / 10	n	2.3×10^{15}	14175

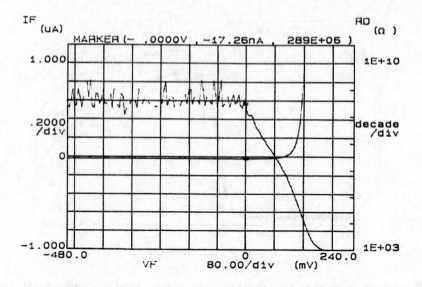

Figure 3. A typical I-V curve of p-n diode obtained by In diffusion.

The p-n junction diodes are fabricated on the p-$Hg_{0.7}Cd_{0.3}Te$ epilayers of $4x10^{15}$ and $1x10^{16}$ cm^{-3} in carrier concentration, which is required to obtain high R_0A products[2]. A typical I-V curve of the diode is shown in figure 3. R_0A of the diodes obtained from the I-V curve and junction area is found to be in the range of 10^3~10^4 ohm-cm^2 at 77 K, which satisfies the condition for back ground limited infrared photodetector(BLIP).

4. Conclusion

In summary, we have grown large area $Hg_{0.7}Cd_{0.3}Te$ films(up to 530 mm^2) on CdZnTe substrates by ISOVPE technique. The most mirror-like surface is obtained on (111)A CdZnTe substrate. The lateral uniformity of composition measured by infrared spectrometer is ±0.003 in x value across the entire surface of HgCdTe epilayer(26 mm in diameter). The hole concentration of the epilayer is in the range of mid 10^{15} to low 10^{16} cm^{-3} at 77 K after Hg vacancy annealing around 370 °C. R_0A of the diodes fabricated by In diffusion is in the range of 10^3~10^4 ohm-cm^2 at 77 K, which satisfies the condition for back ground limited infrared photodetector(BLIP).

References

[1] Hewish M 1993 *International Defense Review* **4** 306-311

[2] Rogalski A 1988 *Infrared Phys.* **28**(3) 139-153

Inst. Phys. Conf. Ser. No 145: Chapter 2
Paper presented at 22nd Int. Symp. Compound Semiconductors, Cheju Island, Korea, 28 August–2 September 1995
ⓒ *1996 IOP Publishing Ltd*

Epitaxial Growth of 3C-SiC(111) Thin Film on Si Wafer by Rapid Thermal Chemical Vapor Deposition using Tetramethylsilane

Y. H. Seo, K. S. Nahm [*], E.-K. Suh [1], Y. H. Lee [1], H. J. Lee [1] and Y. G. Hwang [2]

Department of Chemical Technology, [1] Department of Physics and Semiconductor Physics Research Center, Chonbuk National University, Chonju 560-756, Republic of Korea
[2] Department of Physics, Wonkwang University, Iri 570-749, Republic of Korea

Abstract. We have used rapid thermal chemical vapor deposition (RTCVD) technique to grow epitaxial SiC thin films on Si by pyrolyzing tetramethylsilane (TMS). The growth rate of the SiC film increases with the TMS flow rate and temperature, but it decreases with temperature in the case of higher TMS flow rate. The XRD spectra of the films indicate that the growth direction is along the (111) direction of β-SiC. IR and RBS measurements have been employed to analyze the chemical composition of the films. At 1100°C TMS is almost completely dissociated into Si atoms, CH_4 and C_2H_2 gases. The growth mechanism of the SiC films has been proposed based on the analyses of TEM and QMS.

1. Introduction

Silicon Carbide(SiC) has been drawing much attention as an attractive candidate for the material for the fabrication of electronic devices operating at extreme conditions and blue LED's, owing to its wide band gap, high breakdown voltage, high thermal conductivity, high saturated electron velocity, and good chemical stability. But the development of such devices using SiC has been delayed because single-crystal growth of SiC is technically difficult since it requires extremely high growth temperature. [1]

Recently it had been reported that the high quality SiC film can be grown on Si substrates at relatively low temperatures by pyrolyzing a single organosilane precursor which contains bonds between Si and C atoms, such as methylsilane (CH_3SiH_3), methyltrichlorosilane (CH_3SiCl_3), dimethyldichlorosilane ((CH_3)$_2SiCl_2$), and hexamethyldisilane ((CH_3)$_6Si_2$). [2] Most of the works up to date have focused on the characterization of the grown SiC films, and little attention had been paid to the growth kinetics of SiC films. [3]

In this work, we have used RTCVD technique, capable of an abrupt temperature control, to grow epitaxial β-SiC thin films on Si by pyrolyzing TMS. The structure and chemical composition of SiC films are investigated as a function of the growth condition. Gaseous species in the RTCVD reactor has been also analyzed with a quadrupole mass spectrometer (QMS, Hiden HAL/3F 501) during the deposition. The growth kinetics of SiC films is discussed in detail, based on experimental results.

2. Experimental

The RTCVD reactor, shown in Fig. 1, was specially designed and fabricated for the growth of SiC thin films. High–power tungsten halogen lamps were used to heat a Si substrate. In order to maximize the heating effect, the light of the lamps was focused on the Si substrate in the reactor using specular surfaces set up on backside of the lamps. The temperature of the substrate was monitored using a Pt/Ru thermocouple directly inserted in the substrate holder. TMS and hydrogen gases were introduced into the reactor using mass flow controller(MFC, MKS 247). Rotary vacuum pump was used to evacuate the reactor up to 10^{-4} Torr. The pressure of the deposition reactor was measured with a convectron gauge.

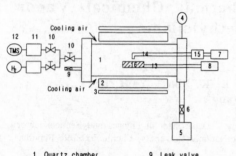

1. Quartz chamber
2. Tungston Halogen lamp
3. Reflector
4. Vacuum guage
5. Rotary pump
6. Gate valve
7. Mass-spectrometer
8. Temp. controller
9. Leak valve
10. Throttle valve
11. MFC
12. Source gas
13. Thermocouple
14. Quartz tube
15. Filter
16. Substrate

Fig. 1. A schematic diagram of rapid thermal CVD apparatus.

P-type Si(111) with the resistivity of 20 Ω·cm was used as a substrate The Si substrate had been cleaned by dipping in acetone and then in HF aqueous solution, and dried with a nitrogen flow, prior to being loaded into the reactor. The cleaned Si substrate was mounted on the substrate holder and kept in H_2 flow at the elevated temperature. The reactor was heated up to the growth temperature at the rate of 20 ℃/s under the flow of hydrogen gas. Experiments of SiC growth had been carried out for 15 minutes at various substrate temperatures (1000~1100 ℃), and H_2 flow rates (20~200 sccm) and TMS flow rates (0.1~5 sccm). After finishing the growth, the reactor was purged with Ar gas, and the sample was taken out of the reactor for characterization.

3. Results and Discussion

3.1. Effects of TMS flow rate and substrate temperature

The growth rate of SiC thin film was measured at varying TMS flow rate (0.1, 0.2, 0.5, 1.0, 2.0 and 5.0 sccm) at the constant temperature of 1100 ℃ and at the constant H_2 flow rate of

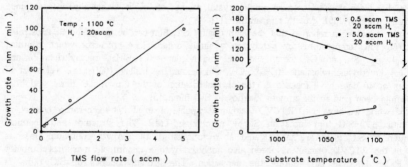

Fig. 2. The growth rates of SiC thin film as functions of (a) TMS flow rate and (b) temperature for 0.5 and 5.0sccm.

20 sccm. Fig. 2(a) shows that the growth rate increases with the TMS flow rate, which is considered to be due to the increase of reactive species for the SiC growth. This will be discussed in more detail later. Fig. 2(b) shows the growth rates of SiC thin films as a function of substrate temperature at the TMS flow rates of 0.5 and 5.0 sccm. With increasing temperature, the growth rate increases for 0.5 sccm, but decreases for 5.0 sccm. The decrease of the growth rate at high TMS flow rate is believed to be caused by the properties of the CVD reactor used in this work. For hot wall type reactor, the reaction at the reactor walls is accelerated with increasing temperature to deplete reactive species on the surface of Si substrate, resulting in the reduction of SiC growth rate.

3.2. The analysis of the structure of SiC films.

XRD is employed to investigate the orientation of the SiC films grown under experimental conditions described in Fig. 2, and the typical XRD spectrum is shown in Fig. 3. It is to be noted from the figure that the films show a preferred orientation of (111) plane of β-SiC. This was also confirmed from Raman spectra for the films, which exhibit TO and LO modes of β -SiC(111) at 795 and 972 RCM^{-1}, respectively. No SiC thin films were deposited on Si substrates below 1000 ℃. It is believed that the degree of TMS decomposition is very low below 1000 ℃

TEM and TED(transmission electron diffraction) measurements have been performed to investigate the structures of the SiC films grown at the pressures of 1.7 Torr and 12 Torr for 15 min at 1100 ℃ when the flow rates of TMS and H_2 are maintained at a constant value of

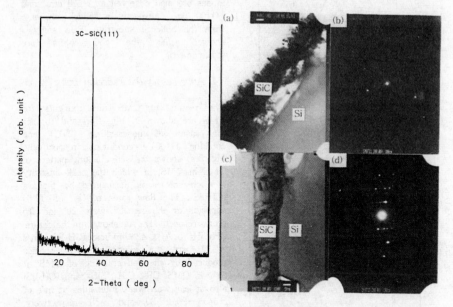

Fig. 3. The typical XRD pattern of 3C-SiC thin films grown for 15 min at 1100℃, 20sccm H_2 and 0.5sccm TMS.

Fig. 4. Cross-sectional TEM micrograph and diffraction patterns of the SiC thin films grown at 1.7Torr ((a) and (b)) and 12Torr ((c) and (d)).

0.5 and 100 sccm, respectively. At 1.7 Torr, a polycrystal SiC film is grown on a Si(111) substrate and large voids are observed in the Si side of the SiC/Si interface. (see Fig. 4(a) and (b)). This void formation is due to the out-diffusion of Si atoms. The same results are obtained from the cross sectional TEM micrographs of the SiC films grown at the H_2 flow rate of 20 sccm at various TMS flow rates between 0.2 and 5.0 sccm and at the substrate temperature between 1000 and 1100 ℃, while maintaining the reactor pressure at a low value of about 2.5 Torr. This indicates that polycrystal SiC films are grown at such a low pressure regardless of the TMS flow rates. On the contrary, at 12 Torr, an intermediate SiC layer of 12 nm thickness is first grown on the Si substrate, and then a single-crystal SiC thin film with twins and defects grows on the layer (see Fig. 4(c) and (d)). Voids are also formed in the Si side at the interface of SiC/Si. As the reactor pressure increases, it is observed that the thickness of the SiC thin film is reduced from 371 to 171 nm and the size of the voids formed in the Si substrate decreases.

3.3. Analyses of the chemical composition of SiC films.

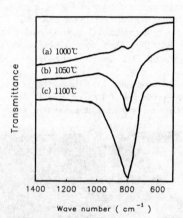

Fig. 5. IR spectra of the SiC thin films grown for 15 min at various temperatures under the condition of 20sccm H_2 and 0.5sccm TMS.

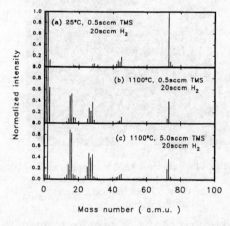

Fig. 6. Quadrupole mass spectra of TMS,

IR was used to analyze the chemical composition of the SiC films. The films were grown at the flow rates of H_2 and TMS of 20 and 0.5 sccm, respectively at various substrate temperatures. Fig. 5 shows that a strong peak of Si–C stretching mode appears at 800 cm^{-1} and the absorption intensity of the peak increases very rapidly with the increasing temperature. Si–Si and Si–O–Si stretching modes are also observed at ~620 cm^{-1} and ~1100 cm^{-1}, respectively, which originate from the bulk Si substrate. RBS analysis showed that the SiC films are stoichiometric.

3.4. Analyses of the reaction gas

The results of QMS measurements for TMS are shown in Fig. 6(a) and (b) for the cases of noncracking (25℃) and cracking (1100℃) conditions, respectively. Fig. 6(a) shows the self-cracking pattern of TMS in QMS, in which the peak intensity of each species is normalized by that of $(CH_3)_3Si$. The flow rates of H_2 and TMS supplied to the reactor were 20 and 0.5 sccm, respectively. As shown in the figure, CH_3SiH_x (m/e = 43~45) and $(CH_3)_3Si$ (m/e = 73) are the major chemical species produced by the self-cracking of TMS at 25℃ in QMS. CH_n, C_2H_n, SiH_n, and $(CH_3)_2Si$ also appear with weak intensities at m/e of ~14, ~26, ~30, and ~58, respectively. This result is in good agreement with that of the previous reports. [4]

Fig, 6(b) indicates the cracking pattern of TMS at 1100 ℃. The peak intensity of each species is normalized by that at 25 ℃. Comparing Fig. 6(a) and (b), remarkable distinctions can be recognized. The peaks related to Si $((CH_3)_3Si$, $(CH_3)_2Si$, CH_3Si, and

SiH$_n$) are disappeared completely at 1100 ℃. According to Avigal and Veintermillas reports [5,6], TMS decomposes into SiH$_n$, CH$_3$, and hydrogen at the SiC growth temperature, and SiH$_n$, in turn, dissociates at this temperature to form Si and hydrogen, and the concentration of the gaseous Si atoms increases with TMS pressure, but the free Si atoms are completely depleted when the SiC decomposition process takes place. This is because SiH$_n$ which is generated by the decomposition of TMS has too short life time to be detected by QMS in this experiment. Meanwhile, the peaks around m/e = 15 and 16, and 25 and 26, which originate from CH$_4$ and C$_2$H$_2$, respectively, increase remarkably at 1100 ℃. Helm and Mark Jr. et. al [7] insisted with the results of their experiment that CH$_3$ radicals generated by thermal decomposition of TMS might united to form ethane but only momentarily, since ethane decomposes rapidly into H$_2$, CH$_4$, C$_2$H$_2$, and C$_2$H$_4$ at this temperature. From our observations, it is plausible that C$_2$H$_6$ generated by the reaction of two methyl radicals is not detected with QMS because it decomposes very rapidly at 1100 ℃ to form mainly CH$_4$ and C$_2$H$_2$.

It is seen from Fig. 6(c) that the tendency of the spectrum obtained at 5 sccm TMS flow rate coincides well with that of Fig. 6(b), but the intensities of the peaks related to hydrocarbon compounds increase very much. This indicates that the amount of gaseous Si atoms and hydrocarbon compounds increase with increasing TMS flow rate.

3.5. Growth Mechanism

According to the analyses using QMS, TMS breaks thermally into gaseous Si atoms and CH$_3$ radicals. CH$_3$ radicals form C$_2$H$_6$ and then quickly breaks into CH$_4$, C$_2$H$_2$, C$_2$H$_4$ and hydrogen. TEM analyses revealed that voids are formed in the Si substrate regardless of experimental conditions, though their sizes are different. Also it was revealed that the intermediate SiC layers above the void are reduced to 18 nm thick when the total pressure increases. These experimental results suggest that the growth reaction occurs in two stages when SiC layer is grown on Si substrates using TMS/H$_2$ without carbonization process: (1) Si atoms on the surface of substrates react with C-containing gases generated by thermal cracking of TMS and form a C-terminated carbonized layer, and (2) gaseous Si atoms and C-containing gases alternatively reacts to form a single crystalline β-SiC layer on top of the carbonized layer.

Density of C atoms is 4 times higher than that of Si atoms in the reaction chamber since TMS breaks into 1 gaseous Si atom and 4 CH$_3$ radicals. These abundant CH$_3$ radicals accelerate the carbonization of surface Si atoms, leading to the depletion of Si atoms from the substrate surface. This depletion causes the outdiffuion of Si atoms from the inside of Si substrate to the surface to maintain stoichiometry as observed by RBS measurements because Si atoms supplied from the cracked TMS are small. The outdiffused Si atoms further react with adsorbed hydrocarbon radicals to form an intermediate SiC layers. When the gas pressure of the reactor is low during this process, the outdiffusion rate of Si atoms is fast due to the high vapor pressure of silicon at low pressure, and the growth reaction at the surface is fast, resulting in the formation of thick intermediate SiC layer. But the crystallinity of thus formed SiC layer is bad, since supply rates of Si and C atoms are higher than the migration rates of the atoms on the surface. Conversely, when the gas pressure of the reaction chamber is high, outdiffusion of Si atoms from the bulk of substrates is difficult and hence even thin intermediate SiC films prevent the outdiffusion of Si atoms from the substrate. Therefore, the higher the gas pressure of the chamber is, the thinner the intermediate SiC layer formed over voids becomes.

Once the intermediate SiC layer becomes thick enough for a given gas pressure to hinder the outdiffusion of Si atoms from the Si substrate, as explained above, the main source of Si atoms needed for the SiC growth reaction is gaseous Si atoms generated by the decomposition of TMS. The gaseous Si atoms adsorb on the C-terminated surface of the intermediate SiC layer and react to form Si-terminated surface. However, the reaction rate is very slow due to the small amount of Si atoms supplied from thermally cracked TMS. Then C-containing gas, for example C$_2$H$_2$, adsorbs on the Si-terminated surface and dissociates into

C atoms and hydrogen gas. C atoms thus formed reacts with surface Si atoms to form C-terminated surface and the reaction continues. In this way, a single crystalline β-SiC layer is grown on the intermediate layer since the amount of Si atoms supplied from TMS is small and then migration times of Si and C atoms are long enough.

At high total pressures, consequently, SiC films with better crystallinity are grown because the supply rate of Si atoms is low since Si atoms are provided only from the dissociation of TMS, but SiC films grown at low pressures have worse crystallinity because a large amount of Si atoms are supplied from the cracked TMS and the bulk Si substrate.

Voids are formed in the place where Si atoms diffuse out to the substrate surface, near the SiC/Si interface regardless of the gas pressure. They are similar in shape as those formed during the carbonization of Si substrates using hydrocarbons such as CH_4 and C_2H_2.[18] The typical size of these voids decreases as the gas pressure increases, as can be seen in Fig. 5. This is believed to be due to the reduction of the vapor pressure of silicon at higher gas pressures, and hence the reduction of the amount of Si atoms outdiffusing from substrate.

4. Conclusions

RTCVD reactor is used to grow epitaxial SiC thin film on Si at various substrate temperatures (1000~1100 ℃), and H_2 flow rates (20~200 sccm) and TMS flow rates (0.1~5 sccm). The grown SiC films show a orientation of (111) plane of β-SiC and free Si and C was not observed in the films. The QMS analysis indicates that TMS is almost completely decomposed at the SiC growth temperature into gaseous Si atoms and CH_3 radicals. Two methyl radicals are united to form C_2H_6, which is dissociated rapidly at this temperature to form mainly CH_4 and C_2H_2. The formation of the voids are observed in the Si side of the SiC/Si interface and the size of the voids decreases with increasing total pressure. The experimental results suggest that the growth reaction occurs in two stages when SiC layer is grown on Si substrate using TMS/H_2 without carbonization process: (1) Si atoms on the surface of substrates reacts with C atoms thermally broken from TMS and form a C-terminated carbonized Si-layer and (2) gaseous Si and C atoms alternatively reacts to form SiC layers on top of the carbonized layer. Enough migration time of the absorbed Si atoms and C-containing species are necessary to grow a single crystal β-SiC(111).

Acknowlegments

This work was supported by Korea Science and Engineering Foundation(KOSEF) through the Semiconductor Physics Research Center at Chonbuk National University.

References

[1] Davis R F, Kelner G, Shur M, Palmour J W and Edmond J A 1991
    Proceedings of the IEEE **79** 677
[2] Steckl A J, Yuan C and Li J P 1993 Appl. Phys. Lett. **63** 3347
[3] Chiu C C, Desu S B and Tai C Y 1993 J. Mater. Res. **8** 2617
[4] Stenhagen E, Abrahamsson S and McLafferty F 1969 *Atlas of mass spectral data*
    (New York: John wiley & Sons)
[5] Avigal Y, Schieber M and Levin R 1974 J. Cryst. Growth **24/25** 188
[6] Veintemillas S, Madigou V, Rodriguez-Clemente R and Figueras A 1995
    J. Crystal Growth **148** 383
[7] Helm D F and Mark E Jr. 1937 J. Am. Chem. Soc. **59** 60
[8] Figueras A, Garelik S, Rodriguez-Clemente R, Armas B, Combescure C
    and Dupuy C 1991 J. Crystal growth **110** 528

Inst. Phys. Conf. Ser. No 145: Chapter 2
Paper presented at 22nd Int. Symp. Compound Semiconductors, Cheju Island, Korea, 28 August–2 September 1995
© *1996 IOP Publishing Ltd*

215

GaN films prepared by hot-wall epitaxy

E.Yamamoto, K.Ishino, M.Ohta[*], M.Kuwabara[*],
S.Sakakibara[**], A.Ishida, and H.Fujiyasu

Faculty of Engineering, Shizuoka University, 3-5-1
Johoku, Hamamatsu 432, Japan

*Hamamatsu Photonics K.K., 5000 Hirakuchi,
Hamakita 434, Japan
**YAMAHA, Matsunokijima, Toyooka-mura, Iwata-gun
438-01, Japan

Abstract. GaN films were prepared by hot-wall epitaxy
(HWE) for the first time. Metallic Ga and NH_3 were used
as sources and sapphire(0001) as substrate. GaN epitaxial films with optically flat surfaces were obtained by
growing GaN films on GaN initial layers prepared by Ga
predeposition on nitrided sapphire substrates and
subsequent nitridation. Formation of epitaxial and
relatively smooth GaN initial layer was confirmed
through reflection high energy electron diffraction
(RHEED) and atomic force microscope (AFM) measurements.
In this paper, growth and characteristics of the GaN
films prepared by the HWE are described.

1.Introduction

GaN related compounds (AlN, GaN, InN, and their alloys) are
promising material for light emitting diodes (LEDs) and laser
diodes (LDs) because of their high emission efficiency due to
direct band gaps. GaN films have been prepared by metalorganic
chemical vapor deposition (MOCVD) or molecular beam epitaxy
(MBE) and so on. However it is difficult to obtain high quality GaN films with optically flat surface because of their
large lattice mismatch and the difference in thermal expansion
coefficients between GaN and the sapphire substrate, usually
used in the GaN growth. Amano et al. [1],[2] and Akasaki et
al. [3] obtained high quality GaN films using AlN buffer
layers grown at a temperature much lower than that of GaN
growth. Nakamura et al. succeeded in obtaining high quality
GaN films using GaN buffer layers grown also at low temperature [4]. Recent success in preparing high quality GaN films
by MOCVD realized bright LEDs in blue and blue green region
[5]. In this paper, preparation of GaN films by hot wall
epitaxy (HWE) is reported. HWE is useful method to prepare
semiconductor films and superlattices, and optical devices and

various superlattices have been prepared, especially in II–VI (e.g. [6],[7]) and IV–VI (e.g. [8]) compound semiconductors.

2. GaN growth by HWE

Schematic diagram of HWE apparatus for GaN growth are shown in Fig.1. GaN films have been grown on sapphire (0001) substrate by HWE using metallic Ga and NH_3 gas as sources. They are put into the vacuum chamber with the background pressure of 1×10^{-6} torr. Two hot wall furnaces (HW–I and HW–II) were used to prepare GaN films. HW–I is for GaN growth and Ga predeposition. Ga was evaporated at about 870°C during the processes and heated wall was kept at the same temperature. HW–II is used for nitridation of the sapphire substrate and the predeposited Ga layer. The flow rate of NH_3 gas was 10sccm during both nitridation and the GaN growth. First, sapphire substrate was moved on the outlet of HW–II furnace and the

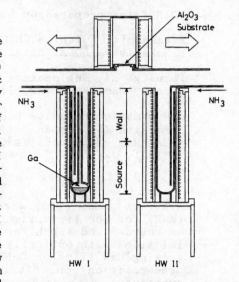

Fig.1. Schematic HWE apparatus used to prepare GaN films.

substrate was nitrided at 1000°C under the NH_3 gas flow for 30 minutes. Then the substrate temperature was lowered to 550°C and the substrate was moved on the outlet of HW–I furnace, where Ga layer was deposited to 100–200Å without introducing NH_3 gas. Next the substrate was moved above HW–II again and the substrate temperature was heated up to 900°C in the NH_3 gas and the predeposited Ga layer was reformed to thin GaN film. The thin GaN film is referred to be GaN initial layer in this paper. After these processes the substrate was moved above HW–I to grow GaN film. The growth rate was 1–3Å/s, the total thickness of the films was about 2 μm. GaN growth without these process was also carried out for comparison.

3. Growth characteristics and effects of initial layers

Scanning electron microscope (SEM) observation and reflection high energy electron diffraction (RHEED) measurement were performed for GaN films. Fig.2 (a)–(c) shows surface morphology and RHEED patterns of GaN films grown directly on sapphire (0001) substrates, grown on Ga predeposited layer, and grown on GaN initial layer, respectively. For the direct growth of the GaN, substrate temperature during growth was 800°C because no film growth was occurred at the temperature higher than 900°C due to difficulty in nucleation of GaN. For

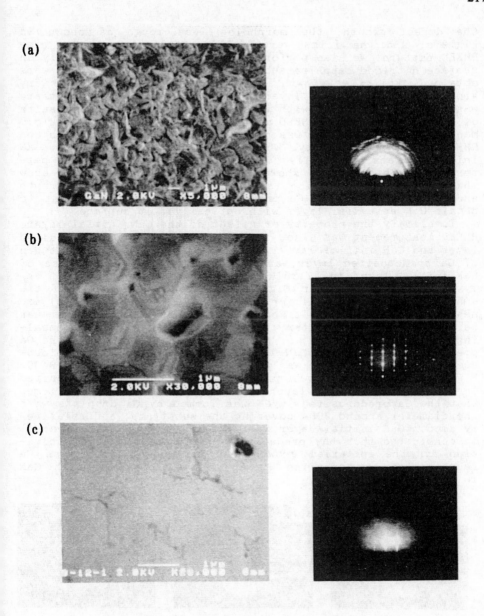

Fig.2. SEM and RHEED photographs of GaN films:
    (a) GaN film grown directly on sapphire substrate.
    (b) GaN film grown on Ga predeposited layer.
    (c) GaN film grown on GaN initial layer.
    Incident direction of electron beam in RHEED is [11$\bar{2}$0].

the direct growth, the morphology was rough as shown in Fig.2(a) and the films were polycrystalline from the ring RHEED pattern. We also performed to grow GaN films directly on surface nitrided sapphire substrates, but similar results were obtained. Epitaxial growth was obtained when the films were grown on Ga redeposited layer, where Ga predeposited substrate was heated up to 900 °C with the supply of Ga and NH_3. Results of the morphology and RHEED measurement are shown in Fig.2(b). Many hexagonal pyramids was observed in the SEM photograph and RHEED pattern was spotty. When GaN films were grown on GaN initial layers, optically flat films with streak RHEED patterns were obtained, as shown in Fig.2(c). These results show that the Ga predeposition is useful for GaN epitaxial growth and nitridation of the Ga predeposited layer is important to obtain GaN epitaxial layer with optically flat surface.

To clarify the property or effect of the GaN initial layer, RHEED measurement was also performed for the Ga predeposited layer and GaN initial layer. Fig.3(a) shows the RHEED pattern of Ga predeposited layer, and Fig.3(b) and (c) show those of GaN initial layer in [11$\bar{2}$0] and [10$\bar{1}$0] incidence, respectively. The halo pattern in Fig.3(a) shows the existence of the amorphous or liquid Ga layer. Patterns in Fig.3(b) and (c) are nearly streak. The result shows that predeposited Ga layer was reformed to the GaN epitaxial initial layer by thermal annealing in the NH_3 gas flow, or nitridation. To investigate the Ga predeposited layer and GaN initial layer, atomic force microscope (AFM) measurement was performed using Shimadzu WET-9400. Fig.4(a) and (b) show AFM photographs of the Ga predeposited layer and GaN initial layer, respectively. It was confirmed that the Ga predeposited layer was formed by Ga droplets with the diameter around 300Å covering the substrate, and relatively smooth GaN initial layer was formed by the nitridation. It is considered that the predeposited Ga droplets gives nucleation for the epitaxial growth and, moreover, the excess Ga reevaporates as annealing in NH_3 flow, forming smooth GaN initial layer.

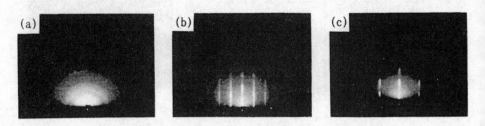

(a) (b) (c)

Fig.3. RHEED patterns of (a) Ga predeposited layer,
(b) GaN initial layer in [11$\bar{2}$0] incidence, and
(c) GaN initial layer in [10$\bar{1}$0] incidence.

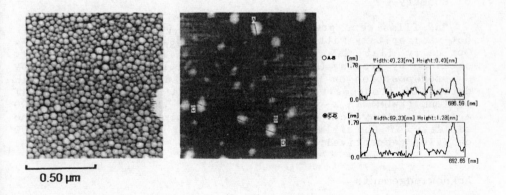

Fig.4. AFM photographs of (a) Ga predeposited layer and
(b) GaN initial layer.

4. Properties of GaN films

X-ray rocking curve (XRC) measurement was also performed for the GaN films grown on GaN initial layer, using four crystal x-ray diffractometer. Typical full width at half maximum (FWHM) of XRC was 400 arcsec.

Photoluminescence (PL) measurements were also performed at 4.2K for the GaN films. A He-Cd laser was used as excitation light. Typical PL spectrum for the films are shown in Fig.5. A sharp I_2 line due to excitons bound to neutral donors corresponding to nitrogen vacancies were observed at 358nm. And D-A pair emission with LO phonon replicas were observed at around 400nm and broad deep level emission at around 560nm considered to be associated with nitrogen vacancies were also observed. Origin of the D-A pair emissions are not identified yet. Hall effect measurement was also performed. Typical carrier concentration, resistivity and mobility of the films were 1×10^{19} cm^{-3}, 0.01 Ohm·cm and 50 cm^2/V·s, respectively.

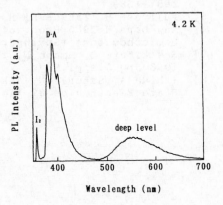

Fig.5. PL spectrum of GaN films prepared by HWE.

220

5. Summary

GaN films were prepared on sapphire (0001) substrates by hot-wall epitaxy with metallic Ga and NH_3 as source materials. Optically flat GaN films with streak RHEED patterns were obtained by growing GaN films on GaN initial layer prepared by Ga predeposition on nitrided substrates and subsequent nitridation of predeposited Ga layer. Typical full width at half maximum (FWHM) in x-ray rocking curve of the GaN films was 400 arcsec, and typical carrier concentration, resistivity, and mobility of the film were 1×10^{19} cm^{-3}, 0.01 Ohm·cm and 50 cm^2/V·s, respectively.

Acknowledgements

This work was partially supported by the Saito Chion Foundation.

References

[1] Amano H, Sawaki N, Akasaki I and Toyoda Y 1986 Appl.Phys. Lett. 48 353-355
[2] Amano H, Akasaki I, Hiramatsu K and Koide N 1988 Thin Solid Films 163 415-420
[3] Akasaki I, Amano H, Koide Y, Hiramatsu K and Sawaki N 1989 J. Cryst. Growth 98 209-219
[4] Nakamura S 1991 Jpn. J. Appl. Phys. 30 L1705-L1707
[5] Nakamura S, Mukai T and Senoh M 1994 Appl. Phys. Lett. 64 1687-1689
[6] Fujiyasu H, Takahashi H, Shimizu H, Sasaki A and Kuwabara H 1985 Proc. of 17th Int. Conf. on Physics of Semiconductors (Springer-Verlag, New York) pp.539-542
[7] Sakakibara S, Amano N, Ishino K, Ishida A and Fujiyasu H 1993 Jpn.J.Appl.Phys. Vol. 32 4703-4708
[8] Ishida A, Sakurai N, Aikawa K and Fujiyasu H 1994 Solid State Electronics 37 1141-1144

Inst. Phys. Conf. Ser. No 145: Chapter 2
Paper presented at 22nd Int. Symp. Compound Semiconductors, Cheju Island, Korea, 28 August–2 September 1995
© *1996 IOP Publishing Ltd*

Strain Relaxation and Crystallographic Tilt of Compositional Graded $In_xGa_{1-x}As$ and $In_xAl_{1-x}As$ (0<x<0.3) Epilayers Grown on GaAs Substrates

J.-L. Shieh, J.-I. Chyi, J.-W. Pan, and R.-M. Lin

Department of Electrical Engineering, National Central University, Chung-Li, Taiwan 32054, R.O.C.

Abstract. The strain relaxation and crystallographic tilt of $In_xGa_{1-x}As$ and $In_xAl_{1-x}As$ (0<x<0.3) epilayers grown on GaAs substrates have been studied. Compositional step-graded or linear-graded buffer layers are used. The residual strain of the $In_xAl_{1-x}As$/graded $In_yAl_{1-y}As$ is found to be larger than that of InGaAs and is strongly dependent on the thickness and the composition of the step-graded buffer layers. The crystallographic tilt of the InGaAs epilayers with respect to GaAs substrate is shown to be strongly dependent on the substrate temperature as well as the layer structure of the underlying buffer layer, while that of InAlAs is insensitive to these two factors. Rough surface due to irregular nucleation is proposed to explain the different behavior between these two material systems.

1. Introduction

Since the introduction of InGaAs/GaAs pseudomorphic heterostructure, the growth of InGaAs with In content within 0.2, has been investigated intensively for the applications in high speed [1,2] and optoelectronic devices [3]. However, the thickness and the In composition have been limited by the critical thickness. On the other hand, the strain-relaxed or so-called metamorphic InGaAs/InAlAs heterostructure grown on GaAs substrate provides much greater freedom than conventional InGaAs/GaAs system for the applications in devices such as resonant tunneling diodes (RTDs) [4], high electron mobility transistors (HEMTs) [5], and heterojunction bipolar transistors (HBTs) [6]. The strain-relaxed InGaAs and InAlAs grown on GaAs, have thus been investigated [7,8] recently. Since the active layers of the above mentioned devices are grown on the strain-relaxed buffer layers, the distribution of the misfit dislocations has to be controlled in order not to degrade the electrical and optical properties of the epitaxial layers. It has been shown that proper buffer layer structure is essential [9] to obtain a high quality strain-relaxed buffer layer for device applications. In this work, a series of InGaAs and InAlAs metamorphic buffers were grown under various conditions. The the residual strain and the crystallographic tilt of the epilayers were investigated in detail.

2. Experiments

The samples studied in this work were grown on (100)-oriented semi-insulating GaAs substrates in a Riber 32P solid-source molecular beam epitaxy (MBE) system. Three types of

(a)		(b)		(c)	
$In_xAl_{1-x}As$	0.75 μm	$In_{0.3}Ga_{0.7}As$ ($In_{0.3}Al_{0.7}As$)	0.5 μm	$In_{0.3}Ga_{0.7}As$ ($In_{0.3}Al_{0.7}As$)	0.5 μm
x = 0.17, 0.26, ...	0.075 or 0.6 μm	$In_{0.2}Ga_{0.8}As$ ($In_{0.2}Al_{0.8}As$)	0.2 μm	x = 0.1 —> 0.3	0.2 μm
$In_{0.1}Al_{0.9}As$	0.075 or 0.6 μm	$In_{0.1}Ga_{0.9}As$ ($In_{0.1}Al_{0.9}As$)	0.2 μm	$In_{0.1}Ga_{0.9}As$ ($In_{0.1}Al_{0.9}As$)	0.2 μm
GaAs Substrate		GaAs Substrate		GaAs Substrate	
(a)		(b)		(c)	

Fig. 1 Schematic layer structures of the metamorphic epilayers grown on (100)-oriented semi-insulating GaAs, (a) step-graded InAlAs epilayers with various thicknesses of step-graded buffer layers, (b) step-graded InGaAs or InAlAs epilayers, and (c) linear-graded InGaAs or InAlAs epilayers.

layer structures were used as illustrated in Fig. 1. Fig. 1(a) shows the structure of a step-graded $In_xAl_{1-x}As$ epilayer, which consists of a multi-stage 0.075 or 0.6 μm-thick step-graded layers with In compositions slowly increasing from 0.1 to 0.26, and finally followed by a 0.75-μm-thick epilayer. Fig. 2(b) shows another step-graded structure, which consists of a 0.2-μm-thick $In_{0.1}Ga_{0.9}As$ ($In_{0.1}Al_{0.9}As$), a 0.2-μm-thick $In_{0.2}Ga_{0.8}As$ ($In_{0.2}Al_{0.8}As$), and a 0.5-μm-thick $In_{0.3}Ga_{0.7}As$ ($In_{0.3}Al_{0.7}As$). The third type of the structure as shown in Fig. 1(c) is similar to that shown in Fig. 1(b), except the second layer is a linear-graded $In_xGa_{1-x}As$ ($In_xAl_{1-x}As$) epilayer with x varying from 0.1 to 0.3. The As_4/III beam equivalent pressure ratio was maintained between 25 and 30 throughout the growth. The growth rate and growth temperature was varied between 0.8 and 1.0 μm/hr, and 420 and 520 °C, respectively, depending on the In composition of the epilayers. Streaky (2x1) reflection high energy electron diffraction (RHEED) patterns, indicating two dimensional growth, were observed throughout the growth.

The strain relaxation of the epilayers were investigated using a double crystal x-ray diffractometer. Because the (400) rocking curves contain the information of both lattice mismatch and epilayer inclination, i.e. crystallographic tilt, four rocking curves along different directions were measured for each sample, i.e. 0°, 90°, 180°, and 270° with respect to [011] direction. The splitting ($\Delta\theta$) due to lattice mismatch and the inclination angle ($\Delta\phi$) between the epilayer and the substrate are therefore deduced from these four rocking curves, respectively. A typical set of rocking curves is shown in Fig. 2. The surface roughness of these epilayers was also analyzed by atomic force microscopy (AFM).

3. Results and discussion

3.1 Residual strain

The splittings between the diffraction peaks of the top $In_xAl_{1-x}As$ (0<x<0.3) layers and the GaAs substrate are shown in Fig. 3, from which the residual strain in the InAlAs layers can be qualitatively analyzed. Since the (400) diffraction peak reflects the lattice constant in the growth direction only, larger splitting means larger residual strain for the samples with the

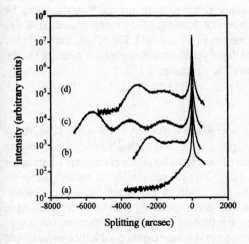

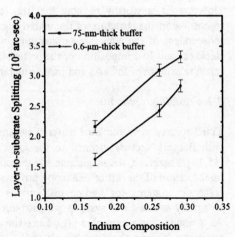

Fig. 2 Four rocking curves measured by double crystal x-ray diffractometer along different directions, i.e. (a) 0°, (b) 90°, (c) 180°, and (d) 270° with respect to [011] direction.

Fig. 3 The splittings of the x-ray diffraction peaks of the top $In_xAl_{1-x}As$ (0<x<0.3) layers with respect to that of GaAs substrates for the thickness of the InAlAs buffer layer being either 75-nm-thick or 0.6-μm-thick.

same In composition. The effect of the thickness of the underlying layer on the residual strain in the top layer can be easily appreciated by comparing the curves in Fig. 3. It can be concluded that thick underlying layer and high In content facilitate the strain relaxation process. Consequently, the top layer has less residual strain as grown on thicker buffer layer. This is consistent with our previous study [10], which indicates that the residual strain of bulk InAlAs epilayers directly grown on GaAs is dependent on the composition as well as the thickness of the epilayer. On the other hand, the strain relaxation of InGaAs grown on GaAs substrates has also been examined and compared with that of InAlAs. The x-ray diffraction peak splittings of the InGaAs and InAlAs epilayers studied are listed in Table I and found to be about 2700 and 3000 arcsec, respectively. It is evident that InGaAs relaxes more completely than InAlAs. The difference in the Peierls force [11], which is related to the atomic bonding, between these two material systems is thought to be the cause of this phenomenon.

Table I

The splitting and the crystallographic tilt of $In_{0.3}Ga_{0.7}As$ and $In_{0.3}Al_{0.7}As$ grown on various metamorphic buffers.

Sample number	Material	Structure	Growth temp. (°C)	Splitting angle (arcsec)	Tilt angle (arcsec)
1	InGaAs	step	520	2750	2800
2	InGaAs	linear	520	2720	80
3	InGaAs	step	420	2700	780
4	InAlAs	step	520	2950	320
5	InAlAs	linear	520	3040	200
6	InAlAs	step	420	2810	750

However, it is worthy to note that the residual strain of both materials is shown to be insensitive to the structure of the underlying layer, i.e. linear-graded or step-graded. It is also independent of the substrate temperature in the range of 420 and 520 °C. In summary, the thickness and In composition of the underlying layer yet its structure play a critical role in the strain relaxation of InGaAs and InAlAs metamorphic structures.

3.2. Crystallographic tilt

Tilted epitaxy in mismatched material systems is generally attributed to the misfit dislocations with Burgers vectors inclined to the growth plane as a mismatched overlayer is growing [12,13]. Therefore, the magnitude and direction of the crystallographic tilt contain information on the nature of the lattice relaxation process. For (100)-oriented zincblende crystals, there are eight slip systems for gliding dislocations to participate in the strain relaxation process. Unequally stressed slip systems will introduce different numbers of dislocations of each type. As a result, crystallographic tilt occurs since each type of dislocation relieves an amount of strain in a specific direction. This is often observed in the heteroepitaxy of lattice-mismatched materials on inclined substrates [14-16]. Surprisingly, there exists a crystallographic tilt between InGaAs epilayer and exact (100)-oriented GaAs substrate in this work. The tilt angle, as shown in Table I, can be as large as 2800 arcsec which is comparable to that reported by K. L. Kavanagh *et. al.* [17] although (100)-oriented 2° off toward [010] GaAs substrates were used in their study.

The effect of the epilayer structure and growth temperature of the underlying buffer layer on the crystallographic tilt is investigated in this work. Comparing the tilt angles of sample 1 and 2 in Table I, it is found that linear-graded structure is capable of reducing the tilt angle of InGaAs epilayer effectively. The different number of slip systems being active in the relaxation process for InGaAs epilayers with different types of underlying structure is thought to be the cause of this behavior. For linear-graded structures, the relaxation process is slower than that for step-graded ones since the strain energy is increased much more gradually. Therefore, the eight slip systems have enough time to participate in the strain relaxation process. As a result, all of the eight active slip systems are competing with each other and the crystallographic tilt is thus minimized in InGaAs epilayers grown on linear-graded buffer. In contrast, only some of the slip systems take part in the relaxation process of the InGaAs epilayers grown on step-graded buffer. It is this unbalanced relaxation that results in the crystallographic tilt along a specific direction. On the other hand, it is also found that lowering the growth temperature from 520 to 420 °C for the underlying $In_xGa_{1-x}As$ buffer layer can also reduce the tilt angle, as shown by sample 3 in Table I. This is again a attributed to the activeness of all the slip systems since low substrate temperature results in slow relaxation.

The $In_{0.3}Al_{0.7}As$ epilayers, however, in general exhibit much less tilt as compared with $In_{0.3}Ga_{0.7}As$ as indicated by the results of sample 4, 5 and 6 shown in Table I. It should also be noted that the crystallographic tilt of $In_{0.3}Al_{0.7}As$ epilayers, as compared to $In_{0.3}Ga_{0.7}As$ epilayers, is pretty much insensitive to the structure of the underlying layer as well as the substrate temperature. The different behavior of these two material systems is believed to be caused by the intrinsic properties of the cations, i.e. In, Al and Ga. It is well known that the surface mobilities of Al and Ga can be very different if the molecular beam epitaxial growth is undertaken at temperatures below 600 °C. Surface roughness develops rapidly due to this reason. For InAlAs epilayers, the difference in surface mobility of In and Al is even greater. More randomized nucleation occurs in the growth of InAlAs. As a result, irregular rough growth front inhibits the existent of preferred orientations for the Burgers vectors of misfit dislocations. Since every type of the slip system participates in the strain relaxation process, the resultant crystallographic tilt is much less significant for InAlAs. To confirm this postulate, the surface topography of these strain-relaxed $In_{0.3}Ga_{0.7}As$ and $In_{0.3}Al_{0.7}As$ grown on various graded buffers was therefore examined by AFM analysis. It is indeed found (not shown here) that the surface morphology of $In_{0.3}Al_{0.7}As$ epilayers appears to be irregular no matter what kind of buffer structure or growth temperature in the range of 420-520 °C is employed, while that of the InGaAs epilayers is rather regular, although both materials have cross-hatching patterns. This is consistent with what discussed above.

4. Conclusion

We have studied the strain relaxation and crystallographic tilt of both $In_xGa_{1-x}As$ and $In_xAl_{1-x}As$ ($0<x<0.3$) metamorphic epilayers grown on GaAs substrate with compositional step-graded and linear-graded buffer layers. The residual strain of the InAlAs epilayer is shown to be strongly dependent on the composition and the thickness of the underlying buffer layers. Besides, $In_{0.3}Ga_{0.7}As$ exhibits less residual strain than $In_{0.3}Al_{0.7}As$. However, the residual strain of both two material systems is independent of the structure of the underlying layer. It is also found that the crystallographic tilt of $In_{0.3}Ga_{0.7}As$ can be as large as 2800 arcsec and is dependent on the substrate temperature as well as the layer structure of the underlying buffer layers. While $In_{0.3}Al_{0.7}As$ reveals less tilt, and the tilt is insensitive to the layer structure and substrate temperature in the range of 420-520 °C. Rough surface induced by random nucleation is proposed to account for the different behavior of InAlAs epilayers.

Acknowledgments

The authors thank Mr. T. E. Nee for the assistance in material growth and the technical support of the MBE laboratory of the Center for Optical Science, National Central University This work was supported by National Science Council, R.O.C. under contract No. NSC-83-0404-E008-002.

References

[1] Kopf R F, Kuo J M, Kovalchick J, Pearton S J, Jones E D and Ourmazd A 1990 *J. Appl. Phys.* **68** 4029-34
[2] Matsumura K, Inoue D, Nakano H, Sawada M and Harada Y 1991 *Jpn. J. Appl. Phys.* **30** L166-9
[3] Yang Y J, Hsieh K Y and Kolbas R M 1987 *Appl. Phys. Lett.* **51** 215-7
[4] Hwang H-P, Shieh J-L, Lin R-M, Chyi J-I, Tu S-L, Peng C-K and Yang S-J 1994 *Electron. Lett.* **30** 826-8
[5] Chyi J-I, Shieh J-L, Wu C-S, Lin R-M, Pan J-W, Chan Y-J and Lin C-H 1994 *Jpn. J. Appl. Phys.* Part 2 **33** L1574-6
[6] Chan Y-J, Chyi J-I, Wu C-S, Hwang H-P, Yang M-T, Lin R-M and Shieh J-L, 1994 *Device Research Conf.* **52nd** VIB-9-10, Boulder, Colorado, U.S.A.
[7] Westwood D I, Woolf D A, Vila A, Cornet A and Morante J R 1993 *J. Appl. Phys.* **74** 1731-5
[8] Chyi J-I, Shieh J-L, Lin R-J, Pan J-W and Lin R-M 1995 *J. Appl. Phys.* **77** 1813-5
[9] Krishnamoorthy V, Ribas P and Park R M 1991 *Appl. Phys. Lett.* **58** 2000-2
[10] Chyi J-I, Shieh J-L, Lin R-M, Nee T-E and Pan J-W 1994 *Appl. Phys. Lett.* **65** 699-701
[11] Johannes Weertman and Julia R. Weertman : Elementary Dislocation Theory, Oxford University, New York, 1992
[12] L. J. Schowalter, E. L. Hall and N. Lewis 1990 *Thim Solid Films* **184** 437-45

226

[13] Ayers J E, Ghandhi S K and Showalter L J 1991 *J. Cryst. Growth* **113** 430-40
[14] Olsen G H and Smith R T 1975 *Phys. Stat. Sol. (a)* **31** 739-47
[15] Kleiman J, Park R M and Mar H A 1988 *J. Appl. Phys.* **64** 1201-5
[16] Ghandhi S K and Ayers J E 1988 *Appl. Phys. Lett.* **53** 1204-6
[17] Kavanagh K L, Chang J C P, Chen J, Fernandez J M and Wieder H H 1992 *J. Vac. Sci. Technol.* **B 10** 1820-3

Inst. Phys. Conf. Ser. No 145: Chapter 2
Paper presented at 22nd Int. Symp. Compound Semiconductors, Cheju Island, Korea, 28 August–2 September 1995

Distribution of As atoms in InP/InPAs/InP and InP/InGaAs/InP hetero-structures measured by X-ray CTR scattering

M. Tabuchi, K. Fujibayashi, N. Yamada, K. Hagiwara, A. Kobashi, H. Kamei† and Y. Takeda

Department of Materials Science and Engineering
School of Engineering, Nagoya University, Nagoya 464-01, Japan

†Opto-electronics R&D Laboratories, Sumitomo Electric Industries, Ltd.
1, Taya-cho, Sakae-ku, Yokohama 244, Japan

Abstract. The distribution of As atoms in InP/InPAs/InP layer and the distributions of As and Ga atoms in InP/InGaAs/InP hetero-epitaxial layer were investigated in the atomic scale by X-ray CTR scattering. The relationsip between PH_3-purge time (after InPAs and InGaAs layers were constructed) and the distributions of As and/or Ga atoms were studied. It was shown that the number of As atoms included in the InP cap layer decreased as the PH_3-purge time increased. However, the As atoms constructing the hetero-layer were also desorbed as the PH_3-purge time increased.

1. Introduction

Hetero-epitaxially grown III-V semiconductor layers are very important to fabricate various electronic and optical devices. For example, AlGaAs/GaAs and InGaAsP/InP systems are used to realize high speed transport devices, such as, HEMT(high erectron mobility transistor) and HBT(heterojunction bipolar transistor), and also utilized for laser diodes[1, 2, 3]. In order to improve characteristics of such devices, techniques to realize well defined low-dimensional structures, i.e., quantum structures, are strongly required. In the quantum structures of a few MLs(monolayers) thickness, quality of hetero-interfaces becomes to play important roles as each layer in the structures becomes thinner. Vagueness of the interface in the scale of only a few Å is no more negligible. Therefore, techniques to characterize the interface structures in the atomic scale are strongly required to control and at least to determine the structure in that scale.

We have demonstrated that X-ray CTR(crystal truncation rod) measurement reveals the interface structure in the atomic scale [4, 5]. X-ray CTR measurement is based on the fact that X-ray diffraction spots are diffused normal to the surface for a finite

228

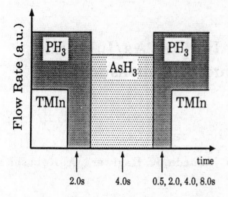

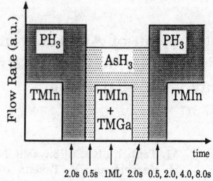

Fig. 1. Flow sequence of source gases to prepare type-A samples.

Fig. 2. Flow sequence of source gases to prepare type-B samples.

crystal and that shapes of the diffused spots are sensitively modified as the structure near surface is changed[6, 7].

In this work, hetero-interfaces of OMVPE grown InP/InPAs(formed by AsH₃-exposure)/InP and InP/InGaAs(1ML-grown)/InP samples are investigated by the X-ray CTR measurement. Since group-V atoms easily exchange during the growth, it is difficult to realize a well defined hetero-interface between materials of different composition of group-V atoms. The interface between InP and InGaAs is one of such cases. Thus, it is important to reveal the real structure of the InP/InGaAs/InP hetero-interface. We have already reported about AsH₃-exposed InP surface and 1ML grown InPAs investigated by the X-ray CTR scattering [8, 9]. Both of them are considered to extract important steps of the growth of InP/InGaAs/InP structures. In this paper, InP/InPAs(formed by AsH₃-exposure)/InP and InP/InGaAs(1ML)/InP samples are discussed to study effects of PH₃-purge time after the InGaAs layer-growth in the InP/InGaAs/InP hetero-epitaxial formation.

2. Sample Preparation

Two types of samples were prepared and measured. One was InP exposed to AsH₃ for 4s (type-A). The other was 1ML-grown $In_xGa_{1-x}As(x=0.53)$ on InP substrate (type-B). All the samples were grown on InP (001) substrates followed by a 1000Å InP buffer layer, and capped by 20Å InP. The growth temperature was 620°C. AsH₃, PH₃, and TMIn were used as the source gases. Figures 1 and 2 show the flow sequence of source gases to prepare type-A and type-B samples, respectively. As shown in Fig. 1 and 2, PH₃-purge time before growing the cap layer was changed from 0.5s to 8s.

3. X-Ray CTR Measurement

The X-ray CTR measurement was conducted at beam lines BL6A₂ and BL18B in the Photon Factory at Tsukuba by Weissenberg camera using imaging plate as a detector.

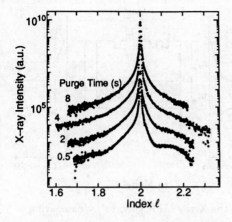

Fig. 3. Measured X-ray CTR spectra for type-A samples.

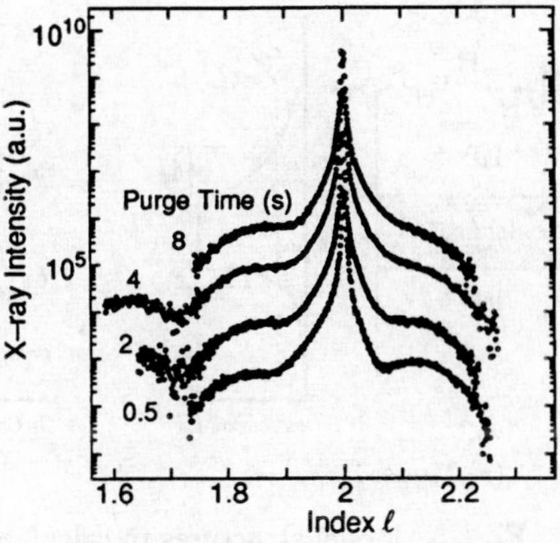

Fig. 4. Measured X-ray CTR spectra for type-B samples.

The wavelength of the X-ray was set at 1.000Å or 0.950Å by a bent Si (111) monochromator. In this work, the X-ray CTR spectrum which appears around the 002 Bragg diffraction spot was analyzed. The CTR spectra were obtained by subtracting the background X-ray diffuse-scattering from the measured X-ray diffraction intensity.

Figures 3 and 4 show the measured X-ray CTR spectra of the type-A and type-B samples, respectively, which spread widely around index $\ell = 2$. In Fig. 3, shoulder is observed at around $\ell = 2.1 \sim 2.2$, which is due to the existence of the InPAs layer in the samples only exposed to AsH_3. The shape of the shoulder reflects the distribution of As atoms in the sample. In Fig. 4, shoulders are observed at both sides of peak. The X-ray CTR spectra of type-B samples are more symmetrical than those of type-A samples. It indicates that the strain in the hetero-layer is less in the type-B samples than in the type-A samples.

4. Data Analysis

The X-ray CTR spectra were analyzed by comparing them with theoretically calculated spectra based on model structures shown in Fig. 5. Model-I contains eight parameters, c/a, I_0 and other parameters which are shown in Fig. 5-(a). In order to express the distributions of both As and Ga atoms, Model-II contains other five parameters as shown in Fig. 5-(b). $n_{c_{(V,III)}}$ and $n_{h_{(V,III)}}$ are the thicknesses of the cap layer and the hetero-layer, respectively. $x_{h_{(V,III)}}$ are the As and Ga compositions in the hetero-layer. $d_{c_{(V,III)}}$ and $d_{b_{(V,III)}}$ represent the distributions of As and Ga atoms in the cap and the InP buffer layer, respectively. The As and Ga compositions, x_V and x_{III}, in the layer away from the upper or lower interface by nML are assumed to distribute as

$$x_{(V,III)} = x_{h_{(V,III)}} exp(-\frac{n}{d})$$

where d is $d_{c_{(V,III)}}$ in the cap layer or is $d_{b_{(V,III)}}$ in the buffer layer. $< \Delta z^2 >$ indicates the surface roughness. c/a and I_0 do not appear in Fig. 5. c/a indicates a tetragonal distortion of the hetero-layer, as the ratio of the lattice constants normal(c) and parallel(a) to

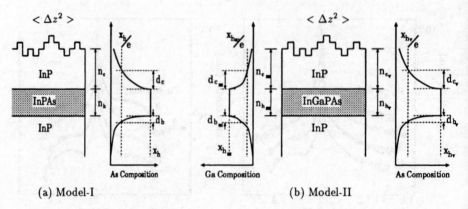

(a) Model-I   (b) Model-II

Fig. 5. Model structures to calculate the X-ray CTR spectra by comparing with the experimental data.

the surface. I_0 is a scaling factor to match the absolute intensity of the measured data to the calculation. A value which is calculated as

$$\frac{\sum |(Measured\ value) - (Calculated\ value)|}{(Number\ of\ Measured\ points)}$$

is usually called 'R-factor'. In this work, we used 'R-factor' to evaluate the curve fitting quality realized by a set of parameters.

5. Results and Discussions

Figures 8, 9 and 10 show the As and Ga distributions, to the 1ML resolution which cannot be obtained by any other characterization techniques. As shown in Fig. 8, in the type-A samples, with the increasing time of PH_3-purge the As/P-exchanged layer and

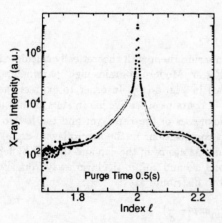

Fig. 6. A result of data fitting for a type-A sample.

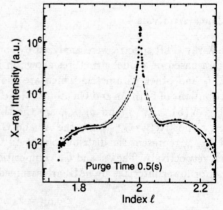

Fig. 7. A result of data fitting for a type-B sample.

the long tail of the distribution into the InP cap layer both of which are formed by the AsH₃-exposure to InP surface are drastically decreased. Therefore, it is expected that the effects of the As/P-exchange in the growth of InP/InGaAs/InP heterostructures can be swept away by the PH₃-purge.

Figure 9 shows the As distributions in InP/InGaAs(1ML)/InP samples with various PH₃-purge time. The peak As composition in 0.5s-purged sample is 1.0, but, the As atoms distributed into the cap layer, which was not intended. The As distribution tail into the cap layer decrease with the increase of PH₃-purge time resulting in an abrupt InP/InGaAs interface. To our surprise the As atoms in the rigidly grown In-GaAs are also partially taken away by the PH₃-purge. In comparison with the type-A and -B samples of the same PH₃-purge time, the amount of desorbed As atoms is almost the same. Figure 10 shows the Ga distributions in InP/InGaAs(1ML)/InP samples. The integrated amount of Ga atoms does not decrease with the increase of the PH₃-purge

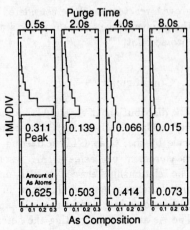

Fig. 8. Distributions of As atoms in type-A samples.

time and corresponds to intended value (0.47ML). However, the Ga atoms are not confined within 1ML. Both $d_{c_{III}}$ and $d_{b_{III}}$ are about 1.0 for all the samples.

It is generally understood that As atoms are distributed and not confined in 1ML because group-V atoms desorb and/or exchange easily and there also exists a memory-effect. On the other hand, Ga atoms had been considered to be confined in 1ML. However, as shown in Fig 10, Ga atoms are not confined in 1ML. It may be caused

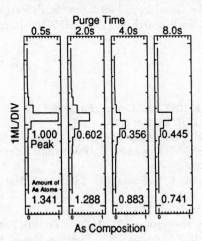

Fig. 9. Distributions of As atoms in type-B samples.

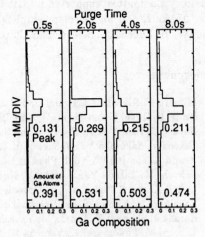

Fig. 10. Distributions of Ga atoms in type-B samples.

by stress between the InP layer and the hetero-layer.

From Fig. 9 it is clear that the PH_3-purge time must be as short as 0.5s to keep the As composition at 1.0. Even at the short purge time of 0.5s, the InP/InGaAs/InP has a few ML transition regions. A further improvement in the growth sequence and/or growth reactors must be considered for the very abrupt interfaces, and the analysis of the quantum levels and electronic states should take such As and Ga distribution profiles into account.

6. Conclusions

The distributions of As and Ga atoms in InP/InPAs(formed by AsH_3-exposure)/InP and InP/InGaAs(1ML-grown)/InP hetero-epitaxial layers were investigated in the atomic scale by the X-ray CTR scattering measurement. It was shown that the X-ray CTR measurement can achieve 1ML resolution in characterization of distribution of atoms. The relationship between the PH_3-purge time after the hetero-layers were grown and the distributions of As and Ga atoms in layers was studied. It was shown that the number of As atoms included in the InP cap layer decreased as the PH_3-purge time increased. The As atoms constructing the hetero-layer were also desorbed as the PH_3-purge time increased. On the other hand, Ga atoms contained in samples did not decrease as the PH_3-purge time increased. Though the integrated amount of Ga atoms corresponded to the intended value, they were not confined in 1ML, contrary to our design.

7. Acknowledgments

Authors would like to thank Prof. J. Harada and Dr. I. Takahashi for their helpful discussion about and assistance to the X-ray CTR measurement. This work was performed as a part of the project (Project No. 93G195) accepted by the Photon Factory Program Advisory Committee, supported in part by the Grant-in-Aid for Scientific Research (B) No. 0745507, and for Developmental Scientific Research (B) No. 07555100 from the Ministry of Education, Science and Culture.

References

[1] Mimura T, Hiyamizu S, Fujii T and Nanbu K 1980 Jpn. J. Appl. Phys. **19** L225
[2] Ito H, Ishibashi T and Sugeta T 1984 IEEE EDL **EDL-5** 214
[3] Hayashi I, Panish M B, Foy P W and Sumski S 1970 Appl. Phys. Lett. **17** 109
[4] Takeda Y, Sakuraba Y, Fujibayashi K, Tabuchi M, Kumamoto T, Takahashi T, Harada J and Kamei H 1995 Appl. Phys. Lett. **66** 332
[5] Tabuchi M, Takeda Y, Sakuraba Y, Kumamoto T, Fujibayashi K, Takahashi I, Harada J and Kamei H 1995 J. Cryst. Growth **146** 148
[6] Kashihara Y, Kimura S and Harada J 1989 Surf. Sci. **214** 477
[7] Harada J, Shimura T, Tanaka M, Yakushiji K and Hoshi K 1990 J. Cryst. Growth **104** 773
[8] Tabuchi M, Fujibayashi K, Yamada N, Takeda Y and Kamei H J. Electron. Mater., to be published
[9] Yamada N, Fujibayashi K, Tabuchi M, Takeda Y and Kamei H J. Electron. Mater., to be published

Inst. Phys. Conf. Ser. No 145: Chapter 2
Paper presented at 22nd Int. Symp. Compound Semiconductors, Cheju Island, Korea, 28 August–2 September 1995
© *1996 IOP Publishing Ltd*

Applications of Liftoff Technology

John C. C. Fan

Kopin Corporation
695 Myles Standish Blvd.
Taunton, MA 02780

Liftoff Technology allows the transfer of semiconductor films of various elements and compounds to dissimilar substrates. Such technology has tremendous potential applications from photovoltaic cells, electronic circuits, optoelectronic devices and circuits, and active matrix liquid crystal displays. Various forms of liftoff technology (for GaAs, InP, Si, etc.) will be reviewed, and examples of applications will be described.

1. Introduction

To enhance device and circuit performance, it has been known for many years to grow epitaxial films using various techniques, such as molecular-beam epitaxy, chemical vapor deposition, organic-metallic vapor phase epitaxy and liquid-phase epitaxy. However there is an evolution of epitaxial techniques, from homoepitaxy, that is growth of materials that are essentially lattice-matched to the substrate materials to heteroepitaxy, where the epitaxial film is not lattice-matched to the substrate. Heteroepitaxy is a very difficult technique since non-lattice matching (usually coupled with other non-matching physical properties) often causes degradation of the epitaxial materials (such as significant increases in defect densities and stresses in the epitaxial layers). These effects can be especially detrimental for optical devices, since the vast majority of optical devices are minority carrier devices, which are very sensitive to material quality. To circumvent the difficulties that arise from heteroepitaxy, a new epitaxial technique is being developed[1], called liftoff epitaxy (see Table 1).

Table1

EPITAXIAL TECHNIQUES	
1980s	Homoepitaxy
1990s	Heteroepitaxy
	Liftoff Epitaxy

As stated in Table 1, the advanced epitaxial techniques for the 1990s are heteroepitaxy and liftoff epitaxy. Both techniques are developed to combine dissimilar materials, which would be the core technologies for optoelectronics.

After all, in this decade, optoelectronics integrated circuits (OEICs) will come to commercial arenas in major ways, ranging from optical communicating, optical interconnects, to optical displays. However, OEICs require interconnecting, electrically as well as optically, various devices with dissimilar structures and materials.

Table 2

OEICs
High Performance
More Functionality
Easier Packaging
Improved Reliability
Low Cost

Table 2 shows the many attributes of OEICs.

Although many different semiconductor materials will be combined together in this decade, it is our belief that the primary combination in the 1990s will be GaAs-based materials and Si materials with various other materials. Silicon is, and will continue to be, the dominant electronic material of choice for the decade and beyond. It has an overwhelmingly established technology and large-scale integration. However, GaAs-based materials have the dominant lead in optical (or photonic and microwave areas). A marriage of these two semiconductors for Optoelectronics is a logical evolution in this field. We will describe the advances in this area. In addition, we will discuss the advantages and applications of transferring either Si or GaAs films to dissimilar substrates such as, glass or plastic, enabling devices that are not possible by other means.

2. Liftoff epitaxy

The basic concept of the liftoff technology is as follows: Thin epitaxial film is grown on a substrate (usually homoepitaxy) and then either by mechanical or chemical means, the film is separated from the substrate. The substrate sometimes can be reused. The thin epitaxial film is supported by other means after separation and processing is usually possible before and after separation. The separated film is then transferred to other substrates or platform, allowing combinations of many materials.

Table 3

THIN-FILM EPITAXIAL LIFTOFF ADVANTAGES
• No substrates
• Double-Sided Processing
• Monolithically-Interconnected Optical Arrays
• Hybrid-Integration of Dissimilar Materials
• Thin High-Quality Epitaxial Structures
• Potentially Low-Cost

Table 3 shows some of the advantages of thin-film epitaxial liftoff technology.

There are three different liftoff techniques, namely mechanical liftoff (CLEFT), chemical liftoff, and etch-back liftoff. Before we describe the features of these liftoff techniques, we will summarize the important issues facing the liftoff epitaxy in general. These issues are shown in Table3.

Table 4

IMPORTANT ISSUES	
ISSUES	**STATUS**
• Large-Area Epitaxial Liftoff	No problem
• Pre- and Post-Liftoff Device Processing	No problem
• Effects on Crystal Quality and Device Performance	None
• Growth of Liftoff Epitaxial Structures	Yes
• Re-Use Substrates	Sometimes
• Thin-Film Handling Procedures	No problem

The liftoff approach overcomes the above problem by utilizing present-day high performance epitaxial layers that are released from the growth substrate and applied to the foreign (Si, LiNbO$_3$, Al$_2$O$_3$, SiO$_2$) substrate without deleterious consequences to device operation. The performance of the lift-off devices is at least equivalent to that obtained in the hybrid approach, but liftoff has several key advantages that can offer superior performance and reliability if fully developed. First, the liftoff technique yields films with precise (within 0.1 μm) thickness control. These properties yield precise vertical registration that is extremely difficult to achieve with hybrid approaches. For this reason, the liftoff approach can yield superior optical coupling between photonic devices, wave guides, and fibers. In addition, the superior vertical registration makes possible the use of photolithography to pattern electrical interconnects, thus enabling precise lateral registration. Of even greater importance is that the use of interconnects with superior vacuum deposited thin-film interconnects; this enhances the reliability and ease of fabrication. For the above reasons, the liftoff approach can be seen to combine the best features of the hybrid and heteroepitaxial approaches. Our view is that the liftoff process is the most promising near-term route to monolithic integration of OEICs.

2.1 Mechanical liftoff (CLEFT)

The mechanical liftoff technique we invented at Massachusetts Institute of Technology and is termed CLEFT, which is an acronym for Cleavage of Lateral Epitaxy for Transfer[2]. This process comprises the growth of thin single-crystal GaAs films by chemical vapor deposition on reusable GaAs substrates, followed by cleaving of the layer from the substrate. Since single-crystal films are used, device performance remains high. The films are thin and their thickness is well controlled, thus permitting precise control over vertical registration. Films with thickness of less than 1 micron have been produced, as well as films up to 20 microns thick. Films composed of different layers of GaAs and AlGaAs are routinely processed. Thus, the use of the CLEFT process yields films with no important constraints on device thickness or composition. Owing to this versatility, a variety of devices may be used, and each may be optimized independently.

The basic steps in the CLEFT process are: growth of a thin single-crystal epilayer on a single-crystal substrate, mechanical separation of the epilayer from the substrate, mounting of the film on a foreign substrate, and reuse of the single-crystal substrate.

The growth is carried out on a substrate which is prepared with a mechanical release layer patterned with parallel openings. These openings expose the substrate, and serve to seed the epitaxial film in its initial phase of growth. The growth nucleates in the openings and, under the proper growth conditions, the film grows laterally across the release layer until a continuous film is formed, in a thickness as small as 1 μm. The desired device structure is then grown upon this laterally overgrown layer. The material quality of overgrown GaAs layers has been examined and is comparable to homoepitaxial layers on GaAs substrates. Solar cells made in these thin-films have been of comparable device quality to cells made in bulk GaAs layers. InP and Si films can also be released by this technique.

2.2 Chemical liftoff

In the past few years, Bellcore has invented and demonstrated what it has termed the chemical epitaxial lift-off technique[3]. Large area (>4 cm^2) multi-layer epitaxial films have been floated off of the original growth substrates and bonded by natural intermolecular attraction (Van der Waals bonding) to foreign substrates such as silicon, sapphire, diamond, and several other glass and metal surfaces. Films as thin as 0.1 μm have been removed from GaAs substrates and have been mounted on glass. At Kopin, we've succeeded in releasing films as large as 4 cm^2. By monitoring minority carrier lifetime, it has been verified that the perfect epitaxial quality of the films has been maintained. The original growth substrate can be re-used.

Chemical epitaxial lift-off is based on the selective etching of $Al_xGa_{1-x}As$ in HF acid. The etch rate varies extremely rapidly, by orders of magnitude, over the composition range of x = .4 to x = .5, and this unusual property can be used to selectively etch $Al_{0.5}Ga_{0.5}As$ and films of higher Al composition, while leaving films of lower composition untouched.

The essential details of the lift-off approach applied to a GaAs-based device, array, or circuit is shown as follows: First, the desired films are grown by OMVPE, including a thin AlAs release layer (usually about 30-50 Å thick) positioned between the active layers and the substrate, as shown in the figure.

Metallization may also be applied after film growth.

The second step consists of the application of an HF-resistant wax to the top surface to protect the metallization and to assist in the removal of the film. The wax film is applied as a liquid in a manner that causes it to dry and to form on the GaAs surface as a solid film with a controlled degree of tension.

The third step comprises the immersion of the wafer in HF, which dissolves the AlAs release layer. The etch starts at the outer exposed edges of the AlAs layer, and proceeds inward through the channel formed by the removal of AlAs. In early work on this type of process, investigators failed to release large film areas, because the transport of reactants through the channel was inadequate. In the Bellcore invention, however, the tension in the wax film assists in the release of the films by bending the released portion of the layers so as to enhance the transport of HF to the etch interface, as well as transport of hydrogen from the interface to the solution. Once completely free, the wax/GaAs composite film floats on the surface of the bath. Since the GaAs thickness is on the order of several microns or less, the wax serves to provide mechanical support to the GaAs film or device until the film can be mounted on a substrate. Similarly, Si films have been successfully released with SiO_2 as a release layer.

2.3 Etch-back liftoff

This etch-back liftoff technology is a simple process without reusing the substrates, as in the previous two techniques. The substrates, in this case, are totally etched off chemically. At Kopin, GaAs films, Si films, and InP films have been liftoff routinely. In the GaAs liftoff, the stop-etch layer is AlAs, in the InP liftoff, the stop-etch is GaInAs.

Table 5

INTEGRATION OF DISSIMILAR MATERIALS			
	Hybrid Bonding	**Heteroepitaxy**	**Liftoff**
Materials Quality	G	F	G
Monolithic Integration	F	G	G
Materials Combinations	G	F	G
Handling	G	G	F

where G = Good
F = Fair

Table 5 shows a comparison of different techniques in integration of dissimilar materials.

3. Example of applications

3.1 Active-Matrix Liquid Crystal Displays (AMLCDs)

Transmissive AMLCDs require transparent integrated circuits on glass or other optically transmissive substrates. We have fabricated, using liftoff technology, AMLCDs with the highest resolution ever. The process is manufacturable and has many advantages over conventional AMLCD approaches.

3.2 The Kopin approach

In contrast to other approaches in the industry, such as amorphous Si, or polycrystalline Si, we have been able to transfer high quality, single crystal, thin film silicon integrated circuits to glass utilizing its proprietary liftoff technology. This capability has enabled development of the SMART SLIDE AMLCD[4]. The SMART SLIDE imaging device includes a thin film single-crystal integrated circuit and liquid crystal, enclosed between two glass panels. As light is projected through a SMART SLIDE imaging device, it is filtered to transform white light into an image, like a photographic slide, except that the SMART SLIDE is computer controlled to display video, data and graphic images.

We have demonstrated and are already manufacturing in limited commercial quantities of AMLCDs formed using Kopin's single crystal thin film silicon wafer engineering technology. An important advantage of its wafer engineering technology is that high speed signal processing and control logic can be integrated directly into the SMART SLIDE imaging device, potentially leading to the development of projectors, monitors, virtual reality game sets and other display products with higher resolution, greater system speed, lower weight, less volume and simplified manufacturing. These SMART SLIDE imaging devices may also provide further cost advantages through the ability to use relatively standard small-format optical components.

Integral to the liftoff approach is the ability to prepare thin film silicon integrated circuits using existing manufacturing facilities. Since integrated circuit manufacturing capacity is in place at multiple sites, this avoids the substantial capital investments, significant process development costs and long lead times typically needed for advanced LCD manufacturing.

238

3.3 OEIC

Integration of Optoelectronic components into OEICs requires interconnecting electrically as well as optically several different types of devices with dissimilar structures and materials. We have successfully integrated arrays of light emitting diodes on silicon CMOS driver circuitry with thin-film photolithographically pattered metalization, thus providing a silicon circuit with an integrated optical output. The process was carried out with full wafer processing[5].

The LED arrays used in this work were formed from $Al_xGa_{1-x}As$ grown by OMCVD on conventional GaAs substrates. A layer of AlAs was interposed between the, substrate and the active layers to enable the use of the chemical liftoff technique. The LED structure consisted of a double heterostructure with a central low band gap active layer into which carriers are injected and within which excess carriers recombine radiatively. The active layer composition was 38% AlAs ($x = 0.38$) and emitted photons with a center wavelength of 660 nm (red). An investigation of the AlGaAs active layer showed strong photoluminescence, indicating that the AlGaAs is of high quality.

The LEDs that comprised the linear array were small area devices with dimensions of 15 μm by 30 μm. The array consisted of 128 emitters with center to center spacing of approximately 60 μm. The total device thickness was about 1.5 μm. The circuit employed complementary metal-oxide-semiconductor logic (approximately 3000 transistors) with an output driver for each of the 128 LEDs. This work suggests that a wide range of materials can be combined monolithically with silicon or with each other, despite lattice or thermal expansion coefficient mismatch, to attain the desired combinations of electrical and optical properties, thus enabling the formation of advanced optoelectronic integrated circuits and other structures.

4. Conclusion

Liftoff technology has been pursued in the laboratories for over one decade. It is now emerging into commercial markets, with enormous potential for electronics, imaging and optoelectronic applications. Its inherent flexibility and its excellent epitaxial material quality provide various possibilities unmatched by other techniques. The handling issues certainly are challenging, but have been resolved.

References

[1]   J.C.C. Fan, ed. M. Razeghi, 1990 *Optoelectronic Materials and Device Concepts* SPIE Optical Engineering Press, p. 202.
[2]   R.W. McClelland, C.O. Bozler, and J.C.C. Fan, 1980 *Applied Physics*, Lett. 37, 560.
[3]   E. Yablonovitch, T. Gmitter, J.P. Harbison, and R. Bhat, 1987 *Applied Physics*, Lett. 51, 2222.
[4]   J. P. Salerno, D .P. Vu, B. D. Dingle, M. W. Batty, A. C. Ipri, R. G. Stewart, D. L. Jose, and M. L. Tilton, 1992 *Society of Information Displays Digest*, 63.
[5]   B. D. Dingle, M. B. Spitzer, R. W. McClelland, J. C .C. Fan, and P. M. Zavracky, 1993 *Applied Physics Letter*. 62, 1.

This work was supported in part by the Defense Advanced Research Projects Agency and the Office of Naval Research.

Inst. Phys. Conf. Ser. No 145: Chapter 2
Paper presented at 22nd Int. Symp. Compound Semiconductors, Cheju Island, Korea, 28 August–2 September 1995
© *1996 IOP Publishing Ltd*

Formation of High Electrical Resistance Region in (InAs)/(AlSb) Superlattice by Ga Focused Ion Beam Implantation

Song Gang Kim*, Hajime Asahi, and Shun-ichi Gonda
The Institute of Scientific and Industrial Research, Osaka University
8-1 Mihogahoka, Ibaraki, Osaka 567 Japan

Soon Jae Yu
Department of Electronics Engineering, Sunmoon University
100 Kalsanri, Tangjeong, Asan, Chungnam 336-840 Korea

Abstract. A region with high electrical resistance as high as 2.7×10^3 $\Omega \cdot cm$ can be formed in $(InAs)_{16}/(AlSb)_{22}$ Superlattice layer by 5×10^{15} cm^{-2} Ga focused ion beam implantation and 400 °C-30 min annealing. The clear photoresponse is also observed in Ga ion implanted diode structures after annealing. Enhancement of the compositional intermixing by Ga ion implantation and the existence of residual defects are confirmed by Raman scattering measurement.

1. Introduction

Focused ion beam (FIB) technology has a wide range of semiconductor processing such as maskless submicron pattern doping, ion beam lithography, ion beam assisted processes and so on [1-4]. As a processing technique to planarize devices, ion mixing appears to be an attractive technology. Formation of a submicron high resistance region in a conductive layer is one of the attractive applications, considering the increasing demands for the formation of a micron size isolation region in smaller size devices in optoelectronic and electronic devices. In the case of optoelectronic and electronic devices, isolation is an important issue. Since Arimoto et. al.[5] reported the formation of the isolation region in n-GaAs by B FIB implantation, many articles on this effect has been reported [6,7,8]. Highly resistive regions can be formed by implantation of B, O, Ga, Cr, and Fe ions into the III-V materials[9-13].

The selective compositional intermixing of GaAs/AlGaAs [14-15] and InGaAs/InP [16,17] systems has been reported as a method for obtaining the region having different physical properties such as band gap and refractive index on the same wafer. Ga ion implantation offers the potential for more abrupt intermixing of III-V compound SL structures without impurity doping,

---

Present address : Semiconductor R&D Lab.2, Hyundai Electronics Industries Co., Ltd.

and can be used to make the microstructures such as quantum well wire and quantum box structure by selective intermixing.

In this paper, we report, for the first time, on the formation of the small area high resistivity region caused by the selective intermixing of InAs/AlSb superlattice by FIB implantation with Ga ions. The intermixing and damage process by Ga ion implantation followed by annealing and the existence residual defects are confirmed by Raman scattering measurement.

2. Experimental

InAs/AlSb SLs in this study were grown on the Cr-O doped (001) GaAs substrates by gas source MBE (Molecular Beam Epitaxy) at 480°C. Elemental Al, In and Sb, and thermally cracked AsH_3 were used as group III and group V sources. The samples consists of 6-periods of $(InAs)_{16}/(AlSb)_{22}$ SLs with GaAs cap layer which is doped with Si ($n=1\times10^{18}$ cm^{-3}) in order to form an ohmic contact. After photolithography, 300x300 µm square patterns were isolated with grave widths of 50 and 20 µm formed by mesa etching process as shown in the insertion of Fig.1. And then single line implantation by Ga FIB were made with 0.3µm beam diameter and at 100 keV ion energy at room temperature. The probe current of Ga FIB was about 50 pA. Implantation doses were varied in the range from 1×10^{14} to 5×10^{15} cm^{-2}. The incident angle for this ion beam was 7°off from the [100] axes. After Ga FIB implantation, samples were annealed at 400 °C for 30 min in a N_2 ambient by face to face method. Resistances between these ohmic contacts were measured without any passivation films at both room temperature and 77K.

3. Results and discussion

Figure 1 shows the annealing temperature dependence of the electrical resistance for Ga ion implanted InAs/AlSb SLs. The sample structure and electrodes are shown in the insertion of Fig.1. The as-grown sample has the two-terminal resistance of 263 Ω. After 1×10^{14} cm^{-2} and 5×10^{15} cm^{-2} Ga-FIB implantation, the two terminal resistance increases to 4.89 kΩ and 6.25 kΩ, respectively, as shown in Fig.1. That is, the sample shows higher resistivity with increasing fluence. This may be come from

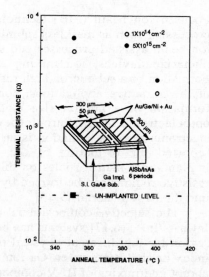

Fig.1 Electrical resistance as a function of annealing temperature and sample structure for I-V measurement.

the formation of higher potential barrier due to the intermixing of InAs and AlSb layers by ion implantation, and the generation of trap centers associated with implantation induced defects. The width of implanted region is 0.3 μm, only this region becomes highly resistive and the distance between two electrodes is 50 μm. In the subsequently annealed (400 °C, 30 min) samples, the resistance decreases from 4.89 kΩ (6.25 kΩ) to 2 kΩ (2.7 kΩ) at the fluence of 1×10^{14} (5×10^{15}) cm^{-2}. This change is attributed to the decrease of trap centers which is due to the recovery of crystalline quality by annealing. However, the resistance value is higher than that of the as-grown sample. This phenomenon is responsible for 1) the enhanced inter-

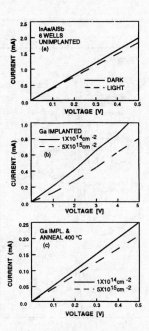

Fig.2 I-V characteristics of InAs/AlSb which was conducted using optical SLs (a) as grown, (b) Ga implanted, and (c) Ga implanted and annealed.

mixing of superlattice layers by annealing, and 2) the incomplete recrystallization of damaged layers.

Figure 2 (a) shows the I-V characteristics for the as-grown InAs/AlSb SLs with (dashed line) and without (solid line) light illumination, from a microscope light (white line, about 10mW/cm^{-2}). For the as-grown sample having two-terminals the resistance value (263 Ω) without light illumination is smaller than that with light illumination (286 Ω), as shown in Fig.2 (a). In general, the resistance decreases due to the increased photocurrent by light illumination. However, in this sample this is contrary. At the present stage, the reason is not clear. Figure 2 (b) and (c) show the I-V characteristics for the unimplanted and the implanted

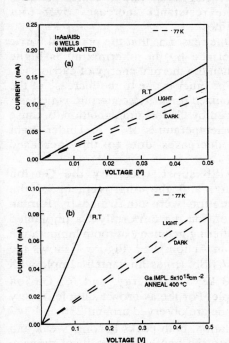

Fig.3 I-V characteristics of InAs/AlSb SLs (a) unimplanted and (b) Ga implanted and annealed as a function of temperature.

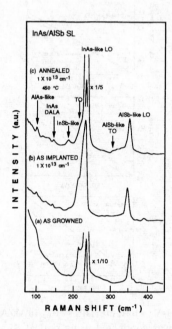

Fig.4 Raman spectra for InAs/AlSb SLs (a) as grown, (b) Ga implanted, and (c) Ga implanted and annealed.

(100 keV) and subsequently annealed (400 °C, 30 min) samples. In the sample subsequently annealed for 400 °C-30 min, the photocurrent increases from 0.2 mA (0.16 mA) to 0.5 mA (0.37 mA) at the fluence of 1×10^{14} (5×10^{15}) cm^{-2} for the 1V bias voltage. This corresponds to the drastic improvement of crystallinity by annealing.

Figure 3 shows the I-V characteristics at 77 K for the same samples of Fig.2. In general, for the lightly doped III-V compound semiconductors, the lattice scattering dominates, and the resistance decreases as the temperature decreases. However, in the case of the as-grown sample the resistance slightly increases at 77 K, and in the ion implanted and annealed samples the resistance increases over four times.

One reason is that the potential barrier formed by the intermixing is higher than the thermal energy of carriers.

The other is due to the increase of the ionized impurity scattering, which is caused by the increase of the defect induced by Ga ions implantation, because the impurity scattering dominates at low temperatures of 77 K. Under light illumination, the resistance slightly decreases due to the increased photocurrent.

The intermixing of the InAs/AlSb superlattices by the Ga ion implantation and subsequent annealing, and the improvement of the crystallinity by annealing after implantation were confirmed by Raman scattering measurement. It was carried out on the uniformly Ga implanted InAs/AlSb superlattice in a quasi back scattering geometry at room temperature by the excitation of the 488 nm line of an Ar ion laser. Figure 4 shows the Raman spectra for the (a) as grown InAs/AlSb, (b) as implanted sample with Ga ion dose of 1×10^{13} cm^{-2} and 100 keV ion energy, and (c) Ga ion implanted, 450 °C-30 min annealed sample. For the as grown sample, a very strong InAs LO mode and a AlSb LO mode are observed around 236 and 347 cm^{-1}, respectively. After Ga ion implantation, both peaks become weak and asymmetric broadening is observed due to the overlap of the local damage regions formed by ion implantation. After 450 °C annealing, however, though the AlSb LO mode intensity remains almost unchanged, the InAs LO mode intensity rapidly increases (not shown). The shift of AlSb LO mode and the increase of the InSb related mode intensity are also observed. These phenomena result from the intermixing of InAs/AlSb SLs by Ga ion implantation and subsequent annealing. Because the InAlAsSb quaternary alloy is formed by the intermixing, and In-Sb and Al-As bonds increase with the progress of the intermixing. Therefore, the increase of InSb related mode intensity is the evidence of the intermixing of InAs/AlSb superlattice. The intermixing is enhanced by the subsequent annealing after Ga ion implantation [18]. The quantitative estimation of the intermixing for InAs/AlSb SLs was also obtained from the result of the blue shift of AlSb LO mode. Its values are given by As mole fraction x for the $InAlAs_xSb_{1-x}$ quaternary alloy. The value x was estimated to be 0.18 after annealing of 450 °C -10 min.

From the above results, we can consider the mechanism of the formation of the high resistive region as follows : First is the formation of the potential barrier due to the intermixing of InAs and AlSb layer by Ga ion implantation and annealing. Second is the formation of the trap centers such as defects induced by ion implantation.

4. Summary

The formation of the small area high resistive region by Ga FIB implantation was studied by means of I-V measurements for the InAs/AlSb superlattices. The resistance increased by more than one order of magnitude by Ga ion implantation and annealing. The resistance increased with increasing fluence and decreased by annealing. At low temperatures, the higher resistance was observed. A high electrical resistive region as high as 2.5×10^3 Ω·cm can be formed in $(InAs)_{16}/(AlSb)_{22}$ superlattice layer by 5×10^{15} cm^{-2} Ga focused ion beam implantation and 400 °C-30 min.annealing. The clear photoresponce is also observed in the Ga ion implanted diode structures after annealing. The formation of high resistive region is attributed to the formation of higher potential barrier due to the intermixing of InAs and AlSb layers by ion implantation and the generation of trap centers associated with implantation induced defects. The intermixing and damage process by Ga ion implantation and annealing and the existence of residual defects are confirmed by Raman scattering study.

References

[1] J. Melngailis, J. Vac. Sci. Technol. B5, 469 (1987)

[2] S. Namba, Proc. Int. Ion Eng. Congr. 111, 1533 (1983)

[3] L. Karapiperis, D. Dubreuil, Ph. David, and D. Dieumegard, J. Vac. Sci. Technol. B3, 353 (1985)

[4] R. L. Seliger, R. L. Kubena, and V. Wang, Jpn. J. Appl. Phys. 24, 965 (1985)

[5] H. Arimoto, A. Takamori, E. Miyauch, and H. Hashimoto, Jpn. J. Appl. Phys. 23, L165 (1984)

[6] C.Vieu, M. Schneider, H. Launois, and B. Descouts, J. Appl. Phys. 71, 4833 (1992).

[7] K. Nakamura, T. Nozaki, T. Shiokawa, K.Toyoda, and S. Namba, Jpn. J. Appl. Phys. 24, L903 (1985)

[8] H. Iguchi, Y. Hirayama, and H. Okamoto, Jpn. J. Appl. Phys. 25, L560 (1986)

[9] E. Miyauchi, H. Arimoto, H. Hashimoto, and T. Utsumi, J. Vac. Sci. Technol. B1,

244

1113 (1983)

[10] Y. Hirayama and H. Okamoto, Jpn J. Appl. Phys. 24, L965 (1985).

[11] H. Asahi, H. Sumida, S. J. Yu, S. Emura, S. Gonda, and M. Komuro, Jpn. J. Appl. Phys. 28, L2119 (1989).

[12] J. P. Donnelly and C. E. Hurwitz, Solid State Electron. 30, 727 (1977).

[13] R. G. Wilson, J. Appl. Phys. 52, 3954 (1981).

[14] Y. Hirayama, S. Tarucha, Y. Suzuki, and H. Okamoto, Phys. Rev. B37, 2274 (1988).

[15] R. J. Blaikie, J. R. A. Cleaver, H. Ahmed, and K. Nakazato, Appl. Phys. Lett. 60, 1618 (1992).

[16] Ch. Lauterbach, D. Römer, and R. Treichler, Appl. Phys. Lett. 57, 481 (1990).

[17] H. Asahi, S. J. Yu, J. Takizawa, S. G. Kim, Y. Okuno, T. Kaneko, S. Emura, S. Gonda, H. Kubo, C. Hamaguchi, and Y. Hirayama, Surface Science 267, 232 (1992).

[18] S. J. Yu, H. Asahi, S. Emura, H. Sumida, S. Gonda, and H. Tanoue, J. Appl. Phys. 66, 856 (1989).

Inst. Phys. Conf. Ser. No 145: Chapter 2
Paper presented at 22nd Int. Symp. Compound Semiconductors, Cheju Island, Korea, 28 August–2 September 1995
© *1996 IOP Publishing Ltd*

Mg-Based Dual Ion Implantation Technique for High Performance P-channel AlGaAs/InGaAs Heterostructure FETs

N Hara, M Shima, H Suehiro, and S Kuroda

Fujitsu Laboratories Ltd., 10-1 Morinosato-Wakamiya, Atsugi 243-01, Japan.

Abstract. We examined a less diffusive and less toxic Mg-based dual ion implantation technique for p^+-layer formation of p-channel heterostructure field effect transistors (p-HFETs). From ion implantation experiments into GaAs/$Al_{0.75}Ga_{0.25}As$/$In_{0.2}Ga_{0.8}As$ heterostructures, P dual implantation is found to be very effective in terms of its low sheet resistance and low Mg redistribution in implanted regions. We then fabricated sub-micron p-HFETs by Mg-based ion implantation and showed the superiority of Mg compared with Be. The device performance was improved by P dual implantation. The peak transconductance and the gate turn-on voltage (defined as the gate-to-source voltage at a gate-to-source current of -1 μA/μm) of 0.5 μm-gate p-HFET fabricated with Mg+P dual implantation are 80 mS/mm and -2.0 V, respectively. This is the best combination among p-HFETs so far reported. We therefore concluded that the Mg+P dual ion implantation technique is promising for forming p^+-layers in p-HFETs with a short gate length.

1. Introduction

The complementary heterostructure field effect transistor (C-HFET) technology is promising for high speed and low power dissipation ICs [1-3]. Since the performance of a C-HFET is limited by that of p-channel HFET (p-HFET), we need to improve the p-HFET's performance by reducing its gate length. P-HFETs with high performance so far have been formed by Be ion implantation in most reports [1, 3-7], because it is easy to form p^+-layers with low sheet resistance. However, Be has some disadvantages. Its lateral diffusion during activation annealing causes a large drain and gate leakage current, which increases the power dissipation. A large drain leakage current particularly limits the performance of sub-micron p-HFETs [3, 4, 6]. Redistribution of the Be depth profile during activation annealing is also a problem when obtaining a shallow profile necessary to reduce the short channel effect. Furthermore, the toxicity of Be is a serious safety problem in producing devices. We therefore used the less diffusive and less toxic Mg as a p-type ion implantation source.

Generally, the activation coefficient in Mg ion implanted AlGaAs is smaller than that in Be ion implanted samples. Mg activation was improved and Mg diffusion was reduced by using the Mg+F, Mg+Ar, Mg+P dual ion implantation technique [8]. In this paper we investigated Mg-based dual ion implantation in order to obtain p^+-GaAs/AlGaAs/InGaAs heterostructures with a low sheet resistance and a shallow profile for ohmic layers in p-HFETs. First, we characterized the electrical properties of Mg-based dual ion implanted heterostructures, then p-HFETs were fabricated and the feasibility of Mg-based dual ion implantation technique was estimated.

2. Mg-based dual ion implantation into heterostructures

We used GaAs (20 nm)/$Al_{0.75}Ga_{0.25}As$ (20 nm)/$In_{0.2}Ga_{0.8}As$ (14 nm) heterostructures grown by metalorganic vapor phase epitaxy. Mg ions were implanted at an energy of 60 KeV with a dose of $1x10^{15}$ cm^{-2}. F, Ar, and P ions were implanted at 20, 70, and 50 KeV, respectively, with a dose of $1x10^{15}$ cm^{-2}. The implanted energies were chosen so that the profiles matched the Mg profile. Samples were annealed in an atmosphere of nitrogen at 750-850°C for 5 seconds. The samples were positioned face-down on a GaAs wafer during annealing to reduce the arsenic evaporation.

The sheet resistance and sheet carrier concentration measured by using the Van der Pauw method are shown in Fig. 1. The sheet resistance is lower and the activation coefficient is higher in Mg+P dual implanted samples, while the electrical properties were degraded in Mg+F and Mg+Ar dual implanted samples. These are generally consistent with results for Mg-based dual ion implanted GaAs, $Al_{0.75}Ga_{0.25}As$, and $In_{0.2}Ga_{0.8}As$ thick single layers. The sheet resistance of the Mg+P dual implanted heterostructure after 800°C annealing, which we used in the p-HFET fabrication process described later, is 72% that of the Mg single implanted sample. The reduction in sheet resistance and the increase in hole concentration by P dual implantation are favorable ways to decrease the source resistance when fabricating a device.

Next, the carrier profiles were characterized by using electrochemical CV measurements (Fig. 2). In the P dual implanted sample, the carrier concentration in all GaAs, AlGaAs, and InGaAs layers was higher than in the Mg single implanted sample, while in the F or Ar dual implanted samples, the carrier concentration was improved only in the AlGaAs layer. This is in good agreement with the Hall measurements described above. The reason for the superiority of P dual implanted samples is that the stoichiometric balance is maintained by the addition of P atoms as reported for $Al_{0.3}Ga_{0.7}As$ [9]. Another purpose of dual ion

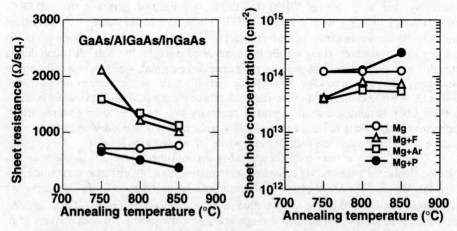

Fig. 1 Sheet resistance and sheet carrier concentration in Mg-based dual ion implanted GaAs/AlGaAs/InGaAs heterostructures measured by using Van der Pauw method.

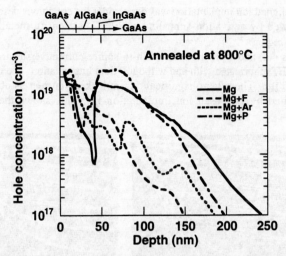

Fig. 2 Carrier profiles in Mg-based dual ion implanted GaAs/AlGaAs/InGaAs heterostructures measured by using electrochemical CV measurements. Activation annealing was done at 800°C.

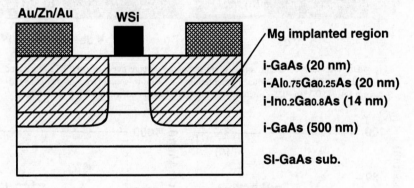

Fig. 3 Cross-section of a p-HFET.

implantation is to reduce the Mg diffusion. As can be seen in Fig. 2, the dual implanted samples have shallower carrier profiles than that in the Mg single implanted sample. It means that the dual ion implantation technique effectively suppresses the Mg diffusion.

3. Fabrication of p-HFETs

We fabricated self-aligned p-HFETs (Fig. 3) by using Mg+P dual ion implantation, and this effectively improved the electrical properties of implanted p+-heterostructures, as described above. The fabrication process was as follows. The epitaxial layers consisted of a GaAs buffer layer (500 nm), an $In_{0.2}Ga_{0.8}As$ pseudomorphic channel layer (14 nm), an $Al_{0.75}Ga_{0.25}As$ barrier layer (20 nm), and a GaAs cap layer (20 nm). All layers were not intentionally doped. After forming the WSi gate, the AlN (20 nm) was deposited and SiON sidewalls were formed. Self-aligned ion implantation was done under the conditions described in the previous section, followed by activation annealing at 800°C. Au/Zn/Au metallization was used to contact the p⁺-region.

Figures 4(a) and 4 (b) show the drain-to-source current versus drain-to-source voltage curve of p-HFETs fabricated with and without P dual implantation. The gate length and width are 0.5 and 10 μm. There is no significant drain leakage current as observed in 0.5 μm p-HFET fabricated by Be-based ion implantation [3]. This shows the superiority of less

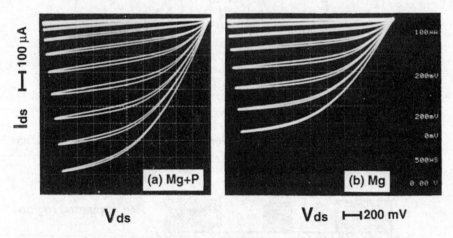

Fig. 4 Drain-to-source current versus drain-to-source voltage curve of p-HFETs fabricated (a) with, and (b) without P dual implantation. The gate-to-source voltage was changed from 0 to -2.0 V in -0.2 V increments. The gate length and width are 0.5 and 10 μm.

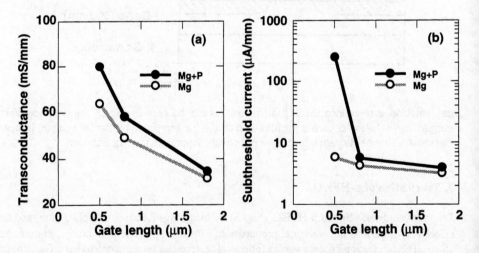

Fig. 5 Gate length dependence of (a) transconductance, and (b) subthreshold current of p-HFETs fabricated with and without P dual implantation.

diffusive Mg-based ion implantation. The gate length dependence of the transconductance is shown in Fig. 5 (a). The transconductance increases with the reduction in gate length, which indicates the feasibility of short gate p-HFETs. Figure 5 (b) shows the subthreshold current as a function of the gate length. Since the devices we fabricated in this article have no buried n-layer to confine carriers, the subthreshold current of the sub-micron devices is higher than the high performance sub-micron p-HFET with a buried n-layer [4]. The increase in the subthreshold current of the 0.5 μm-device with P dual implantation is due to the high activation of Mg in InGaAs and GaAs layers. However, our devices have a smaller subthreshold current than in the p-HFET without a buried n-layer of about 1000 μA/mm [4].

Next, we measured the gate-to-source current (Fig. 6). The gate leakage current not only increases the power dissipation in a C-HFET, but also degrades the p-HFET's

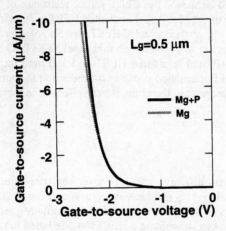

Fig. 6 Gate-to-source current as a function of gate-to-source voltage of p-HFETs with 0.5 μm-gate at a drain-to-source voltage of 0 V.

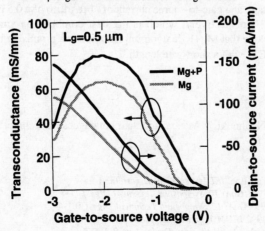

Fig. 7 Transconductance and drain-to-source current versus the gate-to-source voltage curves of p-HFETs with 0.5 μm-gate at a drain-to-source voltage of -1.5 V.

performance by reducing the gate voltage swing. The gate turn-on voltage, defined as the gate-to-source voltage at a gate-to-source current of -1 $\mu A/\mu m$, is about -2.0 V for both HFETs fabricated with and without P dual implantation. The absolute value is large enough for application in C-HFETs. The combination of a 80 mS/mm transconductance and a gate turn-on voltage of -2.0 V for the 0.5 μm device with P dual implantation is the best among p-HFETs so far reported. The large gate turn-on voltage comes from the small lateral diffusion of Mg in the AlGaAs layer as well as the employment of an AlGaAs barrier layer with a high Al fraction.

Finally, we compare the p-HFETs fabricated with and without P dual implantation. The transconductance and the drain current versus the gate-to-source voltage curves are shown in Fig. 7. The peak transconductance of the Mg single implanted HFET is 64 mS/mm, which is not particularly good because it has a high source resistance of 3.2 Ω mm. However, the peak transconductance of the Mg+P dual implanted HFET is 80 mS/mm, which is about 25% higher than that of the Mg single implanted HFET. We also observed the increase in the drain current, which is important to balance with n-channel FETs in C-HFETs. Since the source resistance of the Mg+P dual implanted HFET is 1.2 Ω mm, improvement in the device performance by P dual implantation is due to the reduction in source resistance. The Mg+P dual ion implantation technique, therefore, is very effective in fabricating high performance sub-micron p-HFETs.

4. Summary

We have examined less diffusive and less toxic Mg-based dual ion implantation into GaAs/Al$_{0.75}$Ga$_{0.25}$As/In$_{0.2}$Ga$_{0.8}$As heterostructures and found that P dual implantation is very effective in terms of its low sheet resistance and low Mg redistribution in implanted regions. We then fabricated sub-micron p-HFETs by Mg-based ion implantation and showed the superiority of Mg compared with Be. Finally, the device performance was improved by P dual implantation. The peak transconductance and the gate turn-on voltage (defined as the gate-to-source voltage at a gate-to-source current of -1 $\mu A/\mu m$) of a 0.5 μm-gate device are 80 mS/mm and -2.0 V, respectively, which is the best combination among p-HFETs so far reported. We conclude that Mg+P dual ion implantation is a promising technique for forming p^{+}-layers in p-HFETs with a short gate length.

5. Acknowledgments

We wish to thank N Ohnuki, Y Maeba, and their colleagues for technical support.

References

[1] Akinwande A I et al. 1990 *IEDM Technical Digest* 983-6
[2] Grider D E et al. 1991 *GaAs IC Symposium Technical Digest* 71-4
[3] Kiehl R A et al. 1991 *GaAs IC Symposium Technical Digest* 101-4
[4] Grider D E et al. 1992 *IEDM Technical Digest* 331-4
[5] Daniels R D et al. 1988 *IEEE Electron Device Lett.* **9** 355-7
[6] Ruden P P et al. 1989 *IEEE Trans. on Electron Devices* **36** 2371-9
[7] Abrokwah J K et al.1993 *IEEE Trans. on Electron Devices* **40** 278-84
[8] Hara N et al. 1995 (to be published)
[9] Morizuka K et al. 1986 *Electron. Lett.* **22** 315-6

Inst. Phys. Conf. Ser. No 145: Chapter 2
Paper presented at 22nd Int. Symp. Compound Semiconductors, Cheju Island, Korea, 28 August–2 September 1995
© *1996 IOP Publishing Ltd*

Intermixing Effect of (InAs)/(AlSb) Superlattice by Ga Ion Implantation

Song Gang Kim*, Hajime Asahi, and Shun-ichi Gonda
The Institute of Scientific and Industrial Research, Osaka University
8-1 Mihogahoka, Ibaraki, Osaka 567 Japan

Soon Jae Yu
Department of Electronics Engineering, Sunmoon University
100 Kalsanri, Tangjeong, Asan, Chungnam 336-840 Korea

Abstract. The radiation damage and the intermixing behavior induced by Ga ion implantation and subsequent thermal annealing have been studied for $(InAs)_{16}/(AlSb)_{16}$ SLs using Raman scattering. The compositional intermixing effects by Ga ion implantation are explained in terms of the increase of InSb mode intensity and the shift of AlAsSb-like mode frequency. Raman results show that the compositional intermixing depends sensitively on Ga ion fluence and annealing temperature and consists of two step modes. The first step intermixing occurs by collision effect during Ga ion implantation. The second step intermixing by low temperature annealing recovers the damage states caused by Ga ion implantation which induces the decreasing thermal stability of InAs/AlSb SL structures.

1. Introduction

In III-V compound superlattice structures, compositional disordering has recently received considerable attention due to its potential application to photonic devices [1-3]. A commonly used method to induce compositional disordering in a layered structure is to implant a moderate fluence ($\sim 10^{14}$ cm^{-2}) of ions into the structure at room temperature followed by a high temperature annealing [4,5]. Ga ion implantation offers the potential for more abrupt intermixing of III-V compound SL structures without impurity doping, and can be used to make the microstructures such as quantum well wire and quantum box structure by selective intermixing.

It is important to investigate the behavior of the ion implantation induced defects in order to apply this technique to the actual device fabrication process. Especially, many experimental attempts to exploit ion implantation-induced disordering to device fabrication have been reported [6-9]. In any

Present address : Semiconductor R&D Lab.2, Hyundai Electronics Industries Co., Ltd.,
San 136-1, Ami-ri, Bubal-eub, Ichon-kun, Kyoungki-do, 467-860 Korea

kinds of applications the crystal disorder remaining after thermal annealing is the key parameter which governs the performance of the device.

In epitaxial semiconductor heterostructures, the ion implantation effect is different in each composing layers, for example, the difference of the threshold fluence for amorphization which can strongly influence the recovery behavior. Thus, detailed studies on the damage generation and recovery in ion implanted semiconductor multilayers are necessary.

In this study, we report on the intermixing effects of InAs/AlSb SLs by Ga ion implantation and the subsequent thermal annealing using Raman scattering.

2. Experimental

Gas source MBE growth of InAs/AlSb SLs used in this study was carried out using the elemental Al, In and Sb, and thermally cracked AsH_3 as group III and group V sources. A GaAs buffer layer of 0.1μm was grown on the Cr-O doped SI (001) GaAs substrate at 600°C. And then, 6-periods of $(InAs)_{16}/(AlSb)_{16}$ SLs were grown at 490°C, followed by InAs cap layer (200Å) at 490°C to protect the AlSb layer from the reaction with water vapor in air.

Ga ion implantation was made with a current density of $0.1\mu A/cm^2$ and an ion energy of 100keV at room temperature. Implantation fluences were varied in the range from $1x10^{13}$ to $5x10^{14}$ cm^{-2}. The incident angle was 7°off from the [100] axes. After Ga ion implantation, samples were annealed at various temperatures from 300 to 600 °C in a N_2 ambient by face to face method.

Raman spectra were measured in a quasi- backscattering geometry at room temperature with a 200 mW power level.

3. Results and discussion

Figure 1 shows Ga ion fluence dependence of Raman spectra from InAs/AlSb SLs. In the case of as-grown sample, a very strong InAs LO mode and a AlSb LO mode are observed around 236 and 347 cm^{-1}, respectively. With increasing ion fluence, both peaks become weak and, asymmetric broadening are observed due to the overlap of the local damage regions formed by ion implantation. The damaged states of InAs and AlSb layers by Ga ion implantation is different. In the fluence of $1x10^{13}$ cm^{-2}, the damage effect is much less pronounced in AlSb than in InAs. In the fluence of $1x10^{14}$ cm^{-2}, however, this is reversed. It is noted that the LO mode of AlSb almost disappear in the spectrum with the fluence of $1x10^{14}$ cm^{-2}. This means that AlSb layer becomes amorphized by Ga ion implantation with the fluence of $1x10^{14}$ cm^{-2}. This result is very different from the result of Ga ion implanted AlSb in the ref.[10]. In the case of AlSb implanted with Ga ions, the LO modes are observed even in the implantation with the fluence of $5x10^{14}$ cm^{-2}. That is, AlSb was not completely amorphized even at this fluence.

Furthermore, in Fig.1 InSb like TO mode is also clearly observed in the spectrum with the fluence of 1×10^{14} cm^{-2}. From these results, it is considered that the disappearance of AlSb LO mode in the InAs/AlSb SL sample with the fluence of 1×10^{14} cm^{-2} in Fig.1 is caused by the damage formation and the partial intermixing between InAs and AlSb layers. Therefore, it may be considered that the relatively strong InAs LO mode comes from the 200 Å InAs cap layer. Thus, the intermixing effects are mainly discussed with the behavior of the AlSb LO mode and InSb like TO mode.

Thermal annealing effects were

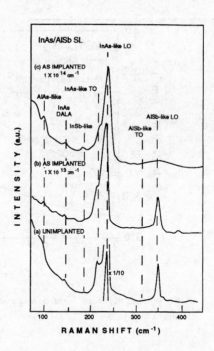

Fig.1 Raman spectra for the InAs/AlSb SLs with InAs cap (a) unimplanted, and (b) and (c) Ga ion implanted with a fluence of 1×10^{13} cm^{-2} and 1×10^{14} cm^{-2}.

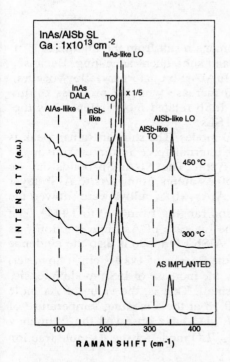

Fig.2 Raman spectra for the InAs/AlSb SLs with InAs cap implanted with 100 keV Ga ions with a fluence of 1×10^{13} cm^{-2} followed by the furnace annealing for 10 min at different temperatures.

investigated on the $(InAs)_{16}/(AlSb)_{16}$ SLs implanted with Ga ion to evaluate the degree of implantation enhanced interdiffusion. Figure 2 shows the annealing temperature dependence of Raman spectrum of InAs/AlSb SL structure after the Ga ion implantation with the fluence of 1×10^{13} cm^{-2}. In annealing of 300 °C, the recovery of the damage state is very lower in both InAs and AlSb. After 450°C annealing, however, though the AlSb LO mode intensity remains almost unchanged, the InAs LO mode intensity rapidly increases. This means that the recovery process between InAs and AlSb is very different. The shift of AlSb LO mode and the increase of the InSb related mode

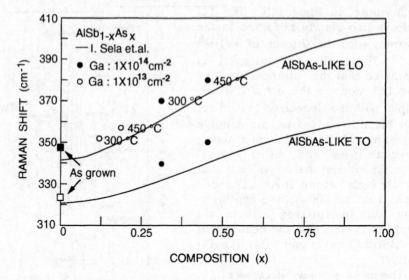

Fig.3 Measured LO and TO mode frequencies of the different $AlSb_{1-x}As_x$ alloys as a function of alloy composition.

intensity are also observed. These phenomena result from the intermixing of InAs/AlSb SLs by Ga ion implantation and subsequent annealing. Because, when the intermixing which forms the InAlAsSb quaternary alloy, occurs, In-Sb and Al-As bonds are formed and increase with the progress of the intermixing. Therefore, the increase of InSb related mode intensity is the evidence of the intermixing of InAs/AlSb SLs.

We also discuss the shift of AlSb LO modes. The shift of Raman peak is generally explained by the variation of internal force in the crystal or the concentration variation of the mixed crystal. The Al-As-Sb bonds can be formed as a result of the interdiffusion of As atoms from InAs to AlSb layer or of Sb atoms from AlSb to InAs layer. $AlAs_xSb_{1-x}$ alloys were showed the single-mode behavior for As mole fraction ranging from x=0 to 1 [11]. The mode shifts to higher frequency with the increase of As mole fraction (as shown in Fig.3). Therefore, the shift of AlSb LO mode is also the evidence of the intermixing of InAs/AlSb SLs. At the fluence of $1x10^{13}$ cm^{-2}, however, the degree of intermixing is low, because the increase of InSb mode intensity and the shift of AlSb LO-like mode is small. That is, the value of As mole fraction x was estimated x = 0.12 and 0.18 at the annealing temperature of 300 and 400 °C, respectively. Sela etal. [11] have observed the frequency shift of 17 and 40 cm $^{-1}$ in the $AlAs_xSb_{1-x}$ LO mode for the As mole fraction of x=0.25 and 0.5, respectively.

The annealing temperature dependence of Raman spectrum on the sample with the fluence of $1x10^{14}$ cm^{-2} is shown in Fig.4. Although the Ga ion fluence increases from $1x10^{13}$ to $1x10^{14}$ cm^{-2}, the recovery behavior of InAs-like mode is similar to that of $1x10^{13}$ cm^{-2} in the whole annealing temperature range. However, at the sample with the $1x10^{14}$ cm^{-2} fluence the recovery behavior of InSb-like and AlSb-like modes is very different from that of 1×10^{13} cm^{-2}. Firstly, the InSb-like mode intensity is increased with increasing annealing temperature. The increase of InSb mode in the 1×10^{14} cm^{-2} fluence is larger than that of the 1×10^{13} cm^{-2} fluence. This is the evidence of Ga ion implantation enhanced interdiffusion in the InAs/AlSb SLs. Secondly, the broad AlSb-like modes are separated to two modes and shift to higher frequency region with increasing annealing temperature due to the intermixing of InAs/AlSb SLs. In the sample with 450 °C annealing, these two new peaks at around 350 and 380 cm^{-1} correspond to Al-As-Sb-related TO and Al-As-Sb-related LO mode, respectively. In the sample with the 1×10^{14} cm^{-2} fluence, the value of As mole fraction x was estimated to be x = 0.31 and 0.45 at

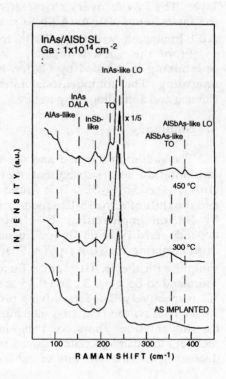

the annealing temperature of 300 and 400 °C, respectively. From the above results, it is found that the intermixing of InAs/AlSb SL structure is enhanced by the subsequent annealing after Ga ion implantation.

Figure 5 shows the fluence dependence of the intensity ratio of LO mode to TO of InAs-like and AlSb-like modes for the Ga ion implanted and annealed InAs/AlSb SLs. The recovery characteristics of InAs and AlSb layers decrease with increasing fluence. Especially, the change of the intensity is much pronounced in AlSb layer than in InAs

Fig.4 Raman spectra for the InAs/AlSb SLs with InAs cap implanted with 100 keV Ga ions with a fluence of 1×10^{14} cm^{-2} followed by the furnace annealing for 10 minutes at different temperatures.

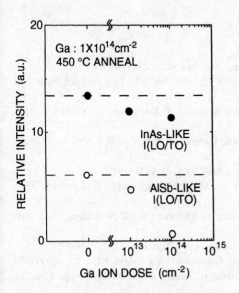

Fig.5 Raman intensity ratios of the AlSb LO/TO and InAs LO/TO modes for Ga ion implanted and annealed InAs/AlSb SLs to the LO/TO mode of AlSb and InAs as a function of fluence.

layer. The low recovery characteristics at the 1×10^{14} cm^{-2} fluence is due to the intermixing of InAs/AlSb SLs, because they are much lower than that of AlSb implanted with the 1×10^{14} cm^{-2} fluence in ref.[10].

From the above results, it is concluded that the compositional intermixing is induced by Ga ion implantation and is enhanced by thermal annealing. The compositional intermixing depends sensitively on Ga ion fluence and annealing temperature.

4. Summary

The radiation damage and the intermixing behavior induced by Ga ion implantation and subsequent thermal annealing have been studied for $(InAs)_{16}/(AlSb)_{16}$ SLs. It is found that the increase of InSb mode intensity and the shift of AlAsSb-like mode are caused by the compositional intermixing by Ga ion implantation. It is shown that the compositional intermixing depends sensitively on Ga ion fluence and annealing temperature. The degree of intermixing for $(InAs)_{16}/(AlSb)_{16}$ SLs is quantitatively estimated : in the sample with the 1×10^{14} cm^{-2} fluence, the value of As mole fraction x was estimated to be x = 0.31 and 0.45 at the annealing temperature of 300 and 400 °C, respectively. The intermixing process of (InAs)/(AlSb) SL consists of two steps. That is, the first step intermixing occurs by collision effect during Ga ion implantation. The second step intermixing by low temperature annealing recovers the damage states caused by Ga ion implantation which induces the decreasing thermal stability of InAs/AlSb SL structures.

References

[1] Y. Hirayama, S. Tarucha, Y. Suzuki, and H. Okamoto, Phys. Rev. B37, 2274 (1988).

[2] J. Cibert, P. M. Petroff, G. J. Dolan, S. J. Pearton, A. C. Gossard, and J. H. English, Appl. Phys. Lett. 49, 1275 (1986)

[3] C.Vieu, M. Schneider, G. Benassayag, R. Planel, L. Birotheau, J. Y. Marzin, and B. Descouts, J. Appl. Phys. 71, 5012 (1992).

[4] F. Laruelle, P. Hu, R. Simes, R. Kubena, W. Robinson, J. Merz, and P. M. Petroff, J. Vac. Sci. Technol. B7, 2034 (1989).

[5] C.Vieu, M. Schneider, H. Launois, and B. Descouts, J. Appl. Phys. 71, 4833 (1992).

[6] P. Gavrilovic, K. Meehan, L. J. Guido, N. Honyak, Jr., V. Eu, M. Feng, and R. D. Burn ham, Appl. Phys. Lett. 47, 903 (1958)

[7] K. Isida, K. Matsui, T. Fukunaga, T. Morita, E. Miyauch, H. Hashimoto, and H. Nakashima, Jpn. J. Appl. Phys., 25, L873 (1986)

[8] J. E. Epler, R. D. Burnham, R. L. Thornton, and T. 1. Paoli, Appl. Phys. Lett. 51, 731 (1987)

[9] K. Isida, K. Matsui, T. Fukunaga, T. Takamori, and H. Nakashima, Jpn. J. Appl. Phys., 25, L690 (1986)

[10] S. G. Kim, H. Asahi, M. Seta, S. Emura, H. Watanabe, H. Tanoue, and S. Gonda, J. Appl. Phys. 74, 2300 (1993)

[11] I. Sela, C. R. Bolognesi, and H. Kroemer, Phys. Rev. B46, 16142 (1992).

Inst. Phys. Conf. Ser. No 145: Chapter 2
Paper presented at 22nd Int. Symp. Compound Semiconductors, Cheju Island, Korea, 28 August–2 September 1995
© *1996 IOP Publishing Ltd*

Evidence for introduction of extra Si from PECVD SiN$_x$ cap during post-implantation annealing of implanted Si due to ion implantation damage in GaAs

Kyung Ho Lee, Jae-Jin Lee, and Kwang Eui Pyun

Electronics and Telecommunications Research Institute, Yusong P.O.Box 106, Taejon, 305-600, Korea

Abstract. The introduction of extra Si from SiN$_x$ cap during 950°C activation annealing of implanted Si in GaAs has been observed in two cases: one was the recoil of Si from SiN$_x$ mask due to through-nitride ion implantation(T-N-II) of $_{28}$Si or $_{40}$Ar in GaAs and subsequent annealing; the other was the gettering of Si from SiN$_x$ cap after post-implantation annealing when SiF or SiF$_2$ was implanted in GaAs as a Si source. For the T-N-II, the implantation condition was 4E13cm^{-2} at 120KeV for Si or Ar with the initial nitride thicknesses in the range of 700 to 1500Å. The total Si dose of T-N-II after activation increased for nitride thicknesses less than 1100Å compared to the direct implantation without initial nitride, according to the secondary ion mass spectrometer (SIMS) depth profiles. However, the sheet resistance increased for all the Si T-N-II samples. Ar ions were implanted through-nitride of 700 to 1500Å thicknesses to find significant amount of Si in GaAs up to approximately 1000Å after annealing. The sheet resistances were not measurable for Ar T-N-II samples. Therefore, we suggest that the recoiled Si atoms were distributed in such a random fashion that the selective ordering redistribution of Si to the Ga sublattice sites was not enhanced for the activation condition of the present study. A dose of 4E12cm^{-2} at the energy range of 30-130KeV has been implanted using SiF or SiF$_2$ to find out that Si concentration increased significantly on the surface side of the projected ranges after activation annealing with SiN$_x$ cap according to SIMS. This is an evidence of the introduction of Si from the nitride during activation annealing. The TRIM-90 calculation was done to elucidate the mechanism of damage enhanced introduction of extra Si in GaAs during activation annealing.

1. Introduction

Although the ion implantation has been one of the major processing technologies used in the fabrication of Si semiconductor device processing for long time, the utilization of this technology for GaAs device processing has not been as extensive as for Si case due to a number of inherent problems in GaAs material. One of the most critical problems for GaAs is its compound nature. The destruction of the lattice structure due to the nuclear stopping of incoming energized ions through the collisions with the sublattice atoms, the implantation damage, is hardly recovered to the entirety during activation annealing. The types of defects created by the damage range from the Frenkel pairs, divacancies, and various complexes depending on the implantation and annealing conditions.[1]-[3] Since those defects can be sources or sinks of point defects, they affect the activation and diffusion behavior of the dopnats during recovery annealing process.[4]

For certain applications, very shallow channel thickness is required. One of the typical examples is in the fabrication of enhancement mode MESFET, where the effective channel thickness should be equal or less than the depletion width of the gate-substrate junction. If one deposits a masking dielectric layer on the substrate before implantation, one can reduce the implant dose and depth compared to direct implantation case. Also, if one uses molecular sources for implantation, the channel thickness can be reduced compared to the elemental source implantation, because the mass of the source is higher for m-Si and the projected range is smaller than e-Si at the same implantation energy. However, due to the higher mass of m-Si than e-Si, it is suspected that the damage effect would be higher for molecular source compared to elemental source. In this study, the activation behavior of shallow channel formed by implanting through-nitride or molecular source of Si and the resulting introduction of extra Si caused by the implantation damage effect were observed.

2. Experimental

The Eaton NV6200 medium current ion implanter was used for implantation. The (100) semi-insulating GaAs substrates were cleaned in O_2 plasma and HCl successively before implantation. Ion implantation was done at room temperature with wafers tilted 7° off the incoming ion beam and rotated 30° to minimize channeling. Silicon nitride was deposited by plasma enhanced chemical vapor deposition (PECVD) at 320°C. The thicknesses of initial nitride deposited after substrate cleaning were varied in the range of 700-1500Å for through-nitride ion implantation (T-N-II). Si or Ar ions were implanted at 120KeV and 1E13cm^{-2} dose. For T-N-II, nitride films were deposited only on the front surface of substrates and removed in HF after implantation and before activation. For molecular source implantation of Si, SiF$^+$ or SiF$_2^+$ was selected for implantation at 4E12cm^{-2} dose for energies of 30-130KeV. For activation, approximately 700Å of nitride films were deposited after implantation on both sides of substrates and were removed after activation for analysis. Ellipsometer was used for thickness and refractive index measurements of nitride. Sheet resistances were obtained by Hall or Lehighton measurements and secondary ion mass spectrometry (SIMS) was used for chemical concentration depth profiles. TRIM calculations were done to explain the increase of Si concentration after activation in terms of abundant vacancy concentration created by implantation damage.

3. Results and discussions

3-1. Through nitride ion implantation (T-N-II)

As the thickness of initial silicon nitride becomes higher, the masking effect of nitride is assumed to become more. In other words, the projected range (R_p) and the total dose in the substrate are expected to become smaller. Figure 1 shows the SIMS profiles of direct (#1 sample) and through-nitride (#'s 3, 5, 7, and 9) implanted profiles of $_{28}$Si of 1E13cm^{-2} dose at

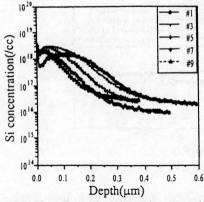
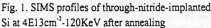
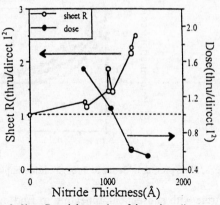

Fig. 1. SIMS profiles of through-nitride-implanted Si at 4E13cm⁻²-120KeV after annealing

Fig. 2. Sheet R and dose ratios of through to direct implanted Si at 4E13cm⁻²-120KeV after annealing

120KeV after activation. As shown in Fig. 1, the peak concentration, N_p, for all through-nitride samples increased compared to the direct implanted one. The nitride thicknesses were 0, 698, 1048, 1315, and 1528Å for sample numbers of 1, 3, 5, 7, and 9, respectively. The total doses calculated from the SIMS profiles in Fig. 1 were normalized to the direct implanted one with respect to the nitride thickness as shown in Fig. 2. The total dose in the substrate decreased as the thickness of the initial nitride increased. An unexpected feature in Fig. 2 is the fact that the total dose in the substrate rather increases for nitride thicknesses up to approximately 1100Å. Therefore, the initial nitride of thicknesses below 1100Å behaves as a source rather than a mask for Si. Also shown in Fig. 2 is the sheet resistance(R) variation with respect to the nitride thickness. Although the Si concentration increased for T-N-II samples with initial nitride thicknesses of 698 and 1048Å, their sheet R values were higher for T-N-II samples than for the direct implanted one. Therefore, the additional Si dose introduced from the capping nitride does not contribute in lowering the sheet resistances. The reason that the extra Si actually lower the activation efficiency is suggested to be due to the fact that extra Si atoms are gettered by the abundant point defects created by the implantation condition of the current experiment and the site occupancy of those extra Si atoms is in such a random fashion that self-compensation occurs and mobility decreases. The recoil of sublattice atoms is dependent on the implantation conditions and the resulting distribution after recoil of sublattice atoms is dependent on their masses. Since the masses of Ga and As are very close, their recoil distribution almost overlap leaving the net vacancy profile shallower to the surface side than the recoil profile on the surface side of the projected range of incoming ions. When thermal energy is provided as a driving force for the redistribution of the vacancies, it seems feasible that Si atoms from the capping nitride are gettered by the vacancies of either Ga or As at random probability during annealing. Therefore, the extra Si becomes self-compensated and the charged impurity scattering effect is increased to lower the electron mobility resulting in the increase of resistance. To confirm if the extra Si dose is actually introduced from the capping nitride during annealing, we implanted only Ar through

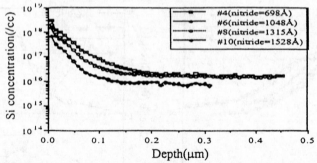

Fig. 3. SIMS profiles of Si due to through-nitride-implanted Ar at $4E13cm^{-2}$-120KeV after annealing

nitride with subsequent annealing to find out significant amount of Si in the substrate by SIMS profiles as shown in Fig. 3. Fig. 3 shows that the thinner the initial nitride, the more extra Si was detected in the substrate. This is expected because the masking role of the initial nitride becomes weaker for the thinner nitride and gettering effect becomes more for the thinner nitride. To check the possibility of introduction of extra Si from the nitride even without implantation damage effect during annealing, silicon nitride was deposited on the GaAs substrate and subsequent annealing was followed without implantation. According to SIMS profiles, Si concentration was found to be negligible(not shown). Therefore, it can be concluded that the introduction of the extra Si from capping nitride during annealing occurs only when the implantation damage exists.

3-2. Molecular source Si implantation (m-Si implantation)

For implantation energies in the range of 30-130KeV, SiF or SiF_2 implantation resulted in significant extent of Si introduction from the capping nitride during activation annealing. Figures 4 and 5 show the typical SIMS profiles before and after post-implantation annealing for SiF and SiF_2 implanted at 70 and 110KeV, respectively, at a dose of $4E12cm^{-2}$. After annealing, Si concentration increased on the surface side of the projected ranges. On the other hand, fluorine profiles dropped to the background level abruptly inside the substrate after activation annealing. The fact that the locations of the extra Si introduced from the capping nitride are on the surface side of the projected ranges is analogous to what has been observed for the case of through-nitride Si implantation discussed in the previous section. The amount of the extra Si increased as the energy increased for one type of m-Si, and was more for the heavier source, SiF_2. Therefore, it can be concluded that the extent of extra Si introduction from the capping nitride during activation annealing is a function of the implantation damage, and it increases as the mass of implantation source or the implantation energy increases.

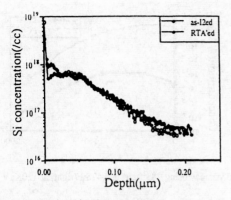

Fig. 4. As-implanted and annealed SIMS profiles with SiF source at 4E12cm^{-2}-70KeV

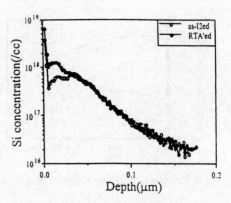

Fig. 5. As-implanted and annealed SIMS profiles with SiF$_2$ source at 4E12cm^{-2}-70KeV

3-3. Defect chemistry and TRIM calculation

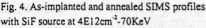

The fact that the gettered extra Si introduced during annealing for T-N-II or m-Si samples did not enhance the electrical activation implies that the redistribution of Si and the vacancies may be limited to the short range ordering so that the probability of the extra Si atoms to occupy either sublattice site is random for the activation condition of the current experiment. If the annealing was processed for longer time, the reordering of atoms and defects would occur more and the probability of Si atoms to occupy Ga sublattice sites is assumed to become higher to yield higher n-type activation efficiency. To check the time dependency of redistribution, we prolonged the annealing time of Si-direct-implanted samples to 20 seconds and found that the sheet resistances decreased significantly compared to the 10 sec-annealed samples for $_{29}$Si at 4E12cm^{-2}-70KeV(1060 to 760Ω/sq.:30% decrease) and $_{28}$Si at 4E13cm^{-2}-120KeV(94 to 74Ω/sq.:20% decrease) conditions. For SiF implanted sample at 4E12cm^{-2}-110KeV, the decrease of sheet resistance for 30" annealed sample with respect to 10" annealed sample was less significant(1080 to 970Ω/sq.:10% decrease). As the results show, the decrease of sheet resistance becomes less significant in the direction of more damage. In other words, the samples with more damage require longer time to obtain identical degree of activation. Although the activation efficiency was enhanced by the longer annealing, the Si concentration profiles for 20 sec-annealed samples showed virtually no change at all compared to 10 sec-annealed profiles within SIMS detection limit. Therefore, the progress of redistribution seems to be still limited to short range order between Si and nearby vacancies to result in the invariant concentration profiles.

To elucidate the gettering phenomenon of Si from the capping nitride to the substrate during activation annealing, TRIM-90 calculation[5] was performed for the as-implanted cases of $_{28}$Si, and $_{47}$SiF in GaAs at 120KeV as shown in Fig. 6. According to the simulation, the vacancy profile for each case lies closer to the surface side than the ion profile. Also, as the peak concentration of the heavier ion is located closer to the surface than that of the lighter ion, the peak concentration of the vacancy corresponding to the heavier ion is located

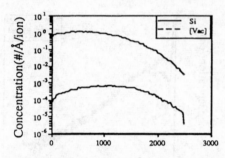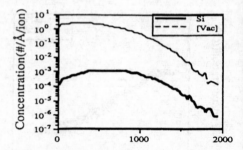

Fig. 6. TRIM-90 calculation profiles of implanted Si and vacancy using 28Si(left) and 47SiF(right) at 120KeV

accordingly closer to the surface than that corresponding to the lighter ion. Moreover, the concentration level of the vacancy is approximately three orders of magnitude
higher at the peak level than the peak ion concentration for each case and is higher for heavier ion source, as previously observed elsewhere.[6] Therefore, such abundant vacancy concentration created by ion implantation located inside the surface is responsible for the gettering of extra Si from the capping nitride during annealing. Naturally, this effect is more prominent for the heavier ion source because of the higher vacancy concentration and its shallower location close to the surface than for the lighter ion confirmed by TRIM-90 simulation.

4. Conclusions

The introduction of extra Si from SiN_x cap during activation annealing of implanted Si sources in GaAs has been observed in two cases: one was the recoil of Si from SiN_x mask due to through-nitride ion implantation (T-N-II) and subsequent annealing of $_{28}Si$ or $_{40}Ar$ in GaAs; the other was the significant increase of Si concentration after post-implantation annealing compared to the as-implanted Si concentration when SiF or SiF_2 was implanted in GaAs as a Si source. For both cases, the extra Si introduced during activation annealing did not lower the electrical resistivity. According to TRIM-90 simulation, the random distribution of the recoiled Si from the initial nitride for the former case and the gettered Si from the capping nitride for the latter case to the abundant vacancy concentration created by the implantation damage is suggested to be responsible.

References

[1] Mader S 1982 *Ion Implantation Techniques* (Berlin:Springer-Verlag) 299-316

[2] Ryssel H and Ruge I 1986 *Ion Implantation* (Chichester: John Wiley & Sons)

[3] Ghandhi S K 1983 *VLSI Fabrication Principles* (New York: John Wiley & Sons)

[4] Orlowski M 1991 *Appl. Phys. Lett.* **58** 1479-1481

[5] Biersack J P and Haggmark L G 1980 *Nucl. Inst. Meth.* **174** 257-269

[6] Christel L A and Gibbons J F 1981 *J. Appl. Phys.* **52** 5080-4

Inst. Phys. Conf. Ser. No 145: Chapter 2
Paper presented at 22nd Int. Symp. Compound Semiconductors, Cheju Island, Korea, 28 August–2 September 1995
© *1996 IOP Publishing Ltd*

Breakdown mechanism of focused-ion-beam-implantation-isolations in a GaAs/AlGaAs heterostructure

S W Hwang[*], H J Lezec[#], T Sakamoto[@] and K Nakamura[@]

[*]Department of Electronics Engineering, Korea University, Sungbukku, Anamdong, Seoul 136-701, Korea
[#]Micrion GmbH, Pilotystrasse 4, 80538 Muchen, Germany
[@]Fundamental Research Laboratories, NEC Corporation, 34 Miyukigaoka, Tsukuba, Ibaraki 305, Japan

Abstract. Breakdown mechanism of focused-ion-beam-implantation-isolations (FIBII's) in a GaAs/AlGaAs heterostructure is systematically studied. The breakdown current-voltage (I-V) characteristics of the FIBII's display the composite of three bias regimes where the current shows three different functional dependences on the bias voltage. The I-V characteristics in those regimes are quantitatively explained by the models of hopping, thermionic emission, and space-charge-limited injection, respectively. Important parameters such as the charged defect density or the barrier height can be obtained from the I-V characteristics.

1. Introduction

Since focused ion beam was shown to be an effective isolation technique in GaAs epi-layers [1,2], it has been a powerful tool in the fabrication of GaAs-based in-plane-gate transistors and many kinds of quantum devices [3]. However, breakdown mechanism across focused-ion-beam-implanted-isolations (FIBII) has not been carefully studied and understood very well. In this paper, we present a systematic study of breakdown mechanism across FIBII's of various implantation doses in a high mobility two-dimensional electron system (2DEG) formed at a GaAs/AlGaAs heterostructure.

2. Experiment

The starting semiconductor substrate was an MBE-grown modulation-doped GaAs/AlGaAs heterostructure wafer with a two-dimensional electron density of 7.56×10^{11} cm^{-2}, and a mobility of 6.32×10^3 cm^2/(Vsec) at a temperature, T = 300 K. Figure 1 shows a schematic sample structure. Several 2DEG channels connecting source and drain were prepared by wet-etching and a single line scan of a focused ion beam of 130 keV Ga$^+$ ions was performed across those channels to create FIBII's. The current (I_{DS}) - voltage (V_{DS}) characteristics of various FIBII's with the width 2~28 μm and the line dose of $2 \sim 5 \times 10^6$ cm^{-1} were measured.

264

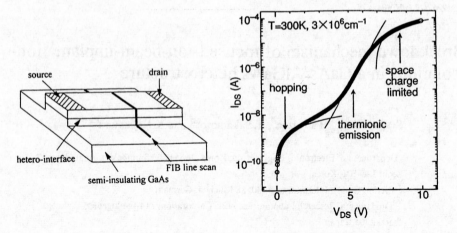

Fig. 1 Schematic sample structure   Fig. 2 A typical breakdown I-V characteristic

3. Data

A typical I_{DS}-V_{DS} characteristic of an FIBII (line dose of 3×10^6 cm^{-1}, measured at T = 300 K) is shown in Fig. 1 plotted in a log-linear scale. A composite of three different functional relationship is shown: power-law increase of I_{DS} in low V_{DS}; exponential increase of I_{DS} in intermediate V_{DS}; linear increase of I_{DS} in high V_{DS}.

      The linear increase of I_{DS} in high V_{DS} is shown more explicitly in the I_{DS}-V_{DS} characteristics plotted in a linear-linear scale. For example, Fig. 3 shows the breakdown characteristics of FIBII's with three different doses wherein perfect linear increases of I_{DS} are observed in high V_{DS} for all three FIBII's. As indicated by the figure, the onsets where the linear relationship occurs increases with increasing line doses. Furthermore, the slopes of the curves are almost the same irrespective of the line dose.

4. Discussion

Three regimes of breakdown observed in Fig. 2 are explained well by the theory of hopping (low V_{DS}), thermionic emission (intermediate V_{DS}), and space-charge-limited injection (high V_{DS}).

      Figure 4 shows schematic band diagrams in those regimes. At low V_{DS}, transport is mainly occurred by hopping [4] along the defect states in the FIBII (Fig. 4 (a)). The characteristics of hopping conduction such as power law dependences of I_{DS} on V_{DS} and on T are identified in our data.

      At intermediate V_{DS}, I_{DS} is determined by the thermionically emitted electrons over the barrier formed by the implantation damage (Fig. 4 (b)). When the zero bias barrier height is Φ_B, applying V_{DS} lowers the barrier and the effective barrier height is given by $\Phi_B - q\alpha V_{DS}$ where q is 1.6×10^{-19} C and αV_{DS} is the voltage drop at the barrier. The current increases exponentially with the effective barrier height, and as a result, I_{DS} is determined by the following formula;

$$I_{DS} = A^*_{2D}WT^{3/2}\exp(-(\Phi_B - q\alpha V_{DS})/(k_B T)),$$

$$A^*_{2D} = 2q(2\pi m^*)^{1/2}k_B^{3/2}/h^2,$$

where A^*_{2D} is the two-dimensional effective Richardson constant, k_B is the Boltzman constant, W is the width of the channel, m^* is the effective mass of the electrons in GaAs, and h is the Planck constant.

At V_{DS} high enough such that no energy barrier exists for the electrons injected into the FIBII ($\Phi_B = q\alpha V_{DS}$), the injected electron density rapidly increases and is comparable to the charged defect density. In such a case, the limiting transport mechanism becomes the space-charge-limited injection and I_{DS} increases linearly with V_{DS} [5]. The slope of I_{DS} is only a function of the saturation velocity and the geometry of the sample. Therefore, almost constant slopes are expected for the samples with different doses. On the other hand, Φ_B increases with increasing the line dose and the onset of the space-charge-limited injection ($\Phi_B = q\alpha V_{DS}$ condition) occurs at a higher V_{DS} for a larger dose.

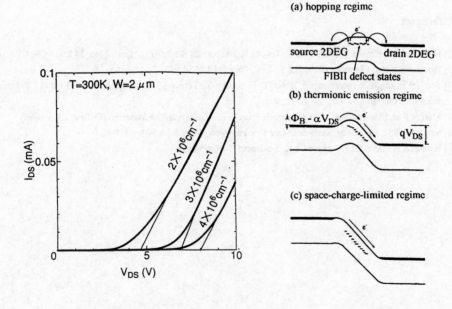

Fig. 3 Breakdown I-V's in a linear-linear scale

Fig. 4 A schematic band diagram

In each bias regime, quantitative fitting of the I_{DS}-V_{DS} characteristics with the models has been performed and the results are shown as solid lines in Fig. 2. The barrier height of 0.52 eV and the charged defect density of 5.0×10^{17} cm^{-3} are obtained. More detailed results such as the dose dependence and the temperature dependences of fitting parameters will be presented in the forthcoming publications [6].

5. Conclusion

The breakdown current-voltage characteristics of FIBII's in a GaAs/AlGaAs heterostructure have been studied. Three regimes showing different functional dependences of the current on the bias voltage have been identified and those regimes are qualitatively explained by the models of hopping (low bias), thermionic emission (intermediate bias), and space-charge-limited injection (high bias), respectively. The barrier height and the charged defect density are obtained from the quantitative fitting of the data with the models.

Acknowledgment

We thank Dr. Rangu, J. Sone, S. Matsui, J. Tsai, T. Baba, Y. Ochiai, and F. Nihey for useful discussions and constant encouragement throughout the work. This work was performed under the management of FED as a part of the MITI R&D program (Quantum Functional Devices project) supported by NEDO.

References

[1] Nakamura K, Nozaki T, Shiokawa T, Toyoda K and Namba S 1985 *J. J. Appl. Phys.* 24 L903-L904

[2] Hirayama Y and Okamoto H 1985 *J. J. Appl. Phys.* 24 L965-L967

[3] See, for example, a recent paper, Bever T, Hirayama Y, Tarucha S, *J. J. Appl. Phys.* 33 L800-L803 and references therein

[4] Mott N F and Davies G A 1979 *Electronic Process in Non-crystalline Materials* (Oxford: Clarendon)

[5] Sze S M 1981 *Physics of Semiconductor Devices* (New York: John Wiley & Sons)

[6] Hwang S W, Lezec H J, Nakamura K, Sakamoto T *unpublished*

Inst. Phys. Conf. Ser. No 145: Chapter 2
Paper presented at 22nd Int. Symp. Compound Semiconductors, Cheju Island, Korea, 28 August–2 September 1995
© *1996 IOP Publishing Ltd*

Progress in Sulfur Passivation of GaAs Surfaces

Xun Wang and Xiaoyuan Hou

Surface Physics Laboratory, Fudan University, 200433 Shanghai, China

Abstract. We develop three new sulfur passivation techniques of GaAs. The first one is the electrochemical sulfur passivation, which can create a thick S-containing layer to prevent the GaAs surface to be reoxidized in air. The second technique is using an oxygen-free sulfur-containing solvent $--S_2Cl_2$. It has been successfully used as a GaAs substrate preparing technique in the heteroepitaxial growth of II--VI wide gap semiconductors to improve the epilayer quality. The third technique is the sulfur vapor glow discharge method. It might be possible to eventually solve the problem of surface passivation for GaAs.

1. Introduction

The lack of effective surface passivation of GaAs has long been a tough problem which prohibits the development of GaAs devices, especially the fabrication of MOS type field effect transistors. Unlike the excellent passivating SiO_2 layer on Si surface, the native oxides of GaAs are soluble in water and thus are very easy to loss. Also, there exists a large amount of interface states between the native oxide and the GaAs substrate. In 1987, Sandroff et al. developed a sulfur-passivation technique by treating the GaAs surface with $Na_2S \cdot 9H_2O$ dipping[1]. The superior effects of S passivation were demonstrated. Since then, the S passivation has been thought to be the most promising technique for GaAs and was widely investigated[2-5]. However, the S passivation by $Na_2S \cdot 9H_2O$ or $(NH_4)S_x$ dipping suggested by previous authors was proved to be very unstable[6]. It degrades in air, especially under the illumination of laser beam. In order to overcome the problem of instability, different approaches have been proposed[7,8].

In this paper, three new passivation techniques are briefly presented. They are the electrochemical sulfur passivation, the S_2Cl_2 solution passivation, and the sulfur glow discharge passivation. The superior passivation effects are demonstrated.

2. Electrochemical Sulfur passivation

2.1. Experimental and Method

N-type Te-doped GaAs(100) single-crystal wafers with the doping concentration of 7×10^{15} cm^{-3} were used in the experiments. The sample was first ultrasonically cleaned in acetone and ethanol, etched by H_2SO_4:H_2O_2:H_2O (5:1:1) solution for 75s, rinsed by deionized water and then fixed on a teflon holder for passivation in $(NH_4)_2S$ solution. A dc voltage was applied between the backside of the wafer and a Ta cathode placed in the solution. During the anodic process, the current density reached $100mA/cm^2$ at the beginning, then decreased gradually and stabilized ultimately at several tenth of mA/cm^2 after about one hour.

In the case of sulfur treatment by using $(NH_4)_2S$ and $Na_2S \cdot H_2O$ dipping, only about one monolayer S atoms bond to the GaAs substrate, while the anodic S passivation can form a thick sulfide layer on GaAs. The XPS measurements showed that a relative thick sulfide layer(about 2nm) was created on GaAs[7].

2.2. Passivation effect

The passivation effect of the anodic sulfurized treatment was checked by the photoluminescence(PL) measurement with an Ar^+ laser as the light source. The power density on the sample surface was about 1.5 kW/cm^2. The PL intensity was recorded as a function of laser illumination time for different samples, as shown in Figure 1(a). The PL intensity not only increases by one and a half orders of magnitude after the anodic S passivation as compared with the as-etched surface, but also remains unchanged under the Ar^+ laser illumination for 30 min. While, for the $(NH_4)_2S$ treated sample, a rapid decay of PL intensity under the same illumination condition is observed although the initial PL intensity is enhanced.

The passivation effect could be maintained for quite a long period. After keeping in atmosphere for seven months, the sample still shows the improvement of PL intensity and its stability under laser illumination, as shown by Figure 1(b). It has been shown that the stability of the S passivation can be greatly improved due to the creation of a thick sulfide layer on GaAs substrate. The sulfide layer prevents the interface from being re-oxidized in air.

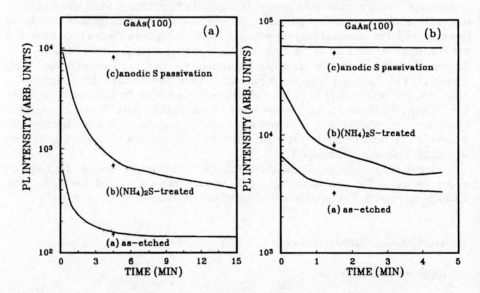

Fig.1 The variation of PL intensity of GaAs(100) versus the laser illumination time for different samples. (a) as-prepared, (b) after seven months.

3. S₂Cl₂ treatment

3.1 Procedure of S₂Cl₂ treatment

GaAs(100) single crystal wafers were chemically cleaned as described in section 2.1. The S passivation was carried out by dipping the wafer in S_2Cl_2 for 5 sec or in the diluted $S_2Cl_2+CCl_4$ solution for longer time. The effect is found to remove the native oxide and form a S-passivation layer. The wafer was then rinsed sequentially by CCl_4, acetone, ethanol and de-ionized water. The sample was then loaded into the ultrahigh vacuum (UHV) chamber of an electron spectrometer to do the characterization. In addition, the surfaces of two GaAs wafers, without chemical treatment, and treated by $(NH_4)_2S_x$ solution dipping at room temperature for 15 hours, were also measured for comparison.

3.2 Composition of passivated surface

Figure 2 shows the Auger electron spectra (AES) of S_2Cl_2-treated, $(NH_4)_2S_x$ treated and as-etched GaAs samples. The most interesting thing is that on the S_2Cl_2-treated sample surface no O KLL AES signal could be detected, even the sample was exposed in atmosphere for several min. after the treatment. It verifies that the S_2Cl_2 solution which is free of oxygen species is much more effective in removing native oxide than the $(NH_4)_2S_x$ aqueous solution. As to other compositions, the S_2Cl_2 treated sample shows about same concentration ratios of S/Ga and As/Ga to that of $(NH_4)_2S_x$ treated surface. But the relative concentration of C is increased by a factor of two for the S_2Cl_2 passivation as compared with the $(NH_4)_2S_x$ dipping. This may arise from the poor purity of the S_2Cl_2 solution.

The results of PL illustrated that the enhancement of PL intensity by S_2Cl_2 treatment is as good as that by $(NH_4)_2S_x$ dipping[9]. XPS study shows that sulfur atoms bond to both Ga and As atoms more effectively on S_2Cl_2 treated surfaces than those passivated by $(NH_4)_2S_x$.

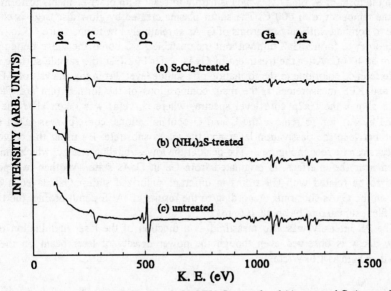

Fig.2 Auger spectra of (a) S_2Cl_2 treated, (b) $(NH_4)_2S_x$ treated and (c) untreated GaAs samples.

3.3 S_2Cl_2 dipping as a pre-treatment of GaAs substrate for epitaxial growth

In order to investigate whether this new technique could be of practical use, the sulfurized GaAs was employed as the substrate for epitaxial growth of ZnSe film[10]. The S_2Cl_2 treated GaAs substrate was annealed in UHV at the temperature of 400°C for 30 min. The residual S would not be the obstacle of growing ZnSe film with good crystalline quality. The epitaxial growth of ZnSe films was performed in a molecular beam epitaxy system. The growth temperature was 350°C with the rate of 1um/h. The thickness of the epilayer is in the range of 1um.

Raman scattering measurements showed the suppression of ZnSe TO/ZnSe LO ratio in S-passivated sample, which is an indication of the improvement of crystalline quality. Besides, the FWHM of ZnSe LO mode was statistically narrower in the case of S-passivated sample as compared with that of unpassivated one. Based on the analysis of the line shape of first order LO phonon peak of ZnSe by using the spatial correlation model of Raman scattering. it is shown that the ZnSe films grown on the GaAs substrates treated by S_2Cl_2 have the large coherence lengths, indicating that their crystalline qualities are better than those passivated by $(NH_4)_2S_x$ solutions. In addition, the barrier heights of ZnSe/GaAs interfaces for different S-passivations have been compared according to the intensity ratios of coupled LO-phonon plasma mode to LO mode of GaAs Raman peaks. The results show that the ZnSe/GaAs samples passivated by S_2Cl_2 solutions have lower density of interfacial states[11].

4. Sulfur glow discharge

The purpose of using SDG is to form a gallium sulfide layer on the GaAs surface by the chemical reaction between activated sulfur and the surface Ga atoms. Sulfur has high vapor pressure, which could reach 10Pa at the temperature of 100°C. However, elemental sulfur exists in the form of S_8 ring[12], which is hardly to react with other elements to form sulfides below the temperature of 600°C. The sulfur plasma created by glow discharge is chemically reactive to combine with surface atoms of GaAs at relatively low temperature.

In order to form a GaS film without the unstable As-S bonds the substrate temperature was kept at 400°C. After the treatment of GaAs wafer by SGD, the sample surface shows a blue interference color due to the formation of a sulfide layer. From the analysis of AES depth profile and XPS measurements, the main compositions of the formed film are Ga and S. Figure 3 shows the Ga2p core level spectra, where curve(a) was taken after the surface sputtered by Ar^+ ion to remove the C and O contaminations, curve(b) was taken near the interface between the passivation layer and the GaAs substrate. By using the curve fitting, three peaks can be seen in Figure 3. The peak at the lowest binding energy, which can be only observed near the interface, is originated from Ga in GaAs state. Another two peaks are believed to be related with Ga atoms in different sulfurized states. GaS layer is certainly formed on the GaAs substrate. According to the results of AES depth profile, the thickness of GaS film estimated to be about 100nm.

The PL intensity was also measured as a function of the laser illumination time. No intensity decay is observed even though the power density of laser beam on the sample surface is as high as $1.6kW/cm^2$.

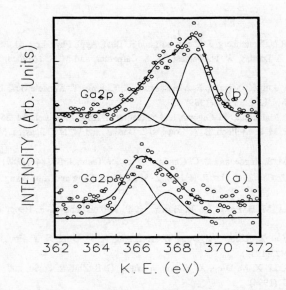

Fig.3 Ga2p core level spectra of (a) GaS film and (b) GaS/GaAs interface between the film.

5. Conclusions

An electrochemical sulfurized passivation technique has been developed. A thick sulfide layer protects the sulfide/substrate interface from being re-oxidized in atmosphere. The improvement of the stability of S passivation is thus achieved.

The advantages of S_2Cl_2 treatment over the ordinary $(NH_4)_2S_x$ treatment are the freedom from oxygen uptake, more effective removal of the native oxide, formation of stronger Ga-S bonds, very short reaction time and the adjustable etching rate of GaAs surface by changing the concentration ratio of S_2Cl_2 to CCl_4. It seems quite promising in preparing the GaAs substrate for the purpose of heteroepitaxial growth of ZnSe.

The S glow discharge technique is a sulfurization process, i. e., the formation of a "native" sulfide on the GaAs surface. In the case of passivation of Si surface, the thermal oxidation which forms a silicon dioxide(native oxide) is proved to be superior than the chemical vapor deposition. This might also be true that a sulfurization process like S glow discharge is a very promising technique. To confirm it, of course, further investigation is necessary.

Acknowledgments

The authors would like to thank many of their colleagues and students who contribute to this work, especially Dr. Z. S. Li, Mr. W. Z. Cai, Miss X. Y. Chen and Q. J. Xu, This work was supported by the National Natural Science Foundation of China.

272

References

[1] C. J. Sandroff, R.N. Nottenberg, J. C. Bischoff and R. Bhat, Appl. Phys. Lett. **51**, 33(1987).

[2] B. A. Cowans, Z. Dardas, W. N. Delgass, M. S. Carpenter, and M. R. Melloch, Appl. Phys. Lett. **54**, 365(1989).

[3] C. J. Spindt, D. Lui, K. Miyano, P. Meissner, T. T. Chiang, T. Kendelewicz, I. Lindan, and W. E. Spicer, Appl. Phys. Lett. **55**, 861(1989).

[4] H. Hirayama, Y. Matsumoto, H. Oigawa, and Y. Nannichi, Appl. Phys. Lett. **54**, 2565(1989).

[5] M. S. Carpenter, M. R. Melloch, B. A. Cowans, Z. Dardas, and W. N. Delgass, J. Vac. Sci. Technol. **B7**, 845(1989).

[6] C. J. Sandroff, M. S. Hegde. and C. C. Chang, J. Vac. Sci. Technol. **B7**, 841(1989).

[7] X. Y. Hou, W. Z. Cai, Z. Q. He , P. H. Hao, Z. S. Li, X. M. Ding and Xun Wang, Appl. Phys. Lett. **60**, 2252(1992).

[8] A. N. MAcInnes, M. B. Power, A. R. Barron, P. P. Jenkins, and A. F. Hepp, Appl. Phys. Lett. **62**, 711(1993).

[9] Z. S. Li, W. Z. Cai, R. Z. Su, G. S. Dong, D. M. Huang, X. M. Ding, X. Y. Hou, and Xun wang, Appl. Phys. Lett **64**, 3425(1994).

[10] W. Z. Cai, Z. S. Li, X. M. Ding, X. Y. hou, J. Wang, C. S. Zhu, R. Z. Su, and Xun Wang, J. Cryst. Growth, **142**, 397(1994).

[11] J. Wang, X. H. Liu, Z. S. Li, R. Z. Su, W. Z. Cai, X. Y. Hou, and Xun Wang, to be Published.

[12] J. E. Bear and M. Carmack, J. Am. Chem. Soc. **71**, 1215(1945).

Inst. Phys. Conf. Ser. No 145: Chapter 2
Paper presented at 22nd Int. Symp. Compound Semiconductors, Cheju Island, Korea, 28 August–2 September 1995
© *1996 IOP Publishing Ltd*

Sulfur passivation for thermal stability enhancement of RuO$_2$ Schottky contact on compound semiconductor

Eun Kyu KIM, Maeng Ho SON, Ho Nyung LEE, Yong Tae KIM, and Suk-Ki MIN

Semiconductor Materials Research Center, Korea Institute of Science and Technology, P.O. Box 131, Cheongryang, Seoul 130-650, Korea

Abstract. We have studied the thermal stability of the electrical properties for RuO$_2$ Schottky contact on sulfur passivated n-type GaAs. The sulfur passivation on the surface of the compound semiconductor was done with (NH$_4$)$_2$S$_x$ solution under 150 watt light illumination at room temperature, and then RuO$_2$ contacts were fabricated by a DC magnetron sputtering method. By the sulfur passivation on semiconductors the RuO$_2$ Schottky contacts showed the good electrical properties and enhancement of its thermal stability during the thermal annealing upto 550℃ for 10 min. From AES(Auger electron spectroscopy) analysis, it was confirmed that the sulfur passivation is responsible for the suppression of an oxidation at interfaces between RuO$_2$ and semiconductors.

1. Introduction

The surface treatments of GaAs and InP using atomic hydrogen and sulfides have been interested because of their remarkable ability to enhance the surface-related characteristics. It has been reported that an improvement of device characteristics from a markable reduction of surface state density in metal-semiconductor(MS) and metal-insulator-semiconductor(MIS) structures by atomic hydrogen or sulfide treatments[1-6]. On the other hand, the RuO$_2$ has a metallic conduction mechanism with the resistivity of about 50 μΩ-cm and good thermal stability at a junction on Si[7,8]. However, the RuO$_2$ shows a rather low barrier height of 0.70 eV as compared to most other metals on n-GaAs contacts[9], where the barrier height can be induced from the effective work function, defects on the interface, metal induced gap states, and reactivity between metal and semiconductor. Recently, the increasing effect for Schottky barrier height on InP by sulfur passivation with supersaturated (NH$_4$)$_2$S$_x$ solution has been reported[4]. The sulfur passivation to GaAs surface expects to be improved the electrical properties of RuO$_2$/GaAs junction.

In this paper, we have studied the effects of surface treatments with sulfur and atomic hydrogen and their thermal stability on RuO$_2$/GaAs junctions. Especially, the enhancement of the thermal stabillity on the electrical property of RuO$_2$/n-GaAs Schottky contact by sulfur passivation through the treatment of (NH$_4$)$_2$S$_x$ solution and the exposure of hydrogen plasma on GaAs surface is mainly treated in this article. The influence of thermal annealing to the interface of these contacts with the surface pre-treatment was investigated also Auger electron spectroscopy(AES) as well as current-voltage(I-V) measurements.

2. Experimental details

The samples used in this study were n-type (100)GaAs wafers with carrier density of 2-3x10^{16} cm^{-3}. The ohmic contacts at one side of the samples were made by alloying indium dots at 200℃ for 5 min in an ambience of N$_2$ gas. After making the ohmic contacts the samples were exposed to hydrogen plasma in a rf (13.56 MHz) plasma chamber. The hydrogenation was done by a power density of 0.08 W/cm^2 with hydrogen pressure of 0.5Torr during 30 min at 200℃. This condition can provide an atomic hydrogen to passivate the eletrically active states of defects and impurities below about 1 μm from the surface of sample[10]. For the sulfur passivation, the GaAs wafer was etched by 5H$_2$SO$_4$: 1H$_2$O$_2$: 1H$_2$O for 1 min to remove the native oxide layer. The wafer was soaked in (NH$_4$)$_2$S$_x$ solution under a white light illumination of 100 W for 1 hr, then followed by wafer rinsing in de-ionized water for 20 s and drying by nitrogen blow.

After the treatments of surface passivation, the GaAs samples were moved into the dc magnetron sputtering system. RuO$_2$ dots were formed about 100 nm thickness on the surfaces of hydrogenated and sulfurized samples with the shadow mask of 0.8 mm diameter in the sputtering system. With the base pressure of 2x10^{-6} Torr and Ru taget purity of 99.9 %, the RuO$_2$ sputtering was carried out[8]. The input power of sputtering was 100 W under 12 mTorr total pressure of mixed gases with argon and 10 % oxygen at room temperature.

Finally, the samples of RuO$_2$ contacts on n-GaAs were annealed at temperature range from 200℃ to 550℃ for 10 min under N$_2$ gas ambient. The electrical characteristics of RuO$_2$/n-GaAs contacts were characterized by the I-V measurement and the thermal reactions existed at the interfaces of these contacts was confirmed by AES.

3. Results and discussion

Fig. 1 represents the dependence of annealing temperature for the ideality factor (n) and the apparent barrier height (ϕ_b) of RuO$_2$ contacts on n-GaAs before and and after the surface passivation with sulfur and atomic hydrogen, respectively.

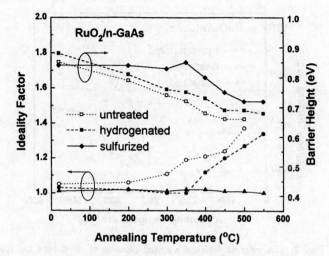

Fig. 1 The variations of the ideality factor(n) and the apparent barrier
height(ϕ_b) of RuO$_2$ contacts on n-GaAs after thermal annealing.

The values of n and ϕ_b were calculated from the typical I-V measurements at
room temperature. The ideality factor of the RuO$_2$/n-GaAs sample with untreated
substrate increases rapidly from 1.05 to 1.37 according to the increasing temperature up to 500℃, showing a severe degradation of Schottky contact above 350℃
annealing temperature. In the sample with hydrogenation, that is maintained a
good value of around 1.01 after thermal annealing at temperature up to 350℃,
but it is severely degraded also with annealing at temperature above 350℃,
showing a similar trend to that of untreated sample. However, in RuO$_2$ contact
on sulfurized GaAs substrate, it shows a prominant stability with the ideality
factor of around 1.01 in whole temperature range.

On the other hand, the barrier heights in these samples decrease gradually
from 0.88 eV to 0.66 eV with the increase of annealing temperature up to 550℃.
The hydrogenated sample shows that the apparent barrier height is about 0.3
eV higher than that of the untreated sample on the whole. However, in the
sulfurized sample it is maintained about 0.84 eV until at 350℃ and decreased
rapidly to 0.71 eV with increasing the annealing temperature.

Fig. 2 shows the results of the reverse leakage current density(J_r) at -3 V
bias for RuO$_2$ contacts on n-GaAs with surface passivation after the thermal
annealing at temperatures up to 550℃. In case of the hydrogen passivation to
GaAs surface, the temperature dependence of J_r is appeared almost the same as
in untreated sample, having the trend of rapid increase at temperature above 350
℃. However, in the sulfur passivated sample, J_r reduced drastically at around
350℃ to about 10 times lower than that of untreated sample. It is also noted
that the low J_r can be maintained up to high temperature range of 550℃.

There are some reports[11,12] that the surface structure of the sulfur treated
sample udergoes a drastic change with thermal annealing and shows a transi-

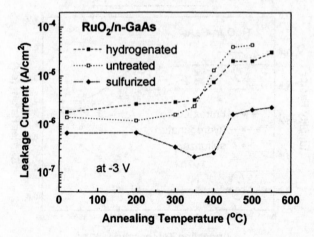

Fig. 2 The reverse leakage current density at -3 V bias for RuO₂ contacts on n-GaAs with surface passivation.

tion temperature of around 250℃. Also, it was reported[4] that the best electrical characteristics in Schottky diode of InP with the surface sulfur treatment appeared after the thermal treatment of 270℃. It is, therefore, believed that the good electrical characteristics of the sulfur passivated sample at around 350℃ shown in figs. 1 and 2 are responsible for the formation of a stable sulfur bond to GaAs surface by thermal annealing above the transition temperature 270℃. On the other hand, it has been generally known that the atomic hydrogen can remove the oxidized layer and neutralize the electrically active impurities in GaAs, which atomic hydrogen bonded to the impurity states in n-GaAs is reactivated by thermal treatment above 400℃ temperature[13]. Also, it was reported that the RuO₂ film was decomposed into Ru by the thermal treatment under oxygen or hydrogen gas ambient[14]. Thus, we can suppose that the interface reaction between RuO₂ and GaAs may be the main causes of the device degradation by thermal treatment, and then the surface passivation of GaAs suppress the reaction. Accordingly, it implies that the sulfur in sulfurized GaAs surface is more hardly decomposed than the hydrogen in the hydrogenated surface, and effectively prevents the thermal reaction at the interface between RuO₂ and GaAs.

Fig. 3 shows the atomic depth-profiles obtained by AES measurements for the RuO₂ contacts on the sulfurized GaAs (a)before and (b)after thermal annealing at 500℃ for 10 min in Ar flow. From the AES signal for the GaAs sulfurized with $(NH_4)_2S_x$ solution during 1 hr under light illumination, the amount of sulfur on GaAs surface was estimated about 20 %. In the AES measurements, the sputtering rate at near interface between RuO₂ and GaAs is about 50 A/min. Unfortunately, the sulfur behavior at interface by the thermal annealing can not be shown here because the most dominant AES signal of sulfur atom is overlapped with one of Ru signals at about 150 eV. Although another signal of the

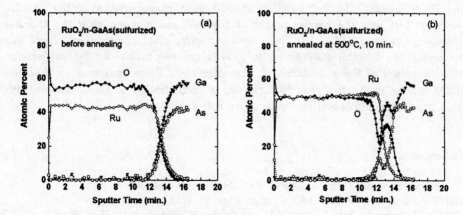

Fig. 3 AES depth profiles for RuO₂ contacts on the sulfurized GaAs (a)before and (b)after the thermal annealing at 500℃ for 10 min.

sulfur atom exists at higher energy energy, it can not detect also because of its weakness. However, this figure shows clearly that GaAs oxidation occurs at the interface between RuO₂ and sulfurized GaAs by thermal annealing at 500℃.

The thermal degradation of the RuO₂ Schottky contact on the GaAs surface passivated with the sulfur or atomic hydrogen may be explained as follows; the sulfur atom or the atomic hydrogen are deactivated and desorbed firstly from the GaAs surface by the thermal annealing, and then the GaAs surface is oxidized by the oxygen atom of RuO₂ film. For instance, the sulfur atom can be deactivated from sulfurized GaAs by the temperature of about 400℃, and then the GaAs oxidation is formed at the interface between RuO₂ and GaAs. From the AES measurement, it can be seen that the thermal annealing above 400℃ for RuO₂/n-GaAs may produce a thin oxide layer such as Ga_2O_3 and As_2O_3 due to the thermal reaction of between RuO₂ and GaAs junction.

Therefore, it is suggested that the sulfur bonding of the GaAs surface may endure to the higher temperature comparing to the hydrogen bonding and then may prevent a thermal reaction of RuO₂/GaAs interface effectively up to the temperature of about 400℃. Conclusively, the sulfur passivation to the compound semiconductors can be one of the most feasible way to enhance the thermal stability of electrical characteristics in the Schottky contact and metalli- zation of a metallic oxide compound such as RuO₂.

4. Conclusion

The thermal stability of electrical property for Schottky contacts of RuO₂ on n-type GaAs was studied with the surface passivation by sulfur and hydrogen. While the hydrogenation of GaAs was not so effective to the thermal stability of

278

the RuO_2 contact, the sulfurization showed good characteristics of the ideality factor of 1.01 and the barrier height of 0.84 eV and endured after the thermal annealing up to 350℃. It was assumed that the sulfurization effects for enhancing the thermal stability of $RuO_2/GaAs$ are responsible for the suppression of a formation of GaAs oxidation on the interface. These results show that the RuO_2 can be used as a good material for a metallization of GaAs devices with high processing and operating temperature below 400℃.

References

[1] Aspnes D E and Heller A 1983 *J. Vac. Sci. Technol. B* **1** 602

[2] Wang Y G and Ashok S 1994 *J. Appl. Phys.* **75** 2447

[3] Van Meirhaeghe R L, Laflere W H and Cardon F 1994 *J. Appl. Phys.* **76** 403

[4] Han I K, Her J, Byun Y T, Lee S, Woo D H, Lee J I, Kim S H, Kang K N and Park H L 1994 *Jpn. J. Appl. Phys.* **33** 6454

[5] Eftekhari G 1994 *J. Vac. Sci. Technol. B* **12** 3214

[6] Silva-Andrade F, Chàvez F and Goméz E 1994 *J. Appl. Phys.* **76** 1946

[7] Takemura K, Sakuma T and Miyasaka Y 1994 *Appl. Phys. Lett.* **64** 2967

[8] Lee J-G, Min S-K and Choh S H 1994 *Jpn. J. Appl. Phys.* **33** 7080

[9] Vandenbroucke D A, Van Meirhaeghe R L, Laflère W H and Cardon F 1985 *J. Phys. D: Appl. Phys.* **18** 731

[10] Kim E K, Cho H Y, Kim H S, Min S-K and Kim T 1992 *Semicond. Sci. Technol.* **7** 695 ; Cho H Y, Kim E K, Min S-K and Lee C 1991 *Phys. Rev. B* **43** 14498

[11] Oigawa H, Fan J, Nannichi Y, Sugahara H and Oshima M 1991 *Jpn. J. Appl. Phys.* **30** L322

[12] Viktovovitch P, Gendry M, Hollinger G, Krawczyk S and Tardy J 1992 *Proc. 4th Int. Conf. Indium Phosphide and Related Materials* (IEEE, New York) 51

[13] Pearton S J, Corbett J W and Shi T S 1987 *Appl. Phys. A* **43** 153

[14] Rard J A 1985 *Chem. Rev.* **85** 1

Inst. Phys. Conf. Ser. No 145: Chapter 2
Paper presented at 22nd Int. Symp. Compound Semiconductors, Cheju Island, Korea, 28 August–2 September 1995
© *1996 IOP Publishing Ltd*

279

Combination of selective etching and AFM imaging for the thickness analysis of AlGaAs/GaAs heterostructures

S. Müller, J.L. Weyher, K. Köhler, W. Jantz
Fraunhofer IAF, 79108 Freiburg, Germany

C. Frigeri
MASPEC-CNR Institute, 43100 Parma, Italy

Abstract. The vertical structure of AlGaAs/GaAs epitaxial layer systems for micro- and optoelectronic device fabrication has been studied by selective etching combined with AFM imaging. Dark etching and photoetching with CrO_3-HF-H_2O solutions were used to transform composition and doping variations into height differences of the cleaved (110) surface. The etching parameters (time, composition and supply of carriers by illumination) were optimized for accurate thickness determination by AFM. The measurements were corroborated by comparison with cross-sectional TEM mapping. The dependence of the etching rate on the composition and the occurence of small, but measurable height variations *without* any etching is discussed.

1. Introduction

Direct measurements of layer thickness in semiconductor heterostructures require nanometer resolution in vertical and horizontal direction combined with high sensitivity for the changes of composition or doping level of the different layers. Up to now qualitative and quantitative information about thickness and/or composition of semiconductor devices have been acquired by secondary ion mass spectroscopy (SIMS) [1], spreading resistance profilometry (SRP) [2], cross-sectional scanning tunneling microscopy (STM) [3] or scanning capacitance microscopy (SCM) [4]. These methods need time consuming sample preparation and/or ultra high vacuum. Hence selective chemical etching of the cross-section of the sample combined with atomic force microscopy (AFM) imaging is interesting as an easy and powerful tool for the investigation of different kinds of heterostructures. It combines the high selectivity of the etching with the advantage of high resolution imaging of the cleavage face by AFM. The localization of the pn-junction and different layers (conductive and isolating) have been investigated at the same time [5]. The high resolution lateral imaging directly shows the geometric parameters of heterostructure layer systems, whereas the equally acurate determination of the depth profile, after calibration of an optimized etching procedure, allows to determine carrier concentrations [6].

Roman and Wilson [7] were the first who demonstrated the effectiveness of Sirtl-type etchants. These etchants are based on a CrO_3-HF-H_2O solution and used with illumination in delineating electronic junctions in semiconductors. The large range of diluted Sirtl-like etchants, used without and with light (DS and DSL methods respectively [8]) to show defects in III-V semiconductors, were also utilized to reveal interfaces in layered structures [9,10].

In this paper we describe the results of cross-sectional AFM studies on AlGaAs/GaAs-heterostructures obtained after selective DS(L) etching. The occurence of height variations on the cleavage face *without* etching is also discussed. The thickness measurements were compared with cross-sectional TEM mapping to verify the AFM measurements.

2. Experimental procedure

The samples used in this experiments are AlGaAs/GaAs heterostructures grown by molecular beam epitaxy (MBE) on GaAs substrates. Fig. 1 shows the composition of the multilayer structure which consists of more than 20 different layers. The thicknesses of the layers anticipated from the control data of the MBE growth are verified by TEM measurements as shown in the right part of Fig. 1. The heterostructure system is composed of a microelectronic and an optoelectronic part. It is designed to fabricate a laser diode with monolithic integration of microelectronic circuitry such as multiplexer and laser driver. Only the laser structure will be analysed in detail. The active area of the laser consists of three GaAs quantum wells (QW) and two AlGaAs superlattices (SL) and has a total thickness of 35.1 nm. The doping concentration is 3×10^{18} Si atoms/cm^{-3} for the n-type layers and ranges from 1×10^{18} to 8×10^{19} Be atoms/cm^{-3} for the p-type layers.

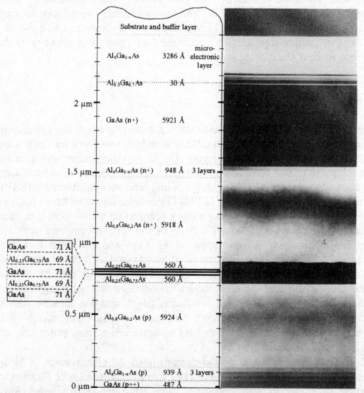

Fig 1: Vertical composition of the heterostructure layer system. The active layer of the laser is shown expanded. The microelectronic layer system at the top is indicated schematically.

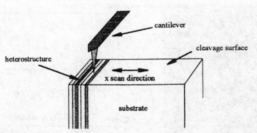

Fig. 2: Experimental setup for the AFM scanning on the cleaved surface. The scan direction is perpendicular to the layers and to the cantilever axis.

All experiments, including cleaving, etching and AFM imaging, were done in normal laboratory environment. For the thickness measurements on the cleavage face we used commercial Si_3N_4 and Si cantilevers with high aspect ratio and a NanoScope III AFM [11] operating in contact and in tapping mode. Both operation modes lead to the same results as described in section 3.

In order to study the layer thickness and the doping profiles, the samples were precisely cleaved along the GaAs (110) plane. The cleaved surface was investigated for reference and then chemically etched with the DS solution either in a dark box or under illumination. When these etchants are applied to the cross-section of the heterostructure, a topography will appear due to the differences in the etch rate for the different compositions and doping concentrations in the layers. The notation $D_{1:x}$ $S_{a/b}$ for the HF-CrO_3-H_2O solution has been adapted from [8]:

$D_{1:x}$ one volume part of the basic mixture was diluted with x volume parts of deionisized water.

$S_{a/b}$ the composition of the basic mixture consists of a volume parts of HF (48 wt%) and b volume parts of CrO_3 (33 wt%) aqueous solution.

L/time designates that a light source has been used for the illumination during etching. If the letter L is missing it indicates that the etching was done in a dark box. The etching time is added at the end of the notation. The etching was done at 20 ± 2 °C.

The DS/DSL methods are commonly used to reveal defects in GaAs. These etchants give reproducible results and do not form precipitates in the etching medium. This is an important advantages in comparison to other mixtures based on the popular AB solution [12]. From the preliminary test-etching of the examined heterostructure, it has been established that the best is $S_{1/5}$ solution diluted at least 1:40 with H_2O.

After cleaving, the heterostructure was immediately etched and measured with AFM to reduce the influence of surface contamination from the ambient atmosphere. Fig. 2 shows the orientation of the cleaved and etched surface with respect to the AFM probe position. The total measurement time including cleaving, etching and AFM imaging was below half an hour. A repetition of the AFM imaging one day later leads to the same results, but an increase of surface contamination, slightly reducing the resolution of the imaging, was discerned.

3. Experimental results and discussion

Fig. 3a shows the two-dimentional AFM image of the laser heterostructure and the corresponding line profile obtained after 30 sec of etching ($D_{1:40}$ $S_{1/5}$ /30sec). The two highly n- and p- doped AlGaAs layers with a thickness of 592 nm are etched much faster than the surrounding GaAs layers and have nearly the same depth. Photoetching ($D_{1:40}$ $S_{1/5}$ L/15 sec) on the same heterostructure produces a much sharper etching profile with a high dependency on the doping type (Fig. 3b).

282

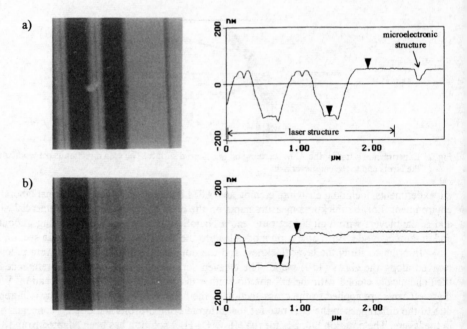

Fig. 3: AFM image and profile of the AlGaAs/GaAs heterostructure shown in Fig. 1 after (a) etching in the dark (DS) and (b) photo-etching (DSL).

For long etching times the DSL mixture leads to sharp, rectangular etching grooves, however such deep trenches are causing measurement problems due to the finite cone angle of the AFM tip.

Selectivity of etching on heterostructures can be influenced by several factors, including a difference in carrier concentration, chemical composition and strains. In agreement with electrochemical studies of the DSL-GaAs system [13] and EBIC measurements [14], we find that the p-type AlGaAs layer was already etched at the highest rate, confirming the importance of holes for surface reactions in the CrO_3-HF containing solution [13]. However, in this study we did not attempt to elucidate these interrelations in detail, but rather to identify empirically the particular etching conditions that are optimal for the subsequent AFM investigations. We

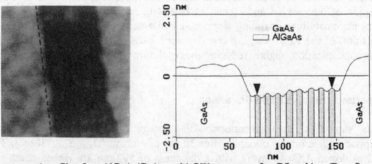

Fig. 4: AFM image and profile of an AlGaAs/GaAs multi QW structure after DS etching ($D_{1:40}$ $S_{1/5}$ /4 sec). The grooves represent the AlGaAs layers, the valleys the GaAs layers. The thickness of each layer is 5 nm. The profile is averaged along the broken line.

find that, in the interest of optimized display of complicated heterostructures with many different compositions and doping levels, it is advantageous to use dark etching rather than photoetching, because there are less pronounced etching rate differences between doped and undoped material.

The best results were obtained using $D_{1:40}$ $S_{1/5}$ solution for 2-4 sec. Under these conditions QW and SL with a thickness down to 5 nm could be resolved (Fig. 4). Other etching conditions, including illumination, might be prefered if different doping levels were to be determined accurately.

Surprisingly a small surface corrugation is already observed after cleavage without any etching. Both sides of the cleaved surface show an identical profile, which is independent of the cleaving procedure. Height differences up to 3-5 nm occur between the GaAs and the AlGaAs layers and increase with the aluminium concentration of the layer. We believe that the individual layers of the heterostructure start to oxidize with different time constants which leads to small height variations on the cleavage surface. Especially the AlGaAs layers of the laser structure described in Fig. 1 quickly develop a significant oxidation layer already after few minutes of exposition to air (Fig 5a). Due to a passivation of the surface this process is self-limiting. High resolution imaging in the active part of the laser structure reveals the three GaAs quantum well layers (69 Å) separated by two AlGaAs superlattices (71 Å) (Fig. 5b). To reduce noise, the line profile has been averaged along the broken line. The typical RMS roughness of a perfectly cleaved substrate is less than 0.7 Å, but in the same heterostructure system the cleavage surface of the epitaxial layers can be rougher than 5 Å.

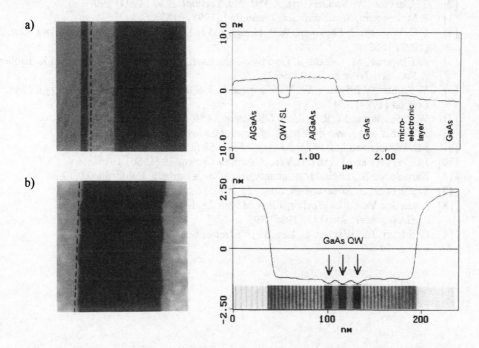

Fig. 5: AFM image and profile of the AlGaAs/GaAs heterostructure shown in Fig. 1 after cleavage without etching. a) The parts of the heterostructure with a high aluminum concentration have the thickest oxidation layer. b) The grooves represent the GaAs layers. The profile is averaged along the broken line.

A comparison to the TEM mapping shows that AFM imaging with and without selective etching can determine layer thickness with an accuracy of better than 5% for layers thicker than 100 nm and less than 10% for SL and QW down to a size of 5 nm. These results can be obtained rapidly to support control and development of epitaxial growth processes.

4. Conclusion

It has been demonstrated that the combined used of DS etching and AFM imaging on the GaAs (110) cleavage allows rapid and precise determination of the layer thickness of the MBE grown AlGaAs/GaAs heterostructures. Different composition and doping concentrations of n-type and p-type layers have been investigated and correlated with cross-sectional TEM measurements. QW and SL down to a thickness of 5 nm can be discerned using this procedure. The conditions of etching (time, composition of etchant and supply of carriers) have been optimized for the highest accuracy of the AFM measurement.

References

[1] S.H. Goodwin-Johansson, Y. Kim, M. Ray, H.Z. Massoud, *J. Vac. Sci. Technol.* B 10 (1992) 369
[2] T. Clarysee ,W. Vandervorst, *J. Vac. Sci. Technol.* B 12 (1994) 290
[3] R.M. Feenstra, *Semicond. Sci. Technol.* 9 (1994) 2157
[4] C.C. Williams, J. Slinkman, W.P. Hough and H.K. Wickramasinghe, *Appl. Phys. Lett.* 55 (1992) 1662
[5] R. Chapman, M. Kellam, S. Goodwin-Johansson, J. Russ, G.E. McGuire and K. Kjoller, *J. Vac. Sci. Technol.* B 10 (1992) 502
[6] V. Raineri, V Privitera, W. Vandervorst, L. Hellemans and J. Snauwaeert, *Appl. Phys. Lett.* 64 (1994) 354
[7] W.C. Roman and L.R. Wilson, *US Patent 3,830,665* (August 20, 1974)
[8] J. Weyher and J. van de Ven, *J. of Crystal Growth* 63 (1983) 285
[9] J.L. Weyher and L.J. Giling, *J. Appl. Phys.* 58 (1985) 219
[10] J.L. Weyher and J. van den Ven, *J. Crystal Growth* 78 (1986) 191
[11] NanoScope III, Digital Instruments, Inc. , Santa Barbara, California 93117
[12] Saitoh et al., *J. Electrochem. Soc.* 122 (1975) 670
[13] J. van den Ven, J.L. Weyher, J.E.A.M. van den Meerakkar and J.J. Kelly, *J. Electrochem. Soc.* 133 (1986) 799
[14] C. Frigeri, J.L. Weyher, L. Lanotti, *J. Electrochem. Soc.* 136 (1989) 262

Inst. Phys. Conf. Ser. No 145: Chapter 2
Paper presented at 22nd Int. Symp. Compound Semiconductors, Cheju Island, Korea, 28 August–2 September 1995
© *1996 IOP Publishing Ltd*

Enhanced Selectivity in GaAs/AlGaAs Selective Dry Etching in $BCl_3 + CF_4$ Plasma by Adsorbed CxFy for Precise Control of HJFET Threshold Voltages

M. Tokushima, H. Hida, and T. Maeda

Microelectronics Research Laboratories, NEC Corporation
34, Miyukigaoka, Tsukuba, Ibaraki 305, Japan

Abstract. An etching selectivity–enhancing effect of fluorocarbon (CxFy) adsorption in GaAs/AlGaAs selective dry etching is reported. Then, a new selective etching technique using non–freon $BCl_3 + CF_4$ plasma based on the selectivity enhancing mechanism is developed and demonstrated. The $BCl_3 + CF_4$ plasma etching technique completely stops etch on an $Al_{0.2}Ga_{0.8}As$ layer as thin as 15 Å, and therefore will be a successful solution to the undesirable threshold voltage variation of HJFET's.

1. Introduction

Selective dry etching of GaAs on AlGaAs is very important for defining the gate recess of AlGaAs/GaAs HJFET's[1]–[3]. The GaAs–AlGaAs etching selectivity has been improved by increasing the aluminum mole fraction in the AlGaAs; the higher concentration of aluminum produces more nonvolatile AlF_3, which prevents etching[3], [4]. However, aluminum mole fractions must generally be restricted to around 0.2 for highly–doped–AlGaAs/GaAs HJFET structures to avoid undesirable effects due to DX centers. Hence, more than several monolayers of the stopper AlGaAs are etched in creating an effective etch barrier of AlF_3 even when a high selectivity etching technique is employed. This loss of the stopper AlGaAs causes undesirable threshold voltage variation, which can reach as high as 0.1 V per nanometer of thickness variation in the etched AlGaAs. Thus, selective etching with the smallest possible AlGaAs stopper loss is essential for precise control of HJFET threshold voltages.

One such etching technique is $CCl_2F_2 + He$ plasma etching, which has been widely used for AlGaAs/GaAs HJFET fabrication[1]. In this technique, the etch can be stopped on an $Al_{0.2}Ga_{0.8}As$ layer as thin as 12 Å, the best result reported so far[5]. However, this etching technique will be discontinued because of CFC regulation. Hence, a new selective etching technique using non–freon gas plasma with the same or smaller loss of AlGaAs is required.

In this paper, the AlGaAs stopper loss in $CCl_2F_2 + He$ plasma etching is compared to that in $BCl_3 + SF_6$ plasma etching. Then, the etch–stopping

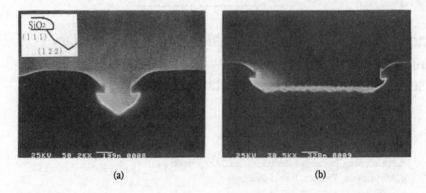

(a)   (b)

Fig. 1. SEM photographs of cross sections for a (100) substrate etched in CCl_2F_2+He plasma with a SiO_2 mask. (a) a 0.3-μm-wide line opening. (b) a 2.5-μm-wide line opening. Both of the openings are parallel to the [011] direction.

mechanism in CCl_2F_2+He plasma etching is investigated with X-ray photoelectron spectroscopy (XPS) and optical emission spectroscopy (OES). Finally, the mechanism is applied to a highly selective plasma etching technique using BCl_3+CF_4 non-freon gas mixture.

2. Etching experiments and results

All etching experiments described below were performed in a Materials Research Corporation MIE760 band-magnetron reactive ion etcher. The total pressure and rf power density were 30 mTorr and 0.07 W/cm^2 at 13.56 MHz. The induced self-biases were set at the same voltage of -12 V by adjusting chamber impedance.

2.1 CCl_2F_2+He plasma etching

The total gas flow rate was 20 sccm, and the gas flow ratio of CCl_2F_2 and He was set at 9:1. First, GaAs etching characteristics were investigated. Figures 1(a) and 1(b) are SEM photographs of cross sections for an etched GaAs surface. Facets of (111) and (122) were observed as in Fig. 1(a), indicating that the etching process of GaAs was mainly a spontaneous radical reaction on the surface and that the self-bias of -12 V was low enough to prevent ions from significant sputtering. In Fig. 1(b), the rough etched-surface is shown. The surface roughness together with an XPS signal of Ga$3d$ for the GaAs etched surface suggests that the GaAs etching was limited by desorption of a low-volatile gallium fluoride on the etched surface. Thus, we presume that dependence of GaAs etching rate on substrate temperature shown in Fig. 2 corresponds to that of etching product desorption rate.

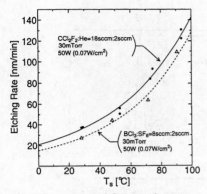

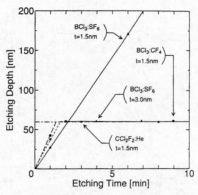

Fig. 2. Dependence of GaAs etching rate on substrate temperature in CCl2F2+He and BCl3+SF6 plasma. Substrate temperatures were measured by a thermocouple set directly on the substrate surface.

Fig. 3. Etching depth versus etching time for a GaAs(600Å)/Al0.2Ga0.8As(t = 15Å or 30Å)/GaAs structure etched in CCl2F2+He, BCl3+SF6 and BCl3+CF4 plasma. The self–bias voltages were all −12V at an rf power density of 0.07W/cm².

Using this etching plasma, the AlGaAs stopper thickness required to stop etch was investigated. Figure 3 shows etching depth versus etching time for a GaAs(600Å)/Al0.2Ga0.8As(15Å)/GaAs structure. A 15–Å layer of Al0.2Ga0.8As was enough to stop the etch.

2.2. BCl3+SF6 plasma etching

The AlGaAs stopper thickness required for CCl2F2+He plasma etching was compared to that of BCl3+SF6 plasma etching. In this experiment, the AlF3 production rate in BCl3+SF6 plasma etching was adjusted to that in CCl2F2+He plasma etching in conformity with the following consideration. The AlF3 production rate and the compositions of the GaAs etching products must vary correlatively because both of them are supposed to depend on the F–Cl ratios in the etching plasma. Therefore, when the compositions of the GaAs etching products are the same, the AlF3 production rates will be also the same. The compositions of the GaAs etching products can be monitored through its desorption rate, which is observed as the GaAs etching rate, as mentioned in the previous section.

When the gas flow rates of BCl3 and SF6 were 8 sccm and 2 sccm, respectively, the GaAs etching rate was nearly the same as that for CCl2F2+He plasma etching throughout the measured range of temperature, as shown in Fig. 2.

In Fig. 3, etching depth versus etching time for GaAs/AlGaAs/GaAs structure for the BCl3+SF6 plasma etching is also shown. A 15 Å–Al0.2Ga0.8As layer was removed, while a 30–Å Al0.2Ga0.8As layer stopped etch. This result indicates the existence of an effect that assists the AlF3 etch–stop reaction in CCl2F2+He plasma etching.

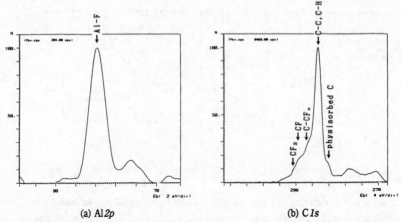

(a) Al2p   (b) C 1s

Fig. 4. XPS signals of an AlGaAs surface on which etch was stopped in CCl₂F₂+He plasma etching. (a) Al2p. (b) C 1s.

3. Selectivity enhancing mechanism

To identify this assisting effect, XPS was employed. Figures 4(a) and 4(b) show XPS signals of an AlGaAs surface on which etch was stopped in the CCl₂F₂+He plasma etching. Al2p and C 1s XPS signals were detected, indicating the existence of AlF_3 and adsorbed fluorocarbon (CxFy). The most probable origin of the CxFy is the CF_2 radical, which was observed in the CCl₂F₂+He plasma by OES as shown in Fig. 5[6]. The production rate of CxFy on the etched stopper surface must be low, because no significant deposition was observed on the etched surface even after long overetching.

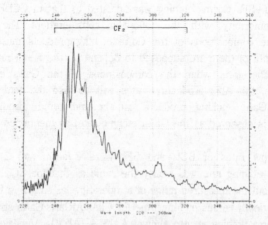

Fig. 5. Optical emission spectrum of CCl₂F₂+He plasma from 220 to 360nm. A static background was subtracted from the original scan.

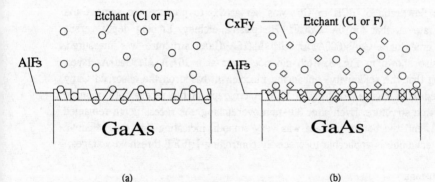

Fig. 6. Selectivity—enhancing mechanisms. (a) A thin barrier layer of AlF3 without fluorocarbon adsorbed. (b) A thin barrier layer of AlF3 with fluorocarbon adsorbed.

Based on these analyses, we propose a selectivity—enhancing mechanism model shown in Figs. 6(a) and 6(b). For a 30—Å Al0.2Ga0.8As stopper, the etched surface is covered with enough AlF3 to prevent etching. However, an AlF3 layer created from a 15—Å Al0.2Ga0.8As layer is too thin and has many leaks through which etchants reach the underlying GaAs, as shown in Fig. 6(a). If there is CxFy adsorbed on the etched surface, those leaks will be sealed by the CxFy, and thus etch of the underlying GaAs does not occur, as shown in Fig. 6(b).

4. BCl3+CF4 plasma etching

Applying the selectivity—enhancing mechanism, we developed a new selective etching technique using non—freon BCl3+CF4 plasma. BCl3 was used as a supplier of the Cl radical, and CF4 was introduced as a supplier of the F radical and CxFy.

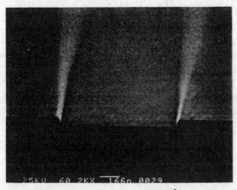

Fig. 7. SEM micrograph of recess pattern on a GaAs(600Å)/Al0.2Ga0.8As(15Å)/GaAs structure etched in BCl3+CF4 plasma. Etch was stopped at the Al0.2Ga0.8As layer. The GaAs etching rate was ~400Å/min at 30°C. Overetching time was 30 min.

The gas flow ratio of BCl_3 to CF_4 was set at 2:3 to produce the same GaAs etching rate as that for the CCl_2F_2+He plasma etching. Etching depth versus etching time for a GaAs(600Å)/$Al_{0.2}Ga_{0.8}As$(15Å)/GaAs structure was measured, and is also shown in Fig. 3. Etch did not occur at a 15−Å $Al_{0.2}Ga_{0.8}As$ layer, indicating that the selectivity−enhancing mechanism based on the adsorbed C_xF_y functioned well. Figure 7 shows an SEM micrograph of the recess pattern made on the same structure. Even after 30−min overetching, the recess depth remained at 600 Å, and the bottom surface was very smooth, indicating that this selective etching technique is applicable to precisely controlling HJFET threshold voltages.

5. Conclusions

Comparing CCl_2F_2+He plasma etching with BCl_3+SF_6 plasma etching, we have shown that adsorbed C_xF_y helps the etch−stop reaction of AlF_3, and consequently enhances etching selectivity. We have also proposed a model of the selectivity−enhancing mechanism, in which adsorbed C_xF_y seals leaks in an AlF_3 barrier. This mechanism was applied to develop a new selective etching technique using non−freon BCl_3+CF_4 plasma. This plasma etching technique completely stops etch on an $Al_{0.2}Ga_{0.8}As$ layer as thin as 15 Å, and therefore will be very useful for precisely controlling threshold voltages.

Acknowledgement

The authors would like to thank Dr. Seiji Samukawa for his useful discussion. They are also grateful to Dr. Tadatoshi Nozaki and Dr. Kazuhiko Honjo for their valuable advice, support, and encouragement throughout the course of this work.

References

[1] Hikosaka K, Mimura T and Joshin K, Jpn. J. Appl. Phys. 20(11), L847−L850 (1981).
[2] Salimian S and Cooper C B, J. Vac. Sci. Technol. B6(6), 1641−1643 (1988).
[3] Guggina W H, Ketterson A A, Andideh E, Hughes J, Adesida I, Caracci S and Kolodzey J, J. Vac. Sci. Technol. B8(6), 1956−1959 (1990).
[4] Knoedler C M and Kuech T F, J. Vac. Sci. Technol. B4(5), 1233−1236 (1986)
[5] Seaward K L, Moll N J and Stickle W F, J. Vac. Sci. Technol. B6(6), 1645−1649 (1988).
[6] Vatus J, Chevrier J, Delescluse P and Rochette J F, IEEE Trans. Electron Devices ED−33(7), 934−937 (1986)

Inst. Phys. Conf. Ser. No 145: Chapter 2
Paper presented at 22nd Int. Symp. Compound Semiconductors, Cheju Island, Korea, 28 August–2 September 1995
© 1996 IOP Publishing Ltd

Annealing effects on low-temperature GaN layer grown by Metalorganic chemical vapor deposition

Bae-Yong Kim, Yeun-Ho Choi, Chang-Hee Hong, Seung-Hee Kim, and Tae-Kyung Yoo

LG Electronics Research Center, 16 Woomyeon-Dong, Seocho-Gu, Seoul 137-140, Korea

Abstract. Low-temperature GaN layers have been grown on (0001) sapphire substrates by metalorganic chemical vapor deposition at 600℃ and the effects of *in-situ* annealing (1000℃) have been investigated using DXRD, SEM, RBS, and SIMS. DXRD measurements indicate that the crystal quality of these low-temperature GaN layers improve with decreasing thickness in the range of 200 Å to 5500 Å. RBS in channeling condition also indicates crystallinity improvement after annealing. The surface of the layers grown at low temperature were rough and upon annealing they exhibited columnar structures. RBS and SIMS revealed considerable interdiffusion between GaN and sapphire.

1. Introduction

The III-nitrides have received considerable interests due to their wide bandgap from 1.9 eV (InN) to 6.2 eV (AlN) for light emitting diodes (LED)[1-3] and laser diodes[4-6] performing in the blue and ultraviolet regions. The lack of substrates having no lattice mismatch and same thermal expansion coefficient makes the development of LEDs or LDs difficult. Sapphire[7], SiC[8], and Si[9] have been the most widely used substrates having considerable mismatch of 13.8%, 3.5%, and 17%, respectively. Recently, Detchprohm et al.[10] attempted to grow high crystal quality GaN film on GaN substrate for the first time. However, GaN and AlN layers deposited at low tempeatures on sapphire are systematically used as buffer layers prior to growing high-quality GaN epilayer growth. Furthermore, there have been few detailed reports on buffer layers in spite of their importance. We report, in this paper, detailed effects of annealing on GaN buffer layers at high-temperature epilayer growth condition.

2. Experiments

Figure 1 shows the vertical MOCVD reactor used in this work. A resistive graphite heater below the susceptor is used to heat the substrate. The temperature is monitored and controlled using an infrared pyrometer. The growth pressure was maintained at 76 Torr and the susceptor was rotated at a speed of 400 rpm. Trimethylgallium (TMG) and ammonia gas

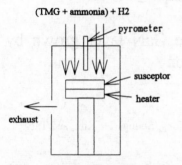

(TMG + ammonia) + H2

pyrometer

susceptor

heater

exhaust

Fig. 1: Schematic of the vertical MOCVD reactor

were used as sources for Ga and N, respectively.

Sapphire substrates were degreased by chemical solvents, etched in a hot H_3PO_4:H_2SO_4(1:3) solution for 10 minutes, rinsed in deionized water and dried with N_2 prior to introducing into the reactor chamber. The substrates were then heated to 1100℃ under H_2 ambient followed by surface nitridation under ammonia flow at 1000℃ for 10 minutes. The buffer growth was carried out at 600℃ using 29 μmol/min TMG, 3.3 slm NH_3, and 15 slm H_2. These as-grown films, with thicknesses ranging from 200Å to 5500Å, were subsequently heated to 1000℃ under ammonia flow to determine the annealing effects.

Rutherford backscattering (RBS) was used to assess the exact thickness and crystallinity of the samples. A Van der Graff particle accelerator was used to accelerate *alpha* particles which were subsequently collected at a detector placed at 165° from the incident beam line after backscattering. Scanning electron microscopy (SEM) and double X-ray diffraction (DXRD) measurements were also utilized to determine film texture and crystallinity, respectively. On selected samples, depth profiles were obtained by secondary ion mass spectrometry (SIMS). A 5 KeV Cesium ion beam which impinged the specimen at 60° from the surface normal was chosen as the sputtering source.

3. Results and discussions

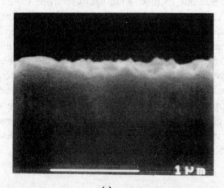 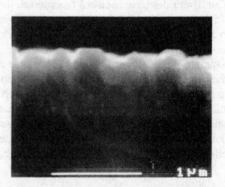

(a)   (b)

Fig. 2: SEM images of GaN/Sapphire (a) as-grown and (b) after annealing

Figure 2 shows SEM images for an as-grown and annealed film. The as-grown film shows very rough surface suggesting small polycrystalline grains as observed in Figure 2(a). The thickness of the film was 3500Å using SEM and 2300Å using RBS. This difference in thickness indicates that a large fraction of the film is composed of voids since RBS is only capable of detecting solid materials and not empty voids. In annealed film, columnar feature

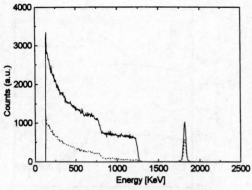

Fig. 3: RBS spectra of GaN(25nm)/Sapphire.
(—) Random and (...) channeling conditions

is observed as shown in Figure 2(b). These features are larger in dimension than those prior to annealing. It is believed that the small grains coalesced to reduce the void space and form larger columnar grains which serve as growth sites for ensuing epitaxial growths. The shape of the grains are consistent with those previously observed in the initial stages of high-temperature GaN epilayers on GaN buffer layers[11]. RBS measurements were also performed to observe the effects of annealing on the crystal quality and to overcome the difficulties of anaylzing very thin films by X-ray diffraction. In the RBS measurement, *alpha* particles of 2.236 MeV were backscattered in random condition and in channeling condition about <0001> crystal axis. The incident charge was fixed at 10 μcoulombs in both cases. In the as-grown film, channeling was not obtained due to the polycrystalline nature of the film. In comparison, channeling condition was realized in the annealed sample with χ_{min} of 57% which indicates improved crystallinity. DXRD measurements (not shown) also supported the trend observed by RBS as shown in Figure 3. The full-width-at-half-maximum (FWHM) of the buffer layers improved after annealing for all the samples investigated in our experiment. These results clearly show crystallization of the as-grown films as a result of annealing.

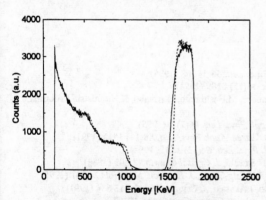

Fig. 4: RBS spectra of GaN(230nm)/sapphire.
(...) as-grown and (—) after annealing

A closer analysis of the RBS spectra shows interdiffusion of Ga and Al species. In Figure 4, which shows the RBS spectra of a 2300 Å-thick as-grown and annealed films, the peaks at 1.778, 0.622 and 1.033 MeV correspond to Ga and N at the surface, and Al at the interface respectively. The as-grown sample shows a relatively abrupt interface compared to the annealed film. This diffusion phenomenum was confirmed by SIMS. The spectra of SIMS are given in figure 5. Considerable difference in the Al profile is evident between as-grown and annealed film due to diffusion. Ga is also found to penetrate into the sapphire substrate. No other significant impurities are observed in both films with the exception of carbon which was likely introduced from the source and graphite susceptor.

294

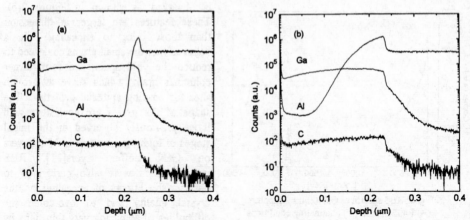

Fig. 5: SIMS profiles of a) as-grown and b) annealed GaN(230nm)/sapphire.

4. Conclusion

In summary, polycrytalline GaN grains were grown on sapphire substrates at buffer layer growth condition and subsequently were crystalized to form columnar structure after annealing as observed by SEM and RBS channeling condition. Prior to annealing, a large fraction of the films were determined to be composed of voids. RBS and SIMS measurements revealed considerable diffusion of Al and Ga as a result of the annealing process.

References

[1] R.J. Molnar, R.Singh, T.D.Moustakas, *Appl. Phys. Lett.* **66**(3) 268 (1995)
[2] S.Nakamura, T.Mukai, M.Senoh, *J.Appl. Phys.* **76**(12) 8189 (1994)
[3] N.Koide, H.Kato, M.Sassa, S.Yamasaki, K.Manabe, M.Hashimoto, H.Amano, K.Hiramatsu, I.Akasaki,
    J. Crystal Growth **115** 639 (1991)
[4] X.H.Yang, T.J.Schmidt, W.Shan, J.J. Song, *Appl. Phys. Lett.* **66**(1) 1 (1995)
[5] M.Asif Khan, D.T.Olson, J.M.Van Hove, J.N. Kuznia, *Appl. Phys. Lett.* **58**(14) 1515 (1991)
[6] S.T.Kim, H.Amano, I.Akasaki, N.Koide, *Appl. Phys. Lett.* **64**(12) 1535 (1994)
[7] I.Akasaki, H.Amano, Y.Koide, K.Hiramastu, N.Sawaki, *J. Crystal Growth* **98** 209 (1989)
[8] T.George, W.T.Pike, M.A.Khan, J.N.Kuznia, P.Chang-Chien, *J. Electron. Materials* **24** 241 (1995)
[9] T.Takeuchi, H. Amano,K.Hiramatsu, N.Sawaki, I.Akasaki, *J. Crystal Growth* **115** 634 (1991)
[10] T.Detchprohm, K.Hiramatsu, N.Sawaki, I.Akasaki, *J. Crystal Growth* **145** 192 (1994)
[11] K.Hiramatsu, S.Itoh, H.Amano, I.Akasaki, N.Kuwano, T.Shiraishi, K.Oki,
    J. Crystal Growth **115** 628 (1991)

Inst. Phys. Conf. Ser. No 145: Chapter 2
Paper presented at 22nd Int. Symp. Compound Semiconductors, Cheju Island, Korea, 28 August–2 September 1995
© *1996 IOP Publishing Ltd*

Comparison of AuGe/Ni/Au, AuGe/Pd/Au, Ti/Pt/Au, and WSi/Ti/Pt/Au ohmic contacts to n-type InGaAs

S.H. Park, M.P. Park, T-W. Lee, K.M. Song, K.E. Pyun, and H-M. Park

Electronics and Telecommunications Research Institute, Yusong P.O. Box 106, Taejon, Korea

Abstract. A comprehensive study of AuGe/Ni/Au, AuGe/Pd/Au, Ti/Pt/Au, and WSi/Ti/Pt/Au ohmic contacts to n-type InGaAs for AlGaAs/GaAs HBT fabrication is presented using various measurements in order to investigate the correlation of the microstructures with the specific contact resistance. As the results, AuGe-based ohmic contact systems with Ni or Pd as the barrier metal show the degradation of specific contact resistance at the relatively low alloying temperature. It is believed because the InGaAs layer near the interface is decomposed due to the out-diffusion of indium. On the other hand, Ti/Pt/Au and WSi/Ti/Pt/Au ohmic systems maintain their excellent electrical resistivities even after alloying at high temperature of 400°C because barrier metals prevent indium and other elements from moving out of interface. Especially for Ti/Pt/Au contact, the specific contact resistance of 8.8×10^{-10} Ω-cm^2, which is one of the best ohmic data reported up to date, is obtained after RTA at 375°C for 10 s.

1. Introduction

Heterojunction bipolar transistors (HBTs) exhibit intrinsically high transconductance, large current drive, and uniform emitter-base voltage (V_{BE}) [1]. It is very critical that the parasitic resistance due to ohmic contact be reduced to maximize these advantageous performances of HBT as a promising device for high speed digital and analog applications. For the conventional AuGe-based ohmic contacts to n$^+$-GaAs, it is difficult to obtain the contact resistivity below low 10^{-6} Ω-cm^2, and easy to cause the poor surface morphology, so called "ball-up", due to alloying treatment. On the other hand, non-alloyed and low-resistance emitter ohmic contacts can be fabricated using the refractory metals on the highly doped InGaAs with smaller bandgap. These ohmic contact schemes have demonstrated the better specific contact resistances between 5×10^{-8} and 2.5×10^{-6} Ω-cm^2 as well as the shallow contact and the smooth metal surface [2][3]. Although in recent years ohmic contacts to n$^+$-InGaAs have been widely investigated for HBT application, most studies are focused mainly on reducing the contact resistivity, and do not analytically explain the effect of metal-semiconductor interfacial reactions on the electrical performances of ohmic contacts.

In this paper we describe the ohmic contacts to n$^+$-InGaAs using the sequential evaporation of AuGe/Ni/Au, AuGe/Pd/Au, Ti/Pt/Au, and WSi/Ti/Pt/Au metals and the RTA (rapid thermal annealing). Also, the electrical properties and the interfacial microstructures of these contact schemes were measured and analyzed utilizing various characterization tools such as HP4145B parameter analyzer, AES (Auger electron spectroscopy), XRD (x-ray diffraction), TEM (transmission electron microscopy). For AuGe/Ni/Au ohmic contact, the specific contact resistance tends to abruptly increase after alloying at the temperature above 250°C for 10 s. The result of AuGe/Pd/Au contact system is similar to that of AuGe/Ni/Au system except for its smoother surface morphology. In contrast with AuGe-based ohmic schemes, it is found that Ti/Pt/Au and WSi/Ti/Pt/Au ohmic systems exhibit the excellent electrical resistivities and the surface morphologies even after alloying at the high temperature of 400°C because barrier metals prevent indium and other elements from moving out of interface.

2. Experiment

The AlGaAs/GaAs heterojunction epi-structure was grown by MOCVD (metal-organic chemical vapor deposition) on 3-inch semi-insulating GaAs substrate. As the emitter cap layer, an n^+-GaAs layer (1300 Å, Si: 4×10^{18} /cm^3), a compositionally graded n^+-In$_x$Ga$_{1-x}$As (x=0→0.5) layer (400 Å, Si: 1×10^{19} /cm^3), and a compositionally uniform n^+-In$_x$Ga$_{1-x}$ As (x=0.5) layer (400 Å, Si: 1×10^{19} /cm^3) were grown sequentially on the Al$_{0.28}$Ga$_{0.72}$As emitter. Prior to metal deposition, the samples were patterned by image reversal process using AZ5214-E photoresist and cleaned in the HCl:H$_2$O (1:1) solution for 1 min. Next, AuGe/Ni/Au (1200/500/1300 Å) and AuGe/Pd/Au (1200/500/1300 Å) multi-layers were evaporated using a thermal evaporator at the pressure of 5×10^{-7} Torr, respectively. Also, Ti/Pt/Au (500/200/1300 Å) and WSi/Ti/Pt/Au (1000/500/200/1300 Å) multi-layers were deposited using a sputter (WSi) and an electron-beam evaporator (Ti/Pt/Au) at the same pressure. After metal deposition, the active area was isolated by chemical mesa etching. Finally, the RTA treatment for ohmic alloying was performed at the temperature ranging from 250 to 425°C for 10 s in N$_2$ atmosphere.

The specific contact resistance was measured using the transmission line method (TLM), in which the pad width was 100 μm, and the spacings between metal pads were 5, 10, 15, 20, 30 μm. The microstructural interaction at the metal/InGaAs interface was investigated by AES, XRD, and TEM.

3. Results and discussions

3.1 Contact resistivity

The change of specific contact resistance (ρ_c) with alloying temperature for AuGe/Ni/Au contacts to n^+-InGaAs is shown in Fig. 1. AuGe/Ni/Au contact scheme shows the minimum specific contact resistance of about 5×10^{-8} Ω-cm^2 after alloying at 200°C for 1 hr. But a rapid increase in the specific contact resistance with increasing RTA temperature above 250°C for 10 s indicates the development of some microstructures at the interface causing to increase the ρ_c. For AuGe/Pd/Au contact system, the change of ρ_c is almost same as that of

AuGe/Ni/Au contact.

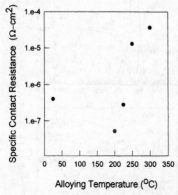

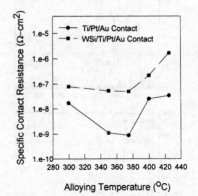

Fig. 1. Dependence of specific contact resistance
of AuGe/Ni/Au contact on alloying temperature.

Fig. 2. Dependence of specific contact resistance
of Ti/Pt/Au and WSi/Ti/Pt/Au contacts on temp.

The specific contact resistance of Ti/Pt/Au and WSi/Ti/Pt/Au metal systems, non-alloyed ohmic contacts, as a function of alloying temperature is shown in Fig. 2. The ρ_c of Ti/Pt/Au contact is below 5.0×10^{-8} Ω-cm^2 through heat treatments from 300 to 425°C for 10 s. Furthermore, ρ_c of 8.8×10^{-10} Ω-cm^2, which is one of the best ohmic data reported up to date, is obtained after alloying of 375°C. This result proves that the metal-to-semiconductor interface for Ti/Pt/Au contact maintains typically its original microstructure in spite of high temperature treatment. For WSi/Ti/Pt/Au contact, its ρ_c is approximately one order of magnitude higher than that of Ti/Pt/Au, although it is superior to that of AuGe-based contacts. It is believed that this difference of ρ_c is due to the stress induced at WSi/InGaAs interface after alloying.

3.2 Interfacial microstructure

Due to the strong reaction of the contact metals with InGaAs for AuGe-based contact schemes, complicated microstructures are expected to be developed at the interface after elevated temperature RTA. The Auger spectroscopy combined with ion beam sputtering was applied in order to analyze the metallurgical composition depth profile in alloyed and non-alloyed ohmic contacts.

The AES depth profiles of all the elements of AuGe/Ni/Au contact as-deposited and subjected to RTA at temperature ranging from 200 to 300°C are shown in Fig. 3. The significant interdiffusion occurred after RTA treatment above 250°C that is a relatively low temperature for ohmic alloying, because a Ni layer with the thickness of 500 Å does not play an appropriate role as the diffusion barrier. The major out-diffusion and in-diffusion elements are In and Au, respectively. It would be believed that the decomposition of InGaAs near the interface due to the out-diffusion of indium, which is very reactive element, is responsible for the abrupt degradation of contact resistivity after alloying at temperature above 250°C for 10 s.

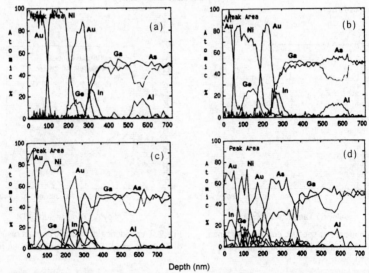

Fig. 3. Auger depth profiles of AuGe/Ni/Au film on n-InGaAs (a) as-deposited, (b) furnace-annealed at 200°C for 1 hr, (c) RTA-treated at 250°C for 10 s, (d) RTA-treated at 300°C for 10 s.

Fig. 4 (a)-(d) show the AES depth profiles of AuGe/Ni/Au, AuGe/Pd/Au, Ti/Pt/Au, and WSi/Ti/Pt/Au contacts after RTA at 400°C for 10 s. For AuGe/Ni/Au contact, the sharpness of metal-semiconductor interface was almost disappeared due to the severe diffu-

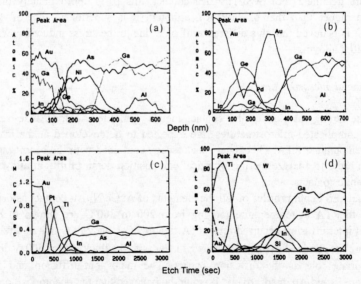

Fig. 4. Auger depth profiles of (a) AuGe/Ni/Au, (b) AuGe/Pd/Au, (c) Ti/Pt/Au, and (d) WSi/Ti/Pt/Au contact RTA-treated at 400°C for 10 s.

sion of In, Ga, As, and Au after alloying. In case of AuGe/Pd/Au contact, the dissolution of

InGaAs was also observed although the inter-diffusion of elements at interface was not so remarkable as in AuGe/Ni/Au contact. As shown in Fig. 4 (c) and (d), however, Ti/Pt/Au and WSi/Ti/Pt/Au contacts exhibit very slight movement of composing elements, because Pt and WSi may prevent effectively the elements from diffusing. So, Ti/Pt/Au and WSi/Ti/Pt/Au contacts do not show such an abrupt degradation of specific contact resistance as AuGe-based contact. These AES results indicate that the specific contact resistances of the above ohmic systems intimately correlate with the diffusion of active elements, especially In, from metal-semiconductor interface.

The compound formation between the metal films and InGaAs layer after alloying was studied by XRD measurement. The diffraction profiles for AuGe/Ni/Au films before and after alloying are shown in Fig 5(a) and (b), respectively. Before alloying only diffraction peaks corresponding to Au, Ni, In, AuGe, and GaAs are observed, while after alloying at 400°C for 10 s [Fig. 5(b)], the Au_xIn_y phases are found to be dominant over wide range of diffraction angle, 2θ. In addition, minor peaks such as β-AuGa and NiAs are also observed. For AuGe/Pd/Au contact, the less Au_xIn_y phases than for AuGe/Ni/Au contact are measured (not shown in this figure). As observed in Fig. 5(c), however, Ti/Pt/Au contact to InGaAs does not exhibit any detectable phase change after RTA treatment ranging from 250 to 400°C. These XRD analyses confirm the previous results concerning the specific contact resistance and AES depth profiles.

The detailed metal/InGaAs ohmic reaction was further confirmed by a cross-sectional TEM (XTEM) analysis, in which the electron energy is 200 keV. Fig. 6 shows the XTEM bright field images of the AuGe/Ni/Au and AuGe/Pd/Au contacts alloyed at the temperature of 300 and 400°C for 10 s, respectively. In the AuGe/Ni/Au contact alloyed at 300°C [Fig. 6(a)], it is observed that the sharp interface dissapears because of the active intermetallic reaction. The compositional nature of these reacted layers is not known accurately at present. For this sample, a phase transformation from

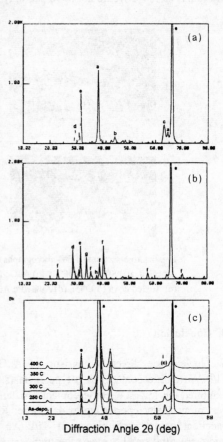

Fig. 5. XRD profiles for AuGe/Ni/Au contact before (a) and after RTA at 400°C (b), and for Ti/Pt/Au contact after RTA at various temperature (c). [a:Au, b: Ni, c:In, d:AuGe, e:GaAs, f: Au_xIn_y, g:β-AuGa, h:NiAs, i:Ti(In).]

poly-crystalline to single-crystalline structure was also confirmed by the transmission diffraction pattern. For the AuGe/Ni/Au contact alloyed at 400°C, it is difficult to distinguish the phase domains (grains) as well as metal layers, This phenomenon can be attributed to the severe interaction between metal films.

Fig. 6(c) and (d) show also the metallurgical behavior of the AuGe/Pd/Au contact alloyed at 300°C and 400°C, respectively. After RTA at 300°C, the XTEM microstructure in the AuGe/Pd/Au contact is almost same as that of 300°C-alloyed AuGe/Ni/Au contact. When the AuGe/Pd/Au contact is alloyed at 400°C, this shows little microstructural change compared with the one annealed at 300°C. This indicates that Pd is a better diffusion barrier than Ni. On the other hand, from the results of the specific contact resistance, AES, and XRD we guess the microstructure for the Ti/Pt/Au and the WSi/Ti/Pt/Au contacts even after RTA at 400°C would not be remarkably transformed.

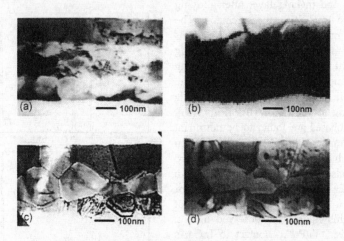

Fig. 6. Cross-sectional TEM micrographs of AuGe/Ni/Au and AuGe/Pd/Au contacts annealed at 300°C and 400°C: (a) AuGe/Ni/Au (300°C), (b) AuGe/Ni/Au (400°C), (c) AuGe/Pd/Au (300°C), (d) AuGe/Pd/Au (400°C).

4. Conclusion

We have made direct comparison of AuGe/Ni/Au, AuGe/Pd/Au, Ti/Pt/Au, and WSi/ Ti/Pt/Au ohmic contacts to n-type InGaAs. Of these contact schemes, Ti/Pt/Au contact exhibited the lowest specific contact resistance of 8.8×10^{-10} Ω-cm^2, which is one of the best ohmic data reported up to date, after RTA at 375°C for 10 s. It was found that this excellent ohmic performance of Ti/Pt/Au contact is due to the maintenance of the microstructural stability even after high temperature alloying.

References

[1] Slater D B, Enquist P M, Hutchby J A, Morris A S and Trew R J 1994 *IEEE Electron Device Lett.* **15** 154-156.

[2] Nittono T, Ito H, Nakajima O and Ishibashi 1986 *Japan. J. Appl. Phys.* **25** L865-L867.

[3] Shanthamara L G, Schumacher H, Leblanc H P, Esagui R, Bhat R and Koza M 1990 *Electronics Lett.* **26** 1127-1129.

Inst. Phys. Conf. Ser. No 145: Chapter 2
Paper presented at 22nd Int. Symp. Compound Semiconductors, Cheju Island, Korea, 28 August–2 September 1995
© *1996 IOP Publishing Ltd*

Direct bonding of lattice-mismatched and orientation-mismatched III-V semiconductor wafers: a step toward establishing "Free-Orientation Integration"

Yae Okuno and Kazuhisa Uomi

Central Research Laboratory, Hitachi, Ltd.
1-280 Higashi-koigakubo, Kokubunji, Tokyo 185, Japan

Abstract

The direct bonding of various III-V wafers, which are mismatched in terms of surface orientations and lattice constants, was systematically investigated for an In-Ga-As-P system. Despite those mismatches, many wafer combinations were successfully united with sufficient mechanical strength. A dislocation-free interface of (001) GaP and (110) InP bonded at the atomic level was observed by transmission electron microscopy. A (001) InP-based long-wavelength laser fabricated on a (110) GaAs substrate exhibited high reliability. These results suggest that a novel concept "free-orientation integration" can be achieved by direct bonding for III-V system.

1. Introduction

Monolithically integrating various device elements will become increasingly important. However, epitaxial growth, a common and basic technique for fabricating compound semiconductor devices, has difficulty in integrating different kinds of materials, because the differences in material properties such as the lattice constant easily degrade crystalline quality of the integrated structure. For example, GaAs heteroepitaxy on Si has been long studied in an effort to integrate photonic and electrical device elements. However, it is difficult to obtain good quality of such a layered structure, so this technique is not yet available for manufacturing integrated devices [1]. Moreover, epitaxial growth has another disadvantage: the crystallographic orientations of epitaxially grown layers are restricted to be the same as that of the substrate. The performance of a device, especially an optical device, greatly depends on its crystallographic axes [2]. Most optical and electrical devices nowadays are fabricated on (001)-based structures, however, their performances may be improved by employing the optimum crystallographic orientations. In that case, if the optimum orientation differs from device to device, the devices can't be integrated monolithically by epitaxial growth.

On the other hand, the direct bonding is a technique that unites two semiconductor materials without interposing any adhesive [3]. It enables integration of different materials, such as InP and GaAs, without degrading their crystalline quality [4,5]. Integrating GaAs layers with different crystallographic orientations is also reported [6]. Furthermore, integration of InP and GaAs with a change in the relation of their crystallographic axes was shown to work well [7,8]. Therefore, the direct bonding can overcome the two difficulties in epitaxial growth: mismatches in material properties and crystallographic orientations. Consequently, direct bonding will extend the possibilities for device integration. "Free-Orientation Integration" represents a new concept of freely integrating various kinds of material with various crystallographic orientations, which should be implemented by the direct bonding. In order to confirm the applicability of this concept, it should be clarified whether the direct bonding is available for any wafer combinations based on III-V materials. In this paper, therefore, we systematically investigate the direct bonding of various combinations of material and crystallographic orientation in III-V wafers.

2. Experimental

The direct bonding procedure used in this study is as follows. First, two wafer surfaces are chemically etched in a mixture of H_2SO_4 and H_2O_2, cleaned in dilute HF, rinsed in deionized water, and dried. Then they are placed face to face, and heated to 600°C for 30 minutes in H_2 flow, while applying a pressure of 30 g/cm^2 by weight. Their horizontal crystallographic axes are usually adjusted by aligning their {110} cleaved facets to be parallel. The wafers used in this study did not have intentional misorientation. Epitaxial growth, when needed, was performed by conventional low-pressure metalorganic vapor phase epitaxy (MOVPE).

3. Results and discussion

3.1. Direct bonding of various wafer combinations

Figure 1 shows the wafer combinations for which direct bonding was performed. In each combination, two wafers were successfully united. Here, the word "successful" means that the wafers were bonded so strongly that they didn't detach even when their united structure was diced into pieces. This means that these

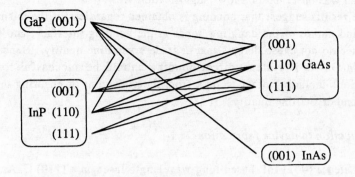

Figure 1   Wafer combinations for which direct bonding was performed
in this experiment, and confirmed as successful.

structures have enough mechanical strength to undergo any device fabrication process. The remaining combinations in Fig. 1 are not "failures", but just they have not been tested yet.

Cross-sectional transmission electron microscopy (TEM) was performed on the united structures. Figure 2 is a high-resolution view of the bonding interface of (001) GaP and (110) InP, which had the largest mismatch in the number of dangling bonds at the interface among the combinations listed in Fig. 1. Two wafers are bonded at the atomic level without occurring any misfit dislocations. Moreover, lattice distortion at the exact interface is very slight on both sides. No threading

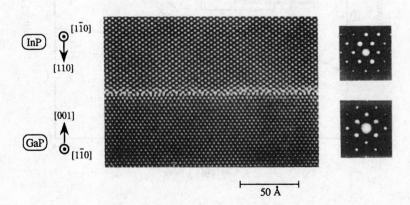

Figure 2   High-resolution TEM view of the bonding interface of (001) GaP
and (110) InP, with diffraction patterns for each material.

dislocation was observed in a low-magnification view.

These results suggest that bonding is obtained regardless of the mismatches in number and nature of the dangling bonds at the bonding interface, and that these mismatches do not cause defects that degrade crystalline quality. Hence, at least for the In-Ga-As-P system, direct bonding should be successful for various combinations including other wafers, from the viewpoint of ensuring mechanical strength and crystalline quality.

3.2. Application to device fabrication

We fabricated a (001) InP-based long-wavelength laser on a (110) GaAs substrate by direct bonding [8]. Figure 3 shows its light-current (I-L) and voltage-current (I-V) characteristics under CW (continuous wave) operation at room temperature. The laser structure, as shown by the insert in Fig.3, is a reversed-mesa ridge-waveguide type buried with polyimide [9], and the active region is a strained multiple quantum well (MQW) emitting at 1.55 μm. Compared with the characteristics of an identical laser fabricated on a (001) GaAs substrate (dashed line), the I-V curve is about 0.4 V higher, while the I-L curve does not differ. This is thought to be due to the mismatch in the number of dangling bonds at the bonding interface [8].

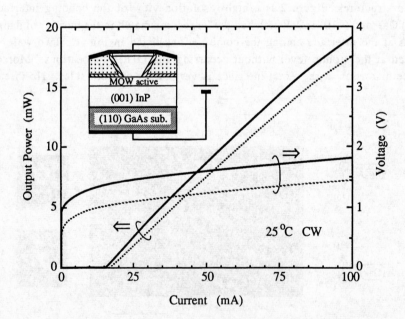

Figure 3 I-L and I-V characteristics of the (001) InP-based 1.55 μm-wavelength laser fabricated on a (110) GaAs substrate, compared with those of the laser on a (001) GaAs substrate (dashed line).

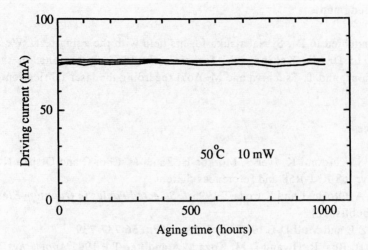

Figure 4   Results of aging test for a (001) InP-based 1.55 μm-wavelength laser
fabricated on a (110) GaAs substrate (3 chips).

Figure 4 is the results of an aging test for three of these laser chips under CW operation at 50°C with a constant light output of 10 mW. They operated for over 1000 hours with no significant degradation. While the orientation-mismatch affected the I-V properties, the effect was small, and it seems that the mismatch did not otherwise degrade laser performance. This fact means that practical device fabrication is thus possible on lattice- and orientation-mismatched wafers. We believe that the other wafer combinations of the In-Ga-As-P system are also feasible for practical device fabrication by direct bonding .

4. Conclusion

We have investigated the use of direct bonding as a tool for free-orientation and free-material integration. We found that various combinations of III-V wafers can be successfully united, in spite of mismatches in lattice constants and surface orientations. TEM showed an excellent interface of (001) GaP and (110) InP, which was dislocation-free and bonded at the atomic level. A laser was fabricated applying a united structure of (110) GaAs and (001) InP. The effect of the orientation mismatch was negligible, and the device showed high reliability. These results indicate that "free-orientation integration" can be conducted for III-V system by direct bonding, which will lead to new types of monolithic integration.

306

Acknowledgments

We are indebted to Dr. S. Nakatsuka for his help with the aging test. We are also grateful to Dr. H. Kakibayashi for his valuable suggestions on the TEM observations, and T. Tsuchiya and M. Aoki regarding the laser fabrication.

References

[1] Fang S F, Adomi K, Iyer S, Morkoç H, Zabel H, Choi C and Otsuka N 1990 *J. Appl. Phys.* **68** R31-R58 and references therein

[2] Niwa A, Otoshi T and Kuroda T 1995 *J. Selected Topics in Quantum Electronics* **1** (to be published)

[3] Liau Z L and Mull D E 1990 *Appl. Phys. Lett.* **56** 737-739

[4] Lo Y H, Bhat R, Hwang D M, Koza M A and Lee T P 1991 *Appl. Phys. Lett.* **58** 1961-1963

[5] Wada H, Ogawa Y and Kamijoh T 1993 *Appl. Phys. Lett.* **62** 738-740

[6] Gordon L, Woods G L, Eckardt R C, Route R R, Feigelson R S, Fejer M M and Byer R L 1993 *Electron. Lett.* **29** 1942-1943

[7] Okuno Y, Uomi K, Aoki M, Taniwatari T, Suzuki M and Kondow M 1995 *Appl. Phys. Lett.* **66** 451-453

[8] Okuno Y, Aoki M, Tsuchiya T and Uomi K 1995 *Appl. Phys. Lett.* **67** (to be published)

[9] Aoki M, Tsuchiya T, Nakahara K, Komori M and Uomi K 1995 *IEEE Photon. Tech. Lett.* **7** 13-15

Inst. Phys. Conf. Ser. No 145: Chapter 3
Paper presented at 22nd Int. Symp. Compound Semiconductors, Cheju Island, Korea, 28 August–2 September 1995
© *1996 IOP Publishing Ltd*

Time-resolved photoluminescence study of radiative transition processes in GaP$_{1-x}$N$_x$ alloys

Hiroyuki Yaguchi†, Seiro Miyoshi†, Hideo Arimoto†, Shiro Saito†, Hidefumi Akiyama‡, Kentaro Onabe†, Yasuhiro Shiraki‡ and Ryoichi Ito†

† Department of Applied Physics, The University of Tokyo, 7-3-1 Hongo, Bunkyo-ku, Tokyo 113, Japan

‡ Research Center for Advanced Science and Technology (RCAST), The University of Tokyo, 4-6-1 Komaba, Meguro-ku, Tokyo 153, Japan

Abstract. Time-resolved photoluminescence (PL) measurements reveal that the radiative transition and carrier relaxation processes in GaP$_{1-x}$N$_x$ alloys with high N concentrations are significantly different from those with low concentrations where NN$_i$ lines are clearly observed. The PL decay curve shows two distinct exponential processes for high concentrations while it is represented by a single exponential for low concentrations. The slow decay in GaP$_{1-x}$N$_x$ alloys with high N concentrations indicates both the long radiative lifetime caused by the weak localization of excitons and the slow relaxation due to the scattered spatial distribution of the states. This is consistent with the fact that the PL occurs at the tails of density of states, which is found from the comparison between absorption and PL spectra.

1. Introduction

GaP$_{1-x}$N$_x$ alloy is one of the candidates for short-wavelength optical device materials since it consists of wide band gap semiconductor materials, GaP ($E_g = 2.3$ eV) and GaN ($E_g = 3.4$ eV). In spite of a large miscibility gap due to the difference in the lattice constant and in the crystal structure, it has recently been possible to grow alloys with N concentrations as high as several % by using molecular beam epitaxy[1] and metalorganic vapor phase epitaxy (MOVPE)[2]. Although the band gap energy of GaP$_{1-x}$N$_x$ is expected to simply increase as the N content increases, the PL peak shifts to lower energy with increasing x[1, 2], which is explained by the reduction in the band gap energy with increasing N concentration. Photoluminescence excitation (PLE) spectroscopy[3] also shows clearly the red-shift in the band gap energy with increasing N concentration. This extremely large band gap bowing is considered to be attributed to the large difference in electronnegativity between N and P atoms.

In the present study, we have carried out time-resolved photoluminescence measurements to examine the radiative transition processes in $GaP_{1-x}N_x$ alloys. We found that the radiative transition processes in $GaP_{1-x}N_x$ alloys with high N concentrations are significantly different from with low concentrations in which NN_i lines are clearly observed. For high nitrogen concentrations ($x > 0.3$ %), the time-decay of the PL intensity shows two distinct exponential processes while it is represented by a single exponential for low nitrogen concentrations.

2. Experimental procedure

The samples used in this study were grown by low-pressure MOVPE using trimethyl-gallium, PH_3 and dimethylhydrazine as the Ga, P and N sources, respectively. Details of the growth procedure have been described elsewhere[2]. The substrates were nominally on-axis (100) GaP. The film thickness was 0.5 μm. The nitrogen concentration was determined by x-ray diffraction on the assumption of Vegard's law between the lattice constant and the N concentration.

Time-resolved PL measurements were performed at 4.2 K. The time-resolved PL intensity was measured with the time-correlated single photon counting method. The excitation source was the frequency-doubled light (315 nm) obtained by a KH_2PO_4 crystal from a dye laser which operates at 630 nm with a repetition rate of 75.4 MHz and a pulse duration of 2 ps.

In cw PL measurements, He-Cd laser (325 nm) was used as an excitation source. The PL was measured with conventional lock-in detection.

We have also carried out low-temperature absorption measurements to examine in more detail the red-shift behavior of the band gap. Monochromatic light obtained from white light of a halogen lamp using a 0.5-m monochromator was used for the absorption measurements. In order to reduce the absorption by the GaP substrate, samples were thinned down to ~100 μm from the back side.

3. Results and discussion

Figure 1 shows 20 K absorption spectra of $GaP_{1-x}N_x$ alloys with various N concentrations. The energy positions of GaP band gap, A line, NN_3 and NN_1 line are shown in this figure. The absorption peak due to A line is distinguished at low N concentrations. With increasing N concentration, it becomes broader and finally disappears. The A-line peak is observed also in GaP because N atoms are only slightly incorporated in the sample. The N concentration dependence of the NN_3 peak is similar to that of the A-line peak though it is not observed in GaP. On the other hand, the peak due to NN_1 pairs becomes larger with increasing N concentration in the range of $x < 1.4$ %. In this way, the absorption lines related to NN_1 pairs can be observed even at $x > 1$ %. This indicates that N atoms in $GaP_{1-x}N_x$ alloys behave like impurities even at such high nitrogen concentrations.

However, the emission related to NN_i pairs is not observed in the PL spectrum of $GaP_{1-x}N_x$ with $x > 1$ %. Thus, the PL does not reflect the density of states (DOS) of NN_i pairs. This agrees with the fact that the emission from NN_1 dominates the other

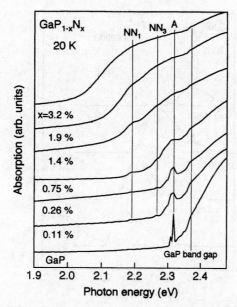

Figure 1. 20 K absorption spectra of $GaP_{1-x}N_x$ alloys with various N concentrations. Absorption edge shifts to lower energy with increasing N concentration. Even for $x > 1$ %, absorption lines related to NN_i pairs can be observed.

emissions for the alloys with $x > 0.2$ %, although the calculation shows that the NN_1 becomes dominant when $x > 3$ %[4].

In addition, as can be seen from this figure, the absorption edge shifts to lower energy as the N concentration increases. This tendency is in agreement with the red-shift of the PL peak[1, 2]. In order to examine in more detail the red-shift behavior, we have compared the absorption spectra with the PL spectra at the same temperature (20 K), as shown in Fig. 2. The PL is found to occur at the DOS tails which are lower in energy than the absorption edge. Namely, the broad emission of $GaP_{1-x}N_x$ alloys with high N concentrations is found to originates not from the band edge but from the tail states, which are closely related to the band edge. This is consistent qualitatively with the stokes shift observed in PLE spectra[3].

Time-resolved PL measurements reveal that the radiative transition and carrier relaxation processes in $GaP_{1-x}N_x$ alloys with high N concentrations ($x > 0.3$ %) are much different from with lower N concentrations where NN_i lines are clearly observed. Figure 3(a) and (b) show the PL spectrum and the PL intensity decay profile of $GaP_{1-x}N_x$ alloy with $x = 0.028$ %, respectively. NN_i lines are dominant in the PL spectrum shown in Fig. 3(a) The decay profile was obtained by detecting NN_5 line. The PL decay curve of $GaP_{1-x}N_x$ alloy with $x = 0.028$ % is expressed by a single-exponential with the decay time of $\tau = 270$ ns, as indicated by the solid line in Fig. 3(b). This decay time is considered to correspond to the radiative recombination lifetime of excitons bound to NN_5 pairs since the lifetime of NN_1 and NN_3 bound exciton is ~100 ns at 4.5 K[5] and

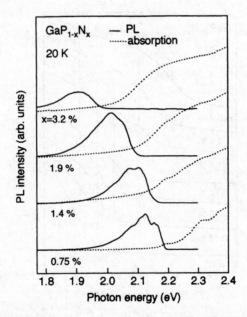

Figure 2. Comparison between absorption and PL spectra for various nitrogen concentrations. The PL is found to occur at the density of states tails.

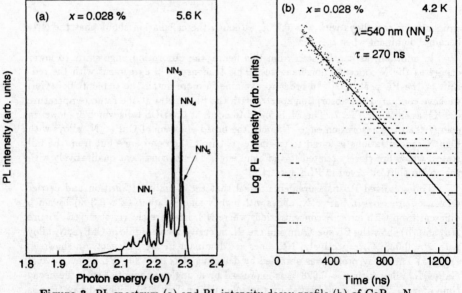

Figure 3. PL spectrum (a) and PL intensity decay profile (b) of $GaP_{1-x}N_x$ alloy with $x = 0.028$ %. The solid line is given by a single exponential with the decay time of 270 ns.

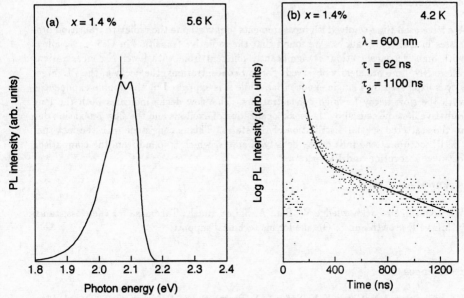

Figure 4. PL spectrum (a) and PL intensity decay profile (b) of GaP$_{1-x}$N$_x$ alloy with $x = 1.4$ %. The solid curve shown in (b) is given by the double exponential with the fast decay time of 62 ns and the slow decay time of 1100 ns.

the radiative lifetime of A, B doublet is ∼500 ns at 4.2 K[6], which are the same order of our result.

Figure 4(a) and (b) show the PL spectrum and the PL intensity decay profile of GaP$_{1-x}$N$_x$ alloy with $x = 1.4$ %, respectively. The emissions related to NN$_i$ pairs are not observed as mentioned before, but a broad emission dominates in the PL spectrum for $x = 1.4$ %. The PL decay curve shown in Fig. 4(b) was obtained by detecting $\lambda = 600$ nm emission, the position of which is indicated by the arrow shown in Fig. 4(a). This decay profile shows two distinct exponential processes and can be expressed by the double exponential $I(t) = I_1 \exp(-t/\tau_1) + I_2 \exp(-t/\tau_2)$ with the fast decay time of $\tau_1 = 62$ ns and the slow decay time of $\tau_2 = 1100$ ns, as indicated by the solid curve.

The fast decay is considered to correspond to the relaxation to nonradiative recombination centers. After the nonradiative recombination centers are saturated, the slow decay becomes dominant. The slow decay represents the radiative recombination lifetime or the relaxation between radiative levels. Therefore, both the radiative lifetime and the relaxation time are found to be as long as or longer than the order of μs. The longer radiative recombination lifetime in comparison with the case of NN$_i$ bound excitons indicates that the localization of excitons becomes weaker as the N concentration increases. Since the tail states are considered to be formed by several N atoms, the excitons may be more weakly localized than the excitons bound to NN$_i$ pairs. On the other hand, the slow relaxation between the radiative levels can be explained by the scattered spatial distribution of the levels. This coincides with the fact that the PL occurs at the DOS tails, which is found from the comparison between absorption and PL spectra.

4. Conclusions

We have used time-resolved PL measurements to investigate the radiative transition processes in $GaP_{1-x}N_x$ alloys. We found that the radiative transition in $GaP_{1-x}N_x$ alloys with high N concentrations is significantly different from with lower N concentrations where NN_i lines are clearly observed. For low concentrations, the decay of the PL intensity is expressed by a single exponential, while it is expressed by the double exponential with the slow decay for high concentrations. The slow decay indicates both the long radiative lifetime caused by the weak localization of excitons and the slow relaxation due to the scattered spatial distribution of the states. This is consistent with the fact that the PL occurs at the tails of the density of states, which is found from the comparison between absorption and PL spectra.

Acknowledgments

We would like to acknowledge W. Pan, A. Shima and K. Takemasa for their assistance in MOVPE growth and S. Ohtake for his technical support.

References

[1] Baillargeon J N, Cheng K Y, Hofler G E, Pearah P J and Hsieh K C 1992 *Appl. Phys. Lett.* **60** 2540–2542

[2] Miyoshi S, Yaguchi H, Onabe K, Ito R and Shiraki Y 1994 *Appl. Phys. Lett.* **63** 3506–3508

[3] Liu X, Bishop S G, Baillargeon J N and Cheng K Y 1993 *Appl. Phys. Lett.* **63** 208–210

[4] Miyoshi S, Yaguchi H, Onabe K, Ito R and Shiraki Y 1994 *J. Cryst. Growth* **145** 87–92

[5] Zheng J and Yen W M 1988 *J Luminescence* **39** 233–237

[6] Cuthbert J D and Thomas D G 1967 *Phys. Rev.* **154** 763–771

Inst. Phys. Conf. Ser. No 145: Chapter 3
Paper presented at 22nd Int. Symp. Compound Semiconductors, Cheju Island, Korea, 28 August–2 September 1995
© *1996 IOP Publishing Ltd*

Reciprocal-Space Analysis of Photoluminescence and Photoluminescence Excitation Spectra

S. D. Yoo and D. E. Aspnes

Department of Physics, North Carolina State University
Raleigh, NC 27695-8202 USA

S. J. Rhee, H. S. Ko, Y. M. Kim, and J. C. Woo

Department of Physics, Seoul National University
Shilim 56-1 Kwan-Ak Seoul, Korea

Abstract. We present an accurate approach for determining energies of features in photoluminescence (PL) and photoluminescence excitation (PLE) spectra that is essentially independent of assumptions about baselines and spectral lineshapes. We Fourier transform these spectra and analyze the phases and amplitudes of the resulting coefficients in reciprocal space. Comparing results with those obtained by conventional real-space analysis, we show that reciprocal-space analysis is more accurate when baseline effects are significant. In addition, our procedure allows the effects of finite monochromator resolution to be corrected, and an optimum monochromator slit width for PL measurements to be defined.

1. Introduction

A primary goal of spectroscopy is the accurate determination of excitation or critical-point energies. It was recently shown that spectroscopic ellipsometric (SE) data, which are characterized by relatively small structures on large backgrounds, could be analyzed more accurately with fewer systematic errors in reciprocal space [1]. In this approach SE data were Fourier transformed and analysis was performed on the amplitudes and phases of the resulting coefficients. This work showed that reciprocal-space analysis offered substantial advantages relative to the more common approach of numerically differentiating or smoothing the data with linear algorithms of the Savitzky-Golay type [2], then fitting the result to standard functional forms [3] by least-squares techniques [4]. Specifically, in reciprocal space, transition energies are defined by phase coherency and are therefore not correlated with the functional forms of the real-space lineshapes, which need not be known a priori. In addition, information is localized in the mid-range Fourier coefficients and is therefore easily separated from the slowly varying baseline and from noise, which dominate the low- and high-index coefficients, respectively.

Deconvolution in reciprocal space has been shown to be useful in separating features of different origins in X-ray photoemission (XPS) [5] and electron energy loss spectroscopies (EELS) [6], which are similar to photoluminescence (PL) spectra in that baseline effects are less prominent than in SE data. Photoluminescence excitation (PLE) spectra appear to represent an intermediate case. For XPS and EELS, procedures have also been given for correcting the data for instrumental broadening, which enters as a convolution and is therefore easy to remove in reciprocal space.

In this work we examine the possibilities of obtaining similar advantages in the analysis of PL and PLE data, which are primary diagnostic techniques for electronic materials and structures. In addition to demonstrating the advantages listed above, we extend previous work by showing that within certain limits these spectral data can be corrected for the effects of finite monochromator resolution, and that this analysis permits an optimal monochromator slit width to be defined.

2. Theory

We provide only a summary, as our basic approach has been described in [1] and more complete details regarding present developments will be published elsewhere [7]. As an example relevant to PL and PLE spectroscopy, we consider the prototypical Lorentzian spectral function

$$L(E) = B(E) + \frac{C\Gamma}{(E - E_g)^2 + \Gamma^2}, \tag{1a}$$

where 2Γ and E_g are the full-width-half-maximum (FWHM) broadening and critical-point energies, respectively, and $B(E)$ is a slowly varying background. The amplitudes and phases of the Fourier coefficients of the Lorentzian term are

$$C_n = \frac{Ce^{-n\Delta}}{E_s}; \tag{1b}$$

$$\xi_n = \pi/2 + n\theta_g; \tag{1c}$$

where

$$\Delta = \Gamma/E_s, \ \theta_g = (E_g - E_0)/E_s, \text{ and } E_s = \frac{M(E_F - E_I)}{2\pi(M-1)}, \tag{1d}$$

and E_0 is the transformation origin. For the Lorentzian lineshape, $\ln(C_n)$ decreases linearly in n. For a Gaussian lineshape $\ln(C_n)$ decreases quadratically in n, allowing Lorentzian and Gaussian lineshapes to be accurately distinguished in reciprocal space. These functional dependencies on n will be observed up to the white-noise limit, which we assume occurs at index n_c. By linear superposition, if multiple structures are present in the real-space region being analyzed, interference patterns will be created in reciprocal space.

Because $B(E)$ is slowly varying, its contribution is expected to be significant only for the lowest few coefficients. On the other hand, white noise (fluctuations) will have a significant effect only for the highest coefficients. The midrange will therefore contain spectral information that is essentially unaffected by either baseline or white-noise effects.

If the data are of sufficiently high quality so that this midrange extends over a reasonable number of coefficients, it follows that accurate line energies can be obtained independent of any assumptions about the baseline, and relatively free of the effects of noise.

A systematic effect that does need to be considered is the finite resolution of the monochromator. This effect can be represented as a convolution with a triangular response function $T(E - E_0)$, where E_0 is the nominal transmitted energy. Since a convolution in real space becomes a product in reciprocal space, the effect of finite monochromator resolution on the data and conversely its correction are easily calculated from the Fourier coefficients of $T(E - E_0)$. Assuming that the monochromator efficiency factor is A and the slit width is W, the Fourier coefficients T_n of $T(E)$ are given by [7]

$$ T_n = \frac{2A}{LW} \left(\frac{L}{2n\pi} \right)^2 \sin^2 \left(\frac{n\pi W}{L} \right) \tag{2} $$

where A is an amplitude, and L and W are the spectral range and monochromator resolution, respectively.

Since A is proportional to W, Eq. (2) shows that for narrow slits the throughput increases as W^2. However, as nW approaches L the amplitudes of the C_n drop off sharply. This behavior shows that there exists an upper limit beyond which the data cannot be corrected for monochromator resolution, and suggests the existence of an optimal slit width depending on system performance and the sample being measured. The effect of increasing the width of very narrow slits is to increase the signal faster than the noise, thus increasing the amount of information available by providing more coefficients for analysis. However, as nW approaches L the C_n near n_c are attenuated below the white-noise threshold, reducing the number of accessible coefficients and therefore resulting in a loss of information. This lost information cannot be recovered by dividing the respective coefficients by those of the monochromator transfer function. A quantitative analysis [7] shows that the optimum resolution for data consisting of a single Lorentzian line is given by the solution of

$$ 0 = \ln\left(\frac{C}{2\pi} \right) - n_c \frac{\Gamma}{E_s} + 2\ln\left(\sin\left(\frac{n_c \pi W}{L} \right) \middle/ n_c \right) - \ln(B_0 W) \tag{3} $$

where C is an amplitude coefficient and $B_0 W$ is the white noise level.

3. Application and discussion

3.1. Determination of line energies

We consider first the PL spectrum of Fig. 1, which was obtained from a single $Al_{0.25}Ga_{0.75}As/GaAs$ $Al_{0.25}Ga_{0.75}As$ quantum well 18 molecular layers (51 Å) thick. In the following, we perform our analyses as a function of λ rather than E to be consistent with the data, which is legitimate because the ranges involved are narrow. The data were obtained at a sample temperature of 14 K and a monochromator resolution (FWHM) of 1.6 Å. The spectrum contains two primary structures. Assuming Lorentzian lineshapes, a simultaneous two-structure real-space fit yields line wavelengths of 7736 ± 1 Å and 7762 ± 1 Å, respectively. The major uncertainty is systematic, arising from the assumption that the baseline is uniform over the spectral region where the fit was performed. This systematic uncertainty does not appear in the statistical uncertainty but can be estimated through false-data calculations. The reciprocal-space equivalent of Fig. 1 is given in Figs. 2(a) and 2(b), which show $\ln(C_n)$ and ξ_n, respectively. Figures 2 show that baseline, information and noise are well separated, occurring in the ranges of 1-5, 6-33, and >34, respectively; and the lineshapes are accurately Lorentzian, thereby justifying the use of this functional form for real-space analysis. An equivalent least-squares analysis in reciprocal space over the index range 6-28 yields line energies of 7735.0 ± 0.3 Å and 7763.4 ± 0.3 Å, respectively, which are similar to the real-space values. This result is expected because the background variation in Fig. 1 is relatively minor, and so the two sets of results should agree. In reciprocal-space analysis, the uncertainty from the baseline is only 0.01 Å. The major uncertainty, ± 0.3 Å, arises from the number of coefficients chosen to represent the information region.

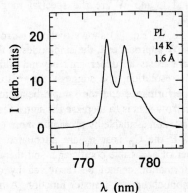

Fig. 1. Photoluminescence (PL) spectrum of a single GaAs/AlGaAs quantum well.

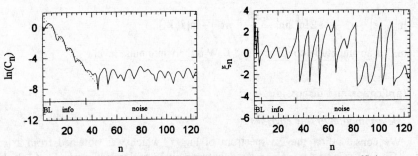

Fig. 2. (a) Natural logarithm of amplitudes and (b) phases of the coefficients.

The situation for the PLE spectrum of Fig. 3 is different. This spectrum was taken on a single $Al_{0.25}Ga_{0.75}As/GaAs$ $Al_{0.25}Ga_{0.75}As$ quantum well at 14 K and a monochromator resolution of 0.8 Å, and the width of the well is 12 molecular layers (34 Å) thick. Here, five lines coexist on a broad background. We consider only the two at the longest wavelengths. In this case the reciprocal-space representation is similar to that shown in Fig. 2 with the information now contained in the range $6 \leq n \leq 40$. Here, we find that the energies of the two highest lines are 7533 ± 2 Å and 7577 ± 2 Å, respectively, from real-space analysis and 7536.4 ± 0.3 Å and 7578.0 ± 0.3 Å, respectively, from reciprocal-space analysis. The differences are about 3.4 and 1 Å, respectively, and can be attributed to real-space asymmetry and baseline effects, as can easily be shown by adding various baselines to the data and repeating the analyses. Because the exact nature of the baseline is not known, it is not possible to obtain a real-space function to accurately mimic this variation. Consequently, we consider the reciprocal-space analysis to be more accurate.

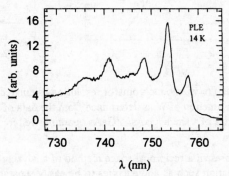

Fig. 3. Photoluminescence excitation (PLE) spectrum of a single GaAs/AlGaAs quantum well.

3.2. Optimum resolution

We now consider Eq. (3), which describes an optimum resolution for a spectrum consisting of a single Lorentzian line. Considering the PL data as an example, the relation $nW = L$ together with the spectral range of 200 Å of Fig. 1 and the maximum coefficient index of 40 from Figs. 2 indicate that the resolution for this case cannot exceed the value 200 Å/40, or 5 Å. Given the 4 Å/mm dispersion of the monochromator, the slit width must therefore be less than 1.25 mm.

The results of a more detailed calculation are shown in Fig. 4. The solid line represents the lowest-order solution of Eq. (3) for n_c as a function of resolution. Although high-order solutions also exist as a result of the oscillatory nature of the sine function, these are not relevant because they are accompanied by zeroes of T_n for $n \leq n_c$, which represent an irretrievable loss of information. It is seen that the optimal resolution for this experiment is 1.93 Å, which corresponds to a slit width of 482 μm. This is consistent with the simple limit obtained above. We recall that the simple limit was based on mathematical consideration alone, where as the optimal resolution depends on the relative width of the dominant line, the total range of the spectrum, and the maximum coefficient index determined by the white-noise level (the performance of the system).

We have investigated this connection experimentally by obtaining PL data at various resolutions as shown in Fig. 5. The resulting values of n_c are shown as points in Fig. 4. The experimental results are in reasonable agreement with the model calculation except for the point at 2.4 Å, which lies far below the model calculation. This deviation can be

318

attributed to the fact that the results depend strongly on the position of the white-noise level, and that this level can be affected drastically by the existence of just one stray point.

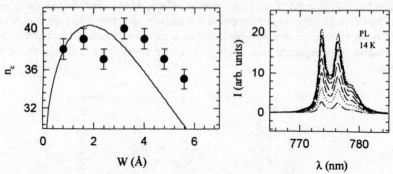

Fig. 4. Solid line : n_c vs. monochromator resolution calculated from Eq. (3). Circles : n_c as a function of W as determined from the data of Fig. 5.

Fig. 5. PL spectra of a single GaAs/AlGaAs quantum well for slit widths from 200 to 1400 μm.

In summary, we present a reciprocal-space method of analyzing PL and PLE spectra, one that allows information such as line energies to be easily separated from baseline and white-noise effects. Where baseline effects are small, both real- and reciprocal-space analyses yield the same results, as expected. However, where baseline effects are nonnegligible, as is the case for PLE, differences occur. We have also shown that this approach allows an optimal resolution to be determined.

Acknowledgements

This work was supported by the Office of Naval Research (ONR) under ONR contract No. N-00014-93-1-0255 and a grant from Texas Instruments.

References

[1] Aspnes, D. E. 1983 Surface Sci. 135 284-306.

[2] Savitzky, A. and Golay, M. J. E. 1964 Anal. Chem. 36 1627-1639; see also corrections given by Steiner, J., Termonia, Y., and Deltour, J. 1972 Anal. Chem. 44 1906-1909.

[3] Aspnes, D. E. 1973 Surface Sci. 37 418-442.

[4] Keeping, E. S. 1962 *Introduction to Statistical Inference*, Van Nostrand, Princeton.

[5] Vasquez, R. P., Klein, J. D., Barton, J. J., and Grunthaner, F. J. 1981 J. Electron Spectrosc. Relat. Phenom. 23 63-81.

[6] Wang, Youqi and Weinberg, W. H. 1992 Surface Sci. 287/288 1102-1105.

[7] Yoo, S. D. and Aspnes, D. E., to be published.

Inst. Phys. Conf. Ser. No 145: Chapter 3
Paper presented at 22nd Int. Symp. Compound Semiconductors, Cheju Island, Korea, 28 August–2 September 1995
© 1996 IOP Publishing Ltd

Photoluminescence, reflectivity, and Raman study of ZnSe, ZnSSe, and ZnMgSSe

Eunsoon Oh, S.D.Lee, H.D.Jung, M.D.Kim, J.R.Kim, H.S.Park, B.J.Kim, and T.I.Kim

Photonics Semiconductor Lab, Samsung Advanced Institute of Technology, Suwon, Korea

Abstract. We have compared photoluminescence (PL) and reflectivity spectra of MBE-grown undoped, Cl-doped, and N-doped ZnSe. Free exciton peak intensity gradually increases with increasing temperature with respect to bound exciton and its energy agrees well with reflectivity below 100 K. At higher temperatures PL peak energies were close to or a few meV below their reflectivity signatures. This probably indicates that the photoluminescence peak is originated either from free exciton or a combination of free exciton and donor-to-free hole (DF), depending upon the temperature regime. We also observe the photoluminescence associated with the first excited excitonic state, whose relative intensity with respect to that of ground state increases with increasing temperature. The temperature dependence of the bandgap in ZnSSe or ZnMgSSe is found to be identical to that of ZnSe within an experimental error.

In the micro-Raman study of ZnSe, ZnSSe, ZnMgSe, and ZnMgSSe, we observe a plasmon-phonon coupling mode as well as their characteristic longitudinal optical phonons. Its intensity is usually stronger for p- or n-type GaAs than for semi-insulating one, as expected.

1. Introduction

Zinc selenide has attracted much attention due to the recent success and development of ZnSe-based laser diodes [1,2]. Successful p-type doping has been achieved by using nitrogen plasma source [3,4] with a net acceptor concentration of about $10^{18}/cm^3$ and optical properties of nitrogen doped ZnSe have been extensively studied [5-8]. As the constituents of ZnSe-based laser diode, electrical and optical properties of ZnSe, ZnSSe, and ZnMgSSe are of interest.

In this paper, we compared photoluminescence (PL) and reflectivity spectra of MBE grown undoped, Cl-doped, and N-doped ZnSe on (100) GaAs substrates at various temperatures. The PL and reflectivity of ZnSSe and ZnMgSSe have been also studied. Micro-Raman measurements were carried out for ZnSe, ZnSSe, ZnMgSe, and ZnMgSSe.

2. Experimental

Samples were grown by Rivere 32P MBE on (001) GaAs substrates. For some samples, ZnS cap layers were grown to enhance the photoluminescence intensity and to reduce the surface effects. Photoluminescence spectra were measured with an excitation wavelength of 325nm from a He-Cd laser. In micro-Raman study, samples were excited by 514.5 nm line from an Ar laser in the backscattering geometry.

3. Results and Discussions

Figure 1 shows photoluminescence spectra of undoped, Cl-doped, N-doped ZnSe at 11 K. The Cl doping concentration and the nitrogen net acceptor concentration are estimated to be 7.5 x 10^{16}/cm³ by Hall measurements and 7x10^{16}/cm³ by C-V measurements, respectively. Undoped ZnSe has a peak at 2.802 eV (labeled Ex), which is attributed to the free exciton, whereas for ZnSe:Cl and for ZnSe:N donor-bound exciton (I_2) and acceptor-bound exciton (I_1) peaks are dominant, respectively. In undoped ZnSe, donor-bound exciton is also observed due to unintentional donors. Also found are two defect-related peaks at 2.774 eV and 2.602 eV, labeled $I_v°$ and Y_o, respectively. Note that a phonon replica of $I_v°$ is observed. For ZnSe:N, donor-to-acceptor pair (DAP) peak and their phonon replicas appear below 2.7 eV. The two DAP peaks have been identified as D^sAP and D^dAP, where D^s and D^d denote shallow and deep donors, respectively [6]. Hauksson et al. suggested a defect complex of (V_{Se}-Zn-N) and several other mechanisms are present such as the formation of N_2 molecules and lattice relaxation associated with DX center.

The temperature dependence of the photoluminescence in ZnSe:N, with net acceptor concentration of 5x10^{16}/cm³ is shown in Fig. 2. Due to the smaller nitrogen concentration, compensation process is not pronounced and thus only D^sAP is seen in the figure. As temperature increases from 11 K to 77 K, the intensity of free exciton becomes relatively stronger with respect to acceptor bound exciton and free electron-to-acceptor (FA) transition dominates over DAP. We also note here the heavy hole and light hole splitting of free exciton mainly due to the thermal strain. Above 100 K, free exciton peak becomes broad and it is difficult to clarify the change of its nature, which will be discuss in detail later. Phonon replicas of DAP, FA, and I_1 are also observed as expected for acceptor-related peaks.

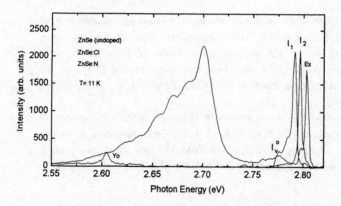

Figure 1. Photoluminescence spectra of undoped, Cl-doped, and N-doped ZnSe at 11 K. Ex, I_1, and I_2 stand for free exciton, acceptor bound exciton, and donor bound exciton, respectively. $I_v°$ and Y_o denote defect related peaks.

We often observe the first excited state of an exciton at 15-17 meV above the ground state exciton energy. This feature is also clearly observed in reflectivity and photo-modulated reflectivity measurements. Its relative intensity with respect to that of the ground state increases with increasing temperature. To enhance the intensity of luminescence and observe the weak signatures more clearly, ZnS cap layer were grown on top of ZnSe epilayers.

In Fig. 3 we show the temperature dependence of luminescence peaks for undoped(solid square), Cl-doped (open triangle), and N-doped ZnSe (solid circle). Bound exciton energies are excluded for clarity. FA energies in the plot are from another ZnSe:N sample with net acceptor concentration of $7x10^{17}/cm^3$. Also shown (cross) is the free exciton energy obtained from reflectivity for undoped ZnSe. By comparing the PL and reflectivity data, we found that the PL peak energies are close to, or a few meV lower than those of reflectivity signatures. It is known that PL peak is expected to be at a higher energy than that of reflectivity if both arise from the free exciton, due to the kinetic energy contribution in PL. Since it is opposite to the experimental results, the PL at room temperature is attributed to donor-to-free hole (DF) transition [9]. Although it is not clear from the figure whether the origin of PL at room temperature is solely due to DF or to a combination of free exciton and DF due to the large uncertainties at high temperatures, it appears that the origin of PL changes from free exciton to a combination of free exciton and DF as temperature increases above 150 K. We found that the luminescence at room temperature is much stronger for ZnSe:Cl than that for undoped ZnSe, also supporting this conclusion.

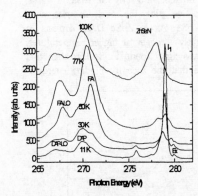

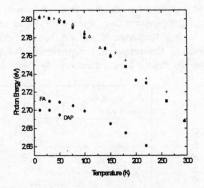

Figure 2. Photoluminescence spectra of ZnSe:N at various temperatures. Free exciton intensity increases with increasing temperature with respect to that of bound exciton and FA gradually dominates over DAP above 50 K

Figure 3. Temperature dependence of photoluminescence for undoped (solid square), Cl- (open triangle), and N-doped (solid circle) ZnSe. Temperature dependence of reflectivity (cross) in undoped ZnSe is shown for comparison.

Another interesting feature is the temperature dependence of FA and DAP. Acceptor binding energy of the nitrogen doped ZnSe has been estimated to be about 110 meV by photoluminescence results at low temperatures [6] and by electronic Raman study at 10 K [10]. Schetzina group reported an acceptor binding energy of 61-82 meV from Hall measurements [11] and later claimed that the acceptor binding energy decreases from 110 meV to 70 meV as temperature increases to room temperature, obtained from the energy difference between the band-gap energy and FA in photoluminescence [12]. On the contrary, Gunshor group reported that the acceptor binding energy remains between 92 -106 meV depending upon the nitrogen concentration, obtained from Hall measurements [13]. From the PL raw data in Ref. 12, the energy shifts of FA, DF (peak at the highest energy), and DAP from 60 K to 260 K are estimated approximately 37 meV, 66 meV, and 55 meV, respectively. The larger shift of DF than that of DAP could be possibly explained if the peak assigned as DF is originated from free exciton at 60 K and mainly from DF at 260 K. The smaller energy shift of FA is partly due to the change of the contribution from the free electron kinetic energy (1/2 kT) of about 8 meV. Thus, if the acceptor binding energy changes, it is 10 meV (55 meV-37 meV-8 meV) rather than about 30 meV.

Our data shown in Fig. 3 also indicate that the energy of FA changes less with temperature than that of free exciton. For a hydrogenated ZnSe:N (after annealing) sample, the temperature dependence of FA agrees with Ref. 12, probably indicating that the character of N acceptor depends upon the growth conditions and post-growth treatments.

Photoluminescence and reflectivity were measured also in ZnSSe and ZnMgSSe. Their bandgap shifts as a function of temperature were identical to that of ZnSe within a few meV, which is the typical experimental error due to the thermal broadening at high temperatures. In Fig. 4 we show the photoluminescence of ZnMgSSe:N at various temperatures. Both DAP and FA peaks are much broader in ZnMgSSe than in ZnSe, and it was difficult to resolve them. However, as we plot the peak energies as a function of temperature they agree with those of ZnSe. As in ZnSe DAP appears to dominate at low temperatures and FA becomes relatively stronger at higher temperatures. In ZnMgSSe:N, the temperature dependence of FA was again found to be less than that of bandgap determined from reflectivity. We note here that the acceptor binding energy

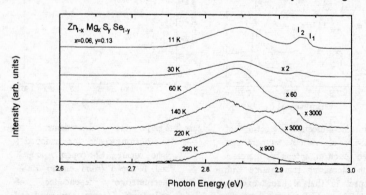

Figure 4. Photoluminescence of N-doped ZnMgSSe at various temperatures. The peak around 2.85 eV at low temperatures is attributed to DAP and the peak around 2.8 eV at high temperatures is to FA.

is expected to increase with increasing band gap due to the decrease of dielectric constant and the increase of the effective mass, which agrees with experimental results obtained from the luminescence intensity *vs.* temperature plot [14]. One would expect that the energy separation between the excitonic peak and the DAP energy should increase with increasing bandgap, due to the increasing acceptor binding energy. However, we were not able to observe any systematic increase of the acceptor binding energy deduced from the luminescence peak energies.

Figure 5 shows the micro-Raman spectra of ZnSe:N, undoped ZnSSe (S: 6%), ZnMgSSe (Mg:3%, S:19%), and ZnMgSe (Mg:14%) on p- or n-type GaAs with an excitation wavelength of 5145 Å from an Ar laser. In ZnSe:N, we observe ZnSe-LO and TO (longitudinal and transverse optical phonons), GaAs-LO, and a lower branch of plasmon-phonon coupling mode near 270 cm⁻¹. The plasmon-phonon coupling mode has been extensively studied [15] and its appearance is considered as an indication of a high quality interface in ZnSe/GaAs. In ZnSSe, ZnMgSSe, and ZnMgSe, we observe additional peaks due to S and/or Mg, depending upon the constituents. Zone-center optical phonons in ZnSSe show "two-mode" behavior, where "ZnSe-like" LO and TO converges into Se gap mode in ZnS and the S local mode in ZnSe (300 cm⁻¹) splits into "ZnS-like" LO and TO modes, which become ZnS-LO and TO in ZnS [16]. "ZnS-like" LO in ZnSSe is labeled S in the figure. As reported in Ref. 17, we observe three LO phonons in ZnMgSSe, two of which are labeled S and Mg. A complete composition dependence of zone-center optical phonons in ZnMgSSe has yet to be studied. For comparison, we also show the Raman spectrum of ZnMgSe, where a mode associated with Mg has been identified.

We would like to note here that the line broadening of DAP luminescence in ZnSSe and ZnMgSSe is partly due to these longitudinal phonon frequencies since DAP is generally accompanied with their phonon replicas. In Ref. 14, authors reported that the phonon replicas were separated by 60-100 meV, the value increasing with increasing bandgap energy. Their result contradicts with our Raman data on ZnMgSSe. We speculate that the luminescence peaks at lower energies of D'A'P could be associated with various donor and/or acceptor levels.

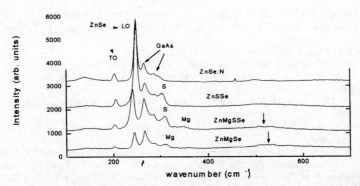

Figure 5 Micro-Raman spectra of ZnSe:N, undoped ZnSSe (S: 6%), ZnMgSSe (Mg: 3%, S: 19%), and   ZnMgSe (Mg: 14%). The longitudinal optical phonons associated with S and Mg, and plasmon-phonon coupling modes are shown in the figure. Arrows near 500 cm⁻¹ indicate the high frequency component of the plasmon-phonon coupling mode.

324

We also observe the high frequency component of the plasmon-phonon coupling mode around 500 cm^{-1}. As power decreases, its energy shifts towards lower wavenumber, whereas the lower frequency component remains almost at the same energy. The relative intensity of the lower frequency component decreases rapidly with respect to GaAs-LO phonon, as carrier density deceases with decreasing power. As expected, we found that the intensity of the plasmon-phonon coupling mode is normally stronger for p- and n-type GaAs substrates than for semi-insulating ones, as expected due to the larger photo-carriers in the former.

4. Concluding remarks

1. Photoluminescence intensity was greatly improved with a cap layer having lager bandgap and weak signatures such as the first excited state of an exciton could be observed.
2. Temperature dependence of FA luminescence energy in ZnSe:N varies with samples, possibly due to the different N acceptor configurations such as interstitials for different growth conditions or post-growth treatments.
3. ZnSSe and ZnMgSSe bandgaps follow the same temperature dependence as ZnSe bandgap, whereas the bandgap of ZnCdSe/ZnSSe quantum wells has smaller shift.
4. The line-shape of DAP in ZnMgSSe is in general complicated by various phonon replicas associated with Mg and S.

References

[1] M.A. Haase, J. Qiu, J.M. Depuydt, and H. Cheng, Appl. Phys. Lett., 59, 127 2 (1991).
[2] H. Jeon, J. Ding, A.V. Nurmikko, W. Xie, D.C. Grillo, M. Kobayashi, R.L. Gunshor, G.C. Hua, and N. Otsuka, Appl. Phys. Lett., 60, 2045 (1992).
[3] K. Ohkawa, T. Karasawa, and T. Mitsuyu, J. Cryst. Growth, 111, 797 (1991).
[4] R.M. Park, M.B. Troffer, C.M.Rouleau, J.M. Depuydt and M.A. Haase, Appl. Phys. Lett. 57, 2127 (1 990).
[5] Z. Zhu, K. Takebayashi, K. Tanaka, T. Ebisutani, J. Kawamata, and T. Yao, Appl. Phys. Lett., 64, 9 1 (1994).
[6] I.S. Hauksson, J. Simpson, S.Y. Wang, K.A. Prior, and B.C. Cavenett, Appl. Phys. Lett., 61, 2208 (1992).
[7] J. Qui, J.M. Depuydt, H. Cheng, and M.A. Haase, Appl. Phys. Lett., 59, 2992 (1991).
[8] B. Hu, A. Yin, G. Karczewski, H. Luo, S.W. Short, N. Samarth, D. Dobrowlska, and J.K. Furdyna, J. Appl. Phys., 74, 4153 (1993).
[9] J. Zheng and J.W. Allen, J. Appl. Phys., 67, 2060 (1990).
[10] D.J. Olego, Intl. Conf. on Solid State Devices and Materials (SSDM '92), Tsukuba, Japan, 1992.
[11] Z. Yang, K.A. Bowers, J. Ren, Y. Lansari, J.W. Cook, Jr. and J.F. Schetzina, Appl. Phys. Lett. vol.61, 2671 (1992).
[12] K.A. Bowers, Z. Yu, K.J. Gossett, J.W. Cook, Jr., and J.F. Schetzina, J. of Electronic Materials, vol. 23, 251 (1994).
[13] Y. Fan, J. Han, L. He, J. Saraie, R.L. Gunshor, M. Hagerott, and A.V. Nurmikko, J. of Electronic Materials, 23, 245 (1994).
[14] H. Okuyama, Y. Kishita, T. Miyajima, A. Ishibashi, K. Akimoto, Appl. Phys. Lett., 64, 904 (1994).
[15] D.J. Olego, J. Vac. Sci. Technol. B 6 1193 (1988); O. Pages, M.A. Renucci, O. Briot, T. Cloitre, and R.L. Aulombard, J. of Cryst. Growth, 117, 569 (1992); A. Krost, W. Richter, and D.R.T. Zahn, Applied Surface Science 56, 691 (1992); O. Pages, M.A. Renucci, O. Briot, and R.L. Aulombard, J. Appl. Phys., 77, 1241 (1995).
[16] K. Hayashi, N. Sawaki, and I. Akasaki, Jpn. J. Appl. Phys., 30, 501 (1991).
[17] E. Oh and A.K. Ramdas, J. Electronic Materials, 23, 307 (1994).

Inst. Phys. Conf. Ser. No 145: Chapter 3
Paper presented at 22nd Int. Symp. Compound Semiconductors, Cheju Island, Korea, 28 August–2 September 1995
© *1996 IOP Publishing Ltd*

Observation of suppressed thermal broadening of photoluminescence linewidth from flow rate modulation epitaxy grown AlGaAs/GaAs quantum wires

Xue-Lun Wang, Mutsuo Ogura, Hirofumi Matsuhata, and Kazuhiro Komori

Electrotechnical Laboratory, 1-1-4 Umezono, Tsukuba 305, Japan

Abstract.   The temperature dependence of photoluminescence (PL) properties of AlGaAs/GaAs quantum wire (QWR) grown by flow rate modulation epitaxy on V-grooved substrates is investigated. PL intensity from QWR is greatly enhanced by properly removing some parts of the AlGaAs barrier layers. PL from a 7.1 nm thick QWR could be observed easily even at room temperature. The full width at half maximum (FWHM) of the ground-state QWR emission peak increases with temperature at low temperatures but becomes almost independent on temperature at high temperatures, a very different behavior from that of a quantum well (QWL) sample in which the FWHM increases with temperature up to room temperature. At high temperatures, the FWHM of QWR is considerably narrower than that of the QWL sample. The suppressed thermal broadening of QWR PL linewidth could be expected easily from the sharp one dimensional density of states but has not been clearly observed experimentally until now.

1. Introduction

The   sharp density of states of quantum wire (QWR) or quantum dot (QD) structures is expected to give rise to various unique and important phenomena which can not be realized using the conventional quantum well (QWL) or bulk structures [1,2]. For example, a weaker thermal broadening of the luminescence linewidth of such low dimensional structures as compared with QWL or bulk structures could be expected, which is considered very useful for high performance optical device applications. Although a variety of fabrication techniques have been developed for the realization of QWRs or QDs [3-5], clear observation of photoluminescence (PL) from such structures at high temperatures is still not so easy at the present time, mainly owing to the structural imperfections (especially for QWRs or QDs fabricated by lithography and etching process) and the influence of PL from neighboring structures (especially for selectively grown QWRs or QDs). Recently, we have proposed the use of flow rate modulation epitaxy (FME) for the growth of QWRs on nonplanar substrates at low growth temperature [6]. A low growth temperature is expected to give QWR structures with low residual impurity concentration and strong lateral quantum confinement (small lateral width). The FME method was first developed by Kobayashi et al. for the low temperature metalorganic vapor phase epitaxial (MOVPE) growth of GaAs [7-8]. The FME growth utilizes   alternative supply of Ga and As sources. Figure 1 shows the typical gas flow

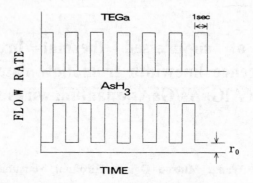

Fig.1. Typical gas flow sequence of FME growth

sequence of this method. To prevent the evaporation of arsenic species and impurity incorporation during Ga flow period, a very small amount of AsH_3 (r_0) is supplied throughout the growth. Due to the extremely low arsenic partial pressure during Ga flow period, Ga species can migrate very rapidly on the substrate surface even at low temperatures, which is very important for the formation of QWRs. The growth selectivity of FME at 600 °C was found even higher than that of conventional MOVPE at 790 °C [6]. We also developed a sample preparation process for PL measurements which can drastically improve the carrier capture efficiency of QWR [9]. These techniques make the investigation of PL from QWR at high temperatures possible [10]. In this paper, we report the first clear observation of suppressed thermal broadening of PL linewidth from AlGaAs/GaAs QWR grown on V-grooved substrate by FME.

2. Experiments

The QWR samples were grown using a low pressure (76 Torr) MOVPE system on V-grooved substrates at a growth temperature of 630 °C. Triethylgallium (TEGa), trimethylaluminum (TMAl) and AsH_3 were used as the source materials. The V-grooves with a pitch of 4.8 μm were formed on (001) semi-insulating GaAs substrate by photolithography and wet chemical etching (NH_4OH:H_2O_2:H_2O=1:3:50). The V-grooves were aligned along [0 $\bar{1}$1] direction.

In FME growth, the total amounts of TEGa and AsH_3 supplied in one flow period (1 sec) were about 1.45×10^{-2} and 2.98 μmol, respectively, which give a growth rate of about 0.8 (monolayer) ML/cycle for GaAs growth on V-groove bottom. The small amount of AsH_3 with a flow rate of about 0.89 μmol/min was supplied with a different AsH_3 line from that for AsH_3 flow period to achieve rapid gas exchange. In conventional MOVPE growth mode, a V/III ratio of about 200 was used.

The optical properties of the grown QWR were characterized by PL measurements with the 5145 Å line of an Ar$^+$ laser as the excitation source. The laser spot diameter was about 190 μm.

3. Results and discussion

Figure 2 shows the layer structure and the cross sectional transmission electron microscopy (TEM) image of the grown QWR sample. This sample consists of the following layers: a 330 nm thick GaAs buffer layer, a 990 nm thick $Al_{0.33}Ga_{0.67}As$ bottom barrier layer, a 5 nm thick

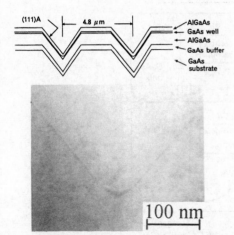

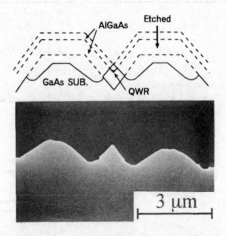

Fig.2. Schematic illustration of the layer structure and the cross sectional TEM image of the grown QWR sample.

Fig.3. Schematic illustration and the SEM image of the sample used for PL measurements. The dashed lines show the cross section before etching.

GaAs well layer, and a 160 nm thick $Al_{0.33}Ga_{0.67}As$ top barrier layer. Here, the layer thicknesses and the Al compositions are all those for layers grown on (001) flat region. Only the GaAs wire layer was grown by FME while all the other layers were grown by the conventional MOVPE. A crescent-shaped QWR with a central thickness of about 7.1 nm and a lateral width of about 31 nm was readily formed on the V-groove bottom from the TEM image of Fig.2. The lateral width obtained here is much smaller than that of QWR grown by conventional MOVPE at high temperatures [4, 11].

In the as-grown sample, luminescence from the QWR region is very difficult to detect probably because the photo-generated carriers redistribute in the AlGaAs barrier layer whose Al composition is nonuniform over the V-groove substrate [9]. Since Ga species migrate more rapidly than Al ones from (111)A side wall plane to (001) flat plane, the Al composition of AlGaAs grown on (111)A side wall will be slightly higher than that of AlGaAs grown on (001) flat region. This will bring a potential discontinuity between the (111)A side wall and (001) flat AlGaAs barrier layers, which in turn causes the carrier redistribution from the (111)A side wall to (001) flat AlGaAs layer. To improve the carrier capture efficiency of QWR region, the (001) flat and part of the (111)A side wall AlGaAs barrier layers were selectively removed by wet chemical etching [9]. Figure 3 shows the schematic illustration and the cross sectional scanning electron microscopy (SEM) image of the etched sample. Wet chemical etching was performed until the original V-grooves barely remained.

The PL spectra of the as-grown and the etched samples at 12 K are shown in Fig.4. All of the PL peaks indicated in Fig.4 were identified using the layer thickness data obtained from TEM observation. The VQWL peak is due to the luminescence from a Ga-rich AlGaAs stripe running through the V-groove bottom, called vertical quantum well [12]. In the spectrum of the as-grown sample, PL from the (001) flat QWL dominates the spectrum and that from QWR could not be resolved as is expected from the carrier redistribution mentioned above. However, PL from QWR was significantly enhanced by the removal of the (001) flat and part of the (111)A side wall AlGaAs barrier layers as shown in Fig.4(a). The QWR peak intensity was found very sensitive to the area of the remained (111)A side wall and the strongest intensity was achieved when the original V-grooves disappeared completely as shown in Fig. 3. The QWR peak in Fig.4 showed a peak intensity of about 50% of that of a 7 nm thick   QWL

328

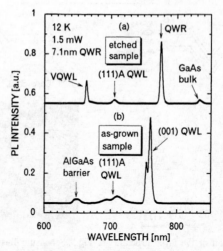

Fig.4. 12 K PL spectra of the as-grown and the etched QWR samples

sample grown on a flat substrate under the same growth conditions. Using an effective QWR lateral width for the ground-state wave function of about 9 nm, the QWR region is estimated to occupy only about 1.87×10^{-3} of the entire substrate surface. Here, the effective lateral width was calculated following the procedure developed by Kapon et al. [13].This means that the equilibrium carrier density in QWR is about 260 times higher than that in QWL with the same excitation power if we assume a similar exciton radiative transition lifetime of QWR with QWL.

Figure 5 shows the temperature dependence of PL spectra of the etched QWR sample. The intensities of the VQWL and (111)A side wall QWL peaks decrease rapidly with the increase of temperature due to the carrier thermalization into QWR [4] and disappeared completely at   temperatures higher than 80 K   beyond which only the QWR emission peak

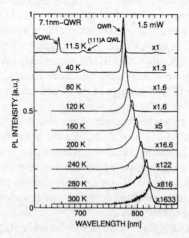

Fig.5. Temperature dependence of the PL spectra of the etched QWR sample

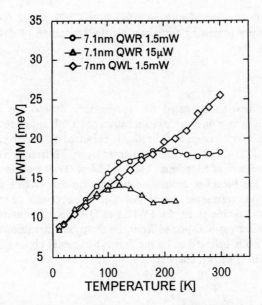

Fig.6. FWHMs of the 1e-1hh emission peaks of the 7.1 nm thick QWR as a function of measurement temperature. FWHM data for a 7 nm thick QWL sample are also shown in the figure for comparison

remained. The QWR peak could be observed easily even at room temperature. A weak shoulder at the short wavelength side begins to appear at high temperatures which is attributed to emission between higher order subbands.

The full width at half maximum (FWHM) of the ground-state electron-heavy hole transition (1e-1hh) peak was calculated from the spectra of Fig.5. The 1e-1hh peaks were separated from the higher order subband peaks for high temperature spectra by assuming two Gaussian distribution functions. The results are shown in Fig.6 as a function of measurement temperature. The FWHM values of the 1e-1hh peaks of a 7 nm thick QWL sample are also given in the figure for comparison. At 12 K, the FWHM of the QWR peak is as narrow as that of the reference QWL sample, suggesting the excellent size uniformity and crystalline quality of the QWR grown by FME. The FWHM of QWL increases linearly with temperature up to room temperature. This is consistent with the theoretical prediction of the 1e-1hh subband-to-subband transition of QWL structure [14] in which the density of states is staircase like. To the contrary, the FWHM of QWR showed very different temperature dependence from QWL. In this case, the FWHM first increases linearly with temperature at low temperatures but becomes almost independent on temperature at high temperatures. The FWHM of QWR also showed strong excitation power dependence at high temperatures. For example, the FWHM of QWR under 15 μW excitation is about 6.5 meV narrower than that under 1.5 mW excitation at temperatures higher than 200 K. As estimated before, the carrier density of QWR under 15 μW excitation is relatively close to that of QWL under 1.5 mW excitation. At room temperature, the FWHMs of QWR with the excitation power of 1.5 mW and 15 W are about 7 and 14 meV narrower than that of QWL, respectively, if we assume that the room temperature FWHM of QWR under 15 μW excitation is equal to that at 220 K. Although the suppression of thermal broadening of PL linewidth of QWR as compared with QWL is easily expected from the sharp density of states of QWR structures, it has not yet been clearly observed experimentally until now due to the structural imperfections. This paper is believed to be the first experimental observation of such an effect from QWR. These results suggest that the FME growth is a very promising technique for the fabrication of device quality QWR on nonplanar substrate.

4. Conclusion

In conclusion, we have investigated the temperature dependence of PL properties of AlGaAs/GaAs QWR grown on V-grooved substrates by FME at low temperature. The carrier capture efficiency of QWR region was drastically enhanced by selectively removing the (001) flat and part of the (111)A side wall AlGaAs barrier layers. PL from QWR was observed easily even at room temperature after etching. The FWHM of QWR increases with temperature at low temperatures, but becomes almost independent on temperature at high temperatures, while that of a QWL reference sample increases linearly with temperature up to room temperature. At room temperature, the FWHM of QWR is much narrower than that of the QWL sample, an effect easily expected from the sharp one dimensional density of states of QWR structures. This is believed to be the first experimental observation of the suppressed thermal broadening of QWR PL linewidth.

Acknowledgments

The authors would like to thank Dr. Tsunenori Sakamoto, director of Electron Devices Division of Electrotechnical Laboratory, and Dr. Keizou Shimizu for their encouragements on this work.

References

[1] Arakawa Y and Sakaki H 1982 Appl. Phys. Lett. **40** 939-941.
[2] Sakaki H 1980 Jpn. J. Appl. Phys. **19** L735-L738.
[3] Tsukamoto S, Nagamune Y, Nishioka M and Arakawa Y 1992 J. Appl. Phys. **71** 533-535.
[4] Walther M, Kapon E, Christen J, Hwang D M and Bhat R 1992 Appl. Phys. Lett. **60** 521-523.
[5] Izrael A, Marzin J Y, Sermage B, Birotheau L, Robein D, Azoulay R, Benchimol J L, Henry L, Thierry-Mieg V, Ladan F R and Taylor L 1991 Jpn. J. Appl. Phys. **30** 3256-3260.
[6] Wang X L, Ogura M and Matsuhata H 1995 Appl. Phys. Lett. **65** 1506-1508.
[7] Kobayashi N, Makimoto T and Horikoshi Y 1985 Jpn. J. Appl. Phys. **24** L962-L964.
[8] Makimoto T, Kobayashi N and Horikoshi Y 1986 Jpn. J. Appl. Phys. **25** L513-515.
[9] Wang X L, Ogura M and Matsuhata H 1995 Appl. Phys. Lett. (in press: 7 August issue).
[10] Wang X L, Ogura M and Matsuhata H 1995 Appl. Phys. Lett. submitted.
[11] Gustafsson A, Samuelson L, Hessman D, Malm J-O, Vermeire G and Demeester P D 1995 J. Vac. Sci. Technol. **B13** 308-317.
[12] Christen J, Kapon E, Colas E, Hwang D M, Schiavone L M, Grandmann M and Bimberg D 1992 Surf. Sci. **267** 257-262.
[13] Kapon E, Hwang D M, Walther M, Bhat R and Stoffel N G 1992 Surf. Sci. **267** 593-600.
[14] Moroni D, Andre J P, Menu E P, Gentric Ph and Patillon J N 1987 J. Appl, Phys. **62** 2003-2008.

Inst. Phys. Conf. Ser. No 145: Chapter 3
Paper presented at 22nd Int. Symp. Compound Semiconductors, Cheju Island, Korea, 28 August–2 September 1995

Spectroscopic characterization for single quantum well structures of compound semiconductors

W Z Shen and S C Shen

National Laboratory for Infrared Physics, Shanghai Institute of Technical Physics, Shanghai 200083, P.R.China

Abstract. The Fourier transform near infrared absorption and photoluminescence spectroscopies have been employed to characterize and investigate some different single quantum well structures of compound semiconductors such as GaAs/InGaAs/GaAs and AlGaAsSb/GaInAsSb/AlGaAsSb. The two-dimensional exciton and light hole subband behavior are emphasized except the characterization of the band structures for the investigated systems.

1. Introduction

Along with the development of both spectroscopic technique and material growth technology, it is now not difficult to characterize different single quantum wells (SQW) of semiconductors by use of different optical spectroscopies. By studying the SQW structure, we can avoid any complexities to the linewidth arising from interwell width variations and also make easy for the mechanism discussion.

Fourier transform (FT) absorption and photoluminescence (FT–PL) measurements are powerful techniques for investigation of the electron states and related characterization in semiconductors. Recently, they have been employed extensively in spectroscopic study of all kinds of materials[1-3]. In this paper, we report a detailed study of the FT absorption in InGaAs/GaAs SQW and FT–PL characteristics in strained GaInAsSb/AlGaAsSb SQW structures. The enhancement of exciton transition due to the decrease of dimension, the influence of the capping layer to the strain relaxation, the transition between the first and second SL states, and PR investigation for different kinds of SQW systems are discussed elsewhere due to the lack of space.

2. Experiment

The strained InGaAs/GaAs SQW sample was grown by MBE on (100)-oriented semi-insulating GaAs substrate. The sample consists of an undoped 25nm $In_{0.20}Ga_{0.80}As$ layer and an undoped GaAs capping layer of thickness of 500nm. The quaternary GaInAsSb/AlGaAsSb SQW structure was also grown by MBE technique on (100)-oriented n^+ GaSb substrates. It was a 10–nm–thick $Ga_{0.67}In_{0.33}As_{0.01}Sb_{0.99}$ well layer structure placed between two 30–nm–thick $Al_{0.25}Ga_{0.75}As_{0.02}Sb_{0.98}$ cladding layers. All layers were undoped.

The FT absorption and FT–PL measurements were performed in a Nicolet 800 Fourier transform infrared spectrometer over the temperature range of 4.0K to room temperature (295.0K), while the absorption spectra were detected by a Si photodiode with a tungsten lamp focused onto the sample and the luminescence spectra were obtained with an Ar–ion (514.5nm) laser as the excitation source and a liquid–nitrogen–cooled InSb detector for receiving the PL signal.

The interband transition energies for the SQW samples were calculated by using a standard finite square–well model. The band gaps for the strained SQW material were calculated from the composition dependence of the band gap of unstrained ones and corrected for the effects of biaxial compressive strain.

3. Results and discussion

3.1. Absorption in InGaAs/GaAs SQW

Temperature-dependent absorption spectra of the $In_{0.20}Ga_{0.80}As$/GaAs SQW sample are shown in Fig.1. At 4.0K, the observed full-width at half-maximum (FWHM) of 1e-1hh peak is only 2.2meV, which is the narrowest peak so far reported on InGaAs/GaAs QW structures, indicative of good optical quality of the highly strained ($\sim$ 1.412%) SQW sample. Five pronounced exciton peaks are observed even at the room temperature absorption spectrum with the FWHM of 1e-1hh peak of 9.6meV. The sharp rise in the highest energy portion of the spectra is due to absorption in GaAs substrate, which was only polished and not removed. The theoretical calculation was employed to identify the origin of the various spectral features observed in the absorption spectra. Q_c=0.70±0.05 provides an excellent fit throughout the observed excitonic transition energies, with all the peaks fitted to experimental results within 2.0meV, except for the highest energy peak 4e-4hh ($\sim$5.0meV). It is resonable that the light-hole-related transitions will not appear in our absorption spectra since the $In_{0.20}Ga_{0.80}As$ potential forms a barrier (type II) for the light holes[4]. An additional feature is observable at the lower energy shoulder of 2e-2hh peak at 200K and well-resloved at 77K, which has been believed to be from a parity-forbidden transition (2e-1hh) due to the increase of the overlap integral of electron and hole wave functions under low temperatures, and the 1e-2hh transition (marked by arrow) appears as a shoulder at 4K due to the same reason. The other small features observed at low temperatures are believed to be longitudinal optical (LO) phonon replicas of the exciton lines from their energy positions.

The measured line shape is a convolution of an inhomogeneous part of FWHM Γ_i and a temperature-dependent homogeneous part (FWHM Γ_h). In order to know the exciton-phonon coupling in InGaAs/GaAs SQW structures, we show the FWHM of the 1e-1hh peak as a function of temperature in Fig.2. We have used the exciton-optical phonon coupling model in which free excitons scatter off LO phonons[5] to describe exciton broadening by the following expression:

$$\Gamma = \Gamma_i + \Gamma_h,$$

$$\Gamma_h = \Gamma_c[\exp(E_{ph}/K_BT) - 1]^{-1}, \tag{1}$$

where Γ_i is the inhomogeneous linewidth due to interface roughness and alloy disorder, Γ_c is a measure of the exciton-phonon coupling, K_BT is the thermal energy, E_{ph} is the LO phonon energy (~ 35.0 meV) in $In_{0.20}Ga_{0.80}As$. A good fit is obtained with $\Gamma_i = 2.2$ meV and $\Gamma_c = 24.0$ meV. The measured value Γ_c in our InGaAs/GaAs SQW sample is much larger than that of InGaAs/GaAs MQW structures (~ 7meV)[6] due to the existence of the periodicity in the growth direction and the reduction in exciton binding energy in MQW structures, based on an overall investigation of temperature–dependent linewidth in both InGaAs/GaAs SQW and MQW structures[7], indicating the exciton-phonon coupling in the $In_{0.20}Ga_{0.80}As$/GaAs SQW sample is quite strong. This result is also consistent with the fact that possible phonon replicas can be seen at low temperatures (see Fig.1) and indicates the linewidth of the InGaAs/GaAs SQW sample above 100K is predominantly homogeneously broadened.

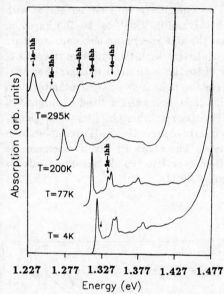

1.227   1.277   1.327   1.377   1.427   1.477
Energy (eV)

Fig.1. Temperature-dependent absorption spectra of $In_{0.20}Ga_{0.80}As$ / GaAs SQW structure.

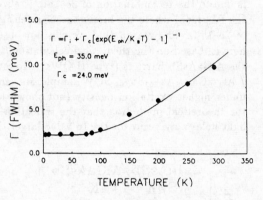

Fig.2. Absorption linewidth (FWHM) of 1e-1hh in $In_{0.20}Ga_{0.80}As$/GaAs SQW as a function of temperature. Solid line is calculated using the exciton-LO phonon coupling model.

3.2. FT–PL characteristics in GaInAsSb/AlGaAsSb SQW

Figure 3 shows the infrared PL spectra from the quaternary $Ga_{0.67}In_{0.33}As_{0.01}Sb_{0.99}$ / $Al_{0.25}Ga_{0.75}As_{0.02}Sb_{0.98}$ SQW at 4K under three different laser excitation density. The integrated intensity of the luminescence peaks (I_{PL}) was found to follow: $I_{PL} \propto I_0^{\alpha}$ with $\alpha = 1.05 \pm 0.10$ in the whole measured temperature region, which reveals the radiative recombination being the prevailing recombination process in GaInAsSb/AlGaAsSb system, and excitonic PL dominating the radiative process up to higher temperatures, which is similar to the quaternary MQW case[8]. It is clearly shown in Fig.3 that a single PL peak is observed, corresponding to the recombination of 1e-1hh, under the lowest excitation density, while an additional luminescence structure gradually appears in the high energy with increasing excitation density due to the band filling effect in quantum well structures (QWs)[9,10], indicating that in GaInAsSb/AlGaAsSb SQWs the photogenerated carriers are efficiently transferred into the wells from the AlGaAsSb barriers before recombination[10]. The experimental fact that we have observed nearly complete suppression of the luminescence from the AlGaAsSb barriers further supports above conclusion. The heating effect of the laser power can be excluded since no detected PL peak shift was observed. The resulting photoluminescence data shown in Fig.3 clearly reveal the stronger band filling in SQW samples than that in MQW samples, as compared with the results of MQW The fact can be explained as follows: the photogenerated carriers are all forced into the single well of the SQW samples, while in MQW samples they might distribute themselves over several wells.

At 4K, the PL spectra show symmetric line shape due to the excitonic transitions, and a good Gaussian fit [Eq.(2)] is observed. Based on this fact, the origin of the luminescence features observed in the PL spectra has been identified by the fitting procedure for the peak energies of 0.6040eV and 0.7242eV, respectively, in good agreement with the theoretically calculation of 0.6046eV for 1e-1hh and 0.7250eV for 2e-1hh based on $Q_c=0.66\pm0.01$. Furthermore, the conclusion that the high energy luminescence structure is due to the recombination of 2e-1hh, is also consistent with their relative strength of the PL signal where the luminescence intensity of 1e-1hh is 10 times stronger than that of 2e-1hh. The reason that we can not observe the light–hole–related recombination here is, based on the theoretical calculation, that light–hole valence band is confined in the AlGaAsSb barriers (type II QWs). However, in our another $Ga_{0.75}In_{0.25}As_{0.05}Sb_{0.95}$ / $Al_{0.25}Ga_{0.75}As_{0.02}Sb_{0.98}$ SQW sample, we can clearly observe the 1e-1lh recombination under higher excitation density (not shown here). The above PL results demonstrate the theoretical prediction that the transition from type I to type II quantum wells for light holes may occur for the In value larger than 0.30[11].

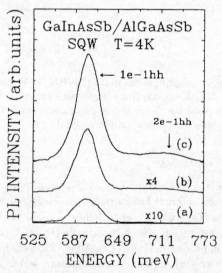

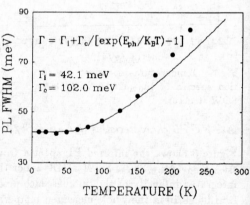

Fig.3. PL spectra of quaternary $Ga_{0.67}In_{0.33}As_{0.01}Sb_{0.99}/Al_{0.25}Ga_{0.75}As_{0.02}-Sb_{0.98}$ strained SQW structure at 4K under different laser excitation density: (a). $100mW/cm^2$, (b). $500mW/cm^2$, (c). $4W/cm^2$. Each spectrum set has been shifted up by a constant for clarity.

Fig.4. PL linewidth (FWHM) in quaternary $Ga_{0.67}In_{0.33}As_{0.01}Sb_{0.99}/Al_{0.25}Ga_{0.75}As_{0.02}-Sb_{0.98}$ strained SQW as a function of temperature under the laser excitation density of $500mW/cm^2$. The solid line is calculated using the exciton–LO phonon coupling model.

The measured luminescence FWHM data below 150K can be well fitted by the above exciton–LO phonon coupling model [Eq.(1)] with the LO phonon energy of $E_{ph}=27.97$meV in $Ga_{0.67}In_{0.33}As_{0.01}Sb_{0.99}$ and $\Gamma_c=102.0$ meV, which is much larger than that in quaternary MQW case[8] (21.2 meV), demonstrating that the exciton–phonon coupling in the GaInAsSb/AlGaAsSb SQW system is quite stronger than that of MQW structure. The LO phonon contribution to the line shape Γ of 25.1 meV at 200K is much larger than the free exciton binding energy ($\sim$ 10meV), and the exciton lifetimes are estimated from the homogeneous linewidth using the uncertainty principle, which implies the mean exciton ionization time of the order of 26fs at 200K. The higher kinetic energy and shorter ionization time reveal the reality of thermal ionization of excitonic states at higher temperatures.

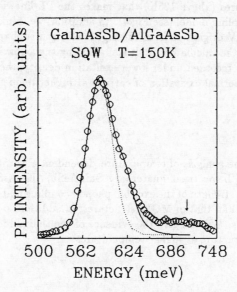

Fig.5. PL spectrum line shape (open circles) and the best fit using Eqs.(2) and (3) with the appropriate parameters (solid curve) of quaternary $Ga_{0.67}In_{0.33}As_{0.01}Sb_{0.99}/Al_{0.25}Ga_{0.75}As_{0.02}Sb_{0.98}$ strained SQW at 150K with the laser excitation density of 500mW/cm². The contribution of exciton recombination is shown by dotted curve. The calculated energy position of 2e–1hh transition is marked by the arrow.

Further evidence of the thermal ionization of excitons is demonstrated in Fig.5 which gives the detailed PL line shape fit. The open circles are the experimental results at 150K under lower excitation density (500mW/cm²). The dotted curve is the calculated exciton recombination (I_{ex}) weighted by the following Gaussian function profiles:

$$I_{ex}(\hbar\omega) = A_{hh}exp\Big[- \frac{[\hbar\omega - E_{hh}(T)]^2}{2\sigma_{hh}^2(T)}\Big]exp\Big(- \frac{\hbar\omega}{K_B T}\Big), \qquad (2)$$

where A_{hh}, E_{hh}, σ_{hh} are the amplitude, energy, and broadening parameters of the 1e–1hh exciton recombination, respectively. We can see that the high energy tail of the PL spectrum extends for several tens of meV from the Gaussian fitting results, in contrast to the excellent fit at low energy side of the PL spectrum. This deviation is due to the thermal ionization of excitons and recombination from free carriers[12]. Therefore, we have also taken into account the contribution of free–carrier recombination[12] (I_{bb}) in the fitting process:

$$I_{bb}(\hbar\omega) = A_{bb}\frac{1}{1 + exp[-\frac{\hbar\omega - E_{bb}(T)}{\sigma_{bb}(T)}]} \times \frac{2}{1 + exp(-2\pi\sqrt{\frac{R}{\hbar\omega - E_{bb}(T)}})} \times exp\left(-\frac{\hbar\omega}{K_B T}\right), \quad (3)$$

where A_{bb}, E_{bb}, σ_{bb} are the amplitude, band–to–band transition energy, and broadening parameters of the free–carrier recombination, respectively. R is the exciton binding energy. The best fit is shown in Fig.5 by solid curve, which clearly shows the presence of band–to–band recombination. Furthermore, it is the partial ionization of excitons at higher temperatures (above 125K) that makes the PL linewidth deviating from the exciton–phonon coupling model (see Fig.4). In addition, we think that the luminescence signal above 0.650 eV originates from the higher electron subbands (2e–1hh), based on its energy position (the calculated result is marked by the arrow in Fig.5). Since the PL spectrum in Fig.5 is recorded under lower excitation density, the appearance of 2e–1hh here is due to the thermal excitation of carriers at higher temperatures rather than the band filling effect.

4. Conclusion

In summary, from the analysis of temperature–dependent absorption spectra of strained InGaAs/GaAs and PL spectra of quaternary GaInAsSb/AlGaAsSb SQW structures, we have determined the strength of the exciton–phonon coupling and shown it is much larger in SQW structures than that of MQW structures, in addition to the characterization of their band structures. We also give the evidence of the thermal ionization of excitons in quaternary SQW structure at higher temperatures.

References

[1] Shen S C 1982 *Progress in Physics* **2** 275 (in Chinese)

[2] Beckett D J S, Nissen M K and Thewalt H L W 1991 *Can. J. Phys.* **69** 427

[3] Storch D R, Schneider R P, Jr. and Wessels B W 1992 *J. Appl. Phys.* **72** 3041

[4] Marzin J Y, Gérard J M, Voision M and Brum J A 1990 *Semiconductors and Semimetals* Vol. **32** 55

[5] Bebbs H B and Williams E H 1972 *Semiconductor and Semimetals* Vol.**8** 256 (Academic, New York)

[6] Huang K F, Tai K, Chu S N J and Cho A Y 1989 *Appl. Phys. Lett.* **54** 2026

[7] Shen W Z, Shen S C, Tang W G, Wang S M and Andersson T G 1995 *J. Phys.: Condens. Matter* **7** L79; *J. Appl. Phys.* **78** 15 July issue (in press)

[8] Shen W Z, Shen S C, Tang W G, Zhao Y and Li A Z 1995 *J. Appl. Phys.* **78** 1 Nov. issue (in press)

[9] Ambrazevičius G, Marcinkevičius S, Lideikis T and Naudžius K 1991 *Semicond. Sci. Technol.* **6** 41

[10] Martelli F, Proietti M G, Simeone M G, Bruni M R and Zugarini M 1992 *J. Appl. Phys.* **71** 539

[11] Shen W Z, Shen S C, Tang W G, Zhao Y and Li A Z *Appl. Phys. Lett.* (to be published)

[12] Colocci M, Gurioli M and Vinattieri A 1990 *J. Appl Phys.* **68** 2809

Inst. Phys. Conf. Ser. No 145: Chapter 3
Paper presented at 22nd Int. Symp. Compound Semiconductors, Cheju Island, Korea, 28 August–2 September 1995
© *1996 IOP Publishing Ltd*

Critical Energies of Photoreflectance and Lineshape Analysis of Photoluminescence of Heavily Si-Doped GaAs

Chul Lee† , Nam-Young Lee† , Kyu-Jang Lee† , Jae-Eun Kim† , Hae Yong Park† , Dong-Hwa Kwak‡ , Hee Chul Lee‡ and H. Lim§

† Department of Physics, KAIST, Taejon 305-701, Korea

‡ Department of Electrical Engineering, KAIST, Taejon 305-701, Korea

§ Department of Electronic Engineering, Ajou University, Suwon 442-749, Korea

Abstract. Room temperature photoluminescence (PL) and photoreflectance (PR) spectra of heavily Si-doped GaAs grown by molecular beam epitaxy (MBE) were investigated as a function of electron concentration and compared with each other. It was found that for highly degenerate semiconductors the critical energy measured by the PR equals to the peak energy of the PL spectrum. When Fermi level lies below the conduction band minimum, the PR spectra revealed the band gap energy as well as the energy E_{max} at which the electron concentration per unit energy in the donor band becomes maximum, and this maximum was observed to merge in the conduction band at about $3 \times 10^{17} cm^{-3}$ electron concentration. And from the line-shape analysis of the PL spectra, it was found that the conduction band tail η_c and the Fermi energy ε_f measured from the conduction band minimum can be expressed as $\eta_c = 2.0 \times 10^{-8} n^{1/3} (eV)$ and $\varepsilon_f = -0.074 + 1.03 \times 10^{-7} n^{1/3} (eV)$, respectively.

1. Introduction

For heavily doped semiconductors, the photoreflectance (PR) can not determine the band gap energy because of the band filling, and the peak energy of photoluminescence (PL) spectrum can not be easily identified with physically meaningful energy such as band gap energy or Fermi energy.[1, 2] We can only obtain these energies from a detailed lineshape analysis of the spectra.[3, 4] In this paper, therefore, we report a detailed line-shape analysis of room temperature PL spectra. And it is demonstrated that the effects of the band tail formation and Fermi energy shift in heavily-doped GaAs can be excellently deduced from this line shape analysis. We also compare the PR with the PL in order to understand a physical meaning of the critical energy determined from the PR spectra.

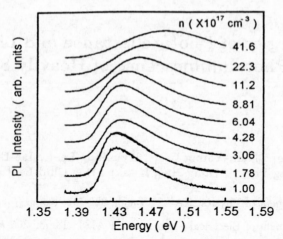

Figure 1. Photoluminescence spectra of Heavily Si-doped GaAs as a function of electron concentration. The smooth curves represent the best fits with Eq. (1) and (2)

2. Experiment

Samples were prepared by molecular-beam epitaxy (MBE). An intrinsic GaAs buffer layer of $1\mu m$ thickness was grown on the Cr-doped GaAs substrate and then followed by a Si-doped layer of $1\mu m$ thickness. The electron concentration of the samples was determined by Hall measurements, and ranged from 1.00×10^{17} to $3.70 \times 10^{18} cm^{-3}$.

3. Results and Discussion

Since the interband transitions are dominant in the room-temperature PL spectrum of heavily doped GaAs, [3] the PL intensity can be described as

$$I(E) \sim E^2 \rho_c(E) f_c(E) |M_{if}|^2, \tag{1}$$

where $\rho_c(E)$ is the density of states in the conduction band ,$f_c(E)$ the well-known Fermi-Dirac distribution function of electrons and $\cdot M_{if}$ the electron transition matrix which is assumed to be constant in energy, since we are dealing with the energy range of about $0.1eV$. We assumed in Eq.(1) that the hole distribution is much narrower in energy adjacent to $k = 0$ states in the valence band than the electron distribution. Since we are dealing with heavily n-doped GaAs, we must include the conduction band tail due to the random fluctuation of the impurity potential. The density of states in the conduction band including this tail is given by [5]

$$\rho_c(E) \sim \int_{-\infty}^{(E-E_0)/\eta_c} \left(\frac{(E - E_c)}{\eta_c} - z \right)^{1/2} exp(-z^2) dz, \tag{2}$$

where E_c is the conduction band edge for parabolic density of states in the absence of the tailing effects. η_c represents the typical well depth of the tail states and at $\sqrt{2}\eta_c$

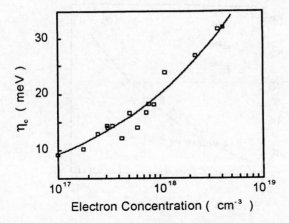

Figure 2. The conduction band tailing η_c as a function of electron concentration n. The best fit is given by $\eta_c = 2.0 \times 10^{-8} n^{1/3} (eV)$.

above the top of the valence band there is the acceptor-like states split-off from the top of the valence band as a result of fluctuations of the spatial distribution of the dopant. [3, 5] Eqs. (1) and (2) were used to fit the observed PL spectra adjusting η_c, E_f and E_c as fitting parameters. The larger value of η_c results in the slower slope of the low-energy side of the PL spectrum. The width of the PL peak is governed by the value of $(E_f - E_c)$ and the energy position of the PL peak is determined by both E_f and E_c in such a way that E_f affects the high-energy side of the PL peak more and E_c the low-energy side. The observed PL spectra of GaAs:Si and the results of the best fits are shown in Fig. 1 as a function of the electron concentration.

Figure 2 shows the band tailing parameter η_c as a function of electron concentration n. The data points were fitted with an equation of $\eta_c = \alpha n^\beta$ with two parameters α and β. And the value of β was found to be 0.351 which is much closer to $1/3$ than $5/12$. We thus fitted with $\beta = 1/3$ and finally deduced $\eta_c = 2.0 \times 10^{-8} n^{1/3}$ with η_c in eV and n in cm^{-3}. The Kane theory, [5] however, predicts the band tailing to be proportional to $n^{5/12}$. And this relation was used for the impurity-induced acceptor-like states in the valence band tail in order to take account of the band gap narrowing in heavily n-doped GaAs. [3, 4, 6] For p-type GaAs, the dependence of the band gap energy at room temperature on the hole concentration p has been reported to be $E_g = 1.424 - 1.6 \times 10^{-8} p^{1/3} (eV)$. [7]

The experimentally determined Fermi energy E_f^{exp} and conduction band minimum E_c^{exp} without the band tailing are depicted in Fig. 3 along with the theoretically calculated E_f^{cal} as a function of the electron concentration. As can be seen, we may consider the conduction band minimum for parabolic density of states in the absence of narrowing effects to be located at about $1\,420 eV$ above the valence band top. The theoretical Fermi level E_f^{cal} was calculated from $n = N_c(T) F_{1/2}[(E_f - E_c)/kT]$ assuming parabolic conduction band. [8] Here, $N_c(T)$ is the effective density of states in the conduction band having the value $N_c(T) = 4.35 \times 10^{17} cm^{-3}$ at $T = 300K$ and $F_{1/2}(\eta)$ is the Fermi-Dirac integral. The values of E_f^{cal} agree fairly well with E_f^{exp} for the lower electron

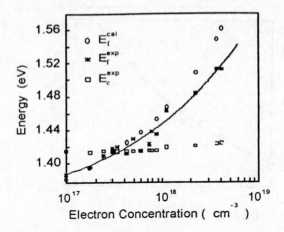

Figure 3. The Fermi energy E_f^{exp}(denoted by $*$), the conduction band minimum $E_c^{exp}(\square)$, and the theoretical value of the Fermi energy $E_f^{cal}(\circ)$ as a function of electron concentration n. The Fermi energy relative to the conduction band minimum $\varepsilon_f = E_f^{exp} - E_c^{exp}$ is given by $\varepsilon_f = -0.074 + 1.03 \times 10^{-7} n^{1/3}$. (eV).

concentrations, but it is larger than E_f^{exp} when the electron concentration becomes relatively larger. Since E_f^{cal} is obtained by neglecting the conduction band tailing, this suggests that the formation of conduction band tail states must be accounted for the exact calculation of E_f for the dopant concentration higher than about $10^{18} cm^{-3}$. The empirical relationship between the Fermi energy relative to the conduction band minimum $\varepsilon_f = E_f^{exp} - E_c^{exp}$ and the electron concentration n was tried to find out. The value of the power of n which gives the best fit through the data points is found to be 0.346. Therefore we express ε_f as $\varepsilon_f = -0.074 + 1.03 \times 10^{-7} n^{1/3}(eV)$ as shown by the solid curve of this figure. At low temperature ($kT \ll \varepsilon_f$), we expect theoretically $\varepsilon_f \sim n^{2/3}$ with parabolic conduction band and neglecting conduction band tail formation. The equation of $\varepsilon_f = 5.209 \times 10^{-14} n^{2/3} - 1.456 \times 10^{-27} n^{4/3}(eV)$, which includes the nonparabolicity of the conduction band, was used to explain the concentration dependence of the PL spectra of heavily Te-doped GaAs at low temperature. [3] Phenomenologically, the deviation of ε_f from $n^{2/3}$ power to $n^{1/3}$ power is quite natural when we note that the E_f^{exp} is deviated from E_f^{cal} due to the nonnegligible electron occupancy in the conduction band tail states for $n \geq 10^{18} cm^{-3}$. We thus believe that ε_f shows empirically $n^{1/3}$ power dependence for the range of electron concentrations $1 \times 10^{17} \sim 3 \times 10^{18} cm^{-3}$ due to the nonnegligible electron occupancy in the conduction band tail states.

On the other hand, the well-known third derivative functional form was used to fit the PR spectra [9] and the results of the best fit are shown in Fig. 4. We should note here that the four less-doped samples have two critical energies and the more-doped samples have one critical energy. Two critical energies in the less-doped samples may be responsible for the band gap energy and the transition energy from the top of the valence band to the energy E_{max} at which the electron concentration per unit energy in the donor band becomes maximum, respectively. With increasing electron concentration

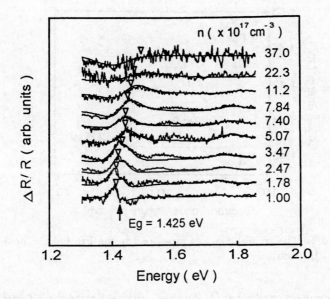

° **Figure 4.** PR spectra of Heavily Si-doped GaAs as a function of electron concentration. The triangles denote the critical energies.

n, however, the donor band merges with the conduction band and eventually becomes the conduction band tail, in which there is no relative maximum. In other words, the donor band may form the conduction band tail with a relative maximum so that this maximum is measured by PR, and then it becomes the tail without the relative maximum as the electron concentration increases. We further note that both the band gap energy and the energy E_{max} were detected for the sample of $n = 3.47 \times 10^{17} cm^{-3}$. When n is larger, the maximum electron concentration in the conduction band is only seen in the PR spectra.

In Fig. 5, we plot the critical energies of the PR and the peak energies of the PL as a function of electron concentration so that their quantitative behavior can be seen clearly. It can be seen that the the critical energy equals to the peak energy for highly degenerate samples. Since the electron concentration per unit energy becomes maximum at the peak energy (see Eq. (1)), we can say that the PR should result from a change in the electron distribution of the conduction band of degenerate semiconductors, so that it determines the energy at which the electron concentration per unit energy becomes maximum.

4. Conclusion

We have demonstrated that the detailed line-shape analysis of the room-temperature PL spectra can determine various physical quantities related to the heavy doping effects, such as band tail formation, the Fermi energy shift and the conduction band edge corresponding to pure GaAs, etc. It was found that both the band tail and the Fermi

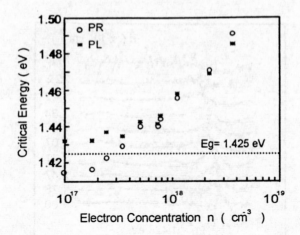

Figure 5. The critical energies measured by the PR and the peak energies of the PL spectra.

energy varies proportionally to $n^{1/3}$. We have also investigated the PR spectra and found that the PR measurement can determine the transition energy of the maximum electron concentration in the conduction band for degenerate semiconductors. And, for nondegenerate semiconductors, it can measure the band gap energy and the transition energy from the top of the valence band to the maximum electron concentration in the donor band, which merges in the conduction band at about $n = 3 \times 10^{17} cm^{-3}$.

References

[1] Peters L and Phaneuf L 1987 *J. Appl. Phys.* **62** 4558–4562

[2] Badakhshan Ali and Glosser R 1991 *Appl. Phys. Lett.* **59** 1218–1220

[3] De-Sheng J and Makita Y 1982 *J. Appl. Phys.* **53** 999–1006

[4] Borghs G and Bhattacharyya K 1989 *J. Appl. Phys.* **66** 4381–4386

[5] Kane E O 1963 *Phys. Rev.* **131** 79–88

[6] Abramov A P, Abramova I N, Verbin S Y, Gerlovin I Y, Grigor'ev S R, Ignat'ev I V, Karimov O Z, Novikov A B and Novikov B N 1993 *Semiconductiors* **27** 647-649

[7] Casey H C and Stern F 1976 *J. Appl. Phys.* **47** 631–643

[8] Blakemore J S 1962 *Semiconductor Statistics* (New York : Pergamon)

[9] Aspnes D E 1973 *Surf. Sci.* **37** 418–442

Inst. Phys. Conf. Ser. No 145: Chapter 3
Paper presented at 22nd Int. Symp. Compound Semiconductors, Cheju Island, Korea, 28 August–2 September 1995
© *1996 IOP Publishing Ltd*

343

Piezoreflectance study of GaAs/AlGaAs digital alloy compositional graded heterostructures

D Y Lin[†], F C Lin[†], Y S Huang[†], H Qiang[‡], F H Pollak[‡], D L Mathine[§] and G N Maracas[§]

[†]Department of Electronic Engineering, National Taiwan Institute of Technology, Taipei 106, Taiwan, Republic of China
[‡]Physics Department and the New York State Center for Advanced Technology in Ultrafast Photonic Materials and Applications, Brooklyn College of CUNY, Brooklyn, NY 11210, USA
[§]Department of Electrical Engineering, Center for Solid State Electronics Research, Arizona State University, Tempe, Arizona 85287, USA

Abstract. We have studied the piezoreflectance (PzR) spectra at 300 K and 80 K related to the intersubband transitions from an asymmetric triangular quantum well (ATQW) and a rectangular quantum well (RQW) fabricated by molecular beam epitaxy using the digital alloy compositional grading (DACG) method. Comparison of the observed intersubband transitions with the envelope function calculations provides a self-consistent verification that the DACG method generated the desired potential profiles.

1. Introduction

Graded band-gap material is of great importance in band-gap engineering and is widely used in high speed optical modulators, photodetectors, and laser diodes, etc. Although rectangular quantum wells (RQWs) have been extensively employed as active regions of opto-electronic devices, they have certain limitations. The symmetrical rectangular quantum wells (SRQW) have an overlap integral between the fundamental conduction and heavy-hole wave functions that are very close to unity when no electric field is applied. However, in most opto-electronic devices the quantum well is subjected to internal and external fields from the device. These fields cause a spatial separation of electron and hole wave functions that decreases the overlap integral between the conduction-band and valence-band wave functions. A reduction in the overlap integral implies a reduction in the emission or absorption processes. To overcome the decrease in the overlap integral, novel quantum well designs such as those containing compositional gradients are deployed. Recently, arbitrary shaped quantum wells have been implemented by molecular beam epitaxy (MBE) in such devices with an emphasis on the study of asymmetric triangular quantum wells (ATQWs) [1-5]. These quantum well structures have demonstrated their potential in increasing the overlap integrals. Thus there is considerable interest in studying the properties of samples fabricated by the digital alloy compositional grading (DACG) method.

In this paper we report a study of the intersubband transitions from two different (001) GaAs/AlGaAs structures, an ATQW and a RQW with 8 unit cells of 12.5 Å, fabricated by MBE using the DACG approach. Measurements were made at 300 K and 80 K using piezoreflectance (PzR). For comparison we also measured the photoreflectance (PR) spectrum of the sample at 300 K. The heavy- or light-hole nature of various transitions were verified by comparing the PzR and PR spectra. The transition energies were extracted from the PzR spectra using a form of the Aspnes equation of the first-derivative Lorentzian line shape (FDLL) [6,7]. The envelope function approximation [8] was used to calculate the bound energy states using the DACG profile determined from the growth conditions. Comparison of the observed transition features with the theoretical calculation provided a self-consistent check that the DACG method generated the desired potential profiles.

2. Experimental

The samples used in this study was fabricated by the DACG method, which approximates the alloy composition profile by using alternating layers of $Al_xGa_{1-x}As$ and GaAs where the ratio of layer thicknesses produces an average alloy composition [5]. The average Al composition in the alloy cell is then approximated by shuttering the Al effusion cell such that the thickness of a layer of AlGaAs centered in the alloy cell provides enough Al for the required average Al composition. For the ATQW the average Al composition in the alloy cell is given by $<x_i> = (d_i/t)x$, where x is the composition of the Al present in the $Al_xGa_{1-x}As$ layer and d_i is the width of the $Al_xGa_{1-x}As$ layer centered in the ith cell of width t. The width of the $Al_xGa_{1-x}As$ layer is given by $d_i = t^2 (i - 0.5)/L_w$, where i ranges from unity to the L_w/t and L_w is the width of the ATQW. For the RQW, the 12.5 Å unit cell RQW regions consisted of 4.2 Å of GaAs, 4.1 Å AlGaAs and 4.2 Å of GaAs repeated 8 times so that the total width of the RQW was 100 Å.

The PzR [6,7] and PR [9] methods have been reported in the literature. The PzR measurement was achieved by gluing the whole sample surface on a 0.15 cm thick lead-zirconate-titanate (PZT) piezoelectric transducer driven by a 200 V_{rms} sinusoidal wave at 230 Hz. The pump beam of the PR measurement was the 6328 Å line of a 5 mW He-Ne laser chopped at 200 Hz. A RMC model 22 closed cycle cryogenic refrigerator equipped with a model 4075 digital thermometer controller was used for low temperature measurements.

3. Results and discussion

Shown by the solid lines in Fig. 1 are the PzR spectra of the ATQW sample at 300 K and 80 K, respectively. The first feature in Fig. 1 corresponds to the direct band gap of GaAs, denoted $E_o(GaAs)$, and originates in the GaAs buffer/substrate of the structure. At 300 K the structure around 1.75 eV is due to $E_o(Al_xGa_{1-x}As)$ corresponding to an Al composition of about 0.23 [10]. The rich spectra between $E_o(GaAs)$ and $E_o(Al_{0.23}Ga_{0.77}As)$ are due to various mnH(L) quantum transitions of the ATQW. The dotted curves are the least-squares fits to the FDLL. The obtained intersubband energies at 300 K and 80 K of the features denoted A-Q are indicated by arrows at the bottom of the figures. The ATQW spectra in Fig. 1 are due to intersubband transitions, similar to the electron beam electroreflectance measurement reported by Kopf et al [2] on a linearly graded well, and are not Franz-Keldysh oscillations as recently reported in the PR study on DACG structures by Xu et al [11].

In order to identify the nature of the large number of transitions observed from the ATQW, we have performed a theoretical calculation based on the envelope function approximation [8] using the intended DACG profile from the growth conditions. Non-parabolic effects were not included. The only adjustable parameter was L_w. We used a

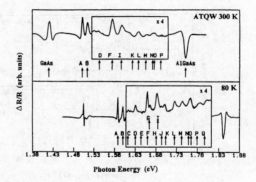

Figure 1. Piezoreflectance (PzR) spectra (solid lines) at 300 K and 80 K from a GaAs/Al$_{0.23}$Ga$_{0.77}$As asymmetric triangular quantum well (ATQW) fabricated by the DACG method. The dotted lines are the least-squares fits to a form of the Aspnes equation of the first-derivative Lorentzian lineshape. The obtained energy values are denoted by arrows.

conduction band offset, Q$_c$=0.65, conduction and valence band masses for GaAs and Al$_{0.23}$Ga$_{0.77}$As listed in the literature [12] and E$_o$(GaAs) and E$_o$(AlGaAs) taken from this experiment. Listed in Table 1 are the results of the calculation for 300 K and 80 K with L$_w$ = 260 Å. An excellent agreement between experiments and theoretical calculations can be achieved, if we identify the feature A as 11H, B as 11L, and C as 12H etc. The numbers in parentheses in Table 1 are for a similar calculation for a linearly graded potential profile with same L$_w$. The close agreement between experiment and theory for this potential profile clearly indicates that the DACG method has produced a desired effective linearly graded potential profile.

Table 1. Comparison of the experimental and theoretical (DACG) values for the mnH(L) transitions of a GaAs/Al$_{0.23}$Ga$_{0.77}$As ATQW at 300 K and 80 K with L$_w$ = 260 Å. The numbers in parentheses are for a similar calculation for a LGW profile using the same L$_w$.

Spectral Feature	mnH(L)	300K Exp. (eV)	300K Theory (eV)	80K Exp. (eV)	80K Theory (eV)
A	11H	1.501	1.501 (1.504)	1.588	1.588 (1.591)
B	11L	1.513	1.514 (1.515)	1.600	1.601 (1.602)
C	12H		1.521 (1.525)	1.612	1.609 (1.613)
D	13H	1.543	1.539 (1.545)	1.630	1.626 (1.632)
E	21H		1.557 (1.561)	1.647	1.644 (1.648)
F	22H	1.573	1.577 (1.582)	1.659	1.664 (1.669)
G	13L		1.579 (1.582)	1.665	1.666 (1.669)
H	23H		1.594 (1.601)	1.681	1.681 (1.688)
I	22L	1.595	1.601 (1.604)	1.683	1.688 (1.691)
J	24H		1.610 (1.617)		1.697 (1.704)
K	32H	1.622	1.623 (1.629)	1.703	1.710 (1.716)
L	33H	1.637	1.640 (1.648)	1.723	1.727 (1.735)
M	34H	1.655	1.656 (1.664)	1.742	1.743 (1.751)
N	35H	1.671	1.672 (1.679)	1.759	1.759 (1.766)
O	33L	1.676	1.678 (1.682)	1.765	1.767 (1.771)
P	44H	1.691	1.696 (1.704)	1.779	1.783 (1.791)
Q	45H		1.712 (1.718)	1.797	1.799 (1.805)

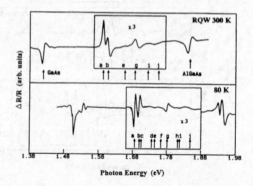

Figure 2. Piezoreflectance (PzR) spectra (solid lines) at 300 K and 80 K from an $Al_{0.1}Ga_{0.9}As/Al_{0.29}Ga_{0.71}As$ rectangular quantum well (RQW) fabricated by the DACG method. The dotted lines are the least-squares fits to a form of the Aspnes equation of the first-derivative Lorentzian lineshape. The obtained energy values are denoted by arrows. The obtained energy values are denoted by arrows.

Shown by the solid lines in Fig. 2 are the PzR spectra from the RQW sample at 300 K and 80 K. Again the first feature in Fig. 2 is related to $E_o(GaAs)$. At 300 K the features in the region of 1.81-1.87 eV is due to $E_o(AlGaAs)$ corresponding to an Al composition of about 29% [10]. The spectra between $E_o(GaAs)$ and $E_o(AlGaAs)$ in Fig. 2 are due to various mnH(L) quantum transitions of the RQW. The dotted curves are the fits to the FDLL. The obtained energies of the features denoted a-j are indicated by arrows and are listed in Table 2. In order to identify the intersubband transitions observed from the RQW samples we have performed a theoretical calculation using the intended DACG profile from the growth conditions. Also the properties of an analog single RQW with an Al composition of 10% and $L_w = 100$ Å were calculated. These results for 300 K and 80 K are listed in Table 2. Exciton binding energy effects were taken into account. There is a very good agreement between experiments and the theoretical calculations. Thus the DACG method has produced the desired RQW profile.

Table 2. Comparison of the experimental and theoretical (DACG) values for the mnH(L) transitions of the RQW at 300 K and 80 K. Also displayed are the calculated values for an analog single RQW with an Al composition of 10% and $L_w = 100$ Å.

Spectral Feature	mnH(L)	300 K		80 K	
		Exp. (eV)	Theory (eV)	Exp. (eV)	Theory (eV)
a	11H	1.594	1.587 (1.594)	1.685	1.676 (1.683)
b	11L	1.608	1.597 (1.605)	1.698	1.686 (1.694)
c	12H		1.613 (1.615)	1.705	1.702 (1.704)
d	13H		1.647 (1.649)	1.736	1.736 (1.738)
e	12L	1.658	(1.654)	1.745	(1.743)
f	21H		1.667 (1.666)	1.763	1.756 (1.755)
g	22H	1.687	1.687 (1.688)	1.781	1.776 (1.777)
h	23H		(1.722)	1.812	(1.811)
i	22L	1.725	1.726 (1.727)	1.818	1.815 (1.816)
j	24H	1.756	1.756 (1.758)	1.849	1.845 (1.847)

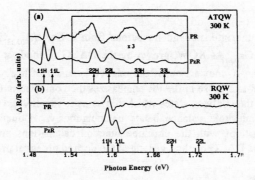

Figure 3. Photoreflectance (PR) and piezoreflectance (PzR) spectra at 300 K for the spectral range of a large number of intersubband transitions from (a) a GaAs/Al$_{0.23}$Ga$_{0.77}$As asymmetric triangular quantum well (ATQW) and (b) an Al$_{0.1}$Ga$_{0.9}$As/Al$_{0.29}$Ga$_{0.71}$As rectangular quantum well (RQW).

In order to verify the heavy- or light-hole nature of the observed mnH(L) transitions from the ATQW and RQW, detailed analysis of the PzR and PR spectra are made and discussed as follows. The intensity of the feature observed in the PzR spectra is proportional to the optical matrix elements and the stress dependence of the energy shift. For the zinc blende band structure the ratio of the optical matrix elements of the heavy hole to that of the light hole is expected to be equal to 3 [13]. The ratio of the intensity between the electron–light-hole and electron–heavy-hole transitions is given by [14]

$$K_{PzR}^{mn} = \frac{1}{3} \frac{[a(2-\lambda)+b(1+\lambda)]}{[a(2-\lambda)-b(1+\lambda)]} \qquad (1)$$

where a and b are the interband hydrostatic and intra-valence band shear deformation potentials, respectively, and $\lambda = -2S_{12}/(S_{11} + S_{12})$, S_{ij} is the elastic compliance constant. For GaAs, using a = -9.8 eV, b = -2.0 eV [15], S_{11} = 1.16 x 10^{-6} bar^{-1} and S_{12} = -0.37 x 10^{-6} bar^{-1} [16], we obtain K_{PzR}^{mn} = 0.765, in reasonable quantitative agreement with the experimental observation. We have also studied the PR spectra of the same samples, in which the relative intensity of the signatures does not involve strain. In Figs. 3(a) and 3(b) the PR and PzR spectra at 300 K of the ATQW and RQW respectively for the spectra range of large number of intersubband transitions are displayed. The PR spectra clearly show that the features associated with heavy hole transitions are about three times stronger than that associated with light hole transitions. A similar behavior has also been previously observed in the results of GaAs/Al$_x$Ga$_{1-x}$As square quantum wells [17] and single parabolic quantum wells [10]. The ratio of the intensity between mnL and mnH of PzR spectra and that of PR is estimated to be 2.0 ± 0.3, which is in good agreement with the ratio of the modulation coefficients between light hole and heavy hole for PzR as compared with that of PR. Thus the combination of the PzR and the PR techniques allows us to unambiguously identify the features associated with the heavy and light holes.

4. Conclusion

We have studied the intersubband transitions from a GaAs/Al$_{0.23}$Ga$_{0.77}$As ATQW and an Al$_{0.1}$Ga$_{0.9}$As/Al$_{0.29}$Ga$_{0.81}$As RQW, grown by the DACG method, at 300 K and 80 K using PzR spectroscopy. A comparison of the relative intensity of heavy- and light-hole related features in the PzR and those in the PR emphasizes the contribution of the strain dependence of the energies of the confined states which allows us to identify the features associated with the heavy- and light-hole valence bands unambiguously. Comparison of the observed intersubband transitions with the envelope function calculations provides a self-consistent verification that the DACG method generated the desired potential profiles.

Acknowledgments

The authors DYL, FCL and YSH would like to acknowledge the support of the National Science Council of the Republic of China under project no. NSC84-2215-E-011-005. The work of HQ and FHP was supported by the US Army Research Office contract no DAAL03-92-G-0189, NSF grant # DMR-9120363 and the New York State Science and Technology Foundation through its Centers for Advanced Technology program. The work of DLM and GNM was supported by US Air Force University Initiative contract AFOSR-90-0118.

References

[1] Cody J G, Mathine D L, Droopad R, Maracas G N, Rajesh R and Carpenter R W 1994 *J. Vac. Sci. Technol.* B **12** 1075-1077

[2] Kopf R F, Herman M H, Schnoes M L, Perley A P, Livescu G and Ohring M 1992 *J. Appl. Phys.* **71** 5004-5011

[3] Qiang H, Huang Y S, Pollak F H, Chi W S, Mathine D L and Maracas G N 1994 *Solid State Electronics* **37** 893-897

[4] Mathine D L, Maracas G N, Gerber D S, Droopad R, Graham R J and McCartney M R 1994 *J. Appl. Phys.* **75** 4551-4556

[5] Gerber D S, Droopad R and Maracas G N 1993 *Appl. Phys. Lett.* **62** 525-527

[6] Lee Y R, Ramdas A K, Chambers F A, Meese J M and Ram-Mohan L R 1987 *Appl. Phys. Lett.* **50** 600-602

[7] Mathieu H Allègre H J and Gil B 1991 *Phys. Rev.* B **43** 2218-2227

[8] Bastard G 1981 *Phys. Rev.* B **24** 5693-5697

[9] Glembocki O J and Shanabrook B V 1992 *Semiconductor and Semimetals* vol 36 ed D G Seiler and C L Littler (New York: Academic) p 221

[10] Kuech T F, Wolford D J, Potemski R, Bradley J A, Kelleher K H, Yan D, Farrell J P, Lesser P M S and Pollak F H 1987 *Appl. Phys. Lett.* **51** 505-507

[11] Xu H, Zhou X, Xu G, Du Q, Wang E, Wang D Zhang L and Chen C 1992 *Appl. Phys. Lett.* **61** 2193-2195

[12] Adachi S 1993 *Properties of Aluminum Gallium Arsenide* ed by S Adachi (London: Inspec) p 58

[13] Mathieu H, Auvergne D and Camassel J 1973 *phys. status solidi* b **58** 227-235

[14] Pollak F H 1990 *Semiconductors and Semimetals* vol 32 ed T P Pearsall (New York: Academic) p 17

[15] Qiang H, Pollak F H and Hickman G 1990 *Solid State Commun.* **76** 1087-1091

[16] Bateman T B, McSkimin H J and Whelan J M 1959 *J. Appl. Phys.* **30** 544-545

[17] Glembocki O J, Shanabrook B V, Bottka N M, Beard W T and Comas J 1985 *Appl. Phys. Lett.* **46** 970-972

Inst. Phys. Conf. Ser. No 145: Chapter 3
Paper presented at 22nd Int. Symp. Compound Semiconductors, Cheju Island, Korea, 28 August–2 September 1995
© *1996 IOP Publishing Ltd*

Birefringence and mode-conversion in ordered GaInP/AlGaInP optical waveguide structures

A. Moritz[1], R. Wirth, C. Geng, F. Scholz and A. Hangleiter

4. Physikalisches Institut, Universität Stuttgart
Pfaffenwaldring 57, D-70550 Stuttgart, Germany

Abstract. Ternary semiconductors like GaInP under certain growth conditions exhibit a (partial) chemical ordering in form of a superlattice of alternate Ga-rich and In-rich planes in $(1\bar{1}1)$ direction. We have studied the polarization properties of light propagating in ordered GaInP/AlGaInP waveguide structures. For the first time, we have observed mode-conversion between transverse electric (TE) and transverse magnetic (TM) modes within typically $60\mu m$ in such structures. We show that the reduced symmetry of the ordered material leads to an optical birefringence with the axis tilted with respect to the growth direction and therefore to a coupling between the TE and the TM mode.

1. Introduction

Recently there has been strong interest in GaInP/AlGaInP as a promising material system for visible laser diodes down to 630 nm. It was found, that like in other ternary semiconductors under special growth conditions the group III constituents are not distributed statistically. Instead, an at least partial chemical ordering is established in form of a superlattice of alternate Ga-rich and In-rich planes in $(1\bar{1}1)$ direction. An ordering parameter η is defined by the composition in the planes $Ga_{x+\eta/2}In_{1-x-\eta/2}P$ and $Ga_{x-\eta/2}In_{1-x+\eta/2}P$. The most prominent consequences of this reduced symmetry are a band gap shrinkage and a splitting of the valence band. In this paper, we will show that ordering also leads to optical birefringence and to coupling of the TE and the TM mode, depending on the propagation direction.

2. Experiment

We have carried out systematic measurements of the polarization properties of light, after it has propagated a variable length L in an optical waveguide of ordered GaInP/AlGaInP.

---

[1] E-mail: a.moritz@physik.uni-stuttgart.de

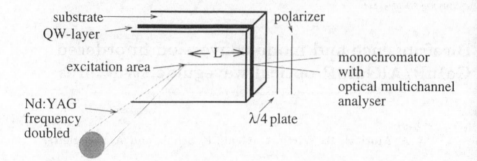

Figure 1. Experimental setup to measure the polarization properties of light propagating in the waveguide.

This was done with laterally unstructured samples consisting of a GaInP double quantum well embedded in a AlGaInP waveguide structure like they are used for lasers. The strain of the quantum wells and the degree of ordering η have been varied. The ordering can be controlled by the growth temperature in metal organic vapor phase epitaxy (MOVPE) and the misorientation of the substrate [1, 2]. It was measured by the shift of the luminescence energy and the valence band splitting [5].

For the measurement a small spot of the sample is optically excited at a variable distance from the sample edge (Fig. 1) where the spontaneous emission is used as a variable light source. The luminescence light of the spot that travels the length L in the waveguide layer to the sample edge can be analyzed after it has been emitted from the sample edge. Using a polarizer and a $\lambda/4$-plate the 4 Stokes parameters are measured and from these the full polarization properties are derived as a function of length L.

Normally, in a planar optical waveguide structure, light can propagate undisturbed in transverse electric (TE) polarization with the electric field vector in the waveguide plane and in transverse magnetic (TM) polarization with the electric field perpendicular to it.

In our experiments with waveguides made of ordered GaInP/AlGaInP, we observe a dramatic difference between light propagating along the (110) direction i.e. parallel to the ordering planes or along the $(1\bar{1}0)$ direction. Whereas for the $(1\bar{1}0)$ direction only the change of intensity expected for absorption close to the band edge is observed, light propagating along the (110) direction exhibits an oscillation of the intensity between TE and TM polarization (Fig.2). The amplitude of the oscillation is strongly dependent on the ordering parameter and on the strain in the quantum wells. The example in Fig.2 shows a (almost) complete TE/TM mode conversion within $60\mu m$.

3. Model

If we neglect the longitudinal parts of the field and assume the deviation from the isotropic material to be so small that the TE and TM modes are still valid, we can describe the electric field in the waveguide for propagation in z-direction in the following manner:

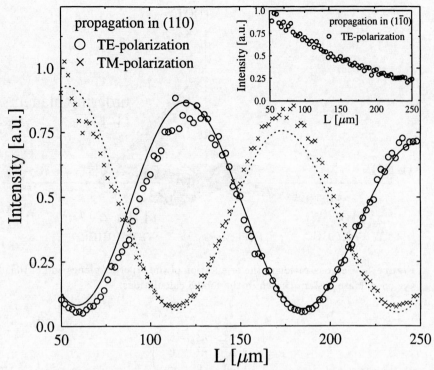

Figure 2. Measured intensity of the TE and the TM mode versus effective waveguide length. The oscillating behavior clearly shows the mode conversion for propagation along (110) . The sample is a $10nm$ double quantum well of highly ordered $G_{0.41}InP/(Al_{0.5}Ga)_{0.51}InP$. The measurement is at $\lambda = 713nm$. The inset shows the behavior for propagation in $(1\bar{1}0)$ direction.

$$\vec{E}(x,y,z) = A_{TE}(z)\hat{E}_{TE}(x,y)exp(-i\beta_{TE}z) + A_{TM}(z)\hat{E}_{TM}(x,y)exp(-i\beta_{TM}z) \qquad (1)$$

With the TE-mode $\hat{E}_{TE} = (0, E_{E_y}, 0)$, the TM-mode $\hat{E}_{TM} = (E_{M_x}, 0, 0)$ and the propagation constants β_{TE} and β_{TM} respectively. $A_{TE}(z)$ and $A_{TM}(z)$ are the amplitudes of the TE and the TM mode.

Chemical ordering in GaInP leads to a reduced symmetry of the crystal structure [4]. The direction within the ordering planes is fundamentally different from that perpendicular to them. Decisive with respect to the mode conversion is the fact, that the ordering direction is not parallel to the growth direction or the quantum well plane but is in the $(1\bar{1}1)$ direction (see fig 3). The consequence is the low crystal symmetry for strained ordered quantum wells C_{1h} [4]. Calculations using a 6-band kp-theory show a strong anisotropy of the gain with that symmetry [6]. Due to the Kramers-Kronig relation one would expect such an anisotropy in the real part of the dielectric tensor, too. This will lead to a birefringence with the main axes tilted towards $(1\bar{1}1)$ and to non-diagonal elements in the dielectric tensor,

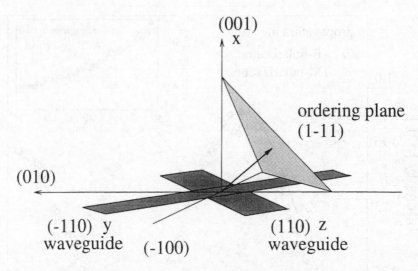

Figure 3. Demonstration of the orientation of the ordering planes and of the xyz coordinate system chosen in the model calculations.

$$\boldsymbol{\varepsilon} = \varepsilon_0 \begin{pmatrix} n_{xx}^2 & n_{xy}^2 & 0 \\ n_{xy}^2 & n_{yy}^2 & 0 \\ 0 & 0 & n_{zz}^2 \end{pmatrix} \tag{2}$$

(see Fig. 3 for the coordinate axes)
which will induce a perturbing polarization $\vec{P}^{pert}$ for propagation in (110) but not for propagation in (1$\bar{1}$0) which also explains the drastic difference between the two directions:

$$\vec{P}^{pert} = \varepsilon_0 n_{xy}^2 \begin{pmatrix} A_{TE} E_{TE_y} exp(-i\beta_{TE} z) \\ A_{TM} E_{TM_x} exp(-i\beta_{TM} z) \\ 0 \end{pmatrix} \tag{3}$$

According to [3] this results in a mode conversion between TE and TM. The conversion rate is determined by:

$$\frac{dA_{TE}}{dz} = i\kappa A_{TM} exp(-i(\beta_{TM} - \beta_{TE})z)$$
$$\frac{dA_{TM}}{dz} = i\kappa A_{TE} exp(-i(\beta_{TE} - \beta_{TM})z) \tag{4}$$

with the coupling constant κ:

$$\kappa = \varepsilon_0 \omega n_{xy}^2 \int \int dx dy \, E_{TE_y}^* E_{TM_x} \tag{5}$$

Within this simple model, equation (4) fully determines the propagation of light in the waveguide. To describe absorption or gain, complex β are used.

It can be seen from (4) that a complete conversion is possible for the phase matching condition $\beta_{TM} = \beta_{TE}$ when the solution is simply a harmonic oscillation of the intensity between TE and TM mode.

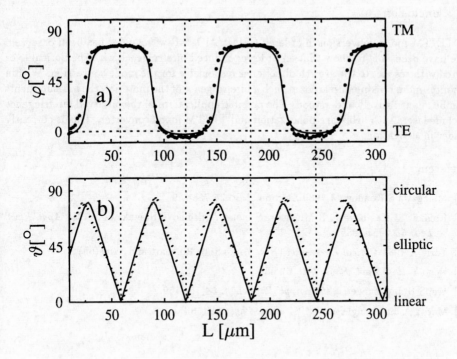

Figure 4. a) Direction of polarization b) polarization type of light versus effective waveguide length L for a ordered waveguide consisting of $G_{0.41}InP/(Al_{0.5}Ga)_{0.51}InP$, measured at $\lambda = 704nm$.

4. Comparison of experiment and model

In Fig. 1 the straight and the dotted lines are fits using eq.4. It can be seen, that the model describes very well the measured intensities of the TE and the TM mode.

Beyond that, in a more detailed investigation, it is possible to explain within the model not only the intensities of the TE and the TM mode, but also the variation of the polarization properties of light, which has traveled in the waveguide for the length L. In Fig. 4 b) the type of polarization ϑ is shown, which is derived from the Stokes parameter. The initially linearly polarized light ($\vartheta = 0°$) oscillates between circular ($\vartheta = 90°$) and linear polarization. The direction of the polarization φ in Fig. 4 a) oscillates between $\varphi \approx -5°$ and $\varphi \approx 80°$. Both angles can be fitted very well. If we extrapolate the measurement to $L = 0$ we find, that the polarization of the luminescence light of the sample, which is coupled into the waveguide is $\varphi \approx -5°$. This is in good agreement with calculations performed with a 6-band kp-model which predict a linearly polarized spontaneous emission with a polarization near TE for ordered and compressively strained quantum wells [6].

5. Conclusion

A TE/TM mode conversion in ordered GaInP/AlGaInP waveguides has been observed. We have been able to show that an ordering induced birefringence with the optical axes tilted with respect to the growth direction is responsible for the mode conversion. Within a simple model taking into account such a birefringence of the material, the measurements can be quantitatively understood. The possible applications of the observed birefringence include lasers with arbitrary polarization and TE/TM mode converter. This is currently under investigation.

References

[1] Minagawa S Kondow M 1989 *Electron. Lett.* **25** 758-759

[2] Kondow M Kakibayashi H Minagawa S Inoue Y Nishino T Hamakawa Y 1988 *Appl. Phys. Lett.* **53** 2053-2055

[3] Yariv A 1985 *Optical Electronics* (New York: Holt, Rinehart and Winston)

[4] Ueno Y 1993 *Appl. Phys. Lett.* **62** 553

[5] Wei S H and Zunger A 1994 *Appl. Phys. Lett.* **64** 757-759

[6] Moritz A and Hangleiter A 1995 *Appl. Phys. Lett.* **66**

Inst. Phys. Conf. Ser. No 145: Chapter 3
Paper presented at 22nd Int. Symp. Compound Semiconductors, Cheju Island, Korea, 28 August–2 September 1995
© *1996 IOP Publishing Ltd*

Suppression of degradation in the crystalline quality of InGaAs/ (Al)GaAs strained quantum-well structures by increasing the barrier thickness

M. Miyashita, S. Karakida, A. Shima, Y. Kajikawa, Y. Mihashi, and
S. Takamiya

Optoelectronic & Microwave Devices Laboratory, Mitsubishi Electric Corporation
4-1 Mizuhara, Itami, Hyogo 664, Japan

Abstract. InGaAs/(Al)GaAs multiple quantum-well (QW) structures with various barrier thicknesses have been grown by metalorganic chemical vapor deposition. Characterization was performed by means of x-ray diffraction measurements as well as microscopic photoluminescence mapping measurements. The characterization results show that the degradation in crystalline quality of the strained QW structure is more suppressed by increasing the barrier thickness. It is pointed out that these results cannot be explained by equilibrium theory concerning the critical thickness in strained layers. It has also been demonstrated that the reliability of InGaAs/AlGaAs double QW lasers is improved by increasing the barrier thickness.

1. Introduction

Recently, InGaAs/(Al)GaAs strained quantum-well (QW) lasers emitting at 0.98 μm have attracted much attention as the pumping source for Er^{3+}-doped fiber amplifiers [1]. High reliability under high-power operation is demanded of this device. In order to meet this demand, it is necessary to suppress any crystalline degradation due to the generation and propagation of misfit dislocations in the strained QW structure. Regarding the critical thickness for dislocation generation in a single strained QW structure, many theoretical models have been proposed [2, 3]. On the other hand, adopting multiple QW structures rather than single QW structures is advantageous for improving the temperature characteristics of high-power lasers. In the case of a multiple QW structure, besides keeping the thickness of each well layer below the critical thickness, it is also important to take into consideration the significant interaction between the individual strained wells. It has been reported that the crystalline quality of strained multiple InGaAs/GaAs QW structures is influenced by the barrier thickness [4, 5]. However, these reports have shown only the characterization results of the crystalline quality, and have not shown the influence of the barrier thickness on the laser characteristics.

In this paper we report on the characterization results of the crystalline quality of InGaAs/ GaAs QW structures grown by metalorganic chemical vapor deposition (MOCVD) as a function of the barrier thickness. We also report for the first time on an improvement in the reliability of InGaAs/AlGaAs double-quantum-well (DQW) lasers by increasing the barrier thickness.

356

2. Experimental Procedure

Epitaxial layers were grown on (100) GaAs substrates by MOCVD in a reactor having a high-speed rotating-disk susceptor [6]. After growing 1-μm-thick GaAs buffer layers on the GaAs substrates, $In_{0.16}Ga_{0.84}As$/GaAs QW structures were grown at a substrate temperature of 675 °C, a V/III ratio of 250, and a growth rate of 1 μm/hour. The thickness of the $In_{0.16}Ga_{0.84}As$ wells was fixed at 3 nm. The thickness of GaAs barriers ranged from 4 to 23 nm. The number of periods of the $In_{0.16}Ga_{0.84}As$/GaAs QW was ten. Finally, 0.1-μm-thick GaAs layers were grown as the uppermost layers of $In_{0.16}Ga_{0.84}As$/GaAs QW structures. These samples were characterized by high-resolution double-crystal x-ray diffraction (HR-DCXRD) and microscopic photoluminescence (PL) mapping measurements at room temperature.

3. Results and Discussion

Figure 1 shows the HR-DCXRD rocking curves obtained for (400) diffraction from samples involving ten $In_{0.16}Ga_{0.84}As$ wells, each separated by 4, 8, 12, and 23-nm-thick GaAs barriers. The satellite peaks of different diffraction orders were observed at regular intervals, reflecting the QW periodicity. The periods of the satellite peaks were changed according to the variation in the GaAs barrier thickness. Furthermore, the widths of the satellite peaks were found to change according to the barrier thickness. Figure 2 shows the full widths at half maximum (FWHM) of the -1st order satellite peaks as a function of the GaAs barrier thickness. In the figures the solid circles indicate the experimental values, and the line shows the theoretical values for ideal QW structures having perfect coherency without any lattice-relaxation. The theoretical calculation was performed on the basis of dynamical x-ray diffraction theory. As can be seen from Fig.2, the experimental value of the FWHM approaches the ideal value upon

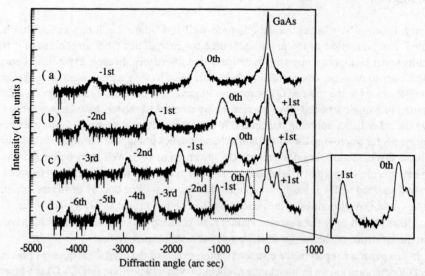

Fig.1   Double-crystal x-ray diffraction rocking curves for (400) diffraction from samples involving ten 3-nm-thick $In_{0.16}Ga_{0.84}As$ wells, each separated by (a) 4, (b) 8, (c) 12, and (d) 23-nm-thick GaAs barriers.

increasing the barrier thickness. It is worth paying attention to the interference fringes between the satellite peaks, as pointed out by Wang *et al.*[4] as well as Takemi *et al.* [7]. The interference fringes can hardly be resolved between any adjacent satellite peaks in the HR-DCXRD rocking curve for a sample having a barrier thickness of 4 nm. On the other hand, the fringes between the GaAs peaks and the 0th order satellite peaks became better resolved along with an increase in the barrier thickness, as can be seen in Fig.1. Moreover, the fringes between the 0th and the -1st order satellite peaks became resolved for a sample having a barrier thickness of 23 nm, when the measurement was performed under a less noisy condition. The broadening of the

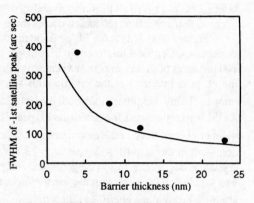

Fig.2 Full widths at half maximum (FWHM) of the -1st satellite peaks in x-ray rocking curves shown in Fig.1 as a function of the barrier thickness. The solid circles (●) indicate the measured values, while the line indicates the theoretical curve of ideal QW structures having perfect coherency without any lattice-relaxation.

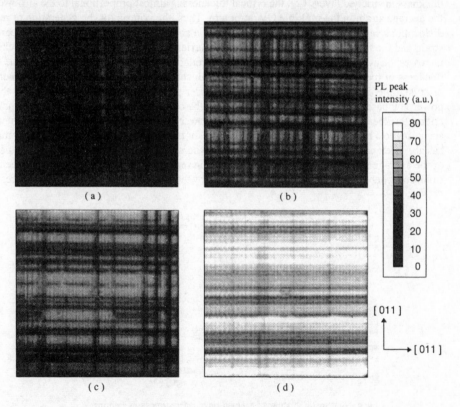

Fig.3 PL peak intensity maps for 100 μm x 100 μm areas of samples involving ten In$_{0.16}$Ga$_{0.84}$As wells, each separated by (a) 4, (b) 8, (c) 12, and (d) 23-nm-thick GaAs barriers.

358

satellite peaks as well as the unresolvability of the fringes in the HR-DCXRD rocking curves suggests that crystalline relaxation occurs more in samples with thinner barrier thicknesses.

Figures 3(a)-3(d) show PL peak-intensity maps for these samples with various barrier thicknesses. Confocal microscopic PL mapping measurements were performed on 100 μm x 100 μm areas of these samples at a scanning interval of 1 μm. As can be seen from the figures, the PL peak intensity of the sample with 4-nm-thick barriers was the weakest among four samples. Many dark lines along both the [0$\bar{1}$1] and [011] directions were observed in the map. As the barrier thickness increased, the PL peak intensity of the sample became stronger, and the number of dark lines in the map decreased. The dark lines along the [0$\bar{1}$1] direction became dominant in the map for a sample with 23-nm-thick barriers. These results indicate that the crystalline degradation due to the generation and propagation of misfit dislocations during growth was suppressed by increasing the barrier thickness. It is also indicated that misfit dislocations are introduced along the [0$\bar{1}$1] directions more easily than along the [011] direction. This preferential direction agrees with a recent report by Beanland et al.[8], but contradicts the results by Kavanagh et al.[9] and Schweizer et al.[10].

We discuss those results, treating a QW structure as a single uniform layer with an averaged alloy composition. According to a mechanical-equilibrium model concerning the critical thickness in strained layers [2], the critical thickness is almost proportional to the -1 power of the average strain in the multiple QW structure. Thus, the criteria for the generation of misfit dislocations can be approximately expressed by a constant value of the product between the strain and the thickness. When increasing the barrier thickness, although the total thickness increases the average strain decreases; thus, the product between the average strain and the total thickness of the multiple QW structure remains unchanged. Therefore, the suppression effect of crystalline degradation by increasing the barrier thickness cannot be explained by the -1 power law based on the equilibrium model. On the other hand, it has been proposed that the critical thickness is almost proportional to the -2 power of the average strain in the QW structure, according to a model which considers the energy for half-loop nucleations [3]. Another model has also been proposed, in which the critical thickness is almost proportional to the -3/2 power of the average strain based on a the consideration regarding the energy balance between the strain energy and the deformation energy reduced by the misfit dislocation [11]. The suppression

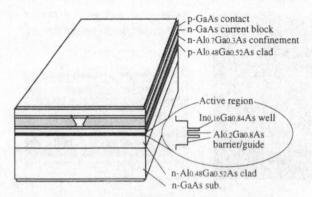

p-GaAs contact
n-GaAs current block
n-Al$_{0.7}$Ga$_{0.3}$As confinement
p-Al$_{0.48}$Ga$_{0.52}$As clad

Active region

In$_{0.16}$Ga$_{0.84}$As well

Al$_{0.2}$Ga$_{0.8}$As barrier/guide

n-Al$_{0.48}$Ga$_{0.52}$As clad
n-GaAs sub.

Fig.4   Schematic structure of a buried-ridge inner-stripe laser emitting at 0.98 μm.

effect of crystalline degradation due to increasing the barrier thickness should be explained by such models that result in higher-power laws than the -1 power.

We fabricated buried-ridge inner-stripe lasers emitting at 0.98μm to examine the influence of the barrier thickness on laser characterictics [12]. A schematic drawing of the laser structure is shown in Fig.4. The active region of the lasers consisted of a DQW structure, i.e., two 8-nm-thick $In_{0.16}Ga_{0.84}As$ strained wells, an $Al_{0.2}Ga_{0.8}As$ barrier between them, and two 30-nm-thick $Al_{0.2}Ga_{0.8}As$ waveguide layers on both outer sides. The barrier thickness ranged from 5 to 20 nm. The cladding and confinement layers consisted of $Al_{0.48}Ga_{0.52}As$ and $Al_{0.7}Ga_{0.3}As$,

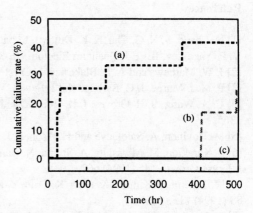

Fig.5 Cumulative failure rate of the DQW lasers with barrier thicknesses of (a) 5, (b) 12.5, and (c) 20 nm under a constant output power of 100 mW at 50°C..

respectively. Figure 5 shows the cumulative failure rate of DQW lasers with barrier thicknesses of (a) 5, (b) 12.5, and (c) 20 nm under a constant output power of 100 mW at 50 °C. While five lasers out of twelve suddenly failed within 500 hours in the case of DQW laser with a 5-nm-thick barrier, all of the tested lasers with 20-nm-thick barriers survived after 1000 hours of operation. It was analyzed by electroluminescence observations from the active regions that these failures of the lasers were caused by the generation of dark-line defects in the strained $In_{0.16}Ga_{0.84}As/Al_{0.2}Ga_{0.8}As$ DQW. It is thought that these dark-line defects are due to the propagation of dislocations, which is caused by stress in the strained DQW, and are suppressed by increasing the $Al_{0.2}Ga_{0.8}As$ barrier thickness. These results indicate that the propagation of dislocations in the InGaAs /AlGaAs DQW structure during laser operation is also suppressed by increasing the barrier thickness.

4. Conclusions

We have shown that increasing the barrier thickness is effective for suppressing the crystalline degradation due to the generation and propagation of misfit dislocations during crystal growth as well as during laser operation. We have found that the FWHM of satellite peaks in the DCXRD rocking curve approaches the ideal value upon increasing the barrier thickness. Moreover, as the barrier thickness increased, the PL peak intensity became stronger and the number of dark lines decreased in the microscopic PL maps. These suppression effects of crystalline degradation cannot be explained by the mechanical-equilibrium theory, which implies that the critical thickness is proportion to the -1 power of the average strain in the QW structure. On the other hand, they can be explained by other models, which result in higher inverse-power laws. Finally, we have demonstrated that the reliability of InGaAs/AlGaAs DQW lasers can be improved by increasing the barrier thickness. All twelve lasers with 20-nm-thick barriers, which were tested under a constant output power of 100 mW at 50 °C, survived after 1000 hours of operation.

References

[1] N. Chand, S. N. G. Chu, N. K. Dutta, J. Lopata, M. Geva, A. V. Syrbu, A. Z. Mereutza and V. P. Yakovlev, IEEE J. Quantum Electon. QE-30 (1994) 424.

[2] J. W. Mathews and A. E. Blakeslee, J. Cryst. Growth 27 (1974) 118.

[3] P. M. J. Marée, J. C. Barbour and J. F. v. d. Veen, J. Appl. Phys. 62 (1987) 4413.

[4] C. A. Wang, S. H. Groves, J. H. Reinold and D. R. Calawa, J. Electron. Mater. 22 (1993) 1365.

[5] M. Hovinen, A. Salokatve and H. Asonen, J. Appl. Phys. 69 (1991) 3378.

[6] S. Karakida, M. Miyashita, A. Shima, N. Kaneno, Y. Mihashi, S. Takamiya and S. Mitsui, J. Cryst. Growth 145 (1994) 662.

[7] M. Takemi, T. Kimura, K. Mori, K. Goto, Y. Mihashi and S. Takamiya, Appl. Surf. Sci. 82/83 (1994) 115.

[8] R. Beanland, M. Aindow, T. B. Joyce, P. Kidd, M. Lourenco and P. J. Goodhew, J. Cryst. Growth 149 (1995) 1.

[9] K. L. Kavanagh, M. A. Capano, L. W. Hobbs, J. C. Barbour, P. M. J. Marée, W. Schaff, J. W. Mayer, D. Pettit, J. M. Woodall, J. A. Stroscio and R. M. Feenstra, J. Appl. Phys. 64 (1988) 4843.

[10] T. Schweizer, K. Kohler, W. Rothemund and P. Ganser, Appl. Phys. Lett. 59 (1991) 2736.

[11] G. Cohen-Solal, F. Bailly and M. Barbe, J. Cryst. Growth 138 (1994) 68.

[12] A. Shima, H. Kizuki, A. Takemoto, S. Karakida, M. Miyashita, Y. Nagai, T. Kamizato, K. Shigihara, A. Adachi, E. Omura and M. Otsubo, in Tech. Dig. IEEE 14th Int. Semiconductor Laser Conf., Hawaii, 1994, p. 131.

Inst. Phys. Conf. Ser. No 145: Chapter 3
Paper presented at 22nd Int. Symp. Compound Semiconductors, Cheju Island, Korea, 28 August–2 September 1995
© 1996 IOP Publishing Ltd

Bound state induced absorption in a QW structure

D. W. Kim

Department of Physics, Sun Moon Univ., Asansi ChungNam 336-840 Korea

S. J. Rhee, Y. M. Kim, H. S. Ko, W. S. Kim, J. C. Woo

Department of Physics, Seoul National Univ., Seoul 151-742 Korea

Abstract. Bound states in a GaAs-Al$_x$Ga$_{1-x}$As quantum well, grown by MBE using phase lock epitaxy were investigated by means of photoluminescence (PL) and PL excitation spectroscopy. An investigation on the localization effects of exciton trapped with defects and/or impurities was performed by means of temperature and excitation intensity dependent PL and PL excitation spectroscopies. When the detection energy was tuned at the bound exciton luminescence, distinct peak was resolved in the exciton absorption region between the two free exciton peaks corresponding to each islands having monolayer well-width difference. This line is prominent in a rather higher temperature range, and is interpreted to be originated from the exciton capture of neutral or ionized impurities.

1. Introduction

The optical properties of semiconductor quantum well (QW) have attracted much attention for the potential of these structures in the optoelectronic applications. And the bound states of the exciton in semiconductor QW structures have been extensively studied. [1-4] Recently, the observation of two hole transitions related with the bound states has been reported. [5] However, in spite of the excellent works, the effects of localization due to the exciton trapping at defects or impurities on the optical properties of QW are still not completely answered. One of the indefinite problems is the correlation between the bound states localized at different islands inevitably formed at heterointerfaces. [6]

In this work, bound states in a QW grown by molecular beam epitaxy (MBE) using phase lock epitaxy were investigated by means of photoluminescence (PL) and PL excitation (PLE) spectroscopy. Through the temperature (T) dependent PL study on a GaAs-Al$_{0.25}$Ga$_{0.75}$As QW with well-width (L$_z$) of 18 monolayer (ML), doublet structures corresponding to free (FE) and bound exciton (BE) recombinations were observed. The relative PL efficiency of BE line with respect to that of FE is observed to vary depending on the excitation photon energy as well as the lattice T and excitation intensity. Below barrier excitation (BBE) generated FEs, which exhibited intrinsic properties like narrow emission peaks with resolved weak BE peaks. While above barrier excitation (ABE) exhibited more intensive BE feature at low T. The feature of excitation spectra obtained with the detection energy tuned at BE luminescence is different from that obtained at FE luminescence. That is, in this case, the bound state induced absorption peak was distinguished below heavy hole (HH) peak of the QW with the most probable L_z in the island distribution. This distinguished line is prominent in a rather higher T range.

2. Experiment

The GaAs-Al$_{0.25}$Ga$_{0.75}$As SQW was grown by MBE using Riber 2300-P system at 600 ℃ on (100) semi-insulating GaAs:Cr substrate in As-rich condition. The QW width is aimed to be 18 ML and controlled by means of counting the number of reflection high energy electron diffraction (RHEED) intensity oscillation. During the growth, the growth interruption of 2 minutes was adopted at each interfaces in order to maximize the flatness of heterointerfaces.

In PL and PLE, He-Ne laser and tunable Ti:Al$_2$O$_3$ laser tuned by a birefringent tuning element were used as excitation sources. Typical linewidths of the excitation sources were less than 1 Å. In PLE, the luminescence was resolved by a monochromator to select the desired photon energy, and the sample T was varied from 15~140 K by temperature controllable cryogenic system.

3. Results and Discussion

In PL spectra of this QW, we were able to observe well resolved peaks originated from the QW islands differing in L_z by 1~2 ML. In addition, there observed some discrepancies between the PL spectra obtained by ABE and BBE. Fig. 1 (a) and (b) are the PL spectra obtained by BBE (hν=1.6256 eV) and by ABE (hν=1.9588 eV), respectively. As shown in Fig. 1, the doublet structures were observed in spectrum (a) and the main peak energies differ by around 1.7 meV, which agrees with the reported values of the energy separation between the FE and BE in GaAs-Al$_x$Ga$_{1-x}$As QWs. [2]

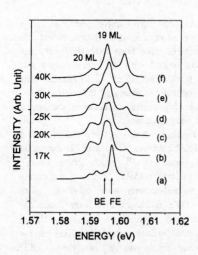

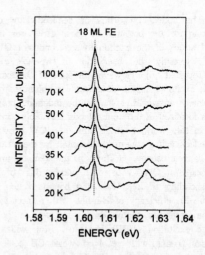

Fig. 1 : PL spectra of a GaAs-Al$_{0.25}$Ga$_{0.75}$As QW obtained (a) by below and (b)-(f) above bandgap excitation at various T.

Fig. 2 : PLE spectra obtained at FE luminescence of 20 ML HH exciton at various T.

This doublet structure is identified as the FE and BE of each QW with L_z=19 and 20 ML. In Fig. 1 (b)~(f), the T dependent ABE PL spectra are shown. Besides the peak energy decrease due to the bandgap shrinkage and the ordinary change of the relative peak intensities upto 70 K due to the carrier thermalization or detrapping effects, [7] it was observed that the FE peak becomes dominant as T increases above 25 K and BE is hardly observable at 40 K. Which is interpreted as the decapturing of BE at this T, and this result is in good agreement with the reports that the ratio of BE/FE starts to be reduced at 20 K and BE peak cannot be observable at the T higher than 40 K. [3, 8]

We have performed the excitation spectroscopy for both BE and FE luminescence by varying T. Fig. 2 and Fig. 3 are the PLE spectra for the FE and BE luminescence of 20 ML heavy hole (HH) exciton, respectively. T was varied from 25 K upto 140 K, and for the easy comparison the spectra are displayed so that the HHFE peaks of the 18 ML QW can align at the same vertical line. As clearly seen in Fig. 2, there appears no critical difference in the spectra with the change of T. On the contrary, there arises a new peak denoted as P below the 18 ML FE peak in the temperature region of 30~60 K in Fig. 3. The position of the peak P as well as the relative intensity changes with T. At 25 K, the magnitude of the separation between peak P and the 18 ML FE peak is 1.7 meV which is same with the value of separation between FE and BE in BBE PL spectrum as shown in Fig. 1. This value varies to 2.5 meV at 30 K, and reaches around 3.4 ± 0.3 meV at 35~60 K. At high T, we can see the peak P even in Fig. 2 though the intensity is much smaller, and the 19 ML HHFE peak energy shifts to the lower energy side as T increases or the detection window is tuned at 21 ML HHFE, which is due to the enhancement of BE peak in PLE spectra. When T is higher than 70 K, it is not possible to observe BE peak because of the breaking of bound exciton.

Preliminary results indicate that the relative intensity of this new peak with respect to the main FE peak is closely related with the incident photon intensity and the areal distribution of the selected QW islands. That is, when we select the BE luminescence of 20 ML QW as a detection window, the intensity ratio of P to 18 ML FE peak reduces rapidly as the excitation intensity was multiplied by 15 times from 5 mW/cm².

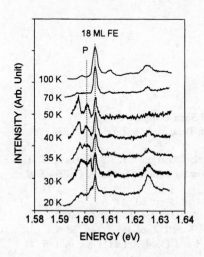

Fig. 3 : PL excitation spectra obtained at BE luminescence of 20 ML HH exciton.

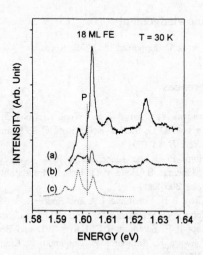

Fig.4 : Typical excitation spectra obtained at (a) FE and (b) BE luminescence at 30 K. (c) is the corresponding PL spectrum.

In this case, peak P is hardly observed and the spectral feature of BE luminescence becomes nearly the same as that of FE. However, when we select the BE luminescence of 19 ML QW as a detection window, the intensity ratio remains nearly the same even though the excitation intensity increased. Considering the high T PL and PLE spectra, the areal distribution of the islands with L_z=19 ML is thought to be larger than that of 20 ML by around 1 order of magnitude. So, we may suggest that the appearance of peak P is originated from the exciton capturing to the traps such as impurities or defects. And the change of the separation is thought to be due to the ionization of the trapping centers with the increase of T. [4]

The spectra in Fig. 4 are the unnormalized excitation spectra of FE and BE luminescence obtained at 30 K. As expected in PL spectrum, the absolute intensity of the excitation spectra is reduced down when the detection window is tuned at BE peak. However, the reduction ratio is different from peak to peak. Considering the background, the reduction ratio of the 19 ML HHFE peak is around 1/2 and that of 18 ML HHFE is around 1/6, on the other hand, that of peak P is nearly 1. Also, at the point of dip between P and 18 ML HHFE peak, the luminescence intensity reduced down to 1/10. This fact indicates that P is not a new absorption center but a relative enhancement of the existing absorption at low excitation power. This phenomenon comes from the transfer efficiency of generated carriers or excitons to the localized states with lower energies. When the generated exciton density is high, most of the excitons transferred to the wider wells recombine in the form of FE, with the assumption of low density of bound states. However, when the generated exciton density is low, they will recombine in the form of both FE and BE.

In summary, through the PL and PLE study on a GaAs–$Al_{0.25}Ga_{0.75}As$ QW with L_z=18 ML, we have observed a bound state induced excitation peak when the detection energy was tuned at the bound exciton luminescence. This peak is observed in the temperature range of 30~60 K and is observed only when the excitation photon intensity is low enough. This line is interpreted to be originated from the exciton capture of neutral or ionized impurities.

Acknowledgement

This work is supported by MOE, Korea.

References

[1] Srinivas V, Chen Y J and Wood C E C 1994 *Solid State Commun.* **89** 611
[2] Reynold D C, Merkel K G, Stutz C E, Evans K R, Bajaj K K and Yu P W 1991 *Phys. Rev. B* **43** 1604
[3] Holtz P O and Monemar B 1994 *J. of Electronic Materials* **23** 513 & Harris C I, Monemar B, Holtz P O, Kalt H, Sundaram M, Merz J L and Gossard A C 1994 *Surf. Sci.* **305** 230
[4] Kirby P B, Constable J A and Smith R S 1991 *J. Appl. Phys.* **69** 517
[5] Holtz P O, Sundaram M, Merz J L and Gossard A C 1990 *Phys. Rev. B* **41** 1489
[6] Kopf R F, Schubert E F, Harris T D and Becker R S 1991 *Appl. Phys. Lett.* **58** 631
[7] Fujiwara K, Katahama H, Kanamoto K, Cingolani R and Ploog K 1991 *Phys. Rev. B* **43** 13987
[8] Ping E X 1993 *J. Appl. Phys.* **73** 7708

Inst. Phys. Conf. Ser. No 145: Chapter 3
Paper presented at 22nd Int. Symp. Compound Semiconductors, Cheju Island, Korea, 28 August–2 September 1995
© 1996 IOP Publishing Ltd

Fast peak quench of the exciton transition in extremely shallow quantum well structures

Sungsoo Park[1] and O'Dae Kwon

Department of Electronics and Electrical engineering, Pohang University of Science & Technology, Pohang P. O. box 125, 790-600 Korea

Abstract. The radius and binding energy of exciton in an extremely shallow quantum well (ESQW) structure are calculated theoretically. As the strength of applied electric field increases, the exciton radius increases and the binding energy decreases rapidly compared with a conventional quantum well, which explains the rapid peak quench in ESQW structures.

1. Introduction

Recently, the room temperature excitons have been observed in $GaAs/Al_xGa_{1-x}As$ multiple quantum wells for x as low as 0.02.[1] The associated electro-absorption in extremely shallow quantum wells (ESQWs) is of special interest because of the rapid sweep-out of the photo-excited carriers in this structure, which allows a high speed modulation at large optical intensities.

When the electric field is applied to the quantum well structures, the absorption peak by exciton shows red shift in transition energy and reduction of oscillation strength, i.e. the so called quantum confined Stark effect. In the ESQW, however, the absorption peak quenches rapidly with increasing electric field, and absorption profile shows a large tail which has not been observed in conventional quantum wells. In an ESQW structure, the distribution of weakly bound particles is strongly affected by an external perturbation. In addition to the change of particle distribution in the quantized direction, the electric field also induces considerable change of distribution in the quantized plane. Thus, the optical absorption in the ESQW shows a strong field dependence compared with the conventional wells.

In this work, we will explain the fast peak quench of the exciton transition in the ESQW by calculating the field dependence of envelope functions and of the binding energy of exciton, and the results are compared with those of a conventional quantum well structure.

2. Theory

The effective mass Hamiltonian for excitons in a quantum well structure with zero

---

[1] E-mail : sspark@jane.postech.ac.kr

center-of-mass momentum can be written as[2]

$$H = -\frac{\hbar^2}{2m_e}\frac{\partial^2}{\partial z_e^2} - \frac{\hbar^2}{2m_h}\frac{\partial^2}{\partial z_e^2} + V_e(z_e) + V_h(z_h) - \frac{\hbar^2}{2\mu}\frac{1}{\rho}\frac{\partial}{\partial\rho}\rho\frac{\partial}{\partial\rho} - \frac{e^2}{\varepsilon(\rho^2 + Z^2)^{1/2}} \tag{1}$$

where μ is defined by $m_e m_h/(m_e + m_h)$, m_e is the conduction band effective mass, m_h is the valence band effective mass, z_e and z_h are the z components (perpendicular to the layers) of the electron and hole position vectors, $V_e(z_e)$ and $V_h(z_h)$ are the additional potentials due to the modulation by the quantum well structure, and ρ, Z are defined as $\mathbf{r}_{e\parallel} - \mathbf{r}_{h\parallel}$ and $z_e - z_h$, repectively. Here, $\mathbf{r}_{e\parallel}$ and $\mathbf{r}_{h\parallel}$ are the parallel components of the particle position.

Introducing the Born-Oppenheimer separation[3]

$$\psi_n(\mathbf{r}_e, \mathbf{r}_h) = \phi(z_e, z_h)g_n(\rho; Z), \tag{2}$$

Leavitt and Little derived an effective Hamiltonian that acts on ϕ only [2] :

$$H_n^{eff} = -\frac{\hbar^2}{2m_e}\frac{\partial^2}{\partial z_e^2} - \frac{\hbar^2}{2m_h}\frac{\partial^2}{\partial z_h^2} + V_e(z_e) + V_h(z_h) - E_n^{(2D)}(Z) + W_n^c(Z). \tag{3}$$

The subscript $n = 1, 2, 3, \cdots$ labels the eigenstates in order of increasing energy, and g_n is an eigenfunction of a two-dimensional Hamiltonian, representing an exciton in which the electron is confined to the plane $z = z_e$ and the hole is confined to the plane $z = z_h$. $E_n^{(2D)}$ is the eigenvalue corresponding to the radial Schrödinger equation and W_n^c is defined by

$$W_n^c(Z) = \frac{\hbar^2}{2\mu}\int_0^\infty |\frac{\partial g_n(\rho; Z)}{\partial Z}|^2 \rho d\rho, \tag{4}$$

Then, the radial Schrödinger equation in a dimensionless form is written by

$$\frac{1}{u}\frac{d}{du}[u\frac{dG}{du}] + \frac{2}{(u^2 + v^2)^{1/2}}G - w(v)G = 0, \tag{5}$$

where $u = \rho/a_0$, $v = Z/a_0$, $E_1^{(2D)} = E_0 w(v)$, and $G(u; v) = a_0 g_1(\rho; Z)$. The parameter a_0 and E_0 are defined by $a_0 = \varepsilon\hbar^2/\mu e^2$ and $E_0 = \mu e^4/2\varepsilon^2\hbar^2$, respectively. The dimensionless binding energy $w(v)$ could be found by introducing a variational wave function,

$$G(u, v) = N(-\lambda[(u^2 + v^2)^{1/2} - v]) \tag{6}$$

where N is a normalization constant and λ is a variational parameter.

Fig. 1 shows the radial wave function $g_1(\rho; Z)$ with various separations, Z. The variational parameter λ is taken as

$$\lambda = \lambda_0(v) = 2/(1 + 2\sqrt{v}), \tag{7}$$

which gives correct results for both $v \to 0$, $(\lambda \to 2)$ and $v \to \infty$, $(\lambda \to 1/\sqrt{v})$. [2]

It clearly shows that the separation in the quantized direction (z-direction) affects the radial distribution function considerably, and the shorter the distance in the z-direction, the shorter the distance in the radial direction (ρ - direction). In consequence, as the overlap between z - directional envelope functions increases, the separation in the ρ - direction decreases.

Due to the v dependence of dimensionless binding energy w, the parameter λ becomes a function of v. The conventional method takes a trial function,

$$\psi_n(\mathbf{r}_e, \mathbf{r}_h) = N' e^{-\lambda'\rho} \phi(z_e, z_h), \tag{8}$$

where N' and λ' is the normalization constant and the variational parameter, respectively. We note that it is equivalent to the removal of the v dependence of radial Schrödinger equation. The conventional method offers a useful parameter, representing the effective separation between electron and hole in the ρ - direction. If we define a exciton radius as $1/\lambda'$ which minimizes the exciton energy,

$$E_{ex}(\lambda') = \frac{<\psi_n|H|\psi_n>}{<\psi_n|\psi_n>}, \tag{9}$$

the exciton radius measures the spacial separation between the electron and the hole in ρ - direction.

Fig. 2 shows the square of the overlap integral between electron and heavy hole ground states, which represents the normalized transition rates of a heavy hole exciton.

We took a $100\mathring{A} - -GaAs/Al_xGa_{1-x}As$ single quantum well as our model structure and assumed that the electric field is applied only on the quantum well region. The real line represents a conventional quantum structure which has 20 % Al fraction in the barrier region, and the dotted line represents an ESQW structure with 4 % Al fraction. As the electric field increases, the conventional well shows a slow decay of the overlap, while the ESQW shows a fast decay associated with the rapid quench. Fig. 3 shows the heavy hole exciton radius in quantum wells with increasing electric field. Because the overlap decreases radpidly with increasing electric field in the ESQW, the spacial distance in ρ - direction increases rapidly. On the other hand, the exciton radius remains almost

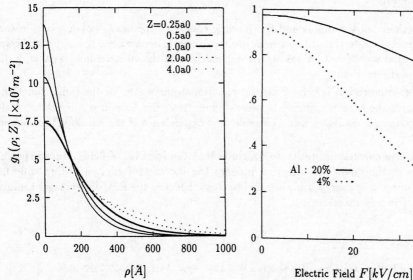

Fig. 1. The radial wave functions with various separations, Z.

Fig. 2. Square of overlap under increasing electric field.

368

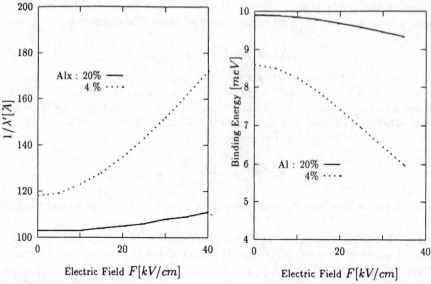

Fig. 3. The heavy hole exciton radius.

Fig. 4. The exciton binding energy under increasing electric field.

unchanged in a conventional structure. As expected, the ESQW has larger exciton radius than the conventional quantum well due to the small overlap.

Fig. 4 shows the binding energy of the heavy hole exciton with increasing electric field. The binding energy in the ESQW also decreases as sharply as the overlap, while the binding energy in a conventional well remains almost unchanged.

3. Conclusion

We have calculated the radius and the binding energy of the heavy hole exciton in the ESQW structure. By the use of the Born-Oppenheimer separation, it has been shown that the radial wave function has strong dependence on the electron-hole separation in the quantized direction.

The exciton radius is defined by the variational parameter for the radial function, and measures the spacial distance between the electron and hole in ρ - direction. Calculated exciton radius shows a considerable field dependence of the envelope function in the ESQW.

From the calculated results we conclude that the increase of field-induced spacial separation in the quantized direction induces the increase of exciton radius and the reduction of binding energy, and thus the excitonic effect in the ESQW decreases rapidly as the electric field increases.

References

[1] Goossen K W, Cunningham J E and Jan W J 1990 *Appl. Phys. Lett.* **57** 2582–2584

[2] Leavitt R P and Little J W 1990 *Phy. Rev. B* **42** 11774–11783

[3] Messiah A 1963 *Quantum Mechanics* (New York: Wiley) **II** 781–793

Inst. Phys. Conf. Ser. No 145: Chapter 3
Paper presented at 22nd Int. Symp. Compound Semiconductors, Cheju Island, Korea, 28 August–2 September 1995
© *1996 US Government*

Electrical and optical characterization of GaSb based diode laser material for 2-4 µm applications

D.K. Johnstone[1], M.A. Marciniak[1], Y.K. Yeo[1], R.L. Hengehold[1], and G.W. Turner[2]

1. Air Force Institute of Technology, Wright-Patterson AFB, OH 45433-7765 USA
2. Lincoln Laboratory, Massachusetts Institute of Technology, Lexington, MA 02173 USA

Abstract. Deep level transient spectroscopy, infrared absorption, or photoluminescence (PL) measurements were made on molecular beam epitaxially grown samples of $Ga_{0.80}In_{0.20}As_{0.12}Sb_{0.88}$, $AlAs_{0.07}Sb_{0.93}$, and $InAs_{1-x}Sb_x$ ($0.031 \leq x \leq 0.192$) materials used to develop 2-4 µm quantum well laser structures lattice-matched to GaSb substrates. In addition to five other hole trap levels, a trap with the greatest effect on nonradiative recombination located at 276 meV above the valence band (E_v) was found in the $Ga_{0.80}In_{0.20}As_{0.12}Sb_{0.88}$ active region material. A DX-like trap was found in the $AlAs_{0.07}Sb_{0.93}$ cladding material at E_c-282 meV. Temperature-dependent absorption measurements made on undoped $InAs_{1-x}Sb_x$ active region material resulted in the closed-form expression for the energy gap. The measured PL linewidth of the band-to-band transition of $InAs_{1-x}Sb_x$ is narrower than any previously reported values, indicating the high quality of this material. Furthermore, two shallow impurity levels were resolved at energies 5-7 meV from the band edge.

1. Introduction

There has been considerable research interest in developing optoelectronic devices which exploit the atmospheric transmission windows in the mid-infrared (IR) for applications such as remote atmospheric gas detection, molecular spectroscopy, and more eye-safe laser radar and communications. Several semiconductor systems have been considered for mid-IR lasers in the past: HgCdTe [1], the Pb-salts [2], and the III-V GaSb-based compositions $(Ga_{1-y}In_yAs_{1-x}Sb_x/Al_{1-y}Ga_yAs_{1-x}Sb_x)$ [3-5]; but of these, the GaSb-based devices have proven to be the most successful.

Since an increase in the Al-composition of the quaternary $Al_{1-y}Ga_yAs_{1-x}Sb_x$ increases the energy gap and decreases the refractive index of the material, $Al_{1-y}Ga_yAs_{1-x}Sb_x$ lattice-matched to GaSb substrate is used for the cladding layers with a $Ga_{1-y}In_yAs_{1-x}Sb_x$ active region. This provides the necessary electrical and optical confinements required for a double heterostructure laser. To develop 2 µm quantum well (QW) lasers, lower Al mole fractions of $Al_{1-y}Ga_yAs_{1-x}Sb_x$ are also used for the barrier material as shown in Fig. 1 (a). For the 4 µm QW lasers, an $In_{1-x}Al_xAs_{1-y}Sb_y$ barrier is used since the band offset of $InAs_{1-x}Sb_x/In_{1-x}Al_xAs_{1-y}Sb_y$ is split between the conduction and valence bands, providing better electron and hole confinement as shown in Fig. 1 (b).

The use of molecular beam epitaxial (MBE) growth techniques for these mid-IR semiconductor lasers has greatly improved the material quality and epitaxial design of the devices such that more advanced laser structures have been successfully implemented using this material. However, even with these reported device successes, further improvements are still necessary. Although it is essential to know the material characteristics of each layer of the $Ga_{1-y}In_yAs_{1-x}Sb_x/Al_{1-y}Ga_yAs_{1-x}Sb_x$ laser system to improve device performance, there are surprisingly few reports of these basic material characterizations in the literature [4,6-8]. Therefore, deep energy levels in the $Ga_{0.80}In_{0.20}As_{0.12}Sb_{0.88}$ and $AlAs_{0.07}Sb_{0.93}$ materials were measured using deep level transient spectroscopy (DLTS). Undoped $InAs_{1-x}Sb_x$ (y=1) was also studied using laser excitation power- and temperature-dependent photoluminescence (PL), and temperature-dependent Fourier transform spectroscopic (FTS) absorption to assess the quality of the material, and to estimate the energy gap.

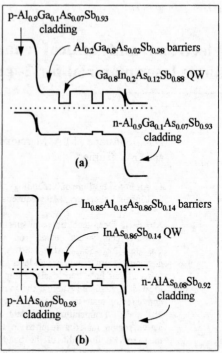

Fig. 1. Energy band diagrams for III-V mid-IR diode laser structures for (a) 2 μm and (b) 4 μm wavelength lasers.

2. Deep level transient spectroscopy

DLTS was used to characterize the deep traps in $AlAs_{0.07}Sb_{0.93}$ (y=0) which is used for the cladding layers of a 4 μm laser structure lattice-matched to GaSb, and in $Ga_{0.8}In_{0.2}As_{0.12}Sb_{0.88}$ which is used for the active region of 2 μm lasers. The $AlAs_{0.07}Sb_{0.93}$ diodes were made by growing Te-doped (8×10^{16} cm^{-3}) n^+-$AlAs_{0.07}Sb_{0.93}$ on the n-type substrate first, followed by unintentionally doped p-$AlAs_{0.07}Sb_{0.93}$ (2×10^{16} cm^{-3}). A mesa etch using H_3PO_4, H_2O_2, H_2O, and L-tartaric acid with a ratio of 1ml:1ml:250ml:0.4g was used to define the diode area of 0.0025 cm^2. The $Ga_{0.8}In_{0.2}As_{0.12}Sb_{0.88}$ diode was grown with background p-type doping of 1×10^{16} cm^{-3}. 460 Å of Si_3N_4 was deposited by plasma enhanced chemical vapor deposition for a metal-insulator-semiconductor (MIS) structure. The capacitance transients measured in the temperature range of 20-475 K were digitized and fit using modulating functions to obtain the emission rate as a function of temperature [9]. Then, the fit for either one or two exponential components was used to determine the activation energy and capture cross section of the trap by plotting the $ln(T^2/e)$ vs. $1/k_BT$ in a standard DLTS Arrhenius plot, where e is the emission rate [10].

The results of DLTS spectra measured from the $AlAs_{0.07}Sb_{0.93}$ p-n junction are shown in Fig. 2 (a) for two rate windows. It shows one very large trap with a well defined emission energy of 282 meV determined from the Arrhenius plot shown in Fig. 2 (b). As the electric field was increased by the reverse bias voltage, the trap energy was reduced. This indicates the presence of the Poole-Frenkel effect for this trap, which identifies it as a donor trap in the n-type or an acceptor trap in the p-type $AlAs_{0.07}Sb_{0.93}$. On the other hand, the reduction of trap energy with an applied electric field was much less than that for a pure Coulombic trap potential. This limited Poole-Frenkel effect is expected from the alteration

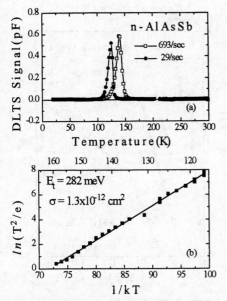

Fig. 2. (a) DLTS spectra obtained from MBE-grown AlAsSb p-n junction. (b) The corresponding Arrhenius plot of T^2/e for a wide range of emission rates.

of the Coulombic potential by the capture barrier [11]. The capture barrier for this trap was determined to be 124 meV using a fit of the amplitude of the temperature-dependent transients to the solution of the capture rate equation. Since the trap shows a Poole-Frenkel effect, and the values of the capture barrier and the trap energy are very close to the previously reported DX-like defects in Te-doped $Al_xGa_{1-x}As_ySb_{1-y}$ [12,13], it is concluded that this trap is a Te-related DX-like center in the Te-doped n-AlAs$_{0.07}$Sb$_{0.93}$. It is believed that the 124 meV capture barrier for this trap is large enough to have significant effects on device performance.

The DLTS spectra for the undoped p-Ga$_{0.8}$In$_{0.2}$As$_{0.12}$Sb$_{0.88}$ measured using an MIS capacitor structure is shown in Fig. 3. The positive going peaks are for hole emission to the valence band, and the negative going peak is for hole emission to the conduction band in the n-type inversion layer. From a multi-exponential fit of the transients, it was found that the low temperature peaks, (a), (b), (c), and (d), from 50–200 K are a superposition of four hole trap energy levels. The position and height of peaks (a), (b), and (c) were found to be very sensitive to the trap filling and measurement biases. That is, shifting the emission window away from the semiconductor-insulator interface resulted in a shift of peaks to lower temperature, and an increase in concentrations. The peak shift to lower temperature with increasing electric field indicates that the defects are acceptor traps in a p-type material. The increase in acceptor trap concentration with distance from the insulator can be a result of a reaction between the Si$_3$N$_4$ and the Ga$_{0.8}$In$_{0.2}$As$_{0.12}$Sb$_{0.88}$. It has been shown for GaAs that Si$_3$N$_4$ deposition causes Ga vacancies near the interface [14]. If a similar phenomenon occurs in Ga$_{0.8}$In$_{0.2}$As$_{0.12}$Sb$_{0.88}$, the Si$_3$N$_4$ reduces the Ga or In concentration near the interface. Since Ga$_x$In$_{1-x}$As$_y$Sb$_{1-y}$ grown by various methods has excess Ga, which makes undoped Ga$_x$In$_{1-x}$As$_y$Sb$_{1-y}$ p-type by forming the Ga$_{Sb}$ antisite double acceptor defect [15], the reduction of the concentration of Ga near the interface by the Si$_3$N$_4$ appears as a higher level of excess Ga away from the interface than that near the insulator. Therefore, the increase in concentration of trap levels (a), (b), and (c) away from the interface suggests that these defects are related to the excess Ga.

The Arrhenius plot depicted in Fig. 4 shows one minority carrier trap level at 320 meV below the conduction band, and six majority carrier trap levels at 24, 76, 108, 122, 224, and 276 meV above the valence band. The six hole traps emit to the valence band and one hole trap emits to the conduction band in the n-inversion layer. The two low energy traps show an interesting behavior, that is, the 24 meV energy level makes a transition to 73 meV as the temperature increases. This can be explained by an emission from a double acceptor defect. At low temperatures, both energy levels are filled, but as the temperature increases, the shallow energy level is emptied. Further increase in temperature is followed

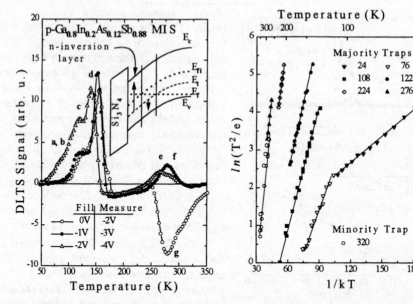

Fig. 3. DLTS spectra for the traps in
MBE-grown p-$Ga_{0.8}In_{0.2}As_{0.12}Sb_{0.88}$.

Fig. 4. The Arrhenius plot for the traps
found in p-$Ga_{0.8}In_{0.2}As_{0.12}Sb_{0.88}$.

by an emission from the deeper energy level to the valence band via the shallow defect level. For GaSb, the theoretical predictions and measurements of the double acceptor by other methods have shown that the two energy levels are 33 and 80 meV [15,16]. Therefore, it is believed that trap levels (a) and (b) are due to a III_V antisite defect in $Ga_{0.8}In_{0.2}As_{0.12}Sb_{0.88}$. However, the origin of traps (c) and (d) are not known at present.

The minority carrier peak (g) shown in Fig. 3 was obtained using a filling pulse of 0.1 sec at 0 V bias and measurement bias of -2 V. Under these conditions, most of the DLTS minority carrier signal originates from the emission of the trapped holes to the n-type inversion layer near the interface as shown in the inset of Fig. 3. Shifting the region of filled traps away from the inversion layer by changing filling and measuring biases decreases the available conduction band states for hole emission, and results in increased competing majority carrier emission. In this case, although the energy levels are difficult to measure accurately due to the spatial dependence of both majority and minority carrier emissions, the energy of the minority carrier signal is estimated to be 320±50 meV. Two energies were also resolved for the majority carrier peak, one (e) at 224 meV, and another (f) at 276 meV. The sum of the minority carrier signal energy (g) and the deeper majority carrier signal energy (f) is very close to the band gap of 594 meV. Therefore, it is believed that both signals (f) and (g) originate from the same trap center. Since this center is located near midgap, it acts as the most efficient non-radiative recombination center in this material.

3. Optical characterization of InAs$_{1-x}$Sb$_x$ active region material

Temperature-dependent FTS absorption, and temperature- and laser excitation power-dependent PL measurements were made on InAs$_{1-x}$Sb$_x$ epilayers (0.40-1.38 μm thick) grown by MBE on GaSb substrates. The Sb-composition ranged between $0.031 \leq x \leq 0.192$, which was calculated using Vegard's law from lattice mismatches of $-0.401\% \leq \Delta a/a \leq +0.708\%$ as determined by double crystal x-ray diffraction. These compositions are candidates for compressively and tensily strained QW lasers which range in wavelength from 3.7 to 4.5 μm at low temperature. The FTS absorption was performed over sample temperatures of 6-300 K. The PL spectra were measured at temperatures as low as 2 K using a grating spectrometer and an InSb detector. The samples were excited using a fiber-coupled 500 mW, 1 μm diode laser, and the beam was chopped at 100 Hz for collection by standard lock-in techniques.

Absorption measurements were performed to establish the band gap energies (E_g) for the $InAs_{1-x}Sb_x$ samples because the reported temperature- and compositional-dependencies of this energy gap differ from one to another in the literature [6-8,17,18]. Typical absorbance spectra for these samples at various temperatures are shown as solid lines in Fig. 5. The absorption edge was typically parabolic in shape and fit well to the standard ($h\nu - E_g)^{1/2}$ form, where $h\nu$ is the photon energy. The temperature dependence of the energy gap for each sample fit well to the Varshni equation, resulting in the empirical relation given by

$$E_g(T,x) = E_g(0,x) - \alpha T^2/(\beta + T),$$

where $E_g(0,x) = 0.3535 - 0.3777x$ eV $(0.031 \leq x \leq 0.192)$, $\alpha = 2.9 \times 10^{-4}$ eV/K, and $\beta = 150$ K. Here, α and β are the mean Varshni fitting parameters taken over all the samples.

PL measurements were also made on the $InAs_{1-x}Sb_x$ samples at various temperatures. The results are plotted in Fig. 5 for the $x = 0.089$ sample together with the absorption data. Only one well resolved PL spectrum has been reported so far for $InAs_{1-x}Sb_x$ grown on InAs [8], and none have yet been reported for $InAs_{1-x}Sb_x$ grown on GaSb [6]. For the present MBE-grown samples, low temperature PL typically resulted in a single main peak, but also included an LO-phonon replica 27-29 meV below it. The low temperature linewidths of these peaks were as narrow as 4.3 meV at low laser excitation powers. Previous narrow linewidths reported are 5.15 meV for $InAs_{0.96}Sb_{0.04}$ on InAs substrate [8] and 22 meV for $InAs_{0.905}Sb_{0.095}$ on GaSb [6]. The observations of the narrow linewidth and the phonon replica indicate the high quality of our material in relation to other $InAs_{1-x}Sb_x$ material. As shown in Fig. 5, a comparison of the PL peak position and the measured band edge for the $InAs_{0.911}Sb_{0.089}$ sample indicates the single PL peak is due to band-to-band recombination.

PL measurements were also made on unintentionally doped $InAs_{0.808}Sb_{0.192}$ (0.708% lattice mismatch with the GaSb substrate) as a function of temperature and laser excitation power. Three resolvable peaks were observed. A comparison of the PL and absorption spectra identified a band-

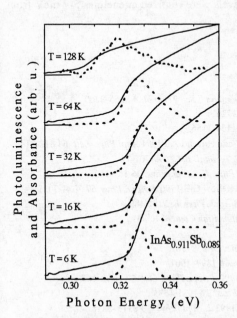

Fig. 5. Comparison of PL (dotted lines) and absorption (solid lines) spectra for MBE-grown $InAs_{0.911}Sb_{0.089}$ on GaSb substrate at five temperatures.

to-band recombination peak at 0.2774 eV. The position of a second PL peak at 0.2714 eV did not change with excitation power, and hence this peak is believed to be due to a free-to-bound (F-B) transition. However, the position of a third peak at 0.2650 eV shifted to higher energies at a rate of 4 meV/decade of laser excitation power until the peak merged with the F-B peak into a single peak. Thus, this peak is attributed to donor-acceptor pair recombination.

4. Summary

From the $Ga_{0.85}In_{0.15}As_{0.12}Sb_{0.88}$ active region material for 2 μm lasers, one minority trap level at 320 meV below the conduction band, and six hole trap levels at 24, 76, 108, 122, 224, and 276 meV above the valence band were found using DLTS measurements. The two low energy levels (24 and 76 meV) are attributed to a Ga_{Sb} double acceptor. It is believed that the minority trap level at 320 meV and the hole trap level at 276 meV originate from the same trap, and it could be the most efficient non-radiative recombination center in this material. From the $AlAs_{0.07}Sb_{0.93}$ cladding layer material for 4 μm lasers, one DX-like center was found at 282 meV below the conduction band, which could limit the minority carrier lifetime. Temperature-dependent IR absorption measurements made on $InAs_{1-x}Sb_x$ active region material for 4 μm lasers resulted in the closed-form expression for the energy gap, $E_g(T,x) = E_g(0,x) - \alpha T^2/(\beta+T)$, where $E_g(0,x) = 0.3535 - 0.3777x$ eV $(0.031 \leq x \leq 0.192)$, $\alpha = 2.9 \times 10^{-4}$ eV/K, and $\beta = 150$ K. PL linewidth of the band-to-band transition narrower than previously reported values for this material indicates the high material quality. Furthermore, two shallow impurity levels were resolved at energies 5-7 meV from the band edge.

References

[1] Arias J M, Zandian M, Zucca R and Singh J 1993 *Semicond. Sci. Technol.* **8** S255-S260.

[2] Partin D L 1988 *IEEE J. Quant. Electron.* **14** 1716-1726.

[3] Choi H K and Eglash S J 1992 *Appl. Phys. Lett.* **61** 1154-1156.

[4] Kurtz S R, Biefeld R M, Dawson L R, Baucom K C and Howard A J 1994 *Appl.Phys. Lett.* **64** 812-814.

[5] Turner G W, Choi H K and Le H Q 1995 *J.Vac. Sci. Technol.* B **13** 699-701.

[6] Elies S, Krier A, Cleverly I R and Singer K 1993 *J. Phys. D: Appl. Phys.* **26** 159-162.

[7] Fang Z M, Ma K Y, Jaw D H, Cohen R M and Stringfellow G B 1990 *J. Appl. Phys.* **67** 7034-7039.

[8] Yen M Y, People R, Wecht K W and Cho A Y 1988 *Appl. Phys. Lett.* **52** 489-491.

[9] Loeb J M and Cohen G M, 1965 *IEEE Trans. on Automation Control* **10** 359.

[10] Lang D 1974 *J. Appl. Phys.* **45** 3023-3032.

[11] Buchwald W R and Johnson N M 1988 *J. Appl. Phys.* **64** 958–961.

[12] Zhu Y, Takeda Y and Sasaki A 1988 *J. Appl. Phys.* **64** 1897–1901.

[13] Nakagawa A, Pekarik J, Kroemer H, and English 1990 *J Appl. Phys. Lett.* **57** 1551–1553.

[14] Dhar S, Mallik K and Mazamdar M 1994 *J. Appl. Phys.* **77** 1531-1535.

[15] Shen J, Dow J, Ren S, Tehrani S, and Goronkin H 1993 *J. Appl. Phys.* **73** 8313–8318.

[16] Nakashima K 1981 *Jap. J. Appl. Phys.* **20** 1085–1094.

[17] Coderre W M and Woolley J C 1968 *Can. J. Phys.* **46** 1207-1214.

[18] Stringfellow G B and Greene P E 1971 *J. Electrochem. Soc.* **118** 805-810.

Inst. Phys. Conf. Ser. No 145: Chapter 3
Paper presented at 22nd Int. Symp. Compound Semiconductors, Cheju Island, Korea, 28 August–2 September 1995
© *1996 IOP Publishing Ltd*

The Variation of the Structures and Excited Subband States for the Thermal Treated GaAs/AlGaAs Multiple Quantum Wells

Y. T. Oh, T. W. Kang[a], and C. Y. Hong

Department of Physics, Dongguk University, 3-26 Pil-dong, Chung-ku, Seoul 100-715, Korea.

T. W. Kim

Department of Physics, Kwangwoon University, 447-1 Wolgye-dong, Nowon-ku, Seoul 139-701, Korea.

Abstract. Photoluminescence (PL) measurements were performed in order to investigate the deformation of the GaAs/$Al_{0.35}Ga_{0.65}As$ multiple quantum wells (MQWs) due to thermal treatment. Rapid thermal annealing was performed at 950℃ for 10, 20, and 30 sec after the Si_3N_4 capping layer was evaporated on the sample. The PL spectra for the as-grown and annealed samples show the transitions from the 1st electronic subband to the 1st heavy hole (E1-HH1) and from the 2nd electronic subband to the 2nd heavy hole (E2-HH2), and the PL signals shift to the high-energy side as the annealing time increases. The potential profile of the quantum well as a function of the Al diffusion length was investigated, and subband energy levels in the GaAs/$Al_{0.35}Ga_{0.65}As$ MQWs were calcualted by a variational method making use of the potential profile. The Al diffusion lengths for the annealed MQWs at 950 ℃ for 10, 20, and 30 sec determined from the (E1-HH1) PL peaks were 2.15, 4.45, and 6.3 nm, respectively. The behavior of the difference between the experimental values of (E2-HH2) and (E1-HH1) as a function of the Al diffusion length is in good agreement with theoretical results.

1.Introduction

Recently, with rapid advances in epitaxial growth technology such as molecular beam epitaxy (MBE) it has become possible to fabricate GaAs/AlGaAs multiple quantum wells (MQWs) with interfacial abruptnesses on the scale of a few lattice constants [1]. Since the GaAs/$Al_xGa_{1-x}As$ MQWs have a large exciton binding energy, they have been very attractive for the investigation of exciton transitions and for applications in optical, electronic, and optoelectronic devices [1]. When the quantum well structures are annealed, the shape of the square potential changes to that of a parabolic

376

potential [2]. The variations of the potential wells, the subband energies [3-5], the Stark shifts of the excitons [6, 7], and the oscillator strengths [8] for the annealed MQWs suggest the possibility of new quantum device applications.

The methods of intermixing quantum wells through thermal treatment are laser-beam-induced disordering [9], impurity-induced disordering [3, 4, 12-14], and capping-layer-induced (SiO$_2$ or Si$_3$N$_4$) disordering [5, 15]. In addition, interdiffusion coefficients have been investigated by photoluminescence (PL) [3-5], double crystal X-ray rocking curves [17, 18], Raman spectroscopy [10, 16], secondary ion mass spectroscopy [14], and tunneling electron microscopy [13]. Although many studies concerning the intermixing behavior in GaAs/AlGaAs MQWs have been performed [2-18], to the best of our knowledge, very little work has been performed on the variations of the excited subband states.

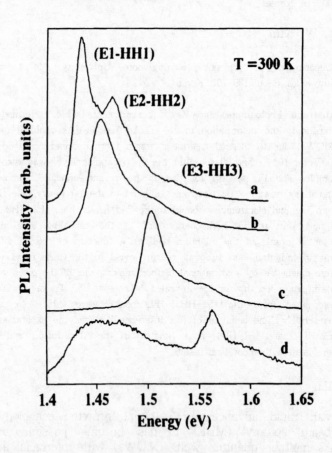

Fig. 1 The PL spectra for the as-grown GaAs/Al$_{0.35}$Ga$_{0.65}$As MQWs (a), and the GaAs/Al$_{0.35}$Ga$_{0.65}$As MQWs annealed at 950 ℃ for 10 sec (b), 20 sec (c), and 30 sec (d). The sample temperature for the PL measurement was 300 K.

This paper PL and DCRC measurements which were performed in order to study the inter-mixed GaAs/Al$_{0.35}$Ga$_{0.65}$As MQWs. The PL spectra show the behavior of the transitions from the 1st electronic subband to the 1st heavy hole (E1-HH1) and from the 2nd electronic subband to the 2nd heavy hole (E2-HH2). The potential profiles of the quantum wells as a function of the Al diffusion length were calculated, and the difference between the (E2-HH2) level and the (E1-HH1) level [(E2-HH2) − (E1-HH1)] as a function of the diffusion length was compared with the PL results.

2. Experimental Details

The samples used in this study were grown by MBE and had the following structure: a 3000-Å Al$_{0.35}$Ga$_{0.65}$As layer, a 25-period GaAs/Al$_{0.35}$Ga$_{0.65}$As superlattice (t = 200 Å for each layer), and a 3000-Å GaAs buffer layer in a semi-insulating GaAs substrate. Every layer was nominally undoped, Rapid thermal annealing (RTA) was performed at 950℃ in a nitrogen atmosphere for 10, 20, and 30 sec. The PL measurements were carried out using a 75-cm monochromator equipped with an RCA C31034 photomultiplier tube. The excitation source was the 4880-Å and 5145-Å lines of a Ar$^+$ laser, and the samples were mounted on a cold finger in a cryostat and were controlled between 10 and 300 K using a helium gas cycling cryostat system. The DCRC measurements were performed using the Rikaku system double-crystal X-ray diffractometer; the reference crystal was an InSb (111) bulk.

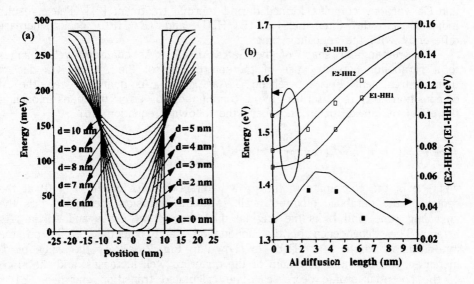

Fig. 2 The calculated potential profile (a) and the energy shift of (E1-HH1), (E2-HH2), (E3-HH3), and [(E1-HH1) − (E2-HH2)] (b) as a function of the interdiffusion length. The filled and empty rectangles represent the experimental values.

3. Results and Discussion

Figure 1 show the PL spectra at 300 K (a), and the GaAs/Al$_{0.35}$Ga$_{0.65}$As MQWs annealed at 950℃ for 10 (b), 20 (c), and 30 sec (d). The PL spectrum for the as-grown GaAs/Al$_{0.35}$Ga$_{0.65}$As MQWs shows peaks at 1.433, 1.465, and 1.530 eV which correspond to (E1-HH1), (E2-HH2), and (E3-HH3), respectively. The PL spectrum for the general quantum well structure do not show the (E2-HH2) signal. However, since the well width of the GaAs/Al$_{0.35}$Ga$_{0.65}$As MQWs used in this study is 200 Å, the electrons and hole are easily excited from the ground state to the excited states because of the small energy difference between the subband states formed in the wells. Thus, the (E2-HH2) and (E3-HH3) peaks were observed at 300 K [20].

These energy values are in good agreement with the theoretical values determined from Barstard's envelope function [19]. These PL peaks shift to the high-energy side as the RTA time increases, as shown in Fig. 1. The shifts of the PL peaks originate from the interdiffusion of the Al and Ga at the GaAs and Al$_{0.35}$Ga$_{0.65}$As heterointerfaces due to thermal treatment, from the change in the shape of the quantum well from a square to a parabolic shape, from the increase of the conduction band minimum, and from the decrease of the valence band maximum. The PL spectrum for the GaAs/Al$_{0.35}$Ga$_{0.65}$As MQWs annealed at 950℃ for 10 sec shows the (E1-HH1) and (E2-HH2) peaks at 1.503 and 1.553 eV, respectively, and that for 30 sec shows the (E1-HH1) and (E2-HH2) peaks at 1.562 and 1.595 eV with a broad peak at 1.45 eV, which is related to the defects due to the Si wdonor and Ga vacancy complex formed during thermal treatment [21]. These results indicate that the intensities of (E1-HH1) and (E2-HH2) peaks increase differently with the annealing time.

The rectangular shape of the GaAs/Al$_{0.35}$Ga$_{0.65}$As quantum well changes to a parabolic shape because of the intermixing of the Al and Ga due to thermal treatment. For a parabolic GaAs/Al$_{0.35}$Ga$_{0.65}$As quantum well, the Al composition x(z) is described by a combination of error functions modeling the effect of interdiffusion and obeys the following equation [15]:

$$x(z) = x_o\left[1 + \frac{1}{2} erf\left(\frac{z-L_z/2}{2\sqrt{Dt}}\right) - \frac{1}{2} erf\left(\frac{z+L_z/2}{2\sqrt{Dt}}\right)\right] \tag{1}$$

where z is the coordinate in the growth direction with origin (z = 0) at the center of the parabolic quantum well, D is the diffusion coefficient, t is the annealing time, and L_z is the original thickness of the GaAs and Al$_x$Ga$_{1-x}$As layers. The dependence of the conduction band potential profile on the interdiffusion length $d_0 = (Dt)^{1/2}$ is shown in Fig. 2-(a). As the value of d_0 increases, the potential minimum of the quantum well increases, and the slope of the potential profile decreases. The calculated transition energies, (E1 – HH1), (E2 – HH2), (E3 – HH3), and [(E1 – HH1) – (E2 – HH2)], as a function of d_0 as determined from the variational method [22] are also shown in Fig. 2-(b). The filled and empty rectangles represent the experimental values. The values of the d_0 corresponding to the (E1 – HH1) peaks obtained from the PL spectra for GaAs/Al$_{0.35}$Ga$_{0.65}$As MQWs annealed at 950℃ for 10, 20, and 30 sec are 2.15, 4.45, and 6.3 nm, respectively. The behavior of the [(E2 - HH2) - (E1 - HH1)] energy values with increasing Al diffusion length shows good agreement between experimental and theoretical values. The difference between the two values might originate from the nonuniformity of the interdiffusion during thermal treatment or from calculation of the energies by a perturbation method.

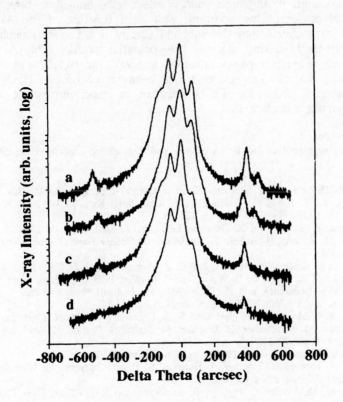

Fig 3 The DCRC spectra for the as-grown GaAs/Al$_{0.35}$Ga$_{0.65}$As MQWs (a) and for the GaAs/Al$_{0.35}$Ga$_{0.65}$As MQWs thermally treated at 950 ℃ for 10 (b), 20 (c), 30 (d), and 50 sec (e).

Figure 3 shows the results of the DCRC spectra for the as-grown GaAs/Al$_{0.35}$Ga$_{0.65}$As MQWs (Fig. 3-(a)) and the GaAs/Al$_{0.35}$Ga$_{0.65}$As MQWs at annealed 950℃ for 10 (3-(b)), 20 (Fig. 3-(c)), and 30 sec (Fig. 3-(d)). The peak originating from the satellite peak of m = 0 was located at origin, and the peaks attributed to m = ± 1 were located at ± 420 arcsec, respectively. The DCRC data for the annealed GaAs/Al$_{0.35}$Ga$_{0.65}$As MQWs show that the intensities of the satellite peaks vary, and that the locations of the satellite peaks do not change in comparison with those for the as-grown GaAs/Al$_{0.35}$Ga$_{0.65}$As MQWs. As the annealing time increases, the peak intensities corresponding to m = ±1 decrease dramatically. This behavior is due to the dectrase of the difference between the structure factor values of the well and the barrier.

4. Summary and Conclusion

In summary, the PL spectra for the as-grown and annealed GaAs/$Al_{0.35}Ga_{0.65}As$ MQWs show that the (E1-HH1) and (E2-HH2) peaks, and the peaks shift to the high-energy side as the annealing time increases. The potential profile of the quantum well as a function of the Al diffusion length was investigated, and the subband energy levels were calculated by a variational method making use of the potential profile. The Al diffusion lengths of the (E1-HH1) peaks annealed at 950℃ for 10, 20, and 30 sec are 2.15, 4.44, and 6.3 nm, respectively. The behavior of [(E2 - HH2) - (E1 - HH1)] determined from the PL spectra is in good agreement with that obtained from the calculation.

Acknowledgement
This work was supported by the Korea Science and Engineering Foundation in 1994.

References
[1] F. Capasso 1990 *Physics of Quantum Electron Devices* (Springer-Verlag, Heidelberg)
[2] S. K. Kim, T. W. Kang, C. Y. Hong, S. H. Cho, J. H. Kim, T. W. Kim, K. S. Chung, and S. J. Yu 1993 *Phys. Stat. Sol. (a)* **136** 99
[3] Y. Hirayama, Y. Suzuki, and H. Okamoto 1985 *Jpn. J. Appl. Phys.* **24** 1498
[4] D. Huang, M. Kallergi, J. Aubel, S. Sundaram, G. DeSalvo, and J. Comas 1991 *J. Appl. Phys.* **70** 4181
[5] B. Elman, E. S. Koteles, P. Melman, and C. A. Armiento 1989 *J. Appl. Phys.* **66** 2104
[6] E. Herbert Li, K. S. Chan, B. L. Weiss, and J. Micallef 1993 *Appl. Phys. Lett.* **63** 533
[7] T. Ishikawa, S. Nishimura, and K. Tada 1990 *Jpn. J. Appl. Phys.* **29** 1466
[8] S. L. Chuang and D. Ahn 1989 *J. Appl. Phys.* **65** 2822
[9] J. Ralston, A. L. Moretti, R. K. Jain, and F. A. Chambers 1987 *Appl. Phys. Lett.* **50** 1817
[10] J. D. Ralston, M. Ramsteiner, B. Dischler, M. Maier, G. Brandt, P. Koidl, and D. J. As 1991 *J. Appl. Phys.* **70** 2195
[11] L. Reynolds, Jr. and M. Geva 1992 *Appl. Phys. Lett.* **61** 165
[12] H. Leier, A. Forchel, G. Horcher, J. Hommel, S. Bayer, Rothfritz, G. Weimann, and W. Schlapp 1990 *J. Appl. Phys.* **67** 1805
[13] E. G. Bithell, M. Stobbs, C. Phillips, R. Eccleston, and R. Gwilliam 1990 *J. Appl. Phys.* **67** 1279
[14] K. B. Kahen and G. Rajeswaran 1989 *J. Appl. Phys.* **66** 545
[15] J. D. Ralston, S. O'Brien, G. W. Wicks, and L. F. Eastman 1988 *Appl. Phys. Lett.* **52** 1511
[16] N. Hara and T. Kotada 1991 *J. Appl. Phys.* **69** 2112
[17] V. S. Sperious and T. Vreeland 1984 *Jr., J. Appl. Phys.* **56** 1591
[18] R. M. Fleming, D. B. McWhan, A. C. Gossard, W. Wiegmann, and R. A. Logan 1980 *J. Appl. Phys.* **51** 357
[19] G. Bastard 1981 *Phys. Rev. B* **24** 5693.
[20] H. H. Dai, M. S. Coi, M. A. Gunderson, H. C. Lee, P. D. Dapkus, and C. W. Myles 1989 *J. Appl. Phys.* **66** 2538.
[21] L. Pavesi, N. H. Ky, J. D. Ganiere, F. K. Reinhart, N. Baba-Ali, I. Harrison, B. Tuck, and I. Henini 1992 *J. Appl. Phys.* **71** 2225
[22] T. E. Schlesinger and Y. Kuech 1986 *Appl. Phys. Lett.* **49** 519

Inst. Phys. Conf. Ser. No 145: Chapter 3
Paper presented at 22nd Int. Symp. Compound Semiconductors, Cheju Island, Korea, 28 August–2 September 1995
© *1996 IOP Publishing Ltd*

The band lineup of AlGaInP / AlInP indirect semiconductors

Y. Ishitani, E. Nomoto, T. Tanaka, and S. Minagawa

Central Research Laboratory, Hitachi Ltd.
1-280 Higashi-koigakubo, Kokubunji, Tokyo 185, Japan

Abstract. A key to achieve better carrier confinement in 0.6 μm laser diodes is to incorporate an AlInP layer into the AlGaInP cladding layer. The band lineup of $(Al_{0.7}Ga_{0.3})_{0.5}In_{0.5}P$ / $Al_xIn_{1-x}P$ (x = 0.47 to 0.61) is studied at both the Γc and Xc points by measuring the optical transition energy of $(Al_{0.7}Ga_{0.3})_{0.5}In_{0.5}P$, $Al_{0.53}In_{0.47}P$, and $(Al_{0.7}Ga_{0.3})_{0.5}In_{0.5}P$ / $Al_xIn_{1-x}P$ super lattices. The energy level of $Al_xIn_{1-x}P$ at the Xc point decreases by 0.09 eV as x increases from 0.47 to 0.61. At x=0.53, the share of the band offset for the conduction band at the Γc point is found to be 78%.

1. Introduction

Although red-light emitting semiconductor lasers made of GaInP/AlGaInP have been established as industrial products, the push towards higher performance continues. The central issue lies in achieving better confinement of injected carriers in the active layer, which will lead to higher operating temperatures, higher output power, and longer life. The band offset at the GaInP / AlGaInP heterobarrier currently used as the basic building block of the lasers, however, is not large enough to the desired level of confinement. Cladding layers composed of GaInP/AlInP super lattices, which utilizes wider AlInP band gap than AlGaInP, have been proposed [1], but problems such as crystal quality and conductivity control limit their practical use. However, strained AlInP layers that are thinner than the critical thickness can be partially incorporated into the AlGaInP cladding layer and this is expected to improve the performance of lasers. To design optimized devices, we must know the exact band energy levels of the semiconductors used and their band lineup at the Γ and X points. In particular, the band lineup of the $(Al_{0.7}Ga_{0.3})_{0.5}In_{0.5}P$ / $Al_xIn_{1-x}P$ heterostructure is important.

The band structures of these phosphides grown on the (100) GaAs substrates are known to be modified by spontaneous long-range ordering (LRO) [2]. The generation of LRO can be suppressed by using of off-angle substrates and high growth temperatures[2]. This present paper examines the band lineup when the disordered crystals are used to avoid the influence of LRO. To determine the band edges of the indirect band, we used photoluminescence (PL) spectra ascribed to the X_c to Γ_v transition [3] and photoreflectance (PR) spectra to observe the Γ_c to Γ_v transitions.

2. Experiments

The samples were grown by low-pressure (100 torr) metal-organic vapor phase epitaxy (MOVPE) on n-type GaAs substrates using trimethylaluminum (TMA), trimethylgallium (TMG), trimethylindium (TMI), arsine, and phosphine as source materials. To suppress spontaneous ordering, (001) substrates 7° off toward [110] were used and the layers were grown at 720°C. The layers were undoped single layers of $Al_xIn_{1-x}P$ and $(Al_yGa_{1-y})_{0.5}In_{0.5}P$, and $(Al_{0.7}Ga_{0.3})_{0.5}In_{0.5}P$ / $Al_xIn_{1-x}P$ super lattices (SLs) with 20 periods. When changing the layer composition, the total concentration of the group III precursors was kept constant. This helped maintain the constant growth rate and thus made controlling the thickness of the super lattice easier.

The layers were characterized by double-crystal x-ray diffraction using (400) planes, PL, and PR. During the PL measurement, the sample temperature was kept at 12 (± 0.5) K by a helium cryostat monitored with an FeAu-Cr thermocouple . The samples were excited by a 476.5 nm line of an Ar+ ion laser or the 325 nm line of a He-Cd laser. The excitation intensity of the Ar laser was varied from 71 mW/cm^2 to 6.4 W/cm^2. The energy gap at the Γ point of indirect semiconductors was measured with a conventional PR system. The spectral resolution of the PL and PR were 2 and 6 meV, respectively.

3. Results and discussion

3.1 x-ray diffraction analysis

The growth rate and the composition of the $Al_xIn_{1-x}P$ layers were obtained from the position of the satellite peaks in the x-ray diffraction patterns of the SLs. The samples were set with the the [110] direction parallel to the outlet slit of the x-ray source, because the substrate orientation was tilted toward [110] from the [001] direction. The SL period of the samples grown on tilted substrates with this setting was found to be nearly equal to that of the (001) on-axis sample. Figure 1(a) shows the relation of the period and the growth time of $Al_xIn_{1-x}P$ while keeping the growth time of the quarternary alloy constant. From this figure, we calculated the growth rates of $Al_xIn_{1-x}P$ and $(Al_{0.7}Ga_{0.3})_{0.5}In_{0.5}P$ to be 0.40 (±0.01) nm /sec and 0.59(±0.01) nm/sec.

The average lattice constant of the SL layer can be written as [4]

$$Aav = (N_1A_1 + N_2A_2) / (N_1 + N_2). \qquad (1)$$

Here, N_1 and N_2 are the number of unit lattices (or thickness) in one cycle, and A_1, and A_2 are the lattice constants of $(Al_{0.7}Ga_{0.3})_{0.5}In_{0.5}P$ and $Al_xIn_{1-x}P$. This equation is rewritten as follows,

$$\Delta A_2 / A_0 = \{ (N_1 + N_2) \Delta Aav/A_0 - N_1 \Delta A_1 /A_0\} / N_2, \qquad (2)$$

where A_0 is the lattice constant of GaAs, and $\Delta Ai /A_0 (i=1, 2$ or av) is the mismatch between the lattice constant A_i of a free layer and A_0. The difference in the diffraction angle between the GaAs and the 0'th satellite peak represents the vertical average lattice mismatch between the SL and GaAs ($\Delta Aav \perp /A_0$). Furthermore, $\Delta Am /A_0$ (m=1, 2, av) can be expressed using elastic stiffness C_{ij} [5] as

$$\Delta Am / A_0 = C_{11} / (C_{11} + C_{12}) \Delta Am \perp / A_0. \qquad (3)$$

From these formulae and the experimental data, we obtained $\Delta A_2/A_0$ and hence the Al composition x. The relation of the intended x and the x obtained by the above method is shown in Fig. 1(b). This figure shows that the composition of the sample was controlled as intended.

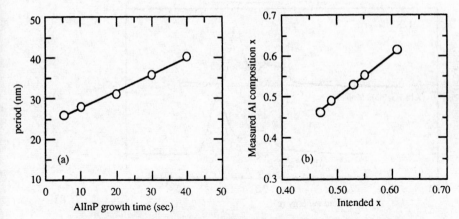

Fig.1 The analytical result of x-ray diffraction of AlGaInP / AlInP super lattices
(a) The dependence of the period of SLs on AlInP growth time
(b) The relation between the intended x and the measured x

3.2 The energy gap of $(Al_{0.7}Ga_{0.3})_{0.5}In_{0.5}P$ and $Al_{0.53}In_{0.47}P$

The PL spectra of $(Al_{0.7}Ga_{0.3})_{0.5}In_{0.5}P$ and $Al_{0.53}In_{0.47}P$ at 12 K is depicted in Fig. 2(a). Since the energy gaps at the Γ point of these alloys were found to be 2.454 eV and 2.709 eV by PR and the Xc point of these crystals is located below the Lc and the Γc point levels, the peak energies of the PL spectra are those for the Xc - Γv transition. Although this transition is forbidden in principle, the violation of the k-selection rule is possible owing to crystal defects or fluctuations in the alloy composition. The small peak shown at the lower energy side is thought to be the phonon replica. Thus, the transition energy of X_c - Γ_v is 2.326 eV for $(Al_{0.7}Ga_{0.3})_{0.5}In_{0.5}P$ and 2.345 eV for $Al_{0.53}In_{0.47}P$. Therefore, the differences in the energy levels between the Γc point and the Xc point at 12 K are estimated to be 128 meV and 364 meV, respectively.

3.3 The photoluminescence of $(Al_{0.7}Ga_{0.3})_{0.5}In_{0.5}P$ / $Al_xIn_{1-x}P$ super lattice

First, we describe the characteristic features of the samples where the thickness of the $(Al_{0.7}Ga_{0.3})_{0.5}In_{0.5}P$ and the $Al_{0.53}In_{0.47}P$ in one period are 24 nm and 8 nm, respectively. Figure 2(b) shows the dependence of the PL spectra of the SLs on the excitation power intensity. When the excitation intensity is very low, only the lower energy peak (peak A) was detected, and the peak position does not move with intensity. The higher energy peak (peak B) appears and increases in relative intensity. It shifts to a higher energy and reaches 2.324 eV. This value agrees fairly well with the Xc-Γv transition energy of $(Al_{0.7}Ga_{0.3})_{0.5}In_{0.5}P$. The small peak located at the lower energy side of the peak A is probably the phonon replica, like the small peak in Fig. 2(a). Figure 2(c) shows the PL profile of $(Al_{0.7}Ga_{0.3})_{0.5}In_{0.5}P/Al_xIn_{1-x}P$ ($0.47 \leq x \leq 0.61$). The peak energy of A decreases remarkably as x increases.

Next, we discuss the band alignment of the $(Al_{0.7}Ga_{0.3})_{0.5}In_{0.5}P/Al_{0.53}In_{0.47}P$ heterostructure. We define the difference in the energy levels of Γ_c, X_c, and Γ_v between $(Al_{0.7}Ga_{0.3})_{0.5}In_{0.5}P$ and $Al_{0.53}In_{0.47}P$ as ΔE_c^{Γ}, ΔE_c^x, ΔE_v^{Γ}, respectively. These values are positive for the conduction band and negative for the valence band when the level of $Al_xIn_{1-x}P$ is higher than that of $(Al_{0.7}Ga_{0.3})_{0.5}In_{0.5}P$.

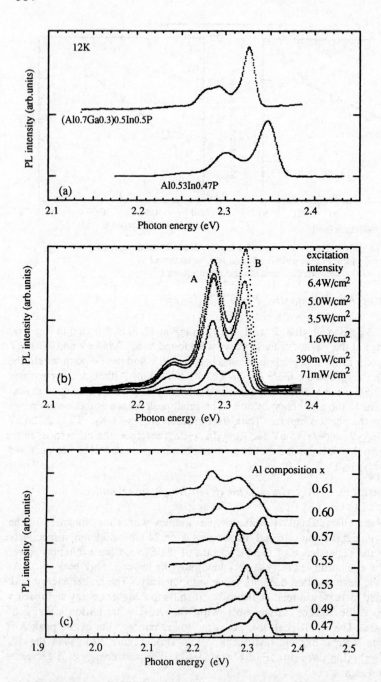

Fig. 2 Photoluminescence spectra at 12K
(a) Photoluminescence of single layers (b) Excitation intensity dependence for the
SL sample ; AlGaInP was 24 nm, AlInP was 8nm . (c) The spectra for various Al composition ; The
excitation intensity was 3.5W/cm²

There are eight possible band lineup schemes, depending on the combination of positive or negative energy differences at each band level. According to Yow et al. [6], in the actual scheme, ΔE_c^{Γ} is positive, ΔE_c^x is negative, and ΔE_v^{Γ} is positive. This leads us to believe peak A originates from the transition of AlInP X_c - AlGaInP Γ_v. The band alignment determined is depicted in Fig. 3. Based on the values obtained from the spectral measurement, we calculated the values of ΔE_c^{Γ}, ΔE_c^x, and ΔE_v^{Γ} to be 200 meV, -37 meV, and 55 meV, respectively, and $\Delta E_c^{\Gamma}/(\Delta E_c^{\Gamma} + \Delta E_v^{\Gamma})$ to be 0.78.

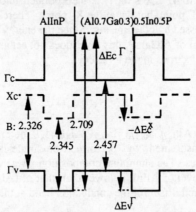

Fig.3 The band lineup of $(Al0.7Ga0.3)0.5InP / Al0.53In0.47P$
$\Delta E_c^x = 200\text{meV}, \Delta E_c^x = -37\text{meV}, \Delta E_v^{\Gamma} = 55\text{meV}$

Last, we will discuss the effect of the composition x. As the energy of peak A represents the transition from AlInP X_c to AlGaInP Γ_v, the energy level of AlInP X_c decreases as x increases, as shown in Fig. 2(c). The dependence of ΔE_c^x on x is plotted in Fig. 4.

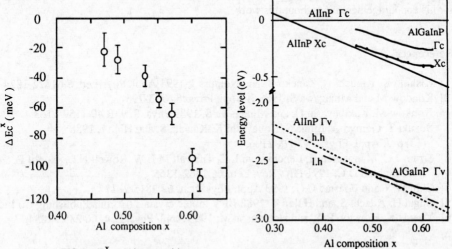

FIG 4. ΔE_c^x vs. Al composition x

Fig.5 $(Al0.7Ga0.3)0.5In0.5P/Al_xIn_{1-x}P$ band lineup
;Xc level of AlGaInP is lower than that of AlInP $(0.47 \lesssim x \lesssim 0.61)$. The Γ_v levels cross each other around x=0.47.

In the range investigated, the X_c level of AlInP never exceeds that of AlGaInP. From these results and the dependence of the energy gap at the Γ point, measured using PR on x, (which will be reported elsewhere), we are able to draw the band alignment of $(Al_{0.7}Ga_{0.3})_{0.5}In_{0.5}P$ / $Al_xIn_{1-x}P$ as shown in Fig. 5. In this figure, the Γ_c level of $Al_xIn_{1-x}P$ is set as the base line and the relative position of each band is shown. The Γ_c levels of these alloys cross at x=0.34 ($\pm$0.01), and the Γ_v levels cross at x=0.47 ($\pm$0.01). As x decreases from 0.53 to 0.47, new spectra must appear in Fig. 2(c), but they are too broad and weak to be identified.

When a layer of AlInP ($0.47 \leq x \leq 0.61$) is incorporated into the AlGaInP cladding it can serve as a barrier at the Γ point, but not at the X_c point. Therefore the carrier leakage via the X valley must be suppressed by other means. Since the energy level of the $Al_xIn_{1-x}P$ layer (x < 0.47) is lower than that of AlGaInP, this layer does not act as a barrier to holes, although it does act as a barrier to electrons in the Γ_c valley.

4. Conclusion

The band lineup of $(Al_{0.7}Ga_{0.3})_{0.5}In_{0.5}P/Al_xIn_{1-x}P$ heterostructure was investigated. Photoluminescence measurement was performed on single layer crystals or super lattices grown on GaAs substrates. The aluminum composition x was varied from 0.47 to 0.61. We found that the Xc level of AlInP is lower than that of $(Al_{0.7}Ga_{0.3})_{0.5}In_{0.5}P$. When x is 0.53, the share of the band offset for the conduction band at the Γ_c point is 78%.

Acknowledgements

We are grateful to prof. Yasuhiro Shiraki and Hiroyuki Yaguchi of the University of Tokyo for providing the PR system and assistance in its use. We also like to thank Hiroshi Hamada of Hitachi C.R.L for his cooperation in crystal growth, and Sin-ichiro Yano of Hitachi C.R.L for his encouragement throughout this work.

References

[1] Kishino K, Kikuchi A, Kaneko Y, and Nomura I, 1991 Appl. Phys.Lett. **58** 1822-1824
[2] Kondow M and Minagawa S, 1988 J. Appl. Phys. 64 793-796
    Konkow M, Kakibayashi H, and Minagawa S, 1989 Phys. Rev. B **40** 1159-1163
    Suzuki T, Gomyo A, Iijima S, Kobayashi K, Kawata S, and Hino I, 1988
        Jap. J. Appl. Phys. 27 2098-2106
[3] Strum J C, Manoharan H, Lenchyshyn L C, Thewalt M L W, Rowell N L, Noel J P, and
        Houghton D C, 1991 Phys Rev. Lett. **66** 1362-1365
[4] Choi W Y and Fonstad C G, 1993 Appl. Phys. Lett. **62** 2815-2817
[5] Nagai H, Adachi S, and Hukui T, 1988 III-V mixed crystals (Corona Publishing Co.Ltd.)
[6] Yow H K, Houston P A, and Hopkinson M, 1995 Appl. Phys. Lett. **66** 2852-2854

Inst. Phys. Conf. Ser. No 145: Chapter 3
Paper presented at 22nd Int. Symp. Compound Semiconductors, Cheju Island, Korea, 28 August–2 September 1995
© *1996 IOP Publishing Ltd*

Temperature Dependence of Type-I Exciton Absorption in Type-II CdSe/ZnTe Heterostructures

E.H. Lee, S. Stoltz, H.C. Chang, W. Y. Yu, I.K. Kang, H. Luo and A. Petrou

Department of Physics and Center for Electronic and Electro-optic Materials, State University of New York, Buffalo, NY 14260, U.S.A.

Abstract. Temperature dependence of type-I exciton absorption in CdSe/ZnTe type-II superlattices is studied as a function of CdSe layer thickness. In a CdSe epilayer, exciton absorption exhibits strong temperature dependence, decreasing in intensity as the temperature is raised, and vanishing at T = 150 K. As the CdSe layer thickness is reduced in a CdSe/ZnTe superlattice, exciton thermal dissociation due to electron-LO-phonon coupling decreases, which is evidenced by the weaker temperature dependence of the exciton absorption in structures with thinner CdSe layers. When the thickness of CdSe is reduced to a nominal value of 125 Å, exciton absorption survives up to room temperature.

1. Introduction

Heterostructures of II-VI semiconductors offer a wide range of opportunities for opto-electronic applications because of their large range of energy gaps, and band offsets. By bandgap engineering, new device configurations can be explored. While most of the current efforts in this area are directed towards light-emitting devices, there are also numerous applications involving absorption of light, such as optical modulators, detectors. Examples of II-VI devices in this area are the fabrication of ZnCdTe/ZnTe, and ZnCdSe/ZnSe quantum wells as modulators [1, 2]. All current quantum-well-based optical modulators consist of type-I quantum wells, which typically have large absorption coefficients.

It was discovered recently that type-I excitons (spatially direct) can also form in type-II superlattices [3]; these excitons involve above-barrier states in the valence (or conduction) band, and confined state in the conduction (or valence) band. Type-I excitons in type-II superlattices result in strong exciton absorption, similar to that observed in type-I structures. While such excitons closely resemble those in type-I quantum wells in the excitation process, their dynamic properties are dramatically different from what has been observed in type-I structures. The lifetime of these excitons are several orders of magnitude shorter because their dissociation process is driven by thermal relaxation of carriers to lower energy states (in the picosecond range), rather than the radiative recombination which is the main mechanism in typical type-I quantum wells (on a nanosecond time scale). The short exciton lifetime is seen qualitatively by the absence of photoluminescence corresponding to the exciton transition observed in the absorption spectra. In contrast, these structures exhibit strong luminescence from the type-II transitions. The fast type-I exciton relaxation can be advantageous for certain devices (e.g., optical modulators) because the saturation power of exciton absorption will be much higher. The combination of large absorption coefficients and short exciton lifetimes in type-II CdSe/ZnTe superlattices--thus high saturation power of exciton absorption--makes these structures interesting candidates for a new type of modulators and other devices involving light absorption. We report in this paper a detailed study of the temperature dependence of type-I exciton absorption in type-II CdSe/ZnTe superlattices, which is an essential step for achieving device applications.

2. Experimental results and analysis

Sample growth using molecular beam epitaxy (MBE) was carried out in a Riber MBE 32P system. All samples were grown on (100) GaAs substrates. Details of the growth condition can be found in Ref. 4. Because of the large lattice mismatch between GaAs and ZnTe (~ 7%), thick ZnTe buffer layers (2 ~ 3 μm) were used to achieve high quality and fully relaxed buffer for the subsequent growth of the closely lattice-matched CdSe/ZnTe superlattices. Since the focus of this study is the type-I excitonic transition occurring in the CdSe layers, all superlattices have the same ZnTe layer thickness (150Å measured by oscillations in reflection high energy electron diffraction) but different CdSe layer thicknesses. A CdSe reference epilayer was also grown on a ZnTe buffer layer for comparison. For the absorption experiments, the GaAs substrates were removed by chemical etching. The samples were placed in variable temperature (10-300K) closed cycle refrigerator.

In the CdSe epilayer a strong exciton absorption peak is observed at 10 K with a full width at half maximun (FWHM) of 8 meV. Because of the strong exciton-LO-phonon coupling, the exciton absorption exhibits strong temperature dependence, similar to other bulk materials, e.g., ZnSe. The intensity drops as the temperature is raised, accompanied by a broadening of the peak. When the temperature reaches $T = 150$ K, the exciton peak vanishes.

We studied the temperature dependence of type-I exciton absorption in type-II CdSe/ZnTe superlattices as a function of CdSe layer thickness. The existence of type-I excitons in type-II heterostructures is the result of the localization of above-barrier states in the barrier region. The valence band offset of CdSe/ZnTe is 0.64 eV and the conduction band offset is 1.35 eV [5]. As a result of the type-II band alignment, shown in the inset of Fig. 1, CdSe acts as a quantum well for the electrons in the conduction band, and as a barrier for the holes in the valence band. The above-barrier holes localized in the barrier region (i.e., CdSe layers) will interact with the confined electrons in the conduction band well (also in the CdSe layers) and form type-I excitons. Because both the hole and the electron are localized in the same layer, such an exciton has type-I behavior, and exhibits a correspondingly strong optical absorption. Absorption coefficents as large as 1.4×10^5 cm^{-1} were obtained in type-II CdSe/ZnTe superlattices.

As the CdSe layer thickness is reduced in a CdSe/ZnTe superlattice, the exciton binding energy increases because of carrier confinement. This results in a decrease of electron-LO-phonon coupling, evidenced by the weaker temperature dependence of the exciton absorption in superlattices with thinner CdSe layers. When the thickness of CdSe is reduced to a nominal

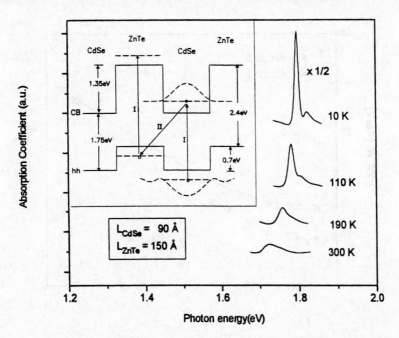

Figure 1. Absorption coefficient of a CdSe/ZnTe superlattice sample with well thickness L_{CdSe} = 90Å as a function of temperature. The inset shows the band alignment of the superlattice. The transition marked in the CdSe layer is responsible for the absorption peak.

value of 125Å, exciton absorption survives up to room temperature. This occurs because the exciton binding energy is higher than the LO-phonon energy. The exciton absorption spectra of a ZnTe/CdSe superlattice, with $L_{CdSe} = 90$Å and $L_{ZnTe} = 150$Å (nominal values), at various temperatures are shown in Fig. 1. The temperature dependence of the exciton absorption peak is very similar to that in a type-I quantum well [6, 7], with an increasing linewidth and a reduced peak value as the temperature is raised. Because of the increased exciton binding energy in thin CdSe layers, exciton absorption can be clearly seen at room temperature.

The FWHM is analyzed with the expression corresponding to one phonon absorption [8]:

$$\Gamma(T) = \Gamma_0 + \Gamma_{LO}/[\exp(\hbar\omega_{LO}/kT) - 1], \tag{1}$$

where Γ_0 represents inhomogeneous linewidth broadening, and Γ_{LO} the strength of exciton-LO-phonon coupling. The value of $\hbar\omega_{LO}$ was obtained by Raman spectroscopy to be 25.4 meV. Figure 2 shows the FWHM for the superlattice shown in Fig. 1. The result for the epilayer is also shown in Fig. 2 for comparison. The lines are the fitted values of $\Gamma(T)$ using Eq. 1. The values of Γ_{LO} are listed in Table 1 for all samples used in this study. The fitted values for Γ_{LO} in Table I show a clear decrease with reduced CdSe layer thickness.

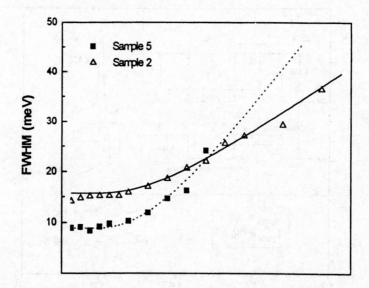

Figure 2. The FWHMs of the exciton absorption peak as a function of temperature for Sample 5 (CdSe epilayer, the squares and the dashed line) and Sample 2 (CdSe/ZnTe superlattice, the triangles and the solid line). The lines are fits using Eq. (1).

Table 1. Sample descriptions, excitonic transition energies, and fitted values of Γ_{LO}.

Sample	L_{CdSe} (Å)	E(T) (eV)	Γ_{LO} (meV)	Type
1	70	1.824	29	SL
2	90	1.797	41	SL
3	125	1.773	53	SL
4	189	1.760	60	SL
5	2700	1.750	82	Epilayer

3. Summary

In summary, we have studied the temperature dependence of type-I exciton absorption in type-II CdSe/ZnTe superlattices with various CdSe layer thicknesses. In the CdSe epilayer, exciton absorption exhibits strong temperature dependence, decreasing in intensity as the temperature is raised, and vanishing at T = 150 K. This is very similar to other bulk materials, in which the process is caused by electron-LO-phonon interaction. As the CdSe layer thickness is reduced in the CdSe/ZnTe superlattices, the exciton binding energy increases due to the enhancement of Coulomb interaction between the electron and the hole, and becomes larger than the LO-phonon energy. This results in a decrease of Γ_{LO}, evidenced by the weaker temperature dependence of the exciton absorption in superlattices with thinner CdSe layers. When the thickness of CdSe is reduced to 125Å, exciton absorption can be observed at room temperature. Although the band alignment is type-II, the type-I excitons confined in the CdSe layers strikingly resemble those in a typical type-I structures, such as ZnSe/ZnCdSe quantum wells, not only in the strong absorption coefficients, but also in the increase of exciton binding energy in thinner layers. The combination of large absorption coefficients and short type-I exciton lifetimes in type-II CdSe/ZnTe superlattices present unique opportunities for a new type of modulators and related optical devices involving light absorption.

Acknowledgment

This work was supported by the Center for Electronic and Electro-optic Materials at SUNY at Buffalo.

392

References

[1]   R.D. Feldman, D. Lee, A. Partovi, R.P. Stanley, A.M. Johnson, J.E. Zucker, A.M. Glass, and J. Hegarty, Critical Reviews in Solid State and Materials Sciences **17**, 477 (1992).

[2]   S.Y. Wang, Y. Kanakami, J. Simpson, H. Stewart, K.A. Prior, and B.C. Cavenett, Appl. Phys. Lett. **62**, 1715 (1993).

[3]   H. Luo, W.C. Chou, N. Samarth, A. Petrou, and J.K. Furdyna, Solid State Commun. **85**, 691 (1993).

[4]   H. Luo, N. Samarth, F. C. Zhang, A. Pareek, M. Dobrowolska, J. K. Furdyna, K. Mahalingam, N. Otsuka, W. C. Chou, A. Petrou, and S. B. Qadri, Appl. Phys. Lett. **58**, 1783 (1991).

[5]   E. T. Yu, M. C. Phillips, J. O. McCaldin, and T. C. McGill, J. Vac. Sci. Technol. B **9**, 2233 (1991).

[6]   D. Lee, A. M. Johnson, J. E. Zucker, R. D. Feldman, and R. F. Austin, J. Appl. Phys. **69** (9), 6722 (1991).

[7]   N. T. Pelekanos, J. Ding, M. Hagerott, A. V. Nurmikko, H. Luo, N. Samarth, and J. K. Furdyna, Phys. Rev. B **42**, 6037 (1992).

[8]   D. Chemla, S. Schmitt-Rink, and D. A. B. Miller, *Optical Nonlinearities and Instabilities in Semiconductors*, edited by H. Haug (Academic, New York, 1988), p. 83.

Inst. Phys. Conf. Ser. No 145: Chapter 3
Paper presented at 22nd Int. Symp. Compound Semiconductors, Cheju Island, Korea, 28 August–2 September 1995
© *1996 IOP Publishing Ltd*

A new method for estimating band parameters in narrow InGaAs/InAlAs quantum wells at room temperature

N. Kotera and K. Tanaka

Kyushu Institute of Technology, Iizuka, Fukuoka 820, Japan

Abstract. A series of interband optical transitions associated with exciton absorptions have been observed in photocurrent spectra of p-i-n junctions including InGaAs/InAlAs multi-quantum wells with well widths of 5 and 10 nm. Eigen energies with consecutive quantum numbers were determined for conduction electrons and heavy holes. The energies were dependent on the square of the respective quantum numbers. The possible ranges of effective masses and band offsets were analyzed using a simple paricle-in-a-box calculation.

1. Introduction

$In_{0.53}Ga_{0.47}As/In_{0.52}Al_{0.48}As$ multi-quantum wells (MQW's) lattice-matched to InP substrates have long been studied for many applications. However, the band parameters within the narrow quantum wells (QW's) have never been estimated experimentally. So far, only the allowed transitions were observed in most optical measurements. Moreover, the estimation of eigen energies or the energy differences between higher subbands were difficult because of the shortage of peaks in the measured spectra. However, photocurrent measurements using p-i-n diodes including undoped MQW layer made us possible to determine the energy differences mainly because of the presence of internal electric field [1]-[3].

In this paper, the eigen energies in 5- and 10-nm wide InGaAs wells were determined using square-law dependence of the quantum numbers. By combining the data of both well widths, the band offsets and the carrier effective masses for conduction electrons and heavy holes (HH's) were deduced consistently.

2. Experimental

The diodes were MBE-grown $In_{0.52}Al_{0.48}As$ p-i-n junctions including undoped $In_{0.53}Ga_{0.47}As/In_{0.52}Al_{0.48}As$ MQW's in the i-layer. Two wafers (V749 and V706) were measured, in which the well width, L_z, were 5 and 10 nm and the numbers of wells were 33 and 25, respectively. The width of InAlAs barrier was commonly 10 nm. Photocarriers flow normally to the QW plane. For the photocurrent measurements, a reverse bias of -1 V was applied. The internal field was 49-50 kV/cm. Between 0 and -1 V bias, the energy at the fundamental absorption edge shifted only by about 1-3 meV for both specimens. The photocurrent spectra were normalized with the used light intensity spectrum.

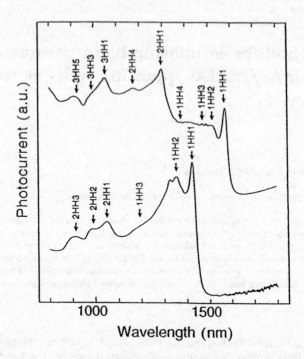

Fig.1 Photocurrent spectra at room temperature. Upper trace; 10-nm well specimen (V706). Lower trace; 5-nm well specimen (V-749). A notation, nHHℓ stands for an optical transition between the ℓ-th heavy-hole subband and the n-th conduction subband.

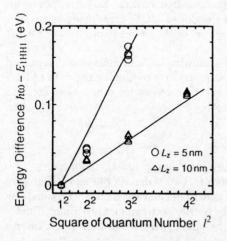

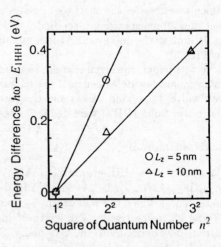

Fig.2 Observed optical transition energy vs. heavy-hole quantum number. Upper straight line; 5-nm well specimen. Lower straight line; 10-nm well specimen.

Fig.3 Observed optical transition energy vs. conduction electron quantum number. Upper straight line; 5-nm well specimen. Lower straight line; 10-nm well specimen.

3. Experimental results

Normalized photocurrent spectra are shown in Fig. 1. The notation, $n\text{HH}\ell$, inserted in the figure stands for a transition between the ℓ-th HH subband and the n-th conduction subband. The peaks due to exciton absorptions and the step-wise current increases were seen at the energy of fundamental transition marked by 1HH1. The energy, $E_{1\text{HH1}}$, was 0.866 eV for 5-nm well specimen and 0.785 eV for 10-nm well specimen.

In 5-nm wide wells (lower trace), all the 6 possible peaks caused by 2 conduction subbands and 3 HH subbands could be observed. Contribution of light hole (LH) was not evident in these experiments [1]. In 10-nm wide wells (upper trace), steplike absorption edges were observed at 1HH1 and 2HH1. Life time broadening possibly coming from LO phonon interactions might cause the dull peaks in due order at 1HH1, 2HH1, and 3HH1. The peak assignment was confirmed later on. All the peaks expected from the consecutive HH quantum number were not observed. At the energies marked by the arrows, tiny peaks were observable when the spectrum was magnified.

Optical transition energies are generally written as $\hbar\omega$ in this paper and the energy differences, $\hbar\omega - E_{1\text{HH1}}$, were plotted as a function of the square of HH quantum number, ℓ, in Fig. 2. In the figure, data of the 1HHℓ transitions ($\ell=1$, 2, 3) were plotted by measuring several p-i-n photodiodes on the same wafer. Using a simple theory, $\hbar\omega - E_{1\text{HH1}}=(\ell^2 - 1)E_{\text{hh}}(1)$ holds for $n=1$, in which HH eigen-energies in a well are written as $E_{\text{hh}}(\ell)$. Within the experimental errors of ±10 meV, the measured points came on the straight lines. The above square law was convincing and the assignment of the quantum numbers to the peaks was validated for both 5- and 10-nm well specimens. The gradients of the two straight lines, $E_{\text{hh}}(1)$, in Fig. 2 were 21 and 7 meV for 5- and 10-nm specimens. The ratio of the gradient was estimated 3.

In a similar manner, the energy differences were plotted as a function of the square of the conduction electron quantum number, n, in Fig. 3. For both specimens, several diodes were measured, though the data coincided as shown in the figure. Using a simple theory, $\hbar\omega - E_{1\text{HH1}}=(n^2 - 1)E_{\text{c}}(1)$ holds for $\ell=1$, in which eigen-energies in a well are written as $E_{\text{c}}(n)$ for conduction electrons. The square law was also applicable in Fig. 3. Therefore, peak assignment in Fig. 1 was validated. The gradients of the straight lines, $E_{\text{c}}(1)$, were 105 and 49 meV for 5- and 10-nm specimens. The ratio of the gradient was 2.1. Knowing those ground-state energies, one can deduce the bandgap energy in the well, $E_{\text{g}}= 0.73\text{-}0.74$ eV, for both specimens.

Since the square law was confirmed, it is considered that the nonparabolicity in conduction subbands or the mixing between HH and LH subbands did not affect much on the energy level determinations. The exciton formation energies are expected 5-15 meV. They are negligible unless they are quite different for each subband.

4. Analysis and discussion

The promising method for obtaining band parameters in QW's is the plot of the band offset vs. effective mass chart. The band parameters were assumed common to both well widths. When the energy-eigen value and the related quantum number are given, one locus can be drawn on the offset vs. mass chart. If the data of different well width are combined, two different loci are drawn and the two intersect. Experimental errors can be treated as the widths of the loci.

According to the process stated above, the following parameters were obtained: (1) The HH effective mass was $0.37\text{-}0.47m_0$ and the band offset of valence subbands was

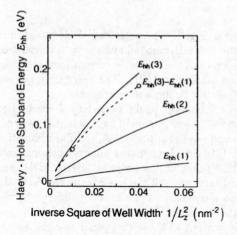

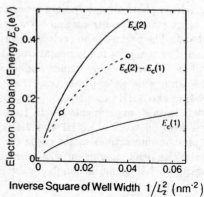

Fig.4 Calculated energies of heavy-hole subbands at the bottom vs. inverse square of the well width.

Fig.5 Calculated energies of conduction subbands at the bottom vs. inverse square of the well width.

0.19-0.31 eV. (2) The effective mass of conduction electrons normal to the QW plane was $0.04m_0$ with 20 % error and the band offset was about 0.5 eV.

Using the above center values, HH subband energies were calculated as a function of the well width, L_z, as shown in Fig.4. The observable energy difference, $E_{hh}(3) - E_{hh}(1)$, could be compared with experiments. The ratio of the values at $L_z=5$ nm and 10 nm shown by open circles in the figure was 3. Conduction subband energies were also shown in Fig. 5. The ratio of the observable energy difference, $E_c(2) - E_c(1)$, at two open circles was 2.2. Those ratios agreed quite well with experiments.

5. Conclusion

Measured eigen-energies for conduction electrons and HH's changed with the square of the respective quantum numbers. The applicability of the simple particle-in-a-box model was exemplified. Band offsets and carrier effective masses in narrow $In_{0.53}Ga_{0.47}As$ /$In_{0.52}Al_{0.48}As$ MQW's were estimated self-consistently.

The authors appreciate Mr. Hitoshi Nakamura of Central Research Laboratory, Hitachi Ltd. for the fabrication of p-i-n photodiodes.

References

[1] Kotera N and Tanaka K 1994 Inst. Phys. Conf. Ser. No 141 (Bristol: IOP Publishing) 233-236

[2] Collins R T, v. Klitzing K, and Ploog K 1986 Phys. Rev. B **33**, 4378-4381

[3] Yamanaka K, Fukunaga T, Tsukada N, Kobayashi K L I and Ishii M 1986 Appl. Phys. Lett. **48** 840-842

Inst. Phys. Conf. Ser. No 145: Chapter 3
Paper presented at 22nd Int. Symp. Compound Semiconductors, Cheju Island, Korea, 28 August–2 September 1995

Determination Of Band Structure Dispersion Curves By Optical Techniques

E. D. Jones, S. K. Lyo, and J. F. Klem

Sandia National Laboratories
Albuquerque, NM 87185-0601, USA

Magnetic field effects upon the photoluminescence spectrum which provide unique information about semiconductor quantum well structures are discussed. Data which provide a simultaneous quantitative measure of both the conduction- and valence-band the energy dispersion curves for an InGaAs/GaAs single-strained-quantum well and a GaAs/AlGaAs lattice-matched single quantum well are presented.

1. Introduction

The semiconductor quantum-well laser is the principal component for optoelectronic applications. The need for custom laser wavelengths, higher laser powers, and better beam quality or control has led to many different designs and configurations. The most common combination for semiconductor devices have been GaAs, InGaAs, and AlGaAs. The requirement for light-hole valence-band masses for optoelectronic devices, lasers, or high-speed electronic devices has been well documented [1-3] and have been mainly achieved in semiconductor structures by compressive biaxial strain in the active quantum layers. This biaxial strain is accomplished by growing layered structures from materials with differing lattice constants, e.g., $In_xGa_{1-x}As$ and GaAs. However, because of level crossing repulsion between the in-plane heavy and light-hole valence-bands, the resulting in-plane valence-band energy dispersion curves are nonparabolic. For lattice-matched quantum well devices, the same conditions hold because of the quantum confinement splitting of the heavy and light-hole valence bands where the in-plane heavy-hole light-hole mixing and hence valence-band nonparabolicity occurs at small values of the wave vector. For devices based upon GaAs and AlGaAs, quantum confinement can be an effective method to alter the energy difference ΔE_{HL} between the heavy and light-hole valence bands. However, for wide (~15 nm) GaAs/AlGaAs structures the heavy-hole light-hole splitting is less than 5meV and thus the mixing is large. For these wide quantum well structures, the in-plane valence-band ground state energy is also "heavy," i.e., for small Fermi energies E_f ~ 2 meV, the valence-band mass $m_v \approx 0.35m_0$. By reducing the quantum-well width, the heavy-hole light-hole energy difference can be increased by quantum confinement and thus reducing the amount of valence-band mixing.

With valence-band nonparabolicity a device which performs at low power levels, i.e, low-carrier densities, may not perform as expected under high power or high-current operation. Device modeling codes predicting the device optical or carrier response should account for any effects which are due to valence-band mixing. Recently, it was demonstrated that a simultaneous determination of the conduction and valence-band energy dispersion curves for modulation doped strained single quantum well (SSQW) [4-7] and lattice-matched single quantum well (LMSQW) [8] structures can be determined from a single sample. These papers discuss the role the energy difference between the heavy and light-hole valence-bands play in determining the degree of mixing. Large values for this energy difference not only give rise to the smallest ground-state in-plane light-hole masses, but also help to reduce the valence-band nonparabolicity.

In this paper we will report on the magnetoluminescence determination of the conduction and valence for two structures, a InGaAs/GaAs SSQW and a LMSQW GaAs/AlGaAS.

2. Experimental

The structures discussed here were prepared using molecular beam epitaxy. The quantum well barrier material was silicon-doped, with a spacing of about 8 nm between the ~3-nm-wide silicon-modulation layer and the quantum well. The SSQW structure (#BC042) consisted of a single 8-nm-wide $In_{0.20}Ga_{0.80}As$ strained quantum-well and unstrained GaAs barriers. The energy difference between the heavy-hole and light-hole valence-bands, which includes contributions from both strain effects and quantum confinement, is about 60 meV. The 4-K two-dimensional carrier concentration N_{2d} and mobility μ were determined by transport measurements and are respectively 5×10^{11} cm^{-2} and 1.2×10^4 cm^2/Vsec. The LMSQW GaAs/AlGaAs structure (#G0260) has a 4.5-nm-wide GaAs quantum well with lattice-matched $Al_{0.25}Ga_{0.75}As$ barriers, and as previously mentioned, the heavy-hole and light-hole valence-band energy difference $\Delta E_{HL} \approx 30$ meV. For the LMSQW and T = 4 K, $N_{2d} = 6.6 \times 10^{11}$ cm^{-2} and $\mu = 2.2 \times 10^4$ cm^2/Vsec. The magnetoluminescence measurements were made in the temperature range of 1.4 and 76 K, and the magnetic fields varied between 0 and 14 T. The luminescence measurements were made with an Argon-ion laser operating at 514.5 nm and CAMAC-based [9] data acquisition system. The direction of the applied magnetic field is parallel to the growth direction with the resulting Landau orbits in the plane of the quantum well. Hence measurements concerning the conduction and valence-band dispersion curves and masses refer to their *in-plane* values.

3. Discussion

A free particle with mass m and charge e moving in a magnetic field B forms quantized states, Landau levels, with an energy $E = (n + 1/2)(e\hbar B/mc) \equiv (n + 1/2)\hbar\omega$, in cgs units, where n is the Landau index, $\hbar$ is Planck's constant over 2π, c is the velocity of light, and $\hbar\omega$ is the cyclotron energy. The distribution function for a degenerate two-dimensional electron gas (conduction-band states for a n-type material) is determined by Fermi-Dirac statistics. However, because of the very small number of photo-induced two-dimensional hole states, the distribution for the valence-band holes are governed by Maxwell-Boltzmann statistics. At high temperatures, where kT is much larger than $\hbar\omega_v$, the n_v = 0, 1, 2, 3, ... valence-band Landau levels are populated and all magnetoluminescence transitions between the n_c and n_v Landau levels obeying the $\delta n_{cv} \equiv (n_c - n_v) = 0$ selection rule are *allowed*. For these high temperatures, the $n_c = n_v \equiv n$ interband luminescence transition energy E(n) is given by

$$E(n) = E_{gap} + \left(n + \frac{1}{2}\right)\left(\frac{e\hbar B}{\mu c}\right),$$ (1)

where E_{gap} is the bandgap energy, here μ is the reduced mass ($\mu^{-1} = m_c^{-1} + m_v^{-1}$) where m_c and m_v are respectively the conduction or valence-band effective masses expressed in terms of the free electron mass m_0.

A schematic showing these *allowed* transitions for a n-type structure is shown in the right side of Fig. 1. The Fermi energy E_f, the bandgap energy E_{gap}, and the Landau level indices n are also indicated in the figure. For large magnetic fields and low temperatures ($\hbar\omega_v \gg$ kT) only the n_v = 0 valence-band Landau level is populated. Here, only the PL transition between the n_c = 0 and n_v = 0 Landau level is *allowed*. Transitions between the higher energy conduction-band Landau levels n_c = 1, 2, 3, ... and the n_v = 0 ground state valence-band Landau

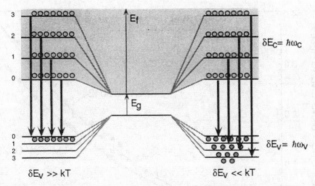

Figure 1. A schematic representation for the Landau levels in a n-type quantum well. The left side is the low temperature case, $\delta E_V \gg kT$, and the right side is the high temperature condition $\delta E_V \ll kT$.

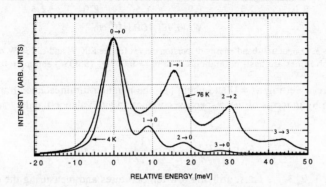

Figure 2. Magnetoluminescence spectra (B = 5.75T) at two temperatures, 4 and 76 K, for an n-type InGaAs/GaAs SSQW showing the *allowed*, $\delta n_{cv} = 0$, and the $\delta n_{cv} \neq 0$ zeroth-order forbidden transitions. The energy axes of the two spectra have been shifted in order to compare them

level are zeroth-order forbidden, but are observable [10, 11] due to higher-order ionized-impurity scattering processes. The left side of Fig. 1 shows the energy level diagram for this case.

Two magnetoluminescence spectra (B = 5.75 T) at 4 and 76 K at are shown in Fig. 2. The origins of the energy axes of the two spectra have been adjusted in order to allow a comparison of the spacing between the magnetoluminescence transitions. For 4 K, E_{gap} = 1330.1 meV while at 76 K, E_{gap} = 1321.1 meV. The indices $n_c \rightarrow n_v$ for each peak are labeled in the figure. As can be seen in Fig. 2, all observed transitions for the 76 K data are *allowed*, i.e, $\delta n_{cv} = 0$, while for the 4.2 K data, the only zeroth-order *allowed* peak is the $0 \rightarrow 0$ transition. Because of the Maxwell-Boltzmann statistics for the holes, the 76 K peak-amplitudes are also governed by Maxwell-Boltzmann distribution function. An analysis of the energy dependence of the peak-amplitudes for this spectrum yields a temperature of about 80 K, in good agreement with the temperature of liquid nitrogen. A theoretical treatment of the energy dependence of the amplitudes of the zeroth-order forbidden transitions has been performed by Lyo [11] and 76-K data shown in Fig. 2 is in good agreement with his calculation.

As can be seen from the left-hand side of Fig. 1, the energy difference δE between the $E(n_c)$ and $E(n_c-1)$ magnetoluminescence peaks depends only on the conduction-band cyclo-

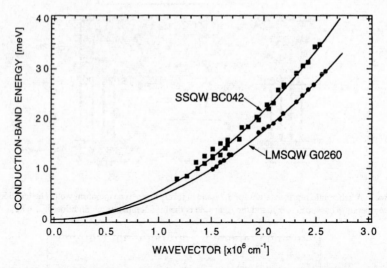

Figure 3. Conduction-band energy dispersion curves for the SSQW and LMSQW samples. The solid lines are best-fit parabolas with $m_c = 0.067m_0$ (SSQW) and $m_c = 0.085m_0$

tron energy $\hbar\omega_c$. Setting $n_v = 0$ and using (1), the magnetoluminescence transition energy $E(n_c)$ is given by in terms of the respective conduction and valence-band cyclotron energies as

$$E(n_c) = E_{gap} + \left(n_c + \frac{1}{2}\right)\hbar\omega_c + \frac{1}{2}\hbar\omega_v, \tag{2}$$

where $n_c = 0, 1, 2, 3, \ldots$ Thus, utilizing low temperatures and measuring the energy differences between the magnetoluminescence peak intensities as a function of magnetic field, (2) provides a method for obtaining the conduction band energy dispersion curves

Figure 3 shows the derived low-temperature-derived conduction-band dispersion curve for the InGaAs/GaAs SSQW. The method (and justification) used relating the magnetic field B to the wavevector k has been adequately discussed [12, 13]. The maximum value of the wavevector k of about 3% of the Brillouin zone is determined by the conduction-band Fermi energy $E_f \sim 35$ meV. The minimum wavevector determinations are limited by our ability to resolve the magnetoluminescence peaks at low magnetic fields. The conduction-band dispersion curve is found to have a small nonparabolic correction with the zone center conduction-band mass $m_c \sim 0.067m_0$ and at the Fermi energy $m_c \sim 0.069m_0$. Nonparabolic effects upon the cyclotron resonance measured conduction-band masses have been discussed in detail [14] and the magnetoluminescence results presented here are in agreement with that derived by conduction-band cyclotron resonance.

The conduction-band dispersion curve for the GaAs/AlGaAs LMSQW sample was also determined by both low temperature magnetoluminescence measurements and the results are also shown in Fig. 3 where a direct comparison with the SSQW data can be made. The conduction-band dispersion curve for the LMSQW structure is again nearly parabolic; however, the zone-center conduction-band mass is $m_c \approx 0.085m_0$. The increase to m_c for the LMSQW conduction-band mass between the bulk GaAs $m_c \approx 0.065m_0$ to $0.085m_0$ is a result [14, 15] of the quantum confinement energy, i.e., increased bandgap energy. In order to verify these masses on our samples, the conduction-band masses for the SSQW and LMSQW structures were measured by far infrared cyclotron resonance techniques with the result that at their

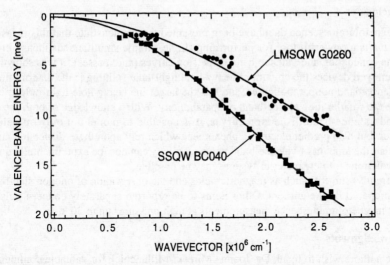

Figure 4. Comparison of the valence-band energy dispersion curves for SSQW and LMSQW samples. The solid lines are a fit of (3) to the data.

respective Fermi energies, $m_c \approx 0.069m_0$ for the SSQW sample and $m_c \approx 0.082m_0$ for the LMSQW structure in excellent agreement with the magnetoluminescence data.

With a knowledge of the conduction-band dispersion curve, (1) can be used to derive the valence-band dispersion curve by performing 76-K magnetoluminescence measurements and analyzing the zeroth-order *allowed* transition peaks as a function of magnetic field. The valence-band data for the SSQW and the LMSQW structures are shown in Fig. 4 and it is evident that the dispersion curves are nonparabolic.

For the range of data shown in the figure, the valence-band effective mass m_v for the SSQW structure varies between a zone-center $0.11m_0$ to about $0.3m_0$ at $E_v(k) \sim 20$ meV. The amount of valence-band nonparabolicity can be quantified by analyzing the valence-band dispersion curve data by a function of the form:

$$E(k) = \left(\frac{\hbar^2 k^2}{2m_v^0} \right) \left[1 - \frac{K}{E_{gap}} \left(\frac{\hbar^2 k^2}{2m_v^0} \right) \right], \tag{3}$$

where m_v^0 is the zone-center mass and K is the nonparabolic term. The solid lines in Fig. 4 are a best-fit of (3) to the data. Using (3), for the SSQW yields $m_v^0 = 0.11m_0$ and $K \sim 16$, a value which is about an order of magnitude larger than measured for the conduction band [14]. For the LMSQW sample ($E_{gap} \sim 1605$ meV) at small wavevectors, the zone-center valence-band mass $m_v \sim 0.15m_0$ and for larger wavevectors, e.g., $E_v \approx 10$ meV, $m_v \approx 0.35m_0$. It is obvious from Fig. 4, that the valence-band dispersion curve for the SSQW structure is more parabolic than that of the LMSQW sample and this result [4, 5] is due to the increased heavy-hole light-hole energy separation in the SSQW sample. The nonparabolic term for the LMSQW $K \sim 35$ is nearly two times larger than found for the SSQW structure, and is due to the smaller heavy-hole light-hole energy separation ~ 30 meV. Magnetoluminescence determined valence-band dispersion curves for other SSQW samples with varying indium concentration, therefore with differing values for the heavy-hole light-hole valence band splitting confirm that as the energy difference increases, the valence-bands become more parabolic and that the ground state in-plane valence-band mass remains relatively constant

4. Conclusions

Magnetoluminescence data have been presented that demonstrate that this measurement technique is a powerful tool for performing single-sample simultaneous measurements of both the conduction and valence-band dispersion curves (and masses). It was shown that for quantum-well devices the amount of heavy-hole light-hole splitting is the determining factor for valence-band nonparabolicities, and that the larger the heavy-hole light-hole energy difference the smaller the valence-band nonparabolicity. With a knowledge of both the conduction and valence-band dispersion curves, it is possible to model not only optoelectronic behavior, but also predict electronic phenomena which rely upon these dispersion curves. In particular, the amount of valence-band nonparabolicity can now be experimentally quantified and predictions for new material systems may be possible.

Currently, subjects such as magnetic fields and the observation of phonon side-bands are being prepared by the authors. Other items to be reported separately concern pulsed field (~60T) magnetoluminescence measurements at high pressures up to 30 kbar.

Acknowledgments

The authors wish to thank Dr. Joanna Mirecki-Millunchick for valuable comments. This work was performed at Sandia National Laboratories and supported by the Division of Material Science, Office of Basic Energy Science, U. S. DOE, No. DE-AC04-94AL8500.

References

[1] Osbourn G C , Gourley P L, Fritz I J , Biefeld R M, Dawson L R, and Zipperian T E 1987 *Principles And Applications of Semiconductor Strained-Layer Superlattices in Semiconductors and Semimetals*, Edited by R. Dingle, Vol. **24** (New York: Academic Press)

[2] Pearsall T P 1990 *Strained-Layer Superlattices in Semiconductors and Semimetals*, Edited by T. P. Pearsall, Vol. **32** (New York: Academic Press)

[3] Morkoc H , Sverdlov B, and Gao Guang-Bo 1993 *Proceedings of the IEEE* **81** 493

[4] Jones E D, Lyo S K , Fritz I J , Klem J F , Schirber J E , Tigges C P , and Drummond T. J 1989 *Appl. Phys. Lett.* **54** 2227

[5] Jones E D, Biefeld R M, Klem J F , and Lyo S K 1990 *Proceedings Int. Symp. GaAs and Related Compounds, Karuizawa, Japan, 1989, Inst. Phys. Conf. Ser. No.* **106** 435

[6] Jones E D , Dawson L R , Klem J F , Lyo S K, Heiman D , and Liu X C 1994 *The Applications of High Magnetic Fields in Semiconductor Physics (SEMIMAG-94), August 8-12, 1994, Boston* (Singapore: World Scientific Publishing)

[7] Jones E D , Dawson L R, Klem J F , Lyo S K , Heiman D, and Liu X C 1995 *Proceedings of the 22nd International Conference of the Physics of Semiconductors*, Edited by D. J. Lockwood, pp. 1492-1494 (Singapore: World Scientific Publishing)

[8] Jones E D , Lyo S K, Klem J F, Schirber J E, and Lin S Y1992 *Proceedings. Int. Symp. GaAs and Related Compounds, Seattle, USA, 1991, Inst. Phys. Conf. Ser. No.* **120** 407

[9] Jones E D and Wickstrom G L 1985 *Proceedings Southwest Conference on Optics, SPIE* **540** 362

[10] Lyo S K, Jones E D, and Klem J F 1988 *Phys. Rev. Lett.* **61** 2265

[11] Lyo S K 1989 *Phys. Rev. B* **40** 8418

[12] Lyo S K and Jones E D 1990 *Electronic, Optical, and Device Properties of Layered Structures*, Edited by J. R. Hayes, M. S. Hybertsen, and C. R. Weber, Fall 1990 Meeting of the Materials Research Society, Boston, MA, pp. 271-274

[13] Lyo S K, and Jones E D 1992 *Proceedings Int. Symp. GaAs and Related Compounds, Seattle, 1991, Inst. Phys. Conf. Ser. No.* **120** 583

[14] Singleton J, Nicholas R J, Rogers D C, and Foxen C T B 1989 *Surface Science* **196** 429

[15] Osório F A P, Degani M H, and Hipólito O 1989 *Superlattices and Microstructures* **6** 107

Inst. Phys. Conf. Ser. No 145: Chapter 3
Paper presented at 22nd Int. Symp. Compound Semiconductors, Cheju Island, Korea, 28 August–2 September 1995
© *1996 IOP Publishing Ltd*

Strain effect on direct- and indirect-gap band lineups of GaAs$_{1-x}$P$_x$/GaP quantum wells

Akio Shima†, Hiroyuki Yaguchi†, Kentaro Onabe†, Yasuhiro Shiraki‡, and Ryoichi Ito†

†Department of Applied Physics, The University of Tokyo, 7-3-1 Hongo, Bunkyo-ku, Tokyo 113, Japan

‡Research Center for Advanced Science and Technology(RCAST), The University of Tokyo, 4-6-1 Komaba, Meguro-ku, Tokyo 153, Japan

Abstract. Strain effect on direct- and indirect-gap band lineups of strained GaAs$_{1-x}$P$_x$/GaP quantum wells with a wide range of x is studied using photoluminescence (PL) and photoreflectance (PR) spectroscopy. By comparing the transition energies obtained by PL and PR measurements with the calculation based on the effective mass approximation, it is found that the band lineups of strained GaAs$_{1-x}$P$_x$/GaP QWs on GaP substrates at any x are of type I at the X and Γ point and that the valence band offset of heavy hole is $506(1-x)$ meV. As a result, it is deduced that biaxial compressive strain makes the energy gap of GaAs$_{1-x}$P$_x$ alloys grown on GaP substrates indirect at any x.

1. Introduction

Recently, there has been considerable interest in heterostructures composed of different semiconductors with unequal lattice constants because the strain due to the difference in the lattice constants is expected to offer semiconductor materials unique electrical and optical properties. The GaAs$_{1-x}$P$_x$ strained-layer system has been widely studied since Osbourn proposed the concept of strained-layer superlattices (SLSs)[1]. In this system, GaAs$_{1-x}$P$_x$/GaP heterostructures are particularly interesting because the band lineup is considered to change depending on the strain configuration. For example, Pistol *et al.*[2] and Hara *et al.*[3] reported that in the strained GaAs$_{1-x}$P$_x$/GaP quantum wells (QWs) grown on GaP (100) substrates, the band lineup at the X point is of type I. On the other hand, Gourley and Biefeld[4] pointed out that in a GaAs$_{1-x}$P$_x$/GaP SLS grown on the GaAs$_{1-x}$P$_x$ layer with average phosphorus composition of SLS, the band lineup at the X point is of type II. Although the strain dependence of the band offset is considered to lead to this difference, it has not yet been investigated experimentally.

In this paper, we report the phosphorus composition dependence of direct- and indirect-gap band lineups of strained $GaAsP_{1-x}P_x/GaP$ QWs grown on GaP (100) substrates using photoluminescence (PL) and photoreflectance (PR) spectroscopy. We studied the strain effect on the band lineups by changing the phosphorus composition x in the well layer. In this structure, only the $GaAs_{1-x}P_x$ well layer is compressively strained to accommodate the lattice mismatch between the $GaAs_{1-x}P_x$ layer and the GaP substrate. In the $GaAs_{1-x}P_x$ layer, this compressive strain splits the valence band into two states, heavy- and light-hole states, which are degenerate at the Γ point without strain. This strain makes the electron–heavy-hole band gap smaller than the electron–light-hole band gap.

2. Experiments

Samples used in this study were strained $GaAs_{1-x}P_x/GaP$ multiple QWs (MQWs) grown on GaP(100) substrates by metalorganic vapor phase epitaxy (MOVPE) using trimethyl-gallium, AsH_3 and PH_3 as sources at 650°C ($0.23 \leq x \leq 0.82$) and 630 °C ($0 \leq x \leq 0.15$). At other temperatures, we could not succeed in the growth of good samples because of the interface roughness caused by islanding to reduce the strain energy at higher temperatures, and because of the insufficient decomposition of PH_3 at lower temperatures. The strained MQW structures consist of 5 periods of GaP barriers and $GaAs_{1-x}P_x$ wells. The barrier width L_b was fixed at 800 Å. The well width L_w and the phosphorus composition x were varied from 11 Å to 31 Å and from 0 to 0.82, respectively. The layer thickness and phosphorus composition x were determined by double-crystal x-ray diffraction. PL and PR measurements were carried out in standard lock-in configuration using He-Cd laser (325 nm) as an excitation source. The samples were cooled down to 6 K (PL) and 120 K (PR) in a closed cycle cryostat.

3. Results

Figure 1 shows the PL spectra of MQWs ($L_w = 22$ Å) with various x. The two peaks are assigned to no-phonon (NP) and longitudinal-acoustic (LA)-phonon-assisted interband transitions. The energy difference between the NP and the LA-phonon-assisted transitions agrees well with the LA phonon energy at the X point obtained by neutron scattering measurements[5]. This indicates that the transitions observed by the PL measurements are indirect-gap (X point) transitions. It can be seen that with decreasing x, the peaks shift toward lower energies, the trend of which is in agreement with the band gap of $GaAs_{1-x}P_x$ alloy at the X point. Also, the feature of the excitation power and temperature dependence of the PL is similar to that of the PL from bulk $GaAs_{1-x}P_x$ alloys[6]. This result suggests that the observed PL transition occurs only in the $GaAs_{1-x}P_x$ layer. Thus, we can suppose the indirect-gap band lineup of strained $GaAs_{1-x}P_x/GaP$ MQWs is of type I at any x, in which both electrons and holes are confined in the $GaAs_{1-x}P_x$ layer.

At the composition from $x = 0.10$ to 0.82, the luminescence of the two transitions are clearly seen and the intensity changes little. On the other hand, at $x = 0$ and 0.06, the luminescence cannot be clearly observed from the MQWs with the well width of 22 Å, though it can be observed from those with narrower well width, $L_w = 11$ Å.

This can be explained in terms of nonradiative recombination centers, such as the misfit dislocations generated by strain relaxation when the layer thickness exceeds the crytical layer thickness.

Figure 2 shows the PR spectra of MQWs ($L_w = 22$ Å) with various x. The signals around 2.89 eV originate from the GaP layer, and the features clearly seen on the lower energy side are the optical transitions related to the MQW levels between the $n = 1$ conduction subband and the $n = 1$ valence subband of heavy hole (1e-1hh) at the Γ point. It can also be seen that with decreasing x, the signals shift toward lower energies, the trend of which is in agreement with the band gap of GaAs$_{1-x}$P$_x$ alloy at the Γ point. This ensures that the signals on the lower energy side originate from the GaAs$_{1-x}$P$_x$ well layer. In order to derive the transition energies from the PR spectra, we used first-derivative Lorentzian functions[7]. The arrows in this figure represent the 1e-1hh transition energies obtained by the fitting.

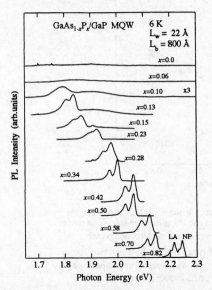

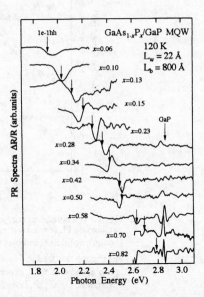

Fig. 1   6 K PL spectra of strained GaAs$_{1-x}$P$_x$/GaP MQWs (L_w=22 Å, L_b=800 Å) with various x. NP and LA-phonon-assisted transitions are clearly seen at $x \geq 0.10$

Fig. 2   120 K PR spectra of strained GaAs$_{1-x}$P$_x$/GaP MQWs (L_w=22 Å, L_b=800 Å) with various x. The signals clearly seen on the lower energy side are the 1e-1hh transitions.

4. Discussion

Composition dependence of the NP transition energy (closed circles) observed by the PL measurements, together with the 1e-1hh transition energy (open circles) observed by the PR measurements, is shown in Fig. 3. In order to examine the band lineups both at the X point and at the Γ point, we have calculated the transition energies based on the effective mass approximation. In this calculation, we have taken the strain effect into account using deformation potentials[8,9]. Solid lines in Fig. 3 are the calculated transition energies assuming that the band lineups are of type I both at the X and Γ point at any x and that the valence band offset of heavy hole (ΔE_{vhh}) depends linearly on the phosphorus composition of the strained $\text{GaAs}_{1-x}\text{P}_x$ layer as follows; $\Delta E_{\text{vhh}} = 506(1-x)$ meV. The theoretical calculation can explain well the PL and PR transition energies, as shown in Fig. 3. For other well widths, $L_w = 11$ Å, 31 Å, the calculation is also in good agreement with experimental results. The offset, $\Delta E_{\text{vhh}} = 506(1-x)$ meV, agrees well with the result reported previously, $\Delta E_{\text{vhh}} = 89$ meV at $x = 0.83$[10].

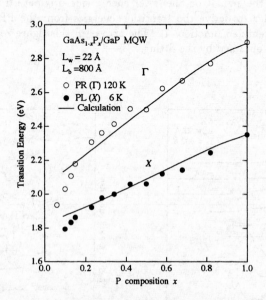

Fig. 3   Phosphorus composition dependence of the transition energies obtained by PL (closed circles, X point) and PR measurements (open circles, Γ point). Solid lines indicate the transition energies calculated on the assumption that the band lineups are of type I both at the X and the Γ point at any x, and that ΔE_{vhh} depends linearly on the phosphorus composition of the strained $\text{GaAs}_{1-x}\text{P}_x$ layer as follows; $\Delta E_{\text{vhh}}=506(1-x)$ meV.

Taking into account that the heavy hole valence band offset is given by $\Delta E_{\text{vhh}} = 506(1-x)$ meV, we have determined the X and Γ point band lineups of strained $\text{GaAs}_{1-x}\text{P}_x/\text{GaP}$ heterointerfaces on GaP substrates using deformation potentials, as shown in Fig. 4. Solid and dashed lines represent the calculated results with and without strain, respectively. As for the X point, the compressive strain splits three equivalent X bulk-band minima into two kinds of band minima, X_{001} and $X_{100,010}$. The $X_{100,010}$ minima of the conduction band in the $\text{GaAs}_{1-x}\text{P}_x$ layer become lower than the X minima in the GaP layer and consequently type-I band lineup is formed. This is different from the Gourley and Biefeld's work[4], in which both the $\text{GaAs}_{1-x}\text{P}_x$ layer and the GaP layer are strained oppositely and the lowest X minima are different between $\text{GaAs}_{1-x}\text{P}_x$ and

GaP, $X_{100,010}$ and X_{001}, respectively.

In addition, the $X_{100,010}$-conduction band minima are always lower in energy than the Γ-conduction band minima, as can be seen from this figure. Thus, the compressive strain is found to make the energy gap of $GaAs_{1-x}P_x$ alloys even at $x < 0.44$ indirect, which is direct without strain. It can also be seen that this compressive strain changes the band lineup into type I at both the X and the Γ point, which would be of type II at the X point if $GaAs_{1-x}P_x/GaP(100)$ QW structure were grown without this strain.

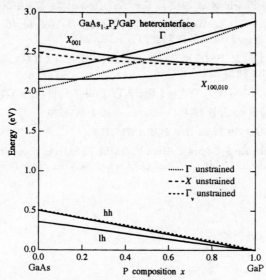

Fig.4 Band lineups of strained $GaAs_{1-x}P_x/GaP(100)$ heterointerfaces on GaP substrates as a function of the phosphorus composition. Solid and dashed lines represent the calculated results with and without strain, respectively.

5. Conclusions

In conclusion, we investigated the phosphorus composition dependence of direct- and indirect-gap band lineups of strained $GaAs_{1-x}P_x/GaP$ QWs grown on GaP (100) substrates using PL and PR spectroscopy. We compared the transition energies obtained by PL and PR measurements with the calculation based on the effective mass approximation, and found that the band lineups of strained $GaAs_{1-x}P_x/GaP$ QWs at any x are of type I at the X and Γ point and that the valence band offset of heavy-hole is $506(1-x)$ meV. As a result, it is deduced that biaxial compressive strain makes the energy gap of $GaAs_{1-x}P_x$ alloy indirect at any x.

Acknowledgements

The authors would like to acknowledge S. Miyoshi, W. Pan, K. Takemasa and S. Hashimoto for assistance in MOVPE growth, K. Ota for useful discussion, S. Ohtake for his technical support. We are also pleased to acknowledge S. Fukatsu and N. Usami for valuable suggestions and advice. This work was partly supported by a Grand-in-Aid for Scientific Research from the Ministry of Education, Science and Culture, Japan.

408

References

[1] Osbourn G C 1982 *J. Appl. Phys.* **53** 1586–1589

[2] Pistol M E , Leys M R and Samuelson L 1988 *Phys. Rev.* **B37** 4664 –4670

[3] Hara Y, Yaguchi H, Onabe K, Shiraki Y and and Ito R 1994 *Inst. Phys. Conf. Ser.* **136** 361–366

[4] Gourley P L and Biefeld R M 1982 *J. Vac. Sci. Technol.* **21** 473–475

[5] Yarnell J L, Warren J L, Wenzel R G and Dean P J 1968 *in U. S. Atomic Energy Commission Report No. LA-DC-9228* (unpublished).

[6] Oueslati M, Zouaghi M, Pistol M E, Samuelson L, Grimmeiss H G and Balkanski M 1985 *Phys. Rev.* **B32** 8220–8227

[7] Shanabrook B V, Glembocki O J and Beard W T 1987 *Phys. Rev.* **B35** 2540–2543

[8] Kleiner W H and Roth L H 1959 *Phys. Rev. Lett.* **2** 334–336

[9] Van de Walle C G 1989 *Phys. Rev.* **B39** 1871–1883

[10] Yaguchi H, Hashimoto S, Sugita T, Hara Y, Onabe K, Shiraki Y and Ito R 1994 *Extended Abstracts of the 1994 International Conference on Solid State Devices and Materials* 108–110

Inst. Phys. Conf. Ser. No 145: Chapter 3
Paper presented at 22nd Int. Symp. Compound Semiconductors, Cheju Island, Korea, 28 August–2 September 1995
© *1996 IOP Publishing Ltd*

Enhancement of carrier confinement in pseudomorphic $In_xGa_{1-x}As/GaAs$ strained quantum wells using interfacial AlAs layers

C. D. Lee, J.Y. Leem, and S. K. Noh

Epitaxial Semiconductor Group, Materials Evaluation Center, Korea Research Institute of Standards and Science, Taedok Science Town, Taejon 305-600, Korea

Abstract. Pseudomorphic $In_xGa_{1-x}As/GaAs$ quantum well structures with one or two monolayer of interfacial AlAs inserted between the $In_xGa_{1-x}As$ layers and the GaAs barriers were studied. The growth parameters were determined using reflection high-energy electron diffraction(RHEED) intensity oscillations in a molecular beam epitaxy(MBE) system. The presence of two monolayer-wide interfacial AlAs was confirmed through high resolution transmission electron microscope(HRTEM). In photoluminescence(PL) spectra a large spectral blue shift of the excitonic transitions was observed.

1. Introduction

Recently the optical and electronic properties of strained $In_xGa_{1-x}As/GaAs$ quantum well (QW) systems have been extensively studied(Reithmaier J.-P 1991). This system is a potentially attractive material from the device prospects in quantum well infrared photodiode(QWIP) fabrication and high electron mobility transistor (HEMT) application.

In $In_xGa_{1-x}As/GaAs$ systems, high quality strained layers can be grown provided that their thicknesses are below the critical layer thickness (CLT) above which the strain is relieved by the formation of misfit dislocations(Price G.L.1991). For thicknesss above CLT, the quality of the epitaxial layers is seriously degraded, affecting the device performance. Therefore, to avoid the generation of misfit dislocations, the well width should be narrowed below the CLT at certain indium compositions. However in the case of thinner well the wavefunctions of electron and hole is easily spread out the barrier region because of the carrier leakage at the heterointerface(Reithmaier J.-P 1991). This delocalization is expected to produce both a reduction of the exciton binding energy and an increase of the radiative time constants. From the point of efficient carrier confinement in a narrow QW our motivation is to enhance the carrier capture efficiency below the CLT by controlling the interface structures.

In this work we examined the nonconventional $In_xGa_{1-x}As/GaAs$ QW heterostructures which is similar to the optical Fabry-Perot interferometer. The growth and optical properties in a nonconventional structures grown by molecular beam epitaxy(MBE) were studied through reflection high-energy electron diffraction (RHEED), high-resolution transmission electron microscope (HRTEM) and photoluminescence (PL) spectroscopy.

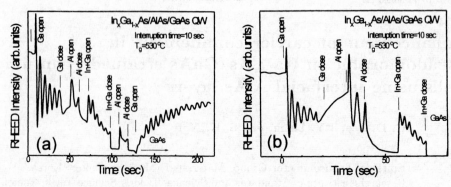

Figure 1. RHEED intensity oscillations:(a) one monolayer AlAs (b) two monolayer AlAs.

2. Experimental

The quantum-well structures investigated in this work were grown in a MBE system on two-inch diameter semi-insulating (001)-oriented GaAs substrates. The substrates were outgassed for 30 min at 200 °C in the entry chamber of the MBE and then moved to the transfer chamber. In the growth chamber, for oxide desorption each sample was heated 590 °C for 10 min. Then the substrate temperature was lowered to the growth temperature of 580 °C. The films were grown using the tetramer arsenic source As_4 and its pressure was high enough to ensure 2x4 As stabilized surface reconstruction. Initially, a 0.5 -μm-thick buffer layer was grown at a growth rate of 1 -μm/h and an As_4 to Ga beam-equivalent pressure of 20. After growing a GaAs buffer layer, four $In_{0.14}Ga_{0.86}As$ QWs of widths 15, 70, 120, and 220 Å were grown between 1000-Å-wide GaAs barriers. One samples is a reference conventional QW's. In the other samples, each well is bordered on each side by ultrathin AlAs layers, of width one monolayer. Cross-sectional specimens for HRTEM studies were prepared in the usual manner by ion milling. For PL measurements, the samples were mounted in a liquid He cryostat and cooled to 4.2 K. The PL spectra were recorded with a computer-controlled data aquisition system with included a double grating spectrometer coupled to a cooled GaAs photomultiplier tube. The light source for excitation, the 488 nm line of an Ar^+ laser was used.

3. Results and discussion

Intensity oscillations of the specular electron beam in the RHEED patterns from (001)-2x4 reconstructed surfaces, [110] azimuth were monitored during all of the work reported here. The oscillations of the specular beem under the constant As_4 pressure at a subtrate temperature of 530 °C are displayed in Fig.1.

The shutters for the Al, Ga and In cells were opened or closed at the times indicated by arrows. A growth interruption of 10 seconds leads to recovery of the RHEED intensity after Al, Ga and In shutters were closed. The damping behavior of RHEED intensity oscillation during the growth of well layer was observed due to the lattice mismatch between the $In_xGa_{1-x}As$ and the GaAs layers(Elman B. 1991). After the growth

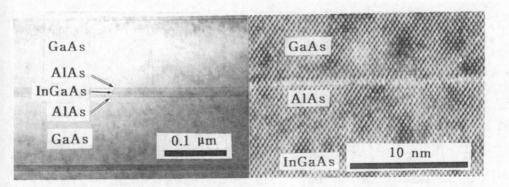

Figure 2. Cross-sectional HRTEM micrograph of $In_{0.14}Ga_{0.86}As$ QWs:(a)with a low magnification (b) with a high magnification.

of $In_xGa_{1-x}As$/GaAs strained-layer QWs with interfacial AlAs layers, the thickness of AlAs layers were investigated through HRTEM. The HRTEM micrograph from $In_xGa_{1-x}As$/GaAs QWs with interfacial AlAs layers is shown in Fig.2. No dislocations or stacking faults in the film and its interface with a low magnifications are displayed. It means that the InGaAs quantum wells were grown pseudomorphically on the GaAs layers. In Fig. 2(b) the sharp contrast between the AlAs and GaAs layers shows that AlAs is well confined. Thus from the image of HRTEM with a high magnification the presence of 2 monolayer AlAs layers are clearly confirmed.

It is well-known that the growth interruption techique seriously affet the interface quality of quantum structures. For 30 second growth interruption the PL spectra of QWs which have the same growth parameters are shown in Fig.3. The PL spectra of the conventional structures shows a broad band due to the longer interruption time. However when the one monolayer-wide AlAs layer was inserted the PL intensities are greatly enhanced about two orders of magnitude. It should be noted that the AlAs layer improve the interface quality. Fig. 4 shows the 4.2 K PL spectra of four $In_{0.14}Ga_{0.86}As$ QWs of width 15, 70, 120, 220 Å with(a) or without AlAs layers(b) for 10 second interruption time. The PL intensities of each QWs show strong and the FWHM(full width at half maximium) is about 2-3meV. The most evident feature of these spectra is the large blue shift of the excitonic recombination energy due to the insertion of the ultrathin AlAs layers. For example the carrier confinement energy of the 70-Å-wide QW with one-monolayer AlAs layers is of the same order as that of a 20-Å-wide single quantum well. Comparing with the sample of 20-Å-wide $In_{0.14}Ga_{0.86}As$/GaAs QW the intensity of 70-Å-wide QW with AlAs layers is strong.

Futhermore it should be noted that the integrated intensitiy is almost of the same order dispite of increasing the confinement energy. It means that the carrier leakage of a nonconventional structure is smaller than that of a conventional one which has the same confinment energy. When the ultrathin AlAs interfacial layers is inserted the quantum mechanical reflectivity at QWs interface is increased. It prevents the carrier capture from being obscured by the other relaxation processes. In summary pseudomorphic $In_xGa_{1-x}As$/GaAs QW structures with two monolayer of interfacial AlAs inserted between the $In_xGa_{1-x}As$ layers and the GaAs barriers were studied to enhance the quantum confinement. Alloy composition and layer thickness of the samples grown by

412

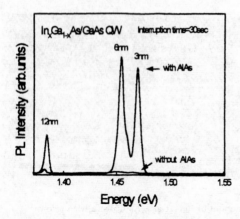

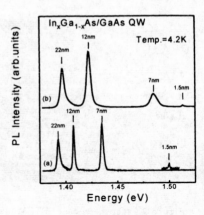

Figure 3. PL spectra of 30sec growth interrupted samples.

Figure 4. Low-temperature PL spectra from $In_{0.14}Ga_{0.86}As/GaAs$ QW's (a) with interfacial AlAs layer (b) without interfacial AlAs layers.

MBE were determined by using RHEED intensity oscillations and the presence of 2 monolayer-wide AlAs were confirmed by HRTEM. In PL spectra of $In_xGa_{1-x}As/GaAs$ QWs with 30 sec interruption the interface quality was greatly improved by inserting the AlAs layers. In PL spectra of $In_xGa_{1-x}As/GaAs$ QWs with interfacial AlAs a large spectral blue shift of the excitonic transition was observed. Results are shown that the interfacial AlAs layer is very effective to increase the carrier confinement energy and enhance the carrier capture than a conventional narrow quantum well.

Acknowledgment

The authors wish to thank Dr. H.J.Lee for HRTEM measurements.

References

Elman B.,Koteles S.,Melman P.,Jagannath C., Lee J., and Dugger D. 1991 J.Appl.Phys **70** 2640

Fujiwara A.,Takahashi Y.,Fukatsu S.,Shiraki Y.,and Ito R. 1995 Phys.Rev.B **51** 2291

Price G.L. 1991 Phys.Rev.Lett. **66** 469

Reithmaier J.-P. R.Hoger R.,H.Riechert H. 1991 Phys.Rev.B **43** 4933

Inst. Phys. Conf. Ser. No 145: Chapter 3
Paper presented at 22nd Int. Symp. Compound Semiconductors, Cheju Island, Korea, 28 August–2 September 1995
© *1996 IOP Publishing Ltd*

Strain and Crystallographic Orientation Effects in (11ℓ) (In,Al,Ga)As/GaAs Heterostructures: Physics and Device Applications

E. Towe, P. A. Ramos, and R. H. Henderson

Laboratory for Optics and Quantum Electronics
University of Virginia, Charlottesville, Virginia 22903-2442 USA

D. S. McCullum, X. R. Huang, and A. L. Smirl

Laboratory for Photonics and Quantum Electronics
University of Iowa, Iowa City, Iowa 52242 USA

Abstract. This paper discusses the influence of crystallographic orientation on some optical properties of III-V compounds. Examples of how some of these orientation-dependent properties can be used to design novel optical devices are presented.

1. Introduction

The III-V alloy compounds have emerged as an important class of semiconductors with a wide range of applications in modern electronics and optoelectronics. These semiconductors have been synthesized both in bulk and epitaxial form. The most important applications of the semiconductors have been in their epitaxial form. Today, most of the epitaxial layers are grown by the advanced crystal growth technique of metalorganic chemical vapor deposition or molecular beam epitaxy. For historical as well as practical reasons, the majority of the layers for both fundamental and device applications have been grown on substrates oriented in the [001] crystallographic axis. There is no fundamental reason for this choice of crystalline orientation other than the fact that the (001) surface is one of the most basic planes in the cubic lattice; the other planes being the {110} and the {111}. All other surfaces with higher Miller indices can be resolved on an atomic scale into stepped structures built up from combinations of these basic planes. Perhaps the most important reason for the widespread use of the (001) surface is the ease of epitaxial growth on it. A departure from this surface is expected to lead to different growth chemistries and kinetics. From the chemical point of view, one expects the chemistry of the important technological processes for III-V semiconductors to depend on the crystallographic orientations. Among these processes are: epitaxial growth, doping, etching, and interaction with the ambient. In this paper, however, we will focus our attention on the effects of orientation that impact the optical properties. At the most elementary level, these effects can be shown to be a consequence of the nature of the chemical bond between the two atoms of the zinc-blende basis.

2. Epitaxial growth

The epitaxy of high quality (In,Ga,Al)As heterostructures on several surfaces of the $\{11\ell\}$ family of planes by molecular beam epitaxy has been demonstrated [1,2]. In general, it has been found that the optimal growth conditions for each non-(001) surface differ from those on the conventional (001) surface. We describe below the conditions that we have used for the growth of heterostructures on the (110) and the (112) surfaces; these surfaces are members of the $\{11\ell\}$ family for $\ell = 0$ and $\ell = 2$, respectively.

Previous work on the growth of (Al,Ga)As layers on exact (110) surfaces was not very successful. In some of the studies conducted, the epitaxial growths reported exhibited instabilities [3] and morphological imperfections [4]. To circumvent some of these problems, we have used vicinal (110) GaAs substrates [5]. We find that GaAs or (In,Al,Ga)As layers with high quality surface morphologies can be grown at the nominal substrate temperature of 625°C with group V to group III beam equivalent pressure (BEP) ratios of 30 or higher. When the layers are to be doped with silicon, a judicious choice of the right combination of substrate temperature and arsenic over-pressure conditions must be used to avoid amphotericity. The details of the silicon doping conditions for the vicinal (110) surface have been reported elsewhere [5].

For the growth of GaAs or (In,Al,Ga)As layers on (112)B GaAs substrates, the nominal substrate temperature is about 620°C and the group V to group III BEP ratio is about 25. These conditions yield epi-layers with reasonably high quality surface morphology and efficient photoluminescence emission.

3. Physics

The optical characteristics of thin epitaxial layers on substrates oriented in directions other than the [001] exhibit properties that are unique to the particular orientations chosen. These properties may be due to the effects of crystallographic orientation, lattice mismatch-induced strain or quantum confinement. The properties may also be due to a combination of these effects. One example of a property that depends on the crystallographic axis is the optical transition matrix element. For the two crystallographic orientations we will consider here— the [110] and the [112]—one can grow quantum well structures that exhibit intrinsic in-plane optical anisotropy. The anisotropy is associated with the reduced symmetry of the structures on these surfaces when compared to the (001) surface. We define the optical anisotropy to be the difference in optical transition strengths for light polarized along two orthogonal in-plane directions. The magnitude of the anisotropy can be dramatically increased by incorporating bi-axial strain in the quantum well layers. With strain, the light- and heavy-hole states are no longer pure eigenstates of the angular momentum as they are in the case for structures on the (001) or the (111) surface. Instead, for the (110) and the (112) surfaces, the light- and heavy-hole states are admixtures of the $J_z = \pm 3/2$ and $J_z = \pm 1/2$ components of the angular momentum eigenvalues. With increasing strain, the admixtures tend to increase the anisotropy.

Another consequence of growing strained thin films on non-(001) surfaces is a manifestation of a strain-induced piezoelectric polarization. All III-V semiconductors are piezoelectric; but the piezoelectricity is only observed in certain symmetry directions. For quantum well heterostructures on certain of the (11ℓ) surfaces, it has been shown that the piezoelectric polarization can lead to an *intrinsic* quantum-confined Stark effect [6]. This effect and the inherent optical anisotropy of quantum well structures grown on these surfaces is used in the design and operation of the polarization rotation modulator discussed in the sequel.

Finally, another physical property which can be altered by simply changing the crystallographic orientation is the second-order nonlinear susceptibility. Since III-V semiconductors belong to the $\bar{4}$3m symmetry class, the polarization of the electric fields which interact to affect the nonlinear polarizability is important. Growth of epitaxial layers in certain crystallographic surfaces allows one to select the pertinent light polarizations for the nonlinear generation of harmonic fields. This orientation-dependent property can be used in the design of blue-green light emitters.

The important theme which emerges from the foregoing discussion is that symmetry plays a fundamental role in the physics of the III-V semiconductors. All the unique orientation-dependent properties exhibited by these compounds can be traced directly to the crystalline symmetry. A discussion of the role of crystallographic orientation in III-V epitaxial layers has been given by Henderson and Towe [7].

4. Devices

The generation of blue or green light by active or passive methods has recently become a very active area of research. Among the several approaches being pursued is the generation of the blue or green light by nonlinear interactions of other color light beams in solid materials.

Second-harmonic generation of the green light has been demonstrated in III-V heterostructures on GaAs substrates oriented in the [001] crystal direction; this process has been effected by the nonlinear interaction of counter-propagating, appropriately polarized fundamental fields coupled into waveguide structures [8]. One drawback of the structures reported, however, was that counter-propagating transverse electric (TE) and transverse magnetic (TM) fields were required for the nonlinear generation of the green, second-harmonic light. This requirement is a consequence of the crystalline symmetry on the (001) surface.

Theoretical considerations show that the requirement for the presence of both TE- and TM-polarized fields in the III-V semiconductor structures can be removed if substrates oriented in crystallographic directions other than the conventional [001] are used [9]. A suitable alternative substrate orientation is the [112]; we have found that structures grown on (112) substrates exhibit large nonlinearities.

We have fabricated special passive waveguide structures on the (112)B GaAs substrate for the nonlinear generation of blue light. The basic structures are of the nonlinear, antiresonant, reflecting optical waveguide type. This type of waveguide has a large core region which can still operate in a single-mode régime [10]. From the GaAs substrate up, the typical structure consists of: a 0.75-μm cladding layer of $Al_{0.80}Ga_{0.20}As$ followed by a 0.3-μm cladding layer of AlAs. These layers satisfy the antiresonance reflection conditions for the fundamental field. Following this, a multi-layered structure composed of 12 periods of $Al_{0.40}Ga_{0.60}As/AlAs$ was grown. Each layer of the period is one-half the wavelength of the second harmonic field. The purpose of the half-wave structure is to provide quasi-phase-matching for the surface-bound second-harmonic field. A typical guide structure is shown in Fig. 1.

The structures were tested by launching ~15 mW of TE-polarized light at $\lambda_0 = 900$ nm into the guide. The optical field reflected from the back facet of the guide provided the second field necessary for the nonlinear interaction in the generation of the second-harmonic field. It is estimated that about 4.5 mW was returned into the guide from the back facet reflection. For equal optical powers coupled into a (112) and a (001) waveguide structure, it was found that higher relative second-harmonic power was generated in the (112) waveguide structure. Specifically, the relative intensity ratio of the second-harmonic power for the (112) waveguide, $P^{2\omega}_{(112)}$, to that of the (001) guide, $P^{2\omega}_{(001)}$, that is: $P^{2\omega}_{(112)}/P^{2\omega}_{(001)}$ was about 2.3; we believe this number is in reasonable agreement with the theoretically calculated value of about 2.7 if allowances are made for losses. Figure 2 shows a photograph of the surface-emitted second-harmonic light from the (112) waveguide structure.

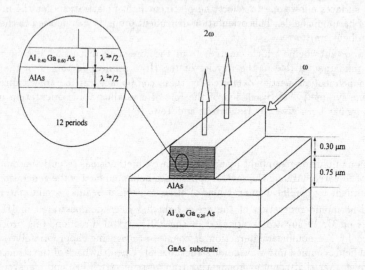

Figure 1. A passive nonlinear, antiresonant, reflecting optical waveguide for second-harmonic blue light generation.

One use of the optical anisotropy found in quantum well structures grown on the (110) surfaces is in optical modulators with high contrast ratios. A typical modulator structure we have studied consists of a 1-μm-thick n^+-GaAs buffer layer grown on a semi-insulating (110) vicinal GaAs substrate; this layer is followed by a 100-nm-thick n-type GaAs layer. An intrinsic region containing 50 periods of 6-nm-thick $In_{0.13}Ga_{0.87}As$ quantum wells separated by 8-nm-wide GaAs barriers is then grown; on top of this region is grown a 100-nm-thick p-type GaAs cap layer.

Figure 2. Photograph of the second-harmonic blue light emitted from a (112) guide structure.

To demonstrate the anisotropy inherent in this structure, the room temperature absorption coefficients for light polarized along two in-plane, orthogonal directions on the (110) surface were measured; the two orthogonal directions are the [$\bar{1}10$] and the [001]. The peak difference in absorption for the two polarizations is about 6.6×10^3 cm^{-1} near the excitonic line center.

The anisotropy of this structure can be used to construct the modulator system shown in the inset of Fig. 3(a). The system consists of the [110]-oriented quantum well structure located between a quarter wave plate and an analyzer on one side and a polarizer on the other. The quantum well structure is located at the intersection of a pump and a probe beam—with the probe beam aligned at 45° to the [$\bar{1}10$] and the [001] axes of the quantum well structure. Both the pump and the probe are pulsed laser beams, with 2 ps pulses at a repetition rate of 500 kHz. During the modulator operation, the probe pulse is set to arrive about 30 ps after the pump in order to allow for the evolution of the optical nonlinearity.

The minimum transmission of the probe beam—the "off" state—is obtained by adjusting the quarter wave plate and the analyzer for minimum light throughput in the absence of the pump. During pump illumination—the "on" state—the photogenerated carriers bleach the exciton, which changes the polarization state of the light transmitted through the quantum well structure. During this period, the fraction of the probe light transmitted to the detector is increased. Note that the quarter wave plate is required in the system because the optical anisotropy in the quantum well structure causes the incident light to be elliptically polarized after passing through it. The quarter wave plate converts the elliptical polarization to linear polarization.

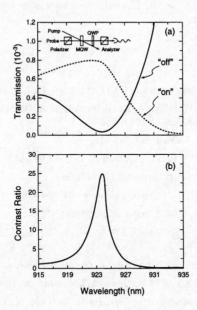

Figure 3. Transmission and contrast ratio characteristics of the polarization rotation modulator.

We define the contrast ratio of the modulator as the amplitude of the transmitted light in the *presence* of the pump beam divided by the transmitted amplitude in the *absence* of the pump as a function of wavelength. The contrast ratio for the device structure measured in this experiment is 25:1 at room temperature. This is shown in Fig. 3(b). A polarization rotation modulator which operates on similar principles like the one discussed here has been reported by Shen *et al.* [11]. The necessary optical anisotropy for its operation, however, was introduced by a uniaxial in-plane strain on (001) (Al,Ga)As/GaAs quantum well structures. The uniaxial strain was created by lifting the quantum wells off the GaAs substrate and re-attaching them to a foreign substrate. The disadvantage of this method is that it is fabrication-intensive.

5. Summary

We have discussed how the crystalline symmetry of III-V compounds affects some optical properties of these semiconductors. Some of the orientation-dependent optical properties can be used in the design of novel optoelectronic devices. In particular, we have presented a novel blue-green light emitter fabricated on the (112) surface with a higher second-harmonic generation efficiency than a similar device on the conventional (001) surface. We have also discussed a polarization rotation modulator with a potential for high contrast ratio.

Acknowledgments

This work was supported by the NSF through a National Science Foundation Young Investigator Award to E. Towe and by NASA. The authors acknowledge contributions to the earlier phases of this work by D. Sun.

References

[1] Beery J G, Laurich B K, Maggiorem and Smith D L 1989 *Appl. Phys. Lett.* **54** 233

[2] Hiyamizu S, Shimomura S, Wakejima A, Kaneko S, Adachi A, Okamato Y, Sano N and Murase J 1994 *J. Vac. Sci. Technol. B* **12** 1043-1046

[3] Wang W 1983 *J. Vac. Sci. Technol. B* **1** 630-636

[4] Ballingall J M and Wood C E C 1982 *Appl. Phys. Lett.* **41** 947-949

[5] Sun D and Towe E 1993 *J. Cryst. Growth* **132** 166-172

[6] Sun D and Towe E 1994 *IEEE J. Quant. Electronics* **30** 466-470

[7] Henderson R H and Towe E 1995 *J. Appl. Phys.* **78** 2447-2455

[8] Vakhshoori D, Walker J, Wang S, Smith J S, Soccolich S E, and Islam M N 1989 *Appl. Phys. Lett.* **54** 1725-1727

[9] Whitbread N D and Robson P N, 1994 *IEEE J. Quantum Electron.* **30** 139-147

[10] Baba T and Kokubun Y 1992 *IEEE J. Quantum Electronic.* **28** 1689-1700

[11] Shen H, Pamulapati J, Wraback M, Taysing-Lara M, Dutta M, Kuo H C and Lu Y 1993 *IEEE Photon. Technol. Lett.* **6** 700-702

Inst. Phys. Conf. Ser. No 145: Chapter 3
Paper presented at 22nd Int. Symp. Compound Semiconductors, Cheju Island, Korea, 28 August–2 September 1995
© *1996 IOP Publishing Ltd*

Characterization of lattice-mismatched InGaAs/AlGaAs heterointerface by modified isothermal capacitance transient spectroscopy

Shigekazu Izumi, Teruyuki Shimura, Norio Hayafuji,

Takuji Sonoda, and Saburo Takamiya

Optoelectronic & Microwave Devices Laboratory, Mitsubishi Electric Corporation,

4-1 Mizuhara, Itami, Hyogo 664 , Japan

Abstract. Modified isothermal capacitance transient spectroscopy is utilized for the characterization of a lattice-mismatched InGaAs/AlGaAs heterointerface, which is inevitable in an AlGaAs/GaAs heterojunction bipolar transistor with a non-alloyed InGaAs ohmic contact layer. Anomalous signals are obtained under an isothermal condition in which the edge of the depletion layer of the p-n junction reaches the InGaAs/AlGaAs interface. It is found that considering another path for the time dependent electron emission is effective for understanding the extremely sharp signal obtained. The temperature dependence of the signals is also well ascertained by using this modified explanation. There are two activation energies (0.62 and 0.26 eV) for multi-level transitions. The modified theory of the isothermal capacitance transient spectroscopy is useful for characterizing the electronic features of the lattice-mismatched heterointerface.

1. Introduction

AlGaAs/GaAs heterojunction bipolar transistors (HBTs) are being developed, for such purposes as the high-power applications [1,2], due to the specific advantages for high power density and high power added-efficiency. A reduction of the parasitic resistance by using InAs or InGaAs contact layers is essential in order to obtain high power and high speed operation as well as reliability. The first work was reported by Woodall et al.[3] concerning an attempt to obtain ohmic contact with low resistivity using InGaAs grown by molecular beam epitaxy (MBE). They realized a low contact resistivity of between 5×10^{-7} and 5×10^{-6} Ωcm^2, which was obtained for a Ag/n-InGaAs/n-GaAs MESFET structure. Peng et al.[4] also showed the non-alloyed ohmic contact with a resistivity of less than 8.5×10^{-8} Ωcm^2 by using n+-InAs/InGaAs and InAs/GaAs strained layer superlattices (SLSs) on GaAs grown by MBE. They also reported excellent performance HBT [5] with the lowest value of about 1.5×10^{-8} Ωcm^2. On the other hand, employing a lattice-mismatched system, such as InGaAs on GaAs or AlGaAs, often brings an inferior surface [6,7] and heterointerface when the layer thickness was over the critical layer thickness for dislocation generation. Although considering the lattice-mismatched systems and their interfaces is inevitable to realize excellent performance of the high-power HBTs, the knowledge is still in short. In this paper, interfacial quality of InGaAs on graded $Al_xGa_{1-x}As$ (x=0.3 to 0) is discussed based on a modified isothermal capacitance transient spectroscopy (ICTS) measurement, which is one of the most useful methods for studying the nature of deep levels in semiconductors.

2. Experimental procedure

Figure 1 shows a schematic drawing of the sample structure studied. The emitter comprised of an $Al_{0.3}Ga_{0.7}As$ flat layer and an $Al_xGa_{1-x}As$ graded layer (x varies from 0.3 to zero), both of which were doped with Si up to 5×10^{17} cm^{-3}. The base comprised of a Be-doped GaAs layer with the carrier concentration of 4×10^{19} cm^{-3}. The thicknesses of the emitter and the base layers were 1500 and 700 Å, respectively. The top layer for non-alloyed ohmic contact comprised of 500 Å thick Si-doped ($N_d = 0.1 \sim 4 \times 10^{19}$ cm^{-3}) graded $In_yGa_{1-y}As$ (y=0~0.5) and 50 nm thick Si-doped $In_{0.5}Ga_{0.5}As$ ($N_d = 4 \times 10^{19}$ cm^{-3}). Ti/Mo/Au non-alloyed ohmic contact electrodes with 300 µm diameter were formed on the InGaAs surface. The

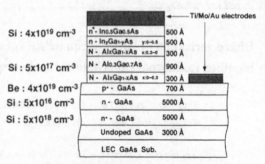

Fig.1. Schematic drawing of the AlGaAs/GaAs HBT structure studied here.

epitaxial layers for the AlGaAs/GaAs HBT structure were grown on (100) oriented LEC semi-insulating GaAs substrates by the conventional MBE at growth rates of 0.3~1.0 µm/h. An arsenic passivation treatment of the GaAs substrates was applied just prior to growth. The arsenic passivation of the GaAs substrate is a more effective pre-treatment than any other wet chemical treatments [8] for reducing accumulated impurities which cause unintentional carrier accumulation. The growth temperature was 630 °C for the base and emitter layers, 450 °C for InGaAs and 600 °C for the other layers, respectively. The V/III ratio was about 20 for n-GaAs layers and 20~40 for the p$^+$-GaAs, AlGaAs and InGaAs layers, respectively. The temperature and duration of the thermal treatment just prior to growth were 640 °C and 10 minutes, respectively. ICTS measurements were performed using a 10 msec filling pulse of 1 V under bias voltage from -1 V to -7 V with a HORIBA DA-155 ICTS system. The capacitance variation followed well from 100 µsec to 100 sec tuning.

3. Basic theory of ICTS

Okushi [9] has proposed a basic theory of ICTS, which suggests the theoretical curve for the single level. The transient response of electron emission from a deep level in the semiconductor, based on conventional theory, is ascribed to be as follows:

$$\frac{dn_T}{dt} = -\frac{n_T}{\tau}. \tag{1}$$

This differential equation can be solved under the boundary conditions at t=0 and $N_T^+ = 0$ as follows:

$$N_T^+(t) = N_T - n_T = N_T\{1 - \exp(-\frac{t}{\tau})\}. \tag{2}$$

where

$$n_T = N_T \exp(-\frac{t}{\tau}). \tag{3}$$

Here, N_T is the deep level concentration; n_T and N_T^+ are the electron concentration which is trapped by the deep level and the ionized deep level concentration, respectively; τ is the electron emission time constant from the deep level. The capacitance ($C(t)$) of the base/emitter p-n junction (E-B junction) is expressed as follows:

$$C(t) = \sqrt{\frac{q\varepsilon_o\varepsilon_s\{N_s + N_T^+(t)\}}{2(V_D + V)}} = C_\infty\sqrt{1 - N\exp(-\frac{t}{\tau})}. \tag{4}$$

where N_S is the shallow level concentrations; C_∞ and N are defined as

$$C_\infty \equiv \sqrt{\frac{q\varepsilon_o\varepsilon_s(N_s + N_T)}{2(V_D + V)}} \qquad (5)$$

and

$$N \equiv \frac{N_T}{N_s + N_T} \qquad (6)$$

The following function in defined to understand the ICTS signals;

$$f(t) = C^2(t) - C_\infty^2 = -C_\infty^2 N \exp(-\frac{t}{\tau}) \qquad (7)$$

Differentiating eq. (8) and multiplying the results by t, the ICTS signal is defined as

$$S(t) = t\frac{df(t)}{dt} = \frac{NtC_\infty^2}{\tau}\exp(-\frac{t}{\tau}) \qquad (8)$$

where S(t) is the ICTS signal. The symbol τ is expressed as

$$\tau = \tau_0 \exp(\frac{\Delta E_A}{kT}) \qquad (9)$$

The features of the ICTS are summarized as follows: (i) S(t) $\propto$ N $\propto$ N_T, (ii) the maximum value appears when t=τ, (iii) only the peak position of the signal changes without changing the shape and height of the spectrum when τ changes dependent on the surrounding temperature, (iv) multi-levels can be recognized as a τ variation.

4. Results and discussion

Figure 2 shows the ICTS signal dependence on the bias voltage for the E-B junction. An anomalous signal was obtained under an isothermal condition for a -7 V bias voltage. The edge of the depletion layer of the p-n junction reached the interface of InGaAs on GaAs at this bias voltage. This signal is thought to have been due to the interfacial feature of InGaAs on AlGaAs. The fitting curve calculated based on the conventional model is also put in fig.2. The experimental ICTS signal was narrower than that conventionally calculated. It is impossible to explain why the full width at half maximum (FWHM)

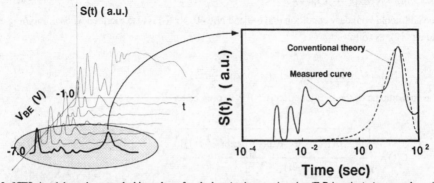

Fig.2. ICTS signal dependence on the bias voltage for the base/emitter p-n junction (E-B junction). An anomalous signal is obtained under an isothermal condition at -7 V bias voltage. The edge of the depletion layer of the E-B junction reaches the interface of InGaAs and AlGaAs at this bias voltage.

of ICTS was narrower than that calculated, though we can explain using conventional theory that the FWHM of ICTS might be wider, since there are many signals (traps). Figure 3 shows the temperature dependence of the ICTS signals along with the basic theoretical curve. Although the conventional ICTS signal also suggests that only the peak position changes without any change in the shape of the spectrum when the surrounding temperature changes, the shape and height of the ICTS signal changed due to the surrounding temperature. In the high temperature range, in particular, the peak intensity of the ICTS signal did not saturate, even when extending the pulse. This phenomenon suggests that dominant transmission occurs in different ways based on the range of the surrounding temperature. Although the ICTS has been demonstrated to be an attractive

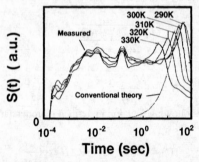

Fig.3. Temperature dependence of the ICTS signal. One of the features of the conventional ICTS signal is that only the peak position changes without changing the shape of the spectrum when the surrounding temperature changes, although the shape and height of the ICTS signal are changed with any surrounding temperature change.

method for deep level detection in semiconductors, the basic theory is not perfect when multi-levels exist. An analysis of the crystal quality of a lattice-mismatched system, such as that used in this study, requires an additional consideration.

In order to understand the narrowing of the FWHM of ICTS , another path for electron emission has been newly considered, which is not constant with time (increasing with time). Since the vacancy ratio of another path depends on the time, the emissive electron concentration per unit time ($[e]_{emt}$) can be written as

$$[e]_{emt} = \gamma \cdot t n_T. \tag{10}$$

where γ is the probability of vacancy generation depending on time. The vacancy behaves as the through path for electrons. This suggestion is well supported by the fact that the peak intensity of the ICTS signal does not saturate even when extending an additional pulse. The ICTS signal is given as follows when a multi-path exists:

$$\frac{dn_T}{dt} = -\frac{n_T}{\tau} - \gamma \cdot t n_T. \tag{11}$$

This differential equation can be solved under the boundary condition that t=0 and $n_T = N_T$, and t=∞ and $n_T = 0$ as follows:

$$n_T = N_T \exp(-\frac{t}{\tau})\exp(-\frac{\gamma}{2}t^2). \tag{12}$$

The additional boundary condition that t=0 and $N_T^+ = 0$. $N_T^+(t)$ is also ascribed when combining eq.(2) and eq.(12), as follows:

$$N_T^+(t) = N_T - n_T = N_T\{1 - \exp(-\frac{t}{\tau})\exp(-\frac{\gamma}{2}t^2)\}. \tag{13}$$

The capacitance of the E-B junction (C(t)) is expressed as

$$C(t) = \sqrt{\frac{q\varepsilon_o\varepsilon_s[N_s + N_T\{1 - \exp(-\frac{t}{\tau})\exp(-\frac{\gamma}{2}t^2)\}]}{2(V_D + V)}}$$

$$= C_\infty \sqrt{1 - N \exp(-\frac{t}{\tau}) \exp(-\frac{\gamma}{2} t^2)}. \tag{14}$$

According to the equations, modified f(t) (fN(t)) and the ICTS signal (SN(t)) are ascribed, respectively as follows:

$$fN(t) = C^2(t) - C_\infty^2 = -C_\infty^2 N \exp(-\frac{t}{\tau}) \exp(-\frac{\gamma}{2} t^2), \tag{15}$$

and

$$SN(t) = t\frac{dfN(t)}{dt} = NC_\infty^2 (\frac{t}{\tau} + \gamma \cdot t^2)[\exp(-\frac{t}{\tau})\exp(-\frac{\gamma}{2} t^2)]. \tag{16}$$

Figure 4 shows the fitting results. In this figure both curves of the conventional and newly developed models are arranged along with the measured one. Our model agrees better with the measured value under even such a lattice-mismatched condition than the conventional one. The γ in eq.(17) is also temperature dependent, as follows :

$$\gamma = \gamma_0 \exp(-\frac{\Delta E_B}{kT}). \tag{17}$$

The temperature dependence of the modified theoretical curves (as in eq.(16)) in account with eqs. (9) and (17) well explains the ICTS signal narrowing and temperature dependence. The anomalous level is ascribed to the additional other emission of the electrons, originated the multi-channeling transmission and resulting part of current flows along this path. This phenomenon is thought be caused by a hopping transposition by the dislocation generated at a lattice-mismatched InGaAs/AlGaAs interface. One schematic of the electron emission paths in this model is shown in fig.5. The transmission paths of A and B show the normal and through path transmission, respectively, assuming that the transmission path of B_1 is reinforced by the vacancies in some levels for the through path. Since the vacancy ratio of the level depends on time, the emissive electron concentration per unit time can be written as eq.(10). The modified ICTS signal calculation results in two activation energies (fig.6): the path B is dominant in the lower temperature range (~310 K), while the path A is dominant in the higher temperature range (310~330 K). The activation energy of the path B (through path transmission, ΔE_B) and the path A (normal electron emission, ΔE_A) are 0.26 and 0.62 eV, respectively. The density of the path A is 6×10^{14} cm^{-3}. Watson et al. [10] showed that the concentration of the deep level traps scaled linearly with dislocation density by deep level transient spectroscopy (DLTS)[11] . Chen N [12] also suggested that the distribution of the DX centers was quite similar to that of

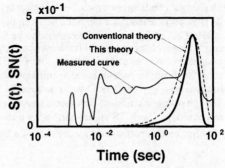

Fig.4. Fitting results of ICTS calculated by both the modified and conventional models along with the measured curve. Our model agrees better with the measured value, even under the lattice- mismatched condition, than the conventional one.

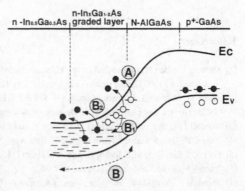

Fig.5. One of the schemes on another electron emission model near the E-B junction to understand the FWHM of ICTS narrowing . Transmission paths A and B_2 show normal emission and multi-channeling electron transmission, respectively. Transmission path B_1 is reinforced by the vacancies in some levels for multi-channeling.

424

dislocations. The value of 0.62 eV obtained in this study is thought to be one of the deep levels in InGaAs layer (path A) affected by a lattice-mismatch. The value of 0.26 eV is thought to be the through path transmission (path B) of a lattice-mismatched system caused by dislocation generation.

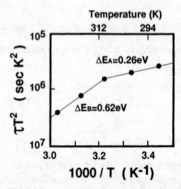

Fig.6 Measured activation energies of path B (through path emission, ΔE_B) and path A (electron emission, ΔE_A). The activation energies for paths A and B are 0.62 and 0.26 eV, respectively.

5. Summary

A modified ICTS has been successfully applied for the characterization of a lattice-mismatched InGaAs/AlGaAs system. An anomalous ICTS signal is obtained under an isothermal condition in which the depletion layer of the E-B junction reaches the interface between the InGaAs and AlGaAs. This signal is ascertained to be due to the concerning with the interfacial features caused by dislocation in a lattice-mismatched InGaAs/AlGaAs system. In order to understand why the FWHM of the ICTS signal becomes narrower, we have considered another path for electron emission which increases with time. The temperature dependence has also been well explained by this new model. Two activation energies have been found: one is the through path transmission; the other is normal electron emission. Their values are 0.26 and 0.62 eV, respectively. The ICTS signal narrowing and temperature dependence are well explained by this modified model, particularly for the case of a lattice-mismatched system, such as InGaAs/AlGaAs.

Acknowledgments

The authors would like to thank Sakai M at Fukami Patent Office, Iuchi Y and Obatake H of HORIBA Ltd. for helpful advice on the ICTS measurements and fruitful discussion.

References

[1] Shimura T, Sakai M, Izumi S,Nakano H, Matsuoka H, Inoue A, Udomoto J, Kosaki K, Kuragaki T, Takano H, Sonoda T, Takamiya S1995 IEEE Trans. on Electron Devices **42** 1-7.

[2] Sakai M, Shimura T, Izumi S, Matsuoka H, Kosaki K, Hattori R, Komaru M, Takano H, and Otsubo M 1993 Proc. Int. Conf. on Solid State Devices and Materials, Chiba Makuhari Messe, p.989.

[3] Woodall J M, Freeouf J L, Pettit G D, Jackson T, and Kirchner P 1981 J.Vac.Sci. & Technol.,**19** 626.

[4] Peng C K, Ji G, S.Kumar N, and Morkoç H 1988 Appl. Phys. Lett. **53**, 900.

[5] Gao G, Selim Unlu M, Morkoç H, and Blackburn D L 1991 Electron Devices, **38** 185.

[6] Yoon S F 1993 J.Vac.Sci. & Technol., **B11** 562.

[7] Shiraishi Y, Yoshida T, Furuhata N, and Okamoto A 1992 Proc. Int. Symp. on GaAs and Related Compounds, Karuizawa, p. 639.

[8] Izumi S, Yoshida N, Takano H, Nishitani K, Otsubo M 1993 J. Cryst. Growth **133** 123.

[9] Okushi H, Tokumal Y 1981 Jpn. J. Appl. Phys. **20** 261.

[10] Watson G P, Ast D G, Anderson T J, Pathangey B, and Hayakawa Y 1993 J Appl. Phys. **71**, 3399.

[11] Lang D V 1974 J. Appl. Phys. **45** 3023.

[12] Chen N, Zhang X, and Den Y 1995 J. Cryst. Growth **148** 219.

Inst. Phys. Conf. Ser. No 145: Chapter 3
Paper presented at 22nd Int. Symp. Compound Semiconductors, Cheju Island, Korea, 28 August–2 September 1995
© *1996 IOP Publishing Ltd*

Interband Transitions in $Cd_xZn_{1-x}Te/ZnTe$ Strained Single Quantum Wells Grown by Double-Well Temperature-Gradient Vapor Deposition

H. L. Park and S. H. Lee

Department of Physics, Yonsei University, Seoul 120-749, Korea

T. W. Kim

Deprtment of Physics, Kwangwoon University447-1 Wolgye-dong, Nowon-ku, Seoul 139-701, Korea

Abstract. The electronic states in a $Cd_xZn_{1-x}Te/ZnTe$ strained single quantum well grown by the simple method of double-well temperature-gradient vapor deposition have been investigated both experimentally and theoretically. X-ray diffractometry measurements were performed to characterize the structural properties of the $Cd_xZn_{1-x}Te/ZnTe$ quantum well. Reflectivity and photoreflectance measurements showed several resonant excitations in the $Cd_xZn_{1-x}Te$ quantum well, and photoluminescence measurements on the strained quantum well structures showed that the sharp excitonic transition from the 1st electron subband to the 1st hevey-hole band. Interband transition energies in the $Cd_xZn_{1-x}Te$ quantum well were calculated by taking into account the strain effects.

1. Introduction

Recently, rapid advancements in several kinds of the epitaxial layer growth technologies have made possible the fabrication of II–VI semiconductor heterostructures, multiple quantum wells, and superlattices with interfacial abruptneses on the scale of a few lattice constants [1-5]. New unique physical phenomena have been investigated [2], and novel quantum devices have been fabricated utilizing them [3]. II–VI semiconductor quantum wells offer the promising possibility of tuning the energy gap to cover the spectrum range from ultra violet to the far infrared [6-8]. Among these nanostructures, even though there are inherent problems due to the different lattice mismatches and thermal coefficients, a $Cd_xZn_{1-x}Te/ZnTe$ quantum well is particularly interesting due to its many possible applications for optoelectronic devices in the blue-green region of the spectrum. Although a few papers concerning $Cd_xZn_{1-x}Te/ZnTe$ quantum wells have been published [2], all these investigations were performed on samples grown by molecular beam epitaxy. Data on a $Cd_xZn_{1-x}Te/ZnTe$ single quantum well (SQW) with high-quality interfaces grown via a simple method have not been reported yet. Furthermore, a clear demonstration of intersubband transitions in the $Cd_xZn_{1-x}Te$ quantum well has not yet succeeded due to the delicate problems of sample growth [2].

This paper presents data for $Cd_{0.027}Zn_{0.973}Te/ZnTe$ and $Cd_{0.053}Zn_{0.947}Te/ZnTe$ SQWs which were grown by the simple technique of double-well temperature-gradient vapor-transport deposition (DWTGVTD) at a chamber pressure of 10^{-6} Torr on semi-insulating (100) GaAs. Even though the growth of the $Cd_xZn_{1-x}Te/ZnTe$ SQWs have been performed in the various Cd mole fractions, only the physical properties of the $Cd_{0.027}Zn_{0.973}Te/ZnTe$ and $Cd_{0.053}Zn_{0.947}Te/ZnTe$ SQWs are reported. X-ray diffraction measurements were carried out to demonstrate the heteroepitaxy of $Cd_xZn_{1-x}Te$ and ZnTe films. Reflectivity, photoreflectance (PR), and photoluminescence (PL) measurements were performed to investigate the interband transitions in the $Cd_xZn_{1-x}Te/ZnTe$ strained quantum wells. Furthermore, the electronic subband energies were calculated by an envelope function approximation method taking into account the strain effects.

2. Experimental Details

Polycrystalline stoichiometric ZnTe and CdTe with purities of 99.9999% grown by the Bridgman method were used as source materials and were precleaned by repeated sublimation. Cr-doped semi-insulating GaAs was degreased in warm trichloroethylene (TCE), rinsed in deionized water thoroughly, etched in a HF solution, and rinsed in TCE again. As soon as the chemical cleaning process was finished, the wafers were mounted onto a molybdenum susceptor. Before $Cd_xZn_{1-x}Te/ZnTe$ SQW growth, the GaAs substrates were thermally cleaned at 600℃ for 5 min in situ in a growth chamber at a pressure of 10^{-6} Torr. The deposition was done at a substrate temperature of 320℃ and a system pressure of 10^{-6} Torr. The ZnTe source temperature for the ZnTe layer was 540℃. The ZnTe and the CdTe source temperatures for the $Cd_{0.027}Zn_{0.973}Te$ layer were 525 and 430℃, respectively, and those for $Cd_{0.053}Zn_{0.947}Te$ were 475 and 430℃, respectively. The growth rate for the ZnTe was 1.4 μm/h, and those of $Cd_{0.027}Zn_{0.973}Te$ and $Cd_{0.053}Zn_{0.947}Te$ layers were 0.96 μm/h and 0.94 μm/h, respectively. A detailed schematic diagram of the single-well TGVTD system was reported previously [4], and the DWTGVTD system is a modified version of the one based on the TGVTD with two effusion cells. The PR measurements were performed using a 75-cm monochromator equipped with a 250-W Xe lamp as a probe source and the 4880-Å line of an Ar^+ ion laser as a modulation source. The PL measurements were carried out using a 75-cm monochromator equipped with an RCA 31034 photomultiplier tube. The samples were mounted on a cold finger in a cryostat and kept either at room temperature or 20 K throughout the experiment.

The typical SQW sample used in this study was grown on a GaAs substrate with a buffer layer of $Cd_xZn_{1-x}Te$ followed by a layer of ZnTe. It consisted of 100-Å $Cd_xZn_{1-x}Te$ well with 255-Å ZnTe barriers. The ZnTe and $Cd_xZn_{1-x}Te$ films grown on GaAs (100) substrates by DWTGVTD had mirrorlike surfaces without any indication of pin holes, which was confirmed by Normarski optical microscopy and scanning electron microscopy. In this structures, the well material was $Cd_xZn_{1-x}Te$ which is a zinc-blende direct-gap semiconductor that has been studied by bulk Raman spectroscopy [9, 10]. The mole fraction of Cd was determined by X-ray diffraction (XRD) and PL measurements, and the thickness of the each layer was determined by cross-sectional transmission electron microscopy.

3. Results and Discussion

The results of the PL, reflectance and the PR spectroscopy measurements on $Cd_{0.027}Zn_{0.973}Te/ZnTe$ and $Cd_{0.053}Zn_{0.947}Te/ZnTe$ SQWs at 20 K are shown in Figs. 1 and 2, respectively. The reflectance spectra for two SQWs in the low-energy region corresponds to the Fabry-Perot oscillation, as shown in Figs. 1 and 2. The peaks observed from the PR measurements on a $Cd_{0.027}Zn_{0.973}Te/ZnTe$ SWQ obtained from the least-square fit to a third-derivative function were 2.3818, 2.3788, 2.3744, 2.3700, 2.3634, 2.3581, and 2.3501 eV, and those on a $Cd_{0.053}Zn_{0.947}Te/ZnTe$ SQW were 2.3818, 2.3706, 2.3633, 2.3509, 2.3420, 2.3241, 2.3162, and 2.3158 eV, respectively. The peaks appeared at the PL spectrum on a $Cd_{0.027}Zn_{0.973}Te/ZnTe$ SQW were 2.3416 and 2.3041 eV, and those on a $Cd_{0.053}Zn_{0.947}Te/ZnTe$ SQW were 2.3171, 2.2922, and 2.2754 eV, respectively.

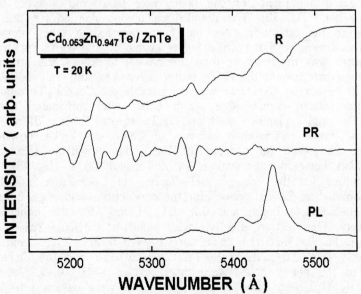

Fig. 1 The results of the photoluminescence, reflectivity and the photoreflectance measurements for a $Cd_{0.053}Zn_{0.947}Te/ZnTe$ single quantum well.

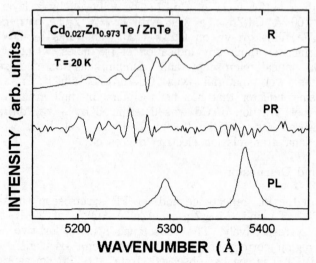

Fig. 2 The results of the photoluminescence, reflectivity and the photoreflectance measurements for a $Cd_{0.027}Zn_{0.973}Te/ZnTe$ single quantum well.

To identify the origin of each peak, the electronic subband energies were calculated taking into account the effect of the strain due to the lattice mismatch between $Cd_xZn_{1-x}Te$ and ZnTe. The residual strain changes the magnitude of the band offset and splits the degenerate valence band into a type- I band alignment for the heavy-hole band and a type II for the light-hole band [11, 12]. The detailed calculation method for the electronic subbands in the quantum wells has been reported in other literature [13]. dSince the thickness of the ZnTe buffer grown on the GaAs substrate to act as a barrier was much larger than the critical thickness, the strain between the GaAs substrate and the ZnTe buffer layer was fully relaxed. Also, since the top ZnTe barrier thickness was comparable the $Cd_xZn_{1-x}Te$ quantum well width, the biaxial compressive strain effect was considered only in the $Cd_xZn_{1-x}Te$ single quantum well and ZnTe layers. Since $Cd_xZn_{1-x}Te$ alloys consist of lattice mismatched binaries of CdTe and ZnTe, bowing effects should be also taken into account [8]. Using the known constants and linear interpolation between their values in ZnTe and CdTe [14, 15], the band discontinuities for the $Cd_{0.027}Zn_{0.973}Te/ZnTe$ and $Cd_{0.053}Zn_{0.947}Te/ZnTe$ SQWs were estimated as 23 and $46meV$ for the conduction band edge, 5 and $11meV$ for the heavy-hole band, and 0.7 and $0.15meV$ for the light-hole band, respectively. The electron effective mass values of the $Cd_{0.027}Zn_{0.973}Te$ and the $Cd_{0.053}Zn_{0.947}Te$ well and the ZnTe barrier were 0.1212, 0.1205, and 0.116 m_e, respectively [15]. The $Cd_{0.027}Zn_{0.973}Te$, $Cd_{0.053}Zn_{0.947}Te$, and the ZnTe effective masses of the heavy holes were taken to be 0.4481, 0.4502, and 0.6 m_e, respectively [15], and those of the light holes were assumed to be 0.1782, 0.1766, and 0.18 m_e, respectively [15].

Since the result of the numerical calculations for the electronic band structures are very complicated, the electronic subband energies and wavefunctions in the conduction and heavy-hole bands in $Cd_{0.027}Zn_{0.973}Te/ZnTe$ and $Cd_{0.053}Zn_{0.947}Te/ZnTe$ SQWs are shown in Figs. 3 and 4. The several peaks observed from the PR measurements for the $Cd_{0.027}Zn_{0.973}Te/ZnTe$ and $Cd_{0.053}Zn_{0.947}Te/ZnTe$ SQWs some peaks correspond to transitions to the conduction band from the 1S light-hole, 1S heavy-hole, 2S light-hole, and 2S heavy-hole bands in the $Cd_xZn_{1-x}Te$ buffer layers. The results of the numerical calculations for the interband transitions indicate that some peaks are interband transitions from the ground, 1st, and 2nd electronic subbands to the ground, 1st, and 2nd heavy-hole levels, respectively, and the other peaks correspond to the interband transitions from the ground, 1st, and 2nd electronic subbands to the ground, 1st, and 2nd light hole levels, respectively. The detailed analyses for the origin of the peaks appeared in PR and PL spectra have been still investigated.

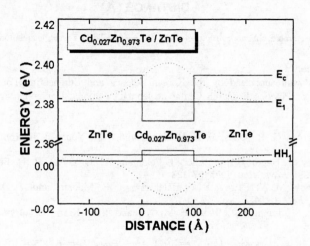

Fig. 3 An electronic subband structure of the $Cd_{0.027}Zn_{0.973}Te/ZnTe$ single quantum well.

4. Summary and Conclusions

High-quality $Cd_xZn_{1-x}Te/ZnTe$ SQWs were grow by the DWTGTVD method. Reflectivity, PR, and PL measurements on the $Cd_xZn_{1-x}Te/ZnTe$ SQWs were performed to investigate the interband transitions. These measurements showed various excitonic interband transitions, and the interband energies of the single quantum wells were calculated taking into account strain effects. The results between the calculated transition energies and the experimental values have been still investigated. Although some detailed investigations remain to be clarified, this growth method with a simple experimental apparatus can be used to farbricate high-quality $Cd_xZn_{1-x}Te/ZnTe$ SQWs with sharp interfaces for use in interesting device implications. Furthermore, high-quality $Cd_xZn_{1-x}Te/ZnTe$ SQWs hold promise for applications such as image processing and electroabsorption modulators.

430

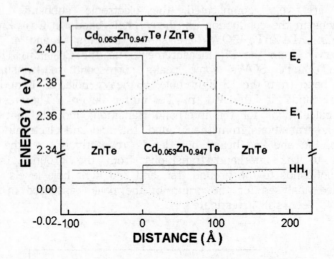

Fig. 4 An electronic subband structure of the Cd$_{0.053}$Zn$_{0.947}$Te/ZnTe single quantum well.

Acknowledgements

This work was supported by the Korean Science and Engineering Foundation through the Semiconductor Physics Research Center at Jeonbuk National University in 1995.

References

[1] R. F. C. Farrow, G. R. Jones, G. H. Williams, and I. M. Young 1981 *Appl. Phys. Lett.* **39** 954

[2] R. P. Stanley, J. Hegarty, R. Fischer, J. Feldmann, E. O. Göbel, R. D. Feldmann, and R. F. Austin 1991 *Phys. Rev. Lett.* **67** 128

[3] I. E. Trofimov, M. V. Petrov, F. F. Balakirev, A. E. Milokin, and V. D. Kuzmin 1992 *Appl. Phys. Lett.* **61** 417

[4] T. W. Kim, M. Jung, H. L. Park, H. K. Na, and J. S. Kim 1992 *Appl. Phys. Lett.* **61** 1101

[5] T. W. Kim, H. L. Park, and J. Y. Lee 1994 *Appl. Phys. Lett.* **64** 2526

[6] J. Allegre, J. Calatayud, B. Gil, H. Mathieu, H. Tuffigo, G. Lentz, N. Magnea, and H. Mariette 1990 *Phys. Rev. B* **41** 8195

[7] T. Li, H. J. Lozykowski, and J. L. Reno 1992 *Phys. Rev. B* **46** 6961

[8] H. Mariette, N. Magnea, and H. Tuffigo 1991 *Physica Scripta T* **39** 204

[9] E. Cohen, R. A. Street, and A. Muranevich 1983 *Phys. Rev. B* **28** 7115

[10] D. J. Olego, J. P. Faurie, S. Sivananthan, and P. M. Racaah 1985 *Appl. Phys. Lett.* **47** 1172

[11] H. Mathieu, J. Allegre, A. Chatt, P. Lefebvre, and J. P. Faurie 1988 *Phys. Rev. B* **38** 7740

[12] H. Tatsuoka, H. Kuwabara, Y. Nakanishi, and H. Fujiyasu 1990 *J. Appl. Phys.* **67** 6860

[13] G. Bastard 1981 *Phys. Rev. B* **24** 5693

[14] P. Peyla, Y. Merle d' Aubigne, A. Wasiela, R. Romestain, H. Mariette, M. D. Sturge, N. Magnea, and H. Tuffigo 1992 *Phys. Rev. B* **46** 1557

[15] H. Mariette, F. Dal'bo, N. Magnea, G. Lentz, and H. Tuffigo 1988 *Phys. Rev. B* **38** 12443

Inst. Phys. Conf. Ser. No 145: Chapter 3
Paper presented at 22nd Int. Symp. Compound Semiconductors, Cheju Island, Korea, 28 August–2 September 1995
© *1996 IOP Publishing Ltd*

Stress in GaAs ridge waveguides integrated with ZnO piezoelectric transducers

Hong Koo Kim, Bandar A. Almashary, Walter Kleemeier, Yabo Li, and Dietrich W. Langer

Department of Electrical Engineering, University of Pittsburgh, Pittsburgh, PA 15261, U.S.A.

Daniel T. Cassidy and Douglas M. Bruce

Department of Engineering Physics, McMaster University, Hamilton, Ontario, CANADA L8S 4L7

Abstract. ZnO thin-film piezoelectric transducers were integrated on GaAs ridge waveguides using RF-magnetron sputtering. Highly c-axis oriented and highly resistive ZnO films were successfully deposited on a GaAs substrate using a SiO_2 thin buffer layer, which was introduced to alleviate a thermal mismatching problem between ZnO and GaAs. Stress on the cleaved facet of the waveguide was imaged with a spatially-resolved and polarization-resolved photoluminescence technique, and was compared with the simulation result. The results showed that the GaAs mesa is stressed up to 1×10^9 dyn/cm^2 (10^{-3} strain) due to residual stress from the ZnO/SiO$_2$/GaAs structure. Potential device applications are discussed that will utilize the intrinsic stress and the piezoelectric stress obtainable from the ZnO transducers.

1. Introduction

ZnO has been recognized as one of the most promising piezoelectric materials for surface acoustic wave transducers due to its high electromechanical coupling factor [1]. ZnO films are also suitable for optical waveguides and can be used as an interaction medium in surface acoustooptic devices [2]. The ZnO single crystal belongs to the (6mm) point group, and its c-axis is a crystallographic polar axis. A c-axis oriented ZnO polycrystalline film has almost the same elastic, piezoelectric, dielectric and other lattice properties as single crystal ZnO.

In this paper, we report sputter deposition of ZnO thin films on a GaAs ridge waveguide structure. Ridge waveguide structures are used commonly in many optoelectronic devices such as semiconductor lasers, waveguide switches/modulators, couplers, etc. In general, optical and/or electrical properties of those devices are sensitive to mechanical stress [4,5,6,7]. Since deposition of thin films usually induces a significant amount of stress as intrinsic or thermal stress, measurement of stress in ridge structures is important for a proper design and control of device characteristics. The effects of stress on device performance have been characterized by many people, using various techniques. However, not many reports are available that dealt with direct measurement of stress in ridge-structure devices [4]. In this paper, we report also our stress measurement results on the fabricated devices obtained with a spatially-resolved and polarization-resolved photoluminescence measurement technique. Potential device applications are discussed that will utilize the intrinsic stress and the piezoelectric stress obtainable from the ZnO transducers.

2. Experimental

GaAs ridge waveguide structures (20 - 60 μm wide, 1 μm high) were fabricated on GaAs substrates with a standard photolithography and a wet chemical etching technique. Stripe geometry mesas were aligned either parallel or 45 degree tilted to the <110>-cleaved-edges of (001)-oriented GaAs substrates, i.e., ridges running along the <110> or <100> directions, as shown in Fig.1. Should there be any stress exerted by films deposited between the mesas, in-plane stress on the (100) plane will be dominated by a uniaxial stress in the <110> direction for the <110> ridges. Similarly, a uniaxial stress in the <100> direction is expected to be dominant for the <100> ridges. After the formation of a mesa structure, a SiO$_2$ buffer layer and a ZnO film were deposited in sequence. Both the ZnO (1-μm-thick) and SiO$_2$ (0.1-μm-thick) films were deposited with an RF planar magnetron sputter system. After film deposition, the ZnO/SiO$_2$ films on the top surface of the mesas were removed selectively to open a window. Please refer to Ref.3 for a detailed description of the fabrication.

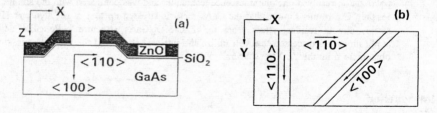

Fig. 1 (a) Schematic drawing of a cross section of the fabricated structure. (b) Plan view of the structure showing orientation of ridges with respect to the <110>-cleaved edges. The GaAs mesas are 1-μm-high and 20-, 30-, 40-, or 60-μm-wide, and are separated with 200-, 400-, or 1000-μm-distance between mesas. The width of the window on top of the mesas is 5, 10, 20, or 30 μm.

Thin-film induced stress in the GaAs mesa was characterized by polarization-resolved photoluminescence. The spatially resolved and polarization-resolved photoluminescence technique has been used for imaging of stress as well-described in Ref.4, and is briefly summarized here. Uniaxial stress destroys the cubic symmetry of GaAs, splitting the heavy- and light-hole bands and causing a reduced (or enhanced) probability of emission of light polarized in the direction of a tensile (or compressive) stress. Thus, by measuring the degree of polarization (DOP) of emitted light, one should be able to determine the local stress or strain. The degree of polarization we define to be DOP = $(L_x - L_z) / (L_x + L_z)$, where L_x and L_z are the measured intensities polarized in the x and z directions, respectively. The x and z directions are indicated in Fig. 1a. For small stresses (below about 2 x 10^9 dyn/cm^2), one can assume a linear relationship between stress and degree of polarization. GaAs is not isotropic in elasticity or photoelasticity, so it is expected that it is not isotropic in the response of DOP to stress. However, due to the present lack of knowledge about the directional dependence of DOP to stress, we assume an isotropic dependence and use the constant measured experimentally for stresses in the (011) plane [4]. This results in the following relationship between DOP and σ, DOP = $K_\sigma (\sigma_{xx} - \sigma_{zz})$, where K_σ is a constant with dimension cm^2/dyn, $K_\sigma = -(5.1 \pm 0.3) \times 10^{-11} cm^2/dyn$ for GaAs [4], and σ_{xx} and σ_{zz} are the (tensile) stresses in the x and z directions, respectively.

3. Results and discussion

Deposited films were examined by $\theta - 2\theta$ X-ray diffraction with Cu-Kα radiation. Fig. 2a shows the X-ray diffraction profile of the ZnO film deposited on a SiO_2-buffered GaAs substrate. Although the as-deposited ZnO film has a strong c-axis orientation, further dramatic enhancement was achieved by the post-deposition heat treatment, i.e., the ohmic contact alloying at 430 °C for 10 min. (See Fig. 2). The intensity of the (0002)-plane peak was increased by more than one order of magnitude, and the full-width at half-maximum (FWHM) of the peak was reduced from 1.23 to 0.37°. It should be noted that the heat treated film shows only c-axis orientation peaks in this wide range scan of 20 - 80°. The c-axis lattice constant of the ZnO films was calculated from the Bragg angle of the (0002)-plane peaks: 2.638 Å for the as-deposited film and 2.603 Å for the post-deposition annealed specimen. Compared with a bulk single-crystal ZnO lattice constant of 2.603 Å, it is obvious that the as-deposited ZnO film has compressive stress along the film plane, thus resulting in tensile strain of 1×10^{-2} normal to the film plane. It is well known that sputter-deposited films usually reveal intrinsic stress along a film plane. This stress, however, was reduced significantly after the brief anneal treatment. Considering the resolution of the X-ray diffractometer used, the residual strain in the heat treated ZnO film is estimated to be less than 10^{-4}. These improvements are attributed to a reduction of grain boundaries and voids with the anneal treatment as supported by the scanning electron microscopy result [9].

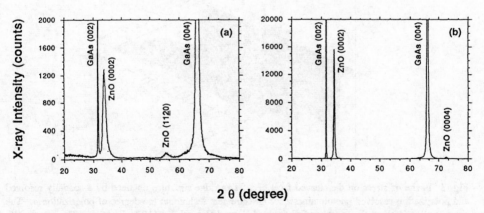

Fig. 2 X-ray diffraction pattern of a ZnO film deposited on a SiO_2-buffered GaAs substrate at room temperature: (a) as-deposited, and (b) after a post-deposition anneal treatment at 430 °C for 10 minutes in forming gas.

Fig. 3 shows an image of stress on the cleaved facet of the fabricated ridge structure obtained by DOP method. The total scanned area was 300 μm wide and 50 μm deep near the mesa area. The stepsize between data points was 0.5 μm in the z direction and 2 μm in the x direction. A spatial resolution of 1.5 - 2.0 μm was estimated since a 20x microscope objective lens was used. The ridge is aligned along the <100> direction and cleaved along the <110> direction. The ridge width is 40 μm and the mesa height is 1 μm. Fig. 3 shows the distribution of the degree of polarization. The height of the graph above the baseline is proportional to DOP; data points above the baseline correspond to σ_{xx} - σ_{zz} < 0, and data points below the baseline correspond to σ_{xx} - σ_{zz} > 0. For $|\sigma_{xx}|$ > $|\sigma_{zz}|$, this corresponds to compressive and tensile stress in the x direction, respectively. Since the structure is open-ended in the z direction, it is expected that $|\sigma_{xx}|$ > $|\sigma_{zz}|$ [8]. The value of DOP is undefined when the scan point is off the facet and no signal is present; DOP has been set to zero in this case. The variations in DOP indicate stressed regions near the ridge structure. The DOP decreases to a minimum value of approximately - 8 % as we approach the surface region near the mesa edges. In the mesa region and along the mesa surface, the DOP becomes positive and reaches a maximum value of + 7 %. Sputter deposited films are usually under compressive stress as a result of the deposition process and/or thermal expansion mismatch. When windows are opened in the deposited films, the edge of the window exerts a force on the GaAs, parallel to the surface of the GaAs and perpendicular to the edge of the window. It produces a complex stress field in the GaAs beneath, which builds up to very high stress values immediately beneath the edge of the films. The result confirms that our sputter-deposited SiO_2 and ZnO films are under considerable compression, thus producing tensile stress beneath the films and compressive stress in the window region. It is interesting to note that the tensile stress beneath the SiO_2/ZnO films remains significant for more than 10 μm depth from the interface. The overall profile of the stress distribution as shown in Fig. 3 is in good agreement with the simulation result obtained with the edge-force assumption [8].

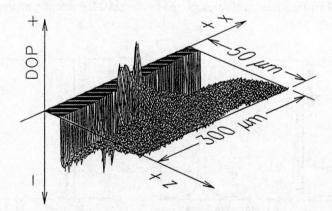

Fig. 3 Image of stress on the cleaved facet of a GaAs ridge structure obtained by a spatially resolved and polarization-resolved photoluminescence measurement technique: the degree of polarization ρ. The ridge is aligned along the <100> direction and cleaved along the <110> direction. The ridge width is 40 μm and the mesa height is 1 μm. The stepsize between data points was 0.5 μm in the z direction and 2 μm in the x direction. The total scanned area was 300 μm wide and 50 μm deep.

Most of the samples characterized showed compressive stress in the mesa region with peak values near edges. In some samples, however, tensile-stress was observed locally in the mesa region as shown in Fig. 3. This is attributed to the SiO_2/ZnO films remaining on top of the mesa. For low levels of DOP (DOP < 5 %) and for GaAs (and depends on direction), 1 % DOP is equivalent to approximately a strain of 3 x 10^{-4} [4]. The peak stress is estimated to be $\sigma_{xx} = 1$ x 10^9 dyn/cm² in the substrate near the ridge, corresponding to 2 x 10^{-3} strain. This number is greater than the value expected from the X-ray diffraction result. X-ray diffraction shows that ZnO is strained in the order of 10^{-4} or less after the anneal treatment. Thus we might expect that the GaAs is strained not more than this value. This difference between the two measurements suggests that there may be another stress component, which is not detected in the X-ray diffraction. The SiO_2 buffer layer may have contributed this additional stress component. This attribution is supported by the fact that a SiO_2 film of thickness ~ 0.1 μm has been reported to produce up to 10^9 dyn/cm² stress[8]. Fig. 3 also shows that the DOP profile is slightly asymmetric, i.e., the profile under the mesa are skewed slightly down to 30 μm depth from the surface. This is attributed to the geometry of this sample, i.e., the cleaved facet is 45° tilted toward the ridge direction as shown in Fig. 1b. Therefore, an asymmetric distribution of edge forces is expected to exist near the cleaved facet.

We have also modeled the stress induced in GaAs waveguides integrated with ZnO. We have used ANSYS to calculate the stress and strain profiles in the mesa. ANSYS is a software package based on a finite element analysis approach.[10] The edge force of 100 N/m is assumed to be applied to the 1-μm-thick ZnO edges. This edge force corresponds to a thin-film stress of 1 x 10^8 N/m². This stress value is within the range estimated from the polarization-resolved photoluminescence measurement result. First, both the horizontal and vertical components of stress were calculated. Then the DOP distribution was calculated. Fig.4 shows the DOP profiles at different depths under the top surface of the mesa. The profile shows that DOP changes abruptly near the mesa edges, and matches well the measurement result. The peak values of DOP are $\pm$ 6 %, which are in close agreement with the measurement result.

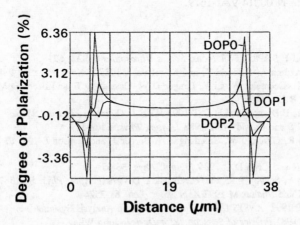

Fig. 4 DOP profiles calculated by ANSYS: DOP0, DOP1, and DOP2 correspond to a 0, 0.5, and 1 μm depth under the top surface of the mesa, respectively.

436

It has been reported that thin-film induced stress is typically in the range of 10^8 - 10^9 N/m^2. Considering the photoelastic coefficients of typical semiconductors[8], this translates into a relative dielectric constant (or refractive index) change of up to 10^{-2} order. This number is considered to be enough to define or modify a waveguide structure.

It is also well known that the bandgap of semiconductor can be changed by stress. In GaAs, for example, the energy/pressure coefficient is 1.29×10^{-10} (eV.m^2/N).[11] Consider a GaAs mesa integrated with a piezoelectric thin film. If 10 V is applied across a 0.5-μm-thick ZnO film, the electric field in the film is calculated to be 2×10^7 (V/m). The pressure applied to the GaAs mesa is, then, estimated to be 1.1×10^7 (N/m^2) (The piezoelectric constant e_{zi} has been reported to be 0.56 (Coul/m^2))[12]. In the PZT or BaTiO$_3$ case, a much higher piezoelectric constant is available (6.5 Coul/m^2 for PZT and 4.4 Coul/m^2 for BaTiO$_3$)[12,13]. This implies that modulation up to 100 Å will be possible.

4. Conclusion

Sputter deposition of ZnO thin films on a GaAs ridge-waveguide structure has been investigated. Highly c-axis oriented and thermally stable ZnO films were deposited on a GaAs ridge structure with a SiO$_2$ thin buffer layer using RF-magnetron sputtering. Stress on the cleaved facet of the waveguide was imaged by a spatially-resolved and polarization-resolved photoluminescence technique. The results showed that the GaAs mesa is stressed up to approximately 10^9 dyn/cm^2 (strain of order 10^{-3}) due to a residual stress from the ZnO/SiO$_2$/GaAs structure, although the post-deposition heat treatment reduced the intrinsic stress of the ZnO film significantly. Device applications of thin-film induced stress are also discussed. For example, the intrinsic stress induced by a thin film for defining and modifying the waveguide structure via a photoelastic effect, and the piezoelectric stress for semiconductor bandgap modulation.

Acknowledgement: This work was supported by the Engineering Foundation / IEEE / Air Force Engineering Research Initiation Grant, and by the Office of Naval Research under Grant No. N 00014-92-J-1679.

References

[1]   Hickernell F S 1985 *IEEE Trans. Sonics Ultrasonics* SU-32, 621
[2]   Wu M, Azuma A, Shiosaki T, and Kawabata A 1987 *J. Appl. Phys.* 62 2482
[3]   Kim H K, Kleemeier W, Li Y, Langer D W, Cassidy D T, and Bruce D M 1994 *J. Vac. Sci. Technol. B* 12 1328
[4]   Colbourne P D and Cassidy D T 1993 *IEEE J. Quantum Electron.* 29 62
[5]   Adams C S and Cassidy D T 1988 *J. Appl. Phys.* 64 6631
[6]   Maciejko R, Glinski J M, Champagne A, Berger J, and Samson L 1989 *J. Quantum Electron.* 25 651
[7]   Dutta N K and Craft D C 1984 *J. Appl. Phys.* 56 65
[8]   Kirkby P A, Selway P R, and Westbrook L D 1979 *J. Appl. Phys.* 50 4567
[9]   Kim H K and Mathur M 1992 *Appl. Phys. Lett.* 61 2524
[10]   Kohnke P 1992 *ANSYS User's Manuals* Swanson Analysis Systems Inc.
[11]   Sze S M 1981 *Physics of Semiconductor Devices* John Wiley
[12]   Ristic V 1983 *Principles of Acoustic Devices* John Wiley
[13]   Jaffe B, Cook W R, and Jaffe H 1971 *Piezoelectric Ceramics* Academic Press 74

Inst. Phys. Conf. Ser. No 145: Chapter 3
Paper presented at 22nd Int. Symp. Compound Semiconductors, Cheju Island, Korea, 28 August–2 September 1995
© *1996 IOP Publishing Ltd*

Confined optical vibrations of (311) GaAs/AlAs superlattices.

A Milekhin, Yu Pusep, D Lubyshev, V Preobrazhenskii, B Semyagin

Institute of Semiconductor Physics, 630090, Novosibirsk, Russia

Abstract. The optical confined modes of the (311)-oriented GaAs/AlAs superlattices were investigated by Infrared Fourier Transform Spectroscopy. The analysis of confined TO and LO vibrational modes observed in experimental IR reflection spectra shows no marked difference between superlattices grown on (311)A and (311)B surfaces. Moreover, no evidence of optical anisotropy caused by corrugation of the (311)A surface was obtained. The comparison of the frequencies of TO confined modes in short-period (311) superlattices with the dispersion of GaAs phonons obtained for long-period superlattices allows to conclude that even confined modes usually forbidden in infrared are active in the case of the (311) short-period superlattices due to different arrangement of normal (AlAs on GaAs) and inverted (GaAs on AlAs) interfaces.

1. Introduction

The optical phonons in (100)-oriented GaAs/AlGaAs superlattices (SL's) have been studied extensively and it's behavior is well understood [1-6]. However, only a few studies of the optical properties of the non-(100)-oriented GaAs/AlGaAs SL's, particular in infrared, have been carried out [7-12]. Recently, the formation of the (311)-oriented GaAs/AlAs SL's with anisotropic electronic and optical properties was demonstrated [13-15]. It was concluded that this anisotropy is probably due to lateral corrugation [16]. However, the recent Raman study of the (311) GaAs/AlAs SL's does not reveal any evidence of corrugation of (311) surface [10]. In our opinion the investigation of optical phonons by Fourier Transform Infrared (FTIR) Spectroscopy can give a new information on the arrangement of the (311) interfaces.

Lowering of symmetry in the (311) SL's comparing with SL's grown along the (100) direction complicates identification of phonon modes in the (311) SL's producing the confined optical modes of pure transverse (A") and mixed longitudinal/transverse (A') character [11]. Moreover, lateral corrugation in (311)A superlattices with a 32 A period can lead to the splitting of the optical confined modes. Due to selection rules all modes (A' and A") can be active in infrared. The wave number of confined modes can be defined as usual

$$q_m = m\,\pi/(n+\delta)d \tag{1}$$

where n is the number of monolayers, $d=a/\sqrt{11}$ is the thickness of one monolayer in the (311) direction, a is the lattice parameter in the (100) direction and m is the number of confined mode. The parameter δ describes the penetration of confined modes into neighbouring layers. The identification of the mixed confined modes with small

438

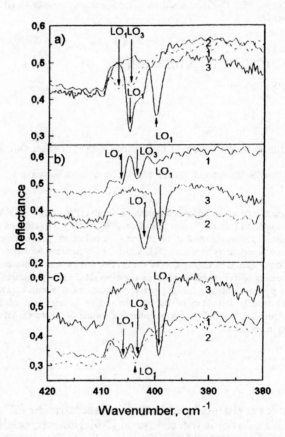

Figure 1. Experimental IR reflection spectra of samples measured with p-polarized light in the spectral range of the LO phonons of AlAs:

a) are (100) SL's: $(GaAs)_{17}/(AlAs)_{15}$ (curve 1).
$(GaAs)_7/(AlAs)_9$ (curve 2),
$(GaAs)_4/(AlAs)_4$ (curve 3),

b) and c) are (311)A and (311)B SL's respectively:
$(GaAs)_{28}/(AlAs)_{24}$ (curve 1),
$(GaAs)_{12}/(AlAs)_{17}$ (curve 2),
$(GaAs)_9/(AlAs)_9$ (curve 3).

wavenumbers becomes easier due to mainly LO or TO polarization of corresponding confined modes [12].

2. Experimental

The samples under investigations were $(GaAs)_n/(AlAs)_m$ SL's (where n=9, 12, 28 and m=9, 17, 24 monolayers) grown on (311)A and (311)B surfaces simultaneously. Thickness of the GaAs and AlAs layers was determined from RHEED experiments using SL's grown in the same process on the (100) GaAs substrates . The layer structure was repeated 100 times for all SL's under investigations.

The Infrared reflection spectra were recorded using FTIR spectrometer Bruker IFS-113V equipped with an Oxford Instruments cryostat at a temperature 79K. The resolution was 0.5 cm^{-1} over the entire spectral range. The reflection spectra of the normal incidence of the light together with p-polarized spectra obtained at oblique incidence

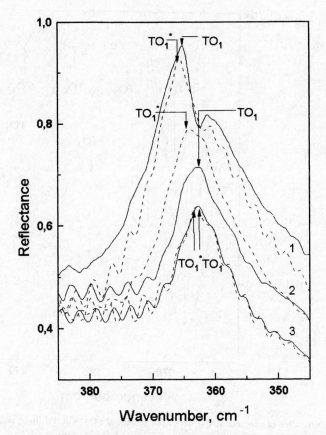

Figure 2. Experimental IR reflection spectra of (311)A (solid lines) and (311)B (dashed lines) GaAs/AlAs
SL's measured at a normal incidence in the spectral range of TO phonons of AlAs:

$(GaAs)_{28}/(AlAs)_{24}$ (curve 1),

$(GaAs)_{12}/(AlAs)_{17}$ (curve 2),

$(GaAs)_9/(AlAs)_9$ (curve 3).

($\theta \approx 70°$) were analyzed. The reflection at the normal incidence of the light allows to observe
the TO confined phonons in SL's while p-polarized spectra reveals in addition the LO
confined phonons owing to the Berreman effect.

3. Results and discussion

The p-polarized reflection spectra of (311)A and (311)B SL's in the spectral range of LO
phonons of AlAs are presented in Fig.1. The structural features in the spectra marked by
arrows here correspond to the LO confined modes localized in the AlAs layers. As can be
seen from Fig.1, the frequencies of fundamental vibrational modes of all SL's are decreasing
with the decrease of the AlAs layer thickness according to the dispersion law of AlAs
phonons. Moreover, the reflection spectra reveal the LO_3 modes that confirms a high
enough crystalline quality of samples. However, neither a marked difference between
reflection spectra of (311)A and (311)B SL's nor an additional splitting of confined modes of

440

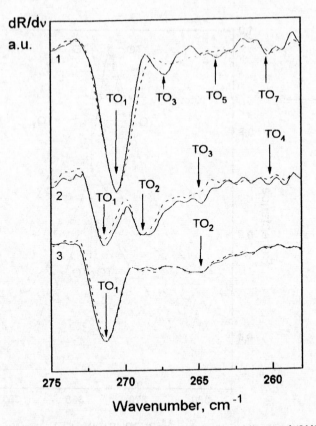

Figure 3. Deriviative of experimental IR reflection spectra of (311)A (solid lines) and (311)B (dashed lin)
SL's measured at normal incidence in the spectral range of GaAs phonons:

$(GaAs)_{28}/(AlAs)_{24}$ (curve 1),

$(GaAs)_{12}/(AlAs)_{17}$ (curve 2),

$(GaAs)_9/(AlAs)_9$ (curve 3).

the (311)A SL's was observed. Alternation of vibrational modes position is caused only by the different localization lenght in SL's. The rotation of the samples around the fixed polarization vector of the light did not show any significant difference between the reflection spectra with the different polarization.

As it was mentioned above, the reflection spectra of SL's recorded at the normal incidence of the light on the sample show the possibility to study the TO confined phonons. Figure 2 presents the reflection spectra of (311)A (solid lines) and (311)B (dashed lines) SL's which reveal only the fundamental TO modes localized in the AlAs layers. Due to the small wave number of the fundamental vibrational modes we can not identify the character (pure transverse or mixed longitudinal-transverse) of confined modes.

The frequencies of the TO modes localized in the GaAs layers of SL's are placed near the "reststrahlen band" of the bulk GaAs. To dertermine the frequencies of these TO modes highly accurately we analyze the derivative of the reflection dR/dv (Fig.3). The minima indicated by arrows correspond to the TO confined modes. As a rule only odd confined optical modes are active in the IR spectra of long-period SL's due to non-zero

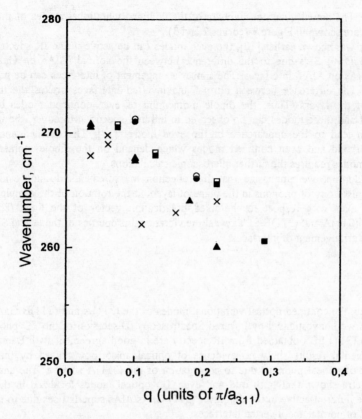

Figure 4. Dispersion of GaAs phonons measured in (311)A (open characters) and (311)B (full characters) (GaAs)/(AlAs) SL's. Experimental data obtained on $(GaAs)_{28}/(AlAs)_{24}$ SL are shown by triangles. Squares and circles are the experimental data obtained on $(GaAs)_{12}/(AlAs)_{17}$ and $(GaAs)_9/(AlAs)_9$, respectively. Raman data (crosses) are taken from reference [10].

dipole momentum of these modes. Influence of interfaces in case of the long-period SL's is insignificant. The experimental rerflection spectra of long-period (311)A and (311)B SL's (solid and dashed curves labeled as 1 in Fig.3) presents the odd confined modes up to TO_7 mode.

It is known that the GaAs dispersion obtained for long-period SL's grown on the (100) surface describes adequate the bulk GaAs dispersion [5]. Dispersion of the TO phonons of GaAs in the (311) SL's obtained using the frequencies of confined modes from the IR reflection spectra and the relation (1) is shown in Fig.4 by triangles. The penetration parameter δ was taken as 1 [12]. Dispersion of the TO phonons of the GaAs in (311) SL's obtained from IR spectra is in good agreement with Raman data presented in Fig.4 by crosses.

The dispersion of the TO phonons in the GaAs obtained for the short-period superlattices was far from one for the long-period SL's. The interface broading can not explain this difference significantly. This disagreement is removed by the assumption that the even confined modes usually forbidden in infrared are active in the case of the (311) short-period SL's. The actual quantum numbers of the confined vibrational modes of the short-period SL's are shown in Figure 3 (curves 2 and 3).

As it was shown earlier [10], the even modes can be active in the IR spectra of the short-period (100) SL's due to the difference between the normal (AlAs on GaAs) and inverted (GaAs on AlAs) interfaces. The similar arrangement of interfaces can be present in (311) SL's. The difference between normal and inverted interfaces brakes the inversion symmerty of the layers. Thus, the dipole momentum of even confined modes becomes nonvanished and these modes can be observed in infrared spectra. Moreover, the interface disorder can lead to the appearence of the even modes [17]. The detailed analysis of intensities of odd and even confined modes which depend on the dipole momentum of vibrational modes requires the further microscopic calculations.

In addition, we emphasize that IR reflection measurements do not evident the significant anisotropy of phonons in the plane of layers at the rotation of the sample around the growth axes with respect to the fixed polarization vector of the light. Thus, the investigated (311)A and (311)B SL's reveal the vibrational properties of the lateral SL's with the different arrangement of interfaces.

Summary

In conclusion, the confined optical vibrational modes of the (311)A and (311)B GaAs/AlAs superlattices were investigated by Infrared Spectroscopy. Dispersion of the TO phonons of the GaAs in (311) SL's obtained from IR spectra are in good agreement with Raman data. The experimental results show no evidence of optical anisotropy caused by the lateral confinement of optical phonons due to corrugation of the (311)A surface. The analysis of infrared spectra shows that both odd and even TO optical modes localized in the GaAs layers can be infrared-active in the thin layer (311) GaAs/AlAs superlattices due to different arrangement of normal and inverted interfaces.

References

[1] Merz J L, Barker A S, Jr, Gossard A C 1977 *Appl. Phys. Lett.* **31** 2, 117-119

[2] Colvard C, Gant T, Klein M, Merlin R, Fischer R, Morkoc H, Gossard A 1985 *Phys.Rev. B* **31** 2080

[3] Cardona M 1990 *Superlatt. and Microstr.* **7** 3 183-191

[4] Jusserand B, Mollot F, Noison J-M, Roux G L 1990 *Appl.Phys. Lett.* **6** 560-562

[5] Milekhin A, Pusep Yu, Preobrazhenskii V, Semyagin B, Lubyshev D 1994 *JETP Lett.* **59** 7   493-496

[6] Pusep Yu A ,Milekhin A G, Toropov A I 1995 *J. Phys.:Cond.Matter* **7** 1493-1498

[7] Popovic Z V, Cardona M, Richter E, Strauch D, Tapfer L, Ploog K 1989 *Phys.Rev. B* **40** 3040

[8] Popovic Z V, Cardona M, Richter E, Strauch D, Tapfer L, Ploog K 1990 *Phys.Rev. B* **41** 5904-5913

[9] Popovic Z V, Cardona M, Richter E, Strauch D, Tapfer L, Ploog K 1991 *Phys.Rev. B* **43** 4925-4932

[10] Pusep Yu, da Silva S W, Galzerani J, Lubyshev D and Basmaji P 1995 *Phys. Rev. B* to be published

[11] Popovich Z, Richter E, Spitzer J, Cardona M, Shield A,Notzel R,Ploog K 1994 *Phys. Rev. B* **49** 11 7577

[12] Castrillo P and Colombo L 1994 *Phys. Rev. B* **49** 15 10362

[13] Notzel R, Daweritz L and Ploog K 1992 *Phys.Rev. B* **46** 4736

[14] Notzel R, Ledentsov N, Daweritz L, Hohenstein M and Ploog K 1991 *Phys. Rev. Lett.* **67** 3812

[15] Notzel R and Ploog K 1992 *J. Vac. Sci. Technol.* **A10** 617

[16] Notzel R, Ledentsov N, Daweritz L, Ploog K and Hohenstein M 1992 *Phys.Rev. B* **45** 3507

[17] Pusep Yu A at all 1995 *Phys.Rev.B* to be published

Inst. Phys. Conf. Ser. No 145: Chapter 3
Paper presented at 22nd Int. Symp. Compound Semiconductors, Cheju Island, Korea, 28 August–2 September 1995
© *1996 IOP Publishing Ltd*

443

Confined AlAs LO phonons in GaAs/AlAs superlattices

D.A.Tenne, V.A.Haisler, N.T.Moshegov, A.I.Toropov,
I.I.Marakhovka, and A.P.Shebanin

Institute of Semiconductor Physics, pr.Lavrenteva, 13,
630090, Novosibirsk, Russia.

Abstract. The phonon spectrum of GaAs/AlAs superlattices in Alas frequency region is studied by Raman spectroscopy. We observed the peaks of confined AlAs LO phonons with mode index 1-5. Experimental frequencies confirm the results of microscopic calculation of bulk AlAs LO phonon dispersion.

1. Introduction.

The vibrational properties of GaAs/AlAs superlattices (SL's) are well understood in principle (for review see [1]). The optical phonons of the SL constituents are confined within their respective layers, leading to a quantization of the phonon wavevector along the SL axis. The series of lines appear in the Raman spectra, whose frequencies correspond to those of optical phonons of bulk material at following wavevectors [1]:

$$q = \frac{m\pi}{(n+\delta)a_0} \tag{1}$$

whefe a_0 is the monolayer thickness, n is the number of monolayers, m is the mode index, and δ is the penetration depth of the phonon into layers of other material. So, the Raman data for confined phonon frequencies allow to determine the optical phonon dispersion of bulk materials that constitute the SL. This opportunity seems to be very essential for AlAs, as the investigation of this material by neutron scattering is difficult because it's unstability.

The confined GaAs optical phonons were studied by many authors [1]. Experimental frequencies of these modes correspond well to the theoretical phonon dispersions of bulk GaAs [2]. But in the AlAs frequency range most groups do not report the observation of successive confined modes. Existing experimental data for AlAs region [3-5] are very scarce and contradictory. The purpose of this paper is the detailed Raman investigation of the LO phonon confinement in AlAs layers of GaAs/AlAs SL's and the analysis of the bulk AlAs LO phonon dispersion.

2. Experimental results.

The characteristic Raman spectrum of GaAs/AlAs SL's in AlAs frequency range is shown in Fig.1a. It contains the LO and TO phonon peaks and the broad feature of interface phonons.

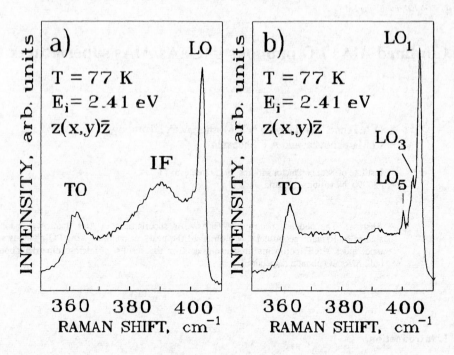

Figure 1. Raman spectra in the AlAs region: a) - for the $(GaAs)_7 (AlAs)_7$ superlattice with rough interfaces; b) - for the $(GaAs)_9(AlAs)_9$ superlattice with more perfect interfaces.

The backscattering spectrum of ideal SL grown in (001) direction should contain only LO phonon peaks. In real SL's the interface roughness may cause the breakdown of momentum conservation in the Raman process, so TO and interface phonon features appear in spectra. This fact causes difficulties in observation and correct interpretation of confined LO phonon lines in spectra

We perform our experiments on the specially selected samples of SL's with perfect interfaces. As the criterion of interface quality we use the absence of interface phonon features in Raman spectra, that allow us to observe the peaks of confined AlAs LO phonons with mode index 1-5 (Fig.1b). We investigated following samples of SL's, satisfying to the criterion of the interface quality: $(GaAs)_{12}(AlAs)_{12}$, $GaAs)_{10}(AlAs)_{10}$, $(GaAs)_9(AlAs)_9$, $(GaAs)_{12}(AlAs)_6$, $(GaAs)_9(AlAs)_5$. All samples were grown by molecular beam epitaxy on GaAs (001) substrates. The Raman spectra were recorded using Jobin Yvon U1000 spectrometer at the temperature 77 K in backscattering geometries $z(x,y)z$ and $z(x,x)z$ (z axis is parallel to the growth direction). For all samples in $z(x,y)z$ geometry we resolved lines of odd confined AlAs phonons LO_1, LO_3 and LO_5 (all spectra are similar to spectrum, plotted in Fig.1b). In parallel geometry $z(x,x)z$ we observed peaks of even confined phonons LO_2, for sample $(GaAs)_{12}(AlAs)_6$ - LO_2 and LO_4.

Experimental data allow us to determine the bulk AlAs LO phonon dispersion. Measured frequency of the confined mode LO_m should be matched by the value of wavevector, derived from Eq.(1). Fig.2 plots the AlAs LO phonon dispersion, determined from our data (triangles) together with experimental data of Ref.3-5. The result of microscopic *ab initio* calculation of bulk AlAs LO phonon dispersion [2] is also shown in Fig.2. It can be seen that our data

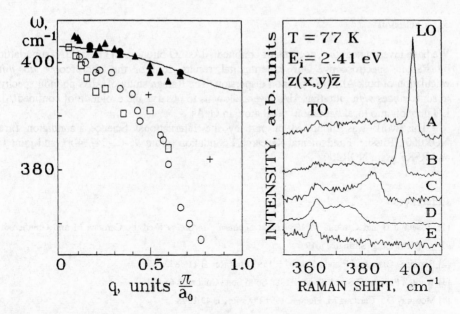

Figure 2. Bulk AlAs LO phonon dispersion. Triangles are the data of this work., full circles are those of Ref.4, open circles and squares show the data of Ref.3 and 5 respectively. Solid line is the result of microscopic calculation [2].

Figure 3. Raman spectra of (GaAs)8(AlAs)n superlattices with n = 3, 2, 1 and 0.5 monolayers for samples A, B, C and D respectively. E is the spectrum of mixed crystal $Al_xGa_{1-x}As$, $x \sim 0.01$.

considerably disagree with data of Ref.3 and 5, but correspond well to data of Ref.4 and to the theoretical dispersion. Small deviation of our experimental points from the calculated dispersion may be explained by some non-ideality of SL's. Interface intermixing leads to decrease of confined AlAs phonon frequencies [6]. It is illustrated in Fig.2, where crosses indicate the calculated confined AlAs LO mode frequencies in SL $(GaAs)_5(AlAs)_5$ with two intermixed $Al_{0.5}Ga_{0.5}As$ monolayers at the interfaces [6]. As can be seen, these values approach to the data of Ref.3 and 5.

Also we studied Raman scattering on AlAs phonons in series of SL's with ultrathin AlAs layers $(GaAs)_8(AlAs)_n$ (n = 3, 2, 1 and 0.5 monolayers) and in the mixed crystal $Al_xGa_{1-x}As$, $x \sim 0.01$. The Raman spectra of these samples are presented in Fig.3. These spectra demonstrate the evolution of the confined AlAs LO phonon into local vibration of Al atom in GaAs. As can be seen from Fig.3, in SL's with 3 and 2 monolayers of AlAs the LO mode frequency lies in the bulk AlAs frequency range, so in these SL's the phonon can be considered as confined LO phonon in AlAs layer. (However it's frequency can not be used for the construction of bulk AlAs dispersion, since the role of interface disorder in SL's with ultrathin layers is very important). As the AlAs layer thickness decreases, the LO mode frequency goes down and approaches to the local phonon frequency of Al atom in GaAs, observed in spectrum of mixed crystal with small Al concentration (~1%).

446

3. Conclusion

We have investigated the spectrum of confined AlAs LO phonons in GaAs/AlAs superlattices by Raman spectroscopy. Our experimental results confirm the microscopic *ab initio* calculation of bulk AlAs LO phonon dispersion. The Raman study of AlAs phonon spectrum in superlattices with ultrathin AlAs layers allow us to observe the evolution of confined AlAs LO phonon into local vibration of Al atom in GaAs.

This work was supported in part by the International Science Foundation (grant NQR000), Russian Fundamental Research Foundation (grant 95-02-04714-a), and joint ISF and RFRF grant NQR300.

References

[1] Jusserand B and Cardona M 1989 *Light Scattering in Solids V* ed. by Cardona M and Guntherodt G
  (Heidelberg: Springer) p 49

[2] Baroni S, Giannozzi P and Molinari E 1990 Phys.Rev.B **41** 3870

[3] Wang Z P, Jiang D S and Ploog K 1988 Solid State Commun. **65** 661

[4] Mowbray D J, Cardona M, Ploog K 1991 Phys.Rev.B **43** 1598

[5] Pusep Yu A, Milekhin A G, Moshegov N T et al 1994 J.Phys.Condens. Matter **6** 93

[6] Molinari E, Baroni S, Giannozzi P, de Gironcoli S 1992 Phys.Rev.B **45** 4280

Inst. Phys. Conf. Ser. No 145: Chapter 3
Paper presented at 22nd Int. Symp. Compound Semiconductors, Cheju Island, Korea, 28 August–2 September 1995
© *1996 IOP Publishing Ltd*

Photoelastic characterization of slip lines generated by thermal processing with ring holder

M. Yamada[†][1]**, M. Fukuzawa**[†]**,**
T. Kawase[‡]**, M. Tatsumi**[‡] **and K. Fujita**[‡]

† Dept. of Electronics and Information Science,
Kyoto Institute of Technology, Matsugasaki, Sakyoku, Kyoto 606, JAPAN

‡ Sumitomo Electric Industries, Ltd.,
1-1-3, Koya-kita, Itami, Hyogo 664, JAPAN

Abstract. Slip lines generated by radiation heating with ring holder have been characterized with the photoelastic technique useful in evaluating residual and/or process-induced strains. It is found that slip lines are strongly reflected in two-dimensional distribution maps of shear strain component, rather than tensile strain component. The slip lines along the ⟨011⟩ directions are clearly observed mainly in the ⟨001⟩ peripheral regions of 4″-diameter (100) wafers processed in the temperature range of 400∼600C.

1. Introduction

GaAs crystals are key materials for developing optoelectonic devices and high-speed integrated circuits. During thermal processes such as epitaxial growth and furnace or rapid thermal annealing after ion implantation, unwanted slip lines are sometimes generated in wafer periphery. Many investigators have known by experience that the slip line generation becomes more severe in larger size of wafer and have to solve this slip line problem to obtain higher yield in fabricating devices. However, not only the origin of slip line generation but also the slip line itself is not so clearly understood. Yamada *et al* (1993) have recently proposed the following possible origins for slip line generation:

(a) Stationary temperature gradient due to inhomogeneous heating,

(b) Transient temperature gradient during heating up and down,

(c) Residual strains caused in crystal growth processes,

(d) Deformation layer caused in all wafer processes from slicing to final polishing.

The former two items depend on the furnace structure, the method of wafer supporting, and the rate of heating up and down; that is, how to anneal the wafer. The latter two items are originating in the wafer itself.

[1] E-mail (SINET): yamada@dj.kit.ac.jp

Kawase *et al* (1992) reported that the slip line generation during epitaxial growth was less in vapor-pressure-controlled Czochralski (VCZ) wafers than in conventional liquid-encapsulated Czochralski (LEC) wafers because the VCZ wafers have a low level of residual strains, compared with the LEC wafers. Yamada *et al* (1993) found that the slip line generation during furnace annealing after ion implantation anomalously increased residual strains and presumed that the transient temperature gradient during heating up and down was a dominant factor in this case.

In this paper, we present photoelastic characterization results on slip lines generated by radiation heating with ring holder, as shown in Fig. 1, whose structure is used for wafer heating in a molecular beam epitaxy (MBE) equipment. In this case, stationary temperature gradient is important rather than transient temperature gradient, because the heating and cooling rate is slow and the wafer temperature distribution is not always homogeneous due to heat flow through ring holder.

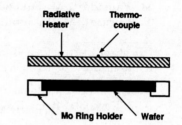

Figure 1. Wafer heating and holding structure used in the present experiment.

2. Experimental procedure

Conventional LEC-grown 4″-diameter (100) GaAs wafers were used in the present experiment. Their front and rear surfaces were mirror-polished, the thickness was $700\mu m$, and the etch pit density was $5 \sim 6 \times 10^4/cm^2$. During thermal processing, the temperature of radiation heater was monitored with a thermocouple embedded in the heater. The wafer temperature relative to the temperature of radiation heater was calibrated with an optical thermometer. In the thermal processing, the actual MBE growth was not made so that the slip line generation might not be effected. The heating and cooling rate was 20C/min and the holding time was 60min in the temperature range of 400~600C.

In the photoelastic measurement, we used a scanning infrared polariscope with a high spatial resolution which was developed as a routine inspection tool to characterize residual strain components quantitatively (Yamada and Fukuzawa, 1994). The beam spot size of the probing light was $200\mu m$ and the wavelength was $1.3\mu m$. The scanning pitch used was typically $500\mu m$. The residual strain components characterized for (100) wafers are $|S_r - S_t|$, that is, the absolute value of the tensile difference between the radial and tangential directions in the cylindrical coordinate system, $|S_{yy} - S_{zz}|$, that is, the absolute value of the tensile difference between the y and z directions in the crystallographic coordinate system, and $2|S_{yz}|$, that is, the absolute value of the shear component in the crystallographic coordinate system. All of these are the in-plane components for (100) wafers.

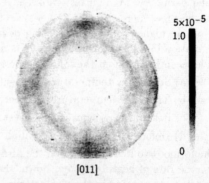

Figure 2. A typical two-dimensional distribution map of $|S_r - S_t|$ measured before thermal processing in LEC-grown 4″-diameter (100) GaAs wafers.

$|S_r - S_t| \ (\equiv [(S_{yy} - S_{zz})^2 + (2S_{yz})^2]^{1/2})$ may be used as a figure of merit for the total in-plane strain component.

3. Results and discussion

The residual strain before thermal processing was characterized by the photoelastic measurement. A typical result of two-dimensional distribution maps of $|S_r - S_t|$ measured in LEC-grown 4″-diameter (100) GaAs wafers is shown in Fig. 2. It is noted here that high residual strain regions with striped patterns are sometimes observed near the $\langle 011 \rangle$ peripheral regions, which may be due to slip-like dislocations caused during crystal growth process. The value averaged over the whole area of wafer was about 10^{-5}.

Figure 3 shows a series of two-dimensional distribution maps of $|S_r - S_t|$ measured after thermal processing at the four different temperatures of 480, 505, 530, and 580C. It should be noticed that the normalized scale is four times larger than that shown in Fig. 2. No change in residual strain distribution is before and after thermal processing at 480C. However, there are drastic changes after thermal processing at 505, 530, and 580C. It is clearly found that there exist many slip lines along the $\langle 011 \rangle$ directions in the $\langle 001 \rangle$ peripheral regions.

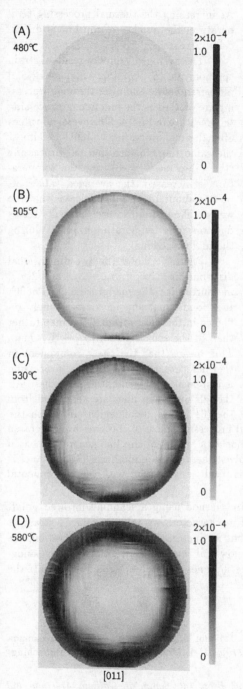

Figure 3. A series of two-dimensional distribution maps of $|S_r - S_t|$ measured after thermal processing at four different temperatures.

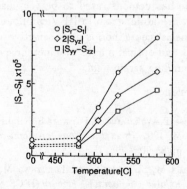

Figure 4. Averaged values of residual strain components; $|S_r - S_t|$, $|S_{yy} - S_{zz}|$, and $2|S_{yz}|$ measured before (T=0) and after thermal processing.

450

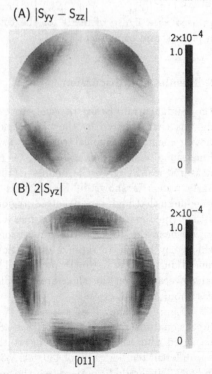

(A) $|S_{yy} - S_{zz}|$

2×10^{-4}
1.0

0

(B) $2|S_{yz}|$

2×10^{-4}
1.0

0

[011]

Figure 5. Two-dimensional distribution maps of $|S_{yy} - S_{zz}|$ and $2|S_{yz}|$ measured after thermal processing at 580C.

As increasing the thermal processing temperature, the number of slip line is drastically increased.

The averaged values of residual strain components: $|S_r - S_t|$, $|S_{yy} - S_{zz}|$, and $2|S_{yz}|$ measured before and after thermal processing are plotted as the function of processing temperature in Fig. 4. The averaged values after thermal processing at 480C are almost the same those before thermal processing (the values at $T = 0$). As the temperature is increased, the averaged values attain to about 10^{-4}. This indicates that the wafer suffered from severe thermal stresses due to strong temperature gradient during thermal processing.

Figure 5 shows the two-dimensional distribution maps of $|S_{yy} - S_{zz}|$ and $2|S_{yz}|$ measured after thermal processing at 580C. It is found that slip lines are strongly reflected in the shear strain $2|S_{yz}|$ map rather than in the tensile strain $|S_{yy} - S_{zz}|$ map. This means that the shear component in thermal stresses during thermal processing is responsible for generating slip lines. Although some slip lines are generated from the $\langle 011 \rangle$ peripheral regions in which the striped patterns are observed before thermal processing, most of slip lines are generated from the $\langle 001 \rangle$ peripheral regions. A thermoelastic model of slip line generation based on resolved shear stress shows that the resolved shear stress gliding the $\{111\}$ planes is strong in the $\langle 001 \rangle$ peripheral regions. This result is consistent with the experimental results observed above.

In concluding, the present photoelastic method using a scanning infrared polariscope has demonstrated to be very useful in characterizing process-induced strain as well as residual strain in III-V compound semiconductors. For slip line observation, the photoelastice method is more sensitive and powerful than the conventional X-ray topography and optical microscope methods, since slip lines themselves strongly modulate the surrounding strain field.

References

Kawase T, Wakamiya T, Fujiwara S, Kimura K, Tatsumi M, Shirakawa T, Tada T, and Yamada M, 1992 *Proc. Int. Conf. on Semi-insulating III-V Materials, Ixtapa* (IOP Publishing) pp.85-90

Yamada M, Shibuya T and Fukuzawa M, 1993 *Proc. Int. Symp. on Gallium Arsenide and Related Compounds, Friburg* (IPCS No.136) pp.505-510

Yamada M and Fukuzawa M, 1994 *Proc. Int. Conf. on Semi-insulating III-V Materials, Warsaw* (World Scientific) pp.95-98

Inst. Phys. Conf. Ser. No 145: Chapter 3
Paper presented at 22nd Int. Symp. Compound Semiconductors, Cheju Island, Korea, 28 August–2 September 1995
© *1996 IOP Publishing Ltd*

Faraday-Stark Electrophotonic Effect

Z.K. Lee and D. Heiman

Massachusetts Institute of Technology, Francis Bitter National Magnet Laboratory, Cambridge, MA 02139 USA

M. Sundaram* and A.C. Gossard

Materials Department, University of California, Santa Barbara, CA 93106 USA

Abstract. We report on a new electrophotonic effect based on resonant Faraday rotation and the quantum confined Stark shift. By applying an electric field to a quantum well structure the exciton energy is red-shifted, thereby tuning the Faraday effect into resonance. We have observed a tunable rotation of 11 deg in a GaAs quantum well structure. This effect could lead to high-speed modulators using magnetic and non-magnetic semiconductor materials.

1. Introduction

Resonant Faraday rotation can be achieved in semiconductor quantum well (QW) structures when the photon energy is close to an excitonic transition [1]. By exploiting the strong dependence of this effect on the exciton energy, important electrically-controlled Faraday rotation devices can be constructed using QW structures in which the exciton energy is electrically tunable due to the quantum confined Stark effect [2]. By applying an electric field to a QW structure the exciton energy is red-shifted, thereby tuning the Faraday effect into resonance, resulting in polarization modulation. Since only a small dc magnetic field is required (typically less than 1 T), permanent magnets can be used. A modulator device based on this effect could operate at very high speed while achieving a modulation depth higher than that obtainable with absorption devices. Similar effects are also expected in Kerr reflection, and in the Voigt geometry appropriate for waveguide implementation.

In addition to red-shifting the exciton energy, applying an electric field across a QW structure also separates the electron and hole wavefunctions spatially, reducing the transition probability [3] and the magnitude of the Faraday rotation. If appropriate operating conditions are met, this effect can be used in conjunction with the Faraday-Stark effect to enable the construction of a modulator requiring a very small modulating electric field.

2. Theory

The excitonic dielectric function of a semiconductor material in a magnetic field can be modeled by a two-level system [1] and is given by:

$$\epsilon_{1\pm} = \epsilon_o + \epsilon_x \frac{\Delta_{\pm}}{1+\Delta_{\pm}^2} \tag{1}$$

$$\epsilon_{2\pm} = \epsilon_x \frac{1}{1+\Delta_{\pm}^2} , \tag{2}$$

where ϵ_1 and ϵ_2 are respectively the real and imaginary parts of the dielectric function for the right and left hand circular polarizations σ_+ and σ_-, ϵ_o is the static dielectric constant, ϵ_x is the excitonic dielectric constant, and Δ is given by

$$\Delta_{\pm} = 2 \frac{(E_x \pm \delta E - \hbar\omega)}{\Gamma} , \tag{3}$$

where E_x is the electric field dependent exciton energy, δE is the half spin splitting between σ_+ and σ_-, and Γ is the exciton linewidth. The Faraday rotation per unit length θ is given by

$$\theta = \frac{\pi}{\lambda} (n_- - n_+) , \tag{4}$$

where λ is the wavelength in the semiconductor, n_+ and n_- are respectively the refractive indices of σ_+ and σ_-, and are given by

$$n_{\pm} = \sqrt{\frac{\epsilon_{1\pm} + (\epsilon_{1\pm}^2 + \epsilon_{2\pm}^2)^{1/2}}{2}} . \tag{5}$$

The excitonic dielectric constant ϵ_x is material dependent and can be shown to be given by [4]

$$\epsilon_x = \frac{2}{\Gamma} \frac{f_{cv} |<e|h>|^2 e^2 \hbar}{m_o \omega \epsilon \pi a_B^3} , \tag{6}$$

where f_{cv} is the conduction-valence band oscillator strength, ϵ is the permitivity of free space, a_B is the exciton Bohr radius, and $|<e|h>|$ is the electric field dependent electron-hole wavefunction overlap integral. It should be noted that $\epsilon_x \propto 1/\Gamma$, indicating that a large Faraday rotation requires a narrow linewidth. Moreover, $\epsilon_x \propto E_g$, since $f_{cv} \propto 1/E_g$, $a_B \propto 1/E_g$, and $\omega \propto E_g$, where E_g is the bandgap.

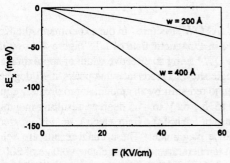

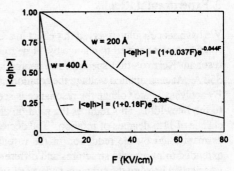

Figure 1. Shift of exciton energy versus electric field for 200Å and 400 Å QWs.

Figure 2. Electric field dependence of electron-hole wavefunction overlap.

Figure 1 shows the calculated change in exciton energy as a function of electric field for both 200 Å and 400 Å QWs in GaAs. Here the Airy function solution [5] and the WKB method [6] are used. Figure 2 shows the corresponding electron-hole wavefunction overlap integrals |<e|h>| calculated by numerically integrating the Airy function solutions. It can be seen that the energy shift for the 400 Å QW is large compared to that of the 200 Å QW. This is desirable since only a small voltage is required to achieve a practical amount of modulation. Moreover, |<e|h>| is almost an exponential function of electric field, which decays much faster in the 400 Å QW than in the 200 Å QW, indicating that the electrons and holes are more spatially confined in the narrower 200 Å QW. This could also be a desirable feature as the sharp reduction of |<e|h>| quenches the Faraday rotation at the same time the electric field tunes the Faraday rotation away from resonance, resulting in the requirement of smaller modulating voltages. Figure 3 shows the Faraday rotation as a function of electric field for various photon energies for a 200 Å QW. The reduction of |<e|h>| at high electric fields is apparent by comparing the magnitudes of the Faraday resonances at various photon energies.

Even though the calculations above are based on GaAs it is applicable to smaller bandgap materials such as InGaAs, and larger bandgap materials such as ZnSe.

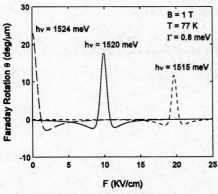

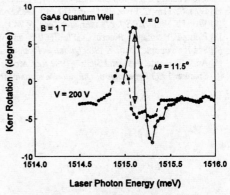

Figure 3. Faraday rotation versus electric field for various photon energies for 200 Å QW.

Figure 4. Kerr rotation spectra of GaAs quantum well at gate voltages of 0 V and 200 V. T = 2 K.

3. Experimental Results

We have successfully observed this effect in a GaAs QW structure. In our experiments the Kerr rotation was observed to change by 11 degrees in a magnetic field of 1 T. Figure 4 shows the resonant Kerr rotation spectra for a parabolic QW having an effective width of approximately 350 Å. At zero applied voltage the rotation angle shows a resonance that peaks at +7 degrees. By applying a gate voltage the resonance is seen to red-shift by an amount approximately equal to the linewidth. This reduces the rotation at 1515.1 meV to –4.5 degrees, resulting in a total change of 11.5 degrees of rotation. The decrease of $|<e|h>|$ at a finite electric field is apparent, as manifested by the reduction of the resonance magnitude. These initial results are being extended to optimized structures and different temperatures. The large gate voltage of 200 V was needed because the structure had a semi-insulating substrate. An n^+ substrate will allow for sizeable electric fields with small voltages of order 1 volt.

4. Conclusion

We have successfully observed the Faraday-Stark electrophotonic effect, which has important potential applications in high-speed modulator devices and optical isolators. This effect is expected to occur in most magnetic and non-magnetic semiconductors. Devices that operate at wavelengths spanning the IR to UV spectral region might be constructed. Work is under way to extend this effect to different materials and structures in order to optimize the operating conditions.

5. Acknowledgements

We thank G. Favrot, F. Plentz, J. Moodera, H. Wang and C. Fonstad for many valuable discussions. Inspiration for this work resulted from National Science Foundation grants DMR–9121452 and DMR-9201614. The quantum well material was grown with support from QUEST, the NSF Science and Technology Center at UCSB. The MIT Magnet Lab is supported by the NSF.

6. References

* Present address: Jet Propulsion Laboratory, Pasadena, CA 91109 USA
[1] Buss C, Frey C, Flytzanis C, and Cibert J 1995 *Solid State Commun.* **94** 543
[2] Wood T 1988 *IEEE J. of Lightwave Tech.* **6**, 743
[3] Plentz F, Heiman D, Pinczuk A, Pfeiffer L N, and West K W 1995 *Phys. Rev. B*
[4] See for example, Datta S 1989 *Quantum Phenomena vol. VIII* (New York:Addison-Wesley)
[5] Ando T, Fowler A B, and Stern F 1982 *Rev. Modern Phys.* **54** 437
[6] Shanker 1993 *Principles of Quantum Mechanics* (New York: Plenum Press)

Inst. Phys. Conf. Ser. No 145: Chapter 3
Paper presented at 22nd Int. Symp. Compound Semiconductors, Cheju Island, Korea, 28 August–2 September 1995
© *1996 IOP Publishing Ltd*

Sensitive observation of eigen-energy in InGaAs/InAlAs quantum wells by means of Stark-effect induced shift of photocurrents

K. Tanaka and N. Kotera

Kyushu Institute of Technology, Iizuka, Fukuoka 820, Japan

Abstract. A novel technique of photocurrent spectroscopy has been studied in In-GaAs/InAlAs multi-quantum wells lattice-matched to InP. Photocurrents in p-i-n junctions had little peaks in the spectra, if the well width was as wide as 20 nm. By subtracting the two spectra under different electric fields, we could observe 6 tiny peaks which could be magnified by a factor of 10^4. The eigen-energies were shown to change with the square of electron quantum number. Electron effective mass within a well as a function of energy from zero to about 0.47 eV in conduction subbands was evaluated. Nonparabolic tendency was observed.

1. Introduction

Multi-quantum wells (MQWs) have a great potential for application to optical and opto-electronic devices. When an electric field is applied normal to the quantum well plane, significant effects on the carrier confinement and the energy states appear within the quantum wells. So far, we observed a series of exciton peaks in photocurrent spectra of MQWs with 5- and 10-nm well widths [1, 2, 3]. The peaks were seen clearly in such narrow well widths, but any peaks didn't appear clearly in case of a 20-nm well width. In this repeort, photocurrent difference spectra produced from the photocurrent spectra taken at two different dc bias voltages were used to analyze eigen-energies in quantum wells. Electron effective mass in the conduction subbands was analyzed and discussed.

2. Experimental

Three specimens of p-i-n photodiodes including undoped $In_{0.53}Ga_{0.47}As/In_{0.52}Al_{0.48}As$ MQWs were grown by MBE with the well widths of 5, 10 and 20 nm. They were made up of a sequence of layers of a 700-nm-thick n-InGaAs buffer layer on a InP (100) substrate, a series of InGaAs/InAlAs MQWs and a p-InGaAs contact layer. Diodes of 0.5-mm diameter were fabricated by mesa etching. Photocurrents were measured as a function of wavelength in electric fields at room temperature. The field in the well was varied by changing a bias voltage between the contact layer and the substrate. Resistance of a p-i-n junction was on the order of 10 MΩ under a typical reverse bias voltage of -1 to -10 V. A halogen lamp filtered by a 600-grid/mm-grating monochrometer was employed as a light source at wavelength range of 300-2000 nm. The resolution was less than 10 meV. Photocurrents were detected with a current preamplifier and a digital lock-in amplifier.

3. Results

Photocurrent spectra of a specimen with 20-nm well width were measured in different applied biases as shown in Fig. 1. As the reverse bias was decreased, photocurrents were drastically decreased. A band edge was near around 1700 nm. These datum were not normalized with a light intensity spectrum. Two peaks near 1000 and 1400 nm were caused by the maximum of used halogen lamp and monochrometer, respectively. Any peaks didn't appear clearly in the spectra except these two.

In order to produce a photocurrent difference sprctrum, the spectra were normalized by the maximum value all over the wavelength range as shown in Fig. 2 (the 1st step). Several discrepancies were clearly seen. A tiny peak possibly caused by the fundamental exciton absorptions could be distinguished by the solid bar in each spectrum. By increasing the reverse bias, the peak became broader and shifted to lower energies due to a quantum-confined Stark effect (QCSE). The shift was 42 meV from +0.2 to –3 V. The one normalized spectrum (–3 V) in Fig. 2 was subtracted from the other (+0.2 V), in order to clarify some differences between two spectra, as shown in Fig. 3 (the 2nd step). Six sharp peaks were clearly extracted from the 20-nm well spectrum (C) as expected from Fig. 2. Positive and negative peaks appeared in pair in the spectra. The negative belonged to the normalized spectrum (–3 V). The positive belonged to the other (+0.2 V). Photocurrent difference spectra of other specimens were produced by the same processing. Two peaks were resolved in the spectrum of 5-nm well width (A). Three sharp peaks were clearly resolved in the spectrum of 10-nm well width (B). The peak positions agreed well with those determined from photocurrent spectra of 5-nm and 10-nm well widths very. In this new spectroscopy, the amplification factor of 3×10^3-1×10^4 times compared to the original normalized spectrum was realized.

4. Analysis and discussion

Based on the theoretical calculation using a finite square well model, energy levels of the n-th conduction subband, $E_\epsilon(n)$, were proportional to the square of the quantum number, n. The square law, $E_\epsilon(n) \simeq n^2 E_\epsilon(1)$ holds generally [3, 4]. The peak energies obtained from the photocurrent difference spectra were fitted with the square-law very well, as shown in Fig. 4. Therefore, it was supposed that these peaks are attributed to excitonic band-to-band transitions between higher conduction subbands and the ground-state heavy-hole subband. The transitions were labeled by using the notation nHHl, where n and l are the quantum numbers of the conduction subband and the heavy-hole subband. The HH denoted the heavy hole. A right-end peak corressponded to a transition, 1HH1. The energies, $E_\epsilon(1)$, determined from a inclination of fitting line on Fig. 4 were 108 meV for 5-nm well, 56 meV for 10-nm well, 16 meV for 20-nm well. Another peak of the 5-nm well width (A) corresponded to 2HH1. Other two peaks of the 10-nm well width (B) corresponded to 2HH1, 3HH1, respectively. A sereis of peaks corresponded to the conduction subband of the consecutive quantum number from right to left. The number of subbands in the conduction band was 2 for 5-nm well width, 3 for 10-nm and 6 for 20-nm. Light-hole-related transition was not observed for the InGaAs/InAlAs MQWs system by photocurrent measurements [5].

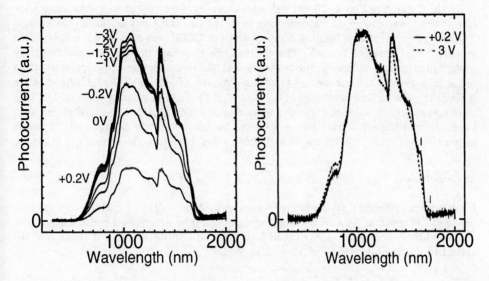

Fig. 1 Photocurrent spectra of 20-nm well width at 300 K. The bias voltage was changed from +0.2 to -3.0 V.

Fig. 2 Photocurrent spectra at two different biases. Each spectrum was normlized by the maximum value.

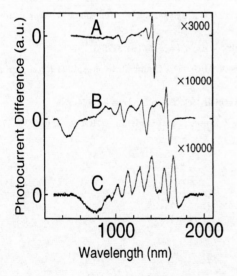

Fig.3 Photocurrent difference spectra produced from the photocurrent spectra taken at 2 different dc biases. A; 5-nm well width. B; 10-nm well width. C; 20-nm well width. The base line for each spectrum is offset for clarity.

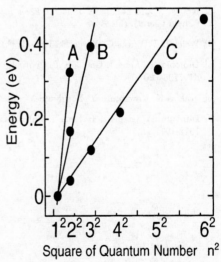

Fig.4 The square-law plot of peak energies on the assigned quantum number, n, of the conduction subband. Diameters of solid circles are 20 meV in energy. Zero of the energy is set at E_{1HH1} for all specimens.

In the C curve in Fig. 4 (20-nm well width), energy levels fitted well with the square law at the lower energy, but a deflection from the law came out at the higher energy. Though errors of exciton binding-energies and the QCSE was considered, a total error was estimated less than 20 meV. The electron effective mass must be increased slightly toward high energy. Nonparabolic tendency was observed. For the numerical estimation of the effective mass, a conduction-band discontinuity of 500 meV, the InGaAs effective mass at the conduction band edge of 0.041 m_0 and the InAlAs effective mass of 0.075 m_0 were assumed [5, 6, 7]. If the effective mass was constant, only five confined states were predicted and the sixth state in the well was not allowed in the 20-nm well. However larger effective mass of 0.052 m_0 at high energy could produce the sixth eigen state.

5. Conclusion

Photocurrent difference spectroscopy was shown to be useful to extract tiny exciton peaks in the spectra. As a result, peak energies could be determined consistently.

The authors appreciate Mr. Hitoshi Nakamura of Central Research Laboratory, Hitachi Ltd. for the fabrication of p-i-n photodiodes.

References

[1] Kotera N and Tanaka K 1994 *the 21st International Symposium on Compound Semiconductors* (San Diego) 233-236

[2] Kotera N, Tanaka K and Nakamura H 1995 *J.Appl. Phys.* (to be published in Oct. 15 issue)

[3] Kotera N and Tanaka K 1995 *the 22nd International Symposium on Compound Semiconductors* (Cheju Island), this issue

[4] Welch D F, Wicks G W and Eastman L F.1984 *J.Appl. Phys.* **55**(8) 3176-3179

[5] Dimoulas A, Leng J. Giapis K P. Georgakilas A. Michelakis C and Christou A 1993 *Phys. Rev. B* **47** 7198-7207

[6] Wakita K, Kawamura Y, Yoshikuni Y and Asahi H 1985 *Electron. Lett.* **21** 574-576

[7] Dimoulas A, Giapis K P, Leng J, Halkias G, Zekentes K and Christou A 1992 *J. Appl. Phys.* **72** 1912-1917

Inst. Phys. Conf. Ser. No 145: Chapter 3
Paper presented at 22nd Int. Symp. Compound Semiconductors, Cheju Island, Korea, 28 August–2 September 1995
© *1996 IOP Publishing Ltd*

Intensity oscillations in the magnetoluminescence of a modulation-doped (Al,Ga)As/GaAs single heterojunction

K.-S. Lee*, E. H. Lee*, C. H. Perry†, and Y. Kim†

* Electronics and Telecommunications Research Institute, Taejon 305-600, Korea

† Northeastern University, Boston, MA 02115 U.S.A. & National High Magnetic Field Laboratory, Los Alamos National Laboratory, Los Alamos, NM 87545, U.S.A.

Abstract.   We report intensity oscillations in the low temperature (T=4.2K) magnetoluminescence of a modulation-doped (Al,Ga)As/GaAs single heterojunction having two occupied subbands at zero field. For B > 1.9T applied in the direction parallel to the growth direction, luminescence peaks from the lowest Landau-level from the second subband are pronounced and show intensity oscillations nearly periodic with the inverse magnetic field. Self-consistent Landau-level calculations indicate that both the population of the second subband and the energy gap between Landau levels from this and the first subband reveal nearly periodic oscillations in inverse magnetic field. Oscillations in the luminescence intensity can be explained by the effect of population oscillations in the second subband.

1. Introduction

Recently, the magnetoluminescence studies of modulation-doped single heterojunctions (MDSH) having two occupied subbands at zero magnetic field (B-field) have shown pronounced oscillations in the photoluminescence (PL) intensity of the interband recombination, as well as in its peak energy and peak width.[1] Oscillations in the PL intensity were explained by the combination of population effects of the second subband (E_2) and a many-body interaction between the E_2 exciton and the Fermi edge resonance of Landau levels from the first subband (E_1).[2-8] The many-body interaction, known as the optical Shubnikov-de Haas (OSdH) effect[8], was found to be largest when the Fermi level lies within the extended states, i.e., at odd-integer filling factors. However, our recent theoretical study[9] showed that the population of E_2 and the energy gap between a Landau level of E_2 and another from E_1 reveal nearly periodic oscillations with the inverse magnetic field. The effect of population oscillation was found to be important for the analysis of magnetotransport and magnetoluminescence spectra from MDSHs having two occupied subbands before other many-body interactions was introduced. In this presentation, focusing only on the transition between E_2 and photogenerated hole states, we report intensity oscillations in the magnetoluminescence of a modulation-doped (Al,Ga)As/GaAs single heterojunction having two occupied subbands and attempt to explain the experimental results in terms of the population oscillation of E_2.

2. Samples and experimental details

The structure of the $Al_{0.35}Ga_{0.65}As$/GaAs MDSH studied in this work consists of an [001]-oriented semi-insulating GaAs substrate, an undoped $2\mu m$ GaAs, an undoped $6nm$ (Al,Ga)As layer as a spacer, a Si-doped (Al,Ga)As layer of thickness $40nm$ with the doping density $2.0 \times 10^{18} cm^{-3}$, and an undoped $5nm$ GaAs cap layer. Low temperature (T=4.2K) PL measurements were carried out with the sample mounted in a strain-free environment. For illumination we used a 632.8nm HeNe laser with the maximum output power of 1 mW. An optical fiber with the core diameter of $400\mu m$ was employed to couple the light in and out of the sample. B-fields up to 15 T were applied in the direction parallel to the growth axis. The PL spectra were dispersed by a $3/4m$ double monochromator with a resolution of 0.3 meV. The PL investigations were performed in two different ways; i) by monitoring PL signals as a function of the emitted photon energy at a fixed magnetic field, and ii) by sweeping the B-field at a fixed photon energy.

3. Results and discussion

Fig. 1 displays B-field dependence of the PL spectra for the MDSH. At zero field (B=0), PL is mainly due to the exciton recombination in the bulk GaAs such as the free exciton at 1.5155 eV, excitons bound to neutral donors at 1.5145 eV, and excitons bound to ionized acceptors at 1.5129 eV. Rather a weak and broad feature which is attributed to the radiative recombination between two-dimensional electrons in E_2 and photogenerated holes in the bulk GaAs layer is observed on the low energy side of bulk exciton peaks. For B>1.9T this feature at B=0 develops into Landau transitions and the lowest(0-0) one becomes well pronounced in the region of the bulk exciton peaks and reveals intensity oscillations. Though not shown in the figure, this peak is pronounced again around $9.5 \pm 0.5T$.

At low fields (B<7T), the intensity oscillations can be clearly resolved using a swept field technique. Fig. 2 shows the intensity of the PL with the spectrometer set at 1.512 eV just below the transition energy of bulk excitons bound to ionized acceptors. At this energy the intensity of the low energy tail of the new peak is monitored. The intensity of PL measured in this way is sensitive to the oscillations in the width of the luminescence of the Landau transition. It is noted that the magnetic fields at the local maxima and minima in Figs. 1 and 2 are in good agreement and both the intensity and width of PL have the same oscillatory behavior. Though not shown in the figure, the locations of the oscillatory peaks are nearly independent of the transition energy. This means that the intensity oscillation depends on the density distribution in the Landau levels of conduction subbands.

In Fig. 3 the lowest Landau transition between electrons in E_2 and photogenerated holes is plotted with dots as a function of B-field. The transition shows a linear behavior with additional oscillations. The slope of the Landau transition is 0.98meV/T, and from this value the reduced mass of an electron-hole pair is determined to be $0.0591m_e$, where m_e is the bare electron mass. Though not shown in the figure, the band-to-acceptor transition shows a linear behavior with the slope of 8.2meV/T, which gives the electron effective mass of $0.0706m_e$. Hence, the effective mass of a photogenerated hole in bulk GaAs layer of the MDSH is determined to be $0.361m_e$.

Fig. 4 displays the inverse B-field at intensity maxima versus the indeces of Landau levels (n=0,1,2,...) from E_1 by those the lowest Landau level of E_2 is crossed. The intensity oscillation is nearly periodic with the inverse B-field. In the figure we also

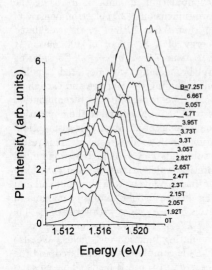

Figure 1. Magnetic field dependence of the PL spectra for the MDSH.

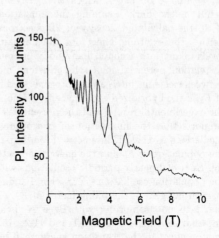

Figure 2. PL intensity versus applied magnetic field with the spectrometer set at 1.512eV.

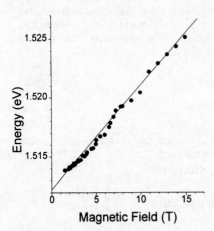

Figure 3. Plot of the energy of the 2DEG peak versus applied magnetic field. The dots denote the experimental results, while the drawn line is a linear fit.

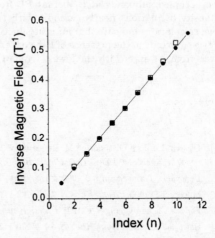

Figure 4. 1/B at intensity maxima (open squares) for the MDSH. The dots are the calculated results using $N_e = 1.05 \times 10^{12} cm^{-2}$.

462

plotted with dots the calculated results of the population maxima in E_2 using the measured electron effective mass and the total two-dimensional(2D) electron density of $N_e = 1.05 \times 10^{12} cm^{-2}$, which is slightly higher than $9.6 \times 10^{11} cm^{-2}$, the calculated result in the dark assuming the ionization energy of a Si donor of 100 meV. At B=0, the calculated density of 2D electrons in E_e was $7.2 \times 10^{10} cm^{-2}$. Considering neither the excitonic effect nor the Landau level broadening, we calculated Landau-levels taking into account many-body interactions including the Hartree and exchange-correlation potential in a self-consistent manner. The probability of recombination between electrons in E_2 and photogenerated holes will be proportional to the population of E_2, if the hybridized excitonic effect is not taken into account. With this assumption, the experimental results and the present calculation are in good agreement. This result supports that the interaction of two occupied subbands can explain the pronounced oscillations in the PL intensity. From our theoretical calculations, we find that both the population of E_2 and the energy gap between Landau levels from this subband and the E_1 reveal nearly periodic oscillations with the inverse B-field. Intensity maxima in the population oscillation occur when Landau levels from E_2 cross with those of E_1, i.e. at even integers of the filling factor of E_1, $\nu_1 = N_1 h/eB$. However, it should be noted that ν_1 is not isomorphic with B, as N_1 is an oscillating function of B-fields. We also find that in B-fields, 7.2 - 7.8T and 11.2 - 16.2T, E_2 is completely depopulated and the exciton effect in the transition energy can be observed in those B-fields. This effect is manifest in the additional oscillations of the Landau transition plotted in Fig. 3.

4. Conclusion

In summary, we have learned from PL investigation of a (Al,Ga)As/GaAs MDSH with two occupied subbands at B=0 that PL from the lowest Landau level of E_2 showing intensity oscillations nearly periodic with the inverse B-field. Self-consistent Landau level calculations indicate that intensity fluctuations in the PL result from the interaction of two subbands in the presence of B-fields, and the intensity is pronounced when Landau levels from E_1 cross with the lowest Landau level of E_2.

References

[1] Drissen F A J M, Olsthoorn S M, Berendschot T T J M, Pen H F, Giling L J, Jones G A C, Ritchie D A, and Forst J E F, 1992 *Phys. Rev. B* **45** 11823-11828

[2] Uenoyama T and Sham L J 1990 *Phys. Rev. Lett.* **65** 1048-1051

[3] Mueller J F 1990 *Phys. Rev. B* **42** 11189-11193

[4] Perakis I E and Chang Y C 1991 *Phys. Rev. B* **43** 12556-12563

[5] Hawrylak P 1991 *Phys. Rev. B* **44** 11236-11240

[6] Bauer G E W 1992 *Phys. Rev. B* **45** 9153-9162

[7] Skolnick M S, Rorison J M, Nash K J, Mowbray D J, Tapster P R, Bass S J, and Pitt A D 1987 *Phys. Rev. Lett.* **58** 2130-2133

[8] Chen W, Fritze M, Nurmikko A V, Ackley D, Colvard C, and Lee H 1990 *Phys. Rev. Lett.* **64** 2434-2437

[9] Lee K S and Lee E H 1995 *Phys. Rev. B* **51** 13315-13319

Inst. Phys. Conf. Ser. No 145: Chapter 3

Paper presented at 22nd Int. Symp. Compound Semiconductors, Cheju Island, Korea, 28 August–2 September 1995

© *1996 IOP Publishing Ltd*

Magnetotransport, Magneto–Optical, and Electronic Subband Studies in In$_x$Ga$_{1-x}$As/In$_y$Al$_{1-y}$As One–Side–Modulation–Doped Asymmetric Double Quantum Wells

T. W. Kim, M. Jung, and D. U. Lee

Department of Physics, Kwangwoon University, 447-1 Wolgye-dong, Nowon-ku, Seoul 139-701, Korea

K. H. Yoo

Department of Physics, Kyung Hee University, Seoul 130-701, Korea

S. Y. Ryu

Department of physics, State University of New York at Buffalo, U. S. A.

Abstract. Shubnikov-de Haas (S-dH) and Van der Pauw Hall-effect, and cyclotron resonance measurements on In$_x$Ga$_{1-x}$As/In$_{0.52}$Al$_{0.48}$As asymmetric double quantum wells grown by metalorganic chemical vapor deposition have been carried out to investigate the electrical properties of a free electron gas and to determine the effective mass of the electron gas, subband energies and wave functions in the quantum wells. Transmission electron microscopy measurements showed that a 100-Å In$_{0.8}$Ga$_{0.2}$As and a 100-Å In$_{0.53}$Ga$_{0.47}$As quantum wells were separated by a 35-Å In$_{0.25}$Ga$_{0.75}$As potential barrier in an active region. The S-dH measurements at 1.5 K demonstrated clearly the existence of a quasi-two-dimensional electron gas in the quantum wells. The results of the cyclotron resonance measurements show that the saturation of the absorption near resonance is a direct evidence of the large electron concentration in the quantum well. The electron effective masses determined from the slopes of the main peak absorption energies as a function of a magnetic field are 0.06171 and 0.05228 m$_e$ for the first exited and ground subbands, respectively. Electronic subband energies and wavefunctions in the quantum wells were calculated by a self-consistent method taking into account exchange-correlation effects together with the strain and nonparabolicity effects. The 1st excited subband wavefunction in the asymmetric quantum well is strongly coupled over both In$_{0.8}$Ga$_{0.2}$As and In$_{0.53}$Ga$_{0.47}$As wells, and the results of the cyclotron resonance measurements satisfy qualitatively the nonparabolicity effect of the double quantum well.

464

1. Introduction

Recently, with rapid advancements in epitaxial growth technologies such as molecular beam epitaxy and metalorganic chemical vapor deposition (MOCVD), the fabrication of new types of strained nanostructures has attracted much attention for both scientific and technological reasons [1]. Potential applications of strained quantum wells in optoelectronic devices have driven an extensive and successful effort to fabricate that system on various system [2-4]. Electrical and optical studies of the two-dimensional electron gas (2DEG) in various strained quantum wells have been attempted for last few years [5-6]. Among these structures, coupled quantum wells have been attractive because of the interest in both an investigation of the fundamental physical properties [7] and tunable coherent light sources for optical communications [8]. Recently, although few works concerning the 2DEG and subband energy studies in the coupled double quantum wells were investigated [9], only the preliminary results for the electrical properties and electroluminescent spectra have been reported.

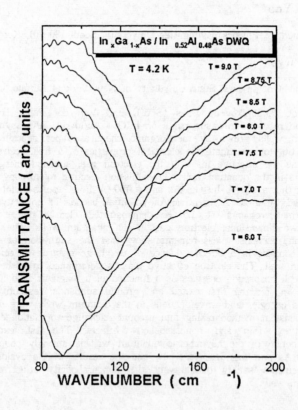

FIG. 1. Transmittance spectra for several different magnetic fields on an $In_xGa_{1-x}As/In_{0.52}Al_{0.48}As$ double quantum well.

This paper reports magnetotransport and magneto-optical data for $In_xGa_{1-x}As/In_{0.52}Al_{0.48}As$ double quantum well. Transmission electron microscopy (TEM) measurements were carried out to investigate the structural properties of the $In_xGa_{1-x}As/In_{0.52}Al_{0.48}As$ asymmetric double quantum wells, and Shubnikiv-de Haas (S-dH) and Van der Pauw Hall-effect measurements were performed in order to investigate the electrical properties of the electronic subbands in the $In_xGa_{1-x}As/In_{0.52}Al_{0.48}As$ double quantum wells grown by MOCVD. Furthermore, cyclotron resonance measurements were carried out to determine the electron effective mass in the quantum wells, and the electronic subband energies and the wavefunctions in the quantum wells have been calculated by a self-consistent method taking into account exchange-correlation effects together with strain and nonparabolicity effects.

2. Experimental details

The samples used in this work were grown on Fe-doped semi-insulating (100)-oriented InP substrates by MOCVD and consisted of the following structures: a 90-Å Si-doped (2×10^{19} cm^{-3}) $In_{0.53}Ga_{0.47}As$ capping layer for ohmic contacts, a 50-Å undoped $In_{0.53}Ga_{0.47}As$ layer, a 150-Å undoped $In_{0.52}Al_{0.48}As$ layer, a 50-Å Si-doped (2×10^{18} cm^{-3}) $In_{0.52}Al_{0.48}As$ modulation layer, a 30-Å undoped $In_{0.52}Al_{0.48}As$ spacer layer, a 100-Å $In_{0.8}Ga_{0.2}As$ quantum well, a 35-Å $In_{0.25}Ga_{0.75}As$ potential barrier, a 100-Å $In_{0.53}Ga_{0.47}As$ quantum well, a 500-Å undoped $In_{0.52}Al_{0.48}As$ buffer layer, and 30 periods of $In_{0.53}Ga_{0.47}As/In_{0.52}Al_{0.48}As$ superlattice buffer layer. The compositions of the samples were measured by using double-crystal X-ray diffraction, and the thickness of the each layer was determined by cross-sectional TEM.

Ohmic contacts to the samples were made by diffusing a small amount of indium through several layers at 450℃ in a hydrogen atmosphere for approximately 10 min. The S-dH and Hall effect measurements were carried out at a temperature of 1.5 K in magnetic fields up to 12 T in an Oxford superconducting magnet system using a Keithley 181 nanovoltmeter. After the ohmic contacts were formed, the $In_{0.53}Ga_{0.47}As$ cap layer was removed to get rid of the parallel conductance for measurements. Cyclotron resonance measurements were made with a Bomen DA-3 far infrared spectrometer in conjunction with a 9 Tesla superconducting magnet system, and the sample substrates were wedged 5° to avoid multiple-reflection interference.

3. Results and discussion

The results of the bright-field TEM image shows that there are no defects, dislocations or stacking faults due to the lattice mismatch among of the each layer. A high-resolution TEM image of a cross-sectional structure shows that 100-Å $In_{0.8}Ga_{0.2}As$ and 100-Å $In_{0.53}Ga_{0.47}As$ quantum well channels were separated by a 35-Å $In_{0.25}Ga_{0.75}As$ potential barrier in its active region. A TEM image shows that there are no misfit dislocations at $In_{0.52}Al_{0.48}As/In_{0.8}Ga_{0.2}As$, $In_{0.8}Ga_{0.2}As/In_{0.52}Al_{0.48}As$, and $In_{0.25}Ga_{0.75}As/In_{0.53}Al_{0.47}As$

heterointerfaces due to the lattice mismatch.

The results of the S–dH oscillations clearly show multiple frequencies, indicative of the occupation of several subbands by electrons in the $In_xGa_{1-x}As/In_{0.52}Al_{0.48}As$ double quantum well. These oscillations vary dramatically with the angle between the magnetic field and the surface normal, indicative of a 2DEG occupation of the quantum well. The quasi–2DEG nature giving rise to the S–dH oscillations was substantiated by studies with the magnetic field oriented 30° and 60° to the normal to the surface.

The transmittance spectra for the $In_xGa_{1-x}As/In_{0.52}Al_{0.48}As$ double quantum wells with different magnetic fields are presented in Fig. 1. Data were taken with the incident far infrared light normal to the sample surface and parallel to the direction of the magnetic field. After the peak absorptions were fitted by a sum of Lorentzian lines, the electron effective masses were obtained from the relationship between the frequency and the magnetic field, as shown in Fig. 2. The corresponding electron effective masses obtained from the slopes are (0.06171 ± 0.00064) and (0.05228 ± 0.00097) m_e for the first excited and ground subbands, repectively. This behavior shows clearly the nonparabolicity of the conduction band in the $In_xGa_{1-x}As/In_{0.52}Al_{0.48}As$ double quantum wells.

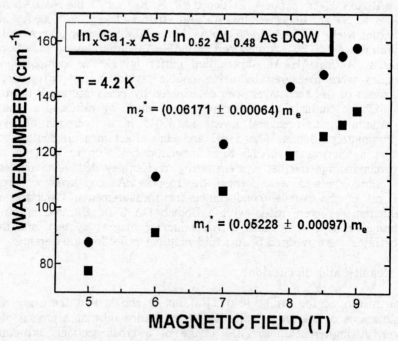

FIG. 2. Cyclotron resonance transition energies as a function of the magnetic field of the transmittance data for an $In_xGa_{1-x}As/In_{0.52}Al_{0.48}As$ double quantum well.

Even though all electrons contribute to the coupled quantum wells with a complicated confinement potential, and the shapes of the potential wells are probably quite complicated, the subband energy levels can be obtained by considering the Fermi level which is assumed to be at the shallow donor level of $In_{0.52}Al_{0.48}As$ modulation doped layer. To determine the subband energies, the calculated carrier density and a self-consistent numerical calculation taking into account the exchange-correlation effects together with the strain and nonparabolicity effects were used [10].

The dielectric constants of all of the $In_{0.52}Al_{0.48}As$ and $In_xGa_{1-x}As$ layers were taken to be 13.5 [11]. Even though the band offsets at the interfaces of a coupled quantum well are not well established yet, the band-edge discontinuities at the $In_{0.52}Al_{0.48}As/In_{0.8}Ga_{0.2}As$, the $In_{0.8}Ga_{0.2}As/In_{0.25}Ga_{0.75}As$, the $In_{0.25}Ga_{0.75}As/In_{0.53}Ga_{0.47}As$, and the $In_{0.53}Ga_{0.47}As/In_{0.52}Al_{0.48}As$ are assumed to be 753, 385, 220, and 588 meV, respectively [11-15]. The electron effective mass values of the $In_{0.52}Al_{0.48}As$ barrier, the $In_{0.53}Ga_{0.47}As$ and $In_{0.8}Ga_{0.2}As$ wells, and the $In_{0.25}Ga_{0.75}As$ barrier were 0.079, 0.029, 0.053, and 0.040 m_e, respectively [11-15]. The results of the numerical calculation for the subbands in an $In_xGa_{1-x}As/In_{0.52}Al_{0.48}As$ double quantum well structure are shown in Fig. 3, and these results show that electrons are occupied at three subbands. In this case, the calculated magnitudes of the two eigenenergies for the potential bottom in the well were 231, 404, and 487 meV corresponding to the ground, first, and second excited electron subbands, respectively. The Fermi level from the potential bottom in the well is 0.491 meV. The electronic subband wave function of the 2nd excited subbands in a double quantum well structure is coupled clearly over both the $In_{0.8}Ga_{0.2}As$ and the $In_{0.53}Ga_{0.47}As$ wells. The energy eigenfunctions are indicated by the dashed lines in Fig. 3.

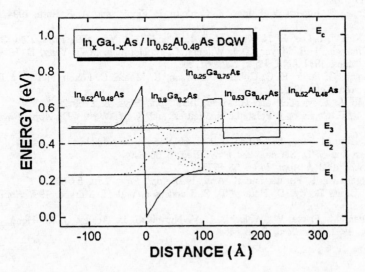

FIG. 3. An electronic subband structure of an $In_xGa_{1-x}As/In_{0.52}Al_{0.48}As$ double quantum well.

468

4. Summary and conclusions

TEM measurements showed that an $In_xGa_{1-x}As/In_{0.52}Al_{0.48}As$ double quantum well was grown on an InP substrate. The results of the S–dH measurements at 1.5 K demonstrate clearly the existence of a 2DEG in the $In_xGa_{1-x}As/In_{0.52}Al_{0.48}As$ double quantum wells grown by MOCVD. The results of the cyclotron resonance showed that the electron effective masses occupied in the ground and first excited subbands are 0.05228 and 0.06171 m_e, and these values are larger than the electron effective masses of the $In_{0.53}Al_{0.47}As$ and $In_{0.8}Ga_{0.2}As$ band edges resulting from the nonparabolicity effects. While the ground subband wavefunction in this structure is localized in the $In_{0.8}Ga_{0.2}As$ well side, the 2nd excited subband wavefunction in the asymmetric double quantum well is coupled over both $In_{0.8}Ga_{0.2}As$ and $In_{0.53}Ga_{0.47}As$ wells.

Acknowledgement

This work was supported in part by the Korean Science and Engineering Foundation (Contract No. 941 - 0200 - 044 - 2). One of the authors (T. W. K) would like to thank Prof. B. D. McCombe for providing the cyclotron resonance facilities and helpful discussion during his stay in SUNY at Buffalo. The authors also would like to thank Drs. K-H. Yoo and W-P. Hong for their useful discussions.

References

[1] F. Capasso 1990 *Physics of Quantum Electron Devices* (Springer–Verlag, Heidelberg)
[2] S. F. Cui, G. M. Wang, Z. H. Mai, W. Feng, and J. M. Chou, 1993 *Phys. Rev. B* **48** 8797
[3] P. W. Yu, B. Jogai, T. J. Rogers, P. A. Martin, and J. M. Ballingall, 1994 *Appl. Phys. Lett.* **65** 3263
[4] C. K. Inoki, E. Ribeiro, V. Lemos, F. Cerdeira, P. Finnie, and A. P. Roth, 1994 *Phys. Rev. B* **49** 2246
[5] T. W. Kim, J. I. Lee, K. N. Kang, K-S. Lee, and K-H. Yoo, 1991 *Phys. Rev. B* **44** 12891
[6] J. P. Omaggio, J. R. Meyer, R. J. Wagner, C. A. Hoffman, M. J. Yang, D. H. Chow, and R. H. Miles, 1992 *Appl. Phys. Lett.* **61** 207
[7] J. Feldman, G. Peter, E. O. Gobel, P. Dawson, K. Moore, C. Foxon, and R. J. Elliot, 1987 *Phys. Rev. Lett.* **59** 2337
[8] L. Y. Liu, E. E. Mendez, and H. Meier, 1992 *Appl. Phys. Lett.* **60** 2971
[9] C. Sirtori, F. Capasso, J. Faist, L. N. Pfeiffer, and K. W. West, 1994 *Appl. Phys. Lett.* **65** 445
[10] T. W. Kim, M. Jung, D. U. Lee, and K. H. Yoo, to be published
[11] F. Stern and S. D. Sarma, 1984 *Phys. Rev. B* **30** 840
[12] S. Adachi 1982 *J. Appl. Phys.* **53** 8775
[13] N. Shigekawa, T. Furuta, and K. Arai, 1990 *Appl. Phys. Lett.* **57** 67
[14] G. Ji, U. K. Reddy, D. Haung, T. S. Henderson, and H. Morkoc, 1987 *Superlatt. and Microstruct.* **3** 539
[15] W. Chen, M. Fritze, W. Walecki, A. V. Nurmikko, D. Ackley, J. M. Hong, and L. L. Chang, 1992 *Phys. Rev. B* **45** 8464

Inst. Phys. Conf. Ser. No 145: Chapter 3
Paper presented at 22nd Int. Symp. Compound Semiconductors, Cheju Island, Korea, 28 August–2 September 1995
© *1996 IOP Publishing Ltd*

469

Exciton formation time measured with Magneto-photoluminescence

J B Zhu, E -K Suh, H J Lee, and Y G Hwang*

Semiconductor Physics Research Center & Department of Physics,
Chonbuk National University, Jeonju, 560-756, R. O. Korea

*Department of Physics, Wonkwang University, Iksan 570-749, Korea

Abstract. Steady state magneto-photoluminescence technique is adopted to measure the dynamical processes of carriers in semiconductors. With the cyclotron periods of carriers as measures of time, the formation time of excitons, and its relation with carrier densities in GaAs/AlGaAs multi-quantum wells can be directly obtained by analysing the photoluminescence intensities under various magnetic fields.

1. Introduction

The radiative recombination time of carriers in semiconductors is the determining factor for the switching times of fast speed devices, and the ratio of the radiative to the nonradiative recombination intensities plays the most important role in determining the quantum efficiency of opto-electronic devices. Although it is evident that at low temperature the radiative recombination of carriers in quantum wells is determined by the formation time and life time of excitons, these two steps still cannot be separated by experimental techniques so far, and the relation between the exciton formation time and the carrier density is still unclear.[1-4]

In this paper, magneto-photoluminescence (MPL) technique has been successfully employed to investigate the dynamical processes of carriers in semiconductors. The radiative recombination time in AlGaAs/GaAs multi-quantum well (MQW) has been separated into exciton formation time and exciton life time. Aided with theoretical analyses, the exciton formation time is directly obtained with the cyclotron periods of carriers in semiconductors as measures of time.

2. Experimental details

The AlGaAs/GaAs MQW sample was grown by molecular beam epitaxy. The 100 Å, 60 Å, 40 Å, 20 Å GaAs quantum wells are grown successively on the 1 μ GaAs buffer layer and covered with 80 Å GaAs cap layer. GaAs quantum wells are separated from adjacent quantum wells and from the buffer and cap layers by 300 Å $Al_{0.3}Ga_{0.7}As$ layers.

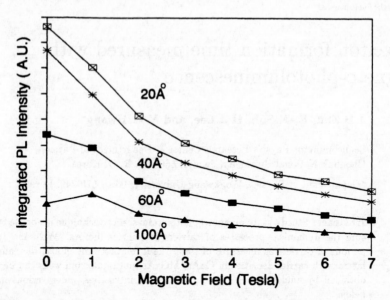

Fig. 1 The integrated PL intensities of AlGaAs/GaAs MQW under various magnetic field along the growth direction. The solid lines are drawn for the guide of eyes.

The high PL intensities and the sharp PL lines demonstrate the high quality of this sample.

MPL was performed at 4.2 K with magnetic field along the growth direction. The 5145 Å line of an Ar⁺ laser was propagated into magnet along a quartz optical fiber with 0.7 mm diameter and illustrated on the sample. Excitation intensity was kept low and unchanged during the measurement. It is found that the PL intensities from the quantum wells change systematically with magnetic field. Except the 100 Å well under small magnetic field, the exciton PL intensities in all the wells decrease as the magnetic field increases, as shown in Fig. 1.

3. Theoretical analyses

To interpret the experimental result, it is necessary to analys the magnetic field effects on all the dynamical processes of photo-generated carriers.

The PL intensity is determined by carrier and exciton densities generated in the corresponding well layers. It is expected that the carrier transport along the growth direction cannot be affected by the perpendicular magnetic field, and the relaxation of carriers, which takes place through the scattering of phonons, is also not changed by external magnetic field. The fact that the PL intensity changes immediately as the magnetic field increases from zero, as shown in Fig. 1, rules out the possibilities of the quadratic magneto-effect as well as the effect of magnetic field on the carrier-phonon interaction. As a result, the carrier relaxation rates from the barriers into the wells do not change with the perpendicular magnetic fields. Or to say, the generation rates of carriers in the wells are constants, and the PL intensity of an individual well layer in the MQW structure can be analysed independently.

At low temperature, the thermal activation of carriers from the wells to the barriers can be neglected. Radiative and nonradiative recombinations are the main carrier losses in well layers. In addition, the exciton PL intensity in every well is much larger than that due to free carrier recombinations, as shown in the PL spectra. This indicates that most of carriers except those recombine nonradiatively combine to form excitons rather than annihilate directly. The balance equations of electrons and holes in an individual well can be written as

$$G_e = n/\tau_e + n/\tau_f$$
$$G_p = p/\tau_p + p/\tau_f \qquad (1)$$

Here G_e and G_p are the electron and hole generation rates in the well, n and p are densities of electrons and holes, τ_e and τ_p are nonradiative loss timess of carriers, respectively, and τ_f is the formation time of exciton. As the nonradiative loss of carriers take place with one electron and one hole annihilate with each other, $\tau_e = \tau_h$ at steat state. And $G_e = G_p = G$, so that the charge density in the well is a constant. As the sample is not intentionally doped, $n = p$ is also a good approximation. Above equations can be simplified as

$$G = n/\tau_e + n/\tau_f. \qquad (2)$$

The exciton PL intensity is determined by the exciton density as well as exciton radiative and nonradiative life times. The radiative life time of exciton is determined by the time in which excitons relax from high energy states to the bottom of exciton band so that radiative recombination can take place with both the energy and momentum conservations. As magnetic field can affect neither the exciton band nor the phonon bands, the exciton-phonon interaction, and hence the radiative life time of excitons are not changed under magnetic field. On the other hand, the nonradiative annihilations of free carriers and excitons must have phonons take part in. With this reason, nonraidative life times of free carriers (τ_e) and the persentage of excitons take part in radiative annihilation (γ) are constants. The PL intensity can be written as

$$I_{PL} = \frac{\gamma G \tau_e}{\tau_e + \tau_f}. \qquad (3)$$

Or to say, the PL intensity affected by the magnetic field implicitly through changing exciton formation time under external magnet field.

When the excitation intensity is very weak, the carrier densities are expected to be very low and the average distance between an electron and a hole is much larger than the mean free path of these carriers. Before the wavefunctions of an electron and a hole overlap with each other and combine to form an exciton, they must move toward each other. Consequently, the formation time of excitons is contributed by the time during which the electron and hole move together under the Coulomb force between them. It is well known that the external magnetic field may affect the average drift velocity of carriers under electric field, as revealed in galvano-magnetic measurements of semiconductors. If the carrier density is very low, the movement of electrons and holes under the Coulomb force between them is expected to be analogous to that of carriers under the external macroscopic electric field. From this point of view, the relative velocity of electron and hole in two dimensional quantum well moving toward each other under perpendicular magnetic field can be written as:

$$\frac{dr}{dt} = \frac{e^2}{8\pi\epsilon\epsilon_0 r^2} \left\{ \frac{\tau_{ef}}{m_e} \frac{1}{1 + \tau_{ef}^2/\tau_{ec}^2} + \frac{\tau_{hf}}{m_h} \frac{1}{1 + \tau_{hf}^2/\tau_{hc}^2} \right\}. \qquad (4)$$

472

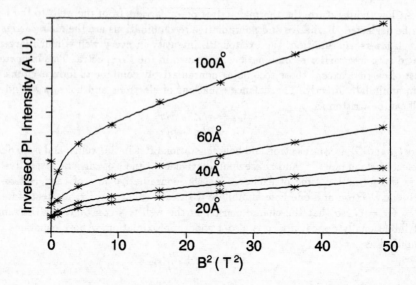

Fig. 2 The inversed PL intensity of AlGaAs/GaAs MQW plotted as the function of B^2. The symbols are experiment results, and The solid lines are the fitting curves with Eq. (3) and (5).

Here, r is the distance between the electron and hole, m_e and m_h are electron and hole effective masses, τ_{ec} and τ_{hc} are the cyclotron periods of electron and hole in quantum well, and τ_{ef} and τ_{hf} are electron and hole free life times, respectively. Assuming the average distance between electron and hole is r_0,

$$\frac{r_0^3}{3} = \frac{e^2}{8\pi\epsilon\epsilon_0} \left\{ \frac{\tau_{ef}}{m_e} \frac{1}{1 + \tau_{ef}^2/\tau_{ec}^2} + \frac{\tau_{hf}}{m_h} \frac{1}{1 + \tau_{hf}^2/\tau_{hc}^2} \right\} \tau_f(B), \tag{5}$$

with $\tau_f(B)$ being the average time during which the electron and hole move toward each other under the magnetic field B. And the carrier density is

$$n = r_0^{-2}/2 . \tag{6}$$

Through fitting the measured PL intensities under various magnetic fields with Eq. (3), (5), and (6), the exciton formation time can be obtained.

4. Discussion

In semiconductors, τ_{ef} and τ_{hf} are in the same order of magnitude, and m_h is usually much larger than m_e. So

$$\frac{\tau_{ef}}{m_e} \gg \frac{\tau_{hf}}{m_h}. \tag{7}$$

As the excitation intensity is extremely low, it is expected that the nonradiative loss of carriers dominates the total carrier loss. With Eq. (5), it is obvious that exciton formation time is dominated by the time of the electron to move toward the hole under

Table 1: The fitting parameters (τ_{hf}/m_h and τ_{ef}/m_e) of experimental results with Eqs. (3) and (5). $\mu_h = e\tau_{hf}/m_h$ and $\mu_e = e\tau_{ef}/m_e$ are electron and hole mobilities when they move toward each other under the Coulomb force between them. $\tau = \alpha n^{5/2}$ is the formation time of exciton with carrier density n, and τ_0 is the exciton formation time with $n = 10^8 cm^{-2}$.

Well Width	τ_{hf}/m_h	τ_{ef}/m_e	μ_h	μ_e	$\tau_e/\tau_f(0)$	α	τ_0
(Å)	$\left(\frac{10^{19}ps}{Kg}\right)$	$\left(\frac{10^{19}ps}{Kg}\right)$	$\left(\frac{10^4 cm^2}{secV}\right)$	$\left(\frac{10^4 cm^2}{secV}\right)$		$(\mu^{-3}ps)$	(ps)
100	0.7047	16.53	1.128	26.45	0.126	57.97	57.97
60	0.6560	11.05	1.050	17.68	0.145	112.7	112.7
40	0.6179	6.824	0.989	10.92	0.172	177.2	177.2
20	0.5075	4.551	0.903	7.282	0.255	260.8	260.8

the very low magnetic field, while under the high magnetic field, the formation time is determined by the hole. At this too extremes, the PL intensity can be written with Eq. (3) and (5) as

$$
I_{PL}^{-1}(B) = \begin{cases} A\dfrac{2\sqrt{2}\pi\epsilon\epsilon_0 m_h}{3e^2 n^{5/2}\tau_{hf}}\left[1 + (\frac{\tau_{hf}e}{2\pi m_h})^2 B^2\right] & \text{for } B \text{ very large;} \\[2ex] A\dfrac{2\sqrt{2}\pi\epsilon\epsilon_0 m_e}{3e^2 n^{5/2}\tau_{ef}}\left[1 + (\frac{\tau_{ef}e}{2\pi m_e})^2 B^2\right] & \text{for } B \text{ very small.} \end{cases} \tag{8}
$$

Here A is a constant determined only by the optical system, $i.e.$, it has no relation with magnetic field and carrier density. The inverse of the measured PL intensities as well as the theoretical fitting curves are plotted as a function of B^2, as shown in Fig. 2. The fitting parameters are listed in Table 1.

As shown in Fig. 2, in every well I_{PL}^{-1} is linear to B^2 when the magnetic field is large. Under small magnetic field I_{PL}^{-1} is not linear to B^2. This is because that under magnetic field near zero the exciton formation time is not much larger than the nonradiative loss time of carriers, as shown in Table 1. In addition, other factors contribute to the formation time should also be considered if $\tau_f(B)$ is not prevailing. With increasing magnetic field, $\tau(B)$ increases very rapidly, and hence dominates the exciton formation time if the magnetic field is sufficiently high. Therefore Eq. (8) turns to be a good approximation.

In Fig. 2, the slop for every well under small magnetic field is larger than that under high magnetic field, and the intersection with I_{PL}^{-1} at small magnetic field is smaller than that of extrapolation of the straight line in high magnetic field range. All these phenomena are consistent with the results described by Eq.(5) and (8).

The excellent fit of the experimental result with Eq. (3) and (5) indicates that the exciton formation time is dominated by the time of electron and hole move toward each other under the Coulomb force between them provided that the carrier densities are very low. In addition, as shown in table 1, the mobility of electron and hole to move toward each other are quite large, suggesting that the excitons are formed after the carrier system are cooled down. This is because the carrier-carrier interaction decreases with decreasing carrier density, while carrier-phonon interaction has almost no relation with the carrier density if the carrier density is not very large. In our case, the carrier density is so low that the carrier-phonon interaction is much stronger than the carrier-carrier interaction, resulting that the carrier system relaxes down its energy before the formation of excitons.

Table 1 also demonstrates that under the small magnetic field, the nonradiative recombination time is comparable with the exciton formation time, *i.e.*, both the radiative and nonradiative losses must be considered. However, as the exciton formation time increases dramatically with increasing magnetic field, the carrier loss is dominated by the nonradiative loss under high magnetic field. This is why I_{PL}^{-1} is linear with B^2 in high magnetic field range while it is not in low magnetic field range.

5. Conclusion

The exciton formation time in quantum wells is directly obtained with the cycrotron periods of carrier as the measures of time. It is found that exciton formation time is a function of carrier density. The higher the carrier density, the smaller the formation time. The high mobility of electrons and holes when they move toward each other under the Coulomb force between them reveals that the excitons are formed after the carrier system is cooled down. In addition, we demonstrated that MPL can be used in measurement of dynamical processes of carriers in semiconductors.

References

[1] Kusano J, Segawa Y, Aoyagi Y, Namba S, and Okamoto H 1989 *Phys. Rev. B* **40** 1685
[2] Damen T C, Shah J, Oberli D Y, Chemla D S, Cunningham J E, and Kuo J M 1990 *Phys. Rev. B* **42**, 7434
[3] Vinatieri A, Shah J, Damen T C, Goosse K W, Pfeiffer L N, Maialle M Z, and Sham L J 1994 *Phys. Rev. B* **50**, 10868
[4] Vinattieri A, Sgah J, Damen T C, Goossen K W, Pfeiffer L N, Maialle M Z, and Sham L J 1193, *Appl. Phys. Lett.* **63**, 3164

Characterization of Boron in 6H–SiC Using Optical Absorption

P. W. Yu, K. Mahalingam[*], W. C. Mitchel, M. D. Roth, and D. W. Fischer

Wright Laboratory, Materials Directorate (WL/MLPO),
Wright–Patterson Air Force Base, OH 45433–7707 U.S.A.

Abstract. Optical absorption measurements were made in the energy range of ~0.41 – 3.5 eV at temperatures from 14 to 300 K for 6H–SiC bulk crystals unintentionally or intentionally doped with boron. The presence of boron is evidenced by two features: (i) the broad photoionization band in the range of 0.5–3.0 eV and (ii) the multiple exciton lines associated with neutral boron in the range of 2.7–3.0 eV. The photoionization band was analyzed for the electronic transition from the valence band to the boron level by taking into account the temperature–dependent phonon broadening effect. The analysis yields the optical activation energy $E^0 = 1.408$ eV and the thermal activation energy $E_{th} = 0.394$ eV, which indicates a large lattice relaxation. The exciton absorption consists of 18 lines with the stronger lines being at 2.791, 2.826, 2.840, 2.862 and 2.889 eV at 14 K. Excellent correlation between the integrated absorption of the two transitions make it possible to obtain the neutral boron concentration of $2 \times 10^{18} - 1.2 \times 10^{19}$ cm^{-3} for the present experiment.

1. Introduction

The intrinsic material advantages[1] of SiC are currently being capitalized in the development of high power and high frequency semiconductor devices for use in high temperature and high radiation environments. In order to improve device performance, it is necessary to thoroughly characterize the material with respect to its electrical and optical properties and to establish a microscopic understanding of impurities and defects which are responsible for these properties. Boron is one of the most important acceptors in SiC and an alternative to aluminum, a common acceptor dopant. However, the microscopic nature of the centers responsible for the energy levels associated with boron is not well understood. Lomikina[2] determined the activation energy of boron in 6H–SiC to be 0.39 eV using temperature dependent Hall effect measurement. Ballandovich et al.[3] determined using the photocapitance effect that the boron acceptor is at Ev + 0.4 eV. Veinger et al.[4] reported two possible levels for boron–doped 6H–SiC from Hall–effect measurements, between either 0.30–0.33 eV or between 0.5–0.6 eV depending on compensation. Suttrop et al.[5] also found, using the deep level transient spectroscopy (DLTS) method, two levels: one at 0.3 eV, which they identified with the isolated substitutional acceptor, and the other at 0.58 eV, known as the boron–related D center. The D level is also observed by admittance spectroscopy.[5] Ikeda et al.[6] also found deep levels of 0.698–0.723 eV for 6H–SiC:B from analysis of donor–acceptor pair and free–to–acceptor luminescence. These experimental facts clearly show that there are at least two energy levels associated with boron.

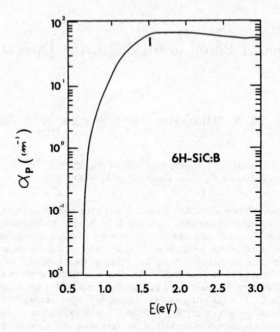

Fig. 1: Experimental a_p (cm^{-1})–vs–E(eV) relationship at T = 14 K for sample #I.

In the present work we have used optical absorption to study the electronic levels of boron and to determine boron concentration in undoped and boron–doped 6H–SiC. The photoionization absorption arising from the transition between a boron energy level and the valence bandedge and the exciton absorption due to neutral boron are analyzed. We find that the integrated absorption intensity due to the photoionization is linearly proportional to that of the exciton lines and that any of the two integrated absorptions can be used to determine the concentration of neutral boron.

2. Experiment

We used bulk 6H–SiC samples grown by the physical vapor transport method and supplied by two manufacturers, Westinghouse Science Center[7] and Advanced Technology Materials, Inc.[8] Optical absorption measurements were made using a Varian Cary 5E Spectrometer over temperatures ranging from 14 to 300 K. The temperature variation was made using a Crydyne cryocooler. Individual samples, usually 0.5 mm thick were wedged (~1°) and polished. All the samples showed p–type conduction. The activation energy related with the p–type conduction, determined from the slope of the hole concentration–vs–10³/T relations, ranges from 0.28 to 0.38 eV for eleven samples used. We believe that the activation energy originates from a shallow boron center.

3. Results and discussion

Figure 1 shows the spectral dependence of the photoionization absorption coefficient (a_p) for a sample #I at T = 14 K in the energy range of 0.5–3.0 eV. The broad absorption feature is attributed to the transition, due to the photoionization of the boron

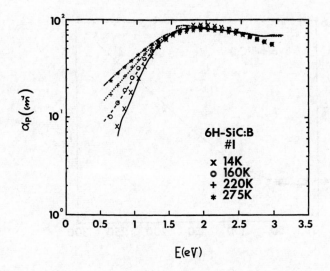

Fig. 2: Experimental α_p (cm^{-1})—vs—E(eV) relationship at 14, 160, 220, and 275 K with solid lines. Symbols are theoretical calculations.

acceptor, from the valence bandedge to the boron acceptor level. The temperature dependence of the photoionization spectra is investigated at temperatures of 14–300 K. Fig. 2 shows the spectra at T = 14, 160, 220, and 275 K. The solid lines represent experimental data, and the symbols are for theoretical calculations, which will be detailed later. From the figure one can easily find the broadening of the spectrum with increasing temperature. The strong temperature dependence of α_p indicates that the photoionization process has to be treated with the boron center–lattice relaxation mechanism. When the phonon broadening is taken into account in addition to the electronic transition, the spectral dependence of the photoionization absorption can be expressed in the following form[9][10]:

$$\alpha_P = \frac{c}{\sqrt{\pi}} \int_0^\infty \left\{ \exp\left[-\left[x - \frac{h\nu - E^0}{\sqrt{2}\,\Gamma}\right]^2 \right] \times \frac{(\sqrt{2}\,\Gamma x)^b}{h\nu(\sqrt{2}\,\Gamma x + E^0 m_T|m^*)} a \right\} dx \quad (1)$$

$$\Gamma = \left[S(\hbar\omega_L)^2 \coth(\hbar\omega_L \mid 2kT) \right]^{\frac{1}{2}}. \quad (2)$$

Here, Γ is the phonon–broadening factor, S is the electron–phonon coupling parameter called the Huang and Rhys factor, and $\hbar\omega_L$ is the energy of the phonon which interacts with the boron level. C, a, and b are constants, E^0 is the optical activation energy (photoionization threshold), m_T is the mass of the trapped carrier, and m^* is the effective mass of the carrier. The constants a and b depend on the form of the defect potential and on the symmetry relation between the initial (trapped) and the final (free) states. We choose the values of a and b, respectively, as 2 and $\frac{1}{2}$ assuming the δ–function potential and allowed photoionization transition.

The symbols in Fig. 2 were calculated with Eq. (1) using E^0 and Γ as fitting parameters. The value of E^0 obtained by the best fit is $E^0 = 1.408$ eV. The fitted values of Γ in the temperature range of 14–300 K are depicted in Fig. 3. The broken line represents the fit of Γ using Eq. (2). The thermal activation energy E_{th} is determined from the Huang–Rhys factor S assuming that the involved phonon energy is

478

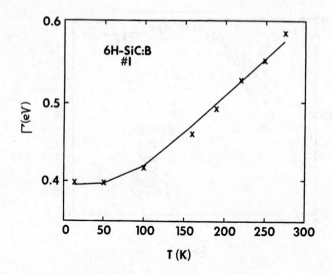

Fig. 3: Γ(eV)–vs–T(K) relationships for sample #I. The broken line is the theoretical fit.

the same as $\hbar\omega_L$. The obtained value of S and E_{th} is 42 and 0.39 eV, respectively. The obtained value of the thermal activation of the boron center is consistent with those obtained using the Hall effect[2],[3] and DLTS[4]. So, it is clear that the shallow boron level plays a dominant role in the photoionization process.

Figure 4 shows multiple absorption lines at 2.7–3.06 eV, with strong lines being at 2.791, 2.826, 2.840, 2.862, and 2.889 eV. Some of these lines persist even at room temperature. The spectra feature, including the line energy and multiplicity, is the same for all investigated samples, across which the boron concentration is estimated to differ by more than ten times. We attribute some of the lines to exciton bound to neutral boron (boron acceptor four particle bound exciton complex 4A). The spectral features are similar to that reported earlier[11] for boron doped 6H–SiC. The multiplicity of the spectral features is also comparable to the aluminum acceptor four particle bound exciton[12] observed from lightly Al doped p–type epitaxial 6H–SiC. We introduce the group theoretical analysis for neutral acceptor bound excitons in covalent zincblende GaP by Dean et al.[13] in order to explain the multiplicity found in the present work. Such a model predicts 12 transitions in GaP and 6 transitions for each of the three ineqivalent silicon sublattice sites for 6H SiC, although specific assignments to each transition is not plausible for the present experimental data.

Figure 5 illustrates the relationship between the intensities of integrated absorption due to the photoionization and bound exciton designated, respectively, as I_P and I_E for eleven 6H SiC samples. Absorption measurements were made at 14 K with identical experimental set–ups in terms of slit, integrated time, and scan time. Samples were named with alphabetic letters and the solid line is the guide for the eye. For simplicity, the strongest absorption line was chosen to calculate I_E. It is evident that the relationship shows an excellent linearity in most of the range. This demonstrates that the boron centers involved in the photoionization and bound exciton processes are the same boron centers, and that this boron center is neutral boron. However, considering the linearity at higher boron concentrations, I_E is lower than the expected value. Samples

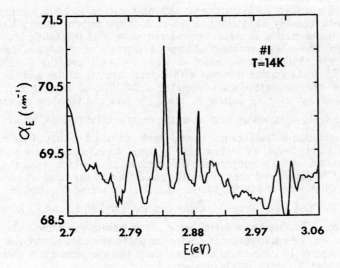

Fig. 4:   Experimental a_E (cm^{-1})–vs–E(eV) relationship for sample #I.

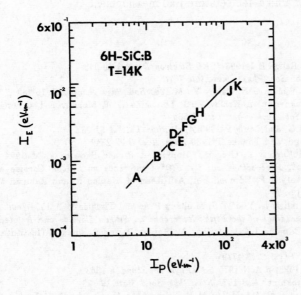

Fig. 5: I_E(eV cm^{-1})–vs–I_P (eV cm^{-1}) relationship obtained at T = 14 K.

J and K show such a tendency. We attribute this phenomenon to boron's complexing with other elements, possibly an intrinsic defect such as the carbon vacancy V_C. It is possible that the boron–related D center is produced more with increasing boron concentrations. Then, the photoionization absorption consists of two photoionization processes due to the shallow boron center at $E_v+0.3$–0.4 eV and the D center at $E_v+0.6$–0.7 eV. Since the exciton involves with isolated neutral borons and the boron complex decreases the concentrations of isolated substitutional neutral boron, the integrated absorption intensity I_E decreases. Since our present data show an excellent linearity of the I_E–vs–I_P relationship in most of the experimental regions, we can safely say that the concentration of the D center is very small compared to that of the isolated boron acceptor at $E_v+(0.3$–$0.4)$ eV. More heavily boron doped samples are needed to clarify this explanation. A high–temperature Hall measurement, which yields the saturation of p–vs–$10^3/T$ relation and thus allows the calculation of the total concentration of donor and acceptor, was made for sample H. The measured N_D and N_A were, respectively, 4.7×10^{16} and 3.1×10^{18} cm^{-3}. Using this we correlate I_P and I_E for neutral boron concentration. The obtained neurtral boron concentration was found to be 2×10^{18} – 1.2×10^{19} cm^{-3} for the present experiment. Since the photoionization absorption can be easily obtained compared to the exciton lines, we prefer the photoionization absorption for correlating the neutral concentration of boron.

Acknowledgements

P. W. Y. acknowledges the support of the National Research Council. We acknowledge R. Perrin and R. V. Berke, for computer work and sample preparation, respectively. We thank J. Jenny for a high–temperature Hall measurement.

References

[1] Bhatnagar M and Baliga B J 1993 *IEEE Electron Devices* **40** 645

[2] Lomakina G A 1965 *Sov. Phys. Solid State* **7** 475

[3] Ballandovich V S, Violina G N, Tairov Yu M 1981 *Sov. Phys. Semicond.* **15** 283

[4] Veinger A I, Vodakov Yu A, Kozlov Yu I, Lomakina G A, Mokhov E I, Odiong V G, and Sokolov V I 1980 *Sov. Tech. Phys. Lett.* **6** 566

[5] Stuttrop W, Pensl G, and Lanig P 1990 *Appl. Phys. Lett. A* **51** 231

[6] Ikeda M, Matsunami, and Tanaka T 1980 *Phys. Rev. B* **22** 2842

[7] Hobgood H M, McHugh J P, Greggi J, Hopkins R H, and Skowronski M 1994 *Inst. Phys. Conf. Ser. No. 137, Proceedings of the Fifth Conference on Silicon Carbide and Related Materials*, eds. Devaty R P, Edmond J A, Asif Khan M, Kaplan R, and Rahman M (Institute of Physics, Bristol) p. 7

[8] Buchan N I, Henshall D N, Yoo W S, Mailloux P A, and Tischler M A 1994 *Inst. Phys. Conf. Ser. No. 137, Proceedings of the Fifth Conference on Silicon Carbide and Related Materials*, eds. Devaty R P, Edmond J A, Asif Khan M, Kaplan R, and Rahman M (Institute of Physics, Bristol) p. 113

[9] Noras J M 1980 *J. Phys. C* **13** 4779

[10] Kopylov A A and Pikhtin A N 1975 *Sov. Phys. Semicond.* **8** 1563

[11] Gorban I S and Krokhmal A P 1977 *Sov. Phys. Solid State* **19** 723

[12] Cleman L L, Devaty R P, MacMillan M F, Yoganathan M, Choyke W L, Larkin D J, Powell J A, Edmond J A, and Kong H S, 1993 *Appl. Phys. Lett.* **62** 2953

[13] Dean P J, Faulkner R A, Kimura S, and Ilegano M 1971 *Phys. Rev.* **B4** 1926

---

* Systran Corporation, 4120 Linden Avenue, Dayton, OH 45432 U.S.A.

Inst. Phys. Conf. Ser. No 145: Chapter 3
Paper presented at 22nd Int. Symp. Compound Semiconductors, Cheju Island, Korea, 28 August–2 September 1995
© *1996 IOP Publishing Ltd*

Photoluminescence Spectra of Heavily Si-doped GaAs at Low Temperature

Nam-Young Lee†, Jae-Eun Kim†, Hae Yong Park†, Dong-Hwa Kwak‡, Hee Chul Lee‡ and H. Lim§

† Department of Physics, KAIST, Taejon 305-701, Korea

‡ Department of Electrical Engineering, KAIST, Taejon 305-701, Korea

§ Department of Electronic Engineering, Ajou University, Suwon 442-749, Korea

Abstract. Photoluminescence spectra of heavily Si-doped GaAs grown by molecular beam epitaxy (MBE) were investigated at $20K$ as a function of electron concentration. We found that the two peaks in the electron populations of the conduction band and the donor band, respectively, are merging as the doping concentration increases, and can not be distinguished at the electron concentration of the order of $10^{18} cm^{-3}$. And we discriminated the artifact peak at $1.49eV$ from the recombination peak by observing its temperature dependent behavior.

1. Introduction

The investigation of photoluminescence (PL) spectrum is a common characterization technique to reveal some heavy doping effects in a semiconductor such as the increase in the interband transition energy [1] and the impurity band merging with the adjacent bands. [2] In this study, we will show the direct evidence of the donor band merging with the conduction band using the PL spectra of MBE-grown Si-doped GaAs samples. We also investigate how the temperature characteristics of the PL peak is related to the physical origin of the $1.49eV$ emission peak of the samples of $n = 7.4 \times 10^{17} cm^{-3}$ and $n = 11.2 \times 10^{17} cm^{-3}$.

2. Experiment

Samples were prepared by molecular beam epitaxy (MBE). An intrinsic GaAs buffer layer of $1\mu m$ thickness was grown on a (1 0 0) oriented semi-insulating GaAs substrate. And then a Si-doped epilayer with thickness of $1\mu m$ was grown on the buffer layer. The

482

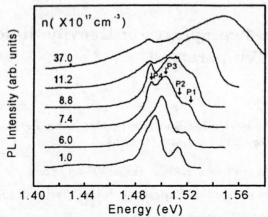

Figure 1. Photoluminescence spectra of Si-doped GaAs at 20K in the region of electron concentration where the donor band merges with the conduction band.

electron concentration of the samples was determined by Hall measurements using van der Pauw technique at room temperature, and it ranges from 1.0×10^{17} to $4.2 \times 10^{18} cm^{-3}$. For the PL measurements, the samples were cooled in a cryogenic system (Air Products 1R02-A displex) to about $20K$ and excited by the 488 nm line of an Ar-ion laser with an intensity of 0.4 W/cm^2. A vibrating mirror attached to a 75 cm monochromator (Spex 750M) enabled us to measure the wavelength-modulated PL and conventional PL spectra, from which the peak energies were read.

3. Results and Discussion

Figure 1 shows some typical PL spectra of Si-doped GaAs samples for various electron concentrations. Four peaks can be seen in all the spectra and we will label them P1, P2, P3 and P4, respectively, in the order of energies. The peak positions observed in all the samples are determined from the wavelength modulated PL spectrum and plotted in Fig. 2 as a function of electron concentration. Note that the peak centers of P1 and P3 shift to higher energy-side as the electron concentration is increased, while those of P2 and P4 remain unchanged. The P2 peaks at $1.514eV$ and $1.511eV$ are known to be due to the bound exciton transitions from the Si-doped epilayer at low doping. [3, 4] This exciton state is difficult to be formed in the samples of rather high electron concentration because of screening effect. [5] Thus, it is clear that this peak is comming from the excition transitions in the buffer layer, since the thickness ($1\mu m$) of the epilayer is comparable to the diffusion length of the photo-generated minority carriers. [6]

The P1 peak is due to the band-band transition as can be easily seen in Fig. 2. In the n-type samples, it is well known that it exhibits the Burstein-Moss shift as the Fermi energy enters the conduction band with increased doping. The energy of P3 peak is also increasing with electron concentration and then the peak disappears by merging with the P1 peak at the doping concentration of the order of $10^{18} cm^{-3}$ in the conventional

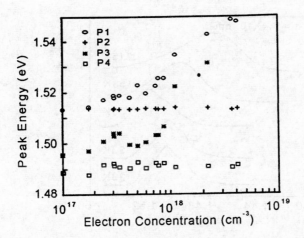

Figure 2. Energy variation of photoluminescence peaks of Si-doped GaAs at 20K. For the doublet P2 only the peak near 1.514 eV is indicated.

PL spectra (Fig. 1). If we look very carefully at the PL spectra of the samples of $n = 1.12 \times 10^{18} cm^{-3}$ (Fig. 2), however, we can still find the very faint trace of the P3 peak and read the peak energies of which value are $1.534eV$ and $1.522eV$, respectively. Using these peak energies, the PL spectrum of this sample was deconvoluted and it was found that the ratio of P1 to P3 peak intensity is about 1 to 3, the same as the value of the sample of $n = 8.8 \times 10^{17} cm^{-3}$. And, in the spectrum of $n = 3.7 \times 10^{18} cm^{-3}$, we cannot read the energy positions separately because the two peaks nearly merge with each other. We should note here that this ratio is about 1 to 3 for the samplte of $n = 1.0 \times 10^{17} cm^{-3}$, too. This clearly shows that the one of the P1 and P3 peaks is not absorbed into the other, but the two peaks merge with each other.

It is well known that the energy level of isolated Si atoms lies at about $6meV$ below the conduction band minimum. [7] The carbon atoms, which are known to be the most common contaminant acceptor in MBE-grown samples, [3, 8, 9] form their energy level at about $27meV$ above the valence band. [7] If P3 peak is due to a DA transition, it must therefore appear at about $33meV$ below the band gap energy at $n = 1.0 \times 10^{17} cm^{-3}$. As can be seen in Fig. 2, however, this peak appears at about $17meV$ below the P1 peak energy. Since the Fermi energy, under the assumption of parabolic band, is located at $12meV$ above the conduction band minimum for this dopant concentration at $20K$, the P1 transition is determined not by the conduction band minimum but by the Fermi energy. Therefore the energy difference between the P1 and P3 peaks should be $18meV$ rather than $6meV$. We thus believe that the P3 peak is due to the transition from a donor band to the valence band or to the acceptor-like states. [8, 5]

The PL intensity of a heavily n-doped sample can be described as follow.

$$I(E) \sim D(E) f_c(E) |M_{if}|^2 \tag{1}$$

where $D(E)$ is the density of states in the conduction band containing the donor band and $f_c(E)$ the well known Fermi-Dirac distribution function of electron given by

$$f_c(E) = \frac{1}{1 + \exp[(E - E_f)/kT]} \tag{2}$$

484

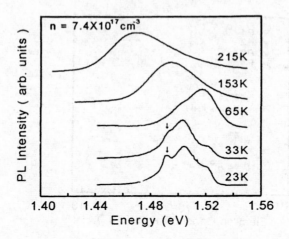

Figure 3. Photoluminescence spectra of Si-doped GaAs with doping concentration of $n = 7.4 \times 10^{17} cm^{-3}$ as a function of temperature.

where E_f is the Fermi energy. M_{if} is the electron transition matrix which may be considered to be constant in energy, since we are dealing with the energy range of a few ten meV. Then $I(E) \sim D(E)f_c(E)$, so that the electron concentration per unit energy becomes maximum at the peak energy of PL spectra. So the merging of P1 and P3 peaks is the direct evidence of the merging of the maxima of the electron populations in the conduction band and the donor band, respectively.

The wavefunction of a donor will overlap with those of neighboring donor atoms if they lie within the Bohr radius of each other, which is of the order of $100 \mathring{A}$ in GaAs. The weak overlapping at relatively low doping will split the isolated donor levels localized at ε_i below the conduction band to form the donor band. But the strong overlapping at heavy doping will cause the donor band to merge with the conduction band. The volume radius, occupied by one donor atom, is about $134 \mathring{A}$ at $n = 10^{17} cm^{-3}$ and $62 \mathring{A}$ at $n = 10^{18} cm^{-3}$. We may, thus, say that the donor band merges with the conduction band at the concentration of $n \sim 10^{17} cm^{-3}$, while the two maxima of the electron distribution in these two bands merge with each other at the concentration of $n \sim 10^{18} cm^{-3}$.

On the other hand, it has not been clear whether the P4 peak at about 1.49 eV can be attributed to a donor-to-acceptor (DA) transition [8, 9, 5] or to a PL signal which traveled to the back surface and was reflected to the front thus adding to the main peak. [10] So it needs to be identified what the peak origin is. Szmyd and Majerfeld [11] proposed one metheod. They showed that the $1.38eV$ peak at room-temperature would be simply due to an artifact induced by the substrate rather than a donor-acceptor recombination in heavily doped GaAs epilayer. By increasing the substrate thickness or by polishing the back surface of the substrate, the peak can be eliminated. But this process would be rather cumbersome and even destructive. Since the two mechanisms responsible for the PL peaks at $1.49eV$ seems to be different, we could expect different dependence of the peaks on the temperature. Hence we examine the PL spectra of the two samples as a function of temperature. And it is not only a non-destructive method, but also a very common practice to observe the temperature-dependence of the PL spectra in order to characterize each peaks in the spectra. The center of the band-

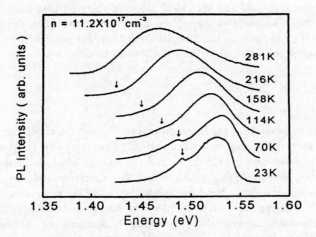

Figure 4. Photoluminescence spectra of Si-doped GaAs with doping concentration of $n = 11.2 \times 10^{17} cm^{-3}$ as a function of temperature. The artifact peak can be seen in all spectra, except for the one at room temperature.

band transition peak shifts to lower-energy-side with increasing temperature in both samples because of the band gap narrowing as shown in Figs. 3 and 4. It can be also seen that the temperature dependence of the $1.49eV$ peak is very different for the two samples. The peak intensity of the sample of $n = 7.4 \times 10^{17} cm^{-3}$ decreases drastically as the temperature increases and disappears at $65K$, while that of $n = 11.2 \times 10^{17} cm^{-3}$ decreases slowly and the peak remains even at $216K$. The disappearance of the peak for the $n = 7.4 \times 10^{17} cm^{-3}$ sample may be related to the thermal excitation of the photo-excited bound hole to the valence band. Since the carbon atoms are apt to be doped with doping concentration of about $10^{14} cm^{-3}$ as the contaminant acceptor in MBE-grown epilayer samples, [13] the peak may be carbon-related. At this doping concentration and at $65K$, 85% of the photo-excited bound hole can be thermally excited to the valence band [14] and, therefore, the acceptor-related peak may disappear above this temperature. If the peak of the sample of $n = 11.2 \times 10^{17} cm^{-3}$ were related to the transition to the acceptor, this peak should also disappear at about $65K$ so that we should not expect to observe it at higher temperature such as $216K$.

Since the artifact results from the transparancy of the substrate, which is directly affected by the bandgap, the bandgap narrowing with increasing temperature will induce the shift of the artifact to lower energy side. So the empirical relationship between the peak energy and temperature was tried with the Varshni's form as follows :

$$E(T) = E(0) - \frac{\alpha T^2}{(\beta + T)} \tag{3}$$

The parameters $E(0), \alpha$ and β, as determined by fitting the peak points with Eq: (3), are $1.49eV$, $8.14 \times 10^{-4} eV/K$ and $326K$, respectively. From the fitting results, the peak energy at $300K$ is obtained to be $1.38eV$ in good agreement with the value $1.38eV$ of the artifact observed by Szmyd and Majerfeld at room temperature. [11] In order to confirm that the $1.49eV$ peak at $23K$ is the artifact, we investigated the PL spectra for the $11.2 \times 10^{17} cm^{-3}$ samples of different substrate thickness and of different roughness of the back surface, as Szmyd and Majerfeld did. We observed that the peak intensity increased when the substrate was lapped down and decreased when the back surface was polished. It is quite natural, therfore, to think that this peak is due to an artifact induced by the substrate rather than the transition to an acceptor.

4. Conclusion

We have studied the behavior of the low temperature PL spectra of MBE-grown heavily Si-doped GaAs varying the doping concentration. The energy of P3 peak, which is related to the transition between the donor and the valence band or the donor and the acceptor-like states, was increasing with electron concentration and the maximum of electron population in the donor band merged with that in the conduction band at the order of $10^{18} cm^{-3}$ electron concentration. And we discriminated the artifact peak at $1.49 eV$ from the recombination peak by observing its temperature behavior. The PL peak at $1.49 eV$ of the sample of $n = 7.4 \times 10^{17} cm^{-3}$ was due to the transition to an acceptor, maybe carbon acceptor, and that of $n = 11.2 \times 10^{17} cm^{-3}$ the substrate-induced artifact. This artifact was observed at much higher temperature than the recombination peak was.

References

[1] Burstein E 1954 *Phys. Rev.* **83** 632-633

[2] Abram R A, Rees G J and Wilson B L H 1978 *Adv. Phys.* **27** 799-892

[3] Fatt Y S 1993 *J. Mater. Sci. Lett.* **12** 609-611

[4] Olego D and Cardona M 1980 *Phys. Rev.* **B22** 886-893

[5] Abramov A P, Abramova I N, Verbin S Y, Gerlovin I Y, Grigor'ev S R, Ignat'ev I V, Karimov O Z, Novikov A B and Novikov B N 1993 *Semiconductiors* **27** 647-649

[6] Castaldini A, Cavallini A, Gombia E, Mosca R, Tarricone L, Motta A and Bora L 1993 *Appl. Surf. Sci.* **63** 208-212

[7] Hellwege K H 1982 *Semiconductors: Physics of Group IV Elements and III-V Compounds* (Landolt-Börnstein New Series vol. 17: Springer-Verlag Berlin: Heidelberg: New York)

[8] De-Sheng J, Makita Y, Ploog K and Queisser H J 1992 *J. Appl. Phys.* **53** 999-1006

[9] Borghs G, Bhattacharyya K, Deneffe K, Mieghem P V and Mertens R 1989 *J. Appl. Phys.* **66** 4381-4386

[10] Yao H and Compaan A 1990 *Appl. Phys. Lett.* **57** 147-149

[11] Szmyd D M and Majerfeld A 1989 *J. Appl. Phys.* **65** 1788-1790

[12] Lu Z H, Hanna M C and Majerfeld A 1994 *Appl. Phys. Lett.* **64** 88-90

[13] Parker E H C 1985 *The Technology and Physics of Molecular Beam Epitaxy* (New York: Plenum Press)

[14] Blakemore J S 1962 *Semiconductor Statistics* (International Series of Monographs on Semiconductors vol. 3: Oxford: Pergamon Press)

Inst. Phys. Conf. Ser. No 145: Chapter 3
Paper presented at 22nd Int. Symp. Compound Semiconductors, Cheju Island, Korea, 28 August–2 September 1995

Characteristics of group VI element DX centers in InGaP

S D Kwon†, Ho Ki Kwon†, Byung-Doo Choe† and H Lim‡

† Department of Physics, Seoul National University, Seoul 151-742, Korea

‡ Department of Electronic Engineering, Ajou University, Suwon 442-749, Korea

Abstract. Deep level properties of S-, Se-, and Te-doped $In_{1-x}Ga_xP$ layers have been studied by DLTS and capacitance-temperature measurements. In $In_{0.32}Ga_{0.68}P$ layers, S, Se, and Te each form deep states whose activation energies are 0.26, 0.23, and 0.14 eV, respectively. In $In_{0.49}Ga_{0.51}P$ layer, only S forms a deep state. From the investigation of defect formation by impurity species, the compositional dependence of binding energy, and the persistent photoconductivity, these deep donors are attributed to the DX centers. The binding energy of S DX center is observed to be the largest in the investigated DX centers.

1. Introduction

The ternary $In_{0.49}Ga_{0.51}P$, lattice-matched to GaAs, is attractive as an alternative to AlGaAs for heterostructure devices. In $In_{0.49}Ga_{0.51}P$, some n-type dopants are expected not to form the deep center. Even in the case of substitutional dopants in alloy semiconductors such as AlGaAs, however, some defect properties can be elucidated only by a detailed study for various alloy compositions.[1] The similarity of energy band structures of InGaP and of AlGaAs suggests the possible presence of DX centers in InGaP. The DX nature of some deep donors, such as persistent photoconductivity (PPC), has been really observed in InGaP.[2, 3] In this work, the evolution of deep centers due to S, Se, and Te is investigated in $In_{0.49}Ga_{0.51}P$ and $In_{0.32}Ga_{0.68}P$. It is found that the S-related DX center has larger thermal activation energy and binding energy than the Se-related and Te-related ones.

2. Experiments

The investigated samples were grown by liquid phase epitaxy. The $In_{0.49}Ga_{0.51}P$ layers were grown directly on n^+- GaAs substrate. To grow $In_{0.32}Ga_{0.68}P$ layers, the n-type

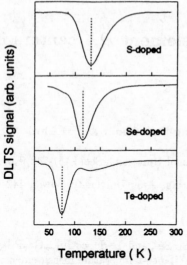

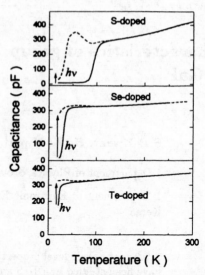

Figure 1. DLTS spectra of $In_{0.32}Ga_{0.68}P$ samples. The rate window is 9.3 s^{-1}.

Figure 2. C-T characteristics of $In_{0.32}Ga_{0.68}P$ Schottky diodes. The solid line is measured in the dark. The dashed line is obtained after the illumination at 20 K.

commercial $GaAs_{0.61}P_{0.39}$ substrates were used to avoid the lattice mismatch. Semitransparent Au Schottky contacts of about 100 Å thickness were also prepared for the study of PPC effects. DLTS and capacitance-vs-temperature (C-T) measurements were performed in a temperature range of 20 – 300 K. To test the PPC effects, the sample was illuminated with photons of $h\nu = 1.3$ eV by using a bright LED.

3. Results and discussion

Typical DLTS spectra obtained from $In_{0.32}Ga_{0.68}P$ layers are shown in Fig. 1. The dopant concentration of these samples was about $5 \times 10^{17} cm^{-3}$. As can be seen in this figure, each dopant gives its characteristic DLTS signals, the Te-related peak appearing at the lowest temperature. The emission activation energies are 0.14, 0.23, and 0.26 eV for the Te-, Se-, and S-related deep levels, respectively. Thus it is evident that the activation energy of these impurity-related deep levels increases as the mass number of impurities decreases. This chemical trend is in contrast with the case of group VI impurities in AlGaAs where the emission activation energy remains constant.[4] The ratios of trap density to dopant density (N_T/N_D) determined from the peak heights in DLTS spectra are 0.58, 0.06, and 0.16 for $In_{0.32}Ga_{0.68}P$ samples doped with S, Se, and Te, respectively.

For the S-doped $In_{0.49}Ga_{0.51}P$ samples, a donor-related deep level is clearly observed. This is consistent with the results obtained in VPE- and MOCVD-grown layers.[2, 3] The activation energy of the S-related deep level in $In_{0.49}Ga_{0.51}P$ was nearly same as that in $In_{0.32}Ga_{0.68}P$. But the ratio N_T/N_D of the former sample was smaller by about one order of magnitude than that of the latter with the same S content. Neither Se- nor Te-related deep level was detected in $In_{0.49}Ga_{0.51}P$ even with the dopant concentrations up to about $2 \times 10^{18} cm^{-3}$. The nonexistence of Se-related deep level in

$In_{0.49}Ga_{0.51}P$ is also consistent with other results.[2, 5] Thus, it is evident that only S donor forms the deep level in $In_{0.49}Ga_{0.51}P$.

The DX-center nature of the observed deep levels was tested by measuring the persistent photocapacitance at low temperature. In the C-T curves of Fig. 2, the solid curves were measured under zero bias in the dark. The dashed curves were obtained after samples were illuminated with photons of $h\nu = 1.3$ eV for about 5 – 10 min at 20 K. For all the samples, the remnant capacitance can return to the dark capacitance only when the sample temperature is increased. This persistent photocapacitance at low temperature indicates the existence of PPC effect aroused from the barrier for electron capturing.[6, 7] The PPC effect is well known as a characteristic feature of the DX centers in AlGaAs and is explained by the large lattice relaxation.[6] Since the PPC effect observed in our samples is certainly concerned with the donor-related deep levels, it is certain that S, Se, and Te donors form DX centers in $In_{1-x}Ga_xP$ over some composition range.

As seen in Fig. 2, the capacitances are largely reduced at low temperature. Since the sample used in this study is quite thick (about 3 μm) and the free electrons freeze out on the DX centers at low temperature, such a drop of capacitance at low temperature results from the increase of series resistance. We can also note that the capacitance drop in S-doped $In_{0.32}Ga_{0.68}P$ sample occurs at a much higher temperature compared with the Se-doped one in Fig. 2. Since the doping level and thickness of the samples were nearly same, this means that the binding energy of S DX center is much larger than that of Se DX center, although the difference of their activation energies is small (about 30 meV). For the S-doped $In_{0.49}Ga_{0.51}P$ samples, however, such a capacitance drop does not appear. This in turn means that the binding energy of S-related DX center in $In_{0.49}Ga_{0.51}P$ is much smaller than that in $In_{0.32}Ga_{0.68}P$. These phenomena are clearer in the admittance measurement. Admittance-vs-temperature characteristic and series resistance R_s deduced from these data are shown in Fig. 3 for S-doped $In_{0.32}Ga_{0.68}P$. No series resistance

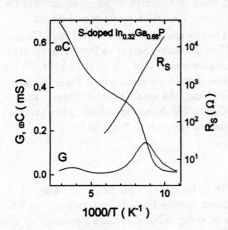

Figure 3. Admittance of S-doped $In_{0.32}Ga_{0.68}P$ Schottky diodes. The frequency of test signal is 1 MHz.

effect in admittance measurement was detected in S-doped $In_{0.49}Ga_{0.51}P$. The binding energy of S DX center in $In_{0.49}Ga_{0.51}P$ was found to be about 13 meV by Hall measurement. The binding energy of S DX center in $In_{0.32}Ga_{0.68}P$ is deduced to be about 135 meV by using the approximate relation between R_s and temperature, $R_s \sim exp(E_t/kT)$. This dependence of binding energy of DX centers on the x value in $In_{1-x}Ga_xP$ is identical to the case of $Al_xGa_{1-x}As$.[1]

The difference of characteristic in the observed DX centers is also confirmable from the dependence of DLTS peak heights on the filling pulse width shown in Fig. 4. For S-doped $In_{0.49}Ga_{0.51}P$ and Se-doped $In_{0.32}Ga_{0.68}P$, the DLTS peak heights have not yet been saturated until the filling pulse of 100 ms. Contrary to this, the DLTS peak of S-doped $In_{0.32}Ga_{0.68}P$ shows nearly saturating behavior as the pulse width gets larger than

490

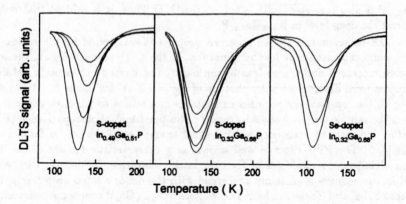

Figure 4. Dependence of DLTS spectra on the filling pulse width. The rate window is 9.3 s^{-1}. The pulse widths are 100, 10, 1, and 0.1 ms, respectively, from the largest peak to the smallest one.

10 ms. This means that, although the capture cross sections determined from Arrhenius plot are similar for these levels in In$_{1-x}$Ga$_x$P, the S DX center in In$_{0.32}$Ga$_{0.68}$P has the largest capture cross section at temperatures where the DLTS peaks are observed.

In Fig. 2, one can note that the persistent photocapacitance of S-doped In$_{0.32}$Ga$_{0.68}$P sample is very small at low temperature and then increases sharply as the temperature is increased. This initially small persistent capacitance means that not all the photo-excited electrons remain in the conduction band. Since the Ga composition of In$_{0.32}$Ga$_{0.68}$P is also near the cross-over point of direct-indirect band, the small remnant capacitance of our sample at low temperature is probably due to the freeze-out of the photo-excited electrons on X-like hydrogenic states as reported in the case of AlGaAs.[7]

4. Conclusion

The formation of DX centers by S, Se, and Te in In$_{1-x}$Ga$_x$P has been investigated. It is found that all these impurities form DX centers in In$_{0.32}$Ga$_{0.68}$P, but only S does it in In$_{0.49}$Ga$_{0.51}$P. The emission activation energy of group VI DX centers decreases as the mass number of impurity atom increases. It is observed that S-related DX center has larger binding energy than Se- and Te-related ones.

References

[1] Mooney P M 1990 *J. Appl. Phys.* **67** R1–26

[2] Kitahara K, Hoshino M and Ozeki M, 1988 *Japan. J. Appl. Phys.* **27** L110–2

[3] Paloura E C, Ginoudi A, Kiriakidis G and Christou A 1991 *Appl. Phys. Lett.* **59** 3127–9

[4] Kumagai O, Kawai H, Mori Y and Kaneko K 1984 *Appl. Phys. Lett.* **45** 1322–3

[5] Fujisawa T, Yoshino J and Kukimoto H 1989 *J. Cryst. Growth* **98** 243–8

[6] Lang D V and Logan R A 1977 *Phys. Rev. Lett.* **39** 635–9

[7] Dobaczewski L and Kaczor P 1991 *Phys. Rev. B* **44** 8621–32

Inst. Phys. Conf. Ser. No 145: Chapter 3
Paper presented at 22nd Int. Symp. Compound Semiconductors, Cheju Island, Korea, 28 August–2 September 1995
© *1996 IOP Publishing Ltd*

Dislocation-related Etch-pits and Deep Level in Strain Relaxed SiGe Layers

H. S. Kim, Y. G. Shin, Y. T. Hwang, J. Y. Kim, H. J. Lee, B. T. Lee*, Y. S. Hwang**, and S. D. Jung***

Department of Physics and Semiconductor Physics Research Center, Jeonbuk National University, Jeonju 560-756, Jeonbuk, Korea
*Department of Metallurgical Engineering, Jeonnam National University, Kwangju 500-757, Korea
**Department of Physics, Korea Advanced Institute of Science and Technology, Taejon 305-701, Korea
***Electronics and Telecommunications Research Institute, Taejon 305-350, Korea

Abstract. Strain relaxed SiGe epitaxial layers were grown on the compositionally graded layers on Si(001) substrates at 730℃, and 900℃ by molecular beam epitaxy. We observed etch-pits by scanning electron microscopy and atomic force microscopy after Schimmel-etch with various time. The analyses show that the etch-pits are related to threading dislocations. The Schimmel-etch seems to follow the kinematic wave theory of dissolution and etch-pit formation. Deep level transient spectroscopy exhibits a hole trap in the present sample. The energy and the capture cross section of the hole trap are $E_H = E_V + 0.62 \pm 0.05\ eV$ and $\sigma_H \sim 5.5 \times 10^{-10}\ cm^2$, respectively. This hole trap is associated with threading dislocations and/or misfit dislocations in compositionally graded layer.

1. Introduction

Strain-relaxed SiGe buffers with low threading dislocation densities (TDDs) have been obtained by compositionally graded growth technique in molecular beam epitaxy (MBE) [1-3]. Reducing TDD in strain relaxed SiGe buffer is important, since electron mobility in strained Si channel on the strain relaxed SiGe layer is strongly dependent on TDD.

In this work, we studied the behavior of etch-pits (EPs) in relaxed SiGe layer as a function of etching time with scanning electron microscopy (SEM) and atomic force microscopy (AFM). The analyses show that the EPs are related to the TDs. The mechanism of TD annihilation and the etching mechanism of the Schimmel etch are discussed also. Deep level transient spectroscopy (DLTS) was used to study the relation between deep level and dislocations in strain-relaxed SiGe/Si heterostructure. We found only one hole trap related to the dislocations.

2. Experiment

The substrates were (001) oriented, p-type Si wafers with resistivity *10-20 Ω-cm*. Wafers were cleaned by modified RCA method in *ex situ*.

Table 1. The growth parameters and etch-pit density of the MBE grown samples.

Sample	Grading rate (%/μm)	Ge composition (%)	Growth temp. (°C)	Etch-pit density (×10⁷ cm⁻²)	Relaxation (%)
2-45	45	45	730	5.6	85
3-2	25	50	900	1.2	100
3-3	15	30	900	0.85	100

The three different samples are described in Table 1. The strain relaxation was deduced from double crystal x-ray diffraction (DCXD) of the (115) asymmetric reflection. Figure 1 shows a schematic diagram of the structures. It is composed of three parts, (A) a 1-2 μm compositionally graded SiGe layer and a 1 μm relaxed $Si_{1-x}Ge_x$ layer, (B) a 50-100 Å strained Si channel and (C) the modulation doped region and a Si cap.

A Schimmel-etch composing of 2:4:3 $0.75mCrO_3:HF:H_2O$ was performed on these samples with stirring as a function of time. In cross sectional TEM image for 50 sec etched sample, the etching rate was estimated to be 43 Å/sec. The etched surfaces were studied with SEM and AFM.

For electrical measurements, ohmic contact was performed on the sample **2-45** by thermal evaporation of Al on both the p-type Si substrate and the Si cap layer. During metalization, Al atom diffuses into the Si cap layer from surface, thus a p-n (Sb doped SiGe layer) junction diode was constructed. I-V and C-V characteristics were measured to ensure device integrity prior to measuring DLTS. While the sample was nominally undoped, C-V dopant profiling revealed a background p-conductivity with a hole concentration 10^{14} cm⁻³. To study reverse bias majority-carrier hole trapping we used quiescent reverse bias $V_R= -1.10$ V and the filling bias pulse $V_F= -0.10$ V. DLTS was measured with various reverse bias voltages and the constant filling pulse of height 1.0 V, for investigating the depth dependence of the trap.

3. Results and discussion

Figure 2 shows a cross sectional TEM image for the sample **2-45**. In the compositionally graded region, very complex dislocation network is shown, but no dislocations is found in the strain relaxed $Si_{1-x}Ge_x$ buffer. This implies that the TDD is much less than 1×10^8 cm⁻² in the buffer [4]. Figure 3 shows the SEM images for the samples in Table 1. etched with various time. The scalers inside indicate 10 μm for (a), (f), (g), and (h) and 1 μm in others. The image of as grown, (a) shows cross hatch similar to the other reports [1]. In the 5 sec etched sample, (b), a lot of bright small etch hillocks are shown. In (c), the 20 sec etched sample shows no more etch hillocks. The etch-pits generated from dislocations are continuous with etching time, but the etch-pits generated from vacancy and impurity clusters do not persist on prolonged etching [5]. So the small bright etch hillocks do not related to dislocations. The presence of impurities may affect the dissolution process in a number of ways. The presence of impurities on surfaces may also modify the nucleation kinetics through lowering of the surface energy. Considering etching rate and etching time (43 Å/sec × 5 sec = 215 Å), the image (b) is from n-$Si_{0.55}Ge_{0.45}$ (Sb doped, 300 Å) layer. So the small bright etch hillocks may be related to the Sb impurities.

In 40 sec etched image (d), we can observe small diamond type EPs. On prolonged etching, EP size increases gradually. (i) shows EPD and the largest EP size as a function of etching time. The EPD increases a little with etching time but not order of magnitude. The EP size increases monotonically with etching time but the EPD is saturated beyond 50 sec. This indicates that the EPs satisfy the condition of continuity which is a feature of TD. So the EPs observed may be related to the TDs.

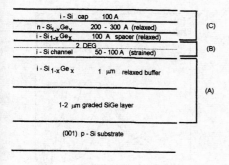

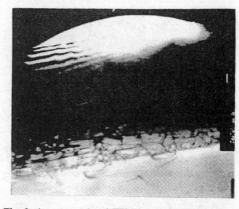

Fig. 1. A schmetic of the structures.

Fig. 2. A cross sectional TEM image for the sample, 2-45.

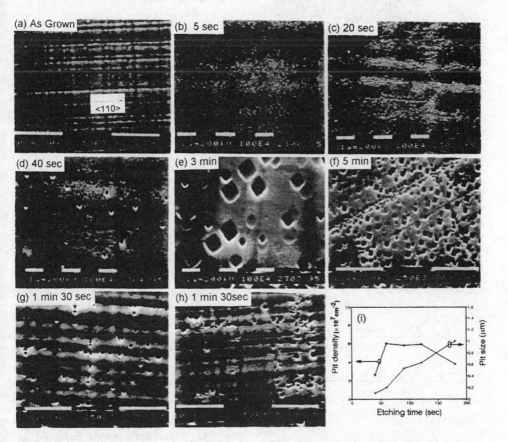

Fig. 3. (a)-(f). The SEM images for the sample 2-45 etched with various time. (g) and (h) are for the samples 3-2 and 3-3 respectively. (i) shows the etch-pit density and etch-pit size with etching time for the sample 2-45.

494

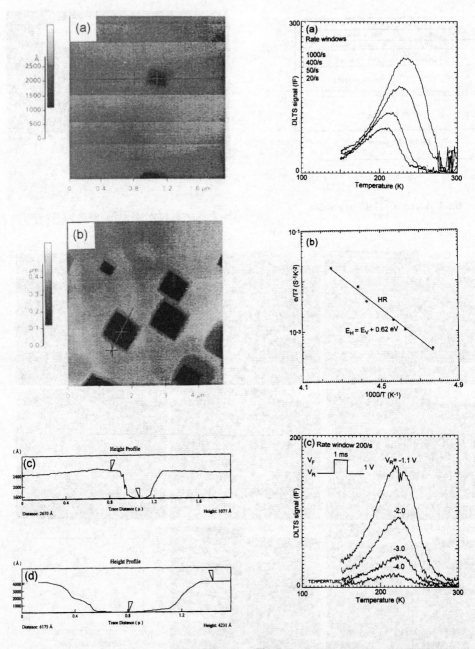

Fig. 4. (a) and (b) AFM images for the 2-45, 1 min and 3 min etched, respectively. (c) and (d) etch-pit profiles for the samples scanned along <100>.

Fig. 5. (a) DLTS spectrum under reverse bias for the 2-45. The different curves correspond to different rate windows indicated inside. (b) Arrhenius plot for the peak. (c) DLTS spectra at a fixed rate window (200/s) for several reverse bias scans. The fill pulse height maintained 1V.

Figure 3 (e) shows very clear EP image. The EPs have a distribution in size and seem to be orthogonal to the growing surface. This means that the EPs are not nucleated at the same time. So the dislocations may be related to partial dislocations and orthogonal to the growing surface.

Figure 3 (f) shows a image for the 6 min. etched sample. Considering etching time, it comes from the compositionally graded layer. The EPs are distributed randomly in the whole image and EP pileups with 3.7×10^5 cm^{-1} line density are observed along <110> direction. While the shape of the randomly distributed EPs is diamond type orthogonal to the growth plane, the shape of the EPs in the pileups is slightly inclined corn type. These pileups seem to a trace of dislocation annihilation by modified Frank-Read (MFR) mechanism [3]. (g) and (h) show SEM images for the 1 min. 30 sec etched for the samples **3-2** and **3-3**, respectively. The EP pileups with 1.2×10^5 cm^{-1} and 2.8×10^5 cm^{-1} line densities are observed along <110> direction, respectively. While EP pileups are observed in the graded layer in the sample **2-45**, the EP pileups are observed in the strain relaxed buffer in the samples **3-2** and **3-3**. This means that the high temperature growth increases not only the TD annihilation probability and the rate of disappearance of the TDs to the substrate edge but also pinning of the TDs along <110> direction.

Figure 4 (a) and (b) show atomic force microscopy (AFM) images for the sample **2-45**, 1 min. and 3 min. etched, respectively. They are similar to the SEM images. Fig. 4. (c) and (d) show depth profiles of the samples (a) and (b) at the positions indicated inside along <100> direction, respectively. The slope of the EPs is larger in the 1 min. etched sample than the 3 min. sample. According to thermodynamic theory [5], the slope of EPs is determined by the nomal and the tangential etch rates, which depend on the value of the chemical potential difference between the etching medium and the crystal, $\Delta \mu$ and ledge retardation factor, β. But the experimental results do not agree to this theory since the $\Delta \mu$ and β were constant in the same etching solution. This fact can be explained by kinematic wave theory of dissolution [5].

Figure 5 (a) shows a typical DLTS spectrum for the hole trap as measured for several rate windows indicated inside as the bias condition described in Section 2. By analyzing the Arrhenius behavior [Fig. 5. (b)] of the trap, the activation energy of the trap and the capture cross section were found to be $E_H = E_V + 0.62 \pm 0.05$ eV and $\sigma_H \sim 5.5 \times 10^{-10}$ cm^2, respectively. (c) shows DLTS peak dependence to various reverse bias voltage. The DLTS peak heights of the trap decreased and finally disappeared with increasing reverse bias. Considering the reverse bias dependency of the depletion region width, it is expected that the hole trap is from the graded layer. As the reverse bias increases , the depletion region is expanded into the substrate so that the trap density decreases. Although there exist two kinds of dislocations such as threading and misfit dislocations in graded layer, we find only a hole trap.

4. Summary

We studied the dislocation-related EPs by SEM and AFM on the strain relaxed SiGe sample after Schimmel-etch. The high temperature growth increases not only the TD annihilation probability and the rate of disappearance of the TDs to the substrate edge but also pinning of the TDs along <110> direction. A hole trap was observed by DLTS measurements and its activation energy and capture cross section were $E_H = E_V + 0.62 \pm 0.05$ eV and $\sigma_H \sim 5.5 \times 10^{-10}$ cm^2, respectively. The observed hole trap may be associated with threading dislocations and/or misfit dislocations in compositionally graded layer.

496

Acknowledgments

This work was supported by Korea Science and Engineering Foundation (KOSEF) through the Semiconductor Physics Research Center (SPRC) of Jeonbuk National University.

References

[1] E.A. Fitzgerald, Y.-H. Xie, D.Monroe, P.J. Silsvsesrman, J.M. Kuo, A.R. Kortan, F.A. Thiel, and B.E. Weir 1992 *J. Vac. Sci. Techno.* **B10**(4) 1807-1819

[2] Y.H. Xie, E.A. Fitzgerald, D. Monroe, P.J. Silverman, and G.P. Watson 1993 *J. Appl. Phys.* **3**(12) 8364-8370

[3] F. K. LeGouses 1994 *Phy. Rev. Lett.* **vol.72** 876-879

[4] G. Patrick Watson, Eugene A. Fitzgerald, Ya-Hong Xie, and Don Monroe 1994 *J. Appl. Phys.* **75**(1) 263-269

[5] Keshra SANGWAL 1987 *Etching of crystals, theory, experiment, and application*, Elservier Science Publishers B. V. (North-Holland Physics Publishing Division)

Inst. Phys. Conf. Ser. No 145: Chapter 3
Paper presented at 22nd Int. Symp. Compound Semiconductors, Cheju Island, Korea, 28 August–2 September 1995
© *1996 IOP Publishing Ltd*

Electrical properties in Si/Si$_{1-x}$Ge$_x$/Si p-type modulation doped heterostructures

Y. T. Hwang, H. S. Kim, S. J. Woo, S. H. Lim and H. J. Lee

Semiconductor Physics Research Center and Department of Physics, Jeonbuk National University, Jeonju 560-756, Korea

Abstract. Si/Si$_{0.79}$Ge$_{0.21}$ p-type modulation-doped single heterostructures were grown by molecular beam epitaxy with different growth temperatures and spacer layer thicknesses. The growth temperature dependence of hole mobility at low temperature is examined in the temperature range from 530 to 730 ℃. The hole mobility increases with increasing growth temperature up to 680 ℃ and then decreases at higher growth temperature. This demonstrates that the dependence of hole mobility on growth temperature may be ascribed to the improvement of interface with increasing temperature.

1. Introduction

Interest has been shown in Si/SiGe p-type modulation-doped heterostructures, not only because of their characteristic energy band structure caused by the large strain in the SiGe layer, but also because of their possible applications in high-performance p-channel field-effect transistors.[1] Several authors have reported enhanced two-dimensional hole mobilities in p-Si/SiGe modulation-doped heterostructures.[1-3]

In this work, the growth temperature dependence of hole mobility of Si/SiGe p-type modulation-doped single heterostructures at low temperature is examined.

2. Experiment

The samples were grown by molecular beam epitaxy in a Vacuum Generators V80S system on n-type Si substrates (resistivity ≈ 10 Ω-cm). The growth temperatures were in the range of 530~730℃. In all the samples 500 Å SiGe layers were grown on Si buffer layers and undoped Si spacer layers of 100 Å thickness, except for the sample (4/26), were grown before capping with B-doped Si layer. The boron concentration of B-doped Si determined by Hall measurements at 300 K was

Table 1. Growth parameters for the single Si/Si$_{0.79}$Ge$_{0.21}$ p-type modulation-doped heterostructures and mobilities and sheet hole concentrations measured by Hall measurements

Sample No.	Growth temp. (℃)	Spacer thickness (A)	Mobility at 5 K (cm^2/V–s)	Sheet hole concentration at 5 K (10^{11} cm^{-2})
4/22	530	100	1,470	5.2
4/12	580	100	1,670	5.5
4/20	680	100	2,450	5.8
4/8	730	100	1,070	22.7
4/26	680	0	550	15.2

3.7×10^{18} cm^{-3}. The Ge content was measured by double crystal x-ray diffraction. To investigate electrical properties, the hole mobility and carrier density were measured as a function of temperature by the van der Pauw method. Ohmic contacts were obtained by evaporating and alloying Al on these samples at 450 ℃ for 20 min. Growth parameters for single Si/Si$_{0.79}$Ge$_{0.21}$ p-type modulation-doped heterostructures and mobilities and sheet hole concentrations measured by Hall measurements are summarized in Table 1.

3. Results and discussion

Temperature dependence of hole mobility of single Si/Si$_{0.79}$Ge$_{0.21}$ p-type modulation-

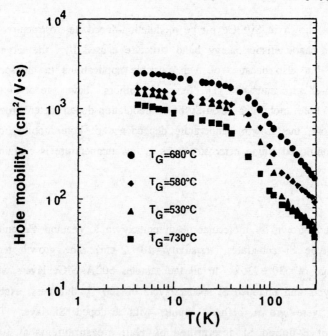

Fig.1 Hole mobility as a function of temperature for single Si/Si$_{0.79}$Ge$_{0.21}$ p-type modulation-doped heterostructures with different growth temperatures.

doped heterostructures with different growth temperatures are shown in Fig. 1. It is readily seen from data in Fig. 1 that all modulation-doped $Si/Si_{0.79}Ge_{0.21}$ heterostructures exhibit enhanced hole mobility at low temperatures. The peak hole mobility of 2450 cm^2/V-s at 5K was obtained for sample (4/20) grown at 680 ℃. As shown in Fig. 1, the hole mobility increases with increasing growth temperature up to 680 ℃ and then decreases at the higher growth temperature. Nagawa and Miyao have asserted that Ge on Si segregation reduces with increasing growth temperature (T_G=450～750 ℃).[5] However, such a segregation dependence in these samples was not identified by x-ray analysis. The dependence of hole mobility on growth temperature has been ascribed to interface charge.[4] Our results indicate that the Si/SiGe interface becomes smoother with increasing growth temperature, with the result of reduction of interface roughness scattering. The hole mobility decreases at growth temperature higher than 680 ℃. This suggests that boron diffusion into spacer layer decreases the mobility. Our results show that a growth temperature of 680 ℃ seems to be optimum for the interfacial quality at the Si/SiGe heterostructures and remote impurity scattering limit.

Figure 2 shows sheet hole concentrations as a function of temperature for single $Si/Si_{0.79}Ge_{0.21}$ p-type modulation-doped heterostructures with different growth temperatures. No carrier freezeout is observed in these heterostructures at low temperatures. This is clear evidence of carrier separation from their ionized boron acceptors due to valence band offset.[3] For the samples except that grown at 730 ℃,

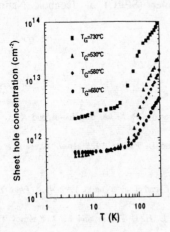

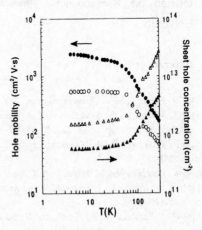

Fig. 2 Sheet hole concentration as a function of temperature for single $Si/Si_{0.79}Ge_{0.21}$ p-type modulation-doped heterostructures with different growth temperatures.

Fig. 3 Hole mobility and sheet hole concentration as a function of temperature for single $Si/Si_{0.79}Ge_{0.21}$ p-type modulation-doped heterostructures. ● ▲; with 100 Å spacer layer, ○ △; without spacer layer.

low-temperature sheet hole concentration increases at temperatures lower than 20 K. The reasons for increased sheet hole concentration at temperatures lower than 20 K may be due to the presence of weak localization and hole-hole interaction.[6]

Figure 3 shows the temperature dependence of Hall mobility and sheet hole concentration for single $Si/Si_{0.79}Ge_{0.21}$ p-type modulation-doped heterostructures with and without spacer layer. The low-temperature hole mobility was affected by the thickness of Si spacer layer. As shown in Fig. 3, the low-temperature hole mobility with spacer layer is about 5 times higher than that without spacer layer. This is due to reduction of remote ionized impurities scattering across the Si spacer layer.

In summary the hole mobility increases with increasing growth temperature up to 680 ℃ and then decreases at higher growth temperature. The peak mobility of 2450 cm^2/V-s at 5 K was obtained in heterostructures with the growth temperature of 680 ℃. This indicates that the Si/SiGe interface becomes smoother with increasing growth temperature, with the result of reduction of interface roughness scattering. The low-temperature hole mobility values of these heterostructures are also affected by the thickness of Si spacer layer. The low-temperature hole mobility is dominantly limited by remote impurity scattering and/or the interfacial quality at the Si/SiGe.

Acknowledgment

This work is supported by the Korea Science and Engineering Foundation (KOSEF) through the Semiconductor Physics Research Center (SPRC) at Jeonbuk National University.

References

[1] R. People, IEEE *J. Quantum Electron.* 1986 **QE-22** 1696-1710
[2] T. Mishma, C. W. Fredriksz, G. F. A. van de Walle, D. J. Gravesteijn, R. A. van den Heuvel, and A. A. van Gorkum 1990 *Appl. Phys. Lett.* **57** 2567-9
[3] P. J. Wang, B. S. Meyerson, F. F. Fang, J. Nocera, and B. Parker 1989 *Appl. Phys. Lett.* **55** 2333-5
[4] K. Nagagawa and M. Miyao 1991 *J. Appl. Phys.* **69** 3058-3062
[5] D. W. Smith, C. J. Emeleus, R. A. Kubiak, E. H. C. Parker, and T. E. Whall 1992 *Appl. Phys. Lett* **61** 1453-5
[6] C. J. Emuleus. T. E. Whall, D. W. Smith, R. A. Kubiak, E. H. C. Parker and M. J. Kearney 1992 *Thin Solid Films,* **222** 24-6

Inst. Phys. Conf. Ser. No 145: Chapter 3
Paper presented at 22nd Int. Symp. Compound Semiconductors, Cheju Island, Korea, 28 August–2 September 1995
© *1996 IOP Publishing Ltd*

Ambipolar diffusion in GaAs/AlGaAs quantum wells with inserted AlAs monolayers

F. Faller, B. Ohnesorge, A. Forchel

Technische Physik, University of Würzburg, Am Hubland, D-97074 Würzburg, Germany

Abstract. By using an optical time of flight technique we have investigated the ambipolar diffusion in GaAs/AlGaAs quantum wells with a varying number of AlAs monolayers (0-5) inserted at the center of the well for different Al-concentrations in the barrier. The AlAs layers push the electron hole wavefunctions into the AlGaAs barriers and result in a reduction of the ambipolar diffusivity particularly in the temperature range between 40 K and 90 K. The decrease of the ambipolar diffusivity can be correlated to the increasing influence of interface roughness scattering at the quantum well interfaces and the AlAs-layers.

1. Introduction

A major part of the activities in the area of thin semiconductor heterostructures aims at high electron mobilities parallel to the interface for fabrication of high-speed electronic devices or at a specific optoelectronic response for applications such as light emitters, heterojunction lasers or detectors [1,2]. Insertion of AlAs or InAs monolayers and submonolayers at the center of e.g. a GaAs/Al_xGa_{1-x}As quantum well (QW) allows both to control the spectral response of the system and to obtain supplementary information on scattering processes relevant for carrier mobility such as interface roughness (IR) scattering.

The influence of AlAs submonolayers (AlAs-coverage of 25%, 50% and 75%) on the mobility of two-dimensional electrons in GaAs/AlAs QWs has been studied resulting in a decrease by a factor of 5-10 as compared with the mobility of a reference QW [3]. In case of InAs submonolayers an increase of electron mobilities has been observed when the InAs coverage reaches 100%, i.e a complete monolayer (ML) is inserted [4]. This is because scattering at isolated InAs islands occurs less frequently.

In this paper, we investigate the ambipolar carrier transport in GaAs/AlGaAs QWs with 1, 3, and 5 MLs of AlAs embedded in the central part of the wells. We find a decreasing mobility with increasing number of AlAs-layers depending on the Al-concentration x in the AlGaAs barrier. For the determination of the ambipolar carrier mobility in our samples we use an optical time of flight method first applied by Hillmer et al.[5].

2. Samples and optical characterization

We have grown two groups of samples with solid source molecular beam epitaxy (MBE), group A with 33% and group B with 12% of Al in the barrier materials. The samples of both groups consist of a 15nm GaAs/Al_xGa_{1-x}As QW, which contains one, three or five MLs of AlAs at the center of the QW. For both groups samples without AlAs-layers were grown as references. We used semi-insulating GaAs (001) substrates, on which we grew a 1 μm-thick undoped GaAs buffer. This was followed by the 15nm undoped GaAs/AlGaAs-SQW with

502

barriers of 500nm thickness on both sides of the QW. To get high quality AlGaAs the barriers were grown at a substrate temperature of 725°C, while the GaAs and the AlAs monolayers were grown at 630°C. The V/III beam flux ratio was 34 in the barriers, the growth rates were 1.0, 0.67, 1.67 μm/h for GaAs, AlAs and $Al_{0.33}Ga_{0.67}As$ for group A samples and 1.0, 0.14, 1.14 μm/h for group B samples.

The samples were analyzed by cw and time resolved spectroscopy. The photoluminescence signals of both series shift to higher energies with increasing number of inserted AlAs monolayers, a behaviour which is well described by model calculations of the wavefunctions and transition energies. The T=2K luminescence line FWHM increase only slightly from 1.2 meV (reference) to 2.4 meV for the sample with 5 MLs (x=12%).

Lifetimes of all samples strongly increase with increasing temperature being almost independent of the number of AlAs-layers in a wide temperature range (50K-100K for group A samples, where lifetimes increase from about 2ns to about 3.2-4ns, 5K-70K for group B where lifetimes increase from 0.4ns to about 3.2ns). This behaviour indicates a high quality of the samples with and without AlAs-layers.

3. Optical time of flight method

We have investigated the ambipolar diffusion by an optical time of flight method using 3ps pulses from a compressed frequency doubled Nd:YAG laser for excitation and a synchroscan streak camera for detection (time resolution of 10ps). The photoluminescence is excited through circular holes in a highly reflective mask (hole radius comparable to diffusion length 1-10μm). The density gradients at the edges of the hole lead to an ambipolar diffusion under the mask which results in an accelerated decay of the luminescence compared to samples investigated without mask (fig.1). Using low excitation densities ($\sim$30W/cm^2) for free exciton formation, the temporal evolution of the photoluminescence directly reflects the evolution of the exciton concentration in the area defined by the hole. It can be modelled numerically by a diffusion equation containing the ambipolar diffusion constant as the only free parameter (fig.1). Details of the fit procedure are given in reference [5].

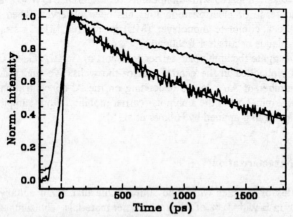

figure 1: Time resolved photoluminescence spectra (5ML sample of series B at T=70K) for excitation through a 100μm-hole (upper curve) and a 10μm-hole including the calculated time evolution for an ambipolar diffusion constant of 65 cm²/s (lower curve)

4. Results

Figure 2 depicts the obtained diffusion constants, figure 3 the deduced ambipolar mobilities $\mu=(e/kT)D$ for the x=33% samples (a) and the x=12% samples (b). Let us first discuss the data for the two samples without inserted AlAs-layers:

At very low temperature T<20K we observe diffusion constants equal to zero due to a localisation of excitons [6]. Photoluminescence rise times as high as 200ps are observed in this regime reflecting a slow relaxation of free excitons to localised states by emission of acoustic phonons. For T>(20-30)K , corresponding to a thermal activation energy of (2-3)meV, excitons move freely in the quantum well layer with diffusion constants increasing up to 60 cm^2/s at 85K for the x=33% sample and up to 85cm^2/s at 70K for the x=12% sample. These values correspond to ambipolar mobilities of about 10^4cm^2/Vs (fig.3). At higher temperatures the diffusion constants decrease due to acoustic and optical phonon scattering.

The temperature dependence and the order of magnitude of the diffusion constants and mobilities in the medium temperature range imply that mainly scattering at well width fluctuations limits the diffusion of electron hole pairs [7,8]. With decreasing Al-content the ground state energy is less sensitive to the roughness of the quantum well interfaces. This is in agreement with the observed higher diffusion constants and smaller linewidths for the sample with a low Al-content (B). Assuming that well width fluctuations give the main contribution to the PL linewidth, we deduce an average terrace height of one monolayer from our PL-spectra.

In order to confirm our interpretation, we calculated the barrier alloy disorder scattering (BAL) caused by a random distribution of Al- and Ga-atoms in the ternary barrier material for our structures following an approach of Hillmer et al.[8]. We obtain theoretical BAL mobilities of the order of 10^7cm^2/Vs with slightly higher values for the x=33% sample. Obviously BAL scattering does not significantly affect the carrier diffusion in case of our relatively wide wells of 15nm.

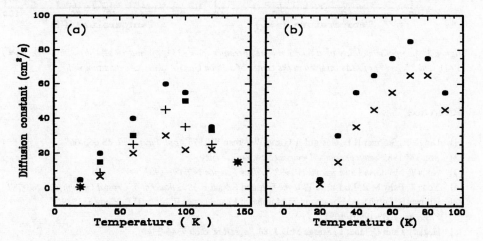

figure 2: Ambipolar diffusion constant as a function of temperature for x=33% (a) and x=12% (b) with O (●), 1 (■), 3 (+), and 5 MLs (×) of AlAs inserted at the center of a 15nm GaAs/Al$_x$Ga$_{1-x}$As quantum well

504

With increasing number of inserted AlAs-layers the diffusion constant is reduced significantly: The maximum diffusion constant in samples with 5 monolayers AlAs reaches only half the value of the reference for x=33% whereas it decreases less remarkably from 80cm²/Vs to 60cm²/Vs for x=12%. We attribute this behaviour to the following two facts: Firstly, the introduction of the AlAs monolayers narrows the quantum well thereby increasing the roughness scattering at the quantum well interfaces. Secondly, an additional scattering process at fluctuations of the AlAs- layers themselves comes into play. Both processes should influence the diffusion constant less for a smaller Al-concentration in the barrier as observed experimentally, since the ground state energy is less sensitive to both, fluctuations of the inserted AlAs-layers and of the quantum well interfaces. However, the systematic decrease of the IR mobility with increasing number of monolayers rather indicates the significance of the first process. An increasing number of AlAs monolayers pushes the electron hole wavefunctions more and more into the AlGaAs barriers thereby decreasing the IR mobility systematically. A high quality of the inserted AlAs-layers can therefore be assumed.

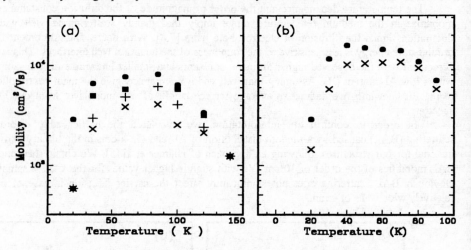

figure 3: *Ambipolar mobility as a function of temperature for x=33% (a) and x=12% (b) with O (•), 1 (■), 3 (+), and 5 MLs (×) of AlAs inserted at the center of a 15nm GaAs/Al$_x$Ga$_{1-x}$As quantum well*

References

[1] Dingle R, Störmer H L, Gossard A C and Wiegman W 1978 *Appl. Phys. Lett.* **33** 665-667
[2] Sze S M 1985 *Semiconductor Devices* (New York: Wiley)
[3] Noda T, Motohisa J and Sakaki H 1992 *Surface Science* **267** 187-190
[4] Noda T, Fahy M R, Matsusue T, Joyce B A and Sakaki H 1993 *Journal of Crystal Growth* **127** 783-787
[5] Hillmer H, Forchel A, Hansmann S, Morohashi M and Lopez E, Meier H P, Ploog K 1989
     Phys. Rev B **39** 10901-10912
[6] Hegarty J and Goldner L, Sturge M D 1984 *Phys. Rev. B* **30** 7346-7348
[7] Sakaki H, Noda T, Hirakawa K, Tanaka M, and Matsusue T 1987 *Appl. Phys. Lett.* **51** 1934-1936
[8] Hillmer H, Forchel A and Tu C W 1993 *J. Phys.: Condens. Matter* **5** 5563-5580

Inst. Phys. Conf. Ser. No 145: Chapter 3
Paper presented at 22nd Int. Symp. Compound Semiconductors, Cheju Island, Korea, 28 August–2 September 1995
© *1996 IOP Publishing Ltd*

Optical investigation of InGaAs/GaAs heterointerfaces

**H. I. Jeon, M. S. Jeong, H. W. Shim, Y. G. Shin, K. Y. Lim,
E. -K. Suh, and H. J. Lee**

Semiconductor Physics Research Center and Department of Physics
Jeonbuk National University, Jeonju 560-756, Republic of Korea

Abstract. Photoluminescence measurements as a function of temperature and excitation intensity were carried out on the strained $In_{0.2}Ga_{0.8}As$/GaAs quantum wells grown by low pressure metalorganic chemical vapor deposition. By analysing the transition energies and linewidths as a function of temperature, and the transition energies versus excitation intensities at various temperatures, we demonstrate the photoluminescence technique which can be used for the evaluation of the structural disorder at the interfaces of the quantum-well heterostructures.

1. Introduction

Quantum-well heterostructures(QWHs) based on InGaAs layers grown on GaAs have recently attracted attention because of interesting physical properties and device applications[1] for optical communication and high-speed electronics. A major limitation for strained layer QW's is the existence of the critical thickness due to the lattice-mismatch. One of the most important problems connected with growth of quantum well heterostructures is the structural disorder on the atomic scale which occurs at interfaces of heterostructures. The variation of confinement energies caused by the well thickness fluctuation in a given heterostructure has a considerable effect on parameters of luminescence signal, i.e. position on energy scale, lineshape, full width at half maximum(FWHM), etc. generated by this hetrostructure.

   In this work we present results of PL measurements on the $In_{0.2}Ga_{0.8}As$/GaAs strained single and double quantum-well heterostructures with various temperatures and excitation intensities. By analysing the transition energies and linewidths as a function of temperature, and the transition energies versus excitation intensities at various temperatures, we demonstrate that the PL technique can be used for the evaluation of the structural disorder at the interface of the QWH.

2. Experiment

Samples used in this study were grown by low-pressure metalorganic chemical vapor deposition(LP-MOCVD) on Cr-doped (100) GaAs substrates. In the case of single QW's 1000 Å-thick GaAs barriers were grown on both sides of $In_{0.2}Ga_{0.8}As$ layers. The well widths of $In_{0.2}Ga_{0.8}As$/GaAs single quantum wells(SQW's) are 18, 20, 35,

506

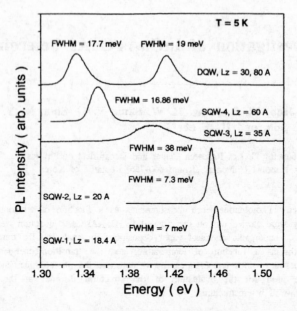

FIG. 1. PL spectra taken at 5 K with excitation power P_L = 30mW.

and 60 Å, for SQW-1, SQW-2, SQW-3, and SQW-4, respectively, and a $In_{0.2}Ga_{0.8}As$/GaAs double quantum-well(DQW) consist of two wells with 30 and 80 Å, respectively. The well layers(InGaAs layers) and confining layers(GaAs layers) of all samples used in this study were not intentionally doped. PL spectra were obtained with the 5145 Å line of an Ar^+ laser and recorded in the temperature range 5~300 K using a variable-temperature optical cryostat operated with liquid helium. A SPEX 1.26 m single spectrometer and a North Coast Ge-detector cooled at liquid nitrogen temperature were used.

3. Results and discussion

Fig. 1 shows the PL spectra taken at 5 K. At a low excitation intensity the full widths at half maximum (FWHM) for the SQW-1 and SQW-2 are 7 and 7.3 meV, respectively, while they are 16.8 and 38 meV for SQW-3 and SQW-4, respectively. Also the FWHM's for DQW are 18 and 19 meV for wells of 30 and 80 Å, respectively. The linewidth of the PL peak is mainly determined by the exciton -phonon interaction, exciton-ionized impurity scattering, and the scattering due to interface roughness in QWH. For samples with low density of residual impurities, the contribution of ionized impurities to the exciton peak linewidth is negligible. Also, the exciton-phonon interactions can be neglected at a low temperature.[2] Therefore, the linewidth of the PL peak at 5 K can provide a direct measure of the degree of the thickness fluctuation of the well. In Fig. 2, we present a plot of the measured FWHM of PL peaks for each sample. In this figure, the solid line represents the calculated fluctuation of the n = 1 quantized energy $\Delta E(\pm(a_o/2))/2$ due to the one monolayer fluctuation in the QW thickness as a function of L_Z, a_o being lattice

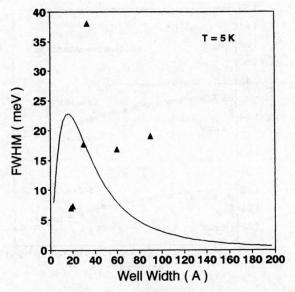

FIG. 2. Plot of FWHM vs well width for $In_{0.2}Ga_{0.8}As/GaAs$ QW's. The solid line represents the calculated fluctuation of the quantized energy $\Delta E(\pm(a_0/2)) /2$ due to a fluctuation of a thickness by one monolayer.

constant to the growth direction in strained $In_{0.2}Ga_{0.8}As$ layer. Based on the results given in Fig. 2, we found that the thickness fluctuation of SQW-1 and SQW-2 are below 1 monolayer and those of other samples are $2\sim5$ monolayers. Hereafter, we will discuss the SQW-1 and SQW-4 with the well thickness fluctuation of 1 monolayer and 2 monolayers, respectively.

In Fig. 3, the transition energies of SQW-1 and SQW-4 are plotted as a function of temperature. For comparison, the values calculated by the standard effective mass model for the QW structures, taking into account the strain and quantum size effects, are marked as a solid line. In the calculation, the variations of GaAs and $In_{0.2}Ga_{0.8}As$ band gaps and GaAs effective masses with temperature are taken into account. The calculated values have been fitted to the experimental QW curves in the high temperature range. In the low temperature range(T<100 K), experimental values deviated from the calculated values. Two possible explanations can be applied for this behavior[2]. One is that the recombination slowly changed from excitonic to free carrier type in the range of $4\sim100$ K, and the shift amounts to the exciton binding energy. The other is that the change taking place from bound exciton to free exciton states in the range of $4\sim100$ K, shifting the PL peak by an amount equivalent to the exciton-defect binding energy. However, results obtained by Delande et al.[3] and Zucker et al.[4] on studies of exciton trapping effects in $Al_xGa_{1-x}As/GaAs$ QW's, demonstrate that the traps involved have their origin in interface defects and typical exciton binding energy on these defects lies in the $2\sim5$ meV range, different from our experimental Stocks shifts at 5 K. In our case, the observed Stokes shift in both samples is approximately $6\sim9$ meV, which is in good agreement with binding energy of free exciton in InGaAs/GaAs QW[5]. Therefore, we conclude that at

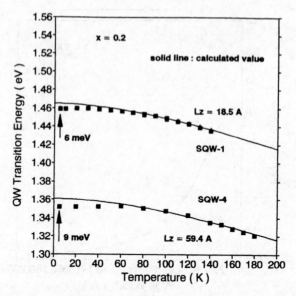

FIG. 3. The transition energy of QW as a function of the temperature for SQW-1 and SQW-4. The solid lines are calculated values.

temperatures above 100 K, the main mechanism of the radiative recombination is the free carrier recombination, while the free exciton recombination dominates at temperatures below 100 K. As shown in Fig. 3, the QW peak energy from SQW-4 undergoes a small blue shift of about 2 meV in the range of 5~40 K. The origin of this observed blue shift can be attributed to the fluctuation of the QW width which produces an effective band-gap modulation in the directions parallel to the QW plane. At low temperatures(4~40 K), the modulation related band-gap minima will act as active recombination sites because photogenerated carriers will drift towards them as drift times are much smaller than radiative recombination times[6]. As the temperature increases, more carriers will be able to occupy states at higher energy in the band-gap modulation scheme and QW peak should broaden and shift toward higher energies. The shift will be approximately proportional to kT. This shift is not usually observed if the thickness fluctuations are smaller than one monolayer due to the competitive red shift induced by the band-gap shrinkage with temperature.

The experimental QW transition energies as a function of excitation intensities are shown in Fig. 4 for SQW-1 and SQW-4. The transition energy of SQW-1 is nearly constant for the excitation intensities investigated at various temperature suggesting that only single confinement energy level exists in the QW. However, the PL peak in SQW-4 shows blue shift up to 9 meV with increase in the excitation intensity at 5 K, while it is nearly constant above 120 K. At 5 K, most of the carriers occupy the lowest level and as the excitation intensity increases carriers start to occupy the higher level leading to the blue shift of the PL peak. At temperatures above 120 K carriers occupied in the high energy level are thermally activated out of well, resulting in nearly constant peak energy with increasing excitation intensity. The QW transition energy in terms of the excitation intensity is consistent with the results

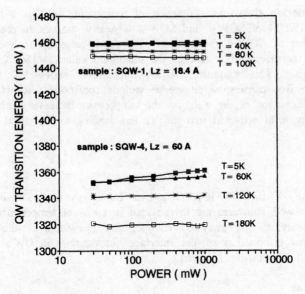

FIG. 4. The transition energy of QW as a function of the excitation intensity for SQW-1 and SQW-4

obtained from the QW transition energies as a function of temperature resulted from well thickness variation as shown in Fig 3.

The experimental values of the half width at half maximum(HWHM) of the

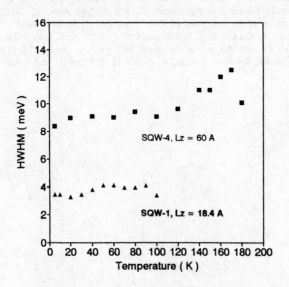

FIG. 5. The experimental values of high energy side of the HWHM as a function of the temperature for SQW-1 and SQW-4.

high energy side are shown as a function of temperature in Fig. 5 for SQW-1, and SQW-4. The HWHM of SQW-1 and SQW-2 is nearly constant in entire temperature range, while that of SQW-4 shows nearly the same value up to about 100 K, and then gradually increase until it reaches the maximum value at 180 K, where it starts to decrease again. This suggests that at temperatures above 100 K excitons are dissociated into free carriers which are at various confinement levels formed due to the thickness fluctuation of the well. As the temperature increases further, free carriers at higher energy level activated into the barriers leading to the slight decrease in the HWHM.

4. Summary

The PL spectra of the LP-MOCVD grown $In_{0.2}Ga_{0.8}As/GaAs$ strained single and double quantum well structures are investigated in terms of temperature and excitation intensity. By analyzing the transition energies and linewidths of the PL peaks, we obtain the direct information of the interface fluctuations of QW's from the peak energy and the linewidth of PL signal.

Acknowledgment

This work is supported by the Korea Science and Engineering Foundation (KOSEF) through the Semiconductor Physics Research Center (SPRC) at Jeonbuk National University.

References

1. Anderson N G, Laiding W D, Kolbas R M and Lo Y C 1986 *J. Appl. Phys.* **60** 2361-2367
2. Doctor M L, Recio M, Golmayo D and Briones F 1992 *J. Appl. Phys.* **72** 5861-5866
3. Delalande C, Meynadier M H, and Voos M 1984 *Phys. Rev. B* **31** 2497-2498
4. Zucker J M, Pinczuk A, Chemla D S, and Gossard A C 1987 *Phys. Rev. B* **35** 2892-2895
5. Atanasov R, Bassani F, D'Andrea A and Tomassini N 1994 *Phys. Rev. B* **50** 14381-14388
6. Bacher G, Schweizer M, Kovac J, Forchel A, Nickel H, Schlapp W and Losch R 1993 *Phys. Rev. B* **47** 9545-9555

511

Paper presented at 22nd Int. Symp. Compound Semiconductors, Cheju Island, Korea, 28 August–2 September 1995

Deep level investigation of bulk and epitaxial 6H-SiC at high temperatures

J.D. Scofield[a], Y.K. Yeo, and R.L. Hengehold

Air Force Institute of Technology, Wright-Patterson AFB, OH 45433-7765 USA

Abstract. Deep level transient spectroscopy measurements from 20 to 800 K were made on n- and p-type sublimation grown bulk SiC and vapor phase epitaxial grown SiC. Several new deep level centers were observed on both the n-type bulk and epitaxial material with ionization energies in the range of 0.58 to 1.4 eV below the conduction band, and on the p-type materials with energies 0.38 to 1.1 eV above the valence band. Field dependency and capture transient measurements on an E_c-0.63 eV center in the bulk n-type material revealed an anomalous emission process, which is dependent upon the depletion region field strength, and the presence of a thermally activated capture cross section. Secondary ion mass spectrometery measurements of the bulk material revealed the presence of several transition metals and other impurities.

1. Introduction

The renewed interest and activity in SiC development in recent years, owing to the availability of mono-polytypic 6H and 4H substrate material, has led to the successful development and demonstration of various devices based on these material systems. Ultraviolet photodiodes [1], junction devices [2], numerous field effect structures [3], and most notably, blue light emitting diodes [4] have all been demonstrated to various levels of efficiency and performance. Fueling this revitalized activity in wide-gap semiconductor development is the promise of significant performance gains for microwave and power devices with the added benefit of increased thermal and radiation hardness. The results of these intensified efforts have been continued improvement to material quality and in device performance. However, several difficulties persist indicating the need for better physical understanding of the electronic properties of this unique material system. Specifically, SiC device technology is hampered by impurity and structurally related crystal flaws, difficulty in processing, relatively deep dopant energy levels, contact metallization problems, and the lack of a controllable process for producing semi-insulating bulk material. Metal-Oxide-Semiconductor (MOS) devices fabricated on p-type 6H-SiC exhibit anomalous transconductance and channel mobility behavior [5] possibly related to the aluminum acceptor and high interface state densities, leakage currents and pinch-off problems in field effect transistors (FETs), and unacceptably low minority-carrier lifetimes in optoelectronic devices are all representative of the difficulties encountered when building devices on these material systems. In many instances, devices built on SiC appear to behave in an unusual fashion. This is due to factors such as the anisotropy of the mobility and conductivity, marginal validity of the effective mass approximation, especially in the valence band, and incomplete ionization of shallow donors and acceptors even at temperatures exceeding

---

[a] Permanent Address: Aeropropulsion and Power Directorate, Wright Laboratory WPAFB OH USA

400 to 500 K. Many of the device related problems are, however, directly related to the high concentrations of defects present in the material with energy levels within the forbidden gap of the semiconductor. These centers act as electron or hole traps and recombination-generation centers. Thus, these centers effectively reduce minority carrier lifetimes, contribute to leakage current, and in general degrade device performance. Of course, controlled introduction of such centers may have the desirable effects of creating semi-insulating material and improving switching speed, but understanding the sources and nature of these centers is required before they can be utilized in this manner.

Therefore, in order for SiC to realize its tremendous potential for high temperature, high power, and high frequency device performance, significant improvements must be made in each of the problem areas as described above, especially in the quality of the starting bulk and epitaxial material. Although significant understanding about the radiative transition centers in SiC has been realized [6,7], little is known regarding device degrading non-radiative deep traps in SiC. Therefore, a deep level study of sublimation grown bulk and vapor phase epitaxial (VPE) grown 6H-SiC was undertaken on Schottky, MOS, and pn junction devices using primarily deep level transient spectroscopy (DLTS) over a wide temperature range up to 800 K.

2. Experimental

The SiC samples used in this work were 6H-polytype n- and p-type bulk material, grown by the vapor transport process, and doped with nitrogen and aluminum, respectively, to a concentration ranging from $1x10^{17}$ to $7x10^{17}$ cm^{-3}. Vapor phase epitaxial grown layers with shallow impurity doping levels ranging from $1x10^{16}$ to $2x10^{17}$ cm^{-3} were also characterized. 500 μm diameter Schottky diodes were formed on the polished side of the wafers using rf sputtered unannealed Al and Ni contacts onto the p- and n-type material, respectively. Ohmic contacts were formed using the same metals processes followed by an 830 °C (Al) or 1050 °C (Ni) tube furnace anneal for 5 minute in an Ar atmosphere. Due to the large diode area, the higher doped n-type samples required an initial deposition of a thin (50 nm) Ti layer prior to Ni coverage in order to eliminate leakage currents in the reverse biased devices. MOS capacitors were obtained by deposition of Mo contacts on thermally grown (1100 °C) native SiO$_2$ layers.

DLTS measurements were taken in the temperature range from 20 to 800 K on a computer controlled system with a fast pulse interface. External pulse generation and software control allow measurements in various DLTS modes including double correlated (DDLTS) and constant-capacitance (CCDLTS). Capacitance or voltage transients are obtained isothermally at pre-defined temperature intervals, typically between 2 K and 5 K. Subsequent determination of trap capture and emission kinetics is made by fitting each transient using the Modulating Functions technique [8]. Use of this method allows superior determination of trap parameters, especially when capacitance transients are truncated and composed of multiple time constants (τ), when compared to the classic rate-window technique originally proposed by Lang [9]. This is exactly the situation that may be encountered for 6H-SiC due to the existence of lattice site inequivalencies. In this case the transient junction capacitance is expressed as $C(t) = C_0 + \sum_i^N C_i \cdot \exp(-t / \tau_i)$, where i=3 for 6H-SiC. Due to the large bandgap (2.86eV) of 6H-SiC, and multiple ionization energies, and relatively large magnitudes for shallow dopant impurities, care has to be taken to ensure that the mathematical expressions, especially simplifying approximations, used to assertain trap parameters are valid.

3. Results and discussions

3.1 Bulk material

Both the epitaxial and bulk material investigated in this work exhibited significant numbers of deep trap levels. However, only those traps which were observed consistently throughout the range of samples are reported. Figure 1 shows a typical rate window plot for the n-type bulk SiC measured by DDLTS at three different rate windows. The spectra show a large majority carrier electron trap observed in all n-type bulk wafers. This trap has an energy level at 0.63 eV below the conduction band (E_c), which is close to the energy level of a reported vanadium (V) related center [10], and the Z_1 level reported by Pensl [11]. Field dependent DDLTS

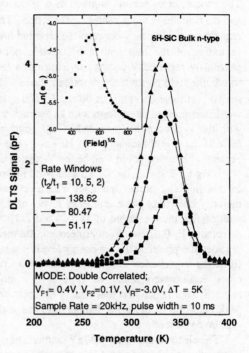

Fig. 1 DLTS rate window plot for the n-type bulk 6H-SiC with the field dependence shown in the inset.

measurements were made to determine acceptor or donor character, limiting the trap emission to a narrow portion of the depletion region of the reverse biased Schottky devices. While this could be the center previously attributed to V, it has been found that this trap shows an unusual field dependency and a thermally activated capture cross section.

Vanadium is known to exist in SiC in three charge states $V_{si}^{5+}(3d^0)$, $V_{si}^{4+}(3d^1)$, and $V_{si}^{3+}(3d^2)$ from electron spin resonance (ESR) data [12], and presumably creates deep donor or acceptor states depending upon the position of the Fermi level. In n-type material, the A^0/A^- electronic configuration should be prevalent, and would not be expected to exhibit a coulombic barrier lowering effect with an applied field (E). However, as shown in the inset of Fig. 1, the emission rate, e_n, plot as a function of $E^{1/2}$ indicates that for fields $\leq 2.5 \times 10^5$ V/cm, the emission rate is linearly dependent on $E^{1/2}$. Also, the measured trap energies decreased from E_c-0.72eV to E_c-0.58eV over this field range. On the other hand, when the applied fields were greater than 3×10^5 V/cm, the emission rate decayed exponentially as seen in the figure. The linear portion of the data was fit using the Poole-Frenkel [13] model given as $e_n = e_{n0} \cdot exp(2q\sqrt{qE/\pi\varepsilon})$, where q is the electronic charge and ε is the dielectric constant of the material. The permitivitty values calculated from the slope differed from the SiC dielectric constant by a factor of three, but this might be expected for localized orbits of tightly bound electronic levels. The large field behavior of this trap potential is not understood at this time, but may be due to a field enhanced screening effect. The capture cross section of this center was also determined to be temperature dependent, and to obey the model commonly attributed to electron or hole capture via multi-phonon emission (MPE), where the electron-phonon coupling is achieved through localized defect oscillations [14].

514

The capture cross section is given as $\sigma = \sigma_\infty \cdot exp(-E_\sigma / kT)$, where E_σ is the capture barrier, and σ_∞ is the high temperature cross section. This strong temperature dependence on σ may explain the discrepancy between the reported lower cross section of $\sigma = 5 \times 10^{-20}$ cm^2 and the saturation of the trap with a 5 µs wide pulse indicating a large cross section for the previously reported V-center. In addition, a thermally activated cross section, will play a role in the field dependence described above. The measured value of σ_∞ was found to be in the 10^{-15} cm^2 range, and $E_\sigma \cong 50$ meV. The spatial distribution of electrically active defects was determined to be uniform with an average concentration of 10^{15} cm^{-3}, which agrees well with the V concentration of 3×10^{15} cm^{-3} obtained from secondary ion mass spectrometery (SIMS) data on these samples. However, the origin of the E_c-0.63 eV trap is unclear at present, although it could well be the V-related center.

Figure 2 shows the DLTS rate window spectra for the Al doped p-type bulk 6H-SiC. In this material, three dominant trap levels are consistently observed. Two new deep majority hole traps were at energy levels of 0.97 and 0.6 eV above the valence band (E_v) with capture cross sections of 3×10^{-13} and 2×10^{-20} cm^2, respectively, and an electron trap was detected at E_c-0.87 eV. Furthermore, careful measurements of the E_v+0.97 eV peak at 400 K revealed the presence of two energy levels instead of one at E_v+1.1 and E_v+0.83 eV from the fit of the capacitance transients at each temperature, as shown in the inset. Their capture cross sections were measured to be 2×10^{-15} and 5×10^{-12} cm^2 for the 1.1 and 0.83 eV traps, respectively. It is noted that widely different energy levels can appear as a single apparent peak emission in the same temperature range due to the emission rate dependence on the capture cross section.

The electron trap at E_c-0.87eV is interesting for several reasons. The existence of such a large minority carrier signal in a majority carrier device of a large bandgap semiconductor would typically not be expected. The large bandgap of SiC removes any possibility of thermally generated minority carriers required for filling this defect in the measured concentration of 10^{16} cm^{-3}. Additionally, filling pulse biases were not large enough to achieve minority carrier injection due to an enhanced drift-field component, thus the minority carrier injection could not be the source of these signals. This behavior was observed on numerous diodes built on both bulk and epitaxial SiC material of various doping densities and both conductivity types. A clue to the source of this behavior may be in the observed temperature dependence of the DLTS signals. In every case, as the temperature increases, the minority carrier emission signal makes a continuous transition to a majority type emission transient, exactly in the manner seen in Figure 2. The details of this deep level assisted generation mechanism will be reported in a subsequent article.

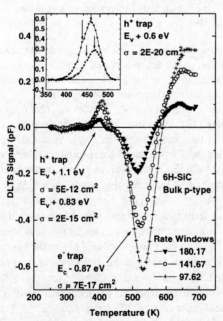

Fig. 2 DLTS spectra for the p-type bulk SiC
showing the resolution of two trap levels for
the low temperature peak in the inset.

3.2 Epitaxial Material

Figure 3 shows the characteristic rate window plots of typical DLTS measurements of the n-type epitaxial SiC layers. The spectra reveal several new majority carrier traps at energies of E_c-0.85, E_c-1.0, and E_c-1.4 eV with capture cross sections of 1×10^{-17}, 1×10^{-16}, 2×10^{-12} cm^2, and average trap concentrations of 5×10^{13}, 3×10^{14}, and 1×10^{14} cm^{-3}, respectively. As seen in the inset of Fig. 3, the DLTS measurements at lower temperature reveal the presence of a shallower level trap. This trap was observed in the majority of samples selected from various wafers. Although the location of the measured energy of E_c-0.59 eV and temperature range of the emission peak correspond to the previously discussed E_c-0.63 eV center observed in bulk samples, this trap did not exhibit the same E-field behavior and cross section dependence. That is, the cross section of this trap was temperature independent, and measured to be 2×10^{-15} cm^2. Also a field dependence was not observed, indicating the electronic charge of the filled trap to be in an ionized acceptor state.

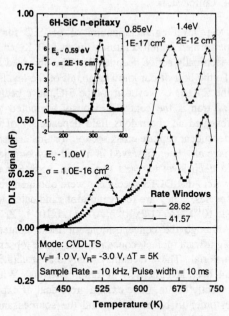

Fig. 3 Rate window spectra for the n-type epitaxial SiC showing lower temperature peak in the inset.

As is observed for the p-type bulk material, the p-epitaxial SiC also exhibits an energetically rich spectra of deep trap levels. The observed positive multiple peaks illustrated in Fig. 4 indicate a series of three deep level energies at 380, 423, and 863 meV above E_V for the corresponding peaks labeled A, B, and C, respectively. The capture cross sections for these traps were measured to be 4×10^{-19}, 2×10^{-20}, and 1×10^{-17} cm^2 in the same order. In this case, the multiple peaks are resolved using carefully chosen rate windows by varying t_1, t_2, and t_2/t_1, as described by [15]. Trap parameters are, however, always determined using the exponential fits to the capacitance transients at each temperature. The average concentrations of the defects, estimated from the heights of the rate window peaks and from Arrhenius data are 10^{14}, 10^{14}, and 10^{13} cm^{-3} for levels A, B, and C, respectively. These concentrations, while lower than those observed in the bulk material, are higher than would be desired especially for optoelectronic and other minority carrier device applications.

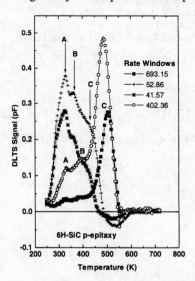

Fig. 4 DLTS plot for in p-type epitaxial 6H-SiC showing 3 deep traps identified

4. Conclusion

Deep level characterization of 6H-SiC for both the bulk and epitaxial materials was performed at temperatures up to 800 K, and resulted in the observation of several new electrically active deep levels. The observed ionization energies range from 0.38 to 1.4 eV. To the best of our knowledge, all of the levels reported here, with the possible exception of the E_c-0.63 eV level in n-type SiC, have been previously unreported. Whether the E_c-0.63 eV trap is the center previously attributed to vanadium or not, our measurements have revealed an anomalous field dependency of the electron emission rate and a temperature dependent capture cross section for this center. High levels of minority carrier trap activity have also been observed in n- and p-type majority carrier devices, and is likely to be related to deep level assisted generation. SIMS measurements, taken from the same bulk SiC wafers that DLTS specimens were obtained, revealed several definitively identifiable atomic species. They are (concentrations indicated in parenthesis in units of cm^{-3}) Ti (5×10^{17}), V (1×10^{16}), Cr (1.2×10^{17}), Cu (1.5×10^{17}), Zr ($\times10^{18}$), Ni (1.6×10^{18}), and Ge (1.1×10^{18}). Although the concentrations of these contaminants in epitaxial SiC are notably lower, significant defect concentrations were observed for several deep traps in the epitaxial material. These facts may imply that structural and native defects are major contributors to the deep level activity in these materials. With the study of deep levels in SiC having only recently begun, there exist a great deal of experimentation and analysis to be accomplished in order to better understand the sources and device related effects of the flaws in this material.

References

[1] Brown D., Downey E., Ghezzo M., Kretchmer J., Saia R., Liu Y., Edmund J., Gati G., Pimbley J., Schneider W., 1993 IEEE Trans. Elec. Dev. **40** 325-331.

[2] Kelner G., Binari S., Shur M., Sleger K., Palmour J., Kong H., 1992 Mat. Sci. and Eng. **B11** 121-124.

[3] Palmour J., Kong H., Carter C., 1992 Proc 2nd Int. Conf. Hi. Temp. Electron 491-494.

[4] Vishnevskaya B., Dmitriev V., Kovalenko I., Kogan L., Morozenko Y., Rodkin V., Syrkin A., Tsarenkov B., Chelenokov V., 1988 Sov. Phys. Semicond. **22** 414-417.

[5] Brown D., Ghezzo M., Kretchmer J., Downey E., Pimbley J., Palmour J., 1994 IEEE Trans. Elec. Dev. **41** 618-620.

[6] Patrick L., Choyke W., 1962 Physical Review **127** 1868-1877.

[7] Dean P., Hartman R., 1972 Phys. Rev **B 5** 4911-4924.

[8] Loeb J., Cohen G., 1965 IEEE Trans. on Automation Control **10** 359-361.

[9] Lang D., 1974 J. Ap. Phy. **45** 3023-3032.

[10] Evwaraye A., Smith S., Mitchell W.,1994 J. Appl. Phys. **76** 5769-5772.

[11] Pensl G., Choyke W., 1993 Physica B **185** 264-283.

[12] Schneider J., Muller H., Maier K., Wilkening W., Fuchs F., Dornen A., Leibenzeder S., Stein R., 1990 Appl. Phys. Lett. **56** 1184-1186.

[13] Frenkel J., 1938 Phys. Rev. **54** 657-662.

[14] Henry C., Lang D.,1977 Phys. Rev. B **15** 989-1016.

[15] Thurber W., Forman R., Phillips W., 1982 J. Appl. Phys. **53** 7397-7400.

Inst. Phys. Conf. Ser. No 145: Chapter 3
Paper presented at 22nd Int. Symp. Compound Semiconductors, Cheju Island, Korea, 28 August–2 September 1995
© *1996 IOP Publishing Ltd*

Negative photoconductivity of $Al_xGa_{1-x}As:Sn$ /GaAs modulation-doped heterostructures

Zhongling Peng, Tadashi Saku and Yoshiji Horikoshi

NTT Basic Research Laboratories, 3-1 Morinosato Wakamiya, Atsugi-shi, Kanagawa Pref., 243-01, Japan

Abstract. Temperature-dependent Hall-effect measurements under different illumination conditions are performed for $Al_xGa_{1-x}As:Sn/GaAs$ modulation doped heterostructures. Two distinct DX center levels are deduced. The two-dimensional electron gas concentration measured under illumination decreases from the value due to persistent photoconductivity to a value less than that measured in the dark when the excitation photon energy is larger than the band gap of the AlGaAs barrier. This decrease occurs only at temperatures below about 90 K. This negative photoconductivity is explained by taking into account the partial freeze-out of electrons into the shallow DX centers and the transfer of holes photo-generated in the barrier into the channel region.

1. Introduction

Recently, Sn-related DX center in AlGaAs systems has received much attention partly due to the need to minimize DX center deleterious effects in III-V device performance. Since device hysteretical behavior I-V collapse and persistent photoconductivity (PPC) at low temperatures are determined by the capture barrier of the DX center[1],, one way to minimize these is the choice of donors with the smallest capture barrier. Sn is believed to be an alternative. While in Te- and Si-doped AlGaAs PPC has been proved at temperatures below about 120 K, in Sn-doped samples the decay of photoconductivity (measured through photocapacitance) has been observed at a quite low temperature[2,3,4], indicating a lower capture barrier. Some authors have even claimed an absence of PPC in Sn-doped $Al_xGa_{1-x}As$ with x=0.39[4].

On the other hand, doublet structures in deep-level transient spectroscopy (DLTS) of Si[5-8], Te[9] and Sn[2, 4, 9-12] have been reported for a wide alloy composition range 0.3<x<0.7, indicating the presence of two deep donor states. In Sn-doped $Al_xGa_{1-x}As,$[11] the second peak was ascribed to photoionization of the DX center which leads to the formation of a localized paramagnetic donor state, $DX^0(Sn)$, according to the reaction $DX^- \rightarrow DX^0 + e^-$. This picture agrees with negative-U models for the DX ground state. The DX^0 state is observed in an alloy range of $0.3 < x < 0.5$, where the $DX^{0/+}$ level is not resonant with the conduction band[11].

PPC in AlGaAs/GaAs heterostructures has been much less thoroughly investigated, in part because the two-dimensional electron gas (2DEG) at the heterojunction (HJ) is itself persistent and this introduces additional complications to the understanding of bulk PPC effects[13]. However, the additional persistent channel is at least useful in preserving the details of the photoexcitation process. For bulk AlGaAs with $x > 0.3$, transport measurement does not give reliable results at low temperatures due to a decrease in mobility. It is possible to avoid these difficulties by studying the heterostructure. Since two-dimensional carriers have high mobility, transport measurement is possible at low temperatures.

In this paper, we report on a photoconductivity study of $Al_xGa_{1-x}As/GaAs$ modulation doped (MD) heterostructures by temperature-variable Hall effect measurement, with emphasis on the above band gap illumination. For a Sn-doped HJ with $x=0.35$, two DX states are deduced and negative photoconductivity is observed.

2. Experimental process

Samples used in this study were AlGaAs/GaAs heterostructures grown by MBE on (100) undoped semi-insulating GaAs substrates. A typical layer structure consists of the following growth sequence: 100 nm undoped GaAs, a 1000 nm AlGaAs/GaAs superlattice buffer, a 1 μm undoped GaAs channel, a 75 nm undoped AlGaAs spacer, Sn delta doping at 1.6×10^{13} cm^{-2}, 500 nm undoped AlGaAs, and a 100 nm GaAs cap. At 6 K, the electron channel mobility is in excess of 10^6 cm^2/V.s after illumination.

The sample was mounted in a variable-temperature continuous-flow He cryostat, and the Hall carrier concentration and mobility were measured using a standard DC Van der Pauw method. Ohmic contacts were fabricated by alloying In balls on the four corners of 5×5 mm^2 portions of the wafers at 400 °C in a N_2 atmosphere. A current of 500 μA was used with an applied magnetic field of 500 gauss. A small monochromator with a tungsten lamp source provided monochromatic light which was transmitted to the cryostat using an optical fiber. Temperature-dependent Hall data was taken without any exposure to light as the samples were cooled down from room temperature to 6 K. The samples were then exposed to light at 6 K until the electron concentration had reached its saturated value. The saturated concentration of the 2DEG measured under illumination with monochromatic light of different

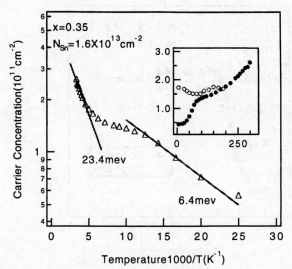

Fig.1 Inverse temperature-dependent 2DEG concentration measured in the dark. The simple relations of n~exp(-E$_{dd}$/KT) yield thermal activation energy of 23.4mev and 6.4mev for deep and shallow DX levels respectively. The inset is the temperature-dependent 2DEG concentration, while the open circle represent 2DEG concentration due to PPC effect.

wavelengths at different temperatures. The stable n$_{2DEG}$ was also measured after the light was turned off.

3. Experimental results and discussion

Figure1 shows the inverse temperature-dependent n$_{2DEG}$ of Al$_x$Ga$_{1-x}$As:Sn/GaAs with x=0.35, measured in the dark. Two distinct regions are identified. In both regions, the Hall electron concentration decreases exponentially as the temperature decreases, due to a partial carrier freeze-out into the DX levels. The simple relations of n~exp(-E$_{dd}$/KT) yield thermal activation energies of 23.4 mev and 6.4 mev for deep and shallow DX levels respectively. Zdansky and Peaker[3] have measured the decay of photocapacitance in Sn-doped Al$_x$Ga$_{1-x}$As with the same composition at low temperature. They obtained the energy E$_D$=30 mev for the deep state and E$_D$=8 mev for the shallow state, to which our data is comparable. However, the shallow DX level is only observed after the samples are illuminated in the previous studies on Si, Te and Sn-doped AlGaAs[2, 4, 9-12]. The direct observation of two DX levels here is probably HJ-related. The open circles in the inset are n$_{2DEG}$ measured in the dark after sufficient white light illumination (PPC effect). In contrast to Si-doped samples, n$_{2DEG}$ due to the PPC effect at 6 K is less than n$_{2DEG}$ at room temperature.

520

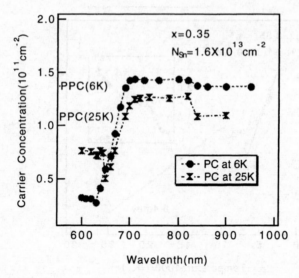

Fig.2 Spectral distribution of 2DEG concentration due to photoconductivity.

This indicates that the capture barrier for the shallow DX level is so low that a fraction of the electrons are trapped out again after they are excited optically from DX centers (deep and shallow) even at the lowest temperature, while others remain in the conduction band. It should be mentioned that saturated n_{2DEG} under illumination (PC) is wavelength dependent. It is inherently related to the photoionization efficiency of both deep and shallow DX centers. With illumination of >1000 nm light, n_{2DEG} can even reach the room temperature value. The details will be discussed elsewhere. After illumination is switched off, the 2DEG concentration experiences a rapid drop initially followed by a slow decay. The decay is so slow that it only causes a small change in carrier concentration with experiment time, so we choose the value at the beginning of the slow decay as the 2DEG concentration due to PPC effect.

Spectral distributions of n_{2DEG} under illumination are shown in Fig. 2 for two different sample temperatures. The n_{2DEG} due to PPC are also labeled. As the photon energy is larger than the band gap energy of GaAs, carrier concentration increases slightly due to the generation of electron-hole pairs in the GaAs channel. However, the carrier concentration drops rapidly to a value even less than that due to PPC if the photon energy is larger than the band gap of AlGaAs barrier. We explain the negative photoconductivity as follows. After electron-hole pairs are generated in the barrier, some of the electrons are recaptured by the shallow DX centers due to the low capture barrier and high Fermi level of photogenerated electrons in the conduction band, while most of the holes are

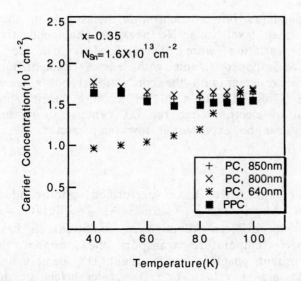

Fig.3 Temperature-dependent 2DEG concentration measured in the dark after illumination(PPC) and under illumination of monochromatic light with three different wavelength(PC).

transferred into the 2DEG channel. This explains the smaller total value of carrier concentration from the Hall-effect measurement. The emission of electrons from shallow DX centers is restrained at low temperature. Actually the thermoequilibrium between the conduction band and a shallow DX center can be maintained only above about 90 K, as shown in Fig.1. The n_{2DEG} in the dark presents a second drop at about 90 K, indicating the beginning of the partial carrier freeze-out into the shallow DX level.

This is further confirmed by the temperature-dependent n_{2DEG} due to the PPC effect and the PC effect under illumination of three different kinds of monochromatic light, as shown in Fig.3. For light of 640 nm (larger than the band gap energy of the AlGaAs barrier), negative photoconductivity occurs at temperatures below 90 K when the freeze-out of a fraction of the electrons into shallow DX centers becomes active. At this wavelength, photoionization efficiency of the shallow DX level is quite low. Part of the electrons excited from the deep DX centers are trapped by the shallow centers due to their low capture and relative high emission barriers, which prevent the establishment of thermoequilibrium between the conduction band and shallow DX level at temperatures below about 90 K. According to the picture of DX^- and DX^0 in Ref. 11, shallow DX centers can be generated by photoionization of deep centers. Thus one need not worry about the saturation of shallow DX centers.

We also measured photoconductivity of Si-MD HJ with x=0.35 and Sn-MD HJ with x=0.3. The above behavior was not observed.

522

Temperature-dependent Hall concentration measured in the dark gives only one deep level state. No negative photoconductivity or its temperature variation were observed. This can be easily understood for a Sn-doped sample with less Al composition if a shallow DX level resonant with the conduction band is taken into account. For a Si-doped HJ, a higher capture barrier prevents partial freeze-out of electrons into the DX centers, so no negative photoconductivity can be expected at low temperatures.

4. Conclusion

In conclusion, we have performed photoconductivity measurement of $Al_{0.35}Ga_{0.65}As:Sn/GaAs$ modulation-doped heterostructures with different illumination conditions. It has been proved that low temperature transport measurement on HJ structures can provide useful details about DX centers in bulk $Al_xGa_{1-x}As$ with large x values. Two DX center levels are deduced from temperature-dependent Hall effect measurement. Negative photo conductivity is observed at temperatures below 90 K if the excitation photon energy is larger than the band gap energy of the AlGaAs barrier. This is explained by taking into account the recapture of photo-excited electrons by shallow DX centers and the transfer of photo-excited holes into the channel region.

The authors wish to thank Dr. K. Muraki for his helpful discussions.

References

[1] Munoz E, Calleja E, Izpura I, Garcla F, Romero A L, SanchezRojas J L, Powell A L and Castagne J 1993 J. Appl. Phys., **73**, 4988
[2] Balland B, Vincent G, Bois D and Hirtz P 1979 Appl. Phys. Lett., **34**, 108
[3] Zdansky K and Peaker A R 1993 Appl. Phys. Lett., **62**, 1393
[4] Hoinkis M, Baranowski J, Dreszer P, Weber E R and Grimmeiss H G 1992 Mater. Sci. Forum, **83-87**, 841
[5] Mohapatra Y N AND Kumar V 1990 J. Appl. Phys., **68**, 3431
[6] Jia Y B, Li M F, Zhou J, Gao J L, Kong M Y, Yu P Y and Chan T K 1989 J. Appl. Phys., **66**, 5632
[7] Seguy P, Yu P Y, Li M F, Leon R and Chan K T 1990 Appl. Phys. Lett., **57**, 2496
[8] Fudamoto M, Tahira K, Tashiro S, Morimono J and Miyakawa T 1989 Jan. J. Appl. Phys., **28**, 2039
[9] Lang D V, Logan R A and Jaros M 1979 Phys. Rev. **B19**, 1015
[10] Kaniewska M and Kaniewski J 1988 MRS Symposia Proceedings **104**, 579
[11] Von.Bardeleben H J, Buyanova I, Belyaev A and Sheinkman M 1992 Phys. Rev., **B45**, 11667
[12] Fockele M, Spaeth J M, Overhof H and Gibart P 1992 Mater. Sci. Forum, **83-87**, 835
[13] Spector M, Pfeiffer L N. Licini J C, West K W, Baraff G A 1994 Mater. Sci. Forum, **143-147**, 1141

Inst. Phys. Conf. Ser. No 145: Chapter 3
Paper presented at 22nd Int. Symp. Compound Semiconductors, Cheju Island, Korea, 28 August–2 September 1995
© *1996 IOP Publishing Ltd*

The Anomalous Photocondutive Decay of Carbon Doped GaAs

Hyunsik Im† Seong-Il Kim† Tae-Geun Kim† Chang-Sik Son† Yong Kim† Moo-Sung Kim† Suk-Ki Min† Yun Chul Chung‡ Jin Ki Hong‡ Bo Ick Chang‡ Sun Ung Kim‡ and Mann Jang Park‡

† Semiconductor Materials Research Centre, Korea Institute of Science and Technology, P.O. Box 131, Cheongryang, Seoul 130-650, Korea

‡ Department of Physics, Korea University, AnamDong 5-1, Seongbukgu, Seoul, 136-701, Korea

Abstract. An anomalous shape of the photoconductive decay(PCD) curve due to the rate of electron captured by traps at 77K is observed for the first time from the carbon doped GaAs . It can be explained by the apparent two different values for the carrier life time in the samples. PCD curves have been calculated also theoretically with the values for the various pulse widths of applied light which determines the initial condition of electron capture rate of trap levels. With the experimental parameters extracted from deep level transient spectroscopy(DLTS), the capture coefficients of the trap levels have been obtained by fitting the calculated PCD curves to the corresponding experimental curves obtained by the various pulse widths of the applied light.

1. Introduction

The effects of heavy impurity doping on the optical and electrical properties of GaAs are very important not only for a fundamental point of view on physics, but also for device applications such as HBT(heterojunction bipolar transistor) and laser diode. An anomalous shape of the photoconductive decay(PCD) curve due to the rate of electrons capture by traps at 77K is observed for the first time from carbon (C) doped GaAs epilayers grown by low-pressure MOCVD. It is confirmed experimentally that the anomalous shape is due to the apparent two different values for carrier life times, the long life time(τ_l) and the short life time(τ_s) [4], in the C-doped GaAs.

PCD curves dominated by Shockley-Read recombination mechanism[1] have been calculated as a function of the various pulse widths of applied light which determine the initial condition of electron concentration of trap levels. It is assumed that there is one deep level, or two levels of which one acts as a recombination center and the other as a shallow level caused by doped carbon. The values for the densities and the energy of the deep levels are extracted from the deep level transient spectroscopy(DLTS) measurements.

As the pulse width becomes longer, the traps are filled more with the electron rather than the hole. Therefore, the traps are relatively unavailable to capture the electron from the conduction band, while the decay occurs. In this way, the inital parts of the transient decay curve which are formed dominantly by electron capture are varied according to the pulse widths. In conclusion, the amount of electron already captured by traps plays an important role in PCD curves.

2. Theoretical background

For homogeneous and nondegenerate semiconductor the recombination mechanism is described by following differential equations, [2]

$$dn/dt = -\sum_i R_{n,i} + g$$
$$dp/dt = -\sum_i R_{p,i} + g \qquad (1)$$
$$dN_i^-/dt = R_{n,i} - R_{p,i}$$

where g is the generation rate of electron-hole pairs due to external light source. The number of filled and empty traps are denoted by N_i^- and N_i^+, respectively. $R_{n,i}$ and $R_{p,i}$ repersent the recombination rate of electron and hole, respectively, by i-th trap level.

$$R_{n,i} = C_{n,i}(nN_i^+ - n_iN_i^-)$$
$$R_{p,i} = C_{p,i}(pN_i^+ - p_iN_i^-) \qquad (2)$$

$C_{n,i}$ and $C_{p,i}$ are the capture coefficient of electron and hole, respectively. The i-th trap level is capable of capturing one electron at average rate $C_{n,i}$ when vacant and one hole at rate $C_{p,i}$ when occupied by an electron. n_i and p_i denote the equilibrium carrier densities when the Fermi level passes through the trap level at E_i [3]. The excess trap occupation densities are related to the excess carrier density.

$$\sum_i \delta N_i^- = \delta p - \delta n \qquad (3)$$

For low carrier injection, the recombination expressions can be linearized. For the case of one trap level case, Eq.(2) and Eq.(3) give the following :

$$\begin{pmatrix} d/dt + \alpha & -\beta \\ -\gamma & d/dt + \eta \end{pmatrix} \cdot \begin{pmatrix} \delta n \\ \delta p \end{pmatrix} = \begin{pmatrix} g \\ g \end{pmatrix} \qquad (4)$$

where α, β, γ, η are the function of trap level(E_i), capture coefficients($C_{n,i}$, $C_{p,i}$) and trap density(N_i).

The calculation of Eq.(4) is done for two parts of generation and decay separately. In the decay part, The solutions to Eq.(4) are the following form

$$\delta n(t) = B\exp(-t/\tau_l) + C\exp(-t/\tau_s)$$
$$\delta p(t) = B\frac{\alpha\tau_l - 1}{\beta\tau_l}\exp(-t/\tau_l) + C\frac{\alpha\tau_s - 1}{\beta\tau_s}\exp(-t/\tau_s) \qquad (5)$$

Here, τ_l and τ_s are long life time and short life time, respectively, which are function of α, β, γ, η. B and C are coefficients determined by initial condition depending on light pulse width. Therefore, various PC decay curves can be obtained, according to the light pulse width.

For low carrier injection, the measured PCD signal is linearly proportional to the photoconductivity transient response as obtained from the following equation :

$$\delta\sigma(t) = q\mu_p(\delta p(t) + b\delta n(t)), \qquad (6)$$

where μ_p is the hole mobility, q is the charge of the electron, $\delta\sigma$ is the photoconductivity, and b is the ratio of electron to hole mobility. b is assumed to be the ratio of hole to electron effective mass.

3. Experimental

The three samples investigated for this study are grown by low-pressure metalorganic chemical vapor deposition(MOCVD) as described elsewhere[6, 7]. The thickness of the samples are 0.4μm. We have used (100) semi-insulaing GaAs wafers with 2° off toward [110] direction as substrates. TMG and AsH$_3$ (20 % diluted in pure hydrogen) have been used as source gases. By reducing the V/III ratios nearly to 1, we have obtained the heavily C doped GaAs epilayers.

For the Photoconductive Decay(PCD) measurements, the light pulse is applied by GaAs laser(wavelength = 780nm) having fall time less than 10 ns. For the case of large injection levels, the decay curves may depend on the density of excess carriers. Therefore, the laser power has been reduced until the amount of the generated excess carriers become less than one hundredth of majority carriers at equillibrium. All measurements are taken at 77K. In order to confirm whether there is any substrate effect, for this measurment, PCD measurements are also done by applying the light pulse on the substrate alone. No substrate effect is found.

4. Results and discussion

From the DLTS measurement, the values for the energy and the density of deep levels acting as a recombination center are found to be 0.44 eV above valence band and 2.5×10^{15}cm^{-3}, respectively.

In general, while the equilibrium carrier concentration is increasing, the recombination occurs faster. According to Eq.(2), this consideration can be easily deduced from the facts that the probability of the recombination of free carriers is proportional to the free carrier concentration and the doping levels in the high dopping materials act on recombination process.

Figure 1(a) and (b) display the experimental PC decay curves for each case of 10μs and 50μs in the applied pulse width. For relatively high dopping specimen (sample(#3)) PC decay curve decays abruptly, meaning that the recombination process is remarkably faster. In this case, the recombination process is thought to be influnced by the dopping level (0.026 eV) of carbon(C) as well as the deep level recombination center found separately by DLTS measurment. The PCD transient behavior will then depend on the

526

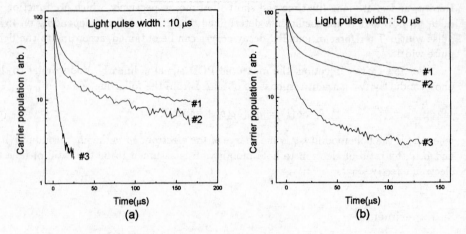

Figure 1. PC decay curves for various carrier concentraion, 1.9×10^{16} (sample(#1)), 1.2×10^{17} (sample(#2)) and $5.7 \times 10^{18} \mathrm{cm}^{-3}$ (sample(#3)) , with light pulse widths (a) $10\mu s$ and (b) $50\mu s$, respectively. As the carrier concentration increases, the recombination becomes faster.

initial ratio of excess carrier densities captured by the traps. This ratio is a function of the light pulse width. The resulting transient curves for the various pulse widths are shown in Figure 2. As the pulse width becomes longer, the traps are filled more with the electron rather than the hole. Therefore, the traps are relatively unavailable to capture the electron from the conduction band, while the decay occurs. In this way, the inital parts of the transient decay curve which are formed dominantly by electron capture are varied according to the pulse widths.

Figure 2(a) displays PCD curves of sample(#1) having 1.9×10^{16} carrier concentration. From the results of DLTS on the sample(#1), the values for the energy and the concentration of deep level acting as the dominant recombination center for sample(#1) are 0.44 eV, $2.5 \times 10^{15} \mathrm{cm}^{-3}$, respectively. In this measurements, each light pulse of $10\mu s$ and $50\mu s$ in widths is applied, separately. In this figure, circular points represent experimental results, and solid lines represent the calculated (using Eq.(5)) lines which are the best fit to the experimental results. Long and short life time, capture coefficients of hole and electron obtained for sample(#1) by fitting are $\tau_l = 1.3 \times 10^{-4}$ s , $\tau_s = 6.45 \times 10^{-6}$ s , $C_p = 4.0 \times 10^{-13} \mathrm{cm}^3/s$ and $C_n = 6.2 \times 10^{-11} \mathrm{cm}^3/s$, respectively.

Figure 2(b) displays PCD curve of the sample(#3) whose carrier concentraion is $5.7 \times 10^{18} \mathrm{cm}^{-3}$ which is relatively high. It is known that Shottky barrier is not formed in the case of such a large free carrier concentration. Although the DLTS measurement is also done for this sample, the deep level of the sample could not be found. In order to fit our calculated results to experimental data, therefore, one trap level with the different values for various fitting parameters is taken at each time for the calculation of PCD decay curve. In spite of the trials for the different one-trap level, it is not possible to get the calculated curves of fitting to experimental data with satisfaction. Two levels, one level(0.026 eV) formed by dopped carbon and the other level(0.44 eV) found in sample(#1), therefore, are used for iterative calculation and thus the values for

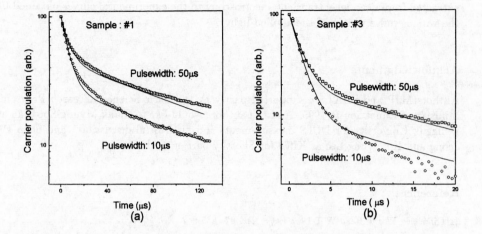

Figure 2. The experimentally measured PCD curves(solid lines) and fitted curves from the caluated results(circular lines). (a) For sample(#1) the carrier concentraion is $1.9 \times 10^{16} \text{cm}^{-3}$, the level and density of trap obtained by DLTS measurment are 0.44 eV and $2.5 \times 10^{15} \text{cm}^{-3}$, respectively. Capture coefficients of hole and electron obtained by fitting are $C_p = 4.0 \times 10^{-13}$ and $C_n = 6.2 \times 10^{-11} \text{cm}^3/\text{s}$, restpectively. (b) For sample(#3) the carrier concentraion is $5.7 \times 10^{18} \text{cm}^{-3}$, the levels and densities of two traps are $E_1 = 0.026 \text{eV}$, $E_2 = 0.44 \text{eV}$, $\text{Nt}_1 = 5.7 \times 10^{18} \text{cm}^{-3}$ and $\text{Nt}_2 = 2.5 \times 10^{15} \text{cm}^{-3}$, respectively. Fitting is done by two levels. Capture coefficients of hole and electron obtained by fitting are $C_{p,1} = 8.9 \times 10^{-15}$, $C_{p,2} = 3.9 \times 10^{-12}$, $C_{n,1} = 2.0 \times 10^{-14}$ and $C_{n,2} = 2.6 \times 10^{-10} \text{cm}^3/\text{s}$, respectively.

the capture coefficient are extracted from this iterative calculation. Capture coefficients of hole and electron obtained by fitting are $C_{p,1} = 8.9 \times 10^{-15}$, $C_{p,2} = 3.9 \times 10^{-12}$, $C_{n,1} = 2.0 \times 10^{-14}$ and $C_{n,2} = 2.6 \times 10^{-10} \text{cm}^3/\text{s}$, respectively. The fitted result are good as shown in Figure 2(b). In the case of relatively low carrier concentration as in the sample(#1), the recombination center by deep level is dominant over all total recombination processes. However, in the case of relatively high carrier concentration as in the sample(#3), it should not be ignored the shallow level which is formed by dopped carbon(C). The discrepancy in the end of th decay curves is ignored as a noise in this work. Besides Shockley-Read recombination mechanism there would be band-to-band Auger, Radiative [5] and surface recombination[8]. However, Shockley-Read recombination mechanism is considerd as a dominant recombination process in this study.

5. Summary

The dependence of applied light pulse widths of the photoconductive decay(PCD) curve due to the rate of electrons capture by traps at 77K is observed for the first time from carbon doped GaAs epilayers grown by low-pressure MOCVD. PCD curves that are dominated by Shockley-Read recombination mechanism have been calculated as a function of the various pulse widths of applied light. The capture coefficients of the trap levels have been obtained by fitting the calculated PCD curves with the experimental data

528

extracted from deep level transient spectroscopy to the experimental curves obtained by the various pulse widths of the applied light.

Acknowledgments

Authors(MJP and SUK) are supported partly by the Basic Science Research Program, Ministry of Education in 1995. And the authors would like to thank Myung-Soo Han at Dongguk University for DLTS measurements and his fruitful discussion , and Soon Pil Jeong and Ho Nyung Lee at KIST for Shottky barrier diode.

References

[1] Shockley W and Read W T 1952 *Phys. Rev.* **87** 835-842

[2] Choo S C 1970 *Phys. Rev.* **1** 687–696

[3] Kao K C and Hwang W *Electrical Transport in Solids* (Oxford : Pergamon Press)

[4] Reichman J 1991 *Appl. Phys. Lett.* **59** 1221–1223

[5] Blakemore J S 1962 *Semiconductor Statistics* (Oxford : Pergamon Press)

[6] Kim S I, Eom K S, Kim Y, Kim M S, Min S K, Lee C, Kwak M H and Ma D S 1993 *J. Cryst. Growth* **126** 441

[7] Kim S I, Kim M S, Kim Y, Eom K S, Min S K and Lee C 1993 *J. Appl. Phys.* **73** 4703

[8] Chung Y C, Hong J K, Lee S B, Chang B I, Kim S U, Park M J and Kim J M 1995 *Proc. of the 7th Int. Conf. on Narrow Gap Semiconductors, Santa Fe*

Inst. Phys. Conf. Ser. No 145: Chapter 3
Paper presented at 22nd Int. Symp. Compound Semiconductors, Cheju Island, Korea, 28 August–2 September 1995
© *1996 IOP Publishing Ltd*

The Effective Species for Nitrogen doping in ZnSe

F. Ito[1], T. Hamada[2] and T. Hariu[3]

Department of Electronic Engineering, Tohoku University, Sendai 980-77, Japan

1)now with Hitachi, Ltd., Kodaira 187

2)now with Oki Electric Industry Co., Ltd., Hachioji 193

3)now with Department of Systems Engineering, Ibaraki University, Hitachi 316

Abstract. The effective species for nitrogen doping have been investigated in N-doped ZnSe grown by using a magnetron-type RF plasma cell. As removing N2+* during growth from nitrogen plasma which contains mainly N2* and N2+* as active species, carrier concentration and PL emission intensity were decreased and PL spectra showed dominant FA emission rather than DdA emission. These results should be caused by reduction of nitrogen incorporation. Taking into consideration the result on ZnSe layers grown in nitrogen-based mixed plasma, higher excited nitrogen species N2+* or atomic nitrogen is more effective for doping.

1. Introduction

ZnSe-related II-VI compound semiconductors have been studied intensively for the realization of practical blue light emitting diodes and laser diodes. Since the first discovery by Yamauchi and Hariu[1] that the plasma-excited nitrogen is effective for shallow acceptors into ZnSe, this dopant source has been employed to grow p-type ZnSe[2-6], and room temperature laser operation was also demonstrated[7]. However net acceptor concentration (N_A-N_D) of N-doped ZnSe films is saturated around $1 \times 10^{18} \mathrm{cm}^{-3}$, despite of incorporating more nitrogen. It is believed that self-compensation due to formation of N-associated deep donors becomes dominant at high nitrogen concentration[8]. In order to overcome this carrier saturation, it is fundamentally important to investigate nitrogen doping mechanism, especially to find out effective excited nitrogen species for doping. In this paper, the effective species for nitrogen doping have been investigated in N-doped ZnSe grown by using a magnetron-type RF plasma cell and also in nitrogen-based mixed plasma.

2. Experimental

ZnSe films were grown on semi-insulating (100)GaAs substrates by molecular beam epitaxy at 320C, otherwise mentioned. The substrates were etched in $5H_2SO_4$-$1H_2O_2$-$1H_2O$ solution and heated up to 500C in the growth chamber before growth. The source materials were elemental Zn, elemental Se both with 6-nine purity and nitrogen gas for doping also with 6-nine purity. Nitrogen doping was achieved by employing a newly-designed magnetron-type RF plasma cell shown in Fig.1. The center electrode is made of zinc in order to prevent contamination from sputtered electrode material, was excited by RF power at 13.56MHz in $4.0x10^{-2}T$ magnetic field. Input RF power and N_2 flow rate were 10W and 1.1sccm, respectively. The growth rate of N-doped ZnSe films ranged from 0.8-1.0 μ m/h and the thickness was between 1.6-2.0 μ m.

Low-temperature photoluminescence (PL) spectra were measured at 4.2K with excitation by UV-light of a 500W Hg(Xe) lamp through a filter for cutting light of wavelength longer than 400nm. The N_A-N_D was evaluated by 10kHz capacitance-voltage (C-V) measurement using Au as a Schottky barrier metal on as-grown surface. Optical

magnetron-type RF plasma cell

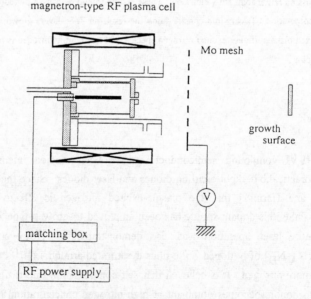

Fig.1 An experimental magnetron-type plasma cell for the growth N-dooped ZnSe.

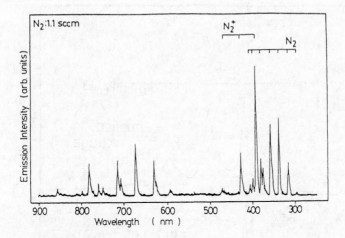

Fig.2 Typical OES spectrum of nitrogen plasma excited in the cell at RF power of 10W, N_2 flow rate of 1.1sccm and total pressure of 1.0×10^{-4}Torr.

emission spectroscopy (OES) measurement was made through the sapphire window by detecting the light reflected by a mirror, because the optical emission from plasma could not be observed directly.

3. Results and discussions

Fig.2 shows the typical OES spectrum of nitrogen plasma excited in the cell under the condition that RF power, nitrogen flow rate and total pressure were 10W, 1.1sccm and 1.0×10^{-4}Torr, respectively. The spectrum was composed of the first negative band emission of N_2^+ ions ($B^2 \Sigma_u^+ \rightarrow X^2 \Sigma_g^+$) and the second negative band emission of N_2^* molecules ($C^3 \Pi_u \rightarrow B^3 \Pi_g$). Although higher emission intensity from N_2^+ ions was obtained compared with other conventional RF plasma sources in MBE system, the optical emission due to atomic nitrogen could not be detected.

In order to compare the doping efficiency between N_2^* molecules and N_2^+ ions, Mo mesh was settled just out of plasma cell and DC bias was applied to the mesh to remove N_2^{+*} ions during growth, as shown in Fig.1. No difference observed in OES spectra by applying both positive and negative DC bias to the mesh in comparison with the spectra without DC bias, indicated no influence of the bias on nitrogen excited species generated in the cell. As increasing positive bias from 0V to 35V to the mesh (removing N_2^{+*} ions from nitrogen plasma), N_A-N_D was decreased from 1.0×10^{16}cm^{-3} to high resistivity and PL

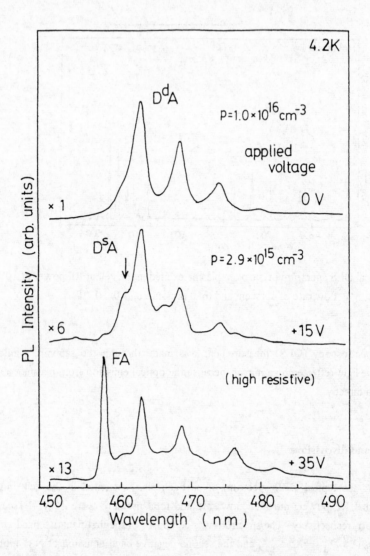

Fig.3 Applied bias dependence of PL spectra at 4.2K and N_A-N_D of N-doped ZnSe.

spectra showed dominant weak FA (free to acceptor) emission rather than strong D^dA (deep donor to acceptor) pair emission or D^sA (shallow donor to acceptor) pair emission, as shown in Fig.3. These results should be caused by reduction of nitrogen incorporation and then indicate that higher excited nitrogen species, N_2^{+*} ions in this case, is more effective for doping.

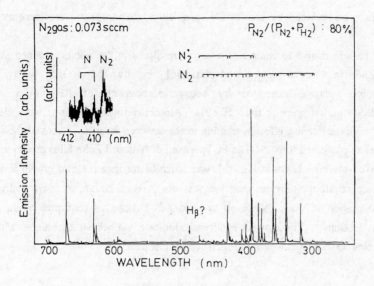

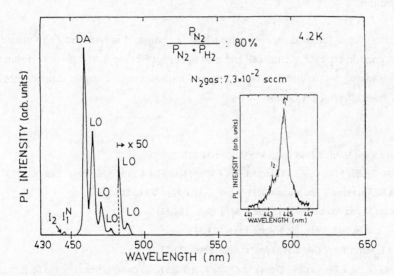

Fig.4 (a) OES spectrum of (H_2+N_2) mixed plasma with $P_{N2}/(P_{H2}+P_{N2})$ partial pressure ratio of 80% and (b) PL spectrum of N-doped ZnSe grown in (H_2+N_2) mixed plasma.

In the case of applying negative bias of -35V to the mesh (accelerating N_2^{+*} ions), PL spectra showed the same strong D^dA pair emission as films grown without DC bias to the mesh. But p-type conductivity was not confirmed in contrast. These results should indicate that damages were induced by accelerated ions, although nitrogens were incorporated effectively.

It would be informative to compare the above results with the results obtained at the growth of ZnSe in nitrogen-based mixed plasma[6]. Fig.4(a) shows OES spectrum of (H_2+N_2) mixed plasma generated by capacitive coupling of RF power with $P_{N2}/(P_{H2}+P_{N2})$ partial pressure ratio of 80%. Atomic nitrogen emissions were clearly observed through the Penning effect in addition to the emission from N_2^{+*} ions of the same intensity level as compared to N_2^*. The PL spectrum of N-doped ZnSe films grown in this (H_2+N_2) mixed plasma, shown in Fig.4(b), was much better than the film grown in pure nitrogen where no atomic nitrogen emission was observed, although excessive hydrogen caused compensation of N-acceptor levels by making N-H bond, as confirmed by infrared absorption at $3192cm^{-1}$. The above overall results indicate, we believe, that higher excited nitrogen species is more effective for nitrogen acceptor doping.

4. Conclusions

The effective species for nitrogen doping have been investigated in N-doped ZnSe grown by using a magnetron-type RF plasma cell and also in nitrogen-based mixed plasma. From the comparison among N_2^* molecules, N_2^{+*} ions and atomic nitrogen, higher excited nitrogen species is more effective for doping.

References

[1] Yamauchi S and Hariu T 1988 Appl. Surf. Sci. **33/34** 862

[2] Park R M, Troffer M B, Rouleau C M, DePuydt J M and Hasse M A 1990 Appl. Phys. Lett. **57** 2127

[3] Ohkawa K, Karasawa T and Mitsuyu T 1991 Jpn. J. App. Phys. **30** L152

[4] Ito S, Ikeda M and Akimoto K 1992 Jpn. J. Appl. Phys. **31** L1316

[5] Ohtuka T and Horie K 1993 Jpn. J. Appl. Phys. **32** L233

[6] Hamada T, Hariu T and Ono S 1993 Jpn. J. Appl. Phys. **32** 674

[7] Nakayama N, Ito S, Okuyama H, Ozawa M, Ohata T, Nakano K, Ikeda M, Ishibashi A and Mori Y 1993 **29** 2194

[8] Hauksson I S, Simpson J, Wang S Y, Prior K A and Cavenett 1992 Appl. Phys. Lett. **61** 2208

Inst. Phys. Conf. Ser. No 145: Chapter 3
Paper presented at 22nd Int. Symp. Compound Semiconductors, Cheju Island, Korea, 28 August–2 September 1995
© *1996 IOP Publishing Ltd*

Effect of Ge Interlayers on Threading Dislocation Behavior in GaAs on Si

M.Tamura and T.Saitoh

Optoelectronics Technology Research Laboratory, 5-5 Tohkodai, Tsukuba, Ibaraki, Japan

Abstract. Threading dislocation morphologies and characteristics have been investigated in 3 μm thick GaAs films with Ge interlayers having various thicknesses above and below the critical thicknesses (h_c) for dislocation generation in Ge on GaAs grown by molecular beam epitaxy on tilted (3° toward [110]) Si (001) substrates using cross sectional transmission electron microscopy. In as-grown samples the running direction of most of the dislocations changes at the lower Ge/GaAs interfaces. However, almost all of the dislocations in the Ge interlayers thread into the upper GaAs films, with no particular change in the running direction at the upper GaAs/Ge interfaces. After annealing at 800 °C for 30 min, the interactions of the dislocations with the interlayers are pronounced in all of the samples, resulting in the enhancement of bent dislocations at the interfaces of both Ge/GaAs and GaAs/Ge. However, in a sample having a 9500 Å thick Ge layer far above h_c, a high density of antiphase domains (APDs) remains in the GaAs film on Ge even after annealing. Interactions of the APDs with the dislocations are observed.

1. Introduction

The heteroepitaxial system of III-V compounds on Si is an essential combination of materials for electronic and photonic device application. In particular, heteroepitaxial GaAs films having a high crystal quality on Si are strongly desired for building active optical devices on Si VLSI chips [1]. However, a significant problem regarding the crystal quality of GaAs on Si is necessarily generated due to the chemical, thermal and lattice mismatchs between GaAs and Si, as well as an antiphase domain problem. Regarding this matter, the generation of a high density of threading dislocations remaining in grown GaAs films is the most serious problem, when we apply GaAs on Si to the above-mentioned type of device fabrication.

We have recently reported that the direction of movement of dislocations in GaAs on Si changes at the position of very thin Si interlayers, resulting in a dislocation density reduction at the film surface [2]. However, the thickness of the Si interlayers was limited to ~10 Å because of the large lattice mismatch between GaAs and Si. Thicker Si layers generated additional misfit and threading dislocations. In this study we used Ge as another material for blocking any upward-propagating dislocations at the interlayer positions. Since the lattice constant of Ge is close to that of GaAs, we should be able to grow Ge layers up to a few thousand Å thick in a GaAs film without forming any new dislocations. The purpose of the present experiments is to observe the interactions of Ge interlayers with dislocations propagating in a GaAs film grown by molecular beam epitaxy (MBE) before and after annealing. All of these Ge insertion experiments are discussed in terms of the results obtained by cross-sectional transmission electron microscope (XTEM) observations.

536

2. Experimental procedures

GaAs, Ge and Si growth was performed in a multi-chamber MBE system. Samples were kept under an ultrahigh vacuum (less than 5×10^{-10} torr) and were transferred through the exchange chamber from the III-V (or Ge and Si) to the Ge and Si (or III-V) MBE chamber. By using this system, GaAs heteroepitaxial growth having Ge interlayers was carried out using Si (001) wafers tilted 3° toward the [110] direction. After surface cleaning of the Si substrates at 900 °C, a 300 Å thick Si buffer layer was deposited at 700 °C by electron-beam evaporation, immediately followed by annealing at 1100 °C. We then adapted a two-step growth procedure in the III-V MBE chamber, which involved of preliminary growth at 400 °C and conventional growth at 600 °C. Just after adequate GaAs growth, the samples were transferred into the Ge and Si MBE chamber, where Ge was deposited at various thicknesses in the range from 0.1 to 1 μm at 400 °C. After growing the Ge layers, the samples were transferred back into the III-V MBE chamber and GaAs growth was again performed to a total thickness of ~3 μm. After growth, the samples were variously annealed by either the rapid thermal annealing or furnace annealing system in face-to-face contact with another GaAs wafer in an N$_2$ ambient.

Both the defect generation and threading dislocation behavior were studied by XTEM observations at 200 keV. The cross sectional specimens for observations were prepared by mechanical griding, followed by 4 kV Ar-ion beam bombardment. Observations were performed on large areas of the films for some samples in order to obtain general information concerning the morphologies of the dislocations. To do this, special care was taken to prepare cross-sectional specimens that were sufficiently thin over a large area.

3. Effects of interlayers on dislocation movement

Si and Ge interlayers affect the movement of threading dislocations in GaAs films in at least two ways, as depicted in Fig. 1. In one case (Fig. 1 (a)), the dislocation disappears from the edge of the specimen after running some distance along the $\pm[110]$ direction at the interlayer. This effect is the same as that observed when a strained-layer superlattice (SLS) structure is grown into a GaAs

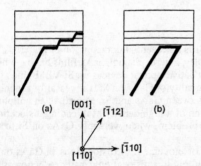

	a (Å)	μ (dyn/cm²)	f (%)	hc (Å)
GaAs	5.653	3.29×10^{11}		
Ge	5.657	5.3×10^{11}	0.07	~ 2500 Å
(Si)	5.431	6.45×10^{11}	4.0	~ 10 Å

Fig. 1 Schematic cross-sectional drawing of two ways in which the movement of dislocations (thick lines) is affected by Si interlayers, and the physical parameters of the lattice constant (a), shear modulus (μ), misfit (f) between the film (GaAs) and interlayers (Ge and/or Si) and the experimentally determined critical thickness (h_c) for Ge/GaAs and Si/GaAs systems [5].

film. In the other case (Fig. 1 (b)), the interlayer changes the dislocation direction twice. First, the dislocation is bent along the $\pm[110]$; then, the dislocation is bent again and forced to return into the substrate. We refer to these two effects as the sweeping out effect and the blocking effect, respectively. The blocking effect is due to a difference in the hardness between the growing GaAs film and the interlayer. Since the dislocation energy is proportional to the product of μ (shear modulus) and the square of the magnitude of the Burgers vector of the dislocation, it is less energetically favorable for dislocations to move into a harder material, i.e. a material having a larger μ. Si and Ge are useful materials for blocking dislocations, because their μ_s are about twice as large as that of GaAs (Fig. 1). In contrast, the blocking effect would be very weak for SLSs grown in GaAs, since the μ of an SLSs is usually close to that of GaAs. From the viewpoint of the interlayer thickness, Ge can be grown in the GaAs layer up to the critical thicknesse (h_c) for dislocation generation in Ge on GaAs of ~2500 Å without forming additional dislocations, in contrast to an h_c of ~10 Å for Si (Fig. 1).

4. As-grown state

Figures 2, 3 and 4 show the effect of Ge interlayers of different thicknesses on the dislocation movement in GaAs. The samples of Figs. 2 and 3 include the Ge interlayers below h_c in between GaAs films; the sample of Fig. 4 has a Ge layer far above h_c. In the two samples with thinner Ge

Fig. 2 Composite XTEM micrographs, observation from the [110] direction showing dislocation morphologies in an as-grown sample with a 1050 Å thick Ge interlayer.

Fig. 3 Composite XTEM micrographs, observation from the [110] direction showing dislocation morphologies in an as-grown sample with a 2300 Å thick Ge interlayer.

interlayers, some dislocations running along the <211> directions in the lower GaAs changed their moving directions into the ±[110] directions and/or the [001] direction at the interface of the Ge/GaAs. However, almost all of these dislocations threaded into the Ge layers and escaped to the upper GaAs films with almost no change in the running direction at the GaAs/Ge interface. We checked the results obtained by carrying out contrast change experiments of the above-mentioned [001] directed dislocations. As a result, they were found to be 30° type dislocations. Therefore,

Fig. 4 Composite XTEM micrographs, observation from the [110] direction showing dislocation morphologies in an as-grown sample with a 9500 Å thick Ge interlayer.

these dislocations were thought to change their moving directions, for example, from the [$\bar{1}$12] to [112] direction and their glide planes, for example, from the (111) to (111) plane at the lower Ge/GaAs interface.

In a sample with a thicker Ge interlayer of 9500 Å, the above-mentioned phenomena of dislocation motion were observed in a more enhanced manner (Fig. 4). Namely, the results clearly show that the dislocation movement is influenced when the dislocations run from a softer (GaAs) to a harder (Ge) material and is hardly affected when they move from a harder to softer material. Although the inserted Ge film of this sample greatly exceeds the h_c of ~2500 Å, newly generated dislocations do not seem to be added to the pre-existing dislocations. This is probably due to the generation of a comparatively low dislocation density caused by a small misfit of 0.07 % between GaAs and Ge. This misfit introduces only two orders of magnitude lower dislocation density along the [110] direction than that in GaAs on Si.

We also note the generation of many antiphase domains (APDs) in the upper GaAs films of all the samples. However, the APDs almost disappear within 3000 Å of the upper GaAs film from the GaAs/Ge interface in the two thinner Ge-inserted samples. While in a sample having a 9500 Å thick Ge layer, we can see that most of the APDs reach the sample surface. In a previous paper [3], we reported that Ge layers thicker than 20 Å which were deposited on vicinal GaAs (001) substrates, showed mixed-RHEED patterns of (1×2) and (2×1), indicating the formation of double domain structures. This resulted in APD formation on GaAs films grown on these Ge layers. However, the Ge films grown on (001) GaAs tilted 2° toward the [110] direction showed (2×1) dominated-mixed RHEED patterns, leading to self-annihilation [4] of the APDs on GaAs films deposited on these Ge during growth. These results will be applied to the present APD formation and annihilation of Figs. 2 and 3, since both GaAs and Ge films were grown on Si (001) substrates tilted toward the [110] direction. Almost no annihilation result of the APDs of Fig. 4 may have some correlation with the GaAs growth on a strain-relaxed Ge layer having a film thickness far above h_c.

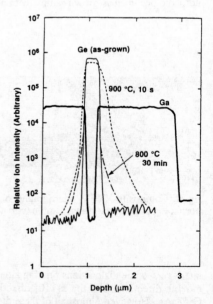

Fig. 5 SIMS depth profiles of the Ge and Ga concentration in a sample with a 2300 Å thick Ge interlayer.

5. Post-growth annealing

Figure 5 shows the Ge depth profiles of a sample having a 2300 Å thick Ge film before and after annealing carried out under the conditions indicated in the figure, together with the Ga profile in the as-grown sample for a reference. The profiles were measured by SIMS (secondary ion mass spectroscopy). In the as-grown sample, a very sharp embedded Ge profile is observed, contrasting to the result of a sharply decreased Ga profile in the region of a Ge layer. This shows that no interdiffusion occurred between Ge and Ga during the growth of GaAs at 600 °C. On the other hand, in the two annealed samples, shoulders are formed in the Ge depth profiles on both sides. In particular, about a 1 μm penetration of Ge is clearly seen on the substrate side in a sample annealed at 900 °C even for a short annealing time of 10 s. In contrast to this, the shape of embedded Ge is clearly maintained in a sample after annealing at 800 °C for 30 min, indicating that Ge hardly diffused into GaAs during annealing at 800 °C. We therefore conducted ex situ thermal annealing on each sample at 800 °C after growth in order to observe the interactions of the threading dislocations with the Ge interlayers during annealing.

Figures 6, 7 and 8 show the effect of the Ge interlayers on the dislocation movement in GaAs for samples corresponding to Figs. 2, 3 and 4 after annealing at 800 °C for 30 min. The interactions of dislocations with Ge as depicted in Fig. 1, are observed to be much more pronounced in every sample than those in as-grown samples. The interactions in this case occur at the interfaces of both Ge/GaAs and GaAs/Ge; this phenomenon is clearly seen in the samples having two thicker Ge layers. For example, in regions X in Fig. 7 of a sample with a 2300 Å thick Ge film, we can recognize that dislocations run not only along the lower Ge/GaAs interface but also along the upper GaAs/Ge interface. This is due to the fact that both interfaces act on dislocations as sources for inducing blocking and sweeping out effects in the case of the motion of dislocations caused by annealing. In a sample of Fig. 8, most of the dislocations change their moving directions along the ±[110] direction at the lower Ge/GaAs interface, although an appreciable number of dislocations thread into a Ge film. These threading dislocations in Ge are mostly bent again at the upper GaAs/Ge interface. However, the remaining dislocation density in the upper GaAs film is approximately one order of magnitude higher than that in the Ge film on this XTEM micrograph. This would be attributed to the residual APDs in the GaAs. In samples of Fig. 6 and 7, the APDs almost completely disappeared during heat treatment, while the developed APDs to the GaAs surface in an as-grown sample having 9500 Å thick Ge still remained after annealing (Fig. 8). The dislocations interacted with these residual APDs during annealing, resulting in the formation of a high density of threading dislocations in localized areas associated with the APDs as seen in Fig. 8. By comparing all of the obtained results, we conclude that the optimum Ge interlayer thickness for reducing the dislocations which can propagate to the film surface, is close to h_c.

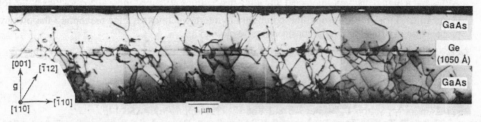

Fig. 6 Composite XTEM micrographs, observation from the [110] direction showing dislocation morphologies in a sample corresponding to Fig. 2 after annealing at 800 °C for 30 min.

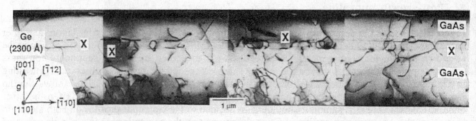

Fig. 7 Composite XTEM micrographs, observation from the [110] direction showing dislocation morphologies in a sample corresponding to Fig. 3 after annealing at 800 °C for 30 min.

Fig. 8 Composite XTEM micrographs, observation from the [110] direction showing dislocation morphologies in a sample corresponding to Fig. 4 after annealing at 800 °C for 30 min.

6. Conclusion

We investigated the interactions of threading dislocations in GaAs on Si with Ge interlayers in between GaAs films before and after ex situ annealing (800 °C for 30 min) by XTEM observations. Most of the dislocations changed their running directions at the lower Ge/GaAs interface, but did not change at the upper GaAs/Ge in all of the as-grown samples having different Ge thicknesses. Also, the formation of antiphase domains (APDs) was observed in the GaAs layers on the Ge interlayers. The APDs were self-annihilated while undergoing GaAs growth in samples having Ge layers below the critical thickness (h_c) for dislocation generation in Ge on GaAs, while most of the APDs developed to the GaAs top surface in the sample with a Ge film above h_c. After annealing, the dislocations interacted with the interfaces of not only the lower Ge/GaAs but also the upper GaAs/Ge, leading to an increase in the number of bent dislocations along the [110] direction. On the other hand, the APDs disappeared in annealed samples having thin Ge films below h_c, although they survived in a sample with a thick Ge film above h_c. The dislocations interacted with these residual APDs and a high density of interacted threading dislocations remained in localized regions of the top

GaAs film. From these results, the optimum Ge interlayer thickness for preventing the upward propagation of dislocations was considered to be close to h_c.

References

[1] Hayashi I 1993 *Jpn. J. Appl. Phys.* **32** 266.
[2] Tamura M, Saitoh T, Palmer J E and Yodo T 1994 *Appl. Phys.* **A58** 145.
[3] Saitoh T, Tamura M and Palmer J E 1994 *Int. Symp. Compound Semicond., San Diego*, p.345.
[4] Kawabe M and Ueda T 1987 *Jpn. J. Appl. Phys.* **26** L944.
[5] Maree P M J, Barbour J C, Van der Veen J F, Kavanagh K L, Bulle-Lieuwma C W T and Viegers M P A 1987 *J. Appl. Phys.* **62** 4413.

Inst. Phys. Conf. Ser. No 145: Chapter 3
Paper presented at 22nd Int. Symp. Compound Semiconductors, Cheju Island, Korea, 28 August–2 September 1995
© *1996 IOP Publishing Ltd*

Properties of GaAs LEC single crystals grown at different inert gas pressures

M Seifert, P Rudolph, M Neubert, W Ulrici *,
J Donecker, J Kluge, E Wolf, D Klinger

Institut für Kristallzüchtung, Rudower Chaussee 6, 12489 Berlin, Germany
* Paul-Drude-Institut für Festkörperelektronik, Hausvogteiplatz 5-7, 10117 Berlin, Germany

Abstract. The structural perfection, surface state, electrical parameters, EL2 concentration and content of important residual impurities (C, B, O) of LEC GaAs single crystals in dependence on the argon pressure between 0.1 and 4.0 MPa have been investigated sytematically. Whereas at pressures > 1.0 MPa the gallium droplets at the crystal periphery disappear completely, the etch pit density, dislocation cell boundaries and EL2 concentration increase with pressure. The incorporated carbon content and the electrical resistivity decrease with increasing pressure, especially, in the region between 0.2 and 1.0 MPa very sensitively. A parallel behaviour has been found for boron. The oxygen concentration behaves opposite to carbon.

1. Introduction

Semi-insulating (SI) GaAs crystals with high structural perfection and exactly controlled residual impurity concentrations are essential for fabrication of integrated high speed devices. It is well known [1,2], that the quality of liquid encapsulated Czochralski (LEC) crystals is considerably affected by preparation and handling of the GaAs starting charge as well as the contamination level, especially, the residual water content in the growth chamber and the B_2O_3 encapsulant. But there is only sporadic information about the influence of the inert gas type and pressure on the quality of as-grown crystals, including residual impurity concentration. After Emori et al. [3] the dislocation density increases in the order Kr < Ar < N_2 due to the increasing heat transfer coefficient leading to an increased temperature gradient in the grown crystal. Katoda et al. [4] reported on a slightly increased EL2 concentration when the argon pressure increases from 2.1 to 3.1 MPa. Recently, Tatsumi et al. [5] attributed the raised carbon contamination in GaAs crystals to the low atmospheric pressure (~ 0.1 MPa) used by them in a modified LEC arrangement.

For the first time we investigated the quality of as-grown GaAs LEC crystals in dependence on the inert gas pressure systematically. An argon ambient was used for the study due to the chemical inertness compared to nitrogen [3]. The value of the pressure was varied from low pressure (LP: 0.2 - 1.0 MPa) up to high pressure (HP: 2.0 - 4.0 MPa) regime in steps keeping all other growth parameters consequently constant.

2. Experimental

Undoped GaAs single crystals, 3 inches in diameter and about 12 cm in length, were grown in <100> direction from pyrolytic boron nitride (PBN) and silica crucibles under different argon pressures of 0.2, 1.0, 2.0 and 4.0 MPa using the Czochralski puller "Mark 3" (LPA Industrie, France). Dry argon (3 vol ppm H_2O) and highly purified GaAs starting material of 6N grade (Freiberger Elektronikwerkstoffe, Germany) were employed. Boron oxide with a constant water content of 200 wtppm (Tomiyama Pure Chemical Industries, Japan) was applied as encapsulant. The following identical growth parameters were used: axial temperature gradient at the melt-solid interface equal to 120 K cm^{-1}, pulling rate, 12 mm h^{-1}, crystal and crucible rotation, 10 and 15 rpm, respectively. Similar heating programs and holding times before the seed contact with the melt has been kept for each growth experiment very accurately.

Thin slices, 1 - 5 mm in thickness, were cut perpendicular to the growth axis from the front and tail regions of the cylindrical part of each as-grown ingot (samples A and E, respectively) and investigated by different analytical methods. For the observation of the etch pit density (EPD) and dislocation cell boundaries (substructure) the (100) wafers were polished and etched by the standard method in KOH at 370 °C. The electrical resistivity and free carrier concentration were obtained from Hall measurements. The content of substitutional carbon atoms (C_{As}) and oxygen centres (O_i, [O_i-V_{As}] and [O_i-X]) were investigated by local vibrational mode (LVM) IR-absorption at liquid nitrogen temperature. For the determination of the carbon concentrations from LVM measurements the calibration factor f_{77K} = 9.2 x 10^{15} cm^{-1} [6] has been used. The conversion relation for [O_i-X] was: $\alpha = 10^{-2}$ (absorption coefficient) corresponds to 8 x 10^{14} cm^{-3} [O_i-X]. Additionally, the residual concentrations of C, O and B were detected by radiofrequency spark source mass spectrometry (rf SSMS) [7]. The EL2 concentration was determined by a near IR-absorption technique.

3. Results and discussion

3.1. Crystal perfection

It is well known the phenomenon of formation of Ga droplets at the growing crystal surface by dissociative evaporation of As species leading to generation of polycrystallinity [8]. We observed microdroplets up to an argon pressure of 1.0 MPa. Under HP growth conditions ($\geq$ 2.0 MPa) the surface of all ingots showed always very shiny without any symptoms of dissociation. Deducing from this result the HP regime seems to be favoured in comparison to the LP one. However, as will be shown below, a markedly deteriorated structural quality of HP crystals has been observed.

Fig. 1 shows the radial distribution of the EPD along the <100> and <110> directions of A-wafers of various crystals grown under different argon pressures. It can be seen that the average EPD increases with increasing gas pressure from about 4 x 10^4 cm^{-2} at < 2.0 MPa to about 1 x 10^5 cm^{-2} at 4.0 MPa. This result correlates with the increasing number of dislocation cell boundaries of the revealed substructure at increasing argon pressure (Figs. 2a,b). Whereas at 1 MPa an average cell diameter of about 800 μm has been found (a) at 4.0 MPa this dimension amounts about 500 μm (b). Taking into account the well known dependence of dislocation density and cell dimension of the substructure on the value of thermomechanical stresses, an increase of the radial and axial temperature gradients within the crystals, grown under HP conditions, can be deduced. Especially, a higher radial temperature gradient promotes the formation of dislocation cell boundaries by polygonization. Such effect is caused by the proportionality of the convective heat transfer coefficient at the gas-crystal interface and radial temperature gradient. After Jordan [9] at higher pres-

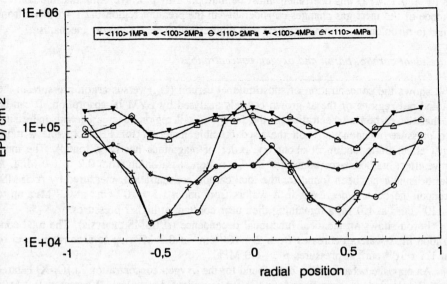

Fig. 1. Radial EPD distribution in GaAs wafers (A-position) along the <100> and
<110> directions grown from PBN crucible at 1.0, 2.0 and 4.0 MPa argon
pressure.

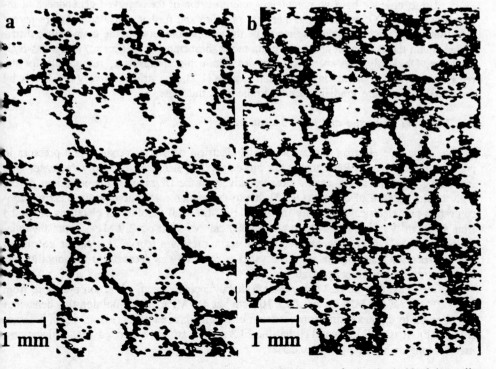

Fig. 2. Etched dislocation cell boundary structure of GaAs crystals (at the half of the radius
of A-wafers) grown from PBN crucible at 1.0 (a) and 4.0 MPa (b) Ar pressure.

sures (p > 0.1 MPa) this coefficient must be multiplied by $p^{1/2}$. We think that the convection mode of the inert gas changes considerably in the pressure region 0.2 - 1.0 MPa from laminar to turbulent, i.e from a modest to a high effective heat transport mechanism.

3.2. Residual carbon, boron and oxygen concentrations

Fig. 3 shows the concentration of substitutional carbon (C_{As}) versus argon pressure in the front and tail regions of the as-grown crystals analysed by LVM IR-absorption. It can be seen that the carbon concentration in the front and tail regions decreases with increasing argon pressure independently on the used crucible material (for PBN crucible a more distinct effect of reduction of about one order of magnitude has been found). The most sensitive effect has been ascertained for the LP region at transition from 0.2 to 1.0 MPa. A similar behaviour has been found for the total carbon concentration, measured by rf SSMS, whereupon the content decreases in A-wafers from about 1.5 x 10^{16} cm^{-3} at 0.2 MPa up to 8.4 x 10^{15} cm^{-3} at 1.0 MPa remaining then near constant at higher pressures.

Boron shows an identical functional dependence (rf SSMS analysis). The total concentration in A-wafers decreases from 2 x 10^{17} cm^{-3} at 0.2 MPa up to a constant level of about 1.5 x 10^{16} cm^{-3} at pressures p > 1.0 MPa.

An opposite behaviour has been found for the oxygen concentration in [Oi-X] centres (not any [O_i-V_{As}] or O_i have been found in the investigated samples). Whereas at 0.2 MPa no oxygen above the detection limit ($\sim 10^{14}$ cm^{-3}) has been observed the related LVM absorption peaks (at 1244 cm^{-1} and 1325 cm^{-1}) of the [O_i-X] centre increase abruptly in samples grown at 1.0 MPa. At higher pressures a constant level of about 8 x 10^{14} cm^{-3} appears.

Up to now we have no completely interpretation of the observed dependence of incorporation of C, O and B on the inert gas pressure and further considerations are necessary. But one of the explanations seems to be the correlation between the temperature distribution within the B_2O_3 melt and convection mode (heat transfer coefficient) at the gas-B_2O_3 interface. Obviously, the chemical reactions within the encapsulant, responsible for the release or association of carbon, boron and oxygen [1,2], are influenced very sensitively by the actual temperature gradient parallel to the B_2O_3 thickness.

3.3. Electrical properties and EL2 concentration

In Fig. 4 the free electron concentration and electrical resistivity versus argon pressure in crystals grown from a PBN crucible are presented. The electrical resistivity decreases with increasing argon pressure, again, most sensitively in the LP range. This behaviour agrees with the dependence of carbon (Fig. 3) due to the known correlation between carbon concentration and electrical resistivity in GaAs [10]. The Hall mobility of electrons increased from 1900 cm^2 V^{-1} s^{-1} at 0.2 MPa to about 4000 cm^2 V^{-1} s^{-1} at p > 2.0 MPa. From the independence of the carrier concentration on the argon presssure (curve 2) it can be concluded that the increased resistivity for the LP crystals is caused by an decrease of the mobility due to the increased carbon concentration.

We found that the EL2 concentration slightly increases with the argon presssure from 1.3 x 10^{16} cm^{-3} at 0.2 MPa up to 1.65 x 10^{16} cm^{-3} at 4.0 MPa. In our opinion this behaviour can correlate to the increasing concentration of cell boundaries at increasing argon pressure (Fig. 2a,b). It is well known [11] that the EL2 defects are agglomerated near dislocations, mainly along dislocation cell boundaries.

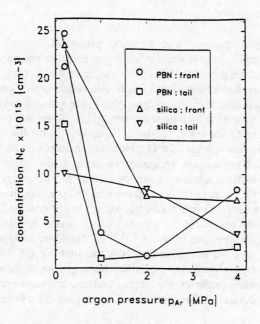

Fig. 3. Concentration of substitutional carbon N_C (C_{As}) from LVM IR-absorption measurements versus argon pressure in the front and tail region (A- and E-wafers, respectively) of LEC GaAs crystals grown from PBN and silica crucibles.

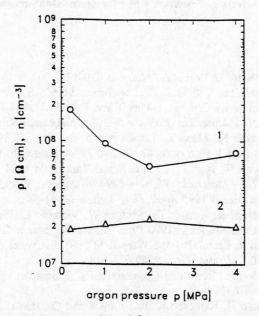

Fig. 4. Electrical resistivity ρ (curve 1) and free electron concentration n (curve 2) in LEC GaAs crystals (A-wafers) versus argon pressure.

4. Conclusions

A correlation between structural and electrical properties as well as concentration of important residual impurities and the applied argon pressure has been found for SI GaAs LEC crystals. At pressures p > 1.0 MPa the formation of Ga droplets at the crystal surface is depressed. But the dislocation density and dislocation cell boundaries increase with pressure due to the increased temperature gradients in the growing crystal which are caused by the increased heat transfer at high pressure conditions. There is a correlation between incorporation of C, B and O and argon pressure. Especially, in the low pressure region the C and B concentrations decrease and the O concentration increases markedly with pressure. More detailed considerations are necessary to explain this phenomenon exactly. The electrical resistivity behaves similar to that of carbon and decreases with pres-sure. The slight increase of the EL2 concentration with the ambient pressure may be attri-buted to the increased concentration of dislocation cell boundaries in HP crystals.

According to our results for the growth of high quality SI GaAs with maximal electrical resistivity an argon pressure of about 1.0 MPa should be achieved for the given LEC conditions. However, in order to reduce the dislocation density and dissociative evaporation at the crystal surface more effectively, a modified inner growth chamber for maintenance of very low axial temperature gradients and arsenic antipressure was developed. Our results of the so-called vapour controlled Czochralski (VCZ) method [9] will be reported soon.

Acknowledgement

The authors are indebted to M. Pietsch and B. Brückner for their help at growth experiments, B. Nippe for the crystal preparation, S. Bergmann for the etching experiments and B. Wiedemann (University Frankfurt a.M.) for the rf SMSS measurements.

References

[1] Lambert U and Wiese U 1991 *Adv. Mater.* **3** 429-435
[2] Nishio J and Terashima K 1989 *J. Crystal Growth* **96** 605-608
[3] Emori H, Terashima K, Orito F, Kikuta T and Fukuda T 1984 *Proc. 3rd Conf. on Semi-insulating III-V Materials, Kah-nee-ta 1984* (Nantwich: Shira) 111-117
[4] Katoda Y, Kakimoto M, Masa Y, Tanaka A and Yokohama T 1984 *Proc. 3rd Conf. on Semi-insulating III-V Materials, Kah-nee-ta 1984* (Nantwich: Shira) 157-159
[5] Tatsumi M, Kawase T, Iguchi Y, Fujita K and Yamada M 1994 *Proc. 8th Conf. on Semi-insulating III-V Materials, Warsaw 1994* (Singapore: World Scientific) 11-18
[6] Alt H Ch and Dischler B 1995 *Appl. Phys. Letters* **66** 61-68
[7] Wiedemann B, Bethge K, Kessler Th, Schütze W, Venzago C, Waldschmidt M, Wolf G, Engelhardt H and Müller D 1990 *Proc. European Vacuum Conf., Trieste 1990*
[8] Tower J P, Tobin R, Pearak P J and Ware R M 1991 *J. Crystal Growth* **114** 665-675
[9] Jordan A S 1980 *J. Crystal Growth* **49** 631-642
[10] Gallium Arsenide for Micro- and Optoelectronics, *Prospectus of Freiberger Elektronikwerkstoffe GmbH* 1994
[11] Oda O, Yamamoto H, Kainosho K, Imaizumi T and Okazaki H 1993 *Proc. 5th Int. Conf. on Defect Recognition and Image Processing in Semiconductors and Devices, Santander 1993*

Inst. Phys. Conf. Ser. No 145: Chapter 3
Paper presented at 22nd Int. Symp. Compound Semiconductors, Cheju Island, Korea, 28 August–2 September 1995
© *1996 IOP Publishing Ltd*

Fermi-level effect on Ga self-diffusion studied using ^{69}GaAs/^{69}Ga^{71}GaAs isotope superlattice

K. Muraki[1] and Y. Horikoshi

NTT Basic Research Laboratories, 3-1 Morinosato-Wakamiya, Atsugi,
Kanagawa 243-01, Japan

Abstract. We have studied the Ga self-diffusion in GaAs and its dependence on Si doping using ^{69}GaAs/^{69}Ga^{71}GaAs isotope superlattices with [Si] = 1×10^{17}- 8×10^{18} cm^{-3} grown by molecular-beam epitaxy. The depth profiles of ^{71}Ga atoms before and after thermal anneals at 900 °C have been investigated via secondary-ion mass spectrometry (SIMS) to obtain the Ga self-diffusion coefficient as a function of the Si concentration. We find that the Fermi-level effect becomes dominant only in the high-doped regime, i.e., [Si] $\geq 1 \times 10^{18}$ cm^{-3}, where the Ga self-diffusion coefficient increased in proportion to the third power of the electron concentration. On the other hand, no change in the diffusion coefficient was observed in the low-doped regime. Our results show that the Ga self-diffusion is driven by the Fermi-level effect and mediated by triply charged Ga vacancies in the high-doped regime, while it is suggested that some other Fermi-level independent mechanisms are operating in the intrinsic and low-doped regime.

1. Introduction

Doping-induced superlattice disordering [1] has attracted considerable interest because of its implications for the fundamental understanding of atom diffusion and defect creation mechanisms in III-V compound semiconductors as well as its potential application for device fabrication [2]. Mei *et al.* [3] have shown that the doping of AlAs/GaAs superlattices with Si significantly enhances the Al-Ga interdiffusion during post-growth annealing. They reported that the Al-Ga interdiffusion coefficient is proportional to the third power of the Si concentration in the range from 2×10^{17} to 5×10^{18} cm^{-3}.

This behavior has later been explained by Tan and Gösele [4] in terms of Fermi-level effect, whose basic idea is that the defect formation energy is partly compensated by dropping electrons from the conduction band to the defect level created in the band gap. Hence, as the doping level increases and the Fermi level moves upwards, creation of defects such as group-III vacancies becomes energetically favorable. The observed third-power dependence has therefore been believed to be the manifestation of the triply

---

[1] E-mail : muraki@will.brl.ntt.jp.

charged group-III vacancies mediating the diffusion, in line with the theory of Baraff and Schläter [5].

Similar diffusion enhancement by grown-in n-doping have been reported for Si [6, 7], Sn [8], Se [9], and Te [10]. On the contrary, Gillin et al. [11], who studied the Al-Ga interdiffusion in Si-doped $Al_{0.2}Ga_{0.8}As/GaAs$ quantum wells (QWs) using photoluminescence (PL), have measured no change in the interdiffusion coefficient for Si concentrations up to $10^{18} cm^{-3}$. Miyazawa et al. [12], on the other hand, have studied the Al-Ga interdiffusion in InGaAs/InAlAs superlattices to find that much higher Si doping levels, in excess of $1.2 \times 10^{19} cm^{-3}$, are required to cause any discernible interdiffusion in this material system. Apparently, these experimental results require satisfactory explanations.

As discussed by Walukiewicz [13, 14], the concentration of Fermi-level induced charged point defects is intimately related to the energy position of the defect level relative to the conduction-band edge, which actually depends on the host material. The larger Si concentration required in the InGaAs/InAlAs system has therefore been interpreted as a consequence of the defect level in InGaAs being located closer to the conduction band edge [14]. If this is the case, however, it is anticipated that the concentration of charged point defects should be different in AlAs and GaAs constituting the superlattices commonly used for the diffusion study. Hence, it is meaningful to investigate the Ga self-diffusion in GaAs and its Fermi-level effect for a more quantitative discussion of the phenomena.

In this paper, we study the Ga self-diffusion and its dependence on Si doping using $^{69}GaAs/^{69}Ga^{71}GaAs$ isotope superlattices grown by molecular-beam epitaxy. The depth profiles of ^{71}Ga atoms before and after thermal anneals at 900 °C are investigated via secondary-ion mass spectrometry (SIMS) to obtain the Ga self-diffusion coefficient as a function of Si concentration. We find that the Fermi-level effect becomes dominant only in the high-doped regime, i.e., $[Si] \geq 1 \times 10^{18} cm^{-3}$.

2. Experiments

The $^{69}GaAs/^{69}Ga^{71}GaAs$ isotope superlattices used in this study were grown by molecular-beam epitaxy at a substrate temperature of 580 °C on semi-insulating undoped GaAs (100) substrates using conventional Ga source (containing 60-% ^{69}Ga and 40-% ^{71}Ga) and ^{69}Ga isotope-purified source. The purity of the ^{69}Ga source was 99.7 %, with 0.3-% residual ^{71}Ga and other impurities [15]. The Hall mobilities of Si-doped samples and the PL of undoped samples confirm that the GaAs layers grown with ^{69}Ga source has crystalline quality comparable to that of GaAs layers grown with conventional Ga sources. Hereafter, $^{69}Ga^{71}GaAs$ layers grown with conventional Ga source will be denoted as GaAs for simplicity.

The sample structure is as follows. First, a 0.7-μm thick undoped GaAs buffer layer was grown. Then, five periods of alternating layers of 20-Å GaAs and 1000-Å $^{69}GaAs$ were grown with conventional Ga and ^{69}Ga isotope sources, respectively. These layers were uniformly doped with $[Si] = 1.3 \times 10^{17}$- $7.9 \times 10^{18} cm^{-3}$, as confirmed by SIMS. The post-growth annealing was carried out using a so-called proximity technique: the samples were placed face-to-face on undoped GaAs substrates, annealed at 900 °C for 1 h in N_2 flow.

The SIMS profile of ^{71}Ga was obtained by detecting ^{71}GaAs$^+$ secondary ions with Cs$^+$ primary ions of 3.0 keV and 0.1 μA. The Si concentration was determined separately by detecting Si$^-$ ions with Cs$^+$ primary ions of 14.5 keV and 0.1 μA and comparing them with those of control samples with known Si concentrations. The accuracy of the Si concentration is estimated to be $\pm$ 10 %. The carrier concentration was obtained by room-temperature Hall measurements with van der Paw geometry.

3. Results and discussions

Figure 1 shows the SIMS profiles of ^{71}Ga atoms in annealed samples with various Si concentrations (undoped, [Si] = 1.3, 3.9, and 7.9$\times$10^{18} cm^{-3}). The as-grown profile of the undoped sample is shown together as a reference. The peak full-width at half maximum (FWHM) of the as-grown profile is 80 Å on the surface side and slightly increases to 90 Å on the substrate side, from which the SIMS resolution can be estimated. After the anneals, the SIMS profiles are broadened. Even in the undoped sample with the sharpest profile, the broadening is discernible, the FWHM increasing to 110 Å on the surface side and 130 Å on the substrate side as a result of Ga self-diffusion. The figure clearly shows that Ga self-diffusion is significantly enhanced with increasing Si concentration.

It is seen in Fig. 1 that the diffusion is somewhat smaller on the surface side. Apparently, this cannot be explained by the surface Fermi-level pinning [16], since the depletion-layer thickness for $n = 5.0\times10^{18}$ cm^{-3} is $\approx$ 100 Å, much shorter than the length scale of the observed gradation. The opposite depth dependence has been reported by Guido *et al.* [17], who explained it by assuming that the sample surface is the source of

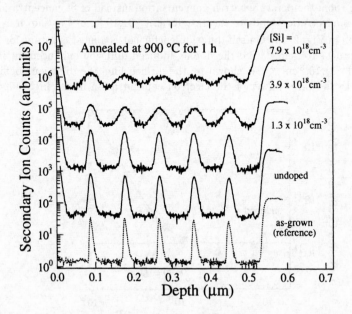

Figure 1. SIMS profiles of ^{71}Ga atoms in isotope superlattices with various Si concentrations. The as-grown profile is shown as a dotted line.

550

point defects. Recalling that As-rich conditions are favored for the group-III vacancies responsible for the Ga self-diffusion [18], our result can be explained as being due to the lower chemical potential of As on the surface side resulting possibly from the incomplete proximity of the sample surface to the GaAs substrate during the annealing. However, this is beyond the scope of this work, and will not be discussed further.

We assume that the diffusion of each Ga atom takes place independently and characterized by a diffusion coefficient D. The SIMS profile of the sample annealed for time t is then given by the convolution of Gaussian function and the as-grown profile $c_0(z)$;

$$c(z,t) = \int_{-\infty}^{\infty} c_0(z')G(z-z',t)dz' \tag{1}$$

$$G(z,t) = \frac{1}{(4\pi Dt)^{\frac{3}{2}}} \exp\left(-\frac{z^2}{4Dt}\right) \tag{2}$$

We have fitted the SIMS profiles of the annealed samples using Eq. (1) with D as an adjustable parameter. This approach is advantageous because the profile broadening due to the SIMS sputtering and its variation with the sputtering depth are naturally incorporated.

In Fig. 2, the Ga self-diffusion coefficients thus obtained are plotted as a function of the electron concentration at 900 °C. To avoid the influence of the sample surface, we plotted the diffusion coefficient averaged over the three peaks on the substrate side. The electron concentration at 900 °C was calculated as $n = \frac{1}{2}n_d + \frac{1}{2}(n_d^2 + 4n_i^2)^{\frac{1}{2}}$ [19], where n_d is the room-temperature carrier concentration and n_i ($= 1\times10^{17}\,\mathrm{cm}^{-3}$) is the intrinsic-carrier concentration at 900 °C calculated from the law of mass action. Here, we used the room-temperature carrier concentration instead of Si concentration because the carrier concentration tends to saturate for [Si] $\geq 3\times10^{18}\,\mathrm{cm}^{-3}$ as shown in Fig. 3.

As seen in Fig. 2, the Ga self-diffusion coefficient is almost constant for Si concentrations up to $1\times10^{18}\,\mathrm{cm}^{-3}$, and the doping-induced diffusion enhancement is observed only for [Si] $\geq 1\times10^{18}\,\mathrm{cm}^{-3}$. As shown by the dashed line, the diffusion coefficient is in proportion to the third power of the electron concentration in the high doped regime,

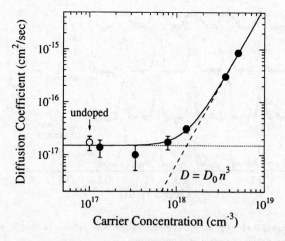

Figure 2. Ga self-diffusion coefficient as a function of carrier concentration at 900 °C.

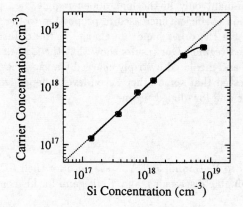

Figure 3. Room-temperature Carrier concentration vs. Si concentration

indicating that triply charged group-III vacancies contribute to the Ga self-diffusion in this regime.

As shown by the solid line in Fig. 2, our data can be fitted by the combination of the Fermi-level induced cubic term (dashed line) and a Fermi-level independent constant term of $1.5 \times 10^{-17} \, cm^2 sec^{-1}$ (dotted line). Tan and Gösele [4], who analyzed the data of Mei *et al.*, concluded that the diffusion is governed by triply charged group-III vacancies in intrinsic as well as *n*-type GaAs. However, extrapolating the cubic line down to the low-doped regime in Fig. 2, we find that the contribution of triply charged group-III vacancies is as small as $10^{-20} \, cm^2 sec^{-1}$ in the intrinsic case. It is therefore suggested that the diffusion is controlled by some Fermi-level independent mechanisms in the intrinsic and low-doped regime.

It is important to point out that the different dependences on the Si concentrations observed by us and Mei *et al.* are due to the different contributions of Fermi-level effect. As a result of the larger Fermi-level effect in their study, the doping-induced diffusion enhancement is observed even for the Si concentrations well below $1 \times 10^{18} \, cm^{-3}$, and the Fermi-level induced cubic term is dominant over the whole doping range [20]. In fact, their 900 °C data agree with ours for the lowest doping concentration where the contribution of Fermi-level effect is thought to be minimum. As we extrapolate their data to the high-doped regime, we find that the Fermi-level effect is about two orders of magnitude larger than in our study.

The origin of the larger Fermi-level effect in their study is not clear at present. Our preliminary SIMS study shows that the diffusion rates of Ga and Al in GaAs are nearly equal [21], and therefore it cannot be ascribed to the difference between Ga self-diffusion and Al-Ga interdiffusion. The larger band gap of AlAs may be of relevance, however, this must be clarified in the future study. Since the $^{69}GaAs/^{69}Ga^{71}GaAs$ isotope superlattice is a more idealized system than the AlAs/GaAs heterostructures, we believe that our Ga self-diffusion study provides better estimate of the Fermi-level effect.

4. Summary

We have studied the Ga self-diffusion and its Si-doping dependence via secondary-ion mass spectrometry of $^{69}GaAs/^{69}Ga^{71}GaAs$ isotope superlattices. We find that the Fermi-

552

level effect becomes dominant only in the high-doped regime, i.e., $[\mathrm{Si}] \geq 1 \times 10^{18}\,\mathrm{cm}^{-3}$, where the Ga self-diffusion coefficient increased in proportion to the third power of the electron concentration. On the other hand, no change in the diffusion coefficient was observed in the low-doped regime. Our results show that the Ga self-diffusion is driven by the Fermi-level effect and mediated by triply charged Ga vacancies in the high-doped regime, while it is suggested that some other Fermi-level independent mechanisms are operating in the intrinsic and low-doped regime.

Acknowledgments

The authors are grateful to Y. Honma and A. Masamoto for their SIMS measurements, H. Yamaguchi for his valuable discussions, and T. Ikegami for his continuous support.

References

[1] Tan T Y, Yu S and Gösele U 1991 *Optical and Quantum Electronics* **23** S863–81

[2] Deppe D G and Holonyak N, Jr. 1988 *J. Appl. Phys.* **64** R93–113

[3] Mei P, Yoon H W, Venkatesan T, Schwarz S A and Harbison J P 1987 *Appl. Phys. Lett.* **50** 1823–5

[4] Tan T Y and U. Gösele U 1988 *Appl. Phys. Lett.* **52** 1240–2

[5] Baraff G A and Schlüer M 1985 *Phys. Rev. Lett.* **55** 1327–30

[6] Mei P, Scwarz S A, Venkatesan T, Schwartz C L, Harbison J P, Florez L, Theodore N D, and Carter C B 1988 *Appl. Phys. Lett.* **53** 2650–2

[7] Beernink K J, Thornton R L, Anderson G B and Emanuel M A 1995 *Appl. Phys. Lett.* **66** 2522–4

[8] Rao E V K, Ossart P, Alexandre F and Thibierge H 1987 *Appl. Phys. Lett.* **50** 588–90

[9] Deppe D G, Holonyak N, Jr., Hsieh K C, Gavrilovic P, Stutius W and Williams J 1987 *Appl. Phys. Lett.* **51** 581–3

[10] Mei P, Schwarz S A, Venkatesan T, Schwartz C L and Colas E 1989 *J. Appl. Phys.* **65** 2165–7

[11] Gillin W P, Bradley I V, Howard L K, Gwilliam R and Homewood K P 1993 *J. Appl. Phys.* **73** 7715–9

[12] Miyazawa T, Kawamura Y and Mikami O 1988 *Japan. J. Appl. Phys.* **27** L1731–3

[13] Walukiewicz W 1989 *Appl. Phys. Lett.* **54** 2094–6

[14] Walukiewicz W 1993 *Proc. 17th Int. Conf. on Defects in Semiconductors, Gmunden, Austria, 1993* Vol 1 (Trans Tech) p 519-29

[15] The main residual impurities of the ^{69}Ga source are K (0.009 %), Si (0.006 %), Fe (0.005 %), Ti (0.004 %), and Sn (0.002 %).

[16] Ogawa K and Kawabe M 1990 *Japan. J. Appl. Phys.* **29** 1240–2

[17] Guido L J, Holonyak N, Jr., Hsieh K C and Baker J E 1989 *Appl. Phys. Lett.* **54** 262–4

[18] Zhang S B and Northrup J E 1991 *Phys. Rev. Lett.* **67** 2339–42

[19] Cohen R M 1991 *J. Electron. Matter.* **20** 425–30

[20] The 700 °C data of Mei *et al.* showing the often cited third-power dependence are doubtful in the low-doped regime. They reported 700 °C diffusion coefficients as small as $3 \times 10^{-21}\,\mathrm{cm}^2\mathrm{sec}^{-1}$ for $[\mathrm{Si}] = 2 \times 10^{17}\,\mathrm{cm}^{-3}$ and $3 \times 10^{-20}\,\mathrm{cm}^2\mathrm{sec}^{-1}$ for $[\mathrm{Si}] = 5 \times 10^{17}\,\mathrm{cm}^{-3}$ [3, 6]. For the annealing time of 3 h they used, these correspond to diffusion lengths of 0.6 and 1.9 Å, which are too small to be measured by SIMS. However, their 900 °C SIMS profile clearly shows the doping-enhanced interdiffusion down to $[\mathrm{Si}] = 2 \times 10^{17}\,\mathrm{cm}^{-3}$.

[21] Muraki K and Horikoshi Y (unpublished)

Inst. Phys. Conf. Ser. No 145: Chapter 3
Paper presented at 22nd Int. Symp. Compound Semiconductors, Cheju Island, Korea, 28 August–2 September 1995
© *1996 IOP Publishing Ltd*

Some anomalies in the electron transport of Se-doped AlAs

B. C. Lee, S. S. Cha, Y. G. Shin, K. Y. Lim, C. J. Youn, C. T. Choi*, and H. J. Lee

Semiconductor Physics Research Center and Department of Physics, Jeonbuk National University, Jeonju 560-756, Korea
*Department of Physics, Suncheon National University, Suncheon 540-742, Korea

Abstract. Se-doped AlAs epitaxial layers are grown by metalorganic chemical vapor deposition. The carrier density increases with increasing [H$_2$Se]/[TMA] and saturates at the ratio of 8×10^{-3}. The samples with carrier densities higher than 10^{18}cm^{-3} exhibit almost constant compensating acceptor density values regardless of carrier density and the free-carrier screening effect is conspicuous in this case. Neutral impurity scattering is dominant for the samples with carrier density in the range of $10^{18} \sim 10^{19}$cm^{-3} at lower temperatures. A two-band model involving the X- and L-band is adopted for the transport calculation.

1. Introduction

The Al$_x$Ga$_{1-x}$As alloys have received much attention because of their application in a variety of sophisticated heterostructure devices. The electrical properties are well known for the most of compositions of Al$_x$Ga$_{1-x}$As [1,2,3], but not for AlAs. In this work, Se-doped AlAs epitaxial layers are grown by the metalorganic chemical vapor deposition. The substrate temperature, reactor pressure, and the V/III ratio are kept constant at 700 ℃, 76 torr, and 45, respectively.

To investigate electrical transport phenomena and doping characteristics of AlAs epitaxial layers, Hall coefficient R$_H$ and conductivity σ are measured as a function of temperature by van der Pauw method. For the analysis of the experiment data, a two-band model involving the X- and L-band is adopted in the calculation, with various scattering mechanisms.

2. Results and discussion.

The doping characteristic at the room temperature is shown in Fig. 1. The present samples show that Hall density increases with increasing [H$_2$Se]/[TMA] ratio and

554

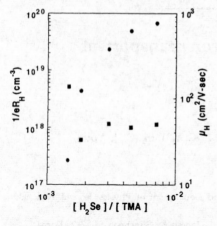

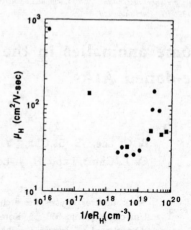

FIG. 1. Variations of Hall density($\bullet$; $1/eR_H$) and Hall mobility($\blacksquare$; μ_H) with [H₂Se]/[TMA] ratio at the room temperature.

FIG. 2. Variation of Hall mobility($\bullet$; μ_H) with Hall density.(Note: $\blacksquare$; μ_H in Fig. 1.)

saturates at the ratio of 8×10^{-3}, but the mobility variation shows the interesting phenomenum contrary to the general belief. That is, the mobility variation shows a minimum value. To check this phenomenum, we examined the variation of mobility as a funtion of Hall density in detail at this region. The result is shown in Fig. 2. The Hall mobility variation exhibits a minimum near $n_H \sim 5 \times 10^{18} cm^{-3}$. It should be noted that the μ_H value increases with n_H above the value showing the minimum. To study this anomalous variation, the Hall coefficient and Hall mobility are

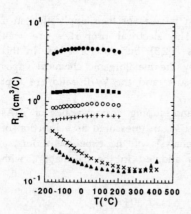

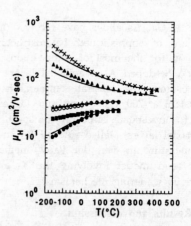

FIG. 3. Variation of Hall coefficient with temperature for various samples.:$\blacktriangle$,SEAA04 ;$\times$,SEAA02;+,SEAA23;$\bigcirc$,SEAA28;$\blacksquare$,SEAA21; $\bullet$,SEAA16.

FIG. 4. Variation of Hall mobility with temperature for various samples with the same symbol in FIG. 3. Solid curves are theoretical values.

Table 1. Some parameters obtained through the analysis at room temperature.

sample name	n (cm^{-3})	μ_H (cm^2/V-sec)	N_{AC} (cm^{-3})	N_{NI} (cm^{-3})	E_D (eV)
SEAA16	1.51×10^{18}	20	3.4×10^{19}	4.5×10^{19}	6.45×10^{-3}
SEAA21	4.65×10^{18}	25	4.2×10^{19}	5.5×10^{19}	4.22×10^{-3}
SEAA28	6.63×10^{18}	35	3.2×10^{19}	6.5×10^{19}	3.42×10^{-3}
SEAA23	8.89×10^{18}	37	3.1×10^{19}	6.5×10^{19}	3.00×10^{-3}
SEAA02	2.49×10^{19}	146	3.2×10^{18}	----	----
SEAA04	3.28×10^{19}	88	3.4×10^{19}	----	----

measured as a function of temperature. This results are shown in Fig. 3 and 4. In general, semiconductors with higher carrier density exhibit strong ionized impurity scattering and thus the mobility increases with temperature at lower region in this case. But the present data are contrary to the general trend as shown in Fig. 4.

We perform the transport calculation to fit the experimental data. A two-band model involving the X- and L-band[1] is adopted in the calculation, with various scattering mechanisms. This results are shown in Fig. 4 ~ 7. In the calculation, we used compensating acceptor density N_{AC} and neutral impurity density N_{NI} for $n<10^{19}$cm^{-3} as adjustable parameter. The compensating acceptor density N_{AC} determined using ionized impurity density, $N_{II}=n+2N_{AC}$, exhibits nearly constant value, $\sim 3.2 \times 10^{19}$cm^{-3}, for all the samples concerned here. This values are given in Table 1.

Therefore, the increasing trend of μ_H with n_H is due to the screening effect of free carriers. Two samples with carrier density higher than 10^{19}cm^{-3} can be reasonably explained in terms of ionized impurity scattering in addition to various lattice scattering mechanisms. But for the samples with lower n_H, impurity scattering is necessary to fit the low temperature data as shown in Fig. 5. This unexpected scattering is found to be dominant at low temperatures for the three samples with lower n_H in Fig. 3.

Figure 6 illustrates the mobility values for the respective scattering mechanisms in the X-band for the sample in Fig. 5. The impurity scattering is dominant at the whole temperature range and in particular, the neutral impurity scattering is prevailing at lower temperature region.

In a single-band transport, the Hall scattering factor is usually assumed to be unity. This assumption is very poor when many bands are involved [1]. From the Fig. 3, Hall coefficient decreases with temperature at lower region and this phenomenum gradually disappears as increasing carrier density. This effect is well explained by

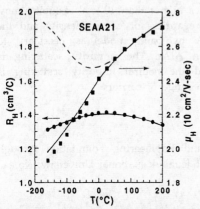

FIG. 5. Hall coefficient R_H and Hall mobility μ_H measured as a function of temperature for a SEAA21 sample. Curves are theoretical values with the assumptions. : —, without neutral impurity ; —, with neutral impurity.

556

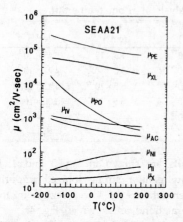

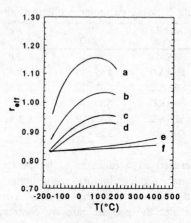

FIG. 6. Variation with temperature of the calculated drift mobility values corresponding to the separate scattering mechanisms in the X-band for the sample of FIG. 5.: μ_X, combined ; μ_{PE}, piezoelectric; μ_{PO}, polar optical; μ_{IV}, intervally; μ_{AC}, acoustic; μ_{XL}, interband; μ_{IL}, ionized impurity; μ_{NL}, neutral impurity.

FIG. 7. Variation of the effective Hall scattering factor with temperature for various samples. : a,SEAA16; b,SEAA21; c,SEAA28; d,SEAA23; e,SEAA02; f,SEAA04.

the effective Hall scattering factor defined by $r_{eff} = neR_H$, where R_H is the measured Hall coefficient and the results are presented in Fig. 7. From this figure, one notices that r_{eff} values are different from unity for the variations of temperature and carrier.

In summary, the samples with carrier densities higher than $10^{18} cm^{-3}$ exhibit almost constant compensating acceptor density values regardless of carrier density and the free-carrier screening effect is conspicuous. A two-band model is necessary to understand the electron transport properties in AlAs. The impurity scattering is dominant at the whole temperature range and the neutral impurity scattering is dominant in lower temperature region for certain range of carriers.

Acknowledgments

This work is suported by Korea Science and Engineering Foundation through Semiconductor Physics Research Center (SPRC) at Jeonbuk National University, Korea.

References

[1] Choon Tae Choi and Hyung Jae Lee 1988 J. Appl. Phys. 64, 4993-4996
Hyung Jae Lee and Choon Tae Choi 1988 J. Appl. Phys. 64, 1906-1909
[2] James J. Yang, Lavada A. Moudy and William I. Simpson 1982 Appl. Phys. Lett. 40, 244-256
[3] T. F. Kuench, K. J. Wolford, E. Venuhoff, V. Deline, P. M. Mooney, R. Potenski, and J. Bradley 1987 J. Appl. Phys. 62, 632-642

Inst. Phys. Conf. Ser. No 145: Chapter 3
Paper presented at 22nd Int. Symp. Compound Semiconductors, Cheju Island, Korea, 28 August–2 September 1995
© *1996 IOP Publishing Ltd*

Temperature coefficients of low-field electron mobility in GaAs/AlGaAs heterostructures

S.K. Noh, J.S. Yuk*, G. Ihm*, K.Y. Lim[†], H.J. Lee[†], and C.T. Choi[‡]

Epitaxial Semiconductor Group, Materials Evaluation Center, Korea Research Institute of Standards and Science, Taedok Science Town, Taejon 305-600, Korea

* Department of Physics, Chungnam National University, Taejon 305-764, Korea

[†] Semiconductor Physics Research Center, Department of Physics, Jeonbuk National University, Jeonju 560-756, Korea

[‡] Department of Physics, Suncheon National University, Suncheon 540-742, Korea

ABSTRACT : The temperature coefficients of 2-dimensional electron mobility have been extensively investigated on 7 different GaAs/AlGaAs heterostructures and an additional GaAs:Si thin film. As the temperature decreases from 400 K to 4 K, the high-mobility samples (≥ 60 Å) show monotonical increase in electron mobility over the entire range, but the low-mobility samples (< 60 Å) have peaks ranging in 50-100 K and positive temperature slopes below the peaks. The temperature dependences satisfy the linear relation in reciprocal mobility due to acoustic phonons, $1/\mu = 1/\mu_{AC}(0) + \alpha T$ ($\alpha > 0$), for the high-mobility samples in the range, but the parabolic contribution ($1/\mu = 1/\mu_{RD}(0) + \beta T^2$, $\beta < 0$) which may be attributed to remote donors appears in the low-mobility samples. From the T^{-k}-dependence in the optical phonon scattering regime, we have obtained an empirical expression as $k = k_0 + k_1 \log[\mu(4.2K)/\mu(300K)]$ ($k_0 = 1.0$, $k_1 = 0.7$), which can be extended to bulk GaAs:Si sample with the mobility ratio much smaller than unity.

1. Introduction

The Hall effect measurement which gives useful informations on transport properties has been widely used in characterization of the 2-dimensional electron gas (2DEG) confined in modulation-doped heterointerfaces. Because of a unique property of high mobility in 2DEG, a variety of theoretical and experimental reports has been continued on scattering mechanisms and quantum phenomena arising in the 2D systems[1-3]. In particular, the temperature dependence of Hall mobility in the modulation-doped heterostructures measured over a wide range is experimentally very valuable for studying the 2DEG transport or its scattering mechanisms due to remote impurities or phonons[4-7]. For more clear picture on transport and scattering in low dimensional systems, it is neccessary to make systematic studies on the temperature coefficients of 2DEG mobility through various layer structures.

In this paper, the evolution of temperature coefficients has been extensively studied when the mobility gradually changes to higher value as the spacer layer becomes thicker. We have systematically investigated the temperature dependence of Hall effect in the range of 4-400 K for various modulation-doped GaAs/AlGaAs heterostructures, and analyzed the temperature coefficients of 2D mobility and the related scattering mechanisms over the wide range of mobility $(2.70 \times 10^3 - 4.67 \times 10^5 \ cm^2/V.s, \ 4.2 \ K)$. Two kinds of temperature coefficients have been determined in consideration of contributions by acoustic phonons and remote ionized impurities in the low temperature range. In the optical phonon scattering regime, we have discussed the temperature coefficients obtained from the T^{-k}-dependence curves.

2. Experimentals

The substrate used in this study is the 7 GaAs/Al$_x$Ga$_{1-x}$As (x=0.3) single-interface heterostructures with different spacer-layer thicknesses ranging from 0 to 170 Å and a GaAs:Si thin film grown by MBE. The heterostructrures have a standard layer configuration known as the high electron-mobility transistor (HEMT)[3], and the thin film is a 1 μm-thick single layer of GaAs:Si ($N_D = 1 \times 10^{17} \ cm^{-3}$) bulk grown on a GaAs buffer layer of 0.5 μm thickness. Mesa-etched Hall devices shaped as a 6-bridged bar were fabricated by photolithographic technique and chemical etching, and have a bridge spacing of 120 μm and a lithographic channel width of 30 μm.

The low-field Hall measurements on the temperature dependence were carried out in the range of 4-400 K by using the automated Hall effect system and ^{4}He cryostat (Variox Plain, Oxford) under the magnetic field of 5 kG. The current applied to the Hall devices was fixed at 10 μA and the temperature sweep rate was kept almost constant as 2 K/min for all the measurements in this experiment.

3. Results and discussion

The temperature dependence curves of low-field Hall mobility taken from different 7 heterostructures (●) and a bulk sample (○) in the range of 4-400 K are given in Fig. 1. As the temperature decreases from 400 K to 4 K, the high-mobility samples whose spacer-layer thicknesses are equal to or larger than 60 Å show monotonical increase in electron mobility over the whole range, but the low-mobility samples with those smaller than

60 Å have the mobility peaks ranging in 50-100 K and positive temperature slopes below the peak temperatures, which shows a similar tendency in bulk mobility. Price[8] suggested that at a temperature low compared to the degeneracy temperature, $T_D=416[10^{-12}n_s]$, the effective lattice-scattering rate is linearly proportional to the temperature. A number of theoretical and experimental researchers[4-12] have generally accepted that at the sufficiently low temperature the acoustic phonon-limited mobility can be effectively simplified to a linear equation expressed as

$$1/\mu(T) = 1/\mu_{AC}(0) + \alpha T, \quad \alpha > 0 \qquad (1)$$

where $\mu_{AC}(0)$ is the zero-temperature mobility and α is the linear temperature coefficient in the reciprocal mobility domain. Figure 2 is plots for $\mu_{AC}(0)$ (■) and α (●) determined by Eq.(1) as a function of electron density. There is a transition in the linear temperature coefficient near the electron density of 10^{12} cm^{-2} obtained from the sample with spacer layer thickness of 60 Å. A comparison between experimental data and theoretical curves (dashed lines) of α for different values of deformation potential is shown in Fig. 3 as a function of electron density. While the results reported by Harakawa and Sasaki[5] and Mendez et al.[6] follow the theoretical curves for U_{DEF}=12-13 eV, the present data are close to that of 14 eV which is a little larger than 13.5 eV known as an optimal value of GaAs[5,6]. This reflects

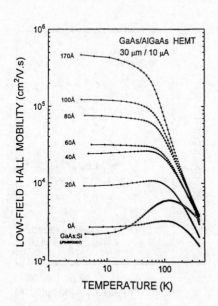

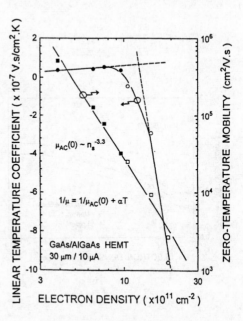

Fig. 1. Temperature dependence curve of low-field Hall mobility for the 7 different GaAs/ AlGaAs heterostructures and a GaAs:Si bulk in the range of 4-400K.

Fig. 2. The linear temperature coefficients and the zero-temperature mobilities plotted as a function of 2D electron density.

that the acoustic phonon scattering is dominant in the low-density samples ($a > 0$) in which a contribution of the ionized impurity is relatively negligible. The large deviation of a positioned at $n_s \sim 10^{12}$ cm^{-2} is probably due to an effect of ionized impurities, since the electron density or the spacer-layer thickness of the sample is just at the transition region changing rapidly from a positive value to a negative one in a as shown in Fig. 2.

It is very interesting that all the present results on the zero-temperature mobility follow an exponential fit, $\mu_{AC}(0) \sim n_s^{-3.3}$, even for the high-density samples ($\square$) in which the ionized impurity scattering is expected to be effective. This means that the dominant scattering near T=0 is not by acoustic phonons but by remote ionized donors. It is well known that the effect of ionized impurities is prevailing over an extension of the whole temperature, and the contribution of remote ionized donors in heterostructures makes the dependence superlinear[5,6,11] at very low temperature. In the present results, the high-density samples have the mobility peaks like bulk, and the nonlinearity and the temperature slope significantly increase below 30 K as shown in Fig. 1. We suppose that the temperature dependence is due to the remote ionized impurities and its functional is parabolic as

$$1/\mu(T) = 1/\mu_{RD}(0) + \beta T^2, \quad \beta < 0. \tag{2}$$

We have found that the zero-temperature mobility, $\mu_{RD}(0)$ and its electron-density

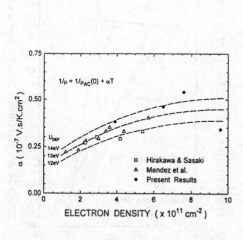

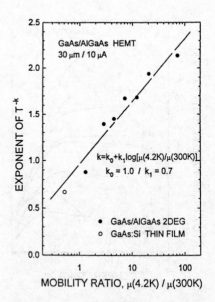

Fig. 3. A comparison between experimental data and thoeretical curves for the linear temperature coefficients as a function of electron density.

Fig. 4. Exponential temperature coefficients plotted as a function of the mobility ratio, $\mu(4.2K)/\mu(300K)$.

dependence are almost the same as $\mu_{AC}(0)$'s, and β gradually increases negatively as the spacer-layer thickness decreases.

Theoretical equations on the optical phonon scattering rate are too complicated to fit for experimental data. Lee et al.[4] proposed a simplified semi-empirical relation on the polar optical phonon mobility, $\mu(T)=A/T^2+B/T^6$, which was deduced from an interpolation between the temperature dependences of mobility near 77 K and 300 K. A plenty of results[5,9] have verified that the optical phonon mobility at an appropriate temperature well obeys a simpler expression of

$$\mu(T) \sim T^{-k}, \tag{3}$$

where k is an exponential temperature coefficient. It is known that the exponent k is 1.0 for pure GaAs bulk and within a range of 2.0-2.6[6,13] for GaAs/AlGaAs heterostructures. Figure 4 is a plot for k ($\bullet$) manipulated by Eq.(3) around room temperature as a function of mobility ratio, $\mu(4.2K)/\mu(300K)$. In the optical phonon regime, we have found that exponent k logarithmically increases with the mobility ratio, and satisfies an expression of $k=k_0+k_1\log[\mu(4.2K)/\mu(300K)]$, where $k_0=1.0$ and $k_1=0.7$. The interesting thing is that the value of $k_0=1.0$ exactly agrees with the k value for pure GaAs, and the curve can be extended to bulk GaAs:Si sample ($\bigcirc$) with the mobility ratio much smaller than unity, as shown in Fig. 4. This implies that the scattering mechanism for polar optical phonon in 2D heterostructures is basically similar to that in 3D bulk. We think this might be an experimental evidence on the argument of Walukiewicz et al.[9] that the optical phonon-limited mobility of 2DEG can be approximated by the mobility calculated for 3D bulks because of smearing effect.

4. Summary and conclusions

We have studied the temperature dependence of Hall mobility in the range of 4-400 K for the 7 modulation-doped GaAs/AlGaAs heterostructures with the wide range of mobility (2.70×10^3-4.67×10^5 cm^2/V.s, 4.2 K) and compared them with that of a GaAs:Si bulk. The high-mobility heterostructures ($\geq3\times10^4$ cm^2/V.s or ≥60 Å) satisfied the linear dependence in reciprocal mobility ($1/\mu=1/\mu_0+\alpha T$, $\alpha>0$) at low temperature, but the parabolic contribution ($1/\mu=1/\mu_0+\beta T^2$, $\beta<0$) appeared in the low-mobility samples ($<3\times10^4$ cm^2/V.s or <60 Å) at very low temperature. We have found the positive α's followed a theoretical curve for deformation-potential constant of 14 eV. According to the present results, while the scattering is dominated by acoustic phonons in the low-density samples in the low temperature range, a contribution of the remote ionized donor becomes more effective as the spacer layer becomes thinner. From the T^{-k}-dependence on the optical phonon scattering, we presented an empirical equation, $k=k_0+k_1\log[\mu(4.2K)/\mu(300K)]$ ($k_0=1.0$, $k_1=0.7$) adaptable to the GaAs:Si bulk sample as an experimental evidence on the argument over

562

the 2D smearing effect.

Acknowlegments

This work was supported by Korea Science and Engineering Foundation (KOSEF) through Semiconductor Physics Research Center (SPRC), Jeonbuk National University.

References

[1] Simserides C D and G.P. Triberis 1993 *Superlatt. Microstr.* **14** 277-282

[2] Schacham S E, Mena R A, Haugland E J and Alterovitz S A 1993 *Appl. Phys. Lett.* **62** 1283-1285

[3] Noh S K, Lee J I, Hwang S J, Ihm G and Yoo K H 1992 *J. Appl. Phys.* **71** 5976-5980

[4] Lee K, Shur M S, Drummond T J and Morkoc H 1983 *J. Appl. Phys.* **54** 6432-6438

[5] Hirakawa K and Sasaki H 1986 *Phys. Rev.* **B33** 8291-8303

[6] Mendez E E, Price P J, and Heiblum M 1984 *Appl. Phys. Lett.* **45**, 294-296

[7] Hess K 1979 *Appl. Phys. Lett.* **35** 484-486

[8] Price P J 1984 *Surf. Sci.* **143** 145-148

[9] Walukiewicz W, Ruda H E, Lagowski J and Gatos H C 1984 *Phys. Rev.* **B30** 4571-4582

[10] Vinter B 1984 *Appl. Phys. Lett.* **45** 581-582; 1986 *Phys. Rev.* **B33** 5904-5905

[11] Lin B J F, Tsui D C, Paalanen M A and Gossard A C 1984 *Appl. Phys. Lett.* **45** 695-697

[12] Lin B J F, Tsui D C and Weimann G 1985 *Solid State Commun.* **56** 287-290

[13] Sano N, Kato H and Chika S 1984 *Solid State Commun.* **49** 123-125

Inst. Phys. Conf. Ser. No 145: Chapter 3
Paper presented at 22nd Int. Symp. Compound Semiconductors, Cheju Island, Korea, 28 August–2 September 1995
© *1996 IOP Publishing Ltd*

Electron and hole multiplication characteristics in short GaAs PINs

S.A. Plimmer, J.P.R. David, T.W. Lee, G.J. Rees, P.A. Houston,

P.N. Robson and R. Grey

Department of Electronic and Electrical Engineering, University of Sheffield, Mappin

Street, Sheffield, S1 3JD, United Kingdom.

D.C. Herbert, A.W. Higgs and D.R. Wight

Defence Research Agency, Electronics Division, St. Andrews Road, Malvern,

Worcestershire, WR14 3PS, United Kingdom.

Abstract. Electron and hole multiplication characteristics have been measured on a series of GaAs homojunction PIN diodes in which the nominal i-region thicknesses, w, range from 1μm to 25nm. Using conventional analysis the effective electron and hole ionization coefficients, α and β respectively, have been deduced. Whereas the results calculated on the w = 1μm and 0.5μm structures agree with previously published data from measurements on thick devices, those observed in the thinner structures show device width dependence. By using a semi-analytical solution of Boltzmann's equation to interpret these results, dead space effects are seen to reduce α and β in short devices at low multiplication values but overshoot effects compensate when the electric field is increased.

1. Introduction

Impact ionization is the high field effect which limits the power performance of many devices, such as bipolar junction transistors (BJTs), and leads to avalanche multiplication and ultimately breakdown. Electron and hole ionization coefficients, α and β respectively, are the parameters used in design work to quantify this process and are defined as the reciprocal distances between electron initiated and hole initiated ionization events.

Previously, investigators have deduced α and β in structures with high-fields regions which are typically several microns long. Under these conditions α and β can be assumed to vary with only the local electric field, E. However, the high field region of modern structures, such as the base collector junction in a BJT, may be much shorter and non-local effects can become important. Although Millidge et. al [1] and Gaul et al. [2] have presented breakdown characteristics in GaAs devices with short avalanche regions, neither of these papers describe multiplication behaviour. In this paper, PIN diodes with submicron i-regions have been used to study non-local effects on ionization.

2. Experimental details

A series of PIN diodes were grown by molecular beam epitaxy (MBE) with high field region lengths, w, from 1μm down to 25nm. Be and Si were the dopants in 1μm p^+ and n^+ cladding regions on n^+ (100) GaAs substrates. The structures included 0.1μm AlAs etch-stop layers. The growth temperature was dropped from 580°C to 520°C during growth of the p^+ layer to ensure abrupt doping. Holes in the substrate were etched to enable light to be coupled onto the n^+ region of the devices.

Current-voltage (I-V) characterization was performed to ensure the samples were suitable for multiplication measurements. In the thicker structures (w ≥ 0.1μm) devices exhibited low dark currents, I_{dark}, and sharp breakdown voltages. In the w = 0.05μm and 0.025μm structures, where the high electric fields caused increased dark currents due to tunneling, characteristics were found to be in good agreement with those of Gaul et. al [2] over the measured voltage range.

To determine the i-region length of the devices, C-V measurements were performed and interpreted using a carrier transport model which incorporates Fermi-Dirac statistics. The p^+ and n^+ doping in the cladding regions, p and n respectively (assumed equal), the unintended n-type doping in the i-region, i and w were all used as adjustable parameters to fit simulated profiles to the experimental results. To corroborate the device parameters obtained in this way and presented in Table I, secondary ion mass spectroscopy (SIMS) measurements were also performed. Modeled values of w were found to be within 5% of those determined experimentally.

To obtain the electron multiplication characteristics, pure electron photocurrents were produced by focusing illumination from a 633nm He-Ne laser on to the top of the mesa. Measurements were taken at both dc and at ac using optically modulated light with the signal detected by a lock-in amplifier. The laser excitation power was varied to give primary photocurrents over the range ~ 10nA - 10μA in the

Table I: Parameters obtained from analysis of C-V profiles.

Nominal w (μm)	Modeled w (μm)	Modeled p=n (cm^{-3})	Modeled i (cm^{-3})
1.0	1.13	2×10^{18}	1×10^{15}
0.5	0.57	2×10^{18}	1×10^{15}
0.1	0.105	1.7×10^{18}	1×10^{17}
0.05	0.053	3.7×10^{18}	1×10^{17}
0.025	0.026	3.5×10^{18}	1×10^{17}

structures with w ≥ 0.1μm and ~ 100nA - 10μA in the w = 0.05μm and 0.025μm structures where dark currents become large as the applied bias is increased. The absense of microplasmas was confirmed by obtaining highly uniform results on several devices which have different diameters and with the laser spot focused at different positions on the top of the mesa. Hole multiplication characteristics were obtained from the same devices by coupling the light via an optical fibre on to the back of the exposed n^+ regions of the samples. Removing part of the substrate on the w = 0.05μm and w = 0.025μm structures, however, caused the devices to degrade as evidenced by an increase in dark currents. Hole multiplication behaviour was therefore investigated by illuminating the edge and floor (i-region and n^+ region) of the mesas producing characteristics in which a mixture of electrons and holes initiate multiplication. Measurements were again carried out at both ac and dc and the laser excitation power was varied. A small baseline slope was seen in the multiplication characteristics due to the electron, hole and mixed carrier photocurrents increasing by ~ 1% between 0V and the onset of multiplication. This is attributed to a widening of the electric field profile toward the contacts resulting in a small increase in the collection efficiency. The correction of Woods et al. [3] was used to give the normalized multiplication coefficients for electron initiated, hole initiated and mixed carrier initiated multiplication, M_e, M_h and M_{mix} respectively. The technique of obtaining M_{mix} from edge illumination was validated on thicker structures (w ≥ 0.1μm) where these normalized coefficients were between M_e and M_h. On the w ≥ 0.1μm structures, normalized multiplication curves exhibited an exponential rise to breakdown after the onset of multiplication. However, in the w = 0.05μm and 0.025μm devices, sub-exponential increases in multiplication with applied bias were seen up to values of ~ 5 before dark currents dominated the photocurrent. For the w = 1μm and 0.5μm PINs results were similar to those predicted by the data of Bulman et al. [5], who deduced α and β on a series of thick PN and PNN structures. Fig. 1 shows the multiplication curves on the structures where w ≤ 0.1μm. M_e and M_h were seen to converge at higher electric fields for w = 0.1μm with M_e and M_h indistinguishable within experimental errors for the two shortest structures.

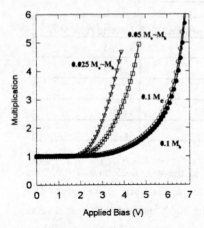

Fig. 1 Normalized multiplication coefficients for the shortest structures. Nominal values of i-region are shown on the graph in microns.

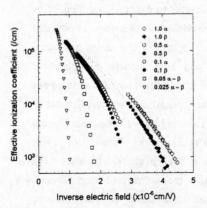

Fig. 2 Effective ionization coefficients plotted against inverse electric field for the range of structures. Numbers refer to the nominal i-region lengths in microns. Lines denote Bulman's data.

3. Discussion.

To illustrate the effect of using short devices on the avalanche process, the conventional analysis described by Stillman & Wolfe [4] was performed to obtain α and β. Because ionization is assumed to be uniform across the device, which is incorrect in these short structures, α and β are now termed "effective ionization coefficients". The associated electric field is calculated from E=V/w where V is the total potential difference across the device including an assumed 1.2V diffusion voltage. The results are shown in Fig. 2 where the data is compared to those of Bulman et al [5]. For the w = 1μm and 0.5μm structures, α and β are seen to be insensitive to w and in excellent agreement with the published data. However, α and β for the w = 0.1μm structure are below the values of Bulman et al. [5] at lower electric fields, but the data converge when the bias is increased. Results from the w = 0.05μm and 0.025μm structures show even greater differences from bulk behaviour. Larger electric fields are required before ionization can be measured and α and β rise more rapidly with electric field.

The semi-analytical trajectory method, which has been explained in detail elsewhere [2,6], has been used to interpret these results. Electron and hole transport is described by a set of energy dependent scattering path lengths and a saturation drift velocity. Separate phonon scattering rates are used for energies greater than 1.3eV to account for the highly non-parabolic band structure. Impact ionization is included as an extras scattering mechanism. The impact ionization rate, I, is assumed to vary with energy, ε, as

$$I = S\left[\frac{\varepsilon - \varepsilon_{th}}{\varepsilon_{th}}\right]^{4.3}$$

where ε_{th} represents the ionization threshold energy. S is a parameter which describes the rate of ionization above the threshold energy. The power 4.3 is obtained from pseudopotential calculations of ionization cross-sections by Kane [7] while S is obtained from a fit to the data of Bulman et. al. [6]. To highlight the discrepancies, theoretical electron and hole ionization integrals, $1-1/M_e$ and $1-1/M_h$ respectively, were then fitted to the experimental results by making slight adjustments to the value of w used in the calculation. This fit is shown for the electron characteristics for structures where $w \leq 0.1\mu m$ in Fig. 3. Local values of ionization coefficients were also calculated which correspond to the steady state ionization coefficient.

The model shows that larger electric fields are required to produce measurable multiplication in short structures because the distance a carrier requires to gain the ionization threshold energy from the electric field, the dead space, becomes a significant fraction of the device. As the electric field is increased, this effect reduces multiplication below values which would be predicted by a local model but becomes less significant because the ballistic dead space distance is inversely proportional to the field. At high electric fields, carriers are driven to energies above the steady state value resulting in energy and thus ionization overshoot. This effect compensates the dead space effect and results in calculations of breakdown voltages from local models producing surprisingly good results.

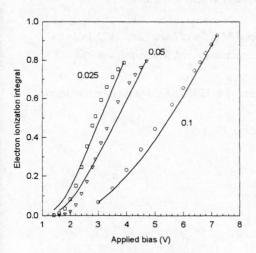

Fig. 3 Theoretical ionization integrals (lines) fitted to experimental results for the thinnest structures. Numbers on the plot denote nominal values of w in microns.

568

4. Conclusion

Electron and hole multiplication data have been obtained from thin GaAs diodes from the analysis of photocurrent measurements. Using a relatively simple model of carrier transport it has been illustrated that the dead space becomes important in structures of this length and that the assumption of uniform ionization is inappropriate. Overshoot of the ionization coefficient is seen to compensate the dead space when the electric fields in the devices are increased.

Acknowledgments

The authors wish to acknowledge M.R. Houlton for supplying SIMs profiles, D.M. Carr for help with the electrical characterization and G. Hill for the device fabrication. This work was supported by DRA (Malvern). SAP is grateful for funding from the University of Sheffield.

References

[1]   Gaul L, Huber S, Freyer J and Claassen M 1991 *Solid State Elec.* **34** 723
[2]   Millidge S, Herbert D C, Kane M, Smith G W and Wight D W
        1995 *Semicond. Sci. Tech* **10** 344
[3]   Woods M H, Johnson W C and Lambert M A *Solid State Elec.* 1973 **16** 381
[4]   Stillman G E and Wolfe C M in *Semiconductors and Semimetals* vol. 12 1977
        (New-York: Academic Press)
[5]   Bulman G E, Robbins V M and Stillman G E 1985 *IEEE Trans. Elec. Devices*
        32 2454
[6]   Herbert D C 1993 *Semicond. Sci. Tech.* **8** 1993
[7]   Kane E O 1967 *Phys. Rev. B* **159** 624

Inst. Phys. Conf. Ser. No 145: Chapter 3
Paper presented at 22nd Int. Symp. Compound Semiconductors, Cheju Island, Korea, 28 August–2 September 1995
© 1996 IOP Publishing Ltd

Impact Ionisation Coefficients in $(Al_xGa_{1-x})_{0.52}In_{0.48}P$

J.P.R. David, R. Ghin, M. Hopkinson, M.A. Pate and P.N. Robson

Department of Electronic and Electrical Engineering,
University of Sheffield, Mappin Street, Sheffield S1 3JD, U.K.

Abstract. The impact ionisation coefficients in $(Al_xGa_{1-x})_{0.52}In_{0.48}P$ have been determined from photomultiplication measurements on p-i-n structures. The results show that the electron and hole ionisation coefficients (α and β respectively) are much lower than in AlGaAs at similar electric fields, and decrease with increasing band-gap, giving rise to higher breakdown voltages. The α/β ratio is found to be close to unity across the composition range.

1. Introduction

One of the main factors limiting the power performance of semiconductor devices is the onset of avalanche multiplication and ultimately breakdown. The multiplication process occurs when a carrier gains enough energy from the electric field to create another electron-hole pair. The minimum energy required to achieve this, the threshold energy (E_{th}), is usually assumed to be a function of the band structure of the material [1] and materials with larger band-gaps have been observed to have higher breakdown voltages [2].

$(Al_xGa_{1-x})_{0.52}In_{0.48}P$ covers a band-gap range of 1.9eV-2.6eV and is becoming increasingly used as a replacement for AlGaAs in devices such as heterojunction bipolar transistors (HBT's). In addition to being latticed matched to GaAs it is thought to have better interfaces and suffer less from deep levels. Since the $(Al_xGa_{1-x})_{0.52}In_{0.48}P$ in many device geometries will experience high electric fields, it is important to know accurately the breakdown behaviour and hence ionisation coefficients. In this paper we report on measurements of the electron and hole ionisation coefficients (α and β) across the $(Al_xGa_{1-x})_{0.52}In_{0.48}P$ alloy range lattice matched to GaAs.

Table I. Details of the measured 'i' region thickness and breakdown voltage in the $(Al_xGa_{1-x})_{0.52}In_{0.48}P$ *diodes.* * *refers to a reach through structure and* † *refers to homojunction diodes.*

x	Band-gap (eV)	V_{bd} (exp.)	'i' (μm)	V_{bd}(GaAs)	ΔV_{bd}(%)
0.0†	1.87	55.0	0.890	30.2	82.1
0.0	1.87	54.0	0.884*	29.0	84.9
0.24	2.07	61.0	0.935	31.4	94.3
0.36	2.12	61.5	0.925	31.2	97.1
0.50	2.25	61.0	0.865	29.6	106.0
0.70	2.38	65.0	0.890	30.2	115.2
0.85	2.46	67.5	0.910	30.7	120.0
1.0	2.58	71.0	0.935	31.4	126.0
1.0†	2.58	43.0	0.495	19.0	126.3

2. Experiment

A series of $Ga_{0.52}In_{0.48}P/(Al_xGa_{1-x})_{0.52}In_{0.48}P$ p-i-n structures with intrinsic region compositions of between x=0.0 and 0.52 as detailed in Table I were grown by solid source molecular beam epitaxy (MBE). The layers comprised a top p^+ $Ga_{0.52}In_{0.48}P$ layer of thickness 0.3μm with intrinsic region thicknesses of nominally 0.9μm and a bottom n^+ layer of GaAs. Two further structures comprising $Ga_{0.52}In_{0.48}P$ and $Al_{0.52}In_{0.48}P$ homojunction p-i-n's were also grown with 'i' region thicknesses of 0.5 and 0.9μm respectively and with a thicker p^+ capping layer.

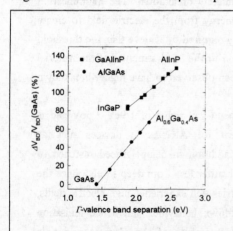

Fig. 1. The normalised breakdown voltage in $(Al_xGa_{1-x})_{0.52}In_{0.48}P$ *is plotted as a function of the direct band-gap. Also shown for comparison is the AlGaAs data from reference [4].*

The composition of the $(Al_xGa_{1-x})_{0.52}In_{0.48}P$ was determined by photoluminescence measurements (in the case of direct band gap material) and by x-ray double crystal rocking curves. Circular mesa diodes of 100-400 μm diameter with optical access were fabricated by wet etching. Devices from all these layers were found to have very low dark currents and sharp, clearly defined breakdown voltages (V_{bd}). Capacitance-voltage measurements on these layers determined the 'i' region thickness accurately and showed that the background doping in all except one deliberately doped structure was below $10^{15}\,cm^{-3}$. V_{bd} was normalised for variations in 'i' region thickness and background doping by comparing it to the calculated V_{bd} of an identical GaAs p-i-n, referred to as $V_{bd}(GaAs)$ in Table I, using the ionisation coefficients of Bulman et al.[3]. The normalised V_{bd}, defined as the ratio between the measured V_{bd} and that calculated, was found to increase linearly with composition from $Ga_{0.52}In_{0.48}P$ to $Al_{0.52}In_{0.48}P$ as shown in Fig. 1 in a similar manner to recent measurements on AlGaAs p-i-n's [4]. No change in gradient was observed as the material became indirect at x>0.25. Comparisons with AlGaAs p-i-ns of similar band-gap show that the $(Al_xGa_{1-x})_{0.52}In_{0.48}P$ structures have a higher breakdown voltage.

In order to understand the V_{bd} behaviour, α and β were determined from photomultiplication measurements. As described in detail elsewhere [5], pure electron and hole injection of carriers was achieved by laser excitation of the p^+ and n^+ cladding layers. The photocurrent was then measured as a function of reverse bias. In order to obtain reliable results, pure electron and hole multiplication characteristics (M_e and M_h respectively) are needed. Pure electron injection was easily achieved in all these structures by focusing the light from a 442nm He-Cd laser spot onto the top of the p^+ $Ga_{0.52}In_{0.48}P$ layer. The high absorption in $Ga_{0.52}In_{0.48}P$ at this wavelength meant that effectively no light was transmitted into the high field region. Pure hole injection was achieved by focusing the laser spot onto the n^+ floor of the mesa structures. No microplasmas or defect related behaviour was observed in any of the devices and several devices from each layer were measured with different laser intensities to ensure reliable results. In addition to this, mixed electron and hole initiated multiplication was also obtained by focusing 633nm onto the top of the p^+ $Ga_{0.52}In_{0.48}P$. Significant amount of light at this wavelength is transmitted through the cap and is absorbed either in the $(Al_xGa_{1-x})_{0.52}In_{0.48}P$ 'i' region or in the bottom n^+ GaAs layer. The 'mixed' multiplication characteristics (M_{mix}) lay between M_e and M_h as expected. Fig. 2 shows the dark current and the photocurrent obtained with different wavelength excitation on the homojunction $Ga_{0.52}In_{0.48}P$ structure. Since these were all either p-i-n or reach through diodes, there was only a small increase in the primary photocurrent prior to the onset of multiplication and this was corrected for[6].

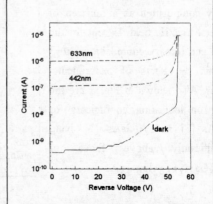

Fig. 2. Dark current and photocurrents for $Ga_{0.52}In_{0.48}P$ homojunction p-i-n with 633nm and 442nm illumination.

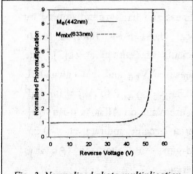

Fig. 3. Normalised photo-multiplication characteristics for a $Ga_{0.52}In_{0.48}P$ structure with M_e and M_{mix} multiplication.

The normalised photomultiplication characteristics, (see Fig. 3), show that $M_e > M_{mix}$ suggesting $\alpha > \beta$. Detailed measurements of M_h on this layer confirmed that M_e was slightly larger than M_h.

Photomultiplication measurements undertaken on the other composition p-i-ns showed that M_e and M_h were very similar across the entire alloy composition range. However for both the $Al_{0.52}In_{0.48}P$ structures M_h was found to be slightly larger than M_e. At electric fields similar to those experienced in these structures, GaAs also shows a similar behaviour with $M_e \approx M_h$ [3]. The multiplication characteristics occur at increasingly larger electric fields as the composition increases in agreement with the breakdown voltage measurements. Fig. 4 shows that large multiplication values are achievable before the onset of catastrophic breakdown.

3. Discussion

α and β were determined from the photomultiplication characteristics using the appropriate p-i-n and reach through diode equations [5]. Fig. 5 shows the α and β across the alloy composition as a function of inverse electric field. Except for the two end ternary compositions, $\alpha \approx \beta$ within the limits of experimental error. Also shown in Fig. 5 are the electron ionisation coefficients for GaAs [3] and AlGaAs [7], lying significantly above those of $(Al_xGa_{1-x})_{0.52}In_{0.48}P$.

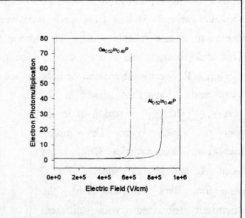

Fig. 4. Normalised M_e as a function of electric field in $Ga_{0.52}In_{0.48}P$ and $Al_{0.52}In_{0.48}P$ structures. Large multiplication values are possible with $Al_{0.52}In_{0.48}P$ breaking down at a higher field as expected.

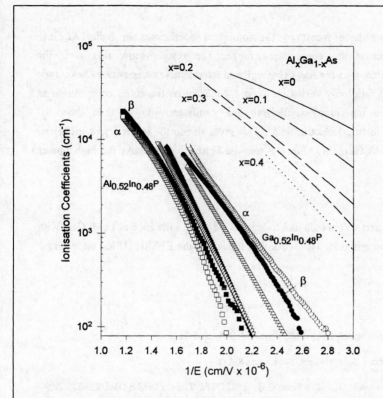

Fig. 5. Impact ionisation coefficients versus inverse electric field for $Al_xGa_{1-x}As$ (x=0.0-0.4)(lines), α only, and $(Al_xGa_{1-x})_{0.52}In_{0.48}P$ (symbols). α>β for $Ga_{0.52}In_{0.48}P$ (circles) but β>α for $Al_{0.52}In_{0.48}P$ (squares). Intermediate compositions have α≈β and lie between the two ternary end points with composition, x, as follows, Δ=0.23, ♦=0.48, and ∇=0.67.

The ionisation coefficients in $(Al_xGa_{1-x})_{0.52}In_{0.48}P$ are found to decrease with increasing band gap as expected from the increasing breakdown voltage. There is very good agreement between the heterojunction and homojunction ternary p-i-n's suggesting that the heterojunction does not affect the multiplication process. The low ionisation coefficients in $(Al_xGa_{1-x})_{0.52}In_{0.48}P$ relative to AlGaAs of equivalent direct band gap suggests that it is not only the Γ-valence band separation that determines the ionisation process and recent pressure studies [8] have confirmed the importance of the satellite valleys. The larger Γ-X separation in $(Al_xGa_{1-x})_{0.52}In_{0.48}P$ relative to AlGaAs therefore probably plays an important part in determining the onset of the ionisation process and accounts for the lower ionisation coefficients and increased breakdown voltages.

4. Conclusions

In conclusion, we have measured the ionisation coefficients in bulk $(Al_xGa_{1-x})_{0.52}In_{0.48}P$ across the composition range. The results show that $\alpha > \beta$ for $Ga_{0.52}In_{0.48}P$ while $\beta < \alpha$ for $Al_{0.52}In_{0.48}P$ and intermediate compositions have $\alpha \approx \beta$. Although the α/β ratio only varied between 1.2-0.8 across the entire composition at higher fields, the ionisation coefficients are significantly lower than those of GaAs/AlGaAs at identical electric fields. This gives rise to the larger V_{bd}'s measured suggesting that $(Al_xGa_{1-x})_{0.52}In_{0.48}P$ may be better than AlGaAs for high power devices.

Acknowledgements: We would like to acknowledge the assistance of D.M. Carr with the material characterisation. This work was funded by the EPSRC (U.K.) under grant GR/J49549.

References

[1]Anderson C.L., and Crowell C.R., 1972, Phys.Rev. B, **5**, 2267-2272

[2]Sze S.M. and Gibbons G.; 1966 Appl. Phys. Lett., **8**, 111-113

[3]Bulman G.E., Robbins V.M. and Stillman G.E., 1985, IEEE Trans. Electron Dev., **ED-32**, 2454-2466

[4]David J.P.R., Allam J., Adams A.R., Roberts J.S., Grey R., Rees G.J. and Robson P.N.; 1995, Appl. Phys. Lett., **66**, 2876-2878

[5]Stillman G.E., and Wolfe C.M., 1977, 'Avalanche Photodiodes', Semiconductors and Semimetals, **Vol.12**, ed. R.K.Willardson and A.C.Beer, Academic Press

[6]Lee C.A., Logan R.A., Baldorf R.A., Kleimack J.J. and Weigman W.; 1964, Phys. Rev., **134**, A761-A773

[7]Robbins V.M., Smith S.C. and Stillman G.E.,1988, Appl. Phys. Lett., **52**, 296-298

[8]Allam J., Adams A.R., Pate M.A. and Roberts J.S., Institute of Physics Conference Series, 1990, No.112, 375-376

Inst. Phys. Conf. Ser. No 145: Chapter 3
Paper presented at 22nd Int. Symp. Compound Semiconductors, Cheju Island, Korea, 28 August–2 September 1995
© *1996 IOP Publishing Ltd*

The reduction of impact ionization and improvement of device reliability in heterostructure doped-channel FETs

Ming-Ta Yang, Yi-Jen Chan and Melvin Chang

Department of Electrical Engineering, National Central University, Chungli, Taiwan, 32054 ROC.

Abstract. Reliability issues of both AlGaAs/$In_{0.2}Ga_{0.8}$As DCFETs and HEMTs were investigated and compared. By placing donors in the conducting channel, this doped-channel approach reduces the field intensity near the heterointerface resulting in a suppression of impact ionization process in the channel. Therefore, a more reliable device characteristics of DCFETs could be expected. The experimental results, through the biasing stress and temperature dependent evaluations, demonstrated a strong correlation between the impact ionization rate and device reliabilities.

1. Introduction

Pseudomorphic heterostructure FETs (HFETs), including a lattice mis-matched channel, have shown state-of-the-art microwave performance [1], primarily as a result of the larger conduction-band discontinuity (ΔE_c) at the AlGaAs/InGaAs heterointerface. Device performance can be enhanced substantially by increasing the indium mole fraction in the InGaAs channel. The increase of In amount in conducting channels leads to an increase of the electron sheet charge density (N_s) in the quantum well, which is a crucial factor for the performance of HEMTs [2]. However, once the occuptance of wave function closes to the deeper area of this triangular potential channel, the possibility of impact ionization process will be triggered due to the associated strong localized electric-field. Chough et al. [3] proposed a graded channel containing various In contents to form a much flatter potential-well channel in HEMTs, which resulted in a higher gate-to-drain and channel breakdown. The avoidness of the triangular potential well can also be obtained in the so-called double heterostructure HEMTs, where a higher breakdown voltage and less short-channel effect have also been demonstrated [4].

Based on the same concept, in this study, we utilized the heterostructure doped channel design to evaluate the impact ionization mechanism in doped-channel FETs (DCFETs). Heterostructure DCFETs with a high N_s have demonstrated a high breakdown voltage, a high linearity and a high current driving capability, which are suitable for the application on large signal microwave power devices [5]. By placing the donors in the heterostructure conducting channel shown in Fig. 1, instead of in the high bandgap region, results in a much flatter potential well profile, which reduces the field intensity near the heterointerface. This will suppress the impact ionization process in conducting channels, and therefore a more reliable device characteristics of DCFETs could be expected. In this study, we compared the device reliability associated with the impact ionization for both HEMT and DCFET built on $Al_{0.3}Ga_{0.7}As/In_{0.2}Ga_{0.8}$As heterostructure system. To monitor the impact ionization process,

576

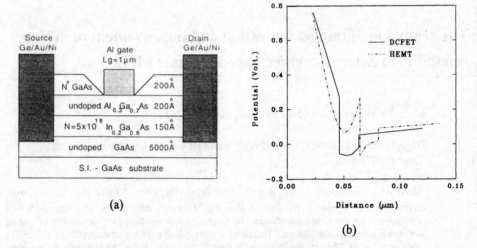

(a)

(b)

Fig. 1 Device cross-section of pseudomorphic $Al_{0.3}Ga_{0.7}As/In_{0.2}Ga_{0.8}As$ doped-channel FETs (a), and schematic band diagram of pseudomorphic HFETs (b).

we utilized the method proposed by Hui et al. [6] to measure the gate current (I_g) under a high gate-to-drain bias (V_{gd}) condition, which provides a convenient way to characterize the impact ionization process in the conducting channel. As to the reliability test, we performed the aged evaluation, where the device was intentionally biased in the high-field regime at room temperature.

2. Strained $Al_{0.3}Ga_{0.7}As/In_{0.2}Ga_{0.8}As$ DCFETs

Pseudomorphic $Al_{0.3}Ga_{0.7}As/In_{0.2}Ga_{0.8}As$ HFETs were grown in a Riber 32-P molecular beam epitaxy (MBE) system on (100)-oriented semi-insulating GaAs substrates. Fig. 1(a) shows the device cross-sections of DCFETs. A 150 Å highly Si-doped (N_D=5x10^{18} cm^{-3}) $In_{0.2}Ga_{0.8}As$ channel was grown by calibrating the growth curves, and verified by double crystal x-ray diffraction afterwards. This doped-channel was sandwiched by the undoped AlGaAs and GaAs layers. The mole fraction of the top undoped 200 Å AlGaAs is 0.3, which was used to improve the Schottky-gate performance, while the bottom GaAs was used as buffer layer. As to the modulation-doped approach, we utilized the conventional single heterostructure HEMT design except for the undoped 150 Å $In_{0.2}Ga_{0.8}As$ pseudomorphic channel to further improve the device transport properties. This channel was followed by a 30 Å undoped $Al_{0.3}Ga_{0.7}As$ spacer and a 150 Å $Al_{0.3}Ga_{0.7}As$ (N_D=5x10^{18} cm^{-3}) donor layer. Both devices were finalized by a 200 Å n$^+$-GaAs cap layer to reduce the ohmic contact resistivities.

Fig. 1(b) shows the conduction band diagram of $Al_{0.3}Ga_{0.7}As/In_{0.2}Ga_{0.8}As$ pseudomorphic DCFETs associated with Fig. 1(a). It illustrates obviously that by using of an uniformly doped-channel scheme modifies the triangular-shaped potential well into an approximately rectangular-shaped one. Since the dopants are uniformly distributed within the channel comparing with HEMTs, the interfacial scattering between $Al_{0.3}Ga_{0.7}As$ and $In_{0.2}Ga_{0.8}As$ layers can be minimized and the localized interfacial electric-field can be reduced as well. In addition, this undoped $Al_{0.3}Ga_{0.7}As$ Schottky layer increases the barrier height, resulting an enhancement of allowing a higher forward bias without a detectable gate leakage. Devices with a 1x50 μm² gate dimension were fabricated by using the conventional optical lithographical and lift-off techniques.

3. Impact ionization in strained $Al_{0.3}Ga_{0.7}As/In_{0.2}Ga_{0.8}As$ DCFETs

Typical drain-source I-V characteristics were observed for DCFETs biased at low V_{ds}. However, by further increasing the V_{ds}, a significant kink of I_{ds} was found [7]. The drain output characteristics of $Al_{0.3}Ga_{0.7}As/In_{0.2}Ga_{0.8}As$ DCFETs are shown in Fig. 2(a). In addition, we also attached the gate current (I_g) curves in the same figure under different reversed gate-source bias (V_{gs}) with a source grounded configuration. The I_g remained almost constant for $V_{gs} \geq -1$ V. By increasing the reversed V_{gs} (≤ -1.5 V), a significant increase of I_g was found at high V_{ds} (>8 V). However, this I_g increase was compensated by the reduction of I_{ds} as further increasing the reversed V_{gs} (-2.5 V). This anomalous I_g change can be easily characterized if we plotted I_g, I_{ds} curves versus V_{gs}, shown in Fig 2(b), under high V_{ds} biasing conditions. At low V_{ds} (<6 V), the I_g curves reflect a simple Schottky-diode behavior over the entire V_{gs} range. For a higher V_{ds} (>6 V), on the other hand, a clear bell-shaped behavior of I_g was observed for DCFETs biased at -2.5 V $\leq V_{gs} \leq$ -1 V regime. The bump peak at V_{gs}=-2 V, and the intensity of this bell-shaped curve was also enhanced by increasing the V_{ds}. In the mean while, I_{ds} also slightly increased, which exactly corresponds to this bell-shaped region. It suggests a related mechanism existed for the both increase of I_g and I_{ds}. The bell-shaped behavior of I_g has been observed in both MESFETs [6] and HEMTs [8], which is associated with the occurrence of impact ionization process in conducting channels. Under such circumstance, holes generated in the channel flow through the gate terminal (V_{gs}<0 V), causing the increase of I_g. While the excess electrons generated in this ionization process will continuously flow into the drain, resulting in an I_{ds} increase. However, by further increasing the reversed V_{gs}, due to a dramatic depletion of channel

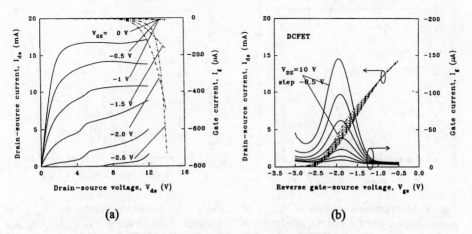

(a)   (b)

Fig. 2 Output characteristics I_{ds}-V_{ds}-I_g (a), and I_{ds}-V_{gs}-I_g (b) of a typical $Al_{0.3}Ga_{0.7}As/In_{0.2}Ga_{0.8}As$ DCFETs measured at room temperature.

carriers near the pinch-off region, impact-ionization rates decreased and therefore a bell-shaped of I_g was formed.

To quantitatively analyze the impact ionization in the channel, we proposed a parameter G_{impact} which is defined as $G_{impact}=\Delta I_g/\Delta V_{ds}$ to represent the degree of impact ionization. Since the V_{ds} dependent I_g was biased at a fixed V_{gs}, the electric field intensity associated with bell-shaped curve in the channel is mainly determined by V_{ds}. Therefore, this paramater, $\Delta I_g/\Delta V_{ds}$, is directly responsible for impact-ionization rates. Fig. 3 shows the measured results of G_{impact} for both $Al_{0.3}Ga_{0.7}As/In_{0.2}Ga_{0.8}As$ DCFETs and HEMTs. The G_{impact} values increased dramatically after $V_{ds}\geq 6$ V, which is corresponding to the threshold of the kink and impact ionization in the channel. In addition, a smaller gate-length of DCFETs resulting in a earlier impact ionization process. It can be clearly found that HEMTs presented a much higher G_{impact} value, which may be responsible from a stronger localized-field near the 2DEG distribution. However, this is not the case for DCFETs. To further confirm that this bell-shaped behavior of I_g is in fact responsible from the impact ionization, we carried out the temperature dependent measurement for both devices with a L_g of 0.8 μm. Due to the phonon scattering, the impact ionization process exhibites a negative temperature behavior [9]. Fig. 4 shows a significant G_{impact} decrease by increasing measurement temperatures; namely, G_{impact} from 0.95 mS/mm at 25 °C to 0.704 mS/mm at 50 °C and 0.59 mS/mm at 75 °C, biasing at $V_{ds}=10$ V, respectively. It therefore provides an another evidence for confirming the occurrence of impact ionization in the channel. Similar temperature behavior was also obtained in HEMT's investigation. In brief, due to a flatter potential well in DCFETs, they present a much lower impact ionization rate comparing with HEMT characteristics.

4. Reliability evaluation

Based on the previous investigation, DCFETs demonstrate a lower impact ionization rate, and it may benefit for DCFETs to be a more reliable device. Aged-testing, where devices

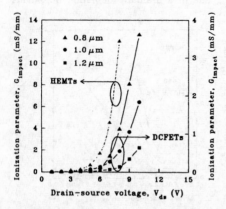
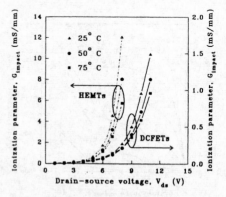

Fig. 3 Measurement results of G_{impact} at room temperature for variousgate-length of $Al_{0.3}Ga_{0.7}As/In_{0.2}Ga_{0.8}As$ HFETs biased at $V_{gs}=-2$ V.

Fig. 4 Measurement results of G_{impact} for various temperature of $Al_{0.3}Ga_{0.7}As/In_{0.2}Ga_{0.8}As$ HFETs as a function of V_{ds} biased at $V_{gs}=-2$ V.

were intentionally biased at the high-field regime, were conducted on both DCFETs and HEMTs at an ambient temperature of 25 °C. Due to a higher operational voltage that can be applied on the DCFETs, we adopted a V_{ds}=5 V for DCFETs and a V_{ds}=3 V for HEMTs for stress condition, respectively. Both devices were surface unpassivated. After a continuous 48 hrs. operation, the channel current of HEMTs dropped 36% from its original value, while this number was only 10% for DCFETs. Both devices all showed a significant initial deterioration. After this initial stage, the devices became stable. Fig. 5 shows the normalized I_{ds} variations of the DCFETs and HEMTs during the 48 hrs. biased-testing. As to the g_m degradation, it also presented the same behavior, namely a 25% drop for HEMTs and a 10% drop for DCFETs, where DCFETs preserved better reliable characteristics after biasing stress. Figs. 6(a) and (b) show the degradation of normalized I_{ds} and g_m characteristics for DCFETs and HEMTs after the stress. Again, the deterioration of device performance is not significant in DCFETs. We also observed an enhancement of device degradation by increasing the reversed gate bias, which resulted in a higher electric-field in the channel. Therefore, it suggests that the deterioration mechanism is associated with the electric-field intensity near the gate to drain region. Since we intentionally biased the devices in the high-field regime, the performance degradation may be associated with Schottky diode. The gate leakages, after 48 hrs. of biasing stress, were 0.45 mA/mm and 2.3 mA/mm for the DCFETs and HEMTs, respectively; however the initial leakage values were similar (0.02 mA/mm). It therefore suggests that, due to different materials underneath the gate, the DCFETs can be survived under a much higher electric-field without the degradation of Schottky performance. This merit together with a flatter potential-well provides DCFETs to achieve a more reliable performance.

Temperature dependent aged-testings were conducted on DCFETs and the results are shown in Fig. 7. DCFETs were found to be more reliable under high temperature operation. The I_{ds} degradation was 10% at 100 °C rather than 16% at 25 °C, which may be due to the reduction of impact ionization rate at high temperatures. Impact ionization coefficiencies

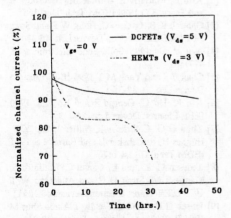

Fig. 5 Normalized channel current percentage variation during aged-testing for HFETs.

were 1.5 mS/mm and 1 mS/mm corresponding to 25 °C and 75 °C at V_{ds}=10 V, respectively. This result proves the existence of strong dependence between the impact ionization and the device reliability.

5. Conclusions

In general, reliability issues of both $Al_{0.3}Ga_{0.7}As/In_{0.2}Ga_{0.8}As$ DCFETs and HEMTs were investigated and compared. Due to the suppression of impact ionization, DCFETs demonstrated better reliable characteristics. The failure mechanism in both devices was associated with the electric-field near the channel and

580

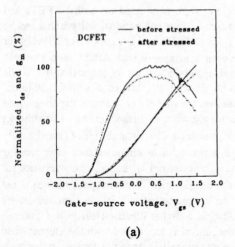

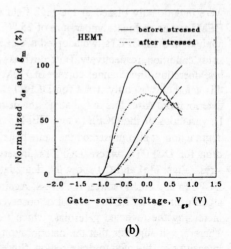

(a)   (b)

Fig. 6 Degradation of normalized I_{ds} and g_m characteristics for DCFETs (a), and HEMTs (b) due to aged-testing.

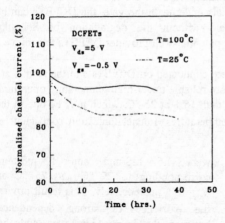

Fig. 7 Normalized channel current percentage variation during aged-testing for DCFETs due to different temperature conditions.

gate-to-drain regions. DCFETs demonstrate not only to be high breakdown, high current density and high linearity devices. Furthermore, DCFETs

are also reliable under large biasing operation, which are very suitable for microwave power device application.

References

[1] Nguyen L D, Tasker P J, Radulescu D C and Eastman L F 1989 IEEE Trans. Electron Devices **36** 2243
[2] Kim B, Matyi R J, Wurtele M, Bradshaw K and Tserg H Q 1988 IEDM Tech. Dig. 168
[3] Chough K B, Caneau C, Hong W P and Song J I 1994 IEEE Electron Device Lett. **15** 33
[4] Wu et al. 1995 Solid-State Electronics **38** 377-381
[5] Chan Y J and Yang M T 1995 IEEE Electron Device Lett. **16** 33-35
[6] Kui K, Hu C, George P and Ko P K 1990 IEEE Electron Device Lett. **11** 113
[7] Zhou G G, Colbrie A F, Miller J, Pao Y C, Hughes B, Studebaker L and Harris J S 1991 IEDM Tech. Digest 247
[8] Tedesco C, Zanoni E, Canali C, Bigliardi S, Manfredi M, Streit D C and Anderson W T 1993 IEEE Trans. Electron Devices **40** 1211
[9] Jensen G U, Lund B, Fjeldly T A and Shur M 1991 IEEE Trans. Electron Devices **38** 840

Inst. Phys. Conf. Ser. No 145: Chapter 3
Paper presented at 22nd Int. Symp. Compound Semiconductors, Cheju Island, Korea, 28 August–2 September 1995
© *1996 IOP Publishing Ltd*

An XPS study of novel wide band gap $Zn_xSr_{1-x}S$ semiconductors

Sang Tae Lee, Masahiko Kitagawa, Kunio Ichino and Hiroshi Kobayashi

Dept. of Electrical and Electronic Engineering, Tottori University, Koyama, Tottori 680, Japan

Abstract. The chemical bonding of Zn and Sr atoms to S atoms for $Zn_xSr_{1-x}S$ thin films has been studied with x-ray photoelectron spectroscopy (XPS). XPS signals from S and Zn in $Zn_xSr_{1-x}S$ made it clear that phase transition occurs at about $x = 0.3$ and 0.85 which is in a qualitative agreement with x-ray diffraction (XRD) results. We also found from the splitting of S $2p_{3/2}$ photoemission lines that the $Zn_xSr_{1-x}S$ for the intermediate composition range have two phases. The chemical shifts are also discussed in conjunction with crystal phases.

1. Introduction

Wide band gap II-VI compounds are promising materials for the light emitting devices which operate in the blue to ultraviolet region. Recent progress in the blue laser diodes (LDs) [1] using Zn-chalcogenide ZnS, ZnSe with zincblende structure and full-color electroluminescent (EL) displays [2] using alkaline-earth chalcogenide CaS, SrS has evoked a strong interest in the study of new optoelectronic materials. Among such II-VI compounds, MgZnS(-Se) solid solution has already been employed as a lattice-matched heteroepitaxial system for the realization of blue LDs [1] since Mg-chalcogenide can crystallize in zincblende structure compatible with Zn-chalcogenide. However, Sr- and Ca-chalcogenide compounds crystallize only in the rocksalt structure. The resultant II_a-II_b-VI compounds including Ca or Sr ions are generally thought to form an immiscible system under the thermal equilibrium preparation condition hence the investigation of the solid solubility for these II_a-II_b-VI compounds is very limited. Up to now, the solubility between II_a-VI and II_b-VI compounds have been studied in the form of sintered-solid solution by several groups [3]. Very recently, we have reported that $Zn_xSr_{1-x}S$ thin films can be deposited by using the thermally nonequilibrium rf sputtering method and found that there exists the solid solution with relatively wide x range. We also reported the characteristics of $Zn_xSr_{1-x}S$ thin films, especially the solid solubility, crystal structure and optical property with composition x [4]. However there is little knowledge about chemical states of this novel $Zn_xSr_{1-x}S$ mixed crystal system.

In this paper, we report on the chemical bonding of the constituent atoms of $Zn_xSr_{1-x}S$ thin films investigated by x-ray photoelectron spectroscopy (XPS). We also discuss the chemical shifts with composition and/or crystal phases as well as the origin of chemical shifts based on ionic character and bond length.

582

2. Experimental

$Zn_xSr_{1-x}S$ thin films were deposited on a quartz glass substrate by a conventional diode-type rf sputtering system. A mixture of ZnS/SrS powder with the required mole fraction was used as a target. The target preparation and the typical deposition conditions of the films have been described elsewhere [4]. Substrate temperature (T_{sub}) was kept at 200°C during deposition.

The composition x of the $Zn_xSr_{1-x}S$ thin films was determined by energy dispersive x-ray (EDX) analysis. An acceleration voltage of 10kV was used. Chemical bonding states of constituent elements were analyzed by XPS (Shimadzu ESCA 750 with ESPAC 100 data processing system, Shimadzu Co.). The x-ray target employed was a Mg anode with a characteristic x-ray energy of 1253.6eV. An energy resolution is ±0.1eV at 368.2eV of Ag $3d_{5/2}$. The XPS signals from the Sr 3d, Zn 2p and S 2p core levels were mainly investigated. Prior to the XPS measurements, *in-situ* Ar$^+$ ion etching with 10-100Å/min was performed to avoid the surface contamination. The XPS measurement was done in a vacuum below 10^{-5}Pa. Shifts of the XPS spectra due to electrical charging of the sample surface were calibrated with reference to the XPS signal from C 1s at 284.6eV.

3. Experimental results

To estimate the effect of the surface contamination, especially the effect of oxygen, XPS spectrum was measured after Ar$^+$ ion etching for 3 minutes. Fig. 1 shows a low-resolution XPS spectrum of $Zn_{0.84}Sr_{0.16}S$ film with scan range of 1100-0eV. XPS spectrum measured before etching is also shown. The wide scanned spectra in Fig. 1 display XPS core level features for each element which is identified with marker. As can be seen in this figure, any other peaks are not detected except Zn, Sr and S for ZnS-SrS mixed system and Ag and C used for charge referencing. Spectrum for sample without etching shows the weak peak from O 1s at about 530eV, while for etched samples O 1s peak almost disappeared and other peak intensities, especially peaks from Zn photoemission lines become stronger. However, energy shifts were not observed between both cases.

Fig. 2(a) shows the typical XPS spectra of S $2p_{1/2}$ and $2p_{3/2}$ core levels for $Zn_xSr_{1-x}S$ thin films. The arrows represent the binding energies of each core level for elemental S and ZnS compound [5], and the vertical broken lines indicate the XPS peaks from S $2p_{3/2}$ bonded in $Zn_xSr_{1-x}S$. The binding energy of S $2p_{3/2}$ in ZnS is about 162.2eV, which agrees well with the reported value. Those of S $2p_{1/2}$ and $2p_{3/2}$ in SrS is about 161.1 and 159.8eV, respectively,

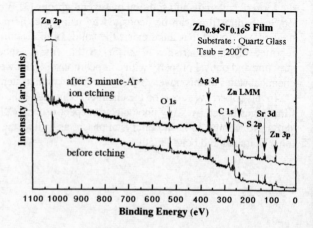

Fig. 1. The low-resolution XPS survey mode spectra of $Zn_{0.84}Sr_{0.16}S$ film. The markers indicate the assignments for various XPS features.

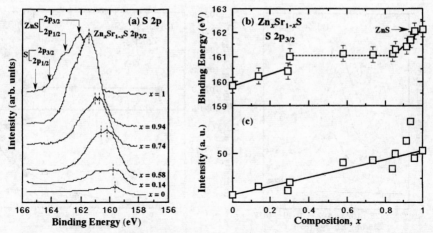

Fig. 2(a) The typical XPS spectra from S $2p_{1/2}$ and $2p_{3/2}$ core levels for $Zn_xSr_{1-x}S$ thin films. The arrows indicate the binding energies of core levels for elemental S and ZnS compound. (b) The composition dependence of binding energy and (c) that of peak intensity of S $2p_{3/2}$ core level.

which differ from the reported values in SrS:Eu, 162.8 and 156.7eV, respectively [6]. The binding energies of S $2p_{3/2}$ in ZnS and SrS are 1.9 and 4.3eV smaller than those in elemental S, respectively. Especially, the XPS signal from S $2p_{3/2}$ bonded in $Zn_{0.74}Sr_{0.26}S$ and $Zn_{0.58}Sr_{0.42}S$ shows two peaks, indicating that S in these films is bound to two components, that is, has two chemical states. These splittings of photoelectron lines were observed in the $Zn_xSr_{1-x}S$ films with $0.3 \leq x \leq 0.84$. Even though XPS were measured before etching, there is no spectroscopically detectable evidence of oxide components such as $-SO_3$ (167.7eV) and SO_4 (169.9eV) in the S 2p photoelectron spectra as shown in the figure. Figs. 2(b) and (c) summarize the composition dependence of binding energy and peak intensity of S $2p_{3/2}$ core level for $Zn_xSr_{1-x}S$ films, respectively. For Sr-rich $Zn_xSr_{1-x}S$ films with $x < 0.3$, binding energy increase almost linearly from 159.8eV for SrS to 160.7eV for $Zn_{0.3}Sr_{0.7}S$ film with composition. For higher Zn composition region from 0.84 to 1, the binding energies from S $2p_{3/2}$ for the films also increase continuously from 161.15eV at $x=0.84$ to 162.15eV of ZnS with increasing x. On the other hand, in the intermediate range of $0.3 \leq x \leq 0.84$, binding energies (Higher peak energies are shown in the figure.) almost keep constant independent of composition, indicating that to some extent, the molar fraction of element related to S bonding is constant. The XPS peak intensities from S $2p_{3/2}$, which are plotted in Fig. 2(c), increase nearly linearly with composition.

Fig. 3(a) shows the typical XPS spectra of Zn $2p_{3/2}$ core level for $Zn_xSr_{1-x}S$ films. The arrows represent the binding energies of Zn $2p_{3/2}$ core level for elemental Zn and ZnS [5], and the vertical broken line indicates the XPS peaks of Zn $2p_{3/2}$ in $Zn_xSr_{1-x}S$. The $2p_{3/2}$ peak energy of the XPS signal from Zn bonded in ZnS is about 1022.25eV which is 0.25eV lower than the reported value. The binding energies of Zn bonded in $Zn_xSr_{1-x}S$ compounds are between that of Zn bonded in ZnS and that of Zn bonded in elemental Zn. XPS peak intensities decrease with decreasing Zn composition. There is no evidence of oxide component such as ZnO (1021.8eV) in Zn $2p_{3/2}$ photoelectron spectra. Figs. 3(b) and (c) summarize the composition dependence of binding energy and peak intensity of Zn $2p_{3/2}$ core level for $Zn_xSr_{1-x}S$ films, respectively. For $Zn_xSr_{1-x}S$ films with $x \leq 0.84$, binding energies are nearly independent of composition by 1021.6eV which is the same as the binding energy of Zn bonded in elemental Zn. For higher Zn composition region from 0.84 to 1, the binding energies from Zn $2p_{3/2}$ for the films also in-

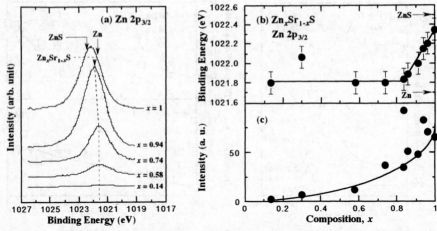

Fig. 3(a). The typical XPS spectra from Zn $2p_{3/2}$ core level for $Zn_xSr_{1-x}S$ thin films. The arrows indicate the binding energies of $2p_{3/2}$ core level for elemental Zn and ZnS compound. (b) The composition dependence of binding energy and (c) that of peak intensity of Zn $2p_{3/2}$ core level.

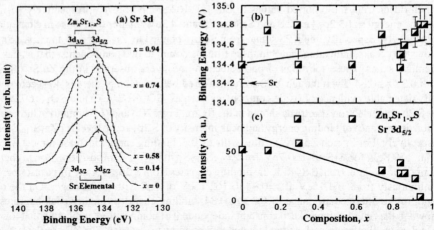

Fig. 4(a). The typical XPS spectra from Sr $3d_{3/2}$ and $3d_{5/2}$ core levels for $Zn_xSr_{1-x}S$ thin films. The arrows indicate the binding energies of $3d_{3/2}$ and $3d_{5/2}$ core levels for elemental Sr. (b) The composition dependence of binding energy and (c) that of peak intensity of Sr $3d_{5/2}$ core level.

crease continuously from 1021.72eV at $x=0.84$ to 1022.25eV of ZnS with increasing x.

Fig. 4 shows the typical XPS spectra of Sr $3d_{3/2}$ and $3d_{5/2}$ core levels for $Zn_xSr_{1-x}S$ films. The arrows represent the binding energies of core levels for elemental Sr [5] and the vertical broken lines indicate the XPS peaks from Sr $3d_{3/2}$ and $3d_{5/2}$ bonded in $Zn_xSr_{1-x}S$ compounds. The spectra show two peaks with energy difference of about 1.6eV except $Zn_{0.14}Sr_{0.86}S$ film, which are attributable to the spin-orbit splitting of Sr $3d_{3/2}$ and $3d_{5/2}$. The binding energy of $3d_{5/2}$ core level of Sr bonded in SrS is about 0.2eV higher than the reported value for elemental Sr. Figs. 4(b) and (c) show the composition dependence of binding energies and peak intensities of Sr $3d_{5/2}$ core level for $Zn_xSr_{1-x}S$ films, respectively. Although the tendency that the binding energies of Sr in $Zn_xSr_{1-x}S$ slightly decrease with decreasing x was observed for higher Zn region, the binding energies with x are scattered on the whole. Considering the error in measurement, the binding energies of Sr are considered to be easily affected by surrounding chemical environment, especially O, Zn and/or S in this system. This result is in discord with the previous report by Kim et al. [6] in which no shift was observed in the binding energy of Sr 3d in SrS_xO_{1-x}:Eu compared with that in SrS:Eu. The XPS peak intensity from Sr $3d_{5/2}$, which are plotted in Fig. 4(c), decreases with increasing x.

4. Discussion

Although ZnS-SrS is considered to be a basically immiscible system, we have shown that solid solutions are formed in limited composition ranges [7]. That is, Zn-rich $Zn_xSr_{1-x}S$, solid solutions with a single-phased zincblende structure were formed for $0.86\sim0.91 \leq x \leq 1$ and Sr-rich $Zn_xSr_{1-x}S$ solid solutions with a single-phased rocksalt structure were formed for $0 \leq x \leq 0.29$. In the intermediate composition range of $0.3 \leq x \leq 0.86\sim0.91$, the miscibility gap including phase separation regions was observed. In conjunction with crystal phases based on these results, we discuss the results of XPS analysis on the ZnS-SrS mixed materials.

First, we discuss the origin of chemical shift based on S $2p_{3/2}$ photoelectron lines as shown in Fig. 2. The chemical shifts are generally known to be due to the bonding changes in the valence-electron charge distribution which arises from the ionic character [6, 7] and the bond length [8, 9] of compound. In our materials, anion S of which electronegativity and the ionic radii are 2.5 and 1.84Å, respectively, is common, hence chemical shifts are considered to be mainly affected by the cation Sr or Zn. The ionic character of ZnS and SrS which are of 0.61 and 0.91 [10], respectively, supports chemical shifts in S bonded in ZnS or SrS which is greater in SrS than in ZnS as plotted in Fig. 2. Furthermore, the energy shifts of Zn with composition in Fig. 3 are probably induced by the increase of bonding length, arising from the partial substitution of Sr atoms with Zn atoms. Second, we discuss the composition dependence of the

Sample No.	Composition x	XPS signal from S $2p_{3/2}$		Crystalline Phase	Structure Model
		Chemical Shift ΔE (eV)	Relative Peak Intensity (a.u.)		
S1	1.000	0.00±0.1	1.00	ZB	ZnS
S3	0.960	-0.07±0.1	0.07	ZB	Zn-rich alloy
S5	0.938	-0.45±0.1	1.58	ZB	
S6	0.914	-0.72±0.1	1.19	ZB+RS	
S8	0.857	-0.83±0.1	0.96	ZB+RS	Zn-rich alloy Sr-rich alloy
S9	0.838	-1.00±0.1	0.66	ZB+RS	
S10	0.737	-1.05±0.1	0.82	ZB+RS	
S13	0.579	-1.00±0.1	0.78	ZB+RS	
S15	0.300	-1.15±0.1	0.40	ZB+RS	
S16	0.292	-1.80±0.1	0.24	RS	Sr-rich alloy
S19	0.137	-1.97±0.1	0.31	RS	
S25	0.000	-2.32±0.1	0.15	RS	SrS

Fig. 5. Summary of experimental results for the $Zn_xSr_{1-x}S$ thin films and structure models with composition. Composition x and structure phase are determined by EDX and XRD, respectively. ZB and RS indicate zincblende and rocksalt structure, respectively. Relative peak intensity ratios are accurate to about ±5%.

586

chemical shifts. As can be seen in Fig. 2, the binding energies of S $2p_{3/2}$ bonded in $Zn_xSr_{1-x}S$ compounds vary with composition for Sr-rich region of $x < 0.3$ and higher Zn composition region of $0.84 \leq x \leq 1$. For these region, a better resolution of the S 2p doublet due to the spin-orbit splitting was clearly observed. For the intermediate range of $0.3 \leq x \leq 0.84$, however, binding energies almost keep constant independent of composition and the splitting of energy with 0.3~0.5eV can be clearly observed at S $2p_{3/2}$ peak instead of the ill-resolved S 2p doublet. This splitting may arise from two chemical states of S, that is, S is bound to Sr-rich $Zn_xSr_{1-x}S$ and to Zn-rich $Zn_xSr_{1-x}S$ compounds. These results is in a qualitative agreement with our results of XRD. In the XPS spectra from the Zn photoelectron lines as shown in Fig. 3, the observation of chemical shifts in Zn-rich films are considered to be due to the influence of cation Sr, not of anion S. This result may also support the phase transition at about $x=0.84$. As for Sr in $Zn_xSr_{1-x}S$, the general tendency of chemical shifts was not observed, which indicates that the bonding states of Sr are easily affected by surrounding chemical environment. Further investigation on the origin of chemical shifts, depth profile and quantitative analysis of these novel films is required.

Film composition by EDX, Structure phases investigated by x-ray diffraction (XRD) and XPS results from S $2p_{3/2}$ core level are summarized in Fig. 5. Based on the above described observations from the XPS spectra as well as the results of XRD, the small structure models are also shown on the right-hand side in this figure.

5. Summary

We have studied the chemical bonding of Zn and Sr atoms to S atoms for $Zn_xSr_{1-x}S$ thin films with XPS in conjunction with crystal phases investigated by XRD. We found that the binding energies of $2p_{3/2}$ core level of S in $Zn_xSr_{1-x}S$ increases nearly linearly with increasing x for Sr-rich region of $x < 0.3$ and Zn-rich region of $0.84 \leq x \leq 1$. For the $Zn_xSr_{1-x}S$ with $0.3 \leq x \leq 0.84$, the splittings of photoelectron lines were observed at S $2p_{3/2}$ photoemission peak. The binding energies of Zn $2p_{3/2}$ shift with composition at $x \geq 0.84$. Above results made it clear that phase transition occurs at about $x = 0.3$ and 0.85 and the $Zn_xSr_{1-x}S$ for the intermediate composition range exist two structure phases, which is in a qualitative agreement with XRD results.

Acknowledgments

The authors wish to thank all members of Industrial Research Institute Tottori Prefecture for EDX measurements. They are also grateful to Professor S. Tanaka for valuable discussion and to Associate Professor S. Kishida of Tottori University for helpful comments on the XPS analysis. One of the authors, S.T.Lee would like to thank Rotary Yoneyama Memorial Foundation, Inc. for the scholarship awarded to him.

References

[1] Okuyama H, Kato E, Itoh S, Nakayama N, Ohata T and Ishibashi A 1995 *Appl. Phys. Lett.* **66** 656-8
[2] Tanaka S 1990 *J. Crystal Growth* **101** 958-66
[3] Viney I V F, Arterton B W, Ray B and Brightwell J W 1994 *J. Crystal Growth* **138** 1055-60
[4] Lee S T, Kitagawa M, Suzukawa R, Ichino K and Kobayashi H, *J. Crystal Growth* (to be published)
[5] Wagner C D, Riggs W M, Davis L E and Moulder J F 1979 *Handbook of X-ray Photoelectron Spectroscopy*, (Perkin-Elmer Corporation, Physical Electronics Division, Minnesota)
[6] Kim T W and Park H L 1993 *Solid State Communications* **85** 635-7
[7] Sporken R, Sivananthan S, Reno J and Faurie J P 1988 *J. Vac. Sci. Technol.* **B6** 1204-7
[8] Berrie C E and Langell M A 1991 *Surf. Interface anal.* **17** 635-40
[9] Yamamoto M, Negishi T, Igarashi J and Ikoma H 1994 *Jpn. J. Appl. Phys.* **33** 4820-8
[10] Phillips J C 1970 *Reviews of Modern Physics* **42** 317-56

Inst. Phys. Conf. Ser. No 145: Chapter 3
Paper presented at 22nd Int. Symp. Compound Semiconductors, Cheju Island, Korea, 28 August–2 September 1995
© *1996 IOP Publishing Ltd*

Novel approach to enhance the optical property in AlGaAs and InGaAlP by natural ordering during growth

A. CHIN[a], B. C. Lin[a], G. L. Gu[b], K. Y. Hsieh[b], M. J. Jou[c], and B. J. Lee[c]

[a]Department of Electronics Engineering, National Chiao Tung University, Hsinchu, Taiwan
[b]Inst. of Materials Science & Engineering, National Sun Yet-Sen Univ., Kaohsiung, Taiwan
[c]Opto-Electronics & Systems Labs., Industrial Tech. Research Institute, Hsinchu, Taiwan

ABSTRACT. Long range alloy ordering is observed by cross-sectional TEM in both (111)A and (111)B AlGaAs. We have measured a 31 meV photoluminescence (PL) peak energy red shift of (111)A and (111)B $Al_{0.30}Ga_{0.70}As$ to that of (100), while the PL integrated intensity in (111)A and (111)B is near an order of magnitude larger than that of (100). The ordering effect is reduced at a high growth temperature of 700 °C and the best PL linewidth of 17 and 3 meV are obtained in (111)A $Al_{0.40}Ga_{0.60}As$ and (111)B $Al_{0.30}Ga_{0.70}As$, respectively. In the $In_{0.6}Ga_{0.4}P$/InGaAlP MQW, similar ordering related PL peak energy red shift of 37 meV and ~2 times enhanced PL integrated intensity are observed for 0° sample to that of 15°.

1. Introduction

The optical properties of AlGaAs and InGaAlP are strongly dependent on the growth conditions such as growth temperature, background moisture, and oxygen contents. This is due to the high reactivity of Al atoms with moisture and oxygen in the reactor, and the slow surface migration velocity of Al adatoms. In spite of reducing the background moisture and oxygen content in the reactor, it is also necessary to growth these compounds at high growth temperatures in order to enhance the surface migration velocity of Al adatoms. However, to growth these compounds at high temperatures are relatively difficult. The re-evaporation rates of As and P are rather high at high growth temperatures that may create group III rich compounds and degrade the optical properties. At high growth temperatures, the group III atoms such as In and Ga will also re-evaporate which makes the control of Al composition difficult. It is therefore necessary to achieve high optical quality of AlGaAs and InGaAlP at low growth temperatures. In this report we have shown a novel approach-growth induced natural ordering [1-5] to achieve this goal. The improved optical property is believed to be due to the increased quantum confinement and recombination of photo-generated carriers by the ordered microstructure.

588

2. Experimental

$Al_xGa_{1-x}As$ (x=0.3 to 0.4) epitaxial layers were grown by MBE on (100), (111)A, and
(111)B. Two growth temperatures of 640 and 700 °C were used to study the material
quality and ordering effect. Typical thickness of the epitaxial $Al_xGa_{1-x}As$ was 1.0 μm.
The growth rates were 1.00, 1.43, and 1.67 μm/hr for GaAs, $Al_{0.3}Ga_{0.7}As$, and
$Al_{0.4}Ga_{0.6}As$ respectively. Strained multiple quantum well (MQW) laser diode structure
of $In_{0.6}Ga_{0.4}P/In_{0.5}(Ga_{0.6}Al_{0.4})_{0.5}P$ was grown by MOCVD on 0°, 2°, 10°, and 15°
(100) GaAs. The active region consists of three periods of 70 Å $In_{0.6}Ga_{0.4}P$ well and 150
Å $In_{0.5}(Ga_{0.6}Al_{0.4})_{0.5}P$ barrier. The top p^+ $In_{0.5}Ga_{0.5}P$ and GaAs contact layers were
etched away before photoluminescence (PL) measurement. The growth temperatures were
700 and 760 °C for these materials. PL was used to characterize the optical properties of
the epitaxial layers. PL was measured at ~15 K using an Ar^+ ion laser at 488 nm as the
excitation source. All the measurements were done under identical conditions in order to
compare the relative PL intensities.

3. Results and discussion

Fig. 1 shows the PL spectra of simultaneously grown (111)A, (111)B, and (100)
$Al_{0.30}Ga_{0.70}As$, at a growth temperature of 640 °C. For (100) AlGaAs, an excitonic
related 1.897 eV transition and an impurity-band or donor-acceptor transition centered at
1.863 eV are observed. However, there is only one broad PL peak centered at 1.866 eV
for both (111)A and (111)B AlGaAs. The PL transitions are not due to impurity and are
confirmed by both Hall and excitation intensity dependent PL measurement. A 31 meV red
shift of PL peak energy is measured for (111)A and (111)B $Al_{0.30}Ga_{0.70}As$ to (100).

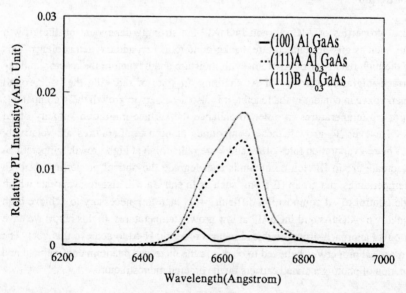

Fig. 1. Low temperature (15K) photoluminescence spectra of 1.0 μm thick 640 °C grown
$Al_{0.30}Ga_{0.70}As$ on (100), (111)A, and (111)B GaAs substrates. The excitation intensity was 1 W/cm².

In both cases, the red shift of PL peak energy is associated with near an order of magnitude enhancement of PL integrated intensity.

We have done cross-sectional Transmission Electron Microscopy (TEM) to understand such PL intensity enhancement and peak energy red shift. Figs. 2(a) and 2(b) show the cross-sectional TEM of (111)A and (111)B AlGaAs, respectively. It is shown that there are superstructures of Al-rich (bright image) and Ga-rich (dark image) AlGaAs in both (111)A and (111)B orientations. In contrast, there is no such superstructure observed in the simultaneously grown (100) AlGaAs (not shown). These naturally ordered AlGaAs microstructures increase the quantum confinement of photo-generated carriers and therefore increase the PL intensity. These Ga-rich domains also make the PL peak energy red shift as compared to the randomly alloyed AlGaAs. The broad PL emission is due to the size fluctuation of ordered domain.

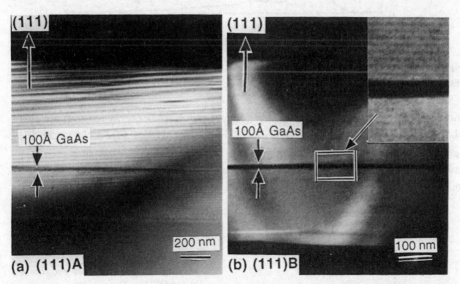

Fig. 2. Cross-sectional view, dark field TEM image of 1.0 μm thick 640 °C grown $Al_{0.30}Ga_{0.70}As$ on (a) (111)A GaAs, (b) (111)B GaAs. A 100Å GaAs quantum well was inserted in AlGaAs in order to enhance the contrast.

The naturally ordered superstructure in both (111)A and (111)B AlGaAs can be disordered or reduced at a higher growth temperature of 700 °C. Fig. 3 shows the PL spectra of 700 °C grown (111)A $Al_{0.40}Ga_{0.60}As$ and (111)B $Al_{0.30}Ga_{0.70}As$. PL linewidth of 17 meV and 3 meV are obtained in (111)A $Al_{0.40}Ga_{0.60}As$ and (111)B $Al_{0.30}Ga_{0.70}As$, respectively. To our best knowledge, these values are the narrowest PL linewidths reported in these orientations. The reduced PL linewidth is related to the increased kinetic energy of adatoms that randomizes the ordered AlGaAs. More detailed TEM analysis will be published elsewhere [6].

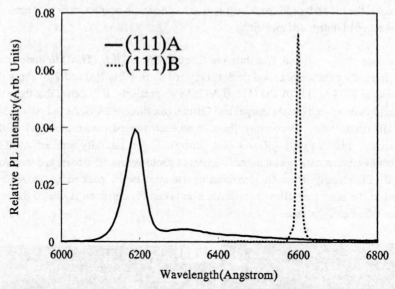

Fig. 3. 15K photoluminescence spectra of 1.0 μm thick 700 °C grown (111)A $Al_{0.40}Ga_{0.60}As$ and (111)B $Al_{0.30}Ga_{0.70}As$.

Similar ordering related PL intensity enhancement and PL peak energy red shift are also observed in $In_{0.6}Ga_{0.4}P/In_{0.5}(Ga_{0.6}Al_{0.4})_{0.5}P$ strained MQWs. Fig. 4 shows the PL spectra of misoriented (100) MQWs grown at 700 °C. The PL peak energy increases monotonically from 1.872, 1.878, 1.937, to 1.967 eV for 0°, 2°, 10°, and 15° misorientations respectively.

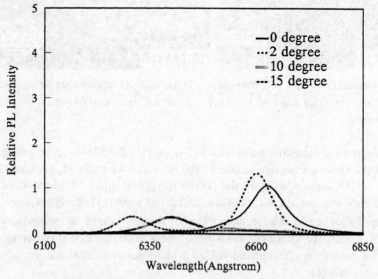

Fig. 4. Orientation dependent 15K PL spectra of $In_{0.6}Ga_{0.4}P$ strained MQWs grown at 700° C. The excitation intensity is at 10 W/cm².

The 95 meV PL peak energy difference among 0° and 15° is related to the ordering effect. [5] The PL integrated intensity is decreased monotonically with the increased degree of misorientation. The 0° on-axis sample shows the highest PL intensity and is ~3 times higher than that of the 15° misorientation.

Fig. 5 shows the PL spectra of the 760 °C grown misoriented (100) $In_{0.6}Ga_{0.4}P$ MQWs. The PL peak energies and relative integrated intensities are summarized in Table 1. The PL intensity is higher for samples grown at 760 °C than those grown at 700 °C. The increased PL integrated intensity with growth temperature is due to the reduced concentration of defect related non-radiative recombination centers. Again, the PL integrated intensity is decreased monotonically with the increased degree of misorientation, and 0° on-axis sample shows the highest PL intensity. The PL peak energies are 1.891, 1.890, and 1.928 eV for 0°, 2°, and 15° misorientations respectively. The amount of red shift among 0° and 15° is reduced to 37 meV as growth temperature is increased to 760 °C.

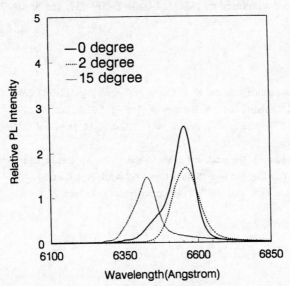

Fig. 5. Orientation dependent 15K PL spectra of $In_{0.6}Ga_{0.4}P$ strained MQWs grown at 760° C. The excitation intensity is at 10 W/cm^2.

Table 1. PL peak energies and relative intensities of misorientated (100) $In_{0.6}Ga_{0.4}P$ MQWs.

	PL peak energy		PL integrated intensity	
miscut	700° C growth	760° C growth	700° C growth	760° C growth
0°(100)	1.872	1.891	2.8	6.1
2°(100)	1.878	1.890	2.6	4.3
10°(100)	1.937	-	1.0	-
15°(100)	1.967	1.928	1.0	3.4

592

4. Conclusions

We have shown alloy ordering related PL peak energy red shift and integrated intensity enhancement both on AlGaAs and $In_{0.6}Ga_{0.4}P$ MQWs. We have measured a 31 meV PL peak energy red shift of (111)A and (111)B $Al_{0.30}Ga_{0.70}As$ to that of (100), which is associated with an order of magnitude enhancement of the PL integrated intensity. The ordering effect is reduced at a high growth temperature and the best PL linewidth of 17 and 3 meV are obtained in (111)A $Al_{0.40}Ga_{0.60}As$ and (111)B $Al_{0.30}Ga_{0.70}As$ respectively. In the $In_{0.6}Ga_{0.4}P$/InGaAlP MQWs, ordering related 95 meV PL peak energy red shift and ~3 times enhanced PL intensity are observed for 0° sample to that of 15°. At a high growth temperature, the red shift of PL peak energy is reduced to 37 meV.

Acknowledgments
We would like to thank the PL support by C. D. Chiang and Y. C. Chien at MRC, CSIST. The work has been supported by NSC (84-0404-E-009-037, and 83-0417-E-009-017) at Taiwan, R. O. C.

References:

[1] K. C. Hsieh, J. N. Baillargeon, and K. Y. Cheng 1990 Appl. Phys. Lett. **57** 2244.

[2] K. Y. Cheng, K. C. Hsieh, and J. N. Baillargeon 1992 Appl. Phys. Lett. **60** 2892.

[3] E. M. Stellini, K. Y. Cheng, P. J. Pearah, A. C. Chen, A. M. Moy, and K. C. Hsieh 1993 Appl. Phys. Lett. **62** 458.

[4] A. Chin T. Y. Chang, A. Ourmazd, and E. M. Monberg 1991 Appl. Phys. Lett. **58** 968.

[5] G. S. Chen, T. Y. Wang, and G. B. Stringfellow 1990 Appl. Phys. Lett. **56** 1463.

[6] A. Chin, B. C. Lin, G. L. Gu, and K. Y. Hsieh, submitted to J. Appl. Phys.

Inst. Phys. Conf. Ser. No 145: Chapter 3
Paper presented at 22nd Int. Symp. Compound Semiconductors, Cheju Island, Korea, 28 August–2 September 1995
© *1996 IOP Publishing Ltd*

Characterization of Semiconductors by Laser-Generated Photocharge Voltage Spectroscopy

Nam-Chun Park

Dept. of Electronics, College of Engineering, Kyungnam University, Masan, Kyungnam 631-701, Korea

A. Abbate

Simulation and Modelling Branch, Technology Division, Benet Labs, Watervliet, NY 12189, USA

P. Das

Electrical, Computers, and System Eng. Dept., Rensselaer Polytechnic Institute, Troy, NY 12180, USA

Abstract. A new technique for evaluating the electrical properties of semiconductor wafers and devices using laser-generated Photo-Charge Voltage(PV) measurements is presented. The technique is based on the measurement of the change in the surface electrical charge induced by a modulated laser beam. This charge is measured capacitivly as a voltage, whose amplitude depends on the surface properties of the sample. In Photocharge Voltage Spectroscopy measurements, the sample is illuminated by both a steady state monochromatic bias light and a pulsed laser. The monochromatic light is used to create a variation in the steady state population of trap levels in the space charge region which does result in a change in the measured voltage. A qualitative analysis of the proposed measurement is presented here along with experimental results performed on GaAs samples passivated with a thin ZnSe film of variable thicknesses. The decrease in surface recombination velocity of GaAs samples as a function of the thickness was measured until a critical thickness is reached.

1. Introduction

Characterization of semiconductor materials and devices plays an important role in the area of solid state device processing. Many techniques have been developed for characterizing defect states at semiconductor surfaces and interfaces, i.e. Capacitance-Voltage[1] and Deep Level Transient Spectroscopy[2] (DLTS). A major drawback of these techniques is that they require the fabrication of contacts on the semiconductor sample. Surface Acoustic Wave techniques have also been useful in characterizing semiconductor films[3]. The development of photonic remote sensing tools has

594

made remarkable progress in the past few years, but their utilization has not been widespread. Various optical techniques have been presented in the literature where a light beam is utilized to excite the sample. Among these, Optical-DLTS[4] and Surface Photovoltage[5] measurements are the most interesting. A new method for the investigation of the surface and interface properties of semiconductor wafers and devices using laser-induced Photocharge Voltage (PV) measurements is presented. The technique is based on the measurement of a small electrical potential difference which appears on any solid body when subjected to illumination by a modulated light beam generated by laser. This voltage is proportional to the induced change in the surface electrical charge and is capacitatively measured. Experiments are easily repeated, and the amplitude of the detected signal depends on the type of material under investigation, and on the surface properties of the sample.

In Photocharge Voltage Spectroscopy measurements, the sample is illuminated by both steady state monochromatic bias light and the pulsed laser. The monochromatic light is used to create a variation in the steady state population of trap levels in the space charge region of semiconductor samples which does result in a change in the measured voltage. This paper is devoted to a description of the Photo-Charge effect and to an experimental demonstration of the PV Spectroscopy technique for the investigation of surface states in semiconductors. In particular results are shown for GaAs samples on which ZnSe films were grown by MOCVD process. For small thicknesses of the film, it is found that ZnSe is a good passivating insulator for GaAs.

2. Experimental Setup

A block diagram of the experimental set-up used in the photo-charge voltage measurement is shown in Fig. 1. The He-Ne laser, used to optically excite the sample, is modulated as a sequence of bursts using either a mechanical chopper or an acousto-optic modulator, with variable pulse length and period. The typical repetition rate for the chopper is 15 Hz with light pulses of approximately 5 m in duration, and the amplitude of the detected signal is of the order of mVolts. The light is incident on the sample, placed in a metal box to shield external electrical disturbance[7]. The output electrical signal is obtained from a metal plate which is pressed against the back surface of the sample. The

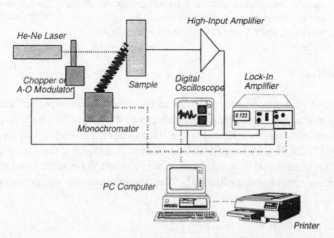

Fig. 1 Experimental set-up used for Photocharge Voltage Spectroscopy Measurements.

ront contact is made by a transparent metal plate with a dielectric spacer. The instrumentation to
nonitor the Photocharge voltage is relatively simple. The output of the sample is connected to a very
ıigh input impedance amplifier with adjustable band pass filter. The output of the amplifier is
:onnected to a digital oscilloscope (Hewlett Packard 54100D) and to a lock-in amplifier (EG&G
Princeton Applied Research 186A). The contacts to the sample, are connected to the input resistance
of the amplifier, so that a variation in the incident photon intensity, as the one induced by the
:hopper, results in a current transient. The lock-in output thus corresponds to the integral over time
of this transient, and thus the total charge induced by the laser illumination. The synchronization to
he oscilloscope and the lock-in amplifier is supplied by the chopper.

A monochromator is also used to vary the surface electrical conditions prior to the laser
ɔulse. The wavelength and intensity of the monochromatic light can be varied and the relative
ımplitude of the PV is measured and stored in the PC computer. The He-Ne laser has a measured
ɔutput power of 8 mWatt, with a calculated photon flux Φ_L of 3×10^{18} cm^{-2}sec^{-1}. A High-Intensity
3aush & Lomb monochromator, with a calculated photon flux Φ_M is 1.6×10^{17} cm^{-2}sec^{-1}, was also
ısed in the presented experiments.

3. Photocharge Effect in Semiconductors

A small electrical potential difference appears on any solid body when subjected to illumination by
ı modulated light beam generated by laser[6]. This voltage was measured on various materials such
ıs conductors, semiconductors, ferrites, ceramics, dielectrics and biological objects. The amplitude
ɔf the detected signal depends on the type of material under investigation, and on the surface
ɔroperties of the sample[7]. The mechanism of generation of the photocharge voltage is different
or the various types of materials, thephotoconductivity and the photovoltaic effect are believed to
ɔe the largest contributors to the photo-charge effect for the case of semiconductor samples. Optical
surface photovoltage has been used in the past to characterize semiconductors[8]. However, the
Photocharge voltage[9] is measured for light of all photon energies and is caused by the
redistribution of charges already present near the surface. Since this voltage is measured
:apacitatively, no net current is present across the sample, thus the photo-induced charge is
:onstrained within the space charge region of the semiconductor sample, resulting in a redistribution
ɔf the total charge compensated by a change in the potential inside the semiconductor space charge
region. In the steady state, the optical generation of free charge carriers is thus balanced by the
recombination at the surface and by diffusion in the bulk.

The response of the semiconductor sample to laser illumination is thus influenced by the
equilibrium conditions existing in the material prior to the laser pulse. The amplitude of the detected
voltage V_L is a function of: (a) the surface space region, through the surface band bending ψ_S, the
width of the space charge region w and its capacitance C_{sc}; (b) the surface states, through the surface
recombination velocity S_p and the density of surface states N_{SS}; (c) the bulk trap levels, through the
lifetime τ_p and average depth L_p. In Photocharge Voltage Spectroscopy measurements, the sample
ıs illuminated by both a steady state monochromatic light, and a pulsed laser. When the sample is
illuminated by a constant monochromatic light of energy $E_M = h \upsilon_M$, and with an intensity $\Phi_M < \Phi_L$,
the conditions of the semiconductor sample, prior to the laser pulse, are varied. In this case the
optical generation is represented by two terms: the first, represented by the term $q\Phi_L/\alpha L$ which
represents the number of free carriers generated per unit time and area by the laser pulse. The second

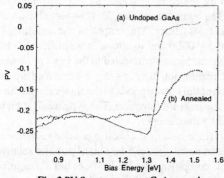

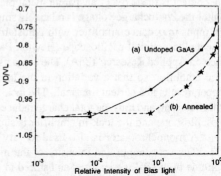

Fig. 2 PV Spectroscopy on GaAs samples   **Fig. 3** PV versus the intensity of the bias beam

term is represented by the product $q\Phi_M \sigma^o N_t$, where σ^o and N_t represent the optical cross section an the density of the trap level involved in the photo-generation process, respectively. A simplifie expression for the Photocharge Voltage is thus given as:

$$V_L = \frac{q}{C_{SC}} \cdot \frac{\dfrac{\phi_l}{\alpha L} \pm \sigma^o\, N_t\, \phi_M}{\left(S_p + \dfrac{L_p}{\tau_B} e^{-q\phi_S/KT} \right)} \tag{1}$$

As it can be seen from eq. 1, the measured voltage is dependent on the surface potential ψ_S throug the exponential factor connected with the recombination current in the bulk, but also through th semiconductor capacitance C_{SC}. It is evident that higher is the value of ψ_S, larger is the amplitude of the PV. This is expected, since the presence of a surface band bending is responsible of th separation of the photogenerated electrons and holes, and of the barrier which impedes diffusion of minority carriers in the bulk.

4. Photocharge Voltage Spectroscopy of GaAs

In the following V_L represents the amplitude of the Photocharge voltage measured using both th monochromator and the laser sources are used, the measured voltage is referred to as V_D. In Fig. the Photocharge voltage V_D is plotted as a function of the energy E_M of the monochromator, fo undoped LEC grown GaAs substrates. Curve (a) represents the PV spectra obtained for a GaA sample as grown, while curve (b) refers to an annealed sample. The rapid change in PV amplitud measured for $E_M \simeq E_g$, indicates transitions from the valence band maximum into the conductio band minimum. The observed reduction in PV is typical of a n-doped behavior of undoped GaA: The sub-bandgap portion of the spectra for curve (a) and (b) is extremely different. For the anneale sample, (curve (b)), no particular change in PV is observed. A large change is instead observed fo the other case, (curve (a)), which suggests the presence of a trap level located at midgap. The bia

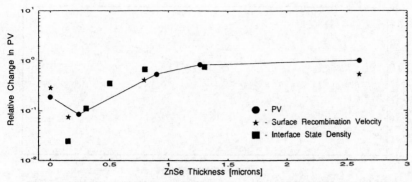

Fig. 4 PV with a bias ligth of 1.0 eV, measured as function of the thickness of the ZnSe film.

monochromatic ligth affects the surface charge balance and thus results in a variation of the measured PV. This effect can thus be used to monitor the change in optically induced charges due to the spatial variation in concentration of impurities in semiconductor samples. For the two samples, the PV has been measured as a function of the bias light intensity (Fig. 3), using a filter in the path of the beam. A value of 1 in the axis corresponds to the full intensity of the monochromatic ligth. It is clear that the effect of the bias light is larger for the undoped sample, and that the change in PV is also more sensitive for a smaller intensity of the bias ligth. These experiments prove our theoretical model of the PV measurement, and the utility of PV measurements for characterizing surface conditions of semiconductor samples.

An increasing interest has been given lately to compound semiconductors such as GaAs for the possibility of obtaining faster electronic devices. Unfortunately the surface properties of GaAs are relatively poor compared to Si, with a detrimental effect on the performance of surface oriented devices. Proper passivation of the unsatisfied bonds at the GaAs surface is increasingly important as the device sizes shrink. Many attempts have been made to improve the surface quality of GaAs using anodic oxides, photochemical treatments and other insulating materials. However these efforts have not yet yielded completely satisfactory results[10]. An alternate method of passivating the GaAs surface is to grow an epilayer of a high quality semiconductor which ties up the dangling bonds at the interface. ZnSe, with its large bandgap, and compatible thermal expansion coefficient and lattice constant (0.25%mismatch), is an attractive choice for use as a passivation layer. It has been shown that thin pseudomorphic ZnSe films, grown by epitaxy on GaAs, reduce the surface recombination velocity (S.R.V.) at the interface. The effect of a ZnSe epitaxial layer on the surface properties of GaAs was studied by measuring the PV signal as a function of the energy of the monochromatic bias beam on samples of different ZnSe epilayer thicknesses. The bandgap of the ZnSe is estimated to be 2.667 eV [11,12], in that range of energies, the V_D is a linear function of the thickness of the ZnSe film. A plot of the measured value of V_D for $E_M \approx 1.0$ eV as a function of the thickness of the ZnSe film is given in Fig. 4. For comparison, the density of the interface defect density[11] and the surface recombination velocity[12], measured as a function of ZnSe thickness, are also plotted in figure. It is interesting to note that the density of surface states, and thus the surface recombination velocity, initially decreases for the pseudomorphic ZnSe layer and increases for thicker samples. This is an expected result, since the small lattice mismatch (0.25%) between ZnSe and GaAs generates a uniform elastic strain between the layer and the substrate. The amount of strain is

598

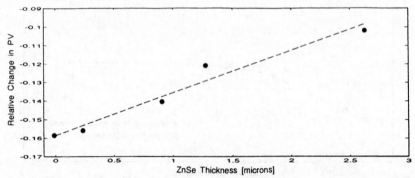

Fig. 5 PV with a bias ligth of 2.7 eV, measured as function of the thickness of the ZnSe film.

proportional to the thickness of the ZnSe film until a "critical thickness" is reached and the strain relieved by the formation of defects such as misfit dislocations [13]. By illuminating the sample wit a monochromatic light beam of energy $E_M \approx 2.7$ eV, a change in PV was observed. This value of V as a function of the thickness of the ZnSe film is given in Fig.5. The straight dotted line was plotte as a visual aid, and was calculated using a linear fit from the experimental data, plotted as circle:

5. Conclusions

A new technique for the characterization of surface properties of GaAs samples using Photocharg Voltage Spectroscopy measurements, was introduced. The change in the surface electrical charg induced by illumination, is capacitatively measured as a function of the wavelength of monochromatic steady-state illumination. GaAs samples passivated with thin ZnSe films wer analyzed using PV technique and an increase in the density of surface states was observed, for ZnS thicknesses above a critical value. Also it is shown that the thickness of the ZnSe film can b evaluated by monitoring the change in PV with a bias ligth of 2.7 eV.

References

[1] C.N. Berglund 1966 *IEEE Trans. Electron Dev.* **13** 701-706
[2] D.V. Lang 1974 *J. Appl. Phys.* **45** 3023-3027
[3] A. Abbate, K.J. Han, I.V. Ostrovskii and P. Das 1993 *Solid State Electron* 321-326
[4] A. Chantre, G. Vincent and D. Bois 1981 *Phys. Rev. B* **23** 3335-3341
[5] J. Lagowski, P. Edelman and M. Dexter 1992 *MRS Symp. Proc.* **261** 217-221
[6] V.I. Pustovoit, M. Borissov and O. Ivanov 1989 *Phys. Lett. A* **135** 59-62
[7] V.I. Pustovoit, M. Borisov and O. Ivanov 1989 *Solid State Communications* **72** 613-617
[8] L. Jastrzebski and J. Lagowski 1981 *Solid-State Sci. Tech.* **128** 1957-1962
[9] P. Das, V. Mihailov, O. Ivanov, V. Gueorgiev, S. Andreev and V.I. Pustovoit, 1992 *IEEE Electron Dev. Lett.* **13** 291-295
[10] C.W. Wilmsen 1989 *Physics and Chemistry of III-V Compound Semiconductor Interfaces* (Plenum New York)
[11] D.J. Olego 1988 *J. Vac. Sci. Technol B* **6** 1193-1200
[12] K.J. Han, A. Abbate, I.B. Bhat, S. Akram and P. Das 1993 *J. Appl. Phys.* **74**, 364-369
[13] D.J. Olego and D. Cammack 1990 *J. Cryst. Growth* **101** 546-551

Inst. Phys. Conf. Ser. No 145: Chapter 4
Paper presented at 22nd Int. Symp. Compound Semiconductors, Cheju Island, Korea, 28 August–2 September 1995

599

SiC High Power Devices

Charles E. Weitzel *, John W. Palmour †, Calvin H. Carter, Jr. †, Karen Moore *,
Kevin J. Nordquist *, Scott Allen †, Chris Thero *, and Mohit Bhatnagar *

* Phoenix Corporate Research Laboratories, Motorola, Inc., Tempe, AZ 85284
† Cree Research, Durham, NC 27713

Abstract. In recent years, silicon carbide has received increased attention because of its potential for high power devices. 4H-SiC Schottky barrier diodes (1200 V) with forward current densities over 732 A/cm^2 have been demonstrated. SiC UMOSFET's (1200 V) are projected to have 15 times the current density of Si IGBT's (1200 V). Sub-micron gate length 4H-SiC MESFET's have achieved f_{max} = 30.5 GHz, f_T = 14.0 GHz, and power density = 2.8 W/mm @ 1.8 GHz. 6H-SiC JFET's have achieved f_{max} = 9.2 GHz, f_T = 7.3 GHz, and 1.3 W/mm power density at 850 MHz.

1. Introduction

The high electric breakdown field of 4 x 10^6 V/cm, high saturated electron drift velocity of 2 x 10^7 cm/sec, and high thermal conductivity of 4.9 W/cm-°K indicate that SiC is a very promising material for high power devices [1]. Until recently the SiC polytype that received the most attention has been 6H-SiC, since it had the best crystal quality of the established polytypes. Another polytype that shows even more potential for high power operation is 4H-SiC, because the electron mobility in 4H-SiC is about twice as high as that in 6H-SiC perpendicular to the c-axis and almost ten times higher in the c-axis direction [2]. Several different power devices - Schottky diodes, UMOSFET's, RF MESFET's, and RF JFET's - will be discussed using simulated and experimental results and compared to similar devices fabricated using the more well established technologies, Si and GaAs.

2. SiC Schottky Diodes

The least complex power device is the Schottky rectifier (Figure 1), which consists of an N$^+$ doped substrate with backside ohmic contact, a lightly doped epitaxial drift layer, and a topside Schottky contact with a high resistivity edge termination [3]. The most important device properties are high reverse breakdown voltage and low forward resistance which yields high forward current density. The lower R_{on} (forward resistance) offered by 4H-SiC Schottky's at breakdown voltages above ≈ 200 V is shown by the modeled results in Figure 2 where R_{on} is plotted versus breakdown voltage. All three curves (Si, GaAs, and 4H-SiC) reach a minimum specific R_{on} below a certain voltage as the thickness of the drift region approaches zero. Experimental results for GaAs and 4H-SiC are included to help validate the modeled results. GaAs diodes (1) and (2) have 200 V and 610 V breakdown voltages, 682 A/cm^2 and 500 A/cm^2 forward current densities at 2 V, and 1.43 mΩ-cm^2 and 2.04 mΩ-cm^2 specific Ron's, respectively. GaAs diode (1) is a Schottky diode and GaAs diode (2) is a

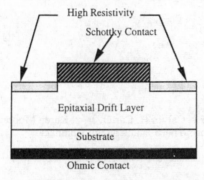

Figure 1. SiC Schottky diode cross section.

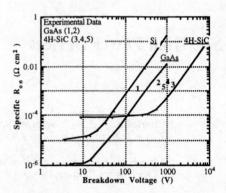

Figure 2. Specific R_{on} for Si, GaAs, and 4H-SiC
Schottky diodes (Modeled and Experimental).

merged pinched Schottky which includes a pn diode in parallel with the Schottky diode [4]. The pn diode reduces R_{on}, but lengthens the reverse recovery time. The forward and reverse I-V curves for a 4H-SiC Schottky diode (3) are plotted in Figure 3. The diode has a 733 A/cm^2 forward current density at 2 V, a 1.5 mΩ-cm^2 specific R_{on}, and a 1200 V breakdown voltage which were achieved with a 10 μm thick drift layer doped 7.5 x 10^{15} cm^{-3}. This particular diode has higher than desired reverse leakage current. Recently, other workers have reported 4H-SiC Schottky diodes (4,5) with 2 mΩ-cm^2 and 1.4 mΩ-cm^2 specific R_{on}'s and 1000 V and 800 V breakdown voltages, respectively [5, 6]. A somewhat more complex device is the pn diode which substitutes a p-type layer and ohmic contact for the Schottky contact. A 2000 V pn diode has been demonstrated with 6H-SiC [7].

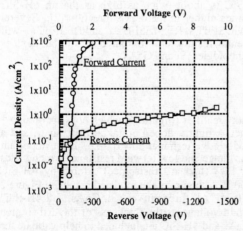

Figure 3. 4H-SiC Schottky diode forward and
reverse current densities.

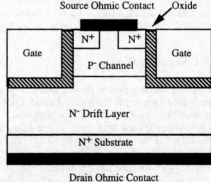

Figure 4. SiC UMOSFET cross section.

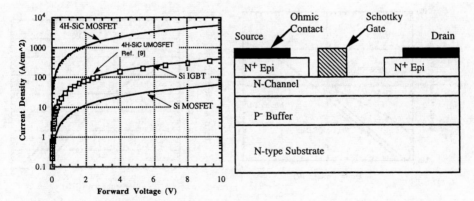

Figure 5. Simulated forward current density of Si MOSFET, Si IGBT, and 4H-SiC UMOSFET with experimental 4H-SiC UMOSFET results [9].

Figure 6. SiC MESFET on conducting substrate.

3. SiC UMOSFET's

The SiC UMOSFET (Figure 4) has a backside drain ohmic contact, N^+ doped substrate, and drift layer similar to the Schottky diode. The upper portion of the UMOSFET contains a p-type channel region and N^+ doped source contacts all of which are enclosed by trenches which contain an SiO_2 gate oxide and polysilicon gate electrode. Current flows from the N^+ source contacts through an inversion layer at both vertical surfaces of the p-type channel region across the drift layer to the N^+ substrate. The forward current density advantage of a 1200 V 4H-SiC UMOSFET over a 1200 V vertical silicon MOSFET and IGBT is shown by the analytical results (solid lines) in Figure 5. The analytical results are based on a design approach that minimizes R_{on} for a given breakdown voltage [8]. The analytically expected current density of a 4H-SiC UMOSFET is 15 times higher than that of a Si IGBT and 100 times higher than that of a Si MOSFET. Experimental data (discrete points) for a 150 V 4H-SiC UMOSFET is also plotted and coincidentally lies on top of the simulated Si IGBT result (Figure 5) [9]. The low current density of the 150 V 4H-SiC UMOSFET is probably a result of very low inversion channel mobility (7-12 cm^2/V-sec) [10], and the immaturity of the technology and device design. Future SiC UMOSFET devices results are expected to show significant improvement.

4. 4H-SiC RF MESFET's

The RF SiC MESFET (Figure 6) is a lateral structure consisting of N^+ source and drain contacts separated by a more lightly doped channel region. The current flow in the channel is controlled by a Schottky gate contact. Device isolation is achieved with a P^- buffer layer on a conducting substrate. The superior power density capability of 4H-SiC RF MESFET's is demonstrated by the simulated and experimental results shown in Figure 7. The device parameters which were important in differentiating the power density of the Si, GaAs, and 4H-SiC MESFET's were low field electron mobility, breakdown field, and electron saturation

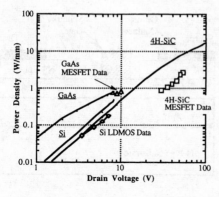

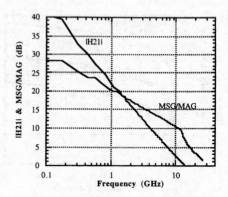

Figure 7. Power density of Si, GaAs, and 4H-SiC MESFET's versus drain voltage.

Figure 8. |H21|, MSG, and MAG of 4H-SiC MESFET on a high resistivity substrate.

velocity. At a doping density of 1×10^{17} cm^{-3} the electron mobility of 4H-SiC is 560 cm^2/V-sec which is slightly lower than that of Si (800 cm^2/V-sec) and significantly lower than that of GaAs (4900 cm^2/V-sec). On the other hand, the breakdown field of 4H-SiC is 10 times that of Si and GaAs, and the saturated drift velocity is 2 times that of Si and GaAs. The simulated results are given credibility by the experimental data [11]. Si LDMOS data was substituted because of the unavailability of Si MESFET data. At low voltages, GaAs MESFET's, which have higher electron mobility, have the highest power density. The higher power density of SiC MESFET's is only achieved at drain voltages higher than those normally used with either Si or GaAs devices. The highest power density reported to date for 4H-SiC MESFET's is 2.8 W/mm at 1.8 GHz with a channel doping of 1.7×10^{17} cm^{-3} and a drain voltage of 54 V (Figure 7) [12]. It is worth noting (Figure 7) that the experimental results of the more mature technologies, Si and GaAs, lie much closer to their simulated curves than do the 4H-SiC experimental results. Lower contact and channel resistances will move experimental SiC results closer to the simulated curve and thereby to higher power densities. In addition, realizing breakdown voltages closer to theoretical limits for a given doping density will extend the experimental data points to higher drain voltages and also to higher power densities.

Recently the use of high resistivity substrates has substantially increased the frequency performance of 4H-SiC MESFET's. Sub-micron gate length 4H-SiC MESFET's fabricated using conducting and high resistivity substrates have achieved f_{max} = 16.3 GHz and 30.5 GHz and f_T = 8.0 GHz and 14.0 GHz, respectively (Figure 8) [13]. Device parameters for the data in Figure 8 are 2×10^{17} cm^{-3} channel doping, 0.45 μm gate length, V_{ds} = 50 V, I_{ds} = 36 mA, and V_{gs} = -6 V. A recently reported 4H-SiC MESFET on a high resistivity substrate achieved a 42 GHz f_{max} [14].

5. 6H-SiC JFET's

High frequency SiC JFET's are of interest for high temperature RF applications because much lower gate leakages should be obtained with a pn junction at high temperature than with a

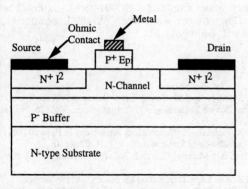

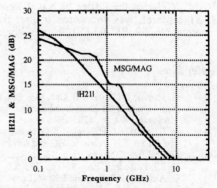

Figure 9. SiC JFET on conducting substrate.　　　Figure 10. |H21|, MSG, and MAG of 6H-SiC JFET.

Schottky gate. The cross section of a SiC RF JFET (Figure 9) is similar to that of the RF MESFET (Figure 6) with the following exceptions [15]. A P$^+$ 6H-SiC epitaxial region with an ohmic contact on top is used in place of a Schottky contact. Ion implanted N$^+$ source and drain contact regions are used in place of the N$^+$ 6H-SiC epitaxial region. The highest frequency performance achieved with a 6H-SiC JFET is f_{max} = 9.2 GHz and f_T = 7.3 GHz (Figure 10). The best power performance is 1.3 W/mm with 40 V drain voltage at 850 MHz. Device parameters for these results are 2.4 x 10^{17} cm^{-3} channel doping, 0.3 μm gate length, 24 mS/mm transconductance, and I_{dss} = 165 mA/mm [16]. SiC JFET performance can be improved significantly by replacing 6H-SiC with higher mobility 4H-SiC and by replacing the conducting substrate with a high resistivity substrate.

6. Conclusions

The simulated and experimental SiC results clearly show the tremendous advantage which this material has over the more mature semiconductor technologies, Si and GaAs, for high power devices. The major challenge facing SiC technology is in the size and defect density of the SiC substrate. In 1995 2" diameter SiC substrates will become commercially available and 3" diameter wafers are on the horizon. This progression to larger and larger substrates is absolutely essential if SiC is to impact the semiconductor market. Substrate defect densities will also have to be reduced. One of the most important defects determined to be fatal to the operation of SiC high voltage devices is the "micropipe" defect, which causes device failures at high fields [17]. However, recent advances in the reduction of micropipe defects has resulted in densities as low as 27 cm^{-2}, which is about 1/10th that of several years ago [18]. Even lower defect densities are needed if SiC is to reach its full potential in the semiconductor market.

7. Acknowledgments

The authors would like to acknowledge the help of S. Macko and C. Capell of Cree Research, processing engineers and technicians at both Motorola and Cree Research, and the management support at both companies. The MESFET research was sponsored in part

604

by M. Yoder of the Office of Naval Research under Contract # N00014-94-C-0293. The JFET research was sponsored in part the Dept. of the Air Force, Wright Laboratories, Contract No. F33615-93-C-1294, monitored by R. Blumgold.

References

[1]   H. Morkoc, S. Strite, G.B. Gao, M.E. Lin, B. Sverdlov, and M. Burns, "Large Bandgap SiC, III-V Nitride, and II-VI ZnSe-based Semiconductor Device Technologies," *J. Appl. Phys.*, Vol. 76, No. 3, 1 August, 1994, pp. 1363-1398.

[2]   W.J. Schaffer, G.H. Negley, K.G. Irvine, and J.W. Palmour, "Conductivity Anisotropy in Epitaxial 6H and 4H SiC," in *Diamond, SiC, and Nitride Wide Bandgap Semiconductors*, C.H. Carter, Jr., G. Gildenblatt, S. Nakamura, and R.J. Nemanich (Eds.), Material Research Society Proceedings, Vol. 339, (MRS, Pittsburgh, PA, 1994) pp. 595-600.

[3]   D. Alok, B. J. Baliga, and P.K. McLarty, "A Simple Edge Termination for Silicon Carbide Devices with Nearly Ideal Breakdown Voltage," *IEEE Electron Device Letters*, Vol. 15, No. 10, October, 1994, p. 394-395.

[4]   L. Tu and B. J. Baliga, "Controlling the Characteristics of MPS Rectifier by Variation of Area of Schottky Region," *IEEE Trans. Electron Devices,* Vol. 40, No. 7, p. 1307-1315, 1993.

[5]   R. Raghunathan, D. Alok, and B.J. Baliga, "High Voltage 4H-SiC Schottky Barrier Diodes," *IEEE Electron Device Letters*, Vol. 16, No. 6, June, pp. 226-227, 1995.

[6]   A. Itoh, T. Kimoto, and H. Matsunami, "High Performance High-Voltage 4H-SiC Schottky Barrier Diodes," *IEEE Electron Device Letters*, Vol. 16, No. 6, June, pp. 280-282, 1995.

[7]   P. G. Neudeck *et al.*, "2000 V 6H-SiC P-N Junction Diodes Grown by Chemical Vapor Deposition," *Appl. Phys. Lett.*, Vol. 64, No. 11, 14 March, 1994 p. 1386-1388.

[8]   M. Bhatnagar, D. Alok and B. J. Baliga "SiC Power UMOSFET : Design, Analysis and Technological Feasibility" in *Silicon Carbide and Related Materials*, M.G. Spencer, R.P. Devaty, J.A. Edmond, M. Asif Khan, R. Kaplan, and M. Rahman (Eds.), Inst. Phys. Pub., Bristol, No. 137, 1993, pp. 703-706.

[9]   J. W. Palmour, J.A. Edmond, H.S. Kong, and C.H. Carter, Jr., "Vertical Power Devices in Silicon Carbide" in *Silicon Carbide and Related Materials*, M.G. Spencer, R.P. Devaty, J.A. Edmond, M. Asif Khan, R. Kaplan, and M. Rahman (Eds.), Inst. Phys. Pub., Bristol, No. 137, 1994, pp. 499-502.

[10]   J.W. Palmour and L.A. Lipkin, "High Temperature Devices in Silicon Carbide," in *Trans. Second International High Temperature Electronics Conference*, D.B. King and F.V. Thome (Eds.), Sandia National Laboratories, June 5-10, 1994, pp. XI-3 - XI-8.

[11]   C.E. Weitzel, "Comparison of Si, GaAs, and SiC RF MESFET Power Densities," to be published in *IEEE Electron Device Letters*, October, 1995.

[12]   C.E. Weitzel, J.W. Palmour, C.H. Carter, Jr., and K. Nordquist, " 4H-SiC MESFET with 2.8 W/mm Power Density at 1.8 GHz," *IEEE Electron Device Letters,* Vol. 15, No. 10, October, 1994, pp. 406-408.

[13]   S.T. Allen, J.W. Palmour, V.F. Tsvetkov, S.J. Macko, C.H. Carter, Jr., C.E. Weitzel, K.E. Moore, K.J. Nordquist, and L.L. Pond, III, "4H-SiC MESFET's on High Resistivity Substrates with 30 GHz f_{max}," *1995 53rd Annual Device Research Conference Digest*, IEEE, pp. 102-103.

[14]   S. Sriram, Westinghouse Science & Technology Center, presented at 1995 Device Research Conference, Charlottesville, Va..

[15]   C.E. Weitzel, J.W. Palmour, C.H. Carter, Jr., K.J. Nordquist, K. Moore, and S. Allen, "SiC Microwave Power MESFET's and JFET's," in *Compound Semiconductors 1994*, H. Goronkin and U. Mishra (Eds.), Inst. Phys. Pub., Bristol, No. 141, 1994, pp. 389-394.

[16]   K. Moore, "DC, RF, and Power Performance of 6H-SiC JFET's," presented at WOCSEMMAD '95, New Orleans, February, 1995

[17]   P.G. Neudeck and J.A. Powell, "Performance Limiting Micropipe Defects in Silicon Carbide Wafers," *IEEE Electron Device Letters*, Vol. 15, No. 2, February, 1994, pp. 63-65.

[18]   C.H. Carter, Jr., V. Tsvetkov, and J.W. Palmour, presented at WOCSEMMAD '95, New Orleans, February, 1995.

Inst. Phys. Conf. Ser. No 145: Chapter 4
Paper presented at 22nd Int. Symp. Compound Semiconductors, Cheju Island, Korea, 28 August–2 September 1995
© *1996 IOP Publishing Ltd*

The fabrication of recessed gate GaN MODFET's

Jinwook Burm, William J. Schaff, and Lester F. Eastman

School of Electrical Engineering and National Nanofabrication Facility, Cornell University, Ithaca, NY 14853

Hiroshi Amano and Isamu Akasaki

Department of Electrical and Electronic Engineering, Meijo University, Nagoya, Japan

Abstract MODFET's were fabricated on GaN/Al$_{0.27}$Ga$_{0.73}$N heterostructure, with and without a gate recess etch. ECR etching was utilized for the gate recess and did not introduce any noticeable damage. After the recess etch, the peak g$_m$ improved from 23mS/mm to 47mS/mm. At the gate lengths of 0.25μm, f$_t$ of 11GHz and f$_{max}$ of 21GHz were measured for non-recessed gates. For 0.4μm recessed gates, f$_t$ of 14GHz and f$_{max}$ of 42.5GHz were achieved. Through the subsequent annealing after the proton bombardment, a severe degradation of the electrical isolation was observed, showing more than factor of 600 reduction of resistance with 450°C 15sec anneal.

1. Introduction

There have been rapid developments and promises on wide bandgap semiconductors for the application of optical devices as well as electrical devices. These developments include p-n junction blue Light Emitting Diodes (LED's) [1,2] and Field Effect Transistors (FET's) [3,4,5]. The employment of wide band gap materials for MODFET's (Modulation Doped Field Effect Transistors) is especially useful for large sheet density from the large conduction band discontinuity, for small gate leakage, and for large breakdown voltage from their wide band gap [6]. With such advantages, the III-nitride MODFET's are an ideal candidate for power amplifiers even at elevated temperature.

An AlGaN/GaN MODFET wafer was processed with and without gate recess etch. This paper deals with the comparison between the two cases and the issues of III-nitride MODFET's fabrication.

2. MODFET fabrication and test

An MOVPE (Metal-Organic Vapor Phase Epitaxy) grown MODFET layer was used for the MODFET fabrication. The layer contains 1600Å unintentionally doped Al$_{0.27}$Ga$_{0.73}$N barrier and GaN channel grown on Sapphire substrate. The electron density of AlGaN layer was about 6×10^{17}cm^{-3}. The sheet density was 1×10^{13}cm^{-2} and the mobility was about

606

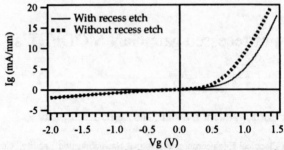

Figure 1. Gate I-V's with recess etch (Solid line) and without recess etch (Dotted line)

$900cm^2/Vs$ at room temperature. The MODFET's were fabricated through mesa isolation, ohmic metal evaporation, and gate metal evaporation. For the mesa isolation, proton bombardment was used. Ti/Au was used for the ohmic contact [3] with very thin (40Å) Ti layer to promote Au adhesion to the wafer. The measured contact resistance of $9.9\Omega mm$ was achieved. Two sets of MODFET's were fabricated, one with, and one without a gate recess etch. For the gate recess, Electron Cyclotron Resonance (ECR) etching was used. The etching gas was the mixture of Ar and CHF_3, with flow rates of 5sccm and 10sccm, respectively. 200W RF power was used with magnetic field confinement of the plasma at the pressure of 5mT [7]. Ti/Pd/Au was used for gate metalization. The gate width of the MODFET's was 100µm. The gate I-V curve showed low leakage Schottky characteristics regardless of the recess etch. (Fig. 1).

The measured DC characteristics (Fig. 2) showed maximum transconductance of 23mS/mm at -2.3V gate bias for a non-recess etched device and 47mS/mm at 1.5V gate bias for a recess etched device. The change of the gate bias of peak g_m (transconductance) was due to the removal of n-doped AlGaN layer and due to the decrease of the distance between the gate and 2DEG (2 Dimensional Electron Gas). As a result, the value of g_m increased at the same time, since g_m is inversely proportional to the distance between the gate and 2DEG [8]. Therefore, the removal of the n-doped layer before depositing the gate metals helped to achieve a better control of 2DEG. For 0.25µm long non-recessed gate, f_t of 11GHz and f_{max} of 21GHz were measured for non-recessed gates (Fig. 3 (a)). For 0.4µm recessed gates, f_t of 14GHz and f_{max} of 42.5GHz were achieved (Fig. 3 (b)). The difference in the gate length was attributed to the resist etching and subsequent resist widening during the ECR recess etch. The ECR recess process did not introduce damage resulting in measurable electrical degradation of the electron supply layer or 2DEG. The improvement of f_t and f_{max} even with a longer gate length after the gate recess etch can be explained by excluding the contribution of carriers in AlGaN barrier. After the gate recess etch, only the high-mobility 2DEG in GaN could contribute to the source-drain current. The estimated resistance of the linear region in I-V characteristics, which is the sum of two contact resistances R_c, and the channel resistance R_{ch}, ($2R_c+R_{ch}$), was $20.6\Omega mm$ for the non-recess etched and $20\Omega mm$ for the recess etched. From these data, any damage to the channel from the ECR recess etch was concluded to be minimal within the error range.

The performance of the MODFET's was mainly hindered by large source resistance [3]. From the contact resistance $9.9\Omega mm$, the intrinsic transconductance, g_{mi}, can be estimated from the equation,

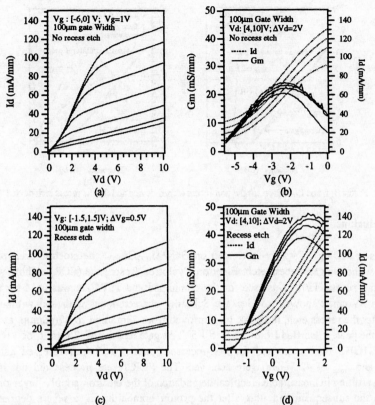

Figure 2. The source-drain current, Id, as a function of drain bias, Vd, and transconductance, G_m as a function of gate bias, Vg, for a non-recess etched device, (a) and (b), and a recess etch device, (c) and (d).

$$g_{mi} = \frac{g_{mo}}{1 - R_s \cdot g_{mo}}, \tag{1}$$

with R_s the source resistance, approximately $(R_c + R_{ch}/2)$ and g_{mo} the extrinsic transconductance measured. The intrinsic transconductance is 30mS/mm for the non-recess etched and 89mS/mm for the recess etched.

3. Isolation degradation

Isolation test patterns were fabricated on another GaN MODFET layer through Ni/AuGe/Ag/Au evaporation and an initial 800°C 10s anneal. The isolation was done by proton bombardment and a good level of isolation, $3 \times 10^6 \Omega$mm, was achieved. After the isolation, the sample was annealed at 450°C for 15s, and showed more than factor of 600 reduction of resistance. Further anneal at 700°C for 10s showed additional resistance degradation by more than factor of 1000. This data indicates that electrical isolation by proton bombardments might be incompatible with high temperature operation of GaN MODFET's.

608

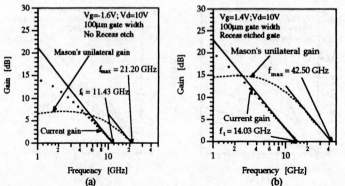

Figure 3. f_t and f_{max} of for (a) the non-recess etched device and (b) the recess etch device

4. Conclusions

Two sets of MODFET's were fabricated on $GaN/Al_{0.27}Ga_{0.73}N$ heterostructures, one with, and one without a gate recess etch. Electron Cyclotron Resonance (ECR) etching was used for the gate recess. Ti/Au was used for the ohmic pad and Ti/Pd/Au was used for the gate. The gate I-V curve showed low leakage Schottky characteristics with and without recess etch. After the recess etch, the peak g_m improved from 23mS/mm to 47mS/mm, as the gate bias for the peak g_m shifted from -2.3V to 1.5V. At gate lengths of .25μm, f_t of 11GHz and f_{max} of 21GHz were measured for non-recessed gates. For 0.4μm recessed gates, f_t of 14GHz and f_{max} of 42.5GHz were achieved. The ECR recess process did not introduce damage resulting in measurable electrical degradation of the electron supply layer or 2DEG. Through the subsequent annealing after the proton bombardment, a severe degradation of the electrical isolation was observed, showing more than factor of 600 reduction of resistance with 450°C 15sec anneal.

Acknowledgment

The work at Cornell University is supported by Office of Naval Research. The work at Meijo University was supported partly by Grant-in Aid for Scientific Research #07505012, #07650025, and #06452114.

References

[1] Amano H, Kito M, Hiramatsu K, and Akasaki I 1989 Jpn. J. Appl. Phys. 73 L212
[2] Nakamura S, Senoh M, and Mukai T 1992 Jpn. J. Appl. Phys. 32 L8-L11
[3] Khan M A, Bhattarai A, Kuznia J N and D. T. Olson 1993 Appl. Phys. Lett. 63 1214-1215
[4] Khan M A, Kuznia J N, Olson D T, Schaff W J, Burm J and Shur M S 1994 Appl. Phys. Lett. 65 1121-1123
[5] Binari S C, Rowland L B, Kelner G, Kruppa W, Dietrich H B, Doverspike K and Gaskill D K to be published in the 1994 ISCS proceedings.
[6] Bykhovski A, Gelmont B, Shur M and Khan A 1995 J. Appl. Phys. 77 1616-1620
[7] Pearton S J, Abernathy C R, Ren F, Lothian J R, Wisk P W, Katz A and Constantine C 1993 Semiconductor Science and Technology 8 310-312
[8] Sze S M 1990 High-Speed Semiconductor Devices (New York:John Wiley & Sons) 302

Inst. Phys. Conf. Ser. No 145: Chapter 4
Paper presented at 22nd Int. Symp. Compound Semiconductors, Cheju Island, Korea, 28 August–2 September 1995
© *1996 IOP Publishing Ltd*

Ion-implantation into α-SiC epilayers and application to high-temperature, high-voltage devices

T.Kimoto, A.Itoh, N.Inoue, S.Yaguchi[1], and H.Matsunami

Department of Electrical Engineering, Kyoto University
Yoshidahonmachi, Sakyo, Kyoto 606-01, Japan

Abstract: N^+ and B^+ implantations into α-SiC epilayers have been systematically investigated. Hot implantation is effective for the reduction of the sheet resistance of N^+-implanted layers, especially in high-dose implantation. 6H-SiC pn diodes formed by N^+ implantation showed a high blocking voltage of 410V at 623K. The reverse characteristics of SiC Schottky rectifiers were significantly improved by employing edge-termination which utilizes highly resistive layers formed by B^+-implantation. Using this technique, 1.4kV Ti/4H-SiC Schottky rectifiers with extremely low power dissipation were realized.

1. Introduction

Silicon carbide (SiC) has been recognized as a vital material for high-power and high-temperature devices, owing to its outstanding properties and the availability of "device-quality" epilayers. Ion implantation is a key technique of selective doping for SiC, in which the diffusion coefficients of impurities are extremely low. However, there have been only a few reports on ion implantation into high-quality α-SiC epilayers [1-4]. In order to clarify the relationship between implantation-induced damages and electrical activation of implanted impurities, systematic studies are required. In this paper, N^+ and B^+ implantations into α-SiC epilayers grown by step-controlled epitaxy [5] are investigated. Using this technique, high-temperature, high-voltage diodes are successfully fabricated with record blocking voltages.

2. Experiments

Samples used in this study were 6H- and 4H-SiC epilayers grown on off-oriented 6H- and 4H-SiC{0001} substrates by atmospheric pressure chemical vapor deposition (CVD) in a SiH_4-C_3H_8-H_2 system [5,6]. N_2 and trimethylaluminum (TMA) were used as n- and p-type dopants, respectively. N^+/B^+ implantations were performed through three steps with different energies of 30~140keV, which forms box profiles of implanted ions. The total dose was varied in the range of 3×10^{13}~2×10^{16}cm^{-2} to investigate the dose effects. Sample temperature during implantation was kept at room temperature or elevated

---

[1] Mitsubishi Heavy Industries, LTD, Kobe, Japan

temperatures of 500~800°C. Post-implantation annealing was carried out in a furnace heated by rf-induction with a gas flow of Ar. The annealing temperature and period were 1000~1500°C and 30min. During the annealing, samples were set on a SiC-coated graphite susceptor. It is known that long-time annealing at high temperatures above 1300°C causes the damage of SiC surface. However, the formation of surface pits or deviation of stoichiometry were not observed in this study.

The lattice damages of implanted layers were monitored by a channeling measurement of Rutherford backscattering spectroscopy (RBS). The electrical properties were characterized by the van der Pauw method. Ni and Al/Ti annealed at 1050°C were used as ohmic contacts for n- and p-type layers, respectively. From SEM observation, the implantation depths were estimated to be 0.45μm and 0.55μm for N$^+$- and B$^+$-implanted samples, respectively.

3. Results and discussion

3.1 N$^+$ and B$^+$ implantations into epilayers

N$^+$ and B$^+$ were implanted into Al-doped (N_a=2×10^{17}cm^{-3}) and N-doped (N_d=2×10^{16} cm^{-3}) epilayers with a thickness of 4~5μm, respectively. Secondary ion mass spectroscopy (SIMS) measurements of N or B in implanted layers revealed that redistribution of implanted ions is negligibly small even if samples are subjected to high-temperature annealing at 1500°C.

Figure 1 shows the total N$^+$ dose dependence of normalized backscattering yield from N$^+$-implanted layers in RBS measurements. Here, the normalized backscattering yield (χ) was defined as the ratio of the yield of an aligned spectrum to that of a random spectrum at the implanted regions (depth=0.1~0.4μm). Open and closed marks denote the data from as-implanted and 1500°C-annealed samples, respectively. In the case of room-temperature (RT) implantation, χ of as-implanted samples reaches 100% when the implant dose exceeds 4×10^{15}cm^{-2}, indicating complete amorphization. Solid phase epitaxial regrowth of this amorphous SiC layer is difficult, which is demonstrated by high χ values of 1500°C-annealed samples (38~87%) in Fig.1. This problem in high-dose implantation could be solved utilizing hot implantation where dynamic annealing takes place during implantation. In fact, crystalline structure was conserved in samples implanted at elevated temperatures even with a 1×10^{16}cm^{-2} dose, and χ was reduced down to about 2%, which almost coincides with a virgin sample.

The authors investigated the annealing temperature dependence of electrical properties of implanted layers, and found that annealing at a higher temperature results in better electrical activation. Thus, the annealing at 1500°C was mainly employed in this study. Figure 2 shows the implant dose dependence of the sheet resistance of N$^+$-implanted layers annealed at 1500°C. Although no significant differences were observed between room-temperature and hot implantations in the low dose region ($\leq$10^{15}cm^{-2}), a reduction of sheet resistance could be achieved in the high dose region ($\geq$10^{15}cm^{-2}) by utilizing hot implantation at 500~800°C. The lowest sheet resistance obtained in this study (550Ω/□) is the best value ever reported in SiC (previously, 843Ω/□ [1]).

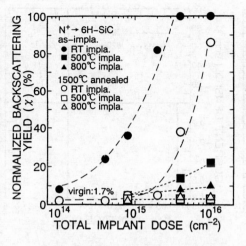

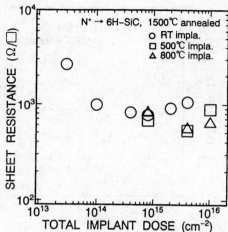

Fig.1 Implant dose dependence of normalized backscattering yield for N$^+$-implanted layers in RBS measurements. The implantation was performed at room temperature, 500°C, and 800°C.

Fig.2 Implant dose dependence of sheet resistance for N$^+$-implanted layers annealed at 1500°C. The implantation was performed at room temperature, 500°C, and 800°C.

The lattice damages of B$^+$-implanted layers exhibited a tendency quite similar to N$^+$ implantation described above. After annealing at 1500°C, the damages were almost removed when the as-implanted layers have crystalline structures. However, the severe damages remain even after 1500°C-annealing, once amorphous layers are formed by implantation. B$^+$-implanted layers were highly resistive which prevented the identification of conduction type (p or n). Figure 3 shows the implant dose dependence of resistivity for B$^+$-implanted layers annealed at 1500°C. The high resistivities (15~120Ωcm) of B$^+$-implanted layers may come from mainly the incomplete ionization of B acceptors because of their rather deep acceptor level (300~400meV) [7,8]. Based on rough estimation, the obtained resistivities turned out to almost coincide with those of B-doped epilayers with the similar B concentrations. The authors believe that these high resistivities are not due to remaining damages but inherent to B-doped SiC. Thus, B$^+$ implantation may be effective to form highly resistive layers rather than to make p$^+$ wells.

3.2 Application to high-temperature, high-voltage diodes

6H-SiC mesa pn junction diodes were fabricated utilizing N$^+$ implantation into p-type epilayers with an acceptor concentration of 4×10^{16}cm^{-3}. The n-layers were formed by N$^+$ implantation (total dose=8×10^{14}cm^{-2}) followed by annealing at 1400~1500°C. The mesa structure was formed by reactive ion etching (RIE) using CF$_4$+O$_2$, and the surfaces were passivated with thermally grown oxides with a thickness of 50nm. The junction area was 7.1×10^{-4}cm^2.

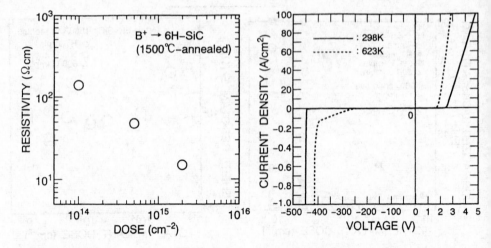

Fig.3 Implant dose dependence of resistivity for B$^+$-implanted layers annealed at 1500°C. The implantation was performed at room temperature.

Fig.4 Current density–voltage characteristics of a 6H-SiC pn junction diode formed by N$^+$ implantation.

Figure 4 shows the current density–voltage characteristics of the diode at room temperature and 623K(350°C). The diode exhibited a high blocking voltage of 450V at room temperature, which is the highest ever reported in SiC pn diodes formed by N$^+$-implantation. The diode demonstrated good rectification with a blocking voltage of 410V at the highest temperature tested, 623K. The mechanism of the negative temperature coefficient of breakdown voltage is not clear at present. The diode exhibited very low reverse leakage current density, 2×10^{-7}A/cm^2 at room temperature and 7×10^{-4}A/cm^2 at 623K at a bias voltage of −100V. The ideality factor under forward bias was 2.1 at room temperature, indicating the recombination current is dominant. The ideality factor decreased down to 1.6 at 623K, which may reflect the increase of diffusion current component owing to the increased intrinsic carrier concentration.

As described in 3.1, B$^+$-implantation resulted in the formation of highly resistive layers. The authors utilized B$^+$ implantation (30keV, 1×10^{15}cm^{-2}) for the edge termination of high-voltage SiC Schottky rectifiers. Figure 5 shows a schematic illustration of an edge-terminated SiC rectifier. In general, electric field enhancement occurs at the edges of Schottky contacts, which leads to a breakdown voltage much lower than the ideal parallel plane breakdown voltage. In Fig.5, the highly resistive layer, which was formed by B$^+$ implantation, promotes the lateral spreading of the potential which reduces electric field at the Schottky periphery.

High-voltage Schottky rectifiers were fabricated using n-type 4H-SiC epilayers, which is the most promising material for efficient power devices [9,10]. The donor concen-

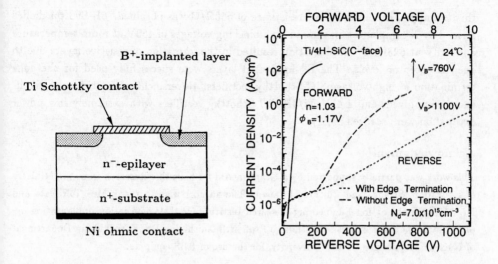

Fig.5   Schematic illustration of an edge-terminated SiC rectifier.

Fig.6   Current density–voltage character-istics of Ti/4H-SiC Schottky rectifiers with and without edge termination.

tration and thickness of n-layers were $7 \times 10^{15} \text{cm}^{-3}$ and $10\mu\text{m}$. The diameter of Schottky contacts was $120\mu\text{m}$.

Figure 6 shows the current density–voltage characteristics of Ti/4H-SiC Schottky rectifiers with and without edge termination. A remarkable improvement in reverse leakage and blocking voltage (≥ 1.1kV) could be achieved, keeping the excellent forward characteristics. Using this technique, a record breakdown voltage of 1.4kV was achieved. Although the edge termination using Ar^+ implantation has been reported [11], the leakage current density of the present rectifier is much lower by two orders of magnitude. Besides, the leakage current did not show significant increase at temperatures as high as 150°C. Although the leakage pass is not clear at present, this study demonstrates a hopeful edge termination utilizing B^+ implantation. Under the forward bias condition, a high current density of 100A/cm^2 can be delivered at a small forward voltage drop of 1.30V. The high breakdown field ($\sim 2 \times 10^6 \text{V/cm}$) and high electron mobility ($\geq 700\text{cm}^2/\text{Vs}$) of 4H-SiC brought about specific on-resistances as low as $2\sim 5\text{m}\Omega\text{cm}^2$. Thus, novel power diodes with high blocking voltages and extremely low power dissipation, which cannot be developed with Si or GaAs, were realized.

4. Conclusions

N^+ and B^+ implantations into α-SiC epilayers have been systematically investigated. Hot implantation at $500\sim 800$°C is effective to suppress amorphization and to promote electrical activation in N^+-implanted layers, especially for high-dose implantation. Using

614

this technique, the lowest sheet resistance of $550\Omega/\square$ was obtained. 6H-SiC pn diodes formed by N^+ implantation showed high blocking voltages of 450V at room temperature and 410V at 623K. B^+ implantation resulted in the formation of resistive layers due to deep B acceptor levels. The B^+-implanted layers were successfully used for the edge termination of high-voltage SiC Schottky rectifiers. Reverse characteristics were significantly improved, and 1.4kV Ti/4H-SiC Schottky rectifiers with extremely low power dissipation were realized.

Acknowledgment

This work was partially supported by a Grant-in-Aid for Scientific Research from the Ministry of Education, Science and Culture of Japan. The authors wish to thank Mr. T.Nakata and Dr. M.Watanabe at Ion Engineering Research Institute for their help in ion implantation experiments. They also express gratitude to Prof. N.Imanishi and Dr. K.Yoshida in Department of Nuclear Engineering, Kyoto University, for the use of RBS equipment.

References

[1] Ghezzo M, Brown D M, Downey E, Kretchmer J, Hennessy W, Polla D L, and Bakhru H, 1992, IEEE Electron Device Lett., **12**, 639-641.

[2] Ghezzo M, Brown D M, Downey E, Kretchmer J, and Kopanski J J, 1993, Appl. Phys. Lett., **63**, 1206-1208.

[3] Rao M V, Griffiths P, Holand O W, Kelner G, Freitas, Jr. J A, Simons D S, Chi P H, Ghezzo M, 1995, J. Appl. Phys., **77**, 2479-2485.

[4] Kimoto T, Itoh A, Matsunami H, Nakata T, Watanabe M, 1995, J. Electron. Mater., **24**, 235-240.

[5] Kuroda N, Shibahara K, Yoo W S, Nishino S, and Matsunami H, 1987, *Ext. Abstr. the 19th Conf. on Solid State Devices and Materials*, p.227-230.

[6] Kimoto T, Nishino H, Yoo W S, Matsunami H, 1993, J. Appl. Phys. **73**, 726-732.

[7] Suttrop W, Pensl G, and Lanig P, Appl. Phys., 1990, **A51**, 231-237.

[8] Vodakov Yu A, Zhumaev N, Zverev B P, Lomakina G A, Mokhov E N, Oding V G, Semenov V V, Simakhin Yu F, 1977, Sov Phys. Semicond. **11**, 214-217.

[9] Palmour J W, Edmond J A, Kong H S, and Carter Jr. C H, 1994, *Silicon Carbide and Related Materials 1994*, Spencer M G, Devaty R P, Edmond J A, Khan M A, Kaplan R, and Rahman M, Eds. (Bristol:Institute of Physics), p.499-502.

[10] Itoh A, Kimoto T, and Matsunami H, 1995, IEEE Electron Device Lett., **16**, 280-282.

[11] Alok D, Baliga B J, McLarty P K, 1994, IEEE Electron Device Lett., **15**, 394-395.

Inst. Phys. Conf. Ser. No 145: Chapter 4
Paper presented at 22nd Int. Symp. Compound Semiconductors, Cheju Island, Korea, 28 August–2 September 1995
© *1996 IOP Publishing Ltd*

Technology of GaAs-Based MMICs for High Temperature Applications

J Würfl, B Janke, S Thierbach

Ferdinand-Braun-Institut für Höchstfrequenztechnik Berlin, Rudower Chaussee 5,
12489 Berlin - Germany,   Tel: +49-30-6392-2690, Fax: +49-30-6392-2642

Abstract: A GaAs MESFET technology for the fabrication of devices specially developed for continuous, reliable operation at high temperatures is presented. The technology is based on highly stable ohmic and Schottky contacts containing WSiN diffusion barriers and is optimized towards minimum temperature induced leakage currents across the substrate or along the semiconductor surface. MESFETs, fabricated by using this technology, have been optimized to match the requirements for continuous operation at high temperatures and have been successfully implemented in high temperature MMICs.

1. Introduction

The high energy bandgap of GaAs enables electronic devices operating at high ambient temperatures of more than 400°C. Furthermore, due to the high electron mobility and the direct band gap, microwave and optoelectronic devices for continuous operation at high temperatures can be realized. This opens new applications for sophisticated sensors and microsystems. At the FBH the process technology for high temperature stable monolithic integrated circuits based on GaAs MESFETs is being developed. The goal is to provide a reproducible process for GaAs devices operating at frequencies up to 30 GHz at temperatures up to 350°C and to establish design rules for high temperature GaAs electronic circuits.

2. High temperature GaAs-MESFET process

Continuous operation at high environmental temperatures means that all degradation processes are considerably enhanced as compared to room temperature. Therefore, special technological processes have to be developed. One most critical issue is the proper design of the metal-semiconductor contacts. Standard GaAs technologies usually employ metallizations that contain highly conductive, chemically inert Au overlayers to reduce parasitic effects. However when being operated at elevated temperatures, the electrical and physical properties of these contacts degrade because of intermetallic interactions with the Au overlayer. This has to be prevented by the incorporation of suitable diffusion barriers. Therefore, for both, the gate Schottky contacts and the drain/source ohmic contacts diffusion barriers are part of the metal layer sequences. The barriers consist of a highly stable, reactively sputtered amorphous WSiN-layer [1]. Figure 1 shows the schematic cross section of Schottky and ohmic metallizations that are used in the high temperature process. The subdivision of the contact structure in layer systems with different electrical and metallurgical functions enables highly temperature stable metal semiconductor contacts without any credits to the electrical performance. The layer system that is placed between the diffusion barrier and the semiconductor determines the electrical properties of the

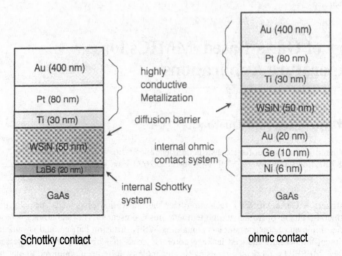

Au (400 nm)

Pt (80 nm)

Ti (30 nm)

WSiN (50 nm)

LaB6 (20 nm)

GaAs

highly
conductive
Metallization

diffusion barrier

internal ohmic
contact system

internal Schottky
system

Au (400 nm)

Pt (80 nm)

Ti (30 nm)

WSiN (50 nm)

Au (20 nm)

Ge (10 nm)

Ni (6 nm)

GaAs

Schottky contact

ohmic contact

Figure 1:   High temperature stable metallizations used for Source/Drain contacts (ohmic) and for the Gate contacts (Schottky)

contacts. The diffusion barrier separates the highly conductive Ti-Pt-Au top layer from the internal layer system to avoid intermetallic interactions and premature degradation at elevated temperatures. For ohmic contacts an optimized, non eutectic Ni-Au-Ge internal layer system is used, the typical contact resistivity is in the 0.1 þmm ranges. The internal layer of the Schottky contacts (gate) consists of a 20 nm thick LaB_6 layer deposited by electron beam evaporation. The LaB_6 layer yields comparably low gate reverse leakage currents since its barrier height on GaAs is around 0,8....0,9 eV depending on the processing and subsequent annealing conditions [2]. This means that the barrier height is much larger as compared to standard Schottky contact systems on GaAs.

A cross section of a typical device, fabricated in high temperature stable technology is shown in Figure 2. The buffer structure of the wafers that are used for high temperature applications contain a p⁻-layer followed by an AlAs/GaAs superlattice system. This arrangement effectively reduces electron scattering from the channel into the substrate at elevated temperatures and thus significantly drops substrate leakage currents. Furthermore it is particularly important to prevent surface corrosion and related detrimental effects. Therefore, an effective passivation of all GaAs and metal surfaces is required. In the high temperature process the free GaAs surface is completely covered by PECVD (Plasma Enhanced Chemical Vapor Deposition) Si_3N_4. All interconnections and passive elements of the MMICs are placed on top of this passivation layer. Metallizations in direct contact with the semiconductor surface are restricted to regions where they are indispensable for device operation (Source, Drain, Gate contacts). This technique enhances device reliability at elevated temperatures and significantly reduces thermally generated leakage currents across the surface [1]. This is especially important if the operation temperature exceeds 250°C. In this case leakage current across the GaAs surface becomes noticeable. Oxygen implantation is used for device insulation on chip. Compared to MESA technology insulation the insulation resistance is increased by one order of magnitude at least. Besides that a quasi planar structure is easily accomplished. Passive components such as NiCr thin film resistors, MIM capacitors and spiral inductors are also implemented.

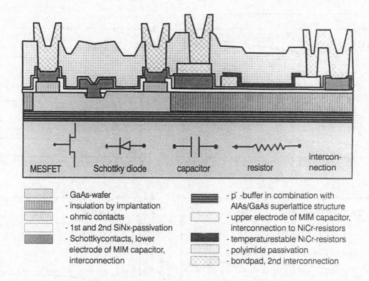

MESFET   Schottky diode   capacitor   resistor   intercon-nection

- GaAs-wafer
- insulation by implantation
- ohmic contacts
- 1st and 2nd SiNx-passivation
- Schottkycontacts, lower electrode of MIM capacitor, interconnection

- p^--buffer in combination with AlAs/GaAs superlattice structure
- upper electrode of MIM capacitor, interconnection to NiCr-resistors
- temperaturestable NiCr-resistors
- polyimide passivation
- bondpad, 2nd interconnection

Figure 2:   Cross section of a typical high temperature MESFET MMIC circuit

3. Devices

The device architecture has been optimized in order to develop high temperature stable MESFET-based monolithic integrated circuits. The optimization primarily concerns the dependence of gate leakage currents, breakdown conditions, transconductance and maximum saturation current on device geometry and doping profile of the active channel. Modelling has been performed by using the physical device simulator SEMICO [4]. A dopant grading in the active channel employing low doped regions under the gate metallization yields best results, however for first technological experiments a rectangular doping profile with a concentration of 5×10^{17} was found to be a good compromise. Based on these simulations MESFETs with 0.5 μm gate lengths have been fabricated. The devices were operated between -196°C (liquid nitrogen) and 500°C (Figure 3). At room temperature the transconductance is about 280 mS/mm: It decreases with increasing temperature to about 100 mS/mm at 500°C. Reliable operation (several hundreds of hours) has been proved for operating temperatures up to 350°C.

The microwave properties of the active devices fabricated in high temperature stable GaAs technology are reflected by the S-parameters. Figure 4 shows the measured S-parameters of a typical MESFET in dependence on temperature and frequency. Gate length and width of the tested device were 0.5 μm and 80 μm respectively. The MESFET was operated at zero gate voltage and 3Volts Drain/Source voltage. Up to temperatures of about 200°C the input and output impedances (S_{11} and S_{22} respectively) of the device remain almost constant. As the temperature increases further to about 250°C the impedances change rapidly. Similarly the transconductance S_{21} gradually decreases as the temperature rises. For the design of MMIC circuits the Maximum Stable Gain (MSG) of the devices is a decisive quantity. At a frequency of 10 GHz the MSG decreases from 9.5 dB to 8.1 dB if the temperature rises from room temperature to 200°C but drops down to only 4.7 dB at 250°C. DC-measurements show that gate leakage becomes noticeable at this temperature and thus plays a decisive role in reducing device performance.

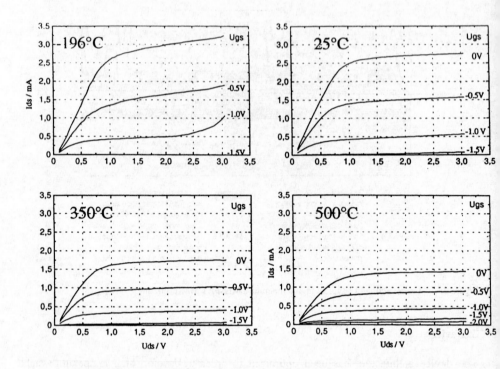

Figure 3 :   DC characteristics of discrete MESFETs fabricated in high temperature stable technology (MESA insulation) at -196°C, 25°C, 350°C and 500°C.

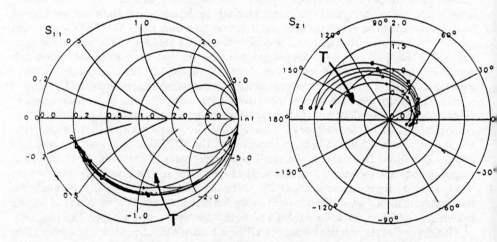

Figure 4:   Scattering parameters (S_{11} and S_{21}) of a typical microwave MESFET in dependence on temperature and frequency. The S-parameters have been determined at 22°C, 100°C, 150°C, 200°C and 250°C. The frequency is scanned from 1 GHz to 50 GHz (The markers are set at 1, 10, 20, 30, 40 and 50 GHz)

To provide the data basis for high temperature MMIC circuit design the temperature dependant equivalent circuit parameters of the active and passive elements were extracted and incorporated in the simulation software. The parameter extraction is based on low frequency electrical data and S-parameters of discrete MESFETs at different temperatures.

A number of microwave and low frequency circuits such as voltage controlled oscillators, mixers, frequency dividers and operational amplifiers have been designed and processed using the high temperature stable technology. Voltage controlled oscillators, mixers and buffer amplifiers have been successfully operated from room temperature to 250°C in some cases up to 300°C. The mechanisms limiting the maximum operation temperature of MMICs are not primarily related with the decreasing transconductance of the MESFETs but are due to the noticeable drift of the input and output impedance of the MESFETs (see figure 4). It is particularly difficult to establish circuit concepts that effectively compensate the drift of input and output impedance of the transistors.

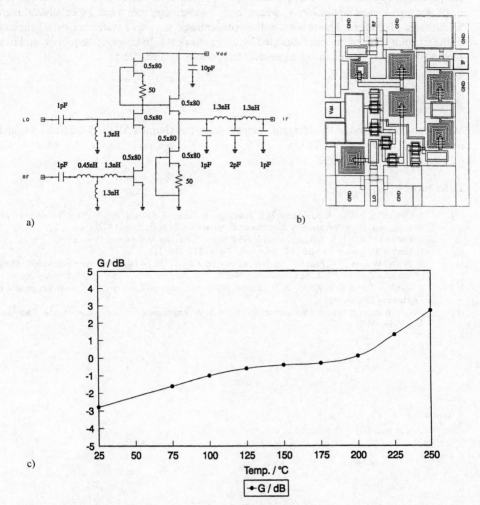

a)

b)

c)

Figure 5: Circuitry of a X-band mixer (a), layout arrangement (b), and conversion gain G versus operating temperature (c)

620

As an example for a MMIC operating at temperatures up to 250°C Figure 5 shows circuitry and layout of a mixer designed for X-band applications. The circuit consists of an cascode type input stage arrangement to enable multiplicative mixing of the RF and the LO signals followed by a source follower output stage and an integrated IF-filter. DC-biasing of the transistors is performed such that maximum conversion efficiency can be obtained. Figure 6 shows the conversion gain G of the mixer as a function of operating temperature. It can be seen that the conversion gain increases with increasing temperature. Simulations show that the observed drift can by related to the variation of input and output impedances of the MESFETs and to a temperature dependant drift of the biasing point for maximum conversion efficiency.

4. Conclusions

At the FBH the process technology for monolithic integrated GaAs devices operating at high temperatures up to 300°C and frequencies up to 30 GHz is available now. This provides the basis for many new applications, where high temperature operation, especially at high frequencies, is required. Future work will be concentrated towards a systematic establishment of the reliability data of devices fabricated by using this new high temperature process and to a method of circuitry design that compensates temperature related effects.

Acknowledgement

We kindly acknowledge the financial support from the German Ministry of Research and Technology (BMFT; 13 MV 0173)

References

[1]   T Böttner K Fricke A. Goldhorn H L Hartnagel A Rappl S Ritter J Würfl 1991 *Proceedings 1st International High Temperature Electronics Conference* Albuquerque NM 77-82
[2]   J Würfl J K Singh H L Hartnagel 1990 *IEEE Reliability Physics 28th Annual Proc.* 87-93
[3]   U Merkel E Nebauer M Mai 1992 *Thin Solid Films* **217** 108-112
[4]   R Stenzel W Klix 1993 *Proceedings 2nd International Workshop on Discrete Time Domain Modelling of Electro-magnetic Fields and Networks* 199-216
[5]   J Würfl B Janke K H Rooch S Thierbach 1993 *1st European Conference on High Temperature Electronics* Proceedings
[6]   J Würfl B Janke K H Rooch S Thierbach 1994 *2nd High Temperature Electronics Conference Charlotte NC* V-33 - V-38.

Inst. Phys. Conf. Ser. No 145: Chapter 4

Paper presented at 22nd Int. Symp. Compound Semiconductors, Cheju Island, Korea, 28 August–2 September 1995

High Power Performance InP/InGaAs Single HBTs

D Sawdai, K Hong, A Samelis, and D Pavlidis

Solid-State Electronics Laboratory, Department of Electrical Engineering and Computer Science, The University of Michigan, Ann Arbor, MI 48109, U.S.A.

Abstract. In this study, single- and multiple-finger HBTs were fabricated from epilayers grown in-house by low-pressure metal organic vapor phase epitaxy (LP-MOVPE). Disilane and DEZn were used for Si and Zn sources for n- and p-type doping. The gas switching sequences were optimized for optimal layers and layer interfaces. The HBTs were fabricated with a double self-aligned wet-etch process designed for reduced device parasitics. Base contacts were Pt/Ti/Pt/Au, which gave a contact resistivity as low as 1.3×10^{-6} Ω-cm^2. High-frequency performance for 1×10-μm^2 emitter HBTs was 93 GHz and 67 GHz for f_T and f_{max}, respectively. DC current densities were as high as 1.4×10^5 A/cm^2. Power and load-pull measurements were carried out at 10 GHz to determine device performance at optimal matching conditions. The maximum power density achieved was 1.37 mW/μm^2 for an HBT with a 4-finger (2×10 μm each) emitter. The maximum output power obtained was 22.58 dBm for a larger device (4 fingers of 5×10 μm each). These characteristics demonstrate good power driving capability for unthinned single HBTs which employ a simple InGaAs collector design.

1. Introduction

InP/InGaAs HBTs offer a number of advantages over AlGaAs/GaAs HBTs resulting from the inherent properties of the InP-based material system. Results up to now on InP/InGaAs single HBTs, however, emphasized primarily high-frequency operation characteristics [1], and only little has been reported on their power performance due to their relatively low breakdown voltages imposed by the low bandgap InGaAs collector [2]. Recent data on power performance of InP-based HBTs have primarily focused on InP/InGaAs/InP or InAlAs/InGaAs/InP double HBTs since their InP collector offers the possibility of higher breakdown voltages [3]. However, the conduction band spike at the base-collector heterojunction of double HBTs tends to degrade the electron transport and subsequently the high speed performance of the device unless special collector designs are used [4]. This paper addresses the design, growth, and fabrication technology of single HBTs with InGaAs collectors and demonstrates good high-frequency and high-power characteristics for such devices.

2. MOVPE growth of InP/InGaAs HBTs

The InP/InGaAs HBTs presented in this work were grown by a modified EMCORE GS3200 low-pressure metal organic vapor phase epitaxy (LP-MOVPE) system at the University of Michigan. Trimethylindium (TMIn) and trimethylgallium (TMGa) were used for In and Ga sources, respectively, and 100% arsine (AsH_3) and phosphine (PH_3) were used for group V sources. The n-type dopant was 1% disilane (Si_2H_6) diluted in hydrogen and controlled by double-dilution lines capable of controlling the dopant flux with a dynamic range over 10^4. The p-type dopant was diethylzinc (DEZn). The susceptor rotation was fixed at a low value of 100 rpm since this offered best thickness and compositional uniformity across the wafer, as well as higher source incorporation efficiencies [5].

All layers were grown on an Fe-doped semi-insulating InP substrate with exact (100) surface. The layer structure used for the HBTs in this study is as follows: 1000 Å undoped InP buffer, 5000 Å n^+ InGaAs subcollector, 5000 Å n^- InGaAs collector, 600 Å p^+ InGaAs base, 100 Å undoped InGaAs spacer, 1500 Å n^- InP emitter, 700 Å n^+ InP contact layer, and 2000 Å n^+ InGaAs cap. The base doping was approximately 1.5×10^{19} cm^{-3}.

Gas switching sequences for normal and inverted InP/InGaAs heterointerfaces were optimized to avoid quaternary formation at the interfaces. The growth cycle employed for this purpose included a transition from InGaAs to InP by (i) switching off the TMIn and TMGa followed by 5 seconds of purging, (ii) exchanging the group V source from AsH_3 to PH_3 with 1 second overlap and (iii) 10 seconds of purging time before introducing TMIn for InP growth. The transition from InP to InGaAs was done following a similar procedure where InGaAs growth started immediately after the group V exchange. This allowed consistent and controlled selective etching during device fabrication since no ternary residuals were formed and led to improved device performance.

In order to minimize zinc diffusion from the heavily zinc-doped base to the emitter, sufficient gas purging time was provided after growing the InGaAs base in conjunction with a 100-Å undoped InGaAs spacer. Silicon encroachment from the heavily-doped subcollector to the lightly-doped collector layer was minimized by purging the dopant dilution line between the two layers. By precisely controlling the dopant profile in the collector regions, the full width of the depletion region was maintained, which greatly improved breakdown characteristics of the devices.

3. Technology of InP/InGaAs HBTs

The devices analyzed in this paper were fabricated using double self-aligned technology with an all wet etching process. After deposition of the emitter ohmic contact (Ti/Pt/Au), the wafer was etched down to the base layer by repeated selective wet etches of the emitter cap, emitter, and spacer layers. The base metal, Pt/Ti/Pt/Au, was deposited self-aligned along two parallel edges of the emitter finger, such that the emitter undercut profile under the emitter metal was identical under all self-aligned edges. The emitter etching times were carefully designed to reliably obtain an emitter undercut so that the lateral distance from the self-aligned base metal to the emitter semiconductor was 0.2-μm. After base deposition, the contacts were annealed at 375° C for 7 seconds, which stabilized the base contacts and lowered their contact resistivities to 1.3×10^{-6} Ω-cm^2.

In order to reduce extrinsic base-collector junction capacitances, the emitter fingers were protected with photoresist patterns while the base semiconductor layer was etched away self-aligned to the base metal contacts. This step effectively reduced the semiconductor junction areas to those directly under the emitter and base metalizations. After the Ti/Pt/Au collector was patterned and lifted off, the HBTs were isolated from one another to the semi-insulating InP level through a wet etch. The same wet etch was used to form trenches in the semiconductor under the emitter and base metalizations, which isolated the semiconductor in the intrinsic device areas from the semiconductor under the airbridge contact pads. The junctions under the airbridge pads are therefore isolated and do not contribute to parasitic capacitances (C_{be} and C_{bc}). Finally, gold airbridges were electroplated to connect the HBTs to interconnects and coplanar testing pads.

The HBTs that were fabricated were all variations on two basic designs: a high-frequency device with thin (2×10 μm^2) emitter fingers and trench-isolated pads for airbridge contacts, and a high-power device with wide (5×10 μm^2) emitter fingers which were directly contacted by the emitter airbridge. Variations on the basic devices were fabricated, including multifinger HBTs. A plan-view photograph of a 10-finger HBT is shown in *Figure 1*.

4. DC and small-signal microwave characterization

The DC characteristics of the fabricated devices proved to be good for single InP/InGaAs HBTs. Breakdown voltages for large devices were as high as $BV_{ce} = 7.2$ V at $I_c = 10$ mA. Maximum collector current densities were also high and about $J_c = 1.4\times10^5$ A/cm^2. The common-emitter I-V plot for a 5×10 μm^2 HBT is shown in *Figure 2*. The collector- and base-current ideality factors from forward Gummel plots were $n_c = 1.3$ and $n_b = 1.4$, respectively, for the same device.

The microwave characteristics were measured on a HP 8510 network analyzer up to 25.5 GHz. Measurements from $|h_{21}|^2$ and U were extrapolated at 20 dB/decade to find f_T and f_{max}, respectively. While showing good breakdown characteristics, these devices also showed good high frequency performance. Maximum values of f_T and f_{max} for 1×10 μm^2, 2×10 μm^2, and 5×10 μm^2 emitters, respectively, were 93 and 67 GHz, 95 and 55 GHz, and 97 and 51 GHz. The degradation of f_{max} with increasing device size indicates that f_{max} is dominated by the $R_b C_{bc}$ time constant. Variations of f_T and f_{max} with respect to bias for a 2×10 μm^2 HBT

Figure 1: Photo of HBT in common-emitter configuration with 10 fingers, each 2×20 μm^2. Electroplated gold forms the airbridge interconnections, soft contact pads of coplanar microwave probes, and heat sinking.

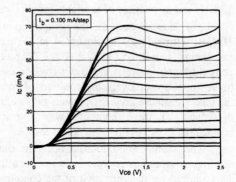

Figure 2: Forward common-emitter I-V plot of HBT with 5×10 μm^2 emitter. $I_b = 0.1$ mA/step.

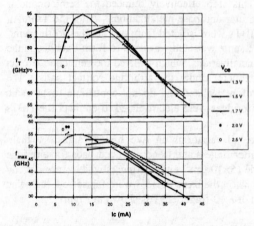

Figure 3: Dependence of f_T and f_{max} on DC bias of a 2×10-μm^2 HBT. Four identical HBTs were measured, two at low current densities and two at high current densities.

Figure 4: Constant load-pull output power contours of $1f\times(5\times10)$ μm^2 HBT at $P_{in} = -1.05$ dBm and $f = 10$ GHz. Values are in dBm.

are shown in *Figure 3*. The drop in performance at high current levels is likely to be due to base push-out and heating effects. A small-signal equivalent circuit [6] was fitted to the 2×10 μm^2 HBT at $I_c = 12.6$ mA and $V_{ce} = 1.5$ V, giving $R_b = 9.8$ Ω, $C_{bc} = 34$ fF, and $\tau_b + \tau_{pcd} \approx 0.6$ ps. The equivalent circuit indicates that the total delay time ($\tau_{ec} = 1 / 2\pi f_T = 1.7$ ps) is also dominated by RC charging time constants.

5. Large-signal power characterization

The HBTs characterized for power performance in this work had emitter areas of $4f\times(5\times10)$ μm^2, $4f\times(2\times10)$ μm^2, and $1f\times(5\times10)$ μm^2. Load-pull measurements were carried out on-wafer using electromechanical tuners in an in-house developed setup. This allowed on-wafer large-signal HBT characterization at the device level and thus extraction of device characteristics under conditions of reduced parasitics. All three transistors were tested at 10 GHz and at a variety of available input power levels and loading conditions.

The dependence of the output power contours on load termination for the $1f\times(5\times10)$ μm^2 device is illustrated in *Figure 4* for input power $P_{in} = -1.05$ dBm. The maximum gain is 9.6 dB at $Z_L = 46.4 + j32.8$ Ω. At lower input power levels, the contours of constant output power were qualitatively quite similar but at lower output power levels. At higher input power levels, the contours of constant output power distorted into ellipses, and the load impedance giving maximum gain shifted closer towards the center. The variation of the shape of these contours was a result of the non-linearities/gain compression characteristics of HBT at high power levels.

The loading conditions at both the input and output of the transistor resulting in maximum output power were evaluated as a function of the power level exciting the device. This allowed for determination of the conditions that provide maximum gain at small-signal power levels as well as those that yield maximum power at larger power levels. To exploit the maximum output power capabilities of all three devices used in this study, a special software routine was employed for controlling the automated tuner system. This routine searched for the termination conditions at a particular input power that resulted in maximum gain and output power. Thus the characterization approach of this work consisted of searching for such conditions at very high power levels for each of the transistors. In this way, the capabilities for maximum power delivery of the devices were analyzed.

Another condition that assured maximum output power delivery was to employ a constant V_{be} in order to bias the base-emitter junction. This allowed the emitter-base junction to stay on at high power levels and led the transistor to high collector current and, consequently, to high output power operation. This also forces Class A operation, however, which limits the power added efficiency of the devices.

The main factor limiting high power performance of single InP/InGaAs HBTs is the use of low V_{ce} values due to the usually low breakdown voltages exhibited by such devices. In the case of our transistors, the breakdown behavior was improved by using 5000-Å collector layers. Another prerequisite ensuring high power driving capabilities is the potential for high output current operation. This was ensured in this work by the design of multi-emitter finger HBTs.

Measurements were carried out at collector current levels between 10 and 200 mA and V_{ce} = 2 V. *Figure 5* shows the output versus input power characteristics for all three transistors. As one observes, the two smaller devices showed a higher small-signal gain of 10 dB versus 5 dB gain of the 4f×(5×10) µm² device at P_{in} = 0 dBm. This is attributed to the higher base-collector capacitance of the latter device.

Moreover, the smallest device in emitter size shows lowest output power driving capabilities. In contrary, the largest device in emitter area shows a maximum output power of 22.58 dBm corresponding to a density of 0.9 mW/µm². This is attributed to the higher current driving capabilities of this transistor compared to the other two devices. On the other hand, the 4f×(2×10) µm² emitter area device showed best output power density of 1.37 mW/µm². Furthermore, a maximum power added efficiency of 33.9% was measured for this transistor, which also is best among the three devices studied in this work. The power density of 1.37 mW/µm² obtained from the four-finger device demonstrates very promising characteristics from unthinned single InP/InGaAs HBTs on 370-µm thick InP substrates.

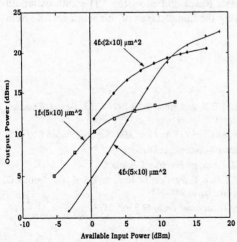

Figure 5: Comparison of power driving capability of HBTs with varying geometries at 10 GHz.

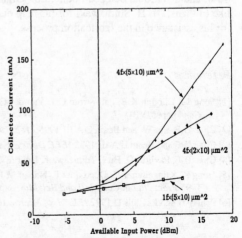

Figure 6: Comparison of current driving capability of HBTs with varying geometries at 10 GHz.

626

Figure 6 shows the dependence of self-biasing effects of the collector current on input power. As expected, the largest emitter area HBT shows maximum current driving capabilities. Moreover, these results indicate significant excursions from the nominal values of DC bias selected for HBT operation and provide further support of their good power driving capability.

6. Conclusion

InP/InGaAs HBTs were fabricated and their power characteristics were evaluated. The studies presented in this paper provide a better insight to the power handling capabilities of InP-based single HBTs, which have not been addressed adequately in the past due to their typically low breakdown voltages. The HBTs were grown in-house by LP-MOVPE and fabricated using a double self-aligned process. Both their small- and large-signal performance was good for InP-based single HBTs, indicating excellent material growth and processing. While still a dominating factor, C_{bc} was significantly reduced through self-aligned etching of the extrinsic base semiconductor and through a special technology involving trench etches to isolate airbridge pads. This allowed for good high-frequency performance of f_T = 93 GHz and f_{max} = 67 GHz from 1×10 μm^2 devices.

Good power performance was demonstrated from the InP/InGaAs HBTs by proper input and output matching selection, which was accurately controlled by means of large-signal on-wafer probing. Breakdown voltages of 7.2 V were demonstrated at 10 μA. Output power densities as high as 1.37 mW/μm^2 were achieved using $4f \times (2 \times 10)$ μm^2 emitter area devices. Moreover, multifinger HBTs were successfully implemented in InP-based HBT technology to increase the total output power handling.

Acknowledgments

This work was supported by URI (DAAL 03-92-G-0109), ARPA COST (MDA 972-94-004), and a National Science Foundation Graduate Research Fellowship. The authors would like to thank Dr. H. Shimawaki for his help during the initial stages of this work and K. Hein for his assistance in the fabrication process.

References

[1] Song J-I, Chough K B, Palmstrom C J, Van der Gaag B P, and Hong W-P 1994 *IEEE Device Research Conf.* paper IVB-5

[2] Chau H-F, Liu W, and Beam E A III 1995 *Proc. 7th Int. Conf. InP and Rel. Mat.* Sapporo, Japan 640-643

[3] Chau H-F and Beam E A III 1993 *IEEE Device Research Conf.* paper IVA-1

[4] Chau H-F, Pavlidis D, Hu J, Tomizawa K 1993 *IEEE Trans. on Electron Devices* 1

[5] Hong K, Klingelhofer C, Ducroquet F, Nuban M F, Bearzi E, Pavlidis D, Krawczyk S K, and Guillot G 1995 *Proc. 7th Int. Conf. InP and Rel. Mat.* Sapporo, Japan 241-244

[6] Pehlke D and Pavlidis D 1992 *IEEE J. of Microwave Theory and Tech.* 12 2367-2373

Inst. Phys. Conf. Ser. No 145: Chapter 5
Paper presented at 22nd Int. Symp. Compound Semiconductors, Cheju Island, Korea, 28 August–2 September 1995
© 1996 IOP Publishing Ltd

AlGaAs/GaAs HBTs with High Fmax

P M Asbeck, M C Ho and R A Johnson

University of California, San Diego, La Jolla, CA USA 92093-0407

Abstract. Approaches to improving the maximum frequency of oscillation of AlGaAs/GaAs HBTs are reviewed. Experimental details are given for an approach based on ion implantation into the sub-collector, and approaches based on multiple epitaxial regrowth. Values of fmax in excess of 200GHz are reported.

1. Introduction

Emerging applications for high speed transistors in millimeter-wave radar, including automotive collision avoidance radar, and lightwave communication systems with bit rates of 40Gb/s and above, require transistors with maximum frequency of oscillation, fmax, reaching above 200GHz. Present production AlGaAs/GaAs HBT fmax values are well below these goals. Fmax is approximately given by the well-known expression

$$f\max = \sqrt{\frac{ft}{8\pi RbCbc}}$$

where Rb is the base resistance, and Cbc the base-collector capacitance. To realize improved fmax, in addition to high ft, requires minimization of Rb and Cbc.

Field effect transistors have demonstrated fmax above 300GHz, obtained using 0.1um gates with T gate structure for very low gate resistance. The input Q of the devices tends to be high, however, making them relatively difficult to match over broad bands. With HBTs there are prospects for maintaining high fmax in a fashion that is both easier to impedance match and easier to manufacture.

2. Strategies for Increased Fmax

A variety of approaches have been pursued for the maximization of fmax, as illustrated in fig.1. A key issue is to eliminate the capacitance associated with the extrinsic base-collector junction, that is, the area underneath the base contacts in the conventional structure.

Transistors with undercut base contacts, in which the extrinsic collector has been removed by etching, have been realized (in SiGe and InP/InGaAs material systems) and shown to have improved fmax. It is regarded as difficult, however, to control the selective etching of the collector to reach the proper position relative to the emitter.

Collector-up devices [1] have been pursued, which eliminate the extrinsic base-collector capacitance associated with the base contact areas. To date their implementation has been difficult because of the problem of restricting the emitter injection to the area underneath the collector.

Transferred substrate devices have recently been reported [2], in which collector size is limited by processing successively both sides of the device (with an intermediate step of transferring the device layers and removing the original substrate). This approach has been demonstrated with InAlAs/InGaAs HBTs.

Deep Implanted HBTs [3]make use of ion implantation damage to render the collector and subcollector semi-insulating in the region underneath the base contact, thereby reducing the extrinsic Cbc to a fringe component only.

Regrown HBTs [4-6], achieved by regrowing the base or the emitter. In these structures, it is possible to make the extrinsic base layer significantly thicker and more heavily doped than the intrinsic base, thereby reducing Rb to achieve high fmax.

Combinations of the above approaches. Very promising structures are HBTs with both regrown bases and extrinsic collector implants, and regrown collector-up HBTs.

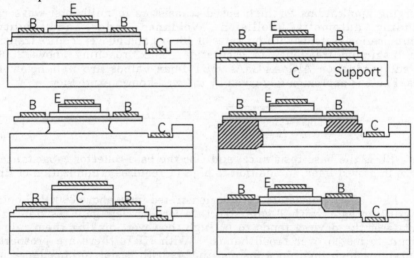

Fig.1 HBT structures for increased fmax: a) conventional emitter-up HBT; b) HBT with undercut base; c) collector-up HBT; d) transferred substrate HBT; e) HBT with deep subcollector implant; f) HBT with regrown base.

In the following, experimental considerations for the realization of deep implanted HBTs and regrown HBTs at UCSD are reported.

3. Deep Collector Implants for Cbc Reduction

Implant damage must reach through the collector and subcollector regions, to a depth greater than 1.5um. With an acceleration energy less than 400KeV, this is attainable with H or He implants. The damage distribution estimated with the TRIM simulation program for the implants used in this work is shown in fig. 2. It has been found possible to tailor the implant to leave the base resistance nearly unchanged (<50% increase) while the subcollector becomes semi-insulating. This tradeoff is possible since damage at the position of the base is 8x lower than that at the subcollector, while the base doping level is 9x higher than that of the subcollector. To account for lateral scattering of the implanted ions and the effects of damage, it was found necessary to space the implant mask 1.2um from the edge of the emitter (done with a non-self-aligned masking operation).

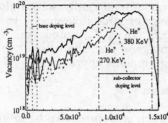

Fig.2 Damage distribution caused by implantation (TRIM simulation)

Using a conventional AlGaAs/GaAs epitaxial layer structure together with the dual-liftoff process technology at Rockwell International, an fmax greater than 200Ghz was obtained, as shown in fig.3. Key to the high fmax was a reduced value of Cbc, which was found through parameter extraction from S-parameter measurements to be lower by 36% than for a baseline device which did not receive the deep collector implant (but which did receive low dose implants of H to compensate the collector only). Fig.4 illustrates current density dependence of Cbc, showing a characteristic drop from electron compensation of the collector doping, followed by increase from Kirk effect. With a properly configured implant, the Kirk effect current density is not increased from that of a baseline device.

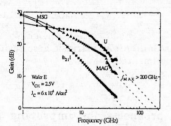

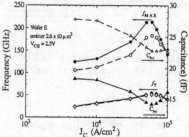

Fig.3 Frequency dependence of microwave gain achieved with deep implanted HBTs

Fig.4 Bias dependence of fmax and Cbc of deep implanted and reference devices.

4. HBTs with Regrown Bases

It was shown by Shimawaki et al that significant advances in fmax (up to 224GHz) could be obtained by regrowing thick heavily doped GaAs base layers on either side of an intrinsic transistor. Regrowth was carried out selectively, with no deposition on SiO2 masking layers, using MOMBE.

Fig.5: Gummel plots of baseline HBT and device with regrown AlGaAs base region.

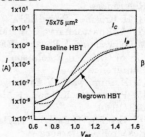

630

In this work, we extend the approach by regrowing layers of AlGaAs. By virtue of the bandgap higher than that of the base, the injection of electrons from the emitter to the regrown base is suppressed. As a result, the emitter edge recombination problem is automatically solved along with the reduction in base resistance. Dc characteristics are shown in fig. 5. An additional extension is to use implants (in self-aligned fashion) to compensate the collector to obtain the advantage of low Cbc without the concern of increasing Rb experienced with the deep implant approach.

5. HBTs with Regrown Emitters

By regrowing the emitter region, structures can be in principle obtained which have very highly scaled emitter dimensions. The approach utilizes sidewall processing similar to double polysilicon self-aligned silicon bipolar devices. Using a three layer base structure in the original growth, it is possible to obtain high conductivity external bases together with thin intrinsic bases with low transit time. The regrown surface constitutes a critical device interface, however. To date, current gain up to 36 has been obtained using selective MOCVD regrowth of a GaInP emitter on GaAs bases.

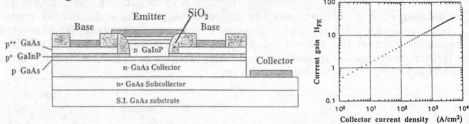

Fig.6: a) Regrown-emitter HBT structure using GaAs/GaInP/GaAs extrinsic base regions; b) current gain vs collector current density achieved experimentally.

6. Conclusions

A variety of HBT structures offer good prospects for improving fmax significantly. An implantation-based approach which is a relatively small departure from the conventional emitter-up design has been used to attain fmax>200GHz. Prospects are better for structures based on multiple growths.

Acknowledgments
This work has resulted from the contributions of many collaborators, chiefly including, from UCSD, M.C.Ho, R.Johnson, N.Y.Li, Y.M.Hsin, S.L.Fu, T.P.Chin, S.Park, C.W.Tu, and P.K.L.Yu; from Rockwell International, W.J.Ho, M.F.Chang and R.L.Pierson; and from Hitachi, T. Nakamura.

References
[1] H. Kroemer, Proc. IEEE 70, 13 (1982).
[2] U.Bhattacharya, M.J.Mondry, G.Hurtz, I.-H.Tan, R.Pullela, M.Reddy, J.Guthrie, M.Rodwell and J. Bowers, IEEE Electr. Dev. Lett. 16, 357 (1995).
[3] M.C.Ho, R.A.Johnson, W.J.Ho, M.F.CHang, and P.Asbeck, 1995 Device research Conference.
[4] H.Shimawaki, Y.Amamiya, N.Furuhata, and K.Honjo, 1993 Device Research Conference.
[5] P.Launay, R.Driad, F.Alexandre, Ph.Legay, AM Duchenois, Inst.Phys.Conf.Ser.No.141, 619 (1995).
[6] S.L.Fu, S.Park, Y.M.Hsin, M.C.Ho, T.P.Chin, C.W.Tu and P.Asbeck, 1994 Device Research Conference

Inst. Phys. Conf. Ser. No 145: Chapter 5
Paper presented at 22nd Int. Symp. Compound Semiconductors, Cheju Island, Korea, 28 August–2 September 1995
© *1996 IOP Publishing Ltd*

Recent Progress in InP-Based HBT Technology

Madjid Hafizi

Hughes Research Lab, 3011 Malibu Canyon Rd, Malibu, CA, 90265

Abstract. Indium Phosphide based heterojunction bipolar transistors have demonstrated advantages over the more widely used GaAs-based HBT's. Recently there has been significant progress in device performance, scaling, reliability and applications of this technology. In this paper I will present a brief review of the device and process and its status to date, this includes a discussion of double heterojunction bipolar transistors for power applications. I will then present examples of key integrated circuit application areas including analog-to-digital converters and optoelectronic IC's. Finally, a brief discussion of reliability and technology qualification for system insertion is presented.

1. Introduction

InP-based heterojunction bipolar transistors (HBT's) have demonstrated performance advantages over other types of HBT's in terms of device RF performance, low-power integrated circuit (IC) applications and optoelectronic receiver IC's. In this material system GaInAs is the material of choice for the base. Generally InP or AlInAs are used as the wide bandgap emitter layer. For the collector there is the choice of GaInAs for single heterojunction bipolar transistors (SHBT) and InP for double heterojunction bipolar transistors (DHBT). Unity gain cutoff frequency f_T and maximum frequency of oscillation f_{max} as high as 200 GHz [1] and 236 GHz [2], respectively, have been reported in this technology for discrete SHBT's. Hughes baseline integrated circuit process offers 2x2 μm^2 emitter transistors with 170 GHz f_T on 3 inch wafers with good uniformity. Using this technology and 130 GHz transistors we have reported record circuit performance [3]. More recently we have evolved this technology into a versatile IC fabrication capability to meet requirements for digital, analog, mixed-signal and optoelectronics applications [4].

For applications requiring high breakdown voltages and low output conductance (such as some analog circuit functions and power amplifiers), InP collector double heterojunction bipolar transistors (DHBT) are used. Record f_T and f_{max} of 160 [5] and 267 [6] GHz, respectively have been reported for discrete InP-based DHBT's. We have applied this technology to analog circuit applications such as analog-to-digital converter (ADC) components [7]. With appropriate device design, we have also achieved [8] base-collector (BV_{CBO}) and collector-emitter (BV_{CEO}) breakdown voltages of 26 V and 32 V. Using this high-breakdown transistor, we demonstrated [9] power cells for C- and X-band applications. In this paper we will present a brief discussion of material and device issues followed by examples of key application areas for InP-based HBT's.

2. Fabrication

SHBT's with AlInAs emitter and GaInAs base and collector can be grown by a conventional solid-source MBE machine. A typical epitaxial profile for the Hughes compositionally graded HBT is shown in Fig. 1. The key features include a GaInAs base thickness of approximately 60 nm doped at 2.5×10^{19} cm^{-3} with Be and a GaInAs collector thickness of 700 nm doped at 5×10^{15} cm^{-3} with Si. The AlInAs emitter was 120 nm thick and doped at 8×10^{17} cm^{-3} with Si. The base-emitter junction was compositionally graded over a distance of 30 nm. This epitaxial structure results in a transistor with a f_T and f_{max} of approximately 100 GHz and a dc current gain of 40. With the compositionally graded emitter-base junction, the transistor exhibits a turn-on voltage, V_{BE}, less than that of the silicon bipolar transistor with nearly ideal base and collector current characteristics.

Devices were fabricated using a triple-mesa process to access the base and collector and to isolate the device. Non-alloyed Ti/Pt/Au was used for the emitter and base ohmic contacts and

AuGe/Ni/Au for the collector contacts. Subsequently, thin-film resistors (TFR) and metal-insulator-metal (MIM) capacitors were fabricated on the substrate. The mesa structure resulting from the device fabrication was planarized by polyimide which was then etched back by reactive ion etching (RIE) to expose the emitter tops. Via holes were subsequently etched in the polyimide using RIE to reach the base and collector and the resistor and capacitor terminals. The second level of metalization was patterned over the polyimide for interconnection. The schematic cross-section of a planarized HBT with resistors is shown in Fig. 2.

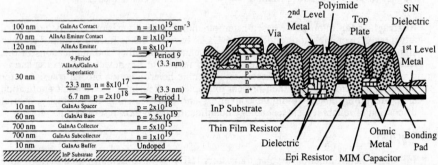

100 nm	GaInAs Contact	$n = 1 \times 10^{19}$ cm^{-3}
70 nm	AlInAs Emitter Contact	$n = 1 \times 10^{19}$
120 nm	AlInAs Emitter	$n = 8 \times 10^{17}$
30 nm	9-Period AlInAs/GaInAs Superlattice	→ Period 9 (3.3 nm)
	23.3 nm $n = 8 \times 10^{17}$	(3.3 nm)
	6.7 nm $p = 2 \times 10^{18}$	→ Period 1
10 nm	GaInAs Spacer	$p = 2 \times 10^{18}$
60 nm	GaInAs Base	$p = 2.5 \times 10^{19}$
700 nm	GaInAs Collector	$n = 5 \times 10^{15}$
700 nm	GaInAs Subcollector	$n = 1 \times 10^{19}$
10 nm	GaInAs Buffer	Undoped
	InP Substrate	

Fig. 1. MBE grown epitaxial profile of a graded single heterojunction transistor.

Fig. 2. Schematic cross-section of HBT IC process, showing the HBT, resistor, capacitor and bond pad.

To increase the f_T of the transistor, we reduce the collector thickness, while for improving the dc current gain reducing the base thickness is very effective. The I_C-V_{CE} characteristic of such a transistor with a collector thickness of 250 nm and a base thickness of 25 nm (doped with Be at 6×10^{19} cm^{-3}) is shown in Fig. 3. The dc current gain was 75 at $J_C = 5 \times 10^4$ A/cm^2. The RF performance of this transistor is shown in Fig. 4. The peak f_T and f_{max} were 170 and 90 GHz respectively. Over a 90% area of a 3 inch wafer, the f_T was in the range of 160 to 170 GHz.

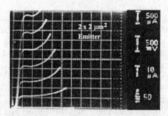

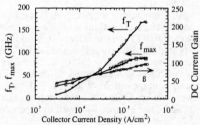

Fig. 3. IC-VCE characteristics of a 2x2 μm2 HBT with 250 nm thick collector and 25 nm thick base.

Fig. 4 Measured f_T, f_{max}, and β for the device of Fig. 3.

3. Device Scaling

For low-power IC applications, InP-based HBT's offer the lowest turn-on voltage of any bipolar transistor (0.6 to 0.7 V). This translates to low-power circuits operating with a 2 to 3 V power supply. We have demonstrated 39.5 GHz D-type flip-flops (in static divider implementation) with 77 mW dc power consumption using our 2x5 μm^2 transistor technology. To further reduce IC power consumption and increase circuit packing density, scaling of device geometries to submicron dimensions is recently being actively pursued.

We have developed [10] a new HBT process to fabricate submicron emitter geometries for applications requiring ultra-low-power consumption and very high-speed performance. We have made devices with an emitter area of approximately 0.3 μm^2 which exhibit a maximum frequency of oscillation, f_{max} of 100 GHz. These transistors are more than an order of magnitude smaller than our current baseline transistors. To achieve these small device geometries we rely heavily on dry etching of semiconductors for defining the device mesas. Figure 5 is a SEM micrograph of a dry etched emitter and base/collector mesa showing the sub-micron circular emitter geometry. This new process is fully self-aligned to minimize the lateral dimensions of the device associated with the base and collector contact regions. In this novel approach the emitter, base and collector ohmic metals are all self-aligned to the emitter mesa. Furthermore, the three ohmic contacts, i.e. emitter, base, and collector are defined and deposited in a single metalization step thereby simplifying the fabrication process. A SEM micrograph of a submicron transistor after polyimide planarization is shown in Fig. 6.

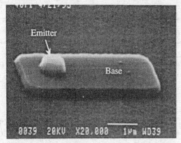

Fig. 5 SEM micrograph of dry etched emitter and base/collector HBT mesas.

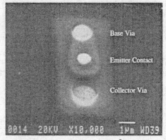

Fig. 6 SEM micrograph of a 0.3 μm^2 emitter HBT after polyimide planarization.

4. DHBT's

We investigated the potential of DHBT's for microwave power applications at X- and C-band frequencies. We fabricated power cells with 480 and 960 μm^2 of emitter area and have characterized them at 9 and 4.5 GHz, respectively. The InP material, which forms the substrate and the collector of our devices, features a higher thermal conductivity and a higher breakdown field than that of GaAs. The epitaxial structure of the DHBT optimized for high breakdown performance is shown in Fig. 7. The main features of the epitaxial structure (compared to the SHBT) is the InP collector of 0.75 μm (n=2.4x10^{16} cm^{-3}) followed by 100 nm of compositional grading and a 40 nm spacer of GaInAs (n=2.4x10^{16} cm^{-3}). The compositional grading at the B-C junction was performed with a 28-period chirped superlattice of InP/GaInAs. Compositional grading at the B-C junction eliminates the potential barrier at the heterointerface between the InP collector and the GaInAs base.

The corresponding I-V characteristic of a DHBT transistor is shown in Fig. 8. This device exhibited collector-emitter breakdown voltage of 26 V and f_{max} of 95 GHz.

100 nm	GaInAs Contact	n = 1x10^{19} cm^{-3}
70 nm	AlInAs Emitter Contact	n = 1x10^{19}
120 nm	AlInAs Emitter	n = 8x10^{17}
30 nm	AlInAs/GaInAs Compositional Grade	n = 8x10^{17}
10 nm	GaInAs Spacer	p = 2x10^{18}
60 nm	GaInAs Base	p = 3x10^{19}
10 nm	GaInAs Spacer	p = 2x10^{18}
40 nm	GaInAs Spacer	n = 2.4x10^{16}
100 nm	Compositional Grade 28-Period Superlattice of GaInAs/InP	n = 2.4x10^{16}
750 nm	InP Collector	n = 2.4x10^{16}
100 nm	InP Subcollector	n = 1x10^{19}
700 nm	GaInAs Subcollector	n = 1x10^{19}
10 nm	GaInAs Buffer	Undoped
	InP Substrate	

Fig. 7 Gas-source MBE grown epitaxial structure of InP-based DHBT for power applications.

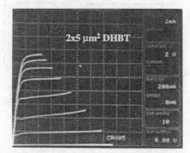

Fig. 8 I_C-V_{CE} characteristics of the DHBT of Fig. 7.

The microwave power performance of large area cells such as the 960 μm^2 multi-finger transistor shown in Fig. 9 is summarized in Table 1.

Size (μm^2)	**240** *(6x2x20)*	**480** *(12x2x20)*	**480** *(12x2x20)*	**960** *(24x2x20)*
Po w *(w/mm)*	0.75 (6)	1.2 (5)	1.37 (5.7)	2.05 (4.3)
PAE %	55	59	65	56
Gain dB	10	9.3	14	10.7
Freq (GHz)	9	9	4.5	4.5

Fig. 9 SEM micrograph of a 24 finger 960 μm^2 emitter power cell.

Table 1 Power performance of different size power cells at 4.5 and 9 GHz.

5. Integrated Circuits

A key application area for InP-based HBT's is high-speed, high-resolution analog-to-digital converters (ADC) which benefit from the high-speed device performance and good device matching (resulting from the low V_{BE} and low R_E) offered by this technology. We are actively pursuing 100 to 200 Ms/s ADC's with 12 to 14 bit accuracy. We have chosen [11] a delta-sigma ($\Delta\Sigma$) ADC architecture as shown in Fig. 10 which includes a modulator and a medium complexity decimation filter implemented in InP HBT technology followed by more extensive and complex filtering performed by a custom silicon IC. The $\Delta\Sigma$ modulator produces a one bit data stream at high frequency. After some initial filtering this data is converted to a lower frequency output for the silicon chip to produce the final digital data. The block diagram of a first order $\Delta\Sigma$ modulator is shown in Fig. 11. In this ADC architecture we rely on over sampling of the input signal to suppress the quantization noise in the signal band. Therefore, the modulator has to operate at a very high frequency (ex. 32 times the input frequency for a 32 time over sampling ratio). We have demonstrated a 2nd order $\Delta\Sigma$ modulator with near ideal performance (for 10-bit resolution) at a sample rate of 3.2 GHz. We have also demonstrated a 1500 transistor accumulator circuit to implement the digital filtering part of the architecture.

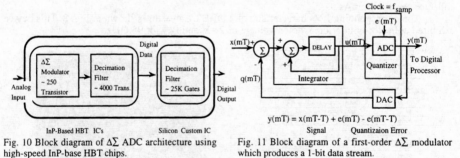

Fig. 10 Block diagram of $\Delta\Sigma$ ADC architecture using high-speed InP-base HBT chips.

Fig. 11 Block diagram of a first-order $\Delta\Sigma$ modulator which produces a 1-bit data stream.

6. Optoelectronics IC's

A unique advantage of InP-based HBT IC technology is the availability of an integrated high-performance photo-detector (PD) for long wavelength (1.3 to 1.55 μm) optical receiver applications. The PIN photo-detector is simply the base-collector (B-C) junction of an ordinary HBT. Our integrated B-C PD's have demonstrated [12] bandwidths of greater than 20 GHz with responsivity of 0.5 A/W for a 1.0 µm thick collector. Our integrated optoelectronics receivers consisting of a PD, transimpedance amplifier and gain block have achieved bandwidth of 20 GHz (showing capability for 40 Gb/s operation). A single transistor receiver is shown in Fig. 12. We are applying our OEIC technology to more complex multi-channel optical front-end receivers for reconfigurable multi-wavelength optical networks. A 1x4 array of front-end receivers was implemented with each channel consisting of a PD with 15 µm diameter, a 2x5 µm² SHBT, a feedback resistor of 200 Ω and an output buffer capable of driving 50 Ω load. The diode had responsivity of 0.5 A/W. The receiver exhibited a transimpedance gain of 180 Ω and a -3 dB bandwidth of 18 GHz.

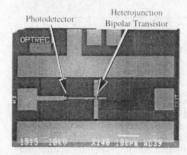

Fig. 12 SEM micrograph of a single transistor receiver circuit.

7. Reliability

The ultimate usefulness of this technology for system applications is determined by its reliability performance. Extensive work in this area have emerged recently with promising results. We have undertaken an extensive and systematic study of reliability of InP-based device and IC technology including physics of device failure and reliability statistics. We have demonstrated [13] that graded junction AlInAs/GaInAs HBT's are free from impurity related instabilities (drift in the turn-on voltage and current gain). This is seen in the plot of I_C and I_B before and after stress shown in Fig. 13. The stability of the turn-on voltage is further evident from the plot of V_{BE} versus stress time (Fig. 14) for a period of approximately 3400 hours.

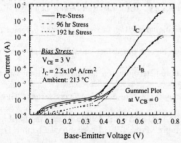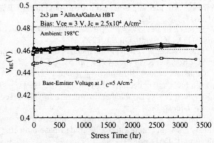

Fig. 13 Gummel plot of collector and base currents before and after stress at 213 C ambient.

Fig. 14 Turn-on voltage as a function of stress time, indicating stability in B-E junction.

The dominant degradation mechanism in these devices is the increase in the leakage current of the base-collector junction which is shown in Fig. 15. Using the increase in the leakage current of the B-C junction as our failure criterion, we have been able to project mean-time-to-failures of mid 10^6 hours at 125°C junction temperatures (Fig. 16) for devices biased with a 3 V collector voltage. Further more we have subjected devices to hydrogen and found no effect. Our tantalum nitride thin-film-resistors also exhibited stability and MTTF exceeding that of discrete devices.

636

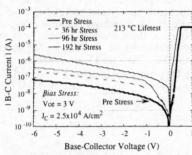

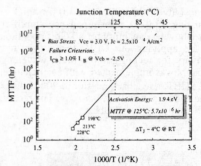

Fig. 15 Increase in collector leakage current of as a function of stress time.

Fig. 16 Projected Mean-time-to-failure (MTTF) of InP-based HBT's.

Conclusions

InP-based HBT technology offers performance advantages for applications in a variety of integrated circuit functions, including digital, analog, mixed-mode, and optoelectronic IC's. The advantages include: inherently higher speed performance, low-power capability, and good reliability. The IC integration complexity is rapidly increasing toward levels required for serious system applications. At the same time process qualification and device/IC reliability is underway to meet system insertion requirements.

Acknowledgments

The Author would like to acknowledge W.E. Stanchina, J.F. Jensen, W.H. Walden, H.C. Sun, and T. Liu for their support. I also acknowledge Y.K. Brown , M.C. Montes, and F. Williams for Their assistance.

References

[1] J.I. Song et al. "Carbon-doped based InP/InGaAs HBTs with f_T= 200 GHz," IEEE Device Research Conf., 1994, paper IVB-5.

[2] H.F. Chau and Y.-C. Kao, "High fmax InAlAs/InGaAs heterojunction bipolar transistors,' IEEE IEDM Tech. Digest, 1993, pp. 783-786.

[3] M. Hafizi, J.F. Jensen, R.A. Metzger, W.E. Stanchina, D.B. Rensch, and Y.K. Allen, "39.5 GHz static frequency divider implemented in AlInAs/GaInAs HBT Technology," IEEE Electron Device Lett., vol. 13, pp. 612-614, Dec. 1992.

[4] W.E. Stanchina et al., "An InP-based HBT Fab for high-speed digital, analog, mixed-signal, and optoelectronic ICs," GaAs IC Symp. Digest, 1995.

[5] Y. Matsuoka et al., "Novel InP/InGaAs double-heterojunction bipolar transistors suitable for high-speed IC's and OEIC's," IPRM Proc., 1994, pp. 555-558.

[6] S.Yamahata et al. "Ultra-high f_{max} and f_T InP/InGaAs double-heterojunction bipolar transistors with step-graded InGaAsP collector," GaAs IC Symp., 1994, pp. 345-348.

[7] J.F. Jensen et al., "Double heterostructure InP HBT Technology for high resolution A/D converters," GaAs IC Symp. Digest, 1994.

[8] M. Hafizi et al., "High-performance microwave power AlInAs/GaInAs/InP double heterojunction bipolar transistors with compositionally graded base-collector junctions," Proc. IEEE International Electron Device Meeting (IEDM), pp. 791-794, 1993.

[9] M. Hafizi, P.A. Macdonald, T. Liu, D.B. Rensch, and T.C. Cisco, "Microwave power performance of InP-based double heterojunction bipolar transistors for C- and X-band applications," Proc. IEEE MTT-S International Microwave Symp., pp 671-674, May, 1994.

[10] M. Hafizi, W.E. Stanchina, and S.C. Sun, "Submicron fully self-aligned AlInAs/GaInAs HBTs for low-power applications," 53rd Device Research Conf., June 1995.

[11] J.F. Jensen et al., "A 3.2 GHz 2nd order $\Delta\Sigma$ modulator implemented in InP HBT technology," IEEE J. Solid-State Circuits, Oct. 1995.

[12] R.H. Walden et al., "An InP-based HBT 1x8 OEIC array for WDM network," 1995 Digest of the LEOS Summer Topical Meeting, pp. 48-49, Aug. 1995.

[13] M. Hafizi et al,. "Reliability of HBT IC Technology for High-Speed, Low-Power Applications," IEEE Trans. Microwave Theory Tech.. Dec. 1995.

Inst. Phys. Conf. Ser. No 145: Chapter 5
Paper presented at 22nd Int. Symp. Compound Semiconductors, Cheju Island, Korea, 28 August–2 September 1995
© *1996 IOP Publishing Ltd*

InGaP/GaAs Drift HBTs with Strained In$_x$Ga$_{1-x}$As Base

Q. J. Hartmann, D. A. Ahmari, M. T. Fresina, P. J. Mares,
J. E. Baker, M. Feng, and G. E. Stillman

Center for Compound Semiconductor Microelectronics, the Department of Materials Science and the Department of Electrical and Computer Engineering, University of Illinois, Urbana, IL 61801

Abstract. In this paper we report on the growth, fabrication and performance of InGaP/GaAs HBTs with a compositionally graded In$_x$Ga$_{1-x}$As base. In characterizing the carbon-doped strained base layer, the interdependence of indium and carbon incorporation during growth was studied. HBTs with the base graded linearly from GaAs at the emitter-base junction to In$_x$Ga$_{1-x}$As at the base-collector junction (where x is the highest mole fraction of In in the base) were fabricated and tested. The dc current gain increased by as much as 50% compared to a device with a standard base design (non-graded base). The unity gain cutoff frequency increased from 69 GHz for the standard device to 83 GHz for the graded base device. The maximum frequency of oscillation of the graded base device also increased compared to the standard device from 183 GHz to 197 GHz. These are the highest f_t and f_{max} reported to date for InGaP/GaAs HBTs.

1. Introduction

Kroemer [1] has suggested that the current gain of npn transistors can be improved by grading the bandgap of the base region from a wide bandgap at the emitter to a narrow bandgap at the collector. The electric field produced in the base adds a drift component to the electron current in addition to the diffusion current normally transporting electrons through the base. By sweeping electrons out of the base, the additional drift component improves the base transport factor, decreases the base transit time (τ_b) and increases the current gain and the device speed. Recent results showing improvements in device performance have increased interest in developing both strained and unstrained graded-base devices. BJTs using strained Si$_x$Ge$_{1-x}$ as the base material have improved the common-emitter current gain up to ten-fold for a given base sheet resistance [2]. GaAs-based HBTs have also been successfully demonstrated using unstrained Al$_x$Ga$_{1-x}$As to grade the base, resulting in increases of 20% in current gain [3].

For the first time, InGaP/GaAs HBTs using strained In$_x$Ga$_{1-x}$As to grade the bandgap of the base have been demonstrated. Although the AlGaAs graded-base technique is relatively easy to implement because Al$_x$Ga$_{1-x}$As is lattice matched to GaAs for all compositions, there are several advantages to using strained InGaAs to create a drift transistor. In the AlGaAs scheme, the base is graded from AlGaAs at the emitter to GaAs at the collector. This reduces the emitter-base discontinuity compared with a non-graded device and lowers the potential barrier that blocks reverse injection of holes into the emitter. In the InGaAs scheme, the base is graded from In$_x$Ga$_{1-x}$As to In$_y$Ga$_{1-y}$As (where x < y) from the emitter to the collector. In this case, the emitter-base discontinuity is at least as large as it is in a non-graded base device, maintaining a large barrier to hole injection. Second, the contact resistance and extrinsic base resistance are higher in the AlGaAs case than in the non-graded case because the bandgap of AlGaAs is larger and the mobility is lower than that of GaAs for a comparable doping level. By using InGaAs, the extrinsic base resistance is smaller for a given doping level compared to GaAs because the majority carrier mobility of InGaAs is higher than GaAs. The mobility is further enhanced for strained InGaAs by the splitting of the valence band and the decrease of the hole effective mass. Finally, AlGaAs presents long term reliability issues due to the high affinity of aluminum for oxygen. Recent reports have suggested that adding indium to the base of carbon-doped GaAs HBTs improves the device reliability[4].

638

2. Experimental Procedure

All of the samples used in this study were grown in a modified Emcore GS-3100 low-pressure metalorganic chemical vapor deposition (LP-MOCVD) reactor. The growth precursors were TEGa, TMGa, TMIn, AsH_3 and PH_3. Diluted CCl_4, 200 ppm in H_2, was used for p-type doping and Si_2H_6 diluted to 18 ppm in H_2 was used for n-type doping. The chamber pressure was maintained at 76 Torr. The base material was grown at a thermocouple temperature of 615° C, using TEGa, TMIn and AsH_3. The TMIn bubbler was held at 1000 Torr and 22° C.

Three samples (referred to as Samples 1, 2 and 3) were grown to investigate the interdependence of the carbon and indium incorporation. In Sample 1, the flow rate of CCl_4 and AsH_3 were held constant at 175 sccm and 20 sccm, respectively, while the TMIn flow rate was stepped from 380 sccm to 0 sccm in 33 twenty-second steps to simulate a linear grade. In Sample 2, the CCl_4 and AsH_3 flow rates were same as in Sample 1, but the indium flow rate was stepped from 101 sccm to 0 cc in 7 two-minute steps. In Sample 3 the indium varied as described in Sample 2, but the AsH_3 and CCl_4 flow rates were adjusted to achieve a constant carbon profile. The flow rate of AsH_3 was evenly stepped from 10 sccm to 20 sccm over the first three steps with a CCl_4 of 200 sccm, then the CCl_4 was decreased from 200 sccm to 175 sccm over the final four steps. The thickness of the samples is approximately 4000 Å. Secondary ion mass spectroscopy (SIMS) measurements of these samples were performed on a Cameca ims5f system using a Cs^+ primary beam and detecting negative secondary ions. The carbon concentration was determined using an ion implant standard and the In was estimated using published relative sensitivity factors.

Four HBTs were grown: one with a standard base design and three with different indium and carbon profiles in the base. The base of the "standard" device is uniformly doped with carbon to a level of 5 x 10^{19} cm^{-3} using a CCl_4 flow rate was 175 cc and an AsH_3 flow rate was 20 cc and does not contain any indium. The base growth conditions for HBTs 1, 2 and 3 are the same as and correspond to the growth conditions for samples 1, 2 and 3, except that the growth time was adjusted to produce a 700 Å base in each device. All of the HBT structures consisted of a 5000 Å GaAs subcollector ($n = 5 \times 10^{18}$ cm^{-3}), a 5000 Å GaAs collector ($n = 3 \times 10^{16}$ cm^{-3}), a 700 Å carbon-doped base, a 700 Å InGaP emitter ($N = 5 \times 10^{17}$ cm^{-3}), a 1300 Å GaAs emitter cap ($n = 5 \times 10^{18}$ cm^{-3}), and 300 Å grade from GaAs to $In_{.05}Ga_{.50}As$ and a 300 Å $In_{.05}Ga_{.50}As$ contacting layer ($n = 1 \times 10^{18}$ cm^{-3}).

Large area devices were fabricated using standard lithographic techniques, selective wet chemical etches, and a single Ti/Pt/Au metallization. Self-aligned high-frequency devices were fabricated using wet chemical etchants for the self-aligned emitter and collector etches and for decvice isolation. The emitter and base contacts were non-alloyed Ti/Pt/Au and the subcollector contact was made of AuGe/Ni/Au. High frequency S-parameters were measured from 1 to 40 GHz using IC-CAP software controlling an HP8510C Network Analyzer and an HP4142B DC Source/Monitor.

3. Results and Analysis

Figure 1 plots the carbon concentration on a log scale and the percentage of indium in sample 1 measured by SIMS. As observed by Stockman et. al. [5], the carbon incorporation decreases as the indium percentage increases, indicating that indium inhibits the incorporation of carbon. In addition, since the carbon concentration decreases linearly on a logarithmic scale, this data shows that the carbon concentration decreases exponentially as the indium concentration increases linearly. The exponential dependence of carbon incorporation on indium flow rate is convenient for device applications because it produces an electric field that is constant as a function of position and is a profile often used to enhance electron transport through the base.

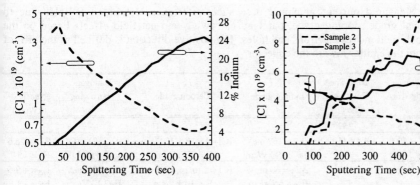

Figure 1. SIMS profile of sample 1.   Figure 2. SIMS of sample 2 and 3.

The doping profiles of Samples 2 and 3 shown in Figure 2 describes the dependence of indium incorporation on CCl_4 and AsH_3 flow rates. In Sample 2, the carbon concentration and indium concentration follow opposite trends, as observed in Figure 1. For Sample 3, adjusting the CCl_4 and AsH_3 flow rates during growth resulted in a nearly flat carbon profile. However, the indium concentration is lower in Sample 3 than in Sample 2. The difference between the indium traces of the two samples is greatest at the high indium concentrations, where low AsH_3 flows and high CCl_4 flows were necessary to achieve $[C] \approx 5 \times 10^{19}$ cm^{-3}. The decrease in indium incorporation is attributed to enhanced etching of InAs by a higher CCl_4 flow rate and a lower AsH_3 flow rate [5].

The profile of the carbon level shown in Figure 1 is important in HBT 1 and HBT 2 which not only have an indium grading, but also a carbon-doping grade that produces an E-field. In addition, the amount of bandgap narrowing decreases as the doping decreases and will produce a third E-field component in the base. The electric field due to the indium grade will move electrons from the emitter to the collector while the field due to changes in bandgap narrowing will move electrons from the collector to the emitter. The field due to the gradient of carbon atoms will move electrons to the collector, but it will also move holes in the opposite direction. The field due to the compositional grade can be calculated from the change in band energy found for strained InGaAs [6]. Assuming the carbon distribution is exponential, the field due to the doping gradient can also be calculated [7]. The magnitude of each field component and the direction electrons and holes will drift in the field (not the direction of the field) is summarized in Table 1 along with the critical parameters used to calculate each component of the field. The positive direction is defined as drift of a carrier from the emitter to the collector. Net fields for electrons and holes are calculated by summing the field components and the sign indicates the direction the electron or hole will drift. The dc current gain and base sheet resistance from large area devices are also included in Table 1.

Drift of both holes and electrons must be considered when analyzing device performance. In all three graded devices, the electrons feel a net field moving them toward the collector. This E-field will decrease the base transit time and increase the current gain. In the case of HBT 3, the field is only due to the compositional grade and electrons are swept toward the collector while holes are not significantly affected by the grading. However, for the devices with a graded carbon profile, the field moves holes toward the emitter and may increase the reverse injection of holes into the emitter. Thus the increase in current gain achieved by decreasing the base transit time may be offset somewhat by the the hole field.

Figure 3 compares the Gummel plots of the standard device and HBT 3 (7% In grade without a doping grade). The collector ideality factor, n_c, is 1.01 for both devices, and the base ideality factor, n_b, is 1.08 and 1.13, respectively, indicating that the strained InGaAs base does not significantly degrade the emitter-base junction.

Table 1. Measured current gain and base sheet resistance, estimated E-fields due to compositional grade, doping grade, bandgap narrowing and net field effects based on the direction each field moves electrons and holes. The positive direction is defined as the drift of a carrier from the emitter toward the collector.

	β @ I_C 1mA	Base Sheet Resistance (Ω/sq.)	Indium grade	Carbon grade	Bandgap Narrowing	Net field (kV/cm)
Standard	65	257	—	—	—	—
HBT 1	50	545	$\Delta E_g = 167$ meV E = 23.9 kV/cm	[C]= 4 to .8 x 10^{19} cm^{-3} E = 5.7 kV/cm	ΔE_g=21 meV E=3.0 kV/cm	E_e = +26.6 E_h = - 5.7
HBT 2	135	375	$\Delta E_g = 60$ meV E = 8.6 kV/cm	[C] = 5 to 2 x 10^{19} cm^{-3} E = 3.1 kV/cm	ΔE_g=20 meV E=2.1 kV/cm	E_e = + 9.6 E_h = - 3.1
HBT 3	97	264	47 meV 6.7 kV/cm			E_e = + 6.7 E_h = —
Direction electrons drift			+	+	−	
Direction holes drift			−			

The theoretical change in base transit time for a graded device compared to a standard device (assuming the base width and doping are the same) is given by the Drift Factor [3],

$$D.F. = \left(\frac{2kT}{\Delta Eg} \right) \left[1 - \frac{kT}{\Delta Eg} + \frac{kT}{\Delta Eg} e^{-\left(\frac{kT}{\Delta Eg} \right)} \right]$$

where ΔE_g is the change in the bandgap across the base. For ΔE_g of 47 meV (7% In grade) D.F. = 0.60 and would result in an increase in the current gain by a factor of 1.67 compared to a standard device, assuming the current gain is dominated by base recombination current. The increase in current gain of HBT 3 compared to the standard device is illustrated in figure 4. The factor of 1.5 (from 65 to 97) increase is close to this theoretical value, indicating that the graded base has significantly shortened the base transit time. Thus, using an InGaAs graded base from 0 to 7% indium and constant carbon profile has resulted in a 50% increase in current gain compared with the standard device with no increase in base sheet resistance.

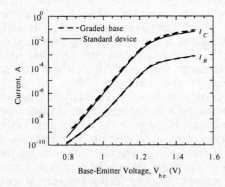

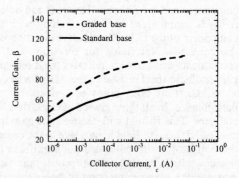

Figure 3 Gummel plot comparing the standard device and HBT 3.

Figure 4. Current gain vs. collector current for the standard device and HBT 3.

Higher indium concentrations were attempted to improve the current gain further. However, the In$_x$Ga$_{1-x}$As/GaAs heterojunction at the base-collector junction affected the transport of electrons out of the base and severely degraded the current gain. In HBT 1, the base was graded to 24% indium resulting in ΔEg $\approx$ 167 meV and the carbon concentration decreased across the base (Figure 1). As expected, the base sheet resistance was much higher than the standard device due to the graded carbon profile, unexpectedly the current gain decreased to 50. However, as Figure 5 shows, the current gain increased when reverse bias was applied to the base-collector junction. A current gain of 150 was obtained for a base-collector voltage of 6 V and a collector current of 1 mA. The variation in current gain with base-collector reverse bias indicates that the heterojunction discontinuity inhibits electron transport through the base and is often observed in double heterojunction bipolar transistors [8]. Figure 6 shows the same data for HBT 2, and indicates that the base-collector heterojunction does not effect base transport characteristics for indium levels of 10% or less.

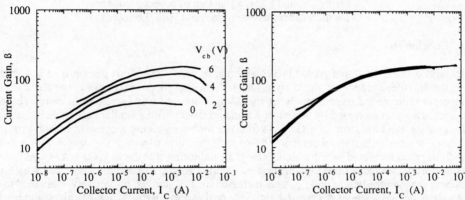

Figure 5. Current gain vs. collector current of HBT 1 for various base-collector voltages.

Figure 6. Current gain vs. collector current of HBT 2 for base-collector voltages stepped from 0 and 6 volts in 2 volt steps.

The high frequency performance of HBT 2 and the standard device are compared in Figure 7. The figure plots the unity current gain cutoff frequncy, f_t, and the maximum frequency of oscillation, f_{max}, as a function of collector current. The collector-emitter voltage, V_{ce}, was 2 V and both devices have an emitter size of 3×10 μm^2. S-parameter data was collected from 1 - 40 GHz using the testing set-up described earlier. The f_t was found by extrapolating H$_{21}$ from 10 GHz at -20 dB/dec and the f_{max} was calculated by extrapolating from Mason's invariant, U, from 20 GHz at -20 dB/dec. The peak f_t of the standard device is 69 GHz. This is very close to the highest reported f_t to date for InGaP/GaAs HBTs of 70 GHz [9], indicating that the material and junction quality of the standard device are high. The peak f_{max} of the standard device is 183 GHz. This is much higher than the previously highest reported value of 116 GHz [10], indicating that the collector-base capacitance and base sheet resistance of the standard device are very low. Using a graded base, the high frequency performance is enhanced further. The f_t increases 20% from 69 GHz to 83 GHz. The f_{max} also increases by 14 GHz from 183 GHz in the standard device to 197 GHz in HBT 2. The increases in f_t and f_{max} are attributed to the decrease in the base transit time caused by the graded base. The improvement in the high frequency performance indicate that the material quality is not severely affected by the strained InGaAs base. These results are the highest reported to date for InGaP/GaAs HBTs and demonstrate the capability of this materials system to compete with AlGaAs/GaAs HBTs.

642

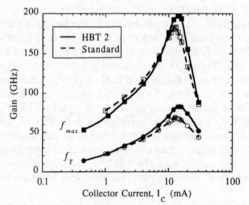

Figure 7. f_t and f_{max} as a function of collector current for the standard device (open) and HBT 2 (closed).

4. Conclusion

InGaP/GaAs HBTs with a graded InGaAs base have been successfully demonstrated. A study of the interdependence of the carbon and indium incorporation showed that the carbon incorporation varied exponentially as the TMIn flow rate was increased linearly. The dc current gain was increased by 50% over a standard device with a similar base sheet resistance by grading the base from $In_{.07}Ga_{.93}As$ to GaAs and maintaining a constant carbon profile. Unlike devices with indium content 10% or less, the current gain of the device with a 24% indium grade increased with base-collector bias, indicating that the $In_{.24}Ga_{.76}As/GaAs$ base-collector junction was impeding electron transport. Using a 10% indium grade and a carbon doping grade in the base, f_t and f_{max} were increased by 20% and 8% respectively, compared to a standard device without a graded base. The peak speed performance of this structure is significantly higher than those previously reported for InGaP/GaAs HBTs.

Acknowledgments

The authors wish to thank Tony Curtis for helpful discussions regarding heavily doped GaAs. This work was supported by Z. Lemnios under ARPA DAAH04-93-G-0172 and the National Science Foundation under NSF ECD 89-43166 and

References

[1] H. Kroemer, *J Vac. Sci. Technol.*B **1**, 126, (1983).

[2] G.L. Patton, D.L Harame, J.M.C Stork, B.S. Meyerson, G.J.Scilla, and E. Ganin, *IEEE Electron Dev. Lett.*, **10**, 534, (1989).

[3] W. Liu, D. Costa, and J.S. Harris, *IEEE Trans. on Electron Dev.* , **39**, 2422, (1992).

[4] T. Nittono, N. Watanabe, H. Ito, H. Sugahara, K. Nagata and O. Nakajima, *Jpn. J. Appl. Phys.* **33**, 6129, (1994).

[5] S.A. Stockman, A.W. Hanson and G.E. Stillman, *Appl. Phys. Lett.*, **60**, 2903, (1992).

[6] M. Ilegems, Optoelectronic Integration: Physics, Technology and Applications, edited by O. Wada, Boston: Kluwer Academic Publishers, 1994.

[7] D. C. Streit, M.E. Hafazi, D.K. Umemoto, J.R. Velebir, L.T. Tran, A.K. Oki, M.E. Kim, S.K. Wang, C.W. Kim, L.P. Sadwick, and R.J. Hwu, *IEEE Electron Dev. Lett.*, **12**, 194, 1991.

[8] A.W. Hanson, S.A. Stockman and G.E. Stillman, *IEEE Electron Dev. Lett.*, **14**, 25, (1993).

[9] W. Liu, S.-K. Fan, T. Henderson and D. Davito, *IEEE Electron Dev. Lett.*, **14**, 176, (1993).

[10] F. Ren, C.R. Abernathy, S.J. Pearton, J.R. Lothian, S.N.G. Chu, P.W. Wisk, T.R. Followan, B.Tseng and Y.K. Chen, *Electon. Lett.*, **28**, 2250, (1992).

Inst. Phys. Conf. Ser. No 145: Chapter 5
Paper presented at 22nd Int. Symp. Compound Semiconductors, Cheju Island, Korea, 28 August–2 September 1995
© *1996 IOP Publishing Ltd*

Thermal Stability and Reliability of Nonalloyed Ohmic Contacts on Thin Base InP/InGaAs/InP HBTs

E F Chor[1], R J Malik, R A Hamm, and R W Ryan

AT & T Bell Laboratories, 600 Mountain Avenue, Murray Hill, NJ 07974, USA.

[1] Present address : Department of Electrical Engineering, National University of Singapore, 10 Kent Ridge Crescent, Singapore 0511, Singapore.

Abstract. Nonalloyed ohmic contacts : Au/Ti, Pt/Ti, and Au/Pt/Ti, have been found to be thermally unstable and unreliable on a 500Å-thick p$^+$-InAsGa base of InP/InGaAs/InP HBTs at 400°C. Pt contact failed at 350°C. Pt or Ti penetrated through the thin base causing a serious leakage short across the base-collector junction and rendering the devices no longer functional. Using Au/Pt/Ti/W as contact, HBTs have been shown to remain functional after a 400°C anneal with no apparent shift in the turn-on voltage for the emitter and collector junctions.

1. Introduction

Nonalloyed contacts such as Pt/Ti and Au/Pt/Ti have commonly been used in InP based heterojunction bipolar transistors (HBTs) [1, 2, 3] as contacts to the thin p$^+$-InGaAs base. These contacts are ohmic as-deposited, unfortunately, the specific contact resistance from 10^{-4} to 10^{-3} Ω-cm^2 [1, 4, 5] is not low enough to satisfy the requirements of high performance circuits. Consequently, some form of thermal anneal is necessary. Most thermal anneal studies on ohmic contacts to InGaAs were carried out on a fairly thick layer with a thickness of at least 0.5 µm [1, 4, 6, 7]. This does not correspond well to the thin base of high performance HBTs, which is around 500Å. In this work, we investigated the thermal stability and reliability of several nonalloyed metallization schemes : Au/Ti, Pt/Ti, and Au/Pt/Ti on a 500Å p$^+$-InGaAs base of triple-mesa InP/InGaAs/InP HBTs. A scheme where only Pt is used as the base contact metal is also examined as Pt is well known as a low-potential barrier height metal to p-type semiconductors, owing to its large work function, and has yielded low ohmic contact resistance without using a highly doped p-layer [8].

In addition, we examined another scheme where a very stable refractory metal tungsten (W) is sandwiched between the above metallization systems and the semiconductor because Ti has been shown [1] to interact with InGaAs beyond a temperature of 350°C and resulted in a complex reaction zone between Ti and InGaAs which can extend up to 800Å [1] into InGaAs. This means serious degradation could occur across a thin base HBT. The metallization scheme tested was Au/Pt/Ti/W, as

Au/Pt/Ti was found to be the most thermally stable and reliable amongst the above metallization schemes.

2. Experiment

By means of MOMBE, the HBTs layers were grown lattice matched on a (100)-oriented Fe-doped InP substrate. Details of the HBT layers are shown in Fig. 1. The 500Å thick InGaAs base layer was heavily carbon doped to a concentration of 2×10^{19} cm^{-3}. All devices were processed identically except the metallization step. The same metallization scheme was adopted to contact the emitter, base and collector regions for simplicity of processing. After the metal liftoff pattern definition, the desired metals, i.e., either Pt, Pt/Ti, Au/Ti or Au/Pt/Ti were electron beam evaporated sequentially at a base pressure better than 10^{-7} Torr onto the sample in a single pump-down process. Shortly before the sample was loaded for metal deposition, it was etched in 10% NH$_4$OH for 30s to remove the native oxide, rinsed in deionized water for a minute and blow-dried by filtered nitrogen gas.

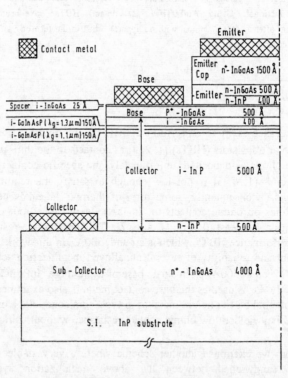

Fig.1 Cross-section and layer details of HBT.

For samples with Au/Pt/Ti/W contact, the metallization step was identical to that described earlier except that prior to the metal liftoff pattern definition and Au/Pt/Ti metal evaporation, a layer of high purity W was evaporated onto the sample in a MBE chamber using an electron gun. Just before loading for W evaporation, the sample was etched in BHF for 30s to remove the native oxide, rinsed in deionized water for a minute and blow-dried by filtered nitrogen gas. W was etched after the metallization liftoff step, where Au/Pt/Ti was used as the etch mask, using 100% CF$_4$ plasma.

During the thermal anneal of metal, the sample was placed on a graphite stripe heater under flowing forming gas at 300°C, 350°C and 400°C up to a duration of 30s in steps of 5s. Temperature higher than 400°C was not used because all HBTs, except those with Au/Pt/Ti/W as contact, became non-functional after a 400°C thermal treatment. Furthermore, higher temperature may also lead to decomposition of InP.

3. Results and discussion

The metallurgical stability and reliability of ohmic contacts : Pt, Au/Ti, Pt/Ti, Au/Pt/Ti, and Au/Pt/Ti/W, on a 500Å p$^+$-InGaAs base of InP/InGaAs/InP triple-mesa HBTs have been investigated as a function of anneal temperature. All contacts exhibited good adhesion to semiconductor and the electrical characteristics of all devices were found to be stable up to a thermal treatment at 300°C for 30s. No degradation in the contact morphology was observed up to a temperature of 400°C, except for the Pt(2000Å) contact which reacted significantly with InGaAs at a relatively low temperature of 350°C. After an anneal at 350°C for 10s, the rectifying characteristic of the base-collector junction greatly deteriorated and became ohmic for HBTs with Pt contact, as shown in Fig. 2. The I_C versus V_{CE} curves have also collapsed into one, as shown in Fig. 3. The base-emitter junction characteristics were intact during the same heat treatment. This seemed to indicate that the reaction between Pt and InGaAs has penetrated through the thin 500Å base layer resulting in a collector leakage short, but not the thicker emitter layer to reach the emitter-base junction.

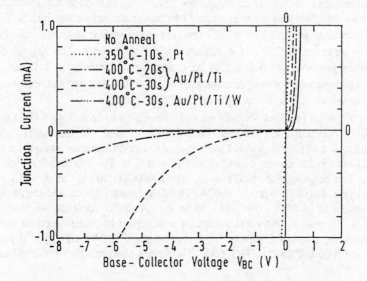

Fig.2 Effects of thermal anneal on the base-collector junction characteristics.

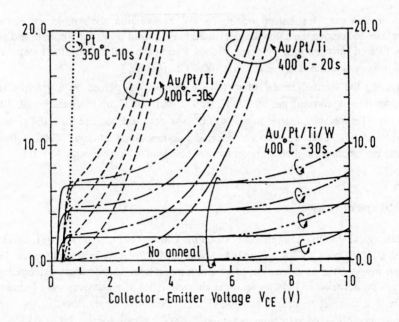

Fig. 3 Effects of thermal anneal on the common emitter characteristics.

The thermal stability of the Pt(2000Å) contact was improved by inserting a thin layer of Ti (~ 400Å) between Pt(2000Å) and InGaAs : there was no degradation in the morphology of the contacts even up to a thermal anneal temperature of 400°C. This showed that Ti has acted as a barrier to Pt diffusion into InGaAs and has effectively prevented the intermixing of Pt and InGaAs. The device characteristics were only marginally altered after an anneal at 350°C for 30s. Unfortunately, the device became non-functional after an anneal at 400°C for 15s. The base-collector junction became almost ohmic then, whereas the base-emitter junction characteristic was intact. The device characteristics after an anneal at 400°C for 15s were similar to those with Pt(2000Å) contact after an anneal at 350°C for 10s shown in Figs. 2 and 3. The degradation of the base-collector junction was deduced to be caused by the interaction between Ti and InGaAs, which occurred at temperatures at or above 350°C [1], and the reaction zone has extended by at least 500Å to reach the base-collector junction.

Au caps are required on both p- and n-contacts to enable wire bonding, Au/Ti and Au/Pt/Ti metallization schemes were therefore tested. In the Au/Ti scheme, Ti served a barrier to Au as it is well known that Au can diffuse into semiconductor and be a potential source for device degradation. In addition, Ti was required for adhesion purposes. The behaviour of HBTs with Au(2000Å)/Ti(200Å) as contacts were very similar to those of HBTs with Pt(2000Å)/Ti(400Å) metallizations and the devices failed after an anneal at 400°C for 10s. With the Au/Pt/Ti contact, we found that the thickness of Ti was an important parameter with regard to contact thermal stability and reliability. A thickness of at least a 500Å was required and we employed the following scheme : Au(2000Å)/Pt(500Å)/Ti(500Å). The device characteristics remained unchanged

A thickness of at least a 500Å was required and we employed the following scheme : Au(2000Å)/Pt(500Å)/Ti(500Å). The device characteristics remained unchanged after an anneal at 350°C for 30s. However, the base-collector junction characteristic continuously degraded with the anneal duration at a temperature of 400°C, as shown in Fig.2. It is seen that the turn-on voltage of the junction decreases to almost zero after an anneal duration of about 20s at 400°C. This reduction is reflected in the I_C versus V_{CE} characteristics as an increase in the V_{CE} offset, as shown in Fig.3, where it is observed that the devices are no longer useful after an anneal duration of 20s at 400°C. At this point, we like to mention that the base-emitter junction characteristic was intact during the 400°C anneal and we concluded that the failure mechanism was similar to that already described - the degradation of the base-collector junction owing to the interaction between Ti and InGaAs.

The Au(2000Å)/Pt(500Å)/Ti(500Å)/W scheme was implemented as an effort to improve the contact thermal stability. Three thicknesses of W were used : 300Å, 600Å and 1500Å. The thickest layer of 1500Å was found to induce too much stress in semiconductor and the thinnest layer of 300Å was found to be inadequate as a refractory barrier to metal interaction with the underlying InGaAs. The latter was also observed in Ref. [9]. The effects of an anneal at on the device characteristics with Au(2000Å)/Pt(500Å)/Ti(500Å)/W(600Å) contacts are shown in Figs. 2 and 3. Note that in contrast to previous contacts, the forward base-collector junction characteristic after a 400°C-30s anneal was almost identical to that prior to thermal anneal, as shown in Fig. 2. It is obvious in Fig. 3 that only HBTs with Au(2000Å)/Pt(500Å)/Ti(500Å)/W(600Å) as contact were still functional after a 400°C anneal and displayed a lower breakdown voltage V_{CE0} owing to a slight increase in the base-collector junction reverse leakage current after a 30s anneal, as shown in Fig. 2. More importantly, there was no apparent shift in the turn-on voltage for the base-emitter and base-collector junctions, as shown in Fig. 3, by the thermally stable V_{CE} offset. The stability of this metallization scheme can be explained by the absence of any intermetallic phases formed between W and In, Ga, or As below a temperature of 800°C [10]

4. Conclusion

It is concluded from this study that the anneal conditions of nonalloyed metallization on thin base InP/InGaAs/InP HBTs must be carefully designed otherwise the device lifetime and reliability will be severely degraded. We have shown that nonalloyed ohmic contacts : Au/Ti, Pt/Ti, and Au/Pt/Ti were not thermally stable at an anneal temperature of 400°C on a 500Å p$^+$-InGaAs base layer of InP/InGaAs HBTs. Pt contact failed at 350°C. Interaction of the contact metal, Ti or Pt, and the thin p$^+$-InGaAs base occurred, leading to a reduction to almost zero of the turn on voltage of the base-collector junction and a significant increase in the reverse junction leakage current, thus rendering the HBTs non-functional. In addition, it has been seen that using a thin layer of thermally stable refractory metal such as W as a contact layer in a multi-layer metallization scheme, e.g., Au/Pt/Ti/W, may be the key to thermally stable and reliable contact. HBTs with Au/Pt/Ti/W as contact have been shown to remain functional after an anneal at 400°C.

Acknowledgement

E.F. Chor, would like to thank Dr. A.Y. Cho for making it possible for her to spend her sabbatical leave at the AT & T Bell Laboratories, Murray Hill. Without which this work would not be possible.

References

[1]   Chu S N G, Katz A *et al* 1990 *J. Appl. Phy.* **67** 3754-3760

[2]   Jensen J F, Stanchina W E *et al* 1991 *IEEE J. Solid-State Cct.* **26** pp.415-420

[3]   Kyono C S, Binari S C *et al* 1992 *Electron. Lett.* **28** 1388-1390

[4]   Katz A, Dautremont-Smith W C and *et al* 1989 *Appl. Phys. Lett.* **54** 2306-2308

[5]   Katz A, Weir B E *et al* 1990 *J. Appl. Phys.* **63** 1123-1128

[6]   Ressel P, Vogel K *et al* 1992 *Electron. Lett.* **28** 2237- 2238

[7]   Stareev G, Künzel H *et al* 1993 *J. Appl. Phys.* **74** 7344-7356

[8]   Okada H, Shikata S-I *et al* 1991 *Jap. J. Appl. Phys.* **30** L558-L560

[9]   Merkel K G, Bright V M *et al* 1994 *Mat. Sci. & Engg.* **B25** 175-178

[10]   Katz A, Weir D M *et al* 1989 *Appl. Phy. Lett.* **44** 2220-2222

Inst. Phys. Conf. Ser. No 145: Chapter 5
Paper presented at 22nd Int. Symp. Compound Semiconductors, Cheju Island, Korea, 28 August–2 September 1995
© *1996 IOP Publishing Ltd*

DC and microwave characteristics of $In_{0.3}(Al_xGa_{1-x})_{0.7}As/$ $In_{0.3}Ga_{0.7}As$ heterojunction bipolar transistors grown on GaAs

H-P Hwang , J-L Shieh, J-W Pan, C-C Chou, and J-I Chyi

Department of Electrical Engineering, National Central University, Chung-Li, Taiwan, 32054,
Republic of China

Abstract. The dc and microwave characteristics of $In_{0.3}(Al_{0.5}Ga_{0.5})_{0.7}As/In_{0.3}Ga_{0.7}As$ and $In_{0.29}Al_{0.71}As/In_{0.3}Ga_{0.7}As$ heterojunction bipolar transistors (HBTs) grown on GaAs are investigated. A step-graded $In_xGa_{1-x}As$ buffer is employed to effectively suppress the threading dislocations resulting from the lattice mismatch between $In_{0.3}Ga_{0.7}As$ and GaAs. These devices exhibit a high collector-emitter breakdown voltage ($BV_{CEO} > 11V$) and a collector offset voltage as small as 66 mV, demonstrating the excellent quality of the base-emitter and base-collector junctions. The typical common-emitter current gain at a collector density of 16 kA/cm^2 is 25 for the $In_{0.3}(Al_{0.5}Ga_{0.5})_{0.7}As/In_{0.3}Ga_{0.7}As$ HBTs with a base doping concentration of $1x10^{19}$ cm^{-3}. The Ft and Fmax for the non-self-aligned device with an emitter size of 4x4 μm^2 is 33 GHz and 20.5 GHz, respectively.

1. Introduction

Heterojunction bipolar transistors (HBTs) are important electronic devices applied in high-speed switching and microwave circuits [1], [2]. Especially for high power applications, HBTs have demonstrated their superior characteristics in current handling capability, power-added efficiency, and linearity [3]. So far, most of the work have been focused on AlGaAs/GaAs material system because of its virtually perfect lattice match and well-established process technology. On the other hand, InAlAs/InGaAs and InP/InGaAs HBTs matched to InP substrate exhibit even higher speed performance. This is attributed to the high electron mobility, large Γ-L energy separation and low contact resistance of InGaAs. The low surface recombination velocity of InGaAs also reduces the emitter size effect. In addition to their outstanding electrical properties, InP-based HBTs can also be integrated with long-wavelength optoelectronic devices to form optoelectronic integrated circuits (OEICs). However, InP-based HBTs have their own deficiencies, such as low breakdown voltage, high output conductance, fragility, and high cost. In order to remedy these problems, we propose the fabrication of $In_{0.3}(Al_xGa_{1-x})_{0.7}As/In_{0.3}Ga_{0.7}As$ HBTs on relatively well developed GaAs substrates. HBTs based on $In_{0.3}(Al_xGa_{1-x})_{0.7}As/In_{0.3}Ga_{0.7}As$ heterostructure are expected to have higher collector breakdown voltage and greater flexibility due to larger bandgap and greater range of conduction band discontinuity (ΔE_C), i.e. 0~0.71 eV [4], compared to InP-based material system. The large ΔE_C of this material system may also be beneficial for p-n-p devices.

The growth of high In-content $In_xGa_{1-x}As$ on GaAs substrates has been pursued for many novel devices such as HBTs [5], [6], resonant tunneling diodes [7], [8] and high electron mobility transistors [9], [10]. However, the large lattice mismatch between InGaAs and GaAs often results in threading dislocations propagating into the active region and adversely affect device performance. Various techniques have been used to decrease the dislocation density below 10^7 cm^{-2}, for example, linear-graded buffer layer [7], [11], step-graded buffer layer [8-10], [12] and superlattice buffer layer [5]. Using step-graded buffer layer, we have successfully obtained device-quality materials for HBT which is much more sensitive to defects than majority carrier device, e.g. field effect transistor. The dc and microwave characteristics of the $In_{0.29}Al_{0.71}As/In_{0.3}Ga_{0.7}As$ and $In_{0.3}(Al_{0.5}Ga_{0.5})_{0.7}As/In_{0.3}Ga_{0.7}As$ HBTs fabricated on these epilayers are presented and discussed. In addition, the current-induced degradation of the $In_{0.3}(Al_{0.5}Ga_{0.5})_{0.7}As/In_{0.3}Ga_{0.7}As$ HBTs is also investigated.

2. Material growth and device fabrication

The metamorphic $In_{0.3}(Al_xGa_{1-x})_{0.7}As/In_{0.3}Ga_{0.7}As$ HBTs were grown on (100) semi-insulating GaAs substrate by molecular beam epitaxy (MBE). Shown in Fig. 1 are the layer structures of the $In_{0.3}(Al_xGa_{1-x})_{0.7}As/In_{0.3}Ga_{0.7}As$ HBT with x values of 1 and 0.5. A step-graded $In_xGa_{1-x}As$ metamorphic buffer layer was first grown for strain relaxation and misfit dislocation confinement. Then, a heavily doped n-type $In_{0.3}Ga_{0.7}As$ subcollector and a 0.5 μm-thick n-$In_{0.3}Ga_{0.7}As$ collector doped to 5×10^{16} cm^{-3} were grown. For the abrupt junction $In_{0.29}Al_{0.71}As/In_{0.3}Ga_{0.7}As$ HBT designated as device A, this was followed by a 0.12 μm p-$In_{0.3}Ga_{0.7}As$ base doped to 8×10^{18} cm^{-3} and a 10 nm undoped $In_{0.3}Ga_{0.7}As$ spacer, then a 0.15 μm n-$In_{0.29}Al_{0.71}As$ emitter was grown. The structure of the graded junction $In_{0.3}(Al_{0.5}Ga_{0.5})_{0.7}As/In_{0.3}Ga_{0.7}As$ HBTs designated as device B is similar to that of device A except that a 0.1 μm p-InGaAs base doped to 1×10^{19} cm^{-3} was used and two 20 nm graded layers were inserted on both sides of the 0.15 μm InAlGaAs emitter. These two graded layers were employed to remove the hetero-interface spike between InAlGaAs and InGaAs and led to reduced offset voltage and emitter dynamic resistance.

The devices were fabricated by conventional wet etching processes and photolithography. The n-type (AuGeNi) and p-type (AuBe) contacts were fabricated by lift-off technique. Non-self-aligned base contacts were formed with a space of 0.8 μm from the emitter mesa. Mesas were etched in $H_3PO_4:H_2O_2:H_2O$ (1:1:20) solution. On these devices,

1000Å n$^+$ $In_{0.3}Ga_{0.7}As$	1×10^{19}	
500Å n$^+$ $In_{0.29}Al_{0.71}As$	$5 \times 10^{17} \rightarrow 1 \times 10^{19}$	
1000Å n $In_{0.29}Al_{0.71}As$	5×10^{17}	
100Å $In_{0.3}Ga_{0.7}As$	undoped	
0.12μm p$^+$ $In_{0.3}Ga_{0.7}As$	8×10^{18}	
0.5μm n $In_{0.3}Ga_{0.7}As$	5×10^{16}	
0.4μm n$^+$ $In_{0.3}Ga_{0.7}As$	5×10^{18}	
0.2μm n$^+$ $In_{0.2}Ga_{0.8}As$	5×10^{18}	
0.2μm n$^+$ $In_{0.1}Ga_{0.9}As$	5×10^{18}	
0.3μm n$^+$ GaAs	5×10^{18}	
(100) S. I. GaAs		

(a)

1000Å n$^+$ $In_{0.3}Ga_{0.7}As$	1×10^{19}	
200Å n$^+$ $In_{0.3}(Al_xGa_{1-x})_{0.7}As$	5×10^{18}	x:0.5->0.1
1500Å n $In_{0.3}(Al_{0.5}Ga_{0.5})_{0.7}As$	5×10^{17}	
200Å n $In_{0.3}(Al_xGa_{1-x})_{0.7}As$	5×10^{17}	x:0.1->0.5
100Å $In_{0.3}Ga_{0.7}As$	undoped	
0.1μm p$^+$ $In_{0.3}Ga_{0.7}As$	1×10^{19}	
0.5μm n $In_{0.3}Ga_{0.7}As$	5×10^{16}	
0.3μm n$^+$ $In_{0.3}Ga_{0.7}As$	5×10^{18}	
0.2μm n$^+$ $In_{0.2}Ga_{0.8}As$	5×10^{18}	
0.2μm $In_{0.1}Ga_{0.9}As$	undoped	
0.25μm GaAs buffer	undoped	
(100) S. I. GaAs		

(b)

Fig. 1 Layer structures of the metamorphic (a) $In_{0.29}Al_{0.71}As/In_{0.3}Ga_{0.7}As$ (device A) and (b) $In_{0.3}(Al_{0.5}Ga_{0.5})_{0.7}As/In_{0.3}Ga_{0.7}As$ (device B) HBTs.

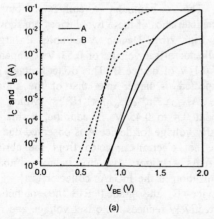
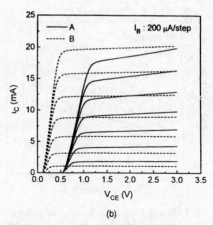

Fig. 2 (a) Gummel plots and (b) common emitter output characteristics for devices A and B. The emitter area is 25x25 µm².

a SiO$_X$ layer was deposited using photo-enhanced chemical vapor deposition for passivation and isolation. In order to investigate the current-induced degradation, the devices were stressed under a fixed collector current density (J$_C$) of 1x10^5 A/cm² for a duration of 10 hours without any cooling. The I-V characteristics were measured every 30 minute. The microwave characteristics of these devices were measured directly by an HF 8510C network analyzer on a Cascade Microtech probe station.

3. Experimental results and discussion

Shown in Fig. 2(a) are the typical Gummel plots of the metamorphic HBTs for device A and B, respectively. It is obvious that device A exhibits a higher turn-on voltage for the collector current than device B. This is attributed to the higher hetero-interface spike at the abrupt In$_{0.29}$Al$_{0.71}$As/In$_{0.3}$Ga$_{0.7}$As heterojunction. The ideality factor of I$_C$ and I$_B$ (η_{IC} and η_{IB}) for device A is 1.57 and 2.0, respectively. Tunneling of electrons through the energy spike at the E-B junction is thought to be the cause of the large ideality factor of I$_C$ [13]. On the other hand, η_{IC} and η_{IB} are 1.02 and 1.8, respectively, for the graded InAlGaAs/InGaAs HBTs (device B). They are comparable to those of the AlGaAs/GaAs HBTs with a graded E-B junction.

Table 1
Summary of the dc characteristics of devices A and B with a 25x25 µm² emitter area. The h$_{fe}$ is measured at J$_C$ = 16 kA/cm² and V$_{CE}$ = 2V.

Device	η_{IC}	η_{IB}	h$_{fe}$	V$_{CE,offset}$	V$_{CBO}$	V$_{CEO}$	V$_{in}$(BC)	V$_{in}$(BE)
					(V)			
A	1.57	2	17	0.53	16.3	11.3	0.82	1.33
B	1.02	1.8	25	0.1	13	11.9	0.82	0.9

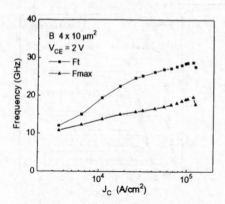

Fig. 3 Ft and Fmax as a function of collector current density for the graded $In_{0.3}(Al_{0.5}Ga_{0.5})_{0.7}As/In_{0.3}Ga_{0.7}As$ HBT.

The common emitter output characteristics of these two devices are shown in Fig. 2(b). Device A exhibits a larger collector offset voltage of 0.53 V, compared to 0.1 V of device B. This offset voltage, as expected, is larger than that of the abrupt $In_{0.52}Al_{0.48}As/In_{0.53}Ga_{0.47}As$ HBTs, which is about 0.4 to 0.45 V. In addition, a smaller Early voltage for device A is observed due to the hot electrons injected from the abrupt InAlAs emitter, which induce impact ionization in the InGaAs collector [14]. For device B, the graded E-B heterojunction effectively reduces the offset voltage and the avalanche effect in the collector region. Compared to InP-based devices, much lower output conductance is observed in this device. Despite the rather large difference in offset voltage and output conductance, both of device A and B have fairly high breakdown voltages ($BV_{CEO} > 11V$), which are comparable to those of the AlGaAs/GaAs HBTs and better than those of the InP-based HBTs. The dc characteristics of these devices are summarized in Table 1.

The measured Ft and Fmax versus collector current density (Jc) of device B are shown in Fig. 3. The device has an emitter size of $4\times10\ \mu m^2$ and exhibits a collector offset voltage as small as 66 mV and a peak current gain of 43 at J_C of 225 kA/cm^2. As shown in Fig. 3, the Ft and Fmax gradually increases with collector current to about 30 GHz and 20 GHz, respectively, and then decreases as Jc is greater than 120 kA/cm^2. This is attributed to the Kirk effect. The device remains at stable operation near 1×10^5 A/cm^2 in spite of a large lattice mismatch between the epilayer and GaAs substrate. This is encouraging for power device applications. For the device with an emitter size of $4\times4\ \mu m^2$, higher Ft and Fmax, i.e. 33 GHz and 20.5 GHz respectively, is measured. The high extrinsic base resistance and base-collector capacitance resulting from the non-self-aligned and non-implanted processes are believed to be the factors that limit the performance of these devices.

Finally, the I-V characteristics of device B after high current stress are measured and discussed. The device used in this study is the same device that gives the microwave characteristics shown in Fig. 3. The Gummel plot of the collector and base currents after stress is shown in Fig. 4. The collector current remains unchanged. While an extra base current is observed and results in a decrease of current gain of about 17% at 100 kA/cm^2. In addition, the degradation of the base-collector junction is more significant than that of the base-emitter junction due to the rapid increase of junction temperature under high current stress. This is mainly attributed to the poor thermal conductivity of InGaAs [15], i.e. the 0.8 μm-thick collector and subcollector layers. The base-collector leakage current (I_{CBO}) at -2 V is increased from about 50 nA up to 153 μA after stress as shown in Fig. 5. Moreover, the turn-on voltage shift of this junction also induces an increment of the collector offset voltage up to 102 mV.

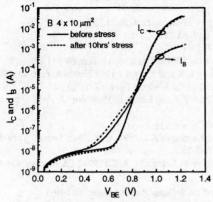

Fig. 4 Gummel plots for device B before and after high current stress.

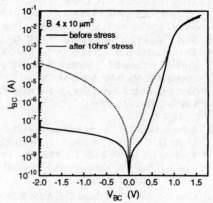

Fig. 5 Measured I-V characteristics of the B-C junction for device B before and after high current stress.

4. Conclusion

In conclusion, we have successfully fabricated n-p-n $In_{0.29}Al_{0.71}As/In_{0.3}Ga_{0.7}As$ and $In_{0.3}(Al_{0.5}Ga_{0.5})_{0.7}As/In_{0.3}Ga_{0.7}As$ HBTs grown on GaAs substrates, to the best of our knowledge, for the first time. The dc and microwave characteristics of these devices are investigated and analyzed. High collector breakdown voltage of about 11 V, which is comparable to that of AlGaAs/GaAs HBTs is obtained for these two structures. The devices with graded $In_{0.3}(Al_{0.5}Ga_{0.5})_{0.7}As/In_{0.3}Ga_{0.7}As$ heterostructure exhibit a collector offset voltage as small as 66 mV and a current gain of 25 at collector current density of 16 kA/cm². An Ft of 33 GHz and an Fmax of 20.5 GHz are obtained for the device with an emitter size of 4x4 µm². Higher device performance is expected as process technology is improved. As to the changes of I-V characteristics after high current stress, a significant increase of I_{CBO} is observed for the graded junction $In_{0.3}(Al_{0.5}Ga_{0.5})_{0.7}As/In_{0.3}Ga_{0.7}As$ HBTs. This is believed to be caused by the poor thermal conductivity of the thick InGaAs collector and subcollector layers.

Acknowledgments. The authors wish to thank R.-M. Lin for the assistance in material growth and the technical support of the MBE Laboratory of the Center for Optical Science, NCU. This work was supported by the National Science Council, R.O.C. under contract NSC 84-2221- E008-011.

References

[1] Asbeck P M, Chang M-C F, Higgins J A, Sheng N H, Sullivan G J and Wang K-C 1989 *IEEE Trans. Electron Devices* **36** 2032-42

[2] Kim M E, Oki A K, Gorman G M, Umemoto D K and Camou J B 1989 *IEEE Trans. Microwave Theory Tech.* **37** 1286-303

654

[3] Wang N L, Sheng N H, Chang M F, Ho W J, Sullivan G J, Sovero E A, Higgins J A and Asbeck P M 1990 *IEEE Trans. Microwave Theory Tech.* **38** 1381-9

[4] Shieh J-L, Chyi J-I, Lin R-J, Lin R-M and Pan J-W 1994 *Electron. Lett.* **30** 2172-3

[5] Won T, Agarwala S and Morkoc H 1988 *Appl. Phys. Lett.* **53** 2311-3

[6] Ito H and Harris, Jr., J S 1993 *Jpn. J. Appl. Phys.* **32** 4923-7

[7] Wolak E, Harmand J C, Matsuno T, Inoue K and Narusawa T 1991 *Appl. Phys. Lett.* **59** 111-3

[8] Hwang H-P, Shieh J-L, Lin R-M, Chyi J-I, Tu S L, Peng C K and Yang S J 1994 *Electron. Lett.* **30** 826-8

[9] Tien N C, Chen J, Fernandez J M and Wieder H H 1992 *IEEE Electron Device Lett.* **13** 621-3

[10] Chyi J-I, Shieh J-L, Wu C-S, Lin R-M, Pan J-W, Chan Y-J and Lin C-H 1994 *Jpn. J. Appl. Phys.* **33** 1574-6

[11] Inoue K, Harmand J C and Matsuno T 1991 *J. Crystal Growth* 313-7

[12] Chang J C P, Chen J, Fernandez J M, Wieder H H and Kavanagh K L 1992 *Appl. Phys. Lett.* **60** 1129-31

[13] Ramberg L P and Ishibashi T 1988 *J. Appl. Phys.* **63** 809-20

[14] Jalali B, Chen Y-K, Nottenburg R N, Sivco D, Humphrey D A and Cho A Y 1990 *IEEE Electron Device Lett.* **11** 400-2

[15] Chau H-F, Liu W and Beam III E A 1995 *7th Int. Conf. on InP and Related Materials* 640-3

Inst. Phys. Conf. Ser. No 145: Chapter 5
Paper presented at 22nd Int. Symp. Compound Semiconductors, Cheju Island, Korea, 28 August–2 September 1995
© *1996 IOP Publishing Ltd*

1/f Noise Characteristics of AlGaAs/GaAs Heterojunction Bipolar Transistor with a Noise Corner Frequency below 1 kHz

Jin-Ho Shin, Joonwoo Lee, Youngseok Suh and Bumman Kim

Department of Electronic and Electrical Engineering
and Microwave Application Research Center,
Pohang University of Science and Technology
Hyoja-dong San-31, Pohang, Kyung-pook, 790-784, Korea

Abstract: The internal low-frequency noise sources of AlGaAs/GaAs HBT's are related to the detailed epi-layer structure as well as surface recombination velocity fluctuation at the extrinsic GaAs base surface. HBT's with a large emitter size of 120 × 120 µm^2 are fabricated on abrupt emitter-base junction materials without undoped spacer, and the HBT's exhibit an internal noise corner frequency of 100 Hz, which is much lower than about 100 kHz of conventional AlGaAs/GaAs HBT's. The existence of resistance fluctuation 1/f noise is clearly verified by the simple comparison of collector current noise spectra with the varying base termination. It is found that, at a high emitter-base bias, the resistance fluctuation 1/f noise becomes dominant for shorted base-emitter termination, but the internal 1/f noise dominant for open base. To improve the low-frequency noise characteristics for a practical small-feature size HBT, device design rules including resistance fluctuation are discussed.

1. Introduction

The low 1/f noise of heterojunction bipolar transistor(HBT) is an important feature for low phase noise microwave oscillator applications [1], [2]. Recently, many experimental data on 1/f noise have been reported for newly emerging material-based HBT's such as GaInP/GaAs [3]-[5], AlInAs/InGaAs [6] and Si/SiGe [7] HBT's. Their noise corner frequencies range from 1 kHz to 100 kHz, while those of AlGaAs/GaAs HBT's [8]-[13] are usually above 100 kHz. In view of the relative maturity of AlGaAs/GaAs HBT technology, it is worthwhile to study the optimized AlGaAs/GaAs HBT structure for reduced low-frequency(LF) noise in more detail.

In order to reduce LF noise, it is essential to identify the dominant noise source. It is generally believed that the noise stems mainly from the recombination at the extrinsic GaAs base surface due to the high surface recombination velocity [9]. A considerable reduction of 1/f noise of AlGaAs/GaAs HBT was achieved by N. Hayama, et al [10]. Their experimental work showed that an HBT with a thin depleted AlGaAs ledge over the extrinsic GaAs base region had lower 1/f noise as compared to a device without the AlGaAs ledge, but still exhibited a large anomalous generation-recombination(g-r) noise bump. It is noteworthy that there have been large deviations of 10-20 dB in 1/f noise data among the surface-passivated HBT's [9]-[12]. This suggests that the major part of the LF noise is closely related to the detailed HBT structure. That is, reduction of LF noise is expected by adopting a proper choice of epitaxial structure. Meanwhile, based on the mobility fluctuation theory, T. G .M. Kleinpenning, et al. suggested that, at a high emitter-base(E-B) forward bias, a significant 1/f noise can be generated by parasitic emitter and base resistances [13]. Since both the noise power spectral densities of the surface recombination and resistance 1/f noises are proportional to bias current square, it is difficult to determine the dominant noise source from the measured bias dependencies. Moreover, the device design rules to reduce the two noise sources are different. This suggests that a direct experimental verification of the existence and amount of resistance 1/f noise is needed to clarify the device design rule. In this paper, we report a great reduction of internal LF noise by selecting a proper epitaxial structure, and a direct verification of resistance fluctuation 1/f noise for AlGaAs/GaAs HBT.

2. Device structure

To reduce the internal 1/f noise, we fabricated AlGaAs/GaAs HBT on an abrupt E-B junction material without spacer. As indicated by comparative studies on abrupt and graded E-B junctions of AlGaAs/GaAs HBT's, HBT with an abrupt E-B junction greatly suppresses the space charge region(SCR) recombination [14]. In addition, an abrupt E-B heterojunction launcher injects electrons into the base with kinetic energy, and the accelerated electrons can transverse the interface state region without recombination. It is also well known that the spacer between emitter and base produces a significant SCR recombination [15]. Materials were grown by MOCVD. The epitaxial structure has a 500 Å 30 % Al emitter layer and 700 Å carbon-doped base layer. The emitter, base, and collector dopant concentrations are 2×10^{17}, 3×10^{19}, and 2×10^{16} cm^{-3}, respectively. We have fabricated MESA-type HBT's and characterized them. To avoid the 1/f noise from extrinsic base surface, we employed a large emitter-size device(120×120 μm^2). The spacing between emitter and base metal is 50 μm, which is large enough to enhance base resistance 1/f noise. DC current gain is typically 67 and the collector current at the unity current gain is 200 pA. Base and collector current ideality factors are 1.31 and 1.12, respectively. These DC characteristics are similar to those of the large emitter-size HBT with abrupt E-B junction without spacer [14]. The base current ideality factor is less than typical values of HBT's with graded E-B junction [14] or with spacer [15], indicating a small SCR recombination current of our HBT. The collector current ideality factor of 1.12 implies that the electron transport mechanism is influenced by the thermionic emission through an abrupt E-B heterojunction discontinuity [16]. Emitter resistance(r_e) is 4 Ω, and base resistance(r_b) ranges from 50 to 250 Ω due to the inherent nonuniformity of wet etching in MESA-type HBT process.

3. Low frequency noise characteristics

Following the generalized small signal noise analysis given in reference [13], we have characterized the LF noise behavior of the HBT. The LF noise equivalent circuit is depicted in Fig. 1. The LF collector noise voltage(S_{Vc}) of HBT was measured from

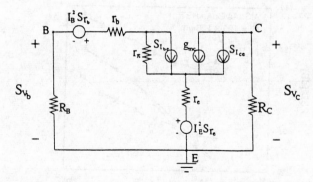

Fig. 1. Low-frequency small-signal noise equivalent circuit. The internal base-emitter current noise S_{Ibe} contains surface recombination 1/f noise. Resistance 1/f noise sources are included in Sr_e and Sr_b.

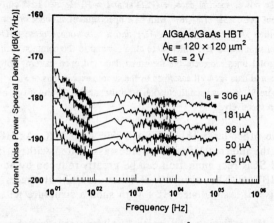

Fig. 2. Low-frequency collector current noise spectra(S_{Vc}/R_C^2) measured in the common emitter configuration, where the base termination resistance(R_B) is much larger than the base input resistance(Z_{in}). In this graph, S_{Vc}/R_C^2 represents only the internal low-frequency noise given by $\beta^2 S_{Ibe} + S_{Ice} + 2qI_B\beta^2 + 2qI_C$, without containing the resistance 1/f noise. The measured spectra exhibit a very low noise corner frequency of about 100 Hz. The white noise level agrees very well with shot noise level given by $2qI_B\beta^2 + 2qI_C$.

10 Hz to 100 kHz in the common emitter configuration. Fig. 2 shows the collector current noise spectra(S_{Vc}/R_c^2) when $R_B \gg Z_{in} = r_b+r_\pi+(1+\beta)r_e$. Here, r_π and β are a base input resistance and a differential current gain of intrinsic transistor, respectively. $R_c(R_B)$ is a collector(base) bias resistance. Here,

$$S_{Vc}/R_c^2 = \beta^2 S_{Ibe} + S_{Ice} + [\beta/(R_B + Z_{in})]^2 S_{Vr} + 2q\,I_B\,\beta^2 + 2q\,I_C \qquad (1)$$

where S_{Ibe} and S_{Ice} represent base-emitter and emitter-collector current 1/f noise, respectively, and S_{Vr} is a resistance 1/f noise given by $I_B^2 Sr_b+I_E^2 Sr_e$. Since $[\beta/(R_B + Z_{in})]^2 S_{Vr}$ is negligible due to large value of R_B, S_{Vc}/R_c^2 represents solely the internal noise, without containing the resistance 1/f noise. The spectra exhibit a very low noise corner frequency of 100 Hz, which is comparable to that of Si BJT, and does not show any E-B junction g-r noise bump, which is common to the reported LF noise data of HBT's. The white noise level agrees very well with shot noise level given by $2qI_B\beta^2+2qI_C$. Our large size AlGaAs/GaAs HBT has at least 20 dB lower

658

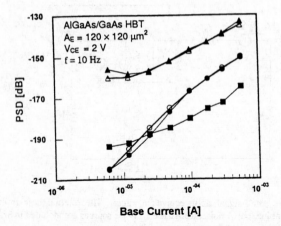

Fig. 3. Low-frequency noise power spectral densities(PSD's) at f = 10 Hz vs. base current. Vertical axis represents a collector current noise PSD of S_{Vc}/R_c^2 with base open(square), a collector current noise PSD of $S_{Ic} = S_{Vc}/R_c^2$ with base-emitter short(circle) in dB(A²/Hz), and a base voltage noise PSD of S_{Vb} with base open(triangle) in dB(V²/Hz) simultaneously, which were all measured in the common emitter configuration. Here, filled patterns with solid lines represent the measured data and open patterns with dotted lines the calculated values. The measured data are well described by the simple expressions of $S_{Ic} = (\beta/Z_{in})^2 S_{Vr} + (S_{Ic})_w$ and $S_{Vb} = S_{Vr} + (S_{Vb})_w$, where S_{Vr} is resistance fluctuation 1/f noise, and $(S_{Ic})_w$ and $(S_{Vb})_w$ are the white noise components of S_{Ic} and S_{Vb}, respectively.

internal noise than the other large [8] and small size AlGaAs/GaAs HBT's [9]-[13]. This data shows that the internal 1/f noise sources associated with surface states, E-B interface states and SCR recombination can be greatly reduced by adopting a large size AlGaAs/GaAs HBT with an abrupt E-B junction without spacer. This implies that the major 1/f noise of a small size HBT with such a structure will be a surface recombination noise.

To investigate the resistance 1/f noise, we also measured S_{Ic}, i.e., S_{Vc}/R_c^2 with $R_B = 0\ \Omega$, and base noise voltage(S_{Vb}) with $R_B \gg Z_{in}$. Fig. 3 shows the base current dependencies of S_{Vc}/R_c^2, S_{Ic} and S_{Vb} at f = 10 Hz. Here,

$$S_{Ic} = [\ \beta\ (r_b + r_e)/\ Z_{in}\ ]^2 S_{Ibe} + [(r_b + r_\pi + r_e)\ /\ Z_{in}\ ]^2 S_{Ice} + (\beta/Z_{in})^2 S_{Vr} + (\ S_{Ic}\ )_w \qquad (2)$$

$$S_{Vb} = (r_\pi + \beta r_e)^2 S_{Ibe} + r_e^2 S_{Ice} + S_{Vr} + (\ S_{Vb}\ )_w \qquad (3)$$

where $(S_{Ic})_w$ and $(S_{Vb})_w$ are white noise components of S_{Ic} and S_{Vb}, respectively. We can observe that $S_{Ic} > S_{Vc}/R_c^2$ at a high E-B bias ($I_B > 50\ \mu A$). Since the contribution of S_{Ibe} to the collector current noise should be reduced with $R_B = 0\ \Omega$, the observation is a clear evidence of the existence of resistance 1/f noise. Our approach of examining resistance 1/f noise differs from that of the previously published work [13] in two aspects. First, we used the HBT with very low internal 1/f noise which does not mask resistance 1/f noise. Second, we directly compared the measured magnitudes of S_{Ic} and S_{Vc}/R_c^2, not relying on the measured bias dependencies. Based on the observation, we assumed $S_{Ic} = (\beta/Z_{in})^2 S_{Vr} + (S_{Ic})_w$ at $I_B > 100\ \mu A$, where S_{Ic} is at least 10 dB larger than S_{Vc}/R_c^2. S_{Vr} can then be expressed as a fitting equation of $S_{Vr} = (7.41 \times 10^{-9}) I_B^{1.65}$, where S_{Vr} is in V²/Hz and I_B is in ampere. In the same way,

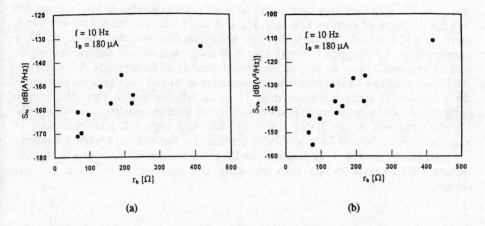

Fig. 4. (a) Collector current noise PSD with base-emitter short(S_{Ic}) vs. base resistance(r_b).
(b) Base voltage noise PSD with base open(S_{Vb}) vs. base resistance(r_b).

$S_{Vb} = S_{Vr} + (S_{Vb})_w$. These expressions for S_{Ic} and S_{Vb} agree very well with the measured data as shown in Fig. 3. This experiment using HBT with ultra low internal noise clearly demonstrates the existence of resistance 1/f noise. Fig. 4 (a) and (b) show S_{Ic} vs. r_b and S_{Vb} vs. r_b, respectively, which were measured from many HBT's on the same wafer. The data show the strong correlation between them, also proving the above statements. As the base resistance(r_b) varies from 50 to 250 Ω, the corresponding deviation of S_{Vb} is about 20 dB, indicating that $S_{Vr} \propto r_b^3$. In view of the large emitter size and large r_b, S_{Vr} is mainly due to the base resistance fluctuation. Then, $S_{Vr} \propto Sr_b$ and $Sr_b = (\alpha_b r_b^2)/(f N_b)$, where α_b is the Hooge coefficient of base layer, f is a frequency and N_b is the effective total number of carriers(holes) within the extrinsic base layer. Since the variation of r_b mainly comes from that of base thickness(W_b), $r_b \propto W_b^{-1} \propto N_b^{-1}$ and $S_{Vr} \propto r_b^3$, agreeing with the above indication. $Sr_b \propto L_{eb}/(W_b P_b W)^3$, where L_{eb} is emitter to base metal spacing, W_b is base thickness, P_b is base doping and W is emitter width. Therefore, to reduce S_{Ic}, base thickness(W_b), base doping(P_b), and emitter width(W) should be kept large and E-B spacing(L_{eb}) small. This device design rule is different from the conventional rule which is concerned with the reduction of surface recombination noise. According to [9], it is desirable to have small W_b for small lateral diffusion, small W for a small periphery/area ratio for the fixed emitter area, and L_{eb} of about 1 μm for surface passivation. Since both resistance and surface recombination 1/f noises should be reduced simultaneously, some trade-off in the above two design rules is needed to have an optimized HBT with a very low 1/f noise.

4. Conclusion

In conclusion, the internal 1/f noise of AlGaAs/GaAs HBT is significantly influenced by using a proper E-B junction structure. Our large size AlGaAs/GaAs HBT with an abrupt E-B junction and no spacer has a very low internal noise corner frequency of 100 Hz. This feature may be attributed to the small surface recombination due to its small emitter periphery/area ratio, small SCR recombination due to the inherent low leakage property of such an E-B junction, and small interface recombination due to more heated electron injection by the heterojunction launcher structure. It is expected that, by adopting our optimized E-B junction structure and minimizing the surface recombination, we can achieve a practical small-feature size AlGaAs/GaAs HBT with very low internal 1/f noise, comparable to Si BJT. A direct verification of the existence of resistance fluctuation 1/f noise was performed through a simple comparison of the noise magnitudes for different base-termination conditions. Our LF noise analyses of the device also show that, at a high E-B bias, resistance 1/f noise becomes dominant for shorted E-B termination, but internal 1/f noise dominant for open base termination. To improve the LF noise characteristics of practical small-feature HBT, device design rules including the resistance fluctuation are also suggested.

Acknowledgment

This work has been partially supported by Korea Agency for Defense Development and Korean Science and Engineering Foundation.

References

[1] H. J. Siweris, B. Schiek, 1985, *IEEE Trans. Microwave Theory Tech.*, **33**, 233-242.

[2] M. N. Tutt, D. Pavlidis, A. Khatibzadeh, B. Bayraktaroglu, 1992, *IEEE MTT-s Int. Microwave Symp. Dig.*, 727-730.

[3] R. Plana, J. Graffeuil, S. L. Delage, H. Blanck, M. A. di Forte-Poisson, C. Brylinski, E. Chartier, 1992, *Electron. Lett.*, **28**, 2354-2356.

[4] W. J. Ho, M. F. Chang, A. Sailer, P. Zampardi, D. Deakin, B. McDermott, R. Pierson, J. A. Higgins, J. Waldrop, 1993, *IEEE Electron Device Lett.*, **14**, 572-574.

[5] J.-H. Shin, J. Lee, Y. Suh, Y. Kim, B. Kim, 1994, *21st Int. Symp. Compound Semicond.*, San Diego, California, 625-628.

[6] Y. K. Chen, L. Fan, D. A. Humphrey, A. Tate, D. Sivco, A. Y. Cho, 1993, *IEEE IEDM Tech. Dig.*, 803-806.

[7] R. Plana, L. Escotte, J. P. Roux, J. Graffeuil, A. Gruhle, H. Kibbel, 1995, *IEEE Electron Device Lett.*, **16**, 58-60.

[8] S. C. Jue, D. J. Day, A. Margittai, M. Svilans, 1989, *IEEE Trans. Electron Devices*, **36**, 1020-1025.

[9] D. Costa, J. S. Harris, 1992, *IEEE Trans. Electron Devices*, **39**, 2383-2394.

[10] N. Hayama, S. R. LeSage, M. Madihian, K. Honjo, 1988, *IEEE MTT-s Int. Microwave Symp. Dig.*, 679-682.

[11] D. Costa, A. Khatibzadeh, 1994, *IEEE Microwave and Guided Wave Lett.*, **4**, 45-47.

[12] J. Cowles, L. Tran, T. Block, D. Streit, C. Grossman, G. Chao, A. Oki, 1995, *IEEE MTT-s Int. Microwave Symp. Dig.*, 689-692.

[13] T. G. M. Kleinpenning, A. J. Holden, 1993, *IEEE Trans. Electron Devices*, **40**, 1148-1153.

[14] W. Liu, 1991, *Electron. Lett.*, **27**, 2115-2116.

[15] H. Ito, 1986, *Japan. J. Appl. Phys.*, **25**, 1400-1404.

[16] W. Liu, S.-k. Fan, T. S. Kim, E. A. Beam III, D. B. Davito, 1993, *IEEE Trans. Electron Devices*, **40**, 1378-1382.

Inst. Phys. Conf. Ser. No 145: Chapter 5
Paper presented at 22nd Int. Symp. Compound Semiconductors, Cheju Island, Korea, 28 August–2 September 1995
© *1996 IOP Publishing Ltd*

AlGaAs/GaAs Heterojunction Bipolar Transistors with InGaAs Etch-Stop Layer

Yosuke Miyoshi[1], Shin'ichi Tanaka, Norio Goto, and Kazuhiko Honjo

Microelectronics Research Laboratories, NEC Corporation 34 Miyukigaoka, Tsukuba, Ibaraki 305, Japan

Abstract. We report the first fabrication of AlGaAs/GaAs HBTs using InGaAs layer as dry-etching stopper. Various HBTs with different location of the etch-stop layer were fabricated to examine the influence of emitter-periphery guardring thickness on surface recombination. We show that it is vital to control the thickness to optimum 50 nm by selective etching technique, when the emitter periphery recombination current can be effectively reduced. We also discuss the current suppression effect caused by charge accumulation in the InGaAs quantum well. Simulation suggests that the undesirable side-effects can be minimized by appropriate epitaxial layer design.

1. Introduction

As the base layer thickness shrinks to less than 100 nm in advanced HBTs, there has been increasing demands for precise dry-etching technique to expose the thin base contact layer. Such technique is also required for delicate fabrication of emitter-guardring (using thinned-emitter) which has been demonstrated to prevent the current gain degradation in modern downscaled emitter HBTs [1]. For HEMTs, selective dry etching techniques using AlGaAs [2] or InGaAs [3] etch-stop layer have been commonly used for gate recess processing step. However, there has yet been few reports on the fabrication of HBTs using etch-stop layer, and little is known how the etch-stop layer affects device performance. In the case of HBTs, InGaAs is a strong candidate for etch-stop layer because the Cl_2, for which the InGaAs etching rate is favorably small, is widely used for mesa etching. The objective of this paper is to demonstrate the applicability of InGaAs based etch-stop layer to HBTs. We show that control of the guardring thickness by selective etching technique is vital to reduce the surface recombination along emitter periphery with high reproductivity.

[1] E-mail : ueda@uhl.cl.nec.co.jp

2. Design for InGaAs etch-stop layer

Prior to fabricating HBTs, we carried out preliminary experiments to determine the optimum In mole fraction for the InGaAs etch-stop layer. The dry-etching system used in this study (ANELVA ECR-6001) takes advantage of electron cyclotron resonance (ECR) plasma source for low damage, high yield GaAs etching. Main and sub-magnetic coils are combined for adjusting collimated magnetic field in ECR chamber. In contrast to conventional reactive ion beam etching (RIBE), no DC bias is applied to extract reactive ions from the plasma source, but samples are located near ECR plasma position to avoid surface damage caused by accelerated ions. In order to assist desorption of chlorides formed on etched surface, the ion energy is controlled by 13.56 MHz rf power which is supplied to the substrate. Throughout the experiment, low gas pressure of 0.4mTorr was used to ensure anisotropic etching profile of emitter mesa.

Figure 1 shows the proposed emitter mesa fabrication process using InGaAs etch-stop layer. For nonalloyed ohmic contact using refractory metal such as WSi, InGaAs is commonly used as an emitter cap layer. However, the low volatility of In chlorides generally leads to rough etched surface morphology unless sufficient rf power is applied. Thus we used high rf power of 60W to etch the InGaAs cap layer, followed by moderate rf power of 15W for GaAs and AlGaAs etching until the etching stopped at the InGaAs

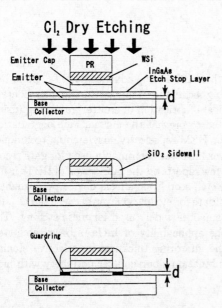

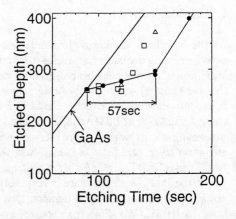

Figure 1. Emitter mesa fabrication process using InGaAs etch-stop layer.

Figure 2. Plots of etched depth vs. etching time for GaAs and the samples with 10 nm thick InGaAs layer. Dots, open rectangles, and open triangles show the data for the samples with In mole fraction of 25, 20, and 15%, respectively.

layer. Next, we formed SiO_2 sidewall (0.2 μm width), which was then used as wet-etching mask to remove the unwanted AlGaAs layer. Note that the thickness of the guardring under the sidewall can be precisely controlled by the location of the etch-stop layer.

The capability of the etch-stop layer to withstand Cl_2 dry-etching is determined by the layer thickness as well as the etching selectivity of GaAs vs. InGaAs. Because the etching selectivity is limited under reduced Cl_2 gas pressure condition, the thickness of the InGaAs etch-stop layer must be traded off against the accuracy of etched depth control. In this study, the thickness was fixed at 10nm, which we believe is the maximum value for meaningful use of selective etching technique. Various GaAs (300nm)/InGaAs (10nm)/GaAs (300nm) samples with different In mole fraction was dry-etched using the aforementioned etching condition (0.4mTorr, rf power of 15W). Figure 2 plots the etched depth vs. etching time for samples with In mole fraction of 15, 20, and 25 %. It can be seen that 25 % InGaAs layer is capable of stopping dry-etching for 57sec (corresponding to etching selectivity of 16.6), which is reasonably long considering that the total etching time required to complete emitter mesa formation is typically 150sec. The experiments as follows use 25 % In mole fraction.

3. Influence of the Etch-Stop Layer on DC Characteristics

Various HBTs with different guardring thickness (**d**) were fabricated in order to investigate the surface passivation effect. All devices had the same 80nm-thick GaAs uniform structure base (Be:2×10^{19}cm^{-3}). In contrast to FETs, the influence of the etch-stop layer on the fundamental dc characteristics is of special interest because the InGaAs layer remain in the transport channel, *i.e.* emitter. For devices with **d**=100, 150 nm, the current gain h_{FE} was typically 15-20 which is comparable to the h_{FE} obtained for a reference device without etch-stop layer, but the **d**=50nm device suffered from substantial current

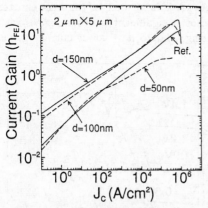

Figure 3. Current gain vs collector current density for the fabricated HBTs with various guardring thickness **d**. "Ref." denotes the reference HBT without etch-stop layer.

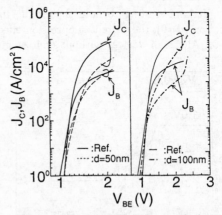

Figure 4. Measured Gummel plots for HBTs with InGaAs etch-stop layer at **d**=50 nm and 100 nm. "Ref." denotes reference HBT without etch-stop layer. V_{BC} was kept constant at 0V.

664

gain reduction (Fig.3). It can be seen from Fig.4 that the ideality factor of the collector current switches from n=1 in the low-injection level to n=2.6 in the medium-injection level (V_{BE} of 1.1 V to 1.6 V). The ideality factor in excess of 2 suggests that a potential barrier exists in the emitter. The current suppression effect of the etch-stop layer was most evident in the $\mathbf{d}$=50 nm device, but was observed for other devices with etch-stop layer as well.

To further examine the effect of the etch-stop layer on carrier transport, we used a device simulator BIUNAP-CT [4] to solve the Poisson equation and current equation. Figure 5 shows the simulated profiles for conduction band edge as well as electron density for $\mathbf{d}$=50 nm device. The simulation suggests that the electron accumulation developed in the InGaAs QW progressively lifts the conduction band with increasing the bias voltage (V_{BE}). Note, however, that the potential barrier has little effect on the carrier transport under low injection level. The onset of the current suppression is determined by the forward bias condition when the potential barrier becomes comparable to the conduction band for the base. Both measurements and simulation show that the critical forward bias voltage V_{BE} is about 1.2 V.

The peculiar I-V characteristics for HBTs with InGaAs etch-stop layer can also be seen by the sudden increase in both base current and collector current for even higher V_{BE} (See Fig.6). As V_{BE} is increased, the potential barrier for the holes in the base is gradually decreased until the holes are back-injected into the InGaAs valence-band QW. Meanwhile the accumulated holes neutralize the net charge in the InGaAs layer. As a result, the potential barrier for the electrons is lowered and thus the collector current is increased, but the current gain is never recovered because the base current is increased as well. Note that the second critical bias voltage (V_{BE}) for the base current anomaly varies for different samples because the potential barrier height in the valence band depends on the location of the InGaAs etch-stop layer (See Figs. 4 and 6). As far as $\mathbf{d}$=50 nm device is concerned, the two critical forward bias voltages were not distinguishable because they were nearly identical.

It is interesting to note that if the InGaAs layer is located at the base-emitter junction ($\mathbf{d}$=0), the current suppression effect is not as significant as other devices with $\mathbf{d}$>0. Although devices without emitter guardring tend to have lower current gains,

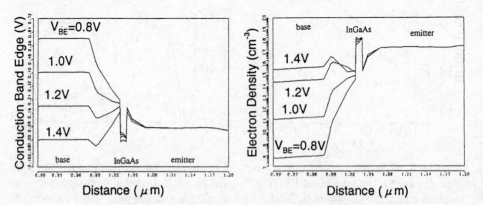

Figure 5. Simulated profiles for conduction band edge and electron density for HBT with InGaAs etch-stop layer at $\mathbf{d}$=50 nm. V_{BE} is varied as 0.8, 1.0, 1.2, and 1.4 V.

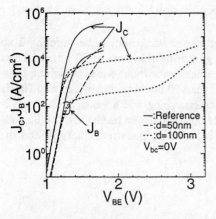

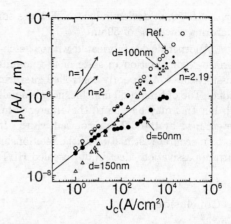

Figure 6. Simulated Gummel plots for HBTs with InGaAs etch-stop layer at **d** =50 nm and 100 nm.

Figure 7. Emitter periphery recombination current I_P v.s. collector current density as a function of **d**.

displacing the InGaAs layer to base-emitter junction would be one possible prescription to avoid the undesirable side-effect of the InGaAs etch-stop layer. On the other hand, the findings obtained in this work lead us to speculation that the side-effect may be perfectly avoided using, for example, quaternary InAlGaAs layer. Within critical thickness, the contents for Al and Ga can be determined so that the layer is bandgap-matched to $Al_{0.25}Ga_{0.75}As$ emitter while using the same 25 % In content as we used in this work. A simple calculation assuming linear interpolation between the bandgaps for InAs and AlAs gives $In_{0.25}(Al_{0.68}Ga_{0.32})_{0.75}As$, which should not induce charge accumulation in the AlGaAs emitter layer.

4. Optimization of Emitter Guardring Thickness

Measurements under low injection level allows us to compare devices with different guardring thickness without being affected by the current suppression effect. To investigate the passivation effect of the emitter guardring, we focused on recombination current along emitter periphery (I_P).

Figure 7 plots I_P versus collector current density J_C for various samples, where I_P was obtained by measuring the emitter size dependence of current gain (emitter size effect). It is well-known that I_P can be decomposed into two terms depending on the location where recombination takes place: The extrinsic base surface recombination component has an ideality factor of $n_P = 1$, because the carriers are once injected into the bulk base region [5]. The second component results from recombination at the SCR (along the emitter periphery) and thus has an ideality factor of $n_P = 2$. Noting that collector current has an ideality factor of $n=1$, the slopes for the plots in Fig.7 directly reflect the ideality factor of I_P (n_P). Under low injection level, it can be seen from Fig. 7 that the **d**=50nm device shows the minimum slope which translates into $n_P = 2.19$, while for other devices we found $n_P = 1.6$. The dominance of the $n_P = 2$ component for the **d**=50nm device indicates that the surface passivation effect is most evident for guardring thickness of 50nm.

Another device fabricated without etch-stop layer turned out to be suffering from surface recombination in spite of the nominal guardring thickness of 50 nm, which is possibly due to the poorly controlled guardring thickness without selective etching technique. The optimum 50 nm thickness is smaller than 90 nm obtained by Liu and Harris [6], but this may be due to the difference in the guardring width which in our case is determined by the 0.2 μm sidewall width. It is noted that our results are the first report on accurate estimation of the relationship between the guardring thickness and the surface passivation effect in downscaled HBTs.

5. Conclusions

We demonstrated the first AlGaAs/GaAs HBT fabrication using InGaAs etch-stop layer for accurate emitter mesa formation. One of the benefit of the selective dry etching technique is that the thinned-AlGaAs guardring can be formed in a precise manner. Various HBTs with different location of InGaAs layer were fabricated in order to investigate the guardring thickness dependence of the passivation effect. Comparing the emitter periphery recombination current among the fabricated HBTs, it is shown the guardring passivate the extrinsic base surface most effectively with 50nm guardring thickness. On the other hand, the narrow-gap InGaAs layer remaining in the bulk emitter was found to induce charge accumulation in the quantum well structure, leading to current suppression effect under high injection level. To counteract the undesirable side-effect, the quaternary InAlGaAs with the same In content should become a strong candidate for ideal etch-stop layer.

Acknowledgments

The authors would like to thank H. Shimawaki and Y. Amamiya for their helpful discussions, M. Tokushima and Y. Takahashi for their helpful advice in dry etching and device simulation, respectively. They also wish to thank J. Yamazaki, M. Mamada, and K. Iioka for their technical support.

References

[1] Ueda Y, Hayama N, Honjo K 1994 *IEEE Electron Device Lett.* **15** 66

[2] Hikosaka K, Mimura T, and Joshin K 1981 *Jpn. J. Appl. Phys.* **20** L847

[3] Kazior T E amd Patel B I 1992 *Mat. Res. Soc. Symp. Proc.* **240** 329

[4] Yano H, Kumashiro S, Goto N, and Ohno Y 1989 *IEDM Tech. Dig.* 151

[5] Liu W and Harris J.S 1992 *IEEE Trans. on Electron Devices* **39** 2726

[6] Liu W and Harris J.S 1993 *J. Vac. Sci. Technol. B* **11** 6

Inst. Phys. Conf. Ser. No 145: Chapter 5
Paper presented at 22nd Int. Symp. Compound Semiconductors, Cheju Island, Korea, 28 August–2 September 1995
© 1996 IOP Publishing Ltd

Investigation of Emitter Degradation in Heterostructure Bipolar Transistors

W A Hagley[1,2], R Rutyna [1], R K Surridge[1] and J M Xu[1,2]

[1]Bell-Northern Research Ltd., P.O. Box 3511, Station C, Ottawa, Ontario, Canada, K1Y 4H7

[2]University of Toronto, Dept. of Electrical and Computer Engineering, 10 King's College Road, Toronto, Ontario, Canada, M5S 1A4

Abstract. The operating characteristics of Heterostructure Bipolar Transistors (HBTs) have been shown to suffer degradation under temperature and current stress, leading to concerns about device reliability. Theoretical and experimental studies of Recombination-Enhanced-Impurity-Diffusion (REID) in AlGaAs/GaAs HBTs were performed in order to determine the cause of the degradation. The theoretical investigation was carried out using a one-dimensional solution of the Schrödinger, Poisson and current continuity equations, including an exact calculation of tunnelling and thermionic field emission at the heterojunctions. Temperature stressing of biased devices was performed at junction temperatures of 200, 215 and 230 C. The results support REID as the cause of the degradation in the AlGaAs/GaAs devices and the activation energy and REID coefficient are calculated.

1. Introduction

Heterostructure Bipolar Transistors (HBTs) are attractive for use in applications such as high speed optical communications, high power microwave circuits and high speed electronics due to their large current gain and fast switching speed. The degradation in the operating characteristics under stress poses a serious problem for applications which require long lifetime devices. Typically the degradation appears as a reduction in β with a corresponding increase in base current for a constant collector current.

Recombination-Enhanced-Impurity-Diffusion (REID) has been postulated as the cause of the degradation in the device characteristics[5].This mechanism involves the removal of dopant ions from the base by highly energetic recombination processes and diffusion of these ions into the emitter of the HBT. The diffused ions then activate in the emitter region, resulting in a net shift in the base-emitter doping profile. As the base emitter doping junction moves further into the emitter region, the gain drops and the base current increases.

REID has been observed in Be-doped GaAs [1] and the diffusion coefficient and activation energy for this process has been calculated from measured data. The increase in the diffusivity of the Be in GaAs was found to be on the order of 10^{15} at room temperature with an activation energy of 0.59meV. The diffusion coefficient was also found to be critically dependant on the current density.

Although the diffusivity of C is lower than that of Be by about 2 orders of magnitude, $(D_c = 6\times10^{-15} cm^2/s$, $D_{Be} = 1\times10^{-12} cm^2/s$ at 900C [6]) it will be shown that the activation

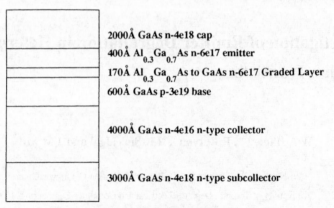

Figure 1 Layer Structure For AlGaAs HBT

energy for the REID is comparable with the values measured for Be in [1].

Simulations and experiments were carried out on AlGaAs/GaAs HBTs grown by Chemical Beam Epitaxy (CBE). The layer structure is shown in Figure 1. The samples were then fabricated using a self-aligned substitutional emitter process [7]. The experimental structures were then packaged and subjected to elevated temperature stressing at a current density of $4\times10^4\ A/cm^2$.

2. Experimental

The experimental portion of the investigation consisted of measuring devices under current and temperature stress. Samples were packaged and stressed at junction temperatures of 200, 215 and 230 C. The current stress was maintained at the collector current density of $4\times10^4 A/cm^2$.

The devices were measured periodically, and the percent change in the gain recorded. The samples consisted of 14 devices from various wafers. The measured data reported is the average of all samples tested.

3. Theoretical

Modelling was carried out using a one dimensional solution of the Schrödinger, Poisson and current continuity equations for both electrons and holes. The equations solved are:

$$\frac{d}{dx}\langle\varepsilon\frac{d\Psi}{dx}\rangle = -q\,(p-n+N_D-N_A) \tag{1}$$

$$\frac{dJ_n}{dx} = q\,(R-G) \tag{2}$$

$$\frac{dJ_p}{dx} = -q\,(R-G) \tag{3}$$

$$J_n = -q\mu_n n\frac{d}{dx}(\Psi_n + \Psi) + kT\mu_n\frac{dn}{dx} \tag{4}$$

$$J_p = -q\mu_p p\frac{d}{dx}(\Psi - \Psi_p) - kT\mu_p\frac{dp}{dx} \tag{5}$$

$$\Psi_n = \delta E_c + \frac{kT}{q} ln \left(\frac{N_c(x)}{N_{cr}} \right) + \frac{kT}{q} ln \left(\frac{F_{1/2}(n_c)}{e^{n_c}} \right) \tag{6}$$

$$\Psi_p = -\delta E_c - \frac{E_g(x) - E_{gr}}{q} + \frac{kT}{q} ln \left(\frac{N_v(x)}{N_{vr}} \right) + \frac{kT}{q} ln \left(\frac{F_{1/2}(n_v)}{e^{n_v}} \right) \tag{7}$$

$$n_c = n_{cr} - (q(\varphi_n - \Psi) + \delta E_c)/kT \tag{8}$$

$$n_{cr} = \frac{1}{2} ln \left(\frac{N_{vr}}{N_{cr}} e^{-E_{gr}/kT} \right)$$

$$n_v = -n_{cr} - (E_g + q(\phi_p - \Psi) + \delta E_c)/kT \tag{9}$$

$$\frac{-h^2}{2m^*} \left[\frac{d^2}{dx^2} + \Psi(x) \right] \psi(x) = E\psi(x) \tag{10}$$

where Ψ is the electrostatic potential, Ψ_n and Ψ_p are the position dependant band parameters which account for the varying material composition and Fermi statistics, δE_c is the conduction band discontinuity calculated as a portion of the total bandgap difference ($\delta E_c = 0.64 \cdot \Delta E_g$ for AlGaAs/GaAs junctions), φ_n and ϕ_p are the electron and hole quasi fermi levels, $\psi(x)$ is the wavefunction and all other symbols have their usual meanings.

At each applied bias, the coupled Poisson and continuity equations are solved using a globally convergent Newton-Rhapson iterative procedure. The potential and carrier profiles are then used in a numerical integration of the Schrödinger equation to calculate the total thermionic and tunnelling currents crossing the heterojunction interfaces. The mobilities of the electrons and holes at the mesh points immediately adjacent to the interface are then adjusted to agree with the thermionic and tunnelling currents. Poisson's equation and the current continuity equations are then solved again with the new value of the mobilities calculated in the previous step. This procedure is iterated until the difference between the potential calculated before and after the Schrödinger solution are within a

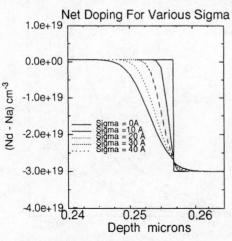

Figure 2 Doping profiles for various values of Sigma

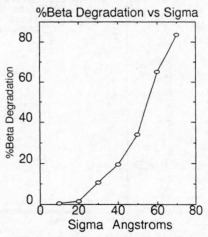

Figure 3 Percent decrease in Beta with Sigma

specified tolerance. At this point, a self-consistent solution has been obtained.

In order to determine the effect of the dopant out diffusion from the base into the emitter of the HBT, the structure shown in Figure 1 was simulated using several dopant diffusion profiles. The expression used to evaluate the carrier profile is

$$N_a(x) = \frac{N_a^0}{2.0}\left(1.0 - erf\left[\frac{x - x_{be}}{\sigma}\right]\right) \tag{11}$$

where N_a^0 is as grown acceptor concentration at the base-emitter junction edge, x_{be} is the location of the base-emitter metallurgical junction and σ is a parameter describing the shape of the out diffusion [2]. σ is related to the time and the diffusion coefficient via

$$\sigma = 2\sqrt{D_c t} \tag{12}$$

where D_c is the diffusion coefficient for C in GaAs.

Figure 2 shows the calculated net doping density used in the calculation of the beta degradation curves. Figure 3 shows the calculated percent decrease in the current gain calculated for the above structure as a function of the out-diffusion parameter σ.

The diffusion coefficient and its associated activation energy can be extracted from the model and the measured data. From the measured data at each temperature, the time for the gain to decrease by 20% can be read from the measured data curves. The value of σ which gives the same gain degradation can then be read off the curve in Figure 3. Substitution of these values into (12) results in a value for the diffusion coefficient at each temperature.

With the values of D_C obtained at the three different temperatures, the activation energy and the value of the REID coefficient can be calculated.

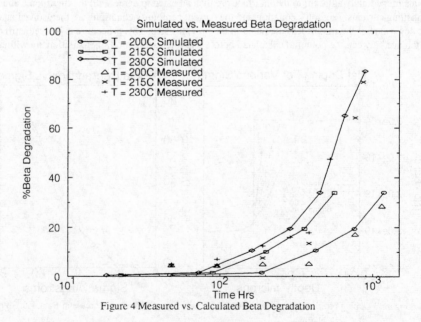

Figure 4 Measured vs. Calculated Beta Degradation

4. Results and discussion

The values of the diffusion coefficient extracted from the simulation and the measured data are

$$D_c(200C) = 1.465 \times 10^{-20} cm^2/s$$

$$D_c(215C) = 3.091 \times 10^{-20} cm^2/s$$

$$D_c(230C) = 3.825 \times 10^{-20} cm^2/s$$

Figure 4 compares the experimentally measured beta degradation with the calculated degradation using the diffusion coefficient extracted from the measurements. The simulated curves and measured data are in good qualitative and reasonable quantitative agreement.

The calculated REID coefficient for GaAs is

$$D_c = 1.58 \times 10^{-13} exp(-0.664eV/kT) cm^2/s \qquad (5)$$

for a current density of $4 \times 10^4 A/cm^2$.

This is consistent with the activation energy calculated for the REID of Be where the activation energy was found to be 0.59eV [1]. This is also consistent with the data on the time to failure of C doped AlGaAs/GaAs HBTs. The workers in [5] measured an activation energy of 0.6eV which is consistent with our measurements of beta degradation and the findings of this paper.

Figure 5 shows the change in the calculated base current at a collector current of $4 \times 10^4 A/cm^2$. This figure clearly demonstrates the increase in the base current occurring in conjunction with the beta degradation, as is seen in experimental measurements. This lends strong support to the theory that the base dopant diffusion is the main cause of the emitter-base junction degradation.

The degradation cannot be ascribed to reduction in the emitter n-type doping alone as would occur if the emitter doping diffused into the base. Experiments using reduced emitter dopings did not show a significant reduction in the gain or increase in base current. In order for the degradation to occur the doping junction must move into the emitter of the HBT and become active there.

Recent work in GaInP/GaAs HBTs doped with carbon in the base do not show the same type of degradation with current and temperature stressing [5]. The reason for the lack of degradation in InGaP has not as yet been resolved. It is known that C does not activate well in InGaP and so the diffusion of C into the emitter does not move the doping junction.

5. Conclusions

The Recombination-Enhanced-Impurity-Diffusion coefficient for GaAs at a current density of $4 \times 10^4 A/cm^2$ has been calculated and is found to be in good agreement with data for the REID of Be. The mechanism of REID has applied to HBT simulations which reproduced the features observed in the measured devices, both the gain degradation and base current increase. This supports the theory that the REID in the base emitter junction of AlGaAs/GaAs HBTs is responsible for the degradation of the transistor characteristics.

There are several modifications to the device structure which can help to reduce the rate of beta degradation in AlGaAs/GaAs HBTs which are currently under investigation. Grading of the base dopant from a lower value at the base-emitter interface will reduce the rate of diffusion by reducing the dopant gradient, as well as reducing the recombination rate at the junction edge, thereby reducing the energy available for REID. Grading of the Al content can also reduce the sensitivity of the device to REID.

Acknowledgments

The authors would like to thank T Moore and P Mandeville for sample growth and useful discussions and J Cook for measurements of devices.

References

[1]Masashi Uematsu and Kazumi Wada 1991 Appl. Phys. Lett. **58** (18) 2015-2017

[2]Y Betser and D Ritter 1995 IEEE Electron Device Letters 16 (3) 97-99

[3]B T Cunningham, L J Guido, J E Baker, J S Major, N Holonyak Jr., and G E Stillman, 1989,Appl. Phys. Lett. **55** (7) 687-689

[4]Byoung-Ho Cheong, K J Chang 1994, Physical Review B, 49 (24), 17436-17439

[5]T Takahashi, S Sasa, A Kawano, T Iwai and T Fujii, 1994, Proc. IEDM, 8.3.1-8.3.4

[6]Koki Saito, Eisuke Tokumitsu, Takeshi Akatsuka, Motoya Miyauchi, Takumi Yamada, Makoto Konagai and K Kiyoshi Takahasi, 1988, J. Appl. Phys. 64 (8), 3975-3979

[7]T Lester, R K Surridge, S Eicher, J Hu, G Este, H Nentwich, B Maclaurin, D Kelly, I Jones, 1993, Gallium Arsenide and Related Compounds Proceedings, Institute of Physics Conference Series 136, 449-454

Inst. Phys. Conf. Ser. No 145: Chapter 5
Paper presented at 22nd Int. Symp. Compound Semiconductors, Cheju Island, Korea, 28 August–2 September 1995
© *1996 IOP Publishing Ltd*

Thermal Analysis of AlGaAs/GaAs Heterojunction Bipolar Transistors Including Base Recombination Current Effect

B. U. Ihn, J. Lee, Y. Suh, and B. Kim

Department of Electrical and Electronic Engineering
and Microwave Application Research Center
Pohang University of Science and Technology
Hyoja-dong San-31, Pohang, Kyung-buk, 790-784, Korea

H. C. Seo, W. Jung, and D. S. Ma

Electronics Division, Kukje Corporation
Daeya-dong San-6, Shihung, Kyunggi, 429-010, Korea

Abstract. One of the major influences on the NDR(negative differential output resistance) effect of AlGaAs/GaAs HBTs is the surface recombination current. It degrades current gain and affects thermal stability. In this work, we have separated the base current into area and edge current components, and have extracted the formula for each current components. The electro-thermal simulation which satisfies the electrical equations and heat transfer equation has been performed. The electrical equations included the temperature-dependent nonuniform base current and current gain. The simulated result shows a good agreement with measured I-V curve, indicating that the model used in this simulation is useful for analyzing exact thermal behaviors of HBTs.

1. Introduction

AlGaAs/GaAs heterojunction bipolar transistors(HBTs) are suitable for high power microwave sources. They have high speed, high breakdown voltage and high current handling capability. However, because of the negative temperature coefficient of the base-emitter turn-on voltage and the poor thermal conductivity of GaAs, the power performance of the HBTs is thermally limited rather than electrically limited. The thermal properties of HBTs have been investigated by many authors[1]~[3].

The thermal properties have been usually characterized from the current variation as a function of junction temperatures. Recently, J.R. Waldrop et al. reported that NDR effect of AlGaAs/GaAs HBTs is related not only to the intrinsic thermal properties but also to the surface recombination current[3]. Therefore, in our study, we separated the base current into two current components, i.e., the area current flowing through the intrinsic base region and the edge current caused by the surface recombination at the emitter periphery. Using the electro-thermal model including nonuniform base current distribution and base edge

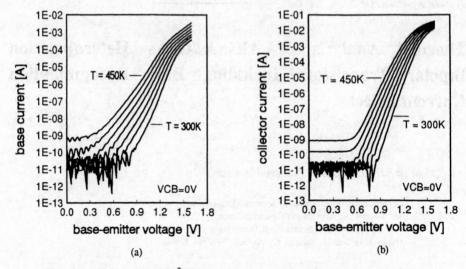

Fig.1 Gummel plots of 2×10 um^2 HBT at various temperatures. (a) I$_B$-V$_{BE}$ plot. (b) I$_C$-V$_{BE}$ plot.

recombination current, we simulated the current variations due to the self-heating effect and compared it with measured I_C-V_{CE} data.

2. Thermal measurements

Devices with three emitter sizes were measured: 2×10, 3×10, and 2×25 μm^2, respectively. The perimeter-to-area ratio(p/A) is 1.2, 0.87, and 1.08 μm^{-1}, respectively. The chip sizes of three devices are $380 \times 440 \times 100$ μm^3.

In this study, Gummel plots of these devices were measured at ambient temperatures

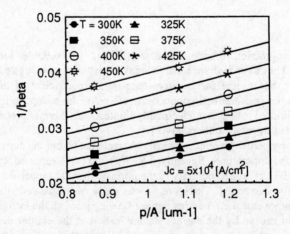

Fig. 2 Current gain versus perimeter-to-area ratio(p/A) at various temperatures.
Lines are least-squares fits.

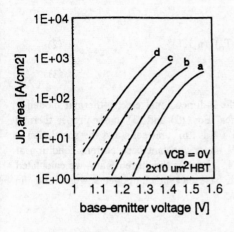
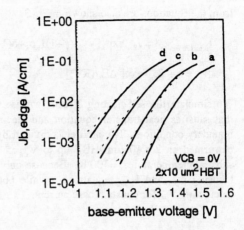

Fig.3 The extracted Gummel plots of base currents for edge and area current components. (a) area base current. (b) edge base current. (———:measured, - - - -:calculated, a:300 K, b:350 K, c:400 K, d:450 K).

between 300 and 450 K. Fig.1 shows the temperature dependence of collector and base currents for $2 \times 10\ \mu m^2$ HBT. In order to minimize the junction temperature rise above the externally applied ambient temperature,the base-collector bias was maintained at 0 V. Fig.2 shows $1/\beta$ vs. p/A over a T_A range of 300 to 450 K at $J_c = 5 \times 10^4\ A/cm^2$. Since the relationship between β and p/A can be expressed as Eq.(1), we can separate the two base current components at different temperatures from the slope of β^{-1} vs. p/A plot.

$$\beta^{-1} = 1/\beta_{area} + (J_{B,edge}/J_C)(p/A) \qquad (1)$$

By measuring such a plot as Fig.2 at various collector current levels, we obtained two Gummel plots for each base current component. Fig.3 shows the Gummel plots for $2 \times 10\ \mu m^2$ HBT.

The HBTs characterized were Si_3N_4 passivated and the surface recombination current($I_{B,edge}$) of the device is about the same as the base current in the intrisic base region($I_{B,area}$) as shown in Fig.4. In addition, we could not see any sharp increase of $I_{B,edge}/I_{B,area}$ ratio with temperature in our Si_3N_4 passivated HBT unlike AlGaAs ledge passivated HBT reported in reference [3].

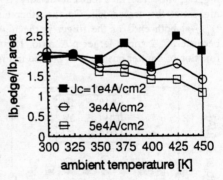

Fig. 4 Edge-to-area base current component ratio of 2x10um² HBT versus ambient temperature

3. Theoretical models

To simplify the calculation of thermal behavior, $J_{B,area}$, $J_{B,edge}$ and β equations given by Eqs. (2) and (3) are extracted from the curve fitting of the measured data. The currents calculated from the equations are also shown in Fig.3.

$$J_{B,area\ or\ B,edge} = J_{BO}.\exp\{[V_{BE}-(\beta+1)J_B\rho_E+\delta(T-T_O)]/(n_BKT)\} \quad (2)$$

$$\beta^{-1} = a + \beta_O^{-1}.\exp[-\Delta E_v/(KT)] \quad (3)$$

To simulate the self-heating effects, we use the 2-dimensional electro-thermal model[1] that satisfies heat transfer equation and electrical Eqs. (2) and (3) under proper thermal boundary conditions. The emitter is divided by 1×1 μm^2 unit cells and the simulation is performed on 2×10 μm^2 HBT up to $V_{CE}=5$ V under constant base current condition at substrate temperature T=300 K. Because only $I_{B,area}$ causes the current gain, we calculated I_C using β_{area} and $I_{B,area}$ equations, while both base current components were used in calculating V_{BE} for constant base current.

4. Results and discussion

Fig.5 shows the calculated common-emitter I_C-V_{CE} curve with base current of 0.1 mA. Due to the surface recombination current which does not contribute to the current amplification, the collector current is reduced by a factor of about 1/2.

Fig.6 shows the variations of $I_{B,area}$ and $I_{B,edge}$ according to emitter positions from the feeding point under constant base current case. The total base current is 0.2 mA. Because the temperature at the center of the finger is the highest, the base area current density is the highest at the center cell. On the other hand, $I_{B,edge}$ at both ends of the finger is the highest because of 2 times larger p/A ratio. I_C as a function of emitter positions is plotted in Fig.7.

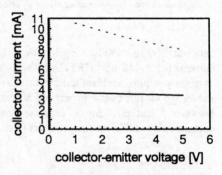

Fig. 5 Simulated IC-VCE plot with and without emitter size effect. (—— : with emitter size effect, - - - : without emitter size effect)

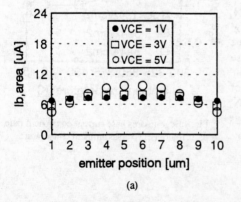

(a)

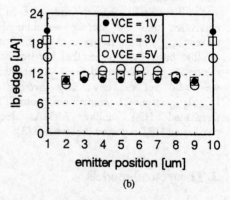

(b)

Fig.6 Base current components at each emitter position. (a) base area current, IB,area. (b) base edge current, IB,edge.

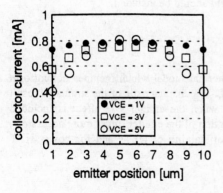 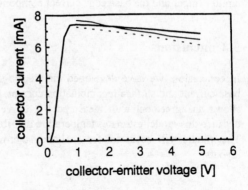

Fig.7 Collector currents at each emitter position

Fig. 8 Ic-Vce characteristics of 2x10 um² HBT
(———: measured Ic
——: simulated Ic with nonuniform base current
- - -: simulated Ic without nonuniform base current)

Because of the negative temperature coefficient of the base-emitter turn-on voltage and higher temperature at the center of the finger, more collector current flows through the center cell. The figure also shows that the electro-thermal feedback causes the current nonuniformity more serious as V_{CE} increases.

Fig.8 shows the I_C-V_{CE} calculation. As shown, the simulation including nonuniform base current produces far more accurate curve than the one with uniform base current. In case of uniform base current, it is assumed that the base edge current does not exist. The calculated I_C with nonuniform base current deviates maximum 9.3 % from measured I_C, while the one with uniform base current deviates maximum 10.6 %. To study the difference, temperature distribution along the emitter finger was calculated (Fig.9). The junction temperature at the center of the nonuniform base current case is lower, producing more collector current for a fixed base current. On the other hand, in case of uniform base current, the base edge current effect is not considered and, therefore, the temperature at the center of the finger is overestimated to produce lower collector current. This study clearly indicates that, in order to study thermal effect, the nonuniform current distribution along the

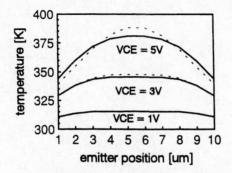

Fig. 9 Temperature profile of nonuniform (solid lines) and uniform (dashed lines) base current case on the emitter finger.

emitter finger and the base edge current component should be included.

5. Conclusions

In conclusion, we have developed the electro-thermal model which includes nonuniform base current and surface recombination current. I_C-V_{CE} characteristic of the simulated result shows an agreement with measured I-V curve when the emitter size effect is included because this model estimates temperature distribution of the emitter finger more accurately. This model will be useful for analyzing thermal stability of HBTs and for power HBT layouts.

References

[1] L.L. Liou, and B. Bayraktaroglu, 1994, *IEEE Trans. Electron Dev.*, **41**, 629-636.
[2] W. Liu, S. Nelson, D. Hill, and A. Khatibzadeh, 1993, *IEEE Trans. Electron Dev.*, **40**, 1917-1927.
[3] J.R. Waldrop and M.F. Chang, 1995, *IEEE Electron Device Lett.*, **16**, 8-10.

Inst. Phys. Conf. Ser. No 145: Chapter 5
Paper presented at 22nd Int. Symp. Compound Semiconductors, Cheju Island, Korea, 28 August–2 September 1995
© *1996 IOP Publishing Ltd*

SiGe HBTs With Very High Germanium Content In The Base

A. Gruhle, A. Schüppen, H. Kibbel, U. König

Daimler-Benz Research, W. Runge Str. 11, 89081 Ulm, Germany

Abstract. A way of improving the performance of SiGe HBTs is to increase the base doping in order to reduce the base resistance. This has to be accompanied by a higher Ge content which in addition will result in a smaller turn-on voltage giving lower power consumption. For more than 50% Ge the collector saturation current of the HBT will be larger than that of a PtSi n-Schottky diode. This opens new fabrication possibilities with self-aligned PtSi base contacts without ion implantation and high-temperature anneal.

This paper investigates SiGe HBTs with very high base Ge contents up to 57% with excellent high frequency performance and with collector currents two orders of magnitude higher than PtSi-diodes. Fabrication of devices with a self-aligned PtSi base, however, is hindered by the formation of PtSiGe alloys which have a smaller barrier height than PtSi.

1. Introduction

The advantage of an HBT is its thin and at the same time highly doped base which reduces base resistance and transit time. Increasing the base doping requires a higher band gap difference, i.e. in the case of SiGe HBTs a higher Ge content in the SiGe base. Typical values [1,2] are $8 \times 10^{19} \text{cm}^{-3}$ doping and 30-40% germanium. Recent experiments show that up to $4 \times 10^{20} \text{cm}^{-3}$ are possible for a further reduction of the base resistance [3]. Even higher Ge contents will therefore be necessary in the future. An additional advantage of these extreme HBTs is the low turn-on voltage which is 0.3V- 0.4V less than in a silicon bipolar transistor thereby reducing the power consumption significantly.

Most important is the possibility of a new low thermal budget fabrication technology: A high Ge content would allow the use of platinum silicide to contact the base without the need for neither mesa formation nor ion implantation. This is because the Schottky saturation current of PtSi on n-Si would be lower than the collector saturation current of the HBT. The absence of ion implantation eliminates problems such as high temperature anneal, end-of-range damage, contact resistance and incomplete activation. In addition a simple self alignment between E and B can be realized with a salicide technology.

2. Experimental results

In this work the growth, fabrication and performance of HBTs with very high Ge contents up to 57% have been investigated. Growth of the complete layer structure was done by MBE as described elsewhere [4].

sample	Ge content	growth temp. °C	base (nm) i_{cb}-w_b-i_{be}	doping (cm^{-3})	J_s (A/cm^2)	f_T/f_{max} (GHz)	remarks
2847	30%	450	10-20-2	8E19	1.5 E-8	74/54	
2844	35%	530	10-15-2	8E19	4.0 E-8	84/59	
3203	40%	530	7-15-3	1E20	5.1 E-8	64/83	BC leak
3230	40%	480	7-15-3	1E20	5.1 E-8	64/80	
3204	45%	530	5-7-2	8E19	9.8 E-8	58/27	BC leak
3234	45%	480	5-7-2	8E19	9.8 E-8	67/37	
2884	45.5%	450	5-5-2	1E20	4.4 E-7	80/37	
2848	46%	450	5-10-2	8E19	4.6 E-7	113/50	
2883	50%	450	3-5-2	1E20	7.8 E-7	68/36	
3263	55%(30%)	450	3-5-2	1E20	2.8 E-6	86/65	BC+BE leak
3383	57%(30%)	400	3-5-2	1E20	5.0 E-6		BE leak
3322	57.5%(30%)	450	3-5-2	1E20	9.0 E-6	36/18	BE leak

Table 1: Data of SiGe HBTs with high Ge contents in the base

The layer parameters of the investigated samples are given in Table 1. Fig.1 shows that depending on the Ge content the maximum base thickness is limited in order to stay within the metastable growth region. Above the indicated line the strained layer relaxes as observed with x-ray rocking experiments. However, dislocations occur much earlier, detectable as leakage currents in fabricated devices. Lowering the growth temperature extends the

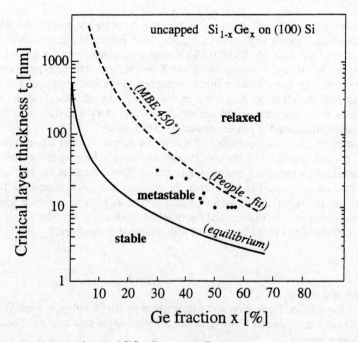

Fig. 1: Critical layer thickness of strained SiGe alloys versus Ge content

metastable region to thicker layers. For this reason most of the samples were grown at 450°C, compared to the standard value of 530°C [1]. However, a further reduction of the growth temperature deteriorates crystal quality leading to neutral base recombination and BE leakage currents [3]. The experimental data points of this paper are indicated in Fig. 1.

The total base thickness includes the doped region (typically 1×10^{20}cm^{-3}) and the undoped spacer layers on both emitter and collector sides that prevent outdiffusion [5]. The necessary BC spacer thickness is always larger than the EB spacer because the collector doping is usually one order of magnitude lower than the emitter impurity level. The necessary total spacer width of about 5 nm determines the smallest possible base width plus the doped region which should have at least a few nm for a decent base resistance. Note that an increase of the base doping requires thicker cladding layers, i.e. for a given base sheet resistance the base cannot be made arbitrarily thin. For this reason in the case of the 4 samples with the highest Ge contents the first 2nm of the CB spacer contained only 30% Ge in order to reduce the total strain.

Table 1 lists the measured collector saturation currents [6] at 294.6K. They were normalized to a base Gummel number of 5×10^{13}cm^{-2}. The given f_t and f_{max} values are the maximum measured frequencies on mesa type devices [1,2]. The numbers do not only reflect the base properties but they also strongly depend on collector design and fabrication technology. Some samples had leakage currents as indicated in the last column. At high collector current densities, however, this did not influence transistor operation.

Fig.2 is a plot of the normalized HBT collector saturation current vs. Ge content in the base. Data points for low Ge values are from reference [6]. The dashed curve is the expected

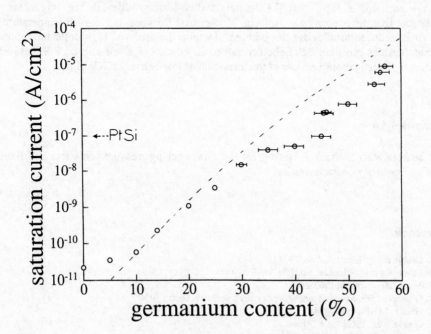

Fig. 2: Measured collector saturation currents of SiGe HBTs at 294.6°C corrected for a base Gummel number of 5×10^{13}cm^{-2}. The dashed line is the People-Bean fit without bandgap narrowing.

value according to the theoretical band gap reduction of strained SiGe alloys [7]. It does not include band gap narrowing which explains the deviation at low Ge contents. However, at high Ge values the experimental points fall below the curve. Possible explanations are a reduction of the band gap narrowing, of the mobility or of the density of states [6]. Note that an uncomplete base dopant activation would have the opposite effect.

The arrow indicates the saturation current density of a PtSi n-Schottky diode with a 0.84eV barrier. A SiGe HBT with more than about 50% of Ge in the base can therefore tolerate the leakage currents of a PtSi Schottky diode at the periphery of a PtSi gate with direct contact to the emitter (unless it touches the n+ emitter contact). First experiments with this type of self-aligned base contact, however, have led to the formation of PtSiGe alloys which showed about 0.7eV barrier height, corresponding to about 10^{-5}A/cm^2. This is probably due to a Ge diffusion during the alloying process leading to a direct contact between PtSiGe and the emitter. A proper choice of emitter and PtSi thickness is therefore crucial. The exact barrier height dependence on Pt thickness and anneal temperature needs further investigation.

3. Conclusion

The investigated samples are close to the physical limits of SiGe HBT growth. Excellent high frequency performance indicate a correct transistor function, i.e. no outdiffusion, complete dopant activation and almost the expected enhancement of the collector saturation current. The latter point needs further investigation, the deviation from the theoretical curve remains unexplained. The DC characteristics of the HBTs with the highest Ge fractions exhibit leakage currents, a proof that the crystal quality suffers: although the layers are still metastable first dislocations start to form. Neither base thickness nor growth temperature can be further reduced to alleviate the problem. Despite the achieved high collector saturation current densities exeeding PtSi Schottky values the concept of a self-aligned PtSi base needs additional investigations because of the formation of low-barrier PtSiGe alloys.

Acknowledgement

The authors wish to thank J. Herzog for the x-ray rocking measurements and U. Erben for the high frequency characterization.

References

[1] A. Gruhle et al., Electron. Lett. 29, 415, 1993
[2] A. Schüppen et al., Electron. Lett. 30, 1187, 1994
[3] F. Beisswanger, unpublished results
[4] A. Gruhle in "Silicon-based mm-wave devices", Springer, Berlin 1994
[5] A. Gruhle, ED-14, 198, 1994
[6] A. Gruhle et al., ISCS 1994, 30
[7] R. People et al., APL-49, 229, 1986

Inst. Phys. Conf. Ser. No 145: Chapter 5
Paper presented at 22nd Int. Symp. Compound Semiconductors, Cheju Island, Korea, 28 August–2 September 1995

Piezoelectric effect suppression in GaAs FETs by using Y-shaped gate structures

M. Fukaishi, M. Tokushima, S. Wada, H. Hida, and T. Maeda

Microelectronics Research Laboratories, NEC Corporation
34 Miyukigaoka, Tsukuba, Ibaraki 305, Japan

Abstract. The overlayer stress dependence and the gate structure dependence of the threshold voltage shift (ΔV_T) caused by the piezoelectric effect are investigated. The reduction of the overlayer stress is effective to reduce ΔV_T in long gate-length FETs, but in short gate-length FETs ΔV_T remains high. In FETs with a Y-shaped gate, ΔV_T is smaller than that of rectangular gate FETs. This is because the Y-shaped gate wings relax the stress under the gate electrode. Experiments and two-dimensional simulations show that ΔV_T decreases as the length (L_w) between the gate edge adjoining the substrate and the Y-shaped gate wing edge increases. They also show that ΔV_T is reduced to almost 0 V when L_w is greater than 0.7 μm.

1. Introduction

The piezoelectric effect, which is caused by the stress of the dielectric overlayer or the gate metal for a GaAs FET, changes the device characteristics, such as the threshold voltage [1]-[4]. This threshold voltage shift (ΔV_T) increases as the gate lengths are reduced in the pursuit of higher-performance FETs. Since ΔV_T limits the IC yield, it must be reduced as much as possible. Although ΔV_T can be reduced by using a substrate orientation vanishing the piezoelectric charges, such as (011) [3], this orientation is not commonly used.

The main purpose of this paper is to find the way to reduce ΔV_T through the piezoelectric effect. The influence of overlayer-stress reduction will be described, and the dependence on a Y-shaped gate structure will be discussed.

2. Experiments

The Y-shaped gate n-$AlGaAs/InGaAs$ heterojuncton FETs were fabricated on (100) GaAs substrate. The fabrication process was the same as that reported in Ref. 5. The gate electrode orientation was changed between 0 and 180 degrees, corresponding to

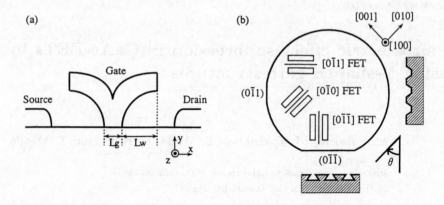

Fig. 1. (a) Cross-sectional view of the Y-shaped gate FET and coordinate axes used in the numerical simulation of the stress. (b) Definition of the FET orientations θ relative to the GaAs crystal, and illustration of the shapes of the etched grooves in the substrate.

the crystal orientations of $[0\bar{1}\bar{1}]$ and $[011]$. Figure 1(a) shows a cross-sectional view of an FET in which the gate wing length (L_w) between gate edge adjoining the substrate and the Y-shaped gate wing edge is typically 0.33 μm. Figure 1(b) shows the gate orientation relative to the substrate and shows the shapes of the etched grooves in the substrate. After fabricating the devices, two types of SiON films were deposited for the dielectric overlayer by controlling process conditions of plasma-enhanced chemical-vapor-deposition (P-CVD). These films had compressive stresses of -3.2×10^9 dyn/cm^2 and -3×10^8 dyn/cm^2. The value of -3.2×10^9 dyn/cm^2 for high-stress film is conventionally used. The value of -3×10^8 dyn/cm^2 for low-stress film is one-tenth of that with the high-stress film and is the minimum value that can be established reproducibly by our equipment. The thickness of both films was 1 μm. The value of ΔV_T was evaluated from the difference in the FET characteristics before and after these films were deposited. By this method, the influence of the gate metal stress and the short-channel effect can be disregarded.

3. Results and discussions

The gate-orientation dependence of the threshold voltage shift (ΔV_T) in typical FETs is shown in Fig. 2, where the gate lengths (L_g) are 0.35 μm and 2.0 μm with the high-stress film and L_w is 0.33 μm. The threshold voltage shift (ΔV_T) varies with a period of 180 degrees; the value of ΔV_T is maximum at 0, 90, and 180 degrees corresponding to the crystal orientation $[0\bar{1}\bar{1}]$, $[0\bar{1}1]$, and $[011]$ and is 0 V at 45 and 135 degrees, or $[0\bar{1}0]$ and $[001]$. This is because the components of piezoelectric tensor vary with a period of 180 degrees in (100) substrate in the case of the zinc-blende structure. These experimental results agree well with those obtained by using two-dimensional device simulator [2].

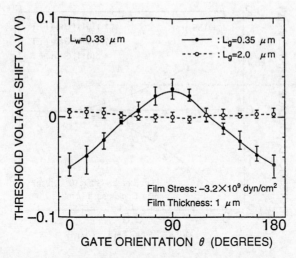

Fig. 2. Threshold voltage shift ΔV_T versus the gate orientation for two gate lengths L_g, where the gate wing length L_w is 0.33 μm with the high-stress film.

Although ΔV_T can be reduced by placing the gate parallel to the orientations at which piezoelectric charges vanish, such as [0$\bar{1}$0], these orientations are not suitable for practical use because of decreasing packing density and increasing chip size. The film stress must, therefore, be reduced in order to reduce ΔV_T.

Figure 3 shows ΔV_T versus L_g for various gate orientations and for two film stresses in typical FETs, where L_w is 0.33 μm. With the high-stress film, ΔV_T increases as L_g decrease below 2.0 μm. The value of ΔV_T is smaller than that for the rectangular gate

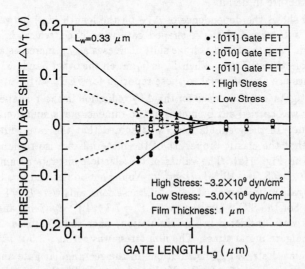

Fig. 3. Threshold voltage shift ΔV_T versus the gate length L_g for various gate orientations and two film stresses, where the gate wing length L_w is 0.33 μm.

686

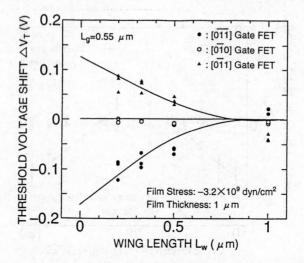

Fig. 4. Threshold voltage shift ΔV_T versus the Y-shaped gate wing length L_w for various gate orientations, where the gate lengths L_g are 0.55 μm with the high-stress film.

FETs [4]. This suggests that ΔV_T is significantly affected by the gate structure. With the low-stress film, ΔV_T is smaller than that with the high-stress film and is negligibly small when L_g is greater than 0.7 μm. However, ΔV_T remains high in short gate-lengths FETs even when the film stress is reduced. Therefore, the stress under the gate electrode must be further reduced in short-gate FETs. Since the Y-shaped gate structure can reduce the piezoelectric effect, we focus on the stress under the gate and investigate the influence of the gate structure in detail.

Figure 4 shows the dependence of ΔV_T on the length (L_w) between the gate edge adjoining the substrate and the Y-shaped gate wing edge, where L_g is 0.55 μm with high-stress film. The threshold voltage shift decreases as L_w increases and is almost 0 V when L_w is more than 0.7 μm. When L_w is 0 μm, on the other hand, ΔV_T is about ± 0.15 V. These values are almost equal to those reported for FETs with rectangular gates [4].

To investigate into the reason for the ΔV_T reduction in the Y-shaped gate FETs, the stress analysis was carried out by means of two-dimensional simulations using the finite-element method. In these simulations, we assumed that the gate width is much larger than L_g and that the elastic properties of the materials are isotropic in the coordinate system shown in Fig. 1(a). The values of the elastic constants assumed for the GaAs substrate were $E = 8.53 \times 10^{11}$ dyn/cm^2 for Young's modules and $\nu = 0.31$ for Poisson's ratio. For the gate metal, assumed to be Au, these constants were $E = 8.1 \times 10^{11}$ dyn/cm^2 and $\nu = 0.42$. For the SiON film they were $E = 7.3 \times 10^{11}$ dyn/cm^2 and $\nu = 0.17$. Only the film stress was considered in this simulation, since the measured ΔV_T excludes the influence of the gate metal stress. The film stress was set at -3.5×10^9 dyn/cm^2.

The simulated stress σ_{yy} distributions in the rectangular gate and Y-shaped gate FETs are shown in Fig. 5. In the Y-shaped gate FET, the stress at the gate edge

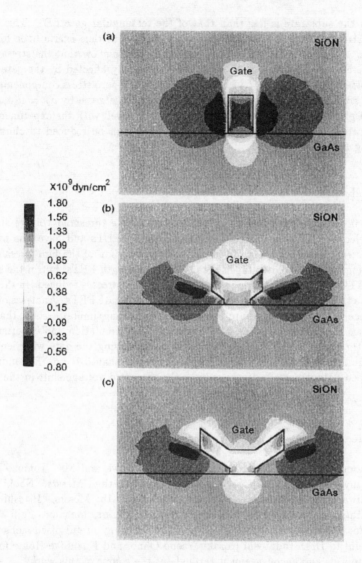

Fig. 5. Numerically simulated stress σ_{yy}, where the gate length is 0.6 μm and the film stress is -3.5×10^9 dyn/cm^2. (a) the rectangular gate FET. (b) $L_w = 0.3\mu$m. (c) $L_w = 0.7\mu$m.

adjoining the substrate is less than that of the rectangular gate FET. This is because the film stress concentrates at the gate wing edges, and thus exerts little force on the substrate. Analyses of the L_w dependence of the stress show that the stress under the total gate electrode adjoining the substrate is markedly affected by the gate structure, and the stress decreases with L_w increasing. Other non-zero stress components (σ_{xx}, σ_{zz}, and σ_{xy}) are also reduced by increasing L_w, where σ_{zz} is given by $\sigma_{zz} = \nu(\sigma_{xx} + \sigma_{yy})$ in a state of plane strain. These simulated results agree well with the experimental results and indicate that ΔV_T for short gate-length FETs can be reduced to almost 0 V by expanding L_w.

4. Conclusion

We have investigated the film-stress dependence of the threshold voltage shift caused by piezoelectric effect due to the film stress in GaAs FETs and have also investigated the gate structure dependence of this shift. The reduction of the film stress is effective for reducing the threshold voltage shift in long gate-length FETs, but in the short gate-length FETs this shift remains high even when the film stress is reduced. In the Y-shaped gate FETs this shift is smaller than that of rectangular gate FETs. Analyses of the stress dependence on the gate structure using two-dimensional simulator show that the gate wings relax the stress under the gate electrode. The simulation and experiments show that the threshold voltage shift is reduced with increasing the gate wing length and is almost equal to 0 V when the wing length is greater than 0.7 μm. Expanding the T- or Y-shaped gate wings effectively reduces the threshold voltage shift in the short-gate FETs.

Acknowledgments

The authors would like to thank Mr. Masaoki Ishikawa and Ms. Tomoko Nakayama for their invaluable contributions to the device fabrication, Messrs. Naoki Furuhata, Takaki Niwa, and Jin Yamazaki for their crystal growth, Messrs. Hitoshi Yano and Noriaki Matsuno for their help with measurements, Messrs. Masahiro Fujii and Keiichi Numata for mask design, and Dr. Junji Matsui for many useful discussions. They are also grateful to Drs. Tadatoshi Nozaki, Yasuo Ohno, and Kazuhiko Honjo for valuable advice, support, and encouragement throughout the course of this work.

References

[1] Asbeck P M, Lee C P, and Chang M F 1984 *IEEE Trans. Elec. Dev.* **ED-31** 1377-1380

[2] Onodera T, and Nishi H 1989 *IEEE Trans. Elec. Dev.* **ED-36** 1580-1585

[3] Onodera T, Kawata H, Nishi H, Futatsugi T, and Yokoyama N 1989 *IEEE Trans. Elec. Dev.* **ED-36** 1586-1590

[4] Kimura T, and Ohshima T 1993 *Jpn. J. Appl. Phys.* **32** L188-L186

[5] Hida H, Tokushima M, Maeda T, Ishikawa M, Fukaishi M, Numata K, and Ohno Y 1993 *GaAs IC Symp. Tech. Dig.* 197-200

Inst. Phys. Conf. Ser. No 145: Chapter 5
Paper presented at 22nd Int. Symp. Compound Semiconductors, Cheju Island, Korea, 28 August–2 September 1995
© *1996 IOP Publishing Ltd*

Design and Fabrication Technologies of a One-chip p-i-n / FET for 155 Mbps Local Loop Applications

Jeong-Woo Park, Sang-Kee Si, *Young-Boo Moon, *Cheol-Soo Sone, Deokho Yeo, Moon-Jung Kim, Sung-June Kim, *Euijoon Yoon ,[#]Chul-Dong Kim and [o]Tae-il Kim

Dept. of electronics engineering, Seoul National University, Seoul, 151-742, Korea.
*Dept. of Inorganic Materials Engineering, Seoul National University, Seoul, 151-742, Korea.
[#] SAMSUNG ELECTRONICS CO., LTD. Chung-Ang Newspaper Bldg., 8-2, Karak-Dong, Songpa-Ku, Seoul, 138-160, Korea.
[o] Samsung Advanced Institute of Technology, P.O.Box 111, Suwon, 440-600, Korea.

Abstract A one-chip p-i-n/FET circuit for 155Mbps local loop application is designed and fabricated. We make emphasis on dynamic range rather than on sensitivity, and on low fabrication cost rather than on high performance, for this application. A new noise analysis technique is used to optimize the circuit. A p-i-n photodiode integrated with a simple source follower circuit consisting of one JFET, one channel resistor, and one NiCr resistor provides a bandwidth of 120MHz, a transimpedance of 1kΩ at 50Ω load, a sensitivity of -34.2dBm and a dynamic range of 34dB. An MOCVD technique is used to grow the whole epi structure without interruption. The InP/InGaAs photodiode layers are stacked on top of the FET layers and the highly doped channel layer of the FET is also the n$^+$ layer for the photodiode. Zn and Cd diffusion processes are used in the simultaneous formation of p-type layers for both photodiode and the junction FET (JFET).

1. Introduction

InP optoelectronic devices are expected to play an important role in the development of fiber optic systems in applications such as long haul and local loop applications. Most of the components used in the present day fiber optic system are hybrid integrated types. Monolithic integration can increase performance by removing some parasitics involved in hybrid packaging. For this reason, many efforts in fabricating Optoelectronic Integrated Circuits(OEIC's) have been devoted to achieving higher sensitivity values operating at higher bit rates than in their hybrid counter parts. However, many are still at performance values below or approaching those of hybrids, primarily because in hybrids, each component can be individually optimized where in monolithics, a trade off is often necessary between electronic and optical devices.

For widespread fiber optics, however, it is essential to have components with lower cost. Monolithic integration can also achieve this objective by increasing packing density of components. The subscriber loop fiber system operating at 155Mbps is expected to provide the first high volume market for the low cost optoelectronic components. In this paper, we present the design and processing technologies used in developing a low cost optical receiver front end.

2. Device design

Our selections for the photodetector and the transistor are p-i-n photodiode and JFET, respectively. These have quantum efficiency and bandwidth high enough for the 155Mbps local loop operation. The JFET has been used widely in the p-i-n/FET fabrications, because of its simple epi structure, relative ease in process, and reliable junction properties[1]. Fig.1 shows the cross sectional view of the epitaxial structure used in the integration. The entire structure is designed for a single growth without interruption. The growth of the JFET layers, consisting of highly doped InGaAs channel layer and InP gate layers, are followed by the p-i-n layers, consisting of undoped InGaAs photoabsorption layer and undoped InP anode layer. The epi layers are grown on semi-insulating InP substrate which provides good isolation among integrated devices.

One cost effective aspect of the structure is in the shared layer approach. The channel layer of the FET also acts as the n+ layer for the p-i-n. The other is that the undoped InP layers with identical thickness are used for both the gate of the FET and the anode of the p-i-n diode. A post-growth selective diffusion is performed to dope these layers, simultaneously. Problems associated with fast diffusion of Zn[2] in-situ doping during the growth, can thus be circumvented.

A disadvantage of this structure is in the height difference of the two devices. Approximately 3 and 5μm differences exist between the top of the two and between the top of the p-i-n mesa and the substrate floor, respectively. We planarize the structure with a polyimide process as will be described below.

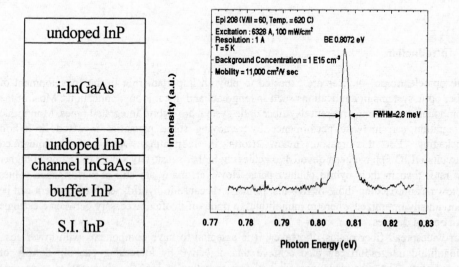

Fig.1 Cross Section View of Epi Structure   Fig.2 PL Spectrum of Undoped InGaAs(5K)

3. Epi growth

We used a low pressure MOCVD technique for the growth of the structure. The growth temperature is 700°C for n-type doped InGaAs channel layer, 600°C for undoped InP and InGaAs layers, and 670°C for other layers. Trimethylindium and trimethylgallium are used as group III sources. Group V gases are 100% arsine and phosphine. Silane(100ppm in H_2) is used as the n-type dopant. For the best sample, undoped InP layer has background carrier concentration of $1.7 \times 10^{14}/cm^3$ and extremely high electron mobility of 138000 $cm^2/Vsec$ and undoped InGaAs samples has background carrier concentration of $1 \times 10^{15}/cm^3$ and electron mobility of 70,000 $cm^2/Vsec$, as measured by van der Pauw method at 77K.

In Fig.2, result of a photoluminescence measurement taken at 5K of the undoped InGaAs layer is shown which shows a bound exciton peak with FWHM of 2.8meV at 0.807eV.

Doping concentration of the n-type InGaAs channel layer is $8 \times 10^{16}/cm^3$. For silane doping of n-type InGaAs, the doping concentration increased exponentially with temperature and decreased linearly with growth rate. These behaviors are related with the fact that silane pyrolysis is thermally activated process and Si incorporation process is controlled by diffusion rate through gas phase boundary layer.

4. Circuit design

Bandwidth, sensitivity, transimpedance and dynamic range are the important parameters that describe a photoreceiver front end. For the photoreceiver to operate at 155Mbps bit rate, more noise comes from the load resistor than from other sources. Use of a high impedance pre-amp configuration can minimize the thermal noise even though it still is the largest noise component. Use of a $2k\Omega$ resistor as the load gives sufficient bandwidth for 155Mbps operation. We used a newly developed method[3] in analyzing noises and in optimizing the circuit. This method allows to optimize the design in terms of bandwidth for a given bit rate. Intersymbol interference(ISI) is included in the consideration. Fig. 3 a) shows the circuit design of the p-i-n/FET device. Increasing the load resistor R_L reduces the bandwidth, and the corresponding thermal noise reduces. The further noise reduction comes from the reduction of effective bandwidth. However, where bandwidth decreases, the ISI increases. Fig. 3 b) shows this trend. Considering both the circuit noise and the ISI, we find an optimum at an R_L value of $2k\Omega$. This gives an ISI at 2.3% of the circuit noise.

Dynamic range is an important parameter in the short haul application. High gain pre-amps such as buffered FET logic amplifier gives higher sensitivity, but dynamic range suffers from the high gain. Using a simple source follower circuit, we can maintain a dynamic range greater than 30dB. Our circuit is designed for a linear detection of input optical power ranging from -34.2dBm to 0dBm as shown in Fig.3 c). The circuit also provides an output impedance of 50Ω, and a transimpedance of $1k\Omega$ from DC to 120MHz.

5. Process and results

Since the junctions are formed by diffusion for the p-i-n diode and the JFET, a reliable diffusion process is needed. We used a diffusion process using evaporated diffusion sources.

692

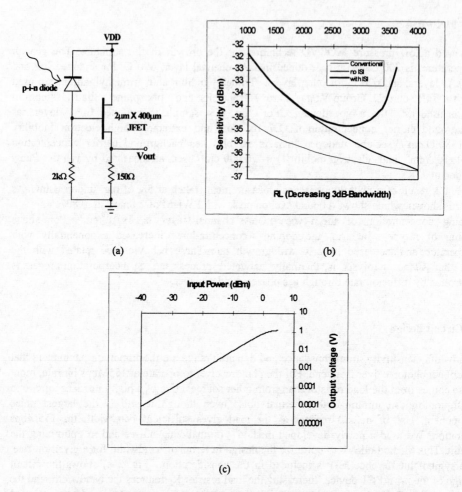

(a)   (b)

(c)

Fig.3   p-i-n/FET Device (a) Circuit (b) Sensitivity Optimized for R_L (c) Dynamic Range

Such method can be advantageous for the large wafer IC process over commonly used ampoule technique[4][5]. We have tried Zn_3P_2 and Cd_3P_2 as the evaporable source. Sample was first cleaned and loaded in a thermal evaporator where a 50 nm thick layer of the diffusion source is deposited. After the sampled is coated with a thin of SiO_2, it is loaded in an rapid thermal annealer with N_2 ambient for a diffusion. After diffusion, remaining SiO_2 and diffusion source can be easily removed by etching in HF and HNO_3 solutions. Fig. 4 a) and b) shows the pn junction profiles for Zn and Cd, respectively. Fig.5 shows the Arrhenius plots for the diffusion process.

The diffusion coefficient for the Zn-based process is larger than that for the Cd-based diffusion. It can be seen that diffusion coefficient of Zn is much larger than that of Cd. Higher doping and slower diffusion are the characteristics of Cd diffusion process[6].

Another important process is the planarization using polyimide as previously mentioned. A step height as large as 5μm should be overcome by this method. We used a photoimagiable polyimide for this process and found that film thickness and develop time are the key parameters in controlling the edge shapes.

Dark current and capacitance measured from fabricated p-i-n photodiode of 75μm diameter are 60nA and 0.3pF at -5V bias voltage, respectively. A responsivity of about 0.7A/W was measured without anti-reflection coating at 1.3μm wavelength. As shown Fig.6, the FET's extrinsic transconductance of 50mS/mm was measured at Vds = 4V. The Idss is 260mA/mm. The gate to source capacitance at 0V gate bias is about 2 pF/mm.

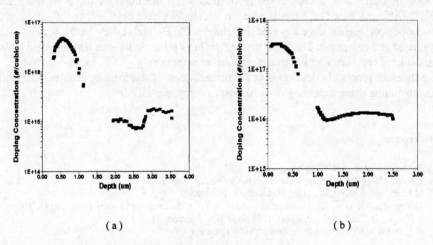

(a)   (b)

Fig 4. pn Junction Doping Profiles using (a) Zn, (b) Cd

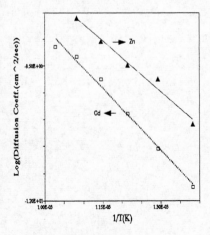

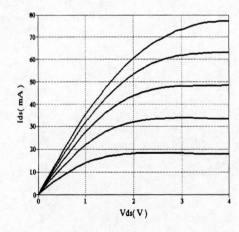

Fig.5 Diffusion Coefficient versus 1/T
       @ InP Substrate

Fig.6 I-V Characteristics of JFET(2μm x 300μm)
       Vgs=-1V step

694

6. Summary

For receiver front-ends used in short-haul fiber-optic application, the emphasis should be made on low cost aspect of design and fabrication technologies. In design, securing large dynamic range is as important as having high sensitivity. Such cost effectiveness is applied in the design of epi structure, circuit, and process technologies. In designing, a new noise analysis technique is introduced in evaluating noise and ISI and thus in optimizing circuit performance. The circuit is expected to have -34.2dBm of sensitivity at 10^{-9} bit error rate for 1.3μm wavelength with a dynamic range greater than 30dB.

Single step diffusion is used in the fabrication of pn junctions for the p-i-n diode and the JFET. Diffusion based on evaporated diffusion sources have been successfully applied in the production of highly doped p-type InP layers. Zn_3P_2 and Cd_3P_2 based technologies are evaluated and compared. The fabricated InP/InGaAs p-i-n photodiode and JFET have 60nA of dark current and 50mS/mm of transconductance, respectively.

A polyimide process is developed in the planarization of the mesa structures. Currently we are integrating these technologies to fabricate a completed device.

References

[1] Y.Akahory, et al., 1989 Electron. Lett., 25, 1, 37
[2] S.Aytac and A.Schlachetzki, 1983 J. Cryst. Growth, 64, 169 -173
[3] M.J.Kim and S.J.Kim, M.D.Das, manuscript in preparation.
[4] M.Yamada, P.K.Tien, R.J. Martin, R.E. Nahory, and A.A. Ballman, 1983 Appl. Phys. Lett.43, 594
[5] A.Hooper, B. Tuck, and A.J. Baker, 1974 Solid State Electron. 17, 531
[6] S.K.Si and J.B.Yoo, S.J.Kim, manuscript in preparation.

Inst. Phys. Conf. Ser. No 145: Chapter 5
Paper presented at 22nd Int. Symp. Compound Semiconductors, Cheju Island, Korea, 28 August–2 September 1995
© *1996 IOP Publishing Ltd*

A New GaAs BiFET Structure using Selective MOCVD Technique

Hyunchol Shin, Jeong-Hwan Son* and Young-Se Kwon

Department of Electrical Engineering, Korea Advanced Institute of Science and Technology, 373-1 Kusong-dong, Yusong-gu, Taejon, 305-701, Korea
*LG Semicon Co., Ltd. 1 Hyangjeong-dong, Hungduk-gu, Cheongju, 360-480, Korea

Abstract : A novel GaAs BiFET technology based on AlGaAs/GaAs heterojunction bipolar transistor and GaAs junction gate floated electron channel field effect transistor (J-FECFET) has been developed. A single selective MOCVD growth is used to fabricate the BiFET structure. The whole process is almost the same as a typical HBT process except for the selective epitaxial growth. Transconductance of fabricated J-FECFET with $1 \times 200 \mu m^2$ gate is 102mS/mm with f_T and f_{MAX} of 10.7 and 27.3 GHz , respectively. DC current gain of HBT is 21 at collector current density of $50KA/cm^2$ with emitter area of $3 \times 2\mu m^2$. The simple process and superior performances of the new BiFET structure could be applicable to various multifunction MMICs.

1. Introduction

Monolithic integration of HBTs and FETs on the same substrate, which is called BiFET technology, is one of the most challenging issues of multifunctional microwave circuit applications. HBTs have high current driving capability, high switching speed performance and low flicker noise. On the other hand, FETs have high input impedance and low noise characteristics. Because of these supplemental relationships of HBTs and FETs, GaAs BiFET technology is expected to be a breakthrough in the fields of GaAs multifunction MMICs.

There have been two approaches for the co-integration of HBTs and FETs. One is based upon stacked epitaxial layers which have FET layers under[1] or over[2] HBT layers. The other is based upon selective area epitaxial regrowth techniques[3]. Both of the approaches are not attractive due to difficult and complex processes.

We have already reported a selectively grown AlGaAs/GaAs HBT[4] and a GaAs junction - gate floated electron channel field effect transistor (J-FECFET) [5] using selective MOCVD. Both devices showed high performance for microwave circuit applications and the co-integration of both devices will imply many advantages such as simple process and superior performance.

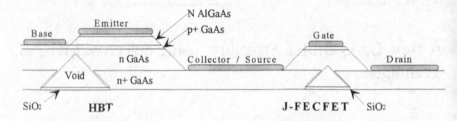

Figure 1. Schematic cross-sectional view of the new BiFET structure

2. Device Structure

Selective MOCVD is a key process for fabrication of the new BiFET structure. Selective MOCVD in this experiment does not mean an etch-and-regrowth process but a selective area growth on patterned substrate which is more simple and stable. When the selective epitaxial growth is performed over patterned (001) GaAs substrate , it has been experimentally proved[6] that the facet angles and shapes of the overgrown layers are determined by several parameters such as mask orientation, AsH_3 mole fraction and growth temperature. We use the reverse mesa shape, which is grown with the SiO_2 mask orientation 10 ° ~ 40 ° off the $[\bar{1}10]$ direction, to fabricate the BiFET structure.

Schematic cross-section of the BiFET structure is shown in fig.1. The BiFET structure is composed of selectively grown HBT and juncticon-gate FECFET. Both devices feature a triangular void over SiO_2 stripe. The collector, sub-collector and base layers of HBT are exactly matched to the active channel, ohmic and gate layers of J-FECFET. The selectively grown HBT is the same as the conventional mesa-type HBT with the exception of a void. The void lies between the intrinsic collector region under the emitter metal and the extrinsic collector region under the base metal so that the extrinsic collector region is electrically isolated from the intrinsic collector region. Therefore the extrinsic base-collector capacitance can be significantly reduced and the microwave performance is improved. J-FECFET has the following advantanges over the conventional MESFETs and JFETs. The effective channel length is very short. The source and the drain resistance are significantly reduced because the thick n+ regions are separated by a short distance underneath the gate. The parasitic gate capacitance is much smaller than that of the conventional JFET. Implementation of the BiFET structure could be possible by directly combining the above two devices without any alterations.

3. Device Fabrication

The SiO_2 stripes are formed on (001)GaAs substrate using e-beam evaporation and lift-off process. The widths of the SiO_2 stripes are 3μm for HBT and 2μm for J-FECFET and the orientation of the SiO_2 stripes is 30 ° off the $[\bar{1}10]$ direction. A single epitaxial growth using atmospheric-pressure MOCVD is followed to form the HBT and J-FECFET layers simultaneously. TMGa, TMAl and AsH_3 are used as source materials. N- and P-type dopant source materials are SiH_4 and CCl_4, respectively. The epitaxial layers are shown in Table 1.

Table 1. Epitaxial layers of the BiFET structure

Material	Thickness (μm)	Doping (cm^{-3})	Growth Temp(°C)	Layers	
				HBT	J-FECFET
n$^+$ - GaAs	0.08	Si 5×10^{18}	700	Emitter Cap	
N - Al$_{0.3}$Ga$_{0.7}$As	0.15	Si 1×10^{18}	700	Emitter	
p$^+$ - GaAs	0.1	C 2×10^{19}	650	Base	Gate
n - GaAs	0.65	Si 5×10^{16}	650	Collector	Channel
n+ - GaAs	0.45	Si 5×10^{18}	650	Sub-Collector	Ohmic

Emitter metal evaporation and emitter edge thinning are performed after selective MOCVD. Collector / source and drain contact-areas are opened and isolation mesa is formed subsequently using conventional chemical etching. Base and collector metals of HBT, source / drain / gate metals of J-FECFET are simultaneously evaporated with AuGe/Ni/Au and alloyed all at once. Gold electroplating is finally performed for interconnection. Fig.2 shows the SEM microphotograph of the fabricated BiFET structure. The photograph clearly exhibits that HBT and J-FECFET are successfully integrated on the same substrate.

4. Measurement Results

Fig.3(a) shows I-V characteristics of J-FECFET. Gate width and length are 200μm and 1μm , respectively. The SiO$_2$ stripe width (2μm) is an important structure parameter determining the saturation current level. Even an enhancement mode J-FECFET can be easily fabricated on the same substrate only by varying the stripe width.[5] A maximum extrinsic transconductance of 102mS/mm is obtained at drain current of 200mA/mm. Threshold voltage is -2.8V. Very low knee voltage of about 0.7V implies low source resistance. Microwave characteristics were measured on wafer using Wiltron 360 Network Analyzer. Current gain cut off frequency f_T and maximum oscillation frequency f_{MAX} are 10.5GHz and 27.3GHz, respectively, at V_{DS} = 3V and I_{DS} = 100% I_{DSS} (35.6mA). Fig.3(b) shows the I-V characteristics of HBT. Emitter area is $3\times2\mu m^2$. The DC current gain is 21 at collector current density of 50KA/cm^2. Offset voltage and ohmic linearity are poor because the base-collector junction and emitter ohmic are not optimized. Specific contact resistivity of the base contact is $1.2\times10^{-5}\Omega cm^2$. Non-alloyed ohmic contact to base and emitter layers can improve the device performance.

Figure 2. SEM photograph of the fabricated BiFET

698

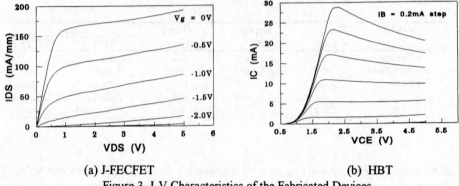

(a) J-FECFET   (b) HBT

Figure 3. I-V Characteristics of the Fabricated Devices

5. Conclusions

We have developed a new BiFET technology which employs selectively grown AlGaAs/GaAs HBT and J-FECFET. A single step of selective area epitaxial growth is used to form the epitaxial layers of HBT and J-FECFET simultaneously. The collector, base and sub-collector layers of HBT and channel, gate, and ohmic layers of J-FECFET are exatctly the same with one another. Therefore any additional stacked layers or any addition process steps are not necessary while the wole process is almost the same as a typical HBT process except for the selective epitaxial growth. The fabricated devices show good performance. The J-FECFET exhibits transconductance of 102mS/mm with f_T and f_{MAX} of 10.7 and 27.3 GHz, respectively, for $1\times200\mu m^2$ gate. HBT shows dc current gain of 21 for $3\times2\mu m^2$ emitter area.

The proposed BiFET structure has many advantages so that it could be widely applicable to various multifunction MMICs.

Acknowledgements

This work has been financially supported by the Korea Telecom Research Laboratories.

References

[1] K.Itakura, Y.Shimamoto, T.Ueda, S.Katsu and D.Ueda, 1989, *IEDM Technical Digest* , 389-392

[2] D.Cheskis, C.E.Chang, W.H.Ku, P.M.Asbeck, M.F.Chang, R.L.Pierson and A.Sailer, 1992, *IEDM Technical Digest* , 91-94

[3] D.C.Streit, D.K.Umemoto, T.R.Block, A.Han, M.Wojtowicz, K.Kobayashi and A.K.Oki ,1994, *Int. Symp. Compound Semiconductor* , 685-688

[4] J.H.Son, C.T.Kim, S.Hong and Y.S.Kwon, 1994, *Inter. Conf. on Solid State Devices and Materials*, 610-612

[5] C.T.Kim, Y.J.Lee and Y.S.Kwon, 1993, *Jpn.J.Appl.Phys*, **32** 1B Jan, 556-559

[6] H.Asai and S.Ando , 1985, *J. Electrochem. Soc.*, **132** Oct., 2445

Inst. Phys. Conf. Ser. No 145: Chapter 5
Paper presented at 22nd Int. Symp. Compound Semiconductors, Cheju Island, Korea, 28 August–2 September 1995
© *1996 IOP Publishing Ltd*

Novel sub-quarter micron GaAs MESFET process with WSi sidewall gate

T Uda, K Nishii, K Fujimoto, and A Tamura

Electronics Research Laboratory, Matsushita Electronics Corporation, 3-1-1, Yagumo-Nakamachi, Moriguchi, Osaka 570, Japan

Abstract. We have developed a novel sub-quarter micron WSi sidewall gate GaAs MESFET (SIG-FET) fabrication process. In this process, using WSi sidewalls as gate electrodes, the gate length is controlled only by the thickness of a WSi thin film deposited by DC sputtering and sub-quarter micron gates can be easily fabricated without using photo-lithography. The 0.15-μm-gate SIG-FET has exhibited gm_{max} = 360 mS/mm, f_t = 50 GHz and f_{max} = 120 GHz with Vth = -0.95 V. This novel process technology is very promising for fabricating high-performance sub-quarter micron gate MESFET's with low costs and high throughput.

1. Introduction

In order to improve the high-frequency performance of GaAs MESFET's, it is important to shorten the gate length of FET's. As it is difficult to form a sub-quarter micron pattern using conventional i-line photo-lithography, direct EB-lithography is usually used to form sub-quarter micron gate FET's [1-4]. However, direct EB-lithography has problems of high costs and limited throughput [5].

In order to improve these problems, we have developed a novel sub-quarter micron WSi sidewall gate GaAs MESFET (SIG-FET) fabrication process. In this process, using WSi sidewalls as gate electrodes, the gate length is controlled only by the thickness of a WSi thin film deposited by DC sputtering and sub-quarter micron gates can be easily fabricated without using photo-lithography.

In this paper, we report the SIG-FET fabrication process and the characterization of a 0.15-μm-gate SIG-FET fabricated by this process.

2. Fabrication process

The substrates used in this work were 3-inch undoped semi-insulating (100)-oriented GaAs crystals grown by LEC method. Fig.1 shows the SIG-FET fabrication process. A buried p-layer lightly doped drain (BP-LDD) structure is used in SIG-FET's to reduce short channel effects. First, the BP layer was formed by 180-keV Mg ion implantation with a dose of 4×

700

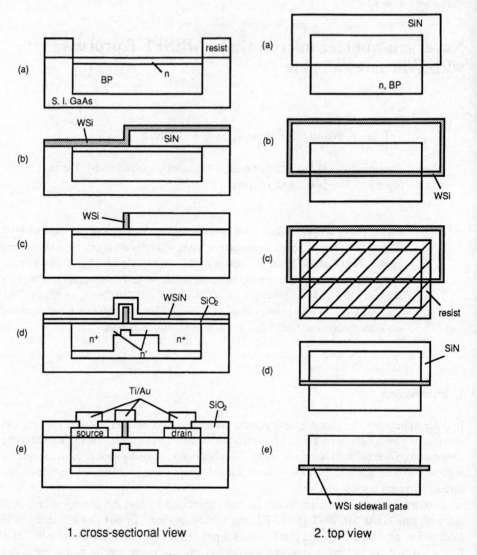

1. cross-sectional view   2. top view

Fig.1. Fabrication process of SIG-FET.

10^{12} cm^{-2} and the channel layer was formed by 25-keV SiF implantation with a dose of 1×10^{13} cm^{-2} (1(a)). Then an SiN film was deposited on the wafer. After patterning a photo-resist layer, the SiN film was etched by reactive ion etching (RIE) using the photo-resist as an etching mask (2(a)). Then a WSi film was deposited by DC sputtering (1(b)). The WSi film was etched by RIE with CF$_4$/SF$_6$ mixed gas without using an etching mask to form a WSi sidewall gate (1(c), 2(b)). Since photo-lithography is not used in gate formation steps and the gate length is controlled only by the thickness of a WSi thin film deposited by DC sputtering, sub-quarter micron gates can be fabricated without limitation of photo-lithography. After patterning a photo-resist layer (2(c)), a part of the WSi sidewall and the SiN film was etched by chemical dry etching (CDE) using the photo-resist as an etching mask (2(d)). In this step, the photo-resist mask was formed to cover all over the BP and n regions in order to protect the regions from getting damage. After removing the SiN film by wet etching (2(e)) and depositing a 40-nm-thick SiO_2 film on the wafer, the Si ion implantations at 50 keV and 120 keV with doses of 2×10^{13} cm^{-2} and 5×10^{13} cm^{-2} were performed for the n' and n$^+$ layers, respectively. An asymmetric LDD structure was formed by n$^+$ layer implantation using a photo-resist mask. All these implantation layers were annealed simultaneously at 820 °C for 15 min in H$_2$ atmosphere with a SiO_2 /WSiN cap [6] (1(d)). Ohmic electrodes were formed by evaporation of AuGe/Ni/Au. Finally, a Ti/Au layer was deposited as gate over-layer or inter-connect metal (1(e)).

3. Results and discussions

3.1. Gate fabrication

Fig.2 shows a cross-sectional SEM photograph of a 0.15-μm WSi sidewall gate using a 200-nm-thick WSi film. From this SEM photograph, it is found that the 0.15-μm gate with a perpendicular cross section was obtained by this process. By further reducing the thickness of the WSi film further, it is possible to shorten the gate length to less than 0.15 μm.

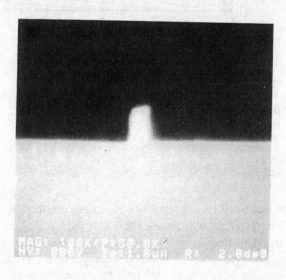

0.15 μm

Fig.2. Cross-sectional SEM photograph of the 0.15-μm WSi sidewall gate.

3.2. DC characteristics

Fig.3 shows the Ids-Vds characteristic of a 0.15-μm-gate SIG-FET fabricated by this process. The gate width (Wg) was 50 μm and the distance between the gate electrode and the source n$^+$ region (Lgs) was 0.25 μm and the distance between the gate electrode and the drain n$^+$ region (Lgd) was 0.6 μm. The good pinch-off characteristics are obtained due to the BP-LDD structure. The threshold voltage was about -0.95 V. The maximum transconductance was 360 mS/mm when the drain voltage (Vds) was 1 V.

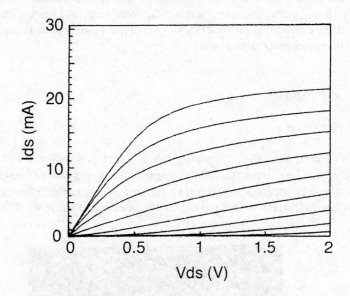

Fig.3 Ids-Vds characteristic of the 0.15-μm-gate SIG-FET with gate width of 50 μm. Maximum gate voltage is 0.6 V and step is 0.2 V.

3.3. RF characteristics

The S-parameters of the 0.15-μm-gate SIG-FET were measured in the frequency range from 1 to 50 GHz. Wg, Lgs, and Lgd were 200 μm, 0.25 μm and 0.6 μm, respectively. The current gain cutoff frequency (f_t) was derived from the extrapolation of calculated current gain $|h_{21}|^2$ to unity using -6 dB/octave decay. In the same way, the maximum frequency of oscillation (f_{max}) was derived from the extrapolation of maximum stable gain MSG at 50 GHz to unity using -6 dB/octave decay. Fig.4 shows the dependences of $|h_{21}|^2$ and MSG on frequency at Vds of 2 V and gate voltage (Vgs) of 0.4 V. The 0.15-μm-gate SIG-FET exhibited f_t of 50 GHz and f_{max} of 120 GHz.

Further improvements in the FET characteristics of the SIG-FET are expected by reducing the gate length or by optimizing the conditions of ion implantations and annealing.

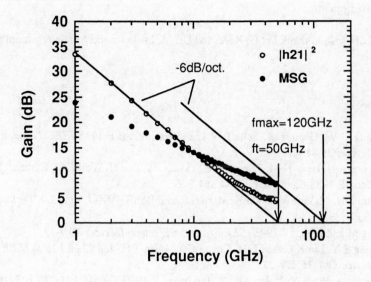

Fig.4 RF characteristics of the 0.15-μm-gate SIG-FET with gate width of 200 μm at drain voltage of 2 V and gate voltage of 0.4 V.

4. Conclusion

A novel sub-quarter micron WSi sidewall gate GaAs MESFET (SIG-FET) fabrication process has been developed. In this process, the gate length is controlled only by the thickness of a WSi thin film deposited by DC sputtering and sub-quarter micron gates can be fabricated without using photo-lithography. The 0.15-μm-gate SIG-FET has exhibited a maximum transconductance of 360 mS/mm with a threshold voltage of -0.95 V, a current gain cut off frequency of 50 GHz and a maximum frequency of oscillation of 80 GHz. Further improvements in the FET characteristics of the SIG-FET are expected by reducing the gate length or by optimizing the conditions of ion implantations and annealing.

This novel process technology is very promising for fabricating high-performance sub-quarter micron gate MESFET's with low costs and high throughput.

Acknowledgment

The authors wish to thank Dr. G. Kano and Dr. M. Inada for their encouragement throughout this work.

References

[1] Wang G W, Chen Y K, Schaff W J, and Eastman L F 1988 *IEEE Trans. on Electron Devices* **35** 818-823

[2] Lepore A N, Levy H M, Tiberio R C, Tasker P J, Lee H, Wolf E D, Eastman L F, and Kohn E 1988 *Electron. Lett.* **24** 364-366

[3] Samoto N, Makino Y, Onda K, Mizuki E, and Itoh T 1990 *J. Vac. Sci. Technol. B* **8** 1335-1338

[4] Feng M and Laskar J 1993 *IEEE Trans. on Electron Devices* **40** 9-17

[5] Chang E Y, Lin K C, Liu E H, Chang C Y, Chen T H, and Chen J 1994 *IEEE Electron Device Lett.* **15** 277-279

[6] Tamura A, Ikeda Y, Yokoyama T, and Inoue K 1990 *J. Appl. Phys.* **67** 6171-6174

Inst. Phys. Conf. Ser. No 145: Chapter 5
Paper presented at 22nd Int. Symp. Compound Semiconductors, Cheju Island, Korea, 28 August–2 September 1995
© *1996 IOP Publishing Ltd*

A 3.3V GaAs Power MESFET for Digital/Analog Dual-mode Hand-held Phone

Jong-Lam Lee, Sung-Jae Maeng, Haecheon Kim, Jae Kyoung Mun,
Chang-Seok Lee, Jae Jin Lee, Kwang-Eui Pyun and Hyung-Moo Park

Semiconductor Division
Electronics and Telecommunications Research Institute
Yusong P. O. Box 106, Taejon, Korea

Abstract. A 3.3V operating GaAs power metal semiconductor field effect transistor for digital/analog dual-mode hand-held phones has been developed with state-of-the-art performance using a high-low doped structure grown by molecular beam epitaxy (MBE). For analog mode, the MESFET tested at a 3.3 V drain bias and a 900 MHz operation frequency displays power-added efficiency of 68% with output power of 32.5 dBm. For digital mode, the device shows that 3rd-order inter-modulation and power-added efficiency at an output power of 28 dBm were -32 dBc and 41%, respectively. A power amplifier operating at 3.3V for the dual-mode hand-held phones are developed by using the high-low MESFET. The amplifier shows an output power of 31.5 dBm and a power-added efficiency of 61% for AMPS mode. The third order intermodulation and the fifth order one are measured to be -32 dBc and -45 dBc at an output power of 26 dBm for CDMA mode. These are good enough for dual-mode requirements.

1. Introduction

The mobile communication has been changed from analog to digital as the enlargement of communication capability is required. It is known that the size and the weight of a phone can be reduced by decreasing the number of battery cells. In order to obtain a high power-added efficiency (PAE) at low biases of 3.0~3.6 V, a low knee voltage in power amplifying device is required, which entails a wide range of voltage swing of the rf output signal, followed by high PAE with high output power. For digital application, the power FET should have a good linearity performance with low third-order inter-modulation, the magnitude of which is dependent upon transconductance uniformity with gate bias in the DC characteristics. Inter-modulation reduces with increasing the gate bias, followed by increase of drain current (class A operation), but PAE decreases because of increasing DC power in operating the power device. A low inter-modulation and a high PAE can be simultaneously achieved by using a power FET with an uniform transconductance with the gate bias, which allow for both a high power gain, subsequent to high PAE, and a low inter-modulation even at a near class B operation. At this point of view, an optimization of channel structure in the MESFET is essentially needed in developing a low-voltage operation power MESFET with a high PAE and a low inter-modulation for digital/analog dual-mode hand-held phones.

Recent advances in material preparation and device fabrication techniques have produced L-band GaAs power FETs operating at drain voltages of 3.0~3.5 V with respectable output power and PAE [1]-[6] for analog hand-held phones. There has been, however, no report on the linearity performance for 900 MHz digital hand-held phones applications. For the digital phones using CDMA mode, power device should have output power higher than 26 dBm with both inter-modulation lower than -30 dBc and PAE higher than 30%. In this work, we developed a power MESFET with a supply voltage of 3.3 V for the CDMA/AMPS dual-mode cellular phone, which simultaneously satisfied the linearity for CDMA mode, and the output power and the PAE for AMPS mode. The MESFETs were used for a dual-mode power amplifier in realizing the dual-mode cellular phones.

2. Design and fabrication

The structure consists of a 1 μm-thick undoped buffer layer including GaAs/AlGaAs superlattices on a C-doped semi-insulating GaAs substrate, a thin active layer doped to mid $10^{17}/cm^3$ (high-doped layer), a thick active layer doped to mid $10^{16}/cm^3$ (low-doped layer) and an undoped GaAs layer for surface passivation. The layer structure was prepared by molecular-beam-epitaxy on 3 inch semi-insulating GaAs wafer. Thickness of the low-doped layer was designed to be two times thicker than depletion layer width formed by Schottky contact. The role of this layer is to decrease gate capacitance and to improve Schottky characteristics of the gate metal, which result in high power gain of MESFET. The high-doped layer acts as a main channel for carriers and provides most of the drain current of MESFET. Thickness and doping concentration of this layer determine DC characteristics, such as pinch-off voltage, gate-to-drain breakdown voltage and transconductance. This layer was made thin to obtain both an uniform transconductance with gate bias and high transconductance at a near class B bias condition.

MESFETs with a total gate width of 16.4 mm having 82 fingers with an unit gate width of 200 μm were fabricated using 3 inch wafer process. In order to decrease source resistance followed by a low knee voltage, gate-to-source spacing was minimized to be 0.5 μm. Details of fabrication processes are described in the previously published article [2]. Figure 1 is a photograph of the process-finished power device chip.

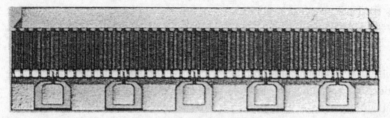

Fig. 1. Photograph of 3.3 V-operation power device chip

3. Performance of deevice

The I-V characteristics were measured by pulse signal with a period of 0.1 msec, the result of which is given in Fig. 2. The maximum drain current, measured at Vgs = +0.5 V, is 5.0 A and its density is 305 mA/mm. The saturated drain current and its density are evaluated to be 4.0 A and 249 mA/mm, respectively. Pinch-off voltage of the device is about ~ -2.4 V. The gate-to-drain breakdown voltage (BV$_{gd}$), determined at a gate-to-drain current density of 1 mA/mm, is evaluated to be 28 V. The source resistance of 0.06 Ω is measured, which results in a very low knee voltage of 1.0V at Vgs = +0.5 V. The transconductance is very uniform around 1.7 S for the gate voltage ranging from -2.0 V to +0.5 V, as shown in Fig. 3. This indicates that the device is good for class B power operation.

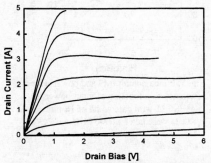

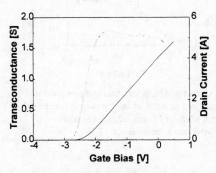

Fig. 2. Typical DC current-voltage characteristics of high-low doped GaAs power MESFET with gate width of 16.4-mm. Top curve corresponds to Vgs = 0.5 V and ΔVgs = 0.5 V.

Fig. 3. Transconductance and drain current as a function of gate voltage at Vds = 2.0 V. Gate width is 16.4-mm.

Power sweep was performed at a range of input power between 0 dBm and 23 dBm under a 3.3 V drain bias and 900 MHz frequency. The device was operated under a near class B condition with a bias current of 0.40 A, corresponding to 10% Idss. The output power of 32.5 dBm and the power-added efficiency of 68% were obtained at an input power of 20 dBm, as shown in Fig. 4. We obtain a very linear power gain with the input power. The linear power gain of 16.5 dB was maintained within +/- 0.10 dB at the output power range of 16~30 dBm. The 1-dB gain compression was measured to be 31.0 dBm with PAE of 60%.

The third-order inter-modulation of the FET with the input power was measured using two tone frequencies, 900.00 MHz and 900.03 MHz. Figure 5 shows the measured IM3, the fundamental output power and PAE as a function of the input power at a drain bias of 3.3 V. The IM3 and PAE at an output power of 28 dBm were measured to be -32 dBc and 41%, respectively, which are evaluated to be enough for CDMA digital applications. Note that third-order intercept point IP3 was determined to be 48.5 dBm. A linearity figure-of-merit (LFOM : IP$_3$/P$_{DC}$) is estimated to be 45.

The power performances in the analog mode are evaluated to be better than the best results reported for both GaAs power MESFET [2] and pseudomorphic heterojunction FET [1]. Especially, the PAE is the highest among the previous results reported at the drain biases of 3.0~3.6 V. The most significant result is the simultaneous achievement of power performances for both analog and digital applications. These results are attributed to simultaneous achievement of low knee voltage and an uniform transconductance with gate biases by using optimized channel structure for high-efficient power operation at 900 MHz frequency.

708

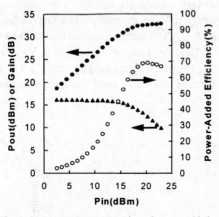

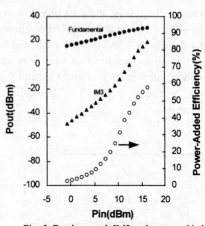

Fig. 4. Output power, power gain and power-added
efficiency as a function of input power for MESFET
with gate width of 16-mm. Data were taken
at 900 MHz and a drain bias of 3.3 V.

Fig. 5. Fundamental, IM3 and power-added
efficiency as a function of input power for
MESFET with gate width of 16-mm.
Data were taken at 900 MHz and 3.3 V.

The optimum impedances for the output power and PAE of the first and second
MESFETs were measured by the load/source-pull method using input and output tuners.
Optimum impedance matching points for both source and load sides were searched by source
pull and load pull methods, followed by power sweeping to measure the power performance
of the device. Figure 6-(a) and -(b) displayed power and efficiency contours for the second
FET using source pull and load pull measurements, respectively. The source pull

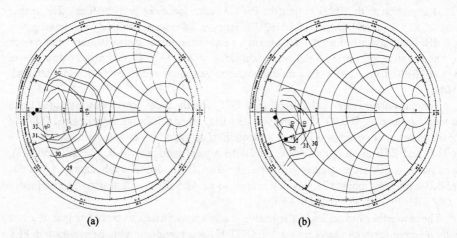

(a)   (b)

Fig. 6. (a) Output power and PAE contours measured by source pull. (b) Output power and PAE contours
measured by load pull.

measurement was done at a load impedance of $\Gamma = 0.793 / -165.8^o$, which is marked as "$\Delta$"
in Fig. 6-(a). The source impedance for both maximum output power (mark of ■) and the
maximum PAE (mark of ♦) coincide with each other, as shown in Fig. 6-(a). The load pull

measurement was performed at a source impedance of $\Gamma = 0.882 / -174.8^O$, which is marked as "$\Delta$" in Fig. 6-(b). In the load pull measurement, we obtained a maximum PAE of 85.4% with an output power of 29.7 dBm at a load impedance point of $\Gamma = 0.825 / -178.6^O$, marked as "$\blacklozenge$". An maximum output power of 33.1 dBm with an PAE of 58% could be obtained a load impedance point of $\Gamma = 0.773 / -159.3^O$.

The power amplifier for the dual-mode cellular phone consists of two stages to obtain a small signal gain more than 31dB. The bias circuits for the first and second stages were designed by near class A and class AB operation, respectively. The gate bias of -3V is supplied through divider resistors with a predetermined voltage. Resistive matching circuits are employed in input and gate bias circuits for more stable operation. The matching networks have high pass characteristics which provide DC bias isolation from the RF input and output. The matching networks were optimized using a linear simulator. In the second stage amplifier, an output matching circuit was designed to have low impedances for the second and third harmonic frequencies to reduce the harmonics. The matching and bias circuits were composed of microstrip transmission lines, chip capacitors and resistors (1005-type). In order to reduce the RF and DC loss, the microstrip lines of 300 um-width were employed in the output matching circuit.

Figure 7 shows a top view photograph of the dual-mode power amplifier with 11.9 x 21.0 mm^2. Glass-based epoxy (FR-4) was used as a substrate, and total cost of the power amplifier can be reduced. Although the low cost substrate with a loss tangent of 0.018 was used, high RF performance was achieved for both AMPS and CDMA modes, simultaneously.

Fig. 7. Photograph of the dual-mode power amplifier with 11.9 x 21.0 mm^2.

Figure 8 shows the output power and PAE of the power amplifier. It was measured at a 3.3V supply voltage and a 7 dBm input power as a function of frequency. At the frequency range between 824 MHz and 849 MHz, the output power of 31.5 dBm was maintained within ± 0.2 dBm, and the minimum PAE of 60% was obtained. These are higher than the requirements for AMPS mode.

The intermodulation of the amplifier with the input power was measured using two-tone frequencies, 836.5000 and 836.9425 MHz. The IM3 and the IM5 at an output power of 26 dBm, as shown in Fig. 9, were measured to be -32 dBc and -45 dBc, respectively, which are good enough for CDMA mode applications.

710

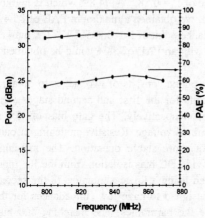

Fig.8. Pout and PAE as a function of frequency

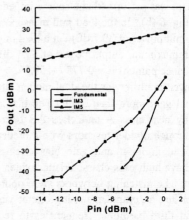

Fig.9. Fundamental, IM3 and IM5 power vs Pin

4. Conclusions

A GaAs power MESFET and a power amplifier for CDMA/AMPS dual-mode hand-held phone, operating at a low voltage of 3.3 V, has been developed. The MESFET has produced output power of 32.5 dBm and power-added efficiency of 68% at an input power of 20 dBm for analog mode, and third-order inter-modulation of -32 dBc and power-added efficiency of 41% for digital mode at an output power of 28 dBm. The power amplifier consists of the linear GaAs power MESFETs and an output matching circuit which reduce the 2nd and the 3rd harmonics. For AMPS mode, the amplifier shows an output power of 31.5 dBm and a PAE of 61%. For CDMA mode, the IM3 of -32 dBc and the IM5 of -45 dBc are obtained at an output power of 26 dBm. These are evaluated to be good enough for dual-mode requirements.

References

[1] N. Iwata, K. Inosako, and M. Kuzuhara, "3V operation L-band power double-doped heterojunction FETs," IEEE MTT-S Digest, June, 1993, pp.1465-1468.

[2] J.-L. Lee, H. Kim, J. K. Mun, H. G. Lee, and H. M. Park, "2.9 V operation GaAs power MESFET with 31.5-dBm output power and 64% power-added efficiency," IEEE Electron Device Letters, vol. 15, pp. 324-326, 1994.

[3] M. Maeda, M. Nishijima, H. Takehara, C. Adachi, H. Fujimoto, Y. Ota, and O. Ishikawa, "A 3.5 V, 1.3 W GaAs Power Multi Chip IC for Cellular Phone", GaAs IC Symp. Tech. Dig., pp. 53-56 (1993).

[4] N. Kuwata, K. Otobe, N. Shiga, S. Nakajima, T. Sekiguchi, T. Hashinaga, R. Sakamoto, K. Matsuzaki, and H. Nishizawa, "High breakdown voltage MESFET with planar gate structure for low distortion power applications," GaAs IC Symp. Tech. Dig. Oct. 1993, pp. 181-184.

[5] J.-L. Lee, H. Kim, J. K. Mun, O. Kwon, J. J. Lee, I. D. Hwang, and H.-M. Park, "A GaAs power MESFET operating at 3.3V drain voltage for digital hand-held phone," ETRI Journal, vol. 16, pp. 1~11, 1995.

[6] D. Ngo, B. Beckwith, P. O'Neil, and N. Camilleri, "Low voltage GaAs power amplifiers for personal communications at 1.9 GHz," in IEEE MTT-S Int. Microwave Symp. Digest, 1993, pp. 1461-1464.

Inst. Phys. Conf. Ser. No 145: Chapter 5
Paper presented at 22nd Int. Symp. Compound Semiconductors, Cheju Island, Korea, 28 August–2 September 1995
© 1996 IOP Publishing Ltd

GaAs MESFET Model for the Temperature Range from 4 K to 625 K

T Ytterdal*+, B J Moon#, T A Fjeldly*+ and M S Shur+

* Dept. of Physical Electronics, Norwegian Institute of Technology, 7034 Trondheim, Norway.
+ Dept. of Electr. Engineering, University of Virginia, Charlottesville, VA 22903-2443, USA.
# Vitesse Semiconductor Corporation, Camarillo, CA 93012, USA.

Abstract. We describe an enhanced GaAs MESFET model which is physics-based and unified, covering all regimes of operation for a wide range of temperatures. Among the effects included are velocity saturation, drain induced barrier lowering, finite output conductance in the saturation region, bias dependent series resistances, bias dependent mobility, gate leakage, non-uniformities in the channel doping, frequency dependent output conductance, backgating and sidegating, and temperature dependent parameters. The output resistance and the trans-conductance are also accurately reproduced, making the model suitable for analog CAD. We implemented this model in SPICE and demonstrated excellent convergence, accuracy and speed.

1. Introduction

One of the advantages of GaAs MESFET technology is its ability to operate in a wide temperature range. The large band gap of GaAs allows GaAs-based FETs to operate at temperatures considerably higher than those for Si MOSFET technology. Also, GaAs MESFET electronics has demonstrated superior performance at cryogenic temperatures, including low noise operation, making it suitable for applications in, for example, photodetector readout circuits even at 4 K [1]. At such low temperatures, Si based technology does not work at all because of freeze-out.

We report on a new GaAs MESFET model which accurately describes both above-threshold and subthreshold characteristics of GaAs MESFETs in the temperature range from 4 K to 625 K. This model has been improved over the existing MESFET models in several areas. The present model accounts for bias dependent series drain and source resistances, velocity saturation in the channel, finite output conductance in the saturation regime, drain induced barrier lowering (DIBL), effects of bulk charge, bias dependent mobility, gate leakage, nonuniformities in the channel doping, frequency dependent output conductance, and sidegating and backgating. The model parameters such as, for example, the low-field mobility and the source and drain resistances, are extractable from experimental data using a direct extraction method [2]. Finally, the model accurately describes the output resistance and the transconductance, making it suitable for analog CAD [3].

2. Temperature Dependence of MESFET characteristics

Characteristics of MESFETs, particularly the threshold voltage and the low-field mobility, depend strongly on the operating temperature. In Fig. 1, we have plotted typical drain current characteristics of a modern sub-micron depletion mode MESFET versus gate-source and drain-source voltages at two temperatures. By comparing the characteristics in Fig. 1a, we observe a negative shift in the threshold voltage, a reduction of the transconductance with increasing temperature, and a "zero temperature coefficient" (ZTC) point. Above threshold, the major temperature effect is a reduction of the saturation current at elevated temperatures as shown in Fig. 1b. The reduced current drive capability lowers the maximum operating frequency of the MESFET. Hence, a major concern is that MESFET circuits may fail to meet speed specifications at elevated temperatures.

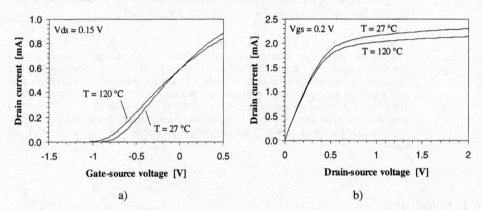

a)   b)

Fig. 1. Typical MESFET drain current characteristics versus a) gate-source voltage and b) drain-source voltage at room temperature and at 120 °C. The effective gate length was 0.5 µm, the gate width was 10 µm, and the nominal threshold voltage was -0.87 V. Experimental data are measured on the samples described in [4].

The above discussion clearly illustrates the importance establishing a reliable model for temperature effects in MESFETs. Our modeling approach incorporates temperature dependent MESFET model parameters, such as the threshold voltage, the low-field mobility, the saturation velocity, the subthreshold ideality factor, the parasitic resistances, and the Schottky barrier height into the basic device model. All these quantities are linked to temperature dependent mechanisms and regions in the MESFET structure as indicated in Fig. 2. In this work, we have concentrated on the temperature modeling of the drain and gate currents.

The threshold voltage of a MESFET is given by (see for example [5])

$$V_T = V_{bi} - V_{po} \tag{1}$$

where V_{bi} and V_{po} are the built-in and pinch-off voltages, respectively. These two voltages are reported to decrease nearly linearly with temperature with slopes of approximately 1 mV/K for V_{bi} [7] and 0.5 mV/K for V_{po} [8]. Hence, the temperature dependence of the threshold voltage can be written as:

$$V_T = V_{T0} - K_{VT}(T - T_0) \tag{2}$$

where V_{T0} is the nominal threshold voltage, K_{VT} is a constant, and T_0 is the nominal temperature.

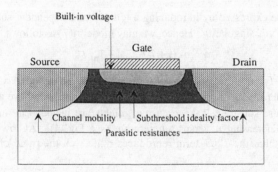

Fig. 2. Temperature dependent parameters of a GaAs MESFET.

The mechanisms behind temperature dependence of the low-field mobility is well understood, and can be expressed analytically through the impurity scattering limited mobility:

$$\mu_{imp} = \mu_0 (T / T_\mu)^{XTM0} \tag{3}$$

and the polar optical scattering limited mobility:

$$\mu_{po} = \mu_1 (T_\mu / T)^{XTM1} + \mu_2 (T_\mu / T)^{XTM2} \tag{4}$$

as follows

$$1 / \mu_{n0} \approx 1 / \mu_{imp} + 1 / \mu_{po}. \tag{5}$$

Here, μ_0, μ_1, μ_2, T_μ, $XTM0$, $XTM1$, and $XTM2$ are parameters.

In a wide temperature range, the dependence of the saturation velocity can be related to the temperature dependence of the low-field mobility. Based on the Monte Carlo simulations reported in [9], the following expression describes the temperature dependence of the saturation velocity:

$$v_s = 0.6 v_{s0} \left(1 + \frac{\mu_{n0}}{\mu_{vs}} - \frac{\mu_{n0}^2}{3\mu_{vs}^2} \right) \tag{6}$$

where v_{s0} and μ_{vs} are parameters. Equation (6) is valid for mobility values up to $1.5\mu_{vs}$.

According to Shoucair *et al.* [7], the Schottky barrier height decreases linearly with temperature. In our model, we use the following equation to describe the temperature dependence:

$$\Phi_b = \Phi_{b0} - \Phi_{b1}(T - T_0) \tag{7}$$

where the parameter Φ_{b1} gives the sensitivity of the barrier height to the temperature.

The subthreshold ideality factor η is found to be practically independent of temperature in the high temperature region where thermionic emission is responsible for the subthreshold current. However, at low temperatures, η increases rapidly with decreasing temperature. This effect may be explained by assuming the presence of a temperature independent trap assisted tunneling current, normally masked by thermionic currents at higher temperatures. At low temperatures, the conventional subthreshold current drops sharply with decreasing gate-source voltage, and soon becomes comparable to the tunneling current. The combined current then can be expressed by introducing a temperature dependent subthreshold ideality factor containing a 1/T singularity. Hence, we may model $\eta(T)$ as follows:

$$\eta = \eta_0(1 + T / T_{\eta 0}) + T_{\eta 1} / T \qquad (8)$$

Here, η_0, $T_{\eta 0}$, and $T_{\eta 1}$ are fitting parameters. Note that even though the subthreshold ideality factor is nearly independent of temperature at high temperatures, we have added a linear term in (8). This because substrate leakage current effects are included in our extraction of η, causing a slightly increasing η versus temperature (see Fig. 4). At low temperatures, we notice from Fig. 3 that the $T_{\eta 1}/T$ term reproduces quite well the trend of the few available measured points.

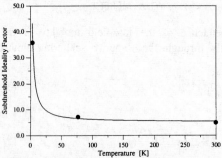

Fig. 3. Experimental (dots) and modeled (line) subthreshold ideality factor versus temperature at low temperatures. (Data from Kirschman *et al.* [1]). Extracted parameters: $\eta_0 = 5$, $T_{\eta 0} = \infty$, and $T_{\eta 1} =$ 115 K (From [4]).

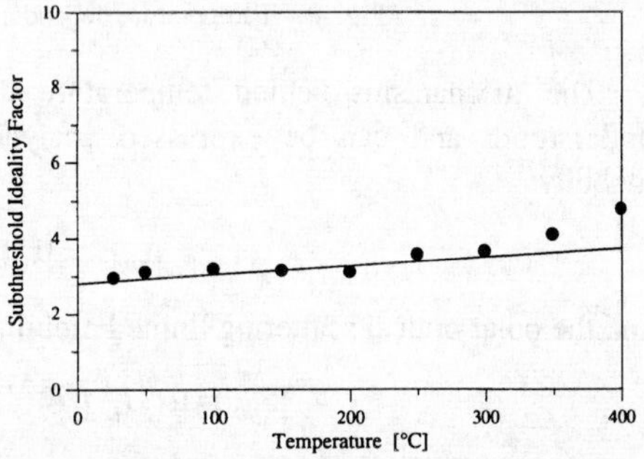

Fig. 4. Experimental (dots) and modeled (line) subthreshold ideality factor versus temperature at high temperatures. (Data from Shoucair *et al.* [7]). Extracted parameters: $\eta_0 = 2.14$, $T_{\eta 0} = 894$ K, and $T_{\eta 1} = 0$ (From [4]).

The standard SPICE equation is used for the parasitic resistances [10]:

$$R = R_0\left[1 + rt_{c1}(T_s - T_0) + rt_{c2}(T_s - T_0)^2\right]. \qquad (9)$$

Here, R_0 is the resistance value at $T = T_0$, and rt_{c1} and rt_{c2} are the first and second order temperature coefficients, respectively.

3. Results

The present model has been implemented in our circuit simulator AIM-Spice [11], [12] and verified extensively against several fabricated MESFETs. To illustrate the accuracy of our model, we considered measured drain and gate current characteristics for two devices fabricated with different technologies, at three different temperatures: 77 K, room temperature and 125 °C. The first device, labeled A, is a depletion mode deep sub-micron MESFET with an effective gate length of 0.515 µm and a gate width of 10 µm. The extracted threshold voltage at room temperature is -0.87 V. The second device, labeled B, is an enhancement mode MESFET with an effective gate length of 1 µm and a gate width of 6 µm. The extracted threshold voltage at room temperature for this device is 0.26 V.

Fig. 5. illustrates the results of comparing measured drain current characteristics and AIM-Spice simulations for device A at different temperatures. Using a direct extraction method similar to the one described in [2] followed by an optimizing step, we obtained nearly perfect fits to measured drain current in the subthreshold regime of operation. Fig. 6 verifies that our model also reproduces accurately the channel conductance, which is very important for accurate simulation of analog circuits.

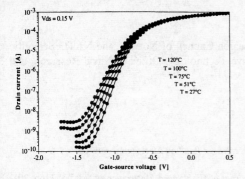

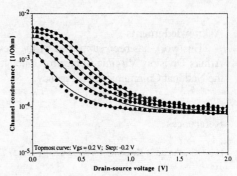

Fig. 5. Measured (symbols) and simulated (lines) subthreshold current characteristics of device A at different temperatures.

Fig. 6. Measured (symbols) and simulated (lines) channel conductance of device A. Operating temperature is 125 °C (From [4]).

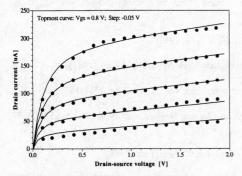

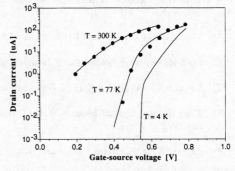

Fig. 7. Measured (symbols) and simulated (lines) above-threshold drain current characteristics at 77 K for device B (From [4]).

Fig. 8. Simulated (lines) subthreshold drain current characteristics at 4, 77 and 300 K Symbols are measured data at 77 and 300 K. Same device as in Fig. 7.

Low temperature results of our model were obtained using the same approach, and are illustrated in Figs. 7 and 8. Here, we have plotted measured and simulated drain current of device B both above and below threshold, operating at room temperature and at 77 K. Simulated results for the drain current at 4 K are also shown in Fig. 8.

Conclusions

We have presented an enhanced GaAs MESFET model which accurately reproduces both above-threshold and subthreshold characteristics of GaAs MESFETs for a wide range of temperatures. The model is physics-based and unified, covering all regimes of operation. The output resistance and the transconductance are also accurately reproduced, making the model suitable for analog CAD. We implemented this model in SPICE and demonstrated excellent convergence, accuracy and speed. The modeling approach used can easily be modified to apply to other GaAs FET technologies such as, for example, HFETs. Also, by including temperature dependence of key model parameters, the model can be extended to account for self-heating.

Acknowledgments

This work has been supported by the Research Council of Norway, the NATO Scientific Affairs Division, Virginia Center for Innovative Technology, Office of Naval Research, and the National Communications Agency.

References

[1]   Kirschman R K, Lemoff S V, and Lipa J A, *SPIE Infrared Readout Electronics*, vol. 1684, 1992.

[2]   Shur M, Fjeldly T A, Ytterdal T, Lee K, *Int. Journal of High Speed Electronics*, vol. 3, no. 2, pp. 201-233, 1992.

[3]   Tsividis Y P and Suyama K, *IEEE J. Solid-State Circuits*, vol. 29, no. 3, pp. 210-216, March 1994.

[4]   Ytterdal T, Moon B J, Fjeldly T A, and Shur M S, *IEEE Trans. on Electron Devices*, vol. 42, no. 10, Oct. 1995.

[5]   Shur M, *GaAs Devices and Circuits*, Plenum Press, New York, 1987.

[6]   Sze S M, *Physics of Semiconductor Devices*, 2nd ed., New York, Wiley, pp. 19-26, (1981).

[7]   Shoucair F S and Ojala P K, *IEEE Trans. on Electron Devices*, vol. 39, no. 7, pp. 1551-1557, July 1992.

[8]   Wong H, Liang C, and Cheung N W, *IEEE Trans. on Electron Devices*, vol. 39, no. 7, pp. 1571- 1577, July 1992.

[9]   Xu J and Shur M, *Applied Physics Lett.*, vol. 52, no. 11, pp. 922-923, Nov. 1988.

[10] Quarles T, Newton A R, Pederson D O and Sangiovanni-Vincentelli A, *SPICE3 Version 3f5 User's Manual*, Berkeley, 1994.

[11] Ytterdal T, Lee K, Shur M, Fjeldly T A, in *Proc. Int. Semiconductor Device Research Symposium, ISDRS'91*, Charlottesville, Virginia, pp. 481-485, Dec. 1991.

[12] Lee K, Shur M, Fjeldly T A, Ytterdal T, *Semiconductor Device Modeling for VLSI*, Prentice Hall, New Jersey (1993).

Inst. Phys. Conf. Ser. No 145: Chapter 5
Paper presented at 22nd Int. Symp. Compound Semiconductors, Cheju Island, Korea, 28 August–2 September 1995
© *1996 IOP Publishing Ltd*

High Device Performance of Ion-Implanted WN 0.25μm Gate MESFET Fabricated Using I-line Photolithography with Application to MMIC

Eung-Gie Oh,   Jeon-Wook Yang, Chul-Soon Park and Kwang-Eui Pyun

Semiconductor Division, Electronics and Telecommunications Research Institute
Yusong P.O.Box 106, Taejeon, Korea, Tel : 82-42-860-5602, Fax : 82-42-860-6200

Abstract.   We developed the WN 0.25μm gate GaAs   MESFET fabrication process using direct ion-implantation and i-line photolithography.   DC current-voltage characteristics does not show short channel effect. The maximum transconductance of 600mS/mm and the K-factor of 450mS/Vmm were obtained.   As high as 65GHz of cut-off frequency has been realized without any de-embedding of parasitic effects. The MESFET shows the minimum noise figure of 0.87dB and the associated gain of 9.97dB at 12GHz.

1. Introduction

Monolithic ICs in the millimeter-wave region are attractive because they can provide the most efficient device circuit interaction at very high frequency. The circuit is fabricated directly adjacent to devices so that loss can be minimum and the size of millimeter-wave circuit is considerably smaller as compared with the equivalent microwave circuit. In spite of the all advantages of MMICs,   the main obstacles to overcome in the MMIC technology are the reduction of price and   the increase of productivity and reliability of chip without the loss of circuit performance.

In this study, to resolve some of the problems discussed above, we developed the WN 0.25μm gate GaAs   MESFET fabrication process using direct ion-implantation and i-line photolithography, which could open a wide opportunity for this device to be used as a building block in millimeter-wave ICs. The refractory metal, WN is used   as the gate metal and the 0.25μm fine gate was formed by using the double exposure process[1] of i-line stepper with 0.4 NA and subsequent reactive ion etching process. With this procedure, the fine 0.25μm photo-resist pattern was formed with larger focus margin than any other lithographic process. Compared to the E-beam lithography and any other conventional photo-lithography,   this process is more competitive in manufacturing cost and also gives better mass productivity and has similar reliability.   High-performance 0.25μm gate GaAs MESFET has been realized by the reduction of source resistance and suppression of short-channel effects due to the self-aligned MESFET structure and the shallow implantation for the channel. The shallow implantation, which gives higher aspect ratio(the gate length over the channel thickness), is possible due to better capping capability of WN gate in the activation process which in turn provides higher free carrier concentration. The self-aligned LDD(Lightly Doped Drain)structure is easily abtained by using WN gate as an ion-implantation mask. Even though without the buried p-type layer, we have obtained excellent

DC current-voltage characteristics which does not show short channel effects by only using the shallow channel implantation and the self-aligned LDD implantation. The maximum transconductance of 600 mS/mm and the K factor of 450 mS/Vmm were obtained and these values are very compatible to the best data obtained from the AlGaAs/GaAs HEMT[2] and MESFET with buried p-layer[3].

2. Device Fabrication

The devices were fabricated on semi-insulating (100) GaAs wafers. First, the n+ source/drain-contact regions was ion implanted using $Si^+(28)$ with a dose of 4×10^{13} cm^{-2} at 100keV. A 30 keV $Si^+(29)$-ion implantation with a dose of 5×10^{12} cm^{-2} followed to form the shallow channel. The refractory gate metal WN was sputtered onto the whole GaAs substrate. Then, the 0.25µm fine photoresist pattern was formed by using the double exposure process of i-line stepper with 0.4 NA and 1.25 µm mask pattern. This process is illustrated in Figure 1. After 1st exposure, the stage of stepper was moved by predetermined distance(1µm) and then, second exposure was performed. With this procedure, the fine 0.25µm photoresist pattern was formed with larger focus margin than the conventional photo-lithographic process. This is because a large pattern has better contrast than a fine pattern and also gives a sharper edge pattern of the developed photoresist pattern. Figure 1(d) shows SEM photograph of the developed 0.25 µm photoresist pattern. In order to form the 0.25µm gate, using fine photoresist pattern as a mask, the WN gate material is reactively ion etched in a fluorocarbon /helium(CF$_4$/He) plasma that gives vertical etch profile. Then, ion implantation of $Si^+(28)$ at an energy of 60keV and a dose of 1×10^{13} cm^{-2} for the lightly doped regions self-aligned to the gate. The channel, n- and n+ regions were activated with rapid thermal annealing at 950°C for 10 sec. The conventional AuGeNi ohmic contacts were deposited and lifted off and alloyed in N$_2$ ambient gas. A schematic cross-sectional view of the WN gate n$^-$-self-aligned MESFET is shown in Figure 2. As a result of 30keV implantation and rapid thermal annealing, the effective channel length a$_{eff}$ was as thin as 55 nm form SIMS measurement. Therefore, the short cannel effect clearly does

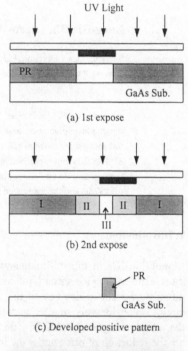

(a) 1st expose

(b) 2nd expose

(c) Developed positive pattern

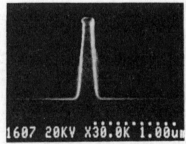

(d) SEM photograph of 0.25µm
Photoresist pattern

Fig. 1 Fine line pattern formation using double exposure process

I : Double exposure area

II : Single exposure area

III : No exposure area

not occur due to the high aspect ratio, $L_g/a_{eff} \approx 4.5$. The gate-to-source spacing was 0.3µm to reduce source resistance and total drain-to-source spacing was 1.5µm. The gate width were 30 µm, 50µm, 100µm, and 200µm. A typical MESFET is shown in Figure 3.

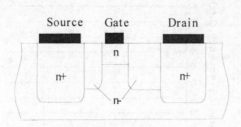

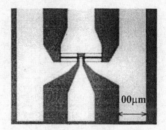

Fig. 2. A schematic cross-sectional view of the WN gate n⁻-self-aligned MESFET

Fig. 3. SEM photograph of a typical 100µm gate width MESFET

3. DC and RF Performance

The Drain I-V characteristics of the 0.25 µm x 30 µm MESFET are shown in Figure 4. The pinch-off characteristic is excellent and short channel effects appears to be suppressed. The device exhibits a very low output conductance g_o of 9mS/mm around zero gate bias. The threshold voltage, Vth = -0.45V was obtained. Figure 5 shows the corresponding plot of transconductance g_m at drain bias of 2V. Maximum transconductance and K value of 600mS/mm and 450mS/V·mm were obtained, respectively. This yields a voltage gain (g_m/g_o) of 66, which is higher than that of 0.25 µm gate pseudomorphic HEMT's [4].

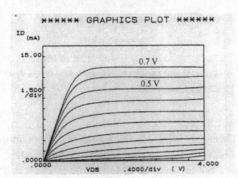

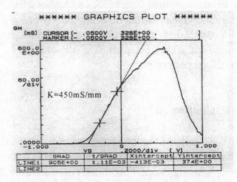

Fig. 4 Drain I-V characteristics of the WN gate MESFET with the gate length Lg= 0.25µm and gate width Wg= 30µm. Vgs is stepped from -0.4 to 0.7 in o.1V intervals. Vth = -0.45 V

Fig. 5 Transconductance versus gate voltage of the WN gate MESFET with the gate length Lg= 0.25µm and gate width Wg= 30µm. The drain- source voltage Vds is 2V.

The S-parameter of the 0.25µm x 200µm gate MESFET was measured from 1GHz to 26GHz with an HP8510B network analyzer and Cascade Microtech microwave probes. The current gain as a function of frequency is calculated from the measured S-parameters and is shown in figure 6. The extrapolation of the current gain to unity with a -20 dB/decade slope yields a cutoff frequency F_t of 65 GHz for 0.25µm gate MESFET. The drain current

720

dependence of the cutoff frequency is shown in figure 7, where the drain voltage is 1.5V. The maximum cutoff frequency was 65GHz at a drain current of 54 mA and in the wide range of drain current, a cutoff frequency of over 60GHz was obtained.

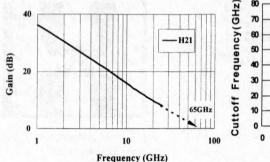

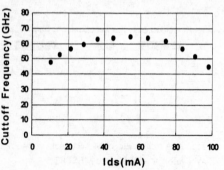

Fig. 6 |H21| versus frequency. The extrapolation of |H21| using a -20dB/decade slope gives an Ft of 65GHz.

Fig. 7 Drain current dependence of current gain cutoff frequency for 0.25μm x 200μm gate MESFET. Drain-source voltage is 1.5V.

Noise figure measurements were performed for MESFET with 0.25μm x 30μm gate at the frequencies from 2GHz to 18GHz with interval of 1GHz by using HP 8510B network analyzer and ATN NP5B noise parameter test set. Figure 8 exhibits minimum noise figure NF_{min} and associated gain G_a as a function of frequency at Vds=2V and Ids=2mA. The measured Nf_{min} and G_a at 12GHz were 0.87 dB and 9.97 dB, respectively.

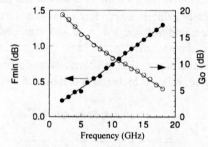

Fig. 8 NF_{min} and G_a as a function of frequency at Vds=2V and Ids=2mA. Nfmin=0.87 dB, Ga=9.97 dB is shown at 12GHz.

4. Conclusion

We devloped the fabrication process for the 0.25μm gate MESFET. This process utilizes only proven technologies, such as ion implantation and conventional i-line photolithography. The demonstration of maximum transconductance of 600mS/mm, K factor of 450mS/mm and 65Ghz cutoff frequency with this process is significant from the view point of practical applications, such as MMICs and digital ICs. The low noise performance (Nf_{min}=0.87 dB, G_a =9.97dB) at high frequency(12GHz) at low power level (4mW)is also important result for low power communication applications.

References

[1] J. W. Yang, B. R. Kim, C. S. Park, and H. M. Park   1994   J. KITE   Vol.31-A   919-924.
[2] Y. Jin, B. Etienne, F. Pardo, G. Faini and H. Launois 1990   Microelectrionic Engineering   11   113-116.
[3] K. Onodera et al. 1988   IEEE   Electron Device Lett.   Vol. 9   no. 9   417-418.
[4] P. C. Pao   et al. 1989   IEEE Trans. on Electron Device   Vol. 36   No. 3   461-473.

Inst. Phys. Conf. Ser. No 145: Chapter 5
Paper presented at 22nd Int. Symp. Compound Semiconductors, Cheju Island, Korea, 28 August–2 September 1995
© *1996 IOP Publishing Ltd*

Non-uniform Light Emission from Parasitic Oscillating GaAs MESFET's

Hidemasa Takahashi, Junko Morikawa, Kazunori Asano and Yasunobu Nashimoto

Compound Semiconductor Development Laboratory
ULSI Device Development Laboratories, NEC Corporation
2-9-1, Seiran, Otsu-shi, Shiga 520, Japan

Abstract: This paper reports on spatially non-uniform light emissions from a GaAs MESFET, undergoing parasitic oscillations. With oscillation, there was a 6-fold increase in emission intensity and a 300K rise in effective electron temperature. In addition, the distribution of emission intensity became spatially non-uniform, as did the electron temperature. The localized electron temperature rise and intensity increase show that the voltage between the drain and the gate was modulated by an electric field standing wave caused by oscillation, consequently a higher electric field accelerates impact ionization. An increase of a few volts of Vds would explain the temperature rise and intensity increase.

1. Introduction

It is well known that power MESFET's often exhibit undesirable parasitic oscillation. For power device applications, in which the gate width is large, oscillation is common. However the oscillation mode is usually not so clear, because there are many factors affecting the parasitic oscillation mode inside and outside the device. In order to analyze the oscillating mode, we observed visible light emissions from such an oscillating device.

The visible light emission phenomena from GaAs MESFET's at high drain-bias voltage condition was studied [1-4]. The dominant emission mechanism of photons with energy greater than the energy gap is thought to be a recombination of channel electrons with holes generated by impact ionization[2,3]. Emission intensity is determined by two factors. One is the number of holes generated by impact ionization, which increases with Vds increase. The other is the number of channel electrons, which increases with Vgs increase. As a consequence, the emission intensity and the electron temperature depend on the bias conditions. Hence, this emission can be used as a probe for analysis of electric field between the gate and the drain.

In this paper we report on the spatially non-uniform light emissions from a GaAs MESFET undergoing such parasitic oscillation.

2. Experiments

The sample device was an ordinary GaAs MESFET with a gate length of 0.45μm, a gate-drain spacing of 0.45μm and a gate width of 1.5mm (125μmx12fingers). Figure 1 shows the Ids-Vds characteristics of the device. The discontinuities in I-V slope show the parasitic oscillation boundary; at drain current higher than this boundary oscillation occurred. The oscillation was observed with a spectrum analyzer. The device was set on microscope a stage in dark ambient at 300K, and DC-biased. Emission was detected with a photon counting camera. Spectroscopy was performed from 1.45 to 1.94eV with 10nm band pass interference filters.

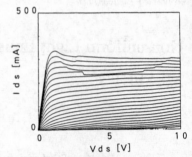

Figure 1   DC drain current characteristics of the device. The maximum gate voltage is 0V and the gate voltage step is 0.1V.

3. Results and discussion

3.1 Light emission comparison

Figure 2 compares the spatial distribution of emissions under oscillating (Vds=8V,Vgs=-0.3V) and non-oscillating (Vds=8V, Vgs=-0.5V) conditions. The light shown here is emitted from the region between each gate finger and the drain electrode. The emission was not spectroscopied in this case, and not corrected with the sensitivity of the detector camera. The emission pattern without oscillation was relatively uniform, even though there was slight unbalance because of gate-drain alignment deviation. On the other hand the pattern became drastically non-uniform with oscillation, and the photon-count rate increased by about 6-fold. Figure 3 shows the energy distribution of photons emitted from the entire gate width for both cases. The electron temperature was calculated from the slope, assuming a Maxwellian distribution. A 300K rise in effective electron temperature was observed with oscillation. Figure 4 and 5 show the measured temperature and emission intensity at the gate-pad side of each gate finger, respectively. The measurement field size was 25μm along the gate by 10μm perpendicular to the gate. The electron temperature distribution was relatively uniform without oscillation, but was non-uniform with oscillation. The maximum increase of temperature was 700K at the outer gate fingers (#1,#12). On the other hand the gate-pad side end of the center gate finger (#6) revealed no temperature rise. Table 1 summarizes these results.

Table 1   Comparison of light emissions between oscillating and non-oscillating state

Vds (V)	Vgs (V)	Parasitic oscillation	Emission pattern	Emission intensity	Electron temperature distribution	Electron temperature
8.0	-0.5	No	Uniform	Weak	Uniform	1700K
8.0	-0.3	Yes	Non-uniform	Strong (6-fold)	Non-uniform	2000K

These results show that the voltage between the gate and the drain was locally modulated by oscillation.

3.2 DC bias dependence of electron temperature and emission intensity

To estimate the bias dependence of electron temperature and emission intensity, we

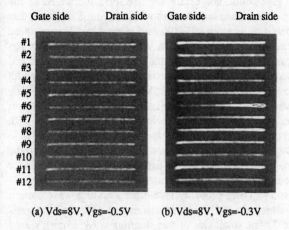

Gate side   Drain side   Gate side   Drain side

#1
#2
#3
#4
#5
#6
#7
#8
#9
#10
#11
#12

(a) Vds=8V, Vgs=-0.5V   (b) Vds=8V, Vgs=-0.3V

Figure 2   Emission images from the MESFET (a) without oscillation and (b) with oscillation.

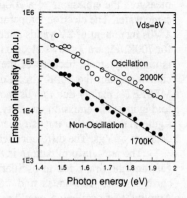

Figure 3   Emitted photon energy distribution at Vds=8V with and without oscillation. The electron temperature is calculated from the energy distribution slopes, assuming a Maxwellian distribution

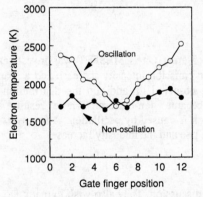

Figure 4   Electron temperature distribution at gate-pad side region.

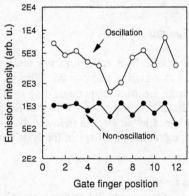

Figure 5   Emission intensity distribution at gate-pad side region.

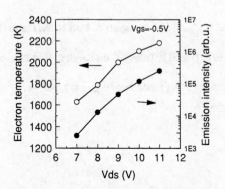

Figure 6   Vds dependence of electron temperature and emission intensity.

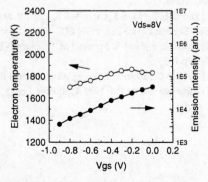

Figure 7   Vgs dependence of electron temperature and emission intensity.

724

measured the emissions of a 200μm gate width, non-oscillating MESFET fabricated on the same wafer. The electron temperature dependence on Vds is shown in Figure 6. It shows that a Vds increment of 2V yields a temperature increase of 300K, and more than 5V is required for 700K. Figure 7 shows electron temperature dependence on Vgs. The dependence on Vgs is relatively small, because of the less sensitivity of Vgs to the electric field intensity, and cannot explain the 300K rise. Consequently the temperature rise seen in large gate width MESFET's mainly due to Vds increase. The Vds increase of 5V corresponds to about 100-fold increase in emission intensity, which is 10 times larger than measured one. As shown in figure 7, the emission intensity depends on Vgs, because the electron number decreases with decrease of Vgs. The difference will be explained by the decrease of Vgs due to oscillation.

The non-uniform pattern is thought to be a standing wave of electric field between the gate and the drain due to oscillation. It seems that the gate-pad side of #1 and #12 finger correspond to anti-nodes and #6 corresponds to a node as shown in figure 4 and 5. Amplitude of oscillation would reach 5V. There probably is another standing wave pattern along the center gate finger (#6), in which the drain-pad side is an anti-node and the gate-pad side is a node as shown in figure 2. The small size of this standing wave pattern shows that the oscillation frequency is more than 10 GHz, and the device will have to be included in the category of distributed elements.

4. Summary

The visible light emission from the parasitic oscillating GaAs MESFET was investigated with an emission microscope. With oscillation an emission intensity increase and a spatial average electron temperature rise of 300K were observed. In addition the distribution of emission intensity and electron temperature became non-uniform. These results are explained by an electric field standing wave which is caused by oscillation. A few volts of Vds increase would explain the 300K temperature rise and the intensity increase.

Acknowledgment
The authors would like to thank Mr. N. Iwata for discussion. They also wish to thank Dr. H. Kohzu and Dr. A. Higashisaka for their support.

References
[1] Zanoni E, Bigliardi S, Cappelletti R, Lugle P, Magistrali F, Manfredi M, Paccagella A, Testa N, and Canali C, IEEE Electron Device Lett. 1990, EDL-11, (11), p487
[2] Zanoni E, Bigliardi S, Manfredi M, Paccagnella P, Pisoni P, Telaroli P, Tedesco C and Canali C, 1991 El. Lett. 27(9), p.770
[3] Tedesco C, Manfred M, Paccagnella A, Zanoni E and Canali C, 1991 IEDM Tech. Dig., p.437, 1991
[4] Zappe H P and As D J, 1990 Appl. Phys. Lett. 57(27). p.2919

Inst. Phys. Conf. Ser. No 145: Chapter 5
Paper presented at 22nd Int. Symp. Compound Semiconductors, Cheju Island, Korea, 28 August–2 September 1995

A Physically Based High Frequency Noise Model of MESFET's Taking Static Feedback Effect into Account

Jong-Hee Han, Jinsu Ko, and Kwyro Lee

Department of Electrical Engineering, KAIST, 373-1, Kusong-dong, Yusong-gu, Taejon, Korea 305-701. Fax. no. 82-42-869-8530.

Abstract. A new physically based thermal noise model for MESFET's has been proposed, which is compatible with small signal equivalent circuit and large signal current-voltage characteristics. Specifically, the static feedback effect is taken into account to model noise characteristics correctly especially in low current regime. The gate and drain bias dependence of the gate noise voltage, the drain noise current, and the correlation coefficient between them has been investigated thoroughly, showing good agreement with experimental results from 0.5μm gate length MESFET. As a result, our formulation is successfully used to model bias dependence of the four noise parameters with reasonably good accuracy. Our model is simple and physical enough for device design and circuit simulation especially for MMIC application.

1. Introduction

The high frequency noise model of MESFET has been widely studied since 1970's[1]. Though Statz's model is physically solid, its formulation was not extensively compared with experimental results and the relationship with large signal current-voltage model is obscure. Moreover, the static feedback effect which is appreciable in modern short channel MESFET was not considered. Though Fukui's model[2] is a good first order estimation for minimum noise figure, the other noise parameters were not investigated. Pospieszalski's[3] model proposed a simple noise equivalent circuit where two noise sources are used at the gate and drain ports. Although this model is simple enough to give physical insight into circuit design, the accuracy is strongly dependent on the noise sources whose values are not physically explained. And the correlation between the noise sources is neglected.

In this paper, we propose a new physically based noise model of MESFET's, which is compatible not only with small signal equivalent circuit, but also with large signal I-V as well as C-V models.

2. Current-voltage and capacitance-voltage relation of MESFET

Our I-V model is derived based on the following assumptions. Firstly, we divide the channel into two regions. One is the so called gradual channel approximation region and the other is the so called velocity saturation region. Secondly, we assume that the channel capacitance per unit area, C_{in}, is constant along the channel for a given gate-source voltage, V_{GS}. Thirdly, the two-piece velocity-electric field relation is assumed, i.e., the electron velocity is proportional to electric field below critical field, F_{crit}, and saturated at v_{sat} above F_{crit}.

The saturation voltage, V_{Dsat}, is calculated as $V_{GST}+V_L-(V_{GST}^2+V_L^2)^{0.5}$ [4]. Here, V_L is $F_{crit}L_G$. As V_D increases over V_{Dsat}, the potential in the velocity saturation region increases exponentially and the effective channel length decreases. As a first order approximation, the potential profile in the velocity saturation region can be written as [5]

$$V(x) = V_{Dsat} + V_\lambda(\exp(x / \lambda) - 1), \tag{1}$$

where λ is the characteristic length of velocity saturation region, and $V_\lambda = F_{crit}\lambda$. Inverting Eq.(1), the length of velocity saturation region, ΔL can be calculated. To take into account the current increase in saturation regime due to the channel length modulation, we use the following expression for the saturation regime drain current[5]

$$I_{DS} \approx I_{sat}\left(1+\frac{V_\lambda}{V_L}\ln\left(1+\frac{V_D-V_{Dsat}}{V_\lambda}\right)\right). \tag{2}$$

Here I_{sat} is the drain current at $V_{DS}=V_{Dsat}$. The static feedback effect can be considered by taking threshold voltage as a function of V_{DS}. We adopt the following dependence[6]

$$V_{GST} = V_{GST0} + \sigma V_{DS}, \tag{3}$$

where V_{GST0} is V_{GST} at $V_{DS}=0V$, σ is the static feedback effect coefficient. Lastly the following expression is used to take σ dependence on V_{GS} into account[7],

$$\sigma = \frac{\sigma_0}{1+\exp\left(\dfrac{V_{GST0}-V_{\sigma T}}{V_\sigma}\right)}. \tag{4}$$

The gate-source capacitance, C_{gs}, and the gate-drain capacitance C_{gd} can be defined by partial derivative of gate charge with respective to source voltage, V_S, and drain voltage, respectively. We assume that the sum of the gate charge and the channel charge is conserved. Based on the gradual channel approximation, the channel current can be written as $\mu C_{in}W(V_{GT}-V(x)) \, dV(x)/dx$. Using this relation, the potential distribution in the region 1 can easily be found as following

$$V(x) = V_{GT} - \sqrt{V_{GST}^2 - 2I_{DS}R_nV_x}, \tag{5}$$

where V_x is $F_{crit}x$ and R_n is the inverse of $v_{sat}C_{in}W$. The total channel charge can be obtained from the following integration,

$$\frac{Q_n}{C_{in}W} = -\int_0^{L_1} dx\sqrt{V_{GST}^2 - 2I_{DS}R_nV_x} - \Delta L(V_{GT} - V_{Dsat}). \tag{6}$$

$x=L_1$ is the point where the potential is equal to V_{Dsat}. Using Eqs.(5)~(6), C_{gs} and C_{gd} can be calculated. Then, C_{gs} can be written as

$$\frac{C_{gs}}{C_{in}W} = L_1\left(\frac{V_{GST}}{V_{GST}-\frac{1}{2}V_{DS}} - \frac{\left(V_{GST}-\frac{1}{3}V_{Dsat}\right)V_{Dsat}}{2I_{DS}R_{B0}\left(V_{GST}-\frac{1}{2}V_{DS}\right)}\right) + \frac{\Delta L V_{GST}}{\sqrt{V_{GST}^2 + V_L^2}}, \tag{7}$$

where the inverse of R_{B0} is defined by $-dI_{DS}/dV_S$ with constant V_{GT} and V_D. And C_{gd} can be written as

$$\frac{C_{gd}}{C_{in}W} = L_1\frac{\left(V_{GST}-V_{Dsat}\right)\left(V_{GST}-\frac{1}{3}V_{Dsat}\right)}{2I_{DS}R_{A0}\left(V_{GST}-\frac{1}{2}V_{Dsat}\right)} - \Delta L\frac{V_{GST}-V_{Dsat}}{I_{DS}R_{A0}}, \tag{8}$$

where the inverse of R_{A0} is defined by dI_{DS}/dV_D with constant V_{GT} and V_S.

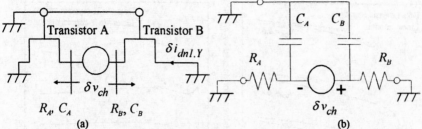

Fig.1. (a) Division of MESFET into two parts. (b) Small signal equivalent circuit of (a).

3. Noise modeling

The infinitesimal thermal noise voltage source, δv_{ch} is distributed along the region 1. In reference to δv_{ch}, we divide MESFET into two parts, i.e., sub-transistors A and B which are located toward the source and drain side of the MESFET, respectively (Fig.1). R_A, R_B, C_A and C_B are the small signal resistance's and capacitance's seen from the drain and source side of the sub-transistors A and B. The values of these parameters are position dependent. $R_A(x)$ and $R_B(x)$ can be derived from partial derivative of I_{DS} of sub-transistors A and B. And $C_A(x)$ and $C_B(x)$ are C_{gd} and C_{gs} of sub-transistors A and B. The infinitesimal noise voltage is thermally generated one at some point of the region 1. Its quantity is proportional to the infinitesimal resistance δR. In addition, we find that the consideration of the mobility degradation by the transverse electric field[8], gives better fit with the experimental results. We introduce the factor, γ, to take the transverse field induced mobility degradation into account. Then, $<\delta v_{ch}{}^2>$ can be written as

$$\left\langle \delta v_{ch}{}^2 \right\rangle = 4kT\Delta f \frac{\gamma \, dV_x}{\mu C_{in} W \sqrt{V_{GT}{}^2 - 2I_{DS} R_n V_x}}, \tag{9}$$

where k is the Boltzmann constant, T the absolute temperature, and Δf is the bandwidth. γ should be equal to 1 at zero drain bias and increase with drain bias. This is taken into account as following equation.

$$\gamma = \frac{2\gamma_0}{\gamma_0 + 1 - (\gamma_0 - 1)\dfrac{V_{DS}{}^3 - 3V_{Dsat}{}^3}{V_{DS}{}^3 + 3V_{Dsat}{}^3}}. \tag{10}$$

The infinitesimal drain noise current in Y-representation induced by δv_{ch} is

$$\delta i_{dn1,Y} = -\frac{\delta v_{ch}}{R_A + R_B}. \tag{11}$$

The infinitesimal gate noise current is also calculated from equivalent circuit in Fig.2 as

$$\delta i_{gn1,Y} = \delta v_{ch} \frac{j\omega (-R_B C_B + R_A C_A)}{R_A + R_B}. \tag{12}$$

The noise sources in H-representation can be easily derived from Eqs.(11)~(12) as

$$\delta v_{gn1,H} = \delta v_{ch} \frac{-R_B C_B + R_A C_A}{(R_A + R_B) C_{gs}}, \text{ and} \tag{13}$$

$$\delta i_{dn1,H} = -\frac{\delta v_{ch}}{R_A + R_B}\left(1 - \frac{g_m (R_B C_B - R_A C_A)}{C_{gs}}\right). \tag{14}$$

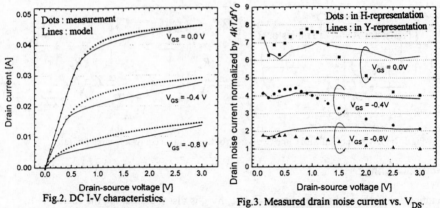

Fig.2. DC I-V characteristics.

Fig.3. Measured drain noise current vs. V_{DS}.

It is obvious from Eqs.(13)~(14) that the correlation between $v_{gn1,H}$ and $i_{dn1,H}$ is not zero.

To calculate the drain noise, we firstly neglect the displacement current to the gate terminal, meaning current continuity along the channel. Under this assumption, we can use the impedance field method to calculate the open circuit drain noise voltage as below[9]

$$\left\langle v_{dn2}{}^2 \right\rangle = \int_{\text{Region II}} 4q^2 \Delta f D_h \left| \frac{dR_{ds}(x')}{dx'} \right|^2 nW dx'. \tag{15}$$

Here, D_h is high field diffusion constant. $R_{ds}(x)$ has two parallel components. One is the resistance due to the channel length modulation, and the other is to the static feedback effect, σg_m. Using the relation $I_{DS}=qnWv_{sat}$ and $qV_T=kT$, $<v_{dn2}{}^2>$ can be written as

$$\left\langle v_{dn2}{}^2 \right\rangle = 4kT\Delta f \frac{I_{DSS}D_h}{v_{sat}V_T} \frac{-1}{6\sigma^2 g_m{}^2 \lambda} \left[\frac{1+3\alpha \exp(x'/\lambda)}{(1+\alpha \exp(x'/\lambda))^3} \right]_0^{\Delta L}. \tag{16}$$

$v_{gn2,H}$ and $i_{dn2,H}$ are easily expressed by v_{dn2} as

$$i_{dn2,H} = \frac{v_{dn2}}{R_{ds}} \left(1 + \frac{g_m R_{ds} C_{gd}}{C_{gs}} \right), \text{ and} \tag{17}$$

$$v_{gn2,H} = -v_{dn2} \frac{C_{gd}}{C_{gs}}. \tag{18}$$

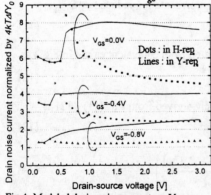

Fig.4. Modeled drain noise current vs. V_{DS}

Table. Measured small signal parameters. MR_{gn} and CR_{gn} are measured and calculated $<v_{gn,H}{}^2>$ normalized by $4kT\Delta f$, respectively.

V_{DS}	1.0			2.0		
V_{GS}	0	-.4	-.8	0	-.4	-.8
r_i [Ω]	5.0	6.4	9.9	5.4	6.9	9.4
g_m[mS]	47	43	32	44	40	32
r_{ds}[Ω]	230	250	250	340	330	310
C_{gs}[pF]	.27	.22	.18	.31	.26	.21
C_{gd}[pF]	.04	.05	.05	.03	.04	.04
MR_{gn}	10	9.0	9.8	12	11	11
CR_{gn}	5.4	6.7	11	5.4	6.4	9.3

4. Measurement and comparison with theory

Both S-parameters and noise characteristics are measured from ion-implanted MESFET whose geometry is 0.5×300 μm^2. The drain saturation current at $V_{GS} = 0.0V$ is 45 mA and the external unit current gain frequency is 22 GHz. The r_g, r_s and r_d values are measured using the conventional method[10]. Since it is not easy to separate C_{gsp} from C_{gs} by s-parameter measurement, we adopt the approximate estimation of $C_{gsp} = 0.18pF/mm$[11].

Fig.2 compares the DC I_{DS}-V_{DS} characteristics between measurement and theory. The overall accuracy is good. The four noise parameters are measured with ATN solid tuner and HP noise figure meter. The intrinsic noise characteristics are Extracted from the measured noise parameters by eliminating the effect of parasitic elements such as r_g, r_s, r_d, C_{gd} and C_{gsp} in Fig.1 using the method in [12].

Fig.3 shows the measured $\langle i_{dn,H}^2 \rangle$ and $\langle i_{dn,Y}^2 \rangle$ versus the drain voltage at several gate bias points. There are several interesting features in Fig.3. Firstly, in linear regime, $\langle i_{dn,H}^2 \rangle$ is almost the same as $\langle i_{dn,Y}^2 \rangle$ for all the gate bias studied in this work. Secondly, in the deep saturation regime, $\langle i_{dn,H}^2 \rangle$ becomes smaller than $\langle i_{dn,Y}^2 \rangle$ and the difference between them increases with increasing V_{GS}. One remarkable feature is that $\langle i_{dn,H}^2 \rangle$ decreases drastically just after the saturation point and becomes constant in the deep saturation regime, while $\langle i_{dn,Y}^2 \rangle$ stays almost constant independently of V_{DS} in saturation regime. Thirdly, near the saturation point, there are bumps both in $\langle i_{dn,Y}^2 \rangle$ and in $\langle i_{dn,H}^2 \rangle$.

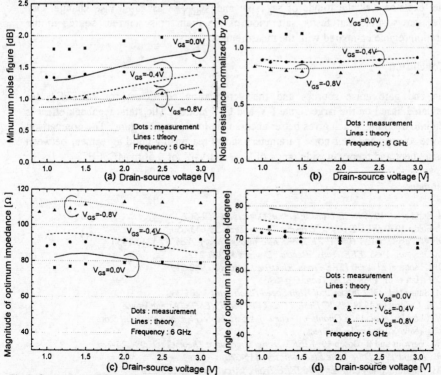

Fig.5. Comparison of four noise parameters between measurement and theory at 6GHz. (a) Minimum noise figure, (b) noise resistance, (c) the magnitude and (d) the angle of optimum impedance.

Moreover, as shown in Table, measured $<v_{gn,H}^2>$ is almost constant independently of V_{DS} as well as V_{GS} in the deep saturation regime. The result of independence on V_{GS} is quite in contrast with the previous results. Both Pospieszalski[3] and Ladbrooke[8] claimed that it is proportional to r_i as a first order approximation. Therefore it is hard to explain why, in our device, $<v_{gn,H}^2>$ is almost constant while the value of r_i changes twice(see Table) as the gate bias varies.

To fit the experimental data, v_{sat}, D_h, γ_0(see Eq.(10))and λ are chosen as 1.2×10^7cm/sec, 11cm^2/sec, 2.3, and 0.055μm, respectively. The values of D_h and λ agree fairly well with the conventionally accepted value of 20 cm^2/sec for D_h[13], $0.1 \sim 0.2$ of the gate length for λ[5]. The value of $\gamma=2.3$ in deep saturation agrees also fairly well with 3 suggested in Ref.[8]. The calculation result of the drain noise current in H- and Y-representation shows good agreement with the measurement as can be seen in Fig.4. The theoretical explanation of the bias dependence of the drain noise current is as follows. The drain noise current from region 2, $<i_{dn2,H}^2>$ begins to be generated after V_{DS} becomes larger than V_{Dsat}. This appears as the increase of the total drain noise current. C_{gd}/C_{gs} is not small near V_{Dsat}, so that $<i_{dn2,H}^2>$ is larger than $<i_{dn2,Y}^2>$(Eq.(17)) This is why total drain noise current in H-representation is larger than in Y-representation near V_{Dsat} at $V_{GS}=0$ and -0.4V. Since C_{gd} over C_{gs} decreases as MESFET operation goes into the deep saturation regime, $<i_{dn1,H}^2>$ also decreases. Meanwhile, the drain noise current from region 1, $<i_{dn1,H}^2>$ also increases with increasing V_{GS}, because total channel conductance becomes higher. As can be seen from the comparison between measured (Fig.3) and calculated(Fig.4) drain noise, our model explains well the trends of $<i_{dn,H}^2>$ and $<i_{dn,Y}^2>$ as V_{GS}, V_{DS} change. Our model shows somewhat faster saturation of the drain noise current source in the saturation regime compared with the measurement.

5. Conclusion

Drain and gate noise sources and their correlation coefficient for MESFET are formulated based on the large signal I-V and C-V models. The static feedback effect is taken into account, which gives better accuracy in low current regime. The comparison of noise sources and four noise parameters shows reasonably good agreement between measurement and theory in wide range of V_{GS} and V_{DS} for 0.5μm MESFET.

References

[1] H. Statz et al. 1974 IEEE Trans. Electron Devices 21 549-562
[2] H. Fukui 1979 IEEE Trans. Microwave Theory Tech. 29 643-650
[3] M. W. Pospieszalski 1989 IEEE Trans. Microwave Theory Tech. 37 1340-1350
[4] K. Lee et al. 1983 IEEE Trans. Electron Devices 30 207-212
[5] B. J. Moon et al. 1990 IEEE Trans. Electron Devices 37 908-919
[6] C. J. Han et al. 1988 IEDM Tech. Dig. 696.
[7] M. Shur et al. 1992 Intern. J of High Speed Electronics 3 201-233
[8] P. H. Ladbrooke 1989 MMIC Design GaAs FETs and HEMTs (Artech House)
[9] W. Shockley et al. 1966 Quantum Theory of Atoms, Molecules, and the Solid State (New York:Academic Press)
[10] M. Berroth and R. Bosch 1991 IEEE Trans. Microwave Theory Tech. 39 224-229
[11] R. Anholt 1991 Solid State Electronics 34 515-520
[12] H. Hillbrand and P. Russer 1976 IEEE Trans. Circuit Syst. 23 235-238
[13] M. de Murcia et al. 1991 IEEE Trans. Electron Devices 38 2531-2539

Inst. Phys. Conf. Ser. No 145: Chapter 5
Paper presented at 22nd Int. Symp. Compound Semiconductors, Cheju Island, Korea, 28 August–2 September 1995
© *1996 IOP Publishing Ltd*

WSi gate GaAs MESFET's with shallow channels fabricated by ion implantation through WSi films

T Uda, H Fujimoto, K Nishii, and A Tamura

Electronics Research Laboratory, Matsushita Electronics Corporation, 3-1-1, Yagumo-Nakamachi, Moriguchi, Osaka 570, Japan

Abstract. We have developed WSi self-aligned gate (SAG) GaAs MESFET's with shallow channels fabricated by ion implantation through WSi films. By using an ion implantation through a 6-nm-thick gate metal thin film formed by DC sputtering, a shallow channel which is thinner by about 40 nm than the conventional process has been obtained. The 0.5-μm-gate buried p-layer lightly doped drain (BP-LDD) SAGFET has exhibited K = 422 mS/Vmm, gm_{max} = 347 mS/mm and f_{max} = 95 GHz with Vth = -0.27 V. These excellent results suggest that this process is very suitable for fabricating low power dissipation GaAs-MMIC's.

1. Introduction

In order to improve the performance of GaAs MESFET's, it is important to form a shallow channel with a high doping concentration using low-energy ion implantation. However, the minimum implantation energy is limited by the apparatus, and a stable implantation generally becomes difficult as the energy decreases. By using the ion implantation through thin films, such as SiO_2 and SiN, a shallower implanted layer than by using the ion implantation into a bare substrate can be formed with the same implantation energy.

Ion implantation through SiO_2 [1] and SiON [2] thin films, which are deposited by plasma-enhanced chemical vapor deposition (P-CVD), have been reported for fabricating GaAs MESFET channels. However, it is difficult to precisely form thin films with less than 100 nm thickness by P-CVD, and the number of process steps increases because SiO_2 and SiON thin films must be removed after the ion implantation.

In order to improve these problems, we have fabricated shallow channels of refractory metal gate GaAs MESFET's by using the ion implantation through gate metal thin films formed by DC sputtering. Compared with conventional P-CVD-SiO_2 and P-CVD-SiON films, the reproducibility of deposited films thickness can be improved by DC sputtering, and the number of process steps decreases because a WSi film is used as a gate metal and this film need not be removed after ion implantation.

In this paper, we report high K-value WSi self-aligned gate (SAG) GaAs MESFET's with shallow channels fabricated by the ion implantation through WSi thin films.

732

2. Fabrication process

The substrates used in this work were 3-inch undoped semi-insulating (100)-oriented GaAs crystals grown by a LEC method. The WSi thin films were deposited by DC sputtering. Fig.1 shows the schematic cross-sectional view of the 0.5-μm WSi gate GaAs SAGFET with a buried p-layer lightly doped drain (BP-LDD) structure [3]. The SAGFET had an off-set gate structure for n$^+$ region (a gate-source spacing Lgs=0.14 μm, and a gate-drain spacing Lgd=1.0 μm). First, the BP layer was formed by 180-keV Mg ion implantation with a dose of 2×10^{12} cm^{-2}. After a 6-nm-thick WSi thin film was deposited by DC sputtering, the channel layer was formed by 25-keV SiF ion implantation through the WSi thin film with a dose of 1.8×10^{13} cm^{-2}. Then 0.5-μm WSi refractory metal gate was formed by DC sputtering and reactive ion etching (RIE). After that, the Si ion implantation for n' layer was performed at 30 keV with a dose of 5.2×10^{12} cm^{-2}. Then a SiO$_2$ film was deposited on the wafer and the n$^+$ layer was formed by Si ion implantation through the SiO$_2$ film at 150 keV with a dose of 5×10^{13} cm^{-2}. An asymmetric LDD structure was formed by n$^+$ layer implantation using a photo-resist mask. All these implantation layers were annealed simultaneously at 820 ℃ for 15 min in a H$_2$ atmosphere with a SiO$_2$/WSiN cap [4]. Ohmic electrodes were formed by evaporation of AuGe/Ni/Au. Finally, a Ti/Au layer was deposited as a gate over-layer or inter-connect metal.

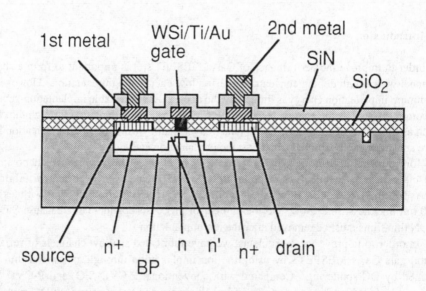

Fig.1. Schematic cross-sectional view of a WSi gate GaAs SAGFET.

3. Results and discussions

3.1. Carrier concentration profiles

Fig.2 shows carrier concentration profiles of channels measured by a C-V method. By using the ion implantation through a 6-nm-thick WSi thin film, a shallow channel which is thinner by about 40 nm than the conventional one was obtained.

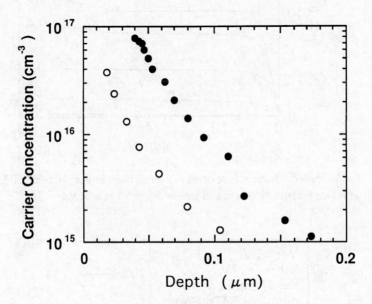

Fig.2. Carrier concentration profiles of channels.
○: n-layer through implantation SAGFET
●: conventional SAGFET

3.2. DC characteristics

Fig.3 shows the typical Id-Vds characteristic of a 0.5-μm-gate SAGFET with a channel formed by the ion implantation through a 6 nm-thick WSi thin film. The gate width (Wg) was 50 μm. The device shows a good static characteristic with a low drain conductance. The threshold voltage was about -0.27 V.

Fig.4 shows the typical $\sqrt{\text{Ids}}$-Vgs characteristic of the 0.5-μm-gate SAGFET. The SAGFET exhibited a K-value of 422 mS/Vmm and a maximum transconductance of 347 mS/mm at the drain voltage (Vds) of 3 V.

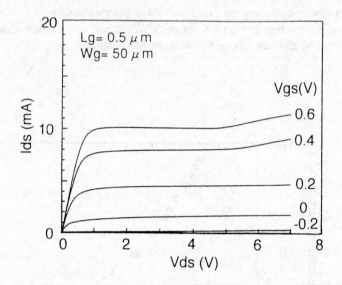

Fig.3. Ids-Vds characteristic of an n-layer through implantation SAGFET with gate length of 0.5 μm and gate width of 50 μm at Vds = 3 V.

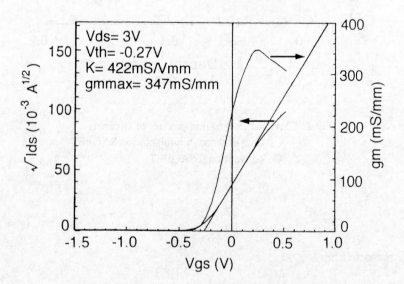

Fig.4. Dependence of $\sqrt{\text{Ids}}$ and gm on Vgs at Vds = 3 V.

3.3. RF characteristics

The S-parameters of the 0.5-μm-gate SAGFET with Wg of 200 μm were measured in the frequency range from 1 to 50 GHz. The maximum frequency of oscillation (f_{max}) was derived from the extrapolation of maximum available gain MAG to unity using -6 dB/octave decay. Fig.5 shows dependences of MSG and MAG on frequency at Vds of 3 V and gate voltage (Vgs) of 0.3 V. The 0.5-μm-gate SAGFET exhibited f_{max} of 95 GHz and MSG of 13.5 dB at 20 GHz.

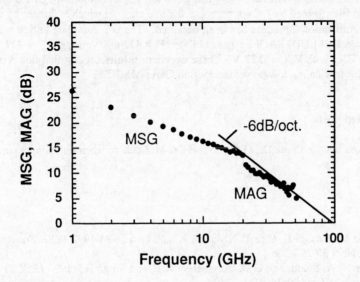

Fig.5. RF characteristic of an n-layer through implantation SAGFET with gate length of 0.5 μm and gate width of 200 μm at Vds = 3 V and Vgs = 0.3 V.

Table 1. FET characteristics of an n-layer through implantation SAGFET and a conventional SAGFET.

	n-layer through implantation SAGFET	conventional SAGFET
Vth (V) [1]	-0.27	-0.26
K (mS/V mm) [1]	422	303
gmmax (mS/mm) [1]	347	231
fmax (GHz) [2]	95	60
MSG (dB) @20GHz [2]	13.5	10.5

[1]: Wg=50μm, Vds=3V

[2]: Wg=200μm, Vds=3V, Vgs=0.3V

736

Table 1 shows the comparison of FET characteristics between the newly-developed SAGFET and conventional SAGFET with the same gate length of 0.5 μm. The DC and RF performances of the newly-developed SAGFET were superior to those of the conventional SAGFET with almost the same threshold voltage.

4. Conclusion

We have developed WSi gate GaAs SAGFET's with shallow channels fabricated by ion implantation through WSi films. By using the ion implantation through a 6-nm-thick gate metal thin film formed by DC sputtering, a shallow channel which is thinner by about 40 nm than the conventional process has been obtained. The 0.5-μm-gate buried p-layer lightly doped drain (BP-LDD) SAGFET has exhibited K = 422 mS/Vmm, gm_{max} = 347 mS/mm and f_{max} = 95 GHz with Vth = -0.27 V. These excellent results suggest that this process is very suitable for fabricating low power dissipation GaAs-MMIC's.

Acknowledgment

The authors wish to thank Dr. G. Kano and Dr. M. Inada for their encouragement throughout this work.

References

[1] Noda M, Hosogi K, Oku T, Nishitani K, and Otsubo M 1992 *IEEE Trans. on Electron Devices* **39** 757-766
[2] Sadler R A, Studtmann G D, Geissberger A E, and Singh H P 1991 *IEEE Trans. on Electron Devices* **38** 1271-1279
[3] Nishitsuji M, Uda T, Yokoyama T, Fujimoto K, Nishii K, Shibuya M, Kitagawa M, and Tamura A *Semicond. Sci. Technol.* (to be published)
[4] Tamura A, Ikeda Y, Yokoyama T, and Inoue K 1990 *J. Appl. Phys.* **67** 6171-6174

Inst. Phys. Conf. Ser. No 145: Chapter 5
Paper presented at 22nd Int. Symp. Compound Semiconductors, Cheju Island, Korea, 28 August–2 September 1995
© *1996 IOP Publishing Ltd*

Novel high power GaAs MESFETs with low distortion and high gate-drain breakdown voltage

H Fujimoto, S Morimoto, H Masato and A Tamura

Electronics Research Laboratory, Matsushita Electronics Corporation
3-1-1, Yagumo-nakamachi, Moriguchi, Osaka 570, JAPAN

Abstract. This paper presents a new approach to a power GaAs MESFET with a planar gate structure, using a MOCVD growth technique to form an undoped GaAs layer with good interface characteristics on an ion-implanted channel layer. The undoped GaAs layer made it possible to increase the gate-drain breakdown voltage, serving as both an ideal passivation layer and an ideal annealing cap of the ion implanted channel. This new simple structure is suitable for high performance power GaAs MESFETs.

1. Introduction

GaAs MESFETs are key devices for high-frequency and high-power applications such as cellular phones[1][2]. For high power GaAs MESFETs, a high gate-drain breakdown voltage is needed. Therefore, conventional power GaAs MESFETs generally have recessed gate structures formed by wet-chemical etching. However, there is a problem of unsatisfactory controllability of wet-chemical etching, which causes poor uniformity and reproducibility in the device characteristics. Moreover, undesired phenomena due to poor interface characteristics between a channel and a passivation film have been reported [3]. In order to solve these problems, some approaches such as forming a structure of an undoped GaAs layer on FET channels have already been reported [4]. For another approach, we have reported a non-recess structure FET using a selectively-doped double-heterojunction[5]. However, these approaches are difficult to mass-produce with a high yield. Rapidly expanding market in cellular phones requires a cost-effective GaAs FET manufacturing technology with a high yield. We have developed a new power GaAs MESFET without a recessed gate structure, combining an ion-implantation technique for a channel formation and an MBE (Molecular Beam Epitaxy) growth technique of an undoped GaAs layer [6].

In this work, we have introduced an undoped GaAs cap layer grown by MOCVD (Metal-Organic Chemical Vapor Deposition) into our new FET fabrication process.

2. Experimental

N-channel layers and n+-source/drain regions of GaAs MESFETs were formed by implanting Si ion directly into semi-insulating (100) GaAs substrates. After the surface treatment of both organic cleaning and thermal treatment in a MOCVD chamber, an undoped GaAs layer with thickness of 50nm was grown by MOCVD. Then, the SiO/WSiN capped furnace annealing [7] at 820°C for 15min was carried out to activate ion-implanted layers. Here, the undoped GaAs layer served as an ideal annealing cap of ion-implanted layers. Ohmic electrodes were formed by alloying AuGe/Ni/Au at 450°C after etching the undoped GaAs layer. Finally an Al/Ti gate electrode was formed on the undoped GaAs layer. Fig.1 shows the cross sectional view of the newly developed GaAs MESFET.

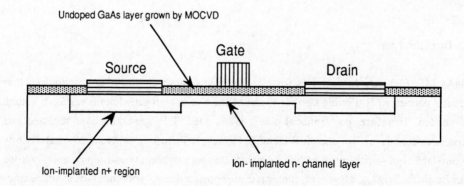

Fig.1 The cross sectional view of the new GaAs MESFET.

3. Results and Discussion

3.1 Characteristics of MOCVD growth interface

In order to investigate the MOCVD growth interface, we analyzed the distribution of atoms around the regrowth interface by SIMS (Secondary Ion Mass Spectroscopy). Fig.2 shows that no distinct pileup of both oxygen and carbon atoms exists in the growth interface. Moreover, it was found by DLTS (Deep Level Transient Spectroscopy) measurements that no trap levels exist in the growth interface[7]. These results indicated that the MOCVD growth interface has good characteristics.

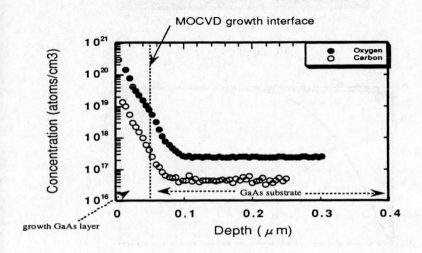

Fig.2 SIMS depth profile of carbon and oxygen atoms around the growth interface by MOCVD.

3.2 DC characteristics of a new MESFET

Typical I-V characteristic of the newly developed MESFET is shown in Fig.3. High Idss current and clear pinch-off characteristics are observed. Fig.4 shows the comparison of typical gate-drain breakdown characteristics. The new type FET shows much higher gate-drain breakdown voltage than that of the conventional FET[1]. The improvement of gate-drain breakdown voltage is due to the undoped GaAs layer under the gate electrode, which plays a role of an insulating layer.

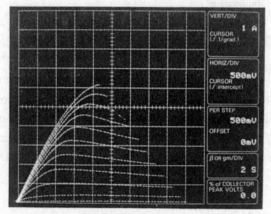

Fig.3 A typical I-V characteristic of a new FET.

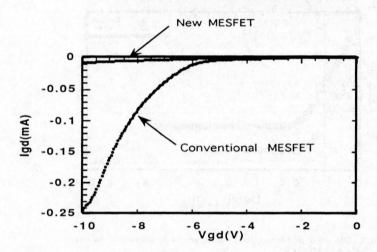

Fig. 4 Comparison of typical gate-drain breakdown characteristics.
(Wg=12mm, Lgd=2 μ m)

3.2 RF characteristics of a new MESFET

Fig.5 shows the RF input-output characteristics at 950MHz. The maximum output power of 1.0W (0.5W/mm power density) at 950MHz and the power added efficiency of 41% at Vds=10V are obtained. Fig.6 shows the output signal spectrum at Pout=31dBm. The lower curve is measured spectrum of output signal and the upper curve is the integrated power envelope of the spectrum. L1, U1, L2 and U2 denote the adjacent channel leakage power at

-50kHz, +50kHz, -100kHz and +100kHz from the center frequency of 948MHz, respectively. It is found that this new type FET gives the low distortion (L1=-51.4dBc, U1=-53.3dBc, L2= -67.3dBc, U2=-66.3dBc). The low distortion characteristics indicate that the undoped GaAs surface layer plays a role of good passivation of the channel layer. This newly developed FET has great potentials for extra-high output power applications with a low distortion and a sufficient power added efficiency.

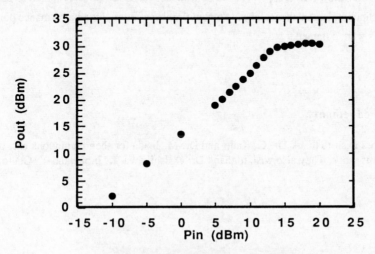

Fig.5 RF input-output characteristics at 950MHz.
(Vds=10V, Wg=2mm)

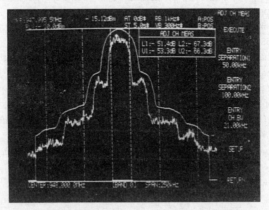

Fig.6 Output signal spectrum at Pout=31 dBm.

4. Summary

We have presented a new approach to power GaAs MESFET with planar gate structures, using the MOCVD growth technique to form an undoped GaAs layer on an ion-implanted channel layer. SIMS analysis revealed that an undoped GaAs layer with good interface characteristics could be formed by MOCVD. The undoped GaAs layer made it possible to increase the gate-drain breakdown voltage, serving as both an ideal passivation layer and an ideal annealing cap of the ion implanted channel. Using a new simple structure, high performance power GaAs MESFETs were obtained.

Acknowledgement

The authors wish to thank Dr. G. Kano and Dr. M. Inada for their continuous encouragement through this work. They also wish to thank Dr. O. Ishikawa, K. Inoue and Y. Ota for valuable discussion.

References

[1] Maeda M, Nishijima M, Takehara H, Adachi C, Fujimoto H, Ota Y and Ishikawa O 1993 GaAs IC Sympo. tech. Digest. pp53-56

[2]Kunihisa T, Yokoyama T, Fujimoto H, Ishida K, Takehara H and Ishikawa O 1994 IEEE MMWMC-S Digest. pp55-58

[3]Iwata N, Takahashi H, Mizutani H, Asano K, Matsunaga K, Hirayama H, Mochizuki A and Kuzuhara M 1992 NEC research & development Vol.33 No.3 pp286-296

[4]Smith F W, Chen C L, Mahoney L J, Manfra M J, Temme D H, Clifton B J and Calawa A R 1991 IEEE MTT-S Dig. pp643-646

[5]Ota Y, Adachi C, Takehara H, Yanagihara M, Fujimoto H, Masato H and Inoue K 1994 IEE Electronics Letters Vol.30 No.11 pp909-907

[6]Fujimoto H, Tanabe M, Maeda M and Tamura A 1995 IEE Electronics Letters Vol.31 No.2 pp137-139

[7]Tamura A, Ikeda Y, Yokoyama T and Inoue K 1990 J. Appl. Phys. 67 pp6171-6174

[8]Fujimoto H, Tanabe M and Tamura A 1995 ISCS-22 to be published

A 3.3 V GaAs MESFET Monolithic Driver Amplifier for Digital/Analog Dual-mode Hand-held Phones

Min-Gun Kim, Choong-Hwan Kim, In-Gab Hwang, Eung-Gie Oh, Jeon-Wook Yang, Chang-Seok Lee, Chul-Soon Park, Jong-Lam Lee, Kwang-Eui Pyun, and Hyung-Moo Park

Compound Semiconductor Department
Electronics and Telecommunications Research Institute
Yusong P. O. Box 106, Taejon, 305-600, Korea

Abstract. A 3.3V GaAs monolithic driver amplifier was developed for Digital/Analog dual-mode hand-held phones. The amplifier was fabricated using a selectively ion-implanted GaAs MESFET process with a gate length of 1μm. The amplifier displayed gain of 11~30dB and noise figure of 6.2~4.3dB at frequency range of 824~849MHz when drain bias is controlled from 0.7V to 2.5V. 1dB compression point of output power was determined to be 8dBm, and the 2nd and 3rd harmonics were -23 and -32 dBc, respectively, at 8dBm output power. Two-tone (Δf=442.5kHz) third-order intermodulation at an output power of 2.5dBm was measured to be -30dBc. Current consumption during power operation was 80mA at a maximum gain of 30dB. Performance of the amplifier is sufficient to be used as a driver amplifier for 3.3V operating Digital/Analog dual-mode hand-held phones.

1. Introduction

Digital hand-held phones are required as communication capability has been increased. The supply voltage of the phone has been decreased to 3.3V and the components are highly integrated in order to reduce its volume and weight [1]. At this point of view, 3.3V-operated RF MMIC components with a small size are important. Until now, there was no report on a monolithic driver amplifier operating at 3.3V for Digital/Analog dual-mode hand-held phones, although the driver amplifiers have been developed for the analog phones. In the present work, a GaAs monolithic driver amplifier operated at 3.3V was developed for Digital/Analog dual-mode hand-held phones.

2. Circuit design

The amplifier was designed with 3-stages using GaAs MESFETs. Each stage was designed in sequence, considering a low noise, a gain control, and a high power. Figure 1 shows the circuit diagram of driver amplifier. The first stage uses reactive serial feedback in the FET source, allowing simultaneous noise and impedance matching [2]. Input reactive matching circuit containing large spiral inductor results in smaller loss and lower noise figure. A resistive loading which has advantages for stability and chip size is used in the 1st stage and reactive ones for higher output power in the 2nd and 3rd stages. A self bias is adopted in all three stages for stable biasing and single positive voltage supply. The operating point of the 1st stage was 20%Idss which has advantages for low noise performance and those of the 2nd and 3rd stage were 50% Idss to obtain high gain and linear output power. The gain was controlled by adjusting the 2nd stage supply voltage, which minimizes the degradation of the input/output VSWR.

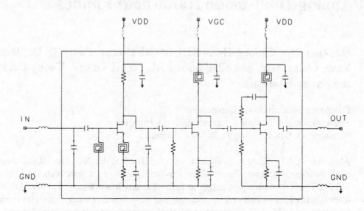

Fig. 1 Circuit diagram of driver amplifier.

To satisfy the unconditional stability of the amplifier, the 1st stage was connected to the 2nd stage with an unmatched structure and the 3rd stage was constructed with shunt-voltage feedback configuration [3]. However, the amplifier was no longer unconditionally stable so far as the package parasitic effects included, especially a common ground parasitic inductance, as shown in Fig. 2 (a). A shunt capacitive reactance inserted between the 1st and 2nd stage improves the stability of the amplifier at the frequency range of about 1~2 GHz. Figure 2 (b) shows the Rollett stability factor of the amplifier, which is greater than 1 in all frequency ranges.

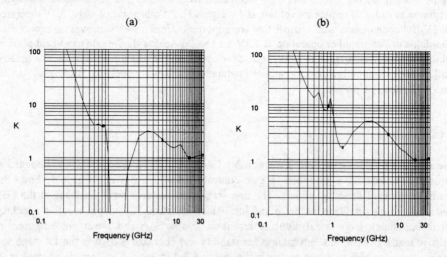

Fig. 2 Rollett stability factor
(a) without shunt capacitive reactance (b) with shunt capacitive reactance

3. Fabrication and performance

Figure 3 shows a top view photograph of the driver amplifier with a size of 1.8 x 1.8 mm^2. It was fabricated using a selectively ion-implanted 1μm gate length GaAs MESFET process which consists of Si$^+$ ion implantation with an implant energy of 70 keV followed by activation using RTA on undoped 3″ semi-insulation GaAs substrates. All of the chip was characterized using chip-on board technique on FR4 epoxy substrate which reduces the cost of driver amplifier.

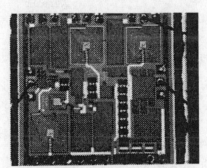

Fig. 3 Photograph of driver amplifier.

Figure 4 shows gain and noise figure of the driver amplifier. The measured gain and noise figure were 30dB and 4.3dB between 824 and 849MHz, respectively, at a high gain mode. As a control voltage varies from 2.5V to 0.7V at 836MHz, the gain changed from 30dB to 11dB and the noise figure from 4.3dB to 6.2dB. Gain control range is determined to be 19dB. The noise figure increases and the gain decreases as the gain control voltage reduces. Since the signal to noise ratio is enough even in a higher noise figure for large input signals, such an amount of degradation of noise figure at a low gain operation is acceptable. The VSWRs for input and output were 1.9:1 and 2.0:1, respectively.

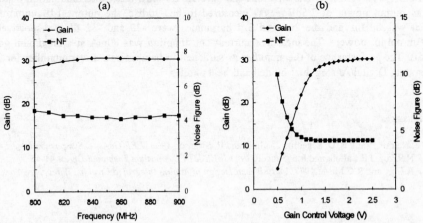

Fig. 4 Gain and noise figure (a) as a function of frequency. (b) as a function of gain control voltage.

The intermodulation of the amplifier was measured using two-tone frequencies, 836.5000 and 836.9425 MHz. The IM3 at an output power of 2.5dBm, as shown in Fig. 5, was measured to be -30dBc. Figure 6 shows that the output 1dB compression point (P$_{1dB}$)

was determined to be 8dBm at 836 MHz. The 2nd and 3rd harmonics at an output power of 8dBm were -23 and -32 dBc, respectively. The operating current was 80mA at a maximum gain of 30dB.

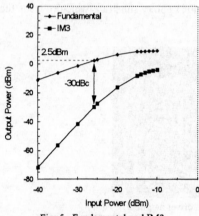

Fig. 5   Fundamental and IM3
as a function of input power.

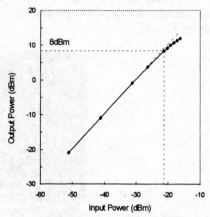

Fig. 6   1dB compression output power.

4. Conclusions

We have developed a GaAs MESFET monolithic driver amplifier. The measured gain and noise figure were 11~30dB and 4.3~6.2dB between 824 and 849MHz, respectively, as a control voltage varies from 2.5V to 0.7V. The VSWRs for both input and output were 1.9:1 and 2.0:1, respectively. The two-tone (Δf=442.5kHz) third-order intermodulation distortion at an output power of 2.5dBm was measured to be -30dBc, the output 1dB compression point was 8dBm, and the 2nd and 3rd harmonics were -23 and -32 dBc, respectively, at 8dBm output power. The measured current consumption was 80mA at a maximum gain of 30dB. The performance of the amplifier is sufficient to be used as a driver amplifier for 3.3V operating Digital/Analog dual-mode hand-held phones.

References

[1] V. Nair, R. Vaitkus, D. Scheitlin, J. Kline, and H. Swanson 1993 *IEEE GaAs IC Symposium Digest* 41-44
[2] P.N.Rigby, J.R.Suffolk and R.S. Pengely 1983 *IEEE MTT Symposium Technical Digest* 41-45
[3] P.R.Gray and R.G.Meyer 1977 *Analysis amd Design of Analog Integrated Circuits, Wiley, New York,* ch.8

Inst. Phys. Conf. Ser. No 145: Chapter 5
Paper presented at 22nd Int. Symp. Compound Semiconductors, Cheju Island, Korea, 28 August–2 September 1995

Fabrication of power MESFETs/Si with a reduced parasitic capacitance

Takashi Aigo, Seiji Takayama, Mitsuhiko Goto, Yasumitsu Ohta,
Aiji Jono, Akiyoshi Tachikawa and Akihiro Moritani
*Toshihiko Ichioka and *Masahiro Akiyama

Electronics Research Laboratories, Nippon Steel Corp.
5-10-1 Fuchinobe, Sagamihara Kanagawa 229, Japan
* Semiconductor Technology Laboratory, Oki Electric Industry Co.
550-5 Higashiasakawa-cho, Hachioji-shi, Tokyo 193, Japan

Abstract. This paper reports on a study of dc and microwave performance for a power MESFET fabricated using a GaAs/Si substrate. In the MESFET, a device pattern layout and a fabrication process are considered in order to reduce a parasitic input capacitance for gate pads originated from a GaAs-Si interfacial conductive layer doped with diffused Si atoms. The MESFET with a gate length and a width of 0.8(μm) and 5.6(mm) demonstrates P1dB of 25.7(dBm), linear gain of 20.5(dB) and ηadd at 1dB compression of 57(%) at a frequency of 0.85(GHz) and a source-drain voltage of 3.6(V). Moreover, almost the same linear gain and ηadd are obtained at a source-drain voltage of 2.4(V). From this result, the GaAs/Si is found to be adequately utilized even in devices driven with a low supply voltage aiming at the application to a mobile communication system.

1. Introduction

In recent years, GaAs electronic devices have been widely utilized with the development of mobile communication. The technology of GaAs/Si, which has the advantages of a large diameter and high thermal conductivity, is very promising for lowering a cost of the devices and, especially for power devices, improving the diffusivity for generated heat by device operation. In the application of the GaAs/Si to the electronic devices, HEMTs or MESFETs are thought to be the most suitable devices because the GaAs/Si is an epitaxially grown material and it is easy to modulate layer structures for fitting in such devices.

One of the disadvantages for the GaAs/Si has been considered as the high density of dislocations. From the evaluation for macroscopic and microscopic distribution of threshold voltage for as-grown and annealed HEMTs/Si, we have reported that the dislocations could not affect the characteristic and its uniformity for the epitaxially grown devices [1,2]. As for the surface morphology which has been taken as another disadvantage and reported to be degraded by a fabrication process

[3], the high uniformity of threshold voltage obtained in [1,2] can prove that the morphology of the present GaAs/Si is satisfactory for device fabrication, which is also supported by the AFM observation on surface roughness [4]. The remaining problem to be overcome for the practical use of the GaAs/Si is to reduce parasitic capacitances and resistances originated from a GaAs-Si interfacial conductive layer doped with diffused Si atoms [5,6,7]. The parasitic parameters degrade microwave performance for GaAs/Si electronic devices, especially for the application to discrete devices.

In this paper, we report on a study of dc and microwave performance for a power MESFET/Si fabricated with a pattern layout and a fabrication process to reduce a major one of the parasitic parameters. The MESFET with a gate length and a width of $0.8(\mu m)$ and 5.6(mm) shows the comparable microwave performance to that for a MESFET/GaAs at a frequency of 0.85(GHz) and a source-drain voltage of 3.6(V). Moreover, the performance for the MESFET/Si is hardly degraded at a source-drain voltage as low as 2.4(V).

2. Characterization of parasitic parameters for GaAs/Si electronic devices

A new equivalent circuit as shown in Fig.1 is used to evaluate the parasitic parameters [8]. We assume in the new model that a parasitic capacitance, Cgp and a resistance, Rgp are induced between a gate pad and a source electrode through the interfacial conductive layer. Similarly, Cdp and Rdp are added for a drain pad. Moreover, an extra output resistance, Rsub is included as the influence of the conductive layer. This model can simulate the characteristic for GaAs/Si electronic devices with the fitting error of 2.1(%), which shows the validity of the model as compared to a conventional model with the error of 16(%) .

Using the model, equivalent circuit parameters are extracted for a HEMT/Si and a HEMT/GaAs with a gate length of $0.8(\mu m)$, a width of $200(\mu m)$ and an area of $80(\mu m)$ sq. for a gate and a drain pad [2]. A large difference is observed for capacitive parameters of Cds, Cgp and Cdp and they are indicated in Table 1. As can be seen in Table 1, the capacitances for the HEMT/Si are at least a few times larger than those for the HEMT/GaAs. This is commonly observed for the present GaAs/Si discrete devices, which causes the inferior microwave performance for GaAs/Si to that for

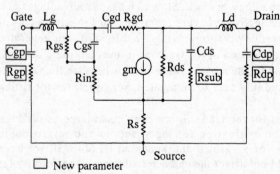

Table1   Comparison of parasitic capacitances between a HEMT/Si and a HEMT/GaAs

	HEMT/ Si	HEMT/ GaAs
Cds (pF)	0.65	0.05
Cgp (pF)	0.75	0.06
Cdp (pF)	0.13	0.03

Fig.1   A new equivalent circuit model for a GaAs/Si electronic device

GaAs/GaAs. As for the new resistive parameters in the model, the values considerably fluctuate for every grown GaAs/Si because carrier density and thickness of the interfacial conductive layer would strongly depend on each growth conditions. The average value of the carrier density and the thickness, however, can be estimated from Laser Raman Spectroscopy and the density of ~10^{18} (cm^{-3}) with the thickness of 500~1000(Å) are verified. Among the parasitic capacitances to affect microwave performance, the simulation indicates that it is the most effective way for obtaining comparable performance to GaAs/GaAs to reduce Cgp [8], namely an extra input capacitance. Considering that it is practical for evading the influence of Cgp to reduce a ratio of Cgp/Cgs rather than to reduce Cgp itself, a large gatewidth device like a power MESFET is concluded to be the most suitable device for the GaAs/Si from the view point of microwave performance. Furthermore, the power MESFET can utilize the high thermal conductivity of a Si substrate.

3. Epitaxial growth and device fabrication

MESFET/Si wafers used in this study are grown by a vertical-type MOCVD system. The growth procedures consist of conventional two-step methods for GaAs growth on a Si substrate with the use of TMG(trimethylgallium), TMA(trimethylaluminum) and AsH3 (arsine). Doping is carried out using disilane (Si2H6) and the total pressure is 60 (Torr). The Si(100) substrate is 3 inches in diameter and misoriented by 3° toward <011>. The substrate is grown by the floating-zone method and its resistivity is 2~6(kΩ·cm).

The structures of a fabricated MESFET/Si are illustrated in Fig.2, which consist of a 2-μm-thick GaAs buffer layer, a 1-μm-thick AlGaAs buffer layer and a 0.3-μm-thick GaAs active layer with carrier concentration of $2x10^{17}(cm^{-3})$. Fabrication processes are carried out by mesa chemical etching for device isolation, AuGe/Ni ohmic contacts metallization and Al gate metallization after recess wet etching. After the gate metallization, a Si3N4 dielectric layer is attached for the surface passivation. Au plating is used for an air-bridge interconnection. As for the backside process of the wafer, Ti/Au multilayer is evaporated after thinning the Si substrate to 300(μm). A SEM photograph for the MESFET is shown in Fig.3. A gate length and a width are 0.8(μm) and 5.6(mm), respectively. The gate width is determined from desirable output power and

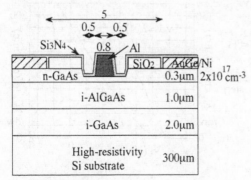

Fig.2 Schematic cross-section for a fabricated MESFET/Si

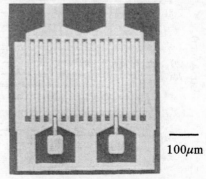

Fig.3 A SEM photograph for a fabricated MESFET/Si

a number of the gate pad is optimized for the width [9]. Each of the pad has an area of 50 (μm) sq., which is determined so as to let the total pad capacitance be less than 10 (%) of Cgs for the MESFET. As for the pad capacitance and Cgs used in the determination, the former is based on the HEMT/Si evaluation [2] and the latter is calculated at the condition of zero gate bias.

A gate bus-line, which connects each of gate fingers, can increase another parasitic capacitance considered as Cgp if it is formed directly on a buffer layer of the wafer. Therefore, a SiO2 dielectric layer with thickness of 0.3 (μm) is inserted under the gate bus-line as indicated in a SEM photograph and an illustrated cross-sectional view in Fig.4(a) and 4(b), respectively.

4. DC and microwave characteristic

The typical current-voltage characteristic for a fabricated MESFET/Si is shown in Fig.5(a) and transconductance and a drain current as a function of a gate voltage is in Fig.5(b). As can be seen in the figures, a source-drain saturation current is ~180(mA/mm) and the maximum transconductance is ~120(mS/mm). A gate-drain breakdown voltage of -12~-14(V) is also obtained as measured at a gate reverse current of

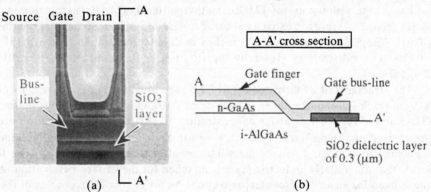

Fig.4 A SEM photograph around a gate bus-line (a) and its schematic cross-section (b)

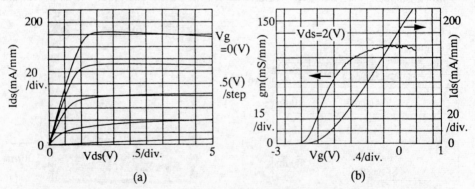

Fig.5 Current-voltage characteristic for a fabricated MESFET/Si (a) and its transconductance and drain current as a function of a gate voltage (b)

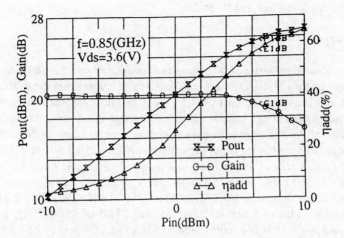

Fig.6 Input-output characteristic for a fabricated power MESFET/Si
(Lg=0.8μm, Wg=5.6mm)

500(μA/mm). These characteristics are almost the same as those for a MESFET/ GaAs fabricated simultaneously.

S-parameters are measured to evaluate the parasitic parameters for the MESFET/ Si at Vds of 2(V) and Vgs of showing the maximum transconductance. From the extraction of equivalent circuit parameters, it is pointed out that the ratio of Cgp including the gate bus-line to Cgs can be reduced to ~10(%), which is almost comparable to that of 5~7(%) for the MESFET/GaAs. The result indicates that the reduction for a ratio of Cgp/Cgs is actually achievable for a large gatewidth device by considering a size for the gate pad and a structure for the gate bus-line.

Input-output characteristic for the power MESFET/Si is shown in Fig.6. The MESFET is mounted in a package and measured at a frequency of 0.85(GHz), a source-drain voltage of 3.6(V) and a source-drain current of ~10(%) of Idss. The measurement condition means the application of the MESFET to a mobile communication system such as a handy phone. After taking impedance matching for input and output using S-parameters, a optimum load impedance is determined by a load-pull and a source-pull method so as to obtain maximum output power for input power of 10(dBm). As can be seen in Fig.6, the MESFET demonstrates output power at 1dB compression (P1dB) of 25.7(dBm), linear gain (GL) of 20.5(dB) and power added efficiency (ηadd) at 1dB compression of 57(%), which are comparable to those for a MESFET/GaAs commercially fabricated. The same evaluation as indicated in Fig.6 is carried out in the case of a lower sorce-drain voltage and shown in Fig.7. It is found in Fig.7 that GL and ηadd can be hardly degraded even

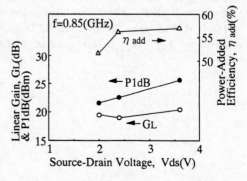

Fig.7 P1dB, GL and ηadd as a function of a source-drain voltage

752

at a source-drain voltage as low as 2.4(V).

5. Summary

We have evaluated parasitic capacitances and resistances peculiar to GaAs/Si electronic devices caused by a Si-GaAs interfacial conductive layer. From the evaluation, a large gatewidth device is expected to be the most suitable application for the GaAs/Si because such a device can reduce a ratio of an extra input capacitance for gate pads, which is one of the parasitic capacitances and is a major cause to degrade microwave performance, to an intrinsic input capacitance. A fabricated power MESFET/Si with a relatively small area of $50(\mu m)$ sq. for a gate pad and a gate bus-line under which a 0.3-μm-thick SiO_2 dielectric layer is inserted can actually reduce the ratio as comparable as to that for a MESFET/GaAs. The MESFET/Si with a gate length of $0.8(\mu m)$ and a width of 5.6(mm) demonstrates P1dB of 25.7(dBm), linear gain of 20.5(dB) and ηadd at 1dB compression of 57(%) at a frequency of 0.85(GHz) and a source-drain voltage of 3.6(V). Moreover, almost the same linear gain and ηadd are obtained even at a source-drain voltage of 2.4(V). The result indicates that the GaAs/Si is adequately substitutable for GaAs/GaAs in the application for mobile communication devices driven by a low supply voltage.

References

[1] Aigo T., Jono A., Tachikawa A., Hiratsuka R. and Moritani A. 1994 *Appl. Phys. Lett.*, **64** 3127-3129

[2] Aigo T. et al. unpublished

[3] Ohori T., Suehiro H., Mitagaki S., Miyagaki T. and Kasai K. 1994 *Proc. 1994 Symp. on Alloy Semiconductor Physics and Electronics*, 57-60

[4] Jono A., Tachikawa A., Aigo T. and Moritani A. 1994 *J. Cryst. Growth*, **145** 353-357

[5] Aigo T., Goto M., Jono A., Tachikawa A. and Moritani A. 1993 *Proc. 1993 Int. Symp. on Gallium Arsenide and Related Compounds*, 87-92

[6] Sriram S., Messham R. L., Smith T. J., Hobgood H. M. and Driver M. C. 1990 *Proc. 1990 IEEE Gallium Arsenide Integrated Circuit Symposium*, 229-232

[7] Georgakilas A., Halkias G., Christou A., Papavassiliou C., Perantinos G., Konstantinidis G. and Panayotatos P. 1993 *IEEE Trans. Electron Devices*, **40** 507-512

[8] Goto M. et al. to be published in *IEEE Trans. Microwave theory and techniques*

[9] Walker J. L.B. 1993 *High-Power GaAs FET Amplifiers* (Boston, Artech House)

Inst. Phys. Conf. Ser. No 145: Chapter 5
Paper presented at 22nd Int. Symp. Compound Semiconductors, Cheju Island, Korea, 28 August–2 September 1995
© *1996 IOP Publishing Ltd*

Effects of Buffer Structures on GaAs MESFET's

Jae-Jin Lee, Dong-Wook Kim *, Hae-Gwon Lee, Kwang-Eui Pyun
and Song-Cheol Hong *, Young-Se Kwon *

Compound Semiconductor Department, Electronics and Telecommunications Research Institute
* Department of Electrical Engineering, Korea Advanced Institute of Science and Technology
373-1 Kusong-dong, Yusong-gu, Taejon, 305-701, Korea

Abstract : To show the effects of buffer structures on GaAs MESFET's, we investigated eight different buffer layers grown on semi-insulating GaAs substrate by Molecular Beam Epitaxy(MBE) and chose four among them to characterize the MESFET performances. We found that AlGaAs/GaAs superlattice layer in the buffer helped the devices to have better performances such as large source-to-drain resistance, small knee voltage and larger gain flatness. Also it was demonstrated that the thickness of intrinsic GaAs spacer layer between AlGaAs/GaAs superlattice and active channel layer must be carefully selected (between 500 Å ~ 1000 Å) to have low knee voltage and better gain flatness.

1. Introduction

Since it was first suggested that the rise beyond the saturation point of the drain-to-source current in GaAs MESFET's was due to the electron injection into the semi-insulating substrate and the related space-charge limited current in the substrate[1], a variety of buffer layers have been studied to eliminate this rise. Also there have been many reports on the buffer layer effects on the device characteristics, especially on kink effect[2], backgating effect[3], channel breakdown voltage[4], microwave output power and efficiency[4], and etc. The buffer layers are thought to have an effect on the knee voltage, transconductance, g_m, near threshold voltage, output resistance and so on. These parameters are known to be very important to determine circuit performances. In this paper, we investigate and compare effects of four different buffer structures, which are chosen among eight different buffers, on GaAs MESFET's. And it is presented that the buffer structure suitable to high efficiency applications should require some thickness of intrinsic GaAs buffer layer.

2. Experimental Procedures and Results

Table 1 shows eight different epi-layer structures grown on semi-insulating GaAs substrate by MBE, measured sheet resistances and their variations. The epi-layers have the same 3000 Å Si doped active layer(N_d=6x10^{17}cm^{-3}) and the different buffer layer structures. The sheet resistance is minimum value when the buffer layer has 1000 Å intrinsic GaAs spacer layer between AlGaAs/GaAs superlattice and 8000 Å i-GaAs. From the trend of the measured results, we find that the electron injection into the buffer layer contributes to the reduction of knee voltage to some extent and optimum value of the effective electron-path length in the i-GaAs spacer layer may be.

Table 1. Summary of eight different buffer structures and sheet resistance as a function of them.

Wafer[***]	Buffer Structure	$R_{sheet}(\Omega/\square)$	Variation	Uniformity
#1	8000Å i-GaAs	75.49	4.48 %	1.20
#2	SL[*] + 8000Å i-GaAs	74.04	4.70 %	1.26
#3	500Å i-GaAs + SL[*] + 8000Å i-GaAs	73.30	4.12 %	1.14
#4	1000Å i-GaAs + SL[*] + 8000Å i-GaAs	71.39	4.13 %	1.15
#5	2000Å i-GaAs + SL[*] + 8000Å i-GaAs	72.52	4.51 %	1.19
#6	3000Å i-GaAs + SL[*] + 8000Å i-GaAs	73.50	4.66 %	1.19
#7	8000Å i-GaAs + SL[*]	78.11	3.85 %	1.16
#8	2000Å i-GaAs + 8000Å LT[**] i-GaAs	72.94	4.33 %	1.15

* The total superlattice thickness is 1800Å and consists of alternating layers of
30Å GaAs and 30Å AlGaAs.

** LT GaAs means low temperature grown GaAs

*** All wafers have the same active channel layer (3000Å 6×10^{17} cm^{-3} Si-doped channel)

Table 2. Summary of epi-layer structures with four different buffer layers
used to characterize GaAs MESFET's

Layer Structure		Epi-Structure
Active Layer		500Å n$^+$ GaAs Nd=1×10^{18} cm^{-3}
		1500Å n$^-$ GaAs Nd=5×10^{16} cm^{-3}
		400Å n^o GaAs Nd=5×10^{17} cm^{-3}
Buffer Structure	Wafer #1	10000Å i-GaAs
	Wafer #2	SL[*] + 5000Å i-GaAs
	Wafer #3	1000Å i-GaAs + SL[*] + 5000Å i-GaAs
	Wafer #4	500Å i-GaAs + SL[*] + 5000Å i-GaAs

* The total superlattice thickness is 1800Å and consists of alternating layers of
30Å GaAs and 30Å AlGaAs.

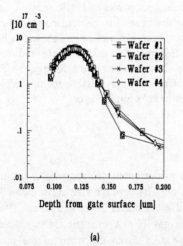

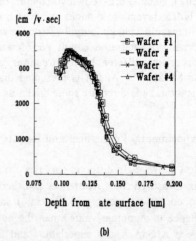

(a)

(b)

Fig. 1 The doping profile and drift mobility profile of each wafer with different buffer layer from FATFET and C-V measurements. FATFET has 150μm gate length and 150μm gate width and 0.5 V drain bias voltage is applied for measurements of transconductance. The respective buffer structures are shown in Table 2. (a) The doping profile from C-V measurement (b) The drift mobility profile from FATFET measurement

To investigate characteristics of MESFET's, we choose four among eight different buffer layers shown in Table 1. They are shown in Table 2 and have the same active layer with low-high doping profile not to be sensitive to gate recess process. The doping profile and drift mobility of each wafer are obtained from FATFET and C-V measurements (Fig. 1). FATFET has 150μm gate length and 150μm gate width. The MESFET on buffer layer with only i-GaAs layer has larger mobility in the low doped region than the others. This is due to larger current contribution of buffer layer. In contrast to that, the MESFET on AlGaAs/GaAs superlattice buffer with no i-GaAs spacer layer has smaller mobility because AlGaAs buffer barrier blocks current injection into the substrate. The deep infiltration of electron into buffer layer causes the slope of mobility curve with gate voltage not to be steep. These are applicable to the other two buffer structures with 500Å i-GaAs spacer layer and 1000Å i-GaAs spacer layer between AlGaAs/GaAs superlattice and doped active layer.

The fabricated MESFET's have 1μm gate length, 100μm gate width and 7μm drain-to-source spacing. AuGe/Ni/Au and Al are used as ohmic metal and Schottky metal, respectively. The drain-source resistance, Rds, of MESFET on buffer layers with AlGaAs/GaAs superlattice(wafer #2,#3,#4 in Table 2) is 1.5~2 times as large as that of MESFET without superlattice. The knee voltage is compared in Fig. 2, which is one of the key parameters for high efficiency applications. The MESFET on the buffer layer with 1000 Å i-GaAs spacer layer has the lowest knee voltage. The knee voltage seems to be affected by the electron mobility and the effective electron-path length in the buffer layer. Since GaAs power MESFET's for digital hand-held phone should have high linearity and efficiency, the transconductance near threshold voltage is important because the steeper increase implicates the better flatness of transconductance curve. Fig. 3 shows the gradient of transconductance with gate bias voltage near threshold voltage. It is found that the transconductance of GaAs MESFET on 500Å i-GaAs spacer layer shows 15 ~ 20 % steeper increase than those of the others.

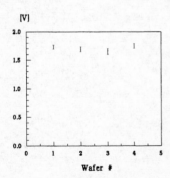 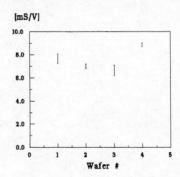

Fig. 2 The knee voltage distribution of each wafer.   Fig. 3 The plot of transconductance gradient of
each wafer near threshold voltage. (Vds=2.5 volt.)
(The measured MESFET's have 1μm gate length, 100μm gate width and 7μm drain-to-source distance.)

Therefore we should carefully determine the thickness of the intrinsic GaAs spacer layer (between 500Å~1000Å) between the superlattice buffer and the active layer to get the low knee voltage and the better gain flatness for high efficiency and high linearity GaAs MESFET.

3. Conclusions

In summary, to show the effects of buffer structures on GaAs MESFET's, we have investigated eight different buffer layers and chosen four among them to characterize the MESFET performances. It is found that the AlGaAs/GaAs superlattice in the buffer layer structure helps the devices to have the better performances such as large Rds, small knee voltage, the larger gain flatness. Also it should be noted that the thickness of the intrinsic GaAs spacer layer between the AlGaAs/GaAs superlattice and the active layer must be carefully selected (between $500\,\text{Å} \sim 1000\,\text{Å}$) to have the better gain flatness and knee voltage.

Acknowledgements

The authors would like to thank S. Jun for his contribution to device fabrication and measurement.

References

[1] L. F. Eastman and M. S. Shur 1979 IEEE Electron Device Lett. vol. 26, No. 9, 1359-1361
[2] Junzi Haruyama, Norio Goto, and Hitoshi Negishi 1992 Appl. Phys. Lett. 61(8), 24, 928-930
[3] F. W. Smith, A. R. Calawa, Chang-Lee Chen, M. J. Manfra, and L. J. Nahoney 1988 IEEE Electron Device Lett. vol. 9, No. 2, 77-80
[4] R. Actis, K. B. Nichols, W. F. Kopp, T. J. Rogers, and F. W. Smith 1995 IEEE MTT-S Digest, 445-448

Inst. Phys. Conf. Ser. No 145: Chapter 5
Paper presented at 22nd Int. Symp. Compound Semiconductors, Cheju Island, Korea, 28 August–2 September 1995
© *1996 IOP Publishing Ltd*

Anomalous Behaviors of Cut-off Frequencies in 0.5 µm GaAs Power MESFET's

J-K Rhee, I-H Lee and S-M Kim

Dept of Electronics Eng.
Dongguk University.
Phil-Dong, Jung-Ku, Seoul, Korea, Tel : 82-02-260-3335

Abstract Anomalous behaviors of the cut-off frequencies in the fabricated 0.5 µm GaAs power MESFET's are possibly detected for the first time. The cut-off frequency of a GaAs power MESFET is supposed to be gradually decreased as the total gate width increases. On the contrary, the gate width increase of a GaAs power MESFET gives rise to increase of the cut-off frequency. The cut-off frequency of a GaAs power MESFET with 0.9 mm gate width is about 13 GHz, and that with 3.0 mm about 16 GHz. This anomalous behaviors in the cut-off frequency of a GaAs power MESFET could be due to the occurrence of resonance phenomena because the stray capacitances are considerably increased in a wider gate width device.

1. Introduction

The cut-off frequency is one of the most important parameters to limit the functional performance of a GaAs power and a low noise MESFET. In general, it is possible that the cut-off frequency of MESFETs will be increased by reducing the gate length. Hence there are many different methods that a sub-0.5 µm gate length can be fabricated using such as electron-beam, laser-beam, X-ray and i-line lithography technologies and so on. But those process equipments are very expensive.

In this paper, GaAs power MESFET's with sub-0.5 µm gates have been fabricated using the conventional UV lithography and the modified evaporation method[1], and then both DC and RF performances are measured and carefully analyzed. The results show that the cut-off frequency increase as the total gate width of a GaAs power MESFET is increased.

These anomalous behaviors of the cut-off frequency, detected for the first time, will be discussed in detail.

2. Experiments

Fig. 1. shows the cross-sectional diagram of the 0.5 μm gate fabrication processes using the conventional UV lithography and modified thermal evaporation method. Fig. 1 (a) depicts a cross-sectional gate pattern with 1 μm gate length after AZ5214E photoresist and image reversal processes for fine line lithography.[2] In Fig. 1 (b) a 5000 Å thick Al layer is evaporated with evaporation angles of 23° onto the entire wafer such that the actual openings of 1 μm gate pattern are supposed to be reduced down to 0.5 μm or less depending on the both evaporation angles. Therefore, a Al Schottky gate is fabricated with 0.5 μm length and 6000 Å thickness as shown in the Fig. 1 (c) and 2. Also the 0.5 μm gate from a SEM photo in fig. 2 is clearly seen between source and drain, and the distance between them is 4 μm.[1]

Epi-layer wafers used for the fabrication of the GaAs power MESFETs have the structures that an undoped GaAs layer is inserted between n-channel and n^+-cap layer in order to improve the microwave characteristics of power devices[3]. The doping concentrations of n^+, undoped, and n-layer are 3×10^{18} cm^{-3}, intrinsic, and 1.5×10^{17} cm^{-3}, respectively. n^+ layer is 500 Å thick, the undoped layer 700 Å, and the channel layer 1200 Å. And these epi-layers are grown by MBE.

For the fabrication of GaAs power devices, a mesa layer is formed by etching away 6000 Å using $H_2SO_4 : H_2O_2 : H_2O(1 : 8 . 160)$ etchant and AuGe/Ni/Au(2300 Å/250 Å/3000 Å) ohmic layer is alloyed at 450 °C for 2 minutes. Air-bridge processes for the interconnection between source electrodes and pads and, furthermore, thinning down to 150 μm of substrates are also very important processes.

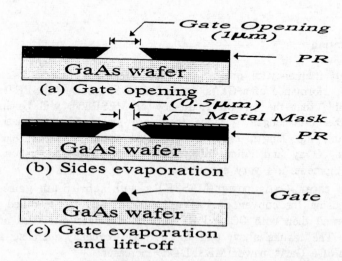

Fig 1. 0.5 μm gate fabrication processes using the conventional
UV photolithography and the modified thermal evaporation

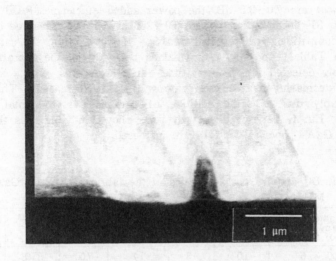

Fig 2. A cross-sectional SEM photo of the fabricated 0.5 μm gate

Briefly the air-bridge is fabricated with combination processes such as 5000 Å thickness of Au as source post metal, source via pattern photolithography, a 100 Å thin layer of Au, image reversal processes and so on.[4] Especially, air-bridged interconnections are very essential to improving RF performances of the power devices in higher frequency bands. Both DC and RF characteristics are measured and carefully analyzed. Anomalous behaviors of the cut-off frequencies in the fabricated 0.5 μm GaAs power MESFET's are detected and analyzed for the first time.

3. Measurements and discussion

S-parameters were measured using a HP8722A vector network analyzer and on-wafer microprobes under the conditions of Ids = 0.5 Idss, Vds = 5 V and frequency ranges from 1 to 18 GHz.

Fig. 3 shows the measured S-parameters from the devices with an unit gate width of 150 μm and 20 fingers. Fig. 4 displays the current gain characteristics as a parameter of the frequency for two different kinds of GaAs power devices, and their total gate widths are 0.9 mm and 3.0 mm, respectively. DC and RF characteristics of various types of the fabricated GaAs power MESFET's are listed in Table 1.

The measured data are carefully analyzed and then the analyzed results show that the cut-off frequencies(f_T) are in the range of 13~17 GHz for the GaAs power MESFET's with total gate widths in the range of 0.6~3.0 mm. Also for the fixed test frequency of 10 GHz and the various total gate widths in the ranges of 0.6~3.0 mm, the maximum unilateral transducer power

760

gains(G_{TUmax}) are 7.0~2.5 dB, the power added efficiencies(PAE) 35.68~30.76 % and the RF linear output powers 60~256 mW. The measured and analyzed data are compatible with reported data[5]. However, from the careful reviews of data in Table 1 and Fig. 4, anomalous behaviors of the cut-off frequencies are possibly detected for the first time. In other words, G_{TUmax} and PAE are gradually decreasing as the total gate widths are increased. This is normal and probably due to some addition of parastics as the total gate width increases. But f_T varies without any rules shown in table 1 as the total gate widths of GaAs power MESFETs are increased.

Table 1. DC and RF characteristics of the fabricated 0.5 μm GaAs power MESFET

Unit gate width (μm) × No. of fingers	Idss (mA)	Vp (V)	f_T (GHz)	G_{TUmax} (dB)	PAE (%)	Pout (mW)
100 × 6	80	−3	14	7.00	35.68	60
110 × 6	90	−3	14	4.12	34.46	66
150 × 6	100	−3	13	4.60	33.91	90
100 × 20	280	−3	17	3.30	32.57	172
110 × 20	300	−3	17	3.01	32.24	192
150 × 20	400	−3	16	2.50	30.76	265

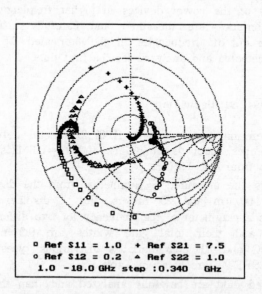

□ Ref S11 = 1.0   + Ref S21 = 7.5
○ Ref S12 = 0.2   ▲ Ref S22 = 1.0
1.0 -18.0 GHz step :0.340   GHz

Fig 3. Measured S-parameter characteristics of
the fabricated 0.5 μm GaAs power MESFET

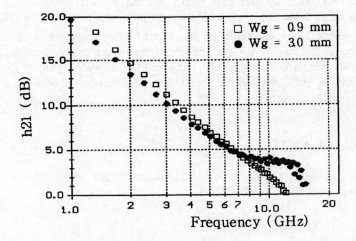

Fig 4. Current gain characteristics as a parameter
of the frequency and total gate width of the
fabricated 0.5 μm GaAs power MESFET

In general, it is very true that the cut-off frequency of a GaAs power MESFET is supposed to be gradually decreased with increasing the total gate width. Fig. 4 tells that h_{21} of a GaAs power MESFET with a smaller gate width of 0.9 mm is normally decreased as increasing test frequency, but, on the other hand, that h_{21} with a wider gate width of 3.0 mm is almost saturated in frequency ranges from 8 to 15 GHz and then drops very rapidly.

This anomalous behaviors in the cut-off frequencies of a GaAs power MESFET could be due to the occurrence of resonance phenomena because the stray capacitances are considerably increased in a wider gate width device. The occurrence of resonance phenomena is easily observed in Fig. 3. Especially two resonance frequencies from S_{11} data in Fig. 3 explain clearly anomalous behaviors of the cut-off frequency in the fabricated GaAs power MESFET's.

4. Conclusion

In this paper, anomalous behaviors of the cut-off frequencies in the 0.5 μm GaAs power MESFET's are possibly detected for the first time. The cut-off frequency of a GaAs power MESFET is supposed to be gradually decreased as increasing the total gate width, i.e., h_{21} of a GaAs power MESFET with a 0.9 mm gate width is normally decreased as increasing test frequency, but, on the other hand, h_{21} with a 3.0 mm gate width is almost saturated in frequency ranges from 8 to 15 GHz and then drops very rapidly. The fabricated GaAs

762

power MESFET with 0.9 mm gate width has the cut-off frequency of 13 GHz, and, on the other hand, the one with 3.0 mm gate width that of 16 GHz.

These anomalous behaviors in the cut-off frequency of a GaAs power MESFET could be due to the occurrence of resonance phenomena because the stray capacitances are considerably increased in a wider gate width device. The occurrence of resonance phenomena is easily observed from the S-parameter measurements.

Acknowledgement

This work has been supported by ADD(contract No. ADD 94-7-1) and by ISRC(contract No. ISRC 94-E-3248)

References

[1] I-H Lee, S-M Kim, E-H Rhee and J-K Rhee 1995 *2nd Proceedings of Semiconductor Conference* 425-426

[2] I-H Lee, S-M Kim, K-K Yoon, S-H Shin and J-K Rhee 1992 *Proceedings of KITE Fall Conference* 356-358

[3] H. Takahashi, K. Asano, K. Matsunaga, N. Lwata, A. Mochizuki and H. Hirayama 1991 *IEEE IEDM* 259-262

[4] I-H Lee, S-M Kim, J-H Lee, K-H Park, Y-M Kim, J-C Woo and J-K Rhee 1994 *1st Proceedings of Semiconductor Conference* 283-284

[5] Harris data book 1990 *GaAs FETs, MMICs & Foundry Services* (Harris Corporation)

Inst. Phys. Conf. Ser. No 145: Chapter 5
Paper presented at 22nd Int. Symp. Compound Semiconductors, Cheju Island, Korea, 28 August–2 September 1995
© *1996 IOP Publishing Ltd*

Microwave Characteristics of *GaAs* MESFET with Optical Illumination

H. J. Kim, S. J. Kim[*], D. M. Kim[*], H. Chung[**], D. H. Woo, S. I. Kim, W. J. Choi, I. K. Han, S. H. Kim, J. I. Lee, K. N. Kang, and K. Cho[***]

Division of Electronics and Information Technology, Korea Institute of Science and Technology, P.O. Box 131, Cheongryang, Seoul 130-650, Korea,
[*]Department of Electronics Engineering, Kookmin University, Jungnung-dong, Sungbuk-gu, Seoul, 136-702, Korea,
[**]Department of Physics, Kyunghee University, Seochunri, Kihung-eup, Yongin-gun, Kyungki-do, 449-701, Korea,
[***]Department of Physics, Sogang University, Sinsu-dong, Mapo-gu, Seoul, 121-742, Korea.

Abstract. We fabricated $1 \times 240 \mu m^2$ *GaAs* MESFET's and investigated their microwave characteristics under optical illumination with varying optical power density from $1.6 mW/cm^2$ to $472 mW/cm^2$. Typical current gain cut-off frequency(f_T) and maximum frequency of oscillation(f_{max}) of fabricated devices were 6.6GHz and 13.6GHz, respectively, at V_{ds}=3.0V and V_{gs}=-0.25V without optical illumination. Under $157 mW/cm^2$ of optical illumination at the same bias, however, f_T and f_{max} were increased to 9.5GHz and 15.3GHz, respectively. Extracted device parasitics with optical illumination are also reported.

1. Introduction

In recent years, considerable attention has been given to optically controlled field effect transistors(OPFET's) with their useful applications to the optoelectronic integrated circuits(OEIC's) for high speed modulation and detection in the optical communication systems. With direct illumination of *GaAs* MESFET's and related devices by an *AlGaAs/GaAs* laser diode, optically controlled switches, gain controlled amplifiers, phase shifters, and optical injection locking oscillators can be implemented for optical communication systems. Even with wide applications of OPFET's and other optically controlled devices, however, there are not enough experimental results reported yet. Characterization of optically controlled microwave devices, based on the experimental results in the high frequency, is indispensible for accurate prediction and implementation of high performance optical communication systems. In this work, we report optical power dependency of microwave characteristics in *GaAs* MESFET's.

2. Device Structure and Fabrication Process

We fabricated *GaAs* MESFET's with $1\mu m$ of gate length(L_g), $240\mu m$ of gate width(W_g), $1\mu m$ of gate-to-source(L_{gs}), and $1\mu m$ of gate-to-drain spacing(L_{gd}). The epitaxial structure is consisted of $3\times10^{18}cm^{-3}$ *Si*-doped layer(1000Å) for ohmic contact, $5\times10^{17}cm^{-3}$ *Si*-doped layer(1200Å) for the active channel, and undoped GaAs buffer layer ($2.5\mu m$) grown by MOCVD on semi-insulating *GaAs* substrate. Diluted ammonium hydroxide was used for isolation mesa etching and recessed gate structure at room temperature. $AuGe/Ni/Au$ was used for ohmic contact followed by 3 minute annealing at $440°C(N_2)$, and Ti/Au was used for Schottky contact for the gate. PECVD-grown silicon nitride(1000Å) was deposited for the passivation of fabricated devices before characterization.

3. Experimental Results and Discussions

We investigated on-wafer microwave characteristics of *GaAs* MESFET's, at room

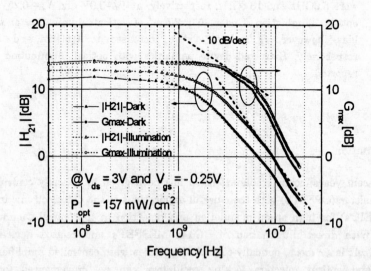

Fig. 1 Typical frequency characteristics of current gain($|H_{21}|$) and maximum available gain($|G_{max}|$) of a fabricated MESFET at V_{ds}=3V, V_{gs}=-0.25V, and P_{opt}=157mW/cm². (open symbol : under illumination, filled symbol : dark condition)

temperature, using the HP8510B network analyzer for frequency sweep from 45MHz~ 18GHz with/without illumination. An *AlGaAs/GaAs* diode-pumped laser module (Spectra-physics, model 7200, λ=0.83μm) was used as an optical power source for the characterization of optically illuminated devices. The optical power source was routed via an optical fiber to the *GaAs* MESFET by bringing the fiber end to a close proximity to the devices' surface such that the light spot covers the entire area of the device under test. We used optical power density from 1.6mW/cm^2 to 472mW/cm^2 for the microwave characterization of *GaAs* MESFET's. While current gain cut-off frequency(f_T) and maximum frequency of oscillation(f_{max}) of fabricated devices were 6.6GHz and 13.6GHz without optical illumination, they were measured to be 9.5GHz and 15.3GHz under 157mW/cm^2 of optical illumination at V_{ds}=3.0V and V_{gs}=-0.25V, as shown in Fig.1

Current gain of the device was improved with illumination, primarily due to increased carriers in the active channel. This also contributes to the enhancement of high frequency figures of merit, f_T and f_{max}, with increased transconductance. Maximum available gain, G_{max}, was also increased in the high frequency region(GHz range) and this improves f_{max}. We plotted measured f_T and f_{max} in Fig.2, as functions of optical power density. This improvement in the microwave performance mainly comes from changes in the channel carrier concentration, gate capacitance, and channel conductance etc[3]. With increasing the optical power density, f_T and f_{max} were improved, but they are saturated at higher incident optical power density.

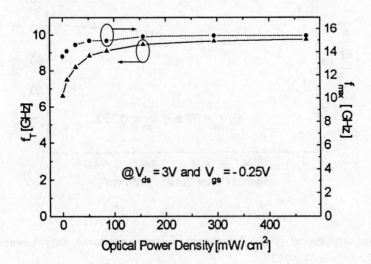

Fig. 2. Current gain cut-off frequency and maximum frequency of oscillation vs. incident optical power density at V_{ds}=3V and V_{gs}=-0.25V.

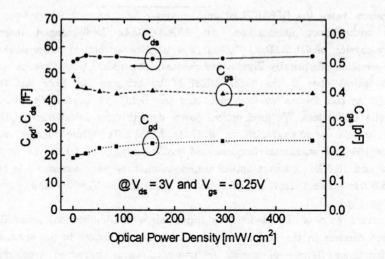

Fig. 3. Incident optical power dependency of extracted gate-to-source, gate-to-drain, drain-to-source capacitance at $V_{ds}=3V$ and $V_{gs}=-0.25V$.

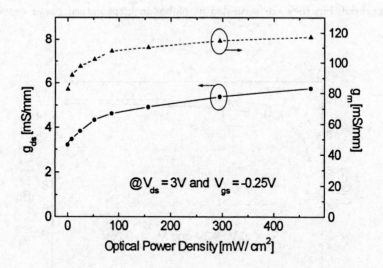

Fig. 4. Incident optical power dependency of transconductance, output conductance at $V_{ds}=3V$ and $V_{gs}=-0.25V$.

In order to investigate physical mechanism of the performance improvement with optical illumination, we extracted model parameters in the small-signal equivalent circuit of GaAs MESFET from S-parameters measured at the active bias region. Fig.3 shows gate-to-source capacitance(C_{gs}), gate-to-drain capacitance(C_{gd}), and drain-to-source capacitance(C_{ds}) depending on the optical power density. While C_{gs} was reduced by about 11% (from 0.46pF to 0.40pF), g_m was improved by about 34% (from 83mS/mm to 111mS/mm) under illumination. Transconductance(g_m) and drain-to-source conductance(g_{ds}) of the device are shown in Fig.4. g_m was improved with illumination by photoconductive and photovoltaic effects.

We obtained smaller C_{gs} under illumination, which is different from previously reported experimental results[5,6]. This has not been fully explained yet, but might come from the difference in the parasitic elements, such as pads and bonding wires, during de-embedding process in the microwave characterization. They used commercially available packaged GaAs MESFET so that they couldn't effectively remove extrinsic capacitances and inductances in the equivalent circuit. Considering analytical relationship of f_T with g_m and C_{gs},

$$f_T = \frac{g_m}{2\pi C_{gs}}, \tag{1}$$

both g_m and C_{gs} contribute to the improvement of f_T. Maximum frequency of oscillation also depends on g_m, C_{gs}, g_{ds}, and parasitic resistances (R_s, R_d, and R_{ch} respectively) through,

$$f_{max} = \frac{f_T}{2\sqrt{(R_s + R_{ch} + R_d)g_{ds} + 2\pi f_T R_g C_{gd}}}. \tag{2}$$

However, the sensitivities of these parameters to f_T and f_{max} are quite different from

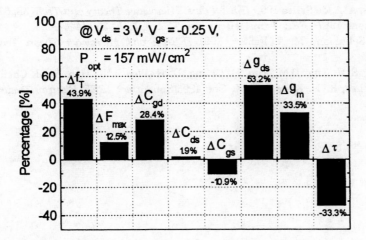

Fig. 5. Variations of device model parameters with/without illumination.

each other. We found that the most dominant model parameters for the improvement in f_{max} are C_{gs} and g_m through f_T. Even though relative change, compared with that of dark condition, in C_{gd} is large with illumination, its contribution to the change in f_{max} is negligible. This is probably due to relatively weak contribution of the second term of the denominator in eq (2). This also apply to the variation of microwave characteristics with g_{ds}. RC time constant(τ) obtained from S-parameters of MESFET's, which is delay time of the carriers passing through the active channel and contributes to the phase difference in the transconductance at high frequency, was decreased with illumination. We summarized relative changes in the key performance parameters of optically illuminated MESFET's in Fig.5.

4. Conclusion

As a conclusion, improved DC and microwave characteristics of optically illuminated *GaAs* MESFET are reported. This is found to be mainly due to photoconductive and photovoltaic effects, which results in improved channel transport property. This shows optically controlled *GaAs* MESFET can be used for optical detection and other applications in the monolithically integrated optoelectronic circuits.

References

[1] R. B. Darling and J. P. Uyemura 1987 *IEEE J. Quantum Electron.* **23**, 1160-1171

[2] A. Madjar, A. Paollela, and P. R. Herczfeld 1995 *IEEE Trans. Microwave Theory and Tech.* **41**, 165-167

[3] P. Chakrabarti, S. K. Shrestha, A. Srivastava, and D. Saxena 1994 *IEEE Trans. Microwave Theory Tech.* **42** 365-375

[4] A. J. Seeds, A. A. de Salles 1990 *IEEE Trans. Microwave Theory and Tech.* **38**, 577-585

[5] R. N. Simons 1987 *IEEE Trans. Microwave Theory Tech.* **35**, 1444-1455

[6] A. A. de Salles and M. A. Romero 1991 *IEEE Trans. Microwave Theory Tech.* **39**, 2010-2017

[7] H. J. Kim, S. J. Kim, D.M. Kim, H. Chung, D. H. Woo, I. K. Han, W. J. Choi, S. H. Kim, J. I. Lee, K. N. Kang, and K. Cho 1995(Submitted) *IEEE Electron Device Lett.*

Inst. Phys. Conf. Ser. No 145: Chapter 5
Paper presented at 22nd Int. Symp. Compound Semiconductors, Cheju Island, Korea, 28 August–2 September 1995
© *1996 IOP Publishing Ltd*

Physical Processes of Effective *V-F* Characteristics for Submicrometer GaAs MESFETs

Y Yamada and T Takahashi

Department of Inf. Sys., Kyushu Tokai University, Japan
Department of Elec. Eng. & Inf. Sci., Kumamoto University, Japan

Abstract. The whole shapes of the effective *V-F* characteristics in the submicron GaAs MESFETs are evaluated using the energy transport drift-diffusion model(ETDDM) including the nonstationary electron transport and the physical processes are studied. The drift velocity is described by a double-valued function of the drift electric field. This property is caused by the inertia motion of the Γ-valley electrons for small drain voltage and by the hot electron effect associated with the large energy relaxation time for large drain voltage. Thus the inertia term should not be neglected in the device simulation. It is confirmed that ETDDM using these effective *V-F* characteristics is effective.

1. Introduction

It is widely known that the nonstationary electron transport, which plays an important role in a submicrometer gate-length GaAs MESFET, cannot be expressed by the bulk *V-F* characteristics and that the maximum velocity in the device exceeds the bulk one[1,2]. Although some studies have been made for an evaluation of the distribution of the mean electron velocity under the gate, there are few reports to evaluate a whole shape of the effective *V-F* characteristics for the nonstationary electron transport both on the source and drain sides of the maximum velocity[3,4]. So the purpose of the present work is to evaluate a whole shape of the effective *V-F* characteristics, to study their physical processes, and to confirm the effectiveness of the drift-diffusion model associated with the characteristics.

2. Basic properties of effective *V-F* characteristics for nonstationary transport

To investigate basic properties of effective *V-F* characteristics for the nonstationary electron transport, it is required to calculate a distribution of the drift velocity(V) for a nonuniform distribution of the drift electric field(F). It is not easy to correctly evaluate V from an exact ensemble Monte Carlo simulation(EMC), because EMC cannot directly separate a drift velocity component from a diffusion velocity component. It has been known that the drift motion in a nonuniform electric field can be well described by the following 1D-balance equations of momentum and energy[4],

770

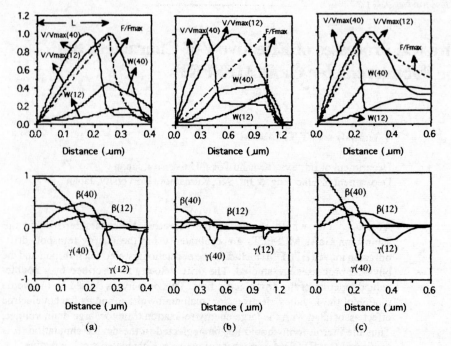

Fig.1 Calculated results for the three types of the distributions of $F(s)$.
The numbers in the parentheses indicate values of F_{max}.

$$m^* \frac{dV(s)}{ds} V(s) = qF(s) - \frac{m^* V(s)}{\tau_p} \quad , \quad (1a)$$

$$\frac{dW(s)}{ds} V(s) = qF(s)V(s) - \frac{W(s) - W_0}{\tau_w} \quad . \quad (1b)$$

Here s is the position coordinate along a main current path, W is the mean energy, and the others have usual meanings. When $V(s)$ is related with $F(s)$ at the same position of s, we can obtain "effective V-F characteristics" depending on $F(s)$ and define an "effective mobility(μ)" by $V(s)/F(s)$. m^*, τ_p and τ_w depend on W. They were calculated for an impurity density of 10^{17}cm^{-3} according to the usual procedure[5-7].

To elucidate physical processes of the characteristics, we here present some results for simple distributions of $F(s)$ modeling a real distribution along a current path under the gate. Fig.1 shows the three types of $F(s)$ which are the combinations of a steep or gentle upward-slope and a steep or gentle downward-slope. $F(s)$ linearly increases toward a maximum(F_{max}) and exponentially decreases behind F_{max}. Fig.1 also shows the distributions of $V(s)/V_{max}$, $W(s)$ in eV unit, $\gamma(\equiv \tau_p dV/ds)$ and $\beta(\equiv m^* V^2/2W)$. The lower figures were obtained for the same conditions as the upper ones. Here V_{max} is a maximum of V, γ means a ratio between the inertia term(the left hand side of (1a)) and the scattering term(the second term on the right hand side of (1a)) and β a ratio between the mean kinetic energy and the mean total energy. A maximum of $V(s)$ occurs where β is maximum or a sign of γ changes. Largeness of $|\gamma|$ or β means that the drift motion is strongly influenced by the inertia term, while their smallness means that it is dominated by the scattering term. Fig.2(a)-(c) show the effective V-F characteristics for the three types of the distribution of $F(s)$ in Fig.1(a)-(c), respectively,

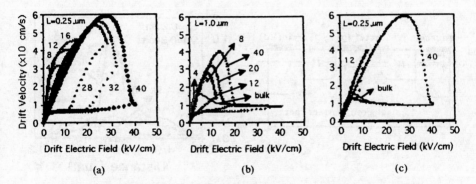

Fig.2 Effective *V-F* characteristics for the three types of the distributions of *F(s)*.
(a), (b) and (c) correspond to Fig.1(a), (b) and (c), respectively. The numbers
associated with the dotted lines indicate values of *Fmax*.

which significantly differ from the bulk one.

Basic properties of the effective *V-F* characteristics behind *Fmax* are summarized as follows:

(1) The effective *V-F* characteristics strongly depend on the distribution of *F*. They are expressed as a double-valued function of *F*.

(2) For a small *Fmax* in Fig.1(a) and 2(a), *V* is larger on the drain side than on the source side at the same *F*. *W* is small and most of the electrons stays in the Γ-valley around *Fmax*. The contribution of the inertia term is fairly large even behind *Fmax*. The inertia motion of the electrons keeps their mean velocity in the region of the steep downward-slope behind *Fmax*. This process is called *inertia effect* hereafter. Thus the double-valued properties for this case are explained by the inertia effect.

(3) For a large *Fmax* in Fig.1(a) and 2(a), *V* is larger on the source side than on the drain side at the same *F*. *W* is fairly large around *Fmax* and most of the electrons stays in the satellite-valleys even in the region of small *F* behind *Fmax*. So the scattering term dominates the electron motion. The electrons cannot immediately release their energies and immediately return to the Γ-valley with the small effective mass in the region of the steep downward-slope due to a large τ_W. This process is called *hot-electron effect* hereafter. Thus the double-valued properties for this case are explained by the hot-electron effect.

(4) In the case of the gentle upward-slope shown in Fig.1(b) and 2(b), the above properties of (1)-(3) are weakened.

(5) For the case of the gentle downward-slope shown in Fig.1(c) and 2(c), the characteristics are close to the bulk one behind *Fmax*.

3. Evaluation of the effective *V-F* characteristics for the GaAs MESFETs

3.1 Evaluation model

A drift-diffusion model associated with the balance equations of (1a) and (1b) was accepted, which is called *Energy Transport Drift Diffusion Model*(ETDDM) hereafter. ETDDM self-consistently solves the Poisson's and current continuity equations. Eqs. (1a) and (1b) are solved only along a selected current path lying at the maximum electron density under the gate. As the effective *V-F* characteristics were expressed as a double-valued function of the drift electric field, the characteristics obtained on the source side of *Fmax* were applied to the source

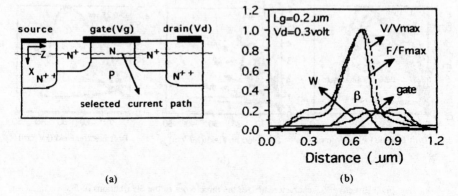

(a) (b)

Fig.3 (a) Device structure of the GaAs MESFETs used in the present work. (b) Distributions of
F/Fmax, V/Vmax, W and β along the selected current path for Vd=0.3volt.

region and the others to the drain region. The drift electric field was obtained by subtracting
the electric field related with a channel formulation from the total electric field. The Einstein
relation was assumed between the effective mobility and the diffusion constant. ETDDM is
more physically meaningful and more correct than the previous models[5,8], because it
doesn't assume the gradual channel and the complete depletion in contrast to the models.

EMC was also carried out to confirm the effectiveness of ETDDM. Details of ETDDM will
be published in [9]. The data of the energy dependence of m^*, τ_p and τ_w were calculated by
EMC for a wide range of impurity density. The energy dependence of m^* is independent of
the impurity density.

Fig.3(a) shows a device structure of GaAs MESFET used in the present work. The first
type is a Gaussian-doped one with $Lg=0.32\mu m$ which was used for the experiment[2] and the
another is a uniform-doped($10^{17}cm^{-3}$) one with $Lg=0.2\mu m$ which was used for EMC[4].

3.2 Numerical results

Fig.3(b) shows the distributions of $F(s)$, $V(s)/Vmax$, $W(s)$ and β in the uniform-doped
MESFET at $Vd=0.3volt$. $F(s)$ has a steep downward-slope behind its maximum. Its
distribution under the gate is close to the first type shown in Fig.1(a). Thus the third type of
Fig.1(c) may not be realized in the MESFETs. Fig.4 shows the effective V-F characteristics
for the uniformly-doped MESFET and the Gaussian-doped one. The arrows indicate an
increasing direction of the z-coordinate from the source to the drain along the selected current
path. The effective V-F characteristics depend on Vd. For large Vd the maximum velocity
exceeds the bulk one(broken line) owing to the near ballistic transport. V is expressed by a
double-valued function of F, as pointed out in the Section 2. For large Vd, V is much smaller
on the drain side than on the source side for the same F. On the other hand the situation
reverses for small Vd. Thus these effective V-F characteristics are basically the same as those
in Fig.2(a). The physical processes of the double-valued properties are also explained as same
as in the Section 2.

To study reasonableness of ETDDM, we compared ETDDM with EMC. The mean velocity
v directly obtained from EMC is a summation of a drift component and a diffusion one. It is
not easy to extract the former from v. In ETDDM, v is defined by

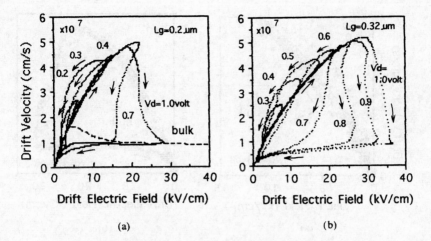

Fig.4 Effective *V-F* characteristics for (a) the uniform-doped GaAs MESFET with $Lg=0.2\mu m$ and
(b) the Gaussian-doped GaAs MESFET with $Lg=0.32\mu m$.

$$v = \frac{J}{qn} = -\mu \text{ grad } \psi + \frac{D}{n}\text{grad } n \qquad (2)$$

Here the first and second terms on the right hand side of (2) are the drift and diffusion
velocities, respectively. We compared the z-component of v between ETDDM and EMC.
Fig.5 shows the relationships between v_z and F_z obtained from ETDDM($\bigcirc$) and EMC($\bullet$),
where v_z is the total mean velocity including both the drift and diffusion components and F_z
the total electric field including both the drift and built-in components. The agreements
between ETDDM and EMC are good. The results indicated by + are as same as those in Fig.4
and so the differences between + and $\bigcirc$ indicate the influence of the electron diffusion on the
characteristics. The influence is clearly observed on the drain side of the peak electric field for
a case of small Vd.

To study the effectiveness of ETDDM further, we compared the drain currents calculated
from ETDDM with those from EMC for the uniform-doped MESFET and with those of the
experiment for the Gaussian-doped one. We confirmed the good agreements between ETDDM
and EMC and between ETDDM and the experiment[2].

4. Conclusions

The basic properties of the effective *V-F* characteristics for the nonstationary electron transport
have been studied and their whole shapes have been evaluated. It has been found that the
characteristics are expressed by a double-valued function of the drift electric field due to the
inertia effect for the small drain voltage and due to the hot-electron effect for the large drain
voltage. The comparisons between ETDDM and EMC and between ETDDM and the
experiment have shown that ETDDM and the effective *V-F* characteristics are reasonable.
Although some models neglecting the inertia term have been proposed previously[10], the
present work tells us that they are not physically reasonable.

774

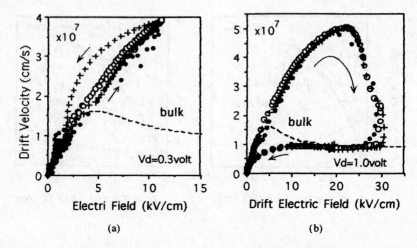

Fig.5 Relationships between Vz and Fz in the uniform-doped GaAs MESFET with $Lg=0.2\mu m$ for (a) $Vd=0.3volt$ and (b) $Vd=1.0volt$.

References

[1] Blotechjaer K : *IEEE Trans. Electron Devices, E.D.* **17**-1,38/47(1970); Woolard D L, Pelouard J-L, Trew R J , Littlejohn M A, and Kelley C T 1989 *Solid-State Electronics* **32**-12 1347-1351

[2] Yamasaki K and Hirayama M 1986 *IEEE Trans. E. D.* **33**-11 1652-1658

[3] Tomizawa M, Furuta T, and Yoshii A 1986 *IEICE Technical Reports CAS***86**-98 93-97

[4] Yamada Y 1994 *Electronics Letters* **30**-4 363-364

[5] Carnez B , Cappy A , Kasynski A, Constant E, and Salmer G 1980 *J. Appl. Phys.* **51**-4 784-790

[6] Fawcett W, Boardman A D, and Swain S 1970 *J. Phys. Chem. Solids* **31** 1963-1990

[7] Ruch J. G. and Fawcett W 1970 *J. Appl. Phys.* **41**-9 3843-3849

[8] Sone J, Kimura T, and Yamada T 1987 *Extended Abstract of the 19th Conf. on Solid State Devices and Materials* Tokyo 379-382

[9] Yamada Y 1995 *20th Int. Conf. on Microelectronics* Serbia Sept. 12-15(to be published)

[10] For example, Cook R K and Frey J 1982 *IEEE Trans. Electron Devices, E.D.* **29**-6 970-977 ; Snowden C S and Pantoja R R 1989 *IEEE Trans. Electron Devices, E.D.* **36**-9 1564-157

Inst. Phys. Conf. Ser. No 145: Chapter 5
Paper presented at 22nd Int. Symp. Compound Semiconductors, Cheju Island, Korea, 28 August–2 September 1995
© *1996 IOP Publishing Ltd*

Effects of Sulfide Treatment on the Gate Voltage Swing of InP MISFETs with Photo-CVD Grown P_3N_5 Gate Insulator

S. K. Jo[†][1], B. H. Lee[†], M. Y. Jeong[†], Y. H. Jeong[†], and T.Sugano[‡]

†Department of Electronics and Electrical Engineering, and Microwave Application Research Center, Pohang University of Science & Technology, Pohang P.O. box 125,790-600 Korea

‡Laboratory for Nano-Electronics Materials, Frontier Materials Research Program,The Institute of Physical and Chemical Research, RIKEN, Saitama, 351-01, Japan

Abstract. The effects of the sulfide treatment on the Al-P_3N_5/InP MIS Devices with a photo-CVD grown P_3N_5 insulating film are investigated. The minimum density of interface trap states is as low as 2.6×10^{10}/cm^2· eV , and has been obtained from the sulfide treated sample at 40 °C for 20min. We have successfully fabricated the depletion mode InP MISFETs for microwave power device applications. The effective channel electron mobility is observed to be 3100 cm^2/V·s at 300K. The extrinsic transconductance of 5.8mS/mm shows a broad plateau region through the 4 V gate voltage swing. The InP MISFET with large gate voltage swing is attractive for power device applications.

1. Introduction

An InP MISFET is an attractive device for high frequency and high power applications because of large current handling capability device and large gate-drain breakdown voltage.

In addition to the potential applications for the power InP based technology, an InP MISFET can be extended to the optoelectronic integrated circuits such as monolithically integrated photoreceiver, laser, or CCDs[1-2]. But high interface traps and drain current degradation have prevented the wide acceptance of InP MISFET as a power device.

In recent years, the sulfide treatments on InP using $(NH_4)_2S_x$ have received much attention because of their possibility of either protecting or at least stabilizing its surface against degradation from subsequent process[3,4]. It is generally accepted that such treatments remove native oxides and replace them with an ultra thin sulfide layer. Because photo-CVD system has a UV light as energy source, it is possible to allow lower temperature deposition which is essential in minimizing thermal degradation of InP substrates.

---

[1] S. K. Jo recently joined the Samsung Electronics Co., Ltd. Korea

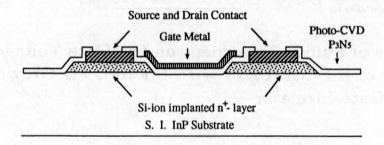

Figure 1. Schematic diagram of the carbino-disc type InP MISFET

Previously, we have successfully demonstrated good interface state properties from the photo-CVD grown P_3N_5 gate insulator on InP and GaAs substrates[5,6].

In this work, we report on the effects of sulfide treatment on depletion mode InP MISFETs with photo-CVD grown P_3N_5 gate insulator for microwave power devices.

2. Experiments and results

In order to examine the effects of sulfide treatment, Al-P_3N_5/InP MIS diodes were first fabricated. Unintentionally doped n-type (100) InP wafers with a carrier concentration of around $3 \times 10^{15} cm^{-3}$ were used as substrates. The samples were cleaned by chemically etching and the ohmic contacts were performed with AuGe/Ni/Au. Following these treatments, they were transferred into $NH_4OH : H_2O(1:15)$ solution for removal of native oxides, and hastly immersed into a beaker containing a $(NH_4)_2S_x : H_2O(1:1)$ solution indirectly heated in water at 40 °C. The treatment time was varied to investigate the time dependency of sulfide treatment on the interface properties of P_3N_5/InP system. The samples were then blown dry with N_2 prior to loading into the photo-CVD chamber for P_3N_5 film deposition. It is generally accepted that $(NH_4)_2S_x$ solution thoroughly removes P and In oxides. $InPS_4$ phase is formed when the sulfidation of InP at low temperature is performed. To convert this layer into more thermodynamically stable layer, In_2S_3 transition annealing step is demanded. Therefore, before P_3N_5 deposition, we performed the sulfide annealing at 250 °C for 30 min. in high purity N_2 ambient to minimize the thermal degradation of InP surface.

The P_3N_5 film deposition was performed using ultraviolet(UV) light with wavelength of 185 nm at 200 °C for 25 min. The in-situ post-deposition annealing was carried out at 300 °C for 10 min. under the flowing N_2/H_2 at the rate of 400 sccm. The measured index of refraction and thickness of the deposited film were 1.75 and 700 Å, respectively. Aluminum metal was finally deposited for gate electrode with shadow mask. The carbino-disc type InP MISFETs were fabricated on Fe-doped semi-insulating InP substrates through the optimal sulfide treatment conditions. The inner and outer radii of devices are 100μm and 200μm, respectively, resulting in 100 μm diameter circle gate length. The source and drain regions were formed with Si^+ ion implantation with a dose $3 \times 10^{15} cm^{-3}$ at 130keV.

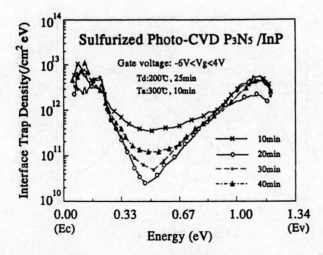

Figure 2. The densities of interface states of the sulfide treated photo-CVD grown InP MIS diode with variation of the sulfur treatment time.

Figure 1 shows a schematic diagram of the carbino-disc type InP MISFET. Figure 2 shows the interface trap state densites deduced from the 1MHz C-V data using Terman method. From Fig. 2, the minimum density of interface trap states of $2.6 \times 10^{10}/\text{cm}^2 \cdot \text{eV}$ was obtained from the sulfide treated sample at 40 °C for 20 min. To our knowledge, the observed minimun density of interface trap states is one of the best reported results. Figure 3 shows a current-voltage characteristics of the depletion mode of InP MISFET. The effective channel electron mobility and extrinsic transconductance was 3100 cm²/V·s and 5.8mS/mm, respectively. Because of large gate size, transconductance is very low. But if gate is scaled down to 1 μm, the expected transconductance can be

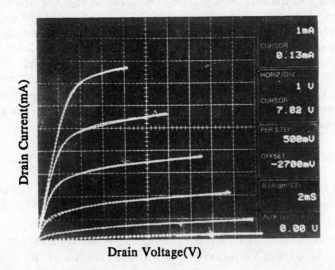

Figure 3. I-V characteristcs of the depletion mode InP MISFET with sulfur treatment at 40 °C for 20 min. and photo-CVD grown P₃N₅ film.

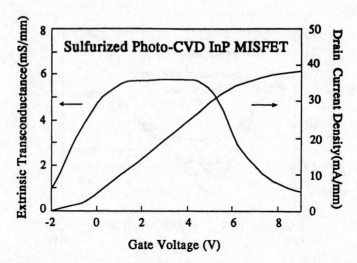

Figure 4. Extrinsic transconductance versus gate voltage profile of the fabricated depletion mode InP MISFETs

400 mS/mm. Figure 4 shows an extrinsic transconductance versus gate voltage profile of a fabricated depletion mode InP MISFET. A broad plateau region extending through a range of nearly 4V gate voltage swing was achieved maintaining maximum extrinsic transconductance of about 5.8mS/mm. Optimization of the device structure should lead to further improvement. Less than 2% change in the measured drain current over 24 h was observed for the device.

3. Conclusions

We have investigated the effect of the sulfide treatment on the Al-P_3N_5/InP MISFET with the photo-CVD grown P_3N_5 gate insulating film. The use of sulfide passivation with photo-CVD grown P_3N_5 gate insulator has led to great improvement in device performance. The minimun density of interface trap states was as low as 2.6×10^{10}/cm^2·eV and was obtained from the sample sulfide treated at 40°C for 20 min. The effective channel electron mobility of the depletion mode InP MIFETs was 3100 cm^2/V·s. The extrinsic transconductance of 5.8 mS/mm showed a broad plateau region through the 4V gate voltage swing. The fabricated depletion mode InP MISFET with large gate voltage swing has a great potential for microwave power device applications.

References

[1] T. Itoh, and K. Ohata, *IEEE Trnas. Electron Devices* Vol. **ED-30**, No. 7, pp 811 -815 (1983)

[2] Y. Iwase, F. Arai, and T. Sugano, *Appl. Phys. Lett.*, Vol. **52**, No. 17, pp. 1437 - 1438 (1988)

[3] R. Iyer et. al., *Appl. Phys. Lett.*, Vol. **53**, No. 2, pp. 134 - 136 (1988)

[4] D. Gallet and G. Hollinger, *Appl. Phys. Lett.*, Vol. **62**, No. 9, pp. 982 - 984 (1993)

[5] Y. H. Jeong, J. H. Lee, Y. H. Bae, and Y. T. Hong, *Appl. Phys. Lett.* vol. **57**, No. 25, pp. 2680-2682 (1990)

[6] Y. H. Jeong, K. H. Choi, and S. K. Jo, *IEEE Electron Devices Lett.* vol. **15**, No. 7, pp. 251-253 (1994)

Inst. Phys. Conf. Ser. No 145: Chapter 5
Paper presented at 22nd Int. Symp. Compound Semiconductors, Cheju Island, Korea, 28 August–2 September 1995
© *1996 IOP Publishing Ltd*

779

Low voltage operated GaAs MISFET using a novel self-alignment technique for power amplifiers in mobile communication system

Y Ota, M Nishitsuji, H Masato, S Morimoto and H Fujimoto

Electronics Research Laboratory, Matsushita Electronics Corporation,
3-1-1 Yagumo-nakamachi, Moriguchi, Osaka 570, JAPAN

Abstract. Ohmic metal self-aligned gate (OMEGA) process has been developed in order to achieve a low knee voltage in GaAs MISFETs. The gate metal is self-aligned to the drain and source ohmic metal which is made of WSi and is formed on a thick n^+-GaAs/n^+-InGaAs cap layer. The distance between the gate metal and the drain/source metal is minimized by using this process, and consequently the parasitic resistances from drain to source can be drastically reduced. The OMEGA FET shows a low knee voltage of 0.8 V with a high breakdown voltage of 12 V. These features indicate that the new FET is highly suitable for power applications under low supply voltage and consequently contributes to minimize the battery cell and the size of microwave mobile communication systems.

1. Introduction

Recent advance in the performance of mobile communication systems such as cellular telephone is remarkable [1], [2]. The technological trend of high frequency devices especially for those handsets is to realize low-voltage operation with high efficiency, because these devices influence the receiving/transmitting time and consequently influence the size of the battery. We have introduced a new AlGaAs/GaAs HBT amplifier with very high gain in microwave band [3]. Since HBTs have a low knee voltage (V_k) around 0.7 V, the amplifier can be operated at 1.5 V which corresponds to one battery cell. On the other hand, the RF properties of conventional GaAs MESFETs degrade at such low voltage, because the FETs have fairly high V_k in comparison with that of HBTs. To reduce the V_k in FETs, the decrease in the parasitic resistances between the drain and the source is most effective. The approach, however, has a limitation in conventional structures, because the pattern alignment using a photo mask makes it difficult to form the gate metal very close to the source and drain ohmic metals.

To solve these problems, we have newly developed Ohmic MEtal self-aligned GAte (OMEGA) process for GaAs MISFETs, where the gate metal is self-aligned to the source and drain metals. As a result that, the parasitic resistances of gate-drain, gate-source, drain contact region and source contact region are minimized. The fabricated OMEGA FET showed a very low V_k and consequently showed a high gain at a low supply voltage in comparison with conventional FETs. In this paper, the fabrication process and the experimental results of the OMEGA FETs are described.

2. Fabrication process

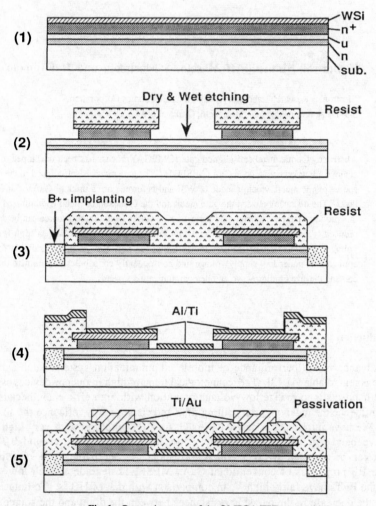

Fig. 1 Processing steps of the OMEGA FET

Figure 1 shows a flow chart of the OMEGA process. The processing steps are as follows, 1) fabricating a multilayered structure comprising a n-GaAs as a channel layer, u-GaAs or AlGaAs as a insulator layer, n^+-GaAs/n^+-InGaAs as a cap layer and WSi as an ohmic metal on a semi-insulating GaAs substrate, 2) forming a drain and a source contact mesa by removing the cap layer with wet etching using the WSi as a mask, 3) implanting B^+ for device isolation, 4) making a gate by evaporating Al/Ti and lifting off it using a patterned resist and the drain/source mesas as a mask, 5) forming a passivation and wirings. Using this OMEGA process, the gate metal can be formed very close to the drain/source metals.

The multilayered structure used in this study is summarized in Table 1. The thick cap layers composed of n^+-$In_{0.5}Ga_{0.5}As$, graded n^+-InGaAs and n^+-GaAs (total thickness = 500 nm) were employed in order to avoid an electrical shortage between the gate metal and the source/drain metals. The OMEGA process can be applied not only to sub-micron gate MISFETs but also to HFETs.

Table. 1 Multilayered structure for the OMEGA FET

Layer	Content (%)	Concentration (cm-3)	Thickness (nm)
n^+-InGaAs	In = 50	2×10^{19}	100
n^+-InGaAs	In = 50-0	grade	50
n^+-GaAs		2×10^{18}	350
u -GaAs		-	150
n -GaAs		2×10^{17}	120
u -AlGaAs	Al = 20	-	
u -GaAs		-	
S. I. Substrate			

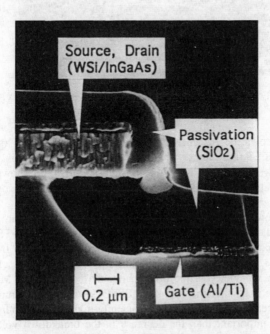

Fig. 2 Cross-sectional SEM view of the OMEGA FET.

The cross sectional SEM view of the fabricated OMEGA FET is shown in Fig. 2.

The shape of the contact mesa depends on the orientation and the etchant. We adopted $H_3PO_4/H_2O_2/H_2O$ etchant with which the mesa had a sufficient under cut for all orientations. Thus, the edge of the gate metal was at least 0.2 µm apart from the edge of the contact mesa.

3. Results and discussion

The fabricated OMEGA FET had a gate length (L_g) of 1.0 µm and a gate-to-drain/source distance (L_{gd}) of 0.2 µm. In order to compare DC characteristics, we prepared a conventional GaAs MISFET which was formed by a simple recess etching process without self-alignment technique. The conventional MISFET had $L_g = 1.0$ µm/$L_{gd} = 0.7$ µm. In addition, four types of conventional GaAs MESFETs formed by an ion-implantation process with Al gate [4] were also prepared; a) $L_g = 0.6$ µm/$L_{gd} = 0.5$ µm, b) $L_g = 0.6$ µm/$L_{gd} = 1.0$ µm, c) $L_g = 1.0$ µm/$L_{gd} = 1.0$ µm and d) $L_g = 1.0$ µm/$L_{gd} = 2.0$ µm.

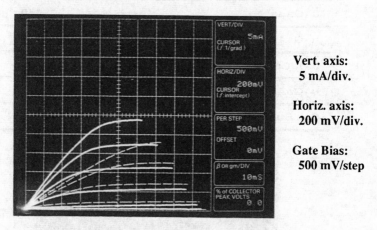

Vert. axis:
5 mA/div.

Horiz. axis:
200 mV/div.

Gate Bias:
500 mV/step

Fig. 3 Comparison of I-V characteristics between the OMEGA FET
(solid lines) and the conventional MISFET (dash lines)

Figure 3 shows the I-V characteristics of the OMEGA FET (solid lines) and the conventional MISFET (dash lines), where the gate width are 100 µm. The OMEGA FET realizes a very low V_k of 0.8 V, while the conventional MISFET shows a V_k of 1.3 V. In addition, the self-aligned process reduces the source resistance and consequently increase the saturation current largely.

The breakdown voltage of the OMEGA FET is 12 V (0.1 mA/mm). The relationships between the V_k and the gate-to-drain breakdown voltage in the OMEGA FET, the conventional MISFET and the conventional MESFETs are summarized in Fig. 4. In case of MESFETs the decrease in the V_k by minimizing the drain-to-source distance resulted in the decrease in the breakdown voltage. On the other hand, the conventional MISFET had a high breakdown voltage in comparison with the MESFETs which had a similar V_k value. Since the value of the breakdown voltage was sufficiently-high for the handsets, the decrease in the V_k was more important than the decrease in the breakdown voltage. Further size minimization, however, is impossible in the conventional MISFET because of the limitation in photo-mask alignment. The OMEGA FET has the lowest V_k with an adequately-high breakdown voltage.

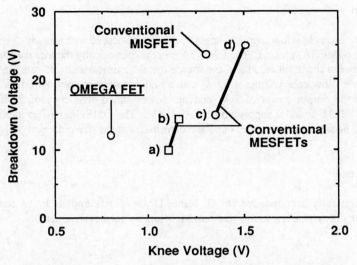

Fig. 4 Relationships between the knee voltage and the breakdown voltage in
the OMEGA FET (L_g = 1.0 µm/L_{gd} = 0.2 µm), the conventional MISFET (1.0/0.7 µm),
the conventional MESFETs (a) 0.6/0.5 µm, b) 0.6/1.0 µm, c) 1.0/1.0 µm, d) 1.0/2.0 µm)

The OMEGA FET also shows good RF properties. The maximum stable gain
(MSG) of 1 GHz are 20 dB, 22 dB and 22.5 dB at a supply voltage of 1.0 V, 2.0 V and 3.0
V, respectively. The output power and the power-added efficiency dependence on the
input power at a frequency of 950 MHz and a supply voltage of 3.0 V are shown in Fig. 5.
The gate length and the gate width are 1.0 µm and 12 mm, respectively. The FET produces
an output power of 30 dBm with an efficiency of 55 %.

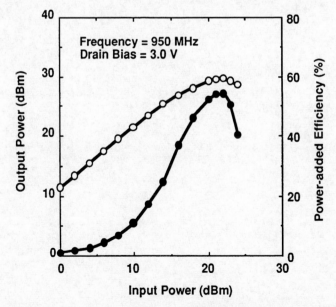

Fig. 5 Output power and power-added efficiency dependence on input power of the OMEGA FET.

4. Conclusion

GaAs MISFETs operable at low supply voltages have been introduced with a newly-developed self-alignment (OMEGA) process. The OMEGA process dramatically reduces the parasitic resistances between the drain metal and the source metal. Fabricated 1 μm-gate OMEGA FET shows a very-low knee voltage of 0.8 V with a high-breakdown voltage of 12 V. The FET achieves the output power of 1 W and the power-added efficiency of 55 % at a frequency of 950 MHz and a supply voltage of 3.0 V. The OMEGA FET is applied to various mobile handsets in microwave band and contributes to the size reduction

Acknowledgment

The authors gratefully acknowledge Dr. G. Kano, Dr. M. Inada and K. Inoue for their encouragement. They would also like to thank M. Miura for the device assembly.

References

[1] Ota Y, Yanagihara M, Yokoyama T, Azuma C, Maeda M and Ishikawa O 1992 IEEE MTT-S 1517- 20

[2] Ota Y, Azuma C, Fujimoto H, Masato H and Inoue K 1994 Electron. Lett. **30** 906-7

[3] Ota Y, Yanagihara M, Tamura A and Ishikawa O 1993 GaAs Related **136** 9-14

[4] Takehara H, Maeda M, Adachi C, Fujimoto H, Ota Y, Konno T and Ishikawa O 1995 Int. Conf. Telecom. (ICT '95, Indonesia) 326-329

Inst. Phys. Conf. Ser. No 145: Chapter 5
Paper presented at 22nd Int. Symp. Compound Semiconductors, Cheju Island, Korea, 28 August–2 September 1995
© *1996 IOP Publishing Ltd*

Detection and Mixing of Terahertz Radiation by Two Dimensional Electronic Fluid

M I Dyakonov and M S Shur*

A. F. Ioffe Physico-Technical Institute, St. Petersburg, 194021, Russia
*Department of Electrical Engineering, University of Virginia,
Charlottesville, VA 22903-2442, USA, shur@virginia.edu

Abstract. We show that a High Electron Mobility Transistor has a resonance response to an incoming electromagnetic radiation at the odd harmonics of the plasma wave frequency of the two dimensional electrons in the device channel (which are in a terahertz range for a 0.1 micron gate device). This response can be used for new types of detectors and mixers that use the plasma waves, which may propagate much faster than electrons. We show that the responsivities of this device may greatly exceed the responsivities of Schottky diodes currently used as detectors and mixers in a terahertz range.

1. Introduction

When the electron mean free path for collisions with impurities and phonons is much greater than the mean free path for electron-electron collisions, electrons behave like a fluid which may be described by hydrodynamic equations. In our papers [1,2], we showed that this condition can be met for two dimensional (2D) electrons in a Field Effect Transistor (FET) and that the hydrodynamic equations describing this electron fluid coincide with those for shallow water. Furthermore, plasma waves in the FET channel are similar to shallow water waves. We also showed that in a short enough device, an instability should occur at a relatively small direct current because of spontaneous plasma wave generation. This provides a new mechanism for the emission of tunable far infrared electromagnetic radiation. We expect that similar waves and a similar instability may occur even if the electron-electron collisions are relatively rare, and the hydrodynamic approach is not justified.

More recently, we discussed the similarity between the plasma waves in 2D systems [3] and sound waves. [4] We showed that the plasma waves in 2D systems are described by the same equations as sound waves in a gas. Therefore, resonant structures, similar to those in musical instruments, may be realized for the plasma waves (i. e. an electronic flute). These waves can be excited by a direct current just like wind musical instruments are excited by air jets. As a consequence of the high plasma wave velocity and small FET dimensions, the plasma wave frequencies are in the terahertz range. The plasma waves are accompanied by a variation of a dipole moment created by charges in the FET channel and the mirror image charges in the gate. This variation should cause the emission of far infrared (terahertz) electromagnetic radiation.

In this paper, we show that a FET has a resonance response at the plasma wave frequency. The width of the resonance curve is determined by the inverse momentum relaxation time. As we discussed in [1], the appropriate boundary conditions for the plasma waves are the short circuited source and the open circuit drain of the FET channel. As we demonstrate below, the asymmetry of these boundary conditions leads to the resonance detection and mixing of electromagnetic radiation at terahertz frequencies.

This new family of solid state devices - a FET emitting a far infrared radiation, an electronic flute, and electron fluid detector and mixer considered in this paper - utilizes the propagation of plasma waves which may be much faster than electron drift velocities. Therefore they should allow us to push a three terminal device operation into a much higher frequency range than has been possible for conventional, transit time limited regimes.

A detector or mixer operating at terahertz frequencies must have an antenna structure in order to collect a weak incoming electromagnetic radiation. Depending on the antenna design, we can represent the FET mixer by one of the two equivalent circuits shown in Fig. 1.

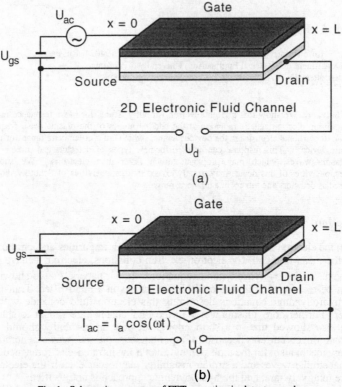

Fig. 1. Schematic geometry of FET operating in detector mode.

Fig. 1 a corresponds to a slot antenna design providing an ac voltage, U_{ac}, between the source and gate. Fig. 1 b corresponds to bow tie antenna design inducing an ac current, i_{ac}, between the source and drain. In both cases, the applied constant gate-to-source bias, U_{gs}, determines the velocity of the plasma waves, $s = [(eU_{gs}-U_T)]/m)^{1/2}$ where m is the electron effective mass, e is the electronic charge, U_T is the threshold voltage, and U_d is the dc output voltage between the drain and source. (As we show below, U_d is proportional to the electromagnetic wave intensity.) As a function of frequency, U_d has a typical resonance response at the resonance frequency $\omega_o = \pi s/2L$ and its odd harmonics, where L is the channel length. The quality factor of the resonance is determined by $Q = s\tau/L$, where s is the plasma wave velocity, τ is the electron collision time with phonons and impurities. For comparison, the parameter that determines whether a FET behaves as a ballistic device is

$v\tau/L$ where v is the drift velocity in a FET. Since v may be much smaller than s, the criterion $v\tau/L \gg 1$ is much harder to meet that the condition of the high quality factor for plasma waves: $s\tau/L \gg 1$. In a High Electron Mobility Transistor (HEMT) structure, the momentum relaxation time at cryogenic temperature may be larger than 10 ps (for GaAs, $\tau \approx 12$ ps corresponds to the electron mobility of 300,000 cm^2/Vs). For $s = 10^8$ cm/s, $s\tau = 10$ micron which corresponds to the quality factor of 50 for $L = 0.5$ micron.

The viscosity of the electronic fluid may also contribute to damping. The kinematic viscosity is on the order of $\hbar/m_e \approx 18$ cm^2/s, for $m_e = 0.063\ m_o$. The corresponding quality factor $Q_v = sLm_e/\hbar$. For $s = 10^8$ cm/s and $L = 0.5$ micron, $Q_v \approx 270$.

2. Basic equations

As discussed in [1], the basic equations describing the two dimensional electronic fluid are the relationship between the surface carrier concentration and gate voltage swing, the equation of motion, and the continuity equation. The surface concentration, n_s in the FET channel is given by

$$n_s = CU/e \tag{1}$$

where C is the gate capacitance per unit area, $U = U_{gc}(x) - U_T$, $U_{gc}(x)$ is the local gate-to-channel voltage. Eq. (1) represents the usual gradual channel approximation [2] which is valid when the characteristic scale of the potential variation in the channel is much greater than the gate-to-channel separation, d.

The equation of motion (the Euler equation) is

$$\frac{\partial v}{\partial t} + v\frac{\partial v}{\partial x} + \frac{e}{m}\frac{\partial U}{\partial x} + \frac{v}{\tau} = 0 \tag{2}$$

where $\partial U/\partial x$ is the longitudinal electric field in the channel and $v(x, t)$ is the local electron velocity, and m is the electron effective mass. (Here, in contrast to [1], we account for electronic collisions with phonons and/or impurities by adding v/τ term.) Eq. (2) has to be solved together with the usual continuity equation which (taking Eq. (1) into account) can be written as :

$$\frac{\partial U}{\partial t} + \frac{\partial (Uv)}{\partial x} = 0 \tag{3}$$

For the equivalent circuit shown in Fig. 1, the boundary conditions for the FET channel (for the detector mode of operation) are

$$U = U_o + U_a \cos\omega t \text{ for } x = 0 \tag{4}$$

$$j = vU = 0 \text{ for } x = L \tag{5}$$

where $U_{ac} = U_a \cos(\omega t)$ is the external ac voltage induced across the gate-to-channel capacitance by the incoming electromagnetic wave and j is the electron current. We will search for the solution of eqs. (2) and (3) in the following form:

$$v = \bar{v} + v_1 \tag{6}$$

$$U = \overline{U} + U_1 \tag{7}$$

where $\overline{v}$, $\overline{U}$ and v_1, U_1 are the average and time-varying components of the electron velocity and channel potential, respectively. Averaging eqs. (1) and (2) with respect to time, we obtain two sets of equations for the time-independent and time-dependent components, respectively: The solution of these equations will be published elsewhere. [5] This solution shows that for the frequencies in the vicinity of the plasma frequency ω_o

$$\omega_o = \frac{\pi s}{2L} \tag{8}$$

and its odd harmonics, that is, for the frequencies such that $\omega - n\omega_o < \gamma = 1/\tau$, where $n = 1$, 3, 5, 7...

$$\frac{U_d}{U_o} = \left(\frac{QU_a}{U_o} \right)^2 \frac{\gamma^2/4}{(\omega - n\omega_o)^2 + \gamma^2/4} \tag{9}$$

where U_d is the dc voltage difference between the drain and source potentials. Hence, an electronic fluid in a FET behaves as a resonant detector of the electromagnetic radiation with the resonance quality factor $s\tau/L$. As we estimated above, this quality factor can easily exceed 200 for 0.1 micron AlGaAs/GaAs HEMTs.

In practical systems, mixing of weak incoming signal, $U_s \cos\omega_s t$ with a strong local oscillator signal, $U_{loc} \cos\omega_{loc} t$, is often more desirable because of a much higher sensitivity. Repeating our derivation for this case, we obtain the equation describing the electron fluid mixer response

$$\frac{U_d}{U_o} = \left(\frac{QU_sU_{loc}}{U_o} \right)^2 \frac{\gamma^2/4}{(\omega_s - n\omega_o)^2 + \gamma^2/4} \tag{10}$$

for the most interesting case when $\Delta\omega = \omega_s - \omega_{loc} \ll \gamma$.

3. Sensitivity estimates.

The amplitude of the ac field in vacuum is related to the electromagnetic field intensity, I, (measured in W/m^2) as follows

$$F_a = \left(\frac{2I}{c\varepsilon_o} \right)^{1/2} \tag{11}$$

where c is the speed of light in vacuum and $\varepsilon_o = 8.854 \times 10^{-12}$ F/m. The ac voltage, U_a, can be estimated as

$$U_a = F_a l_{eff} \approx \left(\frac{2I}{\varepsilon_o c} \right)^{1/2} \frac{\lambda}{4} \tag{12}$$

Here $l_{eff} \approx \lambda/4$ may vary somewhat depending on the antenna design. In the vicinity of the resonant frequency, eq. (24) becomes

$$U_d = \frac{e\tau^2 U_a^2}{L^2 m} = \frac{e\tau^2 I\lambda^2}{8\varepsilon_o c L^2 m} \tag{13}$$

The detector responsivity, $R = U_d/(SI)$ where

$$S = \frac{\lambda^2 G}{4\pi} \tag{14}$$

is the antenna aperture, G is the antenna gain (approximately 1.5 for a slot antenna). Hence, we finally obtain

$$R = \frac{U_d}{SI} = \frac{\pi e \tau^2}{2\varepsilon_o cmL^2} \tag{15}$$

For the device length of 0.1 μm, $\tau = 10$ ps, and $m = 0.063\ m_o$, where m_o is the free electron mass, we obtain $R \approx 1.1 \times 10^7$ V/W, which is far higher than typical values for Schottky diode detectors (typically on the order of 10^3 V/W). Thus, these HEMT detectors and mixers may achieve a far higher sensitivity than Schottky diodes. Similar advantages could achieved with the HEMT mixers. These higher values of sensitivity are caused by a the resonant nature of the proposed HEMT detectors and mixers.

Fig. 2 shows the resonant frequency versus the gate voltage swing for 0.1 μm, 0.25 μm, 0.5 μm, and 1 μm HEMTs. Fig. 3 shows the calculated responsivity as a function of frequency for 0.1 μm, 0.25 μm, and 0.5 μm HEMTs. These figures clearly show that this detector/mixer has the highest responsivity when the device length is small, and the device operates in a terahertz range.

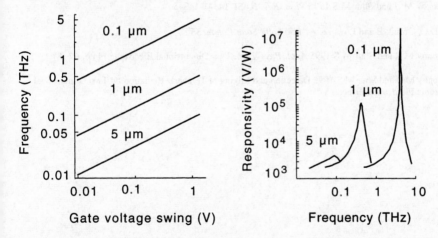

Fig. 2. Resonant frequency versus the gate voltage swing for 0.1 μm, 0.25 μm, 0.5 μm, and 1 μm HEMTs. Parameters used in the calculation: $\tau = 10$ ps, $m = 0.063\ m_o$.

Fig. 3. Responsivity as a function of frequency for 0.1 μm, 0.25 μm, and 0.5 μm HEMT detectors. Parameters used in the calculation: $\tau = 10$ ps, $m = 0.063\ m_o$.

4. Conclusions

We have shown that a HEMT has a resonance response to an incoming electromagnetic radiation at the odd harmonics of the plasma wave frequency of the two dimensional electrons in the device . The plasma frequency varies from approximately 10 GHz to 50 GHz for a 5 μm gate device, from 50 GHz to 500 GHz for a 1 micron device, and from 400 GHz to 5 THz for a 0.1 μm device. This response can be used for new types of detectors and mixers that use the propagation of plasma waves, which may propagate much faster than electrons. The responsivities of these new detectors are larger for shorter devices and may exceed 10^7 V/W for a 0.1 μm gate detector. This greatly exceeds the responsivities of Schottky diodes currently used as detectors and mixers in a terahertz range, which are typically on the order of 1000 V/W.

5. Acknowledgment

The authors are grateful to Professor Robert Weikle for useful discussions and comments. The work at the Ioffe Institute has been partially supported by the US Army through its European Research Office. The work at the University of Virginia has been partially supported by the US Army Research Office (Project Monitor Dr. John Zavada) and by the Office of the Naval Research (Project Monitor Dr. Yoon Soo Park).

References

[1] Dyakonov M I and Shur M S 1993 *Phys. Rev. Lett.* **71** 2465--2468

[2] Dyakonov M I and Shur M S 1995 *Phys. Rev. B.* **51** 14341-14345

[3] Tsui D C, Gornik E and Logan R A 1980 *Solid State Comm.* **35** 875

[4] Dyakonov M I and Shur M S 1995 *Appl. Phys. Lett.*, Two Dimensional Electronic Flute, **67** (8)

[5] Dyakonov M I and Shur M S 1995 Detection and Mixing of Terahertz Radiation by Two Dimensional Electronic Fluid, unpublished

Inst. Phys. Conf. Ser. No 145: Chapter 5
Paper presented at 22nd Int. Symp. Compound Semiconductors, Cheju Island, Korea, 28 August–2 September 1995

Photoelectric measurements of interband transitions in fully fabricated pseudomorphic high electron mobility transistors

F Schuermeyer, J P Loehr, R E Sherriff, C Cerny and M Shur*

Wright Laboratory, Wright Patterson AFB, OH 45433, USA
*University of Virginia, Charlottesville, VA 22903, USA

Abstract. We have performed photoelectric emission and conduction (PEC) studies on fully fabricated $Al_{0.24}Ga_{0.76}As/In_{0.22}Ga_{0.78}As$ pseudomorphic HEMTs and have observed interband transitions between quantum-confined states in the channel at room temperature. The separation between transition energies and the bias dependence of the photoconductivity spectra agree well with a simple model that assumes proportionality between the photocurrent and the interband absorption.

1. Introduction

Optical techniques such as photoluminescence and photoreflectance (Litwin-Staszewska 1994, Svenson 1994, Brierley 1993, Dodabalapur 1989, Yin 1992, Kanata 1992) have been commonly used to measure the energy band profile of modulation-doped quantum well structures and pseudomorphic high electron mobility transistors (PHEMTs). These techniques provide information on the epitaxial layers *before* device processing and sample over areas that are typically much larger than the final device dimensions. Unfortunately, because of processing effects and applied voltages, the band profile of the final *fabricated* device can be dramatically different from that of the virgin epitaxial layer structure. For example, the Schottky gate metal pins the Fermi level at the surface, altering the internal field distribution in the structure. Knowledge of the band profile in fabricated transistors is critical for investigating the effects of the epitaxial layer structure on device performance. Photoelectric techniques such as photoemission and conduction (PEC) sense electrical currents within a device resulting from optical excitation. Because of the localization of the currents and the precision with which they can be measured, these techniques are well-suited to obtain barrier heights and bandgap information on small devices. Previous photoelectric device studies used front side illumination (Spitzer 1964, Okumura 1983, Burrus 1979, Lee 1981, Gilpérez 1992) and required thinned metal gates, entailing extra processing and making the characterization technique destructive. Attempts to use backside illumination were limited by above-bandgap absorption in the substrate. Fortunately, the interband energies in the PHEMT channel are significantly smaller than the substrate bandgap.

Consequently, light incident on the backside can reach the channel, permitting photoelectric studies on fully fabricated devices (Schuermeyer 1995). The ability to measure both optical properties and microwave performance on the same device provides a unique opportunity to connect variations in transistor behavior to changes in material structure and quality.

In this paper we describe the PEC measurement technique and report photoconduction (PC) measurements on fully fabricated PHEMTs with 1 μm gates grown on GaAs substrates. The qualitative shape and bias dependence of the observed PC spectra can be reproduced by a simple model. For samples where all dopants lie between the gate and the well, the experiments and calculations suggest that the $\Delta n = 0$ selection rules for square-well interband transitions are strong for gate biases near pinch-off. Therefore, by comparison with a model the interband transition energies can be identified. At lower biases, where there is significant charge in the well, the selection rules are weak. For samples with dopants both above and beneath the well, however, the selection rules are not strong at any bias.

2. HEMT fabrication

The heterostructure material was grown by MBE. The devices were fabricated on two wafers with distinct layer structures. Wafer 1 nominally consists of the following layers: 35 nm GaAs with 5×10^{18} cm^{-3} Si doping, 3 nm AlAs etch stop, 20 nm Al$_{0.24}$Ga$_{0.76}$As barrier with 5×10^{17} cm^{-3} Si doping, 4×10^{12} cm^{-2} delta doping, 4 nm Al$_{0.24}$Ga$_{0.76}$As spacer, 12.5 nm In$_{0.22}$Ga$_{0.78}$As channel, 489 nm GaAs buffer, and an 11 period superlattice. We refer to Wafer 2 as an underdoped structure since it is nominally identical to Wafer 1 except for an additional 10 nm 1×10^{18} cm^{-3} Si-doped GaAs layer located 3 nm below the well.

The PHEMTs were isolated on the wafer by masking the active areas using standard photolithographic techniques and by forming mesas via wet etching 170 nm down to the undoped GaAs buffer region using HF:H$_2$O$_2$:H$_2$O at (1:1:8). Drain and source ohmic contacts were formed using the metal lift-off technique with standard AuGe/Ni metal evaporation, and were alloyed at 450 °C for 30 seconds using rapid thermal annealing. The ohmic contacts were electrically tested, yielding a contact resistance of 0.060 Ω-mm, a specific contact resistance of 1.73×10^{-7} Ω/cm^2, and a sheet resistance of 191 Ω/square. Next, the gate contact was formed using optical lithography to define the gate region, and the top GaAs ohmic contact layer was removed before gate metallization using a citric acid:H$_2$O$_2$ selective etch (DeSalvo 1992) which stops at the AlAs etch stop layer. Then the AlAs etch stop layer was removed prior to metallization using HCl:H$_2$O (1:10). The gate contact was formed by metal lift-off of evaporated 20 nm Ti/580 nm Au metal, giving 1 μm gate length PHEMT devices with 5 μm source/drain spacing. A final metallization of 20 nm Ti/980 nm Au was then deposited to connect the source and drain to probe pads. Three device configurations were tested. Devices A and B are 2-finger FETs and device C is a 6-finger FET, all with 100 μm unit gate width. Devices B and C have ohmic metal only on the mesa top with final metal connecting the ohmic metal to probe pads. Device A has probe pads and interconnecting metal defined in the ohmic layer to allow testing of this device during gate recess and immediately following gate deposition.

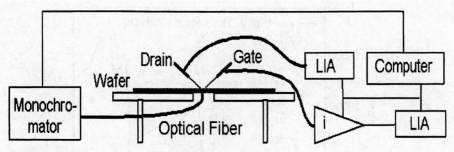

Figure 1 Experimental setup for PEC measurements

3. Experimental techniques

Figure 1 presents a sketch of the experimental setup. The PEC measurements are performed on-wafer at room temperature using either Cascade or shielded Pico probes. The wafer chuck contains a hole in its center. A micromanipulator positions one end of an optical fiber directly underneath a device, and the other end of the optical fiber is connected to the exit slit of a SpectraPro-150 (Acton Research Corporation) monochromator. A 50 W tungsten light source is employed and the light is mechanically chopped. Stanford Research SR530 lockin amplifiers are employed to measure the photocurrents at the drain and gate electrodes. A low noise current preamplifier SR570 is used to amplify the gate photocurrent and to provide the gate voltage.

4. Theory

We have successfully modeled the PC spectra for Wafer 1 by assuming that the photocurrent is proportional to the interband optical absorption coefficient of the channel. The absorption coefficient calculation is performed in three steps: First, the electron charge distribution and band profile are obtained for a given gate bias using a self-consistent Schrödinger-Poisson solution. In the region surrounding the delta-doping and the InGaAs channel, the electron effective mass equation is solved by finite-difference methods and the electron density is obtained from the resulting wave functions; in the remainder of the device the electron density is obtained semi-classically. We calibrate the model for Wafer 1 by adjusting the Schottky barrier on the AlGaAs barrier layer until the calculated pinch-off voltage agrees with experiment. (The underdoping in Wafer 2 created instabilities in the self-consistent Poisson-Schrödinger solution. Since the model would not converge for large voltages it was not possible to model Wafer 2.) Next, using the self-consistent band profile, valence subband energies and wavefunctions are obtained by solution of the 4x4 strain-dependent effective mass equation. Finally, the absorption coefficient for normally incident light is calculated from wave function overlaps with state occupation determined by the self-consistent Fermi level.

In flat (square) quantum wells, the interband absorption coefficient reflects the canonical $\Delta n=0$ selection rules and sharp steps appear at the corresponding energy separations E1-HH1, E2-HH2, etc. Here En represents the n'th electron subband, and HHm

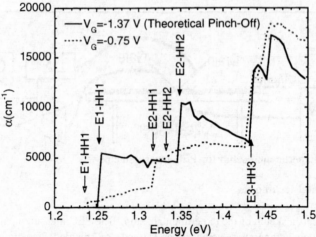

Figure 2   Calculated absorption spectra for interband transitions for pinch-off (solid curve, solid arrows) and "on" (dotted curve, open arrows) conditions

represents the m'th heavy-hole subband. If the band-edge profile of the well is sharply bent or tilted − as when there is significant charge in the well or a large electric field across the well − then the selection "rules" are not as strong, and each hole subband couples strongly to all electron subbands. In this case the absorption curve consists of many small, closely-spaced steps, and the sharp $\Delta n=0$ transitions do not predominate.

In Figure 2 we show the calculated absorption coefficient for the nominal structure of Wafer 1 at room temperature. The solid curve is calculated for the theoretical pinch-off voltage, where the quantum well is essentially devoid of charge. Since Wafer 1 has no underdoping, there is no ancillary space-charge field across the channel. Therefore, the pinched-off well is essentially flat, the selection rules are strong, and the absorption spectrum exhibits a few large, sharp steps only at the $\Delta n=0$ transition energies. The dotted curve, by contrast, is calculated at 0.62 volts above pinch-off. At this gate voltage there *exists* substantial channel charge and there is substantial band-bending in the well. Hence the selection rules are weak, and smaller, more numerous steps are observed corresponding to interband transitions that do not conserve subband number. In addition, the low energy absorption is substantially reduced because the first electron subband lies below the Fermi level.

5. Experimental results

In Figures 3-5 we show several examples of measured photoconduction (PC) spectra for devices A, B and C; we also present differentiated PC spectra to more clearly define the interband transition energies. The photocurrents are all measured with a source-drain voltage of 20 mV and are normalized for constant photon flux. Many devices were tested and the data are reproducible.

In Figure 3, we present PC data for device A fabricated on Wafer 1 at two different

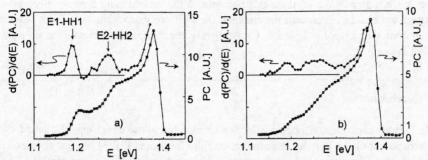

Figure 3 Experimental photoconduction and derivative curves for Wafer 1, Device A, at (a) pinch-off (Vg=-1.5 V) and (b) "on" (Vg=-1 V) conditions

gate biases. Figure 3(a) shows data at pinch-off (-1.5 V for this device), and the spectrum shows sharp steps. The corresponding peaks in the differentiated PC spectrum at 1.19 and 1.28 eV are attributed to the E1-HH1 and E2-HH2 transitions, respectively, since the shape agrees with the model curve in Figure 2. The peak separation also agrees well with the model but the absolute values are lower; this rigid shift may result because the layer thicknesses or compositions differ from the nominal values. The sharp rise in the PC spectrum near 1.35eV may mark the onset of the third step seen in the calculated curve of Figure 2, but a transition energy cannot be clearly estimated because of the nearby GaAs substrate absorption. The Figure 3(b) data was taken at a gate bias of -1 V where the channel contains a significant amount of charge. Here, by contrast, the sharp steps are washed out and the PC spectrum is relatively featureless. This is consistent with the weakening of selection rules discussed in Section 4. Figure 4 presents a PC spectrum for device B immediately adjacent to device A. This data, taken at pinch-off, shows two sharp steps similar to the device A spectrum in Figure 3(a). However, the rise in PC at photon energies above 1.35 eV is missing. The reason for this pronounced difference is not known. Recall that devices A and B differ only in the gate electrode layout and metallization; the epitaxial layer structures are identical.

Figure 5 shows a PC spectrum for device C fabricated on Wafer 2 which contains underdoping. The observed PC spectra, even when measured at pinch-off as shown here,

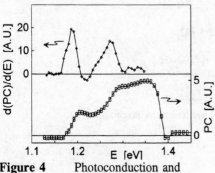

Figure 4 Photoconduction and derivative of Device B on Wafer 1, at Vg=-1.5 V (pinch-off)

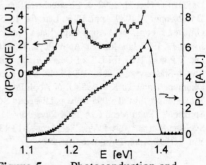

Figure 5 Photoconduction and derivative of Device C on Wafer 2 (underdoping), at Vg=-2.3 V (pinch-off)

lack distinct steps. We attribute this behavior to the underdoping donors. Because of the extra fixed charge underneath the channel, the well profile exhibits significant band tilting even when the channel is depleted. Consequently, the interband selection rules are weakened at all biases.

6. Conclusions

We present experimental photoconduction results that allow direct observation of channel interband transitions at room temperature on fully fabricated PHEMTs. For devices *without* underdoping, the photoconductivity spectra depend dramatically on gate bias. At pinch-off the spectra show strong, sharp steps, while at more positive biases the sharp structure is washed out. This behavior can be qualitatively reproduced with a simple model that assumes proportionality between the photoconductivity and the interband absorption. The bias dependence predicted by the model results from changes in the strength of the $\Delta n=0$ selection rule. Devices *with* underdoping, however, show broad, relatively featureless PC spectra at all biases because the fixed charge under the well tilts the bandedge profile even at pinch-off.

Acknowledgement

The authors would like to express their appreciation to Robert Bacon for setting up the test station and performing measurements, to R Dettmer, Greg Desalvo, Jack Ebel, Dave Via, Tony Quach, Charles Havasy, Chris Bozada, and Ken Nakano for help in device processing and to Keith Evans and Jim Ehret for wafer growth. The high quality of the MBE grown structures and the meticulous device processing that preserved this quality were critical to the success of our experiments.

References

Brierley S K 1993 J. Appl. Phys. 74, 2766 and 1994 J. Appl. Phys. 76, 1343
Burrus C A et al. 1979 Elect. Lett., Vol. 15, pp. 655-657
DeSalvo G C et al. 1992 J. Electrochem. Soc., vol. 139, pp. 831-835
Dodabalapur A 1989 Appl. Phys. Lett, 54, 1675
Gilpérez J. et al. 1992 Appl. Phys. Lett., Vol. 61, pp. 1225-1227
Kanata T 1992 Jap. Journal. Appl. Phys. 31, 756
Lee T P et al.1981 IEEE J. of Quant. Elect. Vol. QE-17, pp. 232-238
Litwin-Staszewska E et al. 1994 Solid State Elect., Vol. 37, pp. 665-667
Okumura T and Tu K N 1983 J. of Appl. Phys., Vol. 54., p. 922
Schuermeyer F et al. 1995 to be published in Solid State Electronics
Spitzer W G and Mead C A 1964 Phys. Rev., 133, A872 and references therein
Svenson S et al. 1994 J. Vac. Sci. Tech. B12 134 and 1086
Yin Y 1992 Appl. Phys. Lett. 61, 1579

Inst. Phys. Conf. Ser. No 145: Chapter 5

Paper presented at 22nd Int. Symp. Compound Semiconductors, Cheju Island, Korea, 28 August–2 September 1995

Metamorphic In(GaAl)As-HEMTs on GaAs-Substrates

S. Kraus, M. Chertouk, H. Heiß, D. Xu, M. Sexl, G. Böhm, G. Tränkle and
G. Weimann

Walter Schottky Institut, Technische Universität München, D-85478 Garching, Germany

Abstract: Metamorphic InGaAs/InAlAs-HEMTs were grown by Molecular
Beam Epitaxy on GaAs substrates using linearly graded $In_xAl_{1-x}As$ buffers to
accomodate lattice misfit. A compositional variation in x of 0.5 per μm in the
buffer layer resulted in practical buffer thicknesses of one μm or less, with moderate cross hatching of 5 nm height for x = 0.32 and 10 nm for x = 0.52. This surface roughness had only little influence on the transport properties of the
metamorphic heterostructures in comparison to lattice matched structures on InP-
substrates. DC- and RF- characteristics of 0.13 μm-HEMTs with an In-content in
the channel of 0.53 were excellent, with maximum drain currents of
800 mA/mm, extrinsic transconductances of 720 mS/mm and cut-off frequencies
of 185 GHz for f_T and 215 GHz for f_{max}, respectively. Metamorphic HEMTs
with a channel of 12 nm $In_{0.53}Ga_{0.47}As$ on top of a 20 nm $In_{0.32}Ga_{0.68}As$ subchannel showed dramatically reduced output conductances of 20 mS/mm, thus
giving very high f_{max}-values of 350 GHz, while still retaining convincing
transit frequencies of 160 GHz.

1. Introduction

Sub-μm HEMTs fabricated from modulation-doped InGaAs/InAlAs-heterostructures grown
lattice matched or strained on InP substrates are well suited for high speed and low noise
applications [1,2]. A successful combination of the superior transport properties of these heterostructures with the advantages of mature, less costly, large area GaAs substrates will give
significant advantages in the application of InGaAs/InAlAs-HEMTs. Metamorphic growth
with molecular beam epitaxy (MBE), on the other hand, allows the arbitrary combination of
III-V semiconductor layers with different lattice constants. A relaxed buffer layer is used to
generate a „new substrate" with a modified lattice constant and moderate defect density [3].
In this paper we compare the layer and device properties of metamorphic InGaAs/InAlAs-
HEMTs (MM-HEMTs) on 3" GaAs substrates to lattice matched ones (LM-HEMTs) on 2"
InP substrates grown by MBE under identical conditions. The lattice misfit in the metamorphic structures was accomodated by linearly graded $In_xAl_{1-x}As$ buffers, keeping the band gap
wider than in the InAlAs supply layers in any case. With industrial applications in mind, the
thickness of the buffer layer was kept at one μm or less. The arbitrary choice of material composition and lattice constant in the metamorphic layers was used as an additional degree of
freedom to tailor the band structure of composite channel HEMTs with reduced output conductances and improved f_{max}-values.

2. MBE growth and transport properties of HEMT layers

The layer sequence of LM- and MM-HEMTs is shown in Figure 1: a 32 nm InGaAs channel, a 12.5 nm InAlAs supply layer Si-doped to 10^{19} cm^{-3}, a 10 nm undoped Schottky barrier layer and a 15 nm InGaAs cap layer were used on a 250 nm InAlAs buffer layer, which was sandwiched between two InGaAs/InAlAs-superlattices. These LM- and MM-HEMTs were grown identically at substrate temperatures of 530 °C and 470 °C, respectively, and at constant As$_4$ beam equivalent pressures of $2 \cdot 10^{-5}$ torr. The individual layers of the structures were mutually lattice matched at In-contents of 0.52 and 0.32. For metamorphic growth on GaAs an additional graded In$_x$Al$_{1-x}$As layer was grown between the substrate and the HEMT structure increasing the In-content linearly to 0.32 or 0.52.

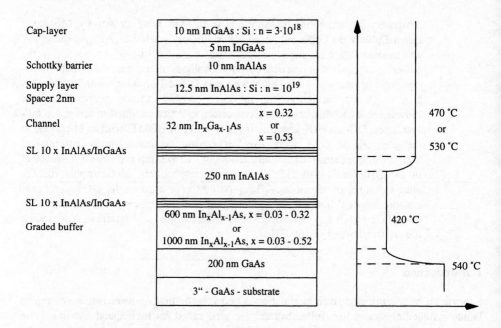

Fig. 1: Layer sequence and growth temperature profile of metamorphic HEMTs

The variation of the substrate temperature in the growth of the graded InAlAs layer yielded lowest defect densities at 420 °C and moderate surface roughness of 5 nm for x = 0.32 and 10 nm for x =0.52, respectively, with lateral cross hatch spacings of ~ 3 µm as shown in Figure 2. Increasing the growth temperature to 530 °C gave a more pronounced cross hatching, while lower temperatures of 350 °C resulted in rough, three-dimensionally grown surfaces. Layer thicknesses of 0.6 µm and 1 µm were optimal for the graded layers at maximum In-contents of 0.32 and 0.52, giving a gradient in In-content of 0.5 per µm. Structures with these low-graded buffers grown at 420 °C showed best morphology, high mobility and high photoluminescence yield. Steeper gradings resulted in a degradation of the transport properties of the layers. The introduction of an additional InAlAs layer of constant composition between the graded layer and the HEMT structure resulted in a pronounced roughening of the surface with no increase in mobility. The optimized buffer structure provided almost complete relaxation of about 90 %, determined by X-ray diffraction. The cross hatching on the surface of MM-HEMTs had only little influence on their electrical transport properties.

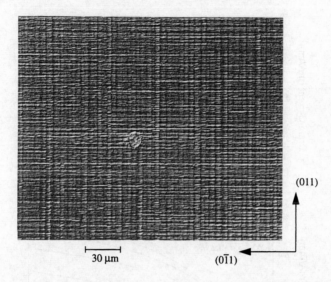

(011)

30 μm   (0Ī1)

Fig. 2: Surface morphology of a metamorphic InGaAs/InAlAs-HEMT with In-content of 0.52

Very high carrier densities and good mobilities were found. Table 1 compares the carrier densities and Hall mobilities measured at room temperatures on samples grown at 530 °C.

In-content x in MM-HEMTs	0.32	0.52	LM-HEMT
Carrier density[10^{12} cm^{-2}]	4.2	5.3	4.7
Hall mobility[cm^2/Vs]	6600	7600	8400

Tab. 1: Room temperature carrier densities and Hall mobilities of MM- and LM-InGaAs/InAlAs-HEMTs grown at 530 °C

3. Metamorphic 0.13 μm-HEMTs : DC- and RF-properties

We fabricated HEMTs from our lattice matched and metamorphic heterostructures. Electron beam lithography was used to define 0.13 μm gates in a two layer resist technology. Gate recess was done by selective wet chemical etching based on succinic acid. Nearly the same source- and drain- resistances, lower than 0.4 Ωmm, were measured for both structures. Inspite of the surface roughness similar yield was obtained for MM-HEMTs and LM-HEMTs, with an increase in threshold voltage variation from $\Delta V_{th} \sim \pm 50$ mV to $\sim \pm 100$ mV, however.

In Figure 4 the I-V-characteristics of LM-HEMTs with gate widths of 150 μm are compared to identical MM-HEMTs with an In-content of 0.52. The layers were grown under identical conditions at 530 °C. The DC- and RF-properties of the HEMTs were similar, although the maximum drain current I_{DSmax} of about 750 mA/mm increased by about 5 % in the MM-HEMTs due to the larger carrier concentration in the channel. The extrinsic transconductance g_m was 750 mS/mm for LM-HEMTs; it decreased slightly to 720 mS/mm for MM-HEMTs. Transit frequencies f_T were 200 GHz in LM-devices, while MM-transistors showed 185 GHz. Thus, DC- and RF-properties of MM-HEMTs are definitely comparable to those of lattice matched devices

800

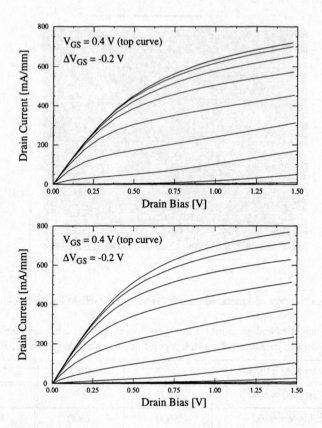

Fig. 3: IV-characteristics of 0.13 μm LM- (top) and MM-In$_{0.53}$Ga$_{0.47}$As/In$_{0.52}$Al$_{0.48}$As-HEMTs (bottom)

Both LM- and MM-HEMTs with an In-content of 0.52 revealed high output conductances g$_0$ with 150 mS/mm and 165 mS/mm for LM-HEMTs and MM-HEMTs, at V$_{DS}$ = 1.5 V. These high output conductances at higher V$_{DS}$ were due to weak impact ionization in the low band gap InGaAs channel (E$_g$ ~ 0.76 eV) induced in the high field region at the drain-sided end of the sub-μm gates [4]. Due to the poor g$_m$/g$_0$-ratio between 4 to 5 reduced f$_{max}$-values were thus restricted to 220 GHz and 185 GHz for LM-HEMTs and MM-HEMTs, respectively.
On the other hand, in MM-HEMTs with a lower In-content of 0.32, the band gap in the InGaAs channel is 1 eV and therefore impact ionization was negligible. Very low output conductances g$_0$ of 10 mS/mm were thus obtained, even for gate lengths of 0.13 μm. Consequently, very high f$_{max}$-values of 310 GHz have been found with corresponding transit frequencies f$_T$ of 105 GHz.

4. Composite channel HEMTs

Based on these results we have fabricated novel composite channel MM-HEMTs , combining the superior transport properties of a 12 nm strained In$_{0.52}$Ga$_{0.48}$As channel with the low output conductance of a 20 nm In$_{0.32}$Ga$_{0.68}$As subchannel. The top layer of high In-content is elastically strained, as the lattice constant of the relaxed metamorphic structure was adjusted to an In-content of 0.32. This resulted in a higher Al-content of the doped InAlAs supply layer under the gate, giving a higher Schottky barrier and high off-state breakdown voltages around 7 V.

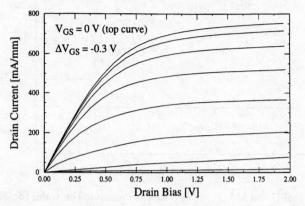

Fig. 4: IV-characteristics of 0.13 μm InGaAs/InAlAs-HEMT with composite channel

In Figure 5 the I-V-characteristics of 0.13 μm HEMTs with composite channel are shown. Maximum drain currents of 750 mA/mm were measured and an extrinsic transconductance g_m of 600 mS/mm at a drain current I_{DS} = 320 mS/mm was achieved. The output conductance g_0 was 20 mS/mm, confirming good carrier confinement and negligible impact ionization [4]. The electrons near the source flow in the narrow gap region, thus having high low field mobility and high carrier density; in the high field region on the drain side of the gate the electrons flow in the wide band gap subchannel unaffected by impact ionization (Fig. 5).

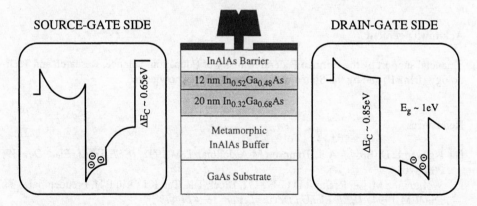

Fig. 5: Schematic band structure on source- and drain side of the gate of composite channel HEMTs

S-parameters, measured on wafer up to 70 GHz, revealed the superior RF-properties of the composite channel 0.13 μm-MM-HEMTs. Convincing f_T-values of 160 GHz and f_{max}-values of 350 GHz were measured for devices with gate widths of 150 μm, biased for maximum extrinsic transconductance (V_{DS} = 1.5 V, V_{GS} = -1.25 V and I_{DS} = 330 mA/mm). The f_{max}-values were extrapolated with a roll-off of -20dB/decade from the measured maximum available gain (MAG) coincidencing with a stability factor k = 1, as indicated in Figure 6.

802

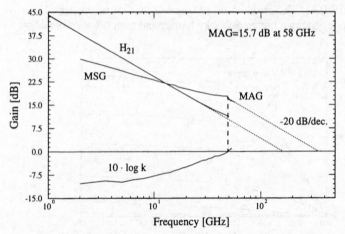

Fig. 6: Power gain (MSG and MAG) versus frequency measured on wafer for 0.13 μm meta-morphic InGaAs/InAlAs-HEMT with composite channel

5. Summary

Metamorphic 0.13 μm InGaAs/InAlAs-HEMTs have been successfully fabricated on 3"-GaAs-substrates with DC- and RF-properties comparable to those of transistors lattice matched on InP.
In composite channel MM-HEMTs very low output conductances were obtained, thus giving very high f_{max}-values of 350 GHz, while still retaining transit frequencies of 160 GHz.

Acknowledgement

Financial support by the German Federal Ministry of Education, Science, Research and Technology (BMBF) and by the Siemens AG is gratefully acknowledged.

References

[1] Nguyen L D, Brown A S, Tompson M A, Jelloian L M 1992, *IEEE Trans. Elec. Dev.* **39**, 2007-2014

[2] Woijtowicz M, Lai R, Streit D C, Ng G I, Block T R, Tan K L, Liu P H, Freudenthal A K, Dia R M 1994, *IEEE Eletron Device Letters* **15**, 477-479

[3] Harmand J C, Matsuno T, Inoue K 1989, *Japan. J. Appl. Phys., Part 2 (Letters)* **28**, L1101-1103

[4] Chertouk M, Heiß H, Xu D, Kraus S, Klein W, Böhm G, Tränkle G, Weimann G 1995, Proc. of the 7th Int. Conf. InP and Related Materials 1995, Sapporo, 737-740

Inst. Phys. Conf. Ser. No 145: Chapter 5
Paper presented at 22nd Int. Symp. Compound Semiconductors, Cheju Island, Korea, 28 August–2 September 1995
© *1996 IOP Publishing Ltd*

Device Structure and Substrate Effects on Drain Lag Phenomena in HJFETs

Masanobu NOGOME, Kazuaki KUNIHIRO and Yasuo OHNO

Microelectronics Research Laboratories, NEC Corporation, 34, Miyukigaoka, Tsukuba, Ibaraki, 305 Japan

Abstract. Rate-limiting processes in drain lag phenomena in HJFETs (Heterojunction Field Effect Transistors) on hole-trap substrates are investigated by using numerical simulation with SRH statistics for deep-level traps. The slow drain-current responses for HJFETs on floating substrates are determined by electron capture and emission even for hole-trap substrates, since total hole concentration is kept constant. In addition, time constants are much smaller than those estimated from SRH statistics since charge variation is enhanced by internal hole movement. On the other hand, drain-lag time constants for HJFETs with substrate electrodes are determined by the hole current travelling through the substrate from the electrode to the trap. These results indicate that special care should be taken in determining the trap parameters causing drain lag in FETs from time constant measurements.

1. Introduction

The drain lag phenomenon is a well-known parasitic effect observed in GaAs FETs. This effect poses serious problems in analog and digital circuit applications. This phenomenon is attributed to the slow response of deep-level traps in semi-insulating (S. I.) substrates. Deep traps in S. I. substrates respond to drain voltage (V_D) variation by capturing or emitting holes or electrons. This phenomenon strongly depends on the type of substrate such as electron-trap and hole-trap substrates.

Recently electron-trap substrates were analyzed by numerical simulation (Kunihiro K and Ohno Y 1994). The results showed that the time constants of drain-current transients for V_D rising steps can be dramatically reduced according to the amplitude of V_D. On the other hand, the time constants for V_D falling steps are almost constant regardless of the amplitude of V_D. Also, the rate-limiting process was shown to be electron capture for V_D rising steps and electron emission for V_D falling steps.

In this work we investigated the drain lag phenomena in HJFETs on hole-trap substrates using numerical simulations. Two cases, with the substrate floating and the substrate grounded through the substrate electrode, were investigated. The actual devices had the substrate electrodes under isolation layers or substrates. Consequently, electrodes could be the indirect source of hole injection or outflow. The effects of trap type difference and the substrate electrode were clarified by these simulations.

Table 1. Semi-insulating substrate trap parameters.

density (cm^{-3})	charge type	$E_C - E_T$ (eV)	σ_n (cm^2)	σ_p (cm^2)
1×10^{16}	donor	0.82	1×10^{-15}	1×10^{-13}
2×10^{15}	acceptor		(shallow level)	

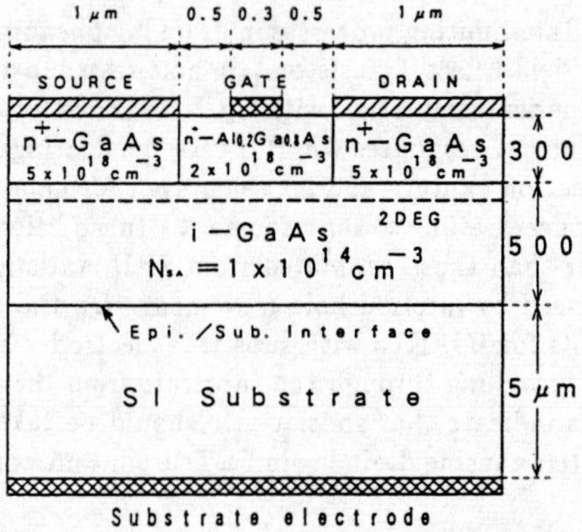

Figure 1. The device structure for simulation with substrate electrode. For floating substrates, the substrate electrode is omitted.

2. Device structure and trap parameters for simulation

For numerical simulations, a two-dimensional device simulator BIUNAP-CT was used (Yano H *et al*. 1989). The deep-level model used in the simulator follows SRH (Shockley-Read-Hall) statistics. Details appear in the ref. (Yano H *et al*. 1989).

The device structure used in simulation is an AlGaAs/GaAs HJFET with a 0.3 μm gate on the semi-insulating substrate as shown in Figure 1. The substrate parameters are shown in Table 1. The trap is a hole trap since the capture cross section for holes is large and trap energy is relatively low. A p-type electrode is placed on the bottom of the substrate for simulation with the substrate electrode.

3. Transients with floating substrate

3.1. Drain current transients and rate-limiting process

Figure 2 shows simulated drain current transients where the horizontal axis is a logarithmic scale of time. The drain voltage is changed for a duration of 10 psec with an amplitude of 0.4, 1.2 and 2.0 V with a center voltage of 2 V, while the source and gate voltage is fixed at 0 V. When the drain voltage is changed, the channel charges respond immediately, reaching a quasi-steady state. After a while the deep traps in the substrate begin to respond and modulate the drain current. Finally, the drain current reaches a steady state. For V_D rising steps, the time constant varies from 10^{-2} sec to 10^{-3} sec as amplitude increases. On the other hand, for V_D falling steps the time constants are always around 10^2 sec regardless of amplitude. This situation is similar to the case of electron-trap substrates (Kunihiro K and Ohno Y 1994).

To investigate the rate-limiting process, the electron capture cross section is varied while that for the hole is constant. Figure 3 shows simulated drain current transients for different electron capture cross sections for V_D falling steps. Since the time constants are different, it is evident that electron emission is the rate-limiting process for V_D falling steps. From similar calculations, electron capture is the rate-limiting process for V_D rising steps. However, the time constants are much smaller than $1/nC_n$ and $1/n_1C_n$, which are the capture and emission time constants, respectively, for the traps estimated from SRH statistics (Table 2.).

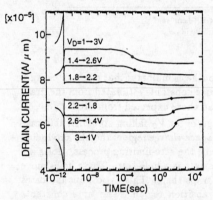

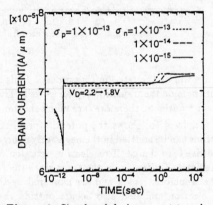

Figure 2. Simulated drain current transients for floating substrates. Upper 3 lines are V_D rising steps and lower 3 lines are V_D falling steps. Dots are the time constants of current variation which is defined as the time when current variation is $1/e$ of total variation.

Figure 3. Simulated drain current transients for various electron capture cross sections. V_D is varied from 2.2 V to 1.8 V.

3.2. Mechanism of rate-enhancing process

Figure 4 (a) shows a potential profile diagram explaining time constant enhancement phenomena for V_D rising steps. The transistor is divided into three regions. These are the space-charge region, transition region and charge neutral region. The boundary between the negatively charged space-charge region and the charge neutral region moves toward the bottom of the substrate. We call this region the "transition region".

Holes are easily emitted from the transition region since the substrate is a hole-trap. When there is no electron, the holes will eventually be captured again by the traps, resulting in no charge variation. However, when there are electrons injected from the channel, the most of the emitted holes recombine with the electrons. This electron

Table 2. Comparison of time constants. Those estimated from SRH statistics with carrier concentration values and those of simulated drain currents.

	V_D rising steps	
	carrier density (cm^{-3})	time constant (sec)
electron capture	$n = 10^{10} \sim 6 \times 10^{7}*$	$1/nC_n = 8.3 \times 10^{-3} \sim 1.4 \times 10^{0}$
hole emission	$p_1 = 3.0 \times 10^{8}$	$1/p_1 C_p = 3.7 \times 10^{-3}$
drain current	——	$10^{-3} \sim 10^{-2}$

	V_D falling steps	
	carrier density (cm^{-3})	time constant (sec)
electron emission	$n_1 = 2.4 \times 10^{3}$	$1/n_1 C_n = 3.4 \times 10^{4}$
hole capture	$p = 8.0 \times 10^{7}*$	$1/pC_p = 1.4 \times 10^{-2}$
drain current	——	10^{2}

$$C_n = 1.2 \times 10^{-8}, C_p = 9.1 \times 10^{-7} \text{cm}^3 \text{sec}^{-1}$$
* estimated from simulation

capture will be the rate-limiting process. This hypothesis indicates that electron capture occurs outside the transition region as well. The time constants estimated from the drain current transients, therefore, are much shorter than those expected from SRH statistics.

Figure 4 (b) shows the potential profile diagram for V_D falling steps. In this case electrons can be emitted in the negatively charged space-charge region, thereby producing a positive trap charge. This electron emission will be the rate-limiting process. Holes are immediately emitted from the traps since the substrate is a hole-trap. These holes move to the transition region and are captured by the traps there. This hypothesis indicates that electron emission also occurs outside the transition region. The time constants estimated from drain current transients are, again, much shorter than those expected from SRH statistics.

4. Transients with electrode placed on bottom of substrate

Figure 5 shows simulated drain current transients when the substrate electrode is placed on the bottom. The time constants are about 10^{-3} sec regardless of amplitude or the step direction of V_D variation. When the hole capture cross section is varied while that for the electron is constant, the curves do not change. So, it can be concluded the rate-limiting process is not hole emission or hole capture.

Figure 6 shows simulated drain current transients for V_D falling steps for substrates with different substrate resistances. Substrate resistance is varied by changing hole density in the substrate. Since time constants of the transients vary, it was found that the rate-limiting process is the hole current travelling through the substrate from and to the substrate electrode. The time constant is almost the same as that estimated from the capacitance between the channel and the trap region, and substrate resistance.

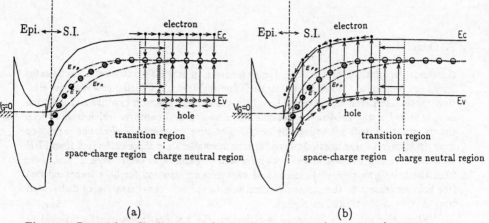

(a)   (b)

Figure 4. Potential profile diagram explaining time constant enhancement phenomena. (a) V_D rising and (b) V_D falling.

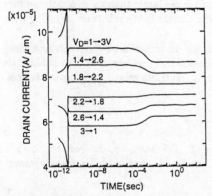

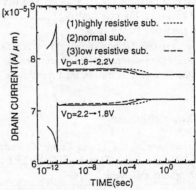

Figure 5. Simulated drain current transients for substrate with substrate electrode. Upper 3 lines are V_D rising steps and lower 3 lines are V_D falling steps.

Figure 6. Simulated drain current transients for different substrate resistances. V_D is varied between 1.8 V and 2.2 V. Hole density in 3 μm thick region of the 5 μm substrate is varied, such as 10^7 cm^{-3} (highly resistive sub.), 10^8 cm^{-3} (normal sub.) and 10^{15} cm^{-3} (low resistive sub.).

In the case of hole-trap substrates, time constants of hole capture and emission are much smaller than those of electron capture and emission, which are competing processes in trap charging and discharging. Consequently, hole capture and emission dominate.

808

However, when the time constant of hole current travelling through the substrate from the external electrode to the trap region is larger than those for hole capture and emission, the former will dominate the process. Thus, it was found that not only the capture and emission in the trap but also carrier movement from the electrode to the trap can be the rate-limiting process for drain current transients.

5. Conclusion

Numerical simulations of the drain lag phenomena in HJFETs on hole-trap substrates have been performed. When the substrate is floating, the rate-limiting processes are electron capture and emission even for hole-trap substrates. Electron capture and emission dominate in floating substrates, because hole emission and capture, which lead to trap charge variation, will not eventually occur. The time constants of electrons estimated from drain current transients, however, are much smaller than those estimated from SRH statistics. This is because electron capture and emission also occur even in places far from the transition region. Efficient hole capture and emission for hole traps and fast free hole movement in the structure contribute to the fast concentration of distributed trap charges in the transition region.

When the substrate electrode is placed on the bottom, holes can be injected or emitted through the electrode. In this case, hole capture or emission may govern trap charge variation. However, if the time constants of hole current travelling through the substrate between the trap region and the electrode are larger than those of hole capture and emission, the rate-limiting process is the hole current in the substrate travelling from the electrode to the trap.

Our simulations show that several factors govern trap charge variation, depending on trap parameters, device structures and their sizes. Therefore, real trap parameters should be carefully determined from drain current transients measured on actual devices.

Acknowledgements

The authors would like to thank K. Kasahara, Y. Takahashi and S. Ohkubo for their helpful discussion. They would also like to thank Dr. K. Honjo for his encouragement throughout this work.

References

Kunihiro K and Ohno Y 1994 *GaAs IC Symp. Tech. Dig.* 267

Yano H, Kumashiro S, Goto N and Ohno Y 1989 *IEDM Dig. Tech. Papers* 151

Inst. Phys. Conf. Ser. No 145: Chapter 5
Paper presented at 22nd Int. Symp. Compound Semiconductors, Cheju Island, Korea, 28 August–2 September 1995
© *1996 IOP Publishing Ltd*

Noise Characteristics of AlGaAs/InGaAs/GaAs Pseudomorphic HEMTs with Wide Head T-shaped Gate Fabricated by Optimized Dose Split E-beam Lithography

Jin-Hee Lee, Hyung-Sup Yoon, Sang-Soo Choi, Chul-Soon Park, Kwang-Eui Pyun and Hyung-Moo Park

Semiconductor Division, Electronics and Telecommunications Research Institute
Yusong P. O. Box 106, Taejeon, Korea, Tel : 82-42-860-5772, Fax : 82-42-860-6200

Abstract. We report the AlGaAs/InGaAs pseudomorphic HEMT with a wide head T-shaped gate fabricated by optimization of dose split electron beam lithography and selective gate recess etching. The device performances were measured for the fully passivated 0.15 μm gate length HEMT. The maximum extrinsic transconductance was 540 mS/mm. The minimum noise figure measured at 12 GHz under the V_{ds} = 2 V and I_{ds} = 22 mA was 0.34 dB with associated gain of 10.22 dB. At 18 GHz, the minimum noise figure was 0.49 dB.

1. Introduction

Recently the low noise HEMTs are widely used in the front end of satellite communications, radio astronomy, and satellite direct broadcasting receiver systems. As these applications progress, extremely low noise devices will be required in the microwave and millimeter wave ranges. The gate resistance is the most important factor in determining the noise performance of the HEMTs if the cut-off frequency and transconductance are highly maintained [1]. In order to improve the gate resistance of the HEMTs, several technologies such as multilayer resists and multiple electron beam lithography has been developed to delineate T-shaped gates to reduce both the gate length and parasitic resistance simultaneously [2]. However, these fabrication techniques adopt trilayer resist which require delicate control of the resist sensitivity and extra processes to form T-shaped gate. Also, there still exists difficulty in fabricating T-shaped gate with a high ratio of the gate head length to gate footprint. The ratio of the gate head length to gate footprint for typical T-shaped gate is around 3 to 5.

In this study, we report the AlGaAs/InGaAs pseudomorphic HEMT with a wide head T-gate fabricated by optimization of dose split electron beam lithography with the high ratio of gate head length to gate footprint larger than 7 and selective gate recess etching. This lithography is so simple, reliable and easy to control the head length that we have successfully fabricated pseudomorphic HEMT with the 0.15 μm gate length and 1.15 μm long head. This device showed extremely low noise figures of 0.34 dB at 12 GHz, and 0.49 dB at 18 GHz.

810

2. Device Fabrication

The AlGaAs/InGaAs pseudomorphic HEMT structures were grown by MBE. They consist of ten AlAs-GaAs superlattices, a GaAs buffer, an $In_{0.15}Ga_{0.85}As$ channel, a planar Si doping layer (5×10^{12} cm^{-2}), an $Al_{0.24}Ga_{0.76}As$ spacer, an $Al_{0.24}Ga_{0.76}As$ Schottky, and n$^+$ GaAs cap layer(5×10^{18} cm^{-3}). The sheet carrier density of 2 DEG and the electron mobility measured at room temperature were 2.21×10^{12} cm^{-2} and 5,870 cm^2/V·s, respectively.

To form the T-shaped gate electrode with wide gate head, a dose split electron beam lithography with two-layer resists systems have been developed. The head length, L_h, is defined by the dimension composed of the patterns and the intervening spaces. The size of central pattern and intervening spaces and the doses for patterns determine the gate footprint. Those patterns were exposed by Leica EBML300 system with 30 kV acceleration voltage.

To reduce the gate resistance, the process of T-shaped gate must be optimized. Fig. 1 shows the variation of gate footprint with the change of spacing and pixel lines. As shown in Fig 1, the optimized spacing was 0.025 μm in case of 2 pixel line for 0.15 μm gate footprint. To obtain a T-shaped gate with footprint of 0.15 μm, we find that optimum dosage for the footprint of 2 pixel (0.050 μm) line is 440 μC/cm^2. Fig. 2 shows the variation of head size with the change of head dosage and spacing. As shown in Fig. 2, it is convenient to control the head size by changing the dosage of head.

To get low source resistance, the n$^+$ GaAs cap layer was selectively etched using

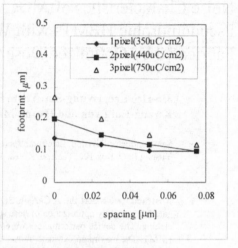

Fig. 1 The variation of gate footprint with the change of spacing and pixel lines.

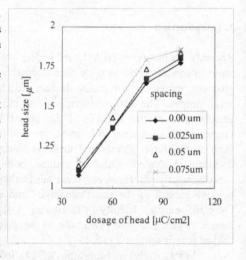

Fig. 2 The head size variation with the change of head dosage and spacing.

citric acid. The etch selectivity of GaAs to AlGaAs was higher than 40. After that, we evaporated Ti, Pt and Au as Schottky gate metal, and evaporated Au on to reduce the gate resistance. Ohmic contacts were made by Ni/Ge/Au/Ti/Au evaporation. Finally, we passivated the device by Si$_x$N$_y$. The T-shaped gate formed by dose split E-beam lithography has the ratio of gate head length (1.15 μm) to gate footprint (0.15 μm) larger than 7, which is higher than twice as that of the conventional T-shaped gate [3]. Fig. 3(a) shows the cross sectional SEM photograph of AlGaAs/InGaAs pseudomorphic HEMT fabricated by optimization of dose split electron beam lithography. The planar view of AlGaAs/InGaAs PHEMT device was shown in Fig. 3(b).

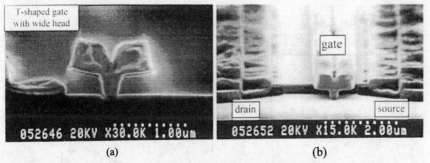

(a)   (b)

Fig. 3. The SEM photographs for the cross sectional view of T-shaped gate fabricated by dose split electron beam lithography(a) and the planar view of AlGaAs/InGaAs PHEMT device(b).

3. Device Performances

The device performances were measured for the fully passivated 0.15 μm gate length HEMT with a total gate width of 140 μm. The threshold voltage, V_{th}, and drain saturation current, I_{dss}, were - 0.8 V and 44 mA, respectively. As shown in Fig. 4, the maximum extrinsic transconductance measured at $V_g = 0$ V and $V_{ds} = 2.0$ V was 540 mS/mm.

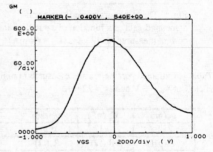

Fig. 4. The transconductance vs. gate bias at $V_{ds} = 2$ V.

The cut-off frequency f_T was obtained from the extrapolation of the current gain, $|h_{21}|$, to unity using a - 6 dB/octave slope, and the maximum frequency of oscillation f_{max} was extracted from small signal parameters. The cut-off frequency and maximum frequency of oscillation in a 0.15 μm gate device were 66 GHz and 168 GHz, respectively.

Noise figure measurements have been carried out in the frequency range between 2 GHz and 18 GHz by using a HP 8510B network analyzer and an ATN NP5 noise parameter test set. The calibration repeatability of the minimum noise figure, NF_{min}, is less than ± 0.02 dB. Fig. 5 shows the dependence of minimum noise figure and associated gain on the percent saturation drain current, which is controlled by the applied gate voltage. Under the fixed drain-source voltage of 2.0 V, NF_{min} can be observed around 50% of the I_{ds}/I_{dss} ratio. At 12 GHz, NF_{min} of 0.34 dB is obtained at 50% of I_{dss}. The NF_{min} of 0.49 dB is obtained at 50% I_{ds}/I_{dss} ratio at 18 GHz. Fig. 6 shows the minimum noise figure and associated gain as a function of frequency at 50% Idss($V_{ds} = 2$ V, $I_{ds} = 22$ mA). The NF_{min} measured at 12 GHz, including passivation loss is 0.34 dB with associated gain G_a of 10.25 dB. At 18 GHz, the NF_{min} is 0.49 dB with G_a of 6.59 dB. These noise figures are the lowest values ever reported for the passivated GaAs-based HEMTs with same gate length. These noise characteristics are attributed to the extremely low gate and source resistances.

Extracted parameters and measured minimum noise figures are listed in Table 1. The minimum noise figure was measured for 0.15 × 140 μm² pseudomorphic HEMT at $V_{ds} = 2$ V

812

and I_{ds} = 22 mA. We believe that the excellent noise characteristics are attributed to the extremely low gate resistance owing to the large cross-sectional area of T-gate fabricated by the dose split E-beam lithography.

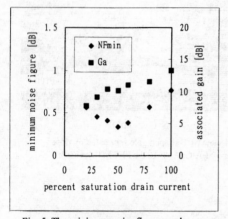

Fig. 5. The minimum noise figures and associated gain as a function of saturation drain current measured at 12 GHz (I_{dss} = 44 mA).

Fig. 6. Noise figure as a function of frequency at V_{ds} = 2 V and I_{ds} = 50 % I_{dss} (I_{dss} = 44 mA).

Table 1. Extracted parameters and measured minimum noise figures for $0.15 \times 140 \ \mu m^2$ pseudomorphic HEMT at V_{ds} = 2 V and I_{ds} = 22 mA.

g_m(mS)	C_{gs}(pF)	τ(psec)	C_{dg}(fF)	C_{ds}(fF)	L_g(pH)	L_d(pH)
75.23	0.12	0.5	40	30	0.10	18.52
L_s(pH)	R_i(Ω)	R_g(Ω)	R_d(Ω)	R_s(Ω)	NF_{m12}(dB)	NF_{m18}(dB)
0.1	16.68	0.30	3.70	0.30	0.34	0.49

NF_{m12} : measured at 12 GHz, NF_{m18} : measured at 18 GHz

4. Conclusion

We report the AlGaAs/InGaAs pseudomorphic HEMT with a wide head T-gate fabricated by optimization of dose split electron beam lithography and selective gate recess etching. The device performances were measured for the fully passivated 0.15 μm gate length HEMT. The minimum noise figure measured at 12 GHz under the V_{ds} = 2 V and I_{ds} = 22 mA was 0.34 dB with associated gain of 10.25 dB. At 18 GHz, the minimum noise figure was 0.49 dB. These data were attributed to the extremely low gate resistance due to the high ratio of gate head length to gate footprint. The proposed technology is adequate for fabricating low noise and millimeter wave devices.

Acknowledgment
This work was financially supported by the Ministry of Information and Communications in Korea.

References
[1] Hwang T et al 1993 *Electron. Lett.* **29** 10-11
[2] Chao P C et al 1989 *Electron. Lett.* **25** 504-505
[3] Lee J H et al 1995 *J. KITE* **32-A** 118-123

Inst. Phys. Conf. Ser. No 145: Chapter 5
Paper presented at 22nd Int. Symp. Compound Semiconductors, Cheju Island, Korea, 28 August–2 September 1995
© *1996 IOP Publishing Ltd*

Single Low-Voltage Operation of Power Heterojunction FETs for Digital Cellular Applications

Naotaka Iwata, Keiko Inosako and Masaaki Kuzuhara

Kansai Electronics Research Laboratory, NEC Corporation
9-1, Seiran 2-Chome, Otsu, Shiga 520, Japan

Abstract. This paper describes 950MHz power performance of a double-doped AlGaAs/ InGaAs/AlGaAs heterojunction FET(HJFET) operated at a single low bias voltage. The developed 1.0 μ m gate-length HJFET exhibited a maximum drain current of 300mA/mm, a transconductance of 240mS/mm, a gate-to-drain breakdown voltage of 13V and a threshold voltage of -0.35V. Operated with a single 3.0V DC bias supply, a 21mm gate periphery HJFET, tuned for maximum power-added efficiency(PAE), showed 1.45W saturated output power and 80.4% maximum PAE. When the device was tuned for minimum adjacent channel leakage power at 50kHz off-center frequency(P_{adj}), a π /4-shifted QPSK output signal of 2.45W was demonstrated with 47.1% PAE and -50.6dBc P_{adj} under a single 3.0V bias operation. The developed HJFET is promising for personal digital cellular power modules operated with a single low-voltage supply.

1. Introduction

Recent personal digital cellular(PDC) phones require small-sized power modules that can be operated at less than 3.6V with more than 1W output power and less than -50dBc adjacent channel leakage power at 50kHz off-center frequency(P_{adj}) while maintaining a high power-added efficiency(PAE). In addition, single DC supply voltage operation is desirable, because it eliminates the need for a negative bias voltage supply, thus reducing size and cost of the cellular phones. In this paper, single low-DC supply voltage operation of a double-doped AlGaAs/InGaAs/AlGaAs heterojunction FET(HJFET) at 950MHz is described.

2. Device Structure and Fabrication

Epitaxial layers were grown by molecular beam epitaxy on a semi-insulating GaAs substrate. Figure 1 shows a cross section of the developed HJFET. The active part of the FET consists of a 13.5nm undoped $In_{0.2}Ga_{0.8}As$ channel layer sandwiched between an upper n-$Al_{0.22}Ga_{0.78}As$ layer($4.5 \times 10^{18}cm^{-3}$/9nm) and a lower n-$Al_{0.22}Ga_{0.78}As$ layer($4 \times 10^{18}cm^{-3}$/4nm). A 3.0 μ m wide recessed undoped $Al_{0.22}Ga_{0.78}As$ Schottky

814

layer was incorporated on the upper n-Al$_{0.22}$Ga$_{0.78}$As layer to achieve a high gate-to-drain breakdown voltage(BV$_{gd}$) and a low knee voltage(V$_K$)[1]. A 1.0 μ m wide narrow recess was then formed in the wide recess using chemical etching to obtain a low threshold voltage(V$_T$). WSi metal was sputter-deposited onto the narrow recess to form a 1.0 μ m long T-shaped gate. With this structure, parallel conduction and surface trapping effects are suppressed, thus reducing V$_K$[1] as well as preventing pre-mature power saturation[2]. After thinning the wafer to 50 μ m, a gold plated heat sink was employed on the back side to ensure low thermal resistance.

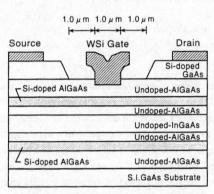

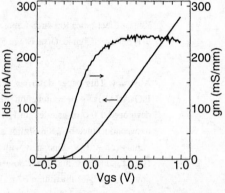

Figure 1 Cross section of the developed HJFET.

Figure 2 g$_m$ and I$_{ds}$ versus V$_{gs}$ for the HJFET.

3. Results and Discussion

Figure 2 shows a transconductance(g$_m$) and a drain-to-source current(I$_{ds}$) versus a gate-to-source voltage(V$_{gs}$) for the fabricated HJFET. The device exhibited a maximum drain current(I$_{max}$) of 300mA/mm and 240mS/mm g$_m$. This high g$_m$ was found to be constant up to V$_{gs}$=1V. The BV$_{gd}$, defined at a reverse gate current of 1mA/mm, was 13V. The high value and flat characteristics of g$_m$ and the high value of BV$_{gd}$ enable high gain and high output power characteristics, as well as low distortion products for the HJFET, as will be discussed later. Since V$_T$ of the fabricated HJFET is -0.35V, it can be operated under class AB condition at V$_{gs}$=0.0V without sacrificing high PAE.

An HJFET with W$_g$=21mm was examined at a source-to-drain voltage(V$_{ds}$) of 3.0V, which would correspond to the minimum DC operating voltage for a PDC system when one Li-ion battery cell or three NiCd(or NiH) battery cells were used. The device was operated under class AB(15% I$_{max}$) condition at V$_{gs}$=0.0V. The optimum load impedance(Z$_L$) was determined through load-pull measurements, where the source impedance(Z$_S$=1.0-j0.5 Ω) was fixed to attain maximum associated gain with 126mW (21dBm) input power(P$_{in}$) at 950MHz. The optimum values of Z$_L$ for maximum P$_{out}$ and for maximum PAE were derived to be 2.5+j1.8 Ω and 4.5-j2.0 Ω , respectively. Figure 3 shows P$_{out}$ and PAE as a function of P$_{in}$ for the HJFET at 950MHz with a single 3.0V DC bias supply under those two Z$_L$ conditions. When the device was tuned for maximum PAE, it exhibited 80.4% maximum PAE(PAE$_{max}$) and 1.45W(31.6dBm) saturated output power(P_{sat}). Tuned for maximum P_{out}, as high as 3.28W(35.2dBm) P_{sat} was obtained with 64.8% PAE_{max}.

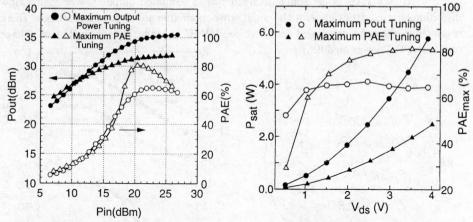

Figure 3 P_{out} and PAE as a function of P_{in} for the HJFET with W_g=21mm at 950MHz.

Figure 4 P_{sat} and PAE_{max} as a function of V_{ds} at 950MHz.

Figure 4 shows P_{sat} and PAE_{max} as a function of V_{ds} under the two Z_L conditions at 950MHz. Tuned for maximum PAE, the device exhibited more than 70% PAE at V_{ds} higher than 1.5V. When the device was tuned for maximum P_{out}, it exhibited more than 1.68W (32.3dBm) P_{sat} at V_{ds} higher than 2.0V.

The device was then characterized with a 950MHz π/4-shifted quadrature phase shift keying(QPSK) signal. Since more than 1.0W(30dBm) P_{out} as well as lower than -50dBc P_{adj} are required for the PDC power application, the optimum Z_L was derived through load-pull measurements with respect to these criteria as well as possible high PAE. Figure 5 shows P_{out}, PAE and P_{adj} contours derived through load-pull measurements. The input conditions(Z_S and P_{in}) were identical to the former ones. The circle with vertical hatching shows the Z_L region that gives rise to more than 1.0W(30dBm) P_{out}, while the region with horizontal hatching corresponds to that ensuring more than 50% PAE. The white circle shows the region for P_{adj} less than -50dBc. The optimum Z_L was determined to be 2.5+j2.8 Ω, taking into account high PAE of more than 45% as well as overcoming the PDC criteria. Figure 6 shows P_{out}, P_{adj} and PAE as a function of P_{in} with a 950MHz π/4-shifted QPSK signal for the HJFET under single 3.0V operation. The HJFET demonstrated 2.45W(33.9dBm) P_{out} and 47.1% PAE with -50.6dBc P_{adj}. The linear gain was 14.8dB. A maximum PAE of 52.2% was obtained with 3.12W(34.9dBm) P_{out}. To the authors' knowledge, this performance is the best ever reported for a PDC power device under single 3.0V operation.

It should be noted that the developed HJFET exhibited relatively flat g_m characteristics with respect to V_{gs}, which is ascribed to the optimum material design of the double-doped structure[1]. Non-linear g_m characteristics with respect to V_{gs} are known to be responsible for odd-order intermodulation distortion(IMD) products[3]. Among the series of odd-order IMDs, the fifth-order IMD products have been found to be closely related to P_{adj}[4]. Thus, the flat g_m characteristics of the developed HJFET are effective to

816

reduce IMD products and hence to suppress P_{adj}. The developed HJFET demonstrated low distortion characteristics of less than -50dBc P_{adj} at a 1.6dB power gain compression point. Since maximum PAE is generally obtained near a saturated output power region, low distortion characteristics under high output power operation are of great importance. These results reveal that the developed double-doped HJFET is promising for single low-voltage operation PDC power modules.

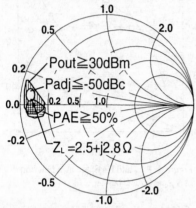

Figure 5 P_{out}, PAE and P_{adj} contours derived through load-pull measurements.

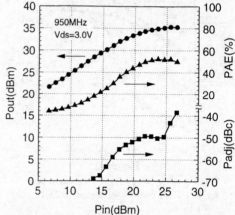

Figure 6 P_{out}, PAE and P_{adj} as a function of P_{in} for the HJFET with a 950MHz π/4-shifted QPSK signal.

4. Summary

We have developed a single low-bias voltage operation power HJFET for PDC phones. The fabricated HJFET exhibited 300mA/mm I_{max}, 240mS/mm g_m, 13V BV_{gd} and -0.35V V_T. The 21mm W_g HJFET, biased at V_{gs}=0.0V and tuned for maximum PAE, showed 1.45W P_{sat} and 80.4% PAE_{max} under 3.0V operation. When the device was tuned for minimum P_{adj}, it demonstrated 2.45W(33.9dBm) P_{out} and 47.1% PAE with -50.6dBc P_{adj}.

Acknowledgment

The authors would like to thank I.Nagasako, Y.Hasegawa, and O.Izumi for useful discussions and suggestions. They are also grateful to Drs. H.Abe, M.Ogawa, T.Noguchi and T.Itoh for their encouragement.

References

[1] Iwata N, Inosako K and Kuzuhara M 1993 IEEE MTT-S Digest 1465-8

[2] Matsunaga K, Okamoto Y and Kuzuhara M 1994 IEEE IEDM Technical Digest 895-8

[3] Higgins J A and Kuvås R L 1980 IEEE Trans. on Microwave Theory and Technique **MTT-28** 9-17

[4] Inosako K, Iwata N and Kuzuhara M submitted to 1995 IEICE Trans.

Inst. Phys. Conf. Ser. No 145: Chapter 5
Paper presented at 22nd Int. Symp. Compound Semiconductors, Cheju Island, Korea, 28 August–2 September 1995
© *1996 IOP Publishing Ltd*

$In_{0.52}(Al_xGa_{1-x})_{0.48}As/In_{0.53}Ga_{0.47}As(0 \leq x \leq 1)$ heterostructure and its application on HEMTs

C.-S. Wu, Y.-J. Chan, C.-H. Chen, J.-L. Shieh and J.-I. Chyi

Department of Electrical Engineering National Central University, Chungli, Taiwan 32054, R.O.C.

Abatract.   Quaternary $In_{0.52}(Al_xGa_{1-x})_{0.48}As(0 \leq x \leq 1)$ compounds lattice-matched InP substrates were grown and characterized. The engery bandgap (E_g) of these $In_{0.52}(Al_xGa_{1-x})_{0.48}As$ compound were $(0.806+0.711x)$ eV. The conduction band discontinuity (ΔE_c) at the $In_{0.52}(Al_xGa_{1-x})_{0.48}As/In_{0.53}Ga_{0.47}As$ (x=0.9 and 0.75) heterojunctions was approximately $(0.68 \pm 0.01)\Delta E_g$. The quaternary $In_{0.52}(Al_{0.9}Ga_{0.1})_{0.48}As/In_{0.53}Ga_{0.47}As$ Q-HEMTs with a gate-length of 0.8μm revealed an extrinsic transconductance (g_m) of 295 mS/mm, an f_T of 35 GHz and an f_{max} of 76 GHz. This Quaternary HEMT was proven to be more reliable, comparing with the conventional InAlAs/InGaAs HEMTs[1].

1. Introduction

InAlAs/InGaAs on InP HEMTs have demonstrated a great potential for the microwave and millimeter wave applications [2],[3]. However, the high quality of InAlAs layer is always hard to be obtained during the MBE growth, where the Al-cluster formation directly impacts the quality of InAlAs layers. The associated abnormal phenomena of the InP-HEMTs include a significant "kink" in the DC I-V curves, a high DC output conductance, and a poor threshold of conducting channels[4]. Furthermore, a high Al content in the InAlAs Schottky layer enhances the oxidation formation, which causes a large gate leakage current. These drawbacks dramatically deteriorate the device performance.

In order to improve the quality of InAlAs layer, one of the possible way is to introduce some Ga atoms in the InAlAs, to form a quaternary InAlGaAs layer. Since the Ga atoms present themselves a high surface mobility than that of Al atoms during the MBE growth , it is possible to eliminate the alloy scattering, which will improve the quality of this quaternary InAlGaAs layer.

In this study, we systematically grew and characterized $In_{0.52}(Al_xGa_{1-x})_{0.48}As$ compounds($0 \leq x \leq 1$), especially focusing on the Schottky diode perforrmance and the conduction-band disccontinuity(ΔE_c), which are essential for HEMTs operation. Finally, we used the quaternary compound as a Schottky layer to build a HEMT with the 0.8μm gate-length and evaluate its DC and RF performance.

2. Characterization of the $In_{0.52}(Al_xGa_{1-x})_{0.48}As(0 \leq x \leq 1)$ layers

2.1. Schottky diode performance

Various $In_{0.52}(Al_xGa_{1-x})_{0.48}As(0 \leq x \leq 1)$ compound layers were grown on semi-insulating [100] InP substrates by a Riber-32P MBE system. The lattice-mismatched $(\Delta a/a_o)$ between the InAlGaAs and InP was within $2x10^{-3}$. Figure 1(a). show Photoluminescence(PL) spectra of the undoped $In_{0.52}(Al_xGa_{1-x})_{0.48}As(0 \leq x \leq 1)$ layers at 20K. The PL peak intensity is associated with the elecrton transition of band-to-band. By increasing the Al content in the InAlGaAs compounds, PL peaks shifted to a short wavelength regime. The energy bandgap (E_g) of the InAlGaAs layers evaluated from the peak position in PL spectra is plotted in Fig. 1(b). A simple linear equations, $Eg(x)=(0.806+0.711x)eV$, can describe the whole quaternary compounds.

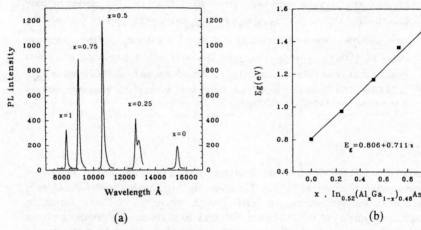

(a)   (b)

Fig.1 (a) PL spectra for various compounds at 20K.

(b) Engery bandgap dependence of Al compositions in $In_{0.52}(Al_xGa_{1-x})_{0.48}As(0 \leq x \leq 1)$.

Al-gate of the $In_{0.52}(Al_xGa_{1-x})_{0.48}As(0 \leq x \leq 1)$ Schottky diodes characteristics are shown in Fig. 2. It was found that the Schottky diode performance could be improved by replacing 10% Al atoms with Ga atoms (x=0.9). However, further increasing the Ga concentration in the $In_{0.52}(Al_xGa_{1-x})_{0.48}As$, a degradation of diode performance was observed due to a decrease of E_g. The ideality factor, Schottky barrier height, and breakdown voltage of Schottky diodes, evaluated from the I-V measurement, were summarized in Tab. 1. The ideality factor was improved from 1.26

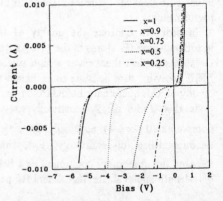

Fig.2 Schottky diode I-V characteristics of different x values in $In_{0.52}(Al_xGa_{1-x})_{0.48}As$ compounds

for x=1 (InAlAs) to 1.15 for x=0.9 ($In_{0.52}(Al_{0.9}Ga_{0.1})_{0.48}As$. However, further increasing the Ga concentration, the ideality factor increased. The Schottky barrier heights slightly dropped from 0.56eV (x=1), 0.54eV (x=0.9), 0.51eV (x=0.75) to 0.54eV (x=0.5) The breakdown voltage was also decreased by adding the Ga atoms.

Table 1: Schottky diode characteristics of $In_{0.52}(Al_xGa_{1-x})_{0.48}As(0.5 \leq x \leq 1.0)$

	Ideality factor	Barrier height(eV)	Breakdown voltage(Volt.)
$In_{0.52}Al_{0.48}As$	1.26	0.56	-4.5
$In_{0.52}(Al_{0.9}Ga_{0.1})_{0.48}As$	1.15	0.54	-4.5
$In_{0.52}(Al_{0.75}Ga_{0.25})_{0.48}As$	1.23	0.51	-3.1
$In_{0.52}(Al_{0.5}Ga_{0.5})_{0.48}As$	1.39	0.54	-3

2.2. Conduction band discontinuities (ΔE_C)

One of the important parameters in InP-HEMT devices is the ΔE_C, which causes the carrier transport along the $In_{0.52}(Al_xGa_{1-x})_{0.48}As/In_{0.53}Ga_{0.47}As$ heterojunction. This ΔE_C can be expressed by the following equation[5],[6]:

$$\Delta \Phi = (q/\varepsilon) \times \{ \int [N(x)-n(x)][x-x_i]dx \} = (\delta_1-\delta_2)+\Delta E_C$$

where N(x) and n(x) are doping concentrations and measured carrier concentrations, respectively, x_i is the heterojunction position, and $\delta_1-\delta_2$ is the difference between the Fermi level and the conduction band in heterointerface. For simplicity, we used the same doping concentration on both sides of heterostructures (~7.5×10^{16} cm^{-3} for x=0.9, 7.8×10^{16} cm^{-3} for x=0.75), and $\delta_1-\delta_2$ can be directly calculated from the individual electron effective mass. Carrier profiles near the heterojunction were measured by capacitance-voltage (C-V) method. To avoid the measurement error caused by the thermal-effect, the C-V measurement was conducted at 77K, and the results for x=0.9 and x=0.75 were shown in Fig. 3. Electrons from the quaternary compounds (higher energy bandgap) layer transfered into the

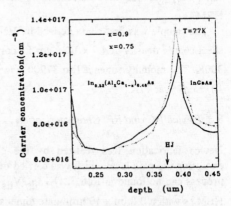

Fig.3 Electron concentration profiles, evaluated by C-V measurements at 77K, near the $In_{0.52}(Al_xGa_{1-x})_{0.48}As/In_{0.53}Ga_{0.47}As$ (x=0.9, 0.75) heterojunctions.

$In_{0.53}Ga_{0.47}As$ (lower energy bandgap) layer due to the ΔE_c . This effect results in a carrier depletion in InAlGaAs and a carrier accumulation in InGaAs, as manifested in Fig.3. By integrating the total amount of these transferred carriers, ΔE_c at the heterojunction can be determined. ΔE_c of 0.443eV and 0.358eV were obtained for x=0.9 and x=0.75, respectively. These two values corresponds to a (0.68 ± 0.01) ΔE_g, which was quite closed to the value of the conventional $In_{0.52}Al_{0.48}As/$ $In_{0.53}Ga_{0.47}As$ heterojunction[6].

3. Quaternary $In_{0.52}(Al_{0.9}Ga_{0.1})_{0.48}As/In_{0.53}Ga_{0.47}As$ HEMTs

3.1. Characterization of the Q-HEMT Material

Based upon the previous characterization of InAlGaAs compounds, we conclude that this quaternary compound with x=0.9 is qualified to be a Schottky layer in HEMT structures. Therefore, in this section we ussed this heterstructure for the HEMT fabrication and evaluation. Fig.4 shows the device cross-section of the $In_{0.52}(Al_{0.9}Ga_{0.1})_{0.48}As/In_{0.53}Ga_{0.47}As$ Q-HEMTs. The $In_{0.52}(Al_{0.9}Ga_{0.1})_{0.48}As$ layers was used as buffer, spacer and Schottky layers, to replace the role of $In_{0.52}Al_{0.48}As$ in the conventional InP-HEMTs.

Si:3×10^{18} $In_{0.53}Ga_{0.47}As$	10 nm
Undoped $In_{0.52}(Al_{0.9}Ga_{0.1})_{0.48}As$	20 nm
Si:4×10^{18} $In_{0.52}(Al_{0.9}Ga_{0.1})_{0.48}As$	20 nm
Undoped $In_{0.52}(Al_{0.9}Ga_{0.1})_{0.48}As$	5 nm
Undoped $In_{0.53}Ga_{0.47}As$	35 nm
Undoped $In_{0.52}(Al_{0.9}Ga_{0.1})_{0.48}As$	250 nm
Undoped $In_{0.52}(Al_{0.9}Ga_{0.1})_{0.48}As/In_{0.53}Ga_{0.47}As$	
S.I InP Substrate	

Fig. 4 Device cross-section of $In_{0.52}(Al_{0.9}Ga_{0.1})$ $_{0.48}As/In_{0.53}Ga_{0.47}As$ Q-HEMTs.

The sample was first characterized by Hall effect measurements. The Q-HEMT had a sheet charge density of 4.0×10^{12} cm^{-2} , corresponding to a mobility 6230 cm^2/V-sec at 300K. The mobility enhanced to 37900 cm^2/V-sec with a sheet charge density of 2.6×10^{12} cm^{-2} at 77K.

3.2. Device DC and RF Characterization

Device fabrication was realized by the conventional optical lithographic techniques excepting the gate level which was carried out by the Deep-UV light source. The detail of process can be found in Ref. 7. DC I_{ds}-V_{ds} characteristics of this InAlGaAs/InGaAs Q-HEMTs with a 0.8μm x 100μm gate dimension are shown in Fig. 5, which revealed good channel threshold characteristics from the I_{ds}-V_{ds} curve. It demonstrates that Q-HEMTs had better Schottky gate performance, which can dramatically enhance device characteristics. The peak extrinsic transconductance (g_m) biased at V_{ds}=2V was 295 mS/mm and output conductance (g_{ds}) was 10 mS/mm with a 0.8μm gate length. The channel threshold voltage was -1.8V. The saturation drain-source current I_{dss} of 360 mA/mm was obtained at V_{gs}=0V.

Microwave S-parmeter measurements from 45 MHz to 50GHz were carried out by a network analyzer in conjunction with a Cascade probe station. Fig. 6 shows the microwave characteristics of Q-HEMTs. The current-gain cut-off frequency f_T of 35GHz was determmined from H_{21} values. The extrapolated maximum oscillation frequency f_{max} was 76 GHz.

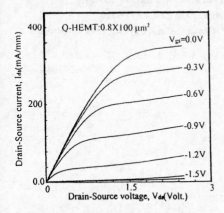

Fig.5 DC I-V characteristics of InAlGaAs/ InGaAs HEMTs(0.8μm x 100μm) at 300 K.

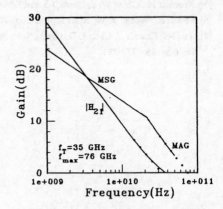

Fig.6 Microwave characteristics of InAlGaAs/ InGaAs HEMTs at (0.8μm x 100μm)300K.

4. Conclusions

Quaternary $In_{0.52}(Al_xGa_{1-x})_{0.48}As$ compounds on InP substrates were grown and characterized. An $E_g(x)=(0.806+0.711x)$ eV of these quaternary compounds was obtained. The ΔE_c of the $In_{0.52}(Al_xGa_{1-x})_{0.48}As/$ $In_{0.53}Ga_{0.47}As$ (x=0.9 and 0.75) heterojunction were (0.68 ± 0.01) ΔE_g. By adding 10% Ga atoms into the InAlAs layers, we observed an improvement of the Schottky performance. The DC peak extrinsic g_m was 295 mS/mm for $In_{0.52}(Al_{0.9}Ga_{0.1})_{0.48}As/In_{0.53}Ga_{0.47}As$ HEMTs. The f_T was 35 GHz, and f_{max} was 76 GHz.

Acknowledgment

This work was financially supported by the National Science Council, Republic of China, under constract no. NSC83-0404-E-006-005.

References

[1] Wu C-S, Chan Y-J, Gan T-H, Shieh J-L and Chyi J-I, 1995 *7th Intl. Conference on InP and Related Material* 412-415

822

[2] Matloubian M., Nguyen L-D, Brown A-S, Larson L-E, Melendes M-A and Thompson M-A, 1991 *IEEE MTT-S Int. Microwave Symp. Dig.* 721-724.

[3] Matloubian M. , Brown A-S, Nguyen L-D, Melendes M-A, Larson L-E, Delaney M-J and Pence J-E, 1993 *IEEE Electron Dev. Lett.*, **14**, no. 4, 188-189

[4] Kuang J-B, Tasker P-J, Wang G-W, Chen Y-K, Eastman L-F, Aina O-A, Hier H. and Fathimulla A., 1988 *IEEE Electron Dev. Lett.*, **9**, 630-632

[5] Kroemer H., Chien W-Y, Harris J-S and Edwall D-D, 1980 *Appl.Phys. Lett.* **36** 295-297

[6] People R., Wecht K-W, Alavi K. and Cho A-Y, 1983 *Appl. Phys. Lett.* **43** 118-210

[7] Wu C-S, Chan Y-J, Chen C-D, Chung T-M, Juang F-Y, Chang C-C and Chyi J-I, 1995 *Solid -State Electron* **38** 377-381

Inst. Phys. Conf. Ser. No 145: Chapter 5
Paper presented at 22nd Int. Symp. Compound Semiconductors, Cheju Island, Korea, 28 August–2 September 1995
© 1996 IOP Publishing Ltd

Submicron PHEMT with a dielectric interface gate

Myeong Kook Gong, Jong Wook Yoon and Jhang Woo Lee

Sammi Technology, Yubang-Ri 205, Yongin, Kyunggi-Do, South Korea

Abstract AlGaAs/InGaAs PHEMTs with ultra-thin dielectric Si_3N_4 layers between the submicron gate metals and the recessed AlGaAs surfaces are fabricated. The low temperature grown Si_3N_4 layer with a thickness range of 5-7 nm is expected to enhance the DC isolation characteristic of the AlGaAs barrier and reduce the gate leakage current. Submicron gate devices with the dielectric layers, fabricated by a simple optical lithography technique, exhibit similar f_t and f_{max} in our RF measurements, but leakage current reduction by two orders of magnitude and 1.5 dB noise improvement compared with the usually processed devices without the dielectric layers.

1. Introduction

Practical need of AlGaAs/InGaAs PHEMTs can be more extended in case of enhancement of gain property of the device as well as the inherited low noise characteristic. It is well known, for this purpose, that thickness reduction of the doped AlGaAs layer is effective to increase the transconductance g_m of the PHEMT, so as to increase its power gain. In this structure the reduced AlGaAs layer needs higher doping concentration in order to maintain the total sheet carrier density at a certain level. The PHEMT, thus in many cases, is likely to suffer from the soft breakdown and sometimes excessive gate leakage. To achieve the high gain low noise PHEMT, it may be desirable to introduce a certain fabrication process which can compromise these two competing parameters.

The insertion of a thin dielectric layer between the gate metal and the AlGaAs layer can be one of the good candidate processes in this purpose, as far as the process does not yield any structure variations and transistor characteristics. To ensure such criterion, we introduced the very low temperature PECVD process for Si_3N_4 deposition in our usual submicron PHEMT fabrication processes. Devices fabricated with this process, called DIG-PHEMTs in our convenience, are proved to have clear advantage over the usually processed devices in our DC and RF measurements. This result indicates that our dielectric layer technique works effectively. In this paper, we

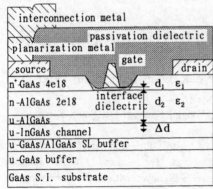

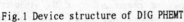

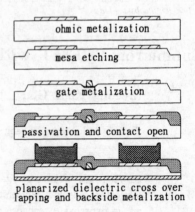

Fig. 1 Device structure of DIG PHEMT | Fig. 2 Fabrication steps of DIG PHEMT

will describe our fabrication processes and device characteristics of the DIG-PHEMT. Comparison of device performances with usual HEMT devices will be discussed more in details.

2. Device Fabrication

The basic structure of our DIG-PHEMT is shown in Fig.1. The PHEMT structures were grown by Riber 32 MBE system on 3 inch GaAs in the usual growth procedure. The active region comprises 15 nm undoped $In_{0.15}Ga_{0.85}As$, 3 nm undoped $Al_{0.22}Ga_{0.78}As$, 30 nm n^+-$Al_{0.22}Ga_{0.78}As$ and 40 nm n^+-GaAs cap layers. Our device fabrication is conventional except the submicron gate formation by optical lithography. Fig.2 describes the basic fabrication steps used in our device processes, consisting of ohmic contact, mesa isolation, gate recess and a 5-7 nm PECVD of Si_3N_4 thin film. The temperature used in PECVD was below 100 ℃ and the gas sources were 2 % SiH_4 diluted in N_2 and NH_3.

The successive steps are submicron optical lithography for gate metal formation, passivation and interconnection processes across over four multi-finger gates. Fig.3 shows the whole view of the completed DIG-PHEMT with four 0.5x50 μm gate fingers. We fabricated many devices using same structured wafers with or without the dielectric layers at the gates. Dielectric strength of the Si_3N_4 layer, estimated from the I-V measurements of the test MIM capacitor formed with 100-200 nm thick Si_3N_4 layer at the equivalent conditions, is approximately $1x10^6$ V/cm. This value is somewaht lower than the ideal case. The interface quality was confirmed by the measurement of threshold voltage shift[1] of a thick DIG-PHEMT. A DIG-PHEMT with 150 nm dielectric layer gives the interface charge of $5x10^{12}$ /cm^2, which is definately larger than the good MIS devices. To avoid such a significant

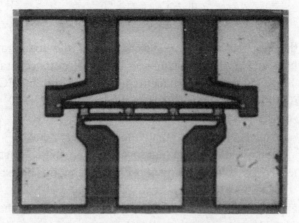

Fig. 3 Photograph of fabricated DIG PHEMT

change of the threshold voltage, we reduced the dielectric thickness below 10 nm.

3. DC characteristics

Fig.4 shows the I-V characteristics (a) and transconductance and saturation current behaviors (b) of our typical DIG-PHEMT with a 0.5x200 μm gate. The overall transistor characteristics are very similar to the usual device. The maximum extrinsic transconductance estimated in Fig.4(b) is 28 mS, which is basically equivalent to the one without the dielectric gate. However, relatively large swing of forward gate bias, greater than 2 V in this device, is observed in Fig.4(a), indicating that it may end up with a higher gain. The estimated maximum intrinsic g_m in this DIG-PHEMT is 32 mS. From the value we can estimate the saturation velocity using the equation[2] of

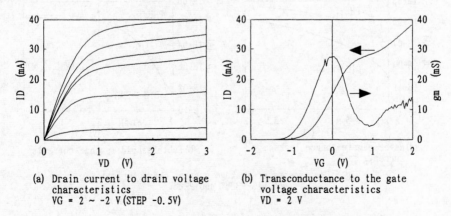

(a) Drain current to drain voltage characteristics
VG = 2 ~ -2 V (STEP -0.5V)

(b) Transconductance to the gate voltage characteristics
VD = 2 V

Fig. 4 DC Characteristics of DIG-PHEMT

$$g^{int}_{m,\,max} = v_s Z \Big/ \left(\frac{d_1}{\varepsilon_1} + \frac{d_2 + \varDelta d}{\varepsilon_2} \right) \qquad (1)$$

where v_s is the electron saturation velocity, Z is the total gate width of the PHEMT(200 μm), $d_{1,2}$, $\varepsilon_{1,2}$ are thicknesses and dielectric constants of epilayers as shown in Fig.1, $\varDelta d$ is the equivalent distance of the 2-DEG peak from the AlGaAs/InGaAs interface. Typical value we obtained is 7×10^6 cm/s, which is comparable to GaAs.

These results indicate that for power device applicaton the DIG-PHEMT is expected to yield improved power characteristics to the level of usual GaAs devices. We attribute those results to the dielectric layer at the gate.

For the comparison of two devices with and without the dielectric gate, the reverse biased gate source current-voltage characteristics are shown together in Fig.5(a). It shows clearly that the gate leakage current is smaller and the breakdown voltage is larger (not shown in the range of our figure) in DIG-PHEMT than in the usual PHEMT approximately two orders of magnitude. According to the Fowler-Nordheim formula for the tunneling[3], the transmission coefficient T is

$$T = \exp\left[-\frac{(2W)^{3/2} m^{1/2} l}{\hbar V} \right] \qquad (2)$$

where W is the work function of metal and m is the effective mass of electron, l is the dielectric thickness, $\hbar$ is the Plank constant, V is the band offset. In equation(2), the expected dielectric thickness is a few angstroms.

Both devices are in the subthreshold region below -1 V of gate voltages, as shown in Fig.5(b). However, the drain current of DIG-PHEMT is much smaller than PHEMT in such region, indicating that the gate leakage current is much lower in the DIG-PHEMT than in the usual device.

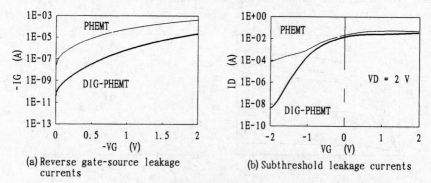

(a) Reverse gate-source leakage currents

(b) Subthreshold leakage currents

Fig.5 Comparisons of leakage current characteristics of DIG-PHEMT and PHEMT

4. RF characteristics

Further comparisons were made in our on-wafer probe RF measurements of S-parameters, current and power gains, and noise figures at 2-18 GHz region. Current gains $|h21|^2$ and the maximum available power gains G_{max} are plotted in Fig.6. It shows the DIG-PHEMT has slightly higher f_t (current cutoff frequency) and f_{max} (maximum frequency of oscillation) than the usual PHEMT. Measured minimum noise figures F_{min} and the associated gains $G_{a,asso}$ are plotted in Fig.7. As shown, $G_{a,asso}$ are almost same in both devices, while F_{min} are much different. At 12 GHz the measured F_{min} is 1.04 dB for the DIG-PHEMT, and 2.60 dB for the usual PHEMT. It shows clearly the 1.5 dB differences over the whole range of frequencies. Since both devices were fabricated on the same epi-wafer and fabricated in equivalent processes except the dielectric gate, these differences are most likely to be caused by the dielectric gate process.

By the Fukui's formula[4] the minimum noise figure F_{min} of HEMT is as

$$F_{min} = 1 + 2\pi K f C_{gs} \sqrt{(R_g + R_s)/g_m} \qquad (3)$$

where f is the measured frequency, C_{gs} is the gate-source capacitance, R_g is the gate resistance, R_s is the source resistance, and K is the fitting parameter. The fitting parameter K estimated from the equation(3) is 2.0 for the DIG-PHEMT and 5.4 for the PHEMT. The K value is normally structure dependent and strongly affected by the material quality. In our comparison, both devices were made on the same epi-wafer, and material qualities are supposed to be identical each other, indicating that the K values are modified simply by introducing the dielectric layer in the gate. The noise figure reduction observed in DIG-PHEMT, therefore, is definitely the outcome of the

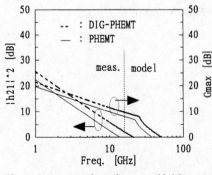

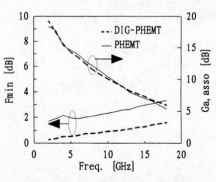

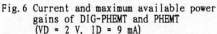

Fig. 6 Current and maximum available power gains of DIG-PHEMT and PHEMT (VD = 2 V, ID = 9 mA)

Fig. 7 Minimum noise figures and associated gains of DIG-PHEMT and PHEMT (VD = 2 V, ID = 9 mA)

dielectric gate process. Although it is not clear, the small leakage current is attributed to the essential reason of the noise reduction.

5. Conclusions

We fabricated so-called DIG-PHEMTs by introducing ultra-thin dielectric layers between the gate metals and the AlGaAs layers in our usual submicron PHEMT processes. The devices demonstrate considerable improvements of device performances in both DC and RF measurements. In our comparison, its gate leakage current decreases by two orders of magnitude of the usual PHEMT. The minimum noise figure is also improved by 1.5 dB from 2.6 dB of PHEMT to 1.04 dB of DIG-PHEMT, while other RF parameters such as $G_{a,asso}$, f_t, f_{max} are basically unchanged. The DIG-PHEMT is likely to be used even for the high power low noise application, since it shows the large forward gate voltage swing beyond 2 V and large breakdown voltage.

6. Acknowledgements

The authors thank Mr. S. K. Cho, Y. D. Joo and S.W. Lee in Sammi for their technical supports, C. S. Lee, S. J. Maeng in ETRI for their RF measurement efforts, and another S.W. Lee for his proof reading of this manuscript

References

[1] Sze S M 1981 *Physics of Semiconductor Devices*(Korea: John Wiley and Sons)
[2] Drummond T J et al 1981 *IEEE Elec. Dev. Lett.* **EDL-3 11** 338-341
[3] Jaros M 1989 *Physics and Application of Semiconductor Microstructures* (New York: Oxford University)
[4] Fukui H 1979 *IEEE Trans. Electron Devices* **ED-26** 1032-1037

Inst. Phys. Conf. Ser. No 145: Chapter 5
Paper presented at 22nd Int. Symp. Compound Semiconductors, Cheju Island, Korea, 28 August–2 September 1995
© *1996 IOP Publishing Ltd*

Al-free GaInP/InGaAs MODFETs on GaAs grown by OMVPE with fT > 100GHz

†Boris Pereiaslavets, ‡Karl Bachem, ‡†Jürgen Braunstein, †Jinwook Burm, and †Lester F. Eastman

†Cornell University, School of Electrical Engineering, Ithaca, NY 14853-5401
‡Fraunhofer Society, IAF, Tullastr. 72, D-79108 Freiburg, Germany

Abstract
Excellent quality $Ga_xIn_{1-x}P/In_yGa_{1-y}As/GaAs$ MODFETs with a pseudomorphic barrier and a pseudomorphic channel were grown by OMVPE. This Al-free material system is the most promising material system for advanced MODFETs on GaAs. Record 2DEG carrier densities of $3.1 \cdot 10^{12} cm^{-2}$ for single sided MODFET were measured. $0.1\mu m$ gate length MODFETs achieved for the first time f_T's over 105GHz and f_{max}'s over 188GHz.

1. Introduction

$Ga_xIn_{1-x}P$ is a very promising barrier material for the MODFET fabrication on GaAs substrates. Several attempts were made in the past [e.g., 1] to fabricate MODFETs with a lattice matched $Ga_{.51}In_{.49}P$ barrier. However the offset in the conduction band depends not only on the composition, but also strongly on the growth conditions. Thus one might have to dope the channel in order to achieve reasonable charge densities. Following the initial simple analytical design by Eastman [2], MODFET structures were grown by OMVPE [3]. The novel concept of grading the barriers was implemented in these structures. Grading the channel results in further improvement [4]. Hall measurements were made on the wafers and compared well with the predictions.

In the second section we present computer simulations that show the advantages of the GaInP system over AlGaAs. The ideas of graded barriers and channels are given in sections three and four. Device fabrication and results are given in section five. Section six summarizes the paper.

2. GaInP vs. AlGaAs

Conduction band profiles and the electron concentrations of the MODFETs were obtained from the computer simulations. Two similar structures were simulated. The first one was with 200Å $Ga_{0.66}In_{0.34}P$ barrier and 90Å channel. The second one was identical to the first one but with the barrier made of $Al_{0.22}Ga_{0.78}As$. The composition of InP in the GaInP was chosen to be 34% to satisfy the Matthews-Blakeslee limit [5]. This is the highest possible strain in this material for a 200Å thick barrier. The composition of AlAs in AlGaAs was chosen to be 22% to avoid DX centers. Thus one can say that these simulations were done for the highest conduction band offsets for AlGaAs and GaInP. The results are shown in fig. 1.

The advantages of the GaInP are obvious. We defined n_{sm} as the concentration of free electrons in the channel at the bias when the concentration of parasitic electrons in the barriers is 10% of the electrons in the channel. n_{sm} is 60% higher for the GaInP case. However, the higher electron sheet density in the channel is not the only advantage of the GaInP material. It is also much easier to process GaInP than AlGaAs. There are highly selective etchants for this material system. Aluminum-free material does not have the disadvantage of fast oxidation. The are no DX centers in this GaInP. Thus the low frequency noise figure should be much better. It is possible to grade the GaInP barriers as will be described in the next section, to further improve the electron sheet densities and the electron confinement in the channel. In the case of AlGaAs the AlAs concentration is limited to approximately 22.5% due to DX centers. Thus the highest possible conduction band offset for AlGaAs/GaAs is limited to 180meV. In this case grading of the barriers would not make an improvement as in the case of GaInP.

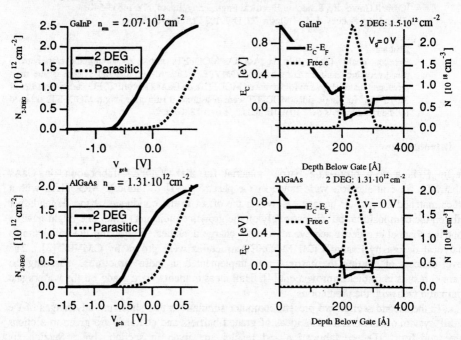

Fig 1.   Uniform MODFETs with 90Å $Ga_{0.76}In_{0.24}As$ channels and 200Å $Ga_{0.66}In_{0.34}P$ or $Al_{0.22}Ga_{0.78}As$ barriers. n_{sm} is 60% higher for the GaInP case.

3. Concept of Graded barriers

A conduction band profile of a conventional MODFET is shown schematically in fig. 2(a). As can be seen from this figure the problem of this structure is the dip in the conduction band due to the δ-doping. Since this region is closest to the Fermi level, a high probability arises to accumulate parasitic charges there. The solution of this problem is to increase the distance of the dip to the Fermi level by increasing the Ga mole fraction in the $Ga_xIn_{1-x}P$ barrier. Therefore the barrier will be pseudomorphic and still be constrained by the Matthews-

Blakeslee limit. However, it is not necessary to strain the whole barrier. It is important to raise the potential of the mainly problematic part of it. In the case of GaInP one needs to increase the GaP concentration at the doped region. The Ga composition in the graded barrier is shown schematically in fig 2(b). The resulting conduction band profile is shown schematically in fig 2(c).

Fig. 2(a) Conventional MODFET Conduction Band

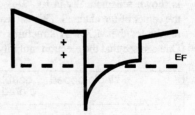

Fig. 2(c) Ideal Conduction Band

after Grading

Fig. 2(b) Grading of the Ga Composition in GaInP

The advantages of the graded barriers are obvious. First of all it leads to the enhanced electron sheet density. The simple analytical model [2] showed 30% increase of the carriers in the channel. Second, material can be strained to a higher value where needed. And the most important is the reduced amount of parasitic electrons in the barrier.

This structure was grown at the Fraunhofer Institute [3]. The Hall measurements are shown in fig.4. As one can see mobility of the single side doped transistors is above $6100 cm^2/Vs$ and for double side doped devices with graded channel it is above $5800 cm^2/Vs$ at room temperature.

4. Concept of Graded Channels

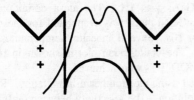

Fig. 3(a) Conventional Channel in the

Double Doped MODFET

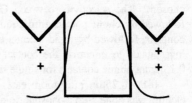

Fig. 3(c) Ideal Conduction Band after Grading

Fig. 3(b) Potential due to the Grading of the In Composition in the Channel

A conventional channel of a double doped MODFET is shown in fig. 3(a). One can note that due to the charge in the channel the conduction band has a parabolic bending in the channel. This leads to a double spike in the electron distribution. Thus electrons are held close to the

832

interfaces. This causes high interface and remote ion scattering. It can also increase the gate leakage current due to the electron injection in the barrier. The obvious solution to this problem is to distribute the electrons more broadly, far from the interfaces. One way to do it is to parabolically grade the channel [6]. The schematic grading of the In composition in the channel is shown in fig. 3(b). The resulting electron distribution and conduction band profile is shown schematically in fig. 3(c). In the graded channel electrons are distributed closer to the center of the channel. This decreases interface and remote ion scattering. The experimental results for the MODFET structures with and without grading channel are shown in fig. 4. One can see that the electron mobility is higher in the structures with graded channels.

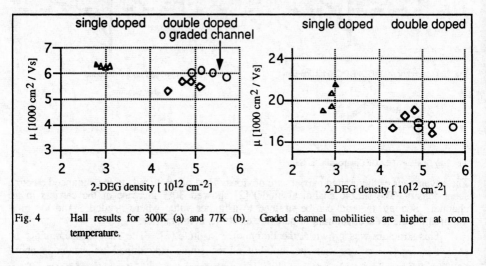

Fig. 4   Hall results for 300K (a) and 77K (b). Graded channel mobilities are higher at room temperature.

5. MODFET fabrication and device results

MODFET fabrication was done in a standard way utilizing the GaInP / GaAs system advantages. Mesa definition was carried out by wet etching. An acetic acid based etchant was chosen. The selectivity between GaInP and GaAs was 1:1. The mesa definition was followed by ohmic contact formation. Au/Ag/AuGe/Ni metal was evaporated for the ohmic contacts, followed by rapid thermal annealing for 10s. Several annealing temperatures were investigated to minimize the contact resistance. The optimization plot is shown in fig. 5. 0.15Ωmm ohmic contacts for single side doped MODFETs were achieved.

0.1 – 0.25μm mushroom gates were defined using electron beam lithography. Recess etching was done with a phosphoric acid based solution. The selectivity between GaInP and GaAs is >300. Such a selectivity cannot be achieved for AlGaAs/GaAs MODFETs.

Device results are shown in fig. 6. The single side doped layer design was optimized for 0.25μm gate length MODFETs which have good DC performance. Fig. 6(a) shows the IV curve for such a device. The saturation current is nearly 700mA/mm with a negligible gate current of less than 1mA/mm. A recent publication reported saturation currents of only 550mA/mm with a doped channel [1]. Our gate drain diode showed a -1mA/mm breakdown voltage of -5.2V. Measurements of the breakdown voltage using a novel method [7] yielded approximately the same result. The barrier height was measured to be 0.9V, and the ideality factor 1.7. The transconductance, as shown in fig. 6(b), is 410mS/mm with a drain bias of 1.5V and a current of 195mA/mm. With the output conductance of 21mS/mm at this bias point a voltage gain of nearly 20 is achieved.

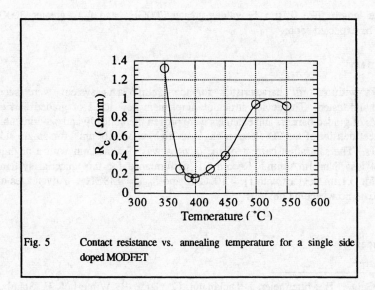

Fig. 5   Contact resistance vs. annealing temperature for a single side doped MODFET

No dispersion was observed between DC and RF data. The 0.25μm gate length MODFETs achieved f_T's about 70GHz. Record RF performance for this material system can be reported here for 0.1μm devices with f_T = 106GHz and f_{max} = 188GHz. With an optimized design for

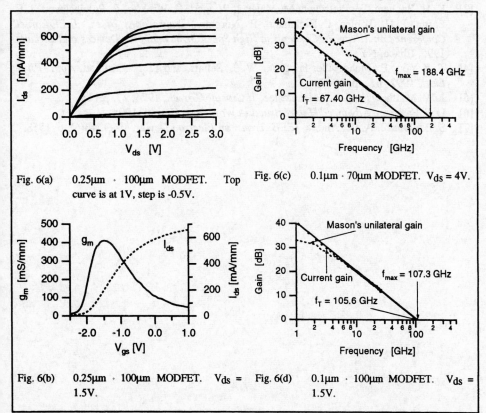

Fig. 6(a)   0.25μm · 100μm MODFET.   Top curve is at 1V, step is -0.5V.

Fig. 6(c)   0.1μm · 70μm MODFET. V_{ds} = 4V.

Fig. 6(b)   0.25μm · 100μm MODFET. V_{ds} = 1.5V.

Fig. 6(d)   0.1μm · 100μm MODFET. V_{ds} = 1.5V.

short gate lengths like 0.1µm f_T values greater 200GHz and f_{max} greater 300GHz can therefore be expected soon.

6. Summary

MODFETs with record characteristics for the GaInP/GaAs system were successfully fabricated and tested. The device structures implement new ideas of graded barriers. The advantages of graded barriers and channels were proven theoretically and experimentally. The carrier sheet density of the 2D electron gas is as high as $3.1 \ 10^{12} cm^{-2}$ for single side doped MODFETs. The saturation current of the devices was 700mA/mm with a negligible gate current of less than 1mA/mm. Record RF performance for this material system can be reported for 0.1µm devices with $f_T = 106GHz$ and $f_{max} = 188GHz$. Advantages of GaInP over AlGaAs now become obvious.

References

[1]   D. Geiger, E. Mittermeier, J. Dickmann, C. Geng, R. Winterhof, F. Scholz, and E. Kohn, IEEE Electron Device Letters, 1995, 16(6), p. 259.

[2]   G. H. Martin, M. Seaford, R. Spencer, J. Braunstein, and L. F. Eastman, *Transactions of the IEEE/Cornell Conference on Advanced Concepts in High Speed Semiconductor Devices and Circuits,* 1995, Ithaca, NY.

[3]   K. H. Bachem, W. Pletschen, M. Maier, I. Wiegert, K. Winkler, B. Pereiaslavets, L.F. Eastman, H. Tobler, I. Dickmann, P. Norozny, *Transactions of the IEEE/Cornell Conference on Advanced Concepts in High Speed Semiconductor Devices and Circuits,* 1995, Ithaca, NY.

[4]   T. K.Yoo, P. Mandeville, H. Park, W. J. Schaff, and L. F. Eastman, *Appl. Phys. Lett.,* 1992, **61** , p. 1942.

[5]   J. W. Matthews and A. E. Blakeslee, *J. Crystal Growth,* 1974, **27**, p.118

[6]   L. F. Eastman, *Advanced Heterostructure Workshop,* Hawaii, 1994

[7]   S. Bahl and J. A. del Alamo, *IEEE Trans. on Electron Devies,* 1993, **40**(6), p. 1558.

Inst. Phys. Conf. Ser. No 145: Chapter 5
Paper presented at 22nd Int. Symp. Compound Semiconductors, Cheju Island, Korea, 28 August–2 September 1995
© *1996 IOP Publishing Ltd*

AlInAs/GaInAs/AlInAs MODFETs fabricated on InP and on GaAs with metamorphic buffer - a comparison

T. Fink, M. Haupt, K. Köhler, B. Raynor, J. Braunstein, H. Massler, and P. J. Tasker

Fraunhofer-Institut für Angewandte Festkörperphysik (IAF)
Tullastr. 72, D-79108 Freiburg, Germany
Phone: +49-761-5159-824, Fax: +49-761-5159-200, e-mail: fink@iaf.fhg.d400.de

Abstract. Employing molecular beam epitaxy, we have grown identical $Al_{0.48}In_{0.52}As/Ga_{0.47}In_{0.53}As/Al_{0.48}In_{0.52}As$ MODFET structures lattice-matched to InP substrates asl as on GaAs substrates using a metamorphic buffer. To fabricate FETs on both substrate types, a fully planar process was developed using ion implantion for device isolation, electron-beam lithography for gate definition, and selective reactive ion etching for gate recess etching. Presently, a yield of 90% for microwave transistors with 0.2 and 0.3 µm gate length and 70% for 0.1-µm devices is achieved. For both substrate types, InP as well as GaAs with metamorphic buffer, very similar transistor performance was observed. Typically measured values are 800-900 mS/mm for $g_{m,max}$, -0.65 V for the threshold voltage, and 90 GHz and 165 GHz for f_T for devices with 0.3 µm and 0.1 µm gate length, respectively. Hence, identical electron transport properties can be assumed for both substrates.

1. Introduction

GaInAs/AlInAs MMICs show clear performance advantages over their GaAs/AlGaAs rivals, especially at higher frequencies where GaAs circuits have higher noise figures and less gain. This is mainly due to the higher values of sheet carrier density, mobility and saturation velocity of the GaInAs/AlInAs/InP system. However, processing of InP substrates is still a difficult task for operators as they are more fragile than GaAs. In addition, at this time, GaAs wafers are available in sizes up to 6 inch diameter whereas for InP the limit is currently 3 inch. Since processing equipment is usually oriented toward the much larger wafer sizes used by the silicon industry, it is increasingly difficult for commercial manufacturers to establish process lines for 2 or 3 inch wafers. A possible solution to these drawbacks for InP based circuit manufacturing could be the epitaxial growth of layer structures lattice-mismatched to the substrate. Originally aimed at combining GaAs and Si, this interesting possibility has been under investigation for some time, and over the last years, also successes in matching the lattices of GaAs and InP by means of metamorphic buffers have been reported [1-4]. Recently, record performance data for GaInAs/AlInAs FETs on metamorphic buffers were obtained [5]. In particular, using metamorphic buffers to grow a substrate with a 'new' lattice constant, transistor structures with higher indium content than the InP based devices can also be developed.

In this paper a direct comparison between the performance of $Al_{0.48}In_{0.52}As/Ga_{0.47}In_{0.53}As/Al_{0.48}In_{0.52}As$ MODFETs with identical layer structure, but on different substrates, InP and metamorphic buffered GaAs substrate, is made. In addition, devices on both substrate types were fabricated employing the same process sequence. The results show that although dislocations are present in the metamorphic substrate type, DC and RF properties of both transistor types are similar.

2. Epitaxial layer structure

The $Al_{0.48}In_{0.52}As/Ga_{0.47}In_{0.53}As/Al_{0.48}In_{0.52}As$ MODFET layers were grown on 2-inch substrates employing solid-source molecular beam epitaxy. For both substrate types, InP and GaAs, the electrically relevant epilayers are identical: a 40 nm thick GaInAs channel covered by 20 nm of AlInAs which contains a δ-doping of 7×10^{12} cm^{-2} at 5 nm distance to the channel, and a 10 nm thick n$^+$ doped cap layer (n = 3×10^{18} cm^{-3}) (Fig.1).

For InP substrates, first a 250 nm thick AlInAs buffer cladded between two superlattices is grown, followed by the MODFET layers (Fig. 1a). For GaAs substrates, a metamorphic buffer is grown at 400° C consisting of 100 nm GaAs, 800 nm $Al_{0.33}Ga_{0.32}In_{0.35}As$, and 800 nm $Al_{0.48}Ga_{0.52}As$. Using this sequence, the transition from the GaAs lattice constant to that of InP is achieved in two steps. After depositing an additional buffer of 250 nm AlInAs topped with a superlattice, the MODFET layers can be grown at the usual temperature of 550 °C (Fig. 1b).

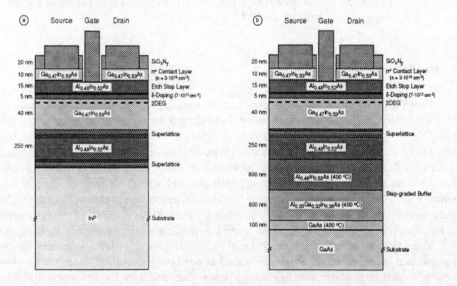

Fig. 1: Layer structure for GaInAs/AlInAs MODFETs on InP (a) and GaAs (b) substrate.

Although the introduction of dislocations during growth of the metamorphic buffer cannot be prevented, we obtained a surface roughness of better than 6 nm. Hall measurements gave a value of 9200 cm^2/Vs at 300 K for the electron mobility which is inferior to the value of 11000 cm^2/Vs for lattice-matched epilayers on InP. However, the sheet carrier density was 3×10^{12} cm^{-2} for both types of epilayer structures.

3. Device fabrication

To fabricate FETs on both substrate types, we developed a fully planar process using ion implantion for device isolation, electron-beam lithography for gate definition, and selective reactive ion etching for gate recess etching [6]. Except for the FET gates, optical contact lithography is used in all other process steps.

The process sequence starts with the deposition of ohmic contacts for source and drain employing metal evaporation and a standard lift-off procedure. The active areas are then isolated from each other by oxygen ion isolation implantation followed by an annealing step for implant activation. For the fabrication of transistor gates, we are using two- or three-layer PMMA resist systems depending on the desired type of gate profile. Direct-write electron-beam exposure of the two-layer resist followed by development, gate-recess etching, metal evaporation, and lift-off results in a trapezoidal shape of the gate profile. The minimum gate length which can be achieved with this process is 0.2 μm. In the case of the three-layer resist, a 'mushroom' gate profile is obtained with a footprint as small as 0.1 μm.

Gate recess etching for removal of the n^+ doped GaInAs cap layer is performed in a reactive ion etching (RIE) reactor employing an HBr/Ar plasma at a bias voltage of -110 V. After gate metal deposition, further metalization levels for device interconnection are processed according to standard procedures used for MMIC and OEIC fabrication in our laboratory. Currently, a yield of 90% for microwave transistors with 0.2 and 0.3 μm gate length and of 70% for 0.1-μm devices is obtained.

Tab. 1: Comparison of extrinsic DC and RF data for 0.3-μm microwave transistors of 2×50 μm width on InP and GaAs substrates, measured at $V_{ds} = 1.0$ V.

	InP Substrate	GaAs Substrate
V_{th} [V]	-0.65	-0.65
$g_{m,max}$ [mS/mm]	900	800
$I_{ds,max}$ [mA/mm]	820	800
R_s [Ωmm]	0.4	0.4
f_T [GHz]	95	90
f_{max} [GHz]	140	130

838

4. Device performance

In DC and RF measurements, transistors of both types of substrates behaved very similar. Hence, identical electron transport properties can be assumed in both cases. Typical data for 0.3-μm microwave transistors with 2×50 μm gate width are given in Tab. 1. These data represent average values; currently, for most parameters, our process reproduces the values given with a standard deviation of approximately 15%. To date, 0.1-μm 'mushroom'-gate FETs have been fabricated only on InP. Threshold voltage, currents and transconductance values for these transistors are similar to the 0.3-μm devices, whereas f_{max} values of 165 GHz were obtained due to the short gate length.

5. Conclusions

In summary, with both types of substrates, InP as well as GaAs with metamorphic buffer, very similar device yield as well as DC and RF transistor performance were found. From the production point of view, this means that the current restriction of fabricating GaInAs/AlInAs based devices and circuits on 2- or 3-inch InP substrates can be overcome by using the larger wafer sizes of GaAs with an additional metamorphic buffer. In addition, although this work was restricted to investigating a MODFET structure with 50% indium for comparison with InP substrates, the metamorphic buffer can easily be grown to match a transistor structure with higher percentages of indium.

Acknowledgments

The authors would like to thank W. Bronner, B. Campillo, S. Emminger, P. Ganser, K. Glorer, R. Haddad, P. Hofmann, A. Hülsmann, G. Kaufel, M. Korobka, M. Krieg, D. Luick, T. Norz, E. Olander, K. Schäuble, G. Schilli, J. Schneider, and B. Weismann for technical assistance in the process laboratory and helpful discussions; N. Grün, J. Rüster, and J. Windscheif for DC measurements; M. Sedler for circuit design; and M. Berroth, T. Jakobus, and H. S. Rupprecht for project management. Financial support by the Bundesminister für Wissenschaft und Bildung (BWB) under grant DFE-C is gratefully acknowledged.

References

[1] P. Win, Y. Druelle, P. Legry, S. Lepilliet, A. Cappy, Y. Cordier, and J. Favre, Electron. Lett. **29**, 169 (1993).
[2] G.-W. Wang, Y.-K. Chen, W. J. Schaff, and L. F. Eastman, IEEE Trans. Electron Devices **ED-35**, 818 (1992).
[3] K. Inoue, J. C. Harmand, and T. Matsuno, J. Crystal Growth **111**, 313 (1991).
[4] M.Chertouk, H. Heiß, D. Xu, S. Kraus, W. Klein, G. Böhm, G. Tränkle, and G. Weimann, Proc. 7th Int. Conf. on InP and Related Materials, Sapporo (1995), p. 737-740.
[5] K. Higuchi, M. Kudo, M. Mori, and T. Mishima, Techn. Digest IEDM '94, 891 (1994).
[6] T. Fink, B. Raynor, M. Haupt, K. Köhler, J. Braunstein, N. Grün, and J. Hornung, J. Vac. Sci. Technol. **B 12**, 3332 (1994).

Inst. Phys. Conf. Ser. No 145: Chapter 6
Paper presented at 22nd Int. Symp. Compound Semiconductors, Cheju Island, Korea, 28 August–2 September 1995
© *1996 IOP Publishing Ltd*

Controlling dissipation in quantum devices

Jean-Pierre Leburton and Yuli Lyanda-Geller

Beckman Institute for Advance Science and Technology and
Department of Electrical and Computer Engineering,
University of Illinois at Urbana-Champaign, Urbana, IL 61801 USA

Abstract. With suitable combinations of material structure design and
geometrical confinement it is possible to tailor efficiently the electronic properties of nanostructures to suppress electron-phonon scattering and considerably reduce the dissipation level in quantum devices. Examples of quenching
of phonon-assisted transport processes are illustrated in resonant tunneling
devices of low dimensionality.

1. Introduction

One of the critical issues for implementing nanoscale devices into practical applications
is the ability to control the level of dissipation during device operation. This issue is
particularly relevant for high density and high speed integrated circuits because it can
imperil the system viability. With advances in nanofabrication techniques, it become
feasible to fabricate quantum nanostructures (QNS) with multiple degrees of confinement which manifest such fundamental phenomena as quantum interference, resonant
tunneling (RT) and single-electron charging effects. It is also well known that the carrier density of states (DOS) which is directly related to the total scattering rate varies
sharply with reduced degrees of freedom and lower dimensionality of quantum systems.
Consequently QNS can be designed with particular geometrical confinement to achieve
arbitrary spectra of electronic states which determine the QNS transport properties,
and enable a control of their level of dissipation. This provides new opportunities for
technological innovation in high speed and high functionality device engineering.

In this paper, we discuss the interplay between appropriate design and carrier
confinement in devices of low dimensionality in order to efficiently reduce phonon-assisted
transport and dissipation in quantum structures. In particular, we demonstrate that
negative differential resistance (NDR) characteristics in Resonant Interband Tunneling
(RIT) devices or RT devices of lower dimensionality such as quantum dots can exhibit
zero-valley current and total quenching of phonon-assisted transitions.

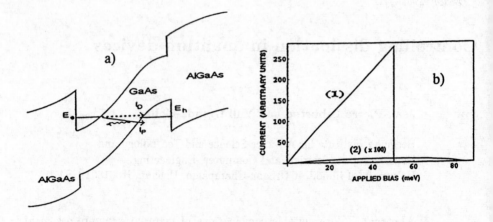

Figure 1. Current-Voltage characteristics of RIT device: a) Schematic representation of direct I_D and POP-assisted I_P transitions in a RIT-BiTFET structure with confined electron and hole states; b) Direct (1) and phonon-assisted (2) tunneling currents in the RIT-BiTFET device described in the text.

2. Quenching of phonon-assisted transitions in resonant tunneling structures

With improved resolution in crystal growth and delta-doping techniques renewed attention has been given to interband tunneling structures [1]. NDR due to RIT with large peak-to-valley current ratios has been demonstrated in two-terminal devices [2]. In the meantime several proposals for RIT-based transistors with important applications in digital and microwave systems have been reported. One of the essential issues for implementing RIT device into practical applications is the ability to control the level of dissipation during the device operation [3]. Specifically, the limit of the valley current in resonant tunneling electronic devices, and, correspondingly, the lowest achievable power dissipation is determined by phonon-assisted transitions.

Let us consider a RIT structure which consists of a heavily-doped p-n junction of a low-gap material (GaAs) sandwiched between a high-gap material (AlGaAs) (Fig 1.a). The homojunction creates a narrow tunneling barrier (similar to a tunnel diode) between the conduction and valence bands. Charge on both sides of the tunnel junction is confined with a 2D density of states at the heterojunction between the wide gap material and the depleted p-n junction. This structure is similar to a BiTFET (bipolar tunneling field-effect transistor) where tunneling is transverse and is controlled by an external gate bias [4]. Heterojunctions form a barrier which efficiently reduces thermionic emission current of minority carriers. In these conditions NDR which occurs when two quantized electron and hole states cease to overlap, is expected to be abrupt with a large peak to valley ratio.

Although, as a result of several overlapping electron and hole channels, multiple negative NDRs with large peak-to-valley ratios have been predicted for this structure we only consider one overlapping electron and (light) hole channel in the present analysis. Moreover, since elastic resonant tunneling between electron and heavy hole channel is negligible compared to the light hole process, we only examine polar optic phonon (POP)-assisted tunneling between an electron and a light hole state, which is expected to be the main contribution to phonon-assisted current. Owing to the high 2D channel conductance, carriers are assumed to be immediately collected in the terminal current flowing parallel to the interface, once they have tunneled.

It can be shown that the ratio between direct tunneling j_D and POP-assisted contributions J_{ph} is given by [5]

$$\frac{j_{ph}}{j_D} \sim \frac{\hbar\Omega e^2 \left(m\omega/\hbar\right)^{1/2} kT \left(\frac{1}{\epsilon_\infty} - \frac{1}{\epsilon_0}\right)}{E_g^2 \hbar\omega}, \tag{1}$$

which is usually small because of the large value of the band gap E_g. This particular expression is due to the electron-phonon interband matrix element between electron and hole states of different $|s>$ and $|p>$ symmetries, in contrast to double quantum barrier RT structures for which intraband tunneling usually occurs between electron states of the same symmetry i.e. $|s>$ states in the conduction band. In our analysis we have assumed that electrons and light holes are characterized by the same effective mass $m = 0.08 m_o$, and the same half-parabolic confinement in the c onduction and valence, with characteristic energy $\hbar\omega =50$ meV. $\hbar\Omega = 36$ meV is the phonon energy and T = 300K. Fig 1.b displays the POP-assisted current and shows the direct and the phonon contributions to the current. We have taken an initial separation $(V = 0)$ between valence band and conduction band $E_v - E_c = 200$ meV across the homojunction. When electron and hole states cease to overlap $(eV = \hbar\omega = 50$ meV) direct tunneling drops to zero (Fig. 1b), while POP-assisted transmission still occurs, until the energy separation between electron and hole levels exceeds the phonon energy $(eV < 86$ meV). It is seen that the POP-assisted tunneling is three orders of magnitude lower than coherent tunneling.

It is important to notice, that in our type of RIT structures electrons are confined in direct semiconductors with the conduction band located at the Γ symmetry point. By contrast, in indirect semiconductors the overlap between the Bloch amplitudes in conduction band, which could be characterized by the X symmetry, and the Bloch amplitudes of the valence band, may not vanish and would lead to significant contribution of POP-assisted tunneling because these transitions are no longer forbidden for symmetry reason at the band edge. This is particularly the case in RIT devices made of type II heterostructures with vectorial (e.g. X or L) symmetry of electron states or with strong $\Gamma - X$ mixing [6]. Let us point out that estimates of the acoustic–phonon–assisted tunneling indicate that their contribution is smaller than that of optical phonons.

In conjunction with the 2D character of the electron and hole states, the suppression of POP-assisted current in RIT structures due to the symmetry difference between Γ-electron and hole wavefunctions is predicted to be responsible for the very low level of dissipation and valley current, which in real NDR structures may be mostly determined by direct tunneling channel between confined electron and heavy hole states [2]. Consequently, POP-assisted tunneling is the lowest achievable limit of the RIT valley current.

3. DOS tailoring and quenching of phonon scattering in quantum structures

In this section, we discuss the possibilities of controlling phonon-assisted processes in nanostructures by using geometrical confinement to limit the number of final scattering states.

3.1. Suppression of polar optic phonon scattering

The possibility of enhancing the transport performances of electronic devices by suppressing POP scattering has been suggested by Sakaki [7] by modulating periodically the semiconductor electronic properties with linear chains of coupled quantum boxes. In the extended electron picture, the widths of the miniband ($\epsilon_b = 4\Delta$) and the minigaps ϵ_g resulting from the periodic structure of the quantum box superlattice (QBSL) can be modified by changing the period and amplitude of the modulating potential. If the miniband width is less than the POP energy ($\epsilon_b < \hbar\Omega$), intra-miniband POP scattering can not occur. Also if the minigaps are greater than the POP energy ($\epsilon_g > \hbar\Omega$), no inter-miniband POP scattering can take place. Consequently, the electron mobility is essentially limited by deformation-potential acoustic-phonon (DAP) scattering which is much waker than POP scattering. Recently, Noguchi [8] investigated theoretical transport in periodically modulated quantum well wires (QWW) by using an iterative technique based on the nearly free-electron model and normal (N) DAP process. In the condition of suppressed POP scattering, they found DAP limited mobilities slightly higher than the GaAs bulk mobility with the structure of the DAP scattering quite similar to the density of states. (DOS). Recent calculations based on a more accurate DAP scattering model have shown a significant enhancement of the transport properties [9].

3.2. Suppression of Acoustic Phonon scattering

The extended electron picture resulting in energy miniband spectrum in QBSL discussed above is valid for relatively small longitudinal electric field F. If the potential drop over the QBSL period d, eFd, exceeds the collisional broadening of the electron levels $\hbar/\tau$, the electronic subband splits into a Wannier-Stark (WS) ladder of localized states

$$\Psi_{nm\alpha} = \sum_{\nu} J_{\nu-\alpha}\psi_{nm\nu}, \tag{2}$$

with the corresponding eigenvalues are

$$E_{nm\alpha} = \epsilon_{nm} - eFd\alpha, \tag{3}$$

where $\psi_{nm\nu}$ is the electron wave functions in QBSL in the Wannier representation, m and n are integer numbers, ν, is the QB label, ϵ_{nm} is the transverse energy, $J_k(2\Delta/eFd)$ are the Bessel functions of order k and α is the Stark diagonal representation index Therefore the band conduction breaks down and electrons move in the z-direction only by hopping from one well to the other due to scattering; because of discrete spectrum ϵ_{mn}, in QBSL electron hopping exists only when the transverse energy spacing is equal to the separation between the WS levels or differs from the latter by the phonon energy (Fig. 2a). It follows then that elastic and optical phonon scattering should manifest in a number of resonant peaks in the current. However, the only background mechanism

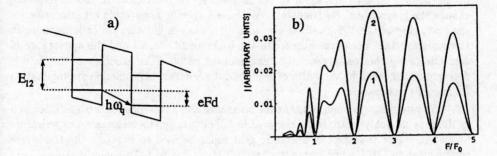

Figure 2. Hopping conductance in quantum box superlattices: a) Schematic representation of phonon-assisted transition in a QBSL with longitudinal WS localization and the first two transverse spatially confined energy states ; b) Current in the conditions of partial occupation of the first level of size quantization in quantum boxes. $F_0 = 2.17$ kV/cm. Curve 1: $E_{12} = 10$ meV; Curve 2: $E_{12} = 20$ meV.

for conduction in QBSL is scattering by acoustic phonons. We describe [10] the hopping conductance between the WS localized states, by the following transparent formula

$$j_z = e \sum_{\alpha,\alpha',N,N'} (z_{\alpha,N} - z_{\alpha',N'}) W_{N,N'}^{\alpha,\alpha'}, \qquad (4)$$

where $z_{\alpha,N} - z_{\alpha',N'}$ is the electron displacement (the hopping length) upon the scattering $(\alpha, N \rightarrow \alpha', N')$, N stands for the set of indices describing the transverse states, $W_{N,N'}^{\alpha,\alpha'}$ is the scattering probability. For POP-assisted hopping

$$W_{NN'}^{\alpha\alpha'} = \sum_{q_\perp q_z} |C_{N,N'}^{\alpha,\alpha'}|^2 \delta(E_{\alpha N} - E_{\alpha'N'} + \hbar\omega_q)[(f_{\alpha N} - f_{\alpha'N'})N_q - f_{\alpha'N'}(1 - f_{\alpha N})], \qquad (5)$$

ω_q and N_q are the acoustic phonon frequency and the occupation number, $f_{\alpha N}$ is the nonequilibrium electron distribution function,

$$| C_{N,N'}^{\alpha,\alpha'} |^2 = | V_{\mathbf{q}} |^2 | \langle N|e^{iq_\perp r_\perp}|N' \rangle |^2 J_{\alpha-\alpha'}^2 \left(\frac{4\Delta}{eFd} \sin \frac{q_z d}{2} \right). \qquad (6)$$

We see that phonons with $q_z = \frac{2\pi n}{d}$ result in vanishing electron transition between different wells α ($J_{\alpha-\alpha'}(0) = 0$ at $\alpha \neq \alpha'$). Assume that the dispersion of acoustical phonons is described by a constant speed of sound s and that phonons propagate in the z−direction. Then the energy conservation law $\hbar s q_z = eFd$ determines q_z. At electric fields $F = nF_0$, n is the integer number,

$$F_0 = 2\pi\hbar s/ed^2, \qquad (7)$$

phonons are ineffective. The scattering probability Eq. (5,6) and the hopping current Eq. (4) vanish (Fig 2.b). This effect takes its origin in Bragg reflection: the transfer of the phonon momentum equal to the momentum of the reciprocal lattice does not change the longitudinal electron state. We have a specific momentum selection rule for the scattering of the WS localized electrons. If there are no other acoustic phonons in the structure this means the absence of the background current and the appearance of zero minima in the conductance, i.e. antiresonant effect. (The accuracy of this zero is determined, as usual, by the width of levels and the accuracy of δ-function approximation for energy conservation)

We would like to point out that the resonance and antiresonance features discussed in the present analysis are to be observed in hopping magnetoconductance experiments at low T, where the effect of magnetic field could be used to simulate the transverse confinement in QBSL. For instance, the resonances due to LO phonon scattering will occur in the magnetoconductance, and the absence of acoustic phonon scattering will manifest in magnetocondactance antiresonances. Earlier claims on zero resistance in the absence of scattering in SL [11] are incorrect because for $\hbar/\tau < eFd$ (and, naturally, in the absence of scattering) the electron states are localized, and consequently cannot carry current irrespective of the absolute value of electric field. LO phonon resonance will result in a resistance maximum only for $\hbar/\tau > eFd$, i.e., in the conditions of miniband conduction (section 3.1).

In summary, we have shown that the influence of quantum confinement on electron states in nanostructures leads to negative differential resistance characterized by valley currents with extremely low dissipation level.

Acknowledgments

This work is supported by NSF under Grant No. ECS 91-08300 and by the U.S. Joint Service Electronic Program under contract No. N00014-90-J-1970.

References

[1] Capasso F, Sen J and Beltram F 1990 *High Speed Semiconductor Devices* 465

[2] Day D J, Yang R Q, Lu J and Xu J M 1993 *Appl. Phys. Lett.* **73** 1542

[3] Bate R T 1989 *Solid State Techn.* **32** 101

[4] Leburton J P, Kolodzey J and Briggs S 1988 *Appl. Phys. Lett.* **52** 1608

[5] Lyanda-Geller Y and Leburton J P 1995 *Appl. Phys. Lett.* **67**

[6] Tehrani S, Shen J Goronkin H, Kramer G, Tsui R and Zhu T X 1994 *Proc. Intl. Symp. Compound Semicond. 855–60*

[7] Sakaki H 1989 *Japan. J. Appl. Phys.* **28** L314

[8] Noguchi H, Leburton J P and Sakaki H 1993 *Phys. Rev.* **B47** 15593

[9] Stocker, J, Leburton J P, Noguchi H and Sakaki H 1994 *J. Appl. Phys.* **76** 4231

[10] Lyanda-Geller Y and Leburton J P 1995 *Phys. Rev.* B **52**

[11] Polyanovskii V M 1983 *Sov. Phys. Semicond.* **17** 1150

Inst. Phys. Conf. Ser. No 145: Chapter 6
Paper presented at 22nd Int. Symp. Compound Semiconductors, Cheju Island, Korea, 28 August–2 September 1995
© 1996 IOP Publishing Ltd

Magnetic-field-induced tunneling and minigap transport in double quantum wells

S K Lyo, J A Simmons, N E Harff, T M Eiles, and J F Klem

Sandia National Laboratories, Albuquerque, NM 87185, U. S. A.

Abstract. We review recent theoretical and experimental results on low-temperature tunneling and in-plane transport properties in double quantum wells (DQWs) in an in-plane magnetic field $B_{\parallel}$. These properties arise from the combined effect of $B_{\parallel}$-induced relative displacement of the wave vectors in the two QWs and the interwell tunneling. In weakly coupled DQWs, the tunneling conductance has two sharp maxima as a function of $B_{\parallel}$. In strongly coupled DQWs, a partial minigap is formed due to the anticrossing of the two QW dispersion curves, yielding sharp $B_{\parallel}$-dependent structures in the density of states and in-plane transport properties. Excellent agreement is obtained between the theory and the data from GaAs/AlGaAs DQWs.

1. Introduction

In this paper, we discuss low-temperature tunneling and in-plane transport properties of double quantum wells (DQWs) in an in-plane magnetic field $B_{\parallel} \parallel x$. DQWs consist of two parallel layers of two-dimensional (2D) degenerate electron-gases (2DEGs) separated by a barrier. Recent surge of interest in DQW structures can be attributed to the fact that DQWs display richer unique physical properties than single QWs due to an extra degree of freedom in the growth direction ($\parallel z$). Some of the examples are the interlayer Coulomb-drag effect [1] in zero field, the tunneling Coulomb gap [2], the interlayer-tunneling excitonic effect [3] in strong perpendicular fields $B_{\perp}$, and 2D-2D interwell tunneling in $B_{\parallel}$ [4-8]. In DQWs, one can control the well-to-well separation and the interwell overlap, both sensitive to the barrier thickness. The charges in the wells can be controlled by gate voltages. We show that $B_{\parallel}$ provides an additional control of the effective interwell coupling by displacing the wave vectors k in the two QWs. As a result, $B_{\parallel}$ can be used to tune the tunneling current and deform the Fermi surface, introducing sharp $B_{\parallel}$-dependent structures in the density of states (DOS) and other in-plane transport properties [9-13]. In contrast to the high $B_{\perp}$ case [2, 3], Coulomb interactions are used in $B_{\parallel}$ only for band bending corrections and can otherwise be neglected: our data are well explained using the combined effect of tunneling and $B_{\parallel}$ within the frame work of noninteracting electrons. Three samples discussed in this paper are symmetric (except for sample 1) GaAs/Al$_{0.3}$Ga$_{0.7}$As DQWs with well widths w and depths V_0. The GaAs wells are separated by an Al$_{0.3}$Ga$_{0.7}$As barrier of thickness t.

The primary effect of $B_{\parallel}$ is to induce a linear transverse displacement $\Delta k_y \propto B_{\parallel}$ in k-space in one QW relative to the other. Here the in-plane wave vector k is a good quantum number. The linear displacement arises from the second term of the Hamiltonian:

$$H = \frac{p_z^2}{2m^*} + \frac{\hbar^2}{2m^*}(k_y - \frac{z}{\ell^2})^2 + V(z),$$

(1)

where $p_z = -i\hbar\partial/\partial z$ and $\ell = (\hbar c/eB_{\parallel})^{1/2}$. The GaAs effective mass $m^* = 0.067$ (in units of the free electron mass) is used in both the QWs and the barriers because the confinement wave functions $\phi_n(z)$ (n = 1, 2) penetrate negligibly into the barriers. The kinetic energy $\varepsilon(k_x) = (\hbar k_x)^2/2m^*$ is to be added to (1). The confinement potential $V(z)$ is a superposition of the potentials $V_1(z)$ and $V_2(z)$ of QW1 and QW2. We assume quasi-2D (i.e., $w < \ell$) thin QW's where only the ground sublevels are populated and are relevant. For the sake of physical argument and for 2D-2D tunneling in Sec. 2, z in the second term of (1) can be replaced by its expectation values $<z>_1$, $<z>_2$ with respect to $\phi_n(z)$ [6]. The net effect of $B_{\parallel}$ is then to shift the origin of k_y of QW2 by $\Delta k_y = d / \ell^2 \propto B_{\parallel}d$ relative to that of QW1, where $d = <z>_2 - <z>_1$. The basic transport properties to be discussed in the following arise from the $B_{\parallel}$-induced shift Δk_y and the tunneling between the two QWs. Spin splitting is neglected in this paper.

2. Resonant 2D-2D tunneling

The tunneling conductance G_{zz} is the steady-state current flowing from QW1 to QW2 per unit linear electric field applied between the QWs. The current flows into QW1 from a source, tunnels into QW2, and then flows out of QW2 to the drain. The source-drain resistance is related to the tunneling conductance and the zero-$B_{\parallel}$ conductances (i.e., mobilities) of the QWs in terms of a transmission-line model [6,7]. The G_{zz} data is obtained from the source-drain resistance data using this relationship [7]. The latter depends on the geometrical structure of the source-drain current paths of the sample [6]. The electrons are rapidly scattered inside the QWs, occasionally tunnelling into the other QW with the tunneling integral J. A weak dependence of J on $\mathbf{k}$ is ignored [6]. For tunneling from $\mathbf{k}$ in QW1 to $\mathbf{k'}$ in QW2 (with $k = |\mathbf{k}|$ and the sublevel energy Δ_n), the initial and final energies are given by

$$\varepsilon_{1\mathbf{k}} = \frac{(\hbar k)^2}{2m^*} + \Delta_1, \quad \varepsilon_{2\mathbf{k'}} = \frac{(\hbar k_x')^2}{2m^*} + \frac{\hbar^2(k_y' - \Delta k_y)^2}{2m^*} + \Delta_2. \tag{2}$$

Only the electrons on the Fermi circles can tunnel when the energy and momentum conservations are satisfied: $\varepsilon_{1\mathbf{k}} = \varepsilon_{2\mathbf{k'}}$ and $\mathbf{k} = \mathbf{k'}$ [5]. For 2D densities $N_1 > N_2$, the Fermi surfaces are concentric circles of radii k_1 and k_2 at $B_{\parallel} = 0$ as illustrated in the inset of Fig. 1(a) and the conservation conditions are not satisfied. As $B_{\parallel}$ is increased, the inner smaller circle slides relative to the outer circle until they touch each other tangentially. The conservation conditions are satisfied at this $B_{\parallel}$, yielding a G_{zz} peak. As $B_{\parallel}$ is increased further, the inner circle begins to move outside the larger circle, intersecting it at two points. G_{zz} begins to decrease as the area of contact decreases. Another G_{zz} maximum is reached at a higher $B_{\parallel}$, when the smaller circle touches the larger circle from outside. G_{zz} begins to drop beyond this field as the two circles are separated as illustrated in Fig 1(b). The G_{zz} maxima diverge in the absence of scattering. However, the data from sample 1 displayed in Fig. 1(a) does not show the divergence due to finite scattering times as will be shown below. Sample 1 is weakly coupled and has $w = 150$ Å, $t = 65$ Å, $N_1 = 1.8\times10^{11}$ cm^{-2}, and $N_2 = 1.0\times10^{11}$ cm^{-2}.

The tunneling conductance is calculated using a linear response theory. The dominant contribution arises from the so-called bubble diagram shown in Fig. 1(b) and is given by [6]

$$G_{zz} = \frac{4\pi e^2}{\hbar} J^2 \sum_{\mathbf{k}} \int_{-\infty}^{\infty} [-f'(\zeta)]\rho_{1\mathbf{k}}(\zeta)\rho_{2\mathbf{k}}(\zeta)d\zeta, \tag{3}$$

where $f'(\zeta)$ is the first derivative of the Fermi function $f(\zeta)$ and

$$\rho_{n\mathbf{k}} = \frac{1}{\pi} \frac{\Gamma_n}{(\zeta - \varepsilon_{n\mathbf{k}})^2 + \Gamma_n^2}. \tag{4}$$

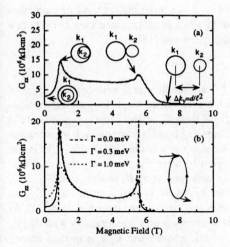

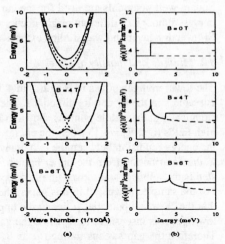

Figure 1. Tunneling conductance per unit area from the data from sample 1 (a) and theory (b). The insets show the relative Fermi surfaces of the QWs (a) and the bubble diagram (b).

Figure 2. (a) Energy dispersion curves from sample 2 with (solid curves) and without (dashed curves) tunneling. (b) The DOS from the lower (dashed curves) and both branches (solid curves).

Here $\tau_n = (2\Gamma_n/\hbar)^{-1}$ is the quantum scattering time of the electrons on the Fermi surface.

The mobilities in the wells are $\mu_1 = 6\times10^4$ cm^2/Vs and $\mu_2 = 6\times10^5$ cm^2/Vs, indicating that $\Gamma_1 \gg \Gamma_2$. In this case, G_{zz} becomes independent of Γ_2. The theoretical G_{zz} displayed in Fig. 1(b) for several values of Γ_1 and $J = 0.04$ meV, show good agreement with the data. This value of J is close to $J = 0.03$ meV estimated from the splitting of the two ground sublevels due to tunneling. For isotropic scattering, the mobility $\mu_1 = 6\times10^4$ cm^2/Vs corresponds to $\Gamma_1 = 0.15$ meV and $\tau_1 = 2.3\times10^{-12}$ s. The G_{zz} maxima diverge for $\Gamma_1 = 0$ at the fields $B_\pm = \hbar c\sqrt{2\pi}(\sqrt{N_>} \pm \sqrt{N_<})/ed$, where $N_> (N_<)$ is the larger (lesser) of N_1 and N_2.

3. In-plane transport properties

3.1. Energy dispersion and anticrossing

DQW structures show interesting in-plane transport properties when interwell tunneling is significant. The $B_\parallel$-induced linear displacement $\Delta k_y \propto B_\parallel$ of the origins of the transverse crystal momenta k_y in the two QWs is shown in Fig. 2(a) by dashed curves for sample 2 which has $w = 150$ Å, $t = 25$ Å, and $N_1 = N_2 = 1.5\times10^{11}$ cm^{-2}. For such a thin barrier (i.e., large overlap) between the QWs, the two energy-dispersion parabolas anticross significantly and a partial minigap is formed as shown by the solid curves therein. For symmetric DQWs, the latter correspond to the upper and lower branches of the eigenvalues of (1) (to the lowest order in the overlap $<\phi_1(z)|\phi_2(z)>$) [9, 13]:

$$\varepsilon_\pm(k_y) = \varepsilon(k_y) + \frac{d_1^2 + (\Delta z_1)^2}{\ell^2}\varepsilon_\ell \pm \left(4\varepsilon_\ell\varepsilon(k_y)d_1^2/\ell^2 + (\frac{E_g}{2})^2\right)^{1/2} + \frac{E_g}{2}, \tag{5}$$

where $\varepsilon(k_y) = (\hbar k_y)^2/2m^*$, $\varepsilon_\ell = (\hbar\ell^{-1})^2/2m^*$, $d_n = |<z>_n|$, and $d_1 = d_2$. Here $(\Delta z_n)^2 = <\phi_n(z)|(z - <z_n>)^2|\phi_n(z)>$ is the mean square deviation of z arising from finite widths of the QWs. In (5), use is made of $\Delta z_1 = \Delta z_2$. The minigap is independent of $B_\parallel$ and is given by

$$E_g = 2|S_{12}<\phi_1(z)|V_2(z)|\phi_1(z)> - <\phi_1(z)|V_1(z)|\phi_2(z)>|. \tag{6}$$

Square-well potentials are used for the curves in Fig. 2. A similar anticrossing of dispersion curves, although of different origin occurs without $B_\parallel$'s in a vicinal surface such as (911) of a Si-inversion layer and was studied extensively many years ago [14].

3.2. Minigap and density of states

The lower minigap edges in Fig. 2(a) at 4 T and 6 T are saddle points with opposite signs of curvatures in the k_x and k_y directions. The DOS has a van Hove singularity diverging logarithmically at the saddle point as shown in Fig. 2(b) at these $B_\parallel$'s. The latter is formed only at high fields (i.e., $B_\parallel$ = 4 T) when the energy of the crossing point is large enough to overcome the energy repulsion between the two branches. The sharp step in the DOS in Fig. 2(b) is due to the contribution from the upper branch. The region between the sharp singularity and the step is the minigap and moves up in energy rigidly with increasing $B_\parallel$ as shown in Fig. 2(b). For the samples discussed in this paper, the 2D density $N_{2D} = N_1 + N_2$ is sufficiently high so that the chemical potential μ lies in the upper branch (i.e., above the step edge) at $B_\parallel = 0$. At low $B_\parallel$'s, μ is insensitive to $B_\parallel$ while the gap rises in energy nearly quadratically with $B_\parallel$ [9]. Therefore the gap sweeps through μ as $B_\parallel$ is increased. Above these $B_\parallel$'s, μ rises in energy eventually together with the bottom of the lower branch. In Fig. 3, we compare the calculated reduced DOS at μ with the capacitive DOS data [15] from sample 3 in units of $m^*A/(2\pi\hbar^2)$, where A is the cross-sectional area of the QWs. Use is made of d = 140 Å, $(\Delta z_1)^2$ = 621 Å^2 obtained by a self-consistent Hartree approximation, and E_g = 1.8 meV determined experimentally from the conductance data [10]. The latter is somewhat smaller than the Hartree value E_g = 2.0 meV. Sample 3 has w = 100 Å, t = 35 Å, and N_{2D} = 2.4×10^{11} cm^{-2}. μ drops off the DOS step suddenly at ~7.2 T and passes through the saddle point at ~ 8.9 T. Sharp edges as well as the singularity peak are rounded by damping [15]. The data are fitted at the theoretical asymptotic value $D(\mu)$ = 2 at high $B_\parallel$'s, where the two QWs are uncoupled. Small oscillations in the data are due to a small $B_\perp$.

3.3 Cyclotron mass

The cyclotron mass m_c is given by $m_c = (\hbar^2/2\pi)\partial S/\partial \varepsilon_F$ where S is the orbit area in k-space and ε_F is the Fermi energy. In 2D structures, m_c can be rewritten as $m_c/m^* = D(\mu)$, where $D(\mu)$ is the reduced DOS at the Fermi level from the orbit under consideration [13]. In Fig. 4, we display m_c/m^* calculated for sample 2 using d = 200 Å, $(\Delta z_1)^2$ = 1150 Å^2 obtained by a self-consistent Hartree approximation, and E_g = 1.2 meV determined experimentally from the conductance data [10]. The latter is somewhat smaller than the Hartree value E_g = 1.4 meV. At low $B_\parallel$'s, μ is above the gap and the cyclotron orbits consist of a large hour-glass orbit in the lower branch and a smaller lens orbit in the upper branch shown in the lower left corner of Fig. 4. The electrons in the hour-glass orbit have a large mass as shown by the dash-dotted curve and are scattered before completing a cycle: the oscillations from the lower-mass inner lens orbit are dominant in this low-$B_\parallel$ two-orbit regime in recent Shubnikov-de Haas measurements [12]. As $B_\parallel$ increases, μ moves toward the gap edge in Fig. 2(b). As a result, the reduced DOS (i.e., m_c) from the lower and upper branch increases (dash-dotted curve) and decreases (solid curve) monotonically as a function of $B_\parallel$, respectively, for the hour-glass orbit and the lens orbit as shown in Fig. 4. At higher $B_\parallel$'s (i.e., above 6.1 T in Fig. 4), μ moves into the gap, depopulating the lens orbit. In this regime of $B_\parallel$ where μ is inside the gap (6.1 T < $B_\parallel$ < 6.9 T in Fig. 4), only the hour-glass orbit is populated and was observed. Therefore m_c increases abruptly to the mass of the hour-glass orbit. At 6.9 T, μ lies at the saddle point. Above this $B_\parallel$, the orbit splits into two separated orbits and m_c drops to half. The mass thus saturates to m^* of the uncoupled QWs. The calculated m_c yields excellent agreement with the data from sample 2 without adjustable parameters, as shown in Fig. 4 .

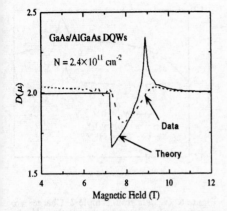

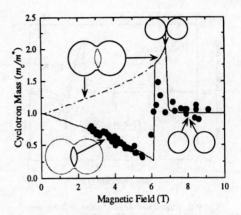

Figure 3. Comparison of the calculated reduced DOS at μ (solid curve) with the capacitive DOS data (dashed curve) from sample 3.

Figure 4. Comparison of the calculated m_c (solid, dash-dotted curves) with the data (black dots) from sample 2 with the relevant cyclotron orbits.

3.3 In-plane conductance

The in-plane conductance $G_\parallel$ in the direction **u** of the electric field is given by [9]

$$G_\parallel = \frac{e^2}{2\pi^2\hbar} \int \tau_\mathbf{k} \frac{(\mathbf{u} \cdot \mathbf{v_k})^2}{v_\mathbf{k}} dk_\parallel, \tag{7}$$

where $\mathbf{v_k}$ is the group velocity, $v_\mathbf{k} = |\mathbf{v_k}|$, $\tau_\mathbf{k}$ is the transport relaxation-time, and the integration is along the orbit. We consider only elastic scattering. In (7), $G_\parallel$ is proportional to $v_\mathbf{k}$ and $\tau_\mathbf{k}$. When μ is above the gap at low $B_\parallel$'s, the contribution to $G_\parallel$ arises from both the hour-glass orbit and the lens orbit in Fig. 4. In contrast to m_c, however, the lens orbit contributes little to $G_\parallel$ because 1) the electrons in the lens orbit have slow velocities due to their small k values and 2) the number of states in the lens orbit is much smaller than in the hour-glass orbit. On the contrary, the lens orbit reduces $\tau_\mathbf{k}$ and therefore $G_\parallel$ by providing states into which the electrons in the hour-glass orbit are scattered rapidly at low $B_\parallel$'s. As $B_\parallel$ is increased, μ falls below the upper gap edge depopulating the lens orbit. In this case, the electrons in the hour-glass orbit cannot be scattered into the upper branch elastically, yielding significantly larger $\tau_\mathbf{k}$ as well as $G_\parallel$ [9]. This behavior is shown by the solid curve (i.e., $\Gamma_o = 0$ meV) in Fig. 5, where we plot $G_\parallel$ calculated for sample 2 by approximating $\tau_\mathbf{k} \propto \rho(\varepsilon)^{-1}$ in (7). Here $\rho(\varepsilon)$ is the total DOS. The maximum in $G_\parallel$ is due to the depopulation of the lens orbit. The $G_\parallel$ minimum, on the other hand, arises when μ lies on the saddle point, where $G_\parallel$ vanishes because $\tau_\mathbf{k} \propto \rho(\varepsilon)^{-1} = 0$ due to the divergence of the DOS [9]. E_g can be determined from the $B_\parallel$'s of the $G_\parallel$ maximum and minimum [10]. The effect of band bending is to increase the effective d from 175 Å to 200 Å by pushing the confinement wave functions away from each other. Since $B_\parallel$ enters the Hamiltonian approximately as $d/\ell^2 \propto B_\parallel d$, the net effect of band bending is to rescale the $B_\parallel$-axis by $175 \times B_\parallel /200$ as shown in the upper axis of Fig. 5. The effect of damping has been treated by a self-consistent linear response theory [16] and is shown in Fig. 5. The calculated results yield good agreement with the data [10] displayed in Fig. 6.

4. Summary

We have discussed low-temperature tunneling and in-plane transport properties in double

850

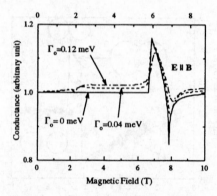

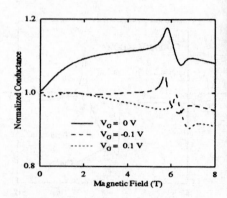

Figure 5. Calculated $G_\parallel$ for sample 2 for several values of damping $\Gamma = 2\Gamma_0$. The upper and lower scales are with and without band bending.

Figure 6. $G_\parallel$ data from sample 2. Charges are approximately balanced at the gate bias $V_G = -0.1$ V.

quantum wells in $B_\parallel$'s. These properties arise from the combined effect of $B_\parallel$-induced relative displacement of the wave vectors in the two QWs and the interwell tunneling. In weakly coupled DQWs, the interwell tunneling conductance has two sharp maxima as a function of $B_\parallel$. In strongly coupled DQWs, a partial minigap is formed due to the anticrossing of the two QW dispersion curves, yielding sharp $B_\parallel$-dependent structures in the density of states, cyclotron mass, and the in-plane conductance. Excellent agreement is obtained between the theory and the data from GaAs/AlGaAs DQWs.

This work was supported by the Office of Basic Energy Sciences, Division of Materials Sciences, U. S. DOE under Contract No. DE-AC04-94AL8500.

References

[1]   Gramila T J, Eisenstein J P, MacDonald A H, Pfeiffer L N, and West K W 1990 Phys. Rev. Lett. **66** 1793-6
[2]   Eisenstein J P, Pfeiffer L N, and West K W 1992 Phys. Rev. Lett. **69** 3804-7
[3]   Eisenstein J P, Pfeiffer L N, and West K W 1995 Phys. Rev. Lett. **74** 1419-22
[4]   Smolinar J, Demmerle W, Berthold G, Gornik E, Weimann G, and Schlapp W 1989 Phys. Rev. Lett. **63** 2116-9
[5]   Eisenstein J P, Gramila T J, Pfeiffer L N, and West K W 1991 Phys. Rev. B **44** 6511
[6]   Lyo S K and Simmons J A 1993 J. Phys.: Condens. Matter **5** L299-L306
[7]   Simmons J A, Lyo S K, Klem J F, Sherwin M E, and Wendt J R 1993 Phys. Rev. B **47** 15741-4
[8]   Zheng L and MacDonald A H 1993 Phys. Rev. B **47** 10619-24
[9]   Lyo S K 1994 Phys. Rev. B **50** 4965-8
[10]   Simmons J A, Lyo S K, Harff N E, and Klem J F 1994 Phys. Rev. Lett. **73** 2256-9
[11]   Kurobe A, Castleton I M, Linfield E H, Grimshaw M P, Brown K M, Ritchie D A, Pepper M, and Jones G A C 1994 Phys. Rev. B **50** 4889-92.
[12]   Simmons J A, Harff N E, and Klem J F 1995 Phys. Rev. B **51** 11156-9
[13]   Lyo S K 1995 Phys. Rev. B **51** 11160-3
[14]   Ando T, Fowler A B, and Stern F 1982 Rev. Mod. Phys. **54** 437-672
[15]   Eiles T M, Simmons J A, Lyo S K, Harff N E, and Klem J F, unpublished.
[16]   Lyo S K, Simmons J A, and Harff N E, 1995 *The Physics of Semiconductors: Proceedings of the 22nd International Conference, Vancouver, 1994, Vol. 1, 843-6,* edited by D. J. Lockwood (World Scientific, Singapore)

Inst. Phys. Conf. Ser. No 145: Chapter 6
Paper presented at 22nd Int. Symp. Compound Semiconductors, Cheju Island, Korea, 28 August–2 September 1995
© *1996 IOP Publishing Ltd*

Resonant tunneling in asymmetric triple barrier diodes

J. Jo

Department of Electronics Engineering, Ajou University, Suwon, 442-749, Korea

K. L. Wang

Department of Electrical Engineering, University of California, Los Angeles, CA 90024, USA

Abstract. We investigated transport properties of triple barrier resonant tunneling diodes. Due to the additional well and barrier inserted, the capacitance in this structure is reduced compared to that of a double barrier resonant tunneling diode. Current-voltage characteristics measured in three different structures show that resonance voltage shifts when the well width is changed. The origin of each resonance can be identified by relating the voltage shift to the width change. Contrary to the generally accepted picture, our data indicate that the energy levels in the neighboring wells are not aligned at resonance. Whenever any quantum well energy level becomes resonant with the emitter level, current shows a resonance.

1. Introduction

The output power and the frequency limit of an oscillator made of a double barrier resonant tunneling diode (RTD) are limited by the capacitance of the RTD. In order to improve performance of the RTD oscillator, there have been efforts to reduce the capacitance of the device. One method is series integration of many RTD's in a single device [1]. Another direction of approach is introducing a multiple barrier structure into an RTD. Due to the additional wells and barriers inserted, the emitter-collector distance is increased, and lower capacitance is expected in this structure. The current-voltage characteristics in multi-barrier RTD structures are not thoroughly understood yet. The existence of transport mechanisms such as high-field domain formation and the Wannier-Stark localization makes it difficult to interpret data. The purpose of this paper is to investigate the details of transport properties in triple barrier RTD's.

2. Experiments

We measured current-voltage (I-V) characteristics and complex admittance in triple barrier RTD's. I-V data were measured by a Hewlett-Packard 4142B dc source/monitor, and the admittance was measured by a Hewlett-Packard 4191A RF impedance analyzer. The RTD capacitance was obtained from the LC resonance due to the wire inductance and the RTD capacitance [2].

The structures used in this study are made of $In_{0.53}Ga_{0.47}As$ wells and AlAs barriers, grown on n+ doped InP substrates by MBE. The triple barrier structure is made of (starting from the substrate side) AlAs barrier, narrow InGaAs well, AlAs barrier, wide InGaAs well, and AlAs barrier. Three different structures of A, B, and C were grown. In these structures the width of the narrow well was fixed to 58 Å. The widths of the wide well and the barrier are 71 Å and 25 Å, respectively for sample A; 65 Å and 25 Å for sample B; 65 Å and 35 Å for sample C. The contacting layers on the top and the substrate side are Si doped InGaAs layers with $N_D = 10^{18}$ cm-3. A 100-μm diameter Au disk was deposited for an ohmic contact, which also serves as a mesa etch mask. All of the measurements reported in this paper were performed at room temperature.

3. Data and discussions

The capacitance values measured in the three structures at zero bias were 20 ± 2 pF. In comparison the capacitance measured in a double barrier RTD at the same condition was 30 pF. We believe that the capacitance reduction in the triple barrier RTD is due to the increased distance between the emitter and the collector.

Figure 1 shows I-V characteristics measured in the triple barrier diodes. Positive polarity means that the top side is positively biased. Samples A and B are different in the wide well width only by 6 Å, but the magnitude of the current differs by a large amount. The

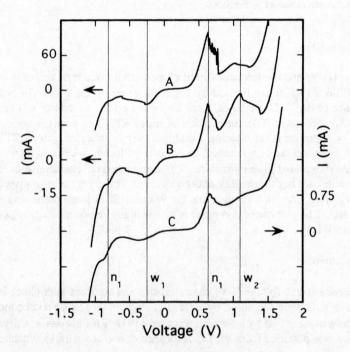

Fig. 1: Current-voltage characteristics of triple barrier diodes. Samples A and B are different in the wide well width, and samples B and C are different in the barrier thickness.

resonance voltages are nearly same for samples A and B, which are marked by the vertical lines at the current peaks.

One of the resonances in Fig. 1 shows a shift when the well width was changed. The current peak at 1.1 V (sample B) moved to 1.0 V (sample A) when the wide well width was increased from 65 Å to 71 Å, while the peak at 0.65 V remained fixed. The shift is related to the well width change, and therefore we assign the peak at 1.1 V as the resonance between the emitter level and the second level of the wide well (w_2). We also assign the peak at 0.65 V as n_1 resonance, which is for the first level in the narrow well. In the negative side it is not easy to tell if there is any shift. The energy change of the second level by the width is about four times larger than that of the first level, and this could be the reason why there is not any noticeable shift in the negative side. When thicker barriers are used (in sample C), the current is reduced, but the resonance positions are almost same as those of sample B.

In Fig. 2 we show I-V data of sample C for a larger bias range up to 4 V. This curve is same as the curve C in Fig. 1, but plotted at different scale. (Samples A and B did not show any structures for biases larger than 2 V). In the negative side four peaks are observed, which are assigned as w_1, n_1, w_2, and n_2 resonances. (The w_1 and n_1 resonances in the negative side are clearly resolved in Fig. 1.) The four energy levels in the quantum wells are schematically shown in Fig. 2.

It is important to note that four resonances are observed in a two-well RTD, in which each well has two energy levels. For multiple barrier RTD structures, it was assumed that the energy levels in the neighboring wells are aligned at resonance [3]. However, our data suggest that this picture is not correct, since if this were true, three instead of four resonances should be observed in our structure. The observation of four resonances indicates that the energy levels are not aligned at resonance, and that the four levels behave independently. It was thought that the levels in the neighboring wells should be aligned at resonance in order to maintain constant

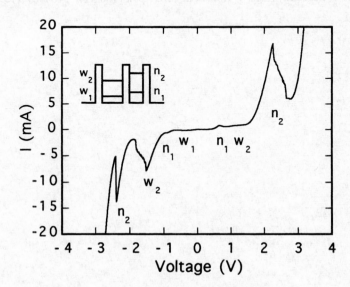

Fig. 2: Current-voltage characteristics of sample C for a larger bias.

current across the structure. This picture does not consider the increasing background current beyond resonance. Because of this current the RTD does not need any level alignment. In an RTD with a single energy level in the well, current beyond resonance also increases due to the increased thermionic current, and this increase is not necessarily related to the higher energy levels. We think that the level alignment picture used in superlattice structures [4] is not directly applicable to multi-barrier RTD's. The level alignment can be achieved with a significant amount of charge build-up in the wells, and resonance voltages will be very different for slightly different structures. In contrast, our data show that for different structures the resonance voltages are nearly fixed, and only the current changed.

4. Summary

In summary, we presented transport data measured in triple barrier resonant tunneling diodes. A current peak was observed when any energy level in the wells became resonant with the emitter level. The energy levels in the neighboring wells were not aligned at resonance. This result is important in understanding the I-V behavior of multiple barrier RTD's.

The work at UCLA was supported by AFOSR/JSEP.

References

[1] Yang C C and Pan D S 1992 *IEEE Trans. Micro. Theo. Tech.* **MTT-40**, 434-441

[2] Jo J, Li H S and Wang K L, to be published.

[3] Nakagawa T, Imamoto H, Kojima T and Ohta K 1986 *Appl. Phys. Lett.* **49** 73-75

[4] Grahn H T, Schneider H and v. Klitzing K 1989 *Appl. Phys. Lett.* **54**, 1757-1759

Inst. Phys. Conf. Ser. No 145: Chapter 6
Paper presented at 22nd Int. Symp. Compound Semiconductors, Cheju Island, Korea, 28 August–2 September 1995
© *1996 IOP Publishing Ltd*

Anomalously Large Stokes and anti-Stokes Real Space Charge Transfer Between Quantum Wells Separated by Thick Alloy Barriers: Beyond the Mean Field Approach

D. S. Kim, H. S. Ko, Y. M. Kim, S. J. Rhee, W. S. Kim, and J. C. Woo

Department of Physics, Seoul National University, Seoul 151-742, Korea

H. J. Choi and J. Ihm
Department of Physics and Center for Theoretical Physics, Seoul National University, Seoul 151, Korea

D. H. Woo and K. N. Kang
Korea Institute of Science and Technology, Cheongryang, Seoul 136-791, Korea

Abstract. Anomalously large real space charge transfer through thick alloy barriers in GaAs asymmetric double quantum wells is studied by photoluminescence excitation. This inter-well excitonic transfer is very large when the barrier is the $Al_{0.3}Ga_{0.7}As$ alloy, but disappears when the barrier is GaAs/AlAs digital alloy with an equivalent x=0.28. Furthermore, for a shallow InGaAs/GaAs/InGaAs ADQW, it largely disappears despite the lower barrier height. These results suggest that the inhomogeneities of the atomic arrangment of Ga and Al in the alloy, which is a strong function of x, might be responsible for this transfer. This picture is qualitatively supported by quantum mechanical calculation in three dimensions whose scopes reach beyond the mean field approach. We, furthermore, presents the existence of an anti-Stokes transfer from the wide well to the narrow well, and discuss its origin.

1. Introduction

Ever since the proposal by Esaki and Tsu [1] and subsequent realization of semiconductor superlattices and quantum wells, the overwhelming majority of the experimental results on alloy superlattices was analyzed using the mean field approach. In this paper, *we present an important problem where the mean field picture fails*: namely, a large real space charge transfer through thick $Al_xGa_{1-x}As$ barriers in GaAs/$Al_xGa_{1-x}As$ quantum wells, where the standard mean field approach predicts prohibittedly low tunneling coefficients [2]. This phenomenon has been observed by many groups over the last 10 years, but its origin has remained a long standing puzzle [3].

In this paper, an extensive study of this "mysterious" transfer in GaAs/$Al_xGa_{1-x}As$ ADQW grown by molecular beam epitaxy (MBE) is presented. Photoluminescence excitation (PLE) was performed in a series of samples with variable x and barrier thickness d. Our results demonstrate the importance of the inhomogeneities and disorders in the alloy

856

potential barrier in charge transfer, which the mean field theory cannot account for.

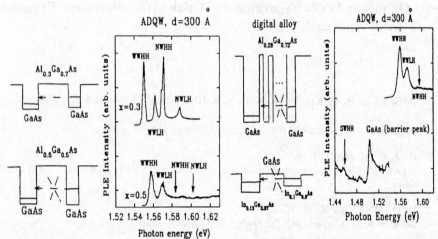

Fig. 1: PLE spectra at 14 K for GaAs/Al$_x$Ga$_{1-x}$As/GaAs ADQW with x=0.3 (top), and x=0.5 (bottom). The schmatics of the sample are shown left to the figure.

Fig. 2: PLE spectra at 14 K for the GaAs/(digital alloy)/GaAs ADQW (top), and In$_{0.1}$Ga$_{0.9}$As/GaAs/In$_{0.13}$Ga$_{0.87}$As/GaAs ADQW (bottom). SWHH denotes for the heavy hole of the shallower well (In$_{0.1}$Ga$_{0.9}$As).

2. Anomalous Stokes Transfer

In Fig. 1, PLE spectra at 14 K obtained with a cw Ti:Sapphire laser, with the PL window on the low energy side of the wide well (WW) exciton, are shown for two ADQW samples. The sample parameters are: (top) GaAs/Al$_{0.3}$Ga$_{0.7}$As/GaAs (75 Å/300 Å/100 Å), and (bottom) GaAs/Al$_{0.5}$Ga$_{0.5}$As/GaAs (75 Å/300 Å/100 Å). For x=0.3, the narrow well (NW) peaks are pronounced, indicating a strong transfer from the NW to the WW. The height of the NW heavy hole (HH) peak is significantly larger than the background due to the continuum excitation. For x=0.5 in Fig. 1 (bottom), there is virtually no sign of the NWHH peak, and the arrows indicate the positions of the NWHH and NWLH peaks obtained from the PLE spectra of the NW. This is of course what one would expect in the mean field approach: namely, virtually no transfer through such thick barrier and therefore no NW peaks for the PLE of the WWHH.

In Fig. 2, striking examples of the complete failure of the mean field approach are found. PLE spectra of a GaAs ADQW with a digital alloy barrier (top), and a shallow In$_x$Ga$_{1-x}$As/GaAs ADQW are shown. The digital alloy sample was chosen so that the effective alloy concentration is 0.28, and d was kept at 300 Å. Within the mean field picture, the transfer of the InGaAs/GaAs ADQW should be much larger than the GaAs ADQW with x=0.3 since the wells are much shallower (by a factor of >5). For the digital alloy barrier, which is roughly equivalent to 100 Å of AlAs, the mean field theory would predict larger transfer coefficient than for the 300 Å Al$_{0.3}$Ga$_{0.7}$As alloy barrier, since the tunneling coefficient exponentially decrease with d. The absence of strong NW peaks in the PLE spectra of the digital alloy ADQW, or the In$_x$Ga$_{1-x}$As/GaAs ADQW tells exactly the opposite story.

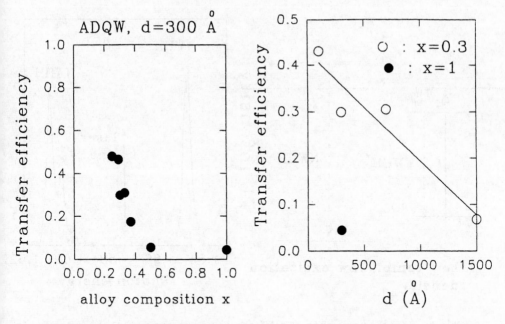

Fig. 3: transfer efficiency, defined as the ratio of the areas under the NW and the WW peaks in the PLE spectra.

Fig. 4: transfer efficiency as defined in Fig. 3, plotted against d for x=0.3 (hollow circles) and x=1 (filled circles).

The absence of the exciton transfer when the barrier is GaAs or AlAs/GaAs digital alloy suggests that this anomalous transfer is a result of the alloy nature of the barrier. Therefore, we contend that the inhomogeneities in the alloy potential barrier is responsible for this puzzling transfer. Since the order parameter or the size of the fluctuations in alloys are often strong functions of x, this picture is consistent with the observed sharp decrease between x=0.3 and x=0.5.

To investigate the x-dependence more systematically, we studied many $GaAs/Al_xGa_{1-x}As$ ADQW samples with varying x. In Fig. 3, the transfer efficiency obtained by normalizing the NWHH peak to the WWHH or to the background is plotted against x. A sharp decrease is observed around x=0.3. From this x-dependence, photon-reabsorption by the WW can be safely ruled out since the barrier region is always transparent at the photon energies used.

Since significant transfer persists up to d=300 Å, it is clear that the d-dependence is much weaker than what the one-dimensional tunneling model would predict. To see this more closely, we studied the d-dependence of the transfer as a function of d for a fixed x=0.3. In Fig. 4, the d-dependence of the normalized transfer efficiency in GaAs/AlGaAs ADQW is shown for x=0.3 (open circle), along with that for x=1 and d=300 Å (closed circle). The d-dependence at x=0.3 is weak, and even at d=1500 Å, it is still greater than that of x=1 and d=300 Å. The weak dependence on d suggests that the apparent penetration depth is of the order of 1000 Å! Since the tunneling coefficient would exponentially decrease with d, with its penetration depth of the order of 10 Å, this dependence is again suggestive of a transfer efficiency that is orders of magnitudes larger than the prediction of tunneling.

858

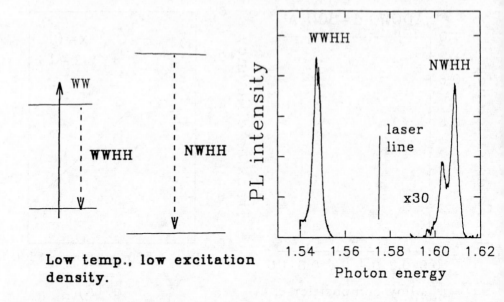

Low temp., low excitation density.

Fig. 5: PL spectrtum when the WW continuum is excited, showing the anti-Stokes transfer from the WW to the NW.

In Fig. 5, yet another surprising experimental result is presented: when the WW continuum was excited for an ADQW with d=200 Å and x=0.37, a fairly strong NWHH luminescence, with an efficiency of >10^{-3} normalized to the WWHH luminescence, was found. This is indicative of an anti-Stokes charge transfer from the WW continuum to the NW, although the experiments were performed at 14 K, so that the Boltzmann coefficient for overcoming the average barrier height is extremely low (~10^{-100}). We believe that this transfer is essentially quantum mechanical, since any coupled two level system can support quantum oscillation of wave functions in real space. Therefore, the real space anti-Stokes charge transfer at low temperature is a natural consequence of any coupled, spatially separated systems, and it does not need any aid from phonons, and does not violate energy conservation. A simple density matrix analysis [4] shows that the transfer efficiency is directly related to the coupling constant of the two wells.

4. Theory

In the mean field approach, the barrier height in GaAs/$Al_xGa_{1-x}As$ superlattice is assumed to be a constant determined by x. On the other hand, recent scanning tunneling microscopic (STM) studies of GaAs/$Al_xGa_{1-x}As$ quantum wells, AlAs/GaAs superlattices, and $Al_xGa_{1-x}As$ alloys [5] show that there exists elongated "clustering" of Ga rich and Al rich regions in the barrier along the growth direction, thus possibly connecting adjacent GaAs wells. We first considered the effect of atomic scale fluctuations on the transmission coefficient with the following model calculation. We divided the barrier into small cubes of atomic scale representing GaAs or AlAs molecules, and randomly assigned either the potential V_0 for AlAs or 0 for GaAs. V_0 is 1.12 eV (0.26 eV) to simulate the band offsets for electrons (holes) (Fig. 6a). We solved the resulting three-dimensional effective mass equation with appropriate boundary conditions to obtain the transmission coefficients. The resulting transmission coefficients are larger than those obtained from the one-dimensional mean field approach, but still far too small to explain the large transmission coefficient of the order of 10^{-3} deduced from experiments. Essentially, the wavelength of the incoming waves (around 100 Å, comparable to the well size) is too large to "see" the low but narrow potential pathways.

We then replaced the cubes in the barrier region with rectangular cylinders (or "wires") long enough to connect the two wells as shown in Fig. 6b, simulating the possible aligning of GaAs or AlAs roughly along the growth direction in light of recent STM studies [5]. We performed the quantum mechanical calculation for various sizes of rectangles to study the cluster size effect on the transmission coefficients. Furthermore, the possible effect of "kinks" was considered (Fig. 6c) in connection with ref. [5] where the GaAs "quantum wires" were shown to "zigzag" their ways through the barrier. In Fig. 6d, the transmission coefficients of holes as a function of x are plotted for several cluster sizes using the model of Fig. 6b. Holes rather than electrons were simulated because it is generally believed that the transport of excitons is determined mostly by the first quantized holes, whose transport is generally slower than electrons. For the grid size of 4 Å, the results are only slightly larger than the prediction of the mean field theory, despite the fact that there exists many low potential "quantum wires" in the barrier. The physics of this is the same as described earlier: the pathways are much narrower than the wavelength. Increasing the cluster size rapidly enhances the transmission coefficient, so that for the cluster size of 30 Å, nearly all holes can pass through the barrier for relatively low x. The strong x-dependence and weak d-dependence can then be explained in this model, albeit somewhat trivially. Finally, the results using the model schematically described in Fig. 4c shows that the effect of the "kink" is to decrease the transmission coefficient only slightly without changing the overall trends.

Our model calculations suggest that the anomalously large transmission coefficient and most features of our experiments can be explained, at least qualitatively, if there were large enough clustering of GaAs or $Al_xGa_{1-x}As$ with very low x, connecting two GaAs quantum wells in a quantum-wire-like fashion. Although a more realistic approach to the detailed mechanism of clustering and the resulting structure and pattern formation would be much desired, our results imply that the clustering of GaAs in the alloy barrier is a likely source of the enormously enhanced interwell coupling. Finally, even without clustering, quantum wells with completely random substitutional alloy as barriers might have extended states below a certain x [6]. This type of approach has not yet shown its full potential in semiconductor physics.

5. Conclusion

In conclusion, we experimentally demonstrated that a mysteriously large transfer of excitons through thick barriers occurs when the barrier is composed of $Al_xGa_{1-x}As$ alloy. Unlike "solid barriers" such as GaAs/AlAs digital alloys, $Al_xGa_{1-x}As$ is essentially leaky, or percolating, especially for x<0.4: there may exist regions of low potential in the barrier connecting two adjacent wells, which allows observed huge charge transfer. Our results show that beyond the widely used mean field approach, a three dimensional approach considering the detailed nature of the barrier such as clustering is needed to understand some of the important dynamics of semiconductor superlattices.

This work was partially supported by Korean Science and Engineering Foundation, Ilju Cultural Foundation, and the Basic Science Research Program of the Ministry of Education.

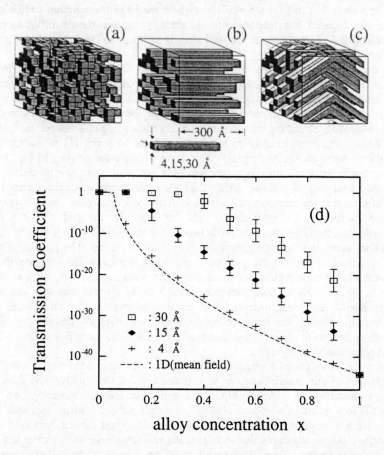

Fig. 6: (a) Schematics for the construction of barriers used in our model calculations assuming completely random, atomic alloy fluctuations. Dark squares represent AlAs "molecules". (b) Schematics of our model taking clustering and the formation of channels into consideration. (c) The same as (b) except for the existence of "kinks". (d) Transmission coefficients using barriers described in (b), plotted against x for several grid sizes. The incident wave simulates holes in the narrow well, with the effective mass of ~0.5m_e, and the wave length of 150 Å.

References

[1] L. Esaki and R. Tsu, IBM J. Res. Develop, 14, 61 (1970).
[2] A. Tomita et al., in Int. Conf. Quantum Electron. Tech. Digest Ser. 9, 116 (1992).
[3] J. J. Song, private communication; T. C. Damen, private communication.
[4] D. S. Kim et al., unpublished.
[5] A. R. Smith et al., Appl. Phys. Lett. 66, 478 (1995).
[6] P. W. Anderson, Phys. Rev. 109, 1492 (1958).

Inst. Phys. Conf. Ser. No 145: Chapter 6
Paper presented at 22nd Int. Symp. Compound Semiconductors, Cheju Island, Korea, 28 August–2 September 1995
© *1996 IOP Publishing Ltd*

Theoretical study of electron tunneling time through a single/double barrier(s), and the effect of wave packet spread

Byoungho Lee, and Wook Lee

Department of Electrical Engineering, Seoul National University,
Kwanak-Ku Shinlim-Dong, Seoul 151-742, Korea

Abstract. Simulated single-barrier tunneling time for various wave packet spread in momentum space is compared with the theoretical phase time. The effect of the shift of the electron momentum peak after the tunneling is discussed. A simple physical formula of the phase time for resonant tunneling structures is given. Simulation results show that the phase time is of the order of decay time in the well between the two barriers. But, the resonant tunneling time is of the order of excitation time of the trapped wave functions and shorter than the phase or decay time.

1. Introduction

The tunneling time through a potential barrier has been controversial for more than 40 years. Recently, Chiao's experiments showed superluminal (but causal) tunneling of photons [1]. The result also showed that the simplest formalism of phase time introduced by Bohm and Wigner [2] is the best model for the tunneling time. For electronic tunneling, there has also been much study on the limitations on the speed of the tunneling device operations [3]. The study has been focused on the modeling of capacitances of tunneling devices such as resonant tunneling diodes because the operation frequency limitation of present devices comes from the RC time constant. But, to probe the physical fundamental limit it is necessary to study the electron tunneling time for those devices.

In this paper, to study the fundamental physical operation speed limit of heterostructure tunneling devices and the wave mixing using tunneling microscopes, the simulated tunneling time for various wave packet spread in momentum space (or real space) is compared with the theoretical phase time. The simulations are performed using the software InterQuanta [4]. The effect of the momentum peak shift after the tunneling (which cannot be seen in photonic tunneling) is discussed. A simple physical formula for the phase time for resonant tunneling structures is derived. The result is compared with simulated well-decay time. Simulated resonant tunneling time and well decay-excitation time are also discussed.

2. Electron tunneling time through a single potential barrier

For an electron wave packet starting from $x = -x_o$ at time $t = 0$ and moving in the positive x direction, assuming the wave packet encounters a rectangular potential barrier of height V_o in

862

the region $0 < x < d$, the phase time approach gives the time t at which the center of the tunneled (transmitted) wave packet reaches position x by the relation of

$$t = \frac{m}{\hbar k}\left[\frac{\partial\phi(k)}{\partial k} + x + x_o\right],$$ (1)

where $\phi(k)$ is the phase of $f(k)$, which is the ratio of the complex coefficient of the transmitted plane wave to that of the input plane wave of momentum $\hbar k$ and mass m. This relies on the fact that the major contribution to the transmitted wave packet is due to the k component with stationary phase [5].

The well-known phase time δt is given by [2]

$$\delta t = \frac{m}{\hbar k}\left[\frac{\partial\phi(k)}{\partial k} + d\right].$$ (2)

Here, the fact that seems not to be paid much attention is that the most probable speed of the electron increases after the tunneling through a single barrier. This is due to the fact that a plane wave component with higher k has a higher transmission coefficient. This is not contradictory to energy conservation. The energy is conserved for each electron. But, incident electrons with higher energies have higher probabilities of tunneling, which results in the increase of the most probable energy of transmitted electrons and the decrease of the most probable energy of the reflected electrons. In Equation (2) it is reasonable to take k as the value corresponding to the most probable k after the tunneling. Equation (2) is derived from Equation (1) assuming that the incident electrons also move with the same k, that is, if we denote the most probable speed of tunneled electrons with v_1, and that of the incident electrons with v_0, and if the transmitted wave packet is located at $x = d + x_1$ at time t, Equation (2) is the same as

$$\delta t = t - \frac{x_o + x_1}{v_1}.$$ (3)

Besides the time δt, we can also consider another physical time given by

$$\Delta t = t - \frac{x_o}{v_0} - \frac{x_1}{v_1},$$ (4)

which involves two different most probable speeds before and after the tunneling. Equation (4) can be reformed as

$$\Delta t = \delta t - x_o\left(\frac{1}{v_0} - \frac{1}{v_1}\right),$$ (5)

which depends on the initial location of the wave packet. This is reasonable because, if the initial wave packet is located much farther from the barrier, larger momentum electrons will arrive much earlier than smaller momentum electrons. Therefore, in some cases, Δt becomes negative because the peak of transmitted wave packets leave the barrier before the peak of incident wave packets arrive at the barrier.

Figure 1 compares tunneling times δt and Δt simulated using InterQuanta [4] with theoretically calculated phase time and the classical time. The classical time is the time required

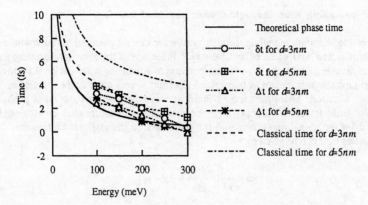

Figure 1. Comparison of simulated single-barrier tunneling time with theoretical phase time and classical time.

for an electron of the same energy pass the potential-free region of the same width d. In Figure 1 the potential barrier height V_o is taken to be $300meV$, and the thickness of the barrier d is taken to be $3nm$ or $5nm$. The theoretical phase times for the two cases are almost the same, that is, they are relatively independent of the barrier thickness. The initial center position of the electron wave packet is taken to be $40nm$ before the barrier. The standard deviation of the wave packet in momentum space is taken to be 10% of the peak momentum. The simulated phase time δt is greater than the theoretical time but smaller than the classical time.

Figure 2 shows simulated times δt and Δt for various wave spread in momentum space. The reference of the percentage of the standard deviation is the peak momentum of the wave packet. Figure 2 shows that the simulated tunneling time strongly depends on the wave packet spread. The dependence is much stronger than the theoretical phase time calculated including the effect of the peak momentum shift after the tunneling. Figure 2 also shows some cases for which Δt is negative, which means that the peak of the transmitted wave packet leaves the barrier before the peak of the incident wave packet arrives at the barrier. This is not contradictory to causality because the wave packet should be treated as a distribution related to the probability of finding the electron. The higher momentum electrons arrive earlier and transmits with higher tunneling probabilities.

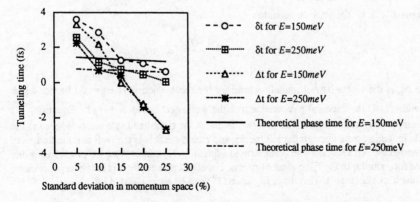

Figure 2. The effect of wave packet spread on tunneling time for the barrier thickness of $3nm$.

3. Electron tunneling time through double potential barriers

Here, we consider resonant tunneling through symmetric double potential barriers with each barrier width of d and well width of w. The major difference of the resonant tunneling from single-barrier tunneling is that the momentum spectrum of a tunneled wave packet is narrower, and the most probable momentum of transmitted electrons is very close to the momentum for the resonant condition. Therefore, for an incident wave packet with the most probable momentum a little lower (or higher) than the resonant momentum, the most probable momentum after the tunneling is increased (or decreased) to the resonant momentum. The phase time in this case can be given by

$$\delta t = \frac{m}{\hbar k} \left[\frac{\partial \phi(k)}{\partial k} + 2d + w \right], \tag{6}$$

where k should be taken as the value corresponding to the resonant condition, and $\phi(k)$ is the phase of $f(k)$ given by

$$f(k) \equiv \frac{t_B^2 \exp(-2ikd)}{1 - r_B^2 \exp(2ikw)}, \tag{7}$$

where [6]

$$t_B = \frac{1}{\cosh(\kappa d) - i \dfrac{k^2 - \kappa^2}{2k\kappa} \sinh(\kappa d)}, \tag{8}$$

$$r_B = -it_B \frac{k^2 + \kappa^2}{2k\kappa} \sinh(\kappa d), \tag{9}$$

and

$$\kappa = \sqrt{\frac{2mV_o}{\hbar^2} - k^2}. \tag{10}$$

Equation (6) can be shown to be equal to

$$\delta t = 2\delta t_b + t_w + \frac{2(1 - T_B)}{T_B}(\delta t_b + t_w), \tag{11}$$

where δt_b is the single-barrier tunneling time for the same electron energy and barrier width and height, t_w is the classical pass time across the well region, and $T_B \equiv |t_B|^2$. Equation (11) has a simple physical meaning: The first two terms on the right-hand side are analogous to the particle tunneling time through the left barrier and pass time across the well and the tunneling time though the right barrier. The other term in Equation (11) comes from the phase factor due to resonance similar to the Fabry-Perot process. Usually, the latter term is much larger because T_B is much smaller than 1. Therefore, Equation (11) can be approximated by

$$\delta t \approx \frac{2}{T_B}\left(\delta t_b + t_w\right).$$ (12)

Figure 3 shows various simulated times, theoretical time, and the classical time. The simulation is performed for the first resonant level tunneling in a symmetric double-barrier structure with barrier height of $300meV$, barrier width of $3nm$, and well width of $5nm$. Excitation time is the time elapse between the moment when the peak of the incident wave packet reaches the front potential barrier and the moment when trapped electron probability becomes maximum. Decay time is the time elapse between the moment when the trapped electron probability becomes maximum and the momentum when it is reduced to half of its maximum. Tunneling time is the time elapse between the moment when the peak of the incident wave packet reaches the front potential barrier and the moment when the peak of the transmitted wave packet leaves the right barrier. Classical time is the pass time of an electron with the same energy across the same region in the absence of the potential barriers. Figure 4 is for the case of $5nm$ barrier thickness with other conditions equal to those for Figure 3.

In Figures 3 and 4 we find that the tunneling time is about the same as (or of the order of) the excitation time, while the decay time is longer. This is contradictory to the usual belief [7] that the transit (tunneling) time is twice the decay time. Our results show that the theoretical phase time is approximately twice the decay time, but the tunneling time is smaller than those times and is about the same as (or of the order of) the excitation time. Therefore, to study the dynamic theory of the resonant tunneling, much attention should be paid on the well-excitation time.

4. Conclusion

In this paper the theoretical phase time is compared with simulation results. For a single potential barrier, it is shown that the phase time gives good estimation of the tunneling time. But, due to the peak momentum shift of electron wave packets after tunneling, in some cases the peaks of the transmitted wave packets leave the barrier before those of the incident wave packets reach the barrier. The effect of wave packet spread is stronger than theoretical expectation.

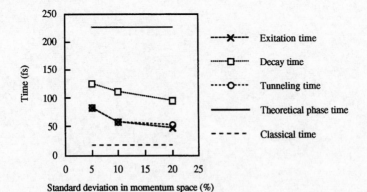

Figure 3. Comparison of simulated times for resonant tunneling with theoretical phase time and classical time. Barrier height is $300meV$, barrier width is $3nm$ each, and well width is $5nm$.

866

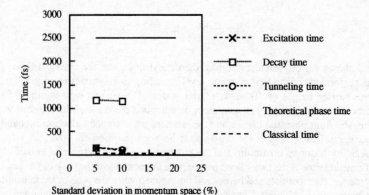

Standard deviation in momentum space (%)

Figure 4. Comparison of simulated times for resonant tunneling with theoretical phase time and classical time. Barrier height is 300meV, barrier width is 5nm each, and well width is 5nm.

For resonant tunneling structures, the decay time of a trapped wave packet in the well is about half of the theoretical phase time. Simulation results show that the tunneling time is about the same as the excitation time of the resonant energy level in the well.

References

[1] Chiao, R. Y., 1993, *Physical Review A*, **48**, R34.
[2] Wigner, E. P., 1955, *Physical Review*, **98**, 145.
[3] For example, Genoe, J. *et al.*, 1991, *IEEE Trans. on Electron Devices*, **38**, 2006.
[4] Brandt, S., and Dahmen, H. D., 1995, *Quantum Mechanics on the Macintosh*, 2nd ed. (Berlin: Springer-Verlag).
[5] Hauge, E. H., and Støvneng, J. A., 1989, *Reviews of Modern Physics*, **61**, 917.
[6] Toombs, G. A., and Sheard, F. W., in the book edited by Chamberlain, J. M. *et al.*, 1990, *Electronic Properties of Multilayers and Low-Dimensional Semiconductor Structures*, (New York: Plenum Press).
[7] Price, P. J., 1988, *Physical Review B*, **38**, 1994.

Inst. Phys. Conf. Ser. No 145: Chapter 6

Paper presented at 22nd Int. Symp. Compound Semiconductors, Cheju Island, Korea, 28 August–2 September 1995

© *1996 IOP Publishing Ltd*

Correlation effects on the single electron tunneling through a quantum dot in magnetic fields

Kang-Hun Ahn, J. H. Oh, and K. J. Chang

Department of Physics, Korea Advanced Institute of Science and Technology, Taejon 305-338, Korea

Abstract. We calculate the spectral weight matrix for quantum dot and find that electron tunneling is strongly spin-related due to electron correlation. We also study the magnetic field dependence of the chemical potential of quantum dot by employing Jastrow correlation factor in an analytic fashion. Our calculational results are in good agreement with the recent experiments.

1. Introduction

In recent years, electron-electron interactions in a quantum dot have been a subject of many theoretical and experimental studies.[1, 2, 3] When the Fermi level of electrodes becomes resonant with a quantum level of a dot, a single electron can tunnel between the electrode and the dot through a tunnel barrier. In transport measurements,[2] the value of the gate voltage at which the conductance peak occurs provides a measure of the electrochemical potential of the droplet, μ_N, which is the difference of the ground state energies of the N and $(N-1)$ electron systems. In this case, the height of the conductance peak is governed by many-body coherence between the N- and $(N-1)$-electron wave functions,[4] thus, electron correlations strongly affect the transport behavior of a quantum dot.

In this work, we calculate the spectral weight matrix elements and find that the electric conduction through a quantum dot sensitively depends on the spin polarization of tunneling electron due to electron correlations. We also calculate the ground state energies of quantum dots containing up to 7 electrons via an exact treatment of electron-electron interactions. We employ a symmetric Jastrow correlation factor so that electron correlations are effectively included through the mixing of higher Landau levels. Our calculated chemical potentials exhibit an unexplained feature, *i.e.*, a ripple in the transition region of the ground state was observed by capacitance spectroscopy measurements.[3]

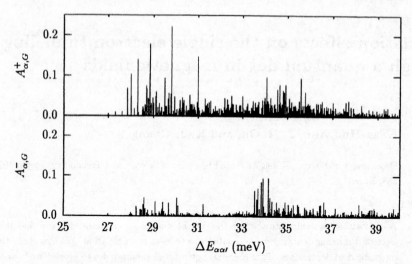

Figure 1. Summed values of the spectral weight matrix elements through a quantum dot containing 4 electrons.

2. Conductance and chemical potential

We consider a droplet of N electrons confined by a parabolic potential $\frac{1}{2}m^*\omega_0^2 r^2$. The conductance through a quantum dot is written as [4]

$$\sigma = \frac{2\pi e^2}{h} \sum_{\alpha,\alpha'} \Gamma_{\alpha\alpha'}[-f'_{FD}(E_{N,\alpha} - E_{N-1,\alpha'} - \mu)](P^{eq}_{N,\alpha} + P^{eq}_{N-1,\alpha'}), \qquad (1)$$

where α and α' represent many-particle states, μ is the chemical potential of a reservoir, f'_{FD} denotes the derivative of the Fermi-Dirac distribution function with respect to energy, and $P^{eq}_{N,\alpha}$ is the probability to find the dot in the N-electron state α. The effective couplings between many-particle states are given by $\Gamma_{\alpha\alpha'}$, which are proportional to the spectral weight matrix elements[4], $A_{\alpha,\alpha'} = A^+_{\alpha,\alpha'} + A^-_{\alpha,\alpha'}$;

$$A^{\pm}_{\alpha,\alpha'} = \sum_n |\langle N, \alpha | c^{\dagger}_{n,s=\pm 1/2} | N-1, \alpha' \rangle|^2 \qquad (2)$$

where $c^{\dagger}_{n,s}$ is the creation operator for the state of single electron orbital (n) and spin (s) quantum numbers. To see the transport behavior of tunneling electron for each spin state, we first calculate the electronic structure for droplets containing 4 and 5 electrons, which are confined by a parabolic potential with the strength of $\hbar\omega_0 = 3.5$ meV. For a magnetic field of 10 Tesla, since the filling factor of the 4-electron droplet in the ground state is $\nu = \frac{1}{3}$, all the electrons are considered to be in the spin-up states. Then, the many-body eigenstates can be obtained by the direct diagonalization of the Hamiltonian matrix with a basis set in the lowest Landau level.[5] The calculated spectral weight matrix elements, $A^+_{\alpha,G}$ and $A^-_{\alpha,G}$, for the up- ($s = 1/2$) and down-spin ($s = -1/2$) states of the tunneling electron, where G denotes the ground state of the droplet, are shown and compared in Fig. 1. We find that the down-spin tunnelings are severely suppressed over a wide range of energy, with the exception near the transition energy of about 34 meV, where selective

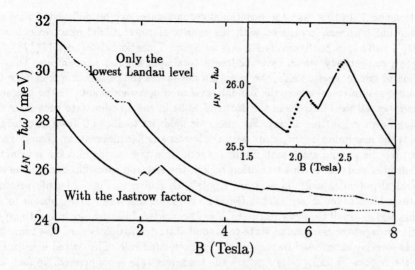

Figure 2. The chemical potential (μ_N) vs magnetic field for $\hbar\omega_0 = 3.5$ meV, with the coupled frequency ω defined as $\omega \equiv (\omega_0^2 + \omega_c^2/4)^{1/2}$, where ω_c is the cyclotron frequency. The results with the Jastrow correlation factor are compared with those from only the lowest Landau level. The dotted lines represent the regions of $|S_z(N) - S_z(N-1)| > 1/2$, where the tunneling rate is suppressed. Inset shows the details near the transition region.

tunnelings occur due to a large tunneling density of states which results from the Coulomb interaction. For the up-spin state, since the same spin polarization of the tunneling electron with the droplet allows for tunneling, we see large enhancements in $A_{\alpha,G}^+$. The selective tunnelings associated with the sharp peaks in $A_{\alpha,G}^+$ at energies between 28 and 31 meV are mainly caused by electron-electron interactions, and these peaks explain the conductance oscillations measured with the variation of gate voltage.[1, 2] Thus, the correlation effect is more important in electron tunneling of a spin-polarized droplet than the well-known spin selection rule, $|\Delta S_z| = |S_z(N) - S_z(N-1)| = 1/2$.[6]

Here we discuss transitions between the ground states of a droplet. Since the lowest Landau-level assumption is only valid in strong magnetic fields, we extend the restricted Hilbert space to evaluate the ground state energy in rather weak magnetic fields. Instead of extending the basis functions to those belonging to higher Landau levels, which are computationally too expensive, we employ a symmetric Jastrow correlation factor,[7, 8] which will be multiplied to the many-body wavefunction, $\Psi_{L,S_z}^{(N)}$, obtained by the direct diagonalization [5] within the restricted Hilbert space spanned by the lowest Landau level. Here L and S_z denote the total orbital and spin angular momentums, respectively. Then, the variational many-body wavefunction in the ground state is $\Phi_{L,S_z}^{(N)}(a) = e^{-a\sum_{ij} r_{ij}^2} \Psi_{L,S_z}^{(N)}$, and the ground state energies can be evaluated in an analytic fashion:[7]

$$E_N(\omega_0, B) = min_{(L,S_z,a)} \langle \Phi_{L,S_z}^{(N)}(a) \mid H \mid \Phi_{L,S_z}^{(N)}(a) \rangle, \tag{3}$$

where H is the Hamiltonian of a quantum dot containing N electrons. For $N=7$, the calculated chemical potential, $\mu_N = E_N - E_{N-1}$, is plotted and compared with the results obtained from only the lowest Landau level in Fig. 2. We find that the chemical potential is significantly reduced at lower magnetic fields ($0 < B < 2$ Tesla), where the Jastrow factor introduces the mixing of higher Landau levels, thus, it includes more effectively the correlation effect. In this case, the variational ground states exhibit different orbital and spin quantum numbers, compared with the results of conventional exact calculations using $\Psi_{L,S_z}^{(N)}$; the spin configurations for $N = 6$ and 7 are found to be $|\uparrow\uparrow\uparrow\downarrow\downarrow\downarrow\rangle$ and $|\uparrow\uparrow\uparrow\uparrow\downarrow\downarrow\downarrow\rangle$, respectively, which have the lowest total spin angular momentum. Thus, in the region of low magnetic fields, the quantum dot favors the equal occupancies for the up- and down-spin states where the Landau level mixing is significant. As the magnetic field increases, all the electrons in the dot tend to be in the up-spin state because of the increasing Zeeman splitting energy. For magnetic fields up to about 6 Tesla, the droplet maintains the maximum density state with the lowest angular momentum. Near 2 Tesla, we find that the ground state with minimum total spin transforms into the maximum-density-droplet state.[9] In this transition region, the electron spins flip one by one and the chemical potential exhibits a clear "ripple", as shown in Fig. 2, while previous calculations have never shown such a feature. We note that the ripple appears in the transition region for all the systems (up to $N = 7$) considered here. More interestingly, at magnetic fields where the ground-state-to-ground-state transition occurs, the tunneling current is severely suppressed because of the spin selection rule. The dotted lines in Fig. 2 show the regions of $|\Delta S_z| > 1/2$, where the tunneling rate is suppressed. In fact, one prominant "ripple" in the $\mu_N - B$ graph was observed, at which the electron tunneling rate is strongly suppressed.[3]

3. Summary

We have investigated the spectral weight matrix elements and find that the electron tunneling through a quantum dot in the fractional quantum Hall regime is strongly spin-related, which is mainly caused by electron correlations. Employing the Jastrow correlation factor, we find a ripple in the transition region where the ground state transitions occur, while the tunneling rate is severely suppressed, in good agreement with experiments.

Acknowledgments

This work is supported by the CMS at KAIST and the SPRC at the Jeonbuk National University.

References

[1] Pfannkuche D and Ulloa S E 1995 *Phys. Rev. Lett.* **74** 1194–7

[2] McEuen P L, Foxman E B, Kastner M A, Meir Y and Wingreen N S 1991 *Phys. Rev. Lett.* **66** 1926–9

[3] Ashoori R C, Stormer J S, Pfeiffer L N, Baldwin K W and West K W 1993 *Phys. Rev. Lett.* **71** 613–6

[4] Kinaret J M, Meir Y, Wingreen N S, Lee P A and Wen X G 1992 *Phys. Rev.* B **46** 4681–92

[5] Yang S R E, Macdonald A H and Johnson M D 1993 *Phys. Rev. Lett.* **71** 3194–7

[6] Palacios J J, Martin-Moreno L, Chiappe G, Louis E and Tejedor C 1994 *Phys. Rev.* B **50** 5760–3

[7] Ahn K H, Oh J H and Chang K J, 1995 *Phys. Rev.* B (submitted)

[8] Fahy S, Wang X W and Louie S G 1990 *Phys. Rev.* B **42** 3503–22

[9] Macdonald A H, Yang S R E and Johnson M D 1993 *Aust.J.Phys.* **46** 345–58

Inst. Phys. Conf. Ser. No 145: Chapter 6
Paper presented at 22nd Int. Symp. Compound Semiconductors, Cheju Island, Korea, 28 August–2 September 1995
© *1996 IOP Publishing Ltd*

A Novel Resonant Tunneling Diode with Single-Peak I-V Characteristics

Kunihiro Arai and Masafumi Yamamoto

NTT LSI Laboratories,
Atsugi-Shi, Kanagawa Pref., 243-01 Japan

Abstract. A novel resonant tunneling diode with single-peak I-V characteristics is proposed. This novel diode utilizes a new mechanism, i.e., coupling between the resonant tunneling effect in a vertical double barrier structure and the pinch-off effect in a lateral collector just beneath the structure. Using a one-dimensional distributed parameter model, the dependence of I-V characteristics on device parameters (emitter width and collector thickness) are investigated systematically and single-peak I-V characteristics are shown to be achieved for appropriate parameter values. The diodes are fabricated using the InGaAs/AlAs/InAlAs material system. The single-peak I-V characteristics are experimentally verified along with their dependence on the device parameters. Good agreement between the theoretical and experimental results confirms the proposed mechanism for the single-peak I-V characteristics. The proposed device could lead not only to reduced power consumption in functional logic circuits but also offer novel functional logic gates.

1. Introduction

Resonant tunneling diodes (RTDs) are widely studied because of their unique negative differential resistance (NDR), which offers the possibility of constructing functional logic gates [1-3]. To reduce power consumption in RTD circuits, RTD current must be suppressed over a wide bias voltage region around the valley voltage. In this study, we propose a novel RTD with single-peak I-V characteristics (SPC) that satisfies this requirement.

To better meet the demand for functional circuit applications, many modifications have been proposed including the construction of three terminal resonant tunneling devices. Some of these concepts are based on the coupling of the resonant tunneling effect with other effects known to occur in semiconductors like the hot electron [3,4], minority carrier injection [5], field effects [2]. Although these approaches all show various possibilities of RTD application, none has shown single-peak I-V characteristics. Others approaches are based on the monolithic integration of an RTD with a conventional device such as a MESFET or HEMT. One can find I-V characteristics analogous to SPC mentioned here in literature [6, 7]. However, insufficient current suppression around the valley voltage and large hysteresis loops in the I-V characteristics make it difficult to apply them for the present purpose. Therefore, in order to achieve a device with SPC, it is necessary to find the appropriate physical effect in a semiconductor that can be coupled with resonant tunneling effect.

In this paper, we describe an RTD in which the pinch-off effect is for the first time to our knowledge coupled with the resonant tunneling effect. It is found that a novel diode specially designed to make the coupling strong shows SPC successfully. We call the novel diode as resonant tunneling pinch-off diode (RTPD). The structure and the fabrication process for RTPD are as simple as those of conventional RTDs. Moreover, essentially no hysteresis appears in the I-V characteristics of RTPD. These excellent features make this approach very attractive for functional logic circuit application.

2. Device structure and modeling

2.1 Basic concept

In the RTPD the resonant tunneling effect in a vertical double barrier structure (DBS) is coupled to the pinch-off effect in a lateral collector just beneath the DBS. The RTPD mainly consists of a barrier, a collector, a DBS and an emitter formed in this order on a semi-insulating substrate as shown schematically in Figure 1 and Table 1. The coupling occurs because the channel opening of the lateral collector, i.e., the neutral layer thickness in the collector is a function of the voltage drop around the vertical DBS (V_{DBS}); the local potential difference between the emitter and collector. In what follows, several terms are defined to make the discussion clearer. DBS peak voltage and DBS valley voltage mean the peak and valley voltage for a virtual RTD, which is the same as the RTPD except that the collector is thick enough to make the coupling week while the collector doping concentration is left unchanged. DBS peak current and DBS valley current also mean the peak and valley current for the virtual RTD.

In the basic RTPD concept, there are two requirements. First, the lateral collector layer should be pinched off for a collector bias voltage, V_{P-O}, at which V_{DBS} is equal or close to the DBS valley voltage. Second, the lateral collector should not be pinched off for a collector bias at which V_{DBS} is equal or close to the DBS peak voltage. Because of the first requirement, the collector current is fixed at a constant value when the collector voltage is larger than V_{P-O}. The second requirement ensures that a current peak occurs at a collector bias less than V_{P-O}. Thus, single-peak I-V characteristics appear.

2.2 Modeling

To test the concept, we first theoretically investigate I-V characteristics using a one-dimensional distributed parameter model. Our main purpose here is to calculate collector currents as a function of collector voltage when emitter voltage is fixed at zero volts. The calculation procedure is as follows. First, electric potential distribution in the collector, $\phi(x)$, is determined by solving coupled equations (1), (2), (3), (4) and (5) below. Then the collector current is derived from the obtained $\phi(x)$ and formula (3).

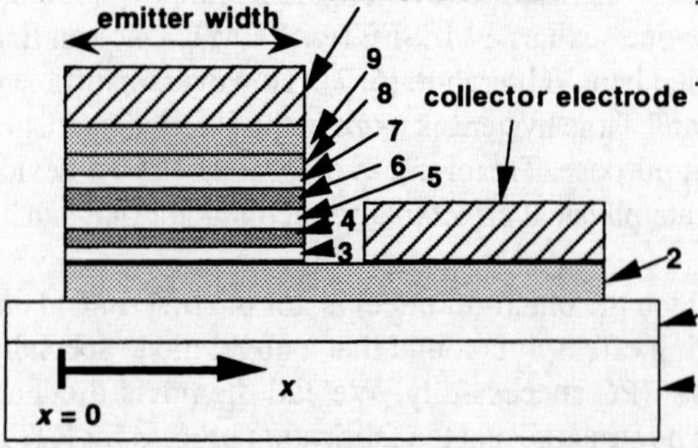

Table 1

layer number	layer name	Si doping (cm^{-3})	material	thickness (nm)
9	cap	1E19	InGaAs	30
8	emitter	1E18	InGaAs	120
7	spacer	undoped	InGaAs	1.4
6	barrier	undoped	AlAs	2
5	well	undoped	InGaAs	4.1
4	barrier	undoped	AlAs	2
3	spacer	undoped	InGaAs	3
2	collector	1E18	InGaAs	60
1	barrier	undoped	InAlAs	200
0	substrate	-	S.I. InP	-

Figure 1. Schematic cross section of an RTPD.

The tunneling current density through DBS at point x, $j_{rtd}(x)$, is assumed to be a function of only the electric potential at x, $\phi(x)$, as shown in (2), where the functional form of J is appropriately assumed taking into account experimental I-V characteristics obtained for a reference RTD, which is shown in section 3. Current flow parallel to the substrate in the collector at position x, $I_c(x)$, is expressed as (3), where the meanings of e, L_E, N_d, T_c, $d(x)$ and $v(x)$ are shown in Table 2. Formula (3) is analogous to the gradual channel approximation for MESFETs where the channel opening is $(T_c - d(x))$. The electron velocity at position x, $v(x)$, is expressed as (4) and the depletion layer thickness in the collector, $d(x)$, is expressed as (5). The meanings of μ and v_s in (4) and ε and d_{DBS} in (5) are also shown in Table 2. Formula (1) is the current continuity equation balancing vertical current density $j_{rtd}(x)$ and lateral current $I_c(x)$ at each position x.

$$I_c'(x) = j_{rtd}(x) \tag{1}$$
$$j_{rtd}(x) = L_E J(\phi(x)) \tag{2}$$
$$I_c(x) = e L_E N_d (T_c - d(x)) v(x) \tag{3}$$
$$v(x) = \mu \phi'(x) / (1 + (\mu / v_s) \phi'(x)) \tag{4}$$
$$d(x) = \{-2 e N_d d_{DBS} + [8 e \varepsilon N_d f(x) + (2 e N_d d_{DBS})^2]^{(1/2)}\} / (2 e N_d) \tag{5}$$

2.3 Simulation results

Example of calculated potential profiles, $\phi(x)$, are shown in Figure 2, where collector thickness and DBS peak current are 50 nm and 7E5 A/cm², respectively. Other parameters are listed in Table 2. The point $x=0$ is assigned to the emitter electrode edge furthest from the collector electrode, as shown in Fig. 1. Because no current flows in the lateral direction at $x=0$, $d\phi(x)/dx$ is equal to zero at $x=0$. The fine lines in Fig. 2 are potential profiles for various values of $\phi(x)$ at $x=0$. The gradient of each fine line increases monotonously with position x and finally reaches a critical point x_c at which $d\phi(x)/dx$ becomes infinite. The critical point x_c is the pinch-off point. In Fig. 2, the pinch-off points are connected to form a solid line. The DBS peak voltage and the DBS valley voltage are also shown by dotted lines in the figure. The differential resistance of the composite DBS is negative if the potential is in a range between the DBS peak and valley voltage. It is positive if the potential is not in that range.

Figure. 2 indicates there are three emitter width regions distinguished by qualitatively different RTPD I-V characteristics. The arrow pointing to C_I in Fig. 2 demonstrates the collector bias dependence of the potential profile for an RTPD in

Table 2

e	elemental charge	-
L_E	emitter length	10 μm
N_d	Si doping concentration	1E18 cm⁻³
T_c	collector thickness	-
$d(x)$	depletion layer thickness	-
$v(x)$	electron velocity	-
μ	electron mobility	5000 cm²/V/s
v_s	electron saturation velocity	2E7 cm/s
ε	dielectric constant	-
d_{DBS}	total thickness of DBS	10 nm

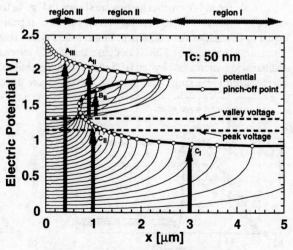

Figure 2. Examples of calculated potential profiles.

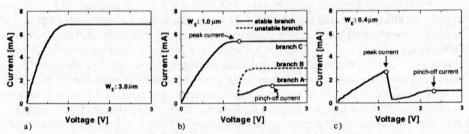

Figure 3. I-V characteristics for typical RTPDs in region I, II and III.

region I with an emitter width of 3 μm. The arrow represents increasing collector bias. Following it upward, one can see several fine lines intersecting it before it reaches C_I. Each fine line shows the potential profile within a range of $0 < x < 3$ μm for a collector bias equal to the respective intersecting points. When the collector bias reaches the pinch-off voltage shown by C_I, the collector current saturates. Figure 3a shows the I-V characteristics for this example.

In region II, it should be noted that three pinch-off points exist for all emitter widths. As an example, RTPD with an emitter width of 1 μm is explained. The arrows pointing to A_{II}, B_{II} and C_{II} in Fig. 2 demonstrate collector bias dependence of the potential profiles corresponding to branches A, B and C plotted in Fig. 3b. Branch A and C are stable, because the corresponding potential profiles essentially locate in a positive differential resistance region of DBS. On the contrary, branch B is unstable because it is related to the NDR region. Although it is stable, branch A will not be observed in a static I-V measurement because there is no path to reach the branch in a static manner.

For region III, again there is one pinch-off point. An RTPD with an emitter width of 0.4 μm is explained as an example. With increasing collector bias from zero to the pinch-off voltage shown by A_{III} in Fig. 2, collector potential increases correspondingly, passing through the *DBS peak voltage* and the *DBS valley voltage*, and resulting in the I-V characteristics shown in Fig. 3c. Thus, SPC are achieved in region III. The ratio of the RTPD peak current to the RTPD pinch-off current (PPR) increases with emitter width. This is because a pinch-off occurs at a collector voltage nearer to an RTPD valley voltage for wider emitters. Thus, ideal SPC with the largest PPR can be obtained at the right edge of region III. It should also be noted that no hysteresis appears in Fig. 3c.

Calculated I-V characteristics are shown in Fig. 4a for several emitter widths with a fixed collector thickness of 40 nm. It is clear that SPC appear for a wide range of emitter widths equal to or smaller than 0.28 μm. For emitters wider than 0.28 μm, a sudden change in I-V characteristics occurs. This is because the RTPD moves from region III to region II when emitter width increases beyond 0.28 μm. When the emitter width is 0.28 μm, hysteresis occurs for a collector bias range near the peak voltage where three current values correspond to single

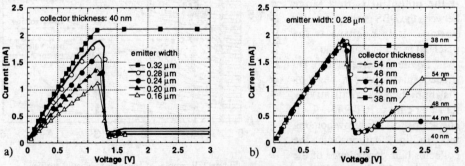

Fugure 4. Dependence of calculated I-V characteristics on emitter width and collector thickness.

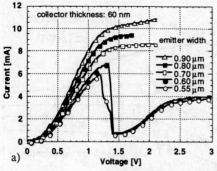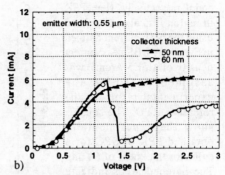

Figure 5. Experimental I-V characteristics for RTPDs.

collector bias. However, the bias range is negligibly small. Collector thickness dependence is shown in Fig. 4b with a fixed emitter width of 0.28 μm. Clear SPC are obtained for collector thicknesses between 40-44 nm. When the collector is less than 40 nm thick, a sudden change in I-V characteristics occurs again. For thicker collectors, the I-V curves deform from the ideal single-peak shape.

3. Experimental

The InGaAs/AlAs/InAlAs material system is used for RTPD fabrication. An example of the epitaxial layer structure is shown in Table 1. The epitaxial layers were grown on semi-insulating InP substrate by conventional MBE and, except for the AlAs layers, are lattice matched to InP. A barrier layer (layer No.1 in Table 1) is inserted in order to enhance the pinch-off effect in the collector by preventing electron injection into the substrate. As a reference, a conventional RTD was also fabricated. Its epitaxial layer structure is the same as Table 1 except the collector thickness and doping concentration are 400 nm and $2E18$ cm^{-3}, respectively and there is no barrier layer (layer No.1 in Table 1).

The RTPD fabrication process is a conventional one including photo lithography for resist patterning, a lift-off method for emitter and collector electrode delineation and wet etching for isolation.

4. Experimental I-V characteristics and discussion

The I-V characteristics of fabricated RTPDs were measured with an HP4156 at 300 K using the pulse mode. The pulse width and period were 1 ms and 50 ms, respectively. Figure 5a shows the results for RTPDs with various emitter widths ranging from 0.55 to 0.9 μm. The collector thickness is fixed at 60 nm. SPC were experimentally verified to occur for emitter widths of 0.55 and 0.60 μm. For larger emitter widths of 0.7, 0.8, 0.9 μm, SPC were not observed. Thus, the boundary between region II and III is experimentally determined to be at an emitter width between 0.6 and 0.7 μm for a collector thickness of 60 nm. The collector thickness dependence was also experimentally investigated. Figure 5b shows the I-V characteristics for collector thicknesses of 50 and 60 nm. Emitter width is fixed at 0.55 μm. SPC was observed only for the collector thickness of 60 nm. This fact also experimentally shows that the boundary between region II and III is at an emitter less than 0.55 μm wide for a collector thickness of 50 nm. It is very interesting to compare these experimental results with the theory developed in section 2. Figure 6 shows PPRs calculated as functions of emitter width for various collector thicknesses. The open and filled circles indicate PPRs for region III

876

and II, respectively. Here, PPR in region II is also calculated using the peak current defined in Fig. 3b. A DBS peak current density of 1.5E5 A/cm^2 was assumed based on the experimental result for the reference RTD. The squares show the experimentally obtained PPR of SPC for a collector thickness of 60 nm. The hatched areas indicate experimentally determined boundary between region II and III. They correspond well with the calculated boundary, which is at a position between the open and filled circles.

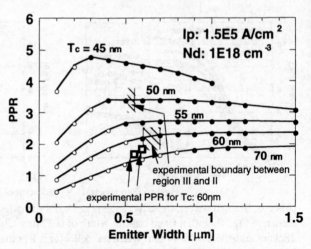

Figure 6. Comparison of PPR obtained by calculation and by experimint.

Good agreement between the theoretical and experimental results confirms the concept and modeling for the RTPD to be correct.

5. Conclusion

A Resonant Tunneling Pinch-off Diode (RTPD) has been proposed and its single-peak I-V characteristics have been systematically demonstrated both theoretically and experimentally along with their parameter dependence. Good agreement between the theory and experiment confirms that the mechanism determining the RTPD characteristics is the coupling between resonant tunneling effect in a vertical double barrier structure and the pinch-off effect in a lateral collector just beneath the double barrier structure. The guidelines for realizing an ideal RTPD have been clarified. RTPD could lead not only to reduced power consumption in functional logic circuits but also offer novel functional logic gates.

Acknowledgment

The authors would like to thank N. Suzuki for growing epitaxial layers. They also would like to thank Y. Imamura and T. Mizutani for their continuous encouragement.

References

[1] Capasso F, Sen S, Beltram F, Lunardi L M, Vengurlekar A S, Smith P R, Shah N J, Malik R J, and Cho A Y, 1989 *IEEE Trans. Electron Devices*, **36**, pp. 2065-2082
[2] Maezawa K, Akeyoshi T, and Mizutani T, 1994 *IEEE Trans. Electron Devices*, **41**, pp. 148-154
[3] Takatsu M, Imamura K, Mori T, Adachihara T, Muto S, and Yokoyama N, *Dig. of Tech. Papers ISSCC94*, pp. 124-125
[4] Yokoyama N, Imamura K, Muto S, Hiyamizu S, and Nishi H, 1985 *Japan. J. Appl. Phys.*, **24**, pp. L853-854
[5] Waho T, Maezawa K, and Mizutani T, 1993 *IEEE Electron Device Lett.*, **14**, pp. 202-204
[6] Woodward T K, McGill T C, Chung H F and Burnham R D , 1987 *Appl. Phys. Lett.* **51**, pp.1542-1544
[7] Tehrani S, Shen J, Goronkin H, Kramer G, Tsui R, and Zhu T X, 1994 *Int. Symp. on Compound Semiconductors, Inst. Phys. Conf. Ser.* **141** (Institute of Physics, Bristol), pp. 855-860

Inst. Phys. Conf. Ser. No 145: Chapter 6
Paper presented at 22nd Int. Symp. Compound Semiconductors, Cheju Island, Korea, 28 August–2 September 1995
© *1996 IOP Publishing Ltd*

GaAs Surface Tunnel Transistors with a symmetric source-to-drain structure

T. Baba and T. Uemura

Fundamental Research Laboratories, NEC Corporation
34, Miyukigaoka, Tsukuba, Ibaraki 305, JAPAN

Abstract. New functional Surface Tunnel Transistors (STTs) with a symmetric source-to-drain structure are proposed. These devices have two p^+-n^+ interband tunnel junctions back to back between a p^+ source and a p^+ drain through an n^+ channel. The symmetric GaAs-STTs were successfully fabricated using MBE regrowth. They exhibit gate-controlled negative-differential-resistance (NDR) characteristics, which are symmetric with drain bias. The peak-voltage increment due to the reverse-biased tunnel junction is negligible, permitting STT operation. These results indicate that functional circuits can be formed by the new STT structure as well as the conventional STT.

1. Introduction

Tunneling transistors are promising candidates for future electronic devices, because of their high speed, low power and functional operability. Operation based on negative-differential-resistance (NDR) helps reduce power consumption by significantly decreasing the number of circuit elements. Resonant tunneling transistors have been extensively investigated and have been used to build circuits with much fewer components than circuits with similar functions containing conventional transistors (Yokoyama 1985, Capasso 1986).

Recently, we proposed a new tunneling transistor, the Surface Tunnel Transistor (STT) (Baba 1992a, b). The STT consists of an n^+-source, a two-dimensional electron-gas (2DEG) channel, a p^+-drain and an insulated gate. The important point in this structure is that the drain must be highly degenerated and have a sharp doping profile. The interband tunneling current at the tunnel junction between the channel and the drain is controlled by the gate. By the analogy of p^+-n^+ tunnel diodes (Esaki diode), for STT operation, we can expect unsaturated transistor characteristics under reverse bias, and can expect NDR characteristics with peak voltages as low as around 0.1 V under forward bias. Therefore, low power (voltage) and functional circuits can be constructed by using these devices. In addition, since the current-voltage characteristics are inherently independent of the gate length or channel length, the gate length can be shrunk as small as the width of the interband tunnel-barrier (about 10 nm). This allows ultra large scale integration of such functional circuits.

The basic operational principle has been exhibited by using simple two-dimensional device simulation (Uemura and Baba 1993). Operation of the STT has been confirmed experimentally by demonstrating gate-controlled negative-differential-resistance (NDR) characteristics using mesa-structure GaAs-STTs as shown in Fig. 1 (Uemura and Baba 1994).

878

In this device, the basic STT structure, i. e., the channel and the gate, is formed at the mesa sidewall. The reason for choosing this structure was that the critical drain region was formed by molecular beam epitaxy (MBE), which affords us very high impurity doping and atomically sharp doping profiles. However, this mesa structure is not adequate for large-scale integration with high operation speed, because large parasitic capacitance is induced at the gate/channel and source/drain overlap regions. In addition, the large step height between the source and the drain prevents reliable wiring. Hence, planar structures are indispensable for overcoming these problems and producing large-scale integrated circuits with STTs. For this purpose, a planar STT, in which the source contact is directly formed on the channel layer, was constructed (Uemura and Baba 1995). However, the number of fabrication process steps or of photo-masks used is larger than in the conventional FET process, due to the asymmetric source-to-drain structure. For further device miniaturization and large-scale integration with a small number of photo-masks, planar and symmetric source-to-drain structures are desirable.

This paper presents preliminary experiments investigating the possibility to obtaining a symmetric structure by fabricating p^+-n^+-p^+ GaAs tunnel diodes, in which two tunnel junctions are connected back to back. Moreover, it describes the first successful fabrication of a new STT with a symmetric source-to-drain structure using molecular beam epitaxy (MBE) regrowth. The electrical properties of the transistor show symmetric NDR characteristics with respect to the polarity of the drain voltage.

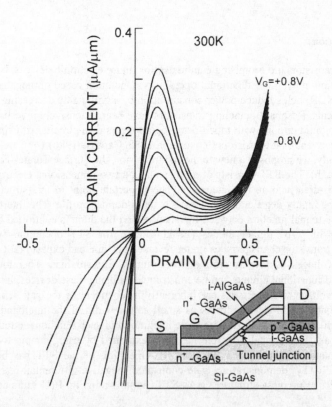

Fig. 1. Transistor characteristics for a conventional GaAs-STT with mesa structure. The inset shows a schematic cross-sectional view of the mesa GaAs-STT.

2. Experiment

The p^+-n^+-p^+ double-junction tunnel-diode structure, which consist of a 400-nm p^+-GaAs (Be=1x10^{20} cm^{-3}), a 50-nm n^+-GaAs (Te=1.0x10^{19} cm^{-3}) and a 150-nm p^+-GaAs (Be=1x10^{20} cm^{-3}), were grown on a semi-insulating (100) GaAs substrate by MBE at 520°C. Relatively low temperature growth was chosen to prevent impurity diffusion and segregation. The middle layer of n^+-GaAs was designed to be thick enough to restrict the interference between the two tunnel junctions. The round-shaped diodes with mesa structure were fabricated (insets of Figs. 3 (a) and (b)). The upper and bottom electrodes were formed by AuGeNi non-alloy contacts. As a reference, conventional p^+-n^+ tunnel diodes with the same doping concentrations were also fabricated. The current-voltage characteristics were measured on a wafer by using a semiconductor parameter analyzer (HP4145B).

Figures 2 (a) and (b) show the surface photograph and the schematic cross-sectional view of the fabricated symmetric STTs. First, a 500-nm i-GaAs buffer layer, a 20-nm p^+-GaAs (Be=1x10^{20} cm^{-3}) drain-source layer, and a 30-nm i-GaAs blocking layer were grown on a (100) GaAs substrate by MBE at 520°C. The source and the drain regions were formed by wet chemical etching with an SiO$_2$ mask. It should be noted that the mesa sidewall of the source and drain regions were designed to minimize the gate overlap area, and its crystal orientation is close to (111)B. This is different from the orientation of (511)B for the mesa-type STTs (Uemura and Baba 1994). After removing the mask, thermal cleaning at 650°C was performed to remove the oxides from the surface. Then, layers of a 13-nm or 14-nm n^+-GaAs (Te=1.4x10^{19} cm^{-3}) channel, a 50-nm i-Al$_{0.5}$Ga$_{0.5}$As gate insulator and a 50-nm p^+-GaAs (Be=1x10^{20} cm^{-3}) gate were successively overgrown by MBE at 520°C. The gate area was formed by wet chemical etching down to the blocking layer, followed by device isolation. An AuZn alloy contacts were used for the source and drain, and Cr/Au non-alloy contacts for the gate. Since source and drain regions are formed simultaneously, reduction of two photo-masks was performed in this fabrication, compared with the process for the asymmetric planar STTs (Uemura and Baba 1995). Due to the structural similarities between symmetric STTs and conventional FETs, self-aligned processes for conventional FETs can be applicable for the fabrication of symmetric STTs, resulting in further reduction of process steps.

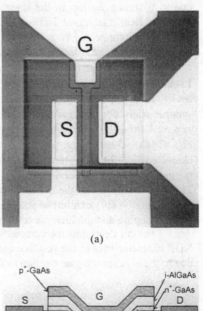

(a)

(b)

Fig. 2. New GaAs-STT with a symmetric source-to-drain structure; (a) surface photograph of fabricated symmetric GaAs-STT, and (b) schematic cross-sectional view. Channel length (source-to-drain spacing), gate length, and gate width are 2 µm, 6 µm and 46 µm, respectively.

880

3. Result and Discussion

Figures 3 (a) and (b) show the current-voltage characteristics of the p^+-n^+ tunnel diode and the p^+-n^+-p^+ double-junction tunnel-diode each 100 μm in diameter, at room temperature. Clear NDR characteristics are observed in both diode structures. Terrace structures in the NDR region are attributed to high-frequency current oscillations. For the double-junction tunnel-diode, NDR characteristics are obtained for positive- and negative-bias conditions. This reason is recognized by considering that one tunnel junction is forward-biased and the other is reverse-biased under both bias directions. The higher resistance of NDR for the forward-biased junction is reflected mainly in the current-voltage characteristics. The peak current and peak-to-valley ratio (PVR) of the double-junction tunnel-diode is almost the same as the conventional tunnel diodes as shown in Figs. 3 (a) and (b). The peak voltage (0.13 V) is slightly higher than that of the conventional tunnel diode (0.09 V). This is mainly due to the voltage drop in the reverse-biased tunnel junction. From these experiments, it becomes clear that good NDR characteristics can be obtained even though two tunnel junctions are connected back to back, and that the symmetric STT structures with the same conduction type of the source and the drain can provide good NDR characteristics.

Figures 4 (a) and (b) show the transistor characteristics of the fabricated new STT with 13-nm and 14-nm thick channels, respectively. Gate voltages were changed from -0.8 V to 0.4 V in 0.4-V increments. Gate-controlled NDR characteristics were clearly observed symmetrically for positive and negative drain voltages. These characteristics are indeed expected from the preliminary results using the double-junction tunnel-diodes as shown in Fig. 3 (b). This result confirms proper STT operation with gate-controlled NDR characteristics for the symmetric source-to-drain structures. The current modification with the gate voltage, however, is relatively weak compared with the conventional STTs (Fig. 1). This is probably due to the channel being thicker than optimal. Actually, the thicker channel device (Fig. 4 (b)) exhibits weaker gate modulation. To improve the gate controllability, channel doping and thickness need to be optimized.

Looking deep into the current-voltage characteristics, the peak and valley current for NDR characteristics in the positive and negative bias conditions differ slightly. This may be due to the asymmetric gate overlap to the source and drain and/or a variation in uniformity of

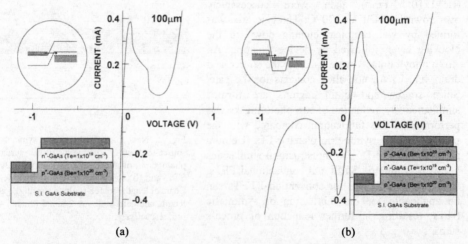

(a)   (b)

Fig. 3. Current-voltage characteristics of tunnel diodes; (a) conventional p^+-n^+ tunnel diode, and (b) p^+-n^+-p^+ double-junction tunnel diode.

surface cleaning before regrowth of the channel layer. Significant increase in the valley current is observed for the negative drain bias with the high gate voltage of 0.4 V. This is due to the gate leakage-current. Where gate bias was larger than 0.4 V, gate leakage-current increased drastically and the NDR characteristics disappeared. The gate breakdown voltages in the fabricated devices were smaller than those for the mesa-type STTs. Although the reason for this is not clear at this stage, the crystal quality of the regrown layers in this fabrication is considered to be poor. This degradation may be caused by the steeper mesa sidewall in the source and the drain, or by high-temperature thermal cleaning before the MBE regrowth.. For improved current-voltage characteristics with higher gate controllability and large peak-to-valley ratio in the NDR characteristics, MBE regrowth conditions and layer structures need to be optimized.

The peak voltage of the observed NDR characteristics is about 0.14 V, which is slightly higher than that of the mesa-type GaAs-STTs (0.10 V), but is close to that of the p^+-n^+-p^+ double-junction tunnel-diodes (0.13 V). Therefore, this voltage shift can be recognized by the voltage drop in the reverse-biased tunnel-junction, as explained for the double-junction tunnel-diodes. The NDR characteristics under positive drain-bias originate from the channel-drain tunnel junction, and under negative bias from the channel-source junction. The other tunnel junction, which is reverse-biased, works as a linear resistor with low impedance. The voltage increase of about 0.04V for the new STT with symmetric source-to-drain structure is small enough compared with the expected operation voltage, which is larger than the valley voltage of about 0.5V, and may not cause severe problems for the application of low-voltage functional-circuits.

Here, we would like to discuss the differences in characteristics between the conventional STTs with the asymmetric source-to-drain structure and the new STTs with the symmetric structure. For the conventional STTs, two specific sets of characteristics, those of

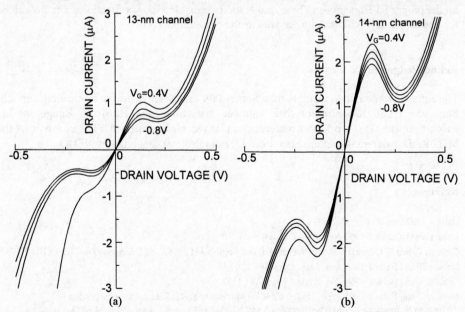

Fig. 4. Transistor characteristics of new GaAs-STT with a symmetric source-to-drain structure; (a) 13-nm thick channel and (b) 14-nm thick channel. The gate voltage is varied in 0.4 V increments from -0.8 V to 0.4 V.

the unsaturated transistor in reverse bias and of the NDR in forward bias, can be used, depending on the application circuits. In contrast, the symmetric STTs can handle only NDR characteristics, because the same characteristics obtain for both bias directions. When we need conventional transistor characteristics with high current density as well as NDR characteristics, conventional asymmetric STTs should be used. If we do not need conventional transistor characteristics and use only NDR characteristics, use of the symmetric STTs instead of the asymmetric STTs is recommended. The symmetric STT configuration allows a reduction in process steps and concomitant reduction in the number of photo-masks in device fabrication. For making the integrated circuits, this is very important in increasing process reliability and in reducing process time and process costs. Furthermore, multiple tunnel-junction structures may be fabricated as an extension of the double tunnel-junction STTs. Such structures enable us to handle the new function with gate-controlled multiple NDR characteristics, and can pave the way for high integration and new functional operation.

4. Conclusion

To investigate the possibility of producing new STTs with a symmetric source-to-drain structure, p^+-n^+-p^+ double-junction tunnel diodes were fabricated. This preliminary experiment showed a negligible increase of peak voltage in NDR characteristics. Based on this result, symmetric GaAs STTs, consisting of a p^+ source, an n^+ channel and a p^+ drain, were successfully fabricated using MBE regrowth. Gate-controlled NDR characteristics, which are symmetric with the drain bias, were obtained for the new STTs. The peak-voltage increment due to the reversely-biased tunnel junction is negligible (0.04 V), and did not hinder proper STT operation. These results indicate that low-power functional circuits can be formed by the new STT structure as well as the conventional STT.

Acknowledgment

The authors would like to thank Hiroko Someya for assistance in device fabrication, Jun'ichi Sone and Hisao Kawaura for their valuable discussion, and Hiroyoshi Rangu for his encouragement. This work was performed under the management of FED as a part of the MITI R&D Program (Quantum Functional Devices Project) supported by NEDO.

References

Baba T 1992 *Jpn. J. Appl. Phys.* **31** L455
Baba T and Uemura T 1992 *NEC Res. & Develop.* **33** 403
Capasso F, Sen S, Gossard A C, Hutchinson A L and English J H , 1986 *IEEE Electron Dev. Lett.* **EDL-7** 573
Uemura T and Baba T 1993 *Inst. Phys. Conf. Ser.* **129** 353
Uemura T and Baba T 1994 *Jpn. J. Appl. Phys.* **33** L1363
Uemura T and Baba T 1995 *Proc. 7th Int. Conf. on Modulated Semiconductor Structures* Madrid
Yokoyama N, Imamura K, Muto S, Hiyamizu S and Nishi H 1985 *Jpn. J. Appl. Phys.* **24** L853

Inst. Phys. Conf. Ser. No 145: Chapter 6
Paper presented at 22nd Int. Symp. Compound Semiconductors, Cheju Island, Korea, 28 August–2 September 1995
© *1996 IOP Publishing Ltd*

Effects of Photogenerated Carriers on Oscillation Frequency in Resonant Tunneling Structure

Hye Yong Chu and El-Hang Lee

Research Department, Electronics & Telecommunications Research Institute
Yusong P.O. Box 106, Taejon, 305-600, Korea

ABSTRACT : The effects of the photogenerated carriers on the oscillation characteristics of the resonant tunneling diode oscillator (RTDO) with a double barrier resonant tunneling structure have been investigated . When the RTDO is illuminated with a laser, the resonant tunneling peak in the dc current-voltage characteristics shifts toward lower voltages. The oscillation frequency decreases linearly as the power of incident light increases at a constant bias voltage.

1. Introduction

Due to its unique device characteristics, the resonant tunneling device can be used as a self-oscillating modulator which operates simultaneously as an oscillator and optical modulator. Thus, the effects of space charge near the heterostructure or inside the quantum well on the resonant tunneling have been investigated to optimize the performance as optoelectronic devices.[1-3] In the microwave oscillator with a double barrier resonant tunneling structure (DBRTS), the optical modulation of frequency and the phase locking between light pulses have been reported.[4,5]

In this report, we present the characteristics of a resonant tunneling diode oscillator (RTDO) with a DBRTS under the illumination of laser. We have investigated the effects of the phtogenerated carriers on the oscillation frequency of the RTDO. Our experimental results suggest that the microwave frequency of the RTDOs can be controlled by the laser illumination.

2. Experimental details

The RTDO used in this work was grown by molecular beam epitaxy on a semi-insulating GaAs substrate. Two-step spacer layers of a 10 nm thick undoped GaAs and a 300 nm thick Si-doped (4×10^{17} cm^{-3}) GaAs were grown at both sides of DBRTS, consisting of two 2.8 nm thick undoped AlAs barriers and a 4.5 nm thick undoped GaAs well layer between them. For ohmic contact, heavily doped (2×10^{18} cm^{-3}) GaAs layers were grown with thickness of 500 nm and 1000 nm, at the contact and buffer sides, respectively. A mesa diode with area of 7×7 μm^2 was fabricated by photolithography and chemical wet etching. In order to measure the characteristics during laser illumination, an electrode with area of 3×3 μm^2 was partially capped on the mesa and a Ti/Au pad for wire bonding was fabricated. The RTDO was diced and bonded to a 50 Ω microstrip line. The dc bias voltage for the RTDO was applied through a bias tee. The RTDO was illuminated with a HeNe laser or a tunable cw Ti:Sapphire laser with 100 μm beam diameter.

3. Results and discussion

Dc current-voltage (I-V) characteristics of the RTDO at a forward bias are shown in Fig. 1. In our experiments, the forward bias is defined as the positive bias applied on the mesa. The solid and dashed lines present the results measured in the dark and under illumination with a HeNe laser of 10 mW power. Since the absorption coefficient of GaAs is larger than 3.0 x 10^4 cm^{-1}, at least 80% of the incident light is expected to be absorbed in the contact layer and the spacer layers above the upper barrier. Thus, the conductivities of the contact layer and the spacer layers increase due to the photogenerated carriers. The electrostatic potential around the DBRTS is also modified because the accumulation of carriers near the DBRTS is enhanced. As a result, the resonant tunneling peak in the dc I-V characteristics shifts toward a lower voltage because the condition for the electron resonant tunneling with laser

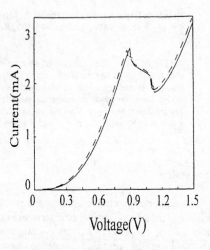

Fig.1 I-V characteristics of RTDO. The solid and dashed lines present the results measured in the dark and under illumination with a HeNe laser of 10 mW power.

illumination is satisfied at the lower voltage than that without the laser illumination. The non-resonant tunneling current near the valley point is found to be more enhanced than that without the laser illumination. This is attributed to the increase of the thermionic emission current that is induced due to the excitation of carriers to the higher energy in the spacer layers and the increase of the temperature by the illumination of the HeNe laser. As a result, the negative resistance is increased even though the series resistance is decreased due to the increased conductivity.

Biasing the RTDO to a negative resistance region using an external voltage source, we have measured the oscillation frequency without the illumination of laser. The frequencies of the RTDO were 5.28 GHz and 5.32 GHz with a constant output power of -26 dBm at the applied voltage of 0.973 V and 1.093 V, respectively. If the frequency dependent spreading resistance is neglected, the frequency of RTDO depends on a capacitance and a negative resistance. Thus, as the applied voltage is increased to the near valley point, the oscillation frequency increases because of the reduced negative resistance and series resistance.[6,7]

The oscillation frequencies of RTDO biased at 1V and illuminated with a HeNe laser of 10 mW power and a 740 nm laser of 100 mW power are shown in Fig. 2(a) and (b), respectively. These are measured at room temperature. In the Fig. 2(a), the oscillation frequency is reduced about 5 MHz with insignificant changes in the

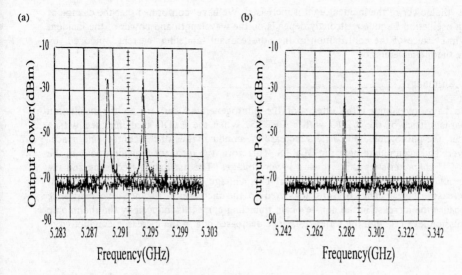

(a)   (b)

Fig 2 Oscillation frequency of RTDO under illumination (a) with 630 nm wavelength and 10 mW light power and (b) 740 nm wavelength and 100 mW light power, repectively. The right trace is without illumination, the left trace is with illumination, respectively.

oscillator output power. However, the frequency spectrum of Fig 2(b) is broadened and shifted to the lower frequency by about 20 MHz from 5.3 GHz under the illumination. The output power of oscillation frequency is also decreased. The decrease in output power and the broadening of frequency spectrum are enhanced as the power of the incident light is increased. We have measured the frequency of the RTDO with 6.5 nm AlAs barriers and 4.5 nm GaAs well layer grown on semiconducting substrate in the condition of the temperature of an RTDO maintained constantly at 80K. When the RTDO is illuminated by the laser with the wavelength of 800 nm and the power of 12 mW, the frequency of the RTDO has been found to enhance from 5.6 MHz to 7.1 MHz. Beacuse the thermionic current is suppressed at 80K, the frequency is increased by the decrease in series resistance due to the photogenerated carriers. However, at room temperature, the decrease in frequency under the illumination of laser is attributed to the increase of negative resistance due to the thermionic emmision current as shown in the Fig. 1. Fig. 3 shows the shift of oscillation frequency with respect to the power of laser with 850 nm wavelength. The open circle in Fig. 3 represent the frequency shift under illumination with a HeNe laser of 10 mW power. The oscillation frequency of the RTDO is decreased linearly

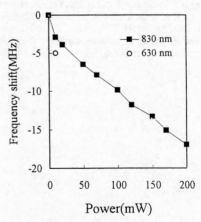

Fig. 3 Shift of oscillation frequency for the power of incident light

as the power of the incident light is increased. We have conformed that the change of the oscillation frequency strongly depends on the wavelength and power of the incident light because of the contribution of the non-resonant tunneling current induced by the thermionic emission.

4. Summary

We have investigated the effects of the photogenerated carriers on the oscillation characteristics of the RTDO with a DBRTS. When the RTDO is illuminated with a laser, the photogenerated carriers increase the conductivities of the contact and spacer layers and accumulate near the BDRTS. Therefore, the resonant tunneling peak in the dc I-V characteristics shifts toward lower voltages. The oscillation frequency and the output power decrease at a constant bias voltage as the power of incident light increases. These behaviors are attributed to the increase in the negative resistance because the increase in the non-resonant tunneling current is induced by the thermionic emission, even though the series resistance decreases due to the increased conductivity.

Acknowledgments

This research was supported by the Ministry of Information and Communications, Korea.

References

[1] Kan S C, Wu S, Sandsers S, Griffel G and Yariv A 1991 *J. Appl. Phys.* **69** 3384-6

[2] Schneider H, Larkins E C, Ralston J D, Schwartz K, Fuchs F and Koidl P 1993 *Appl. Phys. Lett.* **63** 782-4

[3] Chu H Y, Park P W, Han S G and Lee E H 1995 *Jpn. J. Appl. Phys.* **34** 205-7

[4] Higgins T P, Harvey J F, Sturzebecher D J, Paolella A C and Lux R A 1992 *Electronic. Lett.* **28** 1574-6

[5] Lann A F, Grumann E, Gabai A and Golub J E 1993 *Appl. Phys. Lett.* **62** 13-5

[6] Brown E R, Soderstrom J R, Parker C D, Mahoney L J, Molvar K M and Mcgill T C 1991 *Appl. Phys. Lett.* **58** 2991-3

[7] Zarea A, Raven M S, Steenson D P, Chamberlain J M, Henini M and Hughes O H 1992 *Electronic. Lett.* **28** 264-6

Inst. Phys. Conf. Ser. No 145: Chapter 6
Paper presented at 22nd Int. Symp. Compound Semiconductors, Cheju Island, Korea, 28 August–2 September 1995
© *1996 IOP Publishing Ltd*

Quantum Hall Effect Devices of Delta-Doped $Al_{0.25}Ga_{0.75}As$/ $In_{0.25}Ga_{0.75}As$ /GaAs Pseudomorphic Heterostructures grown by LP-MOCVD

J. S. Lee [1], **K. H. Ahn, and Y. H. Jeong**

Department of Electronics and Electrical Engineering and Microwave Application Research Center, Pohang University of Science & Technology, Pohang P.O. box 125,790-600 Korea

Abstract. Quantum Hall effect devices based on delta-doped $Al_{0.25}Ga_{0.75}As$/ $In_{0.25}Ga_{0.75}As$/GaAs pseudomorphic heterostructure materials grown by low-pressure metal organic chemical vapor deposition(LP-MOCVD) are sucessfully fabricated. A high electron mobility of 7200 cm^2/V·s with a sheet carrier density of $2.0 \times 10^{12} cm^{-2}$ has been achieved at room temperature. The temperature coefficient of product sensitivity is -0.1%/K. The minimum detectable magnetic field(B_{min}) of 3 μT at 1 Hz is achieved due to the high electron mobility.

1. Introduction

Recently quantum Hall effect devices using III-V semiconductors using InGaAs/InP and AlGaAs/GaAs heterostructures have been reported[1-5]. Due to the high electron mobility of the two dimentional electron gas(2DEG), these systems showed high quantum magnetic sensitivities with large S/N ratio.

A single heterostructure AlGaAs/GaAs device achieved high quantum magnetic sensitivities of 1000V/AT[1], but has a poor temperature coefficient of 0.7%/K that was ascribed to the DX center in n-$Al_xGa_{1-x}As(x > 0.2)$ layers. To reduce the effects of DX center, AlAs/GaAs superlattice structures were proposed[4]. The temperature coefficients of 0.1%/K with high sensitivities of 1200V/AT was obtained in this structure. However this material system had a low mobilities approximately 5000 cm^2/V·s, due to the interface roughness of the AlAs/GaAs layers and showed large input impedance of 2.2kΩ.

We report high performance delta-doped $Al_{0.25}Ga_{0.75}As$/ $In_{0.25}Ga_{0.75}As$/GaAs pseudomorphic quantum Hall effect devices grown by LP-MOCVD. The delta doping technique is used to suppress the adverse effect of DX center and to achieve a sharp potential curvature in AlGaAs layers for better carrier confinement[6]. In addition, a large AlGaAs/InGaAs conduction band offset is more effective on confining the 2DEG in the channel.

[1] E-mail (J. S. Lee): ljs@hemt.postech.ac.kr

888

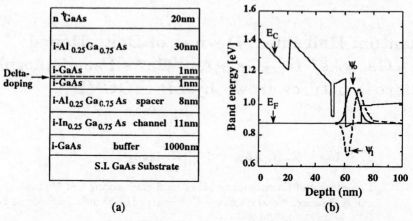

(a)   (b)

Figure 1. Quantum Hall effect device of delta-doped $Al_{0.25}Ga_{0.75}As/In_{0.25}Ga_{0.75}As/$ GaAs pseudomorphic heterostructures; (a)schematic cross section, (b)simulation of the conduction band potential profile(E_C), Fermi-level position(E_F), and electron wavefunction(ψ_0, ψ_1) in the well calculated from one-dimensional self-consistent method

2. Epitaxial Growth and Device Fabrication

The epitaxial layers for quantum Hall effect devices were grown by LP-MOCVD on semi-insulating (100) GaAs substrates at reactor pressure of 76 torr. As the reactant sources, trimethylgallium(TMG), trimethylaluminum(TMA), ethyldimethylindium(EDMIn) and a 10% AsH_3 diluted in H_2 were used. The SiH_4 gas was used as an n-type dopant source for delta doping.

The cross-section of a delta-doped $Al_{0.25}Ga_{0.75}As/In_{0.25}Ga_{0.75}As/GaAs$ pseudomorphic heterostructure is shown in Fig. 1(a). It consists of a 1000-nm undoped GaAs buffer layer, a 11-nm undoped $In_{0.25}Ga_{0.75}As$ channel, a 8-nm undoped $Al_{0.25}Ga_{0.75}As$ spacer layer, a 1-nm undoped GaAs layer, Si-delta doping, a 1-nm undoped GaAs layer, a 30-nm undoped $Al_{0.25}Ga_{0.75}As$ barrier layer, and finally a 20-nm doped GaAs ohmic contact layer. The growth temperature was 700°C for GaAs, 650°C for InGaAs, and 730°C for AlGaAs. The epitaxial growth rate of GaAs and AlGaAs were 1.0 and 1.2μm/h, respectively.

A one-dimensional self-consistent simulator was used to show the conduction band energy profile of the delta-doped $Al_{0.25}Ga_{0.75}As/In_{0.25}Ga_{0.75}As/GaAs$ pseudomorphic heterostructures. Using this method, we extracted the optimized Si-delta-doped GaAs layer thickness of 2nm to prevent parasitic parallel conduction in AlGaAs layer and confirmed electrons were only confined in InGaAs layer. Simulation outputs for the conduction band diagram, electron wavefunctions, and Fermi level position are shown in Fig. 1(b).

The four terminal Greek-cross Hall devices were fabricated by a mesa process with a $H_3PO_4:H_2O_2:H_2O(1:1:25)$ solution. Following this process, an ohmic contact was made by evaporating AuGe/Ni/Au and rapid thermal annealing at 450°C for 30sec. The length and the width of the device are 300 and 100μm, respectively.

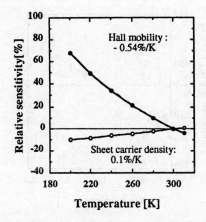

Figure 2. Temperature dependence of a delta-doped $Al_{0.25}Ga_{0.75}As/In_{0.25}Ga_{0.75}As/GaAs$ pseudomorphic quantum Hall effect device for Hall mobility or sheet carrier density

3. Experimental Result and discussion

The $Al_{0.25}Ga_{0.75}As/In_{0.25}Ga_{0.75}As/GaAs$ pseudomorphic structures show a sheet carrier density of $2.0 \times 10^{12}cm^{-2}$ and a Hall mobility of 7200 $cm^2/V{\cdot}s$ at room temperature. The Hall mobility and the sheet carrier density of a delta-doped $Al_{0.25}Ga_{0.75}As/In_{0.25}Ga_{0.75}As/GaAs$ pseudomorphic heterostructures are shown in Fig. 2, with temperature as a parameter. As shown in Fig. 2, the temperature coefficient of the Hall mobility and the sheet carrier density are -0.54 and 0.1%/K, respectively. These values are much smaller than that of AlGaAs/GaAs Hall device grown by MBE technique, because the energy level of the DX centers are located at a higher level by a sharp potential curvature in the AlGaAs layer from the effect of delta doping[1]. The device input impedance is $1.1k\Omega$ at room temperature.

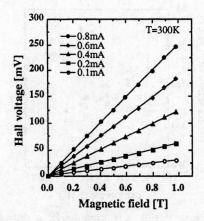

Figure 3. Hall voltage vs. magnetic field for a delta-doped $Al_{0.25}Ga_{0.75}As/In_{0.25}Ga_{0.75}As/$ GaAs pseudomorphic quantum Hall effect device with bias current as a parameter

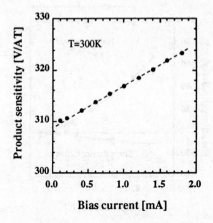

Figure 4. Product sensitivity vs. bias current for a delta-doped $Al_{0.25}Ga_{0.75}As/$ $In_{0.25}Ga_{0.75}As/GaAs$ pseudomorphic quantum Hall effect device

The transverse magnetic-field dependence of the quantum Hall effect device with a length-to-width ratio of 3 is shown in Fig.3, with the input current as a parameter. From the output Hall voltage, the product sensitivity, defined as (V_H/I_SB), was determined as $315 VA^{-1}T^{-1}$. The offset voltage that appeared between Hall output contact in the absence of external transverse magnetic field was also measured. The offset is caused by inhomogeneity in the epitaxial layer and/or by misalignment in fabrication process. The magnitude of offset is defined as the equivalent magnetic field, $B_0 = V_{offset}/(I_s K_H)$. In our devices, B_0 equals 0.8mT, which is much smaller than that of GaAs Hall device grown by MOCVD due to the high electron mobility[5].

The product sensitivity is slightly increased with the bias current. Such behavior known as the backgating effect is shown in Fig.4. The measured slope is

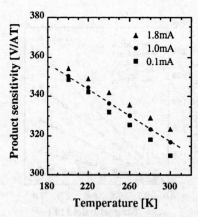

Figure 5. Product sensitivity vs. temperature for a delta-doped $Al_{0.25}Ga_{0.75}As/$ $In_{0.25}Ga_{0.75}As/GaAs$ pseudomorphic quantum Hall effect device with bias current as a parameter

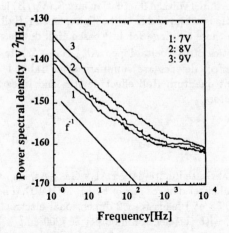

Figure 6. Power spectral density of a delta-doped $Al_{0.25}Ga_{0.75}As/In_{0.25}Ga_{0.75}As/GaAs$ pseudomorphic quantum Hall effect device

$8.4mV(mA)^{-2}T^{-1}$ and is one third of the vertical Si Hall devices[7].

The temperature dependence of the product sensitivity is shown in Fig.5. The sheet carrier variation in III-V heterostructure is mainly due to the activation of the DX centers in a heavily doped AlGaAs layer. In a delta-doped $Al_{0.25}Ga_{0.75}As/In_{0.25}Ga_{0.75}As/GaAs$ pseudomorphic heterostructure, this effect is minimized resulting a small temperature dependence. The temperature coefficient of the product sensitivity which is inversely proportional to sheet the carrier concentration is -0.1%/K.

The noise spectrum of the quantum Hall effect device was measured to determine the minimum detectable magnetic field(B_{min}). The low frequency noise measurements were performed in a brass-shielded room. The output noise voltage fluctuation was amplified by a low-noise preamplifier(ITHACO1201) and fed to a computer-controlled HP3582A spectrum analyzer.

The low-frequency noise spectrum of the Hall terminal voltage at different bias voltages is shown in Fig.6. The generation-recombination noise appears around 10KHz. The noise voltage fluctuation increased as the intput voltage increased. The Hooge parameter, which indicated the material quality, is evaluated to be as low as $\alpha_H=3\times10^{-5}$ in the 1/f region. From the results, we can calculate B_{min} given by[8]

$$B_{min} = [S(f)\Delta f]^{1/2}/S_v$$

Table 1. Characteristics of a delta-doped $Al_{0.25}Ga_{0.75}As/In_{0.25}Ga_{0.75}As/GaAs$ pseudomorphic quantum Hall effect devices

Device parameter	Device performance
Dimensions(L×W)	$300\times100\mu m^2$
Sheet carrier concentration	$2.0\times10^{12}/cm^2$
Carrier mobility	7200 cm²/V·s
Temperature coefficient of Hall mobility	-0.54 %/K
Device impedance	1.1 KΩ
Product sensitivity (K_H)	315 V/AT
Temperature coefficient of product sensitivity	0.1 %/K
Minimum detectable magnetic field	3 μT at 1Hz

892

where $S(f)$ is a Hall terminal voltage fluctuation and $S_v(V_H/B)$ is the absolute sensitivity. The B_{min} of $3\mu T$ at 1Hz is achieved due to the high quantum Hall effect electron mobility and is comparable to those obtained for InP-based Hall device s[9].

The characteristics of the delta-doped $Al_{0.25}Ga_{0.75}As/$ $In_{0.25}Ga_{0.75}As$ pseudomorphic quantum Hall effect devices are summarized in Table 1. This structure is well suitable for intelligent quantum Hall effect devices and integrated circuits fabricated with low noise transistors.

4. Conclusion

We have sucessfully fabricated delta-doped $Al_{0.25}Ga_{0.75}As/In_{0.25}Ga_{0.75}As/GaAs$ pseudomorphic quantum Hall effect devices grown by LP-MOCVD. The large conduction band discontinuity(about 0.35 eV) leads to a 2-dimensional electron gas(2-DEG) density as high as $2.0 \times 10^{12}/cm^2$ with the electron mobility of 7200 $cm^2/V \cdot s$ at 300K. The temperature coefficient at a product sensitivity of $315VA^{-1}T^{-1}$ is -0.1%/K. The resolution of $3\mu T$ at 1 Hz is achieved. Our proposed structure is very promising for high performane quantum Hall effect devices.

References

[1] Y. Sugiyama, T. Taguchi, and M. Tacano, *Proc. 6th Sensor Symp.* , pp. 55-100, 1986

[2] R. Kyburz, J. Schmid, R. S. Popovic, and H. Melchior, *IEEE Trans. Electron Devices* , vol. **41**, pp.315-320, 1994.

[3] Y. Sugiyama, H. Soga, and M. Tacano, *J. Cryst. Growth* , vol. **95**, pp. 394-397, 1989.

[4] Y. Sugiyama, H. Soga, M. Tacano, and H. P. Baltes, *IEEE Trans. Electron Device* , vol. **36**, pp.1639-1650, 1989

[5] R. Campesto, C. Flores, A. Passaseo, and S. Verni, *Sensors and Actuators* , vol. **A32**, pp. 651-660, 1992

[6] B. Etienne and V. Thierry-Mieg, *Appl. Phys. Lett.* , vol. **52**, pp. 1237-1240, 1988

[7] U. Falk and R. S. Popovic, *Tech. Digest, 7th Int. Conf. Solid-State Sensors and Actuators (Transducer'93)*, pp. 902-903, 1993

[8] M. Mathieu, P. Giordano, and A. Chovet, *Sensors and Actuators* , vol. **A32**, pp. 656-660, 1992

[9] Y. Sugiyama, Y. Takeuchi, and M. Tacano, *Sensors and Actuators* , vol. **A34**, pp. 131-135, 1992

Inst. Phys. Conf. Ser. No 145: Chapter 6
Paper presented at 22nd Int. Symp. Compound Semiconductors, Cheju Island, Korea, 28 August–2 September 1995
© *1996 IOP Publishing Ltd*

Quantum Hall effect in GaAs/AlGaAs heterostructures:Its breakdown due to current

S Kawaji

Department of Physics, Gakushuin University, Mejiro, Toshima-ku, Tokyo 171, Japan

Abstract. This paper reviews experimental researches of breakdown of the quantum Hall effect in GaAs/AlGaAs heterostructures due to current carried out in Gakushuin University. We used specially designed Hall bars which eliminate influence of dissipation at current electrodes and found following results: (1) Critical current at the onset of the breakdown is proportional to the sample width. (2) The critical Hall electric field at the breakdown for quantized Hall plateau with plateau quantun numbers $i = 2$ and 4 is proportional to $B^{3/2}$ and does not depend on the electron mobility of the sample where B is the magnetic flux density at the plateau center.

1. Introduction

In the quantum Hall effect (QHE), the Hall resistance in a Hall bar is quantized as $R_{\mathrm{H}}(i) = h/ie^2$ for an integer i in degenerate two-dimensional (2D) electron systems at low temperatures and in strong magnetic fields.[1] The quantized Hall resistance(QHR) has been used as an international resistance standard since 1990.[2] In order to carry out a high precision measurement of the quantized Hall resistance, we have to pass a high electric current through a Hall bar to generate a high Hall voltage across the sample. When the current exceeds a critical value, however, the QHE is broken down.[1]

The QHE state is characterized by non-dissipative current whose direction is orthogonal to the direction of electric field, i.e. Hall electric field, where the diagonal resistance in the Hall bar is zero:i.e. $R_{xx}=0$. Such a nondissipative electronic state appears near the central part of the sample only. Joule heat is always generated at both ends of the sample; i.e. two transition regions between the 2D system and current electrodes.[1, 3, 4] The breakdown of the QHE means appearance of dissipation in a part of a Hall bar other than the parts near current electrodes.

The following is a review of experimental researches of the breakdown of the QHE in GaAs/AlGaAs heterostructures carried out in Gakushuin University.

2. Experiments

Many authors have reported results of experimental researches of the breakdown of the QHE.[5] Our aim is to observe intrinsic breakdown of the QHE. Here, the intrinsic breakdown means the breakdown which occurs without any influence of the dissipation at

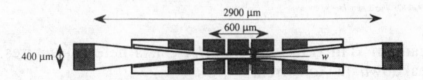

Fig.1. An example of lectrode structure of samples. Shaded areas are ohmic contacts. Breakdown
is measured in the central part of 600 μm long with three pairs of potential probes.[6~8]

the current electrodes. So we designed Hall bar type samples with electrode structures as
shown in Fig. 1.[6] Principles of the design of our samples are as follows. In high precision
measurements of QHRs as a reference standard of electrical resistance, long and wide Hall
bars prepared from GaAs/Al$_x$Ga$_{1-x}$As (x~0.3) heterostructures with a low electron mobility
of order of 10 m^2/Vs have been used in many standards laboratories. We have tried to
simulate those Hall bars in our samples. In order to make small the production of nonequilibrium population of electrons in Landau levels near the current electrodes, we made the
width of current electrodes as large as W = 400 μm. Moreover, in order to thermally
equilibrate the electron population in Landau levels near the center of the samples, we
made the length of the 2D system between two current electrodes L as long as about 3000
μm. These dimensions are close to those of samples used in the reference standards of
resistance. In order to keep conditions for the heat generation at the current electrodes and
thermal equilibration of electron population in Landau levels in the central part, we made all
the samples to have the same width of the current electrodes and the same distance of 2900
μm [6~8] or 2600 μm [9] between two current electrodes. The width of each sample is
linearly narrowed from both current electrodes to the ends of the central part whose width w
ranges from 3 μm to 120 μm. The length of the central part is ℓ = 600 μm in the samples
with L = 2900 μm, and ℓ = 120 μm+12 x w in the samples with L = 2600 μm.

We fabricated samples from six GaAs/Al$_{0.3}$Ga$_{0.7}$As heterostructure wafers by photolithography and wet chemical etching. Electron concentrations and mobilities of these
wafers range from 1.4 x 10^{15} /m^2 to 5.8 x 10^{15} /m^2 and from 13.5 m^2/Vs to 27 m^2/Vs,
respectively. Properties of heterostructure wafers are summarized in Table 1. The sample
width in the central part is reduced from the photolithographic mask width w to effective
sample width w' by chemical etching and formation of space charge layers along the
sample. The effective sample width w' determined by measurements of width dependent

Table 1

Wafer	N_S (10^{15} /m^2)	μ(m^2/Vs)
A0	4.7	21
A	5.8	13.5
B	5.1	17
C	4.5	21
D	2.0	24
E	1.4	27

Electron concentration N_S and electron mobility μ in six GaAs/Ga$_{0.3}$Al$_{0.7}$As wafers.

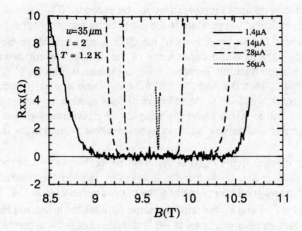

Fig. 2. Diagonal resistance $R_{xx}(2)$ vs change in the magnetic field B
in the sample with the width $w = 35$ μm. [6,8]

sample conductance is in accordance with the effective width determined by breakdown
measurements.   We used the effective width w' in calculations of average current densities
J_{cr} or average Hall electric fields F_{cr}.

The range in the magnetic field ΔB for the quantized Hall-plateau was observed in a
trace of magnetic field dependence of R_{xx} at temperatures between 0.5 K and 1.2 K.
Most measurements were made at 0.5 K. The diagonal resistance was measured by use of a
20 Hz AC constant current. Figure 2 shows an example of R_{xx} vs B traces. The critical
current I_{cr} was determined from the current at which the plateau disappeared.

At low currents, the plateau range in the magnetic field depends on temperature, and the
higher the temperature, the narrower the plateau range. However, the plateau range does not

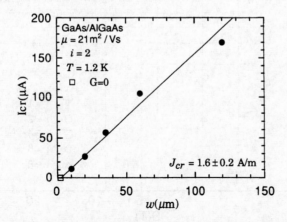

Fig. 3.   Critical current of the breakdown $I_{cr}(2)$ vs sample width w. Open square
shows $w = 2.2$ μm for zero effective width w' where conductance is zero. [6, 8]

896

depend on temperature at high currents close to breakdown. The critical current for the breakdown is independent of temperature in the temperature region in our experiments.[8]

The critical current for the breakdown of the QHE is proportional to the sample width. An example of the sample width dependence of the critical current in samples in Hall plateaus with the plateau quantum number $i = 2$ are shown in Fig. 3. [6, 8] As described above, Fig. 3 shows that the sample width in the central part is reduced from the photolithographic mask width w to the effective sample width w' by chemical etching and formation of space charge layers along the sample. We confirmed similar linear relation between critical current and sample width in samples whose width ranges from 3 μm to 10 μm. [9]

Average Hall electric fields $F_{cr}(i)$ at the onset of the breakdown of the QHE for the plateau quantum number $i = 1, 2, 4, 6$ and 8 observed in samples fabricated from six wafers are shown against the magnetic flux density at the plateau center B in Fig. 4. Figure 4 shows that except $F_{cr}(i)$ for $i = 1$ and 8, the critical electric field for the quantized Hall plateau with a given quantum number is proportional to $\sim B^{3/2}$ and does not show appreciable dependence on electron mobility of the sample. Moreover, the critical fields $F_{cr}(i=2$ and $4)$ are approximately lieing on a line proportional to $B^{3/2}$. The average critical fields for the plateaus with the quantum number $i = 6$ and 8, in particular those with $i = 8$, appear to be appreciablly smaller than the critical fields extraporated from the critical fields for the plateaus with $i = 2$ and 4 in higher magnetic fields. Small Landau level separations at low magnetic fields probably give rise to such smaller critical breakdown fields.

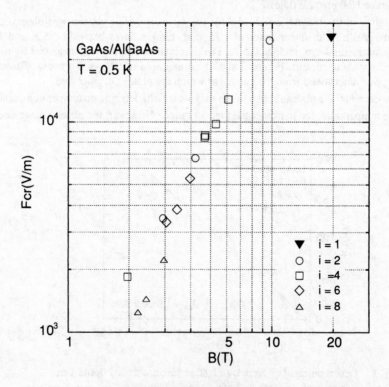

Fig. 4   Critical Hall electric field at the onset of the breakdown of the QHE $F_{cr}(i)$
vs magnetic field at the plateau center B. [9]

3. Discussions

Characteristic features of the breakdown we observed are summarized as follows: (1) The onset of the breakdown of the QHE is caused by the average current density J_{cr} or the average Hall electric field F_{cr} irrespective of electron mobility of the sample in our experiments. (2) The critical electric Hall field F_{cr} for the breakdown does not depend on electron mobility of the sample. (3) Magnetic field dependence of the critical Hall electric field $F_{cr}(i)$ for the breakdown in the Hall plateau with even plateau quantum number i =2 and 4 is approximately given as $F_{cr}(i = 2, 4) \propto B^{3/2}$. In other words, the magnetic field dependence of the critical breakdown field in the Hall plateau with $i = 2$ is not distinguishable from that in the Hall plateau with $i = 4$.

The first characteristic feature of the intrinsic breakdown shows that the edge-state transport does not play a principal role in the quantization of the Hall resistance in the QHE. In other words, the electric current which passes through quantum mechanical edge states only does not result in the quantization of the Hall resistance.

The second and the third characteristic features, i.e. the electron mobility independent nature of the breakdown and the magnetic field dependence of the breakdown, are two keys determining a possible mechanism for the breakdown of the QHE.

Since Ebert et al.[10], several mechanisms of the breakdown of the QHE have been discussed. The mechanism of abrupt phonon emission in analogy to Cerenkov radiation for the breakdown proposed by Stormer et al.[11] does not reconcile with our results. The electron heating mechanism such as discussed by Komiyama et al.[12] leads a dependence of $F_{cr} \propto B$. Eaves and Sheard [13] proposed a mechanism of the breakdown due to production of dissipative carriers by quasi-elastic inter-Landau-level transition of electrons from the highest filled Landau level to the next empty Landau level. They have derived an expression for the critical breakdown field as follows:

$$F_{cr}(\text{Th}) = \frac{\hbar\omega_c}{el_B\left[(2n+1)^{1/2} + (2n+3)^{1/2}\right]} \tag{1}$$

where $l_B = (\hbar/eB)^{1/2}$ is the magnetic length and n is the Landau level quantun number; i.e. $n=0$ for $i=1$ and 2, $n=1$ for $i=3$ and 4, etc.. This equation leads a relation $F_{cr}(i) \propto B^{3/2}$. However, F_{cr} calculated by eq.(1), $F_{cr}(\text{Th})$, is more than one order of magnitude larger than the experimentally observed critical field, $F_{cr}(\text{Exp})$, shown in Fig. 4.[8] Moreover, the third characteristic feature of our results that the breakdown Hall electric field does not depend on the plateau quantum number, $i=2$ or 4, does not reconcile with eq.(1). The enhancement of the inter-Landau-level transition by spacially extended sacatterers proposed by Trugman[12] leads to quantum number independent breakdown fields but its weak magetic field dependence of $F_{cr} \propto B$ does not reconcile with our results. We discussed a possibility that enhancement of the local electric fields in extended states due to electron localization can explain smaller $F_{cr}(\text{Exp})$ than $F_{cr}(\text{Th})$.[8] If this argument be the case, $F_{cr}(\text{Exp})$ in higher mobility samples would be larger than that in lower mobility samples. But, the second feature of our results does not reconcile with any model based on impurity effect or localiztion. However, the second and the third characteristic feature of our experimental results are basic properties of the intrinsic breakdown of the QHE. In particular, the $B^{3/2}$-dependence of the critical field suggests, however, a possibility of the inter-Landau level tunneling. A more realistic approach to the inter-Landau-level transition which occurs in a smaller electric field is required.

In summary we have measured critical Hall electric fields at the onset of the breakdown of the quantum Hall effect in Hall bars fabricated from GaAs/AlGaAs heterostructure

898

wafers with different concentrations and mobilities of two-dimensional electrons. The critical breakdown electric field for the quantized Hall plateau with the quantum number i =2 and 4 in $R_H(i)=h/ie^2$ was proportional to $B^{3/2}$ where B is the magnetic field at the plateau center. The critical electric field has no appreciable dependence on the electron mobility. The magnetic field dependence suggests that the inter-Landau-level transition of electrons is a highly possible mechanism of the breakdown. The mobility independent critical electric field, however, might not be accounted for by simple inter-Landau-level transition mechanism.

Acknowledgments The author would like to thank our colleagues in Gakushuin University, Professor T. Fukase and the staff of HFLSM for the use of the high magnetic field facilities at IMR, Tohoku University and J. Sakai and Y. Kurata of the 2nd Thin Film Engineering Department, ANELVA Corporation for providing GaAs/AlGaAs wafers. This work is supported by the Grant-in-Aid for Scientific Research from the Ministry of Education, Science and Culture.

References

[1] S. Kawaji: *Proc. Int. Symp. Foundation of Quantum Mechanics, 1983, Tokyo,* Eds. S. Kamefuchi et al. (Physical Society of Japan, 1984) p. 327.

[2] T. J. Quinn: Metrologia **26** (1989) 69.

[3] J. Wakabayashi and S. Kawaji: J. Phys. Soc. Jpn. **44** (1978) 1939.

[4] S. Kawaji: Surf. Sci. **73** (1978) 46.

[5] See references cited in Ref. 8.

[6] S. Kawaji, K. Hirakawa and M. Nagata: Physica B **184** (1993) 17.

[7] S. Kawaji, K. Hirakawa, M. Nagata, T. Goto and T. Fukase: Surf. Sci. **305** (1994) 161.

[8] S. Kawaji, K. Hirakawa, M. Nagata, T. Okamoto, T. Goto and T. Fukase: J. Phys. Soc. Jpn. **63** (1994) 2303.

[9] T. Okuno, S. Kawaji, T. Ohrui, T. Okamoto, Y. Kurata and J. Sakai: J. Phys. Soc. Jpn. **64** (1995) 1881.

[10] G. Ebert, K. von Klitzing, K. Ploog and G. Weimann: J. Phys. C: Solid State Phys. **16** (1983) 5441.

[11] H. L. Stormer, A. M. Chang, D. C. Tsui and J. C. M. Hwang: *Proc. 17th Int. Conf. Physics of Semicond., 1984, San Francisco,* Eds. J. D. Chadi and W. A. Harrison (Springer, Berlin, 1985) p. 267.

[12] S. Komiyama and T. Takamasu, S. Hiyamizu and S. Sasa: Solid State Commun. **54** (1985) 479.

[13] L. Eaves and F. W. Sheard: Semicond. Sci. Technol. **1** (1986) 346.

[14] S. A. Trugman and F. R. Waugh: Surf. Sci. **196** (1988) 171, and V. Nikos Nikopoulos and S. A. Trugman: Phys. Rev. Lett. **65**(1990) 779.

Inst. Phys. Conf. Ser. No 145: Chapter 6
Paper presented at 22nd Int. Symp. Compound Semiconductors, Cheju Island, Korea, 28 August–2 September 1995
© 1996 IOP Publishing Ltd

2D-1D crossover behaviors in low-field magnetotransport

S.K. Noh, K.Y. Lim*, G. Ihm[†], and S.J. Lee[‡]

Epitaxial Semiconductor Group, Materials Evaluation Center, Korea Research Institute of
Standards and Science, Taedok Science Town, Taejon 305-600, Korea

* Semiconductor Physics Research Center, Department of Physics, Jeonbuk National
University, Jeonju 560-756, Korea

† Department of Physics, Chungnam National University, Taejon 305-764, Korea

‡ Department of Physics, Korea Military Academy, Seoul 139-799, Korea

ABSTRACT : The 2D-1D crossover behavior has been investigated in narrow
GaAs/AlGaAs heterostructures with a range of aspect ratio (a ratio of length to width of
channel, L/W) of 2-240. In spite of the electron density constant as $(3.9\pm0.3)\times10^{11}$ cm^{-2}, a
rapid change in the electron mobility arises around L/W=18. While the mobility keeps
nearly constant for L/W ⟨18, the exponential reduction has been experimentally clarified by
an empirical equation of $\mu \sim [L/W]^{-0.27}$ for L/W⟩ 18. The 2D interaction parameter
determined in the parabolic regime of negative magnetoresistance is ranging in (0.64 ± 0.05)
for L/W ⟨18, but enhances up to 1.27 for L/W=240 which is close to the calculated 1D
parameter. Similar crossover behaviors appear in changes of the asymmetric and aperiodic
natures of Shubnikov-de Haas oscillation and its onset magnetic field.

1. Introduction

Recently, the advance in lithographic technology has made it possible to realize nanoscale
quantum devices, and a variety of new phenomena on the size effect has been observed in
quasi-1-dimensional (Q1D) systems, such as asymmetric and aperiodic Shubnikov-de Haas
(SdH) oscillation[1-4], positive or negative magnetoresistance[5-8], quenching of the Hall
effect[9], discrepancy between the quantized Hall plateaus and the SdH oscillation minima[10],
which could not be seen in the 2D electrons. These phenomena have revealed that the
electron-electron (e-e) interaction and the boundary scattering become rather dominated as
the conduction channel width approaches 1D from 2D, but an ambiguity still remains in
experimental results.

In disordered 2D systems, two kinds of quantum corrections are unavoidable on the
classical Drude conductivity[11-13]. The first is the localization correction[11] resulting from
the random-potential scattering which leads to negative magnetoresistance (NMR) in weak

magnetic field limit ($\omega_c \tau_0 \ll 1$), where ω_c and τ_0 are the cyclotron frequency given by $\omega_c = eB/m_e^*$ and the scattering time at B=0, respectively. The second is a correction on the e-e interaction[12] due to the Coulomb interaction between diffusive electrons which also leads to NMR under a relatively strong field ($\omega_c \tau_0 \gg 1$). The localization effect is suppressed above the critical field of $B_{C1} = \hbar/4eD\tau_0$, and the interaction effect is affected by the Zeeman splitting which leads to positive magnetoresistance (PMR) above the critical field of $B_{C2} = k_B T/g\mu_B$, where D, g, μ_B are the electron diffusion constant, the conduction-electron g factor, and the Bohr magneton in sequence[13]. An additional correction due to boundary is inevitable in Q1D systems with sufficiently narrow channel. Choi et al.[14] experimentally demonstrated while the samples with channel width larger than 20 μm and inbetween 3 μm and 20 μm agreed fairly well with the 2D and 1D interaction theories, respectively, a correction due to the boundary scattering effect had to be imposed on 1D theory below 2 μm in width. In spite of numerous experimental results on narrow channel heterostructures, the 2D-1D crossover phenomena are not concluded clearly yet.

In this paper, some experimental results on the *sign* of 2D-1D crossover are reported. We have systematically investigated the low-field magnetotransport phenomena in the high-mobility GaAs/AlGaAs heterostructures with a wide range of channel width. We present an empirical relation on the electron mobility as a function of effective channel width and the 2D interaction parameters determined from the parabolic dependence of magnetoresistance, and discuss the crossover behaviors on the SdH oscillation and its onset magnetic field.

2. Experimentals

All of 10 samples used in this experiment are mesa-etched Hall devices fabricated by photolithographic technique and chemical etching on a single substrate of high-mobility GaAs/AlGaAs heterostructure (5×10^5 cm^2/V.s, 1.5 K) with a spacer layer of 170 Å grown by MBE. The channel length of devices is fixed as 120μm and their lithographical channel width is varied in the range of 65-0.9 μm. First of all, prior to main experiment, the thickness of depletion layer, 2d=0.4 μm, was determined by an extrapolation of the sheet conductance plot to apparent channel width[15], G= σ(w-2d)/L, where σ was conductivity of samples, and w and L were the width and length of apparent channel, respectively. Hereafter, we treat the values of apparent width substracted by 0.4 μm as the effective conduction channel widths.

The low-field Hall effect and magnetoresistance measurements were carried out at 1.5 K in the ^{4}He cryostat with a superconducting magnet under the magnetic field of 0.5 T and in the range of 0-1.2 T, respectively. The current density applied to the Hall devices was kept almost constant within (0.2±0.1) A/m, and the field sweep rate was fixed at 10 G/s (3.6 T/hr) for all the measurements. As a couple of samples with a specific channel width were prepared and the same measurements were repeated on two pairs of probe for each sample, we confirmed that a series of data obtained in the identical configuration had no difference within the experimental error range.

3. Results and discusssion

The electron mobility and density for different 10 Hall devices are plotted in Fig. 1 as a function of aspect ratio (L/W=2-240) defined as a ratio of the channel length (L=120 μm) to the channel width (W=65-0.5 μm). Despite the electron density remains constant as (3.9± 0.3)x10^{11} cm^{-2}, a rapid change in the mobility appears at the critical aspect ratio of 18 (W=6.5 μm). While all the samples with L/W ⟨18 keep almost constant as (5.2±0.3)×10^5 cm^2/V.s which holds typical high-mobility characteristics of 2D heterostructure, an exponential reduction of mobility has been experimentally clarified above L/W=18. We have found that the mobility decreases with the aspect ratio and follows an exponential fit as

$$\mu \sim [L/W]^{-0.27}. \tag{1}$$

It is known that the 2D-1D crossover due to the effect of e-e interaction can be observed around W=L$_T$, where L$_T$ is the thermal diffusion length given by L$_T$= π[ℏD/k$_B$T]$^{1/2}$. A transition of the 2D-1D crossover in the present results appears at W = 6.5 μm whose value is quite larger than L$_T$ = 1.4-2.0 μm. This reflects that an additional effect is involved in the mobility reduction mechanism. Considering that the samples used in this work have large aspect ratios and the elastic mean free path (ℓ =(m$^*_e\mu$/e)[2πn$_s$]$^{1/2}$), 2.5-5.0 μm, is comparable to the channel width to induce the boundary scattering, we conclude that the mobility reduction results from a complex effect due to the e-e interaction and the boundary scattering. The crossover behavior in the mobility agrees well with the argument of Choi et al.[14] that the samples with W≥20 μm and 20 μm≥W≥3 μm follow the 2D and 1D interaction theories, respectively.

An additional *sign* on the 2D-1D crossover is in the low-field magnetoresistance curves as given in Fig. 2. All the curves clearly show parabolic NMR signals and typical SdH oscillations. As mentioned in

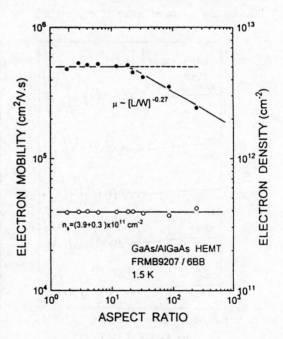

Fig. 1. Plots of the electron mobility and density as a function of aspect ratio for different 10 Hall devices with L/W=2-240.

Introduction, the localization and the e-e interaction are effective in the region of $\omega_c \tau_o \ll 1$ and $\omega_c \tau_o \gg 1$, respectively. Because the value of magnetic field which satisfies $\mu B_o = \omega_c \tau_o = 1$ is low enough as $B_o = 0.01\text{-}0.05$ T and the experimental range (0-1.2 T) satisfies the condition of B $\langle$ B$_{C2}$ (a few teslas for GaAs at 1.5 K), the localization effect has been sufficiently suppressed in the present results.

From the weak-coupling perturbation theory, Altshuler et al.[16] predicted that a correction on the Drude conductivity due to the e-e interaction could be approximated for $\hbar/\tau_o \gg k_B T$ and was independent of magnetic field in the low field limit. Using the relation between conductivity and resistivity, $\delta \rho = -[1-(\omega_c \tau_o)^2] \delta \sigma/\sigma_o^2$ (σ_o is the Drude conductivity), the differential resistivity in 2D and 1D can be written as

$$\Delta \rho(B) = \delta \rho(B) - \delta \rho(0)$$

$$= \begin{cases} -(1/2\pi^2 \hbar n_s^2) \, \alpha_{2D} \ln(\hbar/k_B T \tau_o) B^2 & \text{for 2D} \quad (2) \\ -(1/\pi \, \hbar n_s^2 W) \, \alpha_{1D} \, (\hbar/2k_B T)^{1/2} B^2 & \text{for 1D} \quad (3) \end{cases}$$

where n_s is the sheet electron density, and α_{2D} and α_{1D} are the interaction parameters for 2D and 1D, respectively, which are given by the Hatree factor as an angular average of the screened Coulomb interaction[14]. Figure 3 is a plot of the 2D interaction parameter determined by the parabolic dependence of Eq.(2) as a function of aspect ratio. The dashed lines are the calculated values of α_{2D} and α_{1D} for comparison with the experimental results.

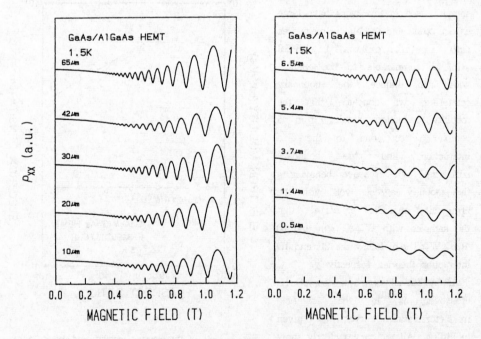

Fig. 2. The low-field magnetoresistance curves which show the parabolic dependence and clear Shubnikov-de Haas oscillation.

While α_{2D}'s for small aspect ratio are ranging in (0.65±0.04), the parameter rapidly becomes more effective and approaches to Q1D system. This 2D-1D crossover behavior is very consistent with that observed in the electron mobiliy.

Similar crossover behaviors appear in the SdH oscillation as shown in Fig. 2. As the channel width becomes narrower, the SdH oscillation shows asymmetric and aperiodic behaviors below 6.5 μm and the strength becomes larger. The onset magnetic field at which the SdH oscillation begins to come out, moves up to high field region, and, eventually, the slight PMR signal appears near B=0 in the near-submicron samples (1.4 μm and 0.5 μm). It is known that the asymmetric and aperiodic behavior is ascribed to the magnetic depopulation of 1D subband[3], and the PMR phenomenon is closely related to the boundary scattering[5-7]. Though it is very difficult to analyze quantitatively these kinds of

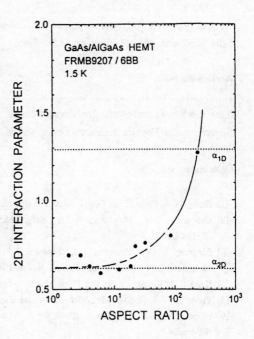

Fig. 3. The 2D interaction parameters as a function function of aspect ratio for 10 different Hall devices with L/W=2-240.

behaviors, all the present magnetotransport results show the 2D-1D crossover behavior which are attributed to the effect of e-e interaction enhanced by squeeze of the conduction channel.

4. Summary and conclusion

We systematically studied the low-field magnetotransport phenomena in the high-mobility GaAs/AlGaAs heterostructures with the wide range of aspect ratio of 2-240. Some experimental results on the *sign* of 2D-1D crossover observed in the low-field electron mobility and magnetoresistance were reported. In spite of the electron density constant as $(3.9\pm0.3)\times10^{11}$ cm^{-2}, an exponential reduction of mobility, $\mu \sim [L/W]^{-0.27}$, was experimentally clarified in the range of L/W>18. We concluded that the crossover behavior in the mobility resulted from a complex effect of the e-e interaction and the boundary scattering dominated as the channel width approaches to the thermal diffusion length and becomes comparable to the mean free path.

The 2D interaction parameters determined by the parabolic relation for magnetoresistance were ranging in (0.65±0.4) for small aspect ratio, but the parameter enhanced up to 1.27 in the range of L/W>18. The enhancement of α_{2D} showed that the effect of e-e interaction became more effective and approached to Q1D system as the channel width decreased.

904

Similar crossover behaviors appeared in changes of the asymmetric and aperiodic nature of SdH oscillation and its onset magnetic field.

Acknowlegments

This work was supported by Korea Science and Engineering Foundation (KOSEF) through Semiconductor Physics Research Center (SPRC), Jeonbuk National University.

References

[1] Zheng H Z, Choi K K, Tsui D C and Weimann G 1985 *Phys. Rev. Lett.* **55** 1144-1147

[2] Grassie A D C, Hutchings K M, Lakrimi M, Faxon C T and Harris J J 1987 *Phys. Rev.* **B36** 4551-4554

[3] Berggren K -F, Thornton T J, Newson D J and Pepper M 1986 *Phys. Rev. Lett.* **57** 1769-1772 ; Berggren K -F, Roos G and Hauten H van 1989 *Phys. Rev.* **B37** 10118-10124

[4] Haug R J, Klitzing K v, and Ploog K 1987 *Phys. Rev.* **B35** 5933-5935

[5] Thornton T J, Roukes M L, Scherer A and Gaag B.P. Van de 1989 *Phys. Rev. Lett.* **63** 2128-2131

[6] Block S, Suhrke M, Wilke S, Menschig A, Schweizer H and Gruentzmacher D 1993 *Phys. Rev.* **B47** 6524-6528

[7] Cheeks T L, Roukes M L, Scherer A and Craighead H G 1988 *Appl. Phys. Lett.* **53** 1964-1966

[8] Cumming D R S, Blaikie R J and Ahmed H 1993 *Appl. Phys. Lett.* **62** 870-872

[9] Roukes M L, Scherer A, Allen, Jr. S J, Craighead H J, Rutten R M, Beebe E D and Harbison M P 1987 *Phys. Rev. Lett.* **59** 3011-3014

[10] Noh S K, Lee J I, Ihm G, Oh J H and Chang K J 1995 *J. Phys. : Condens. Matter* **7** 4517-4523.

[11] Abrahams E, Anderson P W, Licciardello D C and Ramakrishnan T V 1979 *Phys. Rev. Lett.* **42** 673-676

[12] Altshuler B L, Aronov A G and Lee P A 1980 *Phys. Rev. Lett.* **44** 1288-1291

[13] Paalanen M A, Tsui D C and Hwang J C M 1983 *Phys. Rev. Lett.* **51** 2226-2229

[14] Choi K K, Tsui D C and Palmateer S C 1986 *Phys. Rev.* **B33** 8216-8227

[15] Takagaki Y, Gamo K, Namba S, Ishida S, Takaoka S and Murase S 1990 *J. Appl. Phys.* **67** 340-343

[16] Altshuler B L, Khmel'nizkii D, Larkin A I and Lee P A 1980 *Phys. Rev.* **B22** 5142-5153

Inst. Phys. Conf. Ser. No 145: Chapter 6
Paper presented at 22nd Int. Symp. Compound Semiconductors, Cheju Island, Korea, 28 August–2 September 1995
© *1996 IOP Publishing Ltd*

Two Different Types of the *h/e* Oscillation in an AlGaAs/GaAs-based Mesoscopic Ring Structure

Seongjae Lee, KyoungWan Park, Mincheol Shin, El-Hang Lee, and Hyuk Chan Kwon[(*)]

Research Department, ETRI, Yusong P.O.Box 106, Taejon 305 - 600, Korea
*KRISS, Yusong P.O.Box 102, Taejon 305-606, Korea

We have investigated the electrical properties of a GaAs/AlGaAs-based mesoscopic ring structure in the presence of magnetic flux and electrostatic potential. In a certain range of the electrostatic potential, new *h/e* magnetoresistance oscillations have been observed at near zero magnetic field, where the new peaks of the oscillation are interspersed between the conventional ones. This phenomenon is attributed to the combined behavior of the magnetostatic and electrostatic Aharonov-Bohm effects simultaneously occuring in two different configurations of electron paths: one is the Mach-Zender type path and the other is a localization path which is composed of one turn of the ring.

1. Introduction

The phase of electrons in an electrical device can be controlled by the electrostatic potential and the magnetic vector potential that the traveling electrons experience in the conducting path. Changes in magnetic fluxes in a ring structure lead to the magnetostatic Aharonov-Bohm (AB) oscillations[1], whereas changes in gate potentials lead to the electrostatic quantum interference effects[2-4]. Recently, theoretical studies on one dimensional ring structures predicted two types of conductance minima in the electrostatic conductance spectra[5]. The primary minina are caused by the Mach-Zender type interference, and the other ones by the interference by electron's circulation around the ring, that is, "localization type interference". In addition, it has been demonstrated that the two types of the magnetostatic conductance minima are formed in the presence of small electrostatic potential, and that the positions of the minima do not change as the electrostatic potential varies[6]. Experiments have been performed in an attempt to realize such an electron interferometer device[7,8]. However, the applied electrostatic potentials disturbs carrier densities, carrier trajectories[4], and number of the conduction mode[3,7] as well as the elecrostatic AB phase, the results are far from clear to be conclusive. We present experimental results on the formation of new *h/e* oscillations in a small magnetic field range with the variation of the gate potential and the electrostatic AB effect in a corrugated gate AlGaAs/GaAs-based mesoscopic ring structure.

2. Experimental

The conducting path of the ring structure was defined by the electron beam lithography and subsequent chemical etching on a modulation-doped AlGaAs/GaAs heterostructure; the average diameter is 1.9 μm, the width of the conducting path is 0.3 μm, and distance between the measurement probes across the ring is 3.8 μm. Lateral depletion further reduces the conducting width. The second step defined the corrugated Au/Ni gate; the gate length is 0.2 μm on one arm of the ring and 1 μm on the other. Fig.1(a) shows the schematic diagram of our sample. The reason for the corrugated gate is to induce the phase difference with equal transmittance coefficients[2] between electrons passing through upper and lower arms The carrier concentration and mobility of the substrate at 1.5K, as deduced from the Shubnikov-de Haas oscillation, are $n = 3.2 \times 10^{11} cm^{-2}$ and $\mu = 5.5 \times 10^5 cm^2 V^{-1} s^{-1}$, respectively. The lock-in technique was employed for the resistance measurements with 10 nA driving current at 17mK. We found the h/e oscillations at zero gate voltage which are in direct correspondence with the penetration of the flux $\Phi_o = h/e$ through the average area of the annulus. It demonstrates that the conducting path of the ring was well formed.

(a)   (b)

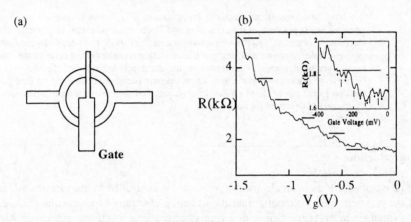

Fig.1. (a) Schematic diagram of the sample geometry. (b) The resistance as a function of gate voltage at B = 0. Finer measurement is shown in the inset.

3. Results and Discussions

The measured resistances as a function of the gate voltage (V_g) are depicted in Fig.1(b). The overall resistance increases as the gate voltage negatively increases. The resistance curve exhibits oscillatory behaviors, which originate possibly from the lateral quantum interference effect between two paths and/or the multiple reflection effect[2]. Since the electrons are confined in the narrow wires, the transverse mode depopulations by the negative gate voltage are also apparent in the resistance curve (indicated by the horizontal bars in the figure). Finer measurements for observation of the electrostatic AB oscillations have been done in a range of low gate voltages without any mode depopulations in between, as shown in the inset of Fig.1(b). Two independent (almost) periodic oscillations (represented by dots and bars, respectively) are seen. The resistance curve with the electrostatic potential sweep alone cannot, however, identify the source of the two types of the oscillations, which leads us to turn to the magnetoresistance measurements at constant gate voltages.

The magnetic AB oscillations was monitored as we varied the gate valtage. As illustrated in Fig.2(a) for $V_g = -80.4 \sim -89.6$ mV, a new peak around B = 0 appears between

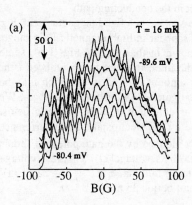

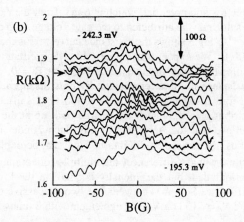

Fig. 2. (a) Generation of a new peak in the usual background of the AB oscillations near B = 0 in the gate voltage range of V_g = -80.4 ~ -89.6 mV. Each magnetoresistance curve is offset for clear viewing. (b) Generation of new set of h/e AB oscillations. Difference in the applied gate voltage between the successive curves is ~ 4 mV. Magnetoresistance curves indicated by the upper and lower arrows are at V_g = - 230.3 mV and V_g = -206.2 mV, respectively.

the usual AB oscillation peaks as the gate voltage decreases. As we lowers the voltage further to -200 mV range, a new set of oscillations with h/e period are clearly seen as shown in Fig.2(b), (see the curve at V_g = - 206.2 mV, for example), in the field range of - 40 G < B < + 40 G, with a phase shifted by π with respect to the background h/e AB oscillations, where the peak positions of both of the h/e oscillations do not change as the gate voltage varies, but only their amplitudes alternate. At V_g = -195.3 mV, the newly formed oscillations begin to appear at the troughs of the background oscillations. As we lowers the gate voltage, the new peaks gradually grow while the background peaks shrink as much. At V_g = -206.2mV, the two set of peaks are almost equal in magnitudes, so that the period of the resistance oscillations as a whole appears to be $h/2e$. If the gate voltage further decreases, the new oscillations continue to grow and they finally take over the background oscillations at - 230.3mV. Therefore, as the gate voltage changes by 35mV, we have half cycle with respect to the alternation between the new and the background oscillations. The estimated period of 35mV is in good agreement with the period of the oscillations in the electrostatic measurements, as estimated from the resistance minima marked by bars in the inset of Fig.1(b). We thus conclude that the electrostatic phase shift between the successive resistance minima in the electrostatic measurement is π.

The reflection probability at the junction points of the ring with the source/drain leads is sufficiently large and we should consider the interference effect around localization path as well as the Mach-Zender type path. Takai and Ohta have investigated the influence of electrostatic potential on the magnetic AB effect in the one-dimensional ring structure[6]. With a sufficient potential, the conductance minima of the usual AB oscillations grow shallower, while new minima appear between the original neighboring minima. These new minima grow and the original minima fade away. The positions of the phase of these new minima differ by π from those of the original minima. The experimental behaviors of h/e oscillations near zero magnetic field in the presence of electrostatic potential, as shown in Fig. 2(b), are qualitatively in good agreement with the calculation results of the magnetoresistance by Takai and Ohta. The discrepancy in the half period along the electrostatic potential, which is 6 mV from their calculation and 35 mV in the experimental results, may be attributed to the gate voltage drop effect in the conducting path.

Fig.3 shows the resistance curve versus the gate voltage at a fixed magnetic field, B = 800 G. We expect that the new h/e AB oscillations disappear in high magnetic field range. Compared with the resistance curve at B = 0, as shown in the inset of Fig. 1(b), some resistance minima represented by dots disappear or the intensities are much reduced (the overall resistance curve at B = 800 G is shifted by amount of - 15 mV, which may be attributed to the interfacial charge built up at the metal gate/GaAs top layer interface). We believe that the AB effect of the Mach-Zender type is dominant in high magnetic fields. Thus, the resistance minima in Fig.3 can be explained by the electrostatic AB interference with 2π phase difference(the multiple reflection effect induced by the gate potential should be included, but the intensity is small in the low gate bias voltage[2]). The gate voltage difference 86.5 mV between the two successive resistance minima, e. g., V_g = - 220.1 mV and V_g = - 133.6mV, is in agreement with 2 times the half period 35 mV at B = 0.

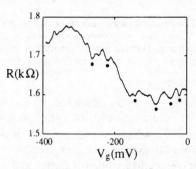

Fig. 3. The resistance as a function of gate voltage at B = 800 G. See text for details.

4. Conclusion

In conclusion, we have carefully investigated the electrical transport properties of a GaAs/AlGaAs-based mesoscopic ring structure in the presence of magnetic flux and electrostatic potential. New h/e AB oscillations near zero magnetic field have been observed. We find that these phenomena are well explained by the combined behavior of the magnetic and electrostatic Aharonov-Bohm effects simultaneously occuring in various configurations of electron paths.

References

[1]Aharonov Y and Bohm D 1959 Phys. Rev. **115** 485 ; Webb R A, Washburn S, Umbach C P, and Laibowitz
       R B 1985 Phys. Rev. Lett. **54** 2696; Timp G, Chang A M, Cunningham J M, Chang T Y,
       Mankiewich P, Behringer R, and Howard R E 1987 Phys. Rev. Lett. **58** 2814.
[2] Park K, Lee S, Shin M, Lee E-H, and Kwon H C 1995 Phys. Rev.B **51** 13805.
[3] Ford C J B, Fowler A B, Hong J M, Knoedler C M, Laux S E, Wainer J J, and Washburn S 1990 Surf. Sci.
       229 307.
[4] Washburn S, Schmid H, Kern D, and Webb R A 1987 Phys. Rev. Lett. **59** 1791.
[5] Cahay M, Bandyopadhyay S, and Grubin H L 1989 Phys. Rev.B **39** 12989.
[6] Takai D and Ohta K 1993 Phys. Rev.B **48** 1537.
[7] de Vegvar P G N, Timp G, Mankiewich P M, Behringer R, and Cinningham J 1989 Phys. Rev.B **40** 3491.
[8] Okuda M, Miyazawa S, Fujii K, and Shimizu A 1993 Phys. Rev. B **47** 4103.
[9] Buttiker M 1986 Phys. Rev. Lett. **57** 1761.

Inst. Phys. Conf. Ser. No 145: Chapter 6
Paper presented at 22nd Int. Symp. Compound Semiconductors, Cheju Island, Korea, 28 August–2 September 1995
© 1996 IOP Publishing Ltd

Aharonov-Bohm oscillations beating and universal conductance fluctuations in a single-mode quantum interferometer

A A Bykov, Z D Kvon, E B Ol'shanetskii

Institute of Semiconductor Physics, Russian Academy of Sciences, Siberian Branch, Novosibirsk, 630090, RUSSIA

Abstract. This paper presents the results of the study of the beating of Aharonov-Bohm oscillations and universal conductance fluctuations in an electron ring interferometer fabricated on the basis of a 2D electron gas in AlGaAs/GaAs heterostructure. It is shown that the behavior of both the beating and universal conductance fluctuations is determined by the width of the conducting channels of an interferometer.

1. Introduction

Beginning with [1] there have been intensive investigations of electron ring interferometers especially after the appearance of quasiballistic interferometers fabricated on the basis of a high mobility 2D electron gas in AlGaAs/GaAs heterostructures [2-4]. This interest has originated mainly from two reasons: in the first place electron interferometer is an excellent tool to study quantum interference in solid state systems and, secondly, it may become a basic component of a so called quantum interference transistor whose switching power is expected to be several orders of magnitude less as compared to conventional devices.

Recently [5,6] first results have been obtained on interferometers operating in a single mode regime i.e. when only the lowest quantum subband in the conducting channels of an interferometer is occupied. It was found that a single mode operation leads to a higher amplitude of Aharonov-Bohm oscillations (up to 35%) and, which is more important, a new feature was observed in the behavior of these interferometers - the beating of Aharonov-Bohm oscillations. No comprehensive explanation of this phenomenon has been given in [5,6] but it was supposed in [5] that the beatings may result from the finite width of the conducting channels of an interferometer.

In the present work we carry out a further experimental investigation of the Aharonov-Bohm oscillations beatings in a quasiballistic electron interferometer and compare our results with a recently developed theory of these beatings in which they are shown to result from the conducting channels' finite width. We have also observed a number of new features of the same nature as beatings, in particular in the $1/B$-spectral region where the universal conductance fluctuations dominate.

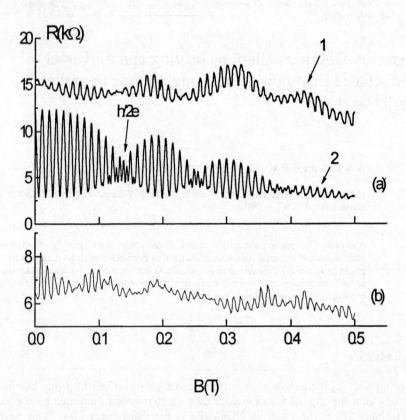

Fig.1 Magnetoresistance dependences: a- before illumination (1-experiment; 2-numerical calculation); b-after illumination.

2. Samples and results

Our experimental devices were interferometers fabricated on the basis of a 2D electron gas in AlGaAs/GaAs heterostructure with the following parameters: the electron mobility $\mu = 10^5$ cm^2/V·s and the electron density $n = (7 \div 9) \times 10^{11}$ cm^{-2}. The ring dimensions were as follows: the inner diameter $d_{in} = 0.4$ µm and the outer diameter $d_{out} = 0.8$ µm. The fabrication technology has been described in detail in [5]. We performed four-terminal magnetoresistance (MR) measurements using a standard ac lock-in technique in magnetic fields up to 10 T and at temperatures 20 mK ÷ 4.2 K. Since, however, each interferometer was positioned in the center of a Hall bar 50 µm wide and with voltage probes 100 µm apart, the resistance thus measured was actually a two-terminal resistance of an interferometer.

Fig.1a shows a magnetoresistance curve measured at 20 mK for one of our interferometers. This curve exhibits well defined high amplitude Aharonov-Bohm oscillations whose period in magnetic field is $\Delta B = 0.012$ T corresponding to a change of $\Phi_o = h/e$ in the magnetic flux through a ring with an effective diameter $d_{eff} = 660$ nm. One can also see that there is also a beating of Aharonov Bohm oscillations with a period approximately an order

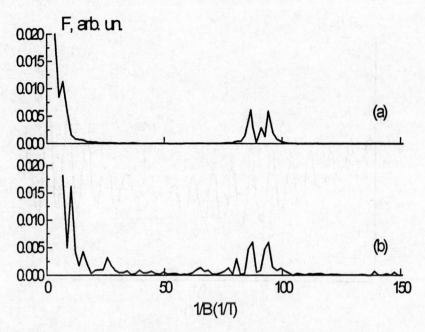

Fig.2 Fourier transforms: a- of the MR dependence in Fig.1; b- of the MR dependence in Fig.4.

of magnitude higher than the period of the latter. Fourier transform of the oscillations shows two closely spaced peaks which account for the observation of the beating (Fig. 2a).

Recently there appeared a theory [7] in which an attempt was made to give a quantitative explanation of such beatings. The model comprises an isolated ring with conducting channels having a certain width. This ring is weakly coupled on both sides to the source and drain contacts. The beating of Aharonov-Bohm oscillations in such a ring results from the lifting of degeneracy of two electronic states corresponding to the clock-wise and anticlock-wise electron wave propagation around the ring. This theory yields for the interferometer conductance:

$$G(B) = \frac{2e^2}{h} \sum_{n,m} \frac{\Gamma^e_{n,m} \Gamma^c_{n,m}}{(E_F - E(B)_{n,m})^2 + ((\Gamma^e_{n,m} + \Gamma^c_{n,m})/2 + \Gamma^i_{n,m})^2},$$

where E_F. Fermi energy, $\Gamma^e_{n,m}$ ($\Gamma^c_{n,m}$) - emitter (collector) binding energy, $\Gamma^i_{n,m}$ - energy of inelastic scattering $E_{n,m} = (n + \frac{1}{2} + \frac{M}{2})\hbar\omega + \frac{m}{2}\hbar\omega_c - V_0$ - energy of (n, m) state,

$$\omega = \sqrt{\omega_c^2 + W_0^2}, \ M = \sqrt{m^2 + \frac{2a_1 m^*}{\hbar^2}}$$

The lower curve in Fig.1a is the result of numerical calculation using the expressions for $G(B)$. Two parameters were varied to fit the calculated curve to the experimental data - the coefficient w_0 which determines the potential across the conducting channels and the Fermi

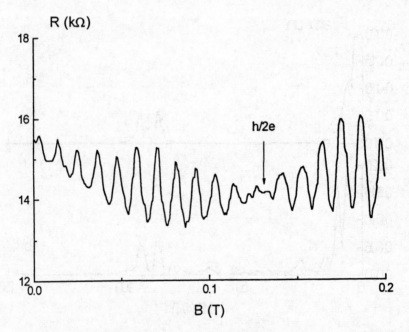

Fig.3 MR dependence in the region of beating

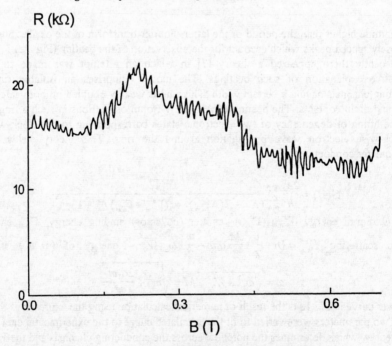

Fig.4 MR dependence in an interferometer with a strong role of disorder

energy of electrons E_F. The knowledge of these two parameters allows one to evaluate the width W of conducting channels. At first glance there is a good agreement between the calculated and experimental curves. However the parameters $W = 73$nm and $E_F = 0.8$ meV determined from the fitting procedure differ significantly from those derived from an independent self-consistent calculation of the energy spectrum in the ring which yielded $W = 40$ nm and $E_F = 5$ meV. Yet another estimate of W was made from the magnetic field value at which total separation of edge current states in the ring leads to suppression of Aharonov-Bohm oscillations (to our opinion this is a more direct estimate). It gives an even lower value for the width of conducting channels - $W \leq 20$ nm. Thus we come to the conclusion that the theory [7] provides only a qualitative description of experiment. That is not surprising since this theory is founded on a model of an isolated ring only weakly coupled to current probes while the experimental structures are basically open systems.

.Apart from the beating of Aharonov-Bohm oscillations there are some other predictions in [7] concerning the behavior of interferometers that have been observed experimentally. It was shown in [7] that the beating is most pronounced in the case of a single-mode operation while the occupation of more then one quantum subbands in the channels leads to a proportional deterioration of the beating pattern. Accordingly, almost entire disappearance of the beating is observed in Fig.1b showing a MR dependence obtained after the interferometer was exposed to a strong illumination that resulted in an increase of electron density in the channels, in a higher value of W and, therefore, in a greater number of occupied subbands.

It was also shown in [7] that low amplitude $h/2e$ - oscillations are expected to arise in the nodes of the beating pattern where the amplitude of Aharonov-Bohm oscillations falls down (see the calculated curve in Fig.1). As seen in Fig.3 there are indeed such oscillations. Originally observed in [5] they were incorrectly interpreted there as resulting from backscattering effects.

Fig.4 shows a MR dependence for an interferometer in a state characterized by a greater role of random potential. This role is evidenced by the appearance of low-frequency irregular fluctuations which are in fact the universal conductance fluctuations resulting from the electron scattering and interference in each of the channels of the interferometer. Fig.2b shows the Fourier transform of the MR dependence in Fig.4. It can be seen that the presence of random potential does not destroy the beating of Aharonov-Bohm oscillations. As far as the single-mode operation is retained, the beating remains well-defined.

Close inspection of the Fourier transforms in Fig.2 reveals one more feature: both curves have a peak corresponding to a periodical component of a frequency ten times lower than the frequency of Aharonov-Bohm oscillations, i.e. lying in the frequency range of universal conductance fluctuations. It appears plausible that these oscillations have the same nature as the beating of Aharonov-Bohm oscillations and, judging from their frequency, correspond to quantization of magnetic flux through each of the interferometer's channels.

3. Summary

We have experimentaly investigated the beatings of Aharonov-Bohm oscillations and universal conductance fluctuations in a single-mode quantum interferometer fabricated on the basis of a high mobility 2D electron gas in AlGaAs/GaAs heterostructure. It is shown that both the beating of Aharonov-Bohm oscillations and universal conductance fluctuations are due to the lifting of the degeneracy by magnetic fields of the clockwize and anticlockwize electron state in ring interferometer.

Acknowledgements

The authors thank M.M.Voronin for numerical calculation. This work was supported by the state program "Physics of solid state nanostructures" and by grant No U84300 from ISF.

References

[1] Webb R A, Washburn S, Umbach S P and Laibowitz R B 1985 Phys.Rev.Lett. **54** 2696

[2] Timp G, Chang A M, Cunningham J E et al. 1987 Phys.Rev.Lett. **58** 2814

[3] Ford C J B, Thornton T J, Newbury R et al. 1989 Appl.Phys.Lett. **54** 21

[4] Ismail K, Washburn S and Lee K Y 1991 Appl.Phys.Lett 59 1998

[5] Bykov A A, Kvon Z D, Olshanetsky E B et al. 1993 JETP Lett **58** 543

[6] Liu J, Gao W X, Ismail K et al 1993 Phys. Rev. **B 48** 15148

[7] Tan W C and Inkson J C 1995 Phys. Rev. (to be published)

Inst. Phys. Conf. Ser. No 145: Chapter 6
Paper presented at 22nd Int. Symp. Compound Semiconductors, Cheju Island, Korea, 28 August–2 September 1995
© *1996 IOP Publishing Ltd*

Optical-Microwave Signal Mixing in a GaAs Uniplanar Ring Resonator

Jong-Chul Lee*, Henry F. Taylor, and Kai Chang

Dept. of Electrical Eng., Texas A&M University, College Station, TX 77843, USA
*Photonic Devices Lab., Semicond. R&D Div. II, Hyundai Electronics Ind. Co., Ltd.,
San 136-1, Ami-ri, Bubal-eub, Ichon-kun, Kyoungki-do, 467-860 Korea.

Abstract. In this paper, a new type of uniplanar ring resonator is introduced and fabricated on semi-insulating GaAs substrate for the first time. A Shottky photodetector is monolithically integrated as a coupling gap between the coplanar waveguide (CPW) feed lines and the slotline ring resonator. Interaction between optical and microwave signals is accomplished using optical excitaion.

1. Introduction

Recent progress in optoelectronic integrated circuit (OEIC) technology has led to the proliferation of high frequency fiber-optic communication systems [1,2]. The availability of high speed light sources and the immunity of lightwaves to electromagnetic interference coupled with the excellent transmission properties of optical fibers have promoted an increased interest in the field of microwave optoelectronics. Monolithic integration of high speed optical and electronic components in the same semiconductor substrate has been an important research fields [3,4].

This paper describes the design and fabrication of a photodetector integrated with a slotline ring resonator which is fed by a coplanar waveguide. Experiments on the interaction between an optically modulated RF signal and an independently applied local oscillator (LO) microwave signal in this structure are described.

2. Device Description

The structure of the CPW-to-Slotline ring resonator is shown in Fig. 1. Here, the CPW feed lines have 703 μm wide signal line and 305 μm wide gap between signal and ground. The Slotline ring has inner and outer radius of 5.555 mm and 5.615 mm, respectively. The photo-sensitive coupling gap is designed to be 5 μm for optical excitation. The rectangular pad of the bottom of the structure is used for biasing the device.

The CPW-to-Slotline ring resonator is fabricated by AZ 5214 double-spin image

916

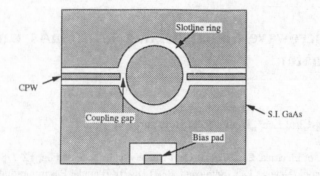

Fig. 1. The structure of the CPW-to-Slotline Ring Resonator.

reversal photolithography [5] and lift-off process. Metal layers with total thickness of about 2.3 μm are deposited by e-beam evaporation through the slots in the photolithographic patterns. Au/Ge and Ni are evaporated as a first layer followed by thin Au. Cu is chosen as a conducting metal layer because of its high conductivity and low cost compared to Au. A thin Au layer is deposited on top of the Cu layer for gold wire bonding and to prevent oxidation in air. In addition, Ti is used as a diffusion barrier between Cu and Au at high temperature. This device can be a strong candidate for OEICs because of its simple uniplanar structure and ease of fabrication.

The microwave response of this device was measured using a HP 8510B network analyzer and its first three resonant peaks are observed to be at 4.19, 7.95, and 11.4 GHz, respectively.

3. Experiments and Results

The experimental set-up for optical excitation is shown in Fig. 2. Optical excitation is accomplished by focusing the light from a current-modulated Ortel laser diode ($\lambda_p = 0.84$ μm and $I_{th} = 6.6$ mA) into one of the gaps between a feed line and the ring. The laser diode is biased at 15 mA and directly modulated by a HP 8340A microwave synthesized sweeper

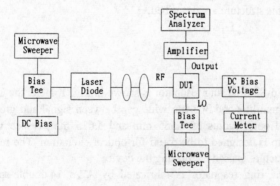

Fig. 2. The experimental set-up.

with -10 dBm of input power. The local microwave signal produced by a HP 8673G synthesized sweeper with 13 dBm of input power is applied to the device via bias-T . If both the RF and LO signals are at near the ring's resonant frequencies, a mixing signal between these two can be generated in this ring structure [6] and the resulting intermediate frequency (IF) signal, defined as $|\omega_{LO} \pm \omega_{RF}|$, is then extracted from the bias pad which is also used as a low pass filter, for the difference signal or from the other feed line for the sum signal via the bias-T. Spectra of sum and difference signals in the CPW-to-Slotline ring resonator are shown in Fig. 3. Here, the RF is at the first resonant frequency, 4.09 GHz, and the LO is at 3.974 GHz. Thedifference frequency is observed at 116 MHz and the sum frequency is at 8.064 GHz.

The IF power levels are investigated as functions of the modulated RF input power, independently supplied LO input power, applied bias voltage for the device, and LO frequency with fixed RF frequency. IF response in each case is shown in Fig. 4.(a) - (d), respectively. As can be seen in the figures, the IF power level is proportional to the RF and LO powers. It increases from 7 V of bias and saturates at 22 V. Also, IF power is a maximum at 3.98 GHz, which is close to the resonant frequency, and decreases when the LO frequency is tuned away from the resonance.

4. Conclusions

A new type of uniplanar ring resonator with monolithically integrated photodetector has been introduced, fabricated on semi-insulating GaAs substrate, and characterized by optical excitation. Sum and difference frequency mixing between an optically modulated RF input and a microwave LO have been successfully demonstrated.

This type of optoelectronic resonator can be used for generating IF signals in receivers for analog fiber-optic communication, and in OEICs for optically controlled microwave oscillators.

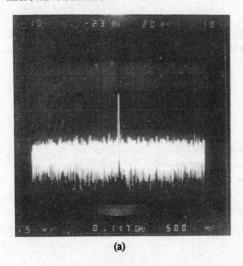 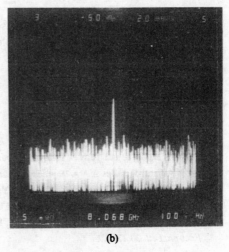

(a) (b)

Fig. 3. Spectra of mixing in the CPW-to-Slotline ring resonator. RF = 4.09 GHz, LO = 3.974 GHz. (a) Difference. IF = 116 MHz, (b) Sum. IF = 8.064 GHz.

918

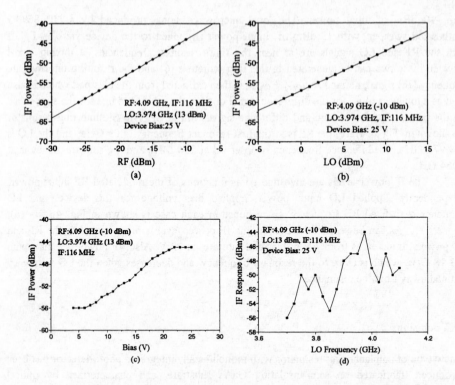

Fig. 4. IF response of the CPW-to-Slotline ring resonator as a function of (a) RF input power, (b) LO pumping power, (c) Applied device bias, and (d) LO frequency.

5. Acknowledgment

The authors would like to thank Vic Swenson for his technical support and C. H. Ho, C. L. Yeh, and J. A. Navarro for their valuable discussions and suggestions. This work was supported by the Office of Naval Research.

References

[1] Horimatsu T and Sasaki M 1989 *J. Lightwave Tech.* **7** 1612-1622

[2] Chaim N B, Ury I, and Yariv A 1982 *IEEE Spectrum* **19** 38-45

[3] Seeds A J and de Salles A A A 1990 *IEEE Trans. Microwave Theory Tech.* **MTT-38** 577-585

[4] Chen T R, Utaka K, Zhuang Y, Liu Y Y, and Yariv A 1987 *IEEE J. Quantum Electron.* **23** 919-924

[5] Lee J C 1994 *Ph.D. Dissertation,* Texas A&M University

[6] Gopalakrishnan G K, Fairchild B W, Yeh C L, Park C S, Chang K, Weichold M H, and Taylor H F 1991 *Eectron. Lett.* **27** 121-123

Inst. Phys. Conf. Ser. No 145: Chapter 7
Paper presented at 22nd Int. Symp. Compound Semiconductors, Cheju Island, Korea, 28 August–2 September 1995
© *1996 IOP Publishing Ltd*

Coherent multiatomic step formation on GaAs (001) vicinal surfaces by MOVPE and its application to quantum well wires

T. Fukui, S. Hara, J. Ishizaki, K. Ohkuri and J. Motohisa

Research Center for Interface Quantum Electronics, Hokkaido University, Sapporo 060, Japan

Abstract.   Coherent multiatomic steps on vicinal (001) GaAs surfaces during metalorganic vapor phase epitaxial (MOVPE) growth are investigated by atomic force microscopy (AFM). AFM images show coherent multiatomic steps with extremely straight edges over a several micron scale. The average spacing of multiatomic steps depends on growth temperature, growth rate and AsH_3 partial pressure. Similar multiatomic steps also appear on GaAs substrate surfaces after thermal treatment under AsH_3/H_2 atmosphere at temperatures higher than 700°C. Furthermore, we fabricate GaAs quantum well wires (QWWs) on these coherent multiatomic steps.   Locally thick GaAs, that is, QWWs are formed at corners of multiatomic steps. Photoluminescence spectra of QWWs show energy shift caused by the QWWs formation. These results suggest that self-organized QWWs can be formed uniformly on coherent multiatomic steps.

1. Introduction

Recently, a novel quantum well wire (QWW) fabrication method using *multiatomic* steps has been demonstrated by metalorganic vapor phase epitaxy (MOVPE) on GaAs (001) vicinal surfaces [1], and molecular beam epitaxy (MBE) on GaAs (110) vicinal surfaces [2]. Similar self-organized QWW structures were also fabricated by using *monoatomic* steps on vicinal (001) GaAs surfaces during MOVPE growth [3] and MBE growth [4]. These *in situ* self-organizing fabrication methods are very promising because high density nanometer-size QWWs can be formed without any damage-introducing processes such as lithography and dry etching, and the size of QWW can be controlled by only adjusting the crystal growth conditions. Therefore, from this point of view, understanding of the behavior and mechanism of the multiatmoic step formation on vicinal surface is very important for application to QWWs using multiatomic steps. Recently, multiatomic step formation phenomena on GaAs (001) vicinal surfaces have been investigated after the MOVPE growth [5-7] and the thermal treatment in both ultra high vacuum (UHV) [8] and AsH_3/H_2 atmosphere [9].

In this paper, we report on the formation of coherent multiatomic steps during MOVPE growth and thermal treatment under AsH_3/H_2 atmosphere. Furthermore, we report on the successful formation and optical characterization of self-organized QWWs using coherent multiatomic steps on GaAs (001) vicinal substrates.

2. Experimental Procedure

The substrates were on GaAs (001) misoriented by 1.0°-5.0° towards [110] and [$\bar{1}$10] directions and singular (001) GaAs as a reference. A horizontal low-pressure MOVPE system with triethylaluminum (TEAl), triethylgallium (TEGa) and AsH_3 as source materials was used. The total gas flow rate was 3.0 l/min, and the working pressure during crystal growth was 76 Torr.

First, a 40-period AlAs/GaAs superlattice buffer layer was grown on the vicinal substrate in order to form the initial surface with *monoatomic* steps. Then, a thick GaAs buffer layer was grown at 600°C-750°C to form the *multiatomic* steps with equal spacing terraces. Multiatomic steps were characterized by atomic force microscopy (AFM) in air.

AlGaAs/GaAs/AlAs single quantum well (SQW) structures were also grown on the GaAs buffer layer with coherent multiatomic steps at 650°C. Multiatomic step edges on the GaAs buffer layer surface became straight after 30-min thermal treatment at 600°C under AsH_3/H_2 atmosphere. A similar SQW structure was simultaneously grown on a singular (001) GaAs substrate as a reference. AsH_3 partial pressures were 4.2 x 10^{-4} atm for GaAs buffer layer growth and 6.7 x 10^{-5} atm for GaAs quantum well layer growth. Growth rates for GaAs buffer layer, AlAs lower barrier layer and GaAs quantum well layer were 1.14, 0.067 and 0.046 nm/s, respectively. Photoluminescence (PL) was measured at 20K using an Ar^+ laser.

3. Results and Discussion

3.1. Multiatomic step formation during MOVPE growth and thermal treatment

During GaAs growth on a vicinal substrate, the surface steps bunch each other to form multiatomic steps with atomically flat terraces, and this behavior rapidly saturates at a growth thickness of less than 100 nm. This tendency was already reported by several authors [6,7]. Figure 1 shows the relation between the terrace width and the growth thickness at the growth temperature of 600°C. At the beginning of the growth, the terrace widths increase linearly as the growth thickness, and are independent of the growth rate. This means that the growth time is not main factor for the step bunching. We also tried to reproduce the step bunching process by using a Monte Carlo simulation. The step bunching phenomenon can be observed only when the activation energy for the migration adatom to the up side step is higher than that to the down side step. From the fitting of the Monte Carlo simulation to the experimental data, we estimated that the activation energy to the up step and down step sites compared with that to the terrace site are 0.47 and 0.07 eV, respectively. Detail about the simulation is reported elsewhere [10]. At the growth temperature of 700°C, the terrace widths have no longer linear development process for the growth thickness, and depend on the growth rate. These results suggest that, at higher growth temperature, the atoms once incorporated in the step sites are easily detached and contribute to the step bunching process as well as the atoms from the vapor phase.

Next, we investigated the effect of thermal treatment under AsH_3/H_2 atmosphere on GaAs vicinal substrates. The substrate misorientation angle is 2.0˚. Figure 2 shows AFM images of (a) a chemically etched GaAs surface, and GaAs surface after thermal treatment (b) under low AsH_3 partial pressure (1.3×10^{-4} atm) and (c) under high AsH_3 partial pressure (1.3×10^{-3} atm) conditions, respectively. After thermal treatment, the clear multiatomic steps appeared. For more detailed investigation about the step bunching mechanism during thermal treatment, AlGaAs/GaAs single quantum well(SQW) structure with thermal treatment process to GaAs quantum well surface were grown by MOVPE. GaAs quantum well layer surface was annealed at 800˚C for 5 min under AsH_3/H_2 atmosphere. The photoluminescence spectra of the SQWs were almost unchanged, compared with the reference sample without thermal treatment process which has no bunching steps on GaAs quantum well layer surface [11]. These results suggest that the atoms detached from the step edge migrate on the terrace and are re-adsorbed to the step sites to form the multiatomic steps, and the amount of the evaporation of the adatoms on the terraces is negligible.

3.2. Fabrication and characterization of quantum well wires using coherent multiatomic steps

A schematic illustration of GaAs QWW structure is shown in Fig. 3. Average thicknesses of AlAs lower barrier layer, GaAs quantum well layer (L_w), and $Al_{0.35}Ga_{0.65}As$ upper barrier layer were 17, 5 and 150 nm, respectively.

First, in order to form coherent multiatomic steps with extremely straight edges prior to QWWs formation, we investigated multiatomic steps on GaAs buffer layer after thermal treatment. Figure 4(a) shows an AFM image of GaAs buffer layer surface after 30-min thermal treatment at 600˚C under AsH_3/H_2 atmosphere and the distribution of the step spacing. AsH_3 partial pressure was 6.7×10^{-5} atm. Coherent multiatomic steps with extremely straight edges were observed over a several micron area. Average height and spacing of multiatomic steps were 5.5nm and 63nm, respectively, and the fluctuation of the step spacing is ±16%. Next, AlAs lower barrier layer surface grown on GaAs buffer layer were observed by AFM. Figure 4(b) shows AFM image of the AlAs lower barrier layer surface on the 5.0˚-misoriented substrate grown at 650˚C and the distribution of the step spacing within the observed area. In this sample, the 3-nm-thick GaAs cap layers were grown on the lower barrier layer for AFM observations in air. Average height and spacing of the multiatomic steps were 5.7nm and 66nm, respectively, and the uniformity of the step spacing was ±14%. The multiatomic step edges on AlAs surface were straight, and the step spacing was almost the same as that on the underlying GaAs surface. In our previous study, it was found that the step edges tended to undulate on the AlGaAs surface, and that the step spacing of multiatomic steps was smaller than that on the underlying GaAs buffer layer surface.[1] Therefore, this result indicates that the QWW structures can be uniformly fabricated over a wide area by using AlAs as the lower barrier layer rather than AlGaAs.

Next, we measured PL spectrum from QWWs and compared with that from quantum well (QW) on singular (001) GaAs substrate. The PL spectra at 20K are shown in Fig. 5. The average quantum well widths (L_w) were 5.5 nm (1.617eV). The PL peak position of the QWWs on the 5.0°-misoriented substrate was 1.593eV. It was found that the PL peak energy of the QWWs was smaller than that of the QW formed on the singular (001) GaAs substrate. Although the peak energy should shift toward the high energy region due to two-dimensional quantum confinement, these energy shifts indicate that locally thick quantum-wire-like structures are formed at the corners of the multiatomic steps.

In these PL spectra at 20K, the full width at half-maximum (FWHM) of the QWWs was 25meV. The result shows that the size uniformity of the QWWs with AlAs lower barrier layer was much improved to the previous report.[1] From the AFM observations, this size uniformity of the QWWs is probably due to the spacing uniformity of the multiatomic steps on AlAs lower barrier layer surfaces. Moreover, in the PL spectra of the sample with AlAs as lower barrier layer, we observed an additional weak spectrum at about 1.748eV. This peak probably corresponds to the quantum wells formed on the (001) terraces connected with QWWs. The result indicates that the most of the photo-excited electrons and holes diffuse to lower energy regions, that is, the QWW regions.

Next, the polarization dependence of PL intensity (I) from the QWWs using AlAs as lower barrier layer was also measured at 20K, and the results are shown in Fig. 6. Since the misorientation direction is $[\bar{1}10]$ and the direction along the QWWs is [110], we defined the degree of polarization as $(I[110]-I[\bar{1}10])/(I[110]+I[\bar{1}10])$. The degree of polarization for the QWWs was found to be 0.12. For normal QW on the singular (001) GaAs substrate, this value was about 0.04, and no large effect due to internal stress is expected in this material system. Therefore, the observed polarization anisotropy supports the successful formation of QWWs at the corners of the multiatomic steps.

Furthermore, we observed fine-area PL spectra for QWWs using Au/Ge patterned mask. Open area is 2μm wide and 800μm long; thus we can obtain PL spectra from about thirty QWWs. It was found that the FWHM of PL spectra from about thirty QWWs is 17meV, which is 8meV smaller than that from the whole area. This result suggests that the size fluctuation of QWWs still remain for a wide area.

4. Conclusions

In this paper, we investigated the behavior and mechanism of the coherent multiatomic step formation during MOVPE growth and thermal treatment under AsH_3/H_2 atmosphere, and also fabricated self-organized QWWs using these coherent multiatomic steps on GaAs (001) vicinal surfaces. Coherent GaAs multiatomic steps with extremely straight edges were formed over a several micron area. It was also found that the underlying coherent GaAs multiatomic steps were *well traced* by AlAs as lower barrier layer of SQW structure. Furthermore, we fabricated self-organized AlGaAs/GaAs/AlAs QWWs on these coherent multiatomic steps, and measured PL spectra at 20K. The peak energies of PL spectra from the QWWs were smaller than those from the QW formed on singular (001) GaAs substrates, and the polarization anisotropy of the QWWs was also observed. These energy shifts and polarization anisotropy of PL spectra indicate that locally thick QWW structures were successfully formed at the corners of the multiatomic steps. Since the FWHM of PL spectrum from about thirty QWWs was 17meV, which is 8meV smaller than that from a whole area, the size fluctuation of QWWs still remain in a several micron area.

Acknowledgment

The authors are grateful to Prof. H. Hasegawa for fruitful discussions.

References

[1] S. Hara, J. Ishizaki, J. Motohisa, T. Fukui and H. Hasegawa: J. Cryst. Growth **145**(1994)692.

[2] K. Inoue, K. Kimura, K. Maehashi, S. Hasegawa, H. Nakashima, M. Iwane, O. Matsuda and K.Murase: J. Cryst. Growth **127**(1993)1041.

[3] T. Fukui and H. Saito: Appl. Phys. Lett. **50**(1987)824.

[4] P. M. Petroff, A. C. Gossard and W. Wiegmann: Appl. Phys. Lett. **45**(1984)620.

[5] T. Fukui and H. Saito: Jpn. J. Appl. Phys. **29**(1990)L483.

[6] M. Kasu and T. Fukui: Jpn. J. Appl. Phys. **31**(1992)L864.

[7] J. Ishizaki, S. Goto, M. Kishida, T. Fukui and H. Hasegawa: Jpn. J. Appl.Phys. **33**(1994)721.

[8] S. L. Skala, S. T. Chou, K. -Y. Chen, J. R. Tucker and J. W. Lyding: Appl. Phys. Lett. **65**(1994)722.

[9] K. Hata, A. Kawazu, T. Okano, T. Ueda and M. Akiyama: Appl. Phys. Lett. **63**(1993)1625.

[10] J. Ishizaki, K. Ohkuri and T. Fukui: to be submitted.

[11] K. Ohkuri, J. Ishizaki, S. Hara and T. Fukui: J. Cryst. Growth, to be published.

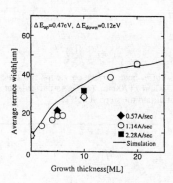

Fig. 1. The relation between the terrace width and the growth thickness at the temperature of 600°C. For fitting the simulation value to the experimental data, the fitting parameters of the activation energy to the up site and down step sites compared with that to the terrace site of 0.47 and 0.07 eV were used.

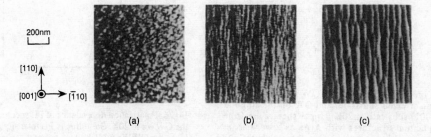

Fig. 2. AFM images of (a) a chemically etched GaAs surface, and GaAs surface after thermal treatment under (b)low and (c) high AsH₃ partial pressure conditions.

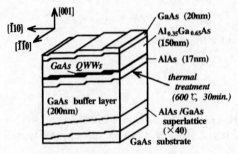

Fig. 3. Schematic illustration of QWW structures on the GaAs vicinal substrate.

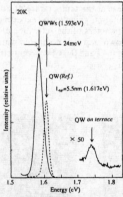

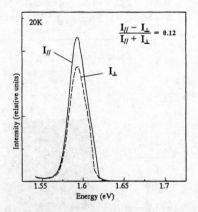

Fig. 4. AFM images and a cross-sectional illustration of GaAs buffer layer surface on vicinal substrate misoriented by 5.0° towards the [Ī10] direction. Observed area is 1700nm x1700nm. (a) GaAs buffer layer surface, (b) AlAs lower barrier layer surface grown on coherent GaAs multiatomic steps at 650°C.

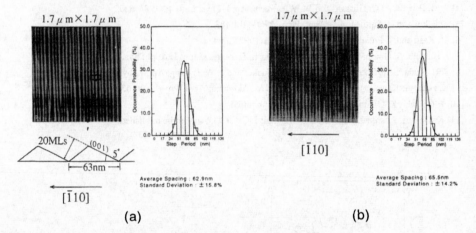

Fig. 5. PL spectra of QWWs on the 5.0°-misoriented substrate, and QW on the singular (001) substrate at 20K. Upper spectra are from quantum structures with AlAs as lower barrier layer, and lower ones are from those with AlGaAs as lower barrier layer. Thicknesses of well layer on the singular (001) substrates, L_w, are also shown.

Fig. 6. Polarization dependence of PL spectra from the QWWs at 20K. Solid line is PL intensity parallel to the QWWs ([110] direction) and dashed line is that perpendicular to the QWWs ([Ī10] direction). Degree of polarization is about 0.12.

Inst. Phys. Conf. Ser. No 145: Chapter 7
Paper presented at 22nd Int. Symp. Compound Semiconductors, Cheju Island, Korea, 28 August–2 September 1995
© 1996 IOP Publishing Ltd

Polarization properties of GaAsP/AlGaAs tensilely strained quantum wire structures grown on V-grooved GaAs substrates

Wugen Pan†, Hiroyuki Yaguchi†, Kentaro Onabe†, Ryoichi Ito†, Noritaka Usami‡, Yasuhiro Shiraki‡

† Department of Applied Physics, The University of Tokyo, 7-3-1 Hongo, Bunkyo-ku, Tokyo 113, Japan

‡ Research Center for Advanced Science and Technology (RCAST), The University of Tokyo, 4-6-1 Komaba, Meguro-ku, Tokyo 153, Japan

Abstract. Crescent-shaped tensilely strained $GaAs_{1-x}P_x/Al_yGa_{1-y}As$ quantum wires ($x = 0.11$, $y = 0.33$) have been grown on 3-μm-period V-grooved GaAs substrates by low-pressure metalorganic vapor phase epitaxy and characterized by cross-sectional transmission electron microscope observation, low temperature cross-sectional photoluminescence and cathodoluminescence. A drastic polarization transition from TM to TE with decreasing quantum wire size has been found in cross-sectional polarized photoluminescence.

1. Introduction

The fabrication of semiconductor quantum wires (QWRs) on patterned substrates using the growth characteristics, such as surface diffusion of atoms, has received extensive attention since the *in situ* formed QWR exhibits high luminescence efficiency[1–3]. The unstrained GaAs/AlGaAs QWR lasers fabricated with this method exhibit reduced threshold currents at room temperature[3, 4].

The QWR laser performance is expected to improve by introducing strain into the active region due to the modified band structure, as well as strained quantum well (QW) structures[5]. The compressive biaxial strain splits the valence band maximum, shifts the heavy-hole band closer to the conduction band than the light-hole band, and causes a considerable reduction of the in-plane heavy-hole mass. The compressively strained QW lasers have a strong polarization-sensitivity of TE gain compared to unstrained QW lasers. Compressively strained InGaAs/GaAs QWR lasers grown by atmospheric pressure metalorganic vapor phase epitaxy (MOVPE) on nonplanar substrates have been reported[6].

In the case of tensilely strained QWs, the strain shifts the light-hole band closer to the conduction band than the heavy-hole band. Therefore, introducing tensile strain to the active region of QW lasers can significantly suppress the spontaneous emission polarized in the growth plane of the laser structure, and consequently enhance TM gain over TE gain. In this paper, we report tensilely strained GaAsP/AlGaAs quantum wires grown by low pressure MOVPE on V-grooved GaAs substrates, demonstrated by transmission electron microscopy (TEM), low temperature cross-sectional photoluminescence (PL), cathodoluminescence (CL), and low temperature polarized cross-sectional PL.

2. MOVPE growth

The samples were grown by low pressure (100 Torr) MOVPE on semi-insulating (100) GaAs substrates patterned with 3-μm-period V-grooves oriented along the direction. The V-grooved substrates were prepared by using conventional photolithography and chemical wet etching in $8H_2SO_4 : H_2O_2 : H_2O$ solution. The etched V-grooves in the substrates are approximately 2.1 μm wide in the top and 1.5 μm in depth. Prior to the growth, the V-grooved substrate was treated with a $2NH_4OH : H_2O_2 : 96H_2O$ solution for 30 s, rinsed in deionized water, dried and finally loaded into the reactor. The epitaxial growth was performed with an RF-heated horizontal reactor. Trimethylgallium, trimethylaluminum, 10 % AsH_3, and 10 % PH_3 were used as Ga, Al, As and P sources, respectively. The growth temperature was 650°C. The typical structure grown on the V-grooved substrate consists of 20-nm GaAs buffer layer, 0.6-μm $Al_{0.33}Ga_{0.67}As$ barrier layer, 5.0–12.5-nm $GaAs_{0.89}P_{0.11}$ layer, 0.6-μm $Al_{0.33}Ga_{0.67}As$ barrier layer, and 10-nm GaAs cap layer. All the compositions and the thicknesses mentioned above correspond to the growth on planar (100) substrates.

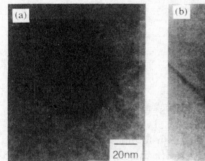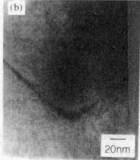

Fig. 1   Cross-sectional TEM images of QWR structures. (a) $t_g = 20$ s, (b) $t_g = 15$ s, (c) $t_g = 10$ s.

Four samples have been grown with different growth time (t_g) of 20, 15, 10 and 5 s for the GaAsP layer, respectively. The crescent-shaped QWRs were observed by cross-sectional TEM. As shown in Fig. 1(a)–(c), the center thickness of the crescent-shaped QWR is 108, 95 and 63 Å, for the growth time of 20, 15, and 10, respectively. The center thickness of QWR for the sample with $t_g = 5$ s is expected to be 40 Å according to the growth time and growth rate of GaAsP layer.

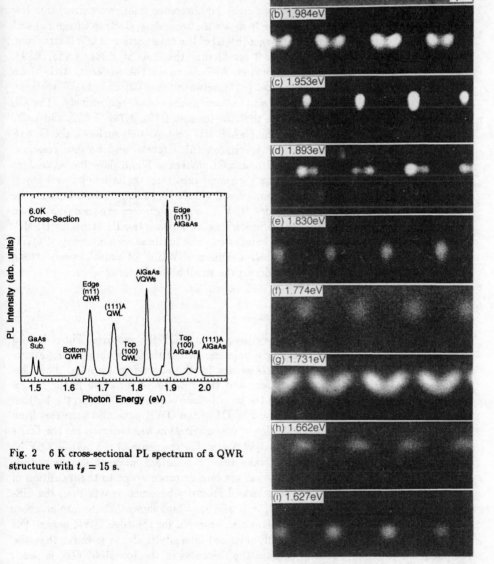

Fig. 2   6 K cross-sectional PL spectrum of a QWR structure with $t_g = 15$ s.

Fig. 3   6 K cross-sectional CL images of a QWR structure with $t_g = 15$ s. (a) secondary electron image, (b) AlGaAs on sidewall (111)A, (c) AlGaAs on top (100), (d) AlGaAs on edge (n11), (e) AlGaAs VQWs, (f) QW on top (100), (g) QW on sidewall (111)A, (h) QWR on edge facet (n11), (i) QWR at bottom of the V-groove.

3. Photoluminescence and cathodoluminescence

Figure 2 shows a 6 K PL spectrum from the cross section of the QWR sample with t_g of 15 s for the GaAsP layer, pumped by 488 nm Ar^+ laser. There are eight luminescence lines in the PL spectrum except for the luminescence lines from GaAs substrate (1.495 and 1.515 eV), centered at 1.627, 1.662, 1.731, 1.774, 1.830, 1.893, 1.953, and 1.984 eV. To make clear the origins of the above eight luminescence lines, we carried out low temperature CL measurements. Figure 3 shows the secondary electron image (a) and spectrally and spatially resolved CL images (b)–(i) of the cross section of QWR structure. The CL images (b)–(e) shown in Fig. 3 reveal that the lines at 1.984, 1.953, 1.893, 1.830 eV are due to the luminescence from AlGaAs on (111)A surfaces, AlGaAs on top (100) surfaces, AlGaAs on edge facet (n11) between top (100) and (111)A surfaces, and spontaneous AlGaAs vertical QWs in bottom intersections, respectively. The CL images (f)–(i) shown in Fig. 3 indicate that the lines at 1.774, 1.731, 1.662, and 1.627 eV are due to the luminescence from the GaAsP QW on top (100) surfaces, the GaAsP QW on (111)A surfaces, edge GaAsP QWRs on (n11) facets, and bottom crescent-shaped GaAsP QWRs, respectively. Although CL images in Fig. 3 show the Al content distribution of AlGaAs layer grown on a V-grooved substrate, we have not found any P content distribution of GaAsP layer grown on a V-grooved substrate. This is because the V/III ratio is much larger than unity for III-V-V' semiconductors grown on patterned substrates, while there exist different diffusion lengths between two III atoms for III-III'-V semiconductors grown on patterned substrates. The luminescence intensity of QWR peak at 1.627 eV, with a full width at half maximum (FWHM) of 5 meV, is very strong compared to that of (111)A QW considering the small filling factor of~ 0.02.

4. Polarized photoluminescence

The PL of QWRs was detected from the cross-section of QWR structures. The polarized PL spectra were obtained by orienting a polarizer before the monochromator. Figure 4 shows the cross sectional polarized PL spectra for the above four samples. The TM-labeled direction is the growth direction, and the TE-labeled direction is the lateral direction of the QWR. With decreasing the growth time of the GaAsP layer, the bottom QWR peak shifts to higher energy. The FWHM of the QWR peak also increases from 6.4 to 15.0 meV. As shown in Fig. 4, almost no polarization was observed for the GaAs substrate-related peaks and the (111)A QW peaks. For the peaks of the bottom QWRs, the TM-polarized intensity is stronger than the TE-polarized intensity for the sample with $t_g = 20$ and 15 s, which supports that the luminescence is due to the transition of 1e–1lh. On the ohter hand, the TM-polarized intensity becomes weaker than the TE-polarized intensity for the sample with $t_g = 10$ and 5 s, which shows that the luminescence is due to the transition of 1e–1hh. This is also observed for the edge QWR peaks. For the peaks of the top (100) QWs, the TE-polarized intensity is always stronger than the TM-polarized intensity. This is because the thickness of the top (100) QW is much thinner.

The polarization of the bottom QWR peak is plotted as a function of the center thickness of the QWRs in Fig. 5, in which the polarization is defined as,

$$P = \frac{I_{TE} - I_{TM}}{I_{TE} + I_{TM}}. \tag{1}$$

The polarization shows the TM character when the center thickness of the QWR is larger than 65 Å, while it shows the TE character when the center thickness of the QWR is smaller than 65 Å.

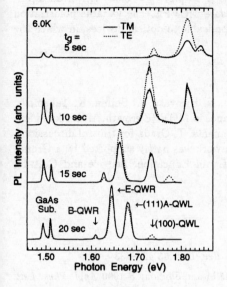

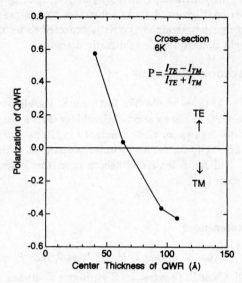

Fig. 4   6 K cross-sectional polarized PL spectra of WR structures.

Fig. 5   Polarization as a function of center thickness of crescent-shaped QWRs.

As shown in Fig. 1 (a)–(c), the lateral width of the crescent-shaped QWR is about 300 Å. Using the finite-element method, we have calculated the energy levels in our crescent-shaped QWR structure, and found that there is ~10 meV splitting between the ground state and the excited state resulting from the lateral confinement. Hence, the lateral quantum confinement effect is weaker than the vertical quantum confinement. As the first approximation, the polarization transition from TM to TE in tensilely strained QWRs is expected to be similar to that in tensilely strained QWs. It is well-known that the dipole moment matrix element of the light hole and the heavy hole for TE and TM polarization[7] are written as

$$M_{lh} = \begin{cases} M_0^2/6 & \text{for TE polarization} \\ 2M_0^2/3 & \text{for TM polarization} \end{cases} \tag{2}$$

and

$$M_{hh} = \begin{cases} M_0^2/2 & \text{for TE polarization} \\ 0 & \text{for TM polarization} \end{cases} . \tag{3}$$

For tensilely strained QWs, when the well thickness is large enough, the luminescence is given by the 1e–1lh transition at low temperatures. Thus, TM polarization becomes dominant over TE polarization. On the other hand, when the well thickness is small enough, the luminescence is given by the 1e–1hh transition at low temperatures, and TE polarization tends to be dominant over TM polarization. This tendency is in good agreement with our experimental results.

930

5. Conclusion

We have observed a drastic polarization transition from TM to TE in crescent-shaped tensilely strained $GaAs_{1-x}P_x/Al_yGa_{1-y}As$ QWs ($x = 0.11$, $y = 0.33$) grown on 3-μm-period V-grooved GaAs substrates by low-pressure MOVPE. The variable polarization of tensilely strained QWRs is expected to be applicable to optical devices, in which the light polarization is arbitrarily designed.

Acknowledgements

The authors would like to thank S. Hashimoto, N. Hiroyasu, A. Shima, K. Takemasa, K. Ota, X. Zhang and S. Miyoshi for their assistance in MOVPE growth, M. Ishikawa and T. Hasegawa for their assistance in CL measurements, T. Osada for fruitful discussions, and S. Ohtake for his technical assistance. This work was partly supported by a Grand-in-Aid for Scientific Research from the Ministry of Education, Science and Culture, Japan.

References

[1] Kapon E, Hwang D M and Bhat R 1989 *Phys. Rev. Lett.* **63** 430–433

[2] Colas E, Clausen, Jr. E M, Kapon E, Hwang D M and Simhony S 1990 *Appl. Phys. Lett.* **57** 2472–2474

[3] Simhony S, Kapon E, Colas E, Hwang D M, Stoffel N G and Worland P 1991 *Appl. Phys. Lett.* **59** 2225–2227

[4] Kapon E, Simhony S, Bhat R, and Hwang D M 1989 *Appl. Phys. Lett.* **55** 2715–2717

[5] Adams A R 1986 *Electron. Lett.* **22** 249–250

[6] Walther M, Kapon E, Caneau C, Hwang D M and Schiavone L M 1993 *Appl. Phys. Lett.* **62** 2170–2172

[7] Chong T C and Fonstad C G 1989 *IEEE J. Quantum Electron.* **QE-25** 171–178

Inst. Phys. Conf. Ser. No 145: Chapter 7

Paper presented at 22nd Int. Symp. Compound Semiconductors, Cheju Island, Korea, 28 August–2 September 1995

© *1996 IOP Publishing Ltd*

Magneto-excitons and Landau level states in quantum wire superlattices

H. Weman,[1] M. Potemski,[2] M.S. Miller,[3,a]
M.E. Lazzouni,[4] and J.L. Merz[3,b]

[1]Department of Physics, Linköping University, S-581 83, Linköping, SWEDEN
[2]High Magnetic Field Laboratory, MPI/FKF and CNRS, F 38042 Grenoble, FRANCE
[3]Center for Quantized Electronic Structures, UCSB, Santa Barbara, CA 93106, USA
[4]Department of Physics, UCSD, La Jolla, CA 92 093, USA

Abstract. Photoluminescence excitation spectra of serpentine superlattice quantum wire arrays with lateral periods near 10 nm have been measured in magnetic fields up to 20 Tesla. The binding energies of coupled 1D excitons have been directly determined from the analysis of Landau-level-like excited magneto-exciton transitions observed in the low-field limit. The enhancement of 1D excitonic binding is observed compared to 2D, with a reduction in the exciton binding energy for very narrow quantum wires.

1. Introduction

Recent progress in the preparation technology of one-dimensional (1D) semiconductor structures makes it now possible to investigate various physical consequences of the 1D confinement [1]. In optical experiments on undoped structures 1D confinement has been observed by the anisotropy in the linear polarization and/or in the magnetic field dependence of the luminescence spectra [2-4]. Until now however, little distinction has been made between different classes of "1D" structures. By analogy, both conventional quantum wells (QWs) and superlattices (SLs) can be referred to as "2D" systems, but they are clearly different due to the distinct differences in electronic structure and, consequently, their properties. In the same way an isolated quantum wire is one example of electronic structure in the class of "1D" objects, and a coupled quantum wire array is another. The expected physical properties for these two different classes should be significantly different. Therefore measuring the magneto-optical properties of an isolated quantum wire would obviously result in different results than for a quantum wire superlattice. The high- and the low-magnetic field limits should also be different.

As we will show in this paper the binding energy of excitons in coupled quantum wires can be directly determined from the analysis of Landau-level-like magneto excitons observed in the low-field limit (where the cyclotron orbit diameter is larger than the wire width), but *not* for isolated quantum wires as was reported by Kohl et al. [5], where Landau level transitions only can be observed in the high-field limit. Our model system consists of (Al,Ga)As serpentine superlattice (SSL) quantum wire arrays grown on vicinal GaAs substrates. The electronic structure for this system has been shown to have strongly coupled electron states and nearly uncoupled hole states [2].

2. Experimental details

The SSL structures used in our experiments have been grown on 0.5° and 1.5° misoriented vicinal (100) GaAs substrates. The deposition sequence corresponded to $Al_xGa_{1-x}As$ barriers

---

a) Present address: Dept. of Solid State Physics, Lund University, S-221 00 Lund, Sweden.

b) Present address: Dept. of Electrical Engineering, University of Notre Dame, Notre Dame, IN 46556, USA.

of $x_{barrier}$ = 20 %, that are 1/2-terrace wide. The SSLs have a thickness of 34 nm with a ramping constant z_0 = 125 nm [6]. A reference (2D) sample of the same average composition though with no lateral composition modulation has also been investigated.

PL and PLE spectra have been measured at 1.6 K in a 10 MW resistive Bitter magnet in fields up to 20 Tesla. For the PL measurements we used the 488 nm line from an Ar⁺ laser and for the PLE measurements a Ti:Sapphire laser. The emitted light from the sample was dispersed through a 1.5 m Jobin Yvon spectrometer and detected with a GaAs photomultiplier. The magnetic field was applied parallel to the x-, y- and z-axes, defined as parallel to the wires, in the periodic direction and normal to the array. In the PL (PLE) experiments the direction of the emitted (exciting) light wavevector, k, was always normal to the vicinal surface (z-direction).

3. Experimental results and discussion

The presence of lateral anisotropy in our structures can already be deduced on the basis of linear polarized luminescence experiments. The representative PL spectrum (solid line) together with the linear polarization of the luminescence (dashed-dotted line) for the 0.5°-SSL structure is shown in Fig. 1. The PL emission is identified as a localized electron (e) heavy-hole (hh) exciton (e1hh1) recombining in the SSL layer. Introducing a model that allows for intermixing between the lateral barrier and well, we can determine the height of the lateral potential from the measured polarization at the PL peak [7]. Here we estimate the actual Al composition in the barrier (wire) to be 15% (5 %) instead of the intended 20 % (0%), giving a lateral potential modulation in the conduction band of 80 meV.

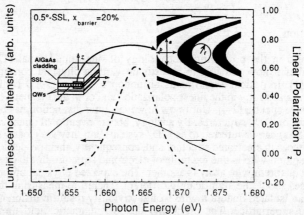

FIG. 1. PL (dash-dotted line) and linear polarization spectrum (solid line) of the luminescence emitted in the z-direction. A schematic figure of a SSL structure with a parabolic cross section is shown in the inset.

In Fig. 2 left (σ⁻) and right (σ⁺) handed circularly polarized PLE spectra of the 0.5°-SSL sample are shown with the magnetic field parallel to the z-direction. At zero magnetic fields four peaks can be resolved that are due to the ground state heavy-hole exciton (e1hh1(1s)) and excited heavy-hole excitons from higher subbands ((e2hh2, e3hh3, e4hh4). By increasing the magnetic field we see several new Landau-level like magneto-exciton transitions, (e1hh1(n = 1, 2, 3, 4 and 5), from fields as low as 3 Tesla. The cyclotron orbit diameter is given by, $d = 2 (h/eB)^{1/2} (2n +1)^{1/2}$, where n is the Landau level quantum number. This means that in the low field regime where the higher Landau-level like magneto-excitons start to appear the orbits are actually tunneling through a few lateral barriers. This situation is distinctly different from isolated quantum wires where Landau levels have been found to be quenched when the cyclotron diameter is wider than the wire width [8].

In Fig. 3 a fan chart is plotted of all the observed transitions in PLE for the 0.5°-SSL sample. For the SSL quantum wire we can determine the band edge by a linear extrapolation of the higher (n = 2, 3, and 4) Landau-level-like transitions to zero field. One can therefore

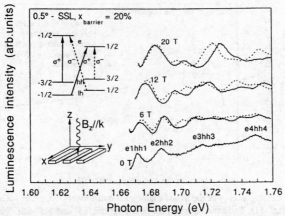

FIG. 2. Left (σ^-) and right (σ^+) handed circularly polarized photoluminescence excitation (PLE) spectra for a 0.5°-SSL quantum wire in the Faraday configuration. Optical selection rules for allowed PLE transitions between electron heavy-hole and electron light-hole states in a magnetic field ($B//k//z$) are shown in the inset.

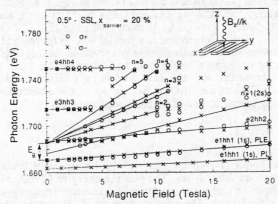

FIG. 3. Fan chart plot of the electron heavy-hole subbands and Landau levels for the 0.5°-SSL sample, with arrows indicating the experimentally determined exciton binding energy (here 13.5 ± 0.5 meV). Solid lines connecting the Landau level like magneto exciton transitions ($n = 2, 3,$ and 4) are linear fits to the data points.

directly determine the binding energy of the delocalized ground state exciton by subtracting the zero-field (e1hh1(1s)) peak position from the Landau level band edge. It should be noted that this determination of the band edge by fitting the higher Landau levels is done in the low field limit where the electrostatic confinement is dominating over the magnetic field induced. The excitonic binding energy has previously been determined in a similar way for quantum wells as well as for superlattices [9,10]. By fitting the Landau levels we determine the exciton binding energy to be 13.5 ± 0.5 meV for the 0.5°-SSL sample. We have performed similar measurements for the 1.5°-SSL sample and for the 2D reference alloy sample. We find the exciton binding energy to be 11.5 ± 0.5 meV in the 1.5°-SSL sample and 8.0 ± 0.5 meV in the 2D reference sample.

By using a fractional dimensional approach the e1hh1 exciton binding energy, in units of the Rydberg energy, E_0, has recently been calculated for AlGaAs cylindrical quantum wires as a function of the wire radius in units of the Bohr radius, a_0, by Christol et al. [11], shown as the solid line in Fig. 4. Our experimental values for the two SSL samples, are shown by the (x) marks in Fig. 4. Here we have assumed that the exciton binding energy is related to the cross sectional area rather than to its geometrical shape, and estimated the wire radius that gives the same cross sectional area as we have in the SSL wires [11]. The decreased binding energy for the narrowest wire is due to the strong lateral penetration of the exciton wavefunction into the barrier region and is in good agreement with the calculations by

934

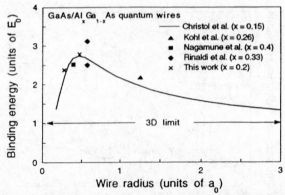

FIG. 4 Calculated binding energy of the e1hh1 exciton for cylindrical GaAs/Al$_{0.15}$Ga$_{0.85}$As quantum wires by Christol et al. (Ref. 11). The experimental data points for the 0.5° and 1.5°-SSL sample determined in this work are shown with the data from Kohl et al. (Ref. 5), Nagamune et al. (Ref. 3), and Rinaldi et al. (Ref. 12).

Christol et al. For comparison we have also included other published values on the exciton binding energy in GaAs/AlGaAs quantum wires in Fig. 4. Kohl et al. determined the excitonic binding energy for 700 Å wide etched GaAs quantum wires, and Nagamune et al. and Rinaldi et al. for 200 Å wide GaAs V-groove wires [3, 5, 11]. The binding energy determined by Kohl et al. was determined in a similar way as for the SSLs, however in the former case the Landau-level transitions were observed in the high-field limit and does therfore not correctly correspond to the binding energy of the wire. All values are in surprisingly good agreement with the calculations by Christol et al., considering the differences in the x values of the Al$_x$Ga$_{1-x}$As barrier of the quantum wires and that the comparison is made by approximation of the cross-sectional areas.

4. Conclusions

In conclusion we have reported on polarization resolved magneto-PLE studies of (Al,Ga)As serpentine superlattice quantum wires. The binding energies of 1D excitons have been directly determined from the analysis of Landau-level-like excited magneto-exciton transitions observed in the low-field limit. The enhancement of 1D excitonic binding is observed compared to 2D, with a reduction in the binding energy for very narrow quantum wires.

Acknowledgments

We want to thank P. Christol, B. Monemar, L.J. Sham, and P. Wyder for fruitful discussions and interest in this work.

References

[1]   Cingolani R and Rinaldi R 1993, *Rivista del Nuevo Cimento* **16** 1
[2]   Miller M S et al. 1992 *Phys. Rev. Lett.* **68** 3464
[3]   Nagamune Y et al. 1992 *Phys. Rev. Lett.* **69** 2963
[4]   Weman H et al. 1993 *Superlattices and Microstructures* **13** 5
[5]   Kohl M, Heitman D, Grambow P, and Ploog K 1989 *Phys. Rev. Lett.* **63** 2124
[6]   Weman H et al. 1993 *Phys. Rev.* **B 48** 8047
[7]   Pryor C 1991 *Phys. Rev.* **B 44** 12 912
[8]   Hammersberg J, Weman H, Notomi M, and Tamamura T 1994 *Superlattice and Microstructures* **16** 143
[9]   Rogers D C et al. 1986 *Phys. Rev.* **B 34** 4002
[10]   Belle G, Maan J C, and Weimann G 1985 *Solid State Commun.* **56** 65
[11]   Christol P, Lefebvre P, and Mathieu H 1993 *J. Appl. Phys.* **74** 5626
[12]   Rinaldi R et al. 1994 *Phys. Rev. Lett.* **73** 2899

Inst. Phys. Conf. Ser. No 145: Chapter 7
Paper presented at 22nd Int. Symp. Compound Semiconductors, Cheju Island, Korea, 28 August–2 September 1995
© *1996 IOP Publishing Ltd*

Fabrication and Optical Characteristics of GaAs Double-Coupled Quantum-Wires on V-grooved Substrate

Kazuhiro KOMORI, Xue-Lun WANG, Mutsuo OGURA, and Hirofumi MATUHATA

Electrotechnical Laboratory, Agency of Industrial Science and Technology, Ministry of International Trade and Industry, 1-1-4 Umezono, Tsukuba, Ibaraki, 305, JAPAN

Abstract. We have fabricated double-coupled quantum-wires, consisting of two GaAs quantum-wires as small as 5nm thick and 30nm in width (~15nm in effective width) separated by a 1 - 3nm thick AlGaAs barrier, using flow rate modulation epitaxy (FME) on V-grooved substrate, and observed its optical characteristics. The room temperature photoluminescence of double-coupled quantum-wires shows the clear energy separation between the fundamental coupled-state and the higher-order coupled-state. Also the separation energy increases as the thickness of barrier decreases from 3nm to 1nm, which result from the coupling effect between the two quantum-wires.

1. Introduction

Superior optical and electrical characteristics of lower-dimensional structures, so-called quantum-wires and quantum-dots, have been predicted theoretically and the fabrication of lower-dimensional structures have been actively studied for the application to high performance devices. By using the coherent characteristics of electron-wave, the coupled low-dimensional structures, in which low-dimensional structures are quantum mechanically coupled, are a more attractive candidate for the future new functional devices with both the high optical nonlinearity and the feature of electron-wave interference.

The coupled electron waveguide structures, which are the weakly coupled quantum-wires, were proposed[1][2] for the use in electronic devices and the fabrication has been reported using the split gate method or the FIB implantation method[3][4]. However, the confinement of the electron-wave by depletion is not sufficient for the application to opto-electronic devices with optical nonlinearities or intersubband-transitions. The conditions of both the ultra-fine size for the electron-wave confinement and the strong coupling are essentially required in the coupled low-dimensional structures. We have fabricated GaAs double-coupled quantum-wires with very small size and strong coupling by MOCVD growth on V-grooved substrate and observed optical characteristics including coupling effects between the two quantum-wires.

2. Fabrication

In the fabrication, V-grooves are formed on (100) oriented semi-insulating GaAs substrate along <01$\bar{1}$> direction by photolithography and wet chemical etching. The pitch of V-grooves is 4.8 μm. Then a 300nm thick GaAs and an 700nm thick Al$_{0.34}$Ga$_{0.66}$As buffer layers, a GaAs quantum-wire, an AlGaAs barrier layer, a GaAs quantum-wire and a 100nm thick Al$_{0.34}$Ga$_{0.66}$As protection layer are successively grown on the V-grooved GaAs substrate using the metalorganic chemical vapor deposition (MOCVD) growth method. In order to obtain very small size and high quality quantum wires, we use the flow rate modulation epitaxy (FME) method during the growth of GaAs quantum wires[5], while other layers except quantum-wires are grown by conventional MOCVD growth methods. The growth temperature was kept 630°C for the FME growth. The gas flow sequence and the flow rate in the FME growth is same as those shown in ref.[5] with a H2 purge period of 1s duration between TEGa and AsH3 flow periods. The growth periods are chosen to grow 5nm thick quantum-wires on the bottom of AlGaAs V-grooves. During the growth of the quantum-wires, the (111) quantum-films, and the (100) quantum-films are formed on the (111) side walls of V-grooves, and the (100) mesa top, respectively. After the masks are formed on the quantum-wires, the (100) quantum-films on the mesa top and the (111) quantum-films around the mesa top are remove using chemical etching. The etching time is chosen to perfectly remove the (100) quantum film by measuring the photoluminescence from the quantum-wires with different etching time[6]. Figure 1 shows the structure of the double-coupled quantum-wires after the etching, where only the quantum-wires and a part of (111) quantum-films are remained on the V-grooves.

The cross-sectional TEM picture of the double-coupled quantum-wire are shown in fig.2. Very small crescent shaped quantum-wires with a central thickness of 5nm and a lateral width of 30nm (effective width of 15nm) separated by a 2nm thick AlGaAs barrier layer are clearly observed at the bottom of the V-groove. While very thin (111) quantum films with a thickness of 1nm are formed on the side walls of the V-groove. The large growth selectivity between the quantum-wires and the (111) quantum-films is attributed to the enhancement of migration of Ga species in the FME growth. The top and the bottom quantum-wires has almost the same crescent shape, while the central thickness and the width of the top quantum-wire is slightly thinner and slightly wider than those of the bottom quantum wire, respectively, which results from the difference of the angle of the V-shaped AlGaAs groove below the top and the bottom quantum-wires .

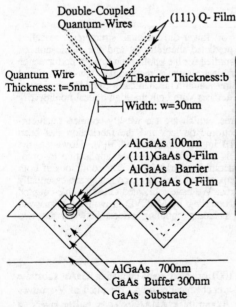

Fig.1 Structure of double-coupled quantum-wires.

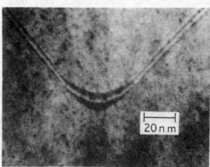

Fig.2 TEM photograph of GaAs/AlGaAs double-coupled quantum-wires.

3. Results and discussion

Figure 3-(a) shows the energy level of isolated two quantum wires with the enough thick barrier, while fig.3-(b) shows those of double-coupled quantum-wires separated by a barrier narrow enough that the wavefunction of the electron and the hole states in the conduction and the valence bands in the adjacent wells overlap. For the simplicity, symmetric type double-coupled quantum-wires, in which both the energy level of the two quantum-wires are the same, are assumed. Also, we can estimate the wave functions of holes has almost the same shape as those of heavy hole (hh) and light hole (lh) of quantum film, since the cross sectional ratio of the width to the thickness of the quantum-wires shown in fig.2 is more than 3. These hh and lh levels are shown in fig.3. Four transitions, symmetric type transitions Echh1, Eclh1, and antisymmetric type transitions Echh2, Eclh2, allowed in the double-coupled quantum-wires are shown in fig.3.

In the case of fig.3-(b), the energy level splitting occurs due to the coupling. We can call the coupled lower energy state, the coupled higher energy state, and the energy difference between the two coupled-states as the fundamental coupled-state, the higher-order coupled-state, and splitting energy respectively. Also, in the case of asymmetric type double-coupled quantum-wires, where the energy level of the two quantum-wires are different, the energy splitting due to the coupling occurs when the energy level difference between the two quantum wires is smaller than the splitting energy.

We measured the photoluminescence (PL) characteristics of the double-coupled quantum-wires at low temperature and at room temperature. Figure 4 shows the low temperature (12K) PL characteristics of the double-coupled quantum-wires with GaAs quantum-wire thickness of 5nm and width of 30nm, separated by a 2nm AlGaAs barrier layer. The sharp single peak around 760nm is the PL from the double-coupled quantum-wires, while those around 660nm and 670nm are the PL from the vertical quantum-film[7] and the (111) quantum-film since the thickness of the (111) quantum-films is as thin as 1nm. The single peak from the double-coupled quantum-wires at low temperature is due to the transition of the fundamental coupled-states.

The PL peak from the coupled quantum-wires shows the shoulder on the high energy side of the main peak as the temperature increase and it changes into two peaks at room temperature. Figure 5(a) shows the room temperature PL characteristics of three double-coupled quantum-wires samples, in which the barrier thickness varies from 1nm to 3nm, with the polarization parallel and perpendicular to the quantum-wires. The PL are measured from <100> direction as shown in Fig.5(b). The PL spectra with the polarization parallel to the quantum wires give the information of the hh transition while that with the polarization perpendicular to the quantum wire give the information of the lh transition.

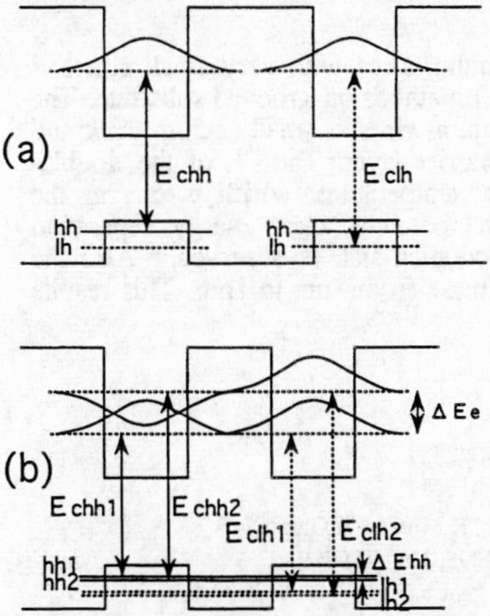

Fig.3 Schematic diagram of the energy levels of
(a) isolated quantum-wires and
(b) double-coupled quantum-wires.

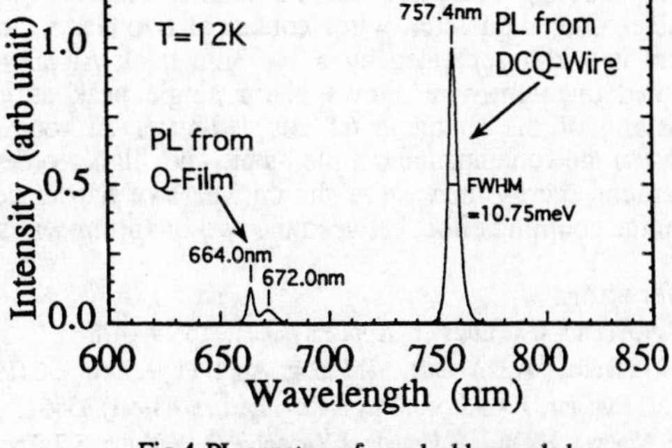

Fig.4 PL spectrum from double-coupled
quantum-wires at low temperature.

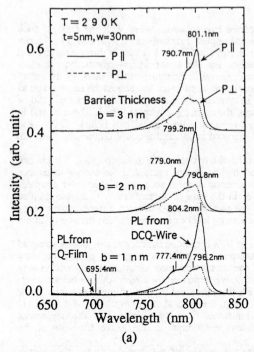

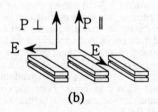

(b)

Fig.5
(a) Room temperature PL spectra from double-coupled quantum-wires with polarization parallel (P ∥) and perpendicular (P⊥) to the quantum-wires. Barrier thickness varies from 1nm to 3nm.
(b) Schematic diagram of polarization measurement. PL are measured from <100> direction with two polarization of P ∥ and P⊥.

(a)

The main peak and the sub-peak with polarization parallel to the quantum-wires are due to the transition of fundamental coupled-state (Echh1) and that of higher-order coupled-state (Echh2), respectively. All the sample show the clear energy separation between the fundamental coupled-state and the higher-order coupled-state. Also, the separation energy between the main-peak and the sub-peak increase as the thickness of a barrier decrease. This is due to the coupling effect between the two quantum-wires.

4. Conclusions

In conclusions, we have fabricated double-coupled quantum-wires with very small size and strong coupling using flow rate modulation epitaxy (FME) methods on grooved substrate. The double-coupled quantum-wires consist of two GaAs quantum-wires as small as 5nm thick and 30nm in width separated by a 1 - 3nm thick AlGaAs barrier layer. The PL of the double-coupled quantum-wire shows sharp single peak at low temperature, which is due to the transition of the fundamental coupled-state. At room temperature, clear energy separation between the fundamental coupled-state and higher-order coupled-state is observed. Also the separation energy increase as the thickness of barrier decrease from 3nm to 1nm. This results from the coupling effect between the two quantum-wires.

References
[1] J.A.Alamo, C.C.Eugster, Appl.Phys.Lett.,56(1990)78
[2] N.Tsukada, A.D.Wieck, K.Ploog, Appl. Phys. Lett., 56 (1990) 2527
[3] C.C.Eugster, J.A.Alamo, Phys. Rev. Lett., 67(1991) 3586
[4] F.Wakaya, H.Otoi, K.Umeda, J.Yanagisawa, Y.Yuba, S.Takaoka, K.Murase, K.Gamo, D.G.Hasko, G.A.C.Jones, and H.Ahmed. Int. Workshop on MPE'95, H6, (1995) 93
[5] X.L.Wang, M.Ogura, Appl.Phys.Lett., 66 (1995) 1506
[6] X.L.Wang, M.Ogura, H.Matsuhata, Appl.Phys.Lett. in press August 7 issue.
[7] M.Walter, E.Kapon, J.Christen, D.M.Hwang, and R.Bhat, Appl.Phys.Lett. 60 (1992) 521

Inst. Phys. Conf. Ser. No 145: Chapter 7
Paper presented at 22nd Int. Symp. Compound Semiconductors, Cheju Island, Korea, 28 August–2 September 1995
© *1996 IOP Publishing Ltd*

Of Serpents, SADs and Cells --
Recent Progress on Quantum Structures

James L. Merz

Department of Electrical Engineering, University of Notre Dame, Notre Dame, IN 46556, USA

Abstract. This paper reviews recent results involving two examples of self-organizing growth of quantum structures: serpentine superlattice quantum wires, and self-assembled quantum dots. Experimental and theoretical advances have been made in both cases which provide hope for their eventual application. One such application, a revolutionary approach to computational hardware (so-called Quantum Cellular Automata) is described in light of the quantum dot results presented earlier in the paper.

1. Introduction

Research activity in the field of low-dimensional quantum structures, or nanostructures, continues to intensify, driven by the requirement for ever-shrinking microelectronic circuits, and the consequent development of techniques for the manipulation and measurement of material systems at the near-molecular level. Many schemes have been proposed and reported for the fabrication of quantum wires and dots, and a considerable body of fundamental information about the behavior of electronic particles in these systems is being accumulated. Common to most of this research is the ever-present difficulty of obtaining collections of these quantum structures whose size and shape distributions are sufficiently uniform that the desired quantum properties do not disappear as a result of averaging. Thus, the "brute force" approach of fashioning nanostructures by assembling atoms or molecules appears to be extremely difficult, and the difficulties multiply if a linear fabrication technique (i.e., assembly of structures one-at-a-time; writing of patterns in real time) is employed. More promising, at least to this author, is the possibility of growing or otherwise fabricating these structures by using natural processes that are self-organizing and self-limiting.

In this paper, recent results involving two different examples of the self-organizing growth of quantum structures are presented, based on work carried out by researchers at the Center for Quantized Electronic Structures (QUEST) at the University of California, Santa Barbara, and by their collaborators. The first of these, the growth of *Serpentine Superlattice* (SSL) quantum wires by molecular beam epitaxy (MBE), involves the surface migration and preferential bonding of atomic species during growth; this is clearly self-organizing but not at all self-limiting. In the second example, the strain-induced growth of *self-assembled dots (SADs)* is also partially self-limiting, at least to the extent that natural forces appear to improve the uniformity of these dots, if (and only if) growth is terminated appropriately. Finally, the question of the *application* of quantum wires and dots will be briefly addressed, with emphasis on a truly revolutionary approach to computation, developed at the University of Notre Dame, utilizing arrays of ordered quantum dots which form so-called *Quantum Cellular Automata*.

2. Of Serpents,....

The growth of serpentine superlattices on vicinal substrates, and their optical characterization, is by now well documented.[1,2] By depositing alternating sub-monolayers of GaAs and AlGaAs on substrates that are slightly (1-2°) misoriented in the (110) direction, under appropriate flux and growth-temperature conditions, the Ga and Al species separate and nucleate successively along the step edges, forming quantum wires oriented parallel with the step edges. Using transmission electron microscopy (TEM), these wires have been shown to be approximately straight for a few hundred nanometers. Detailed polarized photoluminescence (PL) and absorption studies clearly demonstrated their one-dimensional nature, and the expected anisotropy of the luminescence spectra was observed for magneto-optics measurements.[3]

More recently, the character of 1-D confinement achieved in SSLs has been tested with polarization-resolved magneto-optical experiments at magnetic fields up to 20 T.[4] In emission, the diamagnetic shift is critically dependent on the direction of the magnetic field, again giving a clear indication that the ground state exciton is 1-D like, The binding energies of these 1-D excitons was directly determined from Landau-like excited magneto-exciton transitions observed in photoluminescence excitation (PLE) experiments, which are analogous to optical absorption experiments. These binding energies were found to increase from 8 meV for a quantum well, to 11.5 meV and 13 meV (for a 54 Å and 160 Å wide wire region, respectively). This increase agrees with theoretical expectations for 1-D wires. Experiments performed for different orientations of the magnetic field relative to the wire direction show that the carrier confinement achieved in these structures is quantum-well-like (i.e., 2-D) in the direction parallel to the growth, and superlattice-like in the lateral direction.

3. .....SADs....

The situation becomes even more interesting for the strain-induced self-assembled (quantum) dots (SADs) that have been grown by MBE[5] and by metal-organic chemical vapor deposition (MOCVD)[6] in the last two years. In this growth regime, non lattice-matched films (e.g., InGaAs or InAs on GaAs) are grown under conditions for the occurrence of Stranski-Krastanow growth; that is, 3-D islands begin to form well below the critical thickness for the formation of misfit dislocations. The resulting dots are self-organizing in the sense that they nucleate and grow spontaneously (transition from 2-D film growth to 3-D island growth), and they are somewhat self-limiting in the sense that strain causes the dot sizes to become relatively uniform (e.g., 20-30 nm diameters, with a variance of less than 10%). They are not completely self-limiting, because continued growth causes the islands to dislocate and coalesce into highly-damaged 3-D bulk material. Before this onerous condition sets in, however, the dot sizes are uniform, they occur in concentrations of 10^9 to 10^{11} cm^{-2}, have extremely high radiative efficiency, and show discrete quantum states. A detailed PL and PLE investigation of these dots[7] suggested an unusually large Stokes shift between the ground state of these dots observed in absorption and emission, and selective excitation of the PL allowed the excitation of subsets of these dots to be clearly observed. Subsequent PL measurements of small numbers of dots isolated by wet chemical etching[8], or by cathodoluminescence (CL) excitation[9,10] show extremely narrow (<0.15 meV) emission lines emanating from *individual quantum dots*. Such measurements are without precedence, and illustrate the precision and control currently achievable for both the growth and characterization of these structures. In addition, a recent report[11] of capacitance experiments on these dots has demonstrated the controlled occupation of these dots by small numbers of electrons, so that applications requiring the controlled charging of uniform arrays of quantum dots may be possible in the foreseeable future.

To accomplish this, however, several major challenges must first be overcome. Foremost of these is the ability to form uniformly-space arrays of dots; that is, it will be necessary to control the nucleation sites of the SADs, which currently tend to nucleate in random positions. Progress has been made recently on this difficult problem, both theoretically and experimentally. Regular arrays of self-aligned dots have been observed to form naturally on GaAs (100) mesa structures,[12] and self-ordered dots have also been observed to form on GaAs (311)B faces.[6(a)] Understanding the mechanisms responsible for this self-ordering may provide the key to controlling the growth and nucleation of these dots. Theoretical progress has also been made in this regard. Ngo *et al* have recently described a nucleation mechanism which permits ordering of the island nucleation positions.[13] The nucleation of a molecular cluster at a particular site on the sample surface creates an effective zone surrounding the cluster, in which additional adatoms are drawn to the original cluster rather than forming a new cluster at an alternate nucleation site. Thus, a filled nucleation site is surrounded by an "exclusion zone" in which no additional nucleation events occur. This process of itself leads to a certain degree of ordering of the exclusion zones, and hence of the nucleation sites. In this model, the exclusion zone width is independent of the radius of the cluster; as the width is increased, the effective island density is reduced.

Theoretical progress has also been made recently in understanding the radiative recombination mechanisms for these SADs. An alternative explanation of the data presented in ref. 7 has been offered by Tsiper.[14] Using a model based on statistical fluctuations of alloy disorder of the quantum dots, Tsiper proposes that the large apparent Stokes shift between PL and PLE observed in ref. 7 is instead the average distance between the ground and excited states of the dots. That is, the peaks observed by Fafard *et al* in PLE result from absorption by the first excited state, not the ground state as originally believed. The ground state cannot be observed in PLE because it is exactly resonant with the PL emission energy used to monitor the PLE. This result is consistent with direct absorption experiments reported in ref. 10. Furthermore, double peaks observed in ref. 7 are attributed by Tsiper to splittings of the ground state and first excited state as a result of alloy fluctuations which lift the degeneracies of these states. The results of Tsiper's calculations reproduce the PL/PLE data reported in ref. 7 almost exactly.

4. ....and Cells.

Despite the intense research in the field of quantum structures, progress toward the realization of novel applications and substantially improved devices has been slow. Recently, however, an entirely new idea has been proposed by Lent *et al* for performing logic functions in computers.[15] In this scheme individual cells, containing either 4 or 5 quantum dots, are loaded with 2 electrons. The dots are arranged on the corners of a square cell (the fifth dot, if present, is in the center), and the electrons are able to tunnel from dot to dot as a result of the Coulomb repulsion between them. There are therefore two stable configurations of the cells, as shown in Fig. 1, and these configurations are labeled as the polarization states P=+1 and -1. If the two cells of opposite polarization shown in Fig. 1 were brought close together, and one of the cells was held fixed (say, for example, the P=-1 cell was fixed), then the other cell would be

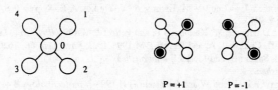

$P = +1$ $P = -1$

Figure 1. The quantum cell consisting of five quantum dots which are occupied by two electrons. The mutual Coulombic repulsion between the electrons results in bistability between the $P = +1$ and $P = -1$ states.

forced to switch to the opposite state (e.g., in the example given, the P=+1 cell would switch to P=-1.) Information can then be made to propagate along a line of such cells, and that propagation occurs in very short times (10s of psec). This is the basic concept of the so-called Quantum Cellular Automata (QCA) paradigm for computing. Lent *et al* have shown that, based on this very simple idea, "and", "or", and "nor" gates can be constructed, leading to such devices as an inverter and a full adder. In fact, all necessary logic operations can be performed with such a scheme. The fundamental property of these cells, which makes simple logic circuits possible, is the large nonlinearity of the switching process. That is, the result of a full scale calculation to solve the Schrödinger Equation for this configuration shows that a slight polarization of the "input" cell causes nearly complete polarization of the "output" cell. This robust character of the polarization switching mechanism provides considerable latitude in the fabrication tolerances of the dots comprising the QCA circuit, making it a promising approach for a radically new device architecture. Not only are the speeds expected to be fast, but the chip real estate should be very small; for example, a full adder should fit inside an area smaller than that expected to be required for a single transistor by the VLSI road map into the 21st Century.

Although considerable progress in improving the technology of quantum dots will be required to realize QCA circuits and architectures, some of the recent results described above auger well. Crucial to this concept will be the ability to control the location and size of dots, and to control the numbers of electrons in them. In both of these areas, significant advances have been reported.

5. Acknowledgments

This paper has relied heavily on the results of the author's colleagues both at Santa Barbara and Notre Dame. Notable among these colleagues are A. Efros, S. Fafard, E.L Hu, C.S. Lent, R. Leon, D. Leonard, M. Miller, T. Ngo, P.M. Petroff, D. Tougaw, E. Tsiper, and H. Weman.

References

[1] Miller M S, Weman H., Pryor C E, Krishnamurthy M, Petroff P M, Kroemer H, and Merz J L, 1992 *Phys. Rev. Lett.* **65** 3464.

[2] Weman H, Miller M S, Pryor C E, Li Y J, Bergman, P, Petroff P M, Merz J L 1993 *Phys. Rev* **B 50** 7719.

[3] Weman H, Jones E D, McIntyre C R, Miller M S, Petroff P M, and Merz J L 1993 *Superlattice and Microstructures* **13** 5.

[4] Weman H, Potemski M, Lazzouni M E, Miller M S, and Merz J L 1993 *unpublished.*

[5] (a) Leonard D, Krishnamurthy M, Reaves C M, DenBaars S P and Petroff P M 1993 *Appl. Phys. Lett.* **63** 3203; (b) Leonard D, Krishnamurthy M, Fafard S, Merz J L, and Petroff P M 1994 *J. Vac. Sci. Technol.* **b 12** 1063; (c) Leonard D, Pond K and Petroff P M 1994 *Phys. Rev.* **B 50** 11687.

[6] (a) Nötzel R and Ploog K 1993 *Adv. Mater.* **5** 22; (b) Nötzel R, Temmyo J and Tamamura T 1994 *Japn. J. Appl. Phys.* **33** 275.

[7] Fafard S, Leonard D, Merz J L. and Petroff P M 1994 *Appl. Phys. Lett.* **65** 1388

[8] Fafard S, Leon R, Leonard D, Merz J L. and Petroff P M 1994 *Phys. Rev.* **B 50** 8086.

[9] Leon R, Fafard S, Leonard D, Petroff P M, and Merz J L. 1995 *Science* **267** 1966.

[10] Grundmann M, Christen J, Ledentsov N N, Böhrer J, Bimberg D, Ruvimov S S, Werner P, Richter U, Gösele U, Heydenreich J, Ustinov V M, Egorov A. Y, Zhukov A E, Kop'ev P S, and Alferov Z I 1995 *Phys. Rev. Lett.* **74** 4043.

[11] Drexler H, Leonard D, Hansen W, Kotthaus J P, and Petroff P M 1994 *Phys. Rev. Lett.* **73** 2252.

[12] Mui D L, Leonard D, Coldren L A, and Petroff P M 1995 *Appl. Phys. Lett.* **66**, 1620.

[13] Ngo T, Petroff P M, Sakaki H, and Merz J L *unpublished.*

[14] Tsiper E V *unpublished.*

[15] (a) Lent C S, Tougaw P D, Porod W and Bernstein G H 1993 *Nanotechnology* **4** 49; (b) Tougaw P D and Lent C S 1994 *J. Appl. Phys.* **75** 1818; (c) Lent C S and Tougaw P D 1994 *J. Appl. Phys.* **75** 4077.

Inst. Phys. Conf. Ser. No 145: Chapter 7   943
Paper presented at 22nd Int. Symp. Compound Semiconductors, Cheju Island, Korea, 28 August–2 September 1995
© *1996 IOP Publishing Ltd*

Polarization dependence of optical gain in unstrained and strained InAlGaAs quantum wire arrays

Jong Chang Yi and Nadir Dagli*

Department of Electrical Engineering, Hong Ik University, Seoul, Korea
*ECE Department, University of California, Santa Barbara, CA93106, U.S.

Abstract. Four types of InGaAs quantum wire lateral superlattices, which are unstrained, or strained laterally, vertically, or uniformly, are investigated, respectively. Their optical properties such as optical gain and polarization dependence are quantitatively analyzed including the effects of the nonuniform strain distribution and the valence band intermixing using a two-dimensional finite-element method. The results show unique polarization dependence characteristics, which can be exploited either for the single polarization optoelectronic devices or for the polarization independent devices. Such results are compared with each other as well as the quantum wells.

1. Introduction

Strained quantum wire structures, which provide additional quantum confinement due to the deformation potential, have undergone extensive study and development for their interesting physical phenomena and potential device applications[1-3]. Among them, the lateral superlattice is one of the most promising structure for its small size and higher wire density[4]. One drawback in it, however, is the relatively weak lateral confinement due to the lateral material intermixing[5]. In this paper, the possibility of additional lateral confinement will be investigated using various types of stress in the lateral superlattice quantum wires. Particularly, their unique polarization dependence characteristics, which can not be seen in typical quantum well or bulk structures, will be quantitatively analyzed including the effects of inhomogeneous strain distribution and valence band intermixing using a two-dimensional finite-element method[6].

2. Stress profiles in various InAlGaAs quantum wires

In quantum well structures, there is only one type of strain profile: That is the uniaxial strain. Whereas, in the quantum wire structures, the strain profile can be uniaxial along either the lateral direction or the vertical direction. The strain can also be hydrostatic when surrounded by one material.

Fig.1 shows some of the possible strain types for the InGaAs quantum wires embedded in the InAlGaAs materials. The wire material is taken as $In_{0.3}Ga_{0.7}As$ and the surrounding materials are $Al_{0.4}Ga_{0.6}As$ *and/or* $In_{0.3}Al_{0.5}Ga_{0.2}As$. Table 1 shows the relevant material parameters for these three materials. The $Al_{0.4}Ga_{0.6}As$ material has a smaller lattice constant and a larger bandgap energy than $In_{0.3}Ga_{0.7}As$ hence it provides both the compressive stress and the potential band offset. On the other hand, the $In_{0.3}Al_{0.5}Ga_{0.2}As$ material has a higher bandgap energy but almost the same lattice constant as the quantum wire material, so it provides only the potential band offset.

Typically, an InGaAs/InAlGaAs lateral superlattice grown on AlGaAs substrate as shown in type B is considered as the strained quantum wire array[3]. In that case the stress is applied uniaxially along the vertical direction, as commonly happening in the strained quantum well structures. However, if one grows an InGaAs/AlGaAs lateral superlattice on AlGaAs substrate, the stress is from both directions as in type A [2]. Furthermore, if one uses InAlGaAs as the substrate material and grows InGaAs/AlGaAs lateral superlattice, the stress is along the lateral direction [1]. Therefore, one can expect laterally uniaxial strain in type C.

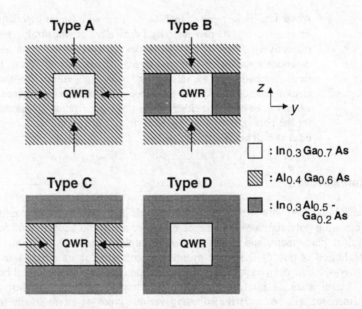

Fig.1 Cross sectional profile of strained quantum wires. Arrows indicate the direction of stress. Type A has the hydrostatic stress, type B has the uniaxial stress along the vertical direction, type C has the lateral stress, and type D has no stress.

Table 1 Material parameters for InAlGaAs materials under investigation. The conduction and valence band offsets are estimated with respect to $In_{0.3}Ga_{0.7}As$ quantum wire material.

	$In_{0.3}Ga_{0.7}As$	$Al_{0.4}Ga_{0.6}As$	$In_{0.3}Al_{0.5}Ga_{0.2}As$
Bandgap, E_g (eV)	1.019	1.966	1.948
lattice constant, a_o (Å) (lattice mismatch)	5.7748	5.6574 (-2.1%)	5.7787 (0.006%)
CB offset, ΔE_c (eV)	0	0.663	0.576
VB offset, ΔE_v (eV)	0	0.285	0.352

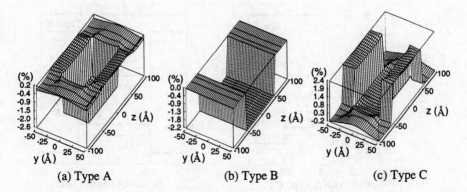

(a) Type A   (b) Type B   (c) Type C

Fig.2 Volume change in the various types of the strained quantum wire arrays shown in Fig.1. Wire width and separation are 50Å and wire height is 100Å. The volume change reflects the hydrostatic strain profile as $\Delta V/V = \varepsilon_{xx} + \varepsilon_{yy} + \varepsilon_{zz}$.

The strain profile in each of the quantum wire arrays is analyzed[7] and its actual lattice volume change profile is plotted in Fig.2. The wire width and the wire separation are taken as 50Å and the wire height is taken as 100Å, which are close to the optimum quantum wire lateral dimensions. The volume change reflects the hydrostatic strain profile as $\Delta V/V = \varepsilon_{xx} + \varepsilon_{yy} + \varepsilon_{zz}$.

In type A, the wire material is compressed approximately 2.6 %. In type B, both the wires and barriers are compressed about 2.2 % and the strain is uniform along the lateral direction since the wire and barrier materials have almost the same lattice constants. In type C, the wire material is very slightly strained but the barrier material is expanded by approximately 2.4 %. In both type A and type C, there is a strong discontinuity in the strain profile along the lateral direction. Type D has almost no lattice mismatch so it is an unstrained quantum wire array and its obvious strain profile is omitted in Fig.2.

3. Band structures of the InAlGaAs quantum wires

The effects of the strain on the band structure are estimated by the deformation potential[7]. Fig. 3 shows the corresponding band profiles along the horizontal direction at the center of the quantum wire. In the figures, C denotes the conduction band, and HH and LH denote the heavy hole and light hole bands, respectively. The dashed lines indicate the bulk band positions with no strain effect.

In type A, the conduction band shifts upwards, and the HH and LH bands split as they shift downwards, as in most compressively strained materials. In type B, the band change shows a similar trend with type A except that the band deformation is almost uniform inside and outside the quantum wires. This is because the lattice constants of the quantum wire and barrier regions are almost the same.

Fig. 3.(c) shows the resulting band deformation in type C. In this case, the net volume change inside the quantum wire is very small hence the conduction band shift inside the quantum wire region is almost negligible. However, the lattice constant of the barrier region is smaller than that of the quantum wire region in type C. Thus, its quantum wire regions

946

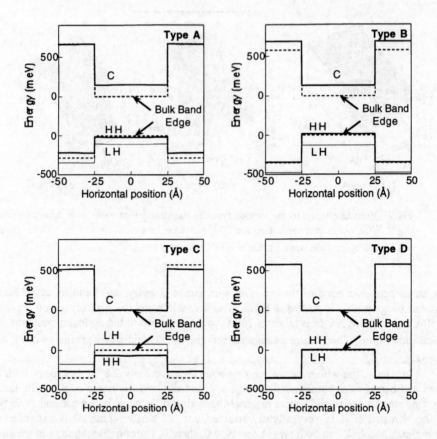

Fig.3 Band profile along the horizontal direction at the center of the quantum wire in various types of the strained(A-C) and unstrained(D) quantum wire arrays shown in Fig.1. C denotes the conduction band, and HH and LH denote the heavy hole and light hole bands, respectively. The dashed lines indicate the bulk band positions with no strain effect.

will be suppressed vertically to match the smaller lattice constant of the barrier regions. The vertically suppressed quantum wire regions will eventually expand along the lateral direction to fill the barrier regions. As a result, the quantum wire regions will be suppressed vertically and expand along the lateral direction, which is quite similar to the quantum well with a tensile strain. Therefore, the LH band moves upwards and the HH band is suppressed downwards as shown in Fig.3.(c). When one compares type C to type A, one can find that the wires in type C has a geometric similarity to the barriers in type A, except that its lattice constant is larger than those of surrounding regions. The quantum wires of type C are lattice matched to the substrate and the barriers between them have a smaller lattice constant, while the barriers of type A are lattice matched to the substrate and the wires between them have a larger lattice constant. Thus, the strain coefficients in the quantum wires of type C would have the opposite signs with those in the barriers of type A. In Fig.3.(a), the HH band is higher than the LH band in the barrier regions of type A. Therefore, in type C, the LH band becomes higher than the HH band since the signs of the strain coefficients are reversed. The largest split energy between HH and LH bands among the four types is observed in type C and the second is in type B. It is almost twice as large as that in type A, not to mention the unstrained type D where the HH and LH bands are degenerate.

4. Optical properties of the InAlGaAs quantum wires

Fig.4 shows the optical gain in various types of quantum wire arrays for three different polarizations. The material gains for three polarizations along the x, y, and z axis are denoted by TE.x, TE.y, and TM polarizations, respectively.

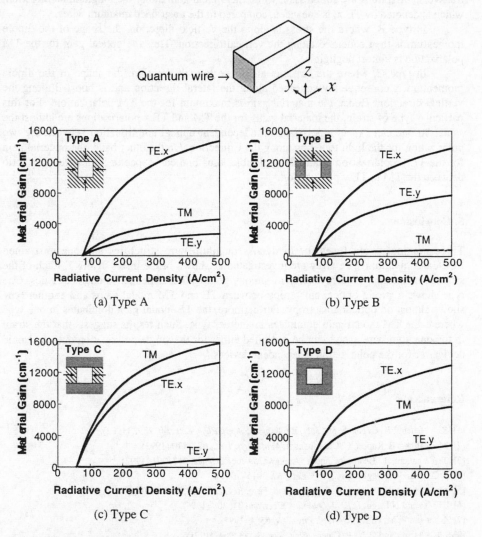

Fig.4 . Polarization dependence of the optical gain in various types of quantum wire arrays shown on the insets. The arrows indicate the directions of stress due to the lattice mismatch.

948

Again, the x-axis is chosen to be parallel to the quantum wire axis, the y-axis along the lateral direction, and the z-axis along the growth direction, as shown in the diagram on the top of Fig.4. The dimensions of the lateral superlattice quantum wires are 50 Å wide, 100Å thick with 50Å separation. In the gain calculation, the energy broadening is taken as 7 meV and temperature is 300 K.

In the unstrained quantum wire array, the top valence band is mostly HH-like one. Thus, the optical gain is maximum for the x- polarization denoted by TE.x, while it is minimum for he y-polarization denoted by TE.y polarization shown on Fig.4.(d).

In case of type A, the dipole momentum is maximum along the TE.x polarization and so is the material gain. However, the overall shape of the dipole momentum is squeezed along the lateral and the vertical directions isotropically. Thus the optical gain along the transverse directions are suppressed, while the optical gain along the longitudinal direction, which is denoted by TE.x, is enhanced, compared to the unstrained quantum wires.

In type B, where the stress is along the vertical direction, the shape of the dipole momentum is then squeezed along the vertical direction. Thus the optical gain for the TM polarization is almost negligible.

In type C, where the stress is along the lateral direction, the shape of the dipole momentum is consequently squeezed along the lateral direction and is boosted along the vertical direction. Hence the material gain is maximum for the TM polarization. For this particular type of strain, the material gains for the TM and TE.x polarizations are almost the same. So one can expect a polarization independent optical amplification between the two polarization for the light traveling along the y direction. On the other hand, the material gain for the TE.y polarization is almost negligible, thus one can expect a high extinction ratio between the TM to TE.y polarization.

5. Conclusion

Three nominal strain types for InAlGaAs quantum wire structures and one unstrained InAlGaAs quantum wire array are investigated as shown in the insets of Fig.1. Each of the strained quantum wire arrays shows a unique polarization dependence characteristics. One type shows a strong optical anisotropy between TE and TM modal gains and another type shows almost no optical anisotropy. Furthermore, the TE modal gain dominates in one type whereas the TM modal gain dominates in another type. Such results suggests that the strain in the quantum wire arrays can be exploited either for the single polarization optoelectronic devices or for the polarization independent devices.

References

[1] K. Y. Cheng, K. C. Hsieh, and J. N. Baillargeon, *Appl. Phys. Lett.*, **60**, 2892 (1992).
[2] M. Walther, E. Kapon, C. Caneau, et al, *Appl. Phys. Lett.*, **62**, 2170 (1993).
[3] T. Yamauchi, T. Takahashi, and Y. Arakawa, *Surface Science*, **267**, 291 (1992).
[4] J. C. Yi and N. Dagli, *Appl. Phys. Lett.*, **59**, 3015 (1991).
[5] M. S. Miller, H. Weman, C. E. Pryor, et al, *Phys. Rev. Lett.*, **68**, 3464 (1992).
[6] J. C. Yi and N Dagli, *IEEE J. Quantum Electron.*, **31**, 208 (1995).
[7] Z. Xu and P. M. Petroff, *J. Appl. Phys.*, **69**, 6564 (1991).
[8] S. L. Chuang and C. Y. P. Chao, *Phys. Rev. B*, **43**, 9649(1991).

Inst. Phys. Conf. Ser. No 145: Chapter 7
Paper presented at 22nd Int. Symp. Compound Semiconductors, Cheju Island, Korea, 28 August–2 September 1995
© *1996 IOP Publishing Ltd*

Photoluminescence from A Single Quantum Dot

Y. Arakawa and Y. Nagamune

Institute of Industrial Science, University of Tokyo
Roppongi, Minato-ku, Tokyo 106 Japan

Abstract. We observed a single photoluminescence peak from a single GaAs/AlGaAs quantum dot, of which the effective size and period are 60 x 60 x 14 nm and 2μm, respectively, using μ-photoluminescence measurement system. The broadening of spectral linewidth is suppressed with the decrease of the excitation laser power, which is mainly due to reduced quantized levels' filling effect. The minimal spectral linewidth observed is 0.9 meV with a low excitation laser power. The time-resolved measurement was also carried out with a streak camera. The use of the near-field optical microscope was also investigated. Finally, using the μ-photoluminescence technique, the two-dimensional electrons' motion in a point contact structure successfully visualized. Time- and spatial resolved optical characterization is essential for nano- and mesocsopic structures.

1. Introduction

Three-dimensional (3D) confinement of carriers is an important phenomenon for fundamental study in physics as well as for quantum device applications using the particularly sharp density of states[1]. In order to realize the predicted quantum effects, large efforts have been made for fabricating quantum dot structures by using etching technique, selective growth, self-organized growth, and other techniques.

However, in many reports the quantum dots show much broader photoluminescence (PL) peak than that of bulk or conventional quantum wells (QWLs), which is mainly due to size fluctuation of the dot structures. In order to clarify the intrinsic effect of the luminescence broadening, it is necessary to eliminate inhomogenous broadening and overlap of luminescence of various sizes of dots by illuminating an individual dot structure precisely . In this paper, we discuss micro-PL (μ-PL) measurement of a single quantum dot structure at low temperature. The result indicates that broadening of the spectral linewidth is suppressed with the decrease of the excitation laser power, which is mainly due to reduced quantized level's filling effect. The minimal spectral linewidth observed is 0.9 meV with a low excitation laser power. In addition, time-resolved measurement was also carried out using the μ-PL system with a streak camera. Finally, we investigated visualization of motion of two-dimensional electrons in a point contact under the electric fields.

2. Photoluminescence from a single quantum dot [2,3]

Samples studied in this work were prepared by the selective growth technique using metal-organic chemical vapor deposition (MOCVD) on SiO2 patterned substrate[1]. The period between the dots is 1 μm. The effective size estimated by magneto-photoluminescence and SEM observation is 60 x 60 x 14 nm. The lateral confinement energy is about 8 meV.

The μ-PL measurements were carried out with a cw Ar$^+$ laser light which was focused onto the sample by an objective lens for an optical microscope. PL spectra from the sample were transferred to a spectrometer or a conventional charge coupled device (CCD) camera by using the same objective lens. Here, the spectrometer was installed with a photomultipular or an optical microchannel analyzer which can be operated by photon counting techniquuueee. The objective lens was placed out of the cryostat. The wavelength of the laser light was 514.5 nm and the beam diameter was 0.7 μm which is equal to the spatial resolution of the present measurement system and smaller than the period of the dot array, so that a single dot can be illuminated by the laser beam. The sample was cooled down at about 15 K in the cryostat by a closed He gas type refrigerator of which the vibration was carefully reduced. Spatial image of the PL spectra from the dots was also detected with the CCD camera through a band-pass filter passing only the PL spectral line of the dots.

Figure 1 shows PL spectra of the sample with a laser power of sub-mW, where the spectra were obtained by illuminating about 20 dots, 4 dots, and 1 dot. In this figure, the PL peaks around 1.54 eV originate from GaAs dots, and those at 1.51 eV and 1.49 eV correspond to bulk transition probably (D_0, X) and that at carbon impurities, respectively. Here, fine peak structures were observed in the quantum dot PL in the middle spectra, of which the number is corresponding to that of illuminated dots. When a single dot was illuminated, a single sharp peak was observed as shown in the lowest spectra.

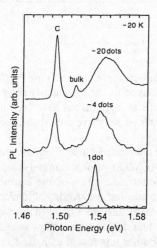

Fig. 1.: Photoluminescence spectra of 20 dots, 4 dots, and 1 dot.

Figure 2 shows spatial PL image of a single dot at an excitation power of 10 mW and the full width of half maximum (FWHM) of the PL spectral line, plotted as a function of laser power. As shown in this figure, FWHM of the quantum dot significantly is decreased with decreasing the laser power. At the lowest power, in which about one electron-hole pair on time average exists in the dot, FWHM was as low as 0.9 meV as shown in Fig. 3. This value may be restricted by the present experimental resolution, but at least smaller than both $k_B T$ and the 1 ML fluctuation, where k_B is the Boltzmann constant and $T = 15$ K. This much sharper PL peak results from the radiative recombination of excitons at the lowest energy state in the quantum dot.

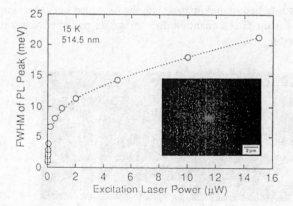

Fig. 2. The FWHM of the photoluminescence peak of the quantum dot and spatial photoluminescence image of a single dot and

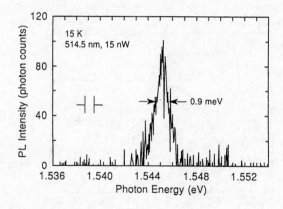

Fig. 3   The photoluminescence spectra of the quantum dot at the lowest laser power

4. Time- and spatial- resolved photoluminescence [4]

We investigated of time-resolved photoluminescence from a single quantum dot structure. For the measurement, we developed spatial and time resolved measurement system using the micro-photoluminescence (PL) technique and a streak camera. The sample is InGaAs/GaAs quantum dots with a period of 2μm. The PL image from the sample was focused onto the streak camera through a bandpass filter corresponding to the measured PL wavelength.

Figure 4(b) shows the spatial and time resolved image of PL under low excitation power, demonstrating time-resolved PL from a single quantum dot. On the other hand, under high excitation power, luminescence also arises with a slight delay from other vicinal quantum dots around the dot excited by laser pulse, as shown in Fig. 1(c) and Fig. 5. This result is the direct observation of the lateral carrier diffusion from the center region into the vicinal quantum dots.

We have also studied spatial images with the near filed optical microscope, obtaining spatial PL images for the GaAs quantum dots grown on patterned substrates. Details are described elsewhere[5].

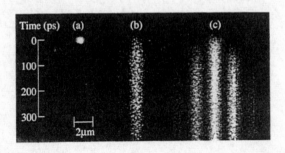

Fig.4: Spatial- and time-resolved images of (a) excitation laser pulse (b) PL from InGaAs quantum dots at 0.1.μW excitation (c) PL from InGaAs quantum dots at 50 uW excitation

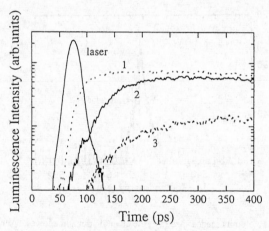

Fig.5: PL time-trace of each InGaAs/GaAs quantum dot shown in Fig.1 (c).

5. Visualization of two-dimensional electrons in mesoscopic systems [6,7]

Direct observation of electrons' motion in quantum nanostructures is quite important for the study of physics of the two-dimensional transport as well as for the characterization of quantum devices. Recently, there have been many reports on transport of electrons in point contact device in which conductance quantization, electron focusing, or electron interference effect have been observed. In these measurements, however, only information on current and voltage in the point devices can be extracted. The energy or spatial distribution of electrons has been so far studied only by computer simulations. Here, we demonstrate a novel method to visualize electron flow b using micro-photoluminescence (μ -PL) measurements.

A sample including a point contact was prepared by molecular beam epitaxy (MBE) grow of an InGaAs/GaAs quantum well with a thickness of 10 mm. A slit was formed by wet chemica etching and re-growth of GaAs. Carrier density and mobility of the unetched quantum well at liquid nitrogen temperature were 4.6 x 10 15 m-2 and 3.0 m2/Vs, respectively. The μ -PL measurement at 18 K was carried out by illuminating a He-Ne laser light in the region around the point contact. The PL image excited by the illumination of the laser light was detected by a charge coupled device (CCD) camera with the various bias voltages of the point contact structures.

Majority electrons accelerated by the electric field at the point contact have large wave number, while minority electrons and holes generated by the laser light have small wave number. Therefore, in the region where electrons have large velocity, probability of electron-hole recombination becomes low and PL intensity at the region decreases. By using this system, electron flow from the point contact with a bias voltage of 0.75V was successfully visualized as a two-dimensional image. Figure 6 shows computer-processed image which was obtained by subtracting the image without illumination of the laser light from the image with the laser light. The bright region is corresponding to "electron jet " region where the wave number of majority electrons is much larger than minority holes.

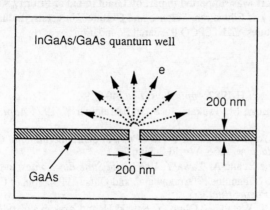

Fig. 6: A schematic illustration of a point contact structures

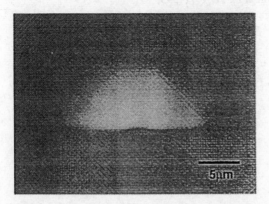

Fig. 7: A plain-view image obtained the PL image with a bias voltage of 0.75V from
        that with a zero bias voltage.

954

Conclusions

We observed a single photoluminescence peak from a single GaAs/AlGaAs quantum dot, of which the effective size and period are 60 x 60 x 14 nm and 2μm, respectively, using μ-photoluminescence measurement system. The broadening of spectral linewidth is suppressed with the decrease of the excitation laser power, which is mainly due to reduced quantized levels' filling effect. The minimal spectral linewidth observed is 0.9 meV with a low excitation laser power. The time-resolved measurement was also carried out with a streak camera. The use of the near-field optical microscope was also investigated. Finally, using the μ-photoluminescence technique, the two-dimensional electrons' motion in a point contact structure successfully visualized. Time- and spatial resolved optical characterization is essential for nano- and mesocsopic structures.

Acknowledgments

We acknowledge Prof. H. Sakaki, H. Watabe, M. Nishioka, Noda, and Y. Ohno for their collaborations. This work was supported in part by Grant in-aid of Priority Area on "Coherent Electronics" by Ministry of Education, Science and Culture, University-Industry Joint Project on Quantum Nanostructures, and TEPCO Research Foundation.

References

[1] Arakawa Y and Sakaki H 1982 *Appl. Phys. Lett.* **40** 939 .

[2] Nagamune Y, Nishioka M, Tsukamoto S, and Arakawa Y, 1994 *Appl. Phys. Lett.* **64**, 2495 ()

[3] Nagamune Y, Nishioka, and Arakawa Y 1994 *Proc. of 22nd International Conference on the Physics of Semiconductors*, 1835, World Scientific

[4] Watabe H, Nagamune Y, and Arakawa Y. 1996 *Solid State Electronics*, to be published

[5] Toda Y, Kourogi M, Nagamune N, Arakawa Y, and Ohtsu 1995 *Proc. of Confference on Las and Electro-Optics / Pacific Rim*, 165

[6] Nagamune Y, Noda T, Watabe H, Ohno, Y, Sakaki H, and Arakawa Y 1995 *Extend Abstrac. the 1995 International Conference on Solid State Devices and Matrerials*, 1083, to be publishee *Jpn. J. of Appl. Phys.*

[7] Nagamune Y, Watabe H, Ohno, Y, Sakaki H, and Arakawa Y 1995 *Confference on Laser anc Electro-Optics / Pacific Rim*, PD-1.4 (1995)

Inst. Phys. Conf. Ser. No 145: Chapter 7
Paper presented at 22nd Int. Symp. Compound Semiconductors, Cheju Island, Korea, 28 August–2 September 1995
© *1996 IOP Publishing Ltd*

Nanometer-scale GaAs dot structures fabricated using in-situ gas etching technique with InAs dots as mask

G. Yusa[1], H. Noge[2], Y. Kadoya[2], T. Someya[1] T. Suga[1], P. Petroff[3] and H. Sakaki[1,2]

1) Research Center for Advanced Science and Technology, Univ. of Tokyo,
4-6-1 Komaba, Meguro-ku, Tokyo 153, Japan
2) Quantum Transition Project, JRDC,
Park Bldg. 4F, 4-7-6 Komaba, Meguro-ku, Tokyo 153, Japan
3) Materials Department and QUEST, Univ. of California,
Santa Babara, California 93106 U.S.A.

Abstract. We propose a novel method to fabricate nano-meter scale GaAs dot structures by using molecular beam epitaxy and *in-situ* gas etching using self-assembled nanometer-scale InAs dots on GaAs (100) are used as masks. Our method takes advantage of the preferential etching of GaAs and InAs by Cl_2 gas and HCl gas, respectively. The average height and base diameter of resultant GaAs dots are 10 nm and 30 nm, respectively, when measured by atomic force microscopy. These GaAs dots are found to dissociate easily at elevated temperatures, indicating that the overgrowth of a barrier layer must be made at a rather low temperatures.

1. Introduction

Semiconductor quantum structures have been widely studied owing to their uniqueness in electronic and optical properties. Especially 10-nm scale InAs self-assembled dot on lattice-mismatched GaAs [1, 2, 3, 4, 5] prepared by the Stranski-Krastanov mode of growth is widely studied for its easiness and contamination-free nature. In this mode of growth, the first monolayer (ML) of InAs grows in the form of fully strained two-dimensional (2D) layer, but three-dimensional (3D) dots are formed automatically InAs with thicker than 1.5 ML is deposited. Although this self-assembled dots approach is attractive, strains between InAs dots and GaAs may lead to complications in the confinement potential and also may result in the material degradation through strain-induced dislocations.

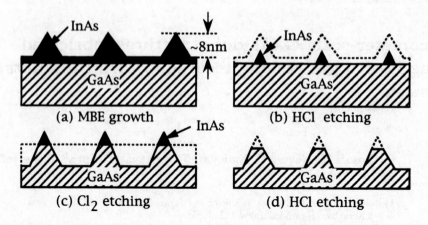

Figure 1. Schematic illustration of the process steps for the fabrication of GaAs dots.

2. Experimental

In this paper, we propose and demonstrate a novel method to form nano-meter scale strain-free GaAs dot structures by using self-assembled InAs dots as etching masks. In this method, a sample is processed entirely in an ultra high vacuum (UHV) multichamber processing system consisting of a molecular beam epitaxy (MBE) chamber and two etching chambers which are all connected via UHV tunnel. UHV etchings of GaAs by Cl_2 and HCl gas lead to clean and damage-free surface with active impurities less than 2×10^{10} cm^{-2} [6, 7, 8]. The cleanliness of this process is also proved by the sharp photoluminescense from an AlGaAs/GaAs quantum well, which was prepared by etching down the freshly grown 20 nm thick GaAs well down to 10 nm with Cl_2 gas and then about 1 nm by HCl gas before overgrowing an AlGaAs barrier layer. In this work, We make use of the selectivity of Cl_2 and HCl gas etching for GaAs and InAs: (1) GaAs can be etched by Cl_2 gas 500 times faster than InAs at substrate temperature T_s below 130 °C [9] and (2) InAs can be etched by HCl gas 60 times faster than GaAs for $T_s \leq 300$ °C [10].

The process steps used in this work are schematically shown in Fig. 1. First, InAs self-assembled dots are formed by MBE on freshly grown GaAs. Then the sample is transferred to etching chamber A, where an InAs wetting layer is removed selectively by HCl gas. The sample is then transferred to etching chamber B where unmasked part of the GaAs layer is selectively etched by Cl_2 gas to form dot structures. The sample is transferred to etching chamber A to remove the remaining InAs masks completely by the second HCl etching, leaving 10 nm scale GaAs dot structures. Note that HCl and Cl_2 etchings are performed in different chambers to minimize the cross-contamination, as will be discussed later. The use of epitaxially grown InAs dots as etching masks seems somewhat advantageous our previous approaches, where 10-nm scale sphere of foreign materials are introduced [11]

We now describe details of our experiment to demonstrate the effectiveness of this

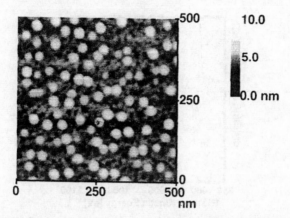

Figure 2. AFM image of the 1.5 ML InAs deposited sample after being exposed to Cl_2 gas. Note that the whole surface of the sample, including InAs dots is unchanged because the InAs wetting layer prevented the etching of GaAs.

approach. First we grew by MBE a 0.3–0.6 μm-thick GaAs layer on a (100) GaAs substrate. We then formed on its surface InAs dot structures with an InAs wetting layer by depositing 1.5 ML InAs at a substrate temperature of 450 °C as reported earlier [3, 4]. InAs dots are typically 8 nm in average height and 10–20 nm in average diameter at the bottom and have an average density of 8×10^{10} cm^{-2}.

Prior to the main experiment an experiment to etch the sample by Cl_2 gas to examine whether or not the 1 ML-thick InAs wetting layer prevents the Cl_2 gas etching. Since the etching rate of InAs by Cl_2 is very low when T_s is lowered [9]. For example, at $Ts = 77 °C$, the expected etch rates are 0.02 nm/min for InAs and 20 nm/min or higher for GaAs of Cl_2 with the pressure at 4×10^{-4} Torr. On the basis of these facts, we etched the sample by Cl_2 gas for one minute at Ts=77 °C. and studied the surface morphology by contact mode atomic force microscope (AFM) as shown in Fig. 2. Note that the size and shape of the dots are almost identical to those of freshly grown InAs dots, which indicates clearly that an InAs wetting layer prevents the Cl_2 etching.

Hence the removal the InAs wetting layer is quite important. Hence, in the main experiment, we sent the InAs-dot-covered sample to etching chamber A and performed at the HCl etching at a pressure of 3×10^{-4} Torr for 30 sec at T_s of 210–230 °C. The etch rate of InAs under this condition is 1–2 nm/min and that of GaAs is 100 times lower [10]. Hence, we expect that the wetting layer and the first (0.5–1) nm surface of InAs dots were removed. Subsequently, the sample was transferred to etching chamber B, where Cl_2 gas etching was carried out at 75 °C for 15 sec. In this step, we expect that unmasked part of GaAs was etched to the depth of 5–10 nm while InAs dots remain intact, since the etch rate of GaAs by HCl is 20–40 nm/min. The surface in this stage was studied by AES (Auger electron spectroscopy), which indicates InAs dot masks as shown in Fig. 3. The sample was transferred back to etching chamber A, where the surface was further etched by HCl gas at T_s of about 290 °C for 10 min to remove the InAs dot masks. Note that the etch rates are 5 nm/min for InAs and 0.1 nm/min for GaAs, respectively.

Figure 4(a) shows the AFM image of the resultant structures. Note that GaAs

958

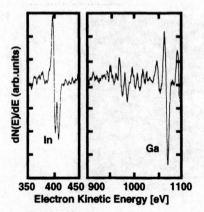

Figure 3. Kinetic energy spectrum of the sample surface after the GaAs etching without an InAs wetting layer by AES.

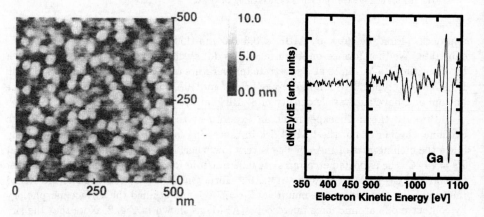

Figure 4. (a) AFM image of the 10-nm scale GaAs dots. (b) Kinetic energy spectrum of the sample surface by AES. Note that the InAs dot masks were etched off by HCl gas.

dots are successfully formed with the average height of 10 nm and the base diameter of 30 nm. The density of GaAs dots was 9×10^{10} cm^{-2}, unchanged from the initial InAs dots. The lateral size of GaAs dots is about 2 times larger than that of initial InAs dots. This is probably due to the surface orientation dependence of thermally-activated Cl$_2$ gas etching. We examined by AES the GaAs dot structure and found no In signal as shown in Fig. 4(b), confirming that InAs was completely removed. Thus the fabrication of nanometer scale GaAs dots free from strains are demonstrated.

In our approach, the prevention of contamination and the control of dot shapes are the two most important factors. To prevent the contamination in etching chamber the Cl$_2$ etching (done typically at T_s=75°C) and the HCl etching (done typically at 220oC) should be performed in separate chambers. This is because the substrate holder and other components in etching chambers may degrade and become sources of contamination when they are heated to high temperatures (> 200 °C) in the presence of residual Cl$_2$ gas. Secondly, we list a few points of importance to control dot shapes. One must avoid

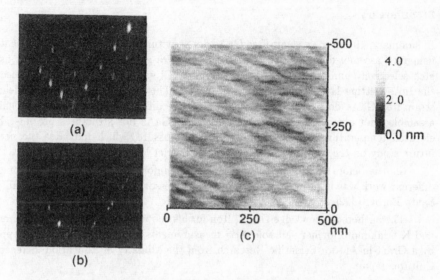

Figure 5. (a) RHEED pattern of the sample with GaAs dots at room temperature. (b) RHEED pattern when the sample was heated up to 580° C under As pressure. (c) AFM image of the sample corresponding to (b). The surface roughness is within 1 nm.

the excessive etching of GaAs by Cl_2 gas, since it results in the undercut which spoils the mask action of InAs dots. Indeed, when the Cl_2 etching was done at $Ts=77$ °C for 1 minute to etch the whole structure to the depth of 30 nm, the number of resultant GaAs dots decreased dramatically to 6×10^9 cm^{-2} and only very large GaAs dots remained. Hence it is quite important to optimize the etching time. In addition, in order to control the shape of resultant GaAs dots, one must clarify the etching process more in detail. In particular, the surface orientation dependence of etching needs to be studied and controlled. In this regard, the application of this method to GaAs surfaces other than (100) plane may be of some use.

In order to form quantum dots surrounded by barrier layers, an AlGaAs layer has to be overgrown on GaAs dots. In this process, we have found that the GaAs dot structure can be rather unstable at elevated temperatures as discussed below.

After the removal of InAs dots by HCl etching, the sample was transferred to the MBE chamber again. The reflection high-energy electron diffraction (RHEED) pattern immediately after the transfer was spotty (Fig. 5(a)), which means the existence of 3D islands on the surface. This spotty pattern remained when T_s was raised to 350-400 °C, where an As beam was turned on to suppress the As deficiency on the GaAs surface. As the sample was heated further, the RHEED pattern was found to change gradually to a streaky pattern and at 580 °C a completely streaky 2x4 pattern was observed (Fig. 5(b)). The AFM image of this sample is shown in Fig. 5(c). Indeed, the GaAs surface was flattened and dot structures disappeared. In another experiment, where As the RHEED pattern was found to become streaky even at $T_s=\sim100$ °C. Such an instability of the GaAs dot structure under As pressure may hinder the overgrowth of a good quality AlGaAs layer. We have attempted to solve this problem by using the low temperature MBE with the alternate beam supply (Migration enhanced epitaxy), on which we report elsewhere.

3. Summary

In summary, 10-nm scale strain-free GaAs dot structures were successfully fabricated by using self-assembled InAs dots as masks for *in-situ* gas etching. Cl_2 gas was used to etch selectively unmasked part of GaAs, while HCl gas was used to remove selectively the InAs wetting layer and the final dot masks. This method provides very clean and strain-free GaAs dots, which can eliminate some difficulties in the conventional self-assembled dot structures. We have found that GaAs dot structures are unstable at elevated temperature especially under the As pressure, which indicates the need for future study on the overgrowth of an AlGaAs barrier layer.

Shortly before submitting this paper, we were informed that a similar but somewhat different work was underway in the group of Professor E. Hu, University of California, Santa Barbara [12].

The authors acknowledge Dr. T. Itoh for his cooperation with AFM measurements and K. Makimoto for her valuable AES measurements. Part of this works is supported by a Grant-in-Aid for Scientific Research from the Ministry of Education, Science, and Culture, Japan.

References

[1] Guha S, Madhukar A, and Rajkumar K C 1990 *Appl. Phys. Lett.* **57**, 2110

[2] Snyder C W, Orr B G, Kessler D, and Sander L M 1991 *Phys. Rev. Lett.* **66** 3032

[3] Leonard D, Krishnamurthy M, Reaves C M, Denbaars S P, and Petroff P M 1993 *Appl. Phys. Lett.* **63** 3203; Leonard D, Pond K, and Petroff P M 1994 *Phys. Rev.* **B 50** 11687

[4] Moison J M, Houzay F, Barthe F, and Leprince L, Andre E and Vatel O 1994 *Appl. Phys. Lett.* **64** 196

[5] Oshinowo J, Nishioka M, Ishida S, and Arakawa Y 1994 *Appl. Phys. Lett.* **65** 1421

[6] Kadoya Y, Noge H, Kano H, Sakaki H, Ikoma N, and Nishiyama N 1992 *Appl. Phys. Lett.* **61** 1658; Kadoya Y, Noge H, Kano H, and Sakaki H 1993 *J. Cryst. Growth* **127** 877

[7] Kadoya Y, Noge H, and Sakaki H: unpublished.

[8] Kadoya Y, Yoshida T, Noge H, and Sakaki H *Nanostructures and Quantum Effects*, ed. Sakaki H and Noge H (Springer Verlag, Heidelberg, 1994), p. 189

[9] Miya S, Yoshida T, Kadoya Y, Akamatsu B, Noge H, Kano H, and Sakaki H 1993 *Appl. Phys. Lett.* **63** 1789

[10] Kadoya Y, Yoshida T, Someya T, Akiyama H, Noge H, and Sakaki H 1993 *Jpn. J. Appl. Phys.* **32** L1496

[11] N. Carlsson, W. Seifert, A. Petersson, P. Castrillo, M. E. Pistol, and L. Samuelson 1994 *Appl. Phys. Lett.* **65** 3093

[12] Strand T A, Mui D S L, Thibeault B J, Coldren L A, Petroff P M, and Hu E L 1995 private communication: presented at QUEST Workshop on Physics of Nanostructure, March 27–28, Santa Barbara

Inst. Phys. Conf. Ser. No 145: Chapter 7
Paper presented at 22nd Int. Symp. Compound Semiconductors, Cheju Island, Korea, 28 August–2 September 1995
© 1996 IOP Publishing Ltd

The electronic structure of coupled quantum dots in magnetic fields

J. H. Oh†, K. J. Chang†, G. Ihm‡, and S. J. Lee§

† Department of Physics, Korea Advanced Institute of Science and Technology, Taejon 305-338, Korea

‡ Department of Physics, Chungnam National University, Taejon 305-764, Korea

§ Department of Physics, Korea Military Academy, Seoul 139-799, Korea

Abstract. We calculate the electronic structure of quantum dots coupled along the growth direction with one or two electrons in magnetic fields and examine the spin transitions and the electronic charge distributions of the ground states. Because of the dot-dot and electron-electron interactions, the coupled-quantum dots exhibit rich electronic structures. We suggest that the effects of these interactions on the energy spectra can be resolved through optical measurements because the transition energies between the energy levels increase for a vertically polarized light as the magnetic field increases.

1. Introduction

If the dimension of a quantum dot is comparable to the effective Bohr radius of a host semiconductor, the quantum dot referring to as an artificial atom gives rise to discrete energy levels, where the number of electrons and confinement potential are controlled artificially. Usually, quantum dots are considered to be two-dimensional or disk-like systems because the lateral sizes are much larger than the extents in the growth direction. Then, the electronic energy levels of the disk-like quantum dot are mainly determined by the lateral motion.[1, 2] A coupled-quantum dot which could be considered as an artificial molecule has attracted much attention, recently.[3, 4, 5] In contrast to a single disk-like quantum dot, a vertically coupled-quantum dot must consider the other degree-of-freedom along the growth direction. Thus, the vertical confinement as well as the lateral electric and magnetic confinements are important for the electronic structure. Several theoretical[3, 4] and experimental[5] studies were performed for a coupled quantum dot, however, the relation of the electronic structure to the dot-dot and electron-electron interactions is still not understood.

In this work we study the electronic structure of a vertically coupled-quantum dot with one or two electrons in magnetic fields and calculate the oscillator strengths for optical transitions. We find that the dot-dot and electron-electron interactions strongly affect the ground states of coupled-quantum dots, which can be measured by optical experiments. For a polarized light along the growth direction, we find one resonance frequency, which exhibits a blue-shift with the increasing of magnetic field.

2. Results and discussions

A coupled-quantum dot is characterized by a parabolic potential with the confinement frequency ω_0 on the lateral plane and by $V(z)$ along the growth direction, where $V(z)$ consists of square wells with an equal width of $w_w = 150$ Å, barriers with a width of $w_b = 50$ Å, and two buffer layers with a thickness of 350 Å. We use the barrier height of 147 meV and the effective mass $m^* = 0.0665m_0$. For an external magnetic field $\vec{B} = B\hat{z}$, the Hamiltonian for the quantum dot with a single electron is written in cylindrical coordinates;

$$H^0 = \frac{1}{2m^*}(\vec{p} + e\vec{A})^2 + \frac{1}{2}m^*\omega_0^2\rho^2 + V(z) \tag{1}$$

where $\vec{A} = \vec{B} \times \vec{\rho}/2$ is the vector potential and $\omega_c = eB/m^*$ is the cyclotron frequency. Since the vertical and lateral motions are decoupled, the eigenvalues and eigenfunctions of the single particle Schrödinger equation $H^0\psi^0(\rho,\varphi,z) = E^0\psi^0(\rho,\varphi,z)$ are easily calculated,

$$E^0_{N,L,k} = E^R_{N,L} + \varepsilon^z_k = \hbar\omega(N + \frac{|L|}{2}) + \frac{L}{2}\hbar\omega_c + \varepsilon^z_k, \tag{2}$$

$$\psi^0(\rho,\varphi,z) = \chi_{N,L}(\rho,\varphi)\phi_k(z),$$

$$\chi_{N,L}(\rho,\varphi) = \frac{1}{\sqrt{2\pi}\lambda}\sqrt{\frac{N!}{(N+|L|)!}}e^{iL\varphi}e^{-\xi/2}\xi^{|L|/2}L_N^{|L|}(\xi) \tag{3}$$

where $\phi_k(z)$ and $\chi_{N,L}$ denote the solutions for the vertical and lateral motions, respectively, $\lambda = \hbar/m^*\omega$, $\xi = \rho^2/\lambda$, and L_N^L is a Laguerre polynomial. Here, the frequency $\omega = \sqrt{\omega_c^2 + 4\omega_0^2}$ indicates a measure of hybrid effects between the magnetic and electric confinements. It is noted that the energy levels $E^R_{N,L}$ for the lateral motion depend on both the radial ($N=0, 1, 2, \ldots$) and azimuthal ($L = 0, \pm1, \pm2, \ldots$) quantum numbers because our system is symmetric about the z-axis. The effect of the dot-dot interaction appears only in the energy levels (ε^z_k) associated with the z-motion. As the separation between quantum dots decreases, the energy of the single-quantum dot, ε^0, is splitted into two states, i.e., the bonding ($\varepsilon^z_a = \varepsilon^0 - v$) and antibonding ($\varepsilon^z_b = \varepsilon^0 + v$) states with a coupling strength v. For dipole transitions which are associated with far-infrared absorptions, two resonance frequencies, $\omega_+ = (\omega+\omega_c)/2$ and $\omega_- = (\omega-\omega_c)/2$, are found for a laterally polarized light due to parabolic lateral confinement.[2, 6] For the z-polarized light, the dipole transitions are allowed between the bonding and antibonding states, i.e., between the ε^z_a and ε^z_b states, however, the transition energy is independent of magnetic field.

For two electrons confined by the same potentials as those used for the single electron case, the Hamiltonian of two electrons is written as,

$$H = H^0(1) + H^0(2) + \frac{e^2}{4\pi\epsilon_0\epsilon\,|\vec{r_1} - \vec{r_2}|} + \frac{g^*\mu_B}{\hbar}\vec{B} \cdot \vec{S} \tag{4}$$

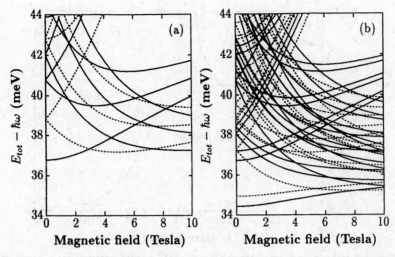

Figure 1. The total energies of two electrons are drawn as a function of magnetic field for the (a) single- and (b) coupled-quantum dots with $\hbar\omega_0 = 4$ meV. The spin singlet and triplet states are described by solid and dotted lines, respectively.

where $H^0(i)$ is the Hamiltonian of a single particle in Eq. (1), g^* is the effective g-factor, μ_B is the Bohr magneton, and ϵ is the dielectric constant of GaAs. Using the center-of-mass and relative coordinates for the lateral motion, the Hamiltonian becomes

$$H = H_R + H_r + \frac{g^*\mu_B}{\hbar}\vec{B} \cdot \vec{S} \tag{5}$$

where H_R and H_r represent the center-of-mass motion and relative Hamiltonians, respectively, on the lateral plane. Then, the eigenstates and eigenfunctions of H_R are $E_{N,L}^R$ and $\chi_{N,L}$, respectively. For H_r, we calculate numerically the eigenstates $E_{n,l}$, with the eigenstates $R_{n,l}(\eta, z_1, z_2)$ expanded in terms of $\chi_{m,l}$ and ϕ_i in Eq. (3) for a given angular momentum l, where η is the radial coordinate of the relative motion on the lateral plane.

For a single-quantum dot, the energy levels $E_{n,l}$ are found to be very similar to those of the disk-like quantum dot.[2, 7] In this case, since the energy difference between the first and second lowest states for the z-motion is larger than that for the radial motion, lower levels $E_{n,l}$ depend on only the radial motion. Thus, the symmetry for particle permutations is determined mainly by angular momentum, i.e., the wavefunctions with even (odd) quantum number l have a singlet (triplet) spin state. For a coupled-quantum dot, however, since the energy difference between the lowest two states for the z-motion is comparable to that for the radial motion, the symmetry of the wavefunctions for H_r depends on both the z- and lateral motions. If electron-electron interactions are excluded, the lowest energy level for a given orbital angular momentum is described by $| b, b >$, that is, two electrons occupy the bonding state of the z-motion. The second lowest levels ($| a, b >$ and $| b, a >$) are degenerate, while the third level is denoted as $| a, a >$. When electron-electron interactions turn on, the energy levels are shifted to higher energies and the degenerated second levels split into the symmetric and antisymmetric states for particle permutations, $(| a, b > + | b, a >)/2$ and $(| a, b > - | b, a >)/2$. Because of the Pauli exclusion principle, the $(| a, b > - | b, a >)/2$ state is regarded as a spin triplet

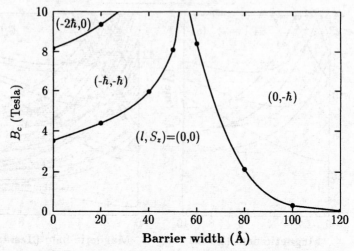

Figure 2. The ground-state phase boundary of the coupled-quantum dot is drawn as a function of barrier width for $w_w = 150$ Å and $g^* = 0.44$.

state, while the other state as a spin singlet for even angular momentums. For odd angular momentums, the $(\mid a, b > - \mid b, a >)/2$ becomes a spin singlet state.

The magnetic field dependences of the total energies, $E_{tot} = E_{n,l} + E_{N,L}^R$ are plotted for single- and coupled-quantum dots in Fig. 2. For the single-quantum dot, the ground state exhibits the orbital angular momentum change at a certain magnetic field B_c, similar to the disk-like quantum dot.[1, 2, 7] These ground state transitions are understood in terms of the Coulomb and kinetic energies. As the magnetic field increases, both the kinetic and Coulomb energies for low angular momentum states increase more rapidly than for higher angular momentum states. Thus, the angular momentums of the ground state follow the sequence of $l = 0, -1, -2, \ldots$ as the magnetic field increases. However, compared with the disk-like quantum dot, the magnetic field B_c at which the ground state transition occurs is larger because the finite size of the vertical motion reduces the Coulomb energy.[7] For the coupled-quantum dot, the ground-state transitions exhibit somewhat different behavior. For barrier widths $w_b < 55$ Å, the transition of the ground state occurs in the same way as the single-quantum dot. However, for large barrier widths ($w_b \geq 60$ Å), a different mechanism for the ground state transition is found, where the radial, vertical, and spin wave functions of the ground state change abruptly while the orbital angular momentum remains the same. Since the second minimum state is a triplet state with $S_z = -\hbar$ for the coupled-quantum dot, its energy is lowered by $g^* \mu_B B$ as the magnetic field increases, finally, being the ground state. Thus, the transition field B_c is strongly influenced by the value of g^* and the energy difference between the first and second states which reflect the dot-dot and electron-electron interactions. If the barrier width increases to reduce the dot-dot interaction, the energy difference between the first and second lowest states for $g^* = 0$ becomes smaller, giving a smaller value of B_c. For $g^* = 0.44$, the ground state transition fields are plotted as a function of barrier width in Fig. 2. As the barrier width increases, two electrons are less correlated and easily spin-polarized, $i.e.$, $S_z = -\hbar$, for low magnetic fields.

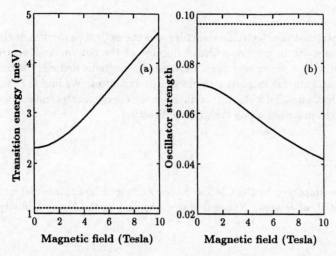

Figure 3. The transition energies and oscillator strengths for the coupled-quantum dot are plotted as a function of magnetic field with (solid line) and without (dotted line) electron-electron interactions.

The optical response in the coupled-quantum dot with two electrons is different from that of the one electron case for the vertical polarization because of the electron-electron interaction. From the generalized Kohn theorem, dipole radiations of a laterally polarized light can only probe the center-of-mass motion in a strictly parabolic potential, thus, it is inadequate to see any effect due to electron-electron interactions. However, for the vertically polarized light, it may be possible to observe the effect of electron-electron interactions on the electronic structure. In Fig. 2, we show the calculated oscillator strength and the resonance frequencies for the coupled-quantum dot. As the magnetic field increases, the resonance energy is blue-shifted while independently of the magnetic field this shift can not be seen in the absence of the Coulomb interaction. Since this resonance energy is attributed to the transition between the $| b, b >$ and $(| a, b > + | b, a >)/2$ states, it strongly depends on the dot-dot and electron-electron interactions.

The effects of the dot-dot and electron-electron interactions can also be found in the electronic charge distribution. For a symmetrically coupled-quantum dot with equal widths, electrons are equally distributed in the dots. In an asymmetrically coupled-quantum dot with different widths, electrons are initially localized in a wider well at zero magnetic field because the Coulomb energy between two electrons in the wider well is smaller than the kinetic energy of these electrons in the narrower and wider wells. As the magnetic field increases, since the reduction of the lateral size increases the Coulomb energy, electrons tunnel into the narrower well and are eventually equally distributed over the wells. A similar distribution of electrons can be also generated by applying an electric field.[4]

966

3. Summary

We have investigated the electronic structure and the optical properties of the vertically coupled-quantum dot in magnetic fields. Because of the dot-dot and electron-electron interactions as well as the hybrid effect between the magnetic and electric confinements, the coupled-quantum dot exhibits rich electronic structures. We find that the effects of the electron-electron and dot-dot interactions on the energy levels can be detectable with far-infrared light polarized along the growth direction.

Acknowledgments

This work was supported by the CMS at Korea Advanced Institute of Science and Technology, the SPRC of Jeonbuk National University, and by the Korea Ministry of Education.

References

[1] Maksym P A and Chakraborty T 1990 *Phys. Rev. Lett.* **65** 108–11

[2] Merkt U, Huser J and Wagner M 1991 *Phys. Rev.* B **43** 7320–3

[3] Chakraborty T, Halonen V and Pietilainen P 1991 *Phys. Rev.* B **43** 14289–92
    Fong C Y, Zhong H and Klein B M 1994 *Phys. Rev.* B **49** 7466–73
    Palacios J J and Hawrylak P 1995 *Phys. Rev.* B **51** 1769–77

[4] Bryant G 1993 *Phys. Rev.* B **48** 8024–34

[5] Tewordt M, Hughes R J F, Martin-Moreno L, Nicholls J T, Asahi H, Kelly M J, Law V J, Ritchie D A, Frost J E F, Jones G A C and Pepper M 1994 *Phys. Rev.* B **49** 8071–5

[6] Liu C T, Nakamura K, Tsui D C, Ismail K, Antoniadis D A and Smith H I 1989 *Appl. Phys. Lett.* **55** 168–70

[7] Oh J H, Chang K J, Ihm G and Lee S J 1994 *Phys. Rev.* B **50** 15397–400

Inst. Phys. Conf. Ser. No 145: Chapter 8
Paper presented at 22nd Int. Symp. Compound Semiconductors, Cheju Island, Korea, 28 August–2 September 1995

Prospects of surface emitting lasers

K. Iga

Tokyo Institute of Technology
4259 Nagatsuta, Midoriku, Yokohama 226
Japan

Abstract. In this paper we review the progress of vertical cavity surface emitting lasers (vcsels) by featuring the materials, process technologies, performances, and so on. Advanced technology for aiming at spontaneous emission control, photon recycling, polarization control, wavelength tuning, integration etc. will be introduced. Also, we discuss on integration issues and applications.

1. Introduction

Large scale networks and computing are now introducing the optical technology, i. e., a new technical area such as optical computing, optical interconnects, and parallel lightwave systems, is being actually opened up[1]. On the other hand, the recent progress of surface emitting (SE) lasers or vertical cavity surface emitting lasers (vcsels) is very fast and it will be a good time to consider, in particular, the technology of parallel optoelectronics including optical interconnects.

We review the performance of SE lasers, i.e., GaInAsP, InGaAs, GaAlAs, and AlGaInP, ZnSe, and GaN. Lastly, we consider some possible applications forwarding to optical interconnects, parallel optical fiber network systems, by featuring ultra-low threshold, high power capability and largely extending 2-D arrays, and so on.

2. Materials, device technology and performances

2.1 GaInAsP/InP SE lasers

The importance of 1.3 or 1.55 μm devices is currently increasing, since parallel lightwave systems are really needed to meet rapid increase of information transmission. However, the GaInAsP/InP system has some substantial difficulties for making SE lasers due to such reasons that the Auger recombination and inter-valence band absorption (IVBA) are noticeable, the index difference between GaInAsP and InP is relatively small, valence band offset is large, and so on. As already introduced in ref.[1], we meet a consistent progress in the reduction of threshold. Pulsed operation was obtained at near room temperature, at room temperature , and at 66 °C.

Recently , a hybrid mirror technology is being developed. One is to use a semiconductor/dielectric reflector, which is demonstrated by chemical beam epitaxy (CBE) as shown in Fig. 1[2]. The other is epitaxial bonding of quaternary/GaAs-AlAs mirror, where 144 °C pulsed operation was achieved by optical pumping. Epitaxially bonded mirror made of GaAs/AlAs was introduced into surface emitting lasers operating at 1.3 μm providing 3 mA of room temperature pulsed threshold. Very recently, record data including the cw threshold of 2.3mA as well as the maximum cw operating temperature up to 32 ℃ have been reported for 1.5μm SE lasers with epitaxially bonded mirrors [3].

Thermal problems for CW operation are now extensively studied. A MgO/Si mirror with

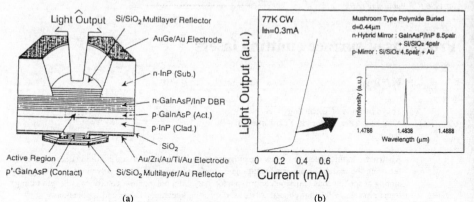

(a)   (b)

Fig. 1   A CBE grown GaInAsP/InP 1.55 μm SE laser using a hybrid mirror.
(a) Structure (b) Its I-L characteristic.

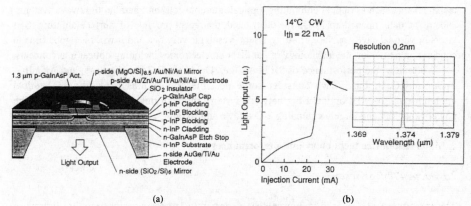

(a)   (b)

Fig. 2   A buried heterostructure GaInAsP/InP 1.3 μm SE laser.
(a) Structure   (b) CW I-L characteristic

good thermal conductivity is demonstrated. For realizing a reliable devices, the buried heterostructure (BH) is crucial. We have fabricated a BH SE laser exhibiting a relatively low threshold at room temperature pulsed operation.

By improving the heat sinking using a diamond submount and highly reflective mirror, we have achieved room temperature CW operation. We show its device structure in Fig. 2 with a typical current vs light power output (I-L) characteristic. The minimum threshold obtained was 22 mA at 14 °C for 12 μm diameter device size[4].

2.2 GaAlAs/GaAs VCSELs

In the GaAlAs/GaAs system, a VCSEL of 5μm long and 6μm in diameter and room temperature CW operation was first realized among other systems. At present, devices exhibiting $I_{th} \cong 2 \sim$

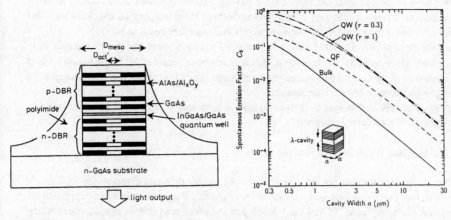

Fig. 3 An oxide-defined GaInAs/GaAs SE laser with 70 μA threshold current.

Fig. 4 Spontaneous emission coefficients for micro-cavity lasers with various active media.

15 mA and 10 mW of output power are available in the laboratory level. A very high coupling efficiency to a single mode fiber ($\cong$ 90 %) was reported. The spectral linewidth of 50 MHz is obtained with an output power of 1.4 mW.

2.3 GaInAs/GaAs VCSELs

The GaInAs/GaAs strained pseudo-morphic system grown on a GaAs substrate emitting 0.98 μm exhibits a high laser gain and has been introduced into surface emitting lasers together with using GaAlAs/AlAs multi-layer reflectors. A low threshold (= 1 mA at CW) has been demonstrated. Recently, the threshold current of vertical cavity surface emitting lasers have been reduced down to sub-milliampere orders in various institutions in the world. The record thresholds reported so far is 0.7 mA, 0.65 mA, $\cong$ 0.2mA. The material which exhibits the best performance for vcsel is pseudo-morphic GaInAs/GaAs system emitting 0.98 micron of wavelength. The minimum reported threshold is 91 micro-amperes at room temperature cw operation. The theoretical expectation was less than 10 micro-ampere or less, if some good current and optical confinement structure can be introduced.

Recently, we have developed a novel laser structure employing selective oxidizing process applied to AlAs which is one of the members of multi-layer Bragg reflector. The schematic structure of the device now developed is shown in Fig. 3. The active region is three quantum wells consisting of 80 Angstrom GaInAs strained layers. The Bragg reflector consists of GaAs/AlAs quarter lamda stacks of 24.5 pairs. After etching the epitaxial layers including active layer and two Bragg reflectors, the sample was treated in the high temperature oven with water vapor which is bubbled by nitrogen gas. The AlAs layers are oxidized preferentially with this process and native oxide of Aluminum is formed at the periphery of etched mesas. The typical size is 20 micron core starting from 30 micron mesa diameter. We have achieved about 1 mW of power output and submicro-ampere threshold. The nominal lasing wavelength is 0.98 microns.

We have made a smaller diameter device having 5 microns started from 20 micron mesa. The minimum threshold achieved is 70 micro-amperes at room temperature cw operation. This value is the world record at this moment for vcsels and any other lasers as well[5].

As the theoretically predicted threshold is about 1 micro-Ampere and this record may soon be cleared. More recently, the threshold of 8.5μA has been reported[6]. But it would be said that surface emitting lasers are now meeting a new stage exhibiting ultra-low power consumption and real application to optical interconnects.

Also, relatively high power as high as 50 mW is becoming possible. The power conversion efficiency 50 % is recently reported[7].

2.4 GaInAlP and visible surface emitting lasers

Visible surface emitting lasers are extremely important for disk and display applications, in particular, red, green and blue surface emitters may provide much wider technical applications, if realized. GaInAlP/GaAs VCSELs are developed and room temperature operation exhibiting submiliampere threshold has been obtained. Blue and green lasers are much more difficult than any other materials for considering SE lasers. Some design consideration and fundamental process technology for ZnSe and GaN systems are now being attempted in the author's laboratory.

3. Ultimate characteristics

By overcoming technical problems to make tiny structures and to improve the thermal resistance, we believe that we can obtain a 1 μA threshold device. A lot of efforts toward improving the characteristics of surface emitting lasers have been made, including surface passivation in the regrowth process for buried heterostructure, micro-fabrication, and fine epitaxies. The ohmic resistance of semiconductor DBR's is reduced down to the order of low 10^{-5} Ωcm^2 .

Spontaneous emission control is considered by taking the advantage of micro-cavity structures. The spontaneous emission factor has been estimated on the basis of 3-D mode density analysis[8]. The result is shown in Fig. 4. The possibility of no distinct threshold devices is suggested.

One another interesting topic for micro-cavity SE lasers is photon recycling. By covering the side-bounding surfaces of the cavity with a highly reflective materials, some amount of spontaneously wasted photons can be recycled. It has been demonstrated that the SE laser device appears to have no distinct threshold. The efficiency of photon recycling has been estimated. The quantum noise characteristics are being studied, such as relative intensity noise (RIN) and linewidths.

4. Photonic integration based on VCSELs

A wide variety of functions, such as frequency tuning, amplification, and filtering can be integrated along with surface emitting lasers by stacking. Moreover, a 2-D parallel optical logic system can deal with a large amount of image information with high speed. To this demand, a surface emitting (SE) laser will be a key device. Optical neural chips have been investigated for the purpose of making optical neuro-computers and VSTEP integrated device.

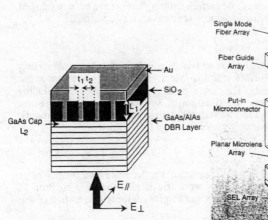

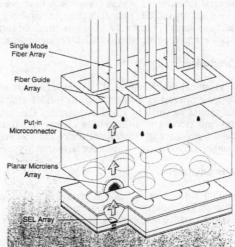

Fig. 5 Polarization control of surface emitting lasers using a resonant grating terminator.

Fig. 6 A 2-D module with VCSELs and pu t-in micro-connectors using planar micronizes.

High power capabilities from SE lasers is very interesting by featuring largely extending 2-D arrays. For the purpose of realizing coherent arrays, coherent coupling of these arrayed lasers has been tried by using a Talbot cavity and phase compensation is considered. It is pointed out that 2-D arrays are more suitable to make a coherent array than a linear configuration, since we can take the advantage of 2-D symmetry. The research activity is now forwarded to monolithic integration of SE lasers taking the advantage of small cavity dimensions. A densely packed array has also been demonstrated for the purpose of making high power lasers and coherent arrays.

Into SE lasers, surface operating photonic elements using quantum wells such as an optical switch, frequency tuner, optical filter, and super-lattices are now tried to be integrated. In Fig. 4 we show a 40 Å continuous tuning by the use of an external reflector. Wide variety of functions, such as polarization control, amplification, detecting, and so on can be integrated along with surface emitting lasers by stacking. The polarization control will become very important for SE lasers. One of the methods is shown in Fig. 5, where a grating terminator to a DBR is incorporated.

5. Subsystem applications

Lastly, we consider some possible applications including optical interconnects, parallel fiber-optic subsystems, and so on. In optical fiber communication systems, ultra-fine semiconductor dealing with a scale of 0.1Å or less. On the other hand, longer than 10,000 km of fibers are considered to be directly connected without any electrical repeaters. It would be said that the optoelectronics can treat the dimensions of 10^{11} to 10^7 m. Another important advance is a parallel lightwave systems including more than 4000 fibers, for example, broad band ISDN systems.

By taking the advantage of wide band and small volume transmission capability, the optical interconnect is considered to be inevitable in the computer technology. Some parallel interconnect scheme is wanted and new concepts is being researched as shown in Fig. 6. Vertical optical interconnect of LSI chips and circuit boards may be another interesting issue. In any way, the two-dimensional arrayed configuration of surface emitting lasers and planar optics will open up a new era of optoelectronics. A new architecture for 64 channel interconnect has been proposed and a modeling experiment was performed using GaAlAs VCSEL arrays.

972

Several schemes for optical computings have been considered, but one of the bottle necks may be a lack of suitable optical devices, in particular, 2-D surface emitting lasers and surface operating switches. Fortunately, very low threshold surface emitting lasers have been developed, and stack integration together with 2-D photonic devices are now actually considered.

6. Conclusion

Vertical optical interconnects of LSI chips and circuit boards and multiple fiber systems may be the most interesting fields related to surface emitting lasers. From this point of view, the device should be small as small as possible. The future process technology for *it* including epitaxy and etching will drastically change the situation of Surface emitting lasers. Some optical technologies are already introduced in various subsystems, but the arrayed microoptic technology would be very helpful for advanced systems.

Acknowledgments

The author would like to thank Emeritus Prof. Y. Suematsu, the former President of Tokyo Institute of Technology, for encouragement of this work. He is also thanking to Prof. F. Koyama and the lab members for performing the study and helpful advice to preparation of this paper.

References

[1] Iga K., Koyama F., and Kinoshita S., "Surface emitting semiconductor lasers", IEEE J. Quant. Electron., vol. **QE-24**, 9, p. 1845, Sept. 1988.

[2] Miyamoto T., Uchida T., Yokouchi N., Inaba Y., Koyama F., and Iga K., "A study on gain-resonance matching of CBE grown l=1.5 μm surface emitting lasers", LEOS Annual, DLTA13.2, Nov. 1992.

[3] Babic D.I., Streubel K., Mirin R.P., Margalit N.M., Bowers J.E. , Hu E.L., Mars D.E., Yang L. and Carey K., "Room temperature continuous-wave operation of 1.54μm vertical-cavity lasers", IOOC'95, Post Deadline Papers, PD1-5, Hong Kong, 1995.

[4] Baba T. , Yogo Y., Suzuki K., Koyama F., and Iga K., "Near room temperature continuous wave lasing characteristics of GaInAsP/InP surface emitting laser", Electron. Lett., vol. 29, no. 10, pp.913-914, May 1993.

[5] Hayashi Y., Mukaihara T., Hatori N., Ohnoki N. , Matsutani A. , Koyama F. and Iga K., "Record low-threshold index-guided InGaAs/GaAlAs vertical cavity surface emitting laser with a native oxide confinement structure", Electron. Lett., vol.31, no.7, pp.560-561, March 1995.

[6] Yang G.M., Macdougal, M.H., Dapkus, P.D., "Ultralow threshold VCSELs fabricated by selective oxidation from all epitaxial structure", CLEO'95, Postdeadline Papers, CPD4-1, Baltimore, 1995.

[7] Lear K. L., Schneider Jr. R. P., Choquette K. D., Kilcoyne S. P., and Geib K.M., "Selectively oxidized vertical cavity surface emitting lasers with 50% power conversion efficiency", Electron. Lett., vol. **31**, pp.208-209, 1995.

[8] Baba T., Hamano T., Koyama F., Iga K.:"Spontaneous emission factor of a microcavity DBR surface-emitting laser", IEEE J. of Quant. Electron., **QE-27**, 6, pp.1347-1358, June 1991.

Inst. Phys. Conf. Ser. No 145: Chapter 8
Paper presented at 22nd Int. Symp. Compound Semiconductors, Cheju Island, Korea, 28 August–2 September 1995
© *1996 IOP Publishing Ltd*

Visible (630-650 nm) Vertical Cavity Surface Emitting Lasers With Al-Oxide/AlGaInP/AlGaAs Distributed Bragg Reflectors

J A Lott

Asian Office of Aerospace Research and Development
Air Force Office of Scientific Research, Roppongi, Minato-ku, Tokyo 106, Japan

L V Buydens

University of Gent, Belgium

K J Malloy

The University of New Mexico, Albuquerque, New Mexico, U.S.A. 87131

Ke Kobayashi, S Ishikawa

NEC Opto-Electronics Research Laboratories, Tsukuba-shi, Ibaraki-ken 305, Japan

Abstract. A study of red AlGaInP vertical cavity surface emitting lasers (VCSELs) with Al-oxide current apertures and Al-Oxide/AlGaInP/AlGaAs distributed Bragg reflectors is reported. Continuous wave (CW) operation at room temperature is obtained for emission wavelengths from 630 to 650 nm. Prototype devices with about 10 μm per side, square apertures have CW threshold currents ranging from 4 mA to 1.5 mA, and peak output powers ranging from 0.1 to 2 mW.

1. Introduction

Vertical cavity surface emitting lasers (VCSELs) consist of an optical cavity active region surrounded by upper and lower quarter-wave stack distributed Bragg reflectors (DBRs). The DBRs are composed of conducting semiconductor layers, dielectric layers, or a combination of the two. In conventional VCSELs, current apertures are formed by ion implantation or by pillar etching. The pillar devices typically emit through a nonabsorbing substrate. A new VCSEL design uses a native oxide $Al_xGa_{1-x}As$ DBR layer (x ~1.0) just above the optical cavity to block current, to define the emission aperture, and to provide lateral index guiding [1-3]. Such native Al-oxide layers were originally introduced by Holonyak *et al* [4-5]. Their use in infrared (IR) VCSELs has led to record threshold currents below 100 μA [6-7] and to power conversion efficiencies over 50% [8]. Native Al-oxide layers have also been used as the low index layers in DBRs, in conjunction with higher index $Al_xGa_{1-x}As$ or $(Al_yGa_{1-y})_{0.5}In_{0.5}P$ semiconductor layers [6, 9-11]. Compared to all-semiconductor DBRs, native oxide DBRs have increased reflectance bandwidth and require far fewer DBR periods for a given peak reflectance. Native oxide DBRs and native oxide current apertures are promising for use in AlGaInP visible VCSELs since the required number of $AlAs/Al_xGa_{1-x}As$ DBR periods increases as the emission wavelength is reduced [12]. Recent studies of red (650-690 nm) VCSELs with Al-oxide apertures have resulted in performance rivaling that of IR VCSELs circa 1991. This paper reports the first application of both native Al-oxide current apertures and native Al-oxide DBRs to red AlGaInP VCSELs emitting below 650 nm.

2. Experiment

The devices are prepared by metalorganic vapor phase epitaxy on (100) (n+)GaAs substrates, misoriented $10°$ toward the (111)A. The lower (n+) all-semiconductor DBR, with 60.5 periods, is composed of Si-doped $AlAs/Al_{0.6}Ga_{0.4}As$ quarter-wave layers with graded interfaces over thicknesses of about 12 nm. The one-lambda AlGaInP active region contains four, ~6 nm-thick, compressively strained $Ga_yIn_{1-y}P$ quantum wells surrounded by unintentionally-doped $(Al_{0.6}Ga_{0.4})_{0.5}In_{0.5}P$ barrier layers [13]. The upper (p+) DBR is composed of a 10.5 period $Al_{0.9}Ga_{0.1}As/Al_{0.6}Ga_{0.4}As$ quarter-wave stack, also with graded interfaces, followed by a quarter-wave (p+)GaAs current spreading/ohmic contacting layer. This top doped DBR is followed by an additional 4 period DBR composed of undoped $AlAs/(Al_{0.2}Ga_{0.8})_{0.5}In_{0.5}P$. The first low index quarter-wave layer next to the AlGaInP optical cavity active region is made of AlAs. After growth, a square mesa about 12 μm on a side is etched down to the (p+)GaAs quarter-wave layer using plasma etching. A second square mesa, about 20 μm on a side and centered on the previous mesa, is etched down to the AlGaInP active region. The AlAs layers in the top DBR are completely oxidized using a recently reported technique [2,3,11]. Also partially oxidized is the AlAs quarter-wave layer directly above the AlGaInP active region, forming a confined current path without the need for proton implantation. A square Ti/Au p-type ring contact is placed on the exposed (p+)GaAs mesa, followed by a full back-surface n-type metallization using Ge/Au/Ni/Au. A schematic diagram of a completed device is shown in Fig. 1. A plot of the calculated refractive index profile and the electric field intensity for an example device is shown in Fig. 2.

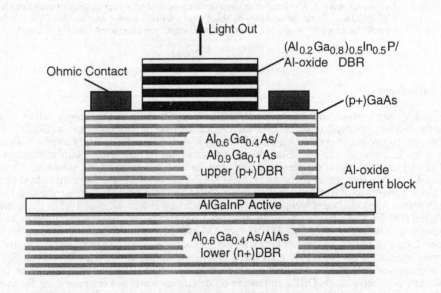

Fig. 1. Schematic diagram of the red VCSEL.

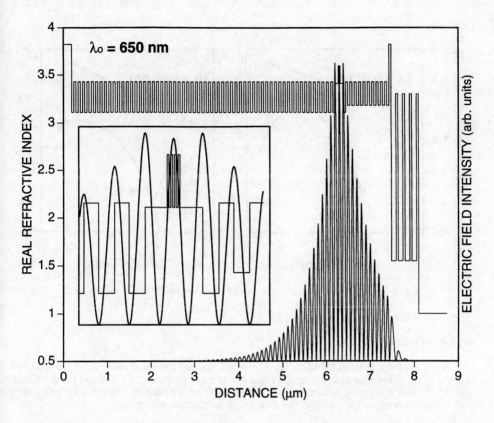

Fig. 2. Plot of the calculated refractive index profile and electric field intensity versus distance for the red VCSEL designed for emission at 650 nm. The DBR interface grading has been neglected for simplicity.

3. Results

Continuous wave (CW) operation at room temperature (20°C) is obtained for emission wavelengths from 630 to 650 nm. The prototype devices have ~10 μm per side, approximately square emitting apertures. Figure 3 shows the L-I-V (output power and voltage vs. current) curves for devices emitting at 630 and 650 nm. The measurements were taken on devices from two separate wafers. Each wafer was unrotated during growth to obtain a range of emission wavelengths. For emission at 630 nm, the CW threshold current is 4 mA, the threshold voltage is 2.5 V, and the peak output power is 0.1 mW (magnified 20 times in Fig. 3.). At 650 nm, the CW threshold current is 1.5 mA, the threshold voltage is 2.2 V, and the peak output power is 2 mW. For the present design, note that the output power is reduced by 10-15% due to absorption in the (p+)GaAs contacting layer. The maximum power conversion efficiency at 650 nm is over 10%, while it drops below 1% at 630 nm. The 650 nm devices operate CW at up to 25°C, while the 630 nm devices are limited to 20°C.

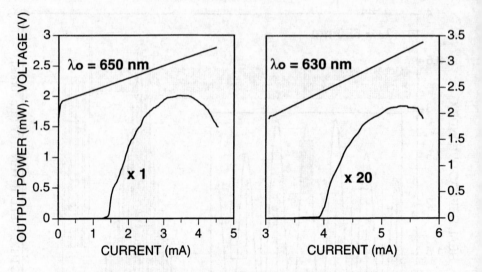

Fig. 3. The L-I-V characteristics of red VCSELs emitting at 650 and 630 nm at 20°C.

4. Conclusion

This work demonstrates the use of native Al-oxide DBR and current aperture layers in AlGaInP red VCSELs. This approach is currently one of the most promising for 630 nm band VCSEL arrays. Further studies of device geometry (e.g. intracavity contacts, wafer fusing techniques with substrate removal), reliability, and processing techniques are required.

References

[1]   Huffaker D L *et al*   1994 *Appl. Phys. Lett.* **65** 1 97-99

[2]   Ries M J *et al*   1994 *Appl. Phys. Lett.* **65** 6 740-742

[3]   Choquette K D *et al.* 1994 *Electr. Lett.* **30** 24 2043-2044

[4]   Dallesasse J M *et al*   1990 *Appl. Phys. Lett.* **57** 26 2844-2846

[5]   Sugg A R *et al*   1990 *Appl. Phys. Lett.* **58** 11 1199-1201

[6]   Hayashi *et al*   1995 *Electr. Lett.* **31** 7 560-562

[7]   Yang G M *et al*   1995 *Electr. Lett.* **31** 11 886-888

[8]   Lear K L *et al*   1995 *Electr. Lett.* **31** 3 208-209

[9]   MacDougal M H *et al*   1994 *Electr. Lett.* **30** 14 1147-1149

[10]   MacDougal M H *et al*   1995 *IEEE Phot. Tech. Lett.* **7** 3 229-231

[11]   MacDougal M H *et al*   1995 *IEEE Phot. Tech. Lett.* **7** 4 385-387

[12]   Lott J A *et al*   1993 *Electr. Lett.* **29** 19 1693-1694

[13]   Bour D P *et al*   1994 *IEEE J. Quant. Electr.* **30** 2 593

Inst. Phys. Conf. Ser. No 145: Chapter 8

Paper presented at 22nd Int. Symp. Compound Semiconductors, Cheju Island, Korea, 28 August–2 September 1995

A Single Transverse Mode Operation of Gain -Guided Top Surface Emitting Laser Diodes with a Very Low Series Resistance.

T. Kim, T. K. Kim, E. K. Lee, J. Y. Kim, and T. I. Kim

Materials & Devices Research Center, Samsung Advanced Institute of Technology, P.O. Box 111, Suwon, Korea

Abstracts. In the 780 nm near infrared spectral region, a fundamental transverse mode vertical cavity top surface emitting laser diode has been fabricated using single step MOCVD grown wafers. Deep proton implantation is employed to make a gain guided device. TEM_{00} mode is maintained up to 1.1 mW with a 5 μm diameter output window. We show that the transverse mode is influenced by implantation power because of the suppression of the thermal lens effect and the spatial hole burning. The threshold current is as low as 4 mA and the slope efficiency is 0.2 mW/mA. Moreover, a record low threshold voltage (1.72 V) is obtained by optimising current path structure.

1. Introduction

Vertical Cavity Surface Emitting Laser diodes are very attractive as devices with a wide range of potential applications including optical communications, optical discs, laser printers and light sensing systems.

In the VCSELs the lasing cavity is perpendicular to the top surface of a laser chip. Therefore, high packing density, compared to the packing density of edge emitting laser diodes with lasing cavity parallel to the surface of the laser chip, is obtainable. This would lead to a promising future in high density laser array, high data transmission in optical communication systems, ultra high parallel processing in optical communication systems, as well as supplying a route for fast and vast data transmission between electronic chips. Furthermore, the circular-like and astigmatism free nature of their beams make them suited for beam-combining for high power applications.

One of the important advantages of VCSELs is its spectral characteristics. Because of the very short cavity length, VCSELs have mode spacings larger than the gain bandwidth of the active material. Thus, they can operate inherently in single longitudinal mode. However, the transverse mode is not the case. The multi-wavelength lateral dimension of the output window facilitates multi-transverse mode operation. Moreover, the thermal lens effect and the spatial hole burning make it difficult to maintain a single transverse mode operation at high power. Although VCSELs operating in single transverse mode with a reasonable output power ($\geq$ 1mW) have been demonstrated, they incorporated a dielectric mirror [1] or used a "native oxide" technique to define the current path [2].

In this paper, we report a 1.1 mW cw single transverse mode operating VCSEL fabricated with conventional all epitaxial structures.

2. Experiments

The epitaxial structure grown by MOCVD consists of a n-GaAs buffer layer, a 40.5 pairs n-$Al_{0.3}Ga_{0.7}As$ / $Al_{0.9}Ga_{0.1}As$ distributed Bragg reflectors with 90 Å graded interface layers, 640 Å $Al_{0.14}Ga_{0.86}As$ active layer sandwiched between 850 Å graded barriers, 30 pairs p-$Al_{0.3}Ga_{0.7}As$ / $Al_{0.9}Ga_{0.1}As$ distributed Bragg reflectors with 180 Å graded interface layers and 90 Å p-GaAs passivation layer. The dopants of p- and n-types are C and Si repectively.

We uses $Al_{0.3}Ga_{0.7}As$ / $Al_{0.9}Ga_{0.1}As$ DBR instead of more conventional AlAs / AlGaAs DBR and insert the graded composition layer between $Al_{0.3}Ga_{0.7}As$ and $Al_{0.9}Ga_{0.1}As$ to minimise the series resistance. These would decrease the series resistance and thus the thermal lens effect. The calculated reflectivity for the top and the bottom mirror are 99.74 % and 99.987 % at 780 nm repectively.

The cross-sectional view of the top surface emitting laser diode is shown in Fig.-1. The fabrication process is as follows ; top p-ohmic metallization, current funnelling proton implantation, rapid thermal annealing, isolation etching and wafer thinning followed by n-ohmic metallization. Cr/Au and AuGe/Ni/Au ohmic contacts are applied to the p- and n-side. The output window size (w) defined by p-metallization is fixed at 5 μm and the implantation mask size (g) is varied. Proton implantation is employed to make a current blocking region near the active layer. To optimise the transverse mode characteristics , we vary the proton implantation power from 410 keV to 470 keV in 20 keV increments. In order to selectively implant H^+ to a depth of ~ 3.7 μm AlGaAs, a blocking structure of ~ 6 μm thickness photo-resist is required [3] instead of normally used ~ 1 μm photo-resist. Following thick photo-resist patterning, hydrogen ions are implanted to a dose of 5×10^{14} cm^{-2}. After the photo-resist stripping , the structure is annealed at 450 ℃ for 45 sec. The anneal serves to shape the resistance profile by recovering the top region while retaing a high resistivity region around active layer.

The results presented in this work are measured under cw condition at room temperature without heat sink.

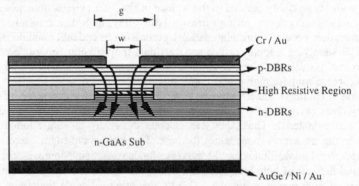

Figure-1. The cross-sectional view of the top surface emitting laser diode.

3. Results and discussion

In order to maintain a single transverse mode operation, the p-metal window diamter is fixed at 5 μm while the implantation mask size (g) is varied to optimise the electrical performance of the device. The larger the gain region size with respect to the output window's, the more efficient the current injection. This will lead to a lower operation voltage and ,thus, smaller the thermal lens effect. To prevent the mode changes with pump level, it is required to reduce the thermal lens effect and the spatial hole burning [4]. But the extreme difference between these two parameters may deteriorate the other performances of laser diodes such as the threshold current. The results are shown in Fig.-2. The lowest resistance is found in w = 5 μm and g = 20 μm pair. But the devices which have a gain region size larger than 10 μm show a relatively high threshold current while the further reduction in the resistance is negligible. Moreover, a relatively larger gain region size could lead to a non-Gaussian carrier distribution and consequently a multi-transverse mode operation.

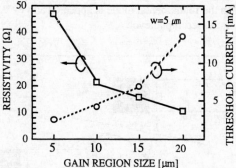

Figure-2. The dependence of the the maximum single mode operation power on the implantation power.

The dependence of the maximum single mode power on the implantation power is shown in Fig.-3. As illustrated in Fig.-3, the mode stability is greatly improved by increasing the proton implantation power from 410 keV and the optimum performance is obtained at 450 keV. The series resistance is also decreased slightly with implantation power. A device fabricated with a implantation power of 450 keV exhibits a stable TEM00 mode upto 1.1 mW. We attribute this increased mode stability with implantation power of 450 keV to the reduced gain profile perturbation by implanted high resistive region on the active layer. At the implantation power of 450 keV, the most resistive region is formed at 3.7 μm below the surface, which is the exactly same depth of the active layer. This can suppress the dispersion of the carrier and ,thus, prevent the mode change with injection level.

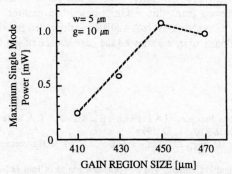

Figure-3 The dependence of the the maximum single mode operation power on the implantation power.

The L-I-V characteristics and lasing spectrum are shown in Fig.-4. The threshold

980

current is 4 mA and the slope efficiency is 0.2 mW/mA. Because this VCSEL has a bulk active layers, the significant further improvements are expected by incorporating a quantum well and strained layer as a active. The series resistance and the threshold voltage are 22 Ω and 1.72 V respectively. These values are lower than any reported 780 nm VCSEL as far as we know. We attribute these record low values partially to the Al0.3Ga0.7As/Al0.9Ga0.1As DBR with graded interface instead of more conventional AlAs/AlGaAs DBR and to the optimised current path structure. The spectrum and near field patterns (NFP) clearly show the single transverse mode operation upto 1.1 mW. Above that maximum single mode operation power, the first higher mode with slightly shorter wavelength than the fundamental mode is appeared and eventually the mode change is accomplished with increased input level.

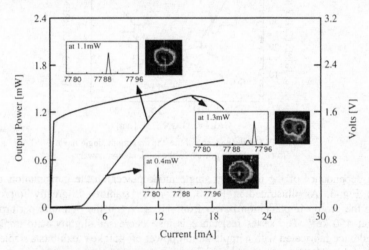

Figure-4. L-I-V characteristics of a VCSEL, with the insets showing the lasing spectrum and near field patterns at various drive levels.

4. Conclusion

We fabricated a 780 nm top surface emitting laser diode which is operated in fundamental transverse mode upto 1.1 mW. Additionally, a record low (1.72 V) threshold voltage in the 780 nm region was achieved from this AlGaAs-based top emitter. Such improvements included low barrier Al0.3Ga0.7As/Al0.9Ga0.1AsDBRs with graded interfaces and optimised current path structure. In particular, we showed the dependence of the mode stability on the implantation power.

References

[1] R.A.Morgan, M.K.Hibbs-Brenner, J.A.Lehman, E.L.Kalweit, R.A.Walterson, T.M.Marta and T.Akinwande, Appl. Phys. Lett. 66(10), 1157, 1995
[2] D.L.Huffaker, J.Shin, H.Deng, C.C.Lin, D.G.Deppe and B.G.Streetman, Appl. Phys. Lett. 65(21),2642,1994
[3] J.F.Ziegler, J.P.Biersack and U.Littmark. The Stopping and Range of Ions in Solids, Vol. 1, Pergamon, New York, 1984.
[4] D. Vakhshoori, J. D. Wynn, G. J. Zydzik, R. E. Leibenguth, M. T. Asom, K. Kojima, and R. A. Morgan, Appl. Phys. Lett. 62(21), 1448, 1993

Inst. Phys. Conf. Ser. No 145: Chapter 8
Paper presented at 22nd Int. Symp. Compound Semiconductors, Cheju Island, Korea, 28 August–2 September 1995
© *1996 IOP Publishing Ltd*

InGaAs Vertical-Cavity Surface-Emitting Lasers Buried in an Amorphous GaAs Passivation Layer

Byueng-Su Yoo, Hye Yong Chu, Min Soo Park*, Hyo-Hoon Park and El-Hang Lee

Electronics and Telecommunications Research Institute, Yusong P.O. Box 106, Taejon, Korea
*Department of Materials Science and Engineering, Korea Advanced Institute of Science and Technology

Abstract. InGaAs vertical-cavity surface-emitting lasers buried in an amorphous GaAs (a-GaAs) layer have been studied to reduce surface recombination and to improve device characteristics. Threshold current and transverse mode characteristics are improved compared to air-post type devices. In particular, smaller devices than 25 μm diameter show more significant reduction of threshold currents. A stable single transverse mode emission for a 10 μm device is also observed. The results suggest that a-GaAs-buried structure is useful for both reducing surface recombination current and obtaining a single transverse mode emission.

1. Introduction

Vertical-cavity surface-emitting lasers (VCSELs) have demonstrated very promising characteristics for several applications in optical interconnections and optical communications.[1] In particular, etched post type index-guided VCSELs show low threshold current density and high output power characteristics due to a strong confinement of current and optical field.[2] However, surface recombinations at the sidewall of active region cause a significant increase of threshold current, especially for small diameter devices. The etched structure also shows multitransverse mode emission characteristics by the strong optical confinement.

In this work, an amorphous GaAs (a-GaAs)-buried surface-emitting laser (SEL) structure has been studied. Since a-GaAs structure has a tetrahedral bonding character similar to crystalline GaAs,[3] we employ the a-GaAs layer for the passivaion of the sidewall defects of the etched laser post. In addition, a-GaAs layer surrounding laser post provides an antiguide effect, which is useful for a stable single mode emission.[4] We report experimental results on threshold current characteristics and modal behaviors for a-GaAs-buried SELs.

2. Experimental

A bottom-emitting InGaAs/GaAs VCSEL structure consisting of a 2λ-cavity periodic gain active structure[5] was used in this work. The top and bottom mirrors are 16 and 23.5 periods of GaAs/AlAs quarter wave stacks, respectively. The back side of the substrate was polished, and then antireflection layers were deposited. Ti/Au and AuGe/Ni/Au were used for top and bottom electrodes, respectively. After laser post was fabricated by ion beam etching through the active region to the top layer of the bottom mirror, highly resistive a-GaAs layer of 2.5 μm thickness was deposited over the post using molecular beam epitaxy technique at 160 °C. Then, the a-GaAs layer on p-metal contact was removed by reactive ion etching. Air-post type VCSELs were also fabricated simultaneously using a nearest piece of the wafer for comparison.

3. Results and Discussion

Fig. 1 shows the CW light output power against current (L-I) characteristics for a-GaAs-buried SELs. The L-I characteristics were measured at room temperature without a heatsink. The lasing wavelength is around 991 nm at threshold current. For 10, 15, 20 and 25 μm diameter circular devices, threshold currents of 0.7, 1.3, 2.1 and 2.9 mA are obtained, respectively. Light output power for the 25 μm diameter device reaches around 4 mW. Threshold current densities for 10 ~ 25 μm diameter devices are typically in the range of 480 ~ 900 A/cm² and differential quantum efficiencies are 14 ~ 28%.

Fig. 2 shows threshold current as a function of the device diameter for both etched air-post and a-GaAs-buried SELs. Threshold current characteristics of the a-GaAs-buried SELs are improved compared to air-post devices. In particular, threshold currents of small devices of 15 μm are significantly reduced. We also observed the enhancement of differential quantum efficiencies measured near threshold for the a-GaAs-buried devices compared to the air-post devices. For the etched post structures, especially small diameter devices, surface recombinations contribute dominantly to the threshold current.[6] Therefore, the improvements of device characteristics obtained in this work indicate the reduction of surface recombinations and leakage currents at sidewall by the deposition of a-GaAs layer. Since the structure of a-GaAs bonding network has a tetrahedral character similar to crystalline GaAs,[3] we expect surface defects at the sidewall of active region would be effectively reduced by the deposition of a-GaAs.

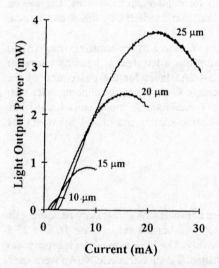

Fig. 1 CW light output power against current characteristics of a-GaAs-buried VCSELs.

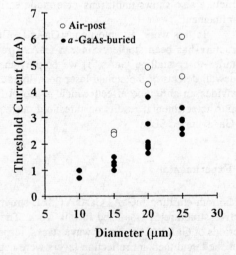

Fig. 2 Threshold currents for etched air-post and a-GaAs-buried VCSELs.

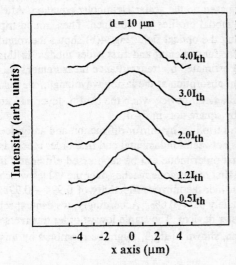

Fig. 3 Near-field intensity profiles of a 10 μm diameter
a-GaAs-buried SEL.

Modal behaviors of the a-GaAs-buried SELs are also investigated. Because a-GaAs has a higher refractive index and a lower band gap than crystalline GaAs and AlAs layers,[7] a-GaAs-buried SEL structure will have an antiguide region with an absorption loss at the clad. Fig. 3 shows the near field intensity profiles measured through the substrate for a 10 μm diameter device. The device has a threshold current I_{th} of 0.7 mA. Below threshold, spontaneous emission is uniformly observed through entire 10 μm diameter, indicating uniform current injection over the area. The field profiles of the device show a stable single mode emission with a lowest order transverse mode up to currents higher than $4I_{th}$ at which light output power reaches to a maximum value. These results suggest that the a-GaAs-buried structure effectively suppresses next higher mode laser operation, resulting from the antiguide effect.

For 15 and 20 μm diameter devices, the lowest order transverse mode is obtained up to the currents around $1.6I_{th}$ and the mode is relatively slowly changed to the next high order mode with increasing currents.

We performed a simple 2-dimensional numerical analysis using a beam propagation method to investigate the transverse mode behavior of the a-GaAs-buried SELs. The structure of the SEL used for the numerical analysis is shown in Fig. 4(a). The active region of 2λ and DBR region are represented by the uniform medium with the effective refractive index of n_1(3.523) and n_2(3.232), respectively. Then, the a-GaAs layer of refractive index n_3 is used as a clad over the active region. The length of DBR region is determined by the penetration depth of DBR mirror. The reflections of n_3/air, n_2/n_3 and n_1/n_2 interfaces were ignored.

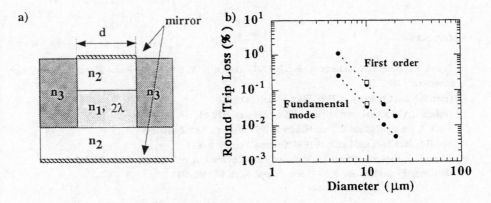

Fig. 4 (a) Schematic of the a-GaAs-buried VCSEL model and (b) the calculated round trip loss as a function of device diameter for a-GaAs-buried VCSELs.

984

The modal profiles for both fundamental and first order modes are found using a finite difference beam propagation method, based on the scalar Helmholtz equation. After several hundreds of round propagation, the modal profiles are obtained. Then, round trip loss for the modes is calculated by integrating the optical field. Fig. 4(b) shows the round trip loss as a function of device diameter for fundamental and first order modes. In this analysis, a-GaAs refractive index n_3 of 3.65 estimated by transmittance measurement was used. Because a-GaAs will have a significant absorption at the lasing wavelength, we also calculated the round trip loss at the 10 μm diameter device when the clad n_3 layer had an absorption coefficient of 10^4 cm^{-1}, indicated by square dots in fig.4(b).

As device diameter reduced, the round trip loss by mirror diffraction and antiguide effect increases and the difference of the loss between fundamental and first order modes is also increased, as shown in fig. 4(b). The first order mode can be suppressed efficiently if the loss difference between fundamental and first order mode reaches over the values which can effect on the total cavity loss. If we consider the mirror reflectivity of 0.994 ~ 0.998, the mirror loss during round trip is approximately 0.2 ~ 0.6%. Accordingly, we can expect a stable single mode emission at small diameter devices. The stable lowest order transverse mode emission at the 10 μm diameter device, shown in fig.3, might be promoted by this antiguide effect of a-GaAs layer.

4. Summary

Lasing characteristics of a-GaAs-buried SEL structure have been investigated. The a-GaAs-buried SELs show low threshold current characteristics compared to etched air-post SELs. A-GaAs-buried antiguide structure also shows a stable single mode emission at the 10 μm diameter device. The results indicate that a-GaAs-buried structures are useful for improving device characteristics.

We gratefully acknowledge Prof. Y. H. Lee and Mr. J.-H. Shin of KAIST for their assistance in laser fabrication and Dr. J. Lee and Mr. H. G. Lee of ETRI for a-GaAs deposition. This work was supported by the Ministry of Information and Communications, Korea.

References

[1] Jewell J L, Lee Y H, Scherer A, McCall S L, Olsson N A, Harbison J P and Florez L T 1990 *Opt. Engineering* **29** 210-4

[2] Geels R S and Coldren L A 1991 *Electron. Lett.* **27** 1984-5

[3] Molteni C, Colombo L and Miglio L 1994 *Phys. Rev.* **B50** 4371-7

[4] Wu Y A, Chang-Hasnain C J and Nabiev R 1993 *Electron. Lett.* **29** 1861-3

[5] Yoo B-S, Park H-H and Lee E-H 1994 *Electron. Lett.* **30** 1060-1

[6] Young D B, Kapila A, Scott J W, Malhotra V and Coldren L A 1994 *Electron. Lett.* **30** 233-5

[7] Matsumoto N and Kumabe K 1980 *Jpn. J. Appl. Phys.* **19** 1583-90

Inst. Phys. Conf. Ser. No 145: Chapter 8
Paper presented at 22nd Int. Symp. Compound Semiconductors, Cheju Island, Korea, 28 August–2 September 1995
© *1996 IOP Publishing Ltd*

High-reflectivity AlAsSb/In(Ga$_{1-x}$Al$_x$)As distributed Bragg reflectors on InP substrates for 1.3 - 1.5 µm wavelengths

H. Asai and H. Iwamura

NTT Opto-electronics Laboratories

3-1 Morinosato-wakamiya, Atsugi, Kanagawa 243-01, JAPAN

Abstract. We demonstrate high-reflectivity semiconductor DBRs for 1.3-1.5 µm wavelengths. The AlAsSb/In(Ga$_{1-x}$Al$_x$)As stacks lattice-matched to InP were grown by molecular beam epitaxy. We determined the refractive indices of the two constituents by reflectance measurement and found that this material system has large refractive index differences of 0.4 at 1.3 µm and of 0.45 at 1.55 µm. The ten-pair AlAsSb/In(Ga$_{1-x}$Al$_x$)As DBRs achieved high reflectivities of 90% in the 1.3 µm wavelength range and of 93% in the 1.5 µm range.

1. Introduction

Semiconductor distributed Bragg reflectors (DBRs) are a key element in a wide variety of opto-electronic devices, such as vertical cavity surface emitting lasers [1]. For high reflectivity there must be a large refractive index difference Δn between the two alternate layers in the DBR, and the two layers must be transparent at operating wavelengths. High-reflectivity DBRs operating at shorter wavelengths have been made using AlAs/GaAs stacks (Δn=0.6). At 1.55 µm wavelengths, InP/InGaAsP DBRs have also been demonstrated [2]. Owing to the small Δn (0.27) of InP/InGaAsP, however, many pairs are required for effective reflectivity in the long wavelength range. Recently, several material systems including the antimonide compounds, such as AlAsSb/GaAsSb [3], AlAsSb/InGaAs [4], and AlPSb/GaPSb [5] have been proposed for high-reflectivity DBRs at 1.55 µm. This is because the AlAsSb and AlPSb can be lattice-matched to InP substrates and have a low refractive index. However, there has been no report for the 1.3-µm DBRs.

In this paper, we demonstrate high-reflectivity semiconductor DBRs at 1.3 µm as well as at 1.55 µm. We used AlAsSb/In(Ga$_{1-x}$Al$_x$)As stacks lattice-matched to InP. We determined their refractive indices by measuring reflectance and found that this material system has large Δn's of 0.4 at 1.3 µm and of 0.45 at 1.55 µm. Ten-pair AlAsSb/In(Ga$_{1-x}$Al$_x$)As DBRs achieved high reflectivities of 90% in the 1.3 µm wavelength range and of 93% in the 1.55 µm range.

2. Experimental

The undoped AlAsSb/In(Ga$_{1-x}$Al$_x$)As layers were grown on Fe-doped (001) InP substrates by solid-source molecular beam epitaxy (MBE). The source materials were Ga, In, Al, As, and Sb metals. Two Al K-cells were used to individually optimize the Al beam fluxes for the AlAsSb and $In(Ga_{1-x}Al_x)As$ layers. The substrate temperature was monitored by a calibrated infrared pyrometer. Growth rates were 1.3 µm/h for both AlAsSb and $In(Ga_{1-x}Al_x)As$. The lattice matching condition of the ternary layers ($AlAs_{0.56}Sb_{0.44}$, $In_{0.53}Ga_{0.47}As$, and $In_{0.52}Al_{0.48}As$) to InP substrates was determined by x-ray diffraction measurement. The composition of the $In(Ga_{1-x}Al_x)As$ quaternary alloy was evaluated from the interpolation between the Ga and Al beam fluxes, for which the InGaAs and InAlAs ternary layers, respectively, were lattice-matched to InP. The As cell shutter was kept open during the $AlAsSb/In(Ga_{1-x}Al_x)As$ DBR growth.

To determine band gaps, we measured the optical absorption and photoluminescence spectra for a single $In(Ga_{1-x}Al_x)As$ layer grown on an InP substrate. We have evaluated the refractive index by measuring the reflectance at normal incidence. For the reflectance measurements, the back sides of the single layer and the DBR samples were roughened to eliminate undesirable reflection.

3. Results and discussion

There are specific difficulties in the MBE growth of $AlAs_{1-y}Sb_y$, because the alloy includes two group-V elements and has a large miscibility gap. To obtain the $AlAs_{1-y}Sb_y$ lattice-matched to InP (y=0.44), it is necessary to stabilize growth temperature (T_g) within ±2 °C as well as to control the As_4 and Sb_4 beam fluxes precisely, as reported previously [6]. Figure 1(a) shows an AlSb mole fraction (y) in the $AlAs_{1-y}Sb_y$ alloy as a function of T_g under the fixed Al, As_4, and Sb_4 beam fluxes. Since the Sb distribution coefficient decreased drastically with increasing T_g, the MBE growth was carried out at the relatively low temperature of 470 °C. In addition, the alloy composition is strongly influenced by the Al beam flux, as shown in Fig. 1(b). With increasing Al flux, the AlSb mole fraction in the antimonide alloy increases steeply. The Al beam flux, therefore, must be precisely controlled not only for the layer thickness but also for the composition in the AlAsSb ternary alloy. For this purpose, the MBE machine used in this work was equipped with two Al K-cells to individually optimize the growth conditions for AlAsSb and $In(Ga_{1-x}Al_x)As$.

The band gaps of the quaternary layers were determined by fundamental absorption

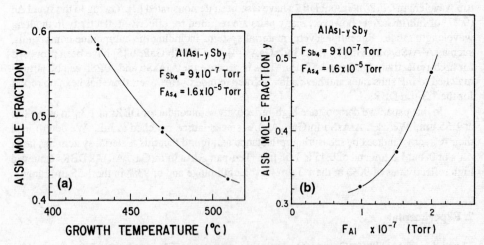

Fig. 1 AlSb mole fraction y in $AlAs_{1-y}Sb_y$ alloy as functions of growth temperature (a) and Al beam flux (b).

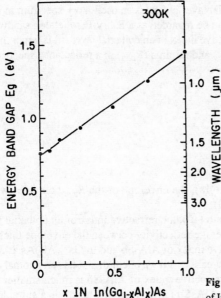

Fig 2 Band gap of In(Ga₁₋ₓAlₓ)As as a function of InAlAs mole fraction.

edges and PL peaks. The results are plotted in Fig. 2 as a function of InAlAs mole fraction (x) in In(Ga$_{1-x}$Al$_x$)As. With increasing InAlAs mole fraction, the quarternary band gap increases linearly from 0.74 eV (InGaAs) to 1.45 eV (InAlAs), corresponding to the wavelength range of 1.68 µm to 0.86 µm. This linear dependence of the band gap agrees with the previous reports [7]. To prevent optical absorption in a DBR, we must use the In(Ga$_{1-x}$Al$_x$)As with a larger band gap than the operating photon energy. For example, the In(Ga$_{1-x}$Al$_x$)As with x>0.43 had a band edge (λ_g) below 1.2 µm and is a suitable constituent material for a 1.3-µm DBR.

The refractive indices of AlAsSb and In(Ga$_{1-x}$Al$_x$)As were determined by a reflectance technique. Figure 3 shows the reflectance spectrum for an In(Ga$_{0.57}$Al$_{0.43}$)As quarternary layer on an InP substrate. The reflectance of a specular InP substrate with no epitaxial layer was also

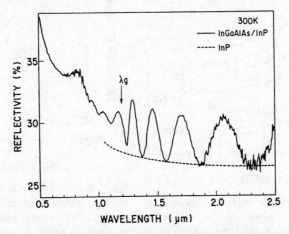

Fig. 3 Reflectance spectra of a single In(Ga$_{0.57}$Al$_{0.43}$)As layer grown on an InP substrate. The λ_g indicates the band gap of the quarternary layer.

988

measured to calibrate the reflectivities using the known refractive index of InP [8]. As shown in Fig. 3, the reflectance of an epitaxial layer indicates an oscillatory spectrum in the transparent region ($>\lambda_g$), because the layer can be regarded as a Fabry-Perot etalon sandwiched by air and an InP substrate. When the refractive index of an epitaxial layer (n) is larger than that of an InP substrate (n_0), the maxima (R_{max}) and minima (R_{min}) in a reflectance spectrum are given by [9]

$$R_{min} = \left(\frac{n_0 - 1}{n_0 + 1}\right)^2, \tag{1a}$$

$$R_{min} = \left(\frac{n^2 - n_0}{n^2 + n_0}\right)^2, \tag{1b}$$

respectively. As expected from Eq. (1a), the envelope of the R_{min} coincides well with the reflectance of an InP substrate (Fig. 3).

According to Suzuki and Okamto [10], the refractive index of an epitaxial layer can be approximately evaluated from the average reflectivity between the envelope lines of the R_{min} and R_{max}. Figure 4 shows the refractive index of AlAsSb and In(Ga$_{1-x}$Al$_x$)As as a function of wavelength. The refractive index decreases with wavelength and has an anomaly around the band gap. It is noteworthy that the refractive index of AlAsSb is much smaller than that of InAlAs or InP. This means that the AlAsSb lattice matched to InP is a suitable low-index material for DBRs in a long wavelength region. From the viewpoint of the application to the DBRs at 1.3 μm and 1.55 μm, the refractive indices at these wavelengths are plotted in Fig. 5 as a function of InAlAs mole fraction in In(Ga$_{1-x}$Al$_x$)As. As is clear from Fig. 5, the refractive index of a quaternary alloy decreases with increasing InAlAs mole fraction. For example, the refractive index of the quaternary alloy at 1.3 μm varied from 3.57 to 3.25.

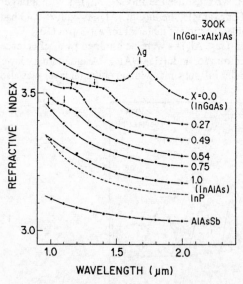

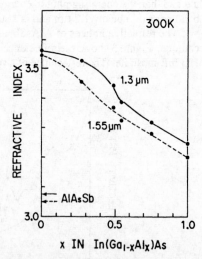

Fig.4 Refractive indices of In(Ga$_{1-x}$Al$_x$)As and AlAsSb as a function of wavelength. The closed circles indicate the measured data. The dashed line indicates the InP data from Ref. [8].

Fig. 5 Refractive indices at 1.3 μm (solid line) and 1.55 μm (dashed line) of In(Ga$_{1-x}$Al$_x$)As as a function of InAlAs mole fraction. The solid and dashed arrows indicate the refractive indices at 1.3 μm and 1.55 μm of AlAsSb, respectively.

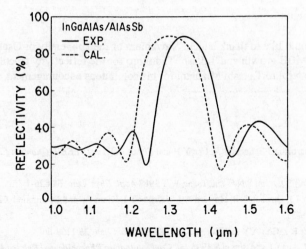

Fig. 6 Measured (solid line) and calculated (dashed line) reflectance spectra of a ten-pair AlAsSb/In(Ga$_{0.57}$Al$_{0.43}$)As DBR.

It is found from Figs. 1 and 5 that the In(Ga$_{0.57}$Al$_{0.43}$)As quaternary alloy is a suitable high-refractive-index material for a 1.3-μm DBR since the alloy is transparent (λ_g=1.2 μm) at the operating wavelength and has a large refractive index (3.47). For a 1.55-μm DBR, In(Ga$_{0.83}$Al$_{0.17}$)As (λ_g=1.45 μm, n=3.50) is an appropriate material for the same reason. On the other hand, the AlAsSb lattice matched to InP was transparent for both operating wavelengths and its refractive index at 1.3 μm and 1.55 μm was 3.07 and 3.05, respectively. By combining the AlAsSb with the In(Ga$_{1-x}$Al$_x$)As, we can obtain Δn's of 0.4 at 1.3 μm and of 0.45 at 1.55 μm.

To demonstrate high-reflectivity AlAsSb/In(Ga$_{1-x}$Al$_x$)As DBRs operating at 1.3 μm and 1.55 μm, we designed the DBRs based on the measured refractive index values. The 1.3-μm DBR consisted of ten pairs of AlAsSb (105.7 nm nominal thickness) and In(Ga$_{0.57}$Al$_{0.43}$)As (93.6 nm) layers. The last layer was the high-index In(Ga$_{0.57}$Al$_{0.43}$)As layer. Figure 6 shows the measured (solid line) and calculated (dotted line) reflectance for the 1.3 μm DBR. A peak reflectivity of 90% was observed near 1.3 μm with a reflectance bandwidth of 140 nm. This is in good agreement with the spectrum of the numerical calculation. We also grew 1.55-μm DBRs consisting of ten pairs of 126.8-nm-thick AlAsSb and 110.0-nm In(Ga$_{0.83}$Al$_{0.17}$)As (λ_g=1.45 μm). A reflectivity of 93% was also observed near 1.55 μm with a reflectance bandwidth of 188 nm, as predicted by the calculation.

4. Summary

We fabricated high-reflectivity AlAsSb/In(Ga$_{1-x}$Al$_x$)As DBRs for 1.3-1.5 μm wavelengths by MBE. We determined the refractive indices of the two constituents by reflectance measurement and found that this material system has large refractive index differences of 0.4 at 1.3 μm and of 0.45 at 1.55 μm. The ten-pair AlAsSb/In(Ga$_{1-x}$Al$_x$)As DBRs achieved high reflectivities of 90% in the 1.3 μm wavelength range and of 93% in the 1.55 μm range. This indicates that the AlAsSb/In(Ga$_{1-x}$Al$_x$)As material system has the potential to be used in making long-wavelength suface-normal photonic devices, such as vertical surface emitting lasers.

990

Acknowledgments

The authors would like to thank Yuichi Kawamura of the University of Osaka Prefecture for his support in MBE growth and Takashi Tadokoro for the reflectivity calculations of DBRs. We are also grateful to Takashi Mizutani for his continuous encouragement.

References

[1] Jewell J L, Harbison J P, Scherer A, Lee Y H and Florez L T 1991 *IEEE J. Quantum Electron.* **QE-27** 1332-46

[2] Tai K, MacCall S L, Chu S N G and Tsang W T 1987 *Appl. Phys. Lett.* **51** 826-7

[3] Blum O, Fritz I J, Dawson L R, Howard A J, Headley T J, Klem J F and Drummond T J 1995 *Appl. Phys. Lett.* **66** 329-31

[4] Tai K, Fischer R J, Cho A Y and Huang K F 1989 *Electron. Lett.* **25** 1159-60

[5] Shinomura H, Anan T and Sugou S 1995 *Int. Conf. on Indium Phosphide and Related Mater.* (Sapporo: IEEE LEOS/EDS) p 801

[6] Asai H and Kawamura Y 1992 *Pro. Int. Conf. on Indium Phosphide and Related Mater.* (Newport: IEEE LEOS/EDS) p 493

[7] Kopf R F, Wei H P, Perley A P and Levescu G 1992 *Appl. Phys. Lett.* **60** 2386-8

[8] Pettit G D and Turner W J 1965 *J. Appl. Phys.* **36** 2081

[9] Nojima S and Asahi H 1988 *J. Appl. Phys.* **63** 479-83

[10] Suzuki Y and Okamoto H 1983 *J. Electron. Mater.* **2** 397-411

Inst. Phys. Conf. Ser. No 145: Chapter 8

Paper presented at 22nd Int. Symp. Compound Semiconductors, Cheju Island, Korea, 28 August–2 September 1995

Anomalous above-threshold spontaneous emission in gain-guided vertical cavity surface emitting lasers.

J H Shin, J K Hwang, K H Ha, and Y H Lee

Department of Physics, Korea Advanced Institute of Science and Technology, Taejon 305-701, Korea

Abstract. Anomalously large decrease of spontaneous emission is observed for the first time in gain-guided vertical-cavity surface-emitting lasers after the onset of lasing. We attribute this unexpected phenomenon to the contraction of lasing mode by thermal lensing. two-dimensional profiles of the lasing mode and carrier density are measured. It is found that, above threshold, the carrier density decreases in contradiction to the general understanding, under the condition of the constant modal gain.

1. Introduction

The clamping of spontaneous emission above threshold is one of the well-known characteristics in semiconductor lasers. We observe anomalous decrease of above-threshold spontaneous emission for the first time in gain-guided vertical-cavity surface-emitting lasers(VCSEL). The decrease of spontaneous emission in edge-emitting semiconductor lasers had been already observed in early seventies[1], and explained from the effect of saturable absorption by some unknown traps[2].

However, the saturable absorption is not the only explanation for the phenomenon, since the decrease of spontaneous emission does not always mean the decrease of gain. It is possible, we think, that the same gain condition can be satisfied with smaller carrier density, since the change of mode size, if exists, results in the change of a transverse confinement factor. Especially, in the case of gain-guided VCSELs, it is well known that the lasing mode shrinks after lasing[3]. To verify the idea, two-dimensional(2-D) profiles of lasing mode and carrier density are measured, and the overlap integration is performed using these 2-D data to obtain the modal gain.

2. Experiments and Analyses

The dramatic decreases of spontaneous emissions after lasing in continuous wave operation are shown in Fig. 1(b). The spontaneous emission is measured from one of the nearest neighbor VCSELs in reverse bias used as a photo–detector(Fig. 1(a)). Only small portion of total spontaneous emission is collected laterally. The down arrows in Fig. 1(b) indicate the points of lasing onset. The tested samples are the MBE–grown proton–implanted 850–nm VCSELs[4]. They consist of a GaAs 4–quantum–well active region, a Si–doped $AlAs/Al_{0.15}Ga_{0.85}As$(26.5 pairs) bottom mirror, and a Be–doped $AlAs/Al_{0.7}Ga_{0.3}As/Al_{0.4}Ga_{0.6}As/Al_{0.15}Ga_{0.85}As$(19 pairs) top mirror. The threshold voltages are about 8.1 V, 5.0 V, and 4.0 V for 10 µm, 15 µm, and 20 µm VCSELs, respectively.

The 2–D profiles of lasing mode($|\Psi(r)|^2$) are measured by a computer–interfaced CCD camera, and the 1–D profiles through the center are shown in Fig. 2. The lasing mode profiles are well fitted to the Gaussian function. The contraction of the lasing mode is noticeable. To obtain the modal gain, the 2–D profiles of spontaneous emission are also measured normally by the CCD camera. To block the intense lasing spectra, a 300–µm–thick undoped GaAs substrate is used as a long–wavelength pass filter. The long wavelength part($\lambda > 875nm$) of spontaneous emission is assumed to represent all spontaneous emission. The 2–D profile of spontaneous emission($P(r)$) is related to the carrier density($N(r)$) by following equation,

$$P(r) = c B N(r)^2 \qquad (1)$$

where, B and c are the radiative recombination coefficient and an arbitrary proportional factor, respectively. Actually, the c factor depends on the injection current level because the spontaneous emission spectra shifts toward long wavelength with temperature. If we assume that the nonradiative recombination is neglected, the c_{th} can be obtained from following equation,

$$I_{th} = e d_{act} \int B N_{th}(r)^2 d^2r = \frac{e d_{act}}{c_{th}} \int P_{th}(r) d^2r \qquad (2)$$

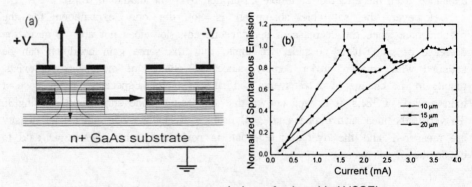

Fig. 1 The lateral spontaneous emissions of gain-guided VCSELs

where, I_{th} and d_{act} are the threshold current and the total thickness of active layers, respectively. The threshold modal gain(G^{th}_{modal}) can be obtained by following overlap integration,

$$G^{th}_{modal} = \frac{\int \Psi^*_{th}(r)\, G_{th}(r)\, \Psi_{th}(r)\, d^2r}{\int |\Psi_{th}(r)|^2\, d^2r} \qquad (3)$$

where, $|\Psi(r)|^2$ and $G(r)$ are the profile of lasing mode intensity and material gain, respectively. $G(r)$ is empirically known to have logarithmic dependence on the carrier density for the 80 Å GaAs quantum wells[5], as

$$G(r) = 3000 \ln\left(\frac{N(r)+1.1}{2.6+1.1}\right) \qquad (4)$$

where, the units of G and N are cm^{-1} and $10^{18}\ cm^{-3}$, respectively. The $N_{th}(r)$ and $G_{th}(r)$ from the 2-D spontaneous emission are plotted in Fig. 3. The threshold modal gains of 10 μm and 15 μm VCSELs are obtained to be 2760 cm^{-1} and 1910 cm^{-1}, respectively.

To verify that the decrease of spontaneous emission is originated mainly from the change of overlap between gain and field, we assume that the modal gain(G_{modal}) remains constant after lasing. The proportionality factors $c(I)$ of Eq.(2) which satisfy this condition of the modal gain constancy are looked after. Then we can calculate the spontaneous emission rate(S) at one current level(I) above threshold by following equation,

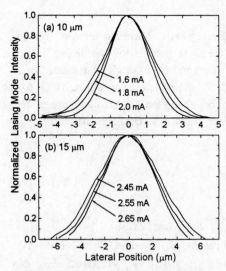

Fig. 2 The lasing mode profiles at several current levels

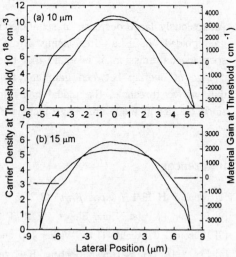

Fig. 3 Carrier density and material gain profiles at threshold

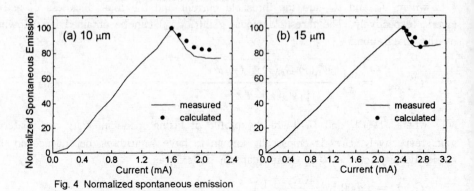

Fig. 4 Normalized spontaneous emission

$$S(I) = \frac{d_{act}}{c(I)} \int P_l(r) \, d^2r \qquad (5)$$

The calculated spontaneous emission rates from Eq.(5) are plotted as solid circles in Fig. 4. The solid lines in Fig. 4 are the laterally measured spontaneous emission as same as that of Fig. 1. The decreasing trend of calculated spontaneous emission is similar to that of measured one, which strongly supports our assumption of constant modal gain. The small discrepancy between calculated and measured spontaneous emission is possibly due to the reduction of some loss mechanisms such as a diffraction loss.

3. Conclusion

Anomalously large decrease of spontaneous emission is observed for the first time in gain-guided VCSELs after lasing. From the 2-D profiles of the lasing mode and spontaneous emission, it is found that this phenomenon is originated mainly from the change of overlap between gain and field by thermal lensing effect. It is possible that, above threshold, the modal gain can be supported by smaller carrier density, since the lasing mode shrinkage increases the transverse confinement factor.

References

[1] Nicoll F H 1971 *J. Appl. Phys.* **42** 2743

[2] Brosson P, Patel N and Ripper J E 1973 *Appl. Phys. Lett.* **23** 94

[3] Wilson G C, Kuchta D M, Walker J D and Smith J S 1994 *Appl. Phys. Lett.* **64** 542

[4] Lee Y H, Tell B, Brown-Goebeler K F, Jewell J L and Hove J V 1990 *Electron. Lett.* **26** 710

[5] Coldren L A and Corzine S W 1994 *Diode Lasers and Photonic Integrated Circuits* (UC-Santa Barbara) Ch.4

Inst. Phys. Conf. Ser. No 145: Chapter 8
Paper presented at 22nd Int. Symp. Compound Semiconductors, Cheju Island, Korea, 28 August–2 September 1995
© *1996 IOP Publishing Ltd*

High Power Tapered InGaAs/GaAs Laser Diodes with Carbon Doped Cladding Layers Grown by Solid Source Molecular Beam Epitaxy

M. Mikulla, W. Benz, P. Chazan, J. Daleiden, J. Fleissner, G. Kaufel,
E.C. Larkins*, M. Maier, J.D. Ralston**, J. Rosenzweig, and A. Wetzel

Fraunhofer Institut für Angewandte Festkörperphysik, Tullastr. 72, 79108 Freiburg, Germany
* Dept. of Electrical and Electronic Engineering, University of Nottingham, Nottingham NG7 2RD, U.K.
** SDL, Inc., 80 Rose Orchard Way, San Jose, CA 95134 USA

Abstract: High power InGaAs/GaAs tapered laser oscillators with a new type of carbon doping in the p-cladding layers grown by MBE are presented. In these devices carbon partially replaces the common beryllium p-dopant near the core region. SIMS depth profiles show, that the carbon doped layer serves as an efficient diffusion barrier for the beryllium in the p-cladding and contact layers. Thus, the previously observed severe beryllium redistribution from the p-cladding layer into the core region of the laser diodes is completely suppressed. In single quantum well (SQW) lasers this doping profile results in low internal losses of 2.6 cm^{-1}. Tapered laser oscillators show output powers as large as 7.2 watts and a maximum power conversion efficiency of 34 %.

1. Introduction

High power operation of semiconductor laser diode devices requires excellent long-time stability of the p-dopant in the cladding layers, both during epitaxial growth and during device operation. In previous investigations of high speed devices [1, 2], we have observed that Be diffusion from the p-type cladding layer during molecular-beam epitaxial (MBE) growth leads to an unintentional p-type background concentration throughout the entire core region, even at reduced growth temperatures [3]. This behavior of the Be dopants has also been observed by other authors [4]. Laser-operation induced migration of Be due to recombination-enhanced diffusion at the mirrors of AlGaAs/GaAs QW lasers has also been identified as a source of device failure [5]. Therefore, carbon has recently been proposed as a high quality p-dopant for MBE grown laser diodes [6, 7], but high power operation has not been presented for this type of laser diodes until today.

Here we report on MBE grown high power tapered laser diode devices in which carbon partially replaces the beryllium dopants near the core region. We show that even low carbon doping levels suppress the Be diffusion out of the highly p-doped cladding and contact layers into the active core region of the laser structures. This results in a substantially improved performance of high power laser diodes with tapered unstable resonators.

996

2. Epitaxial growth

The device structures were grown on 2 inch n-doped GaAs substrates in a Varian GEN II MBE system equipped with a carbon-filament doping source (EPI System). Fig. 1 depicts the InGaAs/GaAs QW laser structures, consisting of $Al_{0.6}Ga_{0.4}As$ cladding layers with one, two or three 7 nm $In_{0.25}Ga_{0.75}As$ quantum wells centered in a 200 nm wide undoped GaAs core region. In the two and three QW structures the quantum wells are separated by 14 nm and 17 nm GaAs barriers, respectively.

Carbon at a doping level of about $5 \times 10^{17}\,cm^{-3}$ was used as p-dopant in the initial 125 nm of the upper $Al_{0.6}Ga_{0.4}As$ cladding region in order to reduce free carrier absorption. The remaining p-doped ($2 \times 10^{18}\,cm^{-3}$) cladding layer and the heavily doped

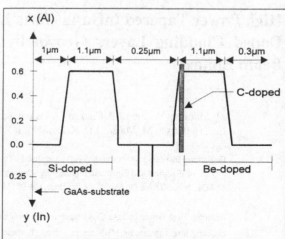

Fig. 1: Schematic epilayer structure of high power lasers with C and Be doped p-cladding layers. Core regions with one , two and three quantum wells were grown by solid source MBE .

($8 \times 10^{19}\,cm^{-3}$) p-contact layer were grown using conventional beryllium as the dopant. Si at a doping level of $1\text{-}4 \times 10^{18}\,cm^{-3}$ was used as n-type dopant throughout. All $Al_xGa_{1-x}As$ layers were implemented using binary GaAs/AlAs short-period superlattices [8]. Growth temperatures were 700°C and 620 °C for the n-cladding layers and the p-cladding layers, respectively. Growth stops were introduced before and after the InGaAs quantum wells in order to facilitate their growth at a substrate temperature of 450 °C.

3. SIMS depth profiles

Fig. 2 shows the SIMS depth profiles of the Be and C dopants. The profiles are not background corrected and the Al profile (arbitrary units) indicates the layer sequence. The In signal from the QWs and the Si signal from the n-cladding layer are not shown. The mass isotope 12 is used for the detection of the C signal in order to suppress any matrix effect arising from the different semiconductor materials of the cladding and core region. In this detection mode the detection limit of our SIMS system for C is about $1 \times 10^{-17}\,cm^{-3}$.

As a result of the MBE growth the Be concentration in the core region of the laser structure remains below the resolutionon limit ($5 \times 10^{16}\,cm^{-3}$) of our SIMS system for beryllium.

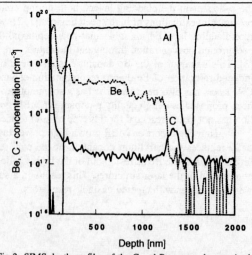

Fig 2: SIMS depth profiles of the C and Be p-type dopants in the cladding layer of the MBE grown epilayer structure.

This behavior of the Be dopants is attributed to the C doping in the cladding layer, which serves as an efficient diffusion barrier for the Be. The previously observed severe Be redistribution which leads to high Be doping levels in the core region is therefore completely avoided by this type of carbon cladding layer doping. Thus, the core region of the laser structure remains completely undoped and internal optical losses due to free carrier absorption are reduced.

4. Broad area laser diodes

Broad area laser diodes were fabricated from these epitaxial layer structures in order to investigate the epitaxial material quality and to compare the performance of the different core designs. The lowest threshold current densities of 154 A/cm^2 for 500 μm long devices together with transparency current densities of about 60 A/cm^2 are obtained from the SQW devices. For these devices internal losses as low as 2.6 cm^{-1} were determined from the linear relationship between the reciprocal of the external quantum efficiency and the cavity length. Compared to these results laser diodes with two or three quantum wells showed significantly larger internal losses of up to 10 cm^{-1} and threshold current densities between 170 A/cm^{-2} and 330 A/cm^{-2}. The internal quantum efficiencies remain at about 0.7 for all different core designs.

5. Tapered laser diodes

Tapered laser oscillators have been fabricated similar to those reported in [9]. The device geometry includes a narrow input waveguide of 7 μm width which acts as a mode selective filter and a tapered gain region of 2 mm length. Tapers with angles between 2° and 5° were fabricated leading to output apertures between 140 μm and 340 μm width. Cavity spoiler grooves etched by CAIBE (chemically assisted ion beam etching) were integrated in this design in order to control the optical beam profile. Fig. 3 shows an SEM micrograph of the narrow input waveguide together with the cavity spoiler grooves. After thinning and cleaving, the output facets are an-tireflexion coated to a residual reflectivity

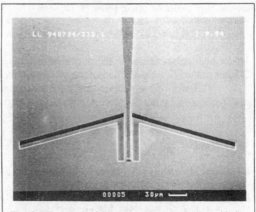

Fig. 3: Micrograph of the waveguide taper together with the CAIBE etched cavity spoiler grooves.

of less than 5 x 10^{-4} by a single λ/4-layer of SiON deposited by PECVD (plasma enhanced chemical vapor deposition). In addition, the narrow facets are coated with two λ/4–pairs of Al$_2$O$_3$ and Si resulting in a reflectivity of about 90 % leading to a maximum power extraction from the broad output facets.

Several devices were mounted p-side down on copper heat sinks with indium solder resulting in a differential resistance of about 50 mΩ. Fig. 4 shows the P-I-characteristic of a two quantum well device with an output power as large as 7.2 watts in a quasi cw operation mode (50 μs pulse width, 200 Hz repetition rate). The threshold current density of this device is about 350 A/cm^2 and an external efficiency of 0.68 W/A is achieved. A maximum electrical to optical power conversion efficiency of 34 % is obtained at a current of 7 A (Fig. 5).

998

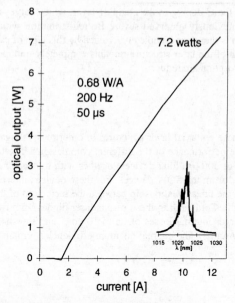

Fig. 4: P - I - characteristic and emissions spectrum of a tapered laser diode. An output power of 7.2 watts is achieved under quasi cw operation conditions.

Fig. 5: Power conversion coefficient of a tapered laser diode with an oscillator length of 2 mm and a taper angle of 4°.

Under high power operation the optical emission spectrum of this device broadens to a FWHM of 2.5 nm and the farfield profile is not transform limited.

This work was supported by the Bundesministerium für Bildung, Wissenschaft, Forschung und Technologie in the framework of the Laser 2000 program.

5. References

[1] Ralston, J.D., Weisser, S., Esquivias, I., Larkins, E.C., Rosenzweig, J., Tasker, P.J., and Fleissner, J., IEEE J. Quant. Electron., 1993, **29**, pp. 1648-1659

[2] Ralston, J. D., Weisser, S., Esquivias, I., Schönfelder, A., Larkins, E. C., Rosenzweig, J., Tasker, P. J., Maier, M., Fleissner, J., Mat. Sci. Eng., 1993, B21, pp. 232-236

[3] Arias, J., Ralston, J.D., Esquivias, I., Larkins, E.C., Weisser, S., Rosenzweig, J., Schönfelder,A., Maier, M., Proc. 1st Int. Conf. Materials for Microelectronics, Barcelona, 17-19 October 1994, pp. 218-219

[4] Chand, N., Chu, S. N. G., Jordan, A. S., Geva, M., Swaminathan, V., J. Vac. Sci. Technol., 1992, B **10**, pp. 807

[5] Jakubowicz, A., Oosenbrug, A., and Horster, Th., Appl. Phys. Lett., 1993, **63**, pp. 1185-1187

[6] Micovic, M., Evaldsson, P., Taylor, G. W., Vang, T., and Malik, R.J., Appl. Phys. Lett., 1994, **64**, pp. 411-413

[7] Ralston, J.D.,Laughton, F. R., Chazan, P., Larkins, E. C., Maier, M., Abd Rahman, M. K., White, I. H., Electron. Lett., 1995, 31, pp. 651-652

[8] Ralston, J.D., Larkins, E. C., Rothemund, W., Esquivias, I., Weisser, S., Rosenzweig, J., and Fleissner, J., J. Cryst. Growth, 1993, **127**, pp. 19-24

[9] Kintzer, E.S., Walpol, J.N., Chinn, S.R., Wang, C.A., and Missaggia, L.J., IEEE Phot. Techn. Lett., 1993, **5**, pp. 605-608

Inst. Phys. Conf. Ser. No 145: Chapter 8
Paper presented at 22nd Int. Symp. Compound Semiconductors, Cheju Island, Korea, 28 August–2 September 1995
© *1996 IOP Publishing Ltd*

MOCVD-grown $Al_{0.07}Ga_{0.93}As$ high-power laser diode array

N.J.Son[1], S.Park, J.C.Ahn, and O.D.Kwon

Department of Electronics and Electrical engineering, Pohang University of Science & Technology, Pohang P. O. box 125, 790-600 Korea

Abstract. A laser diode array structure consisting of a 150Å $Al_{0.07}Ga_{0.93}As$ single quantum well active region operating near 810nm, cladded with an AlGaAs graded-index separate confinement heterostructure, has been grown by MOCVD. 3.1W output power has been obtained from the 500μm aperture, uncoated laser diode array with 460μm cavity length. The internal quantum effciency was found to be 75.8% and the internal loss 4.83cm^{-1}.

1. Introduction

Since the first reports of the use of GaAs-AlGaAs laser diodes(LDs) to pump solid-state lasers over two decades ago[1,2], progress in high-power laser diode technology has led to greatly improved performance of diode-pumped solid-state lasers. For this kind of LD, two characteristics are required. First, the output power is in excess of 1W under CW operation at room temperature. Second, it is necessary to control the lasing wavelength of LDs within the absorption band of a solid-state laser, which is generally very narrow, for example, 805~808nm in the case of Nd:YAG.

To realize high-power operation, one of the methods used for increasing the power from a semiconductor laser is to increase the width of the emitting region. However, as the width is increased, the occurrence of multilateral modes, filaments, and lateral mode instability becomes significant. This produces a far-field pattern which results in reduced brightness. The use of unstable resonators can overcome this problem[3]. However, the most practical method at the present time involves the use of a monolithic array of semiconductor lasers.

For LDs with GaAs single-quantum-well graded-index separate-confinement-heterostructure (SQW-GRIN SCH), one can obtain the lasing wavelength, near 810nm. However, in this case very thin active region($L_z < 40$Å) is needed. On the other hand, for AlGaAs QW, the required lasing wavelength is easily obtained by controlling Al content of QW at various L_z.

In this paper, we report device preparation and laser operating characteristics of AlGaAs SQW-GRIN SCH multi-stripe broad-area LD for Nd:YAG solid-state laser pumping.

[1] E-mail: njson@jane.postech.ac.kr

2. Device preparation

The device structure, consisting of a single-quantum-well graded-index separate confinement-heterostructure(SQW-GRIN SCH), is illustrated in Fig.1. Fig.2 shows the schematic illustration of AlGaAs SQW-GRIN SCH broad-area laser diode array(LDA). Epitaxial layers were grown on an n-GaAs substrate by metalorganic chemical-vapor deposition(MOCVD). The epitaxial growth was carried out at a pressure of 76 Torr in a vertical reactor and V/III ratio was 80, using trimethylgallium(TMGa) and trimethylaluminum(TMAl) as group III precursors, and AsH_3 as group V source. SiH_4 and diethylzinc(DEZn) are used as an n-type and a p-type dopant, respectively. Growth temperature is 750°C with the exception of the p^+-cap layer, in which case the temperature is reduced to 700°C for ohmic contact. The grown layers consist of $1.0\mu m$ GaAs buffer layer(Si doped, $n=5 \times 10^{18} cm^{-3}$), $1.5\mu m$ N-$Al_{0.5}Ga_{0.5}As$ cladding layer(Si doped, $N=6 \times 10^{16} cm^{-3}$), $1,500$Å $Al_{0.5-0.23}Ga_{0.5-0.77}As$ grading layer(undoped),150Å $Al_{0.07}Ga_{0.93}As$ active layer(undoped), $1,500$Å $Al_{0.23-0.5}Ga_{0.77-0.5}As$ grading layer(undoped), $1.5\mu m$ P-$Al_{0.5}Ga_{0.5}As$ cladding layer(Zn doped, $P=1 \times 10^{18} cm^{-3}$), and $0.13\mu m$ heavily doped p^+-cap layer(Zn doped, $p=1.0 \times 10^{19} cm^{-3}$) included for ohmic contact formation (no shallow zinc diffusion or proton implantation before p metalization). The active aperture consists of five emitters separated from each other by means of SiO_2 deposition $(1,500$Å$)$ and stripe formation, which creates insulating regions that channel the current to $100\mu m$ wide stripes placed on $450\mu m$ centers, using Cr-Au p-contact metals. The wafer is thinned to $100\mu m$, the back side is contacted with AuGe-Ni-Au metals(alloyed), and bars with various cavity lengths are cleaved.

3. Laser operating characteristics

At room temperature, pulsed-wave output power(P) of a single-stripe SQW-GRIN SCH LD is plotted as a function of drive current under pulsed operation ($1\mu s$ pulse width, 1kHz repetition rate) in Fig.3. Typical uncoated LD power measurements were made with a calibrated photodiode. The output power of 0.8W has been obtained at

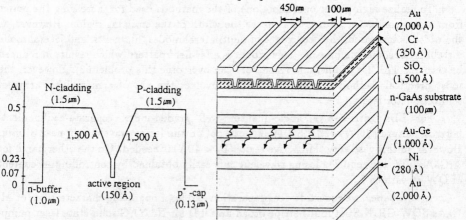

Fig.1 Composition profile of AlGaAs quantum
well laser with graded index separate
confinment heterostructure

Fig.2 Schematic diagram of 5-stripe laser diode array

2.0A, which results in about 64% external quantum efficiency(η_{ext}) and 368mA threshold current for LD with a $500\mu m$ cavity length. Assuming the internal quantum efficiency (η_i) and the internal loss(α_i) to be constant, the dependence of $1/\eta_{ext}$ on the cavity length(L) is given by

$$\eta_{ex}^{-1} = \eta_i^{-1}(1 + \frac{\alpha_i L}{ln(1/R)}).$$

By examining L dependence of η_{ext}, we have found that η_i and α_i are 75.8% and $4.83cm^{-1}$, respectively. This is shown in Fig.4, where the inverse of η_{ext}, measured from a series of uncoated devices of various cavity lengths, is fitted to a linear dependence with L. From Fig.3 and Fig.4 the transparent current density is expected to be $182A/cm^2$ for LD with the $500\mu m$ cavity length. On the other hand, the beam divergence has been measured to be 7° and 34.2° in lateral and transverse directions, respectively.

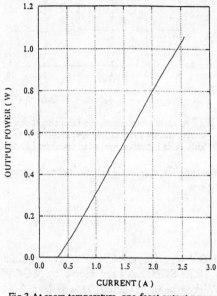

Fig.3 At room temperature, one-facet output power vs. drive current under pulsed operation(1 μs pulse-width, 1kHz repetition rate) of an uncoated LD.

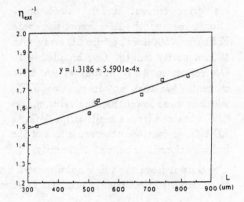

Fig.4 Linear fit of η_{ext}^{-1} vs cavity length(L), used in determination of the internal quantum efficiency (η_i) and the internal loss (α_i)

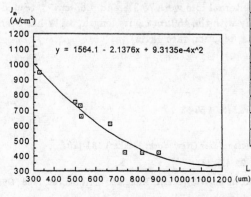

Fig.5 Threshold current density vs cavity length

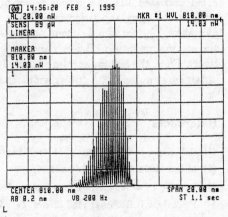

Fig.6 The lasing spectrum of LD with a $500\mu m$ cavity length.

The threshold current density (J_{th}) was measured for cavity lengths ranging from $250\mu m$ to $800\mu m$ and plotted in Fig.5. There is a large increase in J_{th} for short-cavity devices. This has been explained by considering the gain required to overcome the round trip optical loss in the cavity[4].

Fig.6 shows that the peak value of emission wavelength is about $810nm$ at $10°C$. The temperature coefficiency of wavelength is approximately $0.2nm/°C$.

Fig.7 shows the output power versus drive current under pulsed operation(pulse width $1\mu s$, repetition rate 1kHz) for a $500\mu m$ aperture LD array with $460\mu m$ cavity length. Our broad-area LDA results in 3.1W output at 9A drive current. We then obtain $\eta_{ext} = 55.7\%$, which is some lower than the value, $\eta_{ext} = 65\%$, obtained from a multi-stripe AlGaAs LDA[5]. In fact, we observed a lack of the individual emitter output uniformity.

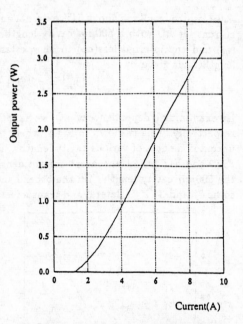

Fig.7 At room temperature, one-facet output power vs. drive current under pulsed operation(1 μs pulse-width, 1kHz repetition rate) of an uncoated LD array.

The output from our single quantumwell broad-area LD($\eta_{ext} = 64\%$) indicates a nonuniform current injections through our LDA emitters.

4. Conclusion

A laser diode(LD) structure consisting of a 150Å $Al_{0.07}Ga_{0.93}As$ SQW active region operating near $810nm$, cladded with an AlGaAs GRIN-SCH, has been grown by MOCVD. From uncoated LD, the output power of 0.8W has been obtained, at a pulsed 2.0A (pulse width $1\mu m$, repetition rate 1kHz), which results in about 64% external quantum efficiency. The threshold current density is $736A/cm^2$ for the $500\mu m$ cavity length, while the internal quantum efficiency and the internal loss were 75.8% and 4.83 cm^{-1}, respectively. From the $500\mu m$ aperture LD array with the $460\mu m$ cavity length, 3.1W has been obtained at pulsed 9.0A (pulse width $1\mu s$, repetition rate 1kHz).

References

[1] R.J. Keyes and T.M.Quist 1964 *Appl. Phys. Lett.* 4 50-52

[2] M.Ross 1968 *Proc. IEEE* 56 196-197

[3] Luis Figueroa 1989 *High-Power Semiconductor Laser* (New York: Dekker) 164-178

[4] P.S.Zory,A.R.ReiSinger 1986 *Electro. Lett.* 22 475-477

[5] M.Sakamoto,G.L.Harnagel,D.F.Welch,C.R.Lennon,W.Streifer,H.Kung,and D.R.Scifres 1988 *Opt. Lett.* 13 378-379

Inst. Phys. Conf. Ser. No 145: Chapter 8
Paper presented at 22nd Int. Symp. Compound Semiconductors, Cheju Island, Korea, 28 August–2 September 1995
© 1996 IOP Publishing Ltd

Picosecond Dynamics of InGaAs Microcavity Lasers -Influence of Carrier Transport/Capture and Gain Flattening

F. Sogawa, A. Hangleiter*, H. Watabe, Y. Nagamune, M. Nishioka, and Y. Arakawa

Institute of Industrial Science, University of Tokyo
7-22-1 Roppongi, Minato-ku, Tokyo 106, JAPAN

Abstract. We have fabricated and studied the dynamics of $In_{0.16}Ga_{0.84}As$ λ-cavity lasers focusing on the influence of carrier transport/capture and gain flattening by using optical resonant pumping and off-resonant pumping. With the off-resonant pumping, shoulders in the lasing pulse shape were observed, while no shoulder appeared with the on-resonant pumping. By solving the rate equations including carrier transport/capture effects, we reproduced such behavior. Saturation of the inverse rise time was observed at high pump levels. This results from gain flattening due to the step-like density of states of 2D carriers.

1. Introduction

Semiconductor microcavity lasers are of interest for high-speed communication due to their higher relaxation frequency than the conventional lasers' one. The relaxation frequency without parasitic limitations is given by $\omega_r = \sqrt{\gamma A(p/p_{th}-1)}$, where γ is damping factor of the cavity, and A is spontaneous emission rate. Therefore microcavity lasers are promising to have high modulation bandwidth because of their short cavity length and high spontaneous emission rate enhanced by the cavity QED effect.[1] In addition, the higher differential gain with quantum well active region should lead to a higher modulation bandwidth.[2] In edge emitting lasers, however, degradation of modulation bandwidth by carrier transport and capture has been discussed.[3, 4] Moreover, to generate a short laser pulse is of great importance. For such a purpose optical switching (i.e. optical pumping) method has a great advantages compared with electrical pumping, because it is free from electrical RC limitations and able to pump with various energies. Generated pulse width is determined mainly by the photon lifetime of the cavity and the gain of the active medium. The observation of the picosecond order laser pulse from an edge emitting quantum well laser has been reported.[5]

Also in microcavity lasers, several experimental results of lasing dynamics have been reported.[6-8] However, effects of carrier transport/capture and gain flattening due to the step-like density of states of 2D carriers[9] on the pulse shape and the width have not been discussed yet. In this paper, we have studied the dynamics of InGaAs λ-cavity lasers focusing on the influence of these two factors.

---

*on leave from Physikalisches Institut, University of Stuttgart, D-70550 Stuttgart, Germany

2. Experimental

2.1. Sample

The microcavity surface emitting laser was prepared with MOCVD growth. The laser has a λ-cavity with 5 $In_{0.16}Ga_{0.84}As$ quantum wells of 50 Å thickness, GaAs barriers of 100 Å and GaAs clads. The top mirror is an $AlAs/Al_{0.4}Ga_{0.6}As$ Bragg reflector consisting of 22 pairs and the bottom mirror consists of 30.5 pairs. Designed reflectivity is 99.3%. Calculated effective cavity length[10] and the photon life time of the cavity are 1.55 μm and 1.3 ps, respectively.

2.2. Optical pumping experiment

The laser was cooled down to 15 K and pumped with a mode locked Ti:sapphire laser with a pulse width of ~ 100 fs. By using a μ-PL setup[11], the pump beam was focused to a spot of about 1 μm diameter in order to reject higher order lateral modes. The size of 1 μm is sufficiently small for a single mode, taking account of the mode radius.[12] Lasing pulses from the microcavity laser were measured using a synchroscan streak camera with a resolution of < 7 ps.

3. Results and discussions

3.1. Picosecond laser pulse generation

Figures 1 and 2 show experimental results of lasing pulse shape of the microcavity laser. Picosecond order laser pulses were successfully generated. The obtained shortest pulse width with deconvolution is 7.7 ps under the condition of 4.4 times the threshold pump power with off-resonant pumping. In Fig. 2, pulse shapes with various off-resonant pump levels are shown.

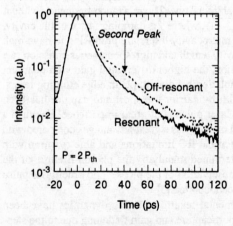

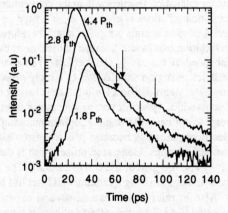

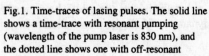

Fig.1. Time-traces of lasing pulses. The solid line shows a time-trace with resonant pumping (wavelength of the pump laser is 830 nm), and the dotted line shows one with off-resonant

Fig. 2. Time-traces of lasing pulses with off-resonant pumping. Shoulders are clearly observed as pointed by the arrows.

3.2. Comparison between off- and on-resonant pumping

In Fig. 1 the solid line shows a pulse shape with resonant pumping which excites only well regions, whereas the dotted line shows one with off-resonant pumping which excites both well and barrier/clad regions. The obtained pulse width was about 14 ps without deconvolution in both cases. Measurement was done with changing pump wavelength from 800 nm to 840 nm in every 5 nm. Pump power was kept at 2 times the threshold pump power. The wavelength of 820 nm corresponds to the energy level of the GaAs barriers/clads. In the range from 800 nm to 820 nm, the laser was resonant pumped and the responses were almost the same. On the other hand, the laser was off-resonant pumped in the range from 825 nm to 835 nm. The pulse shape with 815 nm pump and one with 830 nm pump are shown in Fig. 1. With 840 nm pump, the laser did not lase, for the wavelength of 840 nm corresponds to lower energy than Fermi energy.

By solving the laser rate equations for the case of direct carrier generation in the lasing level (i.e., resonant pumping), it is easy to see that under such conditions only a single lasing pulse is expected. This is in fact observed for the case of resonant pumping as shown by the solid line in Fig. 1. For off-resonant pumping, however, we observed a second peak or a shoulder as shown by the arrows in Figs. 1 and 2. This is only possible if there is a slow supply of carriers into the lasing level. We simulated the lasing response with using the rate equations including carrier transport and capture effects.[13] The rate equations are as follows,

$$\frac{dn_b}{dt} = G_b - \frac{n_b}{\tau_{trans}} - \frac{n_b}{\tau_b}, \tag{1}$$

$$\frac{dn_w}{dt} = G_w + \frac{n_b}{\tau_{trans}} - v_g Sg(n_w, S) - \frac{n_w}{\tau_n}, \tag{2}$$

$$\frac{dS}{dt} = \Gamma v_g Sg(n_w, S) - \frac{S}{\tau_{ph}} + \Gamma \beta B n_w^2. \tag{3}$$

Calculated result is shown in Fig. 3. Therefore the carrier transport and capture should cause the sub peaks of the lasing pulse. From the calculation we estimated the carrier transport/capture time to be of the order of 10 ps.

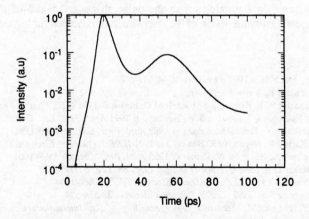

Fig. 3. Calculated results of the pulse shape with off-resonant pumping. Time constant of carrier transport/capture time used is 20 ps and initial carrier density is 2 times the threshold value.

1006

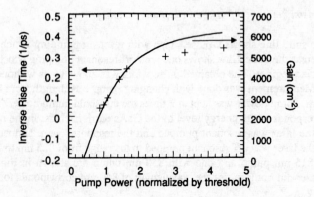

Fig. 4. Inverse rise time of lasing pulse with various pump levels (off-resonant pump) and calculated maximum gain of the laser structure at each pumping power.

3.3. On inverse rise time

Figure 4 shows a plot of the inverse rise time versus pump power. The inverse rise time is expected to be proportional to the gain and therefore to the carrier density minus transparency carrier density in case of no gain flattening. As shown in Fig. 4, the inverse rise time is linearly increased with pump power and saturated at high pump levels. The limit value of inverse rise time from the experimental resolution is 0.51. This saturation is due not only to the experimental resolution but also to gain flattening. This laser has quantum well active region which has step-like density of states of 2D carriers, so gain flattening occurs with increasing carrier density. The solid line shows the calculated material maximum gain against carrier density. Good agreement was obtained and this indicates that the measured saturation of the inverse rise time is due to gain flattening.

4. Conclusion

In conclusion, we experimentally and theoretically demonstrated that carrier transport/capture effects as well as gain flattening affect the pulse shape and width of the gain-switched microcavity lasers. Optimum design of the laser structure is needed for higher performances.

References

[1] Yamamoto Y, Machida S 1991 *Phys. Rev. A* **44** 657-668
[2] Arakawa Y, Vahala K, Yariv A 1984 *Appl. Phys. Lett.* **45** 950-952
[3] Rideout W, Sharfin W F, Koteles E S, Vassell M O, Elman B 1991 *IEEE Photon. Tech. Lett.* **3** 784
[4] Nagarajan R, Fukushima T, Corzine S W, Bowers J E 1991 *Appl. Phys. Lett.* **59** 1835-1837
[5] Sogawa T, Arakawa Y, Tanaka M, Sakaki H 1988 *Appl. Phys. Lett.* **53** 1580-1582
[6] Melcer L G, Karin J R, Nagarajan R, Bowers J E 1991 *IEEE J. Quantum Electron.* **27** 1417-1425
[7] Michler P, Lohner A, Rühle W W, Reiner G 1995 *Appl. Phys. Lett.* **66** 1599-1601
[8] Deng H, Huffaker D L, Shin J, Deppe D G 1995 *Electron. Lett.* **31** 278-279
[9] Arakawa Y, Yariv A 1985 *IEEE J. Quantum Electron.* **QE-21** 1666-1674
[10] Babic D I, Corzine S W 1992 *IEEE J. Quantum Electron.* **28** 514-524
[11] Nagamune Y, Nishioka M, Arakawa Y 1994 *Proceedings of 22nd International Conference on the Physics of Semiconductors* **3** 1835-1838
[12] Ujihara K 1991 *Jpn. J. Appl. Phys.* **30** L901-L903
[13] Grabmaier A, Schöfthaler M, Hangleiter A, Kazmierski C, Blez M, Ougazzaden A 1993 *Appl. Phys. Lett.* **62** 52-54

nst. Phys. Conf. Ser. No 145: Chapter 8
Paper presented at 22nd Int. Symp. Compound Semiconductors, Cheju Island, Korea, 28 August–2 September 1995
© *1996 IOP Publishing Ltd*

Ultralow Laser Threshold Operation of InGaAs-GaAs-InGaP Strained Quantum Well DFB and DBR Lasers

Y. K. Sin

Photonic Devices Lab., Semiconductor R&D Laboratories II,
Hyundai Electronics Industries Co., Ltd.
San 136-1, Ami-ri, Bubal-eub, Ichon-kun, Kyoungki-do, 467-860, Korea

Abstract. Ultralow threshold current characteristics of InGaAs-GaAs-InGaP buried heterostructure (BH) strained quantum well (QW) distributed feedback (DFB) and distributed Bragg reflector (DBR) lasers are reported. Uncoated InGaP DFB lasers show an ultralow CW threshold current of 2.2 mA measured at RT (which is the lowest threshold current ever obtained from DFB lasers), the slope efficiency of 0.36 mW/mA per facet, the sidemode suppression ratio (SMSR) of 30 dB, and the lasing wavelength at 0.98 μm. Also, InGaP DBR lasers show a CW RT threshold current of 7.2 mA with the SMSR of 25 dB and the Bragg wavelength at 0.96 μm.

1. Introduction

Semiconductor lasers with ultralow threshold currents are finding applications in optical interconnects, and these characteristics have been primarily obtained from (In)GaAs-AlGaAs buried heterostructure (BH) quantum well (QW) lasers[1]. However, these devices suffer from poor yield due to the regrowth difficulty of AlGaAs current blocking layers. Recently, owing to lower surface oxidation from InGaP compared to AlGaAs, InGaP Fabry-Perot BH strained QW lasers with as-cleaved facets have shown a high yield with the CW threshold current of 0.6 mA and the slope efficiency of 0.45 mW/mA per facet[2]. In this paper, ultralow threshold current operation of InGaP strained QW distributed feedback (DFB) and distributed Bragg reflector (DBR) lasers is reported.

2. Experimental methods

Both InGaP DFB and DBR lasers were fabricated by three-step low pressure metalorganic vapor phase epitaxy (MOVPE) processes. First, for DFB lasers, second order gratings with a grating period of 3000 Å were formed on a 1-μm thick n-InGaP layer (Si, n = 5 x 10^{18} /cm^3) deposited on an n$^+$-GaAs substrate (Si, n = 5 x 10^{18} /cm^3) by conventional holographic lithography and wet chemical etching using the mixture of HBr, HNO$_3$, and H$_2$O. The corrugated InGaP layer was overgrown by a separate-confinement-heterostructure single-quantum-well (SCH SQW). The SCH SQW consisted of a 70-Å thick In$_{0.18}$Ga$_{0.82}$As quantum well sandwiched between 1000-Å thick GaAs carrier collection layers. The SCH was surrounded by a 1.4 μm-thick Si-doped InGaP cladding layer and a 1.4 μm-thick Zn-doped InGaP cladding layer (p, n = 10^{18} /cm^3) both grown at 700°C. Finally, a 0.15-μm thick p$^+$-GaAs contact layer (p = 5 x 10^{18} /cm^3) was grown. About 1500 Å-thick silicon dioxide was deposited as an etching mask at 300°C by plasma enhanced chemical vapor deposition. Then, 1.5 μm wide mesas with a center-to-center spacing of 350 μm were formed along the <011> direction by standard photolithography and wet chemical etching. The DH structure was etched with a mixture of HBr and H$_2$O$_2$, and the etching was stopped at the lower InGaP cladding layer. Next, the wafer was cleaned with a solution of H$_2$SO$_4$ just before the subsequent MOVPE growth of a current blocking junction. The current blocking junction consisted of a 0.9-μm thick p-InGaP layer (Zn, p = 10^{18} /cm^3) and a 0.6-μm thick n-InGaP layer (Si, n = 10^{18} /cm^3) both grown at 700°C. AuZn-Au and AuGeNi-Au were thermally evaporated for p- and n-ohmic contacts and alloyed at 425°C for 90 sec. The wafer was cleaved into laser diodes and the diodes were bonded for testing. A schematic diagram of the completed InGaP DFB laser is shown in Fig.1.

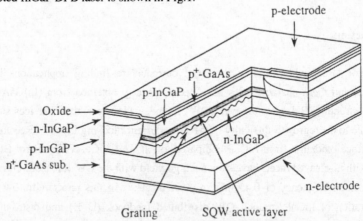

Fig.1　Schematic cross-section of an InGaP DFB laser.

For DBR lasers, second order gratings with a grating period of 3000 Å were only formed on a DBR section of a 1-μm thick n-InGaP layer deposited on the substrate by conventional holographic lithography and wet chemical etching using the mixture of HBr, HNO$_3$, and H$_2$O. The locally corrugated InGaP layer was overgrown by a separate-confinement-heterostructure single-quantum-well that consisted of an InGaAs QW sandwiched between GaAs carrier collection layers. Next, a p-InGaP cladding and a p$^+$-GaAs contact layers were deposited.

After the formation of a mesa, the p-n InGaP current blocking junction was grown. Finally, 500-μm long active sections were electrically isolated from 500-μm long DBR sections by removing p-ohmic contact and GaAs contact layer to form 1-mm long two section DBR lasers. A schematic diagram of the completed InGaAs-GaAs-InGaP DBR BH strained QW laser is shown in Fig.2.

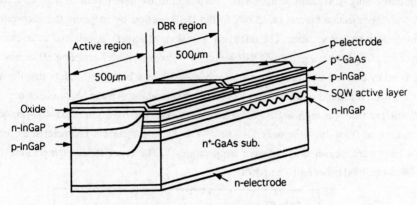

Fig.2. Schematic cross-section of an InGaP DBR laser.

3. Results and discussion

Fig.3 shows the CW light versus current characteristic from an uncoated 500-μm long DFB laser measured at RT showing an ultralow laser threshold of 2.2 mA and a high slope efficiency of 0.36 mW/mA per facet. To the best of our knowledge, this is the lowest laser threshold ever obtained from GaAs or InP based DFB lasers. Previously reported (In)GaAs-AlGaAs DFB QW lasers have shown a limited output power below 10 mW and a high laser threshold over 85 mA. And, these poor device characteristics from AlGaAs DFB lasers can be attributed to fabrication difficulties associated with AlGaAs. Also shown in Fig.3 (insert) is the emission spectra from the same device measured at the injection current of 50 mA (corresponding to the output power of 17 mW) with the SMSR of 30 dB and the lasing wavelength at 0.98 μm. From the stop-band width measured at 0.9 x I_{th}, the coupling coefficient was estimated to be 28 cm^{-1} with the normalized coupling coefficient of 1.4. The temperature sensitivity of the DFB modes was measured to be 0.5 Å/°C between 20 and 50°C. In addition, note that the similar temperature sensitivity of 0.6 Å/°C was obtained between 20 and 40°C using the devices reported in Ref. 3. Also, distributed feedback (DFB) BH lasers show a maximum singlemode output power of 58 mW with the sidemode suppression ratio of 35 dB and the lasing wavelength at 0.98 μm.

Fig.4 shows the output power versus current characteristic measured CW at RT from the two section InGaP DBR laser with the DBR section remaining open. Both of facets were as cleaved. A laser threshold of 7.2 mA and a slope efficiency of 0.14 mW/mA were obtained from the device. To the best of our knowledge, this is the first demonstration of InGaP DBR lasers with the lowest CW threshold current ever obtained from GaAs based DBR lasers reported so far. And, this indicates the advantage of using InGaP DBR lasers over AlGaAs DBR lasers for low-laser-threshold single-mode operation. Previously reported AlGaAs DBR QW lasers have shown high laser thresholds near 30 - 40 mA, attributed to fabrication difficulties using Al containing materials. Furthermore, the laser threshold was reduced to 6.5 mA with the injection current of 15 mA to the DBR section by reducing the absorption loss from the active DBR section. The relatively low slope efficiency is beleived to be due to low power coupling between gain and DBR sections. However, power coupling efficiency can be improved by employing a better coupling scheme such as a butt-joint. Note that the average CW RT laser threshold of 5.0 mA (as low as 3.1 mA) and the average slope efficiency of 0.35 mW/mA per facet (as high as 0.37 mW/mA) were obtained from uncoated 400-μm long FP lasers cleaved from the same wafer. The CW light versus current characteristics using the DBR laser were measured at different temperatures. The characteristic temperature (T_o) of 70K was obtained between 10 and 40°C.

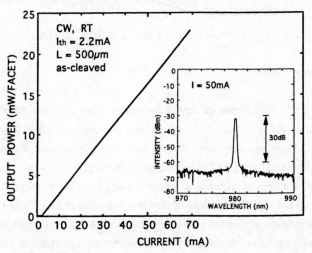

Fig.3 L-I characteristic (CW, RT) and emission spectrum from an uncoated DFB laser.

Also shown in Fig.4 is the emission spectra measured from the uncoated device at the injection current of 20 mA with the Bragg wavelength at 0.959 μm. The monomode oscillation was obtained with the SMSR of 25 dB. The below-laser-threshold-spectrum is usually measured to study characteristics of passive DBR lasers including coupling coefficient (κ), DBR reflectivity, and effective DBR length (the penetration depth into the DBR region). However, the Bragg reflection band was not observed from the below-laser-threshold-spectrum of active InGaP DBR lasers. This is believed to be due to the fact that when biased below the laser threshold optical fields can not easily penetrate deep into the DBR section of active DBR laser that also plays as a saturable absorber. Instead, the coupling coefficient was estimated by studying DFB modes obtained from the same device by injecting currents to the DBR section and with the gain section remaining open. The CW laser threshold of 7.9 mA and the slope efficiency of 0.1 mW/mA per facet were obtained from DFB lasers with the SMSR of 30 dB. From the stop-band width measured at 0.9 x I_{th}, the coupling coefficient was estimated to be 30 cm^{-1} with the product of coupling coefficient and DBR length (κL_{DBR}) of 1.5. In addition, the temperature sensitivity of DBR modes was measured to be 0.53 Å/°C between 20 and 35°C at the injection current of 20 mA from this device. A spectral halfwidth (spectral width at 3 dB down from the peak) of 1.7 Å was obtained at the injection current of 30 mA.

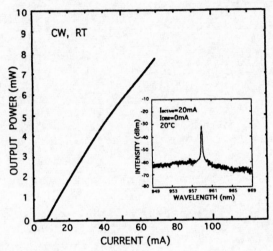

Fig. 4 L-I characteristic (CW, RT) and emission spectra from an uncoated DBR laser.

4. Conclusions

Ultralow threshold current characteristics of InGaAs-GaAs-InGaP DFB and DBR lasers have been reported. The uncoated 500-μm long DFB laser has shown an ultralow laser threshold of 2.2 mA, a high slope efficiency of 0.36 mW/mA per facet and a SMSR of 30 dB at the lasing wavelength of 0.98 μm. In addition, the first demonstration of InGaAs-GaAs-InGaP DBR BH strained QW lasers has been reported. The uncoated 1000 μm-long laser has shown a CW threshold current of 7.2 mA and a slope efficiency of 0.14 mW/mA at RT. The monomode oscillation has been obtained with the SMSR of 25 dB and the Bragg wavelength at 0.96 μm. To the best of our knowledge, this is the lowest CW threshold current ever obtained from GaAs based DBR lasers. These experimental results demonstrate that newly developed InGaP DFB and DBR lasers show excellent characteristics useful for optical interconnects.

Acknowledgment

The author is grateful to D. S. Kim for his assistance in the preparation of the manuscript.

References

[1]   Chen T R, Zhao B, Eng L, Feng J, Zhuang Y H and Yariv A 1995 *Electron. Letters.* **31** 285-287

[2]   Sin Y K and Horikawa H (unpublished)

[3]   Sin Y K and Horikawa H 1993 *Electron. Letters.* **29** 920-922

Inst. Phys. Conf. Ser. No 145: Chapter 8
Paper presented at 22nd Int. Symp. Compound Semiconductors, Cheju Island, Korea, 28 August–2 September 1995
© *1996 IOP Publishing Ltd*

Fabrication of Serpentine shaped Laser Diode using Reactive Ion Beam Etching

Jung-Hwan Choi, Song-Cheol Hong, and Young-Se Kwon

Department of Electrical Engineering, Korea Advanced Institute of Science and Technology, 373-1, Kusong-Dong, Yusong-Gu, Taejon, Korea

Abstract. A serpentine shaped laser diode which has 10 dry etched mirrors and 2 claeved facets is fabricated. Reactive ion beam etching technique with CCl_2F_2 gas is applied to fabricate the total internal reflection mirror. This laser has a stripe width of $10\mu m$ and a cavity length of $960\mu m$. This laser is oscillated under pulsed conditions with a threshold current of 210mA. Up to 20mW of optical power can be obtained from the two facets of the serpentine shaped laser. Total internal reflection mirror loss is evaluated from measurement of the threshold current density for different cavity lengths. The loss of the total internal reflection mirror is about 0.77dB per reflection.

1. Introduction

Cleaving is a common technique used in forming high quality facets for semiconductor lasers. However, cleaving facets is not suitable for ultralong cavity lasers and photonic circuits. A dry etching technique which can reproducibly fabricate laser facet with a quality similar to those obtained through cleaving is very desirable. The fabrication of optoelectronic circuits requires the integration of many optical components, such as semiconductor lasers and waveguides. Obstacles to be overcome in optoelectronic integration include the replacement of the cleaved facet with a reflector made using the etching technology [1] as well as the reduction in the wafer-area required for rounting optical signals from point to point within a single wafer. The mechanism of total internal reflection has been demonstrated as a possible solution for both of these issues. Therefore the dry etched total internal reflection mirror may be a suitable candidate for a high reflectivity integrable laser cavity structure[2]–[10]. In this letter, we described the fabrication of serpentine shaped ridge-waveguide laser which has a long cavity length with 10 dry etched mirrors and 2 cleaved facets. We then evaluate the loss of total internal reflection mirror by plotting the measured threshold current density as a funtion of cavity length.

2. Device Fabrication

The device described in this study was fabricated from the material grown by metalorganic chemical vapor deposition(MOCVD) technique. The structures are grown on the n-type GaAs substrate. The grown epitaxial layer structure for this device is shown in Table 1.

Table 1. Specification of the laser epi wafer.

Layer	Al mole fraction	doping(cm^{-3})	thickness(μm)	function
p-GaAs		10^{18}	0.2	cap layer
p-AlGaAs	0.65	5×10^{17}	0.7	p-cladding
p-AlGaAs	0.3	undoped	0.05	guiding layer
3 quantum wells (w/b)*	0.1/0.3	undoped	80Å / 100Å	active
n-AlGaAs	0.3	undoped	0.05	guiding layer
n-AlGaAs	0.65	10^{18}	1.0	n-cladding
n-GaAs		10^{18}	0.5	buffer layer
n-GaAs		10^{18}	100	substrate

* w: well, b: barrier

First, the p contact metal(AnZn/Au) and RIBE mask material (Cr)[11] were formed using the lift-off process. We used Cr as the impermeable mask for RIBE. The top Cr layer of the contact metal is resistant to CCl_2F_2 and is made thick enough to withstand the etch process. To reduce the tilting effect of mirrors, self-align techique was used. A loadlocked RIBE with a background pressure below 10^{-6} Torr was used to etch the total internal reflecton mirror. The RIBE was performed with CCl_2F_2 at a pressure of 2mTorr and gas flow of 2.5sccm to isotropically etch GaAs and AlGaAs layers. The ridge waveguides and total internal reflection mirrors were etched down to 1μm from the active layer. The resulting etched depth was about 2μm. After lapping the wafer to a thickness of 100μm, AuGe/Ni/Au n contact was evaporated and annealed. Fig. 1 shows a scanning electron micrograph(SEM) of the fabricated serpentine shaped laser diode. This laser has a stripe width of 10μm and a cavity length of 960μm.

The SEM photograph in Fig. 2 shows a close-up view of the etched mirrors in the laser diode. This photograph confirms that a smooth and nearly vertical etch profile was obtained. The surface roughness of the etched mirrors is less than 0.1μm.

3. Results

We tested the laser unbonded and uncoated under pulsed conditions (1μs pulse,1KHz rate), using a broad-area Si photodiode. Input current vs. light output power characteristic of the laser is shown in Fig. 3. The laser was oscillated under pulsed conditions with a threshold current of 210mA. Up to 20mW of optical power can be obtained from the two facets of the serpentine shaped laser.

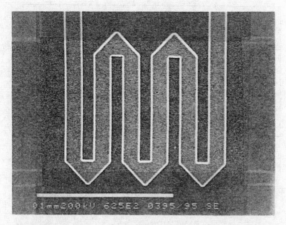

Figure 1. SEM photograph of the fabricated laser diode.

Figure 2. SEM photograph of the etched mirror.

Fig. 4 shows the emission spectra of this laser diode at $I = 1.2I_{th}$. This spectrum shows lasing on many longitudinal cavity modes with about 1.1Å mode spacing.

The total internal reflection mirror quality was evaluated in terms of the threshold current density. The threshold current density of the Fabry-Perot laser $J_{th}(\text{F.P.})$ is expressed[12] as

$$J_{th}(\text{F.P.}) = K_0 + K_1 \cdot \alpha + K_1 \cdot (1/L) \cdot \ln(1/R) \tag{1}$$

where K_0 and K_1 are constant coefficients dependent on the wafer quality and waveguide structure of the lasers, α is the total internal loss, L is the cavity length and R is a cleaved mirror reflectivity of 0.32. In the same manner, the threshold current density of the serpentine laser $J_{th}(\text{S.L.})$ is related to the total internal reflection mirror reflectivity R_m [4], [5], [13] as

$$J_{th}(\text{S.L.}) = K_0 + K_1 \cdot \alpha + K_1 \cdot (1/L) \cdot \ln(1/R) + K_1 \cdot (N/L) \cdot \ln(1/R_m) \tag{2}$$

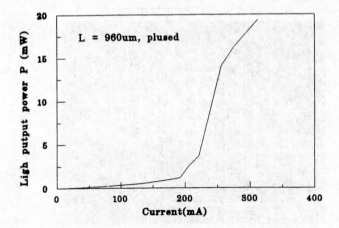

Figure 3. Input current vs. light output power under pulsed operation condition.

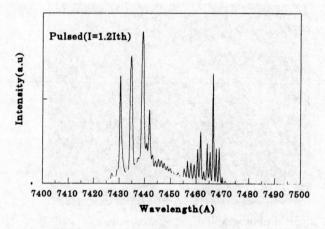

Figure 4. Emission spectrum of a serpentine shaped laser diode.

where N is the number of the etched mirrors. We fabricated a Fabry-Perot laser from the same wafer. Then, K_1 and $K_0 + K_1 \cdot \alpha$ can be determined to be 59A/cm and 388A/cm^2, respectively, from the least-square fitted line of J_{th}(F.P.) versus $1/L$, as shown in Fig. 5.

Using K_1 and $K_0 + K_1 \cdot \alpha$, the R_m is evaluated to be 0.837. The mirror loss β is

$$\beta = -10 \cdot \log(R_m) \tag{3}$$

and is 0.77dB per reflection. Since the mirrors and waveguides are self-aligned, the mirror loss is caused by the mask edge roughness and vetical shape of the mirror. It is thought that the presented mirror loss probably contains both the radiation loss due to structural imperfections on the total internal reflection mirrors and the loss due to field mismatch of the waves around the mirror.

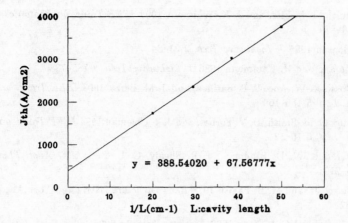

Figure 5. Threshold current density as a function of reciprocal cavity length.

4. Conclusions

Serpentine shaped laser diode which has 10 dry etched mirrors and 2 cleaved facets has been fabricated using the RIBE technique. We have measured up to 20mW output power under pulsed condition. Futhermore, mirror loss of the laser is estimated from the measurement of the threshold current densities for different cavity lengths. The total internal reflecton mirror loss was 0.77dB per reflection, resulting from the mask edge roughness and vertical mirror shape. This cavity geometry is therefore suitable for the monolithic mode locked laser diode which has ultralong cavity with gain modulation section and absorber.

Acknowledgements

The authors would like to thank T. K. You and C. H. Hong of LG Elec. Co. Ltd. for wafer supplyment. This work was financially supported by Korea Science and Engineering Foundation (KOSEF) through OptoElectronic Resaerch Center(OERC).

References

[1] C. J. Chae, and Y. S. Kwon 1987 *Electron. Lett.* **23** 1118–1120

[2] A. Behfar-Red, S. S. Wong, J. M. Ballantyne, B. A. Soltz, and C. M. Harding 1989 *Appl. Phys. Lett.* **54** 493–495

[3] Fusao Shimokawa, Hidenao Tanaka, Renshi Sawada, and Shigeji Hara 1990 *Appl. Phys. Lett.* **56** 1617–1619

[4] Satoshi Oku, Masanobu Okayasu, and Masahiro Ikeda 1991 *IEEE Trans. on Photonics Tech. Lett.* **3** 588–590

[5] John E. Johnson and C. L. Tang 1992 *IEEE Trans. on Photonics Tech. Lett.* **4** 24–26

[6] S. D. Smith, J. L. Fitz, and J. K. Whisnant 1993 *IEEE Trans. on Photonics Tech. Lett.* **5** 876–879

[7] T. M. Benson 1985 *J. Lightwave Tech.* **2** 31–34

[8] P. Buchmann and H. Kaufmann 1986 *J. Lightwave Tech.* **3** 785–788

[9] D. E. Bossi, R. W. Ade, P. P. Basilica, and J. M. Berak 1993 *IEEE Trans. on Photonics Tech. Lett.* **5** 166–169

[10] Z. J. Fang, G. M. Smith, D. V. Forbes, and J. J. Coleman 1994 *IEEE Trans. on Photonics Tech. Lett.* **6** 10–12

[11] W. J. Grande, W. D. Braddock, J. R. Shealy, C. L. Tang 1987 *Appl. Phys. Lett.* **51** 2189–2191

[12] H. C. Casey, Jr. and M. B. Panish 1978 *Heterostructure laser.* (New York:Academic) part A 182

[13] J. E. Johnson and C. L. Tang 1992 *Electron. Lett.* **28** 2025–2026

Inst. Phys. Conf. Ser. No 145: Chapter 8
Paper presented at 22nd Int. Symp. Compound Semiconductors, Cheju Island, Korea, 28 August–2 September 1995
© *1996 IOP Publishing Ltd*

Very low threshold current density 1.3μm InAsP/InP/InGaP/InP/GaInAsP strain-compensated multiple quantum well lasers

A. Kasukawa, N. Yokouchi, N. Yamanaka and N. Iwai

Yokohama R&D Laboratories, The Furukawa Electric Co., Ltd.
2-4-3, Okano, Nishi-ku, Yokohama 220, Japan

Abstract. By newly introducing both InGaP tensile strained layer as barriers of InAsP compressively strained multiple quantum wells, and InP intermediate layer between InAsP and InGaP layers, a high crystalline quality of InAsP/InGaP strain-compensated multiple quantum wells was obtained using metalorganic chemical vapor deposition on (100) InP substrate. A very low threshold current density of 300A/cm^2 was achieved with an emission wavelength of 1.3μm.

1. Introduction

The introduction of strained-layer quantum well (SL-QW) into the active layer has brought us numerous merits. One is the improvement of lasing characteristics (Adams 1986, Yablonovitch et al 1986a). The biaxial strain incorporated into the QWs by an intentional lattice mismatch causes the reduction of hole effective mass, and thus reduces the valence band density of state. Therefore, population inversion occurs at low injection carrier densities. Intervalence band absorption, which is the main cause for temperature sensitive differential quantum efficiency, is also reduced. Auger recombination rate is expected to be reduced due to low injection carrier densities. The other merit is the freedom in choice of material. As for long wavelength lasers, the conventional materials are GaInAsP/InP and GaInAlAs/InP. Using a SL-QW, InAsP/InP system can be a candidate for long wavelength lasers. InAsP/GaInAsP compressively SL-QW lasers emitting at 1.3μm were reported (Kasukawa et al 1993). The use of novel materials might improve temperature characteristics of threshold current for long wavelength lasers, since they could change the band line-up. Actually, InAsP/GaInAsP system has been reported to have a large conduction band offset as compared to GaInAsP/InP system, resulting in a possibility of improved lasing characteristics at a high temperature due to small electron leakage to an optical confinement layer (Yamamoto et al 1994).

In addition, the use of large number of wells is effective for the improvement of temperature characteristics of threshold current. However, SL-QW suffers from critical thickness. In fact, InAsP has a large compressive strain of about 1.5% with thickness of 8nm to realize an emission wavelength of 1.3μm. Therefore, the number of wells is limited to only two due to critical thickness (Kasukawa et al 1993). The idea of strain compensation can relieve this problem. In order to increase the number of wells to improve the temperature characteristics of InAsP/InP SL-QW lasers, tensile strained GaInAsP have been incorporated for strain compensation (Oohashi et al 1994a). Tensile strained InGaP, larger bandgap than that of InP, can make it possible not only to enhance the barrier height but to be used as barrier layer for strain compensation if we can make the best use of freedom of material choice. Moreover, InGaP seems to be easy to control the composition since it is a ternary material. We challenge to grow InAsP/InGaP SC-QWs using metalorganic chemical vapor deposition (MOCVD).

In this paper, we report the growth and characterization of InAsP/InGaP strain-compensated (SC) QWs grown by MOCVD for long wavelength laser. A very low threshold current density of 300A/cm^2 was obtained, which is lower than that of InAsP/GaInAsP SL-QW lasers.

2. MOCVD growth

Low pressure MOCVD is used for the growth with a growth temperature of 550°C (Kasukawa et al 1993). Double heterostructure cladded by InP layer, as shown in Fig. 1, was grown on a (100) InP substrate for sample evaluation. The growth sequence is quite simple as shown in Fig.2. Exact same gas flow rate of trimethylindium (TMIn) and phosphine (PH$_3$) was used for

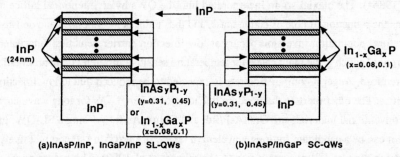

(a)InAsP/InP, InGaP/InP SL-QWs   **(b)InAsP/InGaP SC-QWs**

Figure 1 Schematic diagram of InAsP/InP, InGaP/InP SL-QWs and InAsP/InGaP SC-QWs.

InAsP and InGaP and arsine (AsH$_3$) and triethylgallium (TEGa) were introduced into reactor alternatively for the growth of InAsP and InGaP, respectively, with no growth interruption. The thickness and amount of strain are inferred by X-ray diffraction analysis of MQW consisting of InAsP/InP, InGaP/InP and InAsP/InGaP. The amounts of strain for InAs$_y$P$_{1-y}$ and In$_{1-x}$Ga$_x$P are derived from Vegard's law as 3.23y % (compressive strain) and -7x % (tensile strain), respectively.

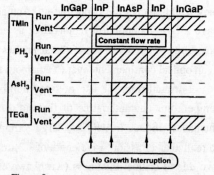

Figure 2
Growth sequence for the growth of InAsP/InP/InGaP strain-compensated QWs.

We confirmed a drastic improvement of the crystalline quality by strain-compensation for a sample with y=0.31 (1.0%) and x=0.08 (-0.57%). In this sample, the net strain was designed to be nearly zero. The good crystalline quality is confirmed by surface morphology, X-ray analysis and photoluminescence (PL) spectrum. Using InAsP/InGaP SC-10QWs, we have reported first lasing operation at a wavelength of 1.2μm (Yokouchi et al 1995). However, in our experiment, degradation of the surface morphology can be observed in a sample with y=0.45 (1.45%) for 1.3μm laser even if net strain was designed to be nearly zero. No laser was obtained. The surface morphology of SC-QWs was worse than those of InAsP/InP and InGaP/InP SL-QWs with same number of wells. From this result, we assumed that transition layer, consisted of quaternary compound (Ga, In, As, P), might be formed at hetero-interfaces. Therefore, we introduced thin InP intermediate layer (IML) between InAsP and InGaP layers.

The importance of InP IML has been investigated in terms of crystal quality. Full width at half maximum (FWHM) of room temperature PL spectrum, excited by Kr+-laser (λ=647.1nm), is plotted in Fig.3 as a function of InP IML thickness. The FWHM of PL spectrum decreases with increase of IML thickness. From this experiment, 5-monolayer (ML) is needed to obtain high crystalline quality. The improvement of crystalline quality of InAsP/InGaP SC-QWs with IML is also supported by X-ray measurement, that is higher order satellite peaks are observed, while no satellite peaks for SC-QWs without IML.

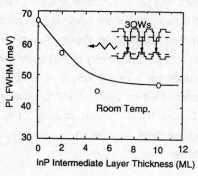

Figure 3
Room temperature PL FWHM as a function of thickness of InP intermediate layer.

3. Laser Structure and lasing characteristics

We have applied InAsP/InP (5MLs)/InGaP SC-QWs to the active layer. Figure 4 shows the schematic diagram of a 1.3μm SC-QW laser. Triple quantum wells is used. A 100nm-thick GaInAsP layer with a bandgap wavelength of 1.1μm is used for optical confinement layer (OCL) in order to increase an optical confinement factor. Figure 5 shows the room temperature PL spectrum. The peak wavelength was 1.3μm with a narrow FWHM of 30meV, which is better than that of GaInAsP/InP compressively SL-QW laser wafer with same number of wells. Sub-peak observed near at 1.21μm is considered to be the transition of second quantized level. The difference of PL linewidth between laser structure and test structure described above arises from the difference of InP upper cladding layer.

1022

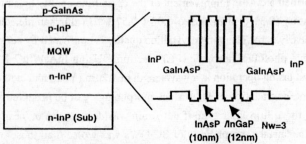

Figure 4. A schematic diagram of 1.3μm InAsP/InP/InGaP SC-QW laser. 100nm-thick GaInAsP is used for optical confinement layer.

Threshold current density of broad area lasers versus inverse cavity length is shown in Fig 6. We have achieved laser oscillation with an emission wavelength of 1.3μm in this material system. A very low threshold current density of 300A/cm^2 was obtained for a cavity length of 1200μm. Assuming a logarithmic gain of the quantum well, threshold current density of the quantum well laser can be expressed as follows:

$$J_{th} = N_w J_0 \exp\{(\alpha_i + \alpha_m)/N_w \Gamma_w G_0\} \quad (1)$$

where N_w is the number of wells, J_0 is the transparent current density, Γ_w is the optical confinement factor, G_0 is the gain coefficient, α_i and α_m are internal loss and mirror loss, respectively. Transparent current density of 88A/cm^2 per well was estimated for infinite cavity length. Lasing wavelength was 1.3μm. The results obtained for InAsP/GaInAsP SL-QW (Kasukawa et al 1993) lasers are also shown in Fig.6. The solid lines are the calculated values using eq.(1). G_0 and J_0 are derived to be 980 cm^{-1} and 88 A/cm^2, 680 cm^{-1} and 95 A/cm^2 for InAsP/InGaP SC-QW and InAsP/GaInAsP SL-QW lasers, respectively.

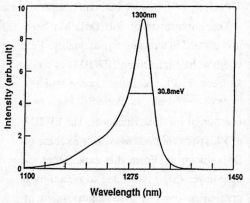

Figure 5
Room temperature PL spectrum of InAsP/InP /InGaP/InP/GaInAsP SC-QW laser structure.

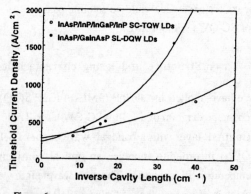

Figure 6
Threshold current density as a function of inverse cavity length for InAsP/InP/InGaP SC-QW and InAsP/GaInAsP SL-QW lasers.

Lower threshold current density was obtained for InAsP/InGaP SC-QW lasers. Especially, large difference of threshold current density was observed for short cavity region. The threshold current densities are 800A/cm^2 and 2500A/cm^2 for 250μm-long SC-QW and SL-QW lasers, respectively. This could be attributed to the fact that carrier overflow into the 1.1μm-Q OCL was suppressed in the latter structure due to the large barrier height of the conduction band. The relatively large barriers in the valence band would not affect the hole injection into the QWs in case of small number of wells. The use of large number of QWs, however, might suffer from nonuniform hole injection. Lower threshold current density can be expected for graded-index separate-confinement-heterostructure (GRIN-SCH) for optical confinement layer.

Using these structural parameters obtain experimentally, threshold current of a narrow stripe laser is calculated. Figure 7 shows the threshold current versus cavity length as a parameter of facet reflectivity. In the calculation, stripe width of 1.5μm is assumed and leakage current is neglected. Submilliampere threshold current can be expected in a LD with both short cavity of about 100μm and high reflectivity coating. Buried heterostructure LDs, as shown in the inset of Fig. 8 , with a 1.5μm wide active layer was fabricated by three-step MOCVD growth. Figure 8 shows the light output power versus injection current characterisitcs of a coated LD with a cavity length of 150μm. A low threshold current of 1.7mA was obtained. However, this is higher than that of the calculated one. This could be attributed to the leakage current in the BH structure.

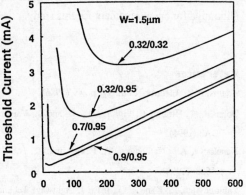

Figure 7   **Cavity Length (μm)**
Calculated threshold current versus cavity length as a parameter of facet reflectivity. Submilliampere threshold current can be obtained by combination of short cavity and high reflectivity.

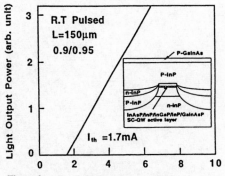

Figure 8   Injection Current (mA)
L/I charcteristics of a coated device with a cavity length of 150μm. Inset shows a schematic diagram of a buried hetero-structure laser grown by MOCVD.

1024

4. Conclusion

In conclusion we have investigated the growth of InAsP/InGaP strain-compensated quantum well for long wavelength laser using low-pressure MOCVD. The introduction of a thin InP intermediate layer can improve the crystalline quality. As a result, a very low threshold current density of $300A/cm^2$ was obtained for InAsP/InP/InGaP/InP/GaInAsP SC-QW lasers emitting at 1.3μm.

Acknowledgment

The authors would like to thank Dr. Y. Hiratani for his valuable advice, S. Kashiwa and T. Ninomiya for encouragement for this study.

References

Adams, A.R., Electron. Lett., 22, p.249 (1986)

Oohashi, H., Hirano, T., Sigiura, H., Nakano, J., Yamamoto, M., Tohmori, Y. and Yokoyama, K., IPRM'94, PDA3 (1994)

Kasukawa, A., Namegaya, T., Fukushima, T., Iwai, N. and Kikuta, T., IEEE J. Quantum Electron., 29, p.1528 (1993)

Yablonovitch, E., Kane, E.O., IEEE J. Lightwave Technol., LT-4, p.961 (1986)

Yamamoto, M., Yamamoto, E. and Nakano, J., IEEE J. Quantum Electron., 30, p.554 (1994)

Yokouchi, N., Yamanaka, N., Iwai, N., Matsuda, T. and Kasukawa, A., IPRM'95, paper TuB-2 (1995)

Inst. Phys. Conf. Ser. No 145: Chapter 8
Paper presented at 22nd Int. Symp. Compound Semiconductors, Cheju Island, Korea, 28 August–2 September 1995
© *1996 IOP Publishing Ltd*

High performance strain-compensated 1.3μm MQW-PBH-LD using two-step etching

H. S. Cho, D. H. Jang, J. K. Lee, J. S. Kim, K. H. Park,

C. S. Park, H. M. Kim, K.-E. Pyun, and H.-M. Park

Compound Semiconductor Research Department
Electronics and Telecommunications Research Institute
Yusong P.O.Box 106, Taejeon, 305-600, KOREA

Abstract. In the fabrication of multiple quantum well planar buried heterostructure laser diodes (MQW-PBH-LDs), a two-step mesa etching process comprising nonselective mesa etching followed by InP selective etching is proposed for obtaining narrow connection width between the p-InP clad layer and the p-InP blocking layer. The leakage current in 1.3μm strain-compensated MQW-PBH-LD has been remarkably reduced by using the two-step etching process for the mesa formation compared to the leakage current of conventional nonselective mesa etching. The threshold current as low as 4.6mA and the maximum slope efficiency of 0.32 mW/mA have been obtained by the using two-step etching process without any coating.

1. Introduction

The 1.3μm GaInAsP/InP laser diodes have been attractive for several applications to optical interconnections[1] and broadband network termination(B-NT) as well as long-haul high-bit rate optical transmission systems.[2] For the laser diodes, low threshold current for low power consumption, stable temperature characteristics for elimination of thermoelectric cooler, and high differential quantum efficiency for elimination of complicated coupling optics are essential from the viewpoint of the cost reduction in the transmitter. Much efforts has been concentrated on the devices those with a high gain active medium and a strong index guiding structure to meet the requirements. The laser diode with the strong index guiding structure, such as PBH-LD has intrinsic low threshold current and stable fundamental transverse mode due to tight confinement of carrier and photon to the active region with p-n-p-n current blocking layers. However, the leakage current flowing outside the active region was increased by increasing the operating temperature and the injection currents. The effective contact area between the p-InP clad layer and the p-InP blocking layer should be minimized in order to decrease leakage currents. Unfortunately there have not been enough investigations on reducing the leakage current via minimizing the connection width, because the controllability of the connection width is poor when the conventional etching method is used for mesa formation. In this letter, we propose the two-step etching method for mesa formation to reduce the effective contact area between the p-InP clad layer and p-InP blocking layer. Low threshold and high efficiency MQW-PBH-LDs were fabricated using proposed etching process and minimum threshold current of 4.6mA and maximum slope efficiency of 0.32mW/mA were obtained without facet coating.

2. Experiments and results

The laser diode structure comprising 6 pairs of 1.4% compressively strained GaInAsP(E_g= 0.905 eV) well and the GaInAsP(E_g=1.107 eV) barrier, surrounded by n- and p- type GaInAsP(E_g=1.107eV) separate confinement layers, was grown in high speed rotating low-pressure metalorganic vapor phase epitaxy(MOVPE) at 630℃. Tensile strain of 0.7% was applied in GaInAsP barrier layer for the strain compensation.[3] Trimethyl indium(TMIn) and triethyl gallium(TEGa) together with phosphine(PH_3) and arsine(AsH_3) were used as precursors. Silane(SiH_4) and diethylzinc(DEZn) were used as n- and p-type dopant, respectively. The narrow stripe mesa structures were prepared by chemical etching. The width of active layer was controlled between 0.8μm and 1.2μm to achieve fundamental transverse mode operation. The blocking layers were grown by MOVPE selective growth for the two types of stripe etched with 3μm SiNx mask along the <011> direction. A wide connection width was obtained after blocking layer growth for conventional nonselective etching. The two-step etching process consisting of nonselective etching followed by a InP selective etching produced the narrow connection width. Cross sectional view of the multiple quantum well planar buried heterostructure laser diode(MQW-PBH-LD) fabricated by two-step etching method is shown in Fig. 1. Owing to the inherent slow growth rate of p-InP blocking layer on the side wall of the active region, the connection width is narrower compared with the thick p-InP blocking layer on the n-InP substrate.

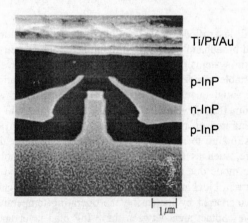

Ti/Pt/Au

p-InP

n-InP

p-InP

1μm

Fig. 1   SEM cross section of a MQW-PBH-LD prepared by two-step etching   process

The temperature dependence of the optical power on the injection current is shown in Fig. 2.   The average threshold current and the slope efficiency of the uncoated MQW-PBH-LD using two-step etching process (Fig. 2(a)) were 5.6mA and 0.27mW/mA , respectively, for the cavity length of 400μm.   The threshold current as low as 4.6mA and the maximum slope efficiency of 0.32mW/mA at the lasing wavelength of 1.30μm were obtained for 320μm-long uncoated laser diode. The internal quantum efficiency was estimated to be 76% with an internal loss of 8.3cm^{-1} from the differential quantum efficiency dependence on cavity length. However, the average threshold current using nonselective etching process was 14.5mA and slope efficiency was 0.22mW/mA for the cavity length of 400μm. The output power of the laser diodes using nonselective etching saturates rapidly at higher temperature due to the increased leakage current passing through the p-p InP (p-InP clad and p-InP blocking layer) as shown in Fig. 2 (b).

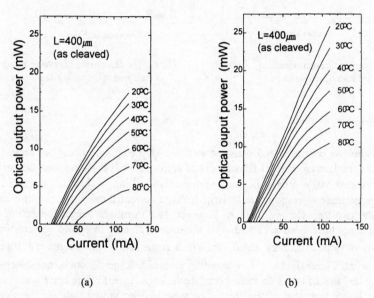

(a)   (b)

Fig. 2   Temperature dependence of optical output power on injection current for
(a) MQW-PBH-LD prepared by two-step etching
(b) MQW-PBH-LD prepared by nonselective etching

The temperature dependence of threshold current is shown in Fig. 3. The characteristic temperature (T$_o$) was 112°K at 20~55℃ and 28°K at 55~80℃ for the nonselective etched laser diode, but that of two-step etching was 50°K through out the temperature range of 20~ 80℃. The anomalous high characteristic temperature at 20~55℃ in the MQW-PBH-LD using nonselective etching was supposed to be owing to the large initial leakage current which is less temperature dependent. The characteristic temperatures of the two type laser diodes show apparently that the laser diode using two-step mesa etching has superior temperature stability to that using the nonselective etching. The dependence of the reverse leakage current on the connection width was measured at a 0.5 volts of the reverse bias to investigate the leakage current dependence on the connection width. The reverse leakage current is increased with the increase of connection width as shown in Fig. 4. The reverse leakage current of the two-step etched laser diode was one order of magnitude lower than that of nonselective etched one owing to the higher resistance of p-p InP leakage current path. From the above results, the two-step etching process of mesa formation was very effective for low threshold and high efficient MQW-PBH-LD.

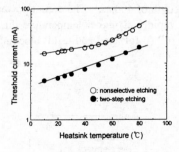

Fig. 3   Temperature dependence of   current
on the connection width

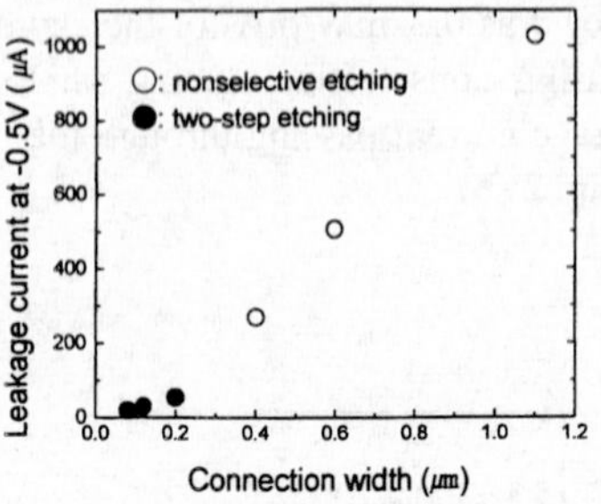

Fig. 4   The dependence of reverse leakage threshold
measured at a reverse bias of 0.5 volts.

3. Conclusion

In conclusion, we proposed and demonstrated two-step etching process for mesa formation that reduces the leakage current flowing out of active layer. It is shown that the performance of the fabricated MQW-PBH-LD using nonselective etching method   is degraded severely at high temperature operation due to large leakage current through the p-p InP connection path. In the two-step etching method, however, low threshold current operation and high slope efficiency have been achieved. The minimum value of threshold current was 4.6mA and maximum slope efficiency was 0.32mW/mA at the lasing wavelength of 1.30μm for 320 μm-long uncoated laser diode.   In comparing reverse leakage current of nonselective etched laser diode and that of two-step etched one, the leakage current of the latter was one order of magnitude lower than that of the former because smaller growth rate on the side wall of active region produced high resistive leakage current path through p-p InP connection.

References

[1] J. W. Goodman, G. J. Leonberger, S. Y. Kung, and R. A. Athale 1984,   Proc. IEEE, **72**, 850-865

[2] Y. Miyamoto, K. Hagimoto, M. Ohhata, T. Kagawa, N. Tsuzuki, H. Tsunetsugu, and I. Nishi 1994, IEEE J. of Lightwave Technol., **12**, 332-341

[3] B. I. Miller, U. Koren, M. G. Young, and M. D. Chien 1991,   Appl. Phys. Lett., **58**, 1952-1954

Inst. Phys. Conf. Ser. No 145: Chapter 8
Paper presented at 22nd Int. Symp. Compound Semiconductors, Cheju Island, Korea, 28 August–2 September 1995

Very low threshold current 630nm band AlGaInP single quantum well laser with strain compensated layers

Won-Jin Choi, Jong-Seok Kim, Meoung-Whan Cho, In-Sung Cho, Shi-Jong Leem, and Tae-Kyung Yoo

LG Electronics Research Center
16 Woomyeon-Dong, Seocho-Gu, Seoul 137-140, KOREA

Abstract: We have introduced strain compensated layers(SCLs) into a tensile strained 630nm band AlGaInP/GaInP single quantum well laser. These SCLs result in lowering the threshold current of the laser due to improving both optical confinement and hole injection efficiency in the well. A buried ridge laser with SCLs has been fabricated by the three-step low pressure vertical metal-organic chemical vapor deposition. The threshold current of this 200 μm long laser without facet coating is 33.3mA, which reflects a 27% reduction in the threshold current compared to the threshold current of 45.4mA for laser without SCLs. Threshold current as low as 25.4mA has been obtained by coating of the facets.

1. Introduction

630nm band AlGaInP lasers have attracted interest for next-generation disc systems. There are two approaches to obtain the 630nm band AlGaInP laser. One is to employ a compressive strained AlGaInP quantum well as an active layer, the other a tensile strained GaInP quantum well. Between the two of these, a tensile strained GaInP quantum well has advantages because GaInP with higher Ga composition can avoid the band-gap reduction by ordering[1] as well as low crystallinity problem due to containing an Al element in the active layer.

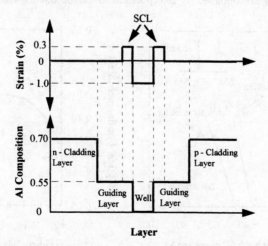

Figure 1. Layer structure of the AlGaInP laser with SCLs.

Recently, a tensile strained GaInP quantum well has been utilized to achieve the 630nm band AlGaInP laser by many groups[2 - 4]. However, 630nm band AlGaInP laser reliabilities have not been able to meet the system's requirements for practical purpose so far. One of the most important factors to enhance laser reliabilities is low threshold current operation. In this paper, we report a considerable improvement in laser threshold currents by introducing strain compensated layers(SCLs) into the conventional strained GaInP single quantum well(SQW) structure.

2. Experimental

We have fabricated a tensile strained 630nm band AlGaInP SQW laser. The laser with a buried ridge structure is realized through the three step growth process using low pressure vertical metal-organic chemical vapor deposition(MOCVD). The layer structure of the laser with SCLs is shown in Figure 1. The degrees of the strain in the well and SCL are -1% and +0.3%, respectively. The thickness of the well is 33nm. The optical guiding layer thickness is determined as the value where the maximum optical confinement factor is obtained. The bottom width of the ridge is about 5 μm. Detail growth condition has been reported elsewhere.[5] The laser is mounted on a Cu heat-sinker with a p-side down configuration. The end facets of the laser are prepared by alternate sputtering of two dielectric materials.

3. Results and discussion

Figure 2 shows current-optical output characteristics of the lasers. The threshold current of the as-cleaved 200 μm long laser with SCLs is 33.3mA, which is much smaller than 45.4mA of the lasers without SCLs. The lasing wavelength is 637nm under the condition of 3mW at room temperature, as shown in the inserted figure.

We calculate optical confinement factor sides of the lasers with and without SCLs. The results are shown in Figure 3. In the calculation, refractive indices of both ternary and quaternary alloy are obtained by interpolation of corresponding values in the reference

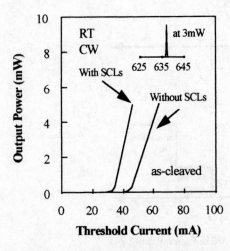

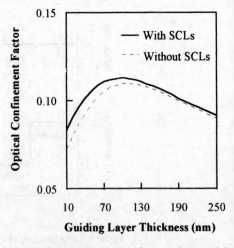

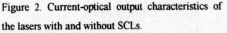

Figure 2. Current-optical output characteristics of the lasers with and without SCLs.

Figure 3. Calculated optical confinement factors of the lasers with and without SCLs.

book[6]. As shown in this figure, it is apparent that optical confinement factor is enhanced by introducing SCLs in the range we calculate.

The relationship between the threshold current(J_{th}) and the optical confinement factor(Γ) is $J_{th} \propto \exp(1/\Gamma)$. Considering this equation, maximizing the optical confinement factor of the active region is important to achieve a low threshold current. Therefore, it is expected that this increase of the optical confinement factor contributes to lowering the threshold current of the laser.

In an unstrained case, the heavy hole and light hole band in the valence band of a semiconductor are degenerated. In the presence of strain, the degeneracy is removed and the dispersion relation is modified according to Luttinger-Kohn Hamiltonian[7]. As a result, the effective hole mass of the valence band becomes lighter. Thus, it is also expected that this reduction of the effective mass resulting from SCLs increases hole mobility and makes hole transport to the well more efficient as shown in the report on the possibility of enhanced hole transport in the InGaAs/GaAs strained layer[8].

Therefore, we believe that SCLs make the threshold current of the laser lower due to both the enhancement of the optical confinement factor and the improvement of the hole injection efficiency.

On the other hand, the threshold current can be reduced further with controlling the reflectivity of the laser facet properly. By coating the laser facets(30% - 90%), we can reduce the threshold current down to the value of 25.4 mA as shown in Figure 4. The 300μm long lasers with SCLs have been operating for longer than 1000 hours maintaining the optical output power of 3mW at 50℃.

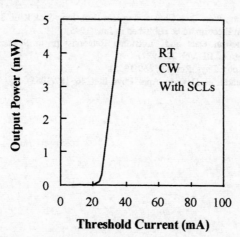

Figure 4. Current-optical output characteristics of the facet coated lasers with SCLs.

4. Summary

We have fabricated a tensile strained 630nm band AlGaInP single quantum well laser with strain compensated layers(SCLs) by three-step low pressure vertical MOCVD. The threshold current of the 200μm long laser without facet coating is 33.3mA. In comparison to the threshold current of 45.4mA for laser without SCLs, the threshold current value is reduced by 27%. The very low threshold current of 25.4mA has been obtained by coating of the facets with alternate sputtering of two dielectric materials. The lasing wavelength is 637nm under the condition of 3mW at room temperature. Under the life time test condition of 3mW at 50℃, the 300μm long lasers with SCLs have been operating for more than 1000 hours.

Acknowledgment

We would like to thank Seung-Chun Choi and Yong-Sung Jin for their helpful discussion, and Hee-Seok Song and Byung-Hun Song for the experimental supports.

References

[1] Won-Jin Choi, Jong-Seok Kim, Hyeun-Chul Ko, Ki-Woong Chung, and Tae-Kyung Yoo, J. Appl. Phys. 77, 3111(1995).
[2] T. Tanaka, H. Yanagisawa, S. Kawanaka, and S. Minagawa, Proc. 14th Int. Semicond. Laser Conf. p.125(1994).
[3] D. P. Bour, D. W. Treat, R. L. Thornton, T. L. Paoli, R. D. Bringans, B. S. Krusor, R. S. Geels, D. F. Welch, and T. Y. Wang, Appl. Phys. Lett. 60, 1927(1992).
[4] A. Valster, C. J. van der Poel, M. N. Finke, and M. J. B. Boermans, Proc. 13th IEEE Int. Semicond. Laser Conf. p.152(1992).
[5] Won-Jin Choi, Ji-Ho Chang, Won-Taek Choi, Seung-Hee Kim, Jong-Seok Kim, Shi-Jong Leem, and Tae-Kyung Yoo, IEEE J. Quantum Electron. to be published in June(1995).
[6] Landolt-Bornstein. Numerical Data and Functional Relationships in Science and Technology (O. Madelung, ed.), New Series group III, Vol. 22(1982).
[7] J. M. Luttinger, and W. Kohn. Phys. Rev. 97, 869(1955).
[8] J. E. Schirber, I. J. Fritz, and L. R. Dawson, Appl. Phys. Lett. 46, 187(1985).

Inst. Phys. Conf. Ser. No 145: Chapter 8
Paper presented at 22nd Int. Symp. Compound Semiconductors, Cheju Island, Korea, 28 August–2 September 1995
© *1996 IOP Publishing Ltd*

High Temperature Operation of 650nm AlGaInP Laser Diode

G. Lim, D. Shin, S. Kang, M. Oh, J. Kim, and T. Kim

New Material & Device Research Center, Samsung Advanced Institute of Technology, P.O.Box 111, Suwon, Korea 440-600

Abstract. High temperature operation of compressively strained GaInP/AlGaInP multiquantum-well(MQW) laser diodes with an emission wavelength of 650nm is reported. The threshold current is 40mA at 25°C and a light output power of 7mW is achieved up to 60°C. The characteristic temperatures (T_0) values for this laser are found to be 110K over a temperature range of 25-40°C, and 76K over the temperature range 45-60°C. Preliminary reliability test results of 400 hours are also presented.

1. Introduction

GaInP/AlGaInP visible-light laser diodes emitting of the 600nm have attracted much interest as light sources for optical information processing systems such as bar-code readers(BCR), laser pointers, and magneto-optical disks [1,2]. In practical applications, the laser diodes must operate in an ambient temperature of 60°C [3]. In the visible diodes, commercial 670nm wavelength products meeting the 60°C specification are available. However, the 670nm laser diodes are not suitable light sources for DVD (Digital Video Disc) systems, because the DVD standard specifies shorter wavelengths of 635nm or 650nm [3]. Recently, the needs for reduction in laser diode power consumption and package size are increasing for battery-operating hand-held BCRs and compact light laser pickup applications. In this letter, we report high temperature operation up to 60°C of compressively strained 650nm GaInP/AlGaInP multiquantum-well(MQW) laser diodes mounted in a 9.6mm-diameter can package. Reliability test of 400 hours are also discussed.

2. Laser structure and fabrication

The devices used in this experiment is a metal clad ridge stripe structure. Epitaxial layers were grown by a low pressure MOCVD method on 7° misoriented substrates. The stripe width is 5μm. The p-cladding layer under the metal clad layer is 0.3μm, and the cavity length is 400μm. Fig.1 shows a schematic cross section of the metal clad ridge stripe structure laser with four quantum well active layers. It consists of 0.3μm Si-doped n-GaAs ($n=2 \times 10^{18} cm^{-3}$) buffer, a 1.1μm Si-doped n-$(Al_{0.7}Ga_{0.3})_{0.5}In_{0.5}P$ ($n=2 \times 10^{18} cm^{-3}$) cladding, a 650Å undoped $(Al_{0.45}Ga_{0.55})_{0.5}In_{0.5}P$ waveguide, an undoped active layers consisting of four 50Å compressively strained $Ga_{0.5}In_{0.5}P$ (+0.5%) quantum wells separated by 70Å $(Al_{0.45}Ga_{0.55})_{0.5}In_{0.5}P$ barriers, a 650Å undoped $(Al_{0.45}Ga_{0.55})_{0.5}In_{0.5}P$ waveguide, a

1.1μm Zn-doped p-$(Al_{0.7}Ga_{0.3})_{0.5}In_{0.5}P$ ($p=7x10^{17}$ cm^{-3}) cladding, a 0.15μm Zn-doped InGaP ($2x10^{18}$ cm^{-3}) contact, and a 0.3μm Zn-doped p-GaAs (10^{19} cm^{-3}) cap layer.

     5μm wide ridge is formed by standard photolithography and wet chemical etching. AuCr/Au and AuGe/Ni/Au were thermally evaporated for p- and n-contacts. Au plating was done on the p-contact for bonding. Passivation coatings were formed by sputtering with a half wavelength thick SiO$_2$ film for both facets. Cleaved laser chips were mounted in a p-side down on silicon submount. The laser with submount is bonded on the Cu heatsink (φ=9.6mm can package) with epoxy, which is capped in a dry nitrogen ambient. Some chips were mounted directly on Cu heatsink with an In solder in a p-side down configuration.

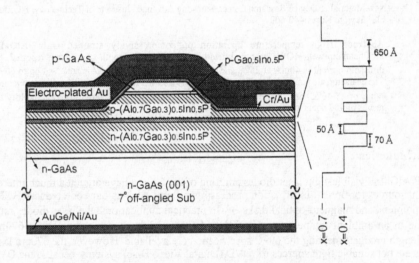

Fig.1. Schematic cross section of the metal clad ridge stripe structure

3. Experimental results and discussion

Fig.2 shows the temperature dependence of the light output against cw current characteristics for the laser diodes. The threshold current is 40mA and the peak wavelength is 650nm at 25°C. The devices operate up to 60°C with an optical output power exceeding 7mW, but thermally saturated to 3mW at 65°C. The maximum cw output power for the passivation coated laser is 19mW, and this is also limited by thermal saturation. Fig.3 shows the temperature dependence of the threshold current in cw and pulse mode (2μs at 2kHz). The characteristic temperature T_0 are 110K between 25 and 40°C, and 76K between 45 and 60°C in pulse mode. This small T_0 above 40°C is due to carrier overflow from the active region. T_0 for the cw operation is smaller than pulse operation due to thermal heating and carrier leakage over the clad layer. Fig.4 shows preliminary reliability test results for the lasers under 5mW cw constant output power operation. Operating current of the lasers with Si submount saturates after 200 hours operation at 40°C, which means that defects begins to saturate. At temperatures above 50°C all the lasers tested failed immediately. Lasers mounted directly on Cu heatsink with an In solder operate reliably at 50°C.

     To investigate the degradation mechanism we compared the threshold current and external quantum efficiency of the lasers before and after reliability test. Results show that threshold current scarcely changes from the initial values and external differential quantum efficiency decreases during aging test, which can be explained by the decrease in internal differential quantum efficiency. Obviously this indicates that defects are generated in the active layer during aging for some reasons [4]. Further experimental work is in progress and will be reported elsewhere.

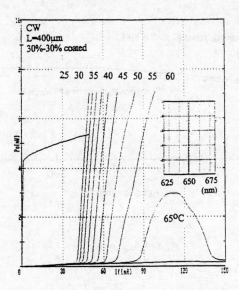

Fig.2. Temperature dependence of the light-current curves

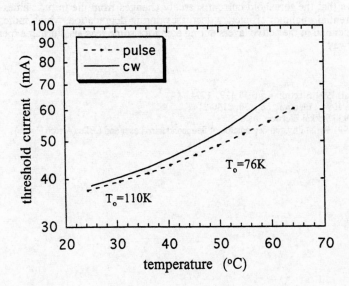

Fig.3. Temperature dependence of threshold current

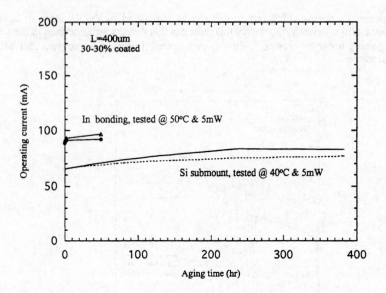

Fig.4. Reliability Test

4. Summary

In summary, high temperature operation of 650nm AlGaInP visible laser diodes with 4 strained QW active layers has been investigated. The devices operate up to 60°C with an optical output power exceeding 7mW, but thermally saturated to 3mW at 65°C. Reliability test has been done under 5mW cw constant output power operation. Operating temperature is limited by thermal saturation due to heating and carrier overflow to the clad layer. A failure analysis shows that the threshold current scarcely changes from the initial values and the external differential quantum efficiency decreases during degradation, which indicates that defects are generated in the active layers during aging for some reasons. Further experimental work is underway.

References

[1] A. Shima et al. 1994 Electron. Lett. **30** 1293-1294
[2] H. Fujii et al. 1994 Electron. Lett. **30**, 2140-2142
[3] K. Tsuda 1995 Nikkkei Electron. Asia 62
[4] M. Fukuda 1991 Reliability and degradation of Semiconductor Laser and LEDs (Artech House)

Inst. Phys. Conf. Ser. No 145: Chapter 8
Paper presented at 22nd Int. Symp. Compound Semiconductors, Cheju Island, Korea, 28 August–2 September 1995
© *1996 IOP Publishing Ltd*

Optimized Characteristics of 1.55 μm MQW DFB Lasers

Sang-Kook Han, Jung-Koo Kang, Bo-Hun Choi, Seung-Jo Jeong, Tae-Jin Kim, You-Ri Jo and Yong-Kun Sin

Photonic Devices Research Laboratory, Semiconductor R&D Laboratory II, Hyundai Electronics Industries Co. Ltd., Ami-ri, Bubal-eub, Ichon-kun, Kyoungki-do, 467-860, Korea

Abstract. We present the power and spectral characteristics of 1.55 μm MQW DFB laser diodes. The device parameters including a coupling coefficient and wavelength detuning are optimized to improve the laser performance. A slope efficiency of ~ 0.3 mW/mA per facet and a single longitudinal mode operation up to 27 mW with a SMSR of larger than 36 dB are achieved.

1. Introduction

High-power, high-speed and single-longitudinal mode long wavelength DFB lasers have attracted great attention for high bit rate optical communication systems. Nevertheless, due to many complicatedly inter-related device parameters, the optimization of DFB laser characteristics has been rather slowly progressed. Considering the relatively low device yield in DFB lasers, the optimization of the device performance is very important. In this paper, we report on the power and spectral characteristics of 1.55 μm MQW DFB laser diodes with the optimum device parameters. The characteristics of DFB lasers have been improved by optimizing the grating coupling condition and the wavelength detuning.

2. Experiment

Planar buried heterostructure(PBH) 1.55 μm DFB lasers with InGaAs/InGaAsP MQW active layer have been fabricated by four step LP MOCVD growth technique. Five of 7 nm thick InGaAs wells sandwiched by 20 nm thick InGaAsP (E_g=1.3 μm) barriers and 50 nm thick confinement layers on both sides of the active layer are grown on n-InP substrate. An InGaAsP (E_g=1.1 μm) layer is used as a waveguide layer on which the gratings are formed. First order gratings (~ 240 nm) are formed by holography and wet chemical etching. P-InP layers are overgrown on top of the gratings and normally 80 % of the grating heights are preserved. The widths of the buried active layer are kept narrower than 1.5 μm (normally 1.0 ~ 1.2 μm) to ensure the single mode operation in lateral direction. To increase the DFB mode selectivity, the front facet is AR coated. Utilizing an up-side

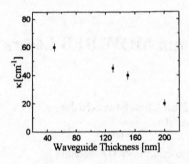

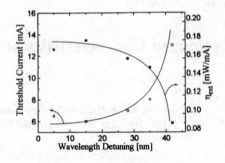

Fig. 1 Variation of κ as a function of waveguide thickness.

Fig. 2 I_{th} and η_{ext} of uncoated laser as a function of wavelength detuning.

grating scheme, the required grating periods can be accurately determined corresponding to the PL peaks measured from the MQW. Different grating periods and the waveguide layer thicknesses are used to see the variation of laser diode characteristics. Keeping the device length at 300 μm, the coupling coefficients (κ) of 15 ~ 65 cm⁻¹ are obtained by varying the waveguide layer thickness. Figure 1 shows the variation of κ as a function of waveguide layer thickness where the κ is determined from stop band measurements[1]. The DFB wavelength is detuned up to 40 nm from the material gain peak by changing the grating period.

3. Results and discussion

Variations of the threshold current (I_{th}) and the external quantum efficiency with different wavelength detunings are shown in Fig. 2. The measured devices are uncoated and the MQW has a PL peak at 1.56 μm. The power characteristics become worse as the wavelength detuning increases. But considering the fact that the linewidth enhancement factor becomes smaller as the wavelength detuning increases, an optimized detuning which facilitates small chirping is essential for high speed modulation characteristics. Figure 3 represents I - L and I - V characteristics of 2 % AR coated DFB laser diodes. The threshold current is in the range of 5 - 7 mA without facet coating and increases to 7 - 10 mA with one facet AR coating. A slope efficiency of 0.3 mW/mA per facet and the maximum output power larger than 20 mW are achieved. The measured DFB lasing wavelengths, which can be easily altered by many device parameters, are quite uniform as 1548 ± 2 nm. Based on this fact, we believe that the thickness and composition uniformity of epitaxial layers and the grating uniformity are precisely well maintained. Temperature dependent I - L characteristics measurements up to 50 °C have been performed and T_0 of 40 K has been obtained from I_{th} variation. The power characteristics variation at the elevated temperature is mainly due to the reduced differential gain possiblely resulted from the electrostatic band-profile deformation [2]. In our DFB-LDs, the differential gain has been decreased up to 30 % when the temperature changes from 20 to 50 °C. For the purpose of comparison, a DFB laser diode which shows severe power saturation at 12 mW

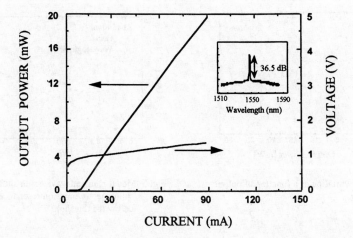

Fig. 3 I-L-V characteristics of DFB laser with an AR facet where the spectrum was measured at 5 mW.

is tested to see the temperature sensitivity of the differential gain. The differential gain has been reduced more than 45 % and this fact may explain why the severe power saturation has been observed. Thus, a well designed QW active layer structure which minimizes the carrier overflow in QWs is imperative to improve the power characteristics in high current injection region.

In the case κ is smaller than 30 cm⁻¹, due to a small threshold gain difference between FP and DFB modes, both FP and DFB modes exist at operating power level of 5 mW. The DFB mode selectivity has been enhanced by either increasing the coupling condition or having AR coating in one facet. A single longitudinal mode operation up to 27 mW with SMSR of larger than 36 dB is observed where κ is around 60 cm⁻¹. The variation of the SMSR as a function of output power is shown in Fig. 4 where the SMSR begins to saturate at the output power of 12 mW.

The small signal frequency response measurements have been performed and resulted in a resonant frequency and 3-dB bandwidth of 3.77 and 5.0 GHz, respectively, at the power level of 3 mW. The differential gain (dg/dn) has been obtained by using a relation between the resonance frequency and square root of the output power [3]. From a D coefficient of ~ 1.5 GHz/(mW)$^{0.5}$, the differential gain turns out to be 2.4 x 10 $^{-16}$ cm^2.

The investigation of the gain characteristic of the MQW laser diodes is important for the observation of the quantum well states. We have carried out the measurements of the injection current dependence of gain and index of MQW DFB laser diodes. Modal gains are obtained by measuring the ratio of the maxima to the minima of the Fabry-Perot resonance in the spontaneous emission spectrum below the threshold current [4]. At the same time, the index variation is obtained by measuring the wavelength shifts of the each resonance peak [5]. Figure 5 shows the variations of modal gain and wavelength shift at 1.54 μm as a function of the injection current below the threshold. The linewidth enhancement factor is determined by using the measured gain variation and the wavelength shifts [6]. The α value of ~ 5 is obtained and it decreases as the DFB lasing wavelength decreases.

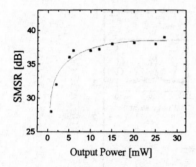

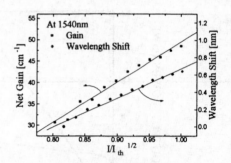

Fig. 4 SMSR variation as a function of output power.

Fig. 5 Modal gain and wavelength shift of spontaneous emission resonance peak below the threshold.

4. Summary

In summary, we have presented the optimized power and spectral characteristics of 1.55 μm MQW DFB laser diodes. The performance comparison with different coupling coefficients and wavelength detunings has been carried out experimentally. A slope efficiency of ~ 0.3 mW/mA per facet and a single longitudinal mode operation up to 27 mW with a SMSR of larger than 36 dB are achieved. Gain spectrum measurements show that our MQW structure has the differential gain of 2.4 x 10^{-16} cm^2 and the linewidth enhancement factor of ~ 5 at the lasing wavelength.

References

[1] Ketelsen L T P, Hoshino I and Ackerman D A 1991 *IEEE J. Quantum Electron.* **27** 965-975

[2] Seki S and Yokoyama K 1995 *IEEE Photonics Tech. Lett.* **3** 251-253

[3] Suemune I 1991 *IEEE J. Quantum Electron.* **27** 1145-1159

[4] Hakki B W and Paoli T L 1975 *J. Appl. Phys.* **3** 1299-1306

[5] Henning I D and Collins J V 1983 *Electron. Lett.* **19** 927-929

[6] Sasai Y, Ohya J and Ogura M 1989 *IEEE J. Quantum Electron.* **25** 662-667

High-Temperature Reliability of Aluminium-free 980 nm and 808 nm Laser diodes

J. Diaz, H. Yi, C. Jelen, S. Kim, S. Slivken, I. Eliashevich, M. Erdtmann,
D. Wu, G. Lukas, and M. Razeghi

Center for Quantum Devices, Northwestern University, Evanston, Il 60208

Abstract. We demonstrate the high-temperature reliability characteristics of uncoated Al-free 980 and 808 nm lasers. Excellent laser characteristics (T_o, J_{th} and η_d), low thermal dissipation (close to the theoretical limit), and low free-carrier loss in the InGaP layers have been measured. We report high power laser arrays (up to 70 W), lasers operating at high-temperatures (100 °C), and lifetime results for Al-free lasers operating over 3000 hours without degradation of the uncoated mirrors.

1. Introduction

980 nm and 808 nm semiconductor lasers are very compact and efficient sources for pumping of Er-doped fiber amplifiers and YAG lasers, respectively. There are several requirements for the practical usage of pump lasers. In most applications, lasers require high output power, a wide range of operating temperatures, and long-term reliability. Low J_{th} (threshold current density) and high η_d (differential efficiency) are other important factors in laser operation. Realization of low J_{th} and high η_d is of utmost importance for high-temperature reliability since it is related to the absence of high non-radiative recombination centers or optical absorption, whose adverse effects will be further enhanced at higher temperatures, leading to gradual or sudden failure of these devices. Minimization of non-radiative recombination and absorption at high temperatures are the essential requisite for high-quality and high-reliability laser devices. Other requirements are a stable, narrow far-field pattern, preferably circularly shaped to optimize the coupling efficiency to optical fibers, and a narrow spectrum width, preferably single longitudinal mode. In this paper, we present detailed studies on Al-free 980 nm and 808 nm lasers operating at high temperatures and high power conditions.

Long recognized advantages of Al-free 980 nm and 808 nm lasers are their resistance against facet degradation due to oxidation and the ease of selective chemical etching between the InGaP and GaAs layers which facilitates multiple-step procedure needed for opto-electronic circuits [1].

2. Experiment and discussion

The separate confinement hetero-structure (SCH) of the 980 nm lasers used in this experiment is composed of two undoped 40 Å-thick InGaAs quantum wells separated by a 100 Å-thick GaAs barrier. The active region is sandwiched between 600 Å-thick GaAs waveguiding layers and 1 μm-thick GaInP cladding layers. The 808 nm lasers have a similar SCH with an InGaAsP (E_g = 1.52 eV) 300 Å-thick quantum well, 0.3 μm-thick InGaAsP (E_g = 1.75 eV) waveguide layers and InGaP cladding layers. These laser structures were grown by low-pressure MOCVD at relatively low temperature (510 °C) compared to the 700 °C commonly used for InGaAsP material growth in order to avoid miscibility problems [2]. Broad-area lasers with 100 μm-wide contact stripes were fabricated using standard photolithography techniques. After ohmic contact deposition and cleaving, the individual diodes were mounted p-side down on copper heat sinks by indium bonding. The mirrors of all the laser diodes used in this experiment are uncoated.

2.1. 980 nm lasers

Threshold current density J_{th} and differential efficiency η_d were measured for uncoated lasers under pulse operation (pulse width 400 ns and frequency 1250 Hz). The dependence of J_{th} on temperature is shown in Fig. 1. High characteristic temperature (T_o) over 350 K was obtained in the temperature range of 20 ~ 40 °C. When considering the overall temperature range, the temperature dependence remains above 170K. This T_o is significantly higher than previously reported value in the same SCH Al-free 980 nm lasers [2-4] and it is comparable with graded index SCH (GRIN-SCH) AlGaAs lasers (T_o ≅ 160 K in the same temperature range) [5]. However, it should be stressed that the laser structure used in this experiment is a simple SCH with GaAs waveguiding layers. According to Ohkubo et al [2], by using GRIN-SCH structure with higher band gap material such as InGaAsP (E_g ~ 1.75 eV) for the waveguide layer, the T_o could be

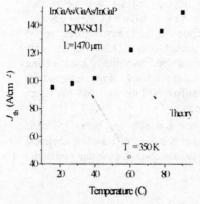

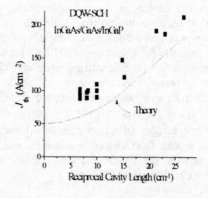

Fig. 1. J_{th} vs. Temperature of 980 nm lasers

Fig. 2. J_{th} vs. 1/L (cavity length) of 980 nm lasers

increased by a factor of 2 or 3 due to the enhanced optical confinement and barrier height which reduces carrier leakage into the waveguide and cladding layers. Along these lines, the laser material system examined in this paper could have a higher T_o if the structure were optimally designed.

Measurements of J_{th} and η_d for various cavity lengths of laser diodes were performed. J_{th} of 80 A/cm^2 (Fig.2) and η_d of 1.0 W/A were obtained in long and short cavity lengths, respectively. Although relatively low band gap GaAs was used for the waveguiding layers, no leakage current was observed, contrary to previous results [2].

$1/\eta_d$ vs. L measurements revealed that the laser structure has a very low internal loss of $\alpha_i \cong 1.8$ cm^{-1}. This is anomalously low because in this structure the light is mainly distributed over the highly doped ($0.5 \sim 1 \cdot 10^{18}$ cm^{-3}) InGaP cladding layers. (Using the refractive indices given in Ref. 6, the optical confinement factor Γ for the InGaP cladding layers is evaluated to be 0.7). Assuming that most of the internal loss comes from free-carrier absorption $\alpha_{f\text{-}c}$, $\alpha_{f\text{-}c}$ is about 2.5 cm^{-1}, which is $3 \sim 4$ times smaller than reported values for GaAs or AlGaAs for the same doping concentrations, as determined by direct transmission or Raman scattering [7]. This result is consistent with a recent experiment on similar 808 nm InGaAsP/InGaP laser structures [8] in which it was found that changing the optical confinement of the highly doped InGaP cladding layers does not affect the threshold condition or the efficiency of the lasers. This strongly supports the above argument that free-carrier absorption in InGaP layers is very small.

Dependence of J_{th} on cavity length is in good agreement with theory (Fig. 2). For the calculation of J_{th}, the heavy hole mass was calculated by the strained-layer QW model within the 4 bands **k·p** perturbation method [9] assuming 20 % indium composition in InGaAs. Gain and radiative recombination rate were calculated without Coulomb-interaction except for intraband scattering [10]. Calculation shows that the shallow GaAs barriers result in relatively high transparent current density (~ 50 A/cm^2) as shown in

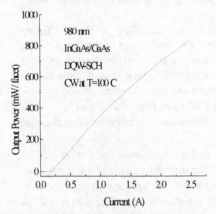

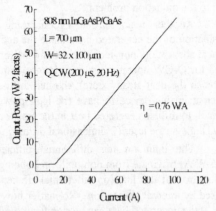

Fig. 3. Light-current characteristics of 980 nm lasers in CW operation at 100 °C

Fig. 4. Light-current characteristics of an 808 nm laser bar in Q-CW operation

Fig.2. Even for low mirror-loss lasers, the quasi-Fermi energy for conduction electrons (~ 95 meV) is close to the conduction band gap discontinuity between $In_{0.2}Ga_{0.8}As$ and GaAs (~ 115 meV), causing significant reduction of both the gain and the injection efficiency. In our calculation, we estimate that the threshold current for long cavity-length lasers could have been lowered to 40 A/cm^2 if higher band gap InGaAsP had been used for barriers, instead of GaAs. The $T_o = 350$ K is also very close to the theoretical limit shown in Fig. 1 by a dotted line. In the temperature dependence calculation, we took into account only the thermal gain broadening effect. Based on the above analysis we conclude that the lasers are free of non-radiative recombination centers or optical absorption which would degrade high-temperature operation of these devices. This suggests high reliability lasers could be operated in such conditions.

Lasers operating at high temperatures directly verify this expectation. Figure 3 shows CW-operation of a single laser chip (100 μm width) with an output of 800mW per facet at a temperature of 100 °C. Further increase in the operating temperature was limited by the available thermo-electric cooling system. Preliminary lifetime testing of laser diodes with a high reflectivity coating on the back-facet and a anti-reflectivity coating on the front-facet was performed at 50 °C with a constant emission power of 1 W. Degradation after aging for 2000 hours was less than 15 %.

2.2. 808 nm lasers

Al-free 808 nm lasers have also shown very high power laser emission due to the high quality of the material and to fabrication technologies such as bonding and metallization [1,11]. Figure 4 shows a 1-cm wide laser bar consisting of 32 stripes with an 100 μm emitting aperture generating 70 W per 2 facets of quasi-continuous (200 μs-20 Hz) power at 100 A of pumping current. Higher power operation was only limited by the available current source. The measured laser array bars had uncoated mirrors. No demonstration of such high power without mirror coating is reported for Al-based lasers due to the oxidation problem.

Another major problem concerning high-power reliability is the inefficient dissipation of the generated heat inside the active layers or other parts of the lasers due to optical absorption, non-radiative recombination, or Joule heating. Since the 200 μs pulse used for Q-CW operation is much longer than the equilibration time (time for the system to reach the near-steady state), thermal diffusion from the heated region should be very efficient in this device to have the high power Q-CW operation. Note that it is more difficult to diffuse generated heat in the laser bar (1-dimensional diffusion) than in the case of a single-stripe laser (2-dimensional diffusion).

The limit on the differential efficiency under high-power operation ($\eta_d \sim 0.75$ W/A) originate from non-uniform bonding or from a defective area of the stripe. We will show that this limit on differential efficiency is mainly due to a temperature increase caused by internal absorption. No matter how small the absorption is (typically $\alpha \cong 2 \sim 4$ cm^{-1}), the generated heat can be significant under high power operation. The temperature inside the active region can be estimated from the thermal conduction equation:

$$K \frac{\partial^2 T}{\partial z^2} = -g \qquad (1)$$

where z is the direction normal to the layers, K is thermal conduction ($= 47$ W/m·K for GaAs [12]), and g is the power density of the heat source. Assuming that the heat is generated mainly due to internal absorption by free-carriers in the SCH region, Eq. (1) gives an approximate solution,

$$\Delta T \cong \frac{P(1 - \eta_d) L_{cl}}{K L_w L_{cav}}. \qquad (2)$$

where L_{cl} is cladding layer thickness ($= 1$ μm), L_w is total aperture width ($= 32 \times 100$ μm), and L_{cav} is cavity length ($= 700$ μm). As shown in Fig. 4, the absorbed power P (1-η_d) is about 30 Watts, giving $\Delta T \cong 40$ °C from Eq. (2). It is worth noting that the thermal resistance $R_{th} \equiv \Delta T / P(1 - \eta_d) \cong 1.3$ °C/W for a laser array bar is consistent with the previous value measured for a single laser diode ($R_{th} \cong 1$ °C/W) [13]. With the heat sink temperature of 25 °C, the active layer temperature (lattice temperature) is about 65 °C. Dependence of differential efficiency on temperature was measured elsewhere [10]: η_d was measured about 0.8 W/A at 70 °C. Thus, this analysis indicates that the limit on differential efficiency entirely originates from temperature increase in the active region, not due to non-uniform bonding or from defective stripes in the array.

The small values of the optical internal absorption, non-radiative recombination, and Joule heating suggests that these laser diodes are highly efficient in converting electrical power to optical power. Figure 5 shows a typical light-current curve for a single diode operating above 6 W per facet under pulse operation with a differential efficiency of 1.3 W/A (86 %) at room temperature. It should be pointed out that uniform results of η_d were obtained (>1.1 W/A) from over 80 % of the samples randomly chosen. This uniformity stems from the high-quality of the grown wafers which is an essential feature of high-power lasers.

For a direct measure of high-power and high-temperature reliability, several uncoated 808 nm lasers were tested at constant powers of 500 mW and 1 W per facet in continuous wave operation. As shown in Fig. 6, no detectable degradation of performance was observed in a 3000 hours aging test. An decrease of current in constant

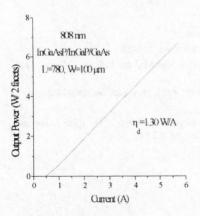

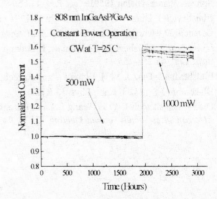

Fig. 5 Light-current characteristics of InGaAsP/ GaAs laser diodes

Fig. 6. 500 mW and 1 W cw aging test at 25 °C for randomly chosen 808 nm InGaAsP/GaAs lasers

optical power mode was observed. This decrease of current may be attributed to current-induced annealing of the indium bond, resulting in a more uniform current distribution across the contact stripe. This lifetime is a few orders of magnitude longer than that of AlGaAs laser systems at the same wavelength [11].

3. Conclusion

In this paper we have demonstrated that Al-free 980 and 808 nm lasers have reliability characteristics in high-power and high-temperature operation superior to the Al-based material system at the same wavelength. Al-free lasers are found to have the following superior characteristics:

(1) Very high T_o (close to limit predicted from thermal gain broadening) due to absence of defects.

(2) 1 Watt of power under CW-operation for 980 nm lasers at 100 °C.

(3) Very small free-carrier absorption in the doped InGaP cladding layer.

(4) Very efficient heat diffusion through the epilayers and the heatsink.

(5) Excellent uniformity and reproducibility of the material grown by MOCVD

(6) A lifetime over 3000 hours for laser with uncoated mirror facets, which is a few orders of magnitude longer compared to Al-based material system.

Reference

[1] Razeghi M 1994 Nature **369** 631

[2] Ohkubo T Ijichi, Iketani A, Kikuta T 1994 IEEE J. Quantum Electron. **QE-30** 408

[3] Vail E, Lim S, Wu Y, Francis D, Chang-Hasnain C, Bhat R and Caneau C 1993 Appl. Phys. Lett. **63** 2183

[4] Wu M, Chen Y, Kuo J, Chin M, Sergent A 1992 IEEE Photon. Technol. Lett. **4** 676

[5] Van der Ziel J and Chand N 1991 Appl. Phys. Lett. **58** 1437

[6] Derry P, Hager H, Chiu L, Booher D, Miao E and Hong C 1992 IEEE Photon. Technol. Lett. **4** 1189

[7] Li Z, Dion M, McAlister S, Willams R, Aers G 1993 IEEE J. Quantum Electron. **QE-29** 346

[8] Hill D 1964 Phys. Rev. **133** A866

[9] Spitzer W and Whelan 1959 J Phys. Rev. **114** 59

[10] Elaishevich I, Diaz J, Yi H and Razeghi M 1994 unpublished

[11] Gershoni D, Henry C, Baraff G 1993 IEEE J. Quantum Electron. **29** 2433

[12] Yi H, Diaz J, Elaishevich I, Stanton M, Erdtmann M, He X, Wang L.J and Razeghi M 1995 Appl. Phys. Lett. **66** 253-255

[13] Elaishevich I, Diaz J, Yi H, Wang L J and Razeghi M 1995 Appl. Phys. Lett. **66** 3087-3089

[14] Blakemore J S 1982 J. Appl. Phys. **53** R123

[15] Diaz J, Elaishevich I, Yi H, Wang L J and Razeghi M 1994 *Proceedings of the IEEE Lasers and Electro Optics Society Annual Meeting*, LEOS '94, Boston, **SL 13.3** 65

Quantum cascade lasers

F Capasso, J Faist, C Sirtori, A L Hutchinson, D L Sivco and A Y Cho

AT&T Bell Laboratories, Murray Hill, NJ 07974

Abstract. Recent advances in quantum cascade lasers are discussed. Using a vertical intersubband transition design with electron Bragg confinement, continuous wave operation has been achieved at 4.6 μm and $\approx$ 8.0 μm wavelength.

1. Introduction

The recently demonstrated quantum cascade (QC) laser is a fundamentally new semiconductor laser [1]-[3]. It relies on only one type of carrier (unipolar laser) and on quantum jumps of electrons between discrete conduction band energy levels of quantum wells. As such the wavelength can be tailored over a very wide range by varying layer thicknesses.

The relevant intersubband radiative transition is between states centered in different neighboring wells to facilitate population inversion, i.e. the transition is *diagonal*. In this design, however, the width of the luminescence transition is relatively broad (FWHM~20 meV) due to the interface roughness since electrons traverse several heterointerfaces in the photon emission process. As a consequence the peak gain is reduced.

To circumvent this problem the structure of Fig. 1 was designed where electrons make a *vertical* radiative transition essentially in the same well [4]. This reduces considerably the width of the gain spectrum (FWHM $\approx$ 10 meV) and therefore the laser threshold current density. To prevent electron escape in the continuum, which is greatly reduced in the case of the diagonal transition, the superlattice of the digitally graded injector is designed to act as a Bragg reflector for electrons in the higher excited state and to simultaneously ensure swift electron escape from the lower states via a miniband facing the latter (Fig. 1). Using this design with AlInAs/GaInAs heterostructures grown by MBE, pulsed operation at 4.6 μm wavelength with $\approx$ 300 mW/A and threshold J_{th} $\approx$ 2.3 kA/cm^2 has been achieved at 80 K [4]. Longer wavelength (λ = 8.4 μm), QC lasers have also been demonstrated in the same material system [5]. These structures were of the diagonal type and operated in a pulsed mode with a threshold of 2.8 kA/cm^2 and peak optical power 30 mW at 100 K heat sink temperature.

In this paper we discuss recent advances in QC lasers, in particular the achievement of single mode continuous wave (cw) operation at 4.6 μm [6] and 7.8 μm [7].

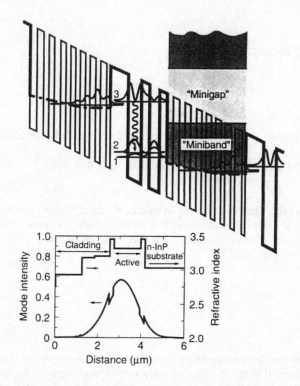

Fig. 1. Schematic conduction band diagram of a portion of the quantum cascade laser under positive bias condition and an electric field of 7.0×10^4 V/cm. The wavy line indicates the transition responsible for laser action. The moduli squared of the relevant wavefunctions are shown. For sample A, the layer sequence of one period of the $Al_{0.48}In_{0.52}As/Ga_{0.47}In_{0.53}As$ structure, in nanometer, left to right and starting from the injection barrier is (6.8/4.8), (2.8/3.9), (2.7/2.2), (2.2/2.1), (2.1/2.0), (2.0/1.8), (1.8/1.7), (2.0/1.6), (2.2/1.6), (2.4/1.4). In sample B, this sequence has one extra period (shown in bold) and is (6.8/4.8), (2.8/3.9), (2.7/2.2), (2.2/2.1), (2.1/2.0), **(2.0/2.0)**, (2.0/1.8), (1.8/1.7), (2.0/1.6), (2.2/1.6), (2.4/1.4). The structures are left undoped, to the exception of layer number twelve to eighteen for sample A and fourteen to eighteen for sample B which are n-type doped to $\cong 3 \times 10^{17}$ cm^{-3}. Inset: refractive index and optical mode profile of the waveguide. The two 200 nm thick GaInAs layers serve the purpose of enhancing the optical confinement.

2. cw Operation of Quantum Cascade Lasers at $\lambda = 4.6 \ \mu m$

In this section we describe the cw operation of QC lasers with a vertical transition at $\lambda = 4.6 \ \mu m$ [6].

Our InGaAs/AlInAs heterostructures, grown lattice-matched to an n^+ InP substrate by MBE, consists of twenty-five stages, each one comprising a coupled-well active region and a graded-gap superlattice injection/relaxation region (see Fig. 1) [4]. The coupled well active region is engineered so that at the threshold field (70 kV/cm), the ground states of the two 4.8 and 3.9 nm quantum wells have anticrossed. As a result, we obtain a short ($\tau_{21} = 0.6$ ps) lifetime between the $n = 1$ and $n = 2$ states since their separation is resonant with the optical phonon. In agreement with the population inversion condition, the lifetime of level 3 $\tau_3 = (\tau_{31}^{-1} + \tau_{32}^{-1})^{-1} = 1.2$ ps is longer, since it involves optical phonon emission associated with a large momentum transfer.

The energy difference Δ between the quasi Fermi-energy in the graded gap region and the lower state ($n = 2$) of the lasing transition is an important parameter controlling the high-temperature behavior of our devices [4], since the thermal population of the $n = 2$ state ($\sim \exp(-\Delta/kT)$) must be compensated by a higher injected current to maintain the population inversion at the threshold value. A higher Δ should lead to lower thresholds at high temperatures, increase the maximum operating temperature and improve the cw slope efficiency. For a threshold electric field (70 kV/cm) fixed by the alignment of the ground state of the injector with the $n = 3$ state of the active region, Δ is proportional to the length of the injector. Two similar structures were grown, differing only by the length of their injectors, longer by one period (4 nm) in sample B, leading to an expected increase from $\Delta \sim 72$ meV for sample A to $\Delta \sim 100$ meV in sample B.

Because of alloy scattering, ternary materials have a thermal resistance fifteen to twenty times larger than binaries such as InP or GaAs. For this reason and to bring the active region closer to the InP substrate, the lower AlInAs cladding layer is eliminated and the first of the two 200 nm InGaAs layers [2]-[4] on both sides of the active region is grown directly on the substrate following a short 35 nm graded gap layer (see inset of Fig. 1). As in Ref. [4] the top cladding layer, grown on top of a 30 nm thick, interface smoothing compositionally graded AlInGaAs layer, consisted of three AlInAs regions of thickness 700, 600 and 1200 nm with n-type doping concentrations of $2 \times 10^{17} \text{cm}^{-3}$, $3 \times 10^{17} \text{cm}^{-3}$ and $7 \times 10^{18} \text{cm}^{-3}$ grown in this sequence. A top 10 nm thick n^{++} contact layer, Sn doped to 10^{20}cm^{-3}, separated from the AlInAs cladding by a 30 nm thick n^+ AlInGaAs graded layer, completes the growth.

The samples were processed into mesa etched ridge waveguides of width $= 7-9 \ \mu m$ by wet chemical etching and SiO_2 (350 nm) insulation [4]. Non alloyed Ti/Au ohmic contacts were provided to the top layer and the substrate. The lasers, 1.3, 2.25 or 3.5 mm long, were then soldered epilayer up to sapphire or Cu holders, wire bonded and mounted on the cold head of a He flow cryostat. Fig. 2 shows the optical power versus drive current obtained using f/0.8 optics and a calibrated, room temperature HgCdTe detector. It is clear that the sample with higher Δ, i.e. B, has significantly better performance in terms of optical power, slope efficiency and maximum operating temperature. At 50 K, the threshold current density is $1.82 \ \text{kA/cm}^2$ with an increased slope efficiency (dP/dI = 80 mW/A). The device delivers maximum powers of 17 mW at 50 K and 2 mW at 85 K.

In pulsed operation (30 ns pulses with 20 kHz repetition rate) the devices of sample B had a peak power as high as 20 mW at the maximum operating temperature (230 K). The measured temperature dependence of J_{th}, obtained from the pulsed data, falls clearly in two regimes: below 140 K where the T_0 is very high (T_0 = 150 K) and above 140 K where its value drops to 96 K. This high T_0 is a fundamental feature of QC lasers [3]. In the vast majority of the temperature and current range, the cw spectrum (measured with an FTIR spectrometer) was found to be monomode with sidelobes below 20 dB, limited by our spectrometer's resolution (inset of Fig. 3). The typical tuning curve of a 2.25 mm long device as a function of holder temperature from sample A is displayed in Fig. 3, at a fixed injection current. The total continuous tuning range is 1.8 cm^{-1}, or 4 nm. Three typical spectra are also reported in the inset. The device is also tunable with current, in which case a total continuous tuning range of 1.2 cm^{-1} at T = 13 K is obtained.

3. cw Operation of a Quantum Cascade Laser at λ = 7.8 μm

The recent demonstration of a QC laser operating at longer wavelengths (λ = 8.4) was achieved by increasing the active region layer thicknesses and by using a plasmon enhanced waveguiding scheme [5]. This design is also used in the devices described in this section [7]. In brief, the TM fundamental guided mode couples to the high loss plasmon mode propagating at the top metal contact semiconductor interface. This coupling can significantly increase waveguide losses and reduce the confinement factor Γ [5]. To minimize this effect, without growing a prohibitively thick top cladding layer, the doping of the contact layer is increased (n^+ = 7×10^{18} cm^{-3}) so that the plasma wavelength (λ_p = 9.4 μm) in this region approaches but does not exceed the wavelength of the guided mode (λ = 7.8 μm). Due to well known behavior of the refractive index near the plasma frequency (for $\omega > \omega_p$), a large refractive index difference is created between the guided mode and that of the contact layer. This considerably reduces the coupling with the interface plasmon mode and therefore also the waveguide losses [5].

The band diagram for the cw 8 μm laser is qualitatively similar to that of Fig. 2. The undoped two GaInAs quantum wells of the 25 active region are separated by a 1.6 nm GaInAs barrier and have thicknesses of 8.6 and 6.4 nm. The AlInAs injection barrier is 5 nm thick. The spatially graded injector relaxation regions separating the active ones consist of four GaInAs wells of thicknesses 4.8 nm, 4.2 nm, 4.0 nm and 3.6 nm, with the thinner well adjacent to the injection barrier, separated by 1 nm AlInAs barriers. The 4.2 nm and 4.0 nm wells and the 1 nm barriers are doped n^+ type to 1.5×10^{17} cm^{-3}. Similarly to the cw 4.6 μm design (Section 2) the InP substrate serves the function of one of the cladding layers while an AlInAs cladding provides optical confinement on the opposite side.

Fig. 4 shows the L-I curve at 80 K; the measured optical power is as high as 10 mW and the threshold is 2 kA/cm^2. The spectrum is shown in the inset.

4. Conclusions

QC lasers are a promising technology for mid-ir lasers, even at this preliminary stage, in light of their significant cw single mode power and the ability to use wider gap technologically mature and more reliable materials such as InP and GaAs based alloys. Lead salt lasers are currently the only commercial mid-ir semiconductor lasers [8]. These small band gap materials are, however, more difficult to process and less reliable than the wider gap III-V compounds. The multimode emission with cw modal power of at most a

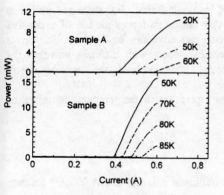

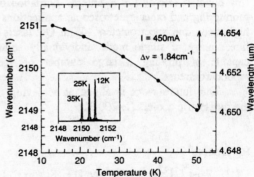

Fig. 2. Continuous optical output power from a single facet versus injection current for various heat sink temperatures, as indicated. Upper curves: sample A (7 µm×3.5 mm). Lower curves, sample B (9 µm×2.25 mm).

Fig. 3. Tuning curve of sample A versus temperature for a fixed injection current I = 450 mA. The spectrum is monomode in the whole temperature range. Inset: three representative spectra at T = 13, 25 and 40 K. The sidelobes on either side of the main peak are artifacts of the Fourier transform used to compute the spectrum.

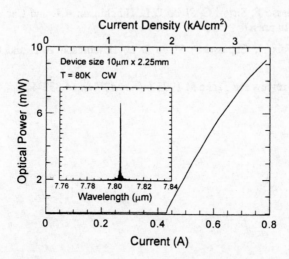

Fig. 4. Continuous optical power from a single facet versus injection current of a λ = 7.8 µm QC laser. The inset shows the spectrum at a current of 450 mA.

few hundred microwatts is another limitation of lead salt lasers. For trace-gas pollution monitoring and other spectroscopic applications this limitation forces the use of expensive high-resolution spectrometers. With QC lasers one can envision for these applications a linear array of single mode, individually addressable lasers with different wavelengths capable of probing a large number of spectral lines without the use of costly monochromators.

Our future work is aimed at achieving cw operation in the temperature range of thermoelectric coolers ($\geq$ 200 K).

References

[1] Faist J, Capasso F, Sivco D L, Sirtori C, Hutchinson A L and Cho A Y 1994 *Science* **264** 553-55

[2] Faist J, Capasso F, Sivco D L, Sirtori C, Hutchinson A L and Cho A Y 1994 *Electron. Lett.* **30** 865-66

[3] Faist J, Capasso F, Sivco D L, Hutchinson A L, Sirtori C, Chu S N G and Cho A Y 1994 *Appl. Phys. Lett.* **65** 2901-3

[4] Faist J, Capasso F, Sirtori C, Sivco D L, Hutchinson A L and Cho A Y 1995 *Appl. Phys. Lett.* **66** 538-40

[5] Sirtori C, Faist J, Capasso F, Sivco D L, Hutchinson A L and Cho A Y 1995 *Appl. Phys. Lett.* **66** 3242-44

[6] Faist J, Capasso F, Sirtori C, Sivco D L, Hutchinson A L and Cho A Y 1995 *Appl. Phys. Lett.* (in press)

[7] Sirtori C, Faist J, Capasso F, Sivco D L, Hutchinson A L and Cho A Y 1995 unpublished

[8] For a recent review see Tacke M 1995 *Infrared Phys.* **64** 833-44

Inst. Phys. Conf. Ser. No 145: Chapter 8
Paper presented at 22nd Int. Symp. Compound Semiconductors, Cheju Island, Korea, 28 August–2 September 1995
© *1996 IOP Publishing Ltd*

Long-Wavelength Lasers and Transmitters: Physics and Technology Roadblocks

P Bhattacharya, H Yoon, A Gutierrez-Aitken, K Kamath, and P Freeman

Department of Electrical Engineering and Computer Science
The University of Michigan
Ann Arbor, Michigan 48109-2122, U.S.A.

Abstract. Our work on high-speed InP-based (1.55μm) lasers and integrated transmitters is described. Issues related to carrier dynamics, device design, epitaxy and fabrication, that limit the performance of these devices and OEICs are discussed.

1. Introduction

A primary requirement of a high-speed optical fiber communication system utilizing wavelength division multiplexing (WDM) is a 1.55μm transmitter with a large modulation bandwidth. The laser diode used in such a transmitter is usually designed and fabricated with InP-based heterostructures. Two schemes of modulation can be employed: internal (or current) modulation, and external modulation. There is therefore a need to develop lasers and modulators which can demonstrate large modulation bandwidths. This has been an elusive goal, particularly in the case of the laser. The underlying physics and technology roadblocks, and some solutions are presented here.

2. Small-signal modulation of discrete 1.55μm lasers

Although the theoretically predicted modulation bandwidth in 1.55μm lasers can be quite high (~35 GHz), practical lasers have demonstrated maximum bandwidths of ~20-25 GHz[1]. The largest bandwidths are obtained only in buried heterostructure (BH) devices with modulation doping measured at high photon densities. For a ridge waveguide laser the maximum reported bandwidth is 17 GHz[2]. We have optimized the design and fabrication of 1.55 μm InP-based ridge and BH lasers with compressively strained and strain compensated MQW separate confinement heterostructures. Our objective was to determine the performance characteristics of optimally designed lasers under low photon density conditions.

Two types of laser structures were grown by MOVPE. In one the active region consists of a 4-period compressively strained (0.38%) MQW with quaternary wells. In the other, the active region consists of a 8-period compressively strained (0.38%) MQW with the strain in the InGaAsP wells *compensated* by tensile strain in the barriers. The larger number of wells was used to increase the differential gain and slightly improve the optical confinement factor. Single-mode $3\mu m$ wide ridge waveguide lasers were made by standard photo lithography and wet and dry etching. Selective hydride VPE growth of undoped InP was used to fabricate single mode BH lasers. An important extrinsic and technology-related factor that affects the ultimate modulation bandwidth of a laser is the nature of the p-type contact and the associated series resistance. The p- and n-ohmic metalizations are Pt/Ti/Pt/Au and Ni/Ge/Au/Ti/Au, respectively. S-parameter measurements on 200 μm length lasers demonstrated electrical impedance bandwidths > 40 GHz, indicating that the parasitics are reduced to a minimum.

Measurements were made at room temperature on 200 μm length devices without any heatsinking. Also, the facets were left uncoated. A maximum -3 dB modulation bandwidth of 20 GHz was measured for the ridge lasers at room temperature. This is the highest bandwidth achieved for a ridge laser operating at $1.55\mu m$. A plot of the damping factor as a function of resonance frequency squared gives a K factor of ~ 0.39 ns and a maximum 3-dB modulation bandwidth of ~ 23 GHz. The photon density (injected current) at this modulation frequency is only $3 \times 10^{16} cm^{-3}$, compared to higher values reported in the literature. From the modulation response, the value of $R(\omega_r)/R(0) = 2.4$ for I=200 mA, clearly indicating that gain-compression effects are small. These results demonstrate that ridge lasers, made by a simple technology, can be used for many high-speed applications.

A very low I_{th}=3.5 mA is measured in the 8-MQW BH lasers. This is the lowest value reported for an uncoated device. The differential gain $\partial g/\partial n$=2.6×10^{-15} cm^2. Figure 1 shows a maximum -3 dB modulation bandwidth of 21 GHz measured in these devices with the response still showing a strong peaking. The value of $R(\omega_r)/R(0) = 3.2$ at I=115 mA, indicating that the device is still operating at a low photon density. The value of the latter parameter is 5.75×10^{15} cm^{-3}. The K-factor derived for these devices is 0.22 ns, which gives a maximum -3 dB modulation bandwidth of 40 GHz. It is important to note that our results are obtained without p-modulation doping of the wells and without device bonding.

3. Hot-carrier effects and the tunneling injection laser

The modulation bandwidths discussed in the last section are limited by device parasitics and/or facet heating. The maximum 3-dB bandwidth of a semiconductor laser is given by

$$f_{-3dB(max)} = \frac{2^{3/2}\pi}{K} \tag{1}$$

and

$$K = 4\pi^2 \left(\tau_p + \frac{\epsilon}{v_g \frac{\partial g}{\partial n}} \right) \tag{2}$$

where τ_p is the photon lifetime, v_g is the photon group velocity, $\partial g/\partial n$ is the differential gain and ϵ is the gain compression factor. With typical values for these parameters, a value of K = 0.11 and $f_{-3dB(max)}$ = 70 GHz is calculated. However the photon density required for this is ~ 100 mW/μm width of cavity, a value which the device would not support. It is important then to lower the photon density and the value of ϵ. In fact gain compression effects will ultimately limit the intrinsic modulation bandwidth of a

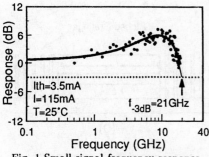

Fig. 1 Small-signal frequency response
of the 8 quantum well BH laser.

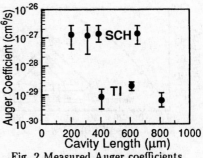

Fig. 2 Measured Auger coefficients
in TI and in MQW-SCH lasers.

laser. We have recently demonstrated a *Tunneling Injection* (TI) Laser[3, 4], in which electrons are injected directly into the lasing subband by tunneling. As a result the carrier distribution remains quasi-Fermi, or "cold", even at high injection levels. In the recent past we have demonstrated[3]-[5] lower dynamic linewidth (almost comparable to DFB lasers), lower gain compression factor (5 times), higher modulation bandwidth (1.5 times), and a higher T_o (1.5 times) in InGaAs/AlGaAs/GaAs tunneling injection (TI) lasers, when compared to equivalent multiquantum well separate confinement heterostructure (MQW-SCH) lasers.

Figure 2 shows the measured Auger coefficients (C_a) in a single-well $In_{0.60}Ga_{0.40}As/In_{0.52}Al_{0.48}As/InP$ tunneling injection laser (λ=1.65μm) and in a conventional InGaAsP/InGaAsP/InP MQW-SCH laser (λ=1.55μm), both grown by MOVPE. The Auger coefficient C_a was derived from the measured turn-on delay times of single mode lasers during large signal modulation. We obtain $C_a = 1.2 \pm 0.6 \times 10^{-29}$ cm^6/s at 10 °C for cavity lengths ranging from 200 to 800 μm. Similar measurements on MQW-SCH lasers yield values of $C_a = 1.3 \pm 1.1 \times 10^{-27}$ cm^6/s. Thus a $\leq 10^2$ times reduction in the value of C_a is obtained in the tunneling injection laser. This drastic reduction is directly attributed to the virtual elimination of hot carrier density resulting from the tunneling injection mechanism. The corresponding Auger recombination rate for an injected carrier density of 5×10^{18} cm^{-3} is 1.5×10^{27} cm^{-3}/s.

Measurement of the temperature dependence of the threshold current (I_{th}) of the tunneling laser yields a maximum value of T_o=71K compared to a value of T_o=40K for the MQW-SCH lasers. Again, a significant enhancement is recorded in the value of T_o resulting from a reduced temperature dependence of I_{th} of the TI laser according to $I_{th}(T) = I_{th}(0)e^{T/T_o}$. It should be remembered that Auger recombination is only one of several factors contributing to the temperature dependence.

4. Lasers with short cavity lengths

The cleaved facet reflectivity of $\cong$0.32 necessitates a cavity length of at least 100μm to maintain a small threshold current. However if higher mirror reflectivities could be attained, cavity lengths can be drastically reduced. This would be the first step towards achieving higher modulation bandwidths. We are exploring two techniques of utilizing *external* integrated mirrors with very high reflectivities. The two schemes are illustrated in Figs. 3(a) and (b). In the first, integrated totally internally reflecting mirrors are utilized in a ring structure, wherein the active gain region can be very

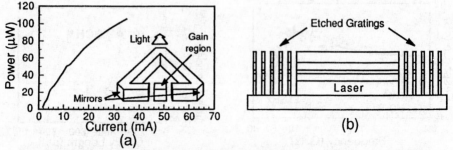

Fig. 3 (a) Light-current characteristic of the ring laser with total internal reflection mirrors, and (b) laser with dry etched Bragg reflector.

small. In the second, a quarter-wave distributed Bragg reflector is formed by very deep dry etching of the heterostructure. The region in between can either be air or filled with a dielectric. A low damage dry etching technique using plasma generated with an electron cyclotron resonance source is used for etching integrated corner mirrors in $In_{0.2}Ga_{0.8}As/Al_{0.3}Ga_{0.7}As/GaAs$ laser structures. A maximum power reflectivity of 92% (0.5 dB) is measured and triangular ridge lasers using such mirrors exhibit threshold currents as low as $\sim$1 mA.

5. External modulators and integrated transmitters

5.1. External modulators

We will describe our recent results from a relatively new type of semiconductor waveguide modulator based on the Blockaded Reservoir and Quantum Well Electron Transfer (BRAQWET) structure[6, 7]. The band diagram of a BRAQWET modulator is shown in Fig. 4(a).

The key regions of the structure are the electron reservoir, the quantum well and the planar doped barrier (PDB). The PDB provides a convenient method of positioning the quantum well energy levels relative to the electron reservoir. By changing the bias across the device, the bound states of the quantum well are moved above and below the Fermi level fixed by the electron reservoir. These states are then filled or emptied by a transfer of electrons from the reservoir region. Optical modulation is achieved as the free carriers screen the Coulombic interaction between the electron and hole in the quantum well. Thus, modulation action is based on turning the oscillator strength on or off, as opposed to merely shifting it in a quantum confined Stark effect (QCSE) modulator. In addition, QCSE modulators which rely on the shifting of the absorption edge exhibit a quadratic electrical to optical transfer function and therefore are not entirely suitable for use as a distortion-free modulator. The BRAQWET device has previously been used only as a phase modulator. We demonstrate its use as a versatile amplitude modulator. When the energy of the quantum well bound state, E_1, is below the Fermi energy E_F, the carrier density in the well can be expressed as

$$n = \frac{m^*}{\pi\hbar^2}(E_1 - E_F). \tag{3}$$

Due to the nature of the planar doped barrier diode structure which is used to shift the height of the quantum well relative to the electron reservoir, the offset in energy

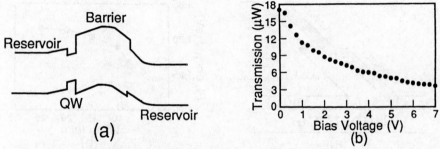

Fig. 4 (a) Band diagram of a BRAQWET modulator, and (b) transmission-voltage characteristics of a GaAs-based BRAQWET modulator.

levels (E_1-E_F) and consequently the electron density n will show linear relationships to the applied voltage. Furthermore, the change in the absorption coefficient due to free carrier screening of excitons in the quantum well has been shown to have a linear relationship to the free carrier concentration[8]. The combination of these effects results in a structure which has a linear relationship between the applied bias voltage and the optical absorption coefficient.

Experimental transmission-voltage characteristics of a GaAs-based BRAQWET modulator are shown in Fig. 4(b). The data agree well with the calculated curve. This curve is not linear, but we believe more linear characteristics can be obtained by using asymmetric quantum wells. Microwave S-parameter measurements on the device yields a device capacitance of $\sim$260 fF. The measured electrical bandwidth (f_{-3dB}) of the device, which is RC-limited, is 20 GHz. These results indicate the enormous potential of the BRAQWET device as an efficient and high-speed amplitude modulator.

5.2. Integrated transmitters

One simple scheme of laser and modulator integration is where both devices are realized by single step epitaxy on a planar substrate. During operation, the bandgap narrowing in the laser provides a photon energy which is 15 - 30 meV smaller than the heavy hole (HH) exciton peak in the absorption spectrum of the quantum well. This is sufficient for the operation of the modulator. The schematic and operation of such an integrated modulator made with $In_{0.53}Ga_{0.47}As/In_{0.52}Al_{0.483}As$ quantum wells are illustrated in Fig. 5(a). The measured bandwidths of the modulator, for three different device dimensions are

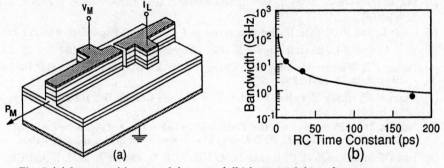

Fig. 5 (a) Integrated laser modulator, and (b) laser modulator frequency response.

1058

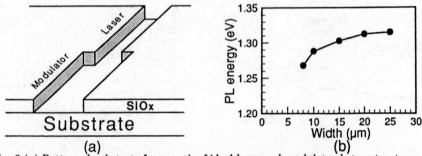

Fig. 6 (a) Patterned substrate for growth of ideal laser and modulator heterostructures in one step, and (b) variation in PL emission as a function of groove width.

also shown in Fig. 5(b). The measured bandwidths are limited by the device capacitance. Similar results have been recently demonstrated by another group.

Another scheme, in which optimized lasers and modulators can be grown by single-step epitaxy is schematically illustrated in Fig.6(a). Due to preferential and nonuniform migration rates of adatoms during molecular beam epitaxy[9], the composition and growth rates of ternary materials are dependent on the width of the groove. Thus quantum wells of different compositions and widths can be grown in the different regions. The variation in the energy of the exciton transition of a single 60Å $In_{0.2}Ga_{0.8}As/GaAs$ quantum well, measured in the photoluminescence spectra, as a function of groove width, is shown in Fig. 6(b). Realization of integrated transmitters on patterned epitaxy will pose considerable technological challenges. Nonetheless, they can be overcome and in the end, better circuit performance will be obtained.

Acknowledgments

This work is supported by ONR under Grant N00014-90-J-1-1831, ARPA under Grants MDA972-94-1-004 (COST) and F30602-92-C-0087, and BMDO and ARO under Grants DAAH04-93-G-0034 and DAAL03-92-G-0109.

References

[1] Morton P A, Logan R A, Tanbun-Ek T, Sciortino Jr. P F, Sergent A M, Montgomery R K and Lee B T 1992 *Electron. Lett.* **28** 2156-2157

[2] Stegmuller B, Borchert B and Gessner R 1993 *IEEE Photon. Technol. Lett.* **5** 597-599

[3] Sun H C, Davis L, Sethi S, and Bhattacharya P 1993 *IEEE Photon. Technol. Lett.* **5** 870-872

[4] Davis L, Sun H C, Yoon H, and Bhattacharya P 1994 *Appl. Phys. Lett.* **64** 3222-3224

[5] Yoon H, Sun H C, Bhattacharya P K 1994 *Electronic Lett.* **30** 1675-1677

[6] Zucker J E, Wegener M, Jones K L, Chang T Y, Sauer N J, and Chemla D S 1990 *Appl. Phys. Lett.* **56** 1951-1953

[7] Wegener M, Chang T Y, Bar-Joseph I, Kuo J M and Chemla D S 1989 *Appl. Phys. Lett.* **55** 583-585

[8] Jaffe M D, 1989 *Studies on the Electronic Properties and Applications of Coherently Strained Semiconductors*, PhD thesis, The University of Michigan, Ann Arbor

[9] Turco F S, Tamargo M C, Hwang D M, Nahory R E, Werner J, Kash K and Kapon E 1990 *Appl. Phys. Lett.* **56** 72-74

Inst. Phys. Conf. Ser. No 145: Chapter 8
Paper presented at 22nd Int. Symp. Compound Semiconductors, Cheju Island, Korea, 28 August–2 September 1995

OPTICAL INTERCONNECTION TECHNOLOGY FOR LARGE COMPUTING AND SWITCHING SYSTEMS

N. K. Dutta

AT&T Bell Laboratories, Murray Hill, New Jersey 07974

The advances in optical interconnection technology for large computing and switching system applications are described. Parallel data lines with 32 channels and optical switches with 64 I/O have been demonstrated.

1. Introduction

Optical interconnection is an important technology for large computing and switching systems. As the clock speed increases, interconnection between processors for a large cluster of workstations becomes increasingly difficult with conventional coax cables. Optical transmission is an important alternative for such systems and is currently implemented in very large systems [1]. As the information superhighway and the multimedia transmission become a reality, large switching systems with several hundred Gb/s transmission capability are needed. Optical interconnection is an important technology for building such a switch [2].

This work describes current directions of our work in the optical interconnection technology. The fabrication and performance characteristics of surface emitting laser array based transmitters and integrated receivers are described. CMOS technology has been used to fabricate laser driver and receiver arrays that operate at 622 Mb/s. An integrated receiver with GaAs MSM photodiode and GaAs based preamplifier, postamplifier, decision circuit and ECL drivers on the same chip has been fabricated. Applications of these subsystems for a clustered workstation environment are described.

The fabrication and performance of optoelectronic integrated circuits used for free space interconnection inside a large switching system are described. The optical devices in these circuits are the optical modulator and the photodetector both of which are fabricated using multiquantum well structures in the AlGaAs material system. The electronic circuits are based on GaAs MESFET technology. Large arrays of switching nodes have been fabricated using these circuits and have been used as a free space interconnection fabric within an experimental high capacity switching system.

2. Computing Systems

The clock speed of a computer system has been increasing steadily over the past decade and it is expected that this trend will continue The number of channels for a parallel interconnection between two elements (e.g. between microprocessors or between memory and microprocessors) of a large computer system is usually given by $2^N + 2$ where N = 5 for a 32-bit system and N = 6 for 64-bit system, etc. The additional two lines are needed for clock and telemetry. Thus for a clock rate of 200 MHz (400 Mb/s), which some of the high end workstations are already approaching, the total throughput for a 64-bit parallel link equals 26.4 Gb/s (($2^N + 2)\times400$ with N = 6).

There are two principal means of transporting this large amount of information. One is a parallel link and the second a serial link. For a parallel link, $2^N + 2$ parallel data path transport the entire information, each channel of the parallel link carrying data at the clock rate. For a serial link, the 2^N data channels are electronically multiplexed to form a single data channel operating at approximately 2^N times the data rate. At the receiver an electronic de-multiplexing is performed to retrieve and retime the information. A key feature of the serial link is the ability to multiplex (MUX) and demultiplex (DEMUX) effectively. The MUX and DEMUX circuits become increasingly difficult to make as the data rates go up.

The optical technology is an alternative to the electronic technology in transporting the information both for the serial and parallel links. Since the optical fiber is much more suited than copper coaxial cable at high data rates (due to both lower loss and lower dispersion), optical data links become the preferred mode of transport for systems operating at high speeds or for very large systems where the individual elements (microprocessors, memory, workstation, etc.) are spread over a long distance (>10 ft.). A horse race will continue to exist between the optical and electronic technologies in this arena and the winner for a particular system will primarily be determined by component availability and cost.

Among the optical components of a serial link which operates at high data rates (>10 Gb/s), are the high speed laser and the high speed photodetector. The laser is part of the transmitter which also includes a laser driver fabricated using GaAs HEMT (high-electron-mobility transistor) based technology. The photodetector is part of the receiver which has a preamplifier, postamplifier and decision circuit fabricated using GaAs HEMT technology. For the parallel link, the data rates are considerably lower and the well established Si-CMOS technology is the preferred technology choice for the electronic drivers and receiver circuits in the optical link.

The main components of a parallel optical link are a laser array and the associated driver array for the transmitter package, and the photodetector array and the receiver electronics array for the receiver package. Among the laser types that have been investigated are the edge emitting laser and the surface emitting laser. The surface emitting laser emits in a narrow circular beam which results in higher optical fiber coupling efficiency over the edge emitting device.

The mirror reflectivity for a surface emitting laser is provided by the multilayer stacks of AlAs and $Al_{0.1}Ga_{0.9}As$ on either side of the active region. The proton implantation is used to confine the current. Various types of laser designs have been fabricated. The reliability results of the proton implanted device is excellent.

The surface emitting laser can be operated at high speed. A transmission experiment at 10 Gb/s using a surface emitting laser source and an optoelectronic integrated photoreceiver has been carried out. Error rates of $< 10^{-9}$ have been obtained.

Laser driver arrays have been fabricated using both GaAs MESFET technology and Si-CMOS technology. The GaAs MESFET driver circuits (fabricated using 1 µm MESFET technology) have been packaged with the laser array to produce an array transmitter. The transmitter has 32 data channels and each channel operated at 500 Mb/s [3]. A 32-channel CMOS driver array has been fabricated using 0.7 µm CMOS technology. Each channel operated at 1.2 Gb/s. For a multichannel transmitter the crosstalk penalty was found to be negligible (<0.1 dB/ch).

For an optical link operating at 0.85 μm wavelength, the optical element in the receiver can be a GaAs p-i-n photodiode, GaAs MSM photodiode or Si photodiode. The key requirement being that the bandwidth and the sensitivity be sufficiently high to detect the signal. The GaAs p-i-n photodiode consists of a low doped GaAs absorbing layer sandwiched between P and N type AlGaAs layers. The detector speed is determined by the size, being higher for the smaller detector. For 25 μm dia. photodiodes, bandwidths and responsivity of 4 GHz and 0.5 A/W, respectively, have been obtained.

The receiver electronics has been fabricated using Si-CMOS technology. The various elements in the receiver are the amplifier, comparator, decision circuit and the ECL driver [4]. The last element is needed to bring the signal out in the form of ECL logic levels typically used in computer systems. Such receivers have operated at 622 Mb/s.

A monolithic receiver chip which incorporates a GaAs MSM photodiode, amplifier, comparator, decision circuit and ECL driver has also been fabricated using 1 μm GaAs MESFET technology. The photodiode has a bandwidth of 3 GHz and responsivity of 0.3 A/W. The receiver operated at 2.4 Gb/s. A similar 32 channel optoelectronic receiver has been fabricated and packaged [3]. This 32-channel receiver along with a 32-channel transmitter has been used to demonstrate a 32-channel optical interconnection system [3].

Thus multi-channel array based subsystems (transmitters and receivers) have been fabricated. These subsystems have been used in several prototype experimental demonstrations. We believe technology choices for manufacturing these subsystems will be dictated by cost effective packaging rather than performance issues.

3. Switching Systems

Another area where optoelectronic array technologies are potentially important is large switching systems [2]. The latter can have 1000 (N) inputs and outputs and therefore if it was to be built as a crossbar, it will have $10^6 (N^2)$ interconnections. Optics is especially suited for providing a very large number of interconnections.

Next generation large switching systems should be capable of routing ATM (Asynchronous Transfer Mode) packets. An ATM packet has 8 routing bytes (8 bit = 1 byte) and 45 data bytes and therefore for a 2.4 Gb/s transmission system, the packet is 176 ns long. The key requirement in the design of a large ATM packet switch is the following: the routing path through the switch of a packet must be completed before the next packet arrives (this implies a path hunt-time of <176 ns for 2.4 Gb/s ATM switches) and the blocking probability of the packets are sufficiently small ($< 10^{-15}$).

Free space optical interconnection is most suited for providing a large number of simultaneous interconnections. Consider a N-input and N-output switch where any one of the N inputs can appear at any of the N outputs. Such a switch has two features [5]: (i) a fan out of each of the N inputs to any of the N outputs and (ii) a MUX circuit (fan-in) which allows any of the 16 signals that arrive at a node to appear at its output. The MUX circuit has a control logic which does the selection process and a memory. The generalized MUX circuit can be represented by a N×1 node (N data inputs with 1 data output). The node also has N control logic inputs (electrical signals) which determines which one of the N inputs appear at the output.

The N×1 node has N optical receivers and DRAMs associated with each receiver to store the output of the photoreceiver before it is called for output at the transmitter. The receiver is a GaAs photodiode in series with an amplifier, the transmitter is a GaAs quantum well modulator.

The N×1 nodes have been fabricated using GaAs/AlGaAs technology [6,7]. The amplifiers are built using 1 μm GaAs MESFETS, the photodiode and the modulator are multilayer stack of GaAs/AlGaAs quantum wells. The modulators and photodiodes are about 10 μm×10 μm in size and the entire 2×1 chip is 200 μm×200 μm in size. Arrays of 16, 2×1 chips in a 4×4 two-dimensional array configuration have been fabricated.

A 16 I/O (16 input, 16 output) switch has been built using these 2×1 switching nodes. The switch is built using a banyon network which needed five 4×4 arrays of the 2×1 nodes. The entire system operates at 155 Mb/s/ch.

This work has been extended recently to 8×1 switching nodes. 8 linear arrays of 8×1 switching nodes have been monolithically configured on a single chip. The entire system has been operated at 700 Mb/s/ch [8]. This system was built using flip-chip bonded technology where the optical devices (modulators and photodiodes) were flip-chip bonded on to Si-CMOS circuits.

4. Summary

The current directions of our work in the optical interconnection technology are described. The fabrication and performance characteristics of surface emitting laser array based transmitters and integrated receivers are described. CMOS technology has been used to fabricate laser driver and receiver arrays that operate at 622 Mb/s. An integrated receiver with GaAs MSM photodiode and GaAs based preamplifier, postamplifier, decision circuit and ECL drivers on the same chip has been fabricated. Multichannel links using these transmitters and receivers have been demonstrated.

The fabrication and performance of optoelectronic integrated circuits used for free space interconnection inside a large switching system are described. The optical devices in these circuits are the optical modulator and the photodetector both of which are fabricated using multiquantum well structures in the AlGaAs material system. The electronic circuits are based on GaAs MESFET technology or Si-CMOS technology. Large arrays of switching nodes have been fabricated using these circuits and have been used as a free space interconnection fabric within an experimental high capacity switching system.

References

[1] Nordin R A, Levi A F J, Nottenburg R N, O'Gorman J, Tanbun-Ek T and Logan R A 1992 *J. Lightwave Tech.* **10** 811

[2] See for example *An Introduction to Photonic Switching Fabrics* 1993 ed. Hinton H S (New York: Plenum Press)

[3] Wang Y M, Muehlner D T, et al. 1995 *Journal of Lightwave Tech.* **13** 995

[4] Yu K. Y., Gabara T J, Levine B F, Wynn J D, Dutta N K and Monteleone K J, Proc. ASIC '95, Austin, TX.

[5] Cloonan T J (unpublished).

[6] D'Asaro L A, Chirovsky L M F, Laskowski E J, Pei S S, Woodward T K, Lentine A L, Leibenguth R E, Focht M W, Freund J M, Guth G and Smith L F 1993 *IEEE JQE* **QE-29** 670

[7] Lentine A L, Chirovsky L M F, D'Asaro L A and Laskowski E J 1994 *Appl. Optics* **33** 2849

[8] Lentine A L, Goossen K W, Walker J, Chirovsky L M F, D'Asaro L A, Hui S P, Tseng B T and Leibenguth R E CLEO'95, Baltimore, MD, Paper CPD11-1.

Inst. Phys. Conf. Ser. No 145: Chapter 8
Paper presented at 22nd Int. Symp. Compound Semiconductors, Cheju Island, Korea, 28 August–2 September 1995
© *1996 IOP Publishing Ltd*

Free-Space Integrated Optics on a Chip

Ming C. Wu, Lih-Yuan Lin, and Shi-Sheng Lee

UCLA, Electrical Engineering Department
405 Hilgard Avenue, Los Angeles, CA 90095-1594

Abstract. A novel scheme for integrating free-space micro-optical systems on a single chip has been demonstrated by surface-micromachining techniques. The *free-space micro-optical bench (FSMOB)* which contains monolithic three-dimensional micro-optical elements as well as active optoelectronic components with self-aligned hybrid integration. The optical axes of these optical elements are parallel to the substrate, which enables the entire free-space optical system to be integrated on a single substrate. Micro-Fresnel lenses, mirrors, beam-splitters, gratings, precision optical mounts, micro-positioners such as rotary stages and linear translational stages, as well as self-aligned semiconductor edge-emitting and surface-emitting lasers have been successfully fabricated and characterized.

1. Introduction

Integrated optics has been an active research area since its proposal in 1969 [1] because it offers many advantages: higher functionality, reduced packaging cost of individual optoelectronic components, improved performance by eliminating parasitics, and more uniform control of the environment (temperature, etc.). To date, most of the research in integrated optics focuses on guided-wave approach. For example, the photonic integrated circuits (PIC) integrates lasers, detectors, and modulators with passive guided-wave components [2]. On the other hand, free-space integrated optics offers further advantages such as high spatial bandwidth (diffraction-limited resolution), non-interfering optical routing, three dimensional optical interconnection, and optical signal processing capability (e.g., Fourier optics). However, it is more difficult to integrate free-space optics on a single substrate since most monolithically fabricated free-space optical elements lie on the surface of the substrate.

2. Approach for free-space integrated optics

Surface micromachining has been used to produce three-dimensional micro-optics [3,4]. Previously, we have proposed a *free-space micro-optical bench* (FSMOB) fabricated by surface micromachining technology for optoelectronic packaging and free-space integrated optics [5,6]. On the micro-optical bench, three-dimensional micro-optical elements such as micro-lenses, mirrors and gratings can be fabricated integrally on a single silicon chip. The fabrication process is compatible with the IC-processing. The micro-optic system can also be pre-aligned in the mask layout stage using computer-aided design. Additional fine adjustment can be achieved by the on-chip micro-actuators and micro-positioners such as rotational and translational stages.

To implement a complete micro-optical system on the micro-optical bench, it is necessary to incorporate active optical devices. Passive alignment of active optical devices with micro-optics is desirable to minimize cost. Hybrid optical packaging on silicon which combines flip-chip mounting and silica waveguide interconnection has been proposed [7]. However, most of these optoelectronic packaging methods confine the optical components on the surface of the substrate, which is two-dimensional in nature and cannot be used for free-space integrated optics. In this paper, we demonstrate the first hybrid integration of semiconductor edge-emitting lasers with micro-optics using novel three-dimensional alignment structures. The optical performance of the edge-emitting laser/micro-Fresnel lens module is reported. Other packaging issues such as heat sinking are also discussed.

3. Fabrication

The three-dimensional micro-optics and the self-alignment structures are fabricated integrally with micro-hinges and micro-spring latches [8] by surface micromachining process. The Si substrate serves as a micro-optical bench. The fabrication process is summarized in the following: First, a 2-μm thick of phosphosilicate glass (PSG) is deposited as the sacrificial material. It is followed by the deposition of a 2-μm-thick polysilicon layer on which the micro-optics patterns and the self-alignment structure are defined by photolithography and dry etching. The hinge-pins holding these three-dimensional structures are also defined on this layer. Then another layer of PSG material with thickness of 0.5 μm is grown uniformly. Before the deposition of second polysilicon layer, contact holes are opened by dry etching through the PSG material to the silicon substrate. Finally the hinge-staples and spring-latches are defined on the second polysilicon layer and their bases are connected to the substrate via the contact holes. The micro-optics plates are released from substrate by selectively removing the PSG material using hydrofluoric acid after fabrication. The polysilicon plates with micro-optics patterns and self-alignment structures can then be rotated out of the substrate plane. The position of the rotated polysilicon plate is fixed by the spring latches which are pushed up by the plate itself.

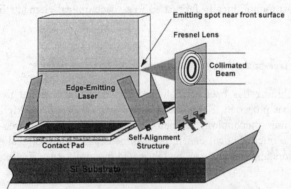

Fig. 1(a) Schematic diagram of the self-aligned hybrid integration of an edge-emitting semiconductor laser and a micro-Fresnel lens.

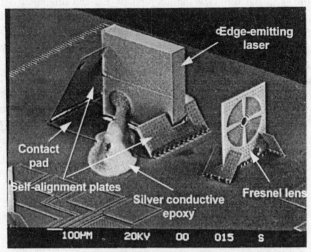

Fig. 1(b) SEM micrograph of the self-aligned hybrid integration of an edge emitting semiconductor laser and a micro Fresnel lens.

Figure 1(a) shows the schematic diagram of the self-aligned hybrid integration of an edge-emitting laser with a micro-Fresnel lens. The Fresnel lens is held by two precision mounting plates fabricated by the same processes to precisely fix its angle and position. The structure of the lens mount has been reported in detail in Ref. [6]. The edge-emitting laser is mounted on its side for accurate position of the active emitting spot. By precise scribing, the optical axis is placed at 254 μm above the Si substrate. The emitting spot of the edge-emitting laser is aligned to the center of the Fresnel lens by the self-alignment structures. Conductive silver epoxy is applied between the laser and the contact pads for the electrical contact. Permanent fixing of the semiconductor laser is achieved by curing the silver epoxy. Potentially, the epoxy can be replaced by other three-dimensional micromechanic structures. Figure 1(b) shows the scanning electron micrograph (SEM) of the laser and the micro-Fresnel lens.

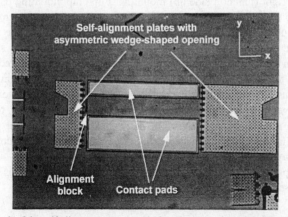

Fig. 2 Photograph of the self-alignment structures before they are released and assembled (top view)

Figure 2 is the top view photograph of the self-alignment structure before it is assembled. The edge-emitting laser is slid into the slot between two electric contact pads until the front facet hits the alignment block built on the MOB, which defines the longitudinal (x-direction, as shown on the picture) position of the emitting spot. The self-alignment plates are then rotated up and the asymmetric wedge-shaped opening on the top gradually guides the active side (waveguide side) of the laser towards the flat edge of the wedges, which defines the transverse (y-direction) position of the emitting spot. This unique design allows us to accommodate lasers with a large variation of substrate thickness (from 100 μm to 140 μm think). The height of the self-alignment structure (400 μm tall for the self-alignment plate on the back facet of the laser) permits more precise alignment.

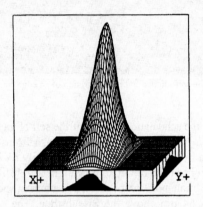

Divisions: top = 15.62 μm
bottom = 100.0 μm

Fig. 3 Beam profile of the semiconductor laser (λ=1.3 μm) after collimated by the micro-Fresnel lens. The profile fits very well with the Gaussian shape

There are other possible schemes for mounting semiconductor lasers: Flip-chip mounting and upright (junction side up) mounting. Flip-chip mounting using indium solder balls can achieve an alignment accuracy of around 1 μm, however, the emitting spot is very close to the Si substrate and is much lower than the optical axis of the free-space optical system. The heat conduction for this mounting scheme might not be sufficient for high power dissipation due to small contact areas. Heat sinking can be improved by increasing the contact area at the expense of alignment accuracy. Upright mounting has good heat conduction because of the large contact area between the semiconductor laser and the Si substrate. However, the height of the optical axis is now defined by the laser substrate thickness, which usually has a tolerance of more than 5 μm and is not suitable for MOB without employing additional adjustable optics. Therefore, side mounting scheme was chosen for the hybrid integration of semiconductor lasers on MOB. The height of the emitting spot is precisely defined by scribing. In our current design, it is placed at 254 μm above the Si substrate, which is suitable for the optical axis of MOB.

4. Experimental results

The collimating performance of the binary amplitude micro-Fresnel lens has been demonstrated successfully using a divergent beam emitted from a single mode fiber as light source [6,9]. Similar experiment has also been conducted using the edge-emitting laser as the light source. Figure 3 shows the beam profile of the collimated beam. The profile is measured at 5 cm after the lens. The laser has a wavelength of 1.3 μm and divergence angles of $18° \times 40°$. After collimation by the micro-Fresnel lens, the Gaussian beam widths become 330 μm × 788 μm, which corresponds to divergence angles of $0.38° \times 0.9°$.

To investigate the heat sinking capability of the side mounting scheme, we compared the temperature dependence of the threshold current I_{th} of the semiconductor lasers with side mounting on MOB and with upright mounting on copper heat sink. The experimental results are shown in Fig. 4. The laser studied is an edge-emitting laser with wavelength of $\lambda = 1.3$ μm. In upright mounting scheme (—●— in the figure), the laser is mounted directly on copper heat sink using indium solder. The characteristic temperature of $T_0 = 40$ K is typical of long wavelength lasers. For side mounting in the hybrid integration (—■— in the figure), the laser is first mounted on the MOB and the MOB is fixed on a copper heat sink using thermal joint compound (whose thermal conductivity is much lower than the indium solder). The threshold currents increase slightly and T_0 becomes 33 K. The threshold current increase at $I_{th} = 12$ mA is equivalent to 7 °C rise in junction temperature. To gain more insight into the thermal conduction of this packaging scheme, we compare the thermal resistivity of these two mounting schemes using the effective thermal resistivity model developed by Joyce and Dixon [10]. From the simulation, it is found that the dominant thermal resistance of the side mounting scheme comes from the thermal joint compound (thermal conductivity = 0.43 $Btu \cdot Ft/Hr \cdot Ft^2 \cdot °F$) between the MOB and the copper heat sink. This thermal joint compound is not used in the upright-mounted laser which is directly soldered to the copper heat sink by indium. Therefore, heat conduction for the hybrid integration can be improved by choosing better thermally conducting material for the mounting of the Si MOB on the copper heat sink.

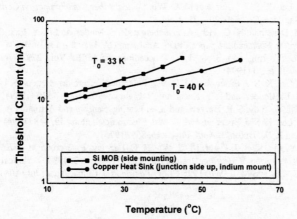

Fig. 4 Temperature dependence of the threshold currents for the semiconductor lasers with side mounting and upright mounting

The free-space MOB technology can also be extended to include vertical cavity surface-emitting lasers (VCSELs) as optical sources, in which case two-dimensional arrays of both micromachined optical elements and VCSELs can be combined for massively parallel operation. An 8×1 array of micro-Fresnel lenses and an 8×1 array of VCSELs have been successfully integrated using similar three-dimensional alignment plates [11]. Individual switching of the VCSEL/lens element was also demonstrated. This combination is particularly attractive for optical interconnect applications.

5. Conclusion

In summary, a novel free-space integrated micro-optical bench technology has been demonstrated. The micro-optical bench integrates three-dimensional free-space optical elements such as lenses, mirrors, gratings, with beam-splitters with self-aligned active optoelectronic components such as edge-emitting lasers and vertical cavity surface-emitting lasers. With this technology, it is possible to integrate the entire micro-optical systems on a chip. A free-space integrated micro-optical pickup head that includes self-aligned semiconductor laser source, collimating and focusing lenses, 45-degree mirrors and rotatable beam splitters have been demonstrated for optical data applications [12]. On-chip tuning or switching can be incorporated into the micro-optical systems. The MOB offers a new approach for optoelectronic packaging, free-space optical interconnect, and single-chip micro-optical systems.

Acknowledgment

This project is supported in part by ARPA ULTRA, NCIPT, and the Packard Foundation.

References

[1] S. T. Miller, Bell Syst. Tech. J., **48**, 2059 (1969)

[2] T. L. Koch and U. Koren, J. Quantum Electronics, **27**, 641 (1991)

[3] L. Y. Lin, S. S. Lee, K. S. J. Pister, and M. C. Wu, *Optical Fiber Communication Conference*, Postdeadline paper PD12, San Jose, CA, Feb. 20 ~ 25, 1994.

[4] O. Solgaard, M. Daneman, N. C. Tien, A. Friedberger, R. S. Muller, and K. Y. Lau, *Conference on Lasers and Electro-Optics*, Postdeadline paper CPD6, Anaheim, CA, May 8 ~ 13, 1994.

[5] M. C. Wu, L. Y. Lin, and S. S. Lee, Proceedings of SPIE, Vol. 2291, *Integrated Optics and Microstructures* II, p. 40 (1994)

[6] L. Y. Lin, S. S. Lee, K. S. J. Pister, and M. C. Wu, Photonics Technology Letters, **6**, 1445 (1994)

[7] C. H. Henry, G. E. Blonder, and R. F. Kazarinov, J. Lightwave Tech., **7**, 1530 (1989)

[8] K. S. J. Pister, M. W. Judy, S. R. Burgett, and R. S. Fearing, Sensors and Actuators A, **33**, 249 (1992)

[9] L. Y. Lin, S. S. Lee, K. S. J. Pister, and M. C. Wu, Electronics Letters, **30**, 448 (1994)

[10] W. B. Joyce and R. W. Dixon, J. Appl. Phys., **46**, 855 (1975)

[11] S. S. Lee, L. Y. Lin, K. S. J. Pister, M. C. Wu, H. C. Lee, and P. Grodzinski, *IEEE Lasers and Electro-Optics Society 1994 Annual Meeting*, Boston, MA, Oct. 31 ~ Nov. 3, 1994, Technical Digest, p. 242.

[12] L. Y. Lin, S. S. Lee, and M. C. Wu, 1995 *CLEO Pacific Rim*, Chiba, Japan, July 10~14, 1995.

Inst. Phys. Conf. Ser. No 145: Chapter 8
Paper presented at 22nd Int. Symp. Compound Semiconductors, Cheju Island, Korea, 28 August–2 September 1995
© *1996 IOP Publishing Ltd*

Performance enhancement of non-biased symmetric self electro-optic effect device with extremely shallow multiple quantum wells using an impedance-mismatched asymmetric Fabry-Perot structure

K. Kim, O.K. Kwon, Y.W. Choi*, and E.-H. Lee

Research Department, Electronics and Telecommunications Research
Institute, Yusong P. O. Box 106, Taejon, Korea 305-600
*Present address: Electronics Engineering Department, Choong-Ang
University, Seoul, Korea 156-756

Abstract. We achieved large non-biased optical bistability (NOB) and significant enhancement in the performance of symmetric self electro-optic effect device (S-SEED) by adopting both extremely shallow quantum wells (ESQWs) and impedance-mismatched asymmetric Fabry-Perot (AFP) cavity structure. From the structure we obtained values as high as 17% of reflectivity change, and 3.7 of contrast ratio, which indicate that ESQWs and impedance-mismatched normally-off AFP improves NOB comparable to conventional S-SEEDs with large external bias. Furthermore, the AFP cavity structure reduces the minimum switching energy and switching time.

1. Introduction

The development of novel semiconductor material epitaxial growth techniques made it possible to realize the multiple quantum wells (MQWs) structure. The excitons due to the discrete energy levels in MQWs make strong optical absorption, which is very susceptible to the applied electric field. For the easy manipulation of the applied electric field intensity, a p-i(MQWs)-n diode structure, the so called self electro-optic effect device (SEED), was developed.[1]

Symmetric SEED (S-SEED)[1] consists of two serially connected identical SEEDs and shows optical bistability at the non-biased exciton absorption peak range with proper external bias. As a result, S-SEED can be used as an optical switching device. In reflection type S-SEED, the reflectivity's of each SEED are different and they can be switched and/or read by light.

For a practical optical switching function, fast switching speed, simple-layout for array operation, large values of reflectivity change, ΔR, and contrast ratio, CR, are essential among many requirements. To satisfy them, we employed extremely shallow quantum wells (ESQWs) and asymmetric Fabry-Perot (AFP) cavity structure simultaneously.

2. Background

When electric field is applied to $Al_xGa_{1-x}As$/GaAs MQWs, the exciton absorption peak of the conventional deep ($x > 0.2$) quantum wells shows red-shift due to the quantum-confined Stark effect (QCSE). However, in ESQWs ($x < 0.04$), because of the low barrier, LO phonon assisted field ionization occurs at room temperature and the exciton absorption peak simply reduces.[2] Furthermore, the carrier sweep-out velocity reaches the saturated drift velocity even at small forward bias.[2] The advantages of ESQWs over the conventional deep quantum wells include stronger low-field electroabsorption and higher saturation intensities.[3]

To get the optical bistability, large external bias is essential in conventional S-SEEDs based on QCSE. Since S-SEED application is mainly directed to parallel processing, this implies complicated electrical wiring in S-SEED array. The diode structure for the supply of external electric field makes built-in potential.[4] The optical bistability without external bias, or non-biased optical bistability (NOB), was demonstrated using p-i(ESQWs)-n structure, where 60.5 periods of $Al_{0.04}Ga_{0.96}As$/GaAs (60 Å/100 Å) ESQWs were employed and the air/semiconductor interface was antireflection coated (E-SEED).[3] The negative photoconductance and the maximum photocurrent at forward bias that are the necessary conditions for NOB. The removal of the external wiring gives many advantages including compact packing and the reduction of crosstalk and array shorts.[3]

When the distance between the two mirrors is adjusted to satisfy the phase-matching condition, the partial mirror of reflectivity R_f naturally formed by the air/semiconductor interface can be used to construct the AFP cavity in conjunction with the bottom BRS mirror of reflectivity R_b. From the optical bistability, the MQWs inside the AFP cavity will have two absorption coefficients α_{off} and α_{on}. Since R_b is larger than

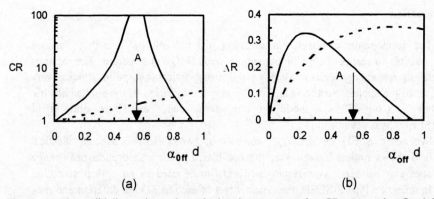

(a)   (b)

Figure 1. The solid lines show the calculated contrast ratio, CR, (a) and reflectivity change, ΔR, (b) of normally-off AFP cavity structure as a function of the total absorption, $\alpha_{off}d$, of the MQWs. Here d is the number of quantum wells times the thickness of a quantum well. Point A is where the impedance-matching condition is satisfied. The dotted lines are antireflection coated results for comparison.

R_f, if the total absorption or the number of quantum wells inside the AFP cavity is adjusted so that the impedance-matching condition is satisfied or the light amplitudes reflected by the two mirrors are same at the larger absorption coefficient α_{off}, the off-state SEED will show virtually no reflection giving very large CR. The observation of CR > 30 and Δ = 44 % from the impedance-matched AFP ESQWs S-SEED (AE-SEED) was reported.[5]

As the calculated results in Fig. 1 clearly show, CR is monotonically increasing in antireflection coated structure as the number of quantum wells increased. However, in AFP cavity structure, CR becomes infinity as the impedance-matching condition is satisfied and it is meaningless to increase the number of quantum wells beyond the point A in the figure. In photonic switching and optical computing systems, Δ and ΔR become the limiting factors in improving beam tolerance and bit-rate, respectively.[6] From Fig. 1, the optimization of these more meaningful parameters can be achieved at an impedance-mismatched condition with a modest CR.

3. Experimental

We constructed a SEED structure incorporating ESQWs and impedance-mismatched normally-off AFP simultaneously. 14 pairs of undoped $Al_{0.1}Ga_{0.9}As/GaAs$ (616 Å/725 Å) BRS were grown on a semi-insulating GaAs substrate, as a bottom mirror of R_b = 0.95. On top of the mirror, an N^+ $Al_{0.1}Ga_{0.9}As$ contact layer, undoped $Al_{0.04}Ga_{0.96}As/GaAs$ (100 Å/100 Å) 20.5 period ESQWs sandwiched by two undoped 200 Å $Al0.1Ga0.9As$ buffers, and a P^+ $Al_{0.1}Ga_{0.9}As$ contact layer were made in order. The number of quantum wells was determined assuming absorption coefficients of α_{off} = ~ 16,000 cm^{-1} and α_{on} = ~ 7,000 cm^{-1}. The air/semiconductor interface gave R_f = 0.32. To satisfy the phase-matching condition of the AFP cavity, the distance between the two mirror was carefully adjusted by monitoring the reflectivity over large spectral range. After electrical isolation of each diode, the NOB S-SEED was accomplished by connecting two 140 x 180 μm^2 mesa diodes.

4. Results and discussion

The NOB was measured by a laser diode tuned to the ESQWs exciton wavelength of 856 nm. The deviation of the exciton peak wavelength from the expected 860 nm may reflect the imperfect growth by MOCVD. The signal beam, P_s, was triangularly modulated from 0 to 110 μW and incident to one diode while a constant bias beam, P_b, of 50 μW was incident to the other. The experimental power range with the focused spot diameter of about 25 μm was low enough to be free from exciton saturation or thermal ohmic heating. The on- and off-state reflectivity's were ~ 23 % and ~ 6 %, respectively, giving ΔR and CR of ~ 17 % and ~ 3.7. Δ of the NOB, defined by 100 x $(P2 - P1)/Pb$, was about 53 %. Here, P1 and P2 are the powers at which the transition between on- and off-state occurs.

The operating voltage was estimated to be ~ 0.9 V from the observation of the I-V characteristics of a SEED under illumination. Since the p-i-n diode built-in potential is ~ 1.5 V and the intrinsic region thickness is 0.45 μm, in the NOB operation, the electric field is changed between ~ 1.3 and ~ 5.3 V/μm in the given structure. In the E-SEED [3], 4.3 V of external bias was needed to get the electric field intensity 5.3 V/μm.

Another advantage of the implementation of the AFP cavity structure comes from the minimum switching energy and the switching speed. The switching characteristics of the 5 V biased E-SEED structure[3] and the non-biased AE-SEED were compared theoretically.[7] When same mesa size of 10x10 μm^2 was assumed, while the E-SEED minimum switching energy was 4.1 fJ/μm^2, the AE-SEED showed minimum switching energy from 1.2 to 3.3 fJ/μm^2, depending on the impedence-matching condition. The switching times in the range of ~ 10 ps for all structures when their own minimum switching energies were used. However, with the minimum switching energy of the E-SEED, the AE-SEED switching times were reduced to less than half of the E-SEED switching time.[7]

In summary, we have shown that Δ and ΔR of NOB can be improved using impedance-mismatched normally-off AFP cavity structure and ESQWs. Though our AE-SEED was slightly deviated in the layer structure, Δ, ΔR, and CR of the NOB were as high as 53 %, 17 %, and 3.7, respectively. Furthermore, the efficient usage of the light in AE-SEED reduces the minimum switching energy and switching time.

The work has been supported by the Ministry of Information and Communications, Korea.

References

[1] Lentine A L, Hinton H S, Miller D A B, Henry J E, Cunningham J E, and Chirovsky L M F 1988 *Appl. Physics. Lett.* **52** 1419-21; 1989 *IEEE J. of Quantum Electron.* **25** 1928-36

[2] Feldmann J, Goossen K W, Miller D A B, Fox A M, Cunningham J E, and Jan W Y 1991 *Appl. Phys. Lett.* **59** 66-8

[3] Morgan R A, Asom M T, Chirovsky L M F, Focht M W, Glogovsky K G, Guth G D, Przybylek G J, Smith L E, and Goossen K W 1991 *Appl. Phys. Lett.*, **59** 1049-51

[4] Sze S M 1985 *Semiconductor Devices Physics and Technology* (New York: John Wiley & Sons)

[5] Choi Y W, Kwon O K, and Lee E-H 1993 *IEEE Photon. Tech. Lett.*, **5** 1406-9

[6] Lentine A L, Miller D A B, Chirovsky L M F, and D'Asaro L A 1991 *IEEE J. of Quantum Electron.*, **27** 2431-9

[7] Choi Y W, Kwon O K, and Lee E-H May 6-10, 1995 *The First Int'l Conference on Low Dim. Structures & Devices* (Singapore)

Inst. Phys. Conf. Ser. No 145: Chapter 8
Paper presented at 22nd Int. Symp. Compound Semiconductors, Cheju Island, Korea, 28 August–2 September 1995
© *1996 IOP Publishing Ltd*

Nonlinearly Chirped Grating for Extended Tuning Range Semiconductor Lasers

Dug-Bong Kim, Tae-Hoon Yoon, Jae Chang Kim, and Sun Ho Kim*

Departmemt of Electronics Engineering, Pusan National University, Pusan, 609-735. Korea
Fax : +82-51- 515-5190   Internet : thyoon @ hyowon. pusan. ac. kr
* Division of Electronics and Information Technology, Korea Institute of Science and Technology.
PO Box 131, Cheongryang, Seoul 130-650, Korea

Abstract. A Superstructure Grating(SSG) Distributed-Bragg-Reflector(DBR) laser has a broad tuning range with a good mode suppression ratio. However, gaps of channel are observed in the wavelength-tuning characteristics of an SSGDBR laser which employs linearly-chirped DBR mirrors. We found by numerical simulation that the gaps may be attributed to the nonuniform reflection-peak heights of a linearly-chirped DBR mirrors. We propose a nonlinearly chirped grating DBR mirror structure that makes reflection-peak heights almost uniform. The uniformity makes it possible to tune the laser without lasing wavelength gaps.

1. Introduction

A tunable laser is an important light source for wavelength division multiplexing(WDM) communication systems. The network capacity in such systems increases monotonously with the number of wavelengths accessible by the tuning range of a semiconductor laser transmitter. Sampled Grating(SG) DBR and Superstructure Grating(SSG) DBR lasers have wide tuning range with good mode suppression ratios[1,2]. For the same refractive index and length of DBR mirror, an SGDBR has lower reflection-peaks. So SGDBR laser needs higher threshold current, which may make its continuous wave(CW) operation difficult. An SSGDBR laser can operate with lower threshold current because its mirrors have higher reflection-peaks, but gaps of channel in the tuning characteristics are observed by experiments.

We found by numerical simulation that the gaps are due to nonuniform reflection-peak heights of an SSGDBR mirror, and the gaps of channel may limit the application of the laser. To overcome this problem, we propose a nonlinearly-chirped DBR structure. This structure can make the reflection-peak heights of the mirror uniform without any decrease of reflectance. This uniformity can make it possible to tune the laser without gaps of channel.

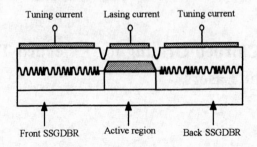

Fig. 1. Schematic structure of a tunable DBR laser with SSG reflectors.

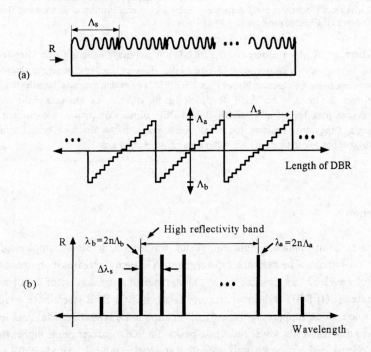

Fig 2. Superstructure grating. (a) grating structure, (b) reflection
characteristics of an SSGDBR mirror.

2. Tuning characteristics of SSGDBR laser

In an SSGDBR laser, chirped DBR mirrors are formed on both sides of the active region as shown in Fig. 1. Those two mirrors are linearly chirped from Λ_a to Λ_b and is repeated with a pitch Λ_s, as shown in Fig. 2(a). This structure gives periodic reflection-peaks with peak spacing $\Delta\lambda_s$ in a broad wavelength range as shown Fig. 2(b). The reflection-peaks are distributed in the wavelength range from $\lambda_b(=2n\Lambda_b)$ to $\lambda_a(=2n\Lambda_a)$ and the peak spacing is[1]

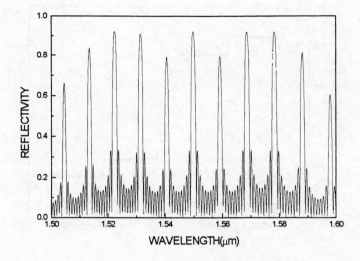

Fig. 3. Reflectivity spectrum of an SSGDBR mirror.

$$\Delta \lambda_s = \frac{\lambda_0^{\,2}}{2n\Lambda_s} \qquad\qquad (1)$$

where λ_0 is the average Bragg wavelength, n the refractive index, and Λ_s the pitch with which a chirped grating structure is repeated. This mirror has a higher reflectance than an SGDBR mirror because any element of the grating on the mirror is not deleted unlike in an SGDBR mirror. This mirror has nonuniform reflection-peak heights as shown Fig. 3. The pitches Λ_s and Λ_s' of those two mirrors in an SSGDBR laser are set to be slightly different. So only one reflection-peak of a mirror is matched to that of the other mirror. Because a cavity mode at the matched peak has the highest reflectance, it has the lowest threshold gain and operates as the lasing mode. The tuning current injected to one mirror of an SSGDBR laser results in the decrease of the refractive index by the plasma effect and reflection-peaks of the mirror shift to shorter wavelength. It causes the matched reflection-peak wavelength to hop to a shorter wavelength. As a result, the lasing mode has a large discontinuous tuning range of $\Delta\lambda_s\times$(Number of peaks in the high reflectivity bandwidth)[2].

We simulated tuning characteristics of an SSGDBR laser by using characteristic matrix method[3] as the injection current to a mirror was increased. The lasing wavelength was tuned to channels with the interval $\Delta\lambda_s$ in a broad wavelength range and gaps of channel occurred as shown Fig. 4. The gaps were observed when we decrease the difference of pitch (Λ_s - Λ_s') between the two mirrors in order to obtain a broader tuning range. Therefore gaps of channel can limit the tuning range of a laser. We found also that an SGDBR laser, which exhibit uniform reflection-peak heights in the reflection spectrum of the mirror, does not have

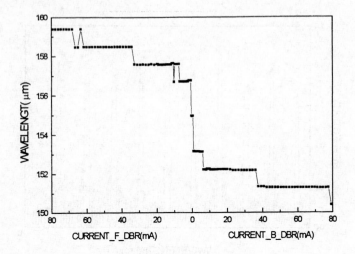

Fig. 4. Tuning characteristics of an SSGDBR laser.

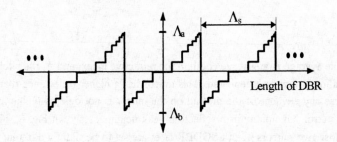

Fig. 5. Grating structure of a nonlinearly chirped grating.

gaps of channel in the tuning characteristics. Therefore we believe that gaps of channel is due to the nonumiformity of reflection-peak heights. Heights of some peaks are smaller than that of neighboring peaks. When these peaks are matched, the main mode may have lower reflectance than that of the side mode. As a result, a laser may not operate at this channel or it may operate in multimodes. Therefore we need a mirror structure that has uniform reflection-peak heights in order to achieve gap-free tuning characteristics.

3. Nonlinearly chirped grating DBR mirror

To make reflection-peak heights uniform, we propose a nonlinearly chirped grating. The grating structure consists of m steps, where the grating period of the ith step is $\Lambda_a + i(\Lambda_b - \Lambda_a)/m$ as shown in Fig. 5. The width of each step is optimized by the try-and-error method for

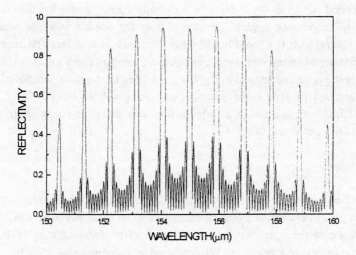

Fig. 6. Reflectivity spectrum of a nonlinearly chirped DBR.

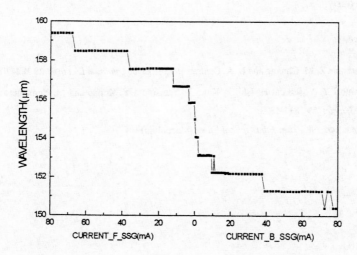

Fig. 7. Tuning characteristics of a nonlinearly chirped DBR laser.

a fixed period Λ_s. As shown in Fig. 6, a nonlinearly chirped grating has high reflectance and uniform reflection-peak heights. Therefore a laser can operate with low threshold current without channel-gaps in a broad tuning range. Moreover, a nonlinearly ch`irped grating does not require any additional complexity in fabrication, compared with a linearly chirped grating.

Under the same condition as in Fig.4, we simulated tuning characteristics of a nonlinearly chirped grating DBR laser as the tuning current is increased. As shown in Fig. 7, a nonlinearly chirped DBR laser operates in gap-free tuning with the same tuning range as a linearly-chirped DBR laser shown in Fig. 4.

4. Conclusions

We found by numerical simulation that the nonuniformity of mirror reflection-peak heights results in gaps of channel in tuning characteristics of an SSGDBR laser, which may limit the tuning range of the laser. We proposed a nonlinearly chirped grating DBR structure to achieve uniform mirror reflection. We confirmed by numerical simulation that a nonlinearly chirped grating DBR laser operates in gap-free tuning. A nonlinearly chirped grating does not require an additional complexity in fabrication, compared with a linearly chirped grating. Therefore a nonlinearly chirped grating structure can be employed in an extended tuning range semiconductor laser to achieve gap-free tuning and low threshold current operation simultaneously.

Reference

[1] V. Jayaraman Z. M. Chuang and L. A. Coldren 1993 *IEEE J. Qunatum Electron.* 29 1824-1834.

[2] Y. Tohmori Y. Yoshikuni H. Ishii F. Kano T. Tamamura Y. Kondo and M. Yamamoto 1993 *IEEE J. Quantum Electron.* 29 1817-1823.

[3] H. A. Macleod 1986 *Thin-Film Optical Filter* (Macmillan).

Inst. Phys. Conf. Ser. No 145: Chapter 8
Paper presented at 22nd Int. Symp. Compound Semiconductors, Cheju Island, Korea, 28 August–2 September 1995
© *1996 IOP Publishing Ltd*

All-optical switching by field-enhancement in MQW structures

C. Knorr, U. Wilhelm, V. Härle, F. Scholz, and A. Hangleiter[1]

4. Physikalisches Institut, Universität Stuttgart
Pfaffenwaldring 57, D-70550 Stuttgart, Germany

Abstract. We propose and demonstrate a novel type of optical switching mechanism which is based on internal field screening effects in pin-SCMQW (separate confinement multiple quantum well) structures. By introducing additional large barriers into conventional SCMQW structures, photogenerated holes can be localized in a way such that the resulting local screening of the internal field in the optical confinement layers leads to an increase of the field in the MQW region.

1. Introduction

Various kinds of optical nonlinearities are presently under investigation for their potential use in all-optical switching devices [1]. Whereas the direct use of the third-order susceptibility of the active material leads to extremely fast switching, rather high power levels are required to take advantage of these effects. Another class of devices like SEED's (self electro-optic effect devices) [2, 3] are based on electrical feedback and the quantum confined Stark effect (QCSE) in low-dimensional semiconductor structures.

Electro-optic modulators based on the QCSE are known to be subject to so-called "power saturation" effects, where photogenerated carriers separated by the internal electric field screen the external bias voltage and therefore reduce the absorption at a given wavelength below the bandgap [4]. In fact, this effect can also be utilized as an optical nonlinearity, if the structure is designed to maximize the field screening [5, 6]. However, since the QCSE is most effective below the fundamental bandgap, this amounts to a negative feedback, i.e. a bistable behaviour can not easily be established.

We have developed a new type of SCMQW structure, where the photogenerated carriers lead to a strong enhancement of the internal field rather than to a screening of the field. In this paper, we will describe the basic idea, report on its first realization, and present experimental results demonstrating its performance.

---

[1] Email: a.hangleiter@physik.uni-stuttgart.de

2. Basic design considerations

Field screening by photogenerated carriers in QCSE modulator structures is controlled mainly by the hole transport properties of the structures. The transport of holes, in turn, is governed by the ability of the holes to overcome hetero barriers by tunneling or thermal emission. We are dealing with the GaInAs/GaInAsP/InP material system, where about 60 % of the total band discontinuity are in the valence band.

Therefore, in order to utilize and maximize field screening effects in a well-defined manner, we introduce additional large barriers at specific places within the structure, in an attempt to get a well-controlled pile-up of holes next to this barrier under optical generation conditions. Since both the barrier height as well as the effective mass are much smaller for electrons, the effect of these barriers on electron transport is negligible.

The basic structure, designed to yield an enhancement of the internal field in the MQW region, is shown in Fig. 1. The rightmost (next to the n-side) of the InGaAsP

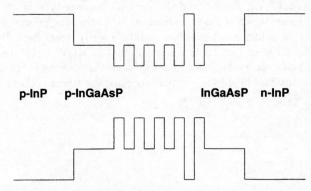

p-InP   p-InGaAsP   InGaAsP   n-InP

Figure 1. Basic layer structure designed to realize field enhancement due to photogenerated carriers.

barriers between the quantum wells of the SCMQW structure has been replaced by an InP barrier. The operation of this structure has first been studied by calculations using a device modeling program for two-contact heterostructure devices, based on a self-consistent solution of Poisson's equation and the continuity equation [7].

The results are shown in Fig. 2, where the band diagrams at a bias voltage of -2 V and varying external generation are plotted. Carriers photogenerated in one of the 4 wells to the left of the InP transport barrier are extracted efficiently by the internal field. The holes generated in the quantum well to the right of the InP transport barrier, however, are trapped and lead to a positive space charge in this quantum well. The electrons generated with these holes are extracted towards the n-doped region and partly compensate the positive space charge there. As a consequence, the electric field to the right of the additional InP transport barrier is screened, whereas the field to the left of this barrier is increased.

As can be seen from Fig. 2, the relative enhancement of the field by the partial screening is determined by the ratio of the total space charge layer width to the width of the region to the left of the InP transport barrier. Therefore it is desirable to make the distance between the p-doped layer and the InP transport barrier as small as possible.

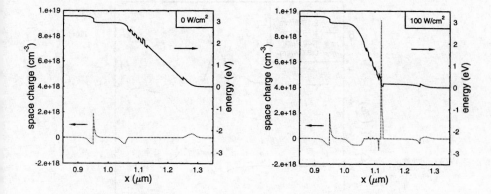

Figure 2. Calculated space charge density and conduction band diagrams for a field-enhancement structure under varying illumination without and with modest external illumination.

As a consequence, we have assumed p-doped InGaAsP up to the quantum well region for these calculations.

3. Experimental results

The layer structures described in the previous section have been realized by LP-MOVPE for the GaInAs/GaInAsP/InP material system. The total width of the waveguiding region was 310 nm, the 5 quantum wells had a thickness of 8 nm, and the barriers were 8 nm thick. The upper waveguiding layer and the upper InP cladding was p-doped at a level of about $1 \cdot 10^{18}$ cm^{-3}.

These structures were studied experimentally by differential transmission spectroscopy, where either the applied bias voltage or the external optical generation rate (by means of a 1.3 μm InGaAsP laser) was modulated, and where the resulting change in transmission was detected using lock-in techniques.

Fig. 3 shows the measured phototransmission spectra at various incident power levels. At lower power, the spectra exhibit a negative change of transmission below the bandgap due to the incident pump light. This corresponds to an increase of absorption in this wavelength region and therefore to an increased internal field across the MQW region of the sample.

At power levels of 2 W/cm^2 and above, the observed behaviour changes drastically. The change of transmission below the bandgap becomes negative, corresponding to a decreasing electric field.

In order to analyse these data quantitatively, we have also measured differential transmission spectra without optical pumping, but with the external bias voltage modulated. These spectra provide us with the energetic position of the bandgap as a function of the internal field. From a comparison of the phototransmission data with these electrotransmission data, we are able to determine the internal field as a function of the pump power level in the phototransmission experiments.

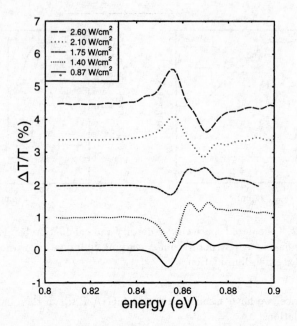

Figure 3. Differential phototransmission spectra for a sample designed for light-induced field enhancement at various incident pump power levels.

The results of this analysis are shown in Fig. 4, where the internal field is plotted versus incident pump power. The figure shows that we get an enhancement of the internal field of about 40 % at a power level of 2 W/cm^2. At higher power levels the field decreases below the dark value.

4. Discussion

Our experimental results clearly demonstrate that field enhancement due to photogenerated carriers can be achieved by introducing an additional larger barrier into a conventional SCMQW structure. As shown above, an enhancement of the electric field in the MQW region of 40 % has been achieved.

At a first glance, it is surprising that at higher power levels the field enhancement is reversed and, finally, the field is completely screened. This is due to the very mechanisms which lead to field screening in MQW modulator structures. As shown in Fig. 5, under these conditions the quantum wells to the left of the transport barrier are flooded with holes, leading to an almost complete screening of the field in the quantum well region. In fact, the "window", where field enhancement exists, is given by ratio of the thermal emission rates over the special transport barrier and over the "normal" barriers.

It is interesting to note that the 40 % field enhancement achieved experimentally is somewhat lower than expected from the calculations, which suggest about a doubling of the field. There are several possible reasons for this discrepancy. Probably, the

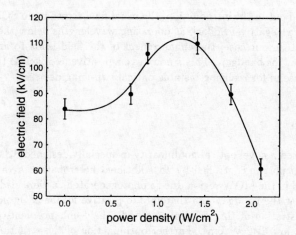

Figure 4. Internal electric field across the MQW region vs. pump power as determined from a comparison of phototransmission and electrotransmission spectra.

most important factor is that the position of the p-doping front is very difficult to control in MOVPE growth. This may lead to some p-doping in the quantum wells which deteriorates the field enhancement. Another possible reason is the early onset of re-screening. This is controlled by the relative thermal emission rates of holes over the respective heterobarriers.

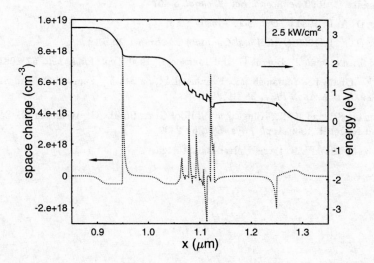

Figure 5. Space charge distribution and conduction band in the field-enhancement structure under strong illumination.

1084

As pointed out in the introduction, the usual field screening in SCMQW structures corresponds to a negative feedback at operating wavelengths below the bandgap of the active material. In our case, the enhancement of the field leads to an <u>increase</u> of the absorption below the bandgap. This amounts to a positive feedback at these wavelengths and might be useful for realizing bistable optical switching devices.

5. Conclusion

We have proposed a novel optical nonlinearity in specially designed SCMQW structures, where hole transport is controlled by an additional heterobarrier. An enhancement of the electric field in the MQW region due to photogenerated carriers leads to an increased absorption below the bandgap and therefore to positive feedback. From our first experimental demonstration of this principle we find a field enhancement of about 40 % at power levels as low as 2 W/cm^2. Further optimization of the structure is currently in progress, with particular emphasis on realizing a bistable behaviour.

Acknowledgement

Support of this work by the Deutsche Forschungsgemeinschaft under contract No. Ha 1670/6 is gratefully acknowledged.

References

[1] Klingshirn C. 1990 *Semicond. Sci. Technol.* **5** 457 .

[2] Miller D. A. B. 1984 *J. Opt. Soc. Am.* **B 1** 857 .

[3] Miller D. A. B. 1990 *Optical and Quantum Electronics* **22** 61 .

[4] Wood T., Pastalan J., Burrus C. and Johnson B. 1990 *Appl. Phys. Lett.* **57** 1081 .

[5] Kan Y., Obata K., Yamanishi M., Funahashi Y., Sakata Y., Yamaoka Y. and Suemune I. 1889 *Jpn. J. Appl. Phys.* **28** L1585 .

[6] Weber C., Schlaad K.-H., Klingshirn C., Hoof C. v., Borghs G., Weimann G., Schlapp W. and Nickel H. 1989 *Appl. Phys. Lett.* **54** 2432 .

[7] Klötzer N. 1991 Ph.D. thesis Universität Stuttgart .

Inst. Phys. Conf. Ser. No 145: Chapter 8
Paper presented at 22nd Int. Symp. Compound Semiconductors, Cheju Island, Korea, 28 August–2 September 1995
© *1996 IOP Publishing Ltd*

New Infrared Materials and Detectors

M. Razeghi
J. D. Kim, S. J. Park, Y. H. Choi, D. Wu, E. Michel, J. Xu, and E. Bigan

Center for Quantum Devices, Department of Electrical Engineering and Computer Science, Northwestern University, Evanston, IL 60208, USA

Abstract. In this paper, we present an overview of the growth of $InAs_{1-x}Sb_x$, $In_{1-x}Tl_xSb$, and $(InP)_{1-x}(TlP_3)_x$ as potential materials for longwavelength (8-12 μm) infrared photodetector applications. Incorporation of thallium into InSb and InP resulted in a bandgap decrease of the alloys. Thallium incorporation has been evidenced by various techniques such as X-ray diffraction, Auger electron spectrometry, and photoconductivity measurements. Preliminary photodetectors fabricated from the grown materials are also reported. InAsSb photodetectors showed room temperature photoresponse up to 13 μm.

1. Introduction

Focal plane arrays of infrared photodetectors operating in the 8-12 μm wavelength range are of great interest for thermal imaging applications. The current standard material system for these applications is based on II-VI semiconductor alloy HgCdTe. However it suffers from severe problems that strongly affect the yield and the cost of these devices. These problems originate from weak bonding characteristic of II-VI semiconductors and from high Hg vapor pressure. Weak bonding results in poor mechanical strength and thermal stability, and high Hg vapor pressure makes the composition control over large area difficult.

A number of III-V semiconductor alloys, benefiting from stronger covalent bonding and advanced growth and device processing technology, have been investigated as alternatives. InAsSb is the III-V semiconductor alloy with the smallest reported bandgap [1,2], whose bandgap is expected to cover the entire 8-12 μm range at room temperature. Some experimental results showed that the bandedge of InAsSb can extend to 12.5 μm [3]. To increase the performance of photodetector at room temperature optimization of the doping and energy gap profiles has been suggested [4]. According to this suggestion, InAsSb with proper bandgap and doping level has been grown to demonstrate the feasibility of III-V long-wavelength (8-12 μm) infrared photodetectors operating at near room temperatures.

Tl containing alloys such as InTlSb and InTlP have been proposed as new III-V semiconductors for 8-12 μm operation at 77 K [5-7]. Band-structure calculation predicted TlSb as a semi-metal having a low-temperature bandgap of -1.5 eV [5]. Hence, InTlSb is expected to have a bandedge of 12 μm at 77 K with only 9 % thallium content. Further theoretical studies indicated that InTlSb would be lattice-matched to InSb within 2 %, thus allowing the growth of device-quality material on InSb substrates. We also propose a new III-V alloy $(InP)_{1-x}(TlP_3)_x$ as a potential infrared material without the problems of large lattice mismatch. Phase diagram shows that TlP_3 is the only stable Tl-P compound with a melting temperature of 418 °C [8]. Furthermore, TlP_3 exhibits a zinc-blende structure with a lattice constant of 5.94 Å, close to that of InP (5.87 Å). This implies that $(InP)_{1-x}(TlP_3)_x$ can have a stable zinc blende structure with a maximum lattice mismatch of 1.28 % with InP. The band calculation predicted the energy gap of TlP to be -0.27 eV [7]. Assuming that the energy gap of TlP_3 is similar to the calculated value of TlP, $(InP)_{1-x}(TlP_3)_x$ has the potential to cover the entire infrared spectral range

In this paper, we report the growth and characterization of InAsSb, InTlSb, and $(InP)_{1-x}(TlP_3)_x$ alloys, as well as the realization of a photodetectors using these materials.

2. Experimental

InAsSb and InTlSb were grown on semi-insulating GaAs substrates by LP-MOCVD at 76 torr. Trimethylindium, trimethylantimony, 0.2 % arsine, and cyclopentadienylthallium were used as the precursors. $(InP)_{1-x}(TlP_3)_x$ was grown on semi-insulating InP substrates using phosphine (PH_3) as a P precursor. P- and n-type doping were achieved by Zn and Sn, respectively. The respective precursors for Zn and Sn were diethylzinc and tetraethyltin. Growths of InAsSb and InTlSb were achieved at a growth temperature of 470 °C while $(InP)_{1-x}(TlP_3)_x$ was grown at 520 °C. The test structure for InAsSb and InTlSb contained the InSb layer which was expected to work as a buffer layer for the growth.

The thicknesses of the epilayers were determined by ball-polishing method. Structural characterization was performed by four crystal high resolution X-ray diffractometer at (400) and (800) crystal orientation. The composition of the InAsSb layer has been determined from X–ray diffraction data and Vegard's law. For InTlSb, (800) orientation was used to clearly resolve the peaks of InSb and InTlSb. Hall measurement was done using Bio-Rad HL 5560 system and Van der Pauw method at both 300 K and 77 K. Infrared transmission and photoconductivity measurements were carried out using Fourier Transform Infrared (FTIR) spectrometer. The absolute responsivity of the photodetectors was determined using a 800 K blackbody.

3. Results and discussion

3.1. InAsSb

The standard structure for photoconductor studies consisted of 1.2 μm p-InSb buffer layer and 2.8 μm $p-InAs_{0.23}Sb_{0.77}$ active layer. P-type doping in $InAs_{0.23}Sb_{0.77}$ ensures the better compromise between optical and thermal generation than n-type doping at near room temperatures [10,11]. Clearly defined x-ray peaks were resolved for InSb and InAsSb with a low FWHM ($\approx$400 arsec for 2.8 μm $InAs_{0.23}Sb_{0.77}$) indicating the high quality of the material. Hall measurements exhibited n-type conductivity at 300 K with electron concentration of 3×10^{16} cm^{-3} and mobility of 36,000 cm^2/Vs. At 77 K, p-type conductivity was observed with a hole concentration of 3.6×10^{16} cm^{-3} and mobility of 923 cm^2/Vs. From absorption spectra bandgap of ≈ 0.103 eV (12.3 μm) was derived for $InAs_{1-x}Sb_x$ layer with x=0.77 and ≈ 0.083 eV (15 μm) for x=0.65.

4×3 mm^2 bar samples have been used for photoconductivity measurements. Au/Ti contacts were deposited by an e-beam evaporator. The sample had a resistance of 23 Ω at

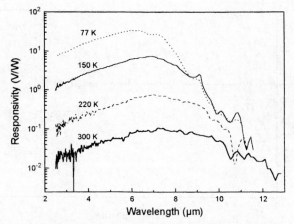

Fig.1. Photoresponse spectra of InAs$_{0.15}$Sb$_{0.85}$ photovoltaic detector.

room temperature and 469 Ω at 77 K. The spectral responsivity measured by FTIR spectrometer showed cutoff wavelength of 13μm at 300 K. The absolute responsivity data was used to determine the effective lifetime. The effective room temperature photoconductivity lifetime determined from the absolute responsivity was calculated to be ≈ 0.14 ns. The Johnson noise limited detectivity of the photoconductor at room temperature was estimated to be ≈ 3.27×10^7 cmHz$^{1/2}$/W at 10.6 μm.

As the photovoltaic devices are more favorable for array applications, 0.5 μm p-InSb/4.8 μm p$^-$ - InAs$_{0.15}$Sb$_{0.85}$/2.0 μm n-InSb double heterostructures have been recently grown on semi-insulating GaAs substarte to fabricate the photovoltaic detectors. The top and bottom InSb layers were doped at the level of 10^{18}cm^{-3} and doping level of p-type InAs$_{0.15}$Sb$_{0.85}$ layer was 3×10^{16}cm^{-3}. 400×400 μm^2 mesa was prepared by photolithography and wet etching. Au/Ti contacts were made on the two InSb layers by an e-beam evaporator. The spectral responsivities were measured with FTIR spectrometer while the illumination of infrared light was on the GaAs substrate side. Fig. 1 shows the absolute spectral responsivities of InAsSb photovoltaic detector. The room temperature responsivity was comparable to that of HgCdTe detectors operating at the same conditions.

3.2. InTlSb

As a first attempt to decrease the bandgap of InSb alloys, InTlSb has been grown. Thallium incorporation into InSb was confirmed by Auger electron spectrometry. When compared to an InSb reference sample, the InTlSb samples showed a noticeable intensity dip near 70 eV, indicating the presence of thallium in the epilayer [9]. The X-ray diffraction measurement of InTlSb/InSb showed two separately resolved peaks around 73°. One peak was attributed to the InSb buffer layer since it was at the correct angular position and the other to the InTlSb epilayer. As the thallium content was increased, X-ray peak corresponding to InTlSb was observed to broaden and shift away from the InSb peak. The broadening of the InTlSb peak has been attributed to an increase in the dislocation density arising from the increased mismatch with InSb as the thallium concentration increases. In contrast to the theoretical prediction, an increase in thallium incorporation led to a decrease in the lattice constant.

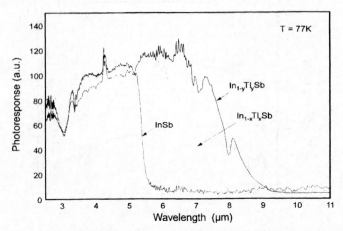

Fig.2. Normalized photoreponse spectra of InTlSb with various Tl compositions.

Infrared transmission measurements exhibited an increase in the absorption edge of the InTlSb samples with increasing thallium concentration [12,13]. As it was difficult to determine the band edge from transmission spectra, photoresponse was used as an alternative method to assess the band edge. Fig. 2 shows the normalized photoresponse spectra for InTlSb/InSb epilayers, along with a photoresponse of a reference InSb epilayer. The onset of the photoresponse was found to increase with increased thallium flow and it was extended up to a wavelength of 9.0 μm. The increase of cut-off wavelength linearly depended on the increase of lattice mismatch of InTlSb with InSb. A 9.0 μm cut-off wavelength corresponded to a small lattice mismatch of -1.3 % with InSb, suggesting the possibility of developing device quality material with reduced dislocation density compared to InAsSb.

The specific detectivity D* of the photoconductors at 77 K and 7 μm increased from 1×10^8 cmHz$^{1/2}$/W up to 3×10^8 cmHz$^{1/2}$/W as the thallium flow increased. This results mainly from the increase in quantum efficiency. To improve the quantum a thicker InTlSb sample was grown under the conditions that yielded a 9.0 μm cut-off wavelength. X-ray rocking curve and photoresponse spectrum were similar to those of the thin structure, indicating good composition control. The room-temperature Hall mobility increased from 2×10^4 cm^2/Vs for the thin structure to 3×10^4 cm^2/Vs for the thick one. This improvement in material quality was confirmed by an increase in the photoconductive gain over bias ratio G/V. The thicker sample presents both a higher absorption quantum efficiency and a higher photoconductive gain, which results in a significantly larger specific detectivity of 1×10^9 cmHz$^{1/2}$/W. This thick InTlSb sample was then used to investigate the effects of thermal annealing on the lattice-mismatch, cut-off wavelength, and magnitude of the photoresponse. No noticeable changes were observed in these parameters even after a 1 hour annealing at a temperature of 350 °C. This is encouraging since the thermal stability of InTlSb has been questioned based on phase-diagram considerations indicating TlSb to be unstable [14].

3.3. $(InP)_{1-x}(TlP_3)_x$

Growth of $(InP)_{1-x}(TlP_3)_x$ was carried out by incorporating Tl into InP. An InP epilayer was also grown under similar conditions as a reference sample. The surface morphology of both

InP and $(InP)_{1-x}(TlP_3)_x$ epilayers was nearly specular, with only a few randomly scattered hillocks. X-ray diffraction measurement of $(InP)_{1-x}(TlP_3)_x$/InP layer showed one peak at the angle corresponding to InP. This is not an unexpected result considering that the lattice-mismatch between the two crystals is small. Despite the similarities in these structural characteristics, a significant difference in the growth rate between InP and $(InP)_{1-x}(TlP_3)_x$ was found. From the measured thicknesses of the epilayers grown for 2 and 1/2 hours, the growth rate has been determined to be 0.60 µm/hr for InP and 0.78 µm/hr for $(InP)_{1-x}(TlP_3)_x$. This increase in the growth rate for $(InP)_{1-x}(TlP_3)_x$ is attributed to the presence of an additional group-III element, which usually determines the growth rate because of their high sticking coefficient under group-V rich environment.

Both InP and $(InP)_{1-x}(TlP_3)_x$ layers exhibited excellent electrical characteristics as determined by Hall measurements. The Hall coefficient sign of the samples were negative, indicating n-type materials. Hall mobility of a 1.7 µm-thick InP sample was as high as 4,200 cm²/Vs at 300 K and 73,000 cm²/Vs at 77 K while the Hall mobility of a 1.6 µm-thick $(InP)_{1-x}(TlP_3)_x$ sample was 4,600 cm²/Vs at 300 K and 77,000 cm²/Vs at 77 K. The higher mobility of $(InP)_{1-x}(TlP_3)_x$ is indicative of a smaller electron mass for the lower bandgap material.

The most interesting results were found from infrared photoconductivity measurements. Simple photoconductors were fabricated from samples grown on semi-insulating InP substrates. Fig. 3 displays the 77 K photoconductivity spectral responses of InP and $(InP)_{1-x}(TlP_3)_x$ samples with different x values. As anticipated, there is no spectral response from the InP sample in the infrared wavelength region. Clear photoresponse is observed from both the Tl-containing samples, with $(InP)_{1-x}(TlP_3)_x$ having an onset wavelength around 5.5 µm and $(InP)_{1-x'}(TlP_3)_{x'}$ around 8.0 µm. This is consistent with our predicted trend since the Tl content in $(InP)_{1-x'}(TlP_3)_{x'}$ is expected to be greater as it was grown under a higher thallium flow than $(InP)_{1-x}(TlP_3)_x$. Schifgaarde et al.[7] have calculated some physical properties of the theoretical TlP compound. They predicted TlP to exhibit a zinc blende structure with a lattice constant of 5.96 Å. This prediction closely resembles the aforementioned properties of TlP_3. Assuming the bandgap energy of TlP_3 to be the same as that of TlP ($E_g = -0.27eV$) [7], a rough estimate of $(InP)_{1-x}(TlP_3)_x$ stoichiometry can be

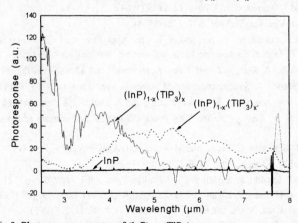

Fig.3. Photoresponse spectra of $(InP)_{1-x}(TlP_3)_x$ with different compositions.

1090

calculated based on our optical measurements and reference 7, the composition for $(InP)_{1-x}$ $(TlP_3)_x$ samples with onset wavelengths of 5.5 μm and 8.0 μm are x=0.6 and 0.64, respectively.

4. Conclusions

InAsSb longwavelength photodetectors operating at near room temperatures have been demonstrated. The photoresponse up to 13 μm has been obtained at 300 K for both photoconductive and photovoltaic detectors. The estimated detectivity of photoconductor at room temperature was 3.27×10^7 cmHz$^{1/2}$/W.

InTlSb alloys of various thallium compositions have been successfully grown and characterized. The photoresponse cut-off was tailored from 5.5 μm to 9.0 μm at 77 K. Experimental observations indicate that the bandgap can be further decreased by increasing the thallium content. A photoconductor fabricated from a thick InTlSb sample had an estimated specific detectivity of 10^9 cmHz$^{1/2}$W^{-1} at 7 μm and 77 K.

$(InP)_{1-x}(TlP_3)_x$ has been grown on InP and some of its physical characteristics have been investigated. X-ray measurements suggest a negligible lattice mismatch with InP. Infrared photoconductivity measurements indicate a reduction of the bandgap energy with increasing Tl incorporation. Further increase in thallium concentration is also expected to extend its wavelength for operation in the 8-12 μm atmospheric window.

Acknowledgement

The authors would like to acknowledge the support of Dr. G. Wright, Dr. Y. S. Park, and Dr. R. Balcerak. This work was supported by the Office of Naval Research under contract No. N00014-92-J1951 and No. N00014-94-1-0902 and the Strategic Defense Initiative Organization under contract No. N0001493-1-0409.

References

[1] J. C. Wooley and J. Warner, Can. J. Phys. **42**, 1879 (1964).
[2] G. C. Osbourn, J. Vac. Sci. Technol. B, **2**, 176, (1984).
[3] C. G. Bethea, B. F. Levine, M. Y. Yen, and A. Y. Cho, Appl. Phys. Lett. **53**, 291 (1988).
[4] A. Rogalski, SPIE Optical Engineering Press, Bellingham, Washington, (1994).
[5] M. van Schilfgaarde, A. Sher, and A.-B. Chen, Appl. Phys. Lett., **62**, 1857 (1993).
[6] Y. H. Choi, C. Besikci, R. Sudharsanan, and M. Razeghi, Appl. Phys. Lett., **63**, 361 (1993).
[7] M. van Schilfgaarde, A.B. Chen, S. Krishnamurthy, and A. Sher, Appl. Phys. Lett. **65**, 2714 (1994).
[8] C. Sharma and Y. A. Chang, "Binary Alloy Phase Diagrams," 2nd edition, edited by T. B. Massalski, 2991 (1990).
[9] Y. H. Choi, P. T. Staveteig, E. Bigan, and M. Razeghi, J. Appl. Phys., **75**, 3196 (1994).
[10] J. Piotrowski and M. Razeghi, Proc. SPIE **2397**, 180 (1995).
[11] C. T. Elliott, N. T. Gordon, R. S. Hall and T. J. Philips, Proc. SPIE, **2269**, 648-657 (1994).
[12] P. T. Staveteig, Y. H. Choi, E. Bigan, and M. Razeghi, Appl. Phys. Lett., **64**, 460 (1994).
[13] M. Razeghi, Y. H. Choi, X. He, and C. J. Sun, Mat. Sci. Technol. **11**, 3, (1995).
[14] B. Predel and W. Schwermann, Z. Naturforsch. A, **25**, 877 (1970).

Inst. Phys. Conf. Ser. No 145: Chapter 8
Paper presented at 22nd Int. Symp. Compound Semiconductors, Cheju Island, Korea, 28 August–2 September 1995
© *1996 IOP Publishing Ltd*

Monolithic Millimeter Wave Optical Receivers using MSM photodetectors and SMODFET's

Jinwook Burm, Kerry I. Litvin, Glenn H. Martin, William J. Schaff, and Lester F. Eastman

School of Electrical Engineering and National Nanofabrication Facility, Cornell University, Ithaca, NY 14853

Abstract Integrating MSM photodetectors and SMODFET's, single stage millimeter wave optical receiver circuits were fabricated monolithically on an MBE grown GaAs wafer. The photodetector layer was designed for the maximum absorption of 770nm light. The SMODFET's utilized 120Å $In_{0.2}Ga_{0.8}As$ channel. The circuits were designed for 44GHz operation. The measurement with beating Ti-Sapphire lasers and a spectrum analyzer on the optical receiver circuits showed 3dB gain over a photodetector at 39GHz, which was the limit of the measurement. The calculated gain of the circuit was 6.6dB over the photodetector and 5.7dB including photodetector signal drop at 44GHz.

1. Introduction

The simple and planar structure of MSM (Metal-Semiconductor-Metal) photodetectors has made the device an ideal candidate for the optical receiver application. For their simple structure, MSM photodetectors have demonstrated outstanding frequency performances to have more than 350GHz of 3dB bandwidth [1,2]. Current developments in making ultrafast MSM photodetectors have been in decreasing the gap between the fingers for short carrier transit time, and decreasing the capacitance by reducing the overall size. In real applications the sensitivity is an important parameter to be considered, and is difficult to optimize for MSM photodetectors. The low sensitivity of MSM photodetectors is mainly due to metal finger shadowing and surface reflectance of the incoming light. To increase the sensitivity of the MSM photodetectors without decreasing the speed, an AlGaAs cap layer minimize the surface recombination of the optically created carriers, a thin GaAs absorption layer to decrease the carrier transit time, and Bragg reflector layers to reflect unabsorbed light back into the absorption layer [3,4] were employed. The idea of employing Bragg reflector layers has been introduced for Schottky photodiodes [5], phototransistors [6] and p-i-n detectors [7].

By integrating the MSM photodetectors with well-developed SMODFET's (Pseudomorphic Modulation Doped Field Effect Transistors), optical receiver circuits were fabricated monolithically. The optical receiver circuit was designed for 44GHz operation. This paper deals with the fabrication and the testing of the optical receivers.

2. Layer Structure

For the high sensitivity of MSM photodetectors, the layer structure was carefully designed for the maximum light absorption. By following the method of layer design for the maximum absorption of the incident light [3,4], the layer structures were determined for the best absorption of the light at 770nm, which can be easily obtained from Ti-Sapphire laser. The optical detector structure includes 16.5 pairs of Bragg reflector layers, 3120Å GaAs absorption layer, and 1906Å $Al_{0.3}Ga_{0.7}$ As cap layer. The thickness of the cap layer was thicker than the calculated optimum thickness 1106Å where the absorption was maximum, to allow process tolerance. The Al mole fraction in the cap layer and in the quarter wavelength Bragg reflector sections were chosen so that these layers were nonabsorbing at the wavelengths of interest (Fig. 1). The GaAs absorption layer was the only absorbing layer. The $Al_xGa_{1-x}As$ cap layer was important to reduced the surface recombination of the carriers and to match the optical resonant impedance of the photodetector structure to that of air, 377Ω. As the result of the impedance matching, the reflectance reduces to nearly negligible levels [3,4].

The photodetector layers were followed by the SMODFET layers. The SMODFET layers were composed of 1×10^{12} cm^{-2} planar p-doped layer, 500Å $Al_{0.3}Ga_{0.7}As$ layer, 3×10^{12} cm^{-2} planar n-doped layer, 50Å $Al_{0.3}Ga_{0.7}As$ spacer, 120Å $In_{0.2}Ga_{0.8}As$ channel sandwiched between two 20Å GaAs layers, 30Å $Al_{0.3}Ga_{0.7}As$ spacer, 6×10^{12} cm^{-2} planar n-doped layer, 250Å $Al_{0.3}Ga_{0.7}As$ barrier and 400Å n-doped GaAs cap layer for metal contacts. The planar p-doped layer was employed to tightly confine the 2-DEG in the pseudomorphic $In_{0.2}Ga_{0.8}As$ channel. The whole optical receiver layers were grown using Varian Gen II MBE (Molecular Beam Epitaxy). Be and Si were used for p and n doping, respectively.

3. SMODFET's (pSeudomorphic Modulation Doped Field Effect Transistors)

SMODFET's were employed as post-detection amplifiers. The transistors were fabricated

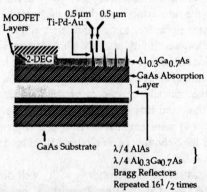

Figure 1. Cross-section of the optical receiver layer. The quarter wavelength Bragg reflector layers ensure the reflection of the light to maximize the absorption. The $Al_{0.3}Ga_{0.7}As$ cap layer on the top of the absorption layer decreases surface recombination rate and minimizes the top surface reflectance from the detector layer. The SMODFET layer on the top of the photodetector layer utilizes strained $In_{0.2}Ga_{0.8}As$ channel and double doping below and above the channel.

using standard processing procedure - mesa isolation, ohmic metal, and gate metal deposition (Fig. 2 (a)). During the mesa isolation, a part of the photodetector cap layer was etched away. As mentioned earlier, a certain thickness of the cap layer was required to match the characteristic impedance of the photodetector layer to that of the air. Thus the surface reflectance was monitored carefully to stop at the optimum place for the minimum reflection from the surface at the desired wavelength. The minimum reflection is important as it implies maximum absorption of incident light. The surface reflectance measurement was measured with a spectrophotometer, Cary 5 by Varian and the desired point could be reached with a good accuracy. Ni/AuGe/Ag/Au were used for ohmic contacts and the contact resistance of $0.1\Omega\cdot mm$ was achieved. Ti/Pd/Au were used for the gate metalization after the Electron-beam lithography and recess etch with citric acid. The gate width was 75µm. The fabricated transistors showed around f_t of 90GHz and f_{max} of 100GHz, however some variations of performance of the devices were observed.

4. Photodetectors

Circular-aperture MSM photodetectors were employed for the optical receiver circuits (Fig. 2 (b)) [8]. 150Å Ti, 150Å Pd, and 600Å Au were used for finger metalization. Before the evaporation of metal fingers, 90% of AlGaAs cap layer had been etched using citric acid. In general the quantum efficiency of MSM photodetectors is difficult to determine due to low frequency gains (more than 100% of quantum efficiency) [9,10]. The internal quantum efficiency of gain-suppressed photodetectors by preserving some of the AlGaAs cap layers, fabricated on about the same layer structure but for different wavelength showed 82% at 5V of bias and 94% at 10V of bias with 54.4µW optical power [4]. No careful quantum efficiency measurement was done for the detectors on the wafer due to low frequency gain, but 0.1A/W (3% of quantum efficiency) of responsivity was obtained for 0.5µm gap and 0.5µm finger width MSM photodetectors at 10mW of optical power. The low quantum efficiency was attributed to excessive carrier-recombinations from high optical power.

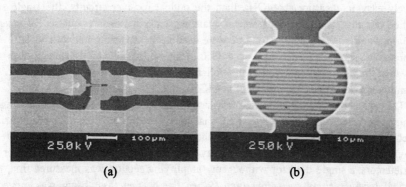

(a) (b)

Figure 2. (a) A SMODFET in a circuit. The gate width is 75µm. The fabricated transistors showed around 90GHz of f_t and 100GHz of f_{max}, however some variations of performance of the devices were observed. (b) A circular MSM photodetector in a circuit. The finger width and the gap between the fingers are 0.5µm. For finger metalization 150Å Ti, 150Å Pd, and 600Å Au was used.

5. Optical receiver design and fabrications

On the fabricated SMODFET's and photodetectors, their S-parameters were measured up to 50GHz using HP8510C network analyzer. Based on the measured S-parameter values at 44GHz and assuming unilateral condition, the S-parameters of adjacent elements were conjugately matched[11], for example, S_{11} of photodetectors and S_{11} of transistors, by way of 50Ω CPW transmission lines and one or two quarter wavelength impedance transformers. The S_{22} of transistors was matched to 50Ω. The design and the matching were further checked with a CAD program, TOUCHSTONE. For this purpose, the frequency response of photodetectors was modeled from both intrinsic carrier response [12] and extrinsic circuit response [13]. For SMODFET's, measured values were used for the simulation. A simulated gain of single stage receiver circuit was 6.6dB over the photodetector at 44GHz. The total gain including the photodetector signal drop was 5.7dB at 44GHz (Fig. 3).

CPW (Coplanar wave guide) transmission lines were employed for matching and filtering network. For DC bias, capacitor-isolated DC bias lines and ground plane capacitor DC isolation were used. The capacitor-isolated DC bias lines had a metal strip as a DC bias line with a resistor connecting between the bias line and center CPW lines where the main signal would flow. The ground plane of the main CPW ran over the metal strip with dielectric isolation in between, forming a capacitor. Thus the metal strip running under the ground plane will be an RF-short due to the capacitor, and through a resistor the desired bias can be provide. A resistor, used to connect the end of bias line and the center of CPW lines should be high enough not to interrupt the signal flow too much. Epitaxial resistors with 600-700Ω were utilized on 50Ω transmission lines. For DC bias isolation, ground plane capacitor isolation [14,15] was employed. The ground plane capacitor isolation had overlay capacitors in ground planes, so that we can bias each section of ground planes separately for desired biases. The each capacitor was 20μm$\times$300μm and expected to have about 3pF of capacitance based on our previous measurement.

Single stage amplifiers, composed of a transistor and matching network but without photodetectors, were fabricated as well (Fig. 4 (a)) with 44GHz design frequency. Being relatively easy to measure at 44GHz than the optical receiver circuits, the single stage amplifier circuits yield the S-parameter matching information. A quarter wavelength transformer and a 50Ω transmission line were used in the matching network to 50Ω. The measurements showed 5.3dB gain at 44GHz (Fig. 4 (b)). Slight mismatch was also observed at the measurements due to the variation of S-parameters of the transistors. The maximum gain was at 40GHz instead of at 44GHz, the design frequency.

The optical receiver circuits were fabricated (Fig. 5) and tested with two beating Ti-Sapphire lasers at 770nm and a spectrum analyzer, Tektronics 2782. The beating signal was fed into a single mode optical fiber to a photodetector. The optical power measured at the end of the fiber was 10mW. As a way to determine the amplification after the photodetectors, a single detector without any amplifying section was measured first, then an optical receiver circuit was measured. When the beating laser beam was moved to another photodetector, the photo-current was maximized and checked to be about the same as before. The beating frequency from two laser beams was set to 39GHz. The RF signal was measured for 2 minutes, registering all the new maximum peaks for more reliable comparison. The resolution bandwidth and sweep time of the spectrum analyzer were

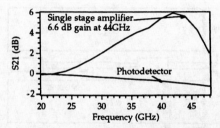

Figure 3. A simulated gain of single stage receiver circuit from Touchstone by EESOF. A measured transistor value was used with the other calculated values. The amplifier showed 6.6dB gain over the photodetector at 44GHz. The total gain including the photodetector signal drop, was 5.7dB at 44GHz.

300kHz and 130ms, respectively. A needle probe was used to monitor the gate bias for the initial bias settings, making a contact on the center conductor of CPW near the gate of SMODFET's. The measured RF results with the needle probe engaged showed little difference with the needle probe up in the air at high frequency (> 20GHz), suggesting a good RF blocking of the needle probe at the frequency range. The signal from a single photodetector was -40dBm. When an optical receiver circuit with a single stage amplifier was measured, the signal was -37dBm showing about 3dB gain at 39GHz. The optical receiver circuits could not be measured at 44GHz, where the circuits were designed to work, due to the limit of the spectrum analyzer. However the result of the single stage amplifier at 44GHz and the gain of the postdetection amplifier of the optical receiver circuit at 39GHz was favorable to expect a working result of optical receiver circuits at 44GHz.

6. Conclusions

The optical receiver circuits tuned at 44GHz were fabricated, integrating MSM photodetectors and SMODFET's. The layers were grown on a GaAs substrate using MBE. The photodetector layers were designed to have maximum absorption for 770nm light. Double doped 120Å $In_{0.2}Ga_{0.8}As$ channel was used in SMODFET layer.

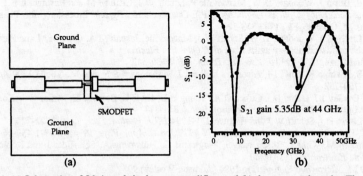

(a)	(b)

Figure 4. (a) Schematics of fabricated single stage amplifiers and (b) the measured results. The measured result showed 5.3dB gain at 44GHz. However the maximum gain occurred at 40GHz instead of at 44GHz where the circuit was design, from a slight mismatch of the circuit. This was largely due to the variation of S-parameters of the transistors.

1096

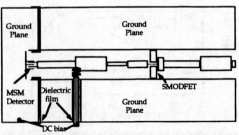

Figure 5. Schematics of fabricated optical receiver circuits. Two sections of quarter wavelength transformers were used to match the S-parameters of the photodetector and the transistor.

CPW waveguide was used for the matching and filtering network. Based on the s-parameters of individual elements, the circuit was designed. To apply DC bias, capacitor isolated bias lines and the ground plane capacitor isolations were employed. The single stage amplifiers without a photodetector were fabricated and measured up to 50GHz with HP8510C network analyzer. The single stage amplifier achieved 5.3dB gain at the design frequency, 44GHz. Two beating Ti-Sapphire lasers and a 40GHz spectrum analyzer were used to measure the fabricated optical receiver circuit. The measured result at 39GHz showed about 3dB gain over photodetectors. However the optical receiver could not be measured at 44GHz due to instrumental limit which was 40GHz. Based on TOUCHSTONE simulation, the receiver circuit was expected to have 6.6dB gain over the photodetector and 5.7dB total gain including photodetector signal drop at 44GHz.

Acknowledgments

This work is supported by Rome Laboratory under contract F30602-91-C-0063 and by Hughes Research Laboratory.

References

[1] Chen Y, Williamson S, Brock T, Smith F W and Calawa A R, 1991 *Appl. Phys. Lett.* **59** 1984-1986
[2] Chou S Y and Liu M Y 1992 *IEEE J. Quantum Electron* **28** 2358-2368
[3] Litvin K I, Burm J, Woodard D W, Mandeville P, Schaff W J, Gitin M M and Eastman L F 1992 *Proc. SPIE-Int. Society of Optical Engineers* (SPIE-Optoelectronic Signal Processing for Phased Array Antennas III, Orlando, FL) **1703** 313-320
[4] Burm J, Litvin K I, Woodard, D W, Schaff W J, Mandeville, Jaspan M A, Gitin M M and Eastman L F 1995 to be published in the August issue of *J. Quantum Electron*
[5] Chin A and Chang T Y 1990 *J. Vac. Sci. Technol.* B **8** 339-342
[6] Ünlü M S, Kishino K, Chyi J-I, Arsenault L, Reed J, Mohammad S N, and Morkoç H 1990 *Appl. Phys. Lett.* **57** 750-752
[7] Tan I-H, Dudley J J, Babic D I, Cohen D A, Young B D, Hu E L, Bower J E, Miller B I, Koren U and Young M G 1994 *IEEE Photonics Technol. Lett.* **6** 811-813
[8] Burm J, Litvin K I, Schaff W J and Eastman L F 1994 *IEEE Photon. Tech. Lett.* **6** 722-724
[9] Sugeta T, Urisu T, Sakata S and Mizushima Y 1979 *Jpn. J. Appl. Phys.* **19** suppl 19-1 459-464
[10] Klingenstein M, Kuhl J, Rosenzweig J, Moglestue C, Hülsmann A, Schneider J and Köhler K 1994 *Solid-State Electronic* **37** 333-340
[11] Pozar D M, *Microwave Engineering* 1990 (Addison-Wesley) 99
[12] Soole J B D and Schumacher H 1990 *IEEE Trans. Electron Device* **37** 2285-2290
[13] Litvin K, Burm J, Woodard D, Schaff W and Eastman L F 1993 *IEEE MTT-S Digest* **2** 1063-1066
[14] Compton R C Private communications.
[15] K. C. Gupta K C 1979 *Microstrip Lines and Slotlines* (Dedham,MA:Artech House) Chapter 7

Inst. Phys. Conf. Ser. No 145: Chapter 8
Paper presented at 22nd Int. Symp. Compound Semiconductors, Cheju Island, Korea, 28 August–2 September 1995
© *1996 IOP Publishing Ltd*

A monolithic ultrahigh-speed InAlAs/InGaAs PIN-HBT photoreceiver with a bandwidth of 18.6 GHz

K Yang, A L Gutierrez-Aitken, X Zhang, P Bhattacharya and G I Haddad

Center for High Frequency Microelectronics, Department of Electrical Engineering and Computer Science, The University of Michigan, Ann Arbor, MI 48109-2122, USA

Abstract. An ultrahigh-speed InAlAs/InGaAs PIN-HBT OEIC photoreceiver was fabricated and characterized. A transimpedance amplifier with a 500Ω feedback resistor demonstrated a -3dB bandwidth of 20GHz and a gain of 46dBΩ. The fabricated photoreceiver integrated with a $83\mu m^2$ p-i-n photodiode demonstrated a -3dB optical bandwidth of 18.6GHz with an input optical dynamic range of 20dB. Clear eye diagrams were obtained up to the 12Gb/s limit of the pattern generator. Large-signal transient simulations indicate that the measured bandwidth of 18.6GHz, the highest reported to date for any integrated front-end photoreceiver, is sufficient for 24Gb/s operation.

1. Introduction

The development of high-speed integrated photoreceivers has been the focus of intense research in recent years as lightwave communication systems progress towards higher data rates. Several optoelectronic integrated circuit (OEIC) photoreceivers have been demonstrated by integrating either an InGaAs p-i-n (PIN) photodiode, or a metal-semiconductor-metal photodetector with transimpedance preamplifiers made of InP-based heterostructure field-effect transistors or heterojunction bipolar transistors (HBT's) [1-6]. Among the variety of integrated designs, a PIN-HBT transimpedance photoreceiver scheme, where the p-i-n photodiode is formed using the base-collector junction of the HBT, has proved to be extremely practical due to its simple epitaxy and fabrication [4-6]. Based on this integration approach, high-speed InP-based photoreceivers have demonstrated -3 dB bandwidths up to 9 GHz [7] and speeds up to 12 Gb/s [8].

In this work, we report the fabrication and characterization of an ultrahigh-speed InAlAs/InGaAs PIN-HBT OEIC photoreceiver with a -3 dB optical bandwidth of 18.6 GHz. The electrical and optical characteristics of individual devices and integrated

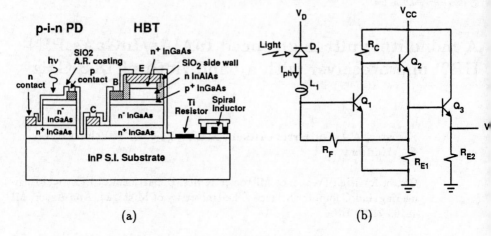

Figure 1. Schematics of (a) fabricated device cross sections and (b) the integrated transimpedance photoreceiver circuit.

circuits were measured and analyzed. In order to perform detailed characterization of the fabricated OEIC photoreceiver, the equivalent-circuit models of devices and passive circuit elements were extracted and used in full-scale circuit simulations.

2. Device structure and OEIC fabrication

The InAlAs/InGaAs HBT epitaxial layers grown by molecular beam epitaxy (MBE) represent a standard InAlAs/InGaAs single HBT structure. The thickness of the lightly-doped InGaAs precollector, 6000 Å, was optimized considering the photodiode responsivity and required HBT high-frequency performance for 20-30 Gb/s operation [7]. The p^+ InGaAs base was grown to be 850Å thick at a doping density of 3×10^{19} cm^{-3}.

A self-aligned emitter-base process was used to fabricate the transistors with nonalloyed Ti/Pt/Au metallization for the emitter and collector ohmic contacts and Pt/Ti/Pt/Au metallization for the base contact. The fabrication of p-i-n photodiodes is completely compatible with the HBT fabrication process [6]. A 7600Å-thick PECVD SiO$_x$ layer was used for isolation of devices and as an antireflecting (AR) coating for the InGaAs photodiode. For the interconnect metal, a 1.3 μm thick Ti/Al/Ti/Au layer was deposited. This interconnect metal layer was also utilized to form spiral inductors used in integrated amplifier circuits. A 300Å thick Ti layer was evaporated and lifted-off to define thin-film resistors. A schematic of fabricated device cross sections is shown in Fig. 1. The schematic circuit diagram of the integrated transimpedance photoreceiver is also shown.

3. Results

The fabricated 5×5 μm^2 emitter HBT's used in the preamplifier circuit demonstrated uniform device characteristics over the wafer such as a dc current gain of 30 and maximum

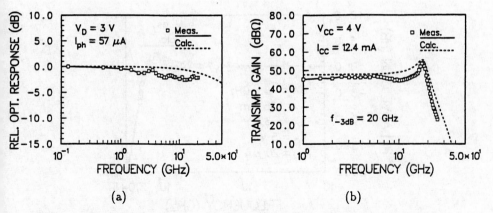

Figure 2. (a) Measured and calculated optical frequency response of the 83 μm^2 p-i-n photodiode and (b) gain-frequency characteristics of the fabricated peaking-inductor transimpedance amplifier.

current and power gain cutoff frequencies, f_T and $f_{max(U)}$, of 67 GHz and 120 GHz at I_C = 10 mA and V_{CE} = 2 V. The dc and small-signal ac characteristics of fabricated HBT's were fully analyzed and modeled using the developed HSPICE HBT macro model [9].

The optical response of the fabricated 83 μm^2 photodiode was measured in a frequency range of 0.13 to 20 GHz using an HP8703 Lightwave Component Analyzer. The photodiode, which has a responsivity of 0.35 A/W at λ = 1.55 μm, showed a -3 dB optical bandwidth larger than 20 GHz (the frequency upper limit of the used Lightwave Component Analyzer). The dark currents of fabricated photodiodes were measured to be less than 10 nA up to a reverse bias of 4 V. The measured optical response of the photodiode is shown in Fig. 2 (a) in comparison with theoretical results based on the extracted equivalent circuit from measured S-parameters of the photodiode without light illumination. The characterization results indicate that the fabricated photodiode has a -3 dB RC-time-constant-limited bandwidth of 45 GHz.

Several different versions of HBT transimpedance amplifiers were designed and fabricated. A schematic circuit diagram of an integrated three-stage peaking-inductor amplifier is shown in Fig. 1 (b). This peaking-inductor scheme includes a spiral inductor (L_1) at the input side of the basic three-stage amplifier to enhance the overall bandwidth. The measured gain-frequency characteristics of a peaking-inductor three-stage amplifier with R_F = 500 Ω and L_1 = 2.2 nH are shown in Fig. 2 (b). The amplifier demonstrated a large transimpedance gain of 46 dBΩ and a -3 dB bandwidth of 20 GHz. The spiral inductor in the input side introduces a gain peak at 16 GHz (-3 dB band edge of the basic amplifier without the inductor) and enhances the overall bandwidth of the amplifier by 60 percent from 14 GHz to 20 GHz. The simulation results based on the developed unified HBT macro model [9] showed good agreement with the experimental results. The circuit model, consisted of individually extracted device models, was implemented in HSPICE and used to characterize the overall performance of integrated circuits including the dc, small-signal ac, noise, and large-signal transient characteristics.

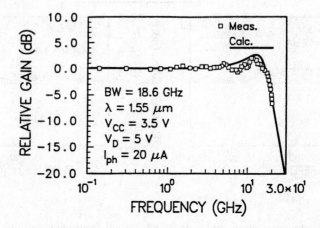

Figure 3. Optical frequency response of the OEIC photoreceiver.

The measured optical frequency response of the OEIC photoreceiver using the Lightwave Component Analyzer is shown in Fig. 3. The integrated photoreceiver consumed a small dc power of 35 mW at an applied bias of $V_{CC} = 3.5$ V and $V_D = 5$ V and demonstrated a -3 dB optical bandwidth of 18.6 GHz, which is the highest reported till date for any integrated front-end photoreceiver. The model predicted the measured optical bandwidth accurately with a slightly reduced inductance value of 2 nH from the value ($L_1 = 2.2$ nH) extracted from the measured S-parameter data of a discrete 3.5-turn spiral inductor.

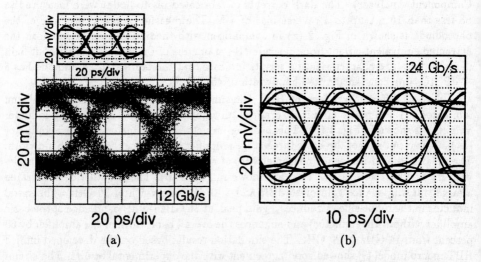

Figure 4. (a) Measured eye diagram of the integrated photoreceiver at 12 Gb/s for a NRZ 2^{31}-1 bit long PRBS (the inset shows a simulated eye diagram at the same bit rate) and (b) simulated eye diagram at 24 Gb/s of the OEIC photoreceiver.

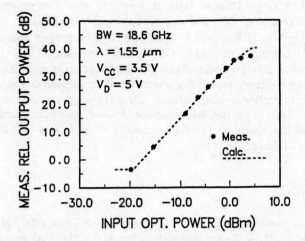

Figure 5. Measured optical-input vs. electrical-output power characteristics of the OEIC photoreceiver. The calculated output-power results showing a -54 dBm minimum noise level are plotted in comparison with the measured output-power results in a relative unit by matching the minimum output noise levels.

The integrated photoreceiver showed clear eye diagrams up to the 12 Gb/s limit of the pattern generator. An on-wafer measured eye diagram of the OEIC for a NRZ 2^{31}-1 bit long pseudo-random binary sequence (PRBS) at 12 Gb/s is shown in Fig. 4 (a). In order to investigate the high-speed performance of the fabricated OEIC photoreceiver with a bandwidth of 18.6 GHz, which can operate at much higher bit rates than the measured 12 Gb/s limit of the pattern generator, large-signal transient simulations were performed. The simulation results confirm the measured OEIC performance at 12 Gb/s and demonstrate that the measured bandwidth of 18.6 GHz of the photoreceiver is sufficient for 24 Gb/s operation as shown in Fig. 4 (b).

The measured optical-input vs. electrical-output power characteristics of the OEIC at 2 GHz using the Lightwave Component Analyzer and an Erbium-doped fiber amplifier are shown in Fig. 5. The calculated results in a unit of dBm are also shown in the figure for comparison. The results demonstrate that the integrated photoreceiver has an input optical dynamic range of 20 dB. From the detailed small-signal noise analysis using the implemented circuit model, it was found that the fabricated photoreceiver has an equivalent-input-noise-current-spectral density less than 11.1 pA/$\sqrt{Hz}$ up to 16 GHz. From the sensitivity calculation based on the noise spectrum, a sensitivity of -18.7 dBm is estimated for OEIC operation at 10 Gb/s for a bit error rate of 10^{-9} and $\lambda = 1.55$ μm.

4. Conclusions

We have fabricated and characterized an ultrahigh-speed OEIC InAlAs/InGaAs PIN-HBT transimpedance photoreceiver. The 83 μm² p-i-n photodiode with a responsivity of 0.35 A/W, formed using the 6000Å thick precollector layer of the HBT, showed an optical bandwidth larger than 20 GHz at $\lambda = 1.55$ μm. The preamplifier showed a transimpedance gain of 46 dBΩ and a -3 dB bandwidth of 20 GHz.

At a supply bias of $V_{CC} = 3.5$ V, the OEIC photoreceiver consumed a small dc power of 35 mW and demonstrated a -3 dB optical bandwidth of 18.6 GHz, which is the highest reported to date for a monolithically integrated front-end photoreceiver. From the measured input-output power characteristics of the photoreceiver, an input optical dynamic range of 20 dB was demonstrated. Clear eye diagrams were measured up to 12 Gb/s, the speed limit of the pattern generator. From large-signal transient simulations, it is found that the measured optical bandwidth of 18.6 GHz is sufficient for OEIC operation up to 24 Gb/s.

Acknowledgments

This research was sponsored by the Advanced Research Projects Agency, MDA 972-94-1-0004, and the U. S. Army Research Office under the URI program, Grant No. DAAL03-92-G-0109. The authors would like to thank Dr. S. Goswami at Bellcore Labs and Dr. L. M. Lunardi at AT&T Bell Labs for optical measurements.

References

[1] Fuji H S, Ray S, Williams T J, Griem H T, Harrang J P, Daniels R R, LaGasse M J and West D L 1991 *IEEE GaAs IC Symp. Dig.* 205–208

[2] Akahori Y, Akatsu Y, Kohzen A and Yoshida J 1992 *IEEE Photon. Technol. Lett.* 4 754–756

[3] Soda M, Nagano N, Takeuchi T, Saito T, Suzaki T, Honjo K and Fujita S 1992 *Electron. Lett.* 28 336–338

[4] Perotti K D, Sheng N H, Pierson Jr R L, Farley C W, Rosker M J and Chang M F 1991 *IEEE J. Quantum Electron.* 27 769–772

[5] Chandrasekhar S, Lunardi L M, Gnauck A H, Hamm R A and Qua G J 1993 *IEEE Photon. Technol. Lett.* 5 1316–1318

[6] Cowles J, Gutierrez-Aitken A L, Bhattacharya P and Haddad G I 1994 *IEEE Photon. Technol. Lett.* 6 963–965

[7] Gutierrez-Aitken A L, Yang K, Zhang X, Haddad G I and Bhattacharya P 1995 *Proc. 7th Int. Conf. on InP and Related Materials* 357–360

[8] Lunardi L M, Chandrasekhar S, Gnauck A H, Burrus C A, Hamm R A, Sulhoff J W and Zyskind J L 1995 *IEEE Photon. Technol. Lett.* 7 182–184

[9] Yang K, Gutierrez-Aitken A L, Zhang X, Haddad G I and Bhattacharya P 1995 *Proc. 7th Int. Conf. on InP and Related Materials* 448–451

High-speed GaAs photodetectors and photodetector arrays for optical processing

Gordon Wood Anderson and Francis J. Kub

Naval Research Laboratory, Washington, DC 20375-5347, USA

Abstract. Optical processors (OPs) can perform parallel signal processing at rates much greater than conventional signal processing and are very promising for signal processing applications. New, high-speed, high-dynamic-range integrated photodetector arrays (PDAs) followed by high-speed microelectronic signal preprocessors are needed with much more demanding parameters than those of conventional modern imaging systems. These include high-speed GaAs and silicon photodetectors and PDAs. GaAs PDAs for spectrum analysis and coherent GaAs photodetectors and PDAs for excision, single-sideband and main-band suppression, and signal synthesis are reviewed and compared with PDAs developed using silicon technology. The PDAs are often the technology pacing components in OP applications.

1. Introduction

Substantially increasing demands are being placed on photodetectors and photodetector arrays (PDAs) for optical processor (OP) applications in areas such as radar and communications. In addition, acousto-optical (AO) signal processing techniques with highly parallel architectures [1] [2] provide significantly greater throughput rates than can be obtained with conventional or anticipated digital signal processing capabilities [3] [4]. Applications of parallel OP techniques include spectrum analysis [2] [3] [4] [5] [6] [7] [8] [9] [10] [11] [12] [13], determination of the direction-of-arrival (DOA) of pulses or precision direction finding (PDF) [3] [4] [8] [9], correlation [2] [7] [8] [9] [14] [15] [16] [17], excision of narrowband signals from wideband backgrounds [3] [4] [8] [9] [12] [15] [16] [17] [18] [19], analog programmable filtering [8] [15] [16] [17] [18] [19], optical signal synthesis [20], and computing [2] [9] [21] [22]. In many applications the performance limits and throughput limits are determined by the PDA. Thus, to obtain the full advantages of OPs, new photodetectors and PDAs are needed with capabilities substantially exceeding those required for modern image sensing applications [3] [4] [10] [12] [23] [24] [25]. The most important performance issues and currently promising solid-state approaches are noted in this paper. More detailed discussions of system signal processing characteristics and requirements [3] [4] [10] [26] [27] [28], photodetector device physics [3] [4] [10] [12] [23] [24] [25], photodetector noise theory [4] [10] [12] [23], photodetector design considerations [4] [10] [12], and imaging arrays [23] [24] [25] are given elsewhere.

A critical OP need arises because of the large number of pixels and high speeds required for some applications. Consequently, the signal reporting rate at the output of the PDAs may be much greater than the processing capabilities of currently available or anticipated digital processors [3] [4] [29]. Thus preprocessing of the pulses from each pixel, preferably on-chip, to optimize the information rate to a host computer is necessary. Preprocessors are used together with PDAs so that all pulses incident at the focal plane are measured but only the desired information is passed onto the host digital processor [3] [4] [29]. Compression is also a critical preprocessing function that occurs because the resolution of available analog-to-digital converters (A/Ds), typically limited to eight to ten bits, is much smaller than the required PDA dynamic range values and because it is difficult to amplify and process voltage signals with > 100-dB ($10^{5.0}$) range [3] [4]. It thus is feasible to integrate the necessary digital processor with the PDA. The digital processor can be operated at possible computation rates because of the signal selection and reduction functions that are performed by the signal preprocessor.

System needs for PDAs include both direct power and heterodyne detectors. Desired PDA parameters for spectrum analysis, depending on the application, include large numbers of detector elements (1 to 12,000 for linear arrays and 128 × 128 to 2,000 × 2,000 for area arrays), high dynamic range (65 dB to 80 dB ($10^{6.5}$ to $10^{8.0}$)), small pixel center-to-center spacing, high sensitivity (noise levels < 100 pW), short-pulse (≤ 100 ns) response capability, high processing rates (> 4 GHz effective sampling rates), short interrogation periods (1 μs to < 100 ns), low crosstalk (−50 dB to −65 dB ($10^{-5.0}$ to $10^{-6.5}$) at the next-nearest-neighbor pixel), precise phase tracking accuracy for intrapulse analysis or PDF, high pixel-to-pixel uniformity (5 %), small bandwidths per pixel or channel (< 25 kHz to ≈ 100 MHz), and low power consumption.

2. Device considerations

Photodetectors are used in AO OPs to convert the optical energy of parts of the laser beam diffracted from the main laser bream in the Bragg cells to electronic energy for further signal processing. This optical-to-electrical transduction occurs at the PDA focal plane [3] [4] [7] [12]. Four types of photodetectors that can be useful for AO OPs are two-dimensional (2-D) vacuum tube sensors, one-element solid-state detectors, and one-dimensional and 2-D PDAs. Solid-state detectors are more useful because of their smaller size, higher reliability, and lower voltage requirements and the possibility of integrating solid-state detectors and some signal preprocessing electronics on the same semiconductor substrate. The detector parameters for OP spectrum analysis applications given above are critical because of the high dynamic range (≈ 80 dB or $10^{8.0}$ in photon flux) and high speed needed. Much of the effort in this area has used silicon technology because of the complexity of the circuits and devices. Nevertheless, because of the direct bandgap and high mobility of GaAs, it has promise, particularly when the circuit requirements are not very complex. The types of detectors that have been developed [3] [4] [10] [12] for OPs include photoconductors [19] [30], avalanche photodiodes [31], charge transfer devices [32] [33], and depletion layer detectors [20] [34] [35] [36] [37] [38] [39] [40] [41].

3. Linear power photodetector arrays

The two functional types of linear PDAs that have been used for real-time spectrum analysis are parallel-addressed (instantaneous) and serial-addressed (integrating-instantaneous). The PDAs were designed for the most part as direct power detectors. Area PDAs for OPs are much more difficult to make because of the very high speeds required along with the other OP requirements. One area PDA development has been initiated for OPs, but these PDAs have not been tested.

An instantaneous PDA is made with parallel video detector-amplifier channels, and the detectors are not reset during the detector-sensing operation. The time-of-arrival (TOA) and frequency of intercepted pulses are given simultaneously. Frequency resolution is limited by the number of package pins and by power dissipation. Silicon 27-element detector-amplifier arrays have been developed [36] [37] [38], where the most successful has a six-stage successive-detection logarithmic compressing amplifier [37]. A 10-element GaAs detector-amplifier that has photoconductors as the light sensing elements has also been developed [30]. The GaAs PDA utilizes a molecular beam epitaxy materials system where the n⁻ active photoactive layer is grown over an undoped buffer layer. An undoped AlGaAs layer is grown over the photoactive layer followed by an n-type electronically active layer in which the metal-semiconductor field effect transistors (MESFETs) and resistors are fabricated followed by an n$^+$ ohmic contact layer to the MESFETs and resistors. The required bandwidth for the GaAs PDA is 100 MHz, and a 350-MHz bandwidth was achieved. Discrete GaAs photoconductors fabricated in the same fashion exhibit bandwidths of $\approx$ 8.8 GHz [42]. In addition, discrete GaAs heterojunction phototransistors have been fabricated for this requirement that also both achieve gain at the focal plane and meet the system bandwidth need. The dynamic range of single channels identical to channels of the GaAs parallel-addressed PDAs is limited to $\approx$ 35 dB, however. Thus for the 100-MHz bandwidth requirement, the GaAs detector-amplifier PDAs with gain at the focal plane have promise provided that increased dynamic range can be achieved. The difficulty is in achieving the high dynamic range and very high uniformity (5 %) required at quite low light levels ($\approx$ 10 pW) for short pulses ($<$ 100 ns) with PDAs of 100 channels or more [3] [4] [10] [12]. For requirements that include 25 or more channels and also only require bandwidths of $\approx$ 15 MHz, the silicon PDAs [37] are preferable.

Serial-addressed (integrating/instantaneous) PDAs, which multiplex the detector signals to one or more serial output ports, are required in OP systems with a large number of channels and in new, wider bandwidth AO systems being developed because of improvements in the bandwidths of Bragg cells to $\approx$ 2 GHz [3] [4]. In these cases optical signal power at the focal plane is integrated for an interrogation period (or frame period). The detector output signals are then transferred to one or more serial output multiplexers just before the detector reset operation. Then the PDA is multiplexed to one or more outputs. The pin-count requirement of serial-addressed PDAs may be substantially lower than that of parallel-addressed PDAs. Thus the total output buffer power dissipation of serial-addressed PDAs may be less than that of parallel-addressed PDAs. Improved frequency resolution can be obtained because of the larger number of pixels. The TOA resolution is limited, however, by the time required for multiplexing. Because of the integrating detector operation, improved detectivity compared to that of instantaneous PDAs can be obtained for long pulses. The four general functions required of serial-addressed PDAs are sensing and integrating the photosignal charge, multiplexing the signal, and resetting the detector. Considerations for achieving these characteristics are described elsewhere [3] [4] [10] [12].

One-hundred-eight-element to 1024-element silicon serial-addressed PDAs have been developed for OP spectrum analysis [32] [33] [39] [40] [41]. The dynamic range values obtained were $\approx$ 60 dB or substantially more for long pulses ($\geq$ 6 μs) [32] [33] [41] and $\approx$ 55 dB for short pulses [40]. Problems in obtaining high uniformity at low light levels as noted above occur for these PDAs. Thirty-two-element and 64-element GaAs serial-addressed PDAs also have been developed for OP spectrum analysis [34] [35]. The sensing elements are Schottky diodes having interdigitated fingers with one set of fingers being Schottky contacts and the other being ohmic contacts. The multiplexers are four phase charge coupled devices (CCDs). Electrical pulses were injected into the CCDs through the parallel inputs at 1-GHz clock rates in this work. The dynamic range of these GaAs PDAs also was limited to $\approx$ 35 dB, however. Thus these serial-addressed GaAs PDAs have potential in that they may permit very short interrogation periods and fast multiplexing and thus make PDAs with very large numbers of channels conceivable. Problems associated with low dynamic range, redevelopment of CCD fabrication processes, and cost will have to be addressed, however.

4. Heterodyne photodetectors and photodetector arrays

Improved dynamic range and sensitivity can be obtained by the use of heterodyne AO spectrum analyzers in comparison to the values obtainable with direct detection (power) AO spectrum analyzers [3] [4] [10]. This is because the induced photocurrent, i_s, for heterodyne detection is approximately proportional to the square root of the product of the optical powers, P_s and P_r, of the Bragg-diffracted signal beam and the reference beam, respectively (Eq. (1)).

$$i_s = S_{pd} \left\{ P_s + P_r + 2\sqrt{P_s P_r} \cos[(\omega_s - \omega_r)t - \theta] \right\}. \tag{1}$$

Here, S_{pd} is the photodetector responsivity; ω_s and ω_r are the angular frequencies of the light in the signal and reference beams, respectively; and θ is the phase difference between the light in the two beams. For the direct-detection approach the induced photocurrent is directly proportional to the optical power of the Bragg-diffracted signal beam. Thus the dynamic range of a heterodyne OP potentially can be the square of that of a direct detection OP. In addition, phase coherence may be maintained for applications such as excision and standard spectrum analysis (Eq. (1)). Thus 2-D OPs are possible, and the DOA of intercepted pulses may be obtained as well as the frequency, amplitude, and TOA. Difficult aspects of the PDA design for heterodyne detection are noted elsewhere [19]. Thus heterodyne photodetectors and PDAs are required. Uses for heterodyne PDAs or discrete photodetectors include spectrum analysis when both coarse and fine frequency information is needed, PDF, correlation, excision, analog programmable filtering, optical signal synthesis, and computing.

Analog programmable OP filtering (excision) has been achieved by using both silicon and GaAs heterodyne PDAs in coherent spectrum analyzers [4] [12] [15] [18] [19]. The PDA was illuminated by both the reference and Bragg-diffracted signal beams, which were mixed at the focal plane. Narrowband frequency components were excised from a wideband RF spectrum by programmably setting the individual photodetector circuits so that they either were or were not responsive to light. A 35-dB excision isolation with a 100-MHz bandwidth was achieved with silicon PDAs [4] [12] [18]. A 40-dB excision isolation over a 6-MHz bandwidth centered at 45 MHz was achieved with a 12-element double-heterojunction GaAs/AlGaAs/GaAs interdigitated-finger metal-semiconductor-metal (MSM) PDA [2] [4] [19]. Six of the detectors were heterojunction photoconductors, and six were heterojunction Schottky diodes. The metals were deposited on an n^+ GaAs layer over an n GaAs layer. The n^+ and n GaAs between the fingers was removed down to a 100-nm AlGaAs layer grown over the n^- GaAs photoactive layer. The Schottky diodes were biased at 0 V during the experiments and therefor were not sources of photocurrent. The excision result thus was obtained with the 6-element photoconductor sub-array. This is the highest excision isolation achieved with PDAs. The response of these GaAs detectors was linear with optical power indicating that these PDAs may be useful as multipliers in optoelectronic circuits such as optical adaptive filter circuits [2] [19].

High bandwidth (> 20 GHz) MSM GaAs Schottky photodiodes also have been used as programmable weights in a 2-tap transversal filter for optical signal synthesis [43]. The bias values on the MSMs were in quadrature, and the optical inputs to the MSMs were also in quadrature. Because the detector response is linearly related to the incident optical power in this case as well, a phase reversal of the output of the detectors occurs when the bias voltage goes from negative to positive. In the case of the 2-tap transversal filter, LO suppression of 40 dB and undesired sideband suppression of 27 dB were obtained when the optical carrier was modulated at 50 MHz. Similar results were obtained when the optical carrier was modulated at 250 MHz, and this result should be obtainable for optical signal synthesis when the optical carrier is modulated up to the bandwidth of the photodetectors, which is > 20 GHz. Silicon technology may not be competitive in this application.

In some coherent AO approaches to excision and signal synthesis, it is desirable to use one large photodetector rather than many small ones to simplify the output circuitry [20]. In this case the detector is required to have a length of up to 6 mm as well as a bandwidth ≥ 5 GHz. The detector thus must have both a very large area and small capacitance. Large-length silicon detectors have been developed for these requirements. They exhibit a bandwidth ≥ 1.3 GHz, and their response is flat over the 0.75 GHz to 1.25 GHz acoustic frequency range of the Bragg cell used [20]. Large-length GaAs detectors are being developed for these applications as well and are expected to perform more satisfactorily than the silicon detectors due to the higher mobility and the direct electronic bandgap of GaAs.

Heterodyne PDAs that sample data in quadrature are also needed for I and Q sampling in spectrum analysis applications. A 32-channel silicon heterodyne detector/bandpass amplifier array has been developed for these requirements. It has a 100-MHz intermediate center frequency, has a 50-ns minimum pulsewidth detection capability, and has achieved a dynamic range of 70 dB ($10^{7.0}$) [44] [45]. A GaAs PDA also has been fabricated for this requirement. The GaAs PDA has not been evaluated, so a comparison between the two approaches cannot be made here.

5. Conclusions

High performance photodetectors and PDAs are critical components for OP applications, which are very promising because they can perform parallel signal processing at rates much greater (≥ 100 times faster) than conventional signal processing. Moreover, the demands on the photodetectors and PDAs for OP applications are substantially greater than those for image sensing applications. In OP applications, such as one-dimensional spectrum analysis, for which the required bandwidth is not extraordinarily high but for which the needed number of channels or pixels is high (≥ 25), silicon technology is and most likely will remain the technology of choice. Increased emphasis is being placed on coherent OP applications requiring bandwidths > 1 GHz, however, and extending up to 40 GHz or more for long term needs. For these cases GaAs technology may be the technology of choice due to the high electronic mobility and direct electronic bandgap of the material.

There are three critical needs for the GaAs detectors and PDAs. The first is greater dynamic range. There are system needs for dynamic range values > 80 dB ($10^{8.0}$) in optical power. To achieve this improvement, more than evolutionary development in existing GaAs detectors and PDAs is required. The second and related need is very uniform detector-to-detector response in PDAs at low light levels (≤ 100 pW). The third need is microelectronic circuits that perform at least a minimum amount of preprocessing functions, such as compression of the output response of the detectors, so the devices may be compatible with subsequent digital signal processing functions. Several promising new areas of solid-state device and solid-state circuit research that may address these needs have been described in this paper.

References

[1] Lee J N 1987 in *Optical Signal Processing*, Horner J L ed. (New York: Academic) pp 165-190
[2] VanderLugt A 1992 *Optical Signal Processing* (New York: Wiley)
[3] Anderson G W *et al* 1991 *Proc. IEEE* **79** 355-388
[4] Anderson G W *et al* 1990 *Opt. Eng.* **29** 58-67
[5] Korpel A 1988 *Acousto-Optics* (New York: Marcel Dekker)
[6] Xu J and Stroud R 1992 *Acousto-Optic Devices: Principles, Design, and Applications* (New York: Wiley)
[7] Hamilton M C and Spezio A E in 1990 *Guided-Wave Acoustooptics* Tsai C T ed Vol 23 Springer Series in Electronics and Photonics (Berlin: Springer-Verlag) pp 235-271
[8] Spezio A E *et al* Feb 1985 *Microwave J.* **28**(2) 155-163
[9] Lee J N and VanderLugt A 1989 *Proc. IEEE* **77** 1528-1588
[10] Anderson G W *et al* 1988 *Appl. Opt.* **27** 2871-2886
[11] Casasent D in 1983 *Acousto-Optic Signal Processing: Theory and Implementation* Berg N J and Lee J N eds (New York: Marcel Dekker) pp 325-367

1108

[12] Borsuk G M 1981 *Proc. IEEE* **69** 100-118
[13] Turpin T M 1981 *Proc. IEEE* **69** 79-92
[14] Sprague 1977 *Opt. Eng.* **16** 467-474
[15] Lafuse J L in 1983 *Bragg Signal Processing and Output Devices*, Markevitch B V and Kooij T eds *Proc. SPIE* **352** 24
[16] Erickson J L 1981 Ph.D. thesis Stanford Univ. Stanford CA USA (Ann Arbor MI USA University Microfilms)
[17] Jackson D W and Erickson J L in 1983 *Acousto-Optic Signal Processing: Theory and Implementation* Berg N J and Lee J N eds (New York: Marcel Dekker) pp 107-137
[18] Roth P J in 1983 *Bragg Signal Processing and Output Devices*, Markevitch B V and Kooij T eds *Proc. SPIE* **352** 17-23
[19] Anderson G W *et al* 1990 *Opt. Eng.* **29** 1243-1248
[20] Alexander E M *et al* 1995 (in press) *NRL Report* (Washington DC: U S Naval Research Laboratory)
[21] Athale R A and Lee J N 1984 *Proc. IEEE* **72** 931-941
[22] Casasent D 1984 *Proc. IEEE* **72** 831-849
[23] Barbe D F 1975 *Proc. IEEE* **63** 38-67
[24] Scribner D A et al 1991 *Proc. IEEE* **79** 66-85
[25] 1979 *Charge Coupled Devices and Systems* Howes M J and Morgan D V eds (New York: Wiley)
[26] Schleher D C 1986 *Introduction to Electronic Warfare* (Dedham MA USA: Artech House)
[27] Tsui J B-Y 1986 *Microwave Receivers with Electronic Warfare Applications* (New York: Wiley)
[28] Lemley L W and Gleason R F 1985 (revised 1986) *NRL Report 8737* (Washington DC: U S Naval Research Laboratory) Defense Technical Information Center AD No. B105737L
[29] Ugarte C A *et al* in 1990 *Optical and Digital GaAs Technologies for Signal Processing Applications* Bendett M P Butler, Jr, D H Prabhakar A and Yang A eds *Proc SPIE* **1291** 11-22
[30] Anderson G W *et al* 1988 *IEEE Electron Device Lett.* **EDL-9** 550-552
[31] Webb P P and McIntyre R J 1984 *IEEE Trans. Electron Devices* **ED-31** 1206-1212
[32] Smith D J *et al* in 1991 *Devices Opt. Processing* Gookin D M ed *Proc SPIE* **1562** 242-250
[33] Satorius D A in 1992 *Opt. Technol. Microwave Appl. VI and Optoelectronic Signal Processing Phased-Array Antennas III* Yao S-K ed *Proc. SPIE* **1703** 43-46
[34] Sahai *et al* in 1984 *Opt. Technol. Microwave Appl.* Yao S-K ed *Proc SPIE* **477** 165-173
[35] Sahai *et al* in 1986 *Proc. IEEE Custom Integrated Circuits Conference* (New York: Institute of Electrical and Electronics Engineers) IEEE No 86CH2258-2 p 521
[36] Van Vonno N W *et al* in 1984 *Proceedings of the 1984 Custom Integrated Circuits Conference* (New York: Institute of Electrical and Electrical Engineers) IEEE No 84CH1987-7 pp 76-80
[37] Boling E J and Dzimianski J W in 1984 *Opt. Technol. Microwave Appl.* Yao S-K ed *Proc. SPIE* **477** 174-177
[38] Allstot D J *et al* in 1985 *IEEE International Solid-State Circuits Conference* Winner L ed (Coral Gables FL USA: Lewis Winner) IEEE No 85CH2122-0 pp 104-105 and 320
[39] Chamberlain S G and Lee J P Y 1984 *IEEE Trans. Electron Devices* **ED-31** 175-182
[40] Hudson L R *et al* in 1988 *Infrared Technology IV* Spiro I J ed *Proc. SPIE* **972** 33-38
[41] Kub F J and Anderson G W 1993 *IEEE Trans. Electron Devices* **ED-40** 1740-1744
[42] Anderson G W *et al* 1988 *Appl. Phys. Lett.* **53** 313-315
[43] Anderson G W *et al* to be published
[44] Koontz M D in 1995 *Transition of Optical Processors into Systems 1995* Casasent D ed *Proc SPIE* **2429** in press
[45] Anderson G W in 1995 *Proceedings of the Wideband RF Science and Technology Workshop* D. Lake ed (Arlington VA USA: Office of Naval Research)

Inst. Phys. Conf. Ser. No 145: Chapter 8
Paper presented at 22nd Int. Symp. Compound Semiconductors, Cheju Island, Korea, 28 August–2 September 1995
© *1996 IOP Publishing Ltd*

Background limited performance in aluminum-free p-doped quantum well intersubband photodetectors

J. Hoff, J. Piotrowski, E. Bigan, and M. Razeghi

Center for Quantum Devices, Department of Electrical Engineering and Computer Science, Northwestern University, Evanston, Illinois 60208

G. J. Brown

Wright Laboratory, Materials Directorate, WL/MLPO, Wright Patterson AFB, Ohio 45433-7707

Abstract. Background limited infrared photodetection has been achieved up to 100 K at normal incidence with two different Aluminum-free p-type quantum well intersubband photodetectors: one GaAs / $Ga_{0.71}In_{0.29}As_{0.39}P_{0.61}$ and the other $Ga_{0.25}In_{0.75}As_{0.13}P_{0.87}/Ga_{0.49}In_{0.51}P$. The samples were grown by low-pressure metalorganic chemical vapor deposition. Both detectors showed extended photoresponse cutoff wavelength over that of similar p-type GaAs/ $Ga_{0.49}In_{0.51}P$ quantum well intersubband photodetectors by virtue of reduced valence band barrier heights. In this paper, the relative merits of both approaches are investigated in order to help determine the most appropriate technique for extending the range of photodetection to 8-12 μm.

1. Introduction

There has been considerable effort directed towards the development of Long Wavelength Infrared (LWIR) photodetectors from multi-quantum well epitaxial structures [1]. These quantum well intersubband photodetectors (QWIPs) offer numerous potential benefits since their photoresponse can be engineered to a particular wavelength range by defining the well width and barrier height through choice of materials and growth parameters. In particular, infrared photodetection, traditionally visible only to narrow bandgap II-VI materials, can be achieved through QWIPs made from robust III-V semiconductor materials.

Most of the QWIP effort thus far has been directed towards n-type GaAs/$Al_xGa_{1-x}As$ heterostructures. However, there are two potential limitations to this approach. First, polarization dependent quantum mechanical selection rules make normal incident detection impossible in n-type QWIPs without external coupling systems such as diffraction gratings or polished facets. Second, $Al_xGa_{1-x}As$ has material problems of its own such as strong Aluminum oxidation. While successful QWIPs have been demonstrated using this approach, performance improvements may be possible through changes in doping type or material system.

Table 1: Well and Barrier Materials in this study.

Sample	Well Material	Well Eg	Barrier Material	Barrier Eg
A	GaAs	1.42 eV	$Ga_{0.87}In_{0.13}As_{0.75}P_{0.25}$	1.53 eV
B	$Ga_{0.71}In_{0.29}As_{0.39}P_{0.61}$	1.75 eV	$Ga_{0.49}In_{0.51}P$	1.87 eV

To research QWIP alternatives, our efforts have been directed towards p-type $In_xGa_{1-x}As_yP_{1-y}$ compounds. p-type QWIPs inherently photorespond to normally incident light due to valence band anisotropy and the band mixing of the conduction band into the valence band quantum well states away from zone center [2]. The Aluminum-free compounds should provide improvements in electrical properties as well as ease of fabrication. In particular, the difference in atomic size for Indium and Gallium impedes the motion of dislocation. This results in improved performance and reliability in the case of laser diodes [3], and should also favor the growth of high quality heterostructures on silicon substrates, which is of major importance for future Focal Plane Arrays monolithically integrated with readout electronics.

In our previous work [4], we have demonstrated normal incidence detection in p-type GaAs/$Ga_{0.49}In_{0.51}P$ QWIPs. However, cutoff wavelengths were short due to the large valence band discontinuity. It was shown that 8-12 μm detection could be achieved only with very narrow quantum wells. In order to widen the wells and improve manufacturability, the barrier height must be lowered. This can be achieved in one of two ways. First, the bandgap of the barrier layer can be reduced by replacing ternary $Ga_{0.49}In_{0.51}P$ with quaternary $Ga_{1-x}In_xAs_yP_{1-y}$ material. Second, the bandgap of the well layer can be increased by replacing the binary GaAs with quaternary $Ga_{1-x}In_xAs_yP_{1-y}$ material. In this paper, we report the realization of both approaches toward longer cutoff wavelength. Each device is capable of Background-Limited Infrared Photodetection (BLIP) up to temperatures of 100K.

2. Growth

The samples were grown by low-pressure metalorganic chemical vapor deposition (MOCVD), and growth conditions have been detailed elsewhere [5]. Both superlattices consist of 50 periods of 30 Å wide quantum wells separated by 280 Å wide barriers. Table 1 shows the well and barrier materials used and quotes each room temperature bandgap. The superlattices were sandwiched between thick GaAs layers for top (0.5 μm) and bottom (1.0 μm) contacts. X-ray diffraction measurements confirmed the nominal thicknesses. All well and contact layers are doped with Zinc to a net acceptor concentration of 3×10^{18} cm^{-3}. The barriers are left undoped. Each sample was then processed into 400 μm x 400 μm mesa structures using standard photolithographic techniques and selective wet chemical etching. For this purpose, we used 1 H_3PO_4 : 1 H_2O_2 : 10 H_2O as the well etchant, and HCl as the barrier etchant. Finally, 100 μm x 100 μm Ti/Pt/Au contacts were formed by electron beam evaporation and lift-off.

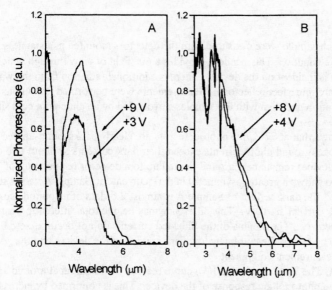

Figure 1: Normalized 77K Photoresponse spectra for A) Sample A and B) Sample B

3. Characterization

3.1 Photoresponse

Photoresponse spectra were measured for both samples at 77K for various biases using a Fourier Transform Infrared spectrometer. Measurements were taken across a load resistor chosen to be 40kΩ making it negligible when compared to both the QWIP differential impedance and the capacitive impedances of cables. Results are shown in Figure 1. Photovoltaic effects were observed in both samples. Dark current measurements are also nonlinear and assymmetric with respect to zero bias (see Figure 2 A and 2 B) indicating asymmetric quantum well potential profiles. Such asymmetry can be linked to either a structural difference in the two quantum well interfaces [6][7] or to dopant migration during the growth [6].

The photoresponse magnitude increases with bias, and in both cases, the photoresponse spectra peak around 4 μm with broad maxima. The shape of the spectral response is determined by both the intersubband absorption spectrum and the dependence of the photo-excited carrier escape probability on energy. Careful examination of the spectra reveal that the cut-off wavelength increases slightly with bias. We believe this cutoff wavelength increase originates in part from the reduction of the effective barrier height (by one-half of the potential drop per well), in part from image force barrier height lowering, and in part from greater tunneling out of the well through the triangular tip of the well made narrower by increased bias.

3.2 Dark Current

Dark currents of each sample were measured with the detectors mounted in a stainless steel cryostat with a ZnSe window. This window provides a 60° field of view through which blackbody radiation is incident on the device. There is additional radiation from the walls of the cryostat which are imperfect reflectors. Measurements were performed first in this unshielded configuration and then with the samples surrounded by an aluminum cold shield at 77 K.

Figure 2 shows that at 77 K, the photodetectors are clearly background-limited as the unshielded, 300 K background photocurrents exceed their dark currents by about two orders of magnitude. The devices remain background-limited up to a detector temperature of 100 K.

Figure 2 also shows a greater asymmetry in dark current for Sample A versus that of Sample B. At +10V, the dark current of Sample A is almost 2 orders of magnitude larger than that of the dark current at -10 V. This difference may be due to a different potential profile for both structures. The minima of the Shielded currents do not occur at zero bias which is believed to be due to the extremely low dark current of the devices and the limitations in our measurement equipment.

In Sample A, The 77 K unshielded I(V) curve reaches its minimum at around +0.2 V. This is a result of the photovoltaic response of the device. This is confirmed by increasing the level of infrared illumination using a blackbody source. At higher illumination, the I(V) curve moves up and reaches a minimum at a voltage that approaches +1.5 V asymptotically.

Shielded dark currents were measured as function of bias for temperatures between 77 K and 250 K for both samples. Figure 2 shows Arrhenius plots of the logarithm of the dark current versus inverse temperature for Samples A and B. Activation energies extracted from these Arrhenius plots closely agree with measured photoresponse cut-off wavelengths. In particular, the cut-off wavelength increases slightly, and the activation energy decreases slightly with increasing bias.

3.3 Blackbody Repsonsivity

Absolute responsivity was measured using a chopped 800 K calibrated blackbody and conventional lock-in detection techniques. Measured blackbody responsivities and photoresponse spectra were combined with Planck's law for spectral exitance, in order to extract the peak responsivities. The corresponding peak responsivities are in the range of a few mA/W for both samples.

4. Conclusion

We have demonstrated normal incidence background limited infrared photodetection up to 100 K for two different aluminum free quantum well intersubband photodetectors. To the best of our knowledge such a high BLIP temperature has only been achieved by one other research group [9]. A cutoff wavelength up to 6. μm has been achieved by reducing the valence band discontinuity two different w5 ays. Both structures yield similar performance. To further extend the cutoff wavelength into the 8-12 μm wavelength range, we propose to use quaternary alloys in both the well and the barrier.

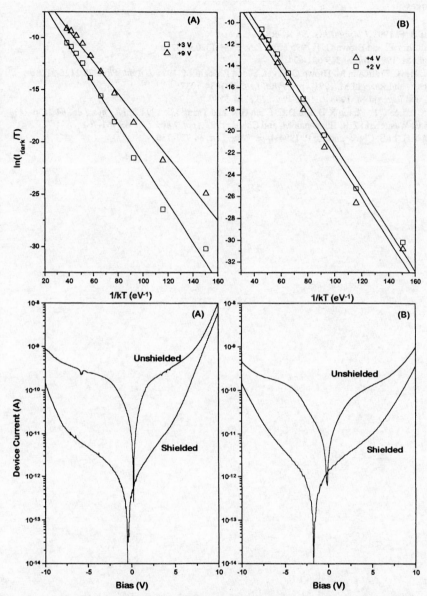

Figure 2: Arrhenius plots and Shielded and Unshielded current plots for Sample A (left) and Sample B (right)

Acknowledgements

The work at Northwestern University was supported by Air Force Contract No. F33615-93-C5382 through Kopin Corporation. The authors would like to acknowledge Dr. Gerald L. Witt from the US Air Force Office of Scientific Research (AFOSR) for his encouragement.

1114

References

[1] Levine B. F. 1993 *J. Appl. Phys.* **50**, R1-R81
[2] Szmulowicz F. and Brown G. J. 1995 *Phys. Rev. B.* **51** 13203-13220
[3] Razeghi M. 1994 *Nature* **369** 631-634
[4] Hoff J. He X. Erdtmann M. Brown G. Bigan E. and Razeghi M. 1995 *J. Appl. Phys.* **78** 2126-2128
[5] Omnes F. and Razeghi M. 1991 *Appl. Phys. Lett.* **59** 1034-1037
[6] He X. and Razeghi M. 1993 *J. Appl. Phys.* **73** 3284-3287
[7] Tsai K.L. Lee C.P. Chang K.H. Liu D.C. Chen H.R. and Tsang J.S. 1994 *Appl. Phys. Lett.* **64** 2436-2439
[8] Liu H.C. Wasilewski Z.R. Buchanan M. and Chu H. 1993 *Appl. Phys. Lett.* **63** 761-764
[8] Wang Y.H. Li S. Chu J. and Ho P. 1994 *Appl. Phys. Lett.* **64** 727-730

Inst. Phys. Conf. Ser. No 145: Chapter 8
Paper presented at 22nd Int. Symp. Compound Semiconductors, Cheju Island, Korea, 28 August–2 September 1995
© *1996 IOP Publishing Ltd*

PIN-like Si(p)/ZnSe(n⁻)/ZnSe(n⁺) visible photodiode

Chung-Cheng Chang, Wen-Shiung Lour and Min-Hsiu Chen

Department of Electrical Engineering , National Ocean University

2 Peining Rd., Keelung , TAIWAN, Republic of China.

Abstract. A PIN-like Si(p)/ZnSe(n⁻)/ZnSe(n⁺) visible photodiode was successfully fabricated by CVD technique. n⁺- ZnSe layer was implemented by driving in the evaporated In metal. The unintentionally n⁻ -ZnSe was employed as an absorption layer which is sensitive on visible light range from red to violet . Under reversed bias condition, the studied device exhibits a breakdown voltage as large as 25V and a dark current of $3.5\,\mu$A/mm. In response to visible light, it is found that the device exhibit strong sensitivity in blue light. We obtained a responsivity of $0.125\,\mu$A/lx for our unoptimized devices. Together with VLSI technology, the development of high-performance short wave-length detector on Si-substrate demonstrate high potential application to photoreceiver.

1. Introduction

II-VI compound semiconductor are some of the many promising materials for the possible fabrication of devices with efficient injection luminesense in the green to blue. ZnSe in particular has an energy gap of 2.67eV at room temperature, and blue emission is expected from band to band transitions. Much effor has been devoted in recent years to study wide -gap II-VI semiconductor heterostructure to realize visible blue and blue-green laser diodes[1]. In addition, a high sensitivity solid-state image sensor using a thin-film $ZnSe - Zn_{1-x}Cd_xTe$ heterojunction photosensor has also been reported[2]. This paper we have successfully fabricated a PIN-like Si(p)/ZnSe(n⁻)/ZnSe(n⁺) visible photodiode. A two-step method was used to prepare the heteroepitaxial ZnSe thin film on Si substrate by vapor phase epitaxy for reducing the lattice mismatch. The achieved device shows strong sensitivity on visible light. High sensitivity of 0.125 μA/lx has been obtained by the structure. Together with Si-based VLSI technology, the development of short wave-length detector provide high potential application to photoreceiver.

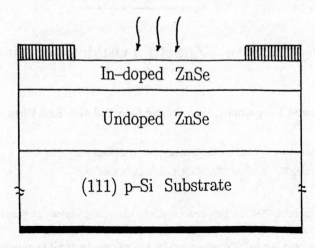

Fig.1 Schematicc cross section of studied device

2. Device fabrication

The PIN-like visible photodiode structure shown in Fig.1 was grown on (111) p-Si substrate using vapor phase epitaxy technique. In this study, two step growth method was employed to overcome the problem of lattice mismatch existing between Si and ZnSe. An epitaxial unintentionally doped ZnSe films is of approx. 2.5 μm. After finishing epitaxial growth, the n^+-ZnSe was achieved by driving-in the evaporated In metal. The measured concentrations are 5×10^{17} and 2×10^{14} cm^{-3} for an In-doped and an undoped ZnSe layers, respectively. In and Al were employed as n and p type ohmic contacts, respectively. Standard photolithography and wet etching techniques were used to implement the device.

3. Results and discussion

3.1. Theoretical consideration

Figure 2 illustrates the corresponding energy band diagram of PIN-like visible photodiode under light illumination. At equilibrium, there is no current flowing in the device, becauce the electron-hole pairs generated by input light are either accumulated in minimum valence band or reflected by potential spike as seen in Fig.2(a).When a negative reverse bias voltage is applied to the PIN diode, holes generated by input light obtain sufficient energy to emit over the heterojunction. As shown in Fig.2(b), this behavior results in conducting photocurrent. Thus the conducting photocurrent is increased with increasing reversed bias and then saturated at some voltage.

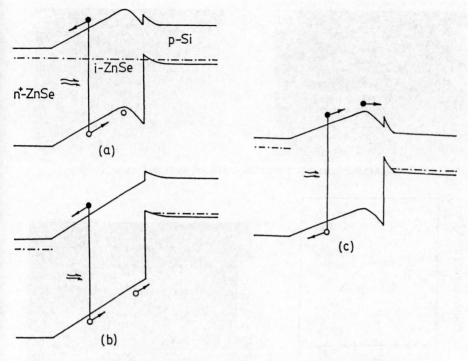

Fig.2 The corresponding energy band diagram of studied photodiode (a) at
equilibrium (b) under reverse bias and (c) under forward bias.

Figure 2 (c) shows the band diagram of the studied structure under forward
bias and illumination. In contrast with reverse bias, electrons instead of holes
contribute to photocurrent. Due to the larger diffusion length for electron,
higher conducting current is expected when   forward bias is applied.

3.2. Experimental results

Figures 3(a) and (b) show the output photocurrent under reverse bias condition
and various light sources, i.e., red and blue. The supplying photo-power is 400
lx per step. The measured breakdown voltage is lager than 25V and the dark
current is of 3.5 uA/mm. In addition, the device exhibits strong sensitivity on
blue but much less on red light. We   obtained a responsivity of 125 μA per
kilo lux when the blue light ia applied. Fig.4 shows the sensitivity as a finction
of light wavelength. Obviously, the studied photodiode exhibits response on
input source ranged in visible region. Figure 5 shows the output photocurrent
measured on ac mode under blue light illumination. The forward photocurrent
of 180 μA was observed and much larger than that obtained in reverse
condition at the same bias voltage. As explained above, the results is due to
the larger diffuion length for electrons.

1118

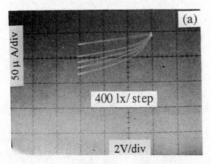
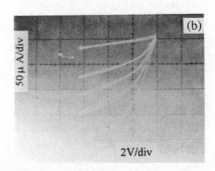

Fig.3 Photocurrent under reverse bias and (a) red, blue illumination.

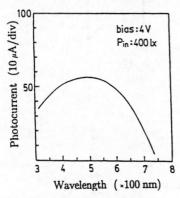

Fig.4 Sensitivity as a function
of light wavelength.

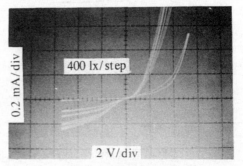

Fig.5 Output photocurrent measured on
ac mode under blue illumination.

4. Conclusions

A new PIN-like Si(p)/ZnSe(n⁻)/ZnSe(n⁺) visible photodiode was proposed and fabricated successfully. Both qualitative analysis and experimental results are described in this study. Experimental results reveal that the studied device shows strong sensitivity on visible light region. A responsivity of 0.125 $\mu A/lx$ was obtained when blue light is applied.

Acknowledgement
Part of this work was supported by the National Science Council of the Republic of China under contract of No. NSC 84-2215-E-019-002.

References

[1] Nishizawa J I 1985 J. Appl. Phys. 57 2210-2216.

Onomura M 1993 Electron. Lett. 29 2114-2115.

[2] Chikamura T 1982 IEEE Trens. Electeron Devices 29 999-1004.

Inst. Phys. Conf. Ser. No 145: Chapter 8
Paper presented at 22nd Int. Symp. Compound Semiconductors, Cheju Island, Korea, 28 August–2 September 1995
© *1996 IOP Publishing Ltd*

High-Responsivity InGaAs Metal-Semiconductor-Metal Photodetectors with Semi-Transparent Schottky Contacts

Rong-Heng Yuang, Jen-Inn Chyi, Yi-Jen Chan,
Wei Lin*, and Yuan-Kuang Tu*

Department of Electrical Engineering, National Central University,
Chung-Li, Taiwan, R. O. C.

*Telecommunication Laboratories, Ministry of Transportation and Communications,
Chung-Li, Taiwan, R. O. C.

Abstract. Novel metal-semiconductor-metal photodetectors (MSM-PDs) with interdigitated semi-transparent Au Schottky contacts have been fabricated on pseudomorphic $In_{0.9}Ga_{0.1}P/InP/InGaAs$ heterostructure grown by low-pressure MOCVD. The responsivity measured at 1.55 µm wavelength is greatly enhanced from 0.4 A/W to 0.7 A/W as the thickness of the Au electrodes are decreased to 10 nm. These devices have an extremely linear photoresponse and show no internal gain. The devices with an active region of 100 µm × 100 µm, a finger width of 3 µm, and a finger spacing of 3 µm exhibit dark currents as low as 16 nA at 10 V. The full-width at half maximum of the temporal response for the device with semi-transparent electrodes is about 85 ps compared to 80 ps for the conventional device with opaque electrodes.

1. Introduction

Planar metal-semiconductor-metal photodetectors (MSM-PDs) have recently shown great promise for monolithic optoelectronic integrated circuit (OEIC) receivers used in lightwave communication systems because of their ease of fabrication and compatibility with field-effect transistor (FET) process technology. On the basis of the lateral interdigitated back-to-back Schottky contact structure, MSM-PDs have inherently very low capacitance, which is advantageous for high-speed operation [1], [2]. However, the main drawback of MSM-PDs with opaque metal electrodes is their low responsivity due to the metal shadowing effect. Several methods have been previously demonstrated to enhance the responsivity of a long-wavelength MSM-PD, e.g., backside illumination and transparent electrodes made of indium tin oxide (ITO) or cadmium tin oxide (CTO). Although an MSM-PD under backside illumination exhibits favorable DC and AC performance [3], it has limited success in practical applications due to its complicated and critical processes on both sides of the wafer. Another straightforward approach is the use of a transparent material for metal electrodes. ITO has been used often in Si solar cells and sometimes in GaAs photodiodes. However, ITO is not appropriate for a long-wavelength photodetector due to its large absorption within this region. Seo *et al.* have reported an improvement in the transmission of ITO at a wavelength of 1.3 µm by adding a forming gas (H_2/N_2) mixture to the Ar sputtering gas during the deposition [4]. Using the modified ITO electrodes, they are able to double the responsivity as compared to conventional devices. However, the resistivity of the modified ITO (1.54×10^{-2} Ω·cm) is one order of magnitude larger than that of the usual one ($1.46 \times$

10^{-3} Ω·cm) at room temperature, which will result in a large RC time constant and severely limit the bandwidth of the detector. Compared with usual ITO, CTO has smaller absorption coefficient in the region of interest for a long-wavelength photodetector and lower resistivity of 4×10^{-4} Ω·cm at room temperature [5]. Recently, Gao *et al.* have successfully fabricated InGaAs MSM-PDs with CTO electrodes [6]. As a whole, the resistivity and transmittance of ITO and CTO strongly depend on deposition conditions. Additionally, if a sputter is used, it usually produces a considerable amount of defects on the semiconductor surface, causing undesirable Schottky barrier lowering and internal gain [6], [7].

Here we present the DC and AC performance of the InGaAs MSM-PDs with semi-transparent Au Schottky contacts for the first time. The comparison between these MSM-PDs and the conventional devices with opaque electrodes is made. The thin Au about 10 nm thick which has been used to fabricated Si photodiodes has shown high optical transmittance over a wide spectral range [8]. The MSM-PDs with the thin Au electrodes exhibit high responsivity and, at the same time, preserve the desirable characteristics of the conventional devices.

2. Device fabrication and measurements

The layer structure consists of a 0.5 μm-thick InP buffer, a 1 μm-thick InGaAs absorption layer, and a 50 nm-thick barrier-enhancement layer which is composed of 40 nm-thick InP and 10 nm-thick $In_{0.9}Ga_{0.1}P$. The epitaxial layers were grown by low-pressure metal-organic chemical vapor deposition (LP-MOCVD) on Fe-doped (100) InP substrate. Detailed growth conditions were described elsewhere [9].

The device structure is schematically shown in Fig. 1. The active mesa regions were defined by using H_3PO_4 : HCL and H_3PO_4 : H_2O_2:H_2O chemical solutions for etching the barrier enhancement and absorption layer, respectively. Without any surface treatment, interdigitated Schottky contacts were formed by thermal evaporation of a 10 nm-thick Au film and a lift-off process. Then a 220 nm-thick silicon nitride layer was deposited by an RF

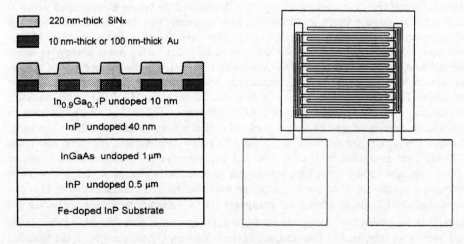

Fig. 1 Schematic device structure of the MSM-PD.

Fig. 2 Device layout of the MSM-PD.

sputter for the purpose of antireflection coating and passivation. Contact holes of 2 μm × 90 μm rectangular size were opened by reactive ion etching the silicon nitride in a $CF_4 : O_2$ plasma. Finally, bonding pads were deposited by thermal evaporation of a 300 nm-thick Ti/Au film. The fabricated devices have an active region of 100 μm × 100 μm and a finger width/spacing of 3 μm, respectively. For comparison, the conventional MSM-PDs with 100 nm-thick opaque Au electrodes were also fabricated together. It is noteworthy that the yield of this process is higher than 90 % since the deposition of a thin Au film is very suitable for lift-off technique and the metallization of the interdigitated finger and bonding pad is accomplished in separate mask levels. The layout of the finished device is shown in Fig. 2.

To carry out DC characterization, a 1.55 μm semiconductor laser diode was used as the light source. The device was illuminated through a 9 μm diameter single mode fiber which was positioned as close as possible to the surface of the active region. The magnitude of the incident optical power was adjusted by an calibrated optical attenuator. For the high-speed characterization, the temporal response and S-parameter measurements were performed. In the temporal response measurements, a 1.55 μm short pulse semiconductor laser diode with 40 ps nominal full-width at half maximum (FWHM) was used as the excitation source and the photocurrent of the device was measured by a Cascade microwave probe and displayed by a 20 GHz digital sampling oscilloscope. In the S-parameter measurements, the S_{11} parameters between 20 MHz and 3 GHz were measured by a lightwave component analyzer. The value of each component in the equivalent circuit of the MSM-PD was determined by the optimization routine in the HP Microwave Design System software.

3. Results and discussion

Fig. 3 shows the typical dark current of the devices. The measured dark current of the MSM-PDs with 10 nm-thick Au electrodes is much the same as that of the MSM-PDs with 100 nm-thick Au electrodes. A dark current of 16 nA was obtained at 10 V, corresponding to a dark current density of 1.6 pA/μm². This value compares favorably to the previous best

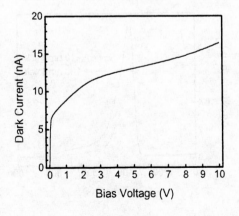

Fig. 3 Typical dark current of the MSM-PDs.

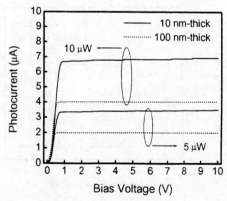

Fig. 4 Photocurrent versus bias voltage under 5 and 10 μW illumination for the MSM-PDs with 10 nm-thick and 100 nm-thick Au electrodes, respectively.

1122

result of 5.6 pA/μm² at 1.5 V for the InGaAs MSM-PDs of the same size with an InP barrier enhancement layer [10]. Our work demonstrates that the pseudomorphic $In_{0.9}Ga_{0.1}P$ cap layer can effectively reduce the dark current of MSM-PDs. Fig. 4 shows the I-V curves under 5 and 10 μW illumination. The photocurrent varies within only 4 % as the bias is increased from the saturation knee voltage to 10 V. Fig. 5 shows a plot of the responsivity versus optical power at 10 V, where the responsivities represented by open symbols and dashed lines are corrected for the dark current. The responsivities are 0.7 A/W and 0.4 A/W for the MSM-PDs with 10 nm-thick and 100 nm-thick Au electrodes, respectively. There is a marked 75 % improvement in responsivity over the conventional MSM-PDs with opaque electrodes. This figure also shows that the responsivities of the devices are nearly constant for illumination levels spanning at least four orders of magnitude, from 0.1 μW to 100 μW. As can been seen in Fig. 4 and Fig. 5, the low frequency internal gain often observed in MSM-PDs, depending on bias voltage and optical power, is fully eliminated in these devices.

Fig. 6 shows the pulse response at 10 V, which is normalized to the peak amplitude of the MSM-PD with 10 nm-thick electrodes. The pulse response of both devices exhibits similar characteristics as observed in the DC responsivity. The MSM-PD with 10 nm-thick Au electrodes exhibits a larger peak amplitude. However, the MSM-PD with 10 nm-thick Au electrodes shows slower response because the photocarriers generated beneath the semi-transparent metal, where the electric field is rather low, take longer transit time to be collected by electrodes. The FWHM of the pulse response for the device with semi-transparent electrodes is 85 ps compared to 80 ps for the conventional device. If we correct for the bandwidth of the measurement system and use the Gaussian approximation (0.312/FWHM), the intrinsic 3 dB bandwidth of the detectors is 6.4 GHz and 7.7 GHz, respectively.

Since RC effect may be a concern of the 3 dB bandwidth as the thin Au electrodes are utilized, we have also examined the parasitic components in the small-signal equivalent circuit of the detectors at 10 V. It is found that the capacitance and series resistance of the MSM-PDs do not significantly increase as the thickness of the Au electrodes is reduced from 100 nm to 10 nm. The obtained capacitance of around 0.25 pF is independent of the

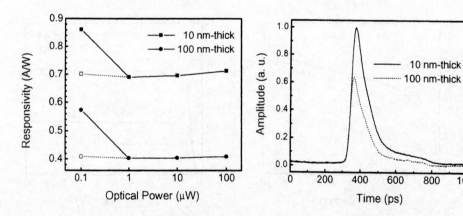

Fig. 5 Responsivity versus optical power for the MSM-PDs with 10 nm-thick and 100 nm-thick Au electrodes, respectively.

Fig. 6 Temporal response at 10 V bias for the MSM-PDs with 10 nm-thick and 100 nm-thick Au electrodes, respectively.

TABLE I

COMPARISON OF VARIOUS InGaAs MSM-PDs WITH TRANSPARENT ELECTRODES.

Authors	This Work	Seo *et al.* [4]	Gao *et al.* [6]
Material of Electrodes	Thin Au	ITO	CTO
Detection Area (μm^2)	100×100	50×50	75×75
Finger Width (μm)	3	3	1
Finger Spacing (μm)	3	3	2
Wavelength (μm)	1.55	1.3	1.3
Dark Current (nA)	16 at 10 V	4 at 5 V	4800 at 10 V
Responsivity (A/W)	0.7	0.76	0.49*
Improvement over the Conventional Devices in Responsivity (%)	75 %	100 %	75 %
Internal Gain	No	Yes	Yes
FWHM (ps)	85 at 10 V	90 at 15 V	—

* Without AR-Coating

thickness of metal electrodes. The measured series resistance of around 40 Ω for both types of devices indicates that this resistance is mainly originated from the metal-semiconductor Schottky contacts rather than the lateral metal finger electrodes. These results reveal that higher speed without sacrificing responsivity can be prospectively achieved using this type of MSM-PDs with smaller finger spacing since their speed is still transit-time limited. Therefore, the problem associated with the high resistivity of ITO electrodes leading to RC-time limited response for finger dimensions below 3 μm can be easily solved [3]. A comparison of the DC and speed performance for the MSM-PDs with previous work using various transparent electrodes as listed in Table I demonstrates that our devices are superior in many aspects.

4. Conclusions

MSM-PDs with 10 nm-thick Au Schottky contacts have been fabricated and investigated. The thin semi-transparent Au electrodes effectively alleviate the metal shadowing effect. The responsivity measured at 1.55 μm wavelength can be significantly increased to 0.7 A/W, which is a 75 % improvement over the conventional devices. This is also close to that of the p-i-n photodetectors. In addition, these devices also have the characteristics of high sensitivity and excellent linearity. It is found that the MSM-PD with semi-transparent electrodes inherently suffer from longer carrier transit time as compared to the conventional device. However, further reduction in finger dimension is expected to improve the speed performance without any expense of responsivity due to its low RC parasitics. These results are encouraging toward the realization of low-cost and high-performance photodetectors used for long-wavelength OEIC receivers.

1124

Acknowledgments

This work was supported by the National Science Council of R. O. C. under contracts No. NSC-83-0417-E-008-015, No. NSC 84-2215-E-008-003 as well as the Telecommunication Laboratories, Ministry of Transportation and Communications of R. O. C. under contract No. TL-NSC-83-5102.

References

[1] M. Ito and O. Wada 1986 *IEEE J. Quantum Electron.* **QE-22** 1073-77

[2] J. B. D. Soole and H. Schumacher 1991 *IEEE J. Quantum Electron.* **27** 737-752

[3] J. H. Kim, H. T. Griem, R. A. Friedman, E. Y. Chan, and S. Ray 1992 *IEEE Photon. Technol. Lett.* **4** 1241-1244

[4] J.-W. Seo, C. Caneau, R. Bhat, and I. Adesida 1993 *IEEE Photon. Technol. Lett.* **5** 1313-1315

[5] P. R. Berger, N. K. Dutta, G. Zydzik, H. M. O'Bryan, U. Keller, P. R. Smith, J. Lopata, D. Sivco, and A. Y. Cuo 1992 *Appl. Phys. Lett.* **61** 1673-1675

[6] W. Gao, A. Khan, P. R. Berger, R. G. Hunsperger, G. Zydzik, H. M. O'Bryan, D. Sivco, and A. Y. Cho 1994 *Appl. Phys. Lett.* **65** 1930-1932

[7] W. C. Dautremont-Smith and L. C. Feldman 1983 *Thin Solid Films* **105** 187-196

[8] A. A. Gutkin, V. M. Dmitriev, and V. M. Khait 1977 *Sov. Phys. Semicond.* **11** 290

[9] J.-I. Chyi, T.-S. Wei, J.-W. Hong, W. Lin, and Y.-K. Tu 1994 *Electron. Lett.* **30** 355-356

[10] C.-X. Shi, D. Grutzmacher, M. Stollenwerk, Q.-K. Wang, and K. Heime 1992 *IEEE Trans. Electron Devices* **39** 1028-1031

Inst. Phys. Conf. Ser. No 145: Chapter 8
Paper presented at 22nd Int. Symp. Compound Semiconductors, Cheju Island, Korea, 28 August–2 September 1995
© *1996 IOP Publishing Ltd*

Analysis of avalanche gain with multiplication layer width and application to floating guard ring avalanche photodiode

Chan-Yong Park, Kyung-Sook Hyun, Jeong Soo Kim, Seung-Goo Kang, Min Kyu Song, Eun Soo Nam, Hong-Man Kim, Kwang-Eui Pyun, Hyung-Moo Park

Optoelectronics Section, Electronics and Telecommunication Research Institute, Yusong P.O.Box 106, Taejon, 305-600, Korea

Abstract The avalanche gain and breakdown voltage are calculated with varying parameters of charge density and multiplication layer width (MLW) in InP/InGaAs avalanche photodiode. As multiplication layer width increases, the breakdown voltage decreases in the range of thinner MLW than w_o while the breakdown voltage increases in the range of thicker MLW than w_o. Here, w_o is a certain MLW where the breakdown voltage has the minimum value for given structure parameters. The value of w_o moves to a larger value as absorption layer thickness increases. On the basis of above result, a floating guard ring avalanche photodiode was fabricated with the dimension of MLW of 0.6 ± 0.1 μm. Two-dimensional gain profile measurement with the aid of focused light beam showed that the avalanche gain at the curved junction was successfully suppressed.

1. Introduction

Since the performance, especially gain-bandwidth (GB) product, of InP/InGaAs avalanche photodiode (APD) is inversely proportional to multiplication layer width (MLW) and it is very difficult to obtain the narrower MLW with a conventional planar guard ring structure, a new structure has been required. Recently, HI-LO structures of avalanche photodiode have been suggested[1] and widely studied[2]-[4]. One of them is floating guard ring (FGR) APD[4] which is simple to fabricate and shows excellent performances. To obtain a high quality APD, the optimum designs of epitaxial layers and guard ring structure are necessary. Especially, multiplication layer width design is very important because it governs the device performance. However, there are few reports on the design of InP/InGaAs APD.

In this paper, we report the experimental results of MLW-dependence of avalanche gain and breakdown voltage in HI-LO structures of InP/InGaAs APD and application of the result to the fabrication of floating guard ring (FGR) APD.

2. Effect of multiplication layer width on breakdown voltage

The APD having epitaxial layer structure as shown in Fig. 1 is fabricated by CH_4/H_2 reactive ion etching (RIE) and metal organic vapor phase epitaxy (OMVPE) regrowth technique. To achieve various MLW, the different depth of zinc diffusion was employed for the same wafer. Details of fabrication process are reported elsewhere[5]. The measured breakdown voltage is defined as the reversely biased voltage where dark current reaches 100 μA. In the gain calculation, we used one dimensional solution of electric field with varying parameters of charge density of layer 2 (defined in Fig.1) and MLW (thickness of layer 1 in Fig. 1). The carrier densities of layer 1 and 3 are assumed to be 2×10^{15} cm^{-3}. In the calculation quaternary layer was regarded as a part of ternary layer. The calculated breakdown voltage is defined as the reversely biased voltage where gain exceeds one hundred. As shown in Fig. 2, all of curves have w_0's where the minimum breakdown voltages occur. As MLW increases, the breakdown voltage decreases in the region of MLW<w_0, while the breakdown voltage increases in the region of MLW>w_0, respectively. The measured data agrees well with the calculated results. When MLW is thinner than 0.2 μm, however, the measured breakdown voltages are lower than the calculated ones. This is attributed to the contribution of InGaAs layer having small bandgap to the avalanche gain because the thinner MLW introduces the high electric field at both of InP multiplication layer region and InP-InGaAs hetero-interface. The w_0 moves as the structure parameters change. According to our calculation, w_0 moves from 0.42 μm to 0.54 μm as the thickness of InGaAs (layer 3) changes from 1.5 μm to 2.0 μm, when 3×10^{12} cm^{-2} of charge density of layer 2 in Fig.1 is employed. The effect of other parameters such as charge density and thickness of layer 2 is negligible.

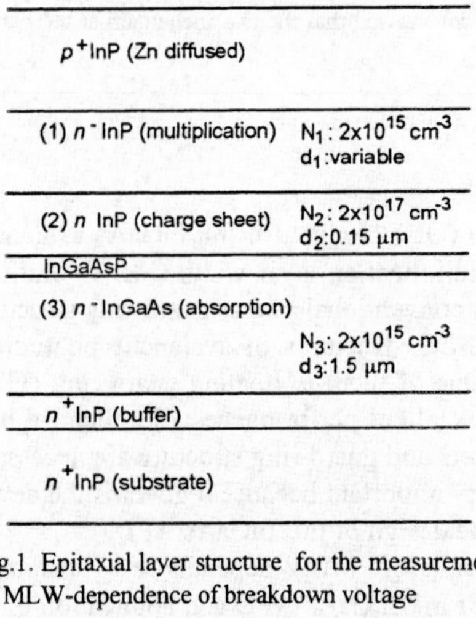

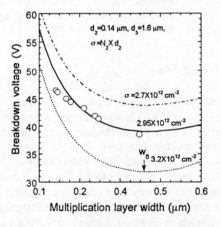

Fig.1. Epitaxial layer structure for the measurement of MLW-dependence of breakdown voltage

Fig. 2. Calculated and measured breakdown voltages. The structure of Fig.1 was used. InGaAsP layer was included into InGaAs.

3. Design of MLW and fabrication of floating guard ring APD

The floating guard ring (FGR) APD[4] has a HI-LO structure and an advantage which is simple to fabricate. As shown in Fig.3, the FGR APD is fabricated by double diffusion. The FGR greatly reduces the electric field (equivalently avalanche gain) at curved junction of diffused edge so that the conventional guard ring is not necessary. However, the avalanche gain at the region of curved junction represented with shadowed line in Fig. 3 is very important and should be suppressed. The curved junction represented shadowed line has varying avalanche gains, i.e., breakdown voltages, according to the varying MLW. Therefore, it is very important to know w_0 for a given structure because MLW should be larger than w_0 for high performance.

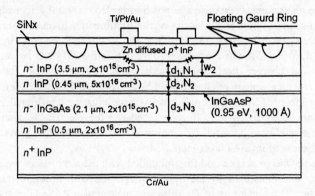

Fig.3. Floating guard ring APD structure

The epitaxial layer is grown by metal organic vapor phase epitaxy (MOVPE) as shown in Fig.3. The first zinc diffusion was performed at 500 °C for 20 minutes by the sealed ampoule method to have 2 μm diffused depth with 1000 Å-thick SiNx mask which has open window having two 2 μm-wide ring pattern. The distance between floating guard rings is 9 μm. The second diffusion is formed at central active region at 500 °C for 25 minutes. During the second diffusion, drive-in diffusion of about 0.4 μm takes place at FGR region. The dimensions of MLW (d_1) and w_2 are 0.6 ± 0.1 μm 1.1 ± 0.1 μm, respectively. The SiNx film was deposited for an antireflection and surface passivation layer. Ti/Pt/Au was used for the formation of p-side contact. The SEM photograph of fabricated device is shown in Fig.4.

Fig.4. SEM photograph of APD

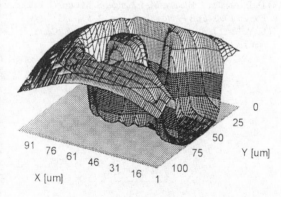

Fig. 5. Two dimensional gain profiles at M=5.

4. Device characteristics and discussions

The breakdown voltages were 80 ~ 85 volts. The 2-dimensional photocurrent response (i.e., gain profile) was measured with the focused light beam diameter of 2 μm and shown in Fig.5. It can be seen that the avalanche gain at the curved junction was successfully suppressed. The photocurrent response at a position far from central active region is due to long diffusion distance of hole and minority carrier lifetime in n^- InGaAs layer.

The measured gain-bandwidth (GB) product was 30 GHz. To enhance the GB product, thinner MLW (d_1 in Fig. 3) and charge sheet layer (d_2 in Fig. 3) must be employed. Since thinner MLW and charge sheet layer introduce the higher electric field at InGaAs absorption layer, charge density of charge sheet layer ($\sigma = N_2 \times d_2$) must be increased. Even in such a case, the effect of MLW on breakdown voltage must be considered and calculated with varying parameters of InGaAs thickness because the electric field and gain at the curved junction should be suppressed. We indicated in section 2 that w_0 moved to the larger value as the thickness of InGaAs layer increased.

The APD chip was packaged and fiber-pigtailed with the aid of laser welder. The APD receiver module was fabricated by the hybrid integration with a commercially available GaAs preamplifier. The sensitivity of -32 dBm was measured at 2.488 Gb/s and 10^{-9} BER.

5. Conclusion

In summary, the FGR APD was designed and fabricated on the basis of gain and breakdown voltage analysis in HI-LO APD. It is shown by 2-dimensional gain measurement that the gain at the curved junction was successfully suppressed.

References

[1] F. Capasso, A. Y. Cho and P. W. Foy 1984 *Electron. Lett.* **20** 635-637
[2] R. Kuchibhotla, J. C. Campbell, C. Tsai, W. T. Tsang and F. S. Choa 1991 *Electron. Lett.* **27** 1361-1362
[3] L. E. Tarof, J. Yu, R. Bruce, D. G. Knight, T. Baird and B. Oosterbrink 1993 *IEEE Photon. Tech. Lett.* **5** 672-674
[4] D. E. Ackley, J. Hladky, M. J. Lange, S. Mason, G. Erickson, G. H. Olsen, V. S. Ban, Y. Liu and S. R. Forrest 1990 *IEEE Photon. Tech. Lett.* **2** 571-573
[5] C. Y. Park, J. B. Yoo, C. Park, K. S. Hyun, D. K. Oh, Y. H. Lee, C. Lee and H. M. Park 1995 *J. Vac. Sci. and Technol. B*, **13**, No.3 , in press.

Inst. Phys. Conf. Ser. No 145: Chapter 8
Paper presented at 22nd Int. Symp. Compound Semiconductors, Cheju Island, Korea, 28 August–2 September 1995
© *1996 IOP Publishing Ltd*

A Monolithic GaAs Photovoltaic Device Array with A Self-Aligned Dielectric Isolation on Sidewall

Young-Gi Kim**, Alec Chen*, Kambiz Alavi and Tsay-Jiu Shieh

Department of Electrical Engineering, University of Texas, Arlington, TX 76019.
* * Young-Gi Kim is with the PCS Division, Korea Telecomm Research Center,
17 Woomyeon-dong, Seocho-gu, Seoul, Korea.
*   Alec Chen is with National Semiconductor, 111 West Bardin Rd., Arlington,
TX 76017..

Abstract.   Monolithic GaAs Schottky photovoltaic device arrays with and without a self-aligned dielectric isolation have been fabricated and characterized to study the possible photo current leak between the Schottky metal and the underlying heavily doped layer. This leakage affects the output power of the array devices used in an electro-optic switch which provides excellent input-output voltage isolation for the application of power controls and telecommunications. In order to reduce the leakage current, the photoresist pattern used for mesa etch was also used in the SiO_2 sputtering deposition process. An improvement of 20% in open circuit voltage and 7% in short circuit current for an array of 25 Schottky (Au) diodes has been observed by using this process.

1. Introduction

Electro-optic solid state switches with good voltage isolation between the input control signal and output load can be used in place of mechanical switches or solid state switches for power control and telecommunication. The electro-optic solid state switch uses a photovoltaic device

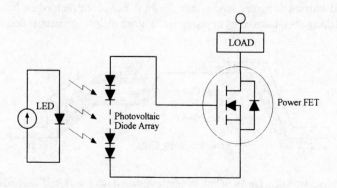

Fig. 1   Schematic representation of the electro-optic solid state switch.

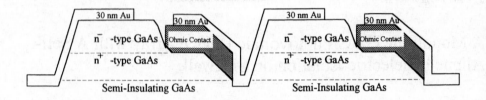

Fig. 2   Cross-section of a 2-cell Schottky barrier photovoltaic device array without any
dielectric isolation.

array to generate the needed voltage to drive a MOSFET transistor from the light generated
by a Light Emitting Diode (LED). This LED is controlled by the input switching signal as
shown in Fig. 1.

Small size, low ON resistance, low leakage current, high shock/vibration resistance, no
arc, no bounce, low sensitivity to magnetic fields and long life time are additional advantages
of the electro-optic switches.[1, 2]   The required MOSFET gate driving voltage which must
be greater than the threshold voltage can be achieved by a series connected photovoltaic
device array. Single photovoltaic device can not provide enough gate to source voltage. A
series connected photovoltaic device array can produce the required high voltage. The power
MOSFET typically has an input capacitance in the range of nano-Farad. The current from the
array needs to be large enough to charge up the gate capacitor. The photovoltaic device is
distinguished from solar cells and photodetectors because the light source used for the array in
a solid state switch is semi-monochromatic light from an LED. A normal photodetector also
requires an external bias in order to increase the sensitivity and speed.

2.   A self-aligned monolithic GaAs photovoltaic device array

Monolithic GaAs Schottky barrier solar cell arrays and p-n junction arrays for consumer
electronic similar to Fig.2 were suggested by Masu *et.al.*[3]   A cross-section of a 2-cell
Schottky barrier photovoltaic device array is shown in Fig. 2.   They reported that a series
array of solar cells has an effect in reducing the series resistance.   Possible photo current leak
between the gold interconnection on the sidewall and the underlying n+ GaAs layer was
speculated.   This part corresponds to reverse biased highly doped Schottky junction in which
reverse biased current becomes larger than forward biased current when bias reaches 0.3
V.[4]   This leakage is improved by applying self-aligned dielectric material deposition

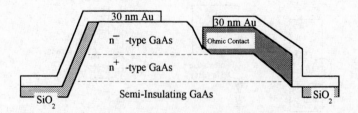

Fig. 3   Cross-section of a GaAs Schottky photovoltaic device with self aligned dielectric
isolation.

isolation at the side wall of the mesa as shown in Fig. 3.

The fabrication process starts with the growth of a 0.5um GaAs buffer layer, a 0.5um n^+ GaAs (Nd=2e18 /cm³), and a 0.2 um n⁻ GaAs (Nd=1e16 /cm³) on a SI GaAs wafer using MBE. After the wafer was etched by mesa etch solution(H_2SO_4 : H_2O_2 : H_2O = 885 : 10 : 1) to define the active cell regions on the semi-insulating layer, the photo-resist pattern was not deleted, in order to reduce the leakage. A 100 nm SiO_2 was deposited and lifted off. Consequently it is self aligned, which has reduced mask alignment error.[5]   An n-type ohmic contact with Au-Ge (Au 88%, Ge 12%, 80nm)/Ni(25nm)/Au(150nm) was made. Then a 30nm Au layer was deposited as a near transparent Schottky electrode and interconnection for the 25 cells connected in series.

After the measurement and comparison, it was found that the open circuit voltage improved from 11.1 V to 13.4 V and that the short circuit current improved from 176 μA to 191 μA   under the same mercury lamp illumination. These results are shown in Fig. 4. and Fig. 5.

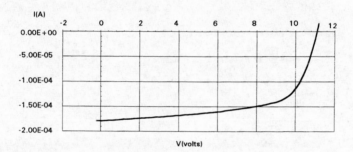

Fig. 4   *I-V* characteristic of a regular 25-cell GaAs Schottky photovoltaic device array
with no dielectric isolation under illumination.

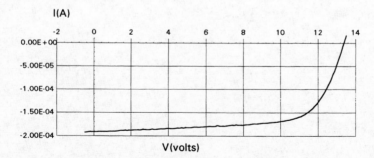

Fig. 5   *I-V* characteristic of a 25-cell GaAs Schottky photovoltaic device array with self-
aligned dielectric isolation under illumination.

3. Conclusion

Monolithic GaAs Schottky photovoltaic device arrays with and without a self-aligned dielectric isolation have been fabricated and characterized to study the possible photo current leak between the Schottky metal and the underlying heavily doped layer.

An improvement of 20% in open circuit voltage and 7% in short circuit current for an array of 25 Schottky (Au) diodes has been observed by using the self-aligned dielectric isolation process. The fabrication process has been simplified by reducing the required mask and the mask alignment error has been improved in the self-aligned process.

References

[1] J. Naples, *Electronic Engineering*, p. 37, August 1992.
[2] *Microelectronic Relay Designer's Manual*, International Rectifier, p. F13, 1990.
[3] K. Masu *et al.*, *Proceeding of the 2nd Photovoltaic Science Engineering Conference*, Japan, p.95, 1980.
[4] E.H. Rhoderick, *Metal-Semiconductor contacts*, Oxford Univ. Press, p.118, 1978.
[5] K. Sato et al., *Jap. J. Appl. Phys*, Vol. 29, p.1946, 1990.

Inst. Phys. Conf. Ser. No 145: Chapter 9
Paper presented at 22nd Int. Symp. Compound Semiconductors, Cheju Island, Korea, 28 August–2 September 1995
© *1996 IOP Publishing Ltd*

GaN based semiconductors for future optoelectronics

Danielle Walker, Patrick Kung, Adam Saxler, Xiaolong Zhang, Manijeh Razeghi

Center for Quanutm Devices, Northwestern University, Department of Electrical Engineering and Computer Science, Evanston, IL 60208, USA

Holger Jürgensen

AIXTRON GmbH, Kackerstr. 15-17, D-52072 Aachen, Germany

Abstract. Ultraviolet photoconductive and photovoltaic detectors were fabricated using n-type GaN and p-n GaN homojunction, respectively. The frequency dependence and the decay of the photoresponse in GaN photoconductors was measured. This revealed the existence of deep level traps in the forbidden gap. A stationary photoconductivity lifetime of about 20 ms was determined. The spectral response of the photovoltaic devices was also measured. A modeling of this response allowed to calculate the hole diffusion length in n-type GaN to be ~0.1 μm.

1. Introduction

There is currently a need for ultraviolet (UV) detectors which have high responsivities and are solar blind. A few potential applications include the detection of spacecraft and the monitoring of combustion chambers, where it is desirable to detect UV radiation in a visible and infrared background. Gallium nitride (GaN) has a direct, wide bandgap (λ_g=365 nm) and is the ideal material for intrinsic UV detectors for wavelengths shorter than 365 nm.

To date, there has been only a limited number of reports on UV photoconductive, photovoltaic, and field effect transistor GaN detectors.[1] Moreover, two theoretical simulations of UV detectors using GaAlN compounds were reported.[2] In this paper, we report the growth and characterization of n-type GaN photoconductive and p-n GaN homojunction photovoltaic detectors.

2. Photoconductivity in n-type GaN

The GaN sample used for photoconductivity experiments was grown on (11•0) sapphire substrates by low pressure metalorganic chemical vapor deposition (MOCVD). It was 1.5 μm thick, and exhibited n-type conductivity with a room temperature n-type carrier concentration of about 10^{17} cm^{-3}. The experimental conditions have been reported earlier.[3] Room temperature photoluminescence from the GaN layer yielded sharp band edge emission at ~365 nm, as well as a broad intense yellow emission at ~560 nm, as shown in figure 1.

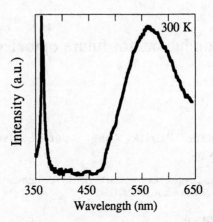

Figure 1. Room temperature photoluminescence of GaN on (11·0) sapphire.

To perform photoconductivity measurements, a van der Pauw clover pattern was realized onto the sample and four indium contacts were annealed on the edges. A bias of 6V was applied onto two adjacent contacts in series with a load of 30 kΩ, while the other two contacts were used for the photoresponse voltage measurements.

A Xe arc lamp was used to measure the spectral responsivity of the detector in a standard synchronous detection scheme. The light was focused into a monochromator, at the exit of which the sample was placed. To measure the responsivity for visible wavelengths, a GaN layer [3] was used as an optical filter between the lamp light and the monochromator to remove any UV radiation due to higher order diffraction and scattering. To account for any fluctuation in the excitation source, the photosignal was calibrated by the lamp power density spectrum. Figure 2 shows the normalized spectral responsivity of the GaN sample. There is a sharp cut-off near the band edge at 365 nm, and the response at longer wavelengths is estimated to be 2-4 orders of magnitude lower than the peak responsivity. For longer wavelengths, the noise level was high and only the upper limit of the photoresponse could be determined. The response in the visible may be due to the deep levels observed in the photoluminescence experiments.

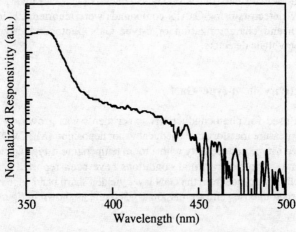

Figure 2. Normalized spectral responsivity of n-type GaN photoconductor.

The frequency dependence of the detector responsivity was measured using a He-Cd laser (325 nm, ~10mW). The frequency of the excitation source was adjusted by varying the chopping frequency, while neutral density filters were used to attenuate the excitation power. Figure 3 shows the detector voltage responsivity as a function of excitation power from 3 to 800 Hz for three excitation power densities. In the whole frequency range and for all three power densities the responsivity remained frequency-dependent. The three curves tend to join at high frequencies where the photogenerated excess carrier concentration becomes dependent only on the excitation period, not the lifetime of the majority carriers which depends on the excitation power density.

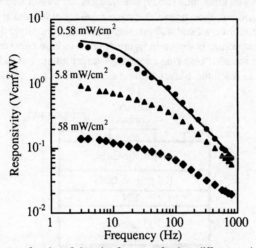

Figure 3. Responsivity as a function of chopping frequency for three different excitation power densities.

The responsivity for the lowest excitation power is the closest to the theoretical responsivity given by:

$$R_v = \frac{R_{v0}}{\left(1+4\pi^2f^2\tau^2\right)^{1/2}}$$

where R_{v0} is the absolute value of stationary voltage responsivity (when $f<<1/2\pi\tau$). The solid line in figure 3 is the best fit of the lowest power density curve to the formula above, which yielded $R_{v0}\approx4.776$ Vcm^2/W. Such a high value would result in a high photoelectric gain and high responsivity photoconductors (tens of kA/W using an optimized geometry).

The photoresponse to a single and intense pulse of He-Cd laser light was measured as a function of time. The signal was amplified and recorded on an X-Y plotter. The decay did not follow a simple exponential law but the slope of the decay decreased with time, suggesting increasing response times. Such phenomenon can be interpreted by the existence of deep level traps in the forbidden gap. When the majority and minority carrier concentrations are high, such as just after the excitation pulse, the deep levels play the role of recombination centers, resulting in a short response time. As the decay goes on, the concentrations of carriers decreases toward their equilibrium values and the deep levels regain their trap character, resulting in a longer response time. To achieve practical applications, these deep levels must be eliminated.

3. Photovoltaic effects in GaN p-n homojunction

The GaN p-n junction sample used for photovoltaic experiments was grown on (00•1) sapphire substrates by atmospheric pressure MOCVD. The experimental conditions have been reported earlier.[4] First, a thin AlN buffer layer was deposited, followed by a 0.1-0.5 μm thick autodoped n-type GaN and a 2 μm thick p-type GaN. The p-type doping was realized by incorporating Mg with bis-cyclopentadienyl-magnesium. The n-type GaN typically yielded a room temperature electron mobility of ~100 cm^2/Vs for a carrier concentration of ~10^{18} cm^{-3}. The as-grown p-type layers were insulating. P-type conduction was achieved after thermal annealing under nitrogen ambient. The typical hole concentration was ~3×10^{16} cm^{-3}. Room temperature photoluminescence from the n-type layers yielded sharp band edge emission, while a broad band emission around 420 nm was observed from p-type layers.

The detector structure is shown in figure 4. Two indium contacts were put on the top, one small contact surrounded by a ring contact. A larger amount of indium was used for the ring contact in order to reach the n-layer without the need to etch a mesa.

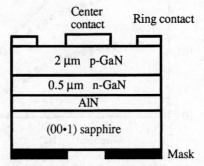

Figure 4. GaN photovoltaic detector structure

The sample was back side illuminated with a Xe arc lamp and a monochromator, as described earlier. A photovoltage was measured accross the two contacts using a standard synchronous detection scheme. The spectral dependence of the photovoltage is shown in figure 5. A strong positive signal can be detected on the central contact at a shorter wavelength (362 nm),while a smaller negative signal was measured just above 365 nm. This spectrum can be interpreted as follows.

Incident light with wavelength < 200 nm (λ_g of AlN) is absorbed by the AlN buffer layer. Radiations between 200 and 365 nm are almost completely absorbed by the n-type GaN layer because of the high absorption coefficient in this wavelength region (>10^5 cm^{-1}). Only light with wavelengths close to 365 nm reaches the p-n junction depletion region and contribute to the signal, thus giving the sharp strong response in figure 5. Longer wavelength radiations can go through the whole sample and reach the p-GaN/metal contact. This contact is rectifying and carriers generated near it contribute to the negative signal observed in figure 5.

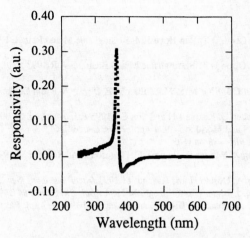

Figure 5. Spectral responsivity of GaN photovoltaic device

The responsivity for shorter wavelengths in figure 5 can be modeled as follows. The current responsivity is given by:

$$R_i = \frac{q\lambda\eta}{hc}$$

where λ is the excitation wavelength, η is the quantum efficiency. We consider three contributions to η: one from the n-type layer, one from the depletion region and one from the p-type layer. Geometry factors, surface recombination velocities, minority carrier diffusion lengths and diffusion coefficients were involved in the computation of the quantum efficiency in each region. The shape of the spectral response was barely affected by the surface recombination velocities, unlike minority carrier diffusion lengths. A diffusion length of 0.1 μm for holes in n-type GaN yielded the best fit.

4. Conclusion

We reported the fabrication and characterization of ultraviolet photodetectors using n-type GaN or p-n GaN homojunction. The n-type GaN photoconductor exhibited a frequency dependent response in the whole range of frequency applied. The photoconductivity decay was also slower than exponential. This suggests the existence of deep level traps in the forbidden gap. The spectral response of the p-n GaN homojunction allowed the estimation of the hole diffusion length in n-type GaN to be ~0.1 μm.

Acknowledgements

The authors wish to acknowledge Max Yoder and Yoon-Soo Park at the Office of Naval Research for their permanent support and interest. The authors would also like to thank Gail Brown and Bill Mitchel at the Wright Laboratory, John Fan at Kopin Corporation. This work is funded by the Office of Naval Research under Grant No. N00014-93-1-0235.

1138

References

[1] Khan M A, Kuznia J N, Olson D T, Van Hove J M, Blasinghame M and Reitz L F 1992 *Appl. Phys. Lett.* **60** 2917-9

Khan M A, Kuznia J N, Olson D T, Blasinghame M and Bhattarai A R 1993 *Appl. Phys. Lett.* **63** 2455-6

Khan M A, Chen Q, Sun C J, Shur M S, Macmillan M F, Devaty R P and Chyke J 1995 *Proc. SPIE* **2397** 283-8

Khan M A, Shur M S, Chen Q, Kuznia J N and Sun C J 1995 *Electron. Lett.* **31** 398-9.

[2] Joshi R P, Dharamsi A N and McAdoo J 1994 *Appl. Phys. Lett.* **64** 3611-3

Joshi R P 1994 *J. Appl. Phys.* **76** 4434-6

[3] Kung P, Saxler A, Zhang X, Walker D, Wang T C, Ferguson I and Razeghi M 1995 *Appl. Phys. Lett.* **66** 2958-60

[4] Sun C J, Kung P, Saxler A, Ohsato H and Razeghi M 1993 *Inst. Phys. Conf. Ser.* **137**   425-8

Saxler A, Kung P, Sun C J, Bigan E and Razeghi M 1994 *Appl. Phys. Lett.* **64** 339-41

Sun C J, Kung P, Saxler A, Ohsato H, Bigan E and Razeghi M 1994 *J. Appl. Phys.* **76**   236-41

Inst. Phys. Conf. Ser. No 145: Chapter 9
Paper presented at 22nd Int. Symp. Compound Semiconductors, Cheju Island, Korea, 28 August–2 September 1995
© *1996 IOP Publishing Ltd*

The 6x6 Luttinger-Kohn model of a cubic GaN quantum well

Doyeol Ahn

LG Electronics Research Center
16 Woomyeon-Dong, Seocho-Gu,Seoul 137-140, Republic of Korea

Abstract. The effects of a very strong spin-orbit (SO) split-off band coupling on the valence-band structure and the optical gain of a 100Å cubic GaN quantum-well are studied within the 6x6 Luttinger-Kohn model. A semi-empirical five level **k·p** model is used to calculate the Luttinger valence-band parameters γ_1, γ_2, and γ_3 for a cubic phase of GaN. Calculated results show that the subbands originated from the "light hole," and the "spin-orbit" are strongly coupled even at the zone center because of the very narrow spin-orbit split off energy. It is found that the optical gains for the TE and the TM polarizations have comparable magnitudes over the wide range of carrier densities and are smaller than the gain calculated using the parabolic model.

1. Introduction

III-V nitrides such as GaN or InGaN are attractive for their potential optoelectronic device applications in the blue-green or the near-ultraviolet spectrum. From the theoretical point of view, one of the most important basic properties of the quantum-well optoelectronic devices is the optical gain and the rigorous calculation of which requires detailed knowledge of the band-structure especially the valence-subband dispersions. Because of its very narrow spin-orbit(SO) split off energy (10 meV), the coupling effects in the valence band of GaN are expected to be quite pronounced. In this study, the effects of a strong spin-orbit (SO) split-off band coupling on the valence-band structure and the optical gain of a 100Å cubic GaN quantum-well are studied within the 6x6 Luttinger-Kohn model [1,2]. We first obtain the Luttinger valence-band parameters γ_1, γ_2, and γ_3 of a cubic phase of GaN, which are not tabulated yet to the best of the author's knowledge, from a semi-empirical five level **k·p** approach [3]. The valence band structure of a cubic GaN quantum well is calculated using the recently developed 6x6 Luttinger-Kohn model [1,2] which takes into account the spin-orbit (SO)split-off coupling. Secondly, the optical gain of a GaN quantum well is calculated from the complex optical susceptibility obtained by the density matrix formalism[4].

2. Theoretical model

Within the five level **k·p** approach, the Luttinger valence-band parameters are given by [3]

$$\gamma_1 = -\frac{1}{3}(F + 2G + 2H_1 + 2H_2) -1 + \frac{1}{2} q,$$
$$\gamma_2 = -\frac{1}{6}(F + 2G - H_1 - H_2) - \frac{1}{2} q, \tag{1}$$
$$\gamma_3 = -\frac{1}{6}(F - G + H_1 - H_2) + \frac{1}{2} q,$$

where F, G, H_1, and H_2 are constants [3,5] originally defined by Dresselhaus, Kip, and Kittel (DKK) [6]. The relative magnitudes are; $|F|$, $|H_1| > |G| \gg |H_2|$; q is always very small and can be neglected [5]. The effective mass m_c^* at the $\Gamma_2^{'}$ band edge is related to F by [6]

$$\frac{m}{m_c^*} = -1 - F (1 - \frac{\Delta}{3(E_G + \Delta)}), \tag{2}$$

where Δ is the spin-orbit splitting, m is the free electron mass, and E_G is the fundamental direct band gap. From (2), F can be calculated from the measured value of m_c^*. Calculation of H_1 requires the knowledge of the momentum matrix element between the Γ_{15} states which is not straightforward. As an ansatz, we propose a simple rule as follows:

$$\frac{m}{m_h^*} = \gamma_1 - 2 \bar{\gamma}$$
$$\approx -1 - \frac{2}{3} H_1, \tag{3}$$

with $\bar{\gamma} = \frac{1}{2}(\gamma_2 + \gamma_3)$.

The Luttinger parameters γ_1, γ_2, and γ_3 can be obtained by substituting the calculated values of F and H_1 into the equation (1). For a comparatively small parameter G, we use G = -0.75 [3]. In order to verify the usefulness of our simple rule, calculated Luttinger parameters γ_1, γ_2, and γ_3 are compared with those obtained by more rigorous approach of Lawaetz [3] for InP, GaAs, ZnSe, and ZnS in table I. Reasonable agreement between two methods can be seen in this table.

Table 1. Comparison of the Luttinger parameters for InP, GaAs, ZnSe, and ZnS.

	γ_1	γ_2	γ_2
InP [a]	6.28	2.08	2.76
(this work)	6.47	2.10	2.81
GaAs [a]	7.65	2.41	3.28
(this work)	7.4	2.24	3.17
ZnSe [a]	3.77	1.24	1.67
(this work)	4.05	1.25	1.73
ZnS [a]	2.54	0.75	1.09
(this work)	2.6	0.63	1.04

[a] Lawaetz P 1971 *Phys. Rev. B* **4** 3460

The calculated Luttinger parameters for GaN are: $\gamma_1 = 3.75$, $\gamma_2 = 0.69$, and $\gamma_3 = 1.44$. One should note that these values are the rough estimates and should be refined with further experimental data in the future. The values of E_G, Δ, m_c^*, and m_h^* for GaN used in the calculations are, respectively,

$$E_G = 3.25 \text{ eV}, \ \Delta = 0.01 \text{ eV}, \ m_c^* = 0.2 \ m, \ m_h^* = 0.8 \ m.$$

We now proceed to carry out the calculation of the valence-band structure of a GaN quantum well surrounded by lattice-matched InGaAlN barriers using the 6x6 Luttinger-Kohn model. We assume that the electrons and holes in a quantum well are confined by the band offset ΔE_c and ΔE_v for electrons and holes, respectively. The value of 250 meV for both ΔE_c and ΔE_v is assumed throughout the computation. Unavailability of more detailed experimental or theoretical data does not allow the rigorous calculations of the band offsets.

The results of our valence-subband calculation for a 100 Å cubic GaN quantum well are customary to label the subbands according to their character at the zone center. It is also well known that the SO band is coupled to the light hole band and is decoupled from the heavy hole band at $k_\parallel = 0$. So far, however, most authors labeled the subbands as "light hole" or "spin-orbit" to describe the subbands which are originated from the spin 1/2 components of the Γ_8 or the Γ_7 bands for bulk semiconductors.

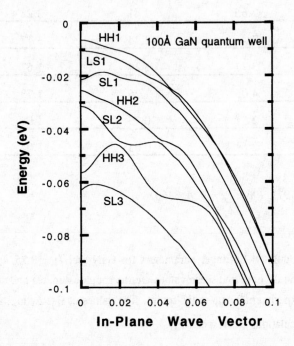

Fig. 1 The valence-subband structures for a 100 Å cubic GaN quantum well surround by lattice-matched barriers.

This is not entirely correct especially for GaN for which the spin-orbit split off energy is very small (10 meV) and as a result there is no pure "light hole" or "spin-orbit" bands even at the zone center when the coupling effects are considered. In this letter, we propose to denote the subbands which are the coupled states of "light hole" and "spin-orbit" bands as "LS" and "SL". Here "LS" or "SL" is the acronym for "light-hole-spin-orbit" and the first letter denotes the dominant component of the wave function. For example, LS1 = LH1 $\oplus$ SO1 where $\oplus$ denotes the tensor addition and the envelope function of LH1 is the dominant mode. Because of the proximity (4 meV) of the HH1 and the LS1 subbands, it is expected that the TE and the TM mode gains have comparable magnitudes.

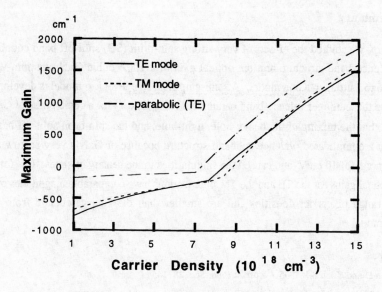

Fig. 2. The peak gain as a function of the injected carrier density for a 100 Å GaN quantum well surrounded by for (i) the TE mode (solid line), (ii) the TM mode (dashed line), and (iii) the parabolic band model for TE polarization (dotted line).

The result of the gain calculations is shown in Fig. 2 for (i) the TE polarization (solid line), (ii) the TM polarization (dashed line), and (iii) the parabolic band model [7] for the TE mode (dotted line). In this study, the strain and the many-body effects are ignored and the inclusion of the many-body effects such as Coulomb enhancement and excitonic gain, and the strain effects will improve our estimation of the optical gain. It is found that the gain is overestimated in the simple parabolic band model and the TE and the TM mode gains have comparable magnitude when the valence band mixing effects are considered. Comparison with our previous result [7] shows that the gain versus carrier density relation for a ZnSe quantum well is much better than that of a GaN quantum well when the strain effects are ignored in both cases. Employing well-engineered strained-layer quantum well seems to be essential to make a GaN quantum-well laser feasible.

1144

3. Summary

In summary, we studied the effects of very strong spin-orbit (SO) split-off band coupling on the valence-band structure and the optical gain of a 100Å cubic GaN quantum-well using the 6x6 Luttinger-Kohn model. A semi-empirical five level $\mathbf{k \cdot p}$ model is developed to calculate the Luttinger valence-band parameters γ_1, γ_2, and γ_3 for a cubic phase of GaN. We found that the mixing of the heavy-hole, light-hole, and the spin-orbit split-off bands resulted in a complicated valence subband structure because of GaN's very narrow SO splitting energy of 10 meV and caused the singularities in the density of states. It is found that the optical gains for the TE and the TM polarizations have comparable magnitudes over the wide range of carrier densities and are smaller than the gain calculated from the parabolic model.

References

[1] Ahn D and Yoon S J 1995 *J. Korean Phys.* Soc. **28** 145

[2] Ahn D, Yoon S J, Chuang S L and Chang C S 1995 *J. Appl. Phys.* accepted

[3] Lawaetz P 1971 *Phys. Rev. B* **4** 3460

[4] Ahn D and Chuang S L 1988 *IEEE J. Quantum Electron.* **24** 2400

[5] Suzuki K and Hensel J C 1974 *Phys. Rev. B* **9** 4184

[6] Dresselhaus G, Kip A F and Kittle C 1955 *Phys. Rev.* **98** 368

[7] Ahn D 1994 *J. Appl. Phys.* **76** 8206

Inst. Phys. Conf. Ser. No 145: Chapter 9
Paper presented at 22nd Int. Symp. Compound Semiconductors, Cheju Island, Korea, 28 August–2 September 1995
© *1996 IOP Publishing Ltd*

Heterojunction GaN Light Emitting Diodes Grown by Plasma-Assisted Ionized Source Beam Epitaxy

Myung C. Yoo*, K.H. Shim, J.M. Myoung, A.T. Ping, K. Kim[†], and I. Adesida

Department of Electrical and Computer Engineering and Center for Compound Semiconductor Microelectronics, University of Illinois at Urbana-Champaign, IL 61801, USA

Abstract. Growth and processing of heterojunction GaN-based light emitting diodes are described. Multilayer LEDs were fabricated out of p-AlGaN/n-GaN/n-AlGaN on sapphire substrate which were grown using plasma-assisted ionized source beam epitaxy. The growth system utilized high-efficiency radio frequency plasma to generate atomic nitrogen from molecular nitrogen. Ga and Al source beams were partially ionized and accelerated to enhance the incorporation with atomic nitrogen. The turn-on voltage of the LEDs was around 2.5 V and soft breakdown occurred at a reverse bias of around 10 V. At a forward bias of 4 V, the current was 10 mA. The LEDs were stable for a sustained period of time. The light intensity increased linearly with increasing current from 17 to 35 mA and started to saturate around 45 mA. The spectrum showed a peak at the wave length of 518 nm (2.4 eV).

1. Introduction

Recently, a great deal of research activities involving GaN-based semiconductors have been reported due to their importance in optoelectronic applications. The resurgence in GaN research is mainly attributed to the significantly improved material quality resulting from the advanced film growth techniques such as MOCVD and MBE. MOCVD is known to be very effective in growing a high-quality GaN epilayer, which has resulted in first introduction of an InGaN-based blue LED in the market [1]. On the other hand, MBE-grown GaN epilayers are still being pursued due to the flexibility and low-temperature capability of the MBE technique. An MBE-grown homojunction blue LED was recently reported by Molnar et. al. [2]. According to this work, post-growth annealing was not necessary due to the elimination of hydrogen ambience during the growth, which is a problem inherent with the MOCVD growth.

* On leave from Samsung Advanced Institute of Technology, Photonics Research Laboratory, Materials & Devices Research Center, P.O. Box 111, Suwon, Korea
† To whom correspondense should be addressed.

The MBE-grown homojunction LED, however, exhibited a small luminous intensity, and the turn-on voltage was too high because of the poor internal quantum efficiency.

In this work, we describe a AlGaN/GaN/AlGaN double heterojunction blue LED grown by the plasma assisted ionized source beam epitaxy (PAISBE). PAISBE is similar to the plasma-assisted MBE in that it employs an atomic radical nitrogen source and the conventional effusion cells. The difference is that it has the capability to ionize and accelerate the source beams and, therefore, can control the electrical charges and the kinetic energies of the individual source beams. The effectiveness of beam ionization and acceleration in conjunction with the growth of GaAs on Si has been previously reported [3-5]. In order to enhance the reaction between atomic nitrogen and Al and Ga, a home-made high-efficiency inductive RF nitrogen source was used. For the fabrication of LEDs, chemically assisted ion beam etching was used to make an etched-junction.

2. Experimental

A detailed description of the PAISBE growth system and the substrate cleaning procedures can be found elsewhere [6]. After the chemical cleaning, the substrate was thermally desorbed for 30 min at 800°C. A 200Å-thick AlN buffer layer was grown on sapphire (0001) at 600°C. For the buffer layer, the temperature of the Al cell, the plasma power, and the flow rate of N_2 were kept at 1070°C, 300W, and 1 sccm, respectively. For the growth of a GaN epilayer, the substrate and the Ga cell were maintained at 700 - 800°C and 920 - 940°C, respectively. The plasma power of the nitrogen source was in the range of 350 - 450 W. For the growth of AlGaN films the substrate temperature and the temperature of the Al cell were elevated to 750°C and 1150°C, respectively, at the same plasma power. To fabricate the p-n junctions, high-purity Mg (6N) and Si (6N) were used in the growth process as a p-and n-type dopant, respectively.

To determine the electrical properties of the grown films, van der Pauw Hall measurements were performed. The quality of the GaN films were evaluated by using the scanning electron microscopy (SEM) and the x-ray diffractometry with Cu Kα radiation using a Ni filter. In particular, rocking curve measurements of x-ray diffractometry were employed to determine the degree of crystallinity of the as-grown films. Finally, optical properties were measured by photoluminescence.

Fudicial cell type LEDs with diameters ranging from 0.1 to 1.0 mm were fabricated. In order to form an etched metal contact on the n-GaN, chemically assisted ion beam etching (CAIBE) was carried out with an Ar ion beam in a Cl_2 environment. The highest etch rate obtained was 300 nm/min with an Ar ion beam energy of 900 eV at 200°C. To form a p-type ohmic contact, a 25-nm Cr/200-nm Au metallic bilayer was deposited on p-GaN using an e-beam evaporator. To improve contact resistance rapid thermal annealing was performed at 360°C for 1 min under an N_2 ambience. For an n-type ohmic contact 25-nm Ti/100-nm Al bilayer was deposited on the n-GaN layer in TLM pattern, and thermal annealing was performed at 900°C for 30 sec under an N_2 ambience.

3. Results and Discussion

A cross sectional view of the n-AlGaN/GaN/p-AlGaN double heterojunction (DH) LED structure is shown in Fig. 1. In order to maximize the quantum efficiency of the LED, namely to confine electrons and holes in the active n-GaN layer, the GaN layer was sandwiched between p-AlGaN and n-AlGaN layers. A heavily doped p-GaN layer is grown on top of the p-AlGaN layer to reduce contact resistance. The n-GaN layer shown below n-AlGaN serves a similar purpose.

Figure 2 shows a representative RHEED pattern of a 1 µm-thick GaN epilayer grown with PAISBE. During the growth the RHEED pattern was observed to be very streaky from the beginning of the growth all the way to the end indicating that the grown GaN film had good crystal quality.

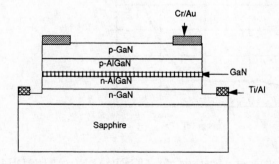

Fig. 1. Cross-sectional view of p-AlGaN/n-GaN/n-AlGaN DH LEDs.

Fig. 2. Representative RHEED pattern of GaN epilayer on sapphire (0001) after 1μm growth.

The full-width at half maximum (FWHM) of the x-ray rocking curve was typically 15 arc min, which is quite comparable to the films grown by MOCVD or MBE [7-8]. The net carrier concentrations of the p-doped GaN and AlGaN layers were high 10^{17} - low 10^{18}/cm^3, whereas those of the n-doped layers were high 10^{18}/cm^3.

The current-voltage curve of the p-AlGaN/n-GaN/n-AlGaN double-heterostructure is shown in Fig.3. Soft breakdown occurred at a reverse bias of around 10 volts. The current was 10 mA at a forward bias of 4 V. This current-voltage characteristic implies that the heterojunction and ohmic metal contacts, as fabricated, were suitable for working devices.

Electroluminescence of the double-heterostructure LED is shown in Fig. 4. This was measured in the wavelength range of 500-600 nm while an ac bias of 16 V(1 kHz) of 50% duty cycle drove 50 mA across the device.

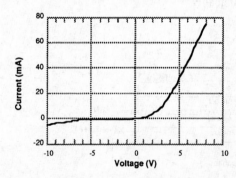

Fig. 3. Typical current-voltage characteristic of p-AlGaN/n-GaN/n-AlGaN DH LED.

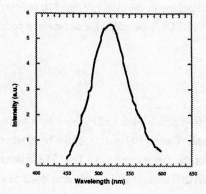

Fig. 4. Electroluminescence spectrum of p-AlGaN/n-GaN/n-AlGaN DH LED.

The FWHM of the peak emission was measured to be 330 nm, which needs to be further reduced. The peak intensity of the spectrum occurs at a wavelength of 518 nm (2.4 eV) with one order of magnitude stronger intensity as compared to the background intensity. Impurity and defect levels in the n-GaN are likely to be responsible for the broad energy distribution of the emitted photons. The doping technique for GaN needs to be improved in order to raise the energy of the photons up to 2.75 eV (450 nm).

Fig. 5. shows a typical spontaneous emission characteristic of the DH LED.

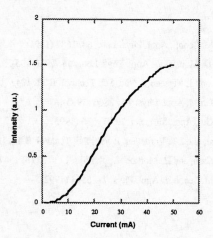

Fig. 5. Typical light intensity vs. current curve of doubleheterojunction LED.

The light intensity increases linearly with the current from 170 to 350 mA, but it starts to saturate around 400 mA. The LEDs were stable for a sustained period of time.

4. Conclusions

A GaN-based double heterojunction blue LED grown by a technique other than MOCVD has been demonstrated. The crystal quality of the heterostructures grown by PAISBE were good enough for the fabrication of LEDs. The bilayer metal contacts that were fabricated and the CAIBE technique that was employed to etch the samples both proved to be effective for device fabrication. Further study, however, is necessary to improve doping and to form better p-type ohmic contacts which will result in desired shorter wavelengths and improved I-V characteristics.

Acknowledgement

Useful exchange of information with Jim van Hoff and Peter Chow of SVT is gratefully acknowledged. This work was supported by Samsung Electronics Co. Ltd.

References

1.   S. Nakamura, T. Mukai, and M. Senoh, Appl. Phys. Lett., **64**, 1687 (1994).

2.   R.J. Molnar, R. Singh, and T.D. Moustakas, Appl. Phys. Lett., **66**, 268 (1995).

3.   M.C. Yoo, S.J. Yun, K. Kim, and J. Rigsbee, J. Vac. Sci. Technol. **B 11,** 1942 (1993).

4.   S.J. Yun, M.C. Yoo, and K. Kim, J. Appl. Phys. **74** , 2866 (1993).

5.   S.J. Yun, M.C. Yoo, and K. Kim, Appl. Surf. Sci. **70/71**, 536 (1993).

6.   K. Kim, M.C. Yoo, K.H. Shim, and J.T. Verdeyen, J. Vac. Sci. Technol. **B 13**, 796 (1995).

7.   M.E. Lin, B. Sverdlov, G.L. Zhou, and H. Morkoc, Appl. Phys. Lett. **62 (6)** , 3479 (1993).

8.   S. Nakamura, T. Mukai, and M. Senoh, J. Appl. Phys. **71**, 5543 (1992).

Inst. Phys. Conf. Ser. No 145: Chapter 9
Paper presented at 22nd Int. Symp. Compound Semiconductors, Cheju Island, Korea, 28 August–2 September 1995
© *1996 IOP Publishing Ltd*

Recent Progress in Optical Studies of Wurtzite GaN Grown by Metalorganic Chemical Vapor Deposition

W. Shan, T. Schmidt, X.H. Yang, and J.J. Song

Center for Laser Research and Department of Physics, Oklahoma State University, Stillwater, Oklahoma 74078, USA

B. Goldenberg

Honeywell Technology Center, Plymouth,, Minnesota 55441, USA

Abstract. We present the recent results of our spectroscopic studies on optical properties of GaN grown by metalorganic chemical vapor deposition, including the issues vital to device applications such as stimulated emission and laser action, as well as carrier relaxation dynamics. By optical pumping, stimulated emission and lasing were investigated over a wide temperature range up to 420 K. Using a picosecond streak camera, free and bound exciton emission decay times were measured. In addition, the effects of pressure on the optical interband transitions and the transitions associated with impurity/defect states were studied using diamond-anvil pressure-cell technique.

1. Introduction

GaN based wide band-gap III-V nitride semiconductors currently attract extensive attention for their potential electronic and optoelectronic device applications such as UV-blue LED's and laser diodes.[1-3] Recent demonstration of superbright high-efficient blue LED's based on nitride heterostructures by the Nichia group and others[4] and the reports on observation of optically pumped stimulated emission in GaN epifilms by a few groups[5,6] have led to much more intense interest in the development of efficient nitride UV-visible light emitters. In this report, we present recent results of our spectroscopic studies on the optical properties of GaN grown by metalorganic chemical vapor deposition (MOCVD).

2. Experimental details

The GaN samples used in this study were all nominally undoped single-crystal films grown on (0001) sapphire substrates by MOCVD. Various optical measurements were carried out on the GaN samples over a temperature range typically from 10 K to room temperature (295 K). Conventional photoluminescence (PL) spectra were measured with a cw HeCd laser (325 nm) as an excitation source and a 1-M double-grating monochromator connected to a photon-counting system. Time-resolved photoluminescence (TRPL) measurements were performed using a frequency tunable pulsed laser (2 ps pulse duration, 82MHz) as an excitation source and a streak camera (2 ps resolution), in conjunction with a 1/4-M monochromator as a detection system. Pressure-dependent PL measurements were conducted using the diamond-anvil pressure-cell technique. Stimulated emission and lasing experiments were carried out employing side-pumping geometry using a nanosecond pulsed laser (10 Hz) as the optical pumping source.

3. Results and discussions

To illustrate the quality and purity of the GaN epifilm samples used in this work, PL spectra taken from the samples are shown in Fig. 1. All GaN samples exhibit strong, predominant near band-edge exciton luminescence lines corresponding to the radiative decay of excitons at low temperatures. The inset shows a broad range of the typical PL spectrum taken at 10 K. The broadband yellow emission and a weak emission band in the blue spectral region could be observed. The strongest emission line marked by BX in Fig. 1 has a full width at half maximum (FWHM) of less than 1.0 meV at 10 K. The second one labeled FX in the figure shows a FWHM of less than 1.5 meV, indicating the high quality, as well as substantial purity of the samples. The intensity of the BX peak was found to decrease much faster than that of the FX as the temperature increased. It became hardly resolvable when the temperature was raised to above 100 K (not shown). Such variations of the luminescence intensity as a function of temperature indicate that the emission line can be attributed to the radiative recombination of excitons bound to neutral donors. The second strongest luminescence line, together with the weak emission feature on the higher energy side, can be assigned to intrinsic free-exciton emissions.

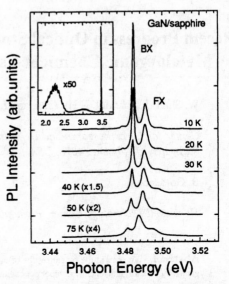

Fig. 1. Exciton luminescence spectra of a MOCVD GaN sample taken at different temperatures. The inset shows the 10 K PL spectrum of the sample over a broad spectral range.

3.1 Optically pumped stimulated emission and lasing

Strong stimulated emission and lasing at near-UV wavelengths could be observed from the GaN samples over a broad temperature range from 10 K to ~ 420 K by optical pumping with a tunable nanosecond laser. In Fig.2 we plot the emission spectra taken at 375 K for pumping power densities below and above the estimated threshold. Under low-excitation conditions, the spectrum is characterized by a very broad spontaneous emission band with the maximum position at 376 nm. With increasing pumping power densities, a new feature with a narrower linewidth appears on the low energy side of the spontaneous emission band and becomes predominant as the pumping power densities are further increased. The emission feature is found to exhibit superlinear increase in intensity and a red shift in the peak position as the pumping power increases. In addition, the longitudinal cavity modes of lasing in some samples with small cavity lengths could be clearly observed with a CCD camera. An example is given in the inset of Fig.2, where the resonant cavity length of the sample is about 200 μm thick. The resonant cavity of the samples was prepared just by cutting the large wafer into bar-like specimens with a diamond saw. The resulting sample edges are far from mirror-like facets.

The threshold pumping power was estimated to be ~500 kW/cm² at 10 K and ~800 kW/cm² at room temperature. Generally, the lasing threshold varies from sample to sample. For a given optical pumping source, the most important influence on the threshold is from the sample itself. This can be classified into two groups: one is associated with material properties, such as impurities, crystallinity, and defects; the other is related to the sample preparation such as the laser cavity length and the quality of sample edge facets. We expect that the lasing threshold in GaN can be lowered substantially by better preparing the sample edge facets, for example, with

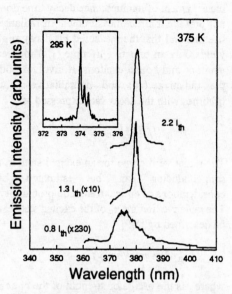

Fig. 2. Emission spectra of GaN at 375 K under different pumping power densities. The inset shows the emission spectrum taken at room temperature (295 K) exhibiting longitudinal mode fringes.

reflective coating. With better designed laser structures, such as double-heterostructures, the lasing threshold is also expected to be significantly lowered. Recently, Zubrilov *et al.* reported the observation of multipass stimulated emission with Fabry-Perot modes from GaN grown on 6H-SiC substrates.[6] In this case, the interference fringes originated from the microcavities. These cavities were formed by microcracks, which were generated during the cleaving process. As better lattice-matched substrates become available for epitaxial growth of GaN with better crsytallinity, the lasing threshold is expected to be lowered further.

We also noted that the threshold is not very sensitive to the change of sample temperature. It is known that high-temperature sensitivity of the lasing threshold usually limits the performance of a laser in high temperature operations.[7] The weak temperature dependence of the lasing threshold suggests that laser operations can be substantially extended to the high temperature range. Thus the results reported here imply that GaN based laser diodes have the potential of operating with a much higher temperature tolerance compared to conventional semiconductor laser diodes.

3.2 Radiative decay of free excitons and bound excitons

The temporal evolution of spectrally integrated exciton luminescence for both free-exciton (FX) and bound-exciton (FX) emissions observed in a GaN sample at 10 K is shown in Fig. 3. The overall time resolution of the experimental system used for conducting the measurements is less than 15 ps. The time evolution for both FX and BX luminescence is dominated by exponential decay. The lifetime of the main PL decay was found to be ~35 ps for the FX emissions and ~55 ps for the BX emissions for the GaN sample at 10 K. It should be pointed out that the

1154

measurement of luminescence decay time does not provide a direct measurement of radiative lifetime and that the measured PL decay time yields only an effective lifetime (τ_{eff}) for free excitons and bound excitons. It involves both the radiative (τ_R) and nonradiative (τ_{NR}) lifetimes with the decay rate expressed as

$$1/\tau_{eff}=1/\tau_R+1/\tau_{NR}. \qquad (1)$$

The radiative lifetime for an excited state in a semiconductor can be estimated by considerations of optical transition probability. The radiative lifetime τ_R of the excited state can be described by:[8,9]

$$\tau_R=2\pi\epsilon_0 m_0 c^3/\tilde{n}e^2\omega^2 f, \qquad (2)$$

where f is the oscillator strength of the optical transition, $\tilde{n}$ is the refractive index, and the other symbols have their usual meanings. By

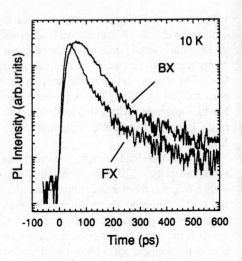

Fig. 3. Temporal variation of spectrally integrated PL for both intrinsic free-exciton and bound-exciton emissions.

using $\tilde{n}=2.4$, and $\omega=\sim5.3\times10^{15}sec^{-1}$ for GaN, one can roughly obtain $\tau_R\sim(800/f)$ ps. The radiative lifetime of bound excitons in GaN is expected to be just slightly shorter than one nanosecond if we take the upper limit with its oscillator strength as unity ($f\sim1$). The oscillator strength of free excitons calculated within the effective-mass approximation is given by $f=E_p/\pi\hbar\omega(V/a_x^3)$, where E_p is the Kane matrix element connecting Bloch states in the valence and conduction bands, V is the volume of unit cell, and a_x is the effective Bohr radius of free exciton. Our result yields $f\sim0.012$ for the free excitons in GaN using $E_p\sim18$ eV and $a_x\sim20$ Å. Thus, the calculated value of the radiative lifetime for free excitons in GaN will be in several tens of nanoseconds domain. The discrepancy between the measured values of PL decay time and the theoretical estimated radiative lifetime can be attributed to nonradiative relaxation processes in competition with the radiative channel. In the case where nonradiative decay rate is larger, the measured decay time is characteristic for the nonradiative processes in accordance with Eq.(1). This situation is typical for recombination from intrinsic states of semiconductors. The nonradiative processes such as multiphonon emission captured by deep centers, Auger effect, etc., give rise to fast relaxation of the excited carriers down to lower states from which they decay radiatively or relax nonradiatively. As a result, the measured PL decay time for a given excited state is an effective lifetime and usually much shorter than a radiative one. This has been observed in a number of semiconductor bulk materials and heterostructures with the measured free-exciton PL decay time decreasing progressively as the density of nonradiative recombination centers increased.[9] The slow rise of bound-exciton luminescence intensity compared to that of the free-exciton PL shown in Fig. 3 is one indicator of such nonradiative relaxation processes for the free excitons relaxing to the bound excitons. We found that the capture of excitons and trapping of carriers by such nonradiative centers as defects and impurities play a major role in the recombination processes responsible for the exciton population decay in the GaN samples studied in this work. The measured PL decay time was found to be directly related to the intensity of broadband emissions lying in the GaN band gap. The broad emission structure referred to as yellow emission in the literature is believed to be associated with the optical transitions between the energy levels involving impurity and/or defect states. The intensity of yellow emission is likely to be proportional to the density of defects and impurities present in samples. We found that the stronger the yellow emission, the shorter the PL decay time in a GaN sample. Fig. 4 shows a comparison of free-exciton PL decays taken from two samples under identical excitation conditions. The yellow emission intensities of the two samples are drastically different, with approximately a 100:1 ratio. The lifetime of free-exciton emission in the sample with stronger yellow emission was deduced to only 15 ps (which is the limit of our instrumental resolution). Therefore, the fast decay behavior of the PL intensity in this sample indicates that the capture of excitons and trapping of carriers at defects and impurities through nonradiative combinations appear to dominate the decay of exciton population. The process of capture is dependent on the density of defects and impurities in the GaN samples.

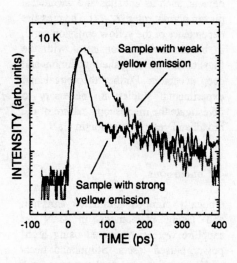

Fig. 4. Comparison of the decay times of free-exciton emission between two GaN samples with the relative intensity ratio of 100:1 for the yellow emission.

3.3 Pressure coefficient and hydrostatic deformation potential for direct Γ band gap

In Fig. 5 we plot the peak energies of exciton emission and yellow emission structures as a function of pressure. The solid lines in the figure are the least-square fits to the experimental data using the quadratic-fit function

$$E(P)=E(0)+\alpha P+\beta P^2. \tag{3}$$

The change of the intense and sharp BX transition with pressure plotted in the figure provides an unmistakable signature of the direct Γ band-gap dependence for wurtzite GaN based on the effective mass approximation. The best fits to the data yield a linear slope of 3.86×10^{-3} eV/kbar with an extremely small sublinear term of -8×10^{-7} eV/kbar^2. Similar results can be obtained from fitting the FX transition as well ($\alpha=3.9\times10^{-3}$ eV/kbar and $\beta=-1.8\times10^{-6}$ eV/kbar^2).

The application of hydrostatic pressure allows a direct estimation of the hydrostatic deformation potential for the direct Γ band gap of wurtzite GaN. The deformation potential is defined as $a=\partial E/\partial lnV$. The hydrostatic deformation potential for the direct Γ band gap of wurtzite GaN could be deduced as $a=-9.2\pm1.2$ eV.

The broad yellow emission band maximum was found to shift with pressure at a rate of 4.0×10^{-3} eV/kbar, almost the same as that of exciton emissions. This spectral structure could be commonly observed in the PL spectra of nominally undoped GaN single crystals regardless of the crystal growth techniques. More strikingly, this band was observed in samples implanted with a variety of atomic species.[10] These results have led to the general belief that the broadband emission in the yellow spectral region involves defects. Recent theoretical studies on the electronic structures of impurities and native defects in GaN have suggested that point defects, such as antisites and vacancies, play important roles.[11-14] The pressure dependence of the yellow emission band, in broad terms, is consistent with the transitions involving shallow donors and deep acceptors. Further theoretical and experimental studies are necessary to investigate the microscopic nature of the broadband yellow emission in GaN.

4. Conclusions

Optically pumped stimulated emission and lasing phenomena in GaN on sapphire were investigated using high-power pulsed lasers. Stimulated near-violet emissions were observed over a wide temperature range from 10 K to ~420 K. The dynamics of photoexcited excess carriers in high quality GaN samples were investigated. This was done

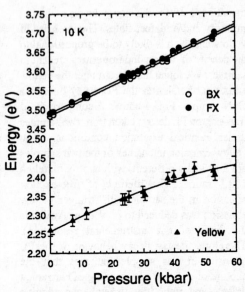

Fig. 5. Dependence of the energy positions on pressure for the various observed PL transitions in the GaN sample.

using transient luminescence spectroscopy in the picosecond regime using a streak camera in the region of near band-edge excitonic emissions. We found that the strong capture of photo-excited carriers in impurities and/or defects through nonradiative recombination processes dominates the decay of carrier population. The capture process depends on the density of impurities and defects in the GaN samples. The effects of pressure on the various optical transitions associated with both intrinsic and extrinsic processes in the GaN samples were examined by photoluminescence spectroscopy. The pressure coefficient of the GaN band gap was determined by studying the shift of exciton emission lines in GaN with applied pressure. Our results yielded the variation of the GaN band gap with pressure to be $\Delta E(P)=3.9 \times 10^{-3}P - 1.0 \times 10^{-6}P^2$ eV. The deformation potential for the direct Γ band was also deduced from the experimental results to be -9.2±1.2 eV.

References

[1]   Strite, S. and H. Morkoç, 1992, J. Vac. Sci. Technol. **B10**, 1237, and references therein.
[2]   Pankove, J.I. 1990, Mater. Res. Soc. Symp. Proc. **162**, 515, and references therein.
[3]   Morkoç, H. *et al.* 1994, J. Appl. Phys. **76**, 1363, and references therein.
[4]   See, for example, Nakamura, S. *et al.* 1994, Appl. Phys. Lett. **64**, 1687.
[5]   Yang, X.H. *et al,* 1995, Appl. Phys. Lett. **66**, 1.
[6]   Zubrilov, A.S. *et al.* 1995, Inst. Phys. Conf. Ser. No.141, 525.
[7]   Agrawal, G.P. and N.K. Dutta, 1993, in *Semiconductor Lasers*, (Van Nostrand Reinhold, New York), p.132.
[8]   Dexter, D.L. 1958, in *Solid State Physics*, ed. F. Seitz and D. Turnbull (Academic, New York), Vol.6.
[9]   't Hooft, G.W. *et al.* 1987, Phys. Rev. **B35**, 8281.
[10]   Pankove, J.I. and J.A. Hutchby, 1976, J. Appl. Phys. **47**, 5387.
[11]   Jenkins, D.W. and J.D. Dow, 1989, Phys. Rev. **B39**, 3317.
[12]   Tansley, T.L. and R.J. Egan, 1992, Phys. Rev. **B45**, 10942; 1993, Physica **B 185**, 190.
[13]   Neugebauer, J. and C.G. Van de Walle, 1994, Phys. Rev. **B50**, 8067.
[14]   Boguslawski, P. *et al.* 1995, Phys. Rev. **B51**, 17255.

Inst. Phys. Conf. Ser. No 145: Chapter 9
Paper presented at 22nd Int. Symp. Compound Semiconductors, Cheju Island, Korea, 28 August–2 September 1995
© *1996 IOP Publishing Ltd*

ZnMgSSe-based Semiconductor Lasers

M Ozawa[1], S Itoh, A Ishibashi and M Ikeda

Sony Corporation Research Center, 174 Fujitsuka-cho, Hodogaya-ku, Yokohama, 240, Japan

Abstract. Most device characteristics of II-VI lasers with ZnMgSSe cladding layers are comparable to those of III-V lasers. The lowest threshold current density we achieved for a gain-guided laser was 460 A/cm^2 under cw operation. A p-ZnTe/p-ZnSe multi-quantum well structure as a p-contact layer reduced the operating voltage to 4.7 V. The use of a GaAs buffer layer can reduce the density of pre-existing defects, resulting in a significant increase in device lifetime.

1. Introduction

ZnSe-based II-VI compounds have long been considered potential materials to realize a light-emitting device with a short wavelength light in blue/green region, because of their band structure of direct transition with an appropriate bandgap energy. Early II-VI materials failed to fulfill their promise as laser materials, primarily because of the difficulty in controlling p- and n-type conductivity using the crystal growth technology of the time.

Low-resistive p-type ZnSe was obtained in 1990 using nitrogen from a plasma source in molecular beam epitaxy[1], followed by the first demonstration of a wide-gap II-VI laser at 77 K in 1991[2]. Shortly after these achievements, many organizations confirmed laser oscillation in similar structures[3][4][5]. Although pulsed operation at room temperature [6] and continuous wave (cw) operation at cryogenic temperature were achieved, there appeared to be no chance for this ZnCdSe/ZnSe/ZnSSe structure to lase under cw mode at room temperature, because both electrical and optical confinement was poor. This poor confinement was caused by the relatively small bandgap and large refractive index of ZnSSe cladding layers. The bandgap energy and the refractive index of ZnSSe cannot be varied because of the requirement of lattice matching to the substrate.

To overcome this difficulty, Okuyama *et al* proposed the ZnMgSSe quaternary alloy for the cladding layers of a II-VI laser[7]. This alloy gives II-VI lasers a high degree of freedom in the design of the laser structure. Making use of this alloy as cladding layers, Nakayama *et al* achieved room temperature cw operation in 1993[8] for the first time,

---

[1] E-mail:mfozawa@src.sony.co.jp

which can be regarded as a milestone for practical II-VI lasers. The lasing wavelength of this laser was in the green region at 524 nm. By increasing the bandgap energy of both the cladding layers and the active layer, they shortened the lasing wavelength under cw mode to 490 nm in 1994[9]. These days the ZnCdSe/ZnSSe/ZnMgSSe strained quantum-well separate-confinement heterostructure (SCH) is widely accepted[10][11][12] and the device properties with this structure are comparable to those of III-V lasers, as is shown in section 2. In this article, we review the current status of II-VI lasers and discuss some remaining problems.

2. General characteristics of ZnMgSSe-based lasers

The physics of laser diodes significantly evolved and the device designing of III-V lasers was established during the 1970's and 80's. When the p-type doping technique and the material for cladding layers were proposed, these physics and designing could immediately be applied to II-VI lasers.

All laser structures in our laboratory are grown by molecular beam epitaxy. (001) n-GaAs is used as a substrate, the buffer layer is composed of ZnSe or ZnSe/GaAs, the cladding layers are N-doped and Cl-doped ZnMgSSe, the active region is a CdZnSe/ZnSSe SCH, the p-contact layer is a ZnSe/ZnTe multi quantum-well structure (MQW) which will be discussed in detail later, the n-type ohmic metal is In, and the p-type is AuPtPd[13]. A stripe is defined by wet etching and the deposition of an insulator. Sophisticated structures such as SCH and MQW were employed even in the early stages. Now, two years after the first cw operation at room temperature[8], most device properties of II-VI lasers are comparable to those of III-V lasers.

Table 1 summarizes the best results published in the literature for important properties as of June, 1995. The lasing wavelength can be controlled by the amount of Cd content and the quantum-well thickness in the active layer. The lasing wavelength under cw mode is in the range from 490 nm to 524 nm[9]. Under the pulsed mode, laser emission of 471 nm from a ZnSe/ZnMgSSe conventional double heterostructure (no quantum-well in the active region) has been obtained by Okuyama et al [14]. This result is interesting in terms of device physics as well as achieving a shorter wavelength. The maximum cw operating temperature is 80 °C[15]. The maximum output power is 840 mW per facet

Table 1. Some best results of II-VI lasers with ZnMgSSe cladding layers reported as of June, 1995.

Items	Condition	Best results	References
Emission wavelength	cw	490 - 524 nm	[9]
Threshold current density	cw gain-guided	460 A/cm^2	[17]
Threshold current density	cw index-guided	310 A/cm^2	[18]
Output power	cw	30 mW	[17]
Output power	pulsed	840 mW	[16]
Operating temperature	cw	80 °C	[15]

under the pulsed mode with a 50 μm wide stripe and without facet coating[16]. It has been reported[16] that this maximum power was limited by the power supply and no catastrophic optical damage (COD) was observed. This indicates that II-VI materials are resistive to COD.

Threshold current density (J_{th}) of a gain-guided laser is a good parameter for evaluating the material quality. The lowest J_{th} of a gain-guided laser under cw mode is 460 A/cm^2[17]. This value is comparable to the J_{th} of III-V lasers. Attempts have been made to reduce J_{th} by employing a index-guided structure. The lowest J_{th} for an index-guided laser under cw mode was reported by Fan et al to be 310 A/cm^2[18]. The effort to fabricate index-guided lasers with a further reduced J_{th} and a controlled beam shape will continue. Furthermore, Ozawa et al indicated that an index-guided structure is necessary to improve the dynamic properties of II-VI lasers[19].

Kondo et al investigated the physical parameters of laser devices[20] and found that the internal loss (α_i) was 4 cm^{-1}, the gain constant (β) was 4×10^{-3} cm-μm/A, the nominal current density at which the gain is zero (J_0) was 2×10^4 A/(cm^2-μm), and the internal quantum efficiency (η_{stim}) was 0.5. These values are comparable to those of III-V lasers, except J_0 and η_{stim}. J_0 of II-VI lasers is larger than that of III-V lasers because the transition probability of electrons from the conduction band to the valence band becomes smaller as the bandgap energy increases. η_{stim} was small probably because of high pre-existing defect density, which is discussed later, in the device and un-optimized growth conditions. Our recent studies show that η_{stim} becomes almost unity when the growth conditions are optimized.

3. Important issues

3.1. Operating voltage

As shown in the previous section, most characteristics of II-VI lasers are comparable to those of III-V lasers. Two important properties, however, are not so good as those of III-V lasers. One is operating voltage and the other is reliability. In the next two sections, these remaining problems are discussed.

The operating voltage (V_{op}) of the first room temperature cw laser was 17 V[8]. This high V_{op} is attributed to the difficulty in obtaining good ohmic contacts to p-ZnSe. Two fundamental problems prevent the formation of an ohmic contact - the low valence band maximum (VBM) and the difficulty in high p-type doping. The VBM of ZnSe is 6.7 eV from the vacuum level[21], which is lower than the Fermi level of any metal. Therefore, contact of any metal to p-ZnSe must be Schottky-like, not ohmic. As for doping, nitrogen doping using a plasma source cannot go far beyond the net acceptor concentration of 10^{18}cm^{-3}. This concentration is not enough to reduce the width of the depletion region so as to get a large tunneling current. Contrary to ZnSe, ZnTe is known to be p-type doped easily, and the VBM of ZnTe is 0.8 eV above that of ZnSe[22]. Making advantage of these properties of ZnTe, Hiei et al proposed the ZnSe/ZnTe multi quantum-well contact layer[23] as shown in Fig. 1. Holes are injected to p-ZnSe by sequential tunneling through a coupled quantum level formed in the ZnSe/ZnTe MQW region. The best result we achieved is V_{op} of 4.7 V in the laser device[15], as shown in Fig. 1. The applied voltage should be reduced to 3.0 - 3.5 V by further optimization, so that no voltage drop occurs in the p-contact region.

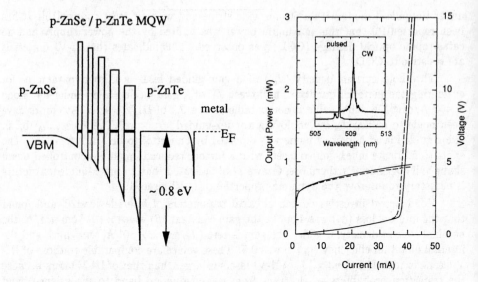

Figure 1. left : Schematic band diagram of a p-contact layer with ZnSe/ZnTe MQW. **right :** Output power and voltage vs. current characteristics of a CdZnSe/ZnSSe/ZnMgSSe SCH laser with the contact layer shown in the left. The emission spectra are also shown as an inset. The lasing wavelength is 509 nm.

3.2. Degradation

The last remaining problem is degradation. The lifetime of II-VI lasers is very short. For our early devices grown directly on a GaAs substrate, the longest lifetime under constant output power of 1 mW at room temperature was only 40 seconds[24].

DePuydt *et al* and Hua *et al* independently investigated the mechanism of degradation in CdZnSe/ZnSSe/ZnMgSSe SCH structures[25][26] and found that the stacking faults play an important role. The stacking faults nucleated at or near the II-VI/GaAs interface extend to the n-cladding layer and become sources for misfit dislocations and threading dislocations near the active layer. During laser operation, these dislocations act as nucleation sites for the formation of dislocation dipoles and eventually form dislocation networks which appear as dark regions under electroluminescence. Similar mechanism was reported by Tomiya *et al* [27], although the details of dislocation, such as Burgers vectors, are different.

Itoh *et al* studied the influence of buffer layers on the density of pre-existing defects, such as stacking faults[28][15]. With a ZnSe buffer layer, the defect density was 10^6 - $10^7 cm^{-2}$. This is the case for our 40 s-lifetime device. The density was significantly reduced to $10^5 cm^{-2}$ by growing the II-VI layers on a homoepitaxial GaAs buffer layer. With the reduced pre-existing defect density, the device lifetimes were significantly improved to the order of 10^0 hours. Further improvement of lifetime is possible by reducing the defect density to the point where there are no pre-existing defects in the vicinity of the stripe region of lasers.

4. Conclusion

Properties of ZnMgSSe lattice-matched to a GaAs substrate grown by MBE are reviewed. By employing this quaternary alloy, both the carrier confinement and the optical confinement are improved compared to the ZnSSe ternary alloy. The current status of ZnMgSSe-based II-VI lasers is summarized. Most device characteristics of II-VI lasers with ZnMgSSe cladding layers are comparable to those of III-V lasers. The physical parameters in laser devices, such as α_i, β and J_0, are also comparable. Two crucial problems, however, still remain. One is the high operating voltage. A p-ZnTe/p-ZnSe multi quantum-well structure as a p-contact layer reduced the operating voltage to 4.7 V. Further reduction is possible by optimizing the structure and doping in the MQW region. The other problem is reliability. Pre-existing defects such as stacking faults play an important role in the degradation of II-VI lasers. The use of a GaAs buffer layer can reduce the density of pre-existing defects, resulting in a significant increase in the device lifetime.

Acknowledgments

The authors would like to acknowledge T. Ohata, K. Nakano, H. Okuyama N. Nakayama, and other members of the II-VI research group at Sony Research Center for their cooperation in this work. The authors also wish to acknowledge Dr. T. Yamada and Dr. J. Seto for their encouragement during the course of this work.

References

[1] Park R M, Troffer M B, Rouleau C M, DePuydt J M and Haase M A 1990 *Appl. Phys. Lett.* **57** 2127–2129

[2] Haase M A, Qiu J, DePuydt J M and Cheng H 1991 *Appl. Phys. Lett.* **59** 1272–1274

[3] Jeon H, Ding J, Patterson W, Nurmikko A V, Xie W, Grillo D C, Kobayashi M and Gunshor R L 1991 *Appl. Phys. Lett.* **59** 3619–3621

[4] Drenten R R, Marshall T, Gaines J and Petruzzello J 1993 *Compact Blue-green Lasers, New Orleans* CTuC3

[5] Yu Z *et al* 1993 *Japan. J. Appl. Phys.* **32** 663–668

[6] Walker C T, DePuydt J M, Haase M A, Qiu J and Cheng H 1993 *Physica B* **185** 27–35

[7] Okuyama H, Nakano K, Miyajima T and Akimoto K 1991 *Japan. J. Appl. Phys.* **30** L1620–L1623

[8] Nakayama N, Itoh S, Ohata T, Nakano K, Okuyama H, Ozawa M, Ishibashi A, Ikeda M and Mori Y 1993 *Electron. Lett.* **29** 1488–1489

[9] Nakayama N, Itoh S, Okuyama H, Ozawa M, Ohata T, Nakano K, Ikeda M, Ishibashi A and Mori Y 1993 *Electron. Lett.* **29** 2194–2195

[10] Gains J M, Drenten R R, Haberern K W, Marshall T, Mensz P and Petruzzello J 1993 *Appl. Phys. Lett.* **62** 2462–2464

[11] Haase M A, Baude P F, Hagedorn M S, Qiu J, DePuydt J M, Cheng H, Guha S, Hoefler G E and Wu B J 1993 *Appl. Phys. Lett.* **63** 2315–2317

[12] Grillo D C *et al* 1993 *Appl. Phys. Lett.* **63** 2723–2725

[13] Ozawa M, Hiei F, Takasu M, Ishibashi A and Akimoto K 1994 *Appl. Phys. Lett.* **64** 1120–1122

1162

[14] Okuyama H, Kato E, Itoh S, Nakayama N, Ohata T, Ishibashi A 1995 *Appl. Phys. Lett.* **66** 656–658

[15] Itoh S *et al* 1994 *Japan. J. Appl. Phys.* **33** L938–L940

[16] Nakayama N *et al* 1994 *Technical Digest of the 5th Optoelectronics Conf.* Makuhari 15A2-6

[17] Nakayama N, Okuyama H, Kato E, Itoh S, Ozawa M, Ohata T, Nakano K, Ikeda M, Ishibashi A and Mori Y 1994 *Electron. Lett.* **30** 568–569

[18] Fan Y *et al* 1994 *J. Vac. Sci. Technol.* B **12** 2480–2483

[19] Ozawa M, Egan A and Ishibashi A 1995 *Solid State Commun.* **94** 87–91

[20] Kondo K, Ukita M, Yoshida H, Kishita Y, Okuyama H, Ito S, Ohata T, Nakano K and Ishibashi A 1994 *J. Appl. Phys.* **76** 2621–2626

[21] Vos M, Xu F, Anderson S G, Weaver J H and Cheng H, 1989 *Phys. Rev.* B **39** 10744–10752

[22] Ukita M, Hiei F, Nakano K and Ishibashi A 1995 *Appl. Phys. Lett.* **66** 209–211

[23] Hiei F, Ikeda M, Ozawa M, Miyajima T, Ishibashi A and Akimoto K 1993 *Electron. Lett.* **29** 878–879

[24] Nakayama N, Okuyama H, Kato E, Itoh S, Ozawa M, Ohata T, Nakano K, Ikeda M, Ishibashi A and Mori Y 1994 *Electron. Lett.* **30** 568–570

[25] DePyudt J M, Haase M A, Guha S, Qiu J, Cheng H, Wu B J, Hoefler G E, Meis-Haugen G, Hagedorn M S and Baude P F 1994 *J. Cryst. Growth* **138** 667–676

[26] Hua G C, Otsuka N, Grillo D C, Fan Y, Han J, Ringle M D, Gunshor R L, Hovinen M and Nurmikko A V 1994 *Appl. Phys. Lett.* **65** 1331–1333

[27] Tomiya S, Morita E, Ukita M, Okuyama H, Itoh S, Nakano K and Ishibashi A 1995 *Appl. Phys. Lett.* **66** 1208–1210

[28] Itoh S, Nakayama N, Ohata T, Ozawa M, Okuyama H, Nakano K, Ikeda M, Ishibashi A and Mori Y 1994 *Japan. J. Appl. Phys.* **33** L639–L642

Inst. Phys. Conf. Ser. No 145: Chapter 9
Paper presented at 22nd Int. Symp. Compound Semiconductors, Cheju Island, Korea, 28 August–2 September 1995

1163

Growth and characterization of II-VI structures for microcavities, distributed Bragg reflectors, and blue-green lasers

M Pessa, K. Rakennus, P. Uusimaa and A. Salokatve

Department of Physics, Tampere University of Technology, P.O. Box 692, 33101 Tampere, Finland

T Aherne, J P Doran and J Hegarty

Department of Physics, Trinity College, Dublin 2, Ireland

Abstract. This paper describes the preparation of ZnSe-based distributed Bragg reflectors (DBRs), microcavities, and edge-emitting lasers using molecular beam epitaxy in conjunction with an *in situ* growth monitoring method. MgZnSSe/ZnSSe DBR's with up to 90 % reflectance have been prepared. A monolithic λ-ZnSSe microcavity with MgZnSSe/ZnSSe DBR and ZnCdSe quantum wells have also been grown. The microcavity exhibits features which are related to the presence of resonant Fabry-Perot cavity and quantum well exciton modes. Edge-emitting MgZnSSe/ZnSSe/ZnCdSe lasers with inverted doping configuration and GaInP/-AlInP barrier reduction layers have been made. They operate at room temperature in pulsed mode, emitting at a wavelength of 520 nm.

1. Introduction

Blue-green ZnSe-based edge-emitting diode lasers grown by molecular beam epitaxy (MBE) have experienced rapid development. The key contributions to this development have been an improvement in ohmic contact to p-ZnSe using a p-ZnSe/ZnTe contacting scheme [1,2], the use of nitrogen species produced from a radio-frequency plasma source as a p-type dopant [3,4], and the use of $Mg_xZn_{1-x}S_ySe_{1-y}$ alloys as cladding layers for a pseudomorphic separate confinement heterostructure [5]. These improvements have increased the *cw* room-temperature lifetime of ZnSe lasers from a few seconds, as first demonstrated by researchers from Sony Co. [6], to the present lifetime of a few hours [7].

There has also been abundant activity in the study of ZnSe vertical cavity surface emitting lasers (VCSEL) and related layer structures. The II-VI microcavity light emitters possess some potential advantages over their III-V counterparts. Besides the fact that they operate in the blue-green spectral region, excitons in ZnCdSe quantum wells can be robust against thermal dissociation and interact more strongly with photons than in III-V quantum wells. Therefore, high-efficiency microcavity visible light emitters operating at room temperature are achievable in ZnSe-based structures. Recently, researchers from Matsushita Electric Co. reported on the first injection blue VCSEL under pulsed drive current at 77 K [8]; other blue VCSELs reported so far are optically pumped [9,10,11].

The topics discussed in this paper concern relatively complicated ZnSe-based layer structures and devices and their physical properties. First, we present the fabrication and characterization of quarter-wavelength ($\lambda/4n$) distributed Bragg reflector (DBR's) of MgZnSSe/-ZnSSe. Secondly, we integrate monolithically a ZnSSe microcavity containing ZnCdSe quantum wells with a MgZnSSe/ZnSSe DBR mirror. Thirdly, in pursuit of finding new ways of preparing edge-emitting ZnSe-based laser diodes and to overcome possible long-term reliability problems of the p-ZnSe/ZnTe contact we consider an "inverted" MgZnSSe/-ZnSSe/ZnCdSe structure having n-on-p doping configuration and interface potential barrier reduction layers.

2. II-VI semiconductor distributed Bragg mirror and microcavity

2.1. Growth of Bragg mirror

Highly reflecting mirrors for microcavities can be fabricated monolithically with Bragg stacks or by hybrid dielectric or metallic layers. Many approaches have been used for the latter.

We have studied growth of $Mg_xZn_{1-x}S_ySe_{1-y}/ZnS_ySe_{1-y}$ DBR's [12,13]. These mirrors are attractive because, in principle, the whole layer structure including the cavity can be prepared by one single growth. A drawback of the monolithic approach is the requirement of extreme stability and close control of MBE growth for many hours. A general feature of MgZnSSe/ZnSSe DBR is the small refractive index step Δn (≤ 0.2) between MgZnSSe and ZnSSe at moderate composition levels. Variations in the band-gap $E_g^{MgZnSSe}(x)$ during long layer growth, and the fact that Δn is small, place serious challenges on the *in situ* epitaxy of MgZnSSe/ZnSSe.

We have studied MgZnSSe layers with different lattice strain and thickness in order to determine to what extent lattice mismatch can be tolerated for pseudomorphic growth of MgZnSSe/ZnSSe. The composition and structural quality of MgZnSSe were determined from room-temperature photoluminescence (PL), X-ray diffraction rocking curves (XRD) and energy dispersive X-ray data. The band-gap wavelength λ of MgZnSSe was taken from the PL measurements and was translated into the alloy composition $(x;y)$ over a range from $(x;y) = (0;0)$ to $(0.3;0.3)$ which is of prime interest to the blue-green light emitters. For practical reasons, lattice mismatch was expressed in terms of $\Delta\Theta$, the peak separation in the XRDs of GaAs and MgZnSSe. The λ *versus* $\Delta\Theta$ relationship obtained is shown in Figure 1. It appears that lattice mismatch must remain within $|\Delta\Theta| < 500$ arc sec for good epitaxial growth. If $\Delta\Theta$ is larger, $|\Delta\Theta| > 500$ arc sec, a critical layer thickness may be exceeded.

Several *in situ* methods for monitoring layer growth in MBE have been developed [14,15,16,17]. We have applied a laser reflectometry method [17] using an Ar-ion 488-nm laser light to control growth of MgZnSSe/ZnSSe in real time. Desired layer thickness can simply be achieved by switching on and off the shutters of the effusion cells of MBE at the extrema of reflected laser beam intensity. Typical reflectance spectra taken at normal incidence are displayed in Fig. 2 for two Bragg mirrors labeled DBR10 and DBR20. These samples consist of 10 and 20 pairs of layers, respectively, having an approximate $Mg_{0.15}Zn_{0.85}S_{0.18}Se_{0.82}/ZnS_{0.07}Se_{0.93}$ composition. The stopbands are centered at $\lambda \approx 473$ nm. The maximum reflectance is 86 % for DBR20 and 70 % for DBR10.

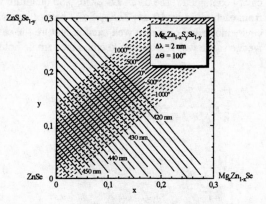

Figure 1. Lattice mismatch $\Delta\Theta$ and band-gap wavelength of MgZnSSe plotted against x and y. The layers are grown on ZnSSe lattice-matched to GaAs. Solid $\Delta\Theta$ lines are for perfect pseudomorphic growth of the alloy at any layer thickness. Dashed $\Delta\Theta$ lines are for growth where excessive lattice strain may generate misfit disloca-tions, depending on layer thickness.

Fig. 2 also shows the theoretical maximum reflectance for these Bragg mirrors at the incident light wavelength $\lambda = 488$ nm, as calculated by the transfer matrix method. Assuming $\Delta n \approx 0.2$ ($0.16 < x < 0.17$) for a stack of 30 pairs of MgZnSSe/ZnSSe one obtains a theoretical re-flectance greater than 99 %, which meets the general requirements of standard plane paral-lel resonator geometry to overcome small gain-length in the microcavity. In our case, theo-retically, $\Delta n = n^{ZnSSe} - n^{MgZnSSe} \approx 0.18$ for which the reflectance should be about 97 % for DBR20 and 90 % for DBR10. The differences between the measured and calculated reflec-tances can be assigned partly to an uncertainty in Δn and also to a residual absorption and scattering at interfaces and diffraction losses, due to mirror roughness.

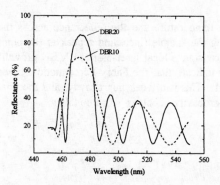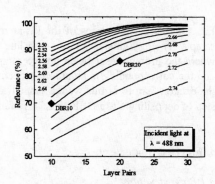

Figure 2. Left panel: Reflectance spectra of DBR10 (dashed curve) and DBR20 (solid curve). Right panel: Theoretical reflectance of $Mg_xZn_{1-x}S_ySe_{1-y}/ZnS_{0.06}Se_{0.94}$ DBR's for incident 488-nm light. Experimental reflectance of DBR10 and DBR20 is also shown. Refractive index of MgZnSSe is varied from 2.50 to 2.74.

2.2. Growth of monolithically integrated microcavity

It is possible to grow a monolithically integrated ZnSSe microcavity sandwiched between two MgZnSSe/ZnSSe DBR mirrors. However, such a structure has yet to be made. We have approached this objective by growing a sample (DBR24C) which contained 24 pairs of $\lambda/4n$-$Mg_{0.15}Zn_{0.85}S_{0.18}Se_{0.82}/ZnS_{0.07}Se_{0.93}$ layers with calculated reflectance of 90 % in air (75 % in the cavity). On top of this mirror a λ/n-$ZnS_{0.07}Se_{0.93}$ cavity (174 nm) was grown. This cavity contained three 6-nm $Zn_{0.90}Cd_{0.10}Se$ quantum wells placed in its center and separated from each other by 10-nm $ZnS_{0.07}Se_{0.93}$ barrier layers. To ensure overlap of the Fabry-Perot cavity mode and quantum well exciton feature the cavity was made slightly wedge-like. The wedge shape was realized by keeping the sample holder stationary during layer growth.

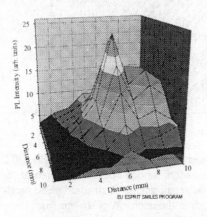 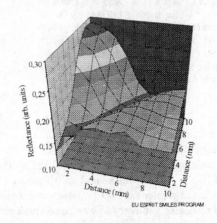

Figure 3. Left panel: (a) Integrated photoluminescence emission intensity over sample DBR24C at room temperature. Enhanced intensity in a central region of the sample indicates an occurrence of overlap of the F-P cavity mode and quantum well exciton feature. Right panel: (b) Reflectance map of DBR24C at room temparature. Minimum reflectance at resonance is reduced on a line across the sample.

Figs. 3a and b show maps of the PL at room temperature and the reflectance across the sample. The high-intensity ridge on a line along the sample (including a peak at the point where the probing laser beam hit the sample causing a local laser-assisted MBE growth) and the corresponding minimum in reflectivity indicate that the Fabry-Perot mode crosses the exciton feature in this wedge shaped sample. This is a week microcavity, and the final step of depositing a top DBR to achieve full microcavity effects is yet to be taken.

3. Inverted diode laser

3.1 Band structure

A blue-green ZnSe laser with n-on-p doping configuration has an advantage over the "usual" laser with p-on-n configuration in that this device has stable, low resistance contacts that can easily be made to both sides of the p/n-junction. Its disadvantage is the presence of a large valence band discontinuity ΔE_v (≥ 1 eV) at the GaAs/ZnSe interface through which the holes from p-GaAs cannot be readily injected into the quantum wells. The problem of large ΔE_v can be alleviated by growing heavily doped p-type wide-gap III-V barrier reduction layers in between the GaAs buffer layer and the ZnSe. Such III-V/II-VI material systems have remained largely uninvestigated, due to technical difficulties associated with their preparation. The layer structure is quite complicated and two MBE reactors are needed to grow it. Because the III-V's with very large band-gaps contain phosphorus one of the reactors should be able to alloy phosphorus with the binaries. This is not usually possible by MBE. In our experiments, we applied either gas-source MBE (with PH$_3$) or all-solid-source MBE equipped with a solid-phosphorus valved cracker cell.

3.2. Laser with n-on-p configuration

Figure 4 shows schematically a separate confinement heterostructures (SCH) with the inverted doping order and GaInP/AlInP barrier reduction layers placed on a p-GaAs substrate. Doping concentrations are N_A-$N_B \approx 10^{17}$ cm^{-3} for MgZnSSe and $\approx 10^{18}$ cm^{-3} for ZnSe.

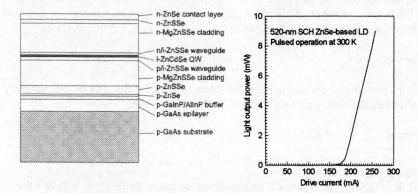

Figure 4. Left panel: Schematic layer structure of ZnSe-based SCH single-quantum-well laser prepared on p-type GaAs. Right panel: Light output power *versus* pulsed drive current of the laser at room temperature.

The ZnSSe waveguides are doped close to the quantum wells; only a region of 10 nm was left undoped on either side of the structure. The lasers were processed into 13-μm-wide oxide stripe structures and cleaved into 800 to 1200-μm bars. The bars were cut into 500-μm wide chips; the mirror facets were left uncoated. The chips were mounted junction-side up on heat sinks. The lasers were characterized at room temperature.

The lasers operated at room temperature in pulsed mode, 1μs pulse duration at a repetition rate of 1 kHz, emitting at a wavelength of 520 nm for a few minutes before they failed (Fig. 4). To the best of our knowledge, these are the first inverted ZnSe lasers that operate at room temperature. The threshold current density is between 1.2 and 1.5 kA/cm^2, varying from one device to another. The threshold voltage is high, over 20 V, despite of the use of barrier reduction layers. The reason for this high threshold voltage is unknown to us at the moment, but we suspect that the main problem is a large voltage drop at the interface between AlInP and ZnSe. Fortunately, this is not an inherent problem of the interface; it can be alleviated by properly preparing the top III-V layer before the growth of the ZnSe layer.

4. Conclusions

The MBE method in conjunction with an *in situ* growth monitoring technique has made it possible to grow complex layer structures of ZnSe-derived semiconductors. With this combination of methods MgZnSSe/ZnSSe distributed Bragg reflectors with reasonable reflectance properties and monolithically integrated ZnSSe microcavities including quantum wells can be prepared by one single growth. This is encouraging from the point of view of vertical cavity surface emitting devices which may be grown, in a straightforward fashion, to a high structural perfection and accuracy. We have also studied edge-emitting lasers. A novel approach of preparing an "inverted" SCH laser structure with barrier reduction layers has been demonstrated to combat the problems related to the preparation and operational reliability of the ohmic contacts to *p*-type ZnSe/ZnTe. The inverted laser exhibit no ohmic contact problem. The laser operated at room temperature in pulsed mode. However, *cw* operation was not possible, likely due to a large voltage drop in the III-V/ZnSe interface region. We believe, however, that a much better performance can be achieved for this kind of laser in the near future when more experience is gained in preparing the interfaces.

Acknowledgements

This work was carried out within *Epimater-Matra #4702-9/9* financed by the Academy of Finland and *ESPRIT Smiles #8447* financed by European Union.

References

[1] Fan Y, Han J, He L, Gunshor R L, Hagerott M, Jeon H, Nurmikko A V, Hua G C and Otsuka N 1992 *Appl. Phys. Lett.* **61** 3160

[2] Hiei F, Ikeda M, Ozawa M, Miyajima T, Ishibashi A and Akimoto K *Electron. Lett.* 1993 **29** 878

[3] Park R M, Troffer M B, Rouleau C M, DePuydt J M and Haase M A 1990 *Appl. Phys. Lett.* **57** 2127

[4] Ohkawa K, Karasawa T and Mitsuyu T 1991 *Jpn. J. Appl. Phys.* **30** L152

[5] Okuyama H, Nakano K, Miyajima T, and Akimoto K 1991 *Jpn. J. Appl. Phys.* **30** L1620

[6] Nakayama N, Itoh M, Ohata T, Nakano K, Okuyama H, Ozawa M, Ishibashi A, Ikeda M and Mori Y 1991 *Electron Lett.* **29** 1488

[7] Ishibashi A 1995 *IEEE J. Selected Topics in Quantum Elecetron.* **1** 741

[8] *Laser Focus World* July 1995 13, and Yoshii S, Yokogawa T, Tsujimura A, Sasai Y and Merz J L 1995 (unpublished, private communication)

[9] Honda T, Yanashima K, Yoshino J, Kukimoto H, Koyama F and Iga K 1994 *Jpn. J. Appl. Phys.* **33** 3960

[10] Jeon H, Kozlov V, Kelkar P Nurmikko A V, Grillo D C, Han J, Ringle M and Gunshor R L 1995 *Electron. Lett.* **31** 106

[11] Floyd P D, Yokogawa T, Merz J L, Luo H, Furdyna J K and Yamada Y 1995, *Appl. Phys. Lett.* **66** 2929

[12] Rakennus K, Uusimaa P, Näppi J, Salokatve A, Pessa M, Aherne T, Doran J P, O'Gorman J and Hegarty J 1996 *J. Crystl Growth* (to be published)

[13] P Uusimaa, K Rakennus, A Salokatve, M Pessa, T Aherne, J P Doran, J O'Gorman and J Hegarty 1995 *Appl. Phys. Lett.* (to be published)

[14] Chalmers S A, Killeen K P and Jones E D 1994 *Appl. Phys. Lett.* **65** 4

[15] Quinn W E, Aspnes D E, Brasil M J S P, Pudensi M A A, Schwarz S A, Tamargo M C, Gregory S and Nahori R E 1992 *J. Vac. Sci. Technol. B* **10** 759

[16] Grothe H and Boebel F G 1993 *J. Cryst. Growth* **127** 1010

[17] Raffle Y, Kuszelewicz R, Azoulay R, Le Roux G, Michel J C, Dugrand L and Toussaere E 1993 *Appl. Phys. Lett* **63** 3479

Inst. Phys. Conf. Ser. No 145: Chapter 9
Paper presented at 22nd Int. Symp. Compound Semiconductors, Cheju Island, Korea, 28 August–2 September 1995
© *1996 IOP Publishing Ltd*

II-VI blue/green laser diodes on ZnSe substrates

Y.S. Park
Office of Naval Research, Arlington, VA 22217

Z. Yu, C. Boney, W.H. Rowland, Jr., W.C. Hughes, J.W. Cook, Jr., and J.F. Schetzina
Department of Physics, North Carolina State University, Raleigh, NC 27695-8202

Gene Cantwell and William C. Harsch
Eagle-Picher Research Laboratories, Miami, OK 74354

Abstract. This paper reports the first blue/green laser diodes grown on ZnSe substrates. The laser structure employed is a p-on-n separate confinement heterostructure (SCH) consisting of 0.8 μm thick ZnMgSSe cladding layers lattice-matched to ZnSe, 0.1 μm thick ZnSe light guiding layers, and a single 60-200 Å thick ZnCdSe quantum well. Green laser emission (507-517 nm; 2.443-2.394 eV) was observed at temperatures from 77-220 K using cw excitation at 77K and pulsed excitation (50 ns; 10^{-1} - 10^{-4} duty cycle) at higher temperatures. Blue laser diodes with outputs at 485 nm (2.553 eV) at 77K have also been fabricated and tested.

1. Introduction

Historically, ZnSe and related II-VI alloys have been grown almost exclusively on GaAs substrates. This is because high-quality GaAs substrates are readily available from commercial sources. In addition, GaAs is closely lattice-matched (to within 0.27%) to ZnSe at room temperature and its pre-growth surface preparation procedure has been well established.

By adding approximately 6% of sulfur to ZnSe to form the ternary alloy ZnSSe, an exact lattice-match to GaAs at typical growth temperatures (240-300 °C) can be achieved. This alloy remains pseudomorphic with GaAs when cooled to room temperature. However, due to the difference in thermal expansion coefficients between the GaAs substrate and ZnSSe epilayer, a tensile strain at the hetero-interface exists at room temperature even when the ZnSSe epilayer is exactly lattice-matched to the GaAs substrate at the growth temperature [1,2]. This interfacial strain may contribute to the generation of stacking faults and dislocations that provide the seed for the production and growth of dark-line-defects. Stacking faults and the dark-line-defects which they seed are believed to be the principal factors which presently limit the useful lifetime of II-VI light emitting diodes and laser diodes grown on GaAs.

In an effort to overcome the above difficulties, ZnSe substrates are being employed for the development of ZnSe-based light-emitting diodes (LEDs) and laser diodes. Significant improvement in the growth of bulk single-crystal ZnSe crystals has been achieved at Eagle-Picher Laboratories using a seeded physical vapor transport (SPVT™) growth technique [3,4]. In this paper, these recent advances in bulk ZnSe crystal growth are discussed, including methods that have been developed for the growth of conducting n-type ZnSe crystals. The successful n-type doping of bulk ZnSe crystals has led us to demonstrations of high-brightness green LEDs using a standard vertical-carrier-transport device structure [5]. Most recently, the world's first blue/green laser diodes on ZnSe have been synthesized and tested. The successful demonstration of laser diodes on ZnSe opens the door to a new (homoepitaxial) approach to develop a variety of ZnSe-based II-VI optoelectronic devices using conducting ZnSe substrates rather than GaAs substrates that have generally been used to date.

2. Experimental details

2.1. Growth of ZnSe bulk crystals

Blue/green laser diode structures were grown on (100) ZnSe substrates produced at Eagle-Picher Research Laboratories by the SPVT^{TM} process. In this technique, a polycrystalline charge is pre-synthesized by reacting vapors of high purity Zn and Se to form ZnSe powder. This ZnSe powder is placed in a quartz ampoule in the hot zone of a crystal growth furnace while seed crystals are placed at opposite ends of the ampoule and maintained at a lower temperature than the charge. The temperature of the polycrystalline charge was maintained at 1200 °C for the ZnSe crystals prepared in this study. The ends of the ampoule containing the seed is usually ~50 °C lower than the charge, as estimated from temperature profiles of the furnace. The temperature of the growth charge is sufficient to establish the vapor pressure necessary to sublime the charge and deposit ZnSe on the seed at the cooler end of the ampoule. By using a single-crystal seed of 50 mm diameter and a straight walled quartz ampoule, a ZnSe ingot can be grown as a single-crystal approximately 50 mm in diameter by 20-40 mm in length. The ZnSe crystals grown by the SPVT^{TM} technique are free from twins and low angle grain boundaries -- defects typically found in ZnSe crystals grown by other methods.

The SPVT^{TM} technique generally produces electrically-insulating ZnSe crystals. For many device applications, however, a conducting substrate is preferred. As a consequence, growth of conducting n-type ZnSe bulk crystals was attempted by systematically employing various potential n-type dopants during the SPVT^{TM} crystal growth process. Hall effect experiments were used to study the doping level achieved in SPVT^{TM} crystals of ZnSe grown using different dopants and crystal growth parameters.

2.2. Growth of laser diode structures by molecular beam epitaxy

Laser diode structures were grown by molecular beam epitaxy (MBE) at North Carolina State University (NCSU) using the conducting ZnSe substrates described above. The laser diode structure employed is a p-on-n separate confinement heterostructure (SCH) as shown in Figure 1. It consists of ~0.8 μm thick ZnMgSSe cladding layers (E_g~3.0 eV) lattice-matched to ZnSe,

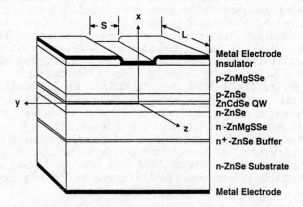

Figure 1. Schematic diagram of laser diode heterostructure on conducting ZnSe substrate.

~0.1μm thick ZnSe light guiding layers, and a single 60-200Å thick pseudomorphically-strained ZnCdSe quantum well. The laser structures were grown at 280 °C using $ZnCl_2$ and plasma-nitrogen for n-type and p-type dopants, respectively. Thin epitaxial layers of p-type ZnSe/ZnTeSe followed by undoped HgSe were deposited onto the top p-type ZnMgSSe layer of the laser structure to improve the p-type contact.

Gain-guided laser devices were fabricated by preparing 500-1000 μm long cleaved-cavity resonator structures with a 10 μm wide Au stripe electrode as the top p-type electrical contact using standard photolithography, metallization, and lift-off techniques [3]. The cleaved facets of the devices were uncoated.

3. Results and discussion

3.1. N-type substitutional doping of bulk ZnSe

By employing a proprietary aluminum-based dopant during the SPVT™ crystal growth process, it was found that controlled substitution doping of ZnSe is possible. Prior to these experiments, conducting bulk crystals of high-quality ZnSe had never been prepared by any technique. Figure 2 shows plots of the carrier concentration versus temperature obtained from Hall effect studies for a series of n-type substitutionally-doped ZnSe crystals prepared using four different dopant fluxes. It is seen that n-type doping up to nearly 10^{18} carriers/cm^3 at room temperature has been achieved using the SPVT™ crystal growth process. The ZnSe crystal doped to 1×10^{17} carriers/cm^3 displays a carrier mobility of about 460 cm^2/V-s. This is

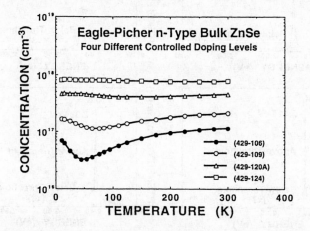

Figure 2. Electron concentration versus temperature for four different n-type ZnSe samples. The data illustrate four different controlled doping levels for different ZnSe crystals.

comparable to the highest mobility ever achieved in n-type ZnSe films grown by MBE [8]. The mobility of the SPVT™ doped crystal increase as the temperature decreases reaching a maximum value of about 800 cm^2/V-s at 125K before decreasing at lower temperatures. As the doping level increases, the carrier mobility progressively decreases but is still impressive in value (~250 cm^2/V-s at 300 K) at the highest doping level achieved (8×10^{17} carriers/cm^3). Note from Figure 2 that for this highest doping level, the carrier concentration is essentially independent of temperature suggesting that a degenerate doping level has been achieved, consistent with theoretical considerations. The data shown in Figure 2 represent an important

1172

achievement in the historical development of wide bandgap II-VI materials -- the demonstration of reproducible and controlled substitutional doping of bulk ZnSe.

3.2. Properties of blue/green laser diodes on ZnSe substrates

Green laser emission spectra (507-517 nm; 2.443-2.394 eV) were recorded using cw electrical excitation at 77K and pulsed excitation (50 ns; 10^{-1}-10^{-4} duty cycle) at higher temperatures. Figure 3 shows emission spectra below and above the lasing threshold for a green laser diode having a cavity length L = 545 μm. Spectra are shown for temperatures of 77K, 115K, 150K, and 220K, respectively. Laser emission near room temperature was also briefly observed but not recorded. Threshold currents ranged from 8.7 mA (77K) to 30.9 mA (220K) for devices having L = 545 μm. For devices with longer cavities, threshold currents were observed to be higher. This is illustrated by the light output versus current (L-I) plot shown in Figure 4 for a laser diode having L = 870 μm. In this case, the threshold current for laser emission at 77K is 22.9 mA as is evident from the data shown in the figure. Threshold voltages for ZnMgSSe-based laser structures that included HgSe/ZnTeSe contact layers to the upper p-type layer of the heterostructures were ≥ 9.0 V.

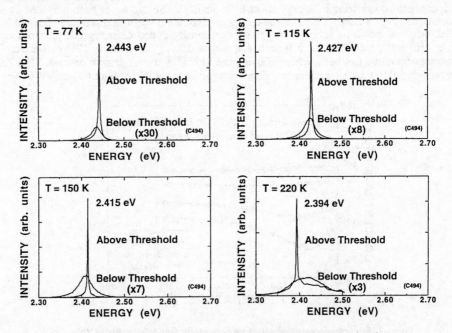

Figure 3. Light emission characteristics of green laser diode on ZnSe operating at temperatures ranging from 77K to 220 K as shown.

Figure 5 shows a high-resolution optical spectrum of a green laser diode on ZnSe operating above threshold at 77K which shows finely spaced cavity modes. The observed mode spacings correspond to an index of refraction of n = 3.83 in the active region of the device under current injection. This is comparable to refractive indices reported for blue/green lasers grown on GaAs substrates [6].

Blue laser diodes with outputs at 485 nm (2.553 eV) at 77K have also been fabricated using the same basic SCH structure, but with less Cd in the ZnCdSe quantum well. An emission spectrum for this type of device is shown in Figure 6. The threshold current for these initial blue-light-emitting devices (L=620 μm) was found to increase to 230 mA at 77K due in part, we believe, to the decrease in carrier confinement associated with the active blue-light-emitting ZnCdSe quantum well.

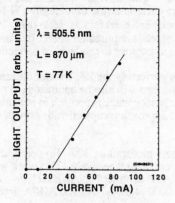

$$\Delta\lambda = \frac{\lambda^2}{2L\left(n - \lambda\frac{dn}{d\lambda}\right)}$$

Figure 4. Light output versus current (L-I) characteristics for green laser diode operating at 77K. The cavity length of this particular device is 870 μm.

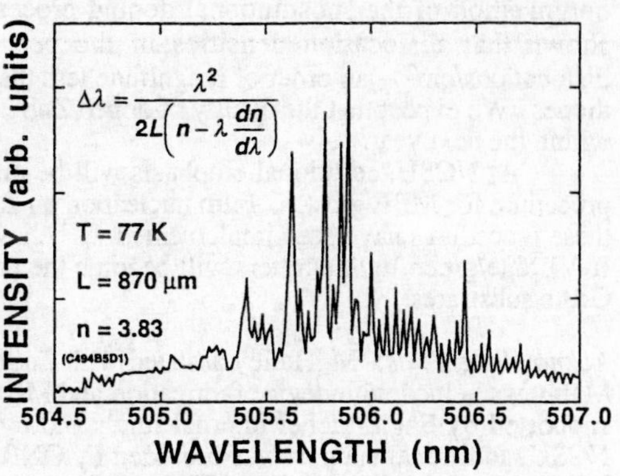

Figure 5. High-resolution optical spectrum of green laser diode at 77K.

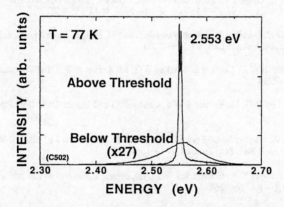

Figure 6. Light emission characteristics of blue laser diode on ZnSe operating at 77K.

4. Summary and Conclusions

In summary, we are reporting the development of device-grade, conducting n-type ZnSe substrates with which the world's first blue-green laser diodes have been successfully synthesized, fabricated, and tested. This work provides the basis for a homoepitaxial approach to II-VI devices in contrast to the use of GaAs substrates.

The key remaining issue that must be addressed before II-VI blue/green light emitters become commercially viable is that of device degradation. Dark-line defects which originate at twins and other imperfections at the epitaxial growth surface are responsible for the rapid degradation of II-VI devices that has been observed in studies throughout the world. Our plan is to circumvent this problem by means of a homoepitaxial growth approach using conducting ZnSe substrates. ZnSe substrate quality at Eagle-Picher Research Laboratories will be improved through better control of the SPVT^{TM} crystal growth technique and through optimization of the substitutional doping process. X-ray transmission topography [5] has shown that dislocation densities in the best undoped ZnSe wafers is less than 10^3 dislocations/cm^2 -- an order of magnitude less than that which is required for defect free laser diodes. We expect that the quality of doped ZnSe crystals to improve to this level of perfection within the next year.

At NCSU, additional emphasis will be placed on perfecting the ZnSe wafer cleaning procedure for MBE growth. Film nucleation on ZnSe substrates will also be optimized. When these procedures have been implemented we believe that a homoepitaxial approach to preparing II-VI blue/green light emitters will become the preferred approach compared with growth on GaAs substrates.

Acknowledgments: M. Hailey assisted with ZnSe substrate preparation. K.A. Bowers and J. Matthews helped with device fabrication and MBE maintenance, respectively. This work was supported by Eagle-Picher internal funds and by NIST ATP contract 70NANB3H1374. At NCSU, additional support was provided by ONR grant N00014-92-J-1644 and ARPA grants N00014-92-J-1893 and DAAL003-91-G-0103.

References

[1] Ren J, Eason D B, Lansari Y, Yu Z, Bowers K A, Boney C, Sneed B, Cook J W Jr., Schetzina J F , Koch M W and Wicks G W 1993 J. Vac. Sci. Technol. **B 11** 955-957

[2] Xie W, Grillo D C, Gunshor R L, Kobayashi M, Jeon H, Ding J, Nurmikko A V, Hua G C and Otsuka N 1992 Appl. Phys. Lett. **60** 1999-2001

[3] Cantwell G, Harsch W C, Cotal H L, Markey B G, McKeever S W S and Thomas S E 1992 J. Appl. Phys. **71** 2931-2934

[4] Cotal H L, Markey B G, McKeever S W S, Cantwell G and Harsch W C 1993 Physica **B 185** 103-106

[5] Eason D B, Yu Z, Hughes W C, Boney C, Cook J W Jr., Schetzina J F, Black D R, Cantwell G and Harsch W C 1995 J. Vac. Sci. Technol. **B 13** 1566-1570

[6] Yu Z, Ren, J, Bowers K A, Gossett K J, Boney, Lansari, Y, Cook, J W, Jr. and Schetzina, J F 1992 Appl. Phys. Lett. **61** 1266-1268

[7] Eason D B, Yu Z, Hughes W C, Roland W H, Boney C, Cook J W Jr., Schetzina J F, Cantwell G and Harsch W C 1995 Appl. Phys. Lett. **66** 115-117

[8] Hwang S, Ren J, Bowers K A, Cook J W, Jr. and Schetzina J F, 1990 Mat. Res. Soc. Symp. Proc. **161**, 133-136

Inst. Phys. Conf. Ser. No 145: Chapter 10
Paper presented at 22nd Int. Symp. Compound Semiconductors, Cheju Island, Korea, 28 August–2 September 1995
© 1996 IOP Publishing Ltd

Device modeling from first principles at the atomic level†

Yia-Chung Chang

Department of Physics and Materials Research Laboratory
University of Illinois at Urbana-Champaign, Urbana, Illinois 61801,USA

Abstract. We demonstrate that device modeling can be done with the use of planar Wannier functions directly constructed from first-principles. Example calculations for GaAs/AlAs heterostructures are illustrated. For a small-size superlattice, it is shown that the results obtained from using the planar Wannier functions are essentially identical to those obtained with the fully convergent plane-wave basis.

1. Introduction

The miniaturization of electronic devices and the development of quantum device structures have reached a point where continuing advances would require improved understanding of the electronic properties of the quantum structure down to the microscopic details. To calculate the electronic properties of a large-scale heterostructure, it would be desirable to use a small set of localized basis functions which can properly describe the interface charge redistribution and at the same time maintain the accuracy needed for device modeling. In this paper, we introduce the "planar Wannier functions" which are directly constructed from first-principles calculations based on the products of two-dimensional plane waves (in x-y plane) and localized functions in the third (z) direction. With the use of these planar Wannier functions, calculation of electronic structures of heterostructures can be made very efficient, while maintain the desired accuracy. We will show that for each in-plane wave vector, $k_{||}$ only eight planar Wannier functions per atomic layer are needed to reproduce the bulk band structure accurately throughout the entire Brillouin zone. With the $k_{||}$-dependent interaction parameters between planar Wannier functions calculated and stored, we can essentially convert a pseudopotential calculation into a simple tight-binding calculation. Example calculation for GaAs/AlAs superlattices within pseudopotential formalism is given. For simplicity, we demonstrate the idea by using empirical local pseudopotentials. The method can be readily extended to self consistent calculations with *ab initio* atomic pseudopotentials.

2. The planar basis method

Recently, we introduced a new planar basis which consists of products of two-dimensional (2D) plane waves (in the x-y plane) and one-dimensional (1D) localized functions along the growth (z) direction for first-principles electronic structure calculations.[1] This basis is particularly suited for describing electronic states of heterostructures including surfaces, interfaces, and superlattices. The new basis has the advantage over the conventional basis in that the layer-like local geometry which appears in surfaces is naturally built in. This would not only make the computation more efficient but also lead to better analysis of interface properties in terms of localized functions, such as layer-resolved local density of states.

For a heterostructure, the in-plane wave vector $k_\parallel$ is conserved. A planar basis function associated with $k_\parallel$, which satisfies the Bloch theorem in the plane and is localized along the z axis, is defined as

$$< r|\nu\alpha, k_\parallel + G_\parallel; z_a > = \frac{1}{\sqrt{A}} e^{i(k_\parallel + G_\parallel) \cdot \rho} f_{\nu\alpha}(z - z_a), \qquad (1)$$

where $G_\parallel$ denotes an in-plane reciprocal lattice vector, ρ is the projection of r in the x-y plane, and A denotes the surface area of the sample. z_a can be either on an atomic site or interstitial site. $f_{\nu\alpha}(z)$ is a localized function which is chosen to have the gaussian form in the present calculation $f_{\nu\alpha}(z) = z^\nu e^{-\alpha z^2}$, where $\nu = 0, 1$ for even and odd functions of z. A more rigorous chose of $f_{\nu\alpha}(z)$ would be one-dimensional wavelets[2], which would satisfy the orthogonality and completeness requirements.

Due to the many different possible reconstructions of semiconductor surfaces, realistic calculations are usually performed with insufficient number of basis functions or with a slab (or supercell) size too small to decouple two end surfaces. With the use of planar Wannier functions for the interior of the slab, while keeping the full set of planar basis functions for the surface layer, one can afford to deal with a thicker slab and much larger energy cutoff. For complicated surfaces such as GaAs (001) (2 × 4) reconstruction[3] or Si (111) (7 × 7) reconstruction[4], such an approach is likely to provide both the efficiency and accuracy needed to fully understand the surface electronic properties.

We have successfully applied the planar-basis method to study the electronic structures and work function of Si (001) 2 × 1 surface with symmetric dimer reconstruction with and without hydrogen passivation.[1] We also calculated the hydrogen dissociation energy and found a pathway for desorption with almost no activation barrier.[5] The result is in accord with previous experimental findings.[6-8] We also studied the Si (001) 2 × 1 surface with As or Sb overlayer.[9] The calculated dispersion of surface states for As/Si(001) is in good agreement with the angle-resolved photoemission measurements.

3. Construction of generalized planar Wannier functions

The generalized Wannier functions are defined as a small number of localized orthonormal functions which form a complete basis for the few low-lying bands of interest. Namely, the exact Bloch states for the few bands can be obtained by linear combinations of these generalized Wannier states. The planar Wannier functions can be defined similarly except they are related to Bloch states of a fixed value of $k_\parallel$. For different $k_\parallel$, different planar Wannier functions are used as the basis for the expansion of the Bloch states. Thus, a planar Wannier function associated with band n and $k_\parallel$ can be formally defined as $< r|W_{n,k_\parallel}(R_z) > = \int dk_z e^{-ik_z R_z} \psi_{n,k}(r)$, where $\psi_{n,k}(r)$ is a Bloch state associated with band n and wave vector $\mathbf{k}$. The planar Wannier functions so defined are usually not well localized. One can construct generalized planar Wannier functions which are some proper linear combinations of $W_{n,\mathbf{k}_\parallel}$ for all bands of interest such that they are well localized. Here, we introduce a simple method to generate localized orthonormal functions which are excellent approximations to the generalized planar Wannier functions. First, we find the expansion coefficients for the Bloch states at $k_z = 0$ and $k_z = k_m$ (the zone boundary) in terms of the orthogonalized planar basis functions for each fixed value of $\mathbf{k}_\parallel$. For the (001) orientation of an f.c.c. lattice considered here, we have $k_m = 2\pi/a$, independent of $\mathbf{k}_\parallel$. Let these expansion coefficients associated with band n be $C_{n,\mathbf{k}_\parallel}^{(0)}(\nu, \alpha, \sigma; \mathbf{G}_\parallel)$ and $C_{n,\mathbf{k}_\parallel}^{(1)}(\nu, \alpha, \sigma; \mathbf{G}_\parallel)$. We then define

$$< \mathbf{r}|W_{n,\mathbf{k}_\parallel}^{(\lambda)}(R_z) >= \sum_{\nu,\alpha,\sigma,\mathbf{G}_\parallel} C_{n,\mathbf{k}_\parallel}^{(\lambda)}(\nu, \alpha, \sigma; \mathbf{G}_\parallel) f'_{\nu\alpha}(z - R_z - z_\sigma) e^{i(\mathbf{k}_\parallel + \mathbf{G}_\parallel)\cdot\rho} e^{-i\mathbf{G}_\parallel\cdot\mathbf{R}_\parallel};$$

$\lambda = 0, 1$, where the orthogonalized $f'_{\nu\alpha}$ functions are constructed from the set of $f_{\nu\alpha}$ functions via the Löwdin orthogonalization procedure.[10] To make the Wannier functions to posses the correct symmetry, z_σ are set at the intersitial positions. Note that elements in the set $\{W_{n,\mathbf{k}_\parallel}^{(0)}; n = 1, ..., N\}$ (N is the number of bands of interest) are in general not orthogonal to elements in $\{W_{n,\mathbf{k}_\parallel}^{(1)}; n = 1, ..., N\}$, although they are orthogonal within the set itself. We define a new set of orthogonal functions $\{W_{m,\mathbf{k}_\parallel}; m = 1, ..., 2N\}$ obtained from an orthonormalization procedure from the above two sets of functions. We can show that the new set of functions are excellent approximations to the generalized planar Wannier functions. First of all, the exact Bloch states at $k_z = 0$ and $k_z = k_m$ can be written in terms of $|W_{n,\mathbf{k}_\parallel}^{(0)} >$ and $|W_{n,\mathbf{k}_\parallel}^{(1)} >$ as $|\psi_{n,\mathbf{k}_\parallel} >= \sum_{R_z} |W_{n,\mathbf{k}_\parallel}^{(0)}(R_z) >$ and $|\psi_{n,\mathbf{k}_\parallel+k_m\hat{z}} >= \sum_{R_z} e^{ik_m R_z}|W_{n,\mathbf{k}_\parallel}^{(1)}(R_z) >$. For k_z between 0 and k_m, the approximate Bloch states are written as linear combinations of $|W_{m,\mathbf{k}_\parallel} >$. The coefficients of expansion are eigenvectors of the Hamiltonian matrix

$$H_{m,m'}(\mathbf{k}) = \sum_{R_z} < W_{m,\mathbf{k}_\parallel}(0)|H|W_{m',\mathbf{k}_\parallel}(R_z) > e^{ik_z R_z}. \tag{2}$$

If $W_{m,\mathbf{k}_\parallel}(R_z)$ are close to the generalized planar Wannier functions, the eigenvalues obtained from diagonalizing $H_{m,m'}$ should be very close to the actual band structure. We will show below that the Wannier functions constructed this way indeed give rise to band structures almost identical to the exact results. In practical calculations, it is sometimes more convenient to use non-orthogonal Wannier functions. In that case, the nonorthogonal Wannier functions are defined as above with the f' functions replaced by the f functions.

The lowest eight bands along [001] for GaAs and AlAs obtained by diagonalizing H in (2) with $N = 8$ (dashed curves) and $N = 16$ (solid curves) are shown in Fig. 1, together with the corresponding band structures obtained by using the 3D plane-wave basis (dotted curves). The results obtained from the Wannier functions with N=16 are found to be almost identical to those obtained by using 3D plane-wave basis. If the reduced set of planar Wannier functions ($N = 8$) is used, noticeable difference can be found only for higher conduction bands. Thus, we conclude that excellent approximations to the generalized planar Wannier functions have been accomplished. We have

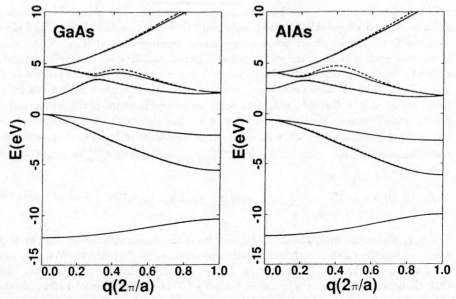

Fig. 1. Band structures of GaAs and AlAs along [001] obtained by using 3D plane-wave basis (dotted), $N = 8$ Wannier basis (dashed), and $N = 16$ Wannier basis (solid).

tried the method with both orthogonal and non-orthogonal Wannier functions. The results are essentially identical.

4. Application to GaAs/AlAs Superlattices

The local pseudopotential is chosen to have the form[11]

$$V_{loc}(\mathbf{r}) = v_1 e^{\alpha_1 r^2} + v_2 r^2 e^{\alpha_1 r^2} + v_3 e^{\alpha_3 r^2},$$

α_1 and α_3 are fixed at 0.61 a.u. and 0.35 a.u. for all atoms of concern (i.e., Ga, As, and Al), while v_1, v_2, v_3 are empirically adjusted to fit the experimental data for conduction band energies at Γ, X, and L (with respect to the valence band top). With proper choice of these parameters, the valence band offset between GaAs and AlAs can be tuned close to the experimental value. For the set of parameters listed in TABLE 1, the valence-band offset is 0.591 eV which corresponds to 37% of the band

TABLE 1. Empirical parameters for atomic local pseudopotentials. All parameters are in units of hartrees.

material	atom	v_1	v_2	v_3
GaAs	Ga	1.63849	-1.60741	-0.05958
	As	2.41947	-1.32375	-1.81770
AlAs	Al	2.46466	-1.61970	-0.52670
	As	2.82374	-1.98034	-1.24894

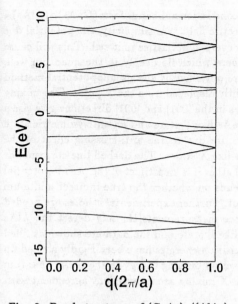

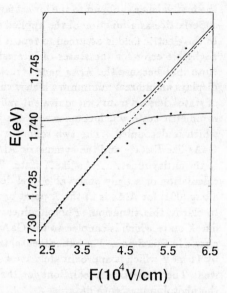

Fig. 2. Band structures of $(GaAs)_1/(AlAs)_1$ superlattice obtained by using full basis (dotted), and Wannier basis (solid).

Fig. 3. Transition energies from the first valence subband to the low-lying conduction subbands of a $(GaAs)_{12}/(AlAs)_{28}$ superlattice obtained by using Wannier basis. Filled circles: data from Ref. [15].

gap difference, in agreement with the experimental observation.[12] The direct band gaps (1.519 eV for GaAs and 3.103 eV for AlAs) and the [001] indirect band gaps (1.975 eV for GaAs 2.259 eV for AlAs) obtained in our calculation are all in good agreement with the corresponding experimental data.[13,14] The energy cutoff used is 20 Ry, which corresponds to 89 2D plane waves at $k_\parallel = 0$. Two gaussian exponents $\alpha = 0.25, 0.75 a.u.$ for symmetric ($\nu = 0$) and $\alpha = 0.5, 1.5 a.u.$ for antisymmetric ($\nu = 1$) $f_{\nu\alpha}$ functions are used in our calculation. Fully convergent results for bulk GaAs and AlAs are obtained. The energy splitting for the three-fold valence band at Γ caused by the non-cubic symmetry of our basis set is found to be less than 0.0004 eV. The potential parameters v_1, v_2, v_3 for As atoms at the interfaces are taken to be the average of the corresponding parameters for GaAs and AlAs. Sixteen wannier functions per diatomic layer are used for all diatomic layers except for those at the interfaces where the number of Wannier functions are doubled. The increase of Wannier functions at the interfaces is necessary in order to better describe the interface electronic properties.

For comparison, we show in Fig. 2 the band structures of the $(GaAs)_1(AlAs)_1$ superlattice calculated by using the planar Wannier functions (solid) and the 3D plane wave basis with the same energy cut-off (dotted). For all the bands shown here, two results are found to be almost identical with the difference barely noticeable in the figure.

Having demonstrated that the results obtained from using the planar Wannier is as good as the full-basis calculation, we can now apply the method to large-size superlattices with confidence. Figure 3 shows the energies for optical transition from the first valence subband to the lowest few conduction subbands for a $(GaAs)_{12}(AlAs)_{28}$ superlattice as a function of the applied electric field. For simplicity, the potential due to the electric field is assumed to repeat in every superlattice unit cell. This will cause negligible error for the states of interest here which lie deeply in the quantum well. Note that because the AlAs band structure obtained by the pseudopotential method displays an indirect minimum which is slightly displaced from the X-point. The mixing of states derived from two equivalent valleys in the [001] and [00$\bar{1}$] directions give rise a symmetric state which couples with the GaAs-like Γ state and an antisymmetric state which is decoupled. The two solid curves are due to the anti-crossing effect of the GaAs-like Γ state and the symmetric AlAs-like X states. The dashed line corresponds to the antisymmetric AlAs-like X state. Whether it is an artifact of the pseudopotential calculation or a truly observable level depends on whether the true indirect minimum along [001] for AlAs is at the X point or not. Further experimental studies are needed to clarify this situation. Photoluminescence measurements[15] only detect the AlAs like X state which is coupled to the GaAs-like Γ state and the data are shown as filled circles in this figure. The experimental transition energies have been rigidly shifted up by 11 meV which is approximately the Γ-exciton binding energy in the GaAs quantum well. The theoretical predictions for the Γ-X mixing are in excellent agreement with the photoluminescence data.

5. Conclusion

We have demonstrated that generalized planar Wannier functions can be constructed directly from first principles. They can be applied to surfaces and interfaces with great efficiency and accuracy. For application to heterostructures, self-consistent calculation is still needed to obtain the correct pseudopotentials at interfaces. Once this is done and the interaction matrix elements for all the $k_\parallel$ values of concern are stored, the remaining calculation is as efficient as an empirical tight-binding method. Thus, Device modeling starting from first principles on the atomic level becomes feasible.

References

† Work supported in part by ONR N00014-90-J-1267
[1] Li G and Chang Y C 1993 Phys. Rev. B **48**, 12032
[2] See, e.g., Ingrid Daubechies 1992 *Ten Lectures on Wavelets* (SIAM, Philadelphia)
[3] Pashley M D *et al* 1988 Phys. Rev. Lett. **60** 2176
[4] Schlier R E and Farnsworth H E 1959 J. Chem. Phys. **30** 917
[5] Li G *et al* 1995 Surf. Sci. **330** 20
[6] Sinniah K *et al* 1989 Phys. Rev. Lett. **62** 567
[7] Wu C J and Carter E A 1991 Chem. Phys. Lett. **185** 172
[8] Shane S F *et al* 1992 J. Chem. Phys. **97** 1520
[9] Li G and Chang Y C 1994, Phys. Rev. B **50** 8675
[10] Löwdin P O 1950 J. Chem. Phys. **18** 365
[11] Kane E O 1976 Phys. Rev. B **13** 3478
[12] Wolford D J et al. 1985 Phys. Rev. **31** 4056
[13] Skromme B J and Stillman G E 1984 Phys. Rev. B29 1982
[14] Monemar B 1973 Phys. Rev. b8 5711
[15] Meynadier M H et al 1988 Phys. Rev. Lett. **60** 1338

Inst. Phys. Conf. Ser. No 145: Chapter 10
Paper presented at 22nd Int. Symp. Compound Semiconductors, Cheju Island, Korea, 28 August–2 September 1995
© *1996 IOP Publishing Ltd*

Wannier-Stark states in semiconductor superlattices

C. Hamaguchi†, M. Yamaguchi†, H. Nagasawa†, K. Murayama†,
M. Morifuji†, A. Di Carlo‡, P. Vogl‡, G. Böhm†, G. Tränkle‡and
G. Weimann‡

† Department of Electronic Engineering, Faculty of Engineering,
Osaka University, Osaka 565, Japan

‡ Walter Schottky Institut, TU München, D-85748 Garching, Germany

Abstract. In order to reveal the behavior of the localized Wannier-Stark states, optical measurements as well as transport measurements are carried out in superlattices with various structures. Experimental results are well interpreted by calculated results based on the realistic tight-binding method. Evidence of the Wannier-Stark oscillation in the Zener tunneling current, which has been recently predicted, is shown.

1. Introduction

When an electric filed F is applied to a crystalline solid with a period d, the energy spectrum disperses into discrete levels expressed as $E_\nu = \nu e F d$ with ν an integer, where e is the electronic charge. These discrete levels are associated with localized wavefunctions and called the Wannier-Stark states or the Stark-ladders. The Wannier-Stark localization can be observed in superlattices much easier than in bulk materials. Thus, the localization has been intensively studied up to now by observing optical transition with various techniques, and explained by using various theoretical methods.[1, 2, 3] In this paper, we report electroreflectance (ER) measurements of the localization and theoretical calculations of the effects by using the tight-binding method. We also show the experimental evidence of a novel phenomenon, Wannier-Stark oscillation, in which resonance between the Wannier-Stark states and tunneling current plays an important roll.

2. Energy spectra of Wannier-Stark states

2.1. GaAs(40Å)/Al$_{0.3}$Ga$_{0.7}$As(20Å) superlattice

First, we show the results of measurements for GaAs(40Å)/Al$_{0.3}$Ga$_{0.7}$As(20Å) superlattice. Figure 1 shows the observed ER spectra for applied bias +0.6 ～ −3.0 V. In the left

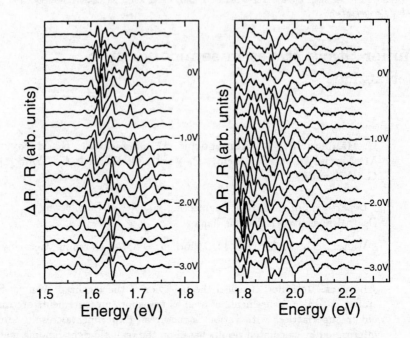

Figure 1. Observed ER spectra of GaAs(40Å)/Al$_{0.3}$Ga$_{0.7}$As(20Å) superlattice.

figure, distinct peaks at the photon energy $1.5 \sim 1.8$ eV are clearly indicating the formation of the Stark-ladders because they show shift in proportion to the applied voltage.[4] In addition to these peaks, we have observed signals in the higher photon energy region ($1.8 \sim 2.1$ eV). Although these signals are weak and complicated, peak positions show shift to lower or higher in proportion to the applied reverse voltage.

In order to analyze the ER signals, we carried out energy spectrum calculation using the tight-binding method.[5, 6] Break down of the translational symmetry due to the electric field prohibits us to apply the Fourier analysis which is usually used. In such a case, energy spectra and wave functions are obtained from the following Hamiltonian:

$$
\mathcal{H}_{j,j'} = \begin{pmatrix} \ddots & & & & 0 \\ & \epsilon + V_j & t(\boldsymbol{k}_{//}) & & \\ & t(\boldsymbol{k}_{//}) & \epsilon + V_{j+1} & t(\boldsymbol{k}_{//}) & \\ & & t(\boldsymbol{k}_{//}) & \epsilon + V_{j+2} & \\ 0 & & & & \ddots \end{pmatrix} . \tag{1}
$$

In eq. (1), $t(\boldsymbol{k}_{//})$ is the transfer integral and ϵ is the site energy, respectively. Both $t(\boldsymbol{k}_{//})$ and ϵ are matrix because multiple orbitals of each atom are considered. $\boldsymbol{k}_{//}$ is a two

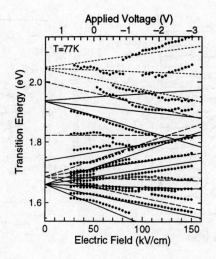

Figure 2. Critical energies obtained from ER spectra (open circles) and calculated transition energies.

dimensional wavevector along the interface and V_j is the applied electrostatic potential at jth atomic layers.

In fig. 2, experimental critical energies and calculated transition energies are shown. In the lower photon energy region, the observed signals are well assigned to the transitions between heavy-hole states and conduction states (denoted by solid lines) and transitions between light-hole states and conduction states (denoted by broken lines). From the calculation, we have found that the spin-orbit split-off (SO) states show the Wannier-Stark localization, although they have energies larger than the AlAs barrier. Signals at 2.0 – 2.1 eV are well assigned to the transition between the SO states to the conduction states (denoted by dot lines).

2.2. Type-II (GaAs)$_8$/(AlAs)$_8$ superlattice

In a short period superlattice, the lowest conduction state is the X-miniband of AlAs. In such a system, we may expect that the Stark-ladders of the X-minibands are observed. The shift of transition energy between the Stark-ladder state of valence states and that of the X-minibands is approximately expressed as νeFd with ν half integer. Thus, we can distinguish the X-related transitions from normal Γ-related transitions by observing shifts of transition energies due to applied electric fields. In order to clarify the existence of the Stark-ladders of the X-miniband of AlAs, we carried out electroreflectance measurement for a (GaAs)$_8$/(AlAs)$_8$ superlattice.[7]

Shown in Figure 3 are the observed ER spectra with reverse voltage $-1.0 \sim -4.0$ V measured at 77 K. The critical energies are shown in Fig. 4 as functions of the applied voltage. In this figure, we estimated the electric field applied to the sample by considering the internal bias associated with the n-i junction 0.9 V. Shifts of the critical energies in

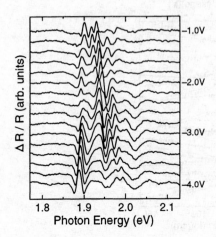

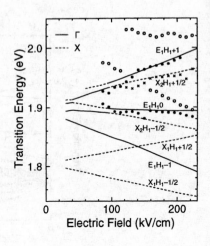

Figure 3. Observed ER spectra of $(GaAs)_8/(AlAs)_8$ superlattice.

Figure 4. Critical energies obtained from ER spectra (open circles and crosses) and calculated transition energies.

proportion to the applied voltage clearly indicate that these signals arise from the Stark-ladders. Although the signals from the Stark-ladders of the X-miniband are expected to be in lower photon energy region (about 1.8 eV), such signals were not observed. However, some critical energies denoted by crosses in Fig. 4 show weaker dependence on voltage than others. This suggest that these signal originate from the Stark-ladder of X-miniband in AlAs.

In order to confirm the assignment of the experimental data, we calculated energy spectra of the superlattice using the second neighbor tight-binding method. Solid and dotted lines in Fig. 6 denote transitions related to the Γ- and X-miniband in the conduction band, respectively. From the comparison between experimental and calculated results, we found that the critical energies denoted by crosses agree well with the calculated transition energies of the second X states $X_2H_1 \pm 1/2$. Other signals are assigned to Γ-related transition $E_1H_1\nu$ ($\nu = 0, \pm 1$). To clarify the reason why $X_2H_1 + 1/2$ is observed while $X_1H_1 + 1/2$ is missing, we calculated the optical matrix elements for those transitions. The matrix elements for $X_2H_1 \pm 1/2$ are larger than that for $X_1H_1 \pm 1/2$ in a wide region of electric fields. From the calculation we found that transition probabilities related to the second X-miniband is largely enhanced due to the mixing with a Γ state. On the other hand, transition probabilities related to the first X-miniband remain small because no such mixing occurs.

3. Wannier-Stark Oscillation in Zener tunneling

Although energy spectra are intensively studied, transport phenomena related to the Wannier-Stark states have rarely studied and have been controversial.[8] Recently, however, Di Carlo *et al.* theoretically studied an interplay between the Zener tunneling current and the Stark-ladder states, and they predicted a novel phenomenon called the Wannier-Stark oscillations.[9] The mechanism of the Wannier-Stark oscillations is as follows: When a strong reverse bias is applied to a highly doped p-i-n diode, electrons can tunnel from the p-region into the n-region through the energy gap which leads to the well-known Zener tunneling. When the intrinsic region of the p-i-n diode consists of a very thin superlattice, localized Stark-ladder states form out of the miniband states due to the high electric field. Consequently, the current-voltage characteristics of the Zener current exhibits an oscillatory structure due to resonant tunneling through these localized Stark-ladder states as schematically shown in Fig. 5. In the figure, only one localized state is related tunneling. If one slightly increases the applied voltage, an energetically higher Stark-ladder state can be reached to a valence electron and an additional resonance contributes to the Zener current. This additional resonance enhances the tunneling current so that another peak occurs in the total current.

Since the Zener tunneling is the tunneling between the valence and conduction bands, the standard transmission theory based on a one-band effective mass approximation is inadequate. Therefore, the transmission coefficient has been evaluated by using a multi-band tight-binding model so as to take the mixing between different bands adequately into account. From eq. (1), we can evaluate a ratio of electronic amplitudes between neighboring atoms at an energy E as [10]

$$T_j(\mathbf{k}_{//}, E) = \begin{pmatrix} 0 & 1 \\ -1 & -\dfrac{\epsilon + V_j - E}{t(\mathbf{k}_{//})} \end{pmatrix}. \tag{2}$$

$T_j(\mathbf{k}_{//}, E)$ is called a transfer matrix. Using the transfer matrix, we can evaluate the transmission coefficients and the Zener tunneling current as

$$j(V) = \frac{-e}{(2\pi)^3\hbar} \int d\mathbf{k}_{//} \int dE\, D(\mathbf{k}_{//}, E)\left[f_p(E) - f_n(E - V)\right], \tag{3}$$

where $-V$ denotes the applied reverse voltage. In Eq. (3), $D(\mathbf{k}_{//}, E)$ is the transmission coefficient between an in-channel state and an out-channel state, and $f_{p(n)}(E)$ is the Fermi-Dirac distribution in the p-region (n-region).

We carried out measurements of the current-voltage characteristics at low temperatures in order to check the predictions concerning Wannier-Stark oscillations. The sample was 7 periods of an undoped $(GaAs)_5/(AlAs)_2$ superlattice sandwiched by p- and n-regions. In fig. 6, the second derivative of the experimental and calculated current density are shown as a function of the applied voltage. Although the magnitude of the experimental current is much larger than that of the calculation, the theoretical and experimental peak positions are in good agreement. The discrepancy in the absolute value of the current may be due to discrepancies between the actual sample parameters (such as width of the buffer layers, doping profiles, and n- and p-concentrations) and the nominal values which the calculations have been based on. It is very important to point out

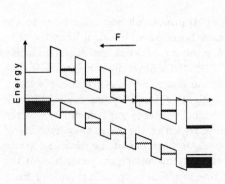

Figure 5. Illustration of Wannier-Stark oscillation in Zener tunneling current.

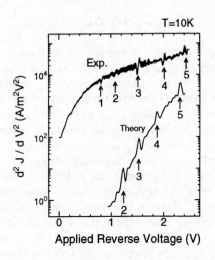

Figure 6. Second derivative of Zener current density in a *p-i-n* diode containing a $(GaAs)_5/(AlAs)_2$ superlattice as function of applied reverse voltage.

that figure 6 shows a good agreement between the observed and predicted peak positions, indicating an increase in the spacing of the peaks with increasing voltage. Since this is a characteristic feature of Wannier-Stark oscillations, this result lends further credence to our interpretation and provides evidence for the existence of Wannier-Stark oscillations.

References

[1] Bleuse J Bastard G and Voisin P 1988 *Phys. Rev. Lett.* **60** 220–223

[2] Mendez E E Agulló-Rueda F and Hong J M 1988 *Phys. Rev. Lett.* **60** 2426–2429

[3] Voisin P Bleuse J Bouche C Gaillard S Alibert C and Regreny A 1988 *Phys. Rev. Lett.* **61** 1639–1642

[4] Yamaguchi M Morifuji M Kubo H Taniguchi K Hamaguchi C Gmachl C and Gornik E 1994 *Solid State Electron.* **37** 839–842

[5] Morifuji M Nishikawa Y Hamaguchi C and Fujii T 1992 *Semicond. Sci. Technol.* **7** 1047–1051

[6] Vogl P Hjalmarson H P and Dow J D 1983 *J. Phys. Chem. Solids* **44** 356–378

[7] Yamaguchi M Nagasawa H Morifuji M Taniguchi K Hamaguchi C Gmachl C and Gornik E 1994 *Semicond. Sci. Technol.* **9** 1810–1814

[8] Argyres P N and Sfiat S 1990 *J. Phys.: Condens. Matter* **2** 7089–7100

[9] Di Carlo A Vogl P and Pötz W 1994 *Phys. Rev.* **B 50** 8358–8377

[10] Schulman J N and Chang Y-C 1983 *Phys. Rev.* **B 27** 2346–2354

Inst. Phys. Conf. Ser. No 145: Chapter 10
Paper presented at 22nd Int. Symp. Compound Semiconductors, Cheju Island, Korea, 28 August–2 September 1995
© *1996 IOP Publishing Ltd*

Intersubband transitions to the above-barrier states controlled by electron Bragg mirrors

B. Sung, H. C. Chui[a], E. L. Martinet[b] and J. S. Harris, Jr.

Solid State Laboratory, Stanford University, Stanford, CA 94305

Abstract. Electron Bragg mirrors grown on both sides of identical quantum wells are used for confining above-barrier states at different energies. By varying the thickness of the mirror layers, the intersubband transition energy between the bound state and the confined above-barrier state (quasi-bound state) changes over 110 meV in 40 Å wide GaAs/$Al_{0.3}Ga_{0.7}As$ quantum wells. The quasi-bound state energy is calculated by considering a superlattice bandstructure and boundary conditions. Theoretical calculations for the absorption spectra with different electron mirrors are compared with experimental results

1. Introduction

Since the pioneering work by Esaki[1], superlattice structures in semiconductor materials have drawn wide interest in application for novel electronic and optical devices[2]-[4]. One of the interesting applications was to trap carriers in a barrier region[5],[6]. Contrary to the conventional method of confining carriers in a potential well, carriers can be confined above a barrier using the optical analogy of Bragg reflectors[7]. Recent experiments of bound to quasi-bound optical absorption have shown clear confinement of the resonant state above the barrier[8]. Spatial confinement using Bragg reflectors was considered for creating population inversion between two above-barrier states[9] and used for improving threshold current in recently invented quantum cascade lasers[10]. Above-barrier states can also be used for intersubband transitions, nonlinear optics and detectors at short wavelengths less than 2 μm when incorporated in deep quantum wells[11],[12].

In this study, above-barrier states at different energies were strongly confined in identical quantum wells by various Bragg mirrors. In addition to the resonant case where the mirrors are quarter wave layers at the quantum well transmission resonance, various dimension of mirror layers were grown to change the quasi-bound state. Considering the Bragg condition and boundary conditions, the quasi-bound state energy of a quantum well with symmetric mirror layers was calculated. Theoretical calculations of dipole moments and absorption spectra are compared with measured spectra.

---

[a] Present address is Sandia National Laboratories, Semiconductor Materials Division (1311), Albuquerque, NM 87185-5800

[b] Present address is Institute for Micro- and Optoelectronics, Ecole Polytechnique Federale de Lausanne, Department of Physics, CH-1015 Lausanne, Switzerland

2. Theoretical Background

If a quantum well is narrow enough to have only one bound state, oscillator strengths between the bound state and above-barrier states become significant. Generally, a broad absorption peak occurs near the ionization threshold for a quantum well cladded by simple barriers because of poor confinement of above-barrier states. Using superlattice structures as Bragg mirrors instead of simple barriers, however, can confine the above-barrier state and produce a narrow absorption peak.

In order to find the quasi-bound state energy of a quantum well with various Bragg mirrors, consider the generic structure in Figure 1. Considering mirror layers as semi-infinite superlattices leads to the Kronig-Penny dispersion relation[13],

$$\cos(KL) = \cos(k_1 d_1)\cos(k_2 d_2) - \frac{1}{2}\left(\frac{m_2^* k_1}{m_1^* k_2} + \frac{m_1^* k_2}{m_2^* k_1}\right)\sin(k_1 d_1)\sin(k_2 d_2) \tag{1}$$

where $k_i = \sqrt{2 m_i^*(E)(E - V_i)/\hbar^2}$ with $i = 1, 2$ and V_i is potential difference relative to the lowest bandedge. Mirror layers behave as Bragg reflectors at energies where the Bragg condition is satisfied, i.e. $K = m\pi/L + i\kappa$ with $m = 1, 2, ...$ and κ given by

$$\kappa = \frac{1}{L}\cosh^{-1}\left|\left[\cos(k_1 d_1)\cos(k_2 d_2) - \frac{1}{2}\left(\frac{m_2^* k_1}{m_1^* k_2} + \frac{m_1^* k_2}{m_2^* k_1}\right)\sin(k_1 d_1)\sin(k_2 d_2)\right]\right|. \tag{2}$$

States exponentially decay and are forbidden in the mirror region at energies which give real κ. Thus a state is localized in the well region if it lies in the superlattice forbidden gap and satisfies boundary conditions[14] at each side of the quantum well which are summarized by the following equation.

$$\tan\delta_3 = \frac{\pm e^{-\kappa L}\cos\delta_2 - \cos 2\delta_2 \cos\delta_1 + \xi_+^{12}\sin 2\delta_2 \sin\delta_1}{\pm \xi_+^{23} e^{-\kappa L}\sin\delta_2 - \xi_-^{12}\xi_-^{23}\sin\delta_1 - \xi_+^{12}\xi_+^{23}\cos 2\delta_2 \sin\delta_1 - \xi_+^{23}\sin 2\delta_2 \cos\delta_1} \tag{3}$$

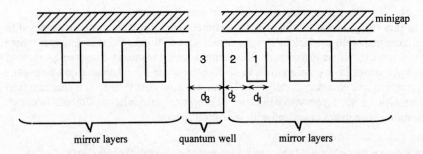

Figure 1. Conduction band diagram of a quantum well and mirror layers. Layer 1 is a mirror well, layer 2 is a mirror barrier and layer 3 is a quantum well. A minigap determined by equation (2) opens up in the superlattice mirror region. .

where + is for (1)>1 and − for (1)<−1, $\delta_i = k_i d_i$ and $\xi_{\pm}^{ij} = \frac{1}{2}\left(\frac{m_j^* k_i}{m_i^* k_j} \pm \frac{m_i^* k_j}{m_j^* k_i} \right)$ with $i, j = 1, 2$ and 3. For the resonant case, the quantum well transmission resonance is the solution of equation (3), where $\delta_1 = \delta_2 = \pi/2$ and $\delta_3 = \pi$. Quasi-bound states for the other mirror layers are also given as solutions of equation (3).

3. Experiments

In this work, we investigated bound to quasi-bound intersubband transitions with varying Bragg mirrors. 40 Å GaAs quantum wells sandwiched by $Al_{0.3}Ga_{0.7}As$/GaAs mirror stacks with different duty cycles were grown by MBE. Thickness of the mirror layers for each sample is presented in Table 1. Only two periods of mirror layers were grown for all of the samples due to the limited coherence at room temperature. Each sample contains 50 quantum wells, n-type doped to $\approx 1e18$ cm^{-3}. As a reference sample, the same quantum wells were grown with 500 Å $Al_{0.3}Ga_{0.7}As$ barriers. The energy of the bound state (E_1) and the transmission resonance (E_{tr}) of the quantum well were calculated to be 93 meV and 307 meV from the bandedge, respectively. Mirror layers of sample #3 were designed to be quarter wave stacks at the transmission resonance of the quantum well (resonant case)[8]. Sample #1, #2 and #4, #5 have mirror layers designed to be quarter wave stacks at energies ($E_{\lambda/4}$) below and above the transmission resonance, respectively as shown in table 1.

In order to study intersubband transitions to quasi-bound states created by various mirror layers, intersubband absorption spectra were taken using a Fourier transform infrared spectrometer (FTIR). The samples were polished at 45° on opposing edges to make a 15-pass waveguides to increase the absorption in quantum wells. Light was incident normal to the polished side of sample. The sample geometry and two different kinds of polarization of incident light ($E_\perp$ and $E_\parallel$) are shown in the inset of Figure 2(a). In order to normalize the intersubband absorption, the transmission ($T_\perp$) for $E_\perp$ polarization was divided by the transmission ($T_\parallel$) for $E_\parallel$ polarization. This excludes the contribution of background and free-carrier absorption. All the measurements were performed at room temperature. Experimental results are compared with the solutions of equation (3) and calculated absorption spectra in the next section.

Table 1. Mirror layer thickness and $E_{\lambda/4}$.

sample	mirror well (layer 1)	mirror barrier (layer 2)	$E_{\lambda/4}$
#1	23.7 Å	134.5 Å	230 meV
#2	22.1 Å	54.9 Å	260 meV
#3	20.0 Å	35.7 Å	307 meV
#4	18.8 Å	29.9 Å	340 meV
#5	17.8 Å	26.5 Å	370 meV

$E_{\lambda/4}$ for #3 corresponds to the transmission resonance of the quantum well and $E_{\lambda/4}$'s are below the resonance for #1 and #2 and above the resonance for #4 and #5

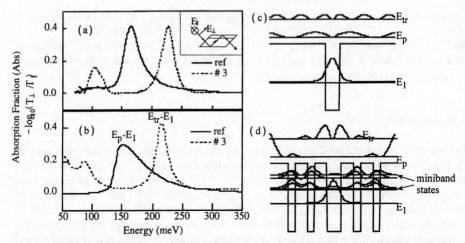

Figure 2. (a) : Measured absorption spectra of reference and resonant case (#3) samples. The inset shows the sample geometry and two kinds of incident light polarization. (b) : Calculated absorption spectra for reference and #3 samples. (c) : Conduction band diagram of reference sample and $|\psi|^2$ of bound and above-barrier states. E_1 is the bound state. E_p and E_{tr} correspond to the absorption peak and transmission resonance energies of reference sample, respectively. (d) : Formation of a quasi-bound state and miniband states by 2 periods of $\lambda/4$ mirror layers. All $|\psi|^2$'s were normalized by the area under themselves.

4. Results and discussion

FTIR spectra taken on the reference and the resonant case (#3) samples are compared with calculated spectra in Figure 2(a) and (b). The bare quantum well (reference sample) yields a broad asymmetric peak below the transition energy (214 meV in calculation) to the transmission resonance as shown in both (a) and (b). This is because above-barrier states of a bare quantum well are just extended states even at the transmission resonance as shown in Figure 2(c). Calculation using the envelope function approximation shows that the dipole matrix element between the bound and the above-barrier states is greater for E_p than for E_{tr}. By putting 2 periods of $\lambda/4$ mirror layers, the absorption peak becomes narrow and is raised to the transmission resonance (#3 in Figure 2(a)). Figure 2(d) shows that the electron state at E_{tr} is localized in the quantum well region by reflection from the mirrors whereas electron states at other energies such as E_p have little probability in the well region because $|\psi|^2$ increases exponentially in the mirror region (actually, those are forbidden states in the super lattice limit of mirror layers). Besides the quasi-bound peak, there is another peak around 100 meV in Figure 2(a). This results from miniband states created by tunnel coupling between quantum wells of mirror stacks as shown in Figure 2(d)[8]. There should be two peaks but the lower peak is out of detection range of the FTIR. These features are presented in Figure 2(b). Theoretical absorption spectra was calculated from Fermi's golden rule using Lorentzian linebroadening[15], which are summarized by the following formula for a 45° multibounce waveguide geometry.

$$\alpha(\omega) = n_p \rho_s \cdot \frac{\pi e^2}{\sqrt{2}\hbar\varepsilon_0 c_0 n} \hbar\omega \int D(E) |\langle 1|z|E \rangle|^2 \delta^\Gamma (E - E_1 - \hbar\omega) dE \tag{4}$$

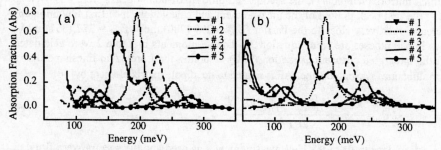

Figure 3. Measured (a) and calculated (b) absorption spectra for various mirror structures.

where n_p is the number of reflection at the waveguide interfaces, ρ_s is the electron sheet density, c_0 is the speed of light in vacuum, n is the refractive index of GaAs, $D(E)$ is the one dimensional density of state at energy E, $|\langle 1|z|E\rangle|^2$ is the dipole matrix element and δ^Γ is the Lorentzian function with FWHM of 2Γ which is used as a fitting parameter.

A study of mirror layer thickness shows that the absorption peak can be tuned by the mirrors. The absorption peak energy changed over 110 meV by varying the mirror layers as shown in Figure 3(a). Thinner mirror layers shift the quasi-bound state to higher energy while less affecting the bound state thus the absorption peak energy increases. But thin mirrors yield stronger miniband absorption as shown in Figure 3(b), so the quasi-bound peak becomes weaker. Linewidth also gets broader for thin mirrors since two periods of mirror stacks do not give strong confinement at high energies above the barrier. FWHM for each sample is shown in Table 2. Calculated absorption spectra using equation (4) are shown in Figure 3(b). Quasi-bound state energies (E_{qb}'s) for various mirror structures are found from the spectra by adding the peak energies to E_1 which ranged 88 meV to 93 meV in theory depending on the mirror layers. E_{qb}'s from the spectra are shown and compared with the solutions of equation (3) in Table 2. Experimental results show good agreement with the theoretical values within 10 %. This discrepancy could be explained by the deviation of sample growth from the design and inaccuracy of material parameters, such as $m_i^*(E)$ and V_i used in equation (3) and (4). In terms of equation (3), the quasi-bound state is a forbidden state in a minigap of the mirror layers and satisfies boundary conditions across the quantum well thus localized in the quantum well region.

Table 2. Comparison of experimental data and theory

| sample | E_{qb} | | | dipole moment | | FWHM(exp.) |
	exp. (Fig. 3a)	calc. (Fig. 3b)	equation (3)	experiment	theory	
#1	257 meV	238 meV	238 meV	13.6 Å	11.2 Å	27 meV
#2	288 meV	271 meV	271 meV	15.4 Å	14.6 Å	20 meV
#3	317 meV	307 meV	307 meV	9.7 Å	9.7 Å	25 meV
#4	341 meV	328 meV	329 meV	7.8 Å	8.0Å	32 meV
#5	366 meV	346 meV	349 meV	5.5 Å	6.7 Å	38 meV

1192

For thick mirrors (sample #1), the second and third quasi-bound peaks were observed as well as the first peak in both of the measurement and calculation since first few minigaps are created relatively close to the barrier (E_{qb2} = 299 (280) meV, E_{qb3} = 377 (351) meV, values in parentheses are from equation 3.). Dipole moments in Table 2 were calculated from the integrated absorption fraction (IAF) assuming the Lorentzian lineshape. For a 45° multibounce waveguide, the IAF is related to the dipole matrix element by[16]

$$IAF = \frac{\pi e^2 n_p \rho_s \left(E_{qb} - E_1\right)\left|\langle 1|z|E_{qb}\rangle\right|^2}{\sqrt{2}\varepsilon_o \hbar c_o n} \tag{5}.$$

The electron density from the Hall measurement was used and the spectra were fitted in the Lorentzian shape to calculate the IAF.

5. Conclusions

In this study, we showed the tuning of bound to quasi-bound intersubband transitions over 110 meV by electron Bragg mirrors. The quasi-bound state energy was calculated theoretically using a single equation which includes the Bragg condition and boundary conditions. Theoretical calculation of absorption spectrum was performed and shows good agreement with experiment.

Acknowledgment

This work was supported by ARPA and ONR through the Center for Non-linear Optics at Stanford, contract # N00014-92-J-1903.

References

[1] L. Esaki and R. Tsu, IBM J. Res. Dev. **14**, 61 (1970)
[2] F. Capasso, K. Mohammed, and A. Y. Cho, IEEE J. Quantum Electron. **QE-22**, 1853 (1986)
[3] K. Furuya and K. Kurishima, IEEE J. Quantum Electron. **QE-24**, 1652 (1988)
[4] E. N. Glytsis, T. K. Gaylord, and K. F. Brennan, J. Appl. Phys. **66**, 6158 (1989)
[5] G. Lenz and J. Salzman, Appl. Phys. Lett. **56**, 871 (1990)
[6] M. Zahler, I. Brener, G. Lenz, J. Salzman, E. Cohen, and L. Pfeiffer, Appl. Phys. Lett. 61, 949 (1992)
[7] F. H. Stillinger, **Physica 85B**, 270 (1977); D. R. Herrick, **Physica 85B**, 44 (1977)
[8] C. Sirtori, F. Capasso, J. Faist, D. L. Sivco, S. N. G. Chu, and A. Y. Cho, Appl. Phys. Lett. **61**, 898 (1992)
[9] G. N. Henderson, L. C. West, T. K. Gaylord, C. W. Roberts, E. N. Glytsis, and M. T. Asom, Appl. Phys. Lett. **62**, 1432 (1993)
[10] J. Faist, F. Capasso, C. Sirtori, D. L. Sivco, A. L. Hutchinson, and A. Y. Cho, Appl. Phys. Lett. **66**, 538 (1995)
[11] J. H. Smet, L. H. Peng, Y. Hirayama, and C. G. Fonstad, Appl. Phys. Lett. **64**, 986 (1994)
[12] H. C. Chui, G. L. Woods, M. M. Fejer, E. L. Martinet, and J. S. Harris, Jr, Appl. Phys. Lett. **66**, 265 (1995)
[13] G. Bastard, *Wave Mechanics Applied to Heterostructures* (Editions de Physique, Paris, 1990), p. 97.
[14] P. Yeh, A. Yariv, and C. S. Hong, J. Opt. Soc. Amer. **67**, 423 (1977)
[15] E. Rosencher, B. Vinter, F. Luc, L. Thibaudeau, P. Bois, and J. Nagle, IEEE J. Quantum Electron. **QE-30**, 2875 (1994)
[16] L. C. West and S. J. Eglash, Appl. Phys. Lett. **46**, 1156 (1985)

Inst. Phys. Conf. Ser. No 145: Chapter 10
Paper presented at 22nd Int. Symp. Compound Semiconductors, Cheju Island, Korea, 28 August–2 September 1995

Windows of full photon-assisted electron transmission via Stark ladder of semiconductor superlattice

O A Tkachenko,† V A Tkachenko,†[1] D G Baksheyev† and
A S Jaroshevich‡

† Novosibirsk State University, Novosibirsk 630090 Russia

‡ Institute of Semiconductor Physics, Novosibirsk 630090 Russia

Abstract. The influence of IR irradiation on resonant tunneling of slow electrons from the contact area through Wannier-Stark quasilevel is modelled numerically for a short-period SL. It is shown that at photon energy close to the spacing of the Stark ladder levels and some optimal amplitude of the alternating field there appear almost rectangular windows of full transparency in the dependencies of transmission coefficient vs. energy, irradiation frequency and voltage. The tunneling electron is delocalized over all SL wells except the outermost ones in that case and the generation of higher harmonics of the external irradiation is possible.

1. Introduction

Under applied bias miniband spectrum of superlattices (SLs) is known to split into Stark ladders where the electrons, which were previously delocalized at miniband states, become localized in separate quantum wells and transport conditions become much more difficult (e.g. [1, 2]). In particular, slow electrons falling from the contact area onto such SL can elastically pass it only with a small probability and only at voltages providing resonant tunneling through separate Wannier-Stark quasilevels [3]. The main contribution into the current is due to inelastic spontaneous electron transitions via the states of Stark ladder, that is due to rather slow process with multiple phase breaks. In this paper we analyze enhancement of resonant transparency of such SLs caused by IR irradiation which induces stimulated electron transitions between neighbouring Stark levels.

Earlier, full photon-assisted inelastic resonant transmission via quasilevels E_1 and E_2 of asymmetric three-barrier structures was predicted for the case the irradiation frequency is such that $\hbar\omega \approx |E_2 - E_1|$ [4]. Recently it was shown by numerical modeling that full transparency of such structures could be achieved only at some optimal amplitude of

---

[1] E-mail (Internet): tkachen@ns.cnit.nsk.su

high frequency (hf) field [5]. High frequency field causes splitting of the quasilevels E_1 and E_2 to two pairs of states $(E_1^{(1)}, E_1^{(2)})$ and $(E_2^{(1)} = E_1^{(1)} + \hbar\omega, E_2^{(2)} = E_1^{(2)} + \hbar\omega)$ with close quasienergies. As the power of irradiation increases, quasilevels interact stronger to each other thus raising transmission coefficient to almost 1. Further, however, the splitting $E_1^{(2)} - E_1^{(1)}$ exceeds the quasilevel width so the tunneling of electrons of energy $E = E_2 \approx (E_2^{(1)} + E_2^{(2)})/2$ becomes non-resonant and the transmission coefficient drops [6]. We suppose that this phenomenon can manifest in cascades of unipolar laser reported recently in [7].

The present paper deals with investigation of analogous effect for semiconductor short-period superlattices in electric field, when the photon energy is equal to the Stark ladder spacing. Note that electron transmission through SL in an hf field is mainly due to multiphoton coherent processes rather than monophotonic ones (as in three-barrier structures). Numerical model is briefly described below in section 1. Results of calculation for SL are discussed in section 2. In the conclusion we discuss use of IR radiation for "lifting" electrons up the Stark ladder of semiconductor SL in p-i-n structures as an example of possible optoelectronic applications.

2. Model of calculations

For modeling of influence of hf field on electron transmission through multibarrier structures we developed a computer program solving the one-dimensional Schroedinger equation with step wise $U(x)$, $V(x)$, $m_{\text{eff}}(x)$ functions:

$$i\hbar \frac{\partial \Psi}{\partial t} = -\frac{\hbar^2}{2} \frac{\partial}{\partial x} \frac{\partial \Psi}{m_{\text{eff}} \partial x} + U(x)\Psi + V(x)\cos\omega t\, \Psi. \tag{1}$$

Here $U(x)$ is the stationary potential of the structure, $V(x)$ is the amplitude of the alternating potential created by monochromatic irradiation of ω frequency polarized along the structure. On assumption that the wave length of irradiation is much larger than dimensions of the structure ($2\pi c/\omega \gg L$, dipole approximation) V(x) is given by:

$$V(x) = \begin{cases} 0, & x < 0 \\ \mathcal{E}x, & 0 < x < L \\ \mathcal{E}L, & L < x. \end{cases}$$

The equation was solved by matching of quasienergetic wave function $\Psi_\varepsilon(x,t)$ ($0 \leq \varepsilon \leq \hbar\omega$) at discontinuity points [8]. Transmission and reflection coefficients of a particle with energy $E = \varepsilon + N\hbar\omega$ through open channels with energies $E_n = \varepsilon + (N+n)\hbar\omega$, $n = 0$, ± 1, ... were calculated using that method. Asymptotically closed channels were also taken into account. The features of our program MULTI-HW are described in [6].

3. Results and discussion

We consider electron transmission through 10-period $(GaAs)_{11}/(AlAs)_3$ SL with bias applied (Fig. 1 (a)) Slow electrons from the contact area of SL with the energy E_2 (quasilevel of second well of SL) are transmitted through SL with small probability (10%) without irradiation and with almost 100% probability if the structure is irradiated by photons of resonant energy $\hbar\omega = 50$ meV and optimal alternating field amplitude 2 kV/cm (Fig. 1 right).

Wave function in these two cases is as follows. Without irradiation an electron is mainly localized in the second well from the SL edge. The appearance of $|\Psi(x)|^2$ is typical for Wannier-Stark quasilevels (Fig. 1 (b)). At the optimal hf field intensity the time averaged probability density is distributed over all the SL wells except outermost ones (Fig. 1 (c)). It resembles delocalized miniband states without external fields. However, the maxima of probability density in six inner wells have approximately equal heights (unlike that for miniband states). Note that analogous distribution of probability density was obtained for full transparency states in graded-barrier structures with maximally flat peaks of transmission coefficient [9, 10].

In our case the repetition of analogous peaks of $\overline{|\Psi|^2}$ is due to summation of coordinate distributions shifted relatively to each other by SL period which correspond to inelastic transmission of electron accompanied with emission of 1, 2, ..., 5 photons. Indeed, Fig. 1 (d) shows contributions of each of 6 strongly coupled channels ($n = 0, -1$, ..., -5) into the full time averaged probability density. These contributions are similar in appearance to electron distributions for Wannier-Stark quasilevels in absence of hf field with electron localization in the second, the third etc. wells. This suggests that in Stark SL we deal with multiple coupling of equidistant levels and multiphoton process of coherent inelastic transmission unlike monophotonic process in the case of two wells. Note that 15 channels ($n = 0, \pm 1, ..., \pm 7$) of expansion of quasienergetical wavefunction were used in the computations for accuracy, though the number of important channels

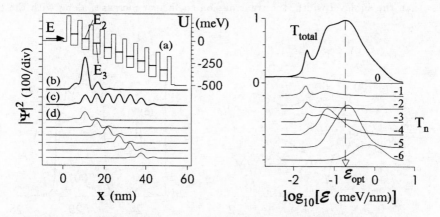

Figure 1. *On the left:* (a) SL conduction band profile with Stark ladder E_1, E_2, ... (b) $|\Psi|^2$ for $E = E_2 = 25$ meV without irradiation. (c) Time averaged total $|\Psi|^2$ for optimal amplitude $\mathcal{E}_{\text{opt}}$ of resonant hf field $\hbar\omega = E_2 - E_3$. (d) Contributions of channels $n = 0, -1, ..., -5$ (from top) into total $|\Psi|^2$. *On the right:* Total transmission coefficient and its constituents by channels $n = 0, -1, ..., -6$ (from top) vs. amplitude of hf field $\mathcal{E}$ for the same E and $\hbar\omega$. The optimal amplitude $\mathcal{E}_{\text{opt}} = 2$ kV/cm is marked.

is less. Stark levels in the outermost wells make no significant contribution in the full wavefunction because of edge effects. Besides that, channels of electron transmission with photon absorption in relatively small alternating field 2 kV/cm appear to be inefficient.

Fig. 1 right shows the full transmission coefficient vs. hf field amplitude. At $\mathcal{E} = 100$ V/cm there is a threshold in $T(\mathcal{E})$ from which T rapidly increases. The narrow peak at 200 V/cm corresponds to maximum transmission through channels $n = -1, -2, -3$, whereas the shoulder at 1 kV/cm is due to contribution of the channel $n = -4$. At the optimal hf intensity there appears the maximum of transmission through the channel $n = -5$, namely, 2/3 of transmitted electrons have energy $E_2 - 5\hbar\omega$. These electrons are emitted mainly from the seventh well (Fig. 1 left) easily overcoming last SL barriers. It means that the trasmission of each electron is accompanied by emission of ≈ 5 photons so the generation of higher harmonics (up to 5th) of the external radiation is possible. Indeed, the 5th harmonic of frequency ω is represented clearly in time oscillations of probability density at the exit from SL (Fig. 2 left (c)). In contrast to this, inside SL $|\Psi(x,t)|^2$ oscillates in time similar to Bloch oscillations. Really, Fig. 2 left curves (a) and (b) shows that probability densities in the second and in the seventh well are in antiphase, i.e., the electron that gets into SL reaches its opposite edge in 1/2 period. As it is easier for it to leave SL in that moment, we should expect a large contribution of $n = -5$ channel in the full transmission.

At small hf field intensity the dependencies of transmission coefficient vs. energy of incident electrons (Fig. 2 right) and vs. voltage (Fig. 3 left) show ~10% maxima corresponding to resonant transmission of electron through Wannie-Stark quasilevel E_2 (curves marked with (a)). Notice that at small $\mathcal{E}$ coefficient T does not depend on frequency. With increasing hf intensity $\mathcal{E}$ from zero Stark quasilevels start to interact, which results in splitting of each of them onto six sublevels. As long as the splitting is less than the sublevel width, the transmission coefficient increases along with the $\mathcal{E}$ as

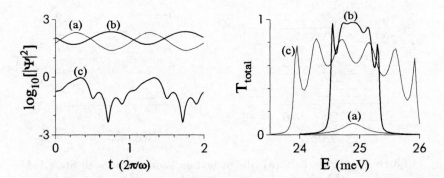

Figure 2. *On the left:* $|\Psi(t)|^2$ for different x: At the center of the 7th **(a)** and 2nd **(b)** well of the SL. **(c)** Beyond the SL, at $x = 60$ nm. *On the right:* T_{total} vs. incident electron energy E for different $\mathcal{E}$: **(a)** Without hf field. **(b)** Optimal amplitude $\mathcal{E}_{opt}$. **(c)** Large amplitude $2.5\mathcal{E}_{opt}$. Other parameters as in **Fig. 1**.

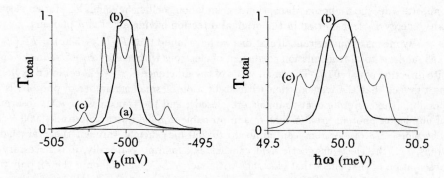

Figure 3. Total transmission coefficient vs. the bias V_b applied to SL (*on the left*) and vs. photon energy $\hbar\omega$ (*on the right*) for different amplitude $\mathcal{E}$ of hf field: **(a)** No hf field. **(b)** $\mathcal{E} = \mathcal{E}_{\text{opt}}$. **(c)** $\mathcal{E} = 2.5\mathcal{E}_{\text{opt}}$.

single peak. At the optimal intensity $\mathcal{E}_{\text{opt}}$ (Fig. 1 right) the peaks become maximally high ($T_{\text{total}} \approx 1$) and flat (Figs. 2 right, 3, curves (b)). These windows of full SL transparency resemble maximally flat peaks of T(E) obtained without external fields as a result of suppression of multiple reflections of electron from edges of short SL by means of linear [9] or other special [10] narrowing of the barriers from center to the ends of SL. In our case repeated reflections of localized electrons from well edges and from edges of the Stark SL are effectively suppressed by resonant IR irradiation of optimal intensity.

Increase of hf power above optimal causes destruction of the above mentioned transparency windows onto series of peaks, because the splitting becomes larger than sublevel width (Figs. 2 right, 3, curves (c)). The dependencies $T_{\text{total}}(E)$ and $T_{\text{total}}(V_b)$ show six peaks, by the number of sublevels of each Stark state, which subsequently come to resonance with energy of incident particles as E or V_b varies.

Deviations of photon energy from Stark level spacing results in different conditions of tunneling via sublevels for different Stark levels. It is observed as both abrupt decrease of the outermost peaks in $T_{\text{total}}(V_b)$ and absence of these peaks in $T_{\text{total}}(\hbar\omega)$ plots (Fig. 3 (c)). Owing to accumulation of detuning this effect is stronger in longer SLs than in shorter ones. However, cojoint varying of $\hbar\omega$ and V_b, such that the chain of Stark levels remains coupled, makes the dependency $T_{\text{total}}(\hbar\omega)$ look similar to $T(E)$.

4. Conclusions

It has been shown above that in optimal hf field Stark SL acts as effective converter of potential energy of electrons into radiation. The effect is caused mainly by suppression of repeated electron reflections from the edges of wells and edges of the SL as whole. Basing on the principle of detailed equilibrium we expect that in the opposite direction the same SL can act as an efficient converter of electromagnetic field energy into potential energy of electrons. Indeed, if an electron with energy $E_2 - n\hbar\omega$ is falling onto SL from the opposite side, the n-photon absorption transmission coefficient equals T_{-n} for an electron with energy E_2 transmitted in the original direction having emitted n photons.

By means of numerical calculations we have found that $T_{+5}(E_2 - 5\hbar\omega) = T_{-5}(E_2) = 0.65$, and the total transmission coefficient of electrons with energy $E_2 - 5\hbar\omega$ in the opposite direction equals 0.67. The small value of the difference $T_{\text{total}} - T_{+5}$ could be explained by the fact that the transmission with absorption of smaller number of photons is impossible (the correspondent channels are closed), and the transmission with absorption of number of photons greater than 5 is improbable in a field with amplitude 2000 V/cm. Therefore, 2/3 of electrons can climb up to 250 meV height through the Stark ladder steps, i.e., the electromagnetic field can act as a pump. Naturally, it is necessary for observation of that effect that electrons with energy $E_2 - 5\hbar\omega$ fall onto the SL only from the lower energetically (left) contact area. In the higher energetically (right) contact area electrons with energy E_2 must be absent. It could be achieved by means of a p-i-n structure which contains the SL in the i-area and a thick layer of $Al_{0.2}Ga_{0.8}As$ in the n-area. When such an element of a closed electric circuit is irradiated by IR light, one should expect very large optical susceptibility at the resonant frequency $\hbar\omega = 50$ meV and direct current of electrons through 250 meV potential barrier without source of constant voltage.

Acknowledgements

We wish to thank V. L. Alperovich and D. Lenstra for interest to work. This work is performed within frames of the program "Universities of Russia".

References

[1] Mendez E E, Agullo-Rueda F and Hong J M 1988 *Phys. Rev. Lett.* **60** 2426

[2] Alperovich V L, Haisler V A, Jaroshevich A S, Moshegov N T, Terekhov A S, Toropov A I and Tkachenko V A 1992 *Surf. Sci.* **267** 75–79

[3] Tkachenko V A, Tkachenko O A, Kotkin G L and V.G.Tupitsin 1991 *Physica B* **175** 75–79

[4] Sumetskii M Yu and Felshtyn M L 1991 *Pis'ma Zh. Eksp. Teor. Fiz.* **53**(1) 24

[5] Baksheyev D G 1994 Diploma thesis (Novosibirsk State University, Novosibirsk, Russia)

[6] Baksheyev D G, Tkachenko V A and Tkachenko O A 1995 in *Physics, chemistry and application of nanostructures* Eds. Borisenko V E *et al* (Minsk, Belarus) 268–270

[7] Faist J, Capasso F, Sirtori C *et al* 1994 *Appl. Phys. Lett.* **64** 1144–46

[8] Coon D D and Liu H C 1985 *J. Appl. Phys.* **58** 2230

[9] Vanbesien O, Leroux H and Lippens D 1992 *Sol. St. Electron.* **35** 665

[10] Tkachenko V A and Tkachenko O A 1993 *Pis'ma Zh. Tekh. Fiz.* **19** 36

Inst. Phys. Conf. Ser. No 145: Chapter 10
Paper presented at 22nd Int. Symp. Compound Semiconductors, Cheju Island, Korea, 28 August–2 September 1995
© *1996 IOP Publishing Ltd*

Electric field induced type-I to type-II switching in GaAs/AlAs quantum wells

M U Erdoğan†, V Sankaran†, K W Kim†, M A Stroscio‡ and G J Iafrate‡

†Department of Electrical and Computer Engineering
North Carolina State University
Raleigh, North Carolina 27695-7911 USA

‡U.S. Army Research Office
Research Triangle Park, North Carolina 27709-2211 USA

Abstract. A theoretical study of the optical properties of GaAs/AlAs quantum well structures in the presence of an electric field is presented near the type-I to type-II transition point. The band structure is calculated using a second nearest neighbor empirical sp^3 tight binding method. Intersubband transition energies and absorption coefficient are given as a function of the electric field. It is shown that the optical properties of these structures can be significantly modified with field near this point.

1. Introduction

Optical properties of layered semiconductor structures depend significantly on whether electrons and holes are confined within the same layer (type-I) or in adjacent layers (type-II) [1]. In GaAs/AlAs heterostructures, holes are localized in the GaAs region. For wide enough GaAs regions, electrons are also confined within the GaAs in the Γ subband. On the other hand, when the GaAs layers are thin, electrons are mainly localized in the AlAs resulting in a type-II alignment. In a heterostructure grown along the z-direction, the momentum in this direction is no longer conserved. Therefore, transitions between the X_z levels in the conduction band and the Γ levels in the valence band become direct (allowed) in k-space. However, since holes at the Γ-point are confined in GaAs, the transition probability associated with the X_z state is small. These are so called pseudo-direct (weakly allowed) transitions. Accordingly, optical transitions in type-II structures are characterized by lower efficiency and slower photoluminescence decay rates. Type-I to type-II transitions have been experimentally observed in GaAs/AlAs heterostructures under an externally applied electric field, with interesting switching properties [2, 3]. In this paper, we present a theoretical study of the optical properties of GaAs/AlAs quantum well structures in the presence of an electric field, particularly when the structure is switched from direct (type-I) to indirect (type-II).

2. Theory

In this study, we calculate the absorption coefficient as the optical property of interest. However, optical matrix elements obtained in calculating the absorption coefficient can be useful for an understanding of other optical properties as well. The absorption coefficient, apart from a constant, is given by

$$\alpha(\hbar\omega) \propto \frac{1}{\omega} \sum_{\mathbf{k}_{||}} \sum_{n,n'} | \hat{\epsilon} \cdot \mathbf{P}_{nn'}(\mathbf{k}_{||}) |^2 \; \delta(E_{n'}(\mathbf{k}_{||}) - E_n(\mathbf{k}_{||}) - \hbar\omega), \qquad (1)$$

where $\hat{\epsilon}$ indicates the direction of polarization, $\mathbf{P}_{nn'}$ is the momentum matrix element between the nth and n'^{th} subbands, and $\mathbf{k}_{||}$ is the carrier momentum component in the plane of the layers. For evaluation of the absorption coefficient, an accurate description of the band structure is needed. In the present work, an empirical tight binding (TB) method is employed with an sp^3 basis including spin-orbit effects and interactions up to second nearest neighbors. This TB model has been demonstrated to yield accurate heterostructure energy levels and wavefunctions in the presence of Γ-X mixing, and is being applied increasingly for investigation of electronic and optical properties of low-dimensional structures [4]. The mixing of the Γ and X valleys in the conduction band, as well as the heavy hole-light hole (HH-LH) interaction in the valence band are intrinsically included in this model. In the tight binding method, the wave function is built up from atomic-like orbitals:

$$\psi_n(\mathbf{k}_{||}, \mathbf{r}) = \sum_i \sum_\beta c_{n,i,\beta}(\mathbf{k}_{||}) \frac{1}{\sqrt{N}} \sum_a e^{i\mathbf{k}_{||} \cdot (\mathbf{R}_a + \mathbf{r}_i)} \zeta_{i,\beta}[\mathbf{r} - (\mathbf{R}_a + \mathbf{r}_i)], \qquad (2)$$

where i indexes the ions in the unit cell, β denotes the orbital (i.e. s, p_x, p_y and p_z for the anion and the cation), $c_{n,i,\beta}$ are expansion coefficients, N is the number of unit cells in the crystal, a indexes the Bravais lattice sites of the crystal, and $\mathbf{R}_a$ is the corresponding lattice translation. In this equation, $\mathbf{r}_i$ and $\zeta_{i,\beta}$ denote the relative position of the ith ion within the unit cell and the atomic-like orbitals centered at the ith ion, respectively. Heterostructure energy levels depend strongly on relevant bulk energies and effective masses. Accordingly, special attention is paid to the fitting of these effective masses. The parameters across an interface are taken to be the average of the bulk values in the two materials. The effect of an electric field on the electronic states is also included within the framework of the TB calculation by addition of diagonal terms in the TB Hamiltonian.

Matrix elements of the momentum operator between the atomic orbitals that make up the TB basis is obtained by comparison of the calculated bulk dielectric constant with the experimental data. Upon examination of the band structure and matrix elements, we find that the energies and the matrix elements essentially depend only on the magnitude of $\mathbf{k}_{||}$ for the portion of the Brillouin zone we are interested in. Therefore, for ease of calculation, we assume that the energies and the matrix elements are a function only of the magnitude of $\mathbf{k}_{||}$. With this simplification, the integration becomes one-dimensional in k-space. Final integration is performed using a modified version of the Gilat-Raubenheimer method [5] with an interpolation scheme.

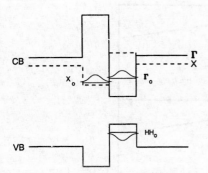

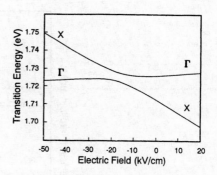

Figure 1. Schematic band diagram of the structure. The middle region is a GaAs/AlAs quantum well contained between two $Al_{0.5}Ga_{0.5}As$ cladding layers.

Figure 2. Transition energies from the first valence subband to the two lowest conduction subbands as a function of electric field. The thicknesses of GaAs and AlAs are 10 and 9 monolayers, respectively.

3. Results and discussion

To facilitate direct comparison with experiment, we choose to work with a structure that was used by Zrenner *et al*, which consists of a GaAs/AlAs quantum well structure contained between two $Al_{0.5}Ga_{0.5}As$ cladding layers. The thicknesses of the GaAs and AlAs regions are 10 monolayers and 9 monolayers, respectively. A schematic drawing of the band alignment is shown in Fig. 1. For this choice of geometry, the lowest Γ level (Γ_0), mostly confined in the GaAs, and the lowest X level (X_0), mostly confined in the AlAs are very close energetically. At zero bias, the X level is lower in energy and the structure is indirect. By application of a small negative bias, the X level can be pushed above the Γ level and the structure becomes direct. Figure 2 shows the intersubband transition energies from the highest valence subband to the lowest two conduction subbands as a function of the electric field. The two subbands anticross for a small negative bias. Away from the anticrossing point, the HH_0-X_0 transition exhibits a linear shift with field.

Figure 3 shows the absorption coefficient for different values of the electric field. Absorption coefficient decreases with increasing energy, mainly because of the reduction of the matrix element due to HH-LH mixing. At zero field, the structure is indirect and X_0 level shows up in the absorption spectrum as a small step due to weak pseudo-direct transitions. The bulk of absorption comes from the Γ_0 level. As the field is increased, the Γ_0 and X_0 levels come closer and interact strongly. This causes the absorption strength to the X_0 level to increase. For F=-20 kV/cm, there is sizable absorption to both levels. This is due to the strong mixing of the Γ and X levels. As field is further increased, the two levels anticross and the structure becomes direct. For F=-40 kV/cm, the absorption threshold is due to the absorption to the Γ_0 level. The X_0 level shows itself as a very small step after the HH_0-Γ_0 transition.

These results show that absorption from the valence subbands to the conduction subbands can be controlled by a bias applied to the structure. The same effect can be observed for other optical properties as well. The results are are in good agreement with the experiment.

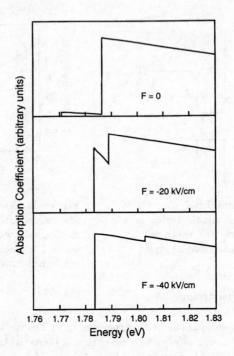

Figure 3. Absorption coefficient for different values of the electric field.

4. Conclusion

In summary, we have calculated transition energies and absorption coefficient for a GaAs/AlAs quantum well structure in the presence of an electric field when the structure is switched from direct (type-I) to indirect (type-II). The results show that optical properties of these structures can be modified to a great extent by application of a bias. This may be useful for designing new optoelectronic devices.

Acknowledgements

This work was, in part, supported by the Office of Naval Research and the U.S. Army Research Office.

References

[1] For a review of work on type-II structures, see Wilson B A 1988 *IEEE J. Quantum Electron.* **QE-24** 1763-1777

[2] Meynadier M H, Nahory R E, Worlock J M, Tamargo M C, de Miguel J L and Sturge M D 1988 *Phys. Rev. Lett.* **60** 1338-1341

[3] Zrenner A, Leeb P, Schäfer J, Böhm G, Weimann G, Worlock J M, Florez L T and Harbison J P 1992 *Surf. Sci.* **263** 496-501

[4] Lu Y -T and Sham L J 1989 *Phys. Rev. B* **40** 5567-5578

[5] Gilat G and Raubenheimer L J 1966 *Phys. Rev.* **144** 390-395

Inst. Phys. Conf. Ser. No 145: Chapter 10
Paper presented at 22nd Int. Symp. Compound Semiconductors, Cheju Island, Korea, 28 August–2 September 1995
© 1996 IOP Publishing Ltd

Effective-Mass Approximation at Heterointerfaces: Intervalley Mixing and Interface Fluctuations

Tsuneya Ando

Institute for Solid State Physics, University of Tokyo

7–22–1 Roppongi, Minato-ku, Tokyo 106, Japan

Abstract. A brief review is given of the interface matrix giving boundary conditions for envelope functions at heterointerfaces. The envelope function approximation works reasonably well for connection of envelopes between Γ valley conduction band minima in III-V semiconductors. Mixing between Γ and X conduction-band valleys can be successfully described by a 6×6 interface matrix. An even-odd oscillation of the optical intensity as a function of the monolayer number for short-period GaAs/AlAs superlattices with lower X states is destroyed almost completely in the presence of interface fluctuations.

1. Introduction

In the effective-mass approximation, the wave function appearing in the Schrödinger equation is not the total wave function but the envelope of the Bloch functions varying rapidly within each unit cell. The conventional effective-mass approximation is not directly applicable to heterostructures, because the potential varies strongly within the distance of the lattice constant in the vicinity of interfaces, and effects of such heterointerfaces can be incorporated only in the form of boundary conditions for envelope functions. In this paper a brief review is given of such boundary conditions at semiconductor heterointerfaces.

One of the most typical semiconductor superlattices is that made of GaAs/AlAs heterostructures. In this system the AlAs layer functions as a quantum well for the X valley, while the energy of the GaAs layer is lower for the Γ valley. Because the effective mass of the Γ valley is smaller than that of the X valleys, electrons in the Γ valley are raised higher in energy compared with those in X valleys for superlattices with a sufficiently thin GaAs layer. This leads to a sudden change of the conduction-band bottom from the character of the Γ valley in GaAs to that of X valleys in AlAs. Within the effective-mass approximation, mixing of different valleys can be incorporated also as boundary conditions.

In Sec. 2 a brief review is given on the basic formulation of the interface matrix and the validity of the envelope function approximation for the Γ valleys. An extension to 6×6 interface matrix describing inter-valley mixing is discussed in Sec. 3. In Sec. 4 effects of interface fluctuations are included in a simple model and the results are compared with recent experiments. A short summary is given in Sec. 5.

2. Interface matrix

Let us consider an interface of a semiconductor A occupying the left $(z < 0)$ half-space and B occupying the right $(z > 0)$ half-space. We first consider the case that both semiconductor A and B possess a single band maximum or minimum which lies close in energy, and discuss energy levels close to the band extrema and corresponding wave functions. In bulk semiconductors, these states are described well by the effective-mass approximation in which the envelope of the Bloch function satisfies a second-order differential equation. In the presence of heterointerfaces, the wave functions of A and B should match smoothly with each other at the interface ($z=0$). This does not necessarily mean that associated envelope functions are continuous across the interface. However, if the envelope in A is given, that of B is uniquely determined. This fact can be described by the following linear relation between envelopes and their derivatives at $z=0$:

$$\begin{pmatrix} \zeta_B \\ \nabla_B \zeta_B \end{pmatrix} = T_{BA} \begin{pmatrix} \zeta_A \\ \nabla_A \zeta_A \end{pmatrix}, \quad \nabla_A = \frac{m_0}{m_A}\nabla, \quad \nabla_B = \frac{m_0}{m_B}\nabla, \quad \nabla = a\frac{\partial}{\partial z}, \quad (2.1)$$

where $T_{BA} = (t_{ij})$ is a 2×2 matrix called interface matrix, a is the lattice constant, m_0 is the free electron mass, and m_A and m_B are the effective masses of A and B, respectively.

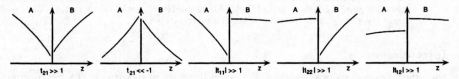

Fig. 1 Various kinds of boundary conditions described by the interface matrix $T_{BA} = (t_{ij})$.

The interface matrix can describe various kinds of boundary conditions as shown in Fig. 1. Among various elements t_{21} is most important, because it is related to the presence of interface states. To see this, we consider a uniform system in which a δ–function potential $V_0\delta(z)$ is present at $z = 0$. The δ–function potential can be incorporated into the interface matrix T_{BA} given by $t_{11} = t_{22} = 1$, $t_{21} = 2maV_0/\hbar^2$, and $t_{12} = 0$. This shows that the positive t_{21} roughly corresponds to the presence of a repulsive δ–function potential and leads to the reduction of the envelope function at the interface. The negative t_{21}, on the other hands, corresponds to an attractive δ–function potential and leads to the increase of the envelope at the interface, often giving rise to a bound interface state.

This interface matrix can be calculated through the matching of the wave functions at the heterointerface. The wave function consists of the Bloch function and many evanescent waves decaying exponentially away from the interface. The Bloch function can be expressed in terms of the envelope $\zeta(z)$ as

$$\psi_\zeta(\mathbf{r}) = \psi_0(\mathbf{r})\zeta(z) + \psi_0'(\mathbf{r})\nabla\zeta(z), \quad (2.2)$$

where $\psi_0(\mathbf{r})$ is the Bloch function a the band extremum and $\psi_0'(\mathbf{r})$ is the first derivative with respect to ak_z.

In a simplest tight-binding in which we consider s orbital of group III material and p_z orbital of group IV material, the interface matrix can be calculated to be $t_{11} = \sqrt{m_B E_g^A/m_A E_g^B}$, $t_{22} = \sqrt{m_A E_g^B/m_B E_g^A}$, and $t_{12} = t_{21} = 0$ for GaAs/AlAs system, where E_g^A and E_g^B are the band gaps of materials A and B. Usually for $Al_x Ga_{1-x}As$, we have $m(x) = m(1+\alpha x)$, $E_g(x) = E_g(1+\beta x)$, $m \approx 0.066m_0$, $E_g \approx 1.519\text{eV}$, $\alpha \approx 0.895$, and $\beta \approx 1.333$. The interface matrix can be rewritten as

$$T_{BA} \approx \begin{pmatrix} 1+\gamma x & 0 \\ 0 & 1-\gamma x \end{pmatrix}, \quad \gamma = (\alpha-\beta)/2 \approx 0.22, \quad (2.3)$$

with $x = x_B - x_A$. This shows that the envelope function approximation ($T_{BA} = 1$) works reasonably well in GaAs/AlGaAs systems.

Calculations in a more realistic empirical pseudo-potential gave the interface matrix approximately given by (2.3) in which γ varies as a function of the position of the matching plane but lies in the range $-0.04 < \gamma < +0.04$. An empirical tight-binding model sps^* also gives (2.3) with $\gamma \approx 0.18$.

This conclusion is more general and applicable to most of the cases in which envelope functions associated with the Γ valley minimum of the conduction bands of III-V semiconductors are connected to each other. For InAs/GaSb, where the envelope function associated with the conduction-band minimum of InAs is connected to that of the valence-band maximum of GaSb, we have the boundary conditions in such a way that the derivative is essentially determined by the value of the envelope on the other side.

3. Intervalley mixing

For the Γ valley we use the conventional envelope $\zeta_\Gamma(z)$ satisfying a single-valley effective mass equation characterized by an effective mass m_Γ. For X valleys we must use a 2×2 multi-components effective-mass equation because of the presence of two bands denoted as u (conventionally X_1) and v (X_3) at the X point close to each other in energy.

The presence of abrupt interface potentials gives rise to mixing between 3 different valleys $\alpha = \Gamma$, u, and v. Such mixing can be included as a form of a 6×6 interface matrix.

$$\begin{pmatrix} \zeta_\beta^B(0) \\ \nabla_\beta^B \zeta_\beta^B(0) \end{pmatrix} = \sum_\alpha T_{BA}^{\beta\alpha} \begin{pmatrix} \zeta_\alpha^A(0) \\ \nabla_\alpha^A \zeta_\alpha^A(0) \end{pmatrix},$$

where $T_{BA}^{\beta\alpha} = (t_{ij}^{\beta\alpha})$ is a 2×2 matrix and $\alpha, \beta = \Gamma$, u, and v.

The interface matrix was first calculated in an empirical sps^* tight-binding model by a direct matching of the wave function [1]. Unfortunately however, this interface matrix violates the current or flux conservation at the interface. Then, it has been determined in such a way that the flux conservation is satisfied and it reproduces the previous result obtained as well as possible [2].

The result is given by

$$T(\text{AlAs} \leftarrow \text{GaAs}) = \begin{array}{c} \zeta_\Gamma \\ \nabla\zeta_\Gamma \\ \zeta_u \\ \nabla\zeta_u \\ \zeta_v \\ \nabla\zeta_v \end{array} \begin{matrix} \zeta_\Gamma & \nabla\zeta_\Gamma & \zeta_u & \nabla\zeta_u & \zeta_v & \nabla\zeta_v \\ \begin{pmatrix} 1.24 & 0 & 0 & 0 & 0 & 0 \\ 0 & 0.81 & 0 & 0 & 0.83p & 0 \\ 0 & 0 & 0.97 & 0 & 0 & 0 \\ 0 & 0 & 0 & 1.03 & 0.26 & 0 \\ 0 & 0 & 0 & 0 & 0.83 & 0 \\ 1.24p & 0 & -0.40 & 0 & 0 & 1.20 \end{pmatrix} \end{matrix}, \tag{3.1}$$

Two elements $t_{21}^{\Gamma v} = 0.83p$ and $t_{21}^{v\Gamma} = 1.24p$ giving the strength of Γ-X mixing contain a parameter p, which is determined to be about 0.5 by the direct comparison of the wave functions with those calculated in short-period superlattices in an empirical sps^* tight-binding model. This interface matrix corresponds roughly to an introduction of a δ-function-like potential connecting only Γ and X valley envelopes at the heterointerface as postulated previously for a model of Γ-X mixing [3-5].

It is straightforward to calculate transmission and reflection probabilities at a single heterointerface and tunneling probabilities for various barrier structures consisting

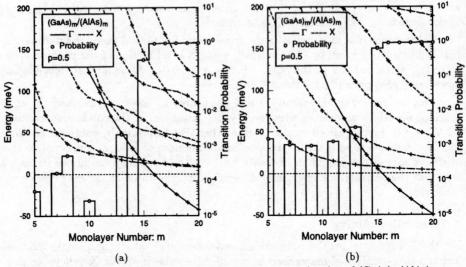

(a)   (b)

Fig. 2 Energy bands calculated in the effective-mass approximation of $(GaAs)_m(AlAs)_m$ superlattices as a function of m (a) for band parameters reproducing an sps^* tight-binding model and giving double minima in the vicinity of the X point and (b) for parameters given a single minimum at the X point. The energy is measured from the bottom of the conduction-band minimum near the X point. The parity of each state is denoted by $+$ or $-$. The optical transition probability determined by the overlap of the wave function of the lowest energy level with that of the valence-band top is denoted by open circles.

of $Al_xGa_{1-x}As$ with varying x [1]. The results show that the mixing between Γ and X valleys is much stronger than that estimated by using the interface matrix which is obtained by a direct matching of the wave functions and violates the flux conservation condition.

Figure 2(a) shows the energy levels (at the Γ point) of $(GaAs)_m(AlAs)_m$ superlattices as a function of m calculated for the band parameters reproducing the empirical tight-binding model [1]. In this model the X valley has minima slightly away from the X point in AlAs, which leads to valley splitting oscillation as a function of m. For $m \geq 15$ the conduction-band minimum comprises mainly the Γ valley minimum and for $m \leq 14$ it comprises the X valley minima.

Because of the small mixing between Γ and X, the transition probability for $m \leq 14$ becomes more than two orders of magnitude smaller than that for $m \geq 15$. It also exhibits an oscillatory behavior as a function of m for $m \leq 14$. This is a result of the characteristic parity change pointed out for the first time by Lu and Sham [6] using group theory and demonstrated in a tight-binding model calculation. In Fig. 2 the parity change is modified by a valley splitting.

Figure 2(b) shows an example for the band parameters giving the X valley minimum just at the X point. It shows a clear even-odd symmetry dependence of the optical intensity. Unfortunately the band structure of AlAs in the vicinity of the X point is not well understood yet and it remains an open question whether it has a single minimum or double minima.

Quite recently, the photoluminescence decay rate in $(GaAs)_m/(AlAs)_m$ short-period superlattices was measured. The experiments of Minami et al. [7] indicated a strong dependence of the decay rate on m, while those of Nakayama et al. [8] did not. One possible origin of such a discrepancy lies in the presence of interface fluctuations.

In the following a strong effect of fluctuations is demonstrated through calculation of optical intensities for $(GaAs)_\alpha/(AlAs)_\beta$ superlattices with different α and β.

4. Interface fluctuations

The above interface matrix should be applied to the band-structure calculation with special care because the X-valley Bloch function changes its sign for each monolayer. We denote a certain interface consisting of GaAs on the left and AlAs on the right as $n = 0$ where we use the interface matrix T given by eq. (3.1). Then, the interface matrix at the interface $n = \beta$ is given by T^{-1} for even β, while it is given by $\bar{T}^{-1}$ for odd β. Here, $\bar{T} = U^{-1}TU$ with $U = (u_{ij})$ where $u_{ij} = 0$ for $i \neq j$, $u_{ii} = 1$ for $i = 1, 2$, and $u_{ii} = -1$ for $i = 3, 4, 5, 6$. Similarly, the interface matrix at $n = \alpha + \beta$ becomes T for even $\alpha + \beta$ and $\bar{T}$ for odd $\alpha + \beta$. This leads to the conclusion that the envelope wave-function associated with the X valley is given by $\exp[i(k + \pi/d)z]u_X(z)$ for odd $\alpha + \beta$ in contrast to $\exp(ikz)u_X(z)$ for even $\alpha + \beta$, where d is the superlattice period given by $d = (\alpha + \beta)a/2$ and $u_X(z)$ is a periodic function with period d.

Results of calculation for the band parameters giving a single X valley minimum in AlAs are given in Fig. 3 where the optical intensity is plotted as a function of $(\alpha + \beta)/2$. Effects of interface fluctuations can be included roughly by taking an average of the results for different α and β. We assume that each interface position is shifted by a monolayer with probability r and remains at the ideal position with probability $1 - 2r$. The solid and dashed lines represent the results for $r = 0$ and 0.25, respectively, for $(GaAs)_m(AlAs)_m$ superlattices. The even odd symmetry is destroyed completely for $r \sim 0.25$.

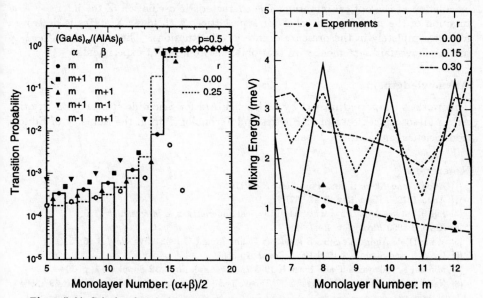

Fig. 3 (left) Calculated optical intensity between the lowest conduction-band state and the highest valence-band state as a function of $(\alpha + \beta)/2$ for $(GaAs)_\alpha/(AlAs)_\beta$ superlattices. Solid lines show the result for ideal $(GaAs)_m/(AlAs)_m$ superlattices and dashed lines that for $(GaAs)_m/(AlAs)_m$ with interface fluctuation $r = 0.25$.

Fig. 4 (right) Calculated Γ-X mixing energy as a function of the monolayer number m in $(GaAs)_m/(AlAs)_m$ superlattices. Experimental results [8] are also shown.

Nakayama et al. [8] determined the mixing energy $(\Gamma|\mathcal{H}|X)$ defined by the condition that the optical intensity is proportional to $|(\Gamma|\mathcal{H}|X)|^2/(E_\Gamma - E_X)^2$, where E_Γ is the energy of the lowest Γ-like conduction-band state and E_X is that of the X-valley state. Figure 4 gives a comparison between the experiments and the present theoretical estimate. The even-odd oscillation diminishes again with the increase of fluctuation r. The result for $r \sim 0.3$ shows a monolayer dependence in qualitative agreement with the experiments. The absolute magnitude of the mixing energy seems to be almost twice as large as that determined in the experiments. We can conclude, however, that the agreement is satisfactory, since the sps^* tight-binding model can be too simple and the mixing parameter may strongly depend on the band structure near the X point.

A similar calculation for the parameters giving double X-valley minima in AlAs shows that the mixing energy is much smaller than that from the experiments and follows an m dependence different from the experiments. Therefore, we can conclude that the band structure in the vicinity of the X point calculated using the sps^* tight-binding model is inaccurate.

5. Summary and conclusions

The presence of heterointerfaces can be incorporated by the interface matrix as a form of boundary conditions of envelope functions and their first derivatives at the interface. Calculations of the interface matrix in various models show that the envelope function approximation works reasonably well for the cases that envelope functions associated with the Γ valley minimum are connected to each other.

An interface matrix incorporating mixing between Γ and X conduction-band valleys is determined in GaAs/AlAs short-period superlattices. It has been used for calculation of the optical intensity. The characteristic oscillation of the intensity as a function of the monolayer number for superlattices with lower X states is destroyed almost completely in the presence of interface fluctuations. The strength of the Γ-X mixing shows fair agreement with that obtained from recent experiments.

Acknowledgments

This work is supported in part by a Grant-in-Aid for Scientific Research on Priority Area "Mesoscopic Electronics: Physics and Technology" from the Ministry of Education, Science and Culture, Japan.

References

[1] Ando T and Akera H 1989 *Phys. Rev. B* **40** 11619
[2] Ando T 1993 *Phys. Rev. B* **47** 9621
[3] Liu H C 1987 *Appl. Phys. Lett.* **51** 1019; 1990 *Superlattices & Microstruct.* **7** 35.
[4] Xia J B 1990 *Phys. Rev. B* **41** 3117
[5] Wang D, de Andrada e Silva E A and da Cunha Lima I C 1992 *Phys. Rev. B* **46** 7304
[6] Lu Y -T and Sham L J 1989 *Phys. Rev. B* **40** 5567
[7] Minami F, Nakayama T and Inoue K 1993 *Jpn. J. Appl. Phys.* **32** Suppl. 32-1, p. 70
[8] Nakayama M, Imazawa K, Suyama K, Tanaka I and Nishimura H 1994 *Phys. Rev. B* **49** 13564

Paper presented at 22nd Int. Symp. Compound Semiconductors, Cheju Island, Korea, 28 August–2 September 1995
© *1996 IOP Publishing Ltd*

Energetics of N-related defects in Zn-based II-VI semiconductors

Byoung-Ho Cheong†and K. J. Chang‡

† Department of Supercomputer Applications, Samsung Advanced Institute of Technology, P. O. Box 111, Suwon, Korea

‡ Department of Physics, Korea Advanced Institute of Science and Technology, 373-1 Kusung-dong, Yusung-ku, Taejon, Korea

Abstract. We study the energetics of N-related defects in Zn-based II-VI semiconductors such as ZnS, ZnSe, and ZnTe through first-principles pseudopotential calculations and find that a substitutional N at an anion site to be most stable at low doping levels. From the calculated formation energies, we determine defect concentrations and find that p-type doping levels are different between ZnSe and ZnTe. The maximum hole density in ZnTe is estimated to be higher by about 5 times than ZnSe, which is in good agreement with the experiments. We also investigate the microscopic mechanism for the compensation of N acceptors in ZnSe.

1. Introduction

Due to wide band gaps, Zn-based II-VI semiconductors such as ZnS, ZnSe, and ZnTe are known to be promising materials for blue-green light emitting and laser devices. However, n- and p-type carrier concentrations sensitively depend on the type of dopants and host semiconductors. Especially, ZnSe can be easily doped n-type but not easily p-type, while ZnTe is easily doped p-type but not easily n-type. Recently, it has been possible to achieve high hole concentrations for device operations with the use of a plasma excited N source in ZnSe.[1] However, the net hole concentration dose not exceed about $10^{18} cm^{-3}$, although N is incoporated into ZnSe over $10^{19} cm^{-3}$, indicating the self compensation of acceptors. Despite several theoretical attempts, the maximum hole concentration and the type of defects which cause compensation are not fully understood yet.[2, 3, 4] In previous first-principles calculations, we reported that most N impurities behave as an acceptor for N concentrations up to $10^{18} cm^{-3}$.[5] However, for higher N concentrations, the N acceptors were shown to be compensated for by $(NN)_{Se}^{2+}$ complexes under Zn-rich conditions, while they are neutralized by interstitial N_2 molecules under Se-rich condition. In the Zn-rich limit, the net hole concentration was suggested to increase up to $10^{19} cm^{-3}$.

In this work, we report the results of theoretical calculations for the energetics of N-related defects in Zn-based II-VI semiconductors, ZnS, ZnSe, and ZnTe. Our calculations are based on the first-principles pseudopotential method within the local-density-functional approximation.[6] For the Zn pseudopotential, instead of including the Zn $3d$ states in the valence shell, we employ partial core corrections, thus, we are able to produce correctly the bulk properties of Zn-based semiconductors. The details of calculational methods were discussed in our previous work.[5]

2. Results

We calculate the formation energies for various N-related defects as well as native point ones and examine the energetics as a function of N chemical potential. The results for substitutionals (N_{anion}^-, where $anion = $ S, Se, and Te), N-N complexes at anion sites [$(NN)_{anion}^{2+}$], and interstitial N_2 molecules are listed and compared in Table 1, for p-type ZnS, ZnSe, and ZnTe grown under stoichiometric conditions. In addition, we set the N chemical potential to be the ground state energy of a N_2 molecule in vaccum.[5] For a N-N complex at an anion site, we find a [100]-split interstitial configuration to be lowest in energy. For all the semiconductors considered here, the substitutional N acceptor is most stable. Because of the small atomic radius of the N atom, the neighboring Zn atoms of the N acceptor undergo large lattice relaxation, with the distortions of 0.37, 0.46, and 0.63 Å toward the N atom in ZnS, ZnSe, and ZnTe, respectively. In this case, the second neighboring anion atoms are also relaxed toward the first neighboring Zn atoms: 0.07, 0.09, and 0.10 Å for ZnS, ZnSe, ZnTe, respectively. Then, the bond lengths between the Zn and second neighboring anion atoms are increased by 0.08, 0.10, and 0.17 Å for ZnS, ZnSe, and ZnTe, respectively. Since the formation energy of the N acceptor in ZnS is lower than for ZnSe and ZnTe, we expect that ZnS has higher hole concentrations. However, it is difficult to compare directly our results with experiments at this point, because of the lack of experimental data for N-doped ZnS.

Table 1. The formation energies for a substitutional acceptor (N_{anion}^-), a N-N complex [$(NN)_{anion}^{2+}$], and an interstitial N_2 molecule are listed for ZnS, ZnSe, and ZnTe. The formation energies are defined in Ref. [5].

	N_{anion}^-	$(NN)_{anion}^{2+}$	N_2
ZnS	1.47	3.03	3.24
ZnSe	2.03	3.38	3.36
ZnTe	1.90	3.64	3.35

From the formation energies and the charge neutrality condition, we estimated the maximum hole carrier concentration for ZnSe to be about $10^{18} cm^{-3}$.[5] In ZnTe, since the formation energy of the N_{Te}^- acceptor is lower by 0.13 eV than for the N_{Se}^- in ZnSe, the maximum hole concentration achievable with N impurities is higher by about 5 times, in good agreement with experiments.[7]

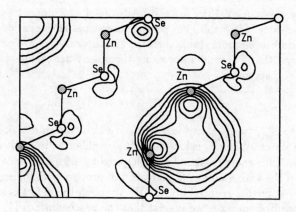

Figure 1. The charge density plot for the V_{Se}° defect level on the [110] plane.

Since an interstitial N_2 molecule is located in the empty space, it behaves as an inert molecule, with almost no interactions with the neighboring anions. Thus, the formation energy of an interstitial N_2 molecule changes very slightly as going from ZnS to ZnTe. However, in the case of the N-N complex at an anion site, although the N atoms are strongly bonded to each other, each N atom is weakly interacting with its two neighboring Zn atoms, giving rise to a donor level. Thus, the formation energy of the $(NN)_{anion}^{2+}$ complex sensitively depends on host semiconductor, as shown in Table 1, because of the change of the lattice constant. Since a substitutional N-N complex at an anion site behaves as a double donor, this complex compensates for N acceptors at high N doping concentratoins.[5]

We also investigate the formation energies of native point defects such as interstitials and vacancies. However, it is difficult to expect the self compensation by native defects, because their formation energies are generally larger than those for N-related defects. This results are consistent with previous theoretical calculations, which suggested that the compensation mechanism in ZnSe can not be explained by native point defects only.[2] Recently, it was reported that the compensation of p-type acceptors in ZnSe is caused by the formation of complexes between the substitutional acceptor and the charged native defects such as V_{Se} and Zn_i.[8] In this case, since a Se vacancy induces large lattice relaxations for the neighboring atoms, the formation energy of V_{Se} was suggested to be significantly lowered, with the lowest energy for the doubly ionized Se vacancy. In fact, we find that four Zn dangling bonds around the V_{Se} vacancy interact strongly each other, with inward relaxations of 0.32 Å, lowering the formation energy. However, our calculations show that despite such large lattice relaxations, the formation energy of V_{Se} is higher than for a N-N complex. The electron charge density for the Se-vacancy in a neutrally charged state is drawn in Fig. 1. The defect energy level for V_{Se}° is found to lie 0.12 eV above the valence band maximum (VBM). However, for the singly ionized state V_{Se}^{+}, the interactions between the dangling bonds become weaker, with the Zn atoms relaxed outward by 0.13 Å from their ideal positions, then, the defect level moves to higher energies by 1.24 eV. For the doubly ionized state, the outward relaxations of the Zn atoms are found to be 0.55 Å, and the interactions between the dangling bonds no longer exit, with the empty defect state at 2.41 eV above VBM.

We find that a N interstitial acts as a single donor. However, because its formation energy is higher by about 5 eV than for a substitutional acceptor, the concentration of N_i is found to be extremely small. For several interstitial sites considered here, the hexagonal site is found to be more stable than the tetrahedral site. We next consider a process that an interstitial N atom diffuses along the interstitial channel in ZnSe, then, it is captured by a N_{Se} acceptor:

$$N_i^+ + N_{Se}^- \rightarrow (N_i + N_{Se})^* \rightarrow (N-N)_{Se}^{2+} + 2e^- \tag{1}$$

As N_i^+ approaches to N_{Se}^-, two N atoms feel attractive Coulomb interactions because of the opposite charge states. When N_i is located at the neighboring interstitial site of N_{Se}, denoted as $(N_i+N_{Se})^*$, the energy is found to be lowered by 4.6 eV because of the strong interactions between the two N atoms. Finally, in the process of changing to a $(NN)_{Se}$ complex, the energy is further reduced by 1.3 eV. Although the defect concentration of N_i is negligible in equilibrium state, we suggest that the process in Eq. (1) is possible in highly N-doped ZnSe and under non-equilibrium state.

3. Conclusion

We have performed first-principles pseudopotential calculations for studying the compensation mechanism of N acceptors in ZnS, ZnSe, and ZnTe. Comparing the formation energies of various N-related defects, we find that the N acceptors are compensated for by N-N complexes or neutralized by interstitial N_2 molecule, depending on the stoichiometric condition. We also show that the N-N complex can be easily formed by combination of a substituional N acceptor and interstitial N atom.

Acknowledgments

This work was supported by the CMS at Korea Advanced Institute of Science and Technology and the Korea Ministry of Science and Technology.

References

[1] Haase M A , Qiu J, DePuydt J M and Cheong H 1991 *Appl. Phys. Lett.* **59** 1272-74

[2] Laks D B, Van de Valle C G, Neumark G F, Blöchl P E and Pantelides S T 1992 *Phys. Rev.* B **45** 10965-78

[3] Chadi D J and Chang K J 1989 *Appl. Phys. Lett.* **55** 575-77

[4] Park C H and Chadi D J 1995 *to be published*

[5] Cheong B H, Park C H and Chang K J 1995 *Phys. Rev.* B **51** 10610-14

[6] Cohen M L 1982 *Phys. Scr.* *T* **1** 5-10

[7] Fan Y, Han J, He L, Gunshor R L, Brandt M S, Walker J, Johnson N M and Nurmikko A V 1994 *Appl. Phys. Lett.* **65** 1001-03

[8] Garcia A and Northrup J E 1995 *Phys. Rev. Lett.* **74** 1131-34

Inst. Phys. Conf. Ser. No 145: Chapter 10
Paper presented at 22nd Int. Symp. Compound Semiconductors, Cheju Island, Korea, 28 August–2 September 1995
© *1996 IOP Publishing Ltd*

Miniband formation in graded-gap superlattices

H T Grahn†[1], L Schrottke†, K H Ploog†, F Agulló-Rueda‡ and U Behn§

† Paul-Drude-Institut für Festkörperelektronik, Hausvogteiplatz 5-7, D-10117 Berlin, Germany

‡ Instituto de Ciencia de Materiales de Madrid, CSIC, Campus de Cantoblanco, E-28049 Madrid, Spain

§ Fachhochschule Schmalkalden, Blechhammer 4 und 9, D-98564 Schmalkalden, Germany

Abstract. Two different graded-gap superlattices (SLs) have been realized and investigated by photocurrent (PC), electroreflectance (ER) and photoluminescence (PL) spectroscopy. The first SL is compositionally graded, while the second system consists of layers with different well and barrier thicknesses. In contrast to the compositionally graded SL, the PC spectra of the second system are dominated by excitonic effects. The formation of a miniband at finite electric fields can be clearly observed in the ER-spectra of the compositionally graded SL.

1. Introduction

Over the last decade semiconductor multiple quantum wells superlattices (SLs) have been used for many interesting optoelectronic and microelectronic applications such as lasers, modulators or high speed transistors.[1] If in a SL adjacent wells are strongly coupled, new electronic and optical effects such as miniband transport [2] and Wannier-Stark localisation [3] can arise. In a conventional, strongly coupled SL the electronic as well as the hole miniband exist at zero electric field. Applying an electric field parallel to the superlattice direction, leads first to a destruction of the heavy hole miniband, while the electronic miniband splits up into Wannier-Stark states. The optical interband transitions between the valence and conduction band exhibit the well-known Wannier-Stark ladder transitions, which consist of one spatially direct and several of spatially indirect transitions.[4, 5] For larger electric fields the electronic miniband is also destroyed resulting in a complete localization of both electronic and heavy hole states. At even larger field strengths a delocalization of the electronic and heavy hole states may

---

[1] on leave at: Tokyo Institute of Technology, 2-12-1 O-okayama, Meguro-ku, Tokyo 152, Japan.

occur again due to resonant tunneling between different subbands.[6] This coupling is accompanied by an anticrossing of the involved optical transitions resulting in non-linear optical properties, which can be exploited for optical switches and modulators.[7]

We have investigated the formation of minibands in two different graded-gap super-lattices using photocurrent (PC), electroreflectance (ER) and photoluminescence (PL) spectroscopy. The advantage of graded-gap SLs in terms of miniband formation is that the miniband forms at finite electric fields. Therefore, in contrast to conventional SLs the miniband in the conduction band and the miniband in the valence band are present at different electric fields. Furthermore, the transition regime between localized and extended states, i.e., the Wannier-Stark regime, can be confined to a much narrower field range than in conventional SLs.

2. Experimental

Both superlattices represent the intrinsic region of a p-i-n diode. Sample 1 consists of $Al_xGa_{1-x}As/Al_{0.3+x}Ga_{0.7-x}As$ layers with constant well and barrier thickness of 4.0 and 2.0 nm, respectively, increasing only the Al mole fraction x from one period to the next by 0.03. The first 8 periods of sample 1 are shown in Table 1 on the left side. The total number of periods is 15. Sample 2 contains GaAs wells and $Al_{0.35}Ga_{0.65}As$ barriers with varying thicknesses as shown in Table 1 on the right side. The number of periods is 8. For both samples the parameters are chosen in order to achieve an electronic miniband

Table 1. Layer sequence of the two graded-gap superlattices labeled sample 1 and 2. d denotes the nominal layer thickness. Sample 1 consists of 15 wells following the recipe shown in the table.

Well	Sample 1 d (nm)	Sample 1 Material	Sample 2 d (nm)	Sample 2 Material
	2.0	$Al_{0.30}Ga_{0.70}As$	1.4	$Al_{0.35}Ga_{0.65}As$
# 1	4.0	GaAs	7.0	GaAs
	2.0	$Al_{0.33}Ga_{0.67}As$	0.9	$Al_{0.35}Ga_{0.65}As$
# 2	4.0	$Al_{0.03}Ga_{0.97}As$	5.0	GaAs
	2.0	$Al_{0.36}Ga_{0.64}As$	2.5	$Al_{0.35}Ga_{0.65}As$
# 3	4.0	$Al_{0.06}Ga_{0.94}As$	3.6	GaAs
	2.0	$Al_{0.39}Ga_{0.61}As$	3.3	$Al_{0.35}Ga_{0.65}As$
# 4	4.0	$Al_{0.09}Ga_{0.91}As$	2.8	GaAs
	2.0	$Al_{0.42}Ga_{0.58}As$	3.9	$Al_{0.35}Ga_{0.65}As$
# 5	4.0	$Al_{0.12}Ga_{0.88}As$	2.2	GaAs
	2.0	$Al_{0.45}Ga_{0.55}As$	4.5	$Al_{0.35}Ga_{0.65}As$
# 6	4.0	$Al_{0.15}Ga_{0.85}As$	1.7	GaAs
	2.0	$Al_{0.48}Ga_{0.52}As$	5.3	$Al_{0.35}Ga_{0.65}As$
# 7	4.0	$Al_{0.18}Ga_{0.82}As$	1.2	GaAs
	2.0	$Al_{0.51}Ga_{0.49}As$	6.1	$Al_{0.35}Ga_{0.65}As$
# 8	4.0	$Al_{0.21}Ga_{0.79}As$	0.8	GaAs
	2.0	$Al_{0.54}Ga_{0.46}As$	6.5	$Al_{0.35}Ga_{0.65}As$

width at finite electric fields of 50-60 meV. The p-i-n diode has a built-in voltage so that the flat band condition is actually realized at about 1.6 V forward bias.

A schematic diagram of the conduction and valence band potential for both samples is shown in Fig. 1. The profile in (a) corresponds to flat band, while (b) shows the formation of the miniband at finite electric fields. The profile in (c) shows the band profile at high fields when the miniband is again destroyed. The absorption spectrum

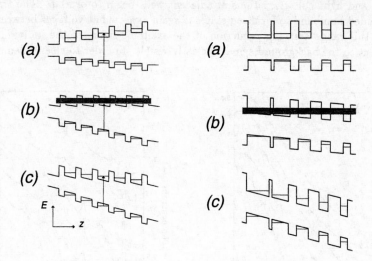

Figure 1. Schematic conduction and valence band potential profile in a SL with a compositionally grading (left) and variation of the well and barrier thicknesses (right) for different values of the applied electric field: (a) flat band, (b) the field range corresponding to miniband formation in the conduction band, and (c) at large electric fields, when the conduction band states are completely localized again.

should therefore display a red-shift of approximately half the miniband width for the absorption edge of each well at the applied electric field at which the miniband is formed. At higher fields the absorption edge should exhibit the typical blue-shift of the Wannier-Stark localisation. This transition is indicated by the arrows on the left side of Fig. 1.

3. Photocurrent spectra

The photocurrent spectra of sample 1 are shown in Fig. 2(a). A clear modulation of the increasing photocurrent by transitions from the highest heavy hole subband to the lowest electronic subband of each individual well is observed. However, at 0.7 V this modulation almost disappears and then becomes visible again at even lower voltage, i.e., larger electric fields. This region corresponds to the transition from extended miniband states to Wannier-Stark states. The first transition, i.e., from localized to extended states cannot be observed with photocurrent spectroscopy, since the current in this field range is very small. Please note that the spectra clearly show the two-dimensional density of states as expected for quantum wells. However, no excitonic effects are present. Furthermore, the light-hole transitions are not resolved.

In contrast to the first sample the photocurrent spectra of the second sample are dominated by excitonic transitions as shown in Fig. 2(b). In addition to the heavy hole transitions, also weaker transitions are present, which are attributed to light hole transitions. At least six of the eight wells are visible. However, for the narrower wells the heavy and light hole transitions of adjacent wells begin to overlap. The transition from extended miniband to localized states is again expected at voltages between 1 and 0.7 V. In this range a strong reduction of the excitonic signal of the individual wells with respect to the background signal is clearly visible. In order to clearly demonstrate

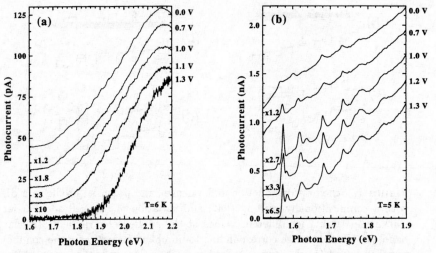

Figure 2. Photocurrent spectra of sample 1 (a) at 6 K and sample 2 (b) at 5 K for different applied voltages as indicated in the figure. The spectra are shifted vertically for better visibility.

the formation of a miniband at finite electric fields, the graded-gap superlattice has to be investigated at all field strengths, even close to flat band. This can be achieved either by performing absorption measurements, which are not easily performed on these structures, or by electroreflectance spectroscopy. Differential photocurrent spectroscopy has been performed on sample 1. However, only the transition from extended miniband to Wannier-Stark states could be observed.[8, 9] ER spectroscopy has the advantage over any photocurrent and aborption related technique that it can be applied at all field strengths and at the same time it is differential. Therefore, the background signal, which would be present in the absorption spectrum, is suppressed.

4. Electroreflectance spectra

The electroreflectance spectra of sample 1 (the compositionally graded SL) have been measured at 80 K in a standard set-up under normal incidence. In Fig. 3 the results of the ER spectra are compiled in an isointensity plot. The darker areas correspond to minima in the ER signal. At low fields (between 1.45 and 1.35 V) a red-shift of the ER-peaks by 17 meV occurs. This regions corresponds to the formation of the miniband, which can be observed until 1.15 V. Between 1.15 and 1.05 V the spectra show a rather complex lineshape again due to the interaction of the Wannier-Stark ladder transitions. For even larger fields the individual lines are blue-shifted by 17 meV. These spectra clearly demonstrate the formation of a miniband at finite electric fields. Numerical simulations of the differential absorption spectra have confirmed this interpretation.[10] The ER spectra for sample 2 have not been recorded yet.

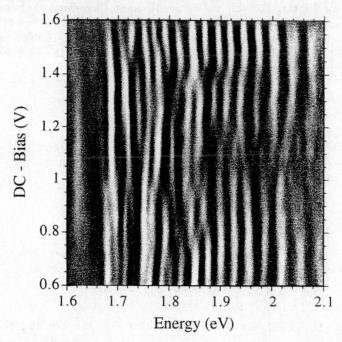

Figure 3. Isointensity plot of the ER spectra at different applied voltages in the compositionally graded SL at 80 K. The darker areas correspond to minima in the ER-signal.

5. Photoluminescence spectra

Sample 1 does not show any PL signal for applied voltages below flat band condition bias or an electroluminescence signal for applied voltages above flat band. This is probably due to the Al content in the wells. Sample 2 exhibits a PL signal, but only for the two widest wells. The coupling between the wells might be too large so that at all fields the transport through the superlattice occurs in a much shorter time interval than the recombination. Very recently, photoluminescence and PL excitation spectra obtained on a similar system as sample 2 have been reported showing the miniband formation at finite electric fields.[11]

1218

6. Summary

The formation of an electronic miniband at finite electric fields has been investigated in two different types of graded-gap superlattices in the material system GaAs/Al$_x$Ga$_{1-x}$As. The first sample consists of a compositionally graded SL in which the formation of the miniband can be observed with electroreflectance spectroscopy. The Wannier-Stark transition regions at small and large fields are confined to a much smaller field range than in conventional superlattices. A photoluminescence signal could not be detected in this system probably because of Al content of the wells. In the second sample the graded gap is achieved by changing the well and barrier thickness throughout the superlattice. In contrast to the compositionally graded system, the absorption spectrum of this system is dominated by excitonic effects. The GaAs wells of this system exhibit a photoluminescence signal. Further investigations are necessary to optimize the second structure so that the optical properties of this system can be improved.

Acknowledgments

We would like to thank A. Fischer for sample growth and N. Linder for the numerical simulations. This work was supported in part by the Bundesministerium für Bildung und Wissenschaft.

References

[1] See, e.g. Weisbuch C and Vinter B 1991 *Quantum Semiconductor Structures* (Boston: Academic)

[2] Esaki L. and Tsu R. 1970 *IBM J. Res. Develp.* **14** 61-64

[3] Wannier G H 1960 *Phys. Rev.* **117** 432-439

[4] Mendez E E, Agulló-Rueda F and Hong J M 1988 *Phys. Rev. Lett.* **60** 2426-2929

[5] Voisin P, Bleuse J, Bouche C, Gaillard S, Alibert C and Regreny A 1988 *Phys. Rev. Lett.* **61** 1639-42

[6] Schneider H, Grahn H T, Klitzing K v and Ploog K 1990 *Phys. Rev. Lett.* **65** 2720-23

[7] See, e.g., Agulló-López F, Cabrera J M and Agulló-Rueda F 1994 *Electrooptics: Phenomena, Materials, and Applications* (New York: Academic)

[8] Agulló-Rueda F, D'Intino A, Schmidt K H, Döhler G H, Grahn H T and Ploog K 1993 *Europhys. Lett.* **23** 283-288

[9] Grahn H T, Agulló-Rueda F, D'Intino A, Schmidt K H, Döhler G H and Ploog K 1994 *Sol. Stat. Electron.* **37** 835-838

[10] Behn U, Linder N, Grahn H T and Ploog K 1995 *Phys. Rev.* B51 17271-74

[11] Cao S M, Willander M, Toropov A A, Shubina T V, Mel'tser B Ya, Shaposhnikov S V, Kop'ev P S, Holtz P O, Bergman J P and Monemar B 1995 *Phys. Rev.* B51 15 June

Inst. Phys. Conf. Ser. No 145: Chapter 10
Paper presented at 22nd Int. Symp. Compound Semiconductors, Cheju Island, Korea, 28 August–2 September 1995
© *1996 IOP Publishing Ltd*

Simple mesoscopic dimensional approach to exciton properties in a quantum well under an applied electric field

B. W. Kim and E. M. Charlson

Department of Electrical and Computer Engineering, University of Missouri at Columbia, Columbia, Missouri 65211, U S A

Abstract. A simple method to describe the electric-field dependent exciton properties in a quantum-well system has been proposed. A mesoscopic dimensional space is defined to satisfy the excitonic virial theorem at any point, and can be described using only one scale factor corresponding not only to the well width but also to the applied electric field in a quasi two-dimensional system. The scale factor that implies the dimensionality of the system is obtained from the preintimated information of confined states for conduction-band and valence-band wells. This preintimated information is essential in the analysis of the quantum-well system, and has independent importance together with the properties of an exciton such as binding energies, orbital radii, and oscillator strengths.

We used the multistep-potential tunneling-resonance method, and considered the band-nonparabolicity effect on the quantized states of a quantum-well system when solving the Schrödinger equations of individual electrons and holes. A GaAs/Al_xGa_{1-x}As quantum-well structure is analyzed and the results of our calculations are compared to those from very sophisticated methods reported previously. Excellent agreement is shown.

1. Introduction

In a quantum-well system, not only is a particle allowed in its quantized energy levels, but also its motion is reduced from three to two degrees of freedom. Even relatively strong external perturbation is not likely to remove these restrictions, so that excitons persist. This persistence of excitons implies that free-exciton binding energies are increased by confinement in a well structure. The optical properties of the quantum-well system are mainly dependent on the extent to which this confinement can be controlled by inducing an electric field in the potential well.

The most common approach to obtain the solution of the exciton problem in quantum wells is the variational method [1]-[4]. The exciton binding energy is calculated from the trial wave-function parameters that give the minimum expectation value of the exciton Hamiltonian. Choosing a trial wave-function is somewhat artificial, and it is a major task to reach successful results. To include the valence-band coupling effect in binding energies of excitons, Ekenberg and Altarelli used the off-diagonal elements of the Luttinger Hamiltonian and treated them with a perturbational approach [5]. The Coulomb Green's-function approach has been applied to the exciton problem recently to show the limitations of the variational method [6].

In recent years, much effort has been devoted to finding a simple method to solve the exciton problem without a lengthy calculation process. By introducing the concept of mesoscopic dimensionality [7] or fractional dimensionality [8], an anisotropic system can be transformed to an isotropic system. The mesoscopic dimension is determined by the degree of anisotropy. Mathematically the dimension is the same as the number of degrees of freedom and it can be any positive real number from three to zero. We will briefly discuss the choice between two and three dimensions, since the dimensions of any quantum-well system fall into this range.

2. Mesoscopic-dimensional space and scale factor

A mesoscopic-dimensional space can be defined such that each coordinate in three-dimensional space is multiplied by an arbitrary scale factor η and the state vector in this space has the form [7]

$$\psi_\eta = \eta^{3/2} \psi(\eta z_e, \eta z_h, \eta r), \tag{1}$$

where the subscripts e and h denote, respectively, the electron and the hole; z is the coordinate perpendicular to the plane of the quantum-well layers; r is the relative distance between electron and hole in the x-y plane. If we choose η to satisfy the minimization condition of the variational principle, the virial theorem of the mesoscopic-dimensional space can be expressed [7]

$$KE_\eta = -\frac{1}{2} PE_\eta, \tag{2}$$

where KE_η and PE_η are the kinetic and potential energy terms of the exciton binding energy respectively.

First consider only the ground state of the exciton binding energy. η varies from 1 for a purely two dimensional system to 1/2 for a three dimensional system, so that the exciton binding energy for the ground state in the space being considered can be written

$$E_\eta = (2\eta)^2 Ry, \tag{3}$$

where Ry is the three-dimensional exciton Rydberg constant. The mesoscopic-dimensional exciton radius is defined to have the following relation to the potential energy term

$$PE_\eta = -\frac{e^2}{\varepsilon a_\eta}, \tag{4}$$

where e is the electron charge, and ε is the permittivity of the medium. The virial theorem gives

$$E_\eta a_\eta = -\frac{e^2}{2\varepsilon}. \tag{5}$$

Thus, the mesoscopic-dimensional exciton radius can be expressed by the scale factor

$$a_\eta = \frac{a_B}{(2\eta)^2}, \tag{6}$$

where a_B is the effective Bohr radius. It should be emphasized that the mesoscopic-dimensional exciton radius does not refer to the Bohr radius in that dimension. The Bohr radius for the two-dimensional system, for example, is $a_B/2$ and can be determined from the exponential term of the corresponding wave function; however, $a_{\eta=1}$ for the two-dimensional system is $a_B/4$ and can be obtained from the position where the virial theorem is satisfied. a_η, then, should be the maximum probability-peak position of a particle in any mesoscopic-dimensional system and represents the 1s orbital radius in a hydrogenic system.

3. Exciton under an electric field

An electric field perpendicular to the quantum-well layers pulls the electron and the hole to opposite sides. The exciton, which is still not broken by the field increases its radius by reducing its binding energy. The quantum-well system is now in another mesoscopic-dimensional space and that space could be defined by the field-dependent scale factor $\eta(F)$.

In most cases, the analysis for an individual particle in a quantum-well system is essential and the results of the analysis probably include its eigenvalues and eigenfunctions. We consider the first confined states of both particles that form an exciton, and solve the Schrödinger equation of an empirical two-band model [9] [10] by using the multistep-potential tunneling-resonance method to include band nonpabolicity effects on the solutions of the effective Hamiltonian [11]. From this preintimated information, obtaining the probability peak position of each particle in the direction perpendicular to the layers and the distance between electron and hole are straightforward.

It is found that the mesoscopic-dimensional exciton radius for the quantum well under an electric field can be well described by

$$a_{\eta(F)} = \sqrt{a_\eta^2 + \left[\eta(F)z_{eh}^*\right]^2}, \tag{7}$$

where z_{eh}^* is the distance between the probability peak position of the electron and that of the hole. It should be noted that $\left[\eta(F)z_{eh}^*\right]$ is just the coordinate transform of z_{eh}^* for the newly-defined mesoscopic-dimensional space. From the definition, one obtains

$$\eta(F) = \frac{1}{2}\sqrt{\frac{a_B}{a_{\eta(F)}}}. \tag{8}$$

Equation (7) and Equation (8) can be solved in self-consistent manner. Once, $\eta(F)$ is known, calculating $E_{\eta(F)}$ from Equation (3) is a simple matter.

Since the heavy hole and light hole masses are different from each other, their effective Bohr radii must be calculated separately and the each radius can be calculated from the relation

$$a_B = \frac{\varepsilon_r m_0}{\mu_t} \cdot a_0, \tag{9}$$

where ε_r is the relative permittivity of the medium, a_0 is the Bohr radius and t stands for hh (heavy hole) or lh (light hole). m_0 is the free electron mass, and μ_t is the exciton reduced

Table 1. Device parameters used in exciton binding-energy calculations of Figure 2 by different groups.

Well width	x (Al)	Band-gap offset	
100 Å	0.30	65:45	Present model
100 Å	0.32	57:43	Nojima [13]
100 Å	a	60:40	Hong and Singh [14]

[a]Unknown.

mass corresponding to t. We do not consider the off-diagonal terms in the Luttinger Hamiltonian when calculating the reduced mass. That means that the reduced mass does not change with any external perturbation and it retains its two-dimensional value. The effective Rydberg constant in the well material has the magnitude

$$R_y = \frac{\mu_t}{\varepsilon_r^2 m_0} R_H, \qquad (10)$$

where R_H is the Rydberg constant.

If we assume that all state vectors in the mesoscopic-dimensional space of an exciton system can be represented by one scale factor $\eta(F)$, it is possible to obtain all of the discrete bound-state energies and their corresponding orbital radii from the following:

$$E_n = \frac{R_y}{\left[n + \dfrac{1}{2\eta(F)}\right]^2}, \qquad (11)$$

$$a_n = \left[n + \frac{1}{2\eta(F)}\right]^2 a_B, \qquad (12)$$

where n=0, 1, 2, 3, ... is the principal quantum number for the mesoscopic-dimensional exciton system. The above relations are essentially the same as the relations proposed by He [12] if we define the mesoscopic dimension as

$$D = 1 + \frac{1}{\eta(F)}. \qquad (13)$$

The oscillator strength of an exciton is simply proportional to $\eta^3(F)$ [7], so that it can be obtained from the strength of the purely two- or three-dimensional exciton system and the above scale factor.

3. Results

We compare the results to those of other calculations [13], [14] that involved very similar structures (see Table 1) to the quantum well analyzed in this work. The zero-field exciton binding energies for heavy and light holes in a 100 Å GaAs/Al$_{0.3}$Ga$_{0.7}$As quantum well are chosen to be 8.5 meV and 9.5 meV, respectively [15], and the dielectric constants for the whole quantum-well system are averaged to obtain the geometric mean.

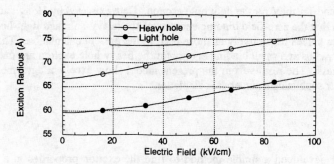

Figure 1. Exciton radius variation under electric fields.

To obtain the confined states of the quantum well, the conduction-band nonparabolicity parameter and the light-hole-band nonparabolicity parameter are assumed to be 4.9×10^{-19} m^2 and 7.35×10^{-19} m^2 respectively. The other material constants relevant to present calculations are: The band offset ratio is 0.65, the electron effective mass is $m_0(0.067+0.083x)$, and the heavy-hole and light-hole effective masses are $m_0(0.34+0.175x)$ and $m_0(0.094+0.069x)$, respectively. Here m_0 refers to the electron rest mass and x denotes Al mole fraction in the barrier.

Figure 1 illustrates the variation of the mesoscopic-dimensional exciton radius for the 1s state of an exciton in a 100 Å GaAs/$Al_xGa_{1-x}As$ quantum well under electric fields. As expected, we can see the gradual increase of the mesoscopic-dimensional exciton radius with increasing electric field from the figure. An electric field perpendicular to the quantum-well layer stretches the exciton in the quantum well, so that the mesoscopic-dimensional space of the exciton approaches three-dimensional space by reducing the scale factor. Finally, the exciton binding energy decreases with increasing electric field as shown in Figure 2. The binding energy drops in a field of 100 kv/cm are 1.0 meV for the heavy

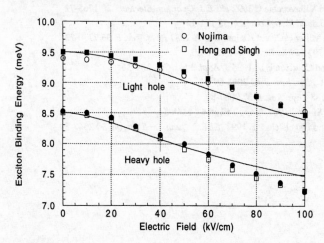

Figure 2. Exciton binding energy variation under electric fields. The solid lines indicate the present results.

hole exciton and 1.1 meV for the light hole exciton. Data extracted by Hong and Singh are adjusted such that the zero-field binding energies of the heavy-hole and light-hole excitons have the same values respectively as those used in the present calculation. The relatively large exciton radius and small binding energy for the heavy hole exciton are mainly due to its heavier mass. The results from the present model show excellent agreement with the values of other sophisticate numerical calculations.

4. Summary

We have developed a simple method to find the exciton properties in a quasi two-dimensional system, even in the presence of an electric field. Despite recent considerable attention given to the mesoscopic-dimensional or fractional-dimensional approach, this approach does not adequately treat external perturbations. The proposed method overcomes this weakness and is useful in modeling any semiconductor quantum-well device that deals with the exciton, such as quantum-well lasers, quantum-well optical modulators, and self-electro-optic effect devices.

References

[1] Miller D A B, Chemla D S, Damen T C, Gossard A C, Wiegmann W, Wood T H, and Burrus C A 1985 *Phys. Rev. B* **32** 1043-60
[2] Miller D A B, Chemla D S, Damen T C, Gossard A C, Wiegmann W, Wood T H, and Burrus C A 1984 *Phys. Rev. Lett.* **53** 2173-76
[3] Miller R C, Kleinman D A, Tsang W T, and Gossard A C 1981 *Phys. Rev. B* **24** 1134-36
[4] Bastard G, Mendez E E, Chang L L, and Esaki L 1982 *Phys. Rev. B* **26** 1974-79
[5] Ekenberg U and Altarelli M 1987 *Phys. Rev. B* **35** 7585-95
[6] Chuang S L, Schmitt-Rink S, Miller D A B, and Chemla D S 1991 *Phys. Rev. B* **43** 1500-9
[7] Campi D and Villavecchia C 1992 *IEEE J. Quantum Electron.* **28** 1765-72
[8] He X F 1990 *Solid State Commun.* **75** 111-14
[9] Nelson D F, Miller R C, and Kleinman D A 1987 *Phys. Rev. B* **35** 7770-73
[10] Leavitt R P 1991 *Phys. Rev. B* **44** 11270-80
[11] Kim B W and Charlson E M 1994 *J. Appl. Phys.* **76** 1334-36
[12] He X F 1991 *Phys. Rev. B* **43** 2063-69
[13] Nojima S 1988 *Phys. Rev. B* **37** 9087-88
[14] Hong S and Singh J 1987 *J. Appl. Phys.* **61** 5346-51
[15] Kaushik S and Hagelstein P L 1994 *IEEE J. Quantum Electron.* **30** 2547-59

Inst. Phys. Conf. Ser. No 145: Chapter 10
Paper presented at 22nd Int. Symp. Compound Semiconductors, Cheju Island, Korea, 28 August–2 September 1995
© *1996 IOP Publishing Ltd*

A simple approach for the valence-band structure of strained-layer quantum wells within the 6x6 Luttinger-Kohn model

Sean J. Yoon and Doyeol Ahn

LG Electronics Research Center, 16 Woomyeon, Seocho-Gu, Seoul 137-140,
Republic of Korea

Abstract : In this paper, we present a simple method for calculating the valence-band structure of strained-layer quantum wells. The method consists of expanding the envelope functions of the Luttinger-Kohn Hamiltonian using Fourier expansion, and would be useful to calculate the confined states in the quantum well for which the envelope function decays very rapidly in the barrier regions so that the homogeneous boundary conditions can be imposed. The effects of the spin-orbit (SO) split-off band coupling on the valence-band structure are studied within the 6x6 Luttinger-Kohn model.

1. Introduction

The calculation of the hole states is essential both for the fundamental research and the development of optoelectronic devices, especially for the strained-layer quantum wells in which the strain and the quantum size effects modify the valence band structure considerably. Up to now, the numerical techniques based on the finite difference method[12,13], variational method[14-17], and the propagation matrix approach[18] have been used to calculate the valence band structure.

In this paper, we present a simple method for making reasonably accurate valence band structure calculations of the strained-layer quantum wells which requires much less computation time than the finite difference method. We shall refer to the method as the Fourier expansion method.

2. The model

To calculate the band structure and electron (or hole) wave functions of a quantum well, we use the multiband effective-mass theory ($\vec{k} \cdot \vec{p}$ method). For the valence band with the spin-orbit (SO) coupling effects, the 6x6 Luttinger-Kohn Hamiltonian[11] is employed. We first make a unitary transformation of the 6x6 Luttinger-Kohn Hamiltonian into two 3x3 blocks H^U and H^L within the axial approximation.

We expand the envelope functions as a linear combination of an orthonormal set of functions u_n

$$g^{(v)}(z) = \sum_{n=1} c_n^{(v)} u_n(z) , \qquad (1)$$

where $c_n^{(v)}$ is the expansion coefficient and $u_n(z)$ is the sinusoidal function. We obtain the associated matrix eigenvalue equations for the expansion coefficient $c_n^{(v)}$:

$$\sum_{m=1}^{N} < u_n | \sum_{v'} ([H^\sigma_{vv'} (k_\parallel, -i\frac{\partial}{\partial z}) + V_v(z) \delta_{vv'} + H^\sigma_{\varepsilon vv'}]) | u_m > c_m^{(v')} = E\, c_n^{(v)}. \qquad (2)$$

Here $V_v(z) = V_h(z)$ for $v = 1,2$ or $5,6$; $V_v(z) = V_{so}(z)$ for $v = 3,4$; H^σ_ε is the Hamiltonian for strain (shear) potential; and $\sigma = U$(or L) refers to the upper (or lower) blocks, respectively. The hole confining potentials $V_h(z)$ and $V_{so}(z)$ are obtained from the calculated valence band offsets from the model-solid theory[19] of Van de Walle and Martin.

The procedure requires the matrix elements of the following two operators

$$\frac{\partial}{\partial z} A(z) \frac{\partial}{\partial z} \quad \text{and} \quad \frac{1}{2}(B(z)\frac{\partial}{\partial z} + \frac{\partial}{\partial z} B(z)) \qquad (3)$$

in addition to the step-like function $C(z)$ to ensure that the Hamiltonian is self-adjoint. Matrix elements are then in analytical form for step-like functions $A(z)$ and $B(z)$ which are given elsewhere.

The drawback of the Fourier expansion method is that only the confined states can be obtained and the procedure becomes inaccurate for higher states near the continuum. Since there are only a finite number of confined states in the quantum well, we can use finite basis set applying the Weierstrass theorem[20]. We can obtain the optimum matrix size by varying the set of values (L,N) for each quantum well width for numerical accuracy, where L is the total width of the structure including the barriers. It is found that the matrix size required for the Fourier expansion method (this paper) is about one-fifth of the matrix size for the finite difference method for comparable numerical accuracy. As a result, an order of magnitude reduction of the computation time can be obtained in the Fourier expansion method.

In Figs. 1(a)~1(c) we compare the valence-band structures obtained by the Fourier expansion method (solid curves) and the Finite difference method (dotted curves) of 60 Å $In_xGa_{1-x}As$ surrounded by InP outer barriers versus in-plane wave vector (in units of $2\pi/a_0$). The matrix sizes for the former and latter methods are 27 and 120, respectively and the total width L of 200 Å is used in the calculation. It is found that the discrepancies between the two methods become larger for higher subbands. For the lowest two subbands, however, the agreement is reasonably good for all three cases.

We show the results for a 60 Å $In_xGa_{1-x}As$ (x=0.53) quantum well for four different set of values of (N,L) in Fig. 2 : (i) N=5, L=120Å, (ii) N=9, L=200Å, (iii) N=13, L=300Å, and (iv) N=50, L=1200Å. Reasonable agreements in the lowest three subband structures for the four cases (i) to (iv) verifies our assumption that N is proportional to L. We note that the matrix size for each block Hamiltonian is 3Nx3N.

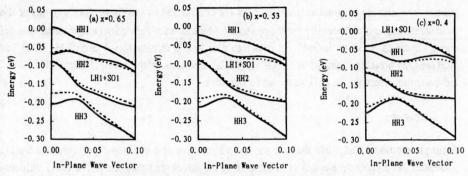

Fig.1 Valence band structure of a 60 Å $In_xGa_{1-x}As/InP$ quantum well is shown for (a) x=0.65 (compressive), (b) x=0.53 (no strain), and (c) x=0.4 (tensile) with the Fourier expansion method (solid curves) and the finite difference method (dashed curves).

The calculated valence band structures obtained by the Fourier expansion method (solid curves) and the finite difference method (dashed curves) for a 60 Å $Si_{0.75}Ge_{0.25}/Si$ quantum well are shown in Fig. 3.

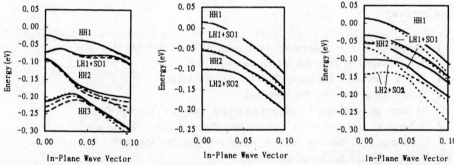

Fig.2 Valence band structure of a 60Å $In_{0.53}Ga_{0.47}As/InP$ quantum well for cases : (i) L=120 Å, N=5 (dotted), (ii) L=200 Å, N=9 (solid), (iii) L=300 Å, N=13 (short-dashed) and (iv) L=1200 Å, N=50 (long-dashed).

Fig.3 Valence band structure of a 60 Å $Si_{0.75}Ge_{0.25}/Si$ quantum well with the Fourier expansion method (solid curves) and the finite difference method (dashed curves).

Fig.4 Valence band structure of a 60 Å $Si_{0.75}Ge_{0.25}/Si$ quantum well with (solid curves) and without (dashed curves) SO coupling.

In order to see the effect of the spin-orbit (SO) split-off band coupling effects on the band structure, we compare the calculated results of the valence-subband structure obtained from the 6x6 (solid curves) and the 4x4 (dotted curves) models in Fig. 4 for a 60 Å $Si_{0.75}Ge_{0.25}/Si$ quantum well. Since the heavy-hole band is decoupled from the rest of the valence bands at the zone center, the SO coupling changes only the subband energies of the light hole states when $k_{||} = 0$. However, both heavy and light holes are affected by the coupling for a finite $k_{||}$. Because of the narrow SO splitting (~0.1eV), the subband structures, especially for the light hole states, of SiGe/Si are strongly affected by the SO coupling as can be seen in the figure.

It has been customary to label the subbands according to their character at the zone center and is well known that the SO band is coupled to the light hold band even at $k_\parallel = 0$. So it is not entirely correct to label the subband as "light hole" or "spin-orbit" to describe the subbands which are originated from the spin 1/2 component of the Γ_8 or the Γ_7 bands of the bulk semiconductors especially for those semiconductors with very narrow spin-orbit split off energy such as SiGe or InGaP. In order to clarify the situation, we use "LH1+SO1" or "LH2+SO2" notations instead of "LH1" or "LH2" notations, respectively. Here "LH1+SO1" indicates the admixture of SO1 state and LH1 state.

3. Summary

A simple method to calculate the valence-band structures of a strained-layer quantum well is presented. The procedure consists of expanding the envelope functions of the 6x6 Luttinger-Kohn Hamiltonian using the sinusoidal components of the Fourier series and is found to be useful in the calculation of the confined states in the quantum well. The effects of the spin orbit (SO) split-off band coupling on the band structure are studied by solving the 6x6 Hamiltonian numerically. Comparison with the results obtained by the finite difference method shows that the method suggested in this work reduces the computation time required to calculate the band structure by an order of magnitude for comparable accuracy.

References

[1]   R. Eppenga, M. F. H. Schuurmans, and S. Colak 1987 *Phys. Rev. B36* 1554

[2]   E. P. O'Reilly 1989 *Semicond. Sci. Technol.* 4 121

[3]   C. Y. P. Chao and S. L. Chuang 1992 *Phys. Rev. B46* 4110

[4]   B. A. Foreman 1993 *Phys. Rev. B48* 4964

[5]   M. Sugawara, N.Okazaki, T. Fujii, and S.Yamazaki 1994 *Phys. Rev. B48* 8102

[6]   A. Baliga, D. Trivedi, and N. G. Anderson 1994 *Phys. Rev. B49* 10402

[7]   G. Edwards, E. G. Valadares, and F. W. Sheard 1994 *Phys. Rev. B50* 8493

[8]   C. S. Chang and S. L. Chuang, unpublished.

[9]   D. Ahn and S. J. Yoon 1995 *J. Korean Phys. Soc.* 28 145

[10]   D. Ahn, S. J. Yoon, S. L. Chuang and C. S. Chang, *J. Appl. Phys.*, Accepted (Aug. 15, 1995).

[11]   J. M. Luttinger and W. Kohn 1955 *Phys. Rev.* 97 869

[12]   D. Ahn, S. L. Chuang, and Y. C. Chang 1988 *J. Appl. Phys.* 64 4056

[13]   D. Ahn and S. L. Chuang 1988 *IEEE J. Quantum Electron.* 24 2400

[14]   M. Altarelli, U. E. Kenberg, and A. Fasolina 1985 *Phys. Rev. B32* 5138

[15]   D. A. Broido and L. J. Sham 1985 *Phys. Rev. B31* 888

[16]   G. D. Sanders and Y. C. Chang 1987 *Phys. Rev. B35* 1300

[17]   A. Twardwski and C. Hermann 1987 *Phys. Rev. B35* 8144

[18]   S. L. Chuang 1991 *Phys. Rev. B43* 9649

[19]   C. G. Van de Walle 1989 *Phys. Rev. B39* 1871 and references therein.

[20]   C. L. DeVito 1990*Functional analysis and linear operator theory* (Addison-Wesley, New York), pp37-44.

[21]   C. H. Henry 1991 *IEEE J. Quantum Electron.* 27 523

[22]   T. Fromherz, E. Koppensteiner, M. Helm, G. Bauer, J. F. Nutzel and G. Abstreiter 1994 *Phys. Rev. B50* 15073

Inst. Phys. Conf. Ser. No 145: Chapter 10
Paper presented at 22nd Int. Symp. Compound Semiconductors, Cheju Island, Korea, 28 August–2 September 1995
© 1996 IOP Publishing Ltd

High-Field Electron Transport of the ZnS Phosphor in AC Thin-Film Electroluminescent Devices

Insook Lee, S. Pennathur, K. Streicher, T. K. Plant, J. F. Wager, P. Vogl†, and S. M. Goodnick

Department of Electrical and Computer Engineering,
Oregon State University, Corvallis, OR 97331-3211, USA

† Walter Schottky Institute, Technical University Munich,
D-85748 Garching, Germany

Abstract. A full-band Monte Carlo simulation of the hot carrier distribution and impact excitation yield for ZnS doped with Mn^{+2} and Tb^{+3} is performed. A nonlocal empirical pseudopotential calculation has been performed for ZnS and used in the full-band Monte Carlo model which includes all the pertinent electron scattering mechanisms. Band-to-band impact ionization and impact excitation of luminescence centers are included, both of which are threshold phenomena crucial to the operation of an AC thin-film electroluminescent (ACTFEL) device. The impact excitation yield (the number of excited centers per traversing electron in the phosphor) exhibits a monotonic increase with increasing phosphor field in agreement with experiment. However, at high fields where impact ionization begins to significantly affect the transport, the impact excitation yield is somewhat suppressed.

1. Introduction

ACTFEL devices are an essential technology for future flat-panel display applications. The basic device structure consists of a phosphor layer sandwiched between two insulating layers and a pair of electrodes. The phosphor consists of a wide bandgap semiconductor such as ZnS (E_g=3.7 eV) heavily doped ($\sim$ 1%) with a luminescent impurity such as Mn^{+2}. The mechanism for light emission is the application of large AC voltages to the layer structure such that electrons are sourced into the high-field phosphor layer from interface states at the insulator-semiconductor interface either due to tunneling or due to field emission. For sufficiently high-fields (typically 1-2 MV/cm), electrons in the phosphor layer may gain sufficient energy to impact excite electrons in the luminescent impurities from the ground to excited states, which subsequently undergo radiative decay emitting photons [1]. The luminescent and power conversion efficiency of such devices critically depend upon the high-field distribution of electrons in the conduction bands

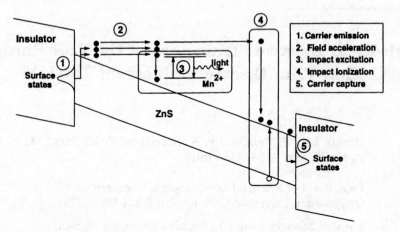

Figure 1. Band diagram of an ACTFEL device showing the important interactions during transport across the phosphor layer including impact excitation of luminescent impurities (3) and band-to-band impact ionization (4).

of the phosphor, and the impact excitation rate for exciting luminescent impurities. Thus, an understanding of high-field carrier transport in the phosphor layer and of the physics of different threshold processes such as band-to-band impact ionization and impact excitation of luminescent impurities is essential for device design, especially when newer phosphors are continually being developed in the quest for a full-color EL display. Further, the ACTFEL device structure provides a unique probe of the high energy hot carrier distribution function in a semiconductor since the output intensity for a particular luminescent impurity is a direct measure of the carrier population at the excitation energy for the impurity, which has a well defined threshold energy as it involves an atomic transition. Some of the key physical processes that are fundamental to the operation of an ACTFEEL device are illustrated in Fig. 1.

2. A full-band Monte Carlo Model

The band structure of ZnS has been calculated using the nonlocal empirical pseudopotential method (EPM) [2], in which the calculated reflectivity spectrum is fit to recent uv spectroscopic ellipsometric data [3] for this material. For the purpose of full-band Monte Carlo simulations, the dispersion relation and density of states for the first four conduction bands have been used to describe the carrier dynamics. The nonlocal correction to the pseudopotential was included, along with a corresponding correction for calculating velocities as well as effective masses [4]. The energy values and energy derivatives (i.e. velocities) are computed and stored so that the carrier energies and velocities during the course of the Monte Carlo simulation can be determined (interpolated, when necessary) from the tabulated full-band dispersion relation. In addition to including all the relevant phonon and impurity scattering mechanisms which includes polar optical, acoustic, and nonpolar optical phonon scattering as well as elastic scattering due to ionized and neutral impurities [5], the full-band Monte Carlo model also includes both band-to-band

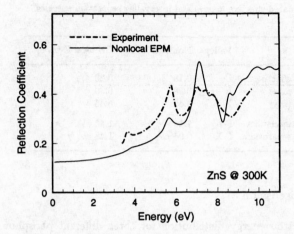

Figure 2. Experimental and theoretical reflectivity spectra for ZnS.

impact ionization and impact excitation of luminescent impurity atoms. We observe that the distribution function is unstable with respect to runaway at the highest fields unless mitigated by impact ionization. While a nonparabolic band model for computation of the excitation cross section for Mn^{2+} [6] is already included in the simulation, a formulation of the full-band impact excitation rates has also been completed. An approximation of the energy-dependent band-to-band impact ionization rate [7] computed using an empirical pseudopotential band structure is incorporated into the Monte Carlo model.

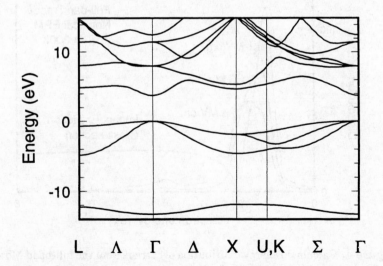

Figure 3. Nonlocal empirical pseudopotential band structure used in the Monte Carlo simulation of transport in the ZnS phosphor layer.

Table 1. Calculated effective masses and intervalley separation energies.

	Valley	Nonlocal EPM	CB local EPM[8]
effective	Γ	0.18	0.20
mass	L	0.28	0.30
(m_0)	X	0.46	0.46
intervalley	Γ-L	1.16 eV	1.48 eV
separation	Γ-X	0.94 eV	1.48 eV

3. Results and discussion

The simulated electron energy distribution for three different phosphor electric field strengths is shown in Fig. 4, compared with the calculated impact excitation cross-section of Mn^{2+}. The distribution becomes hotter (i.e. increasing average energy) with increasing phosphor fields. For manganese centers (used for yellow luminescence) with an excitation threshold energy of about 2.1 eV, it is seen that a considerable number of the electrons in the ensemble are energetic enough to cause impact excitation. From the distributions shown, we estimate that about 2.1 %, 15.3 %, and 34.4 % of the electrons possess energy above 2.1 eV at fields of 1 MV/cm, 1.5 MV/cm and 2.0 MV/cm respectively. The corresponding numbers at the three fields for terbium (which is used for green luminescence and has an excitation threshold of 2.5 eV) are 0.3 %, 7.1 %, and 21.1 %, respectively.

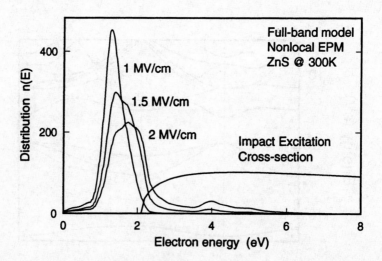

Figure 4. Calculated energy distribution of carriers from the full-band Monte Carlo model for various electric field strengths compared with the calculated impact excitation cross-section of Mn^{2+}.

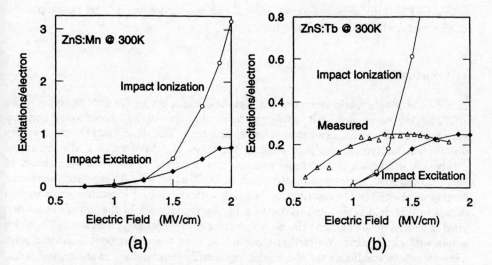

Figure 5. (a) Calculated impact excitation and impact ionization yield (number of events per electron across a 0.5 micron phosphor layer) in a ZnS:Mn ACTFETL device. (b) Calculated impact excitation and impact ionization yield (number of events per electron across a 0.5 micron phosphor layer) along with the field-dependent impact excitation yield estimated from experiments, of a ZnS:Tb ACTFEL device.

The internal luminescence intensity is proportional to a convolution of the electron energy distribution function with the impact excitation cross-section of the luminescent centers in the phosphor [9]. An increasing overlap of the excitation cross-section with the electron energy distribution with increasing phosphor field, therefore, predicts an increasing luminescence from the thin-film EL device, for increasing electric fields. An estimate of the quantum yield (a measure of the per-electron contribution to the luminescence) is obtained by tracking the number of impact excitation events per electron across a 0.5 μm phosphor layer. Figure 5(a) shows a plot of this impact excitation quantum yield as a function of the phosphor field for Manganese luminescent centers. It shows a threshold of approximately 0.7 MV/cm, increases monotonically for increasing fields. This monotonic increase is consistent with experimentally observed trends in the brightness-voltage curves of ACTFEL devices [10]. A comparison of the average number of impact ionization events to that of the impact excitation events in Fig. 5(a) reveals that although the impact ionization process has a larger threshold field of about 1 MV/cm, it increases rapidly and crosses the impact excitation curve at around 1.25 MV/cm, and dominates above 1.25 MV/cm. This establishes the importance of band-to-band impact ionization as one of the important physical processes in typical ACTFEL operating electric fields.

Figure 5(b) plots the computed impact excitation and impact ionization quantum yields for a ZnS phosphor with Terbium lumiescent impurities. Also shown are estimated quantum yield values from measured luminescence data for ZnS:Tb devices [11]. It is observed that both the experimental and simulated impact excitation quantum yields, display similar qualitative trends. It is noted that the simulated data for ZnS:Tb devices shown in Fig. 5(b) were generated using a two conduction-band local pseudoptential band structure in the Monte Carlo model.

4. Conclusions

A full-band Monte Carlo simulation of high-field transport in the ZnS phosphor of an ACTFEL device is presented, using the band structure of ZnS based on a nonlocal empirical pseudopotential calculation for the first time. The steady-state electron energy distribution for typical phosphor fields in the range of 1 MV/cm to 2 MV/cm reveal a substantial fraction of electrons with energies in excess of the excitation threshold energies of typical luminescent centers such as Mn^{2+} and Tb^{3+}. As expected, increasing overlap of the tail of the distribution function with the impact excitation cross section occurs in this field range, giving rise to a monotonic increase in the impact excitation yield of the carriers transiting the phosphor layer with increasing phosphor fields, which agrees with experiment. We further observe that band-to-band impact ionization plays a crucial role in stabilizing the electron energy distribution function at the highest fields. The impact excitation yield is found to be strongly affected by the presence of impact ionization and the associated threshold voltage for this mechanism.

References

[1] D. C. Krupka, J. Appl. Phys. **43**, 476, 1972.

[2] J. R. Chelikovsky and M. L. Cohen, Phys. Rev. B **14** 556, 1976.

[3] J. Barth J, R. L. Johnson and M. Cardona, *Handbook of Optical Constants of Solids II* (Boston: Academic) p 213, 1991.

[4] M. M. Rieger and P. Vogl, Phys. Rev. B **48** 14 276, 1993.

[5] K. Bhattacharyya, S. M. Goodnick, and J. F. Wager, J. Appl. Phys., **73**, 3390, 1993.

[6] E. Bringuier and K. Bhattacharyya, Unpublished work.

[7] M. Reigrotzki, M. Stobbe, R. Redmer, and W. Schattke, To be published in Phys. Rev. B.

[8] M. L. Cohen and T. K. Bergstresser, Phys. Rev., **166**, 789, 1966.

[9] J. W. Allen, J. Lumin., **48/49**, 18, 1991.

[10] D. H. Smith, J. Lumin., **23**, 209, 1981.

[11] K. Streicher, M.S. Thesis, Oregon State University, 1994.

Inst. Phys. Conf. Ser. No 145: Chapter 10
Paper presented at 22nd Int. Symp. Compound Semiconductors, Cheju Island, Korea, 28 August–2 September 1995
© *1996 IOP Publishing Ltd*

Γ-X Electron Tunneling in AlAs/GaAs/AlAs Double-barrier Quantum-well Heterostructure

Gyungock Kim

Research Department, Electronics and Telecommunications Research Institute, Yusong
P. O. Box 106, Taejon, Korea 305-600.

Abstact. Γ-X intervalley interlayer electron tunneling in a AlAs/GaAs/AlAs (001) double-barrier structure has been investigated using the tunneling formalism based on the non-equilibrium Green's function technique. The effects of indirect bandgap, band non-parabolicity, multi-orbitals in the aperiodic layered structure is taken into account by the Green's function formulation of scattering theory incorporated with the multi-orbital tight-binding model. The high external bias effect on the junction is solved exactly within this model. The effect of the X valley originated barrier confined states on the electron tunneling are analyzed with the spectral local density of states of the double-barrier structure in the [001] direction under the external bias. The smaller current peak to valley ratios are found in the double-barrier structures with thicker barriers, which is consistent with the experimental results. The enhancement of Γ-X interlayer intervalley electron transfer and reduction of the Γ-Γ resonant electron interlayer transfer is found with the increase of the phenomenological imaginary potential term.

1. Introduction

A material system of particular interest is that of AlAs/GaAs or $Al_{1-x}Ga_xAs$/GaAs based double-barrier quantum-well (DBQW) heterostructures because of their applicability to modern electrical and optical devices. Due to small conduction band discontinuity between the GaAs Γ valley and the AlAs X valley, experimental studies report the electron tunneling through both QW Γ states and barrier X valley originated states in these systems. The interlayer Γ-X intervalley scattering process in multiple quantum-well structures is also important in the ultrafast relaxation of photoexcited electrons. Various experimental observations have recognized the situations, which go beyond the limit of effective mass approximation[1], and require more complicated pictures and understanding of the layer-structured heterostructure. Some of the examples include the indirect bandgap barrier and the interlayer intervalley transfer process, and the high external electric field problem in the heterojunction. The electron tunneling by the Γ-X transfer includes the change of wave vector, which, not by conserving the wave vector, subsequently complicates the tunneling situation. External electric fields applied to quantum-well structures are expected to produce pronounced effects on the electronic structures. The band nonparabolicity becomes important for electron transport properties under high external field. Ultra-thin heterostructure systems are also expected to deviate from the results obtained with the approach which directly uses the bulk-like treatment for each constituent layer. In the aperiodic layered system, the translational periodicity is broken in the tunneling direction. Therefore instead of the electron wave-vector $\vec{k}$, the consideration of the real space Green's function of the DBQW heterostructure under an external bias in this direction, can describe the electron propagation more adequately. The electron propagation through AlAs/GaAs/AlAs (001) double-barrier structures has been investigated with the tunneling formalism based on the non-equilibrium Green's function developed by Keldysh and Caroli et al., which allows to calculate nonlinear response properties of a non-equilibrium system.[2] The Green's function formulation of scattering theory[3] incorporated with the realistic multi-orbital sp³s* tight-binding model[4] is employed for the description of the electronic structure in order to handle multi-band effect, bulk material specific characteristics and symmetry considerations of constituent materials of a DBQW structure.[5] As a result, the formalism can naturally include the effects of the indirect bandgap, the band non-parabolicity of constituent materials and the interlayer intervalley transfer process through a heterostructure in this calculation. In addition to the aperiodicity of a heterostructure, the externally applied bias further destroys the symmetry of the system. The high electric field effect in the barrier and quantum-well regions is solved to all orders in Dyson's equation within the tight-binding basis.[5]

2. Theory

The Hamiltonian H of a double-barrier quantum-well heterostructure subject to an externally applied bias can be expressed as,

$$H = H^1 + H^2 + H^3 + H^{LE} + H^{RE} + H^{Cp} + V_{ext} = H^{sub} + H^{Cp} + V_{ext} = H^{sub} + H' . \qquad [1]$$

H^1, H^2 and H^3 are three isolated thin film Hamiltonians for barrier 1 (B1), quantum-well (QW) and barrier 2 (B2), respectively. H^{LE} and H^{RE} are the left and the right electrode Hamiltonians. In the Bloch-Wannier (BW) orbital basis $|b,n,\vec{k}_{//}>$, where b indicates the basis orbital, n is the atomic plane index in the z direction and $\vec{k}=(\vec{k}_{//},k_z)$, the coupling term H^{Cp} which combines 5 subsystems is expressed as,

$H^{Cp}(\vec{k}_{//}) = \sum_I \sum_{b,b'} [T^I_{b,b'}(N_I, N_I+1,\vec{k}_{//})|b,N_I,\vec{k}_{//}><b',N_I+1,\vec{k}_{//}|]+H.c.,$ where I indicates interfaces. $V_{ext}(\vec{k}_{//}) = \sum_b \sum_n V(n)|b,n,\vec{k}_{//}><b,n,\vec{k}_{//}|$ represents externally applied bias.

Within the framework of the non-equilibrium Green's function, and assuming a stationary situation, the current density $J(\vec{k}_{//})$ along the tunneling direction at the interface between the left electrode and barrier 1 can be obtained as[2,5]

$$J(\vec{k}_{//}) = \frac{e}{i\hbar} \int_{-\infty}^{\infty} dE \sum_{b,b'} [T^1_{b,b'}(1,0,\vec{k}_{//}) \, G^+_{b,b'}(1,0,E,\vec{k}_{//}) - T^1_{b,b'}(1,0,\vec{k}_{//}) \, G^+_{b',b}(0,1,E,\vec{k}_{//})]. \qquad [2]$$

0 and 1 indicate the interface sites of the left electrode and the barrier 1 (B1), respectively. The non-equilibrium Green's function G^+ of a DBQW heterostructure carries the information on the electron distribution and essentially represents the non-equilibrium distribution function.[2,5,6] The nonequlibrium Green's matrices G^+ satisfy the following symbolic equations,

$G^+(1,0) = G^+(1,1)T^1(1,0)G^{\mu^a}(0,0) + G^r(1,1) \, T^1(1,0) \, G^{\mu^+}(0,0)$

$G^+(0,1) = G^{\mu^+}(0,0)T^1(0,1)G^a(1,1) + G^{\mu^r}(0,0)T^1(0,1) \, G^+(1,1)$

$$G^{+}(1,1)=G^{\mu^{+}}(1,1)+G^{r}(1,n)H^{'}(n,n')G^{\mu^{+}}(n',1)+G^{\mu^{+}}(1,n)H^{'}(n,n')G^{a}(n',1)$$
$$+G^{r}(1,n)H^{'}(n,n')G^{\rho^{+}}(n',n'')H^{'}(n'',m)G^{a}(m,1) \qquad [3]$$

where $G^{\mu r}$ and $G^{\mu a}$ are the equilibrium retarded and advanced Green's function of the isolated left electrode. The index ρ of the equilibrium Green's function $G^{\rho^{+}}$ indicates the isolated subsystems of LE, RE, B1, QW or B2, when both indices n', n" belong to each region, respectively. In evaluating the Green's function G of a DBQW heterostructure, we begin with the perfect bulk Green's functions of constituent materials, and solve a series of Dyson's equations $G^{\kappa}=G^{\mu}+G^{\mu}\,H^{ptb}\,G^{\kappa}$ with the corresponding perturbing term H^{ptb} in the BW orbital basis, in order to produce isolated thin films of barriers and quantum-well, and semi-infinite electrodes, and to combine these subsystems to complete a DBQW heterostructure.[3,5] The perfect bulk Green's function is expressed as,[3,5]

$$G_{b,b'}^{o}(n,n',E,\vec{k}_{//})=<b,n,\vec{k}_{//}|\frac{1}{(E\pm i\delta-H^{o}(\vec{k}))}|b',n',\,\vec{k}_{//}>=\frac{1}{2\pi}\int G_{b,b'}^{o}(E,\vec{k})\cdot e^{ik_{z}\cdot(n-n')d}dk_{z}. \qquad [4]$$

H^{o} is the perfect semiconductor bulk Hamiltonian, for which we employ the empirical Tight-Binding (TB) model with a basis of $sp^{3}s^{*}$ localized quasi-atomic orbitals of anion and cation.[4] AlAs parameters are adjusted to make ΔE_{v}^{max}, the difference between the valence band maximums of bulk GaAs and AlAs for the energy alignment between these constituent materials.[4] The perturbing terms H^{ptb} can be determined from the nearest-neighbor interaction parameters between the anion and the cation, $<b,0,\vec{k}_{//}|H^{o}|b',1,\vec{k}_{//}>$ of the bulk Hamiltonian.[3,5] The Dyson matrix equation in each case is solved exactly by matrix inversions and recursive relations in the computer.[5] The ideal junction, perfect anion and cation matching at interfaces and the confinement of the spatial variation of the external potential to double barriers and quantum-well region are assumed for simplicity. For the left and right electrodes, there is a chemical difference assumed to be eV_{ext}. The spectral local density of state (DOS) is obtained from, $D_{P}(E,\vec{k}_{//})=\pm\frac{1}{\pi}\sum_{n\subset P}\sum_{b}$ Im $G_{b,b}^{r(a)}(n,n,\vec{k}_{//},E)$, where the subscript P indicates either QW or barriers of a DBQW structure. We also investigate the inelastic effect not only within the quantum well but also throughout the heterostructure.[6] The phenomenological imaginary potential energy term which may be a function of energy $\delta_{s}(E)$ is set to a constant for simplicity. We discuss the quasi-one-dimensional aspects of the electron tunneling in the [001] direction based on the $(\vec{k}_{//}=\vec{0})$ component in the AlAs/GaAs/AlAs double-barrier structure, which can explain the prevailing electron tunneling and the interlayer intervalley transfer process.

3. Results and discussion

The tunneling current density $J(\vec{k}_{//}=\vec{0})$ is illustrated and analyzed with the spectral DOS of DBQW heterostructures with and without an externally applied bias. Fig. 1 depicts the calculated current density $J(\vec{k}_{//}=\vec{0})$ versus voltage curve of a 15Å-30Å-15Å AlAs/GaAs/AlAs DBQW structure with increasing imaginary potential energy δ_{s}. The first current density peak feature near 0.3 V (an electric field of 5×10^{5} Vcm⁻¹) in each curve of the figure is due to the electron resonant tunneling through the QW Γ valley

1238

confined state. The spectral LDOS also shows that the first two X confined states of B2 are also between the Fermi energy E_F and the bottom of conduction band E_C^{LE} of the left

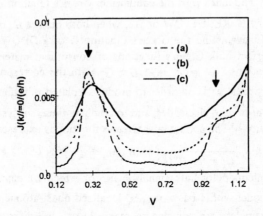

[Fig. 1] The calculated current density $J(\vec{k}_{//} = \vec{0})$ versus voltage curve of a 15Å-30Å-15Å AlAs/GaAs/AlAs DBQW in the (001) direction, shows the dependence of $J(\vec{k}_{//} = \vec{0})$ on the imaginary potential term δ_s. (a) The dashed line is for δ_s=0.005 eV, (b) the dotted line is for δ_s=0.015 eV and (c)the solid line is for δ_s=0.030 eV. The Fermi energy E_F of the electrode is set to 80 meV above the bulk GaAs conduction band minimum and ΔE_v^{max}=0.51 eV.

electrode. The second current peak feature near 1.02V manifests the electron propagation through the first B1 confined barrier X state. As the imaginary energy increases, the resonant peak through the QW confined state decreases (the reduction of the electron Γ-Γ transfer component), whereas the current density through the barrier states (B1) around 1.02 V increases, indicating the enhancement of the electron Γ-X interlayer transfer component. There is also increase in the valley current and the background current. Fig. 2 depicts the calculated average spectral DOS per atomic plane, $\overline{D}_P(E, \vec{k}_{//} = \vec{0})$ at the applied bias of (a) V_{ext}=0V, and (b) V_{ext}=1.02V in the conduction energy range. $\overline{D}_P$ of each region shows the sharp localized energy peaks in the lower conduction energy range for QW Γ states and X valley confined barrier states. The orbital decomposition of spectral local density of states, $\overline{D}_P(E, b, \vec{k}_{//} = \vec{0})$ at zero bias reveals the characteristics which can be related to the Γ valley states of bulk GaAs and X valley states of bulk AlAs.[5] In addition to the layered aperiodicity, the external bias further destroys the symmetry properties of Γ or X valley originated states. The bulk-like characteristics become less meaningful as the external bias increases to a high value. The externally applied bias V_{ext}=1.02 V (an electric field of 1.7×10^6 Vcm^{-1}) provides considerable modification and mixing between barrier states and the quantum-well states in Fig.2(b). The manifestation of the peak structure in current density around 1.02 V of Fig. 1 as the electron propagation through the B1 confined state is evident by the fact that the first B1 confined barrier state (X valley originated state at zero bias) falls well within the energy range between the conduction band minimum and the Fermi level of the left electrode (between two vertical lines in Fig. 2(b)). Also the study of the dependence of the electron tunneling current in the [001] direction on the valence band

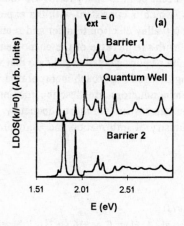 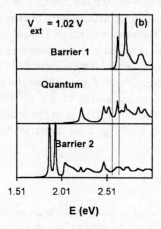

[Fig. 2] The average spectral local DOS $\overline{D}_P(E, \vec{k}_{//} = \vec{0})$ of a AlAs/GaAs/AlAs (15 Å -30 Å -15 Å) DBQW structure in the (001) direction. (a) $\overline{D}_P(E, \vec{k}_{//} = \vec{0})$ without the external bias. (b) $\overline{D}_P(E, \vec{k}_{//} = \vec{0})$ with applied external bias V_{ext}=1.02V. Two vertical lines indicate the conduction band minimum and the Fermi level of the left electrode, respectively, at a given external bias. The top figures are for the barrier 1 (B1), the middle figures are for the quantum-well (QW), and the bottom figures are for the barrier 2 (B2).

offset ΔE_v^{max} shows that not only the shift of the peak but also the shape change of the current density peak, indicating the role of the X valley barrier confined states. The calculated spectral local density of states in the [001] direction shows that in contrast to confined QW Γ valley states, X valley originated barrier states show sensitive behaviors to the value of the ΔE_v^{max}.[5] More complicated features are involved in the current density $J(\vec{k}_{//} = \vec{0})$ versus voltage curve where the barrier is thicker than the quantum well. The thicker barrier in the DBQW structure provides more available local density of states for the electron tunneling in the lower energy range. The current density is enhanced by the electron propagation through these B1 X valley states. Hence, after the Fermi energy of the left electrode catches up to the first barrier confined states, more available states in the barrier contribute to the current density through the interlayer intervalley process. The calculated result that the thicker AlAs barrier (30 Å -15 Å -30 Å) in the DBQW structure (Fig.4) yields the smaller peak to valley ratio (PVR) (~2:1) than that of the thinner barrier case of (15 Å -15 Å -15 Å DBQW) (~10:1) with δ_s=0.005eV is consistent with the experimental report by Mendez et al..[7]

4. Summary

The electron tunneling in the [001] direction is analyzed with the calculated spectral local DOS of the DBQW at finite external bias, which directly exhibits confined electron states originated from the GaAs Γ valley and AlAs X valley originated barrier states as well as their mixed states. The calculated current density $J(\vec{k}_{//} = \vec{0})$ versus voltage curve exhibits the structures for the interlayer intervalley electron transfer, which are manifested as the electron propagation through Γ-valley originated states and through X-valley barrier confined states. The smaller current peak to valley ratios are found in the double-barrier structures with thicker barriers, which is consistent with the experimental results. The enhancement of Γ-X interlayer intervalley electron transfer and reduction of the Γ-Γ resonant electron interlayer transfer with the increase of the phenomenological imaginary potential term is identified. The results of our calculation indicate that the non-equilibrium Green's function with the scattering theoretic approach incorporated with the realistic bandstructures of the constituent semiconductors can describe the intervalley interlayer electron transfer in the double-barrier heterostructure more adequately.

This work has been supported by the Ministry of Information and Communication, Korea.

References

[1] R. Tsu and L. Esaki, Appl. Phys. Lett, 22, 562 (1973)

[2] L. V. Keldysh, JETP 20, 63 (1965), C. Caroli, et. al., J. Phys. C 4, 916 (1971), J. Phys. C 5, 21 (1972); R. Combescot, et al., J. Phys. C 4, 2611 (1971).

[3] D. Kalkstein, et. al., Surf. Sci. 26, 85 (1971), E. Foo, et. al., Phys. Rev. B, 9, 1857 (1974), J. Pollman, et. al., Phys. Rev. B, 18, 5524 (1978), E. Economou, "Green's Functions in Quantum Physics" (Springer-Verlag Berlin Heidelberg, 1983).

[4] T. Boykin, et. al., Phys. Rev. B, 46, 12796 (1992), Phys. Rev. B, 47, 12696 (1993), P. Vogl, et. al., J. Phys. Chem. Solids, 44,810 (1980); The TB parameters in Ref.10 produce the valence band maximum of bulk GaAs Γ_v^8 at 0 eV. The and gaps are $E_{g,direct}^{(GaAs)}$ $[\Gamma_c^6 - \Gamma_v^8]$~1.54 eV, $E_{g,indirect}^{(AlAs)}[X_c^1 - \Gamma_v^{15}]$~2.50 eV, and $E_{g,direct}^{(AlAs)}[\Gamma_c^1 - \Gamma_v^{15}]$~3.14 eV, and bulk effective masses are $m_\Gamma^*(GaAs) = 0.0689$, $m_\Gamma^*(AlAs) = 0.1582$, $m_{x,l}^*(GaAs) = 1.457$ and $m_{x,l}^*(AlAs) = 1.046$.

[5] G. Kim et al., Phys. Rev. B, 38, 3252 (1988): Phys. Rev. B, 50, 7582 (1994).

[6] Y. Isawa, ``Transport Phenomena in Mesoscopic Systems" edited by H. Fukuyama et al., (Springer-Verlag, Berlin Heidelberg, 1992).

[7] E. Mendez, et al., Appl. Phys. Lett, 50, 1263 (1987), Appl. Phys. Lett, 53, 977 (1988).

Inst. Phys. Conf. Ser. No 145: Chapter 10
Paper presented at 22nd Int. Symp. Compound Semiconductors, Cheju Island, Korea, 28 August–2 September 1995
© *1996 IOP Publishing Ltd*

Modeling of strained quantum-well lasers and comparison with experiments

S L Chuang[1], C S Chang, J Minch and W Fang

University of Illinois at Urbana-Champaign
Department of Electrical and Computer Engineering
1406 W. Green Street, Urbana, Illinois 61801, U. S. A.

Abstract. We discuss a general theory for strained quantum-well lasers start-ing from an expression for amplified spontaneous emission. The electronic band structure, optical gain spectrum, refractive index change and the linewidth en-hancement factor in the presence of carrier injection are calculated by taking into account the valence band-mixing effects. We show that our theoretical results com-pare very well with experimental data for strained InGaAsP quantum-well lasers obtained by measuring the amplified spontaneous emission spectrum as a function of the injection current.

1. Introduction

Strained quantum-well lasers have been of great interest in many semiconductor material systems covering from visible to near-infrared regions of the spectrum. The modeling of strained quantum-well lasers [1, 2, 3] poses a challenging task because many im-portant material parameters for the electronic properties such as the effective masses, the band edge discontinuity ratio, and the deformation potentials in strained ternary and quaternary quantum-well systems have not been well documented or experimentally determined. Very often, a linear interpolation scheme from the binary compound semi-conductors has to be adopted. Optical properties such as the refractive index and the optical gain coefficient as a function of the injection current in a semiconductor laser have to be determined taking into account the optical waveguide effects in a realistic quantum-well laser structure using double confinement. To date, little work has been done to compare directly theoretical results with experimental data obtained for the op-tical gain spectrum and the induced change in refractive index in a population inverted medium using strained quantum wells.

We have developed a comprehensive theoretical model [1, 3] including the valence-band mixing effects for strained InGaAsP quantum-well lasers. The optical gain coeffi-cient and the refractive index change as a function of the injection carrier density were

[1] E-mail: s-chuang@uxh.cso.uiuc.edu, Phone: 1-217-333-3359, Fax: 1-217-333-5701

calculated first, taking into account band gap renormalization. Then the linewidth enhancement factor [4], which determines the laser spectral linewidth and also affects the laser chirps in high-speed modulation, was calculated. We next measured the amplified spontaneous emission spectrum from an index-guided InGaAsP quantum-well laser as a function of the injection current. The optical gain spectrum, refractive index profile and the linewidth enhancement factor were obtained directly from the amplified spontaneous spectra at different injection currents. By taking into account the dispersion effects of the refractive index in calculating the optical confinement factor and the modal gain, our theoretical results agree very well with the experimental data.

2. Theory

2.1. Amplified spontaneous emission spectroscopy

The amplified spontaneous emission is modeled as a distribution of uniform electromagnetic (optical) current sheet along the longitudinal direction z. The magnitude of the current sheet is chosen to have a radiation power density to the right $W(\hbar\omega)$ (W/cm^3) equal to half of the spontaneous emission power density that coupled into the waveguide mode,

$$W(\hbar\omega) = \frac{\beta}{2} r_{sp}(\hbar\omega)\Delta E \, \hbar\omega \tag{1}$$

where β is the fraction of the spontaneous emission events that are coupled into the waveguide mode, and $r_{sp}(\hbar\omega)\Delta E$ is the spontaneous emission rate per unit voulme of photons with energy between $\hbar\omega$ and $\hbar\omega + \Delta E$.

The fields due to the current sheet at $z = z_s$ can be written as superpositions of a forward and a backward propagating wave inside and outside the cavity. By matching the boundary conditions for the tangential electric and magnetic fields at the two mirrors and $z = z_s$, the intensity of the output wave (light) from the right facet $I(z_s, \hbar\omega)$ can be determined. Due to the incoherent nature of the spontaneous emission events, the total amplified spontaneous emission intensity $I_{ASE}(\hbar\omega)$ can be found by integrating the intensity $I(z_s, \hbar\omega)$ over the cavity. The final expression $I_{ASE}(\hbar\omega)$ is found to be

$$I_{ASE}(\hbar\omega) = (1-R)\left(\frac{W(\hbar\omega)L}{GL}\right)\frac{(e^{GL}-1)(1+Re^{GL})}{|1-Re^{GL}e^{i2k_o n_e L}|^2} \tag{2}$$

where we have assumed both left and right mirrors have the same field reflection coefficient $\sqrt{R}$, $k_o = 2\pi/\lambda$, λ is the wavelength in free space, n_e is the effective index of the waveguide mode, L is the cavity length, $G = g_M - \alpha_i$ is the net gain, g_M is the modal gain, and α_i is the intrinsic loss.

2.2. Gain and refractive index model with valence band-mixing

The subband structures are solved by the effective mass theory [5] with a 6×6 Hamiltonian which includes the valence band mixing between the heavy-hole, light-hole, and spin-orbit split-off bands. The theory of Bir-Pikus [6, 7] is used to take into account the effects of strain. The spontaneous emission rate $r_{sp}(\hbar\omega)$ for a single quantum-well can be calculated as

$$r_{sp}(\hbar\omega) = \frac{n_e^2\omega^2}{\pi^2\hbar c^2}\left\{\frac{q^2\pi}{n_e c\epsilon_o m_o^2\omega L_z}\sum_{n,m}\int |M_{nm}^{sp}(k_t)|^2\frac{f_n^c(k_t)\left(1-f_m^v(k_t)\right)(\gamma/\pi)}{\left(E_{nm}^{cv}(k_t)-\hbar\omega\right)^2+\gamma^2}\frac{k_t dk_t}{2\pi}\right\} \tag{3}$$

where L_z is the well width, f_n^c (f_m^v) is the Fermi occupation probability for electrons (holes) in the n_{th} (m_{th}) conduction (valence) subband, E_{nm}^{cv} is the transition energy between the n_{th} conduction and m_{th} valence subbands, $M_{nm}^{sp} = (2M_{nm}^{TE} + M_{nm}^{TM})/3$, M_{nm}^{TE} (M_{nm}^{TM}) is the momentum matrix element for the TE (TM) polarization.

The material gain can be determined from the spontaneous emission rate using [8]

$$g^p = g_{sp}^p \left[1 - \exp \left(\frac{\hbar\omega - \Delta F}{k_B T} \right) \right] \tag{4}$$

where the superscript p stands for polarization, $p = TE$ or $p = TM$, ΔF is the seperation between the electron and hole quasi-Fermi level, and g_{sp}^p is the expression inside the curly braket of Equation (3) with M_{nm}^{sp} change to M_{nm}^p. The modal gain is obtained by multiplying the material gain by the optical confinement factor Γ for the well regions. Γ can be estimated by weighting the optical confinement factor for the waveguide mode with $n_w L_z / W_{mode}$, where n_w is the number of wells and W_{mode} is effective width of the waveguide mode.

The induced change in the effective refractive index due to interband transitions is

$$\delta n_e(\hbar\omega) = \Gamma \frac{q^2 \hbar^2}{n_e m_o^2 L_z \epsilon_0} \sum_{n,m} \int \frac{k_t dk_t}{2\pi} |M_{nm}^p|^2 \frac{(E_{nm}^{cv} - \hbar\omega)}{E_{nm}^{cv}(E_{nm}^{cv} + \hbar\omega)} \frac{(f_n^c - f_m^v)}{(E_{nm}^{cv} - \hbar\omega)^2 + \gamma^2} \tag{5}$$

which has been derived in [8].

3. Comparison with experiments

The device tested is an index-guided buried heterostructure laser grown on an InP substrate with a width about 1 μm. The active region consists of five 105 Å In$_{0.55}$Ga$_{0.45}$As$_{0.88}$P$_{0.12}$ wells with 78 Å lattice matched InGaAsP barriers with a bandgap wavelength $\lambda_g = 1.28$ μm. The wells have 0.26% tensile strain. The multiple quantum wells are surrounded by InGaAsP spacer layers with the same composition as the barrier material. The total thickness of the InGaAsP region is about 2200 Å.

Amplified spontaneous emission from one facet of the laser was directed into a 1.25 m spectrometer and detected by a cooled germanium detector. Our measurements cover the the whole range of the spectrum including the transparency (zero modal gain) energy and a flat portion of the modal gain spectrum below the bandgap energy of the well region. The gain spectrum of the laser was measured based on the Hakki-Paoli method [9] for current biases below threshold. The intrinsic loss is determined by using the fact that the modal gain g_M approaches zero in the long wavelength limit.

The peak wavelength of each mode in the amplified spontaneous emission spectrum is a function of the refractive index. By comparing the peaks of two spectra of two different injection currents, the refractive index change due to the injection current difference is derived. The resonance conditions of two longitudinal modes are used to estimate the cavity length L. By using the formula from Adachi [10] at two wavelengths λ_1 and λ_2 for the refractive index of the waveguide core and InP cladding, the effective indices $n_e(\lambda_1)$ and $n_e(\lambda_2)$ can be determined by finding the propagation constant of the lowest waveguide mode. Choosing wavelengths from each end of our measurement range, $\lambda_1 \approx 1460$ nm and $\lambda_2 \approx 1590$ nm, we find the cavity length L to be 398 μm. The extracted effective

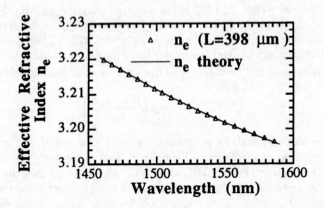

Figure 1. A plot of the measured dispersion of the effective index (triangles) and comparison with the theoretical effective index profile. From the experimental data and the theoretical fit, we obtain a cavity length of 398 μm.

index spectrum is shown in Fig. 1 as triangles. The solid line is the calculated effective index of the lowest (TE$_0$) mode of a slab waveguide, using Adachi's formula [10] for the refractive indices of the waveguide core and InP cladding. We can see from Fig. 1, that the solid line and the triangles match very well.

In Fig. 2, we show the measured and calculated TE polarized modal gain spectra at three different injection currents, 6, 8, and 10 mA. The bulk bandgap E_g and electron effective mass m_n used in our calculation for the In$_{1-x}$Ga$_x$As$_y$P$_{1-y}$ are taken from [11]. The material parameters for GaAs, GaP, InAs, and InP are taken from [8, 11, 12]. The model-solid theory [12] is used for the bandgap offsets. The bandgap renormalization is taken into account by adding ΔE_g^{BGR} to the bulk bandgap E_g. We use [13] $\Delta E_g^{BGR} = -CN^{1/3}$, where C is a fitting parameter, to obtain the transparency energy position. The carrier densities are determined from the transparency energy corresponding to the quasi-Fermi level separation. The carrier densities are $N = 2.41, 2.66$, and 2.82×10^{18} cm^{-3}, for the three injection currents, 6, 8, and 10 mA, respectively. The bandgap renormalization coefficient is found to give a bandgap shift of -30 meV for a 2D carrier density 1×10^{12} cm^{-2}. The Γ for the tested laser is about 0.1, calculated directly from the waveguide theory. The experimental and theoretical results agree very well with each other. Note that only two fitting parameters, C and γ, are used, and the experimental spectra are explained very well with the full valence band-mixing model. The spectra show a clear blue-shift of the peak modal gain wavelength and a reduction of the differential modal gain ($\partial g_M/\partial I$) as the injection current is increasing. The blue-shift shows that the band-filling effect is larger than the bandgap renormalization.

In Fig. 3 we show the dispersions of both measured and calculated data for the induced refractive index change (Fig. 3(a)) and linewidth enhancement factor (Fig. 3(b)). Our model agrees well with the experimental refractive index change and linewidth enhancement spectra. Note that our formula (5) is a direct numerical evaluation of the refractive index profile and it automatically takes into account the Kramers-Kronig re-

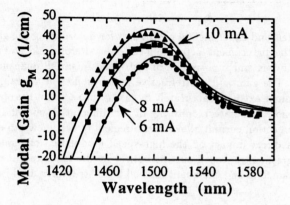

Figure 2. Our calculated (solid curves) and measured (symbols) modal gain spectra below threshold at I= 6, 8 and 10 mA using the Hakki-Paoli method.

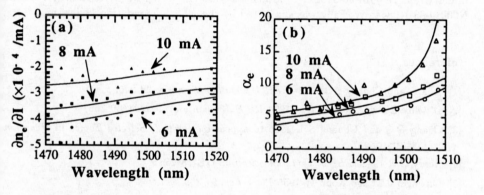

Figure 3. Comparison of our experimental data (symbols) and theoretical calculations (solid curves) for the (a) refractive index change per injection current change and (b) linewidth enhancement factor spectra.

lation and the scattering broadening mechanism. The refractive index decreases as the injection current increases. The magnitude of the refractive index change per unit injection current change is a decreasing function of the injection current.

The linewidth enhancement factor is proportional to the ratio of the refractive index change per injection current change and the differential modal gain profile. Since the refractive index change is roughly constant over the wavelength range of interest, the α_e spectrum is determined mainly by the differential gain profile. α_e tends to decrease at shorter wavelengths because of the increased differential gain caused by the band filling effects.

4. Conclusion

Theoretical models and experimental procedures for determining the gain, refractive index and linewidth enhancement factor from the Fabry-Perot peaks and peak shifts in the amplified spontaneous emission spectra of a strained InGaAsP quantum-well laser have been presented. The gain and the refractive index profiles of a strained quantum-well laser structure can be extracted accurately by taking into account the optical confinement factor. Furthermore, accurate monitoring of the differential gain and the Fabry-Perot peak shift with injection current allows measurement of the linewidth enhancement factor, which has a direct impact on the high-speed modulation of semiconductor lasers. The excellent agreement between our theoretical results and experimental data verifies the material parameters and the quantum-well band structures used in our model.

Acknowledgements

The authors thank Drs. Y. K. Chen and T. Tanbun-Ek at AT&T Bell Laboratories for providing the laser. This work was supported by the U.S. Office of Naval Research under Grant N00014-90-J1821. J. Minch was also supported by ONR AASERT Grant N00014-93-1-0844. W. Fang was supported by the AT&T Foundation Fellowship.

References

[1] Ahn D and Chuang S L 1988 *IEEE J. Quantum Electron.* **24** 2400

[2] O'Reilly E P and Adams A R 1994 *IEEE J. Quantum Electron.* **30** 366

[3] Chang C S and Chuang S L 1995 to appear on *IEEE J. Selected Topics in Quantum Electronic.*

[4] Henry C H 1982 *IEEE J. Quantum Electron.* **QE-18** 259

[5] Luttinger J M and Kohn W 1955 *Phys. Rev.* **97** 869

[6] Bir G L and Pikus G E 1974 *Symmetry and Strain-Induced Effects in Semiconductors* (New York: John Wiley)

[7] Chuang S L 1991 *Phys. Rev. B* **43** 9649

[8] Chuang S L 1995 *Physics of Optoelectronic Devices* (New York: Wiley)

[9] Hakki B M and Paoli T L 1973 *J. Appl. Phys.* **44** 4113

[10] Adachi S 1982 *J. Appl. Phys.* **53** 5863

[11] Hellwege K H 1982 *Landolt-Börnstein Numerical Data and Functional Relationships in Science and Technology* **17a** (Berlin: Springer); 1986 **22a**

[12] Van de Walle C G 1989 *Phys. Rev. B* **39** 1871

[13] Tränkle G, Lach E, Forchel A, Scholz F, Ell C, Haug H and Weimann G 1987 *Phys. Rev. B* **36** 6712

Inst. Phys. Conf. Ser. No 145: Chapter 10
Paper presented at 22nd Int. Symp. Compound Semiconductors, Cheju Island, Korea, 28 August–2 September 1995
© *1996 IOP Publishing Ltd*

Enhanced Carrier and Optical Confinement of Quantum Well Lasers with Graded Multi-Quantum Barriers

J.-I. Chyi, J.-H. Gau, S.-K. Wang, J.-L. Shieh, and J.-W. Pan

Department of Electrical Engineering, National Central University, Chung-Li, Taiwan, 32054, R.O.C.

Abstract   The enhancement of electron barrier height by multi-quantum barrier structure is simulated using transfer matrix method. An effective barrier height as high as 5.5 times the classical potential barrier is designed by using five stacks of GaAs/AlAs superlattices. Based on the simulated results, we construct both 0.78 μm and 1.3 μm graded-index separate confinement heterostructure lasers with enhanced carrier and optical confinements using graded multi-stack multi-quantum barriers. The threshold current densities of the lasers are estimated to be lower than those of the conventional graded-index separate confinement heterostructure lasers. Higher characteristic temperatures are also expected for these lasers.

1. Introduction

For heterostructure lasers, the carrier confinement efficiency is strongly dependent on the barrier height at the heterojunction. The barrier height is the inherent characteristic of a material system since it is determined by the conduction band discontinuity between the active and guiding layer. However, for short wavelength lasers, the built-in barrier height is often not sufficient to confine hot carriers, and therefore leads to low efficiency and characteristic temperature. This difficulty can be alleviated by increasing the effective barrier height using multi-quantum barriers (MQBs) as proposed by Iga [1]. That is, electron wave would be reflected by MQBs even when the energy of electron is higher then the bulk potential barrier. The enhanced carrier reflectivity has been examined by I-V and PL measurements previously [2-4]. The MQBs has also been applied to lasers and show great potential on decreasing threshold current density (J_{th}) and raising characteristic temperature (T_0). [5-7] In this work, transfer matrix method is used to simulate the electron reflectivity of multi-stack GaAs/Al$_x$Ga$_{1-x}$As (x=0.5,1) and In$_{0.53}$Ga$_{0.47}$As/In$_{0.52}$(Al$_x$Ga$_{1-x}$)$_{0.48}$As MQBs. Based on the simulated results, new graded index separate confinement heterostructure single quantum well lasers for 0.78 mm and 1.3 mm with enhanced carrier and optical confinements are proposed.

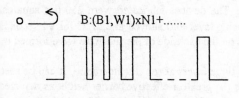

B:(B1,W1)xN1+·······

Fig. 1 Schematic diagram of a MQB

2. Simulation and experiments

The schematic diagram of a basic MQB structure used in this work is shown in Fig. 1 and expressed as B:(B$_n$,W$_n$)xN$_n$, where B, B$_n$, and W$_n$ is the width of the first barrier, and well in the unit of monolayer (ML) repetively, and N represents the number of period. The barrier height is defined as U$_0$.

For the purpose of demonstration , we chose GaAs/AlGaAs material system in our calculation

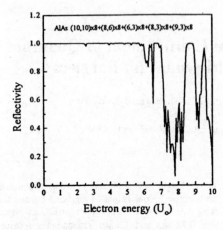

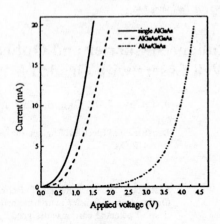

Fig. 2 Electron reflectivity of a GaAs/AlAs multi-stack multi-quantum barrier

Fig. 3 Current-voltage characteristics of three n-i-n diodes with various structures

due to its easy implementation.

In this work, the effective mass approximation is used and intervalley scattering is ignored because of the thin MQB used. Therefore, the transport of electrons in MQB is supposed to be Γ-band to Γ-band. This assumption is confirmed by the experimental results. We examined several MQBs with various combinations of barrier width and well width while keep the first barrier width and period number at fixed values. Due to the interference effect, each of them has different high reflectivity regions with energies larger than the bulk potential barrier. Knowing these characteristics, we were able to construct a superior effective potential barrier using a multi-stack MQB structure to complement the low reflection regions associated with each MQB. Fig. 2 shows the simulated results of a GaAs/AlAs10:(10,10)x8+(8,6)x8+(6,3)x8+(8,3)x8+(9,3)x8 5-stack MQB. An effective barrier height as high as 5.5 U_0 can be obtained. For the multi-stack MQB, it can be shown that the sequence of the stacks does not result in noticeable change on the electron reflectivity. Therefore, one can construct a multi-stack MQB with the desired effective composition grading while maintain the same electron reflectivity. This is extremely useful for its applications in optoelectronic devices since the effective refractive index profile can be tailored as desired.

In order to verify our simulated results, we have grown a series of samples and measured their current-voltage (I-V) characteristics. The layers were grown on (100)-oriented Si-doped GaAs substrates by molecular beam epitaxy. The device structure following the growth sequence consists of 0.2 μm n^+-GaAs buffer, 0.2 μm n-GaAs, 5 nm undoped GaAs spacer, MQB which is listed in Table I, 5 nm undoped GaAs spacer, 0.2 μm n-GaAs, and 0.2 μm n^+-GaAs capping layer. The total thickness of the MQBs in these layers were controlled to be equal in order to compare their performance. The devices fabricated were 250 μm squares. AuGeNi was evaporated on the n^+-GaAs capping layer and alloyed at 320 °C for 90 seconds to form ohmic contacts. Good ohmic contacts on the backside of the substrate were formed by Indium during the MBE growth.

Fig. 3 shows the I-V characteristics of all three layers at room temperature. It can be seen that the GaAs/AlAs multi-stack MQB exhibits the greatest effective barrier height as depicted by the dotted line. On the other hand, the four-stack GaAs/$Al_{0.5}Ga_{0.5}$As MQB, which is

Table I. Layer structures for current-voltage characterization

Layer Number	Structure	Barrier
1	single 70 nm	$Al_{0.5}Ga_{0.5}As$
2	3:(3,5)x8+(3,3)x8+(4,4)x8+(5,3)x8	$Al_{0.5}Ga_{0.5}As$
3	10:(10,10)x8+(8,6)x8+(6,3)x8+(8,3)x8+(9,3)x8	AlAs

supposed to give an even greater effective barrier height if Γ-band to X-band transport mechanism is in effective, apparently doesn't work as good as GaAs/AlAs multi-stack MQB. These experimental results indicate that electron transport via Γ-band to Γ-band is a valid assumption in our calculation. It should also be noted that the effective barrier height varies with external bias because the electric field across the MQB alters the potential profile as the bias voltage changes.

3. Graded-index separate confinement heterostructure lasers with graded MQBs

To be able to be used in graded-index separate confinement heterostructure (GRIN-SCH) lasers, multi-stack MQBs have to give not only good carrier confinement but also proper optical confinement. Therefore, we calculate the optical confinement factor of some 0.78 mm GRIN-SCH laser structures with and without multi-stack GaAs/AlGaAs MQBs. The structure of the multi-stack MQB is 60:(6,12)x8+(6,10)x4+(6,6)x4+(8,4)x4+(4,4)x4+(10,4)x4+(10,6)x 4. Since the multi-stack MQB is most effective in confining electrons when it is next to the quantum well, the Al mole fraction of the AlGaAs barrier is designed to follow a parabolic function as shown in Fig. 4 to form a good guiding layer. The resultant effective barrier height is about 1.5 times that of the conventional graded layer as indicated in Fig. 5 by the dashed line. The optical confinement factor is also enhanced from 2.45% to 2.89%. The threshold current density of a single QW laser can be approximated by [8]

$$J_{th} = (\frac{J_o}{\eta_i})\exp\{[\frac{1}{\Gamma b_o J_o}] \times [\alpha_i + (\frac{1}{L})\ln(\frac{1}{R})]\}$$

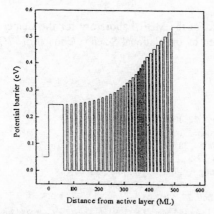

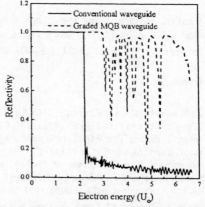

Fig. 4 Schematic diagram of the conduction band profile for a 0.78 μm GRIN-SCH MS MQB laser

Fig. 5 Electron reflectivity for conventional and graded multi-stack MQB GRIN-SCH lasers

1250

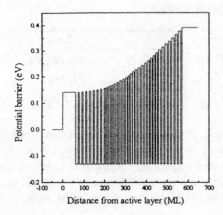

Potential barrier (eV)

Distance from active layer (ML)

Fig. 6 Schematic diagram of the conduction band profile for a 1.3 μm GRIN-SCH MS MQB laser

where J_o and b_o is the threshold current density and gain-current coefficient at transparency, I is confinement factor, η_i is internal quantum efficiency, α_i is internal loss, L is cavity length, and R is facet reflectivity. Using the following typical parameters for GaAs/AlGaAs lasers i.e. $J_o=100$ A/cm^2, $b_o=20$ cm/A, $R=0.32$, $\alpha_i=10$ cm^{-1}, $\eta_i=0.8$, and $L=500$ μm, the threshold current density of the new laser can be reduced by 10 %.

Same principle has been applied to the design of 1.3 μm $In_{0.52}(Al_xGa_{1-x})_{0.48}As/InF$ GRIN-SCH quantum well lasers. The $In_{0.53}Ga_{0.47}As/In_{0.52}(Al_xGa_{1-x})_{0.48}As$ (x=0.5-1.0) multi-stack MQB incoparated in the laser strucutre is 60:(6,12)x8+(6,6)x4+(7,7)x4+ (8,4)x4+(9,6)x4+(8,5)x4+(10,6)x8 and its conduction band energy diagram is shown in Fig. 6. Compared with the conventional laser structure, the effective barrier height in this case is two times higher and the optical confinement factor is also improved from 1.68% to 1.92%. This is encouraging toward uncooled 1.3 μm quantum well lasers for low cost fiber communications.

4. Conclusions

We have examined the enhanced carrier confinement effect of MQBs both theoretically and experimentally. GRIN-SCH multi-stacks MQB waveguide are designed to enhance carrier and optical confinements. Employing those structure for 0.78 and 1.3 mm quantum well lasers threshold current can be reduced. Improvement on the characteristic temperature of these lasers is also expected.

Acknowledgment

The authors wish to thank the technical support of the MBE Laboratory for the Center for Optical Science, NCU. This work was supported by the National Science Council of R.O.C under contract No. NSC-82-0417-E-008-179.

References

[1] Iga K, Uenohara H, and Koyama F, 1986 Electron. Lett. 22, 1008-10.
[2] Takagi T, Koyama F, and Iga K, 1991 IEEE. J. Quantum. Electron. 27, 1511-8.
[3] Takagi T, Koyama F, and Iga K,1990 Appl. Phys. Lett. 59, 2877-9.
[4] Yen T, Tsai C M, Lee C P, and Liu D C, 1993 EDMS, 32-5.
[5] Yokoyama H, Iwata H, Sugimoto M, Onabe K, and Lang R, 1988 J. Appl. Phys, 63, 4755-8.
[6] Kikuchi A, Kaneko Y, Numura I,and Kishino K, 1990 Electron. Lett. 27, 1669-71.
[7] Kikuchi A, Kishino K,and Kaneko Y,1991 Electron. Lett. 27, 1301-03.
[8] Zou W X, Chuang Z M, Law K -K, Dagli N, Coldren L A, and Merz J L, 1991 J. Appl. Lett. 69, 2857-61.

Inst. Phys. Conf. Ser. No 145: Chapter 10
Paper presented at 22nd Int. Symp. Compound Semiconductors, Cheju Island, Korea, 28 August–2 September 1995
© *1996 IOP Publishing Ltd*

Negative transconductance and depletion effects in gated resonant tunneling diodes

Chomsik Lee

Electronics Engineering, Honam University Kwangju, Korea 506-090

Abstract. The characteristics of resonant tunneling devices, under the condition of quantum size effect, are theoretically studied in the AlGaAs/GaAs material system. The electron transport mechanisms and the current-voltage characteristics two-dimensional structures are investigated in double-barrier heterostructure electron devices. In a gated resonant tunneling diode (GRTD), we have seen that different resonances can be observed in forward and reverse collector bias by depletion effects. We also investigate the effect of barrier size on the existence of negative transconductance at a fixed emitter-collector bias with elastic and inelastic scattering mechanism.

1. Introduction

Many theoretical models have been proposed to explain the electron transport properties of semiconductor interfaces. In a previous work, a self-consistent numerical solution of the Poisson equation and Schrödinger equation was made by Vassell *et al.* [1] in a one-dimensional model of the RTD in 1983. Frensley [2] studied the potential and density of the resonant tunneling diode by the self-consistent Wigner function(1987). A gated resonant tunneling diode (GRTD) has been proposed for a single electron transistor. The three-terminal device (GRTD) is designed to control the depletion region and it reduces the dimensionality in the active channel area. Lee and Weichold [3] also predicted the negative transconductance in the GRTD (1993).

In this paper, we solve the two-dimensional Poisson equation and the continuity equation to find the depletion region and initial condition of the conduction band structure by the finite difference method. We calculate the current-voltage characteristics in gated resonant tunneling diodes with reverse gate bias for a weak lateral confinement case (3D to 2D tunneling when the GRTD reaches two-dimensional conditions in the well region). Simulation of the GRTD using a self-consistent quantum model shows depletion effects on current-voltage characteristics. Scattering processes are included to calculate resonant tunneling transmission coefficient. All simulations have been performed on a CRAY Y-MP supercomputer.

2. Fabrication and theory

One way to make the gated resonant tunneling diode(GRTD) involves the integration of a double-barrier $(Al_xGa_{1-x}As/GaAs)$ tunnel structure with a field-effect transistor in which the electrical size of the channel within the vertical resonant tunneling diode (RTD) is directly controlled by a simple, self-aligned rectifying electrode in the low-doped GaAs material (Fig. 1). The ohmic contact of the emitter was used as a shadow mask for nonconformal deposition of chromium on the etched MBE surface. Details of the fabrication of such a device have been previously reported in ref. [3].

There are two different systems: one has weak lateral confinement, the other has strong lateral confinement. The important difference is that the emitter region is not quantized in the former case, but it is quantized in the latter case. Therefore, the quantum transport effects are very much changed in the two cases. In both systems, we have solved the Schrödinger equation and calculated tunneling probabilities differently. The strong confinement system has subband mixing and evanescent modes are considered to calculate transmission probabilities. The weak confinement system is dominated in current-voltage characteristics by the depletion effect. The most voltage drop occurs in the barriers and the undoped area throughout the device.

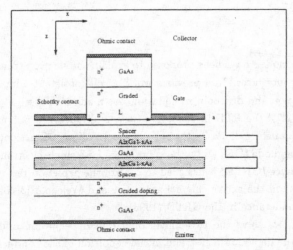

Figure 1: Schematic of the gated resonant tunneling diodes with GaAs/AlGaAs.

The application of a potential difference across the gate will cause lateral depletion and this leads to a reduced cross-section for electron transport. The peak-to-valley current ratio of the negative-differential-resistance (NDR) present in the emitter-collector circuit can be modulated with gate voltage. Several attempts[4] have been made to incorporate the NDR into three-terminal devices using field-effect transistor structures. However the main features of this simulation are controllable NDR in the emitter-collector current and negative transconductance by controlling depletion region. Numerical solutions to the two-dimensional Poisson's equation and the continuity equation have been used to calculate the lateral depletion region and carrier concentrations. We examine the effects of elastic and inelastic scattering by analyzing the phase changing in the GRTD double-barrier structure. Once a self-consistent solution is achieved for a given applied bias, the band structure is saved and the next solution is converge faster than the previous one.

3. Discussion of results

At the gate region, low doping is preferred to increase the depletion layer length under bias and thus decrease the capacitance. Lateral and vertical extension of the depletion region on the gate contact changes the active channel region of heterostructures. In our calculations, the heights of the barriers are V=0.342 eVs, their width 50Ås, and their separation is 50Å, the gate-to-gate space is L=0.7μm, and the mole fraction of Al, x=0.3. The width of the conducting channel between the emitter and the collector is expected to be minimum at the undoped well.

The question to be considered is why the peaks in the I-V occur at such a high bias. The extension of the depletion region on the gate contact changes the resistance of the channel through which electrons must travel. Thus, increasing the resistance in the emitter and collector area shifts the peak voltage to high bias. The peak position will shift to higher bias due to additional gate depletion. Depletion effect dominates when the gate is located very close to the double-barrier structure. Our result is an excellent match with this experimental result. It is clear that depletion effects will therefore dominate the $I - V$ curve from $V_g = -1.5V$ reducing the peak-to-valley current ratio (PVCR) drastically (Fig. 2), and will be significant even for V_g=0, leading to a reduction in PVCR as compared to a large active channel width device.

Negative transconductance is observed in figure 2. The peak current can be subsequently quenched by increasing V_g. The current flow mainly is controlled by the cross section of well region if the gate voltage is small. If the gate bias is large enough to have edge effects, then the current mainly is controlled by the resonant tunneling current. On the other hand, resonant states are more important than the cross section of the active channel area. If the barrier thickness is very large, the PVCR is very large. Then it may happen that the peak current of the higher reverse gate bias is larger than the one at the lower reverse gate bias. Therefore, negative transconductance exists in the sample device at $V_{ce} = 0.2$ V.

4. Conclusions

Different behavior is observed in the common-emitter mode as compared to the common-collector mode. The ratio is affected more dramatically by gate bias (V_g) in common-collector than in common-emitter. As the gate location is made close to barriers, the

1254

depletion region has more of an effect on the collector-emitter current. We have found in the case of large barrier size, current flow is more dependent on the tunneling probability of electrons than on the cross section of active channel area. The peak-to-valley current ratio is very small in narrow barrier case because the resonant transmission coefficient as a function of incident electron energy is broadened. Therefore, the negative transconductance does not exist in this case.

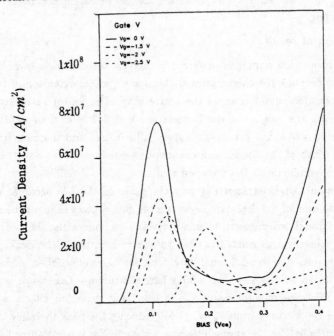

Figure 2: Current-Voltage characteristics at 77 K, barrier width L_b= 50, well width L_w = 50Å with various V_g from 0 to $-2.5V$, $V_{ce} = 0.5V$, in common-collector bias, side contact a= 50Å. The gate is located $0.52\mu m$ from the top mesa structure.

Acknowledgements

The authors would like to thank Dr. M. Weichold for his valuable discussions.

References

[1] M. A. Vassell, J. Lee, and H. F. Lockwood, "Multibarrier Tunneling in GaAlAs/GaAs Heterostructures," *J.Appl.Phys.*, vol. 54, 1983, pp. 5206.

[2] W. R. Frensley, "Wigner Function Model of a Resonant Tunneling Semiconductor Device," *Phys.Rev.B*, vol. 36, 1987, pp. 1570.

[3] C. Lee and M. H. Weichold, *Proc. of the Ninth Intl. Conf. on the NASECODE*, Copper Mountain, CO, April 6-9, 1993.

[4] F. Beltram, F. Capasso, S. Luryi, S. G. Chu, A. Y. Cho, and D. L. Sivco, Appl. Phys. Lett. **53**, 219 (1988)

Inst. Phys. Conf. Ser. No 145: Chapter 10
Paper presented at 22nd Int. Symp. Compound Semiconductors, Cheju Island, Korea, 28 August–2 September 1995
© *1996 IOP Publishing Ltd*

Mobility model for III-V compounds suited for hydrodynamic device simulation

Ch Köpf [1], H Kosina and S Selberherr

Institute for Microelectronics, TU Vienna,
Gusshausstrasse 27–29, A-1040 Vienna, Austria

Abstract. We propose an improved mobility model for hydrodynamic device simulation. The mobility is modeled as function of the total mean energy. Expressions for the individual valley mobilities are combined by means of the relative valley populations estimated by a modified Boltzmann distribution. Our model is based on steady-state Monte Carlo calculations and reflects the situation of intervalley transfer in III-V compounds. As an example the model is applied to $Ga_x In_{1-x} As$.

1. Introduction

Hydrodynamic device simulation is believed to be today's best suited tool for device characterization in the submicron regime in view of accuracy and computational effort. It bridges the gap between the classical drift-diffusion simulation tools, which run into difficulties if the characteristic length dimensions approach the thermal wavelength, and the computationally expensive Monte Carlo codes which are most accurate in describing non-local effects. The drift-diffusion (DD) and the hydrodynamic (HD) methods are approaches to solve the semiclassical Boltzmann equation via balance equations which are solved for the moments of the carrier distribution function. An important physical parameter governing electron transport in devices within the mentioned frameworks is the carrier mobility μ. The Monte Carlo (MC) technique calculates the distribution function by observing particle trajectories, the only input being basic physical quantities and scattering formulae [1].

2. Mobility modeling

Carrier mobility modeling is usually split into the characterization of the mobility at zero field, μ_0, which takes into account all doping and temperature dependences and the deviation from this value at increasing fields. A usual approach in the DD framework is to assume mobility as a function of the local driving force $\mathbf{F}$, $\mu = \mu(\mu_0, \mathbf{F})$, which

---

[1] E-mail: koepf@iue.tuwien.ac.at

consists of the electric field and a diffusion term, $\mathbf{F} = \mathbf{E} - \frac{1}{n}\nabla\frac{nkT}{q}$. A summary of various formulations can be found in [2]. In the HD case a dependence on the local carrier energy w is commonly assumed, $\mu = \mu(\mu_0, w)$. This includes non-local dissipation since the electron energy is a solution of the energy balance equation.

3. Methods of calculation

In calculating μ_0 one usually assumes an equilibrium distribution function (Boltzmann or Fermi-Dirac) and calculates mobilities for the individual scattering mechanisms which implies that momentum relaxation times exist. These contributions are then combined by Mathiessen's rule. This is believed to be reasonably accurate in most cases of elemental and III-V semiconductors.

Different methods for the calculation of high-field mobility have been proposed in the literature [3]. The simplest one is based on the relaxation time approximation. The energy dependent relaxation times are averaged over the assumed distribution function. Though very crude, this gives analytic expressions for isotropic parabolic bands. The method of calculating the distribution function by a variational principle gives more accurate results. The most accurate method is the Monte Carlo technique. It removes the restrictions of the relaxation time approximation and allows full inclusion of the complex band structure of semiconductors [4]. This technique yields a self consistent distribution function, therefore we have used it in this work.

3.1. Monte Carlo calculation

Our simulation program incorporates a many-valley isotropic nonparabolic band structure representation (Γ,L,X valleys) and takes into account scattering by polar and non-polar optic phonons, acoustic phonons, ionized impurities, and alloy disorder. An interpolation routine for the basic material parameters is included both for ternary and quaternary alloys. The band edges are given by quadratic expressions in the composition ratio. In the ternary case we have

$$E_{A_x B_{1-x} C} = x E_{AC} + (1-x) E_{BC} - x(1-x) C_{ABC}. \tag{1}$$

The remaining parameters are interpolated linearly.

Fig. 1 and Fig. 2 show the calculated low-field mobility and velocity-field relation, respectively, with the typical negative differential resistivity, both as function of the composition x for $Ga_x In_{1-x} As$. Table 1 shows the material parameters used in the MC calculation.

4. The hydrodynamic mobility model

4.1. The conventional model

Hänsch [5] obtained the following expression for the momentum relaxation time τ_m from a series expansion of the distribution function into its first four moments,

$$\tau_m^{-1} = A + B\frac{\mathbf{j} \cdot \mathbf{S}}{\mathbf{j}^2}, \tag{2}$$

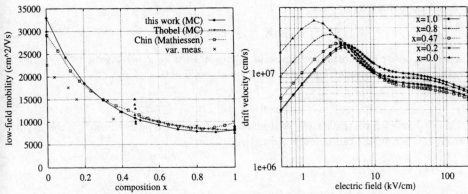

Figure 1. Low-field mobility for $Ga_xIn_{1-x}As$ at 300K as function of the composition x.

Figure 2. Electron velocity for $Ga_xIn_{1-x}As$ at 300K as function of the electric field.

Table 1. Material parameters used in MC calculation.

Quantity		GaAs		InAs		Unit
a		5.653		6.058		Å
v_l		5.23		4.28		10^5cm/s
v_t		2.47		2.65		10^5cm/s
ϵ_s		12.9		15.15		ϵ_0
ϵ_∞		10.92		12.75		ϵ_0
ρ		5.36		5.67		g/cm^3
$\hbar\omega_{LO}$		36.25		30.2		meV

		Γ	L	X	Γ	L	X	Unit
E		1.439	1.769	1.961	0.36	1.442	1.98	eV
m^*		0.063	0.22	0.41	0.023	0.29	0.64	m_0
α		0.61	0.461	0.204	1.39	0.536	0.90	eV^{-1}
D_{ac}		7.0	9.2	9.27	8.0	8.0	8.0	eV
D_o		–	3.0	–	–	3.0	–	10^8 eV/cm
D_{ij}	Γ	–	10.0	10.0	–	10.0	10.0	10^8 eV/cm
	L	10.0	10.0	5.0	10.0	10.0	9.0	10^8 eV/cm
	X	10.0	5.0	7.0	10.0	9.0	9.0	10^8 eV/cm
$\hbar\omega_{ij}$	Γ	–	27.8	29.9	–	27.8	29.9	meV
	L	27.8	29.0	29.3	27.8	29.0	29.3	meV
	X	29.9	29.3	29.9	29.9	29.3	29.9	meV

C_Γ		0.44						eV
C_L		1.10						eV
C_X		2.0						eV
D_{all}		0.5						eV

where **j** denotes the current density and **S** the energy flux. A and B are constants with respect to the moments. The balance equations for momentum and energy read in the stationary homogeneous case

$$\mathbf{j} = q\mu n\mathbf{E} \qquad \mathbf{j} \cdot \mathbf{E} = n\frac{w - w_0}{\tau_w}, \tag{3}$$

where w_0 is the equilibrium energy, and τ_w denotes the energy relaxation time which is assumed to be constant. This gives the mobility as function of w,

$$\mu(w) = \frac{\mu_0}{1 + \eta(w - w_0)} \qquad \eta = \frac{\mu_0}{q\tau_w v_s^2}. \tag{4}$$

This dependence represents the well-known saturation curve in silicon via the explicit $w(\mathbf{E})$ relation which can be obtained from (3).

4.2. The new model

However, because of the different band structure of compound semiconductors a different behavior is observed from MC results (Fig. 4). A much steeper decay is clearly evident, which can be related to intervalley transfer.

In principle two different approaches for dealing with the many-valley bands of compounds can be thought of. Firstly, one could introduce a carrier concentration for each valley and find models as functions of the average energy of that carrier types (measured from the valley minimum). The intervalley processes must be accounted for via energy dependent generation terms in the balance equations. This approach increases the number of PDEs, both the number of variables and the model complexity are increased too. Keeping in mind the rather poor convergence of the HD method itself, it is not often used. Secondly, one retains the single electron model in the calculation, where the different valleys are lumped together into the physical models, thus using a single energy variable. This approach is pursued in our work.

4.2.1. The valley mobilities.
We propose a functional relation for each valley ($i = \Gamma, L, X$) which is a modified version of (4)

$$\mu_i(w) = \frac{\mu_{0,i}\left(\frac{w}{w_{0,i}}\right)^{\gamma_i}}{1 + \left(\frac{w - w_{0,i}}{\delta_i}\right)^{\beta_i}}. \tag{5}$$

The equilibrium energy $w_{0,i}$ in the nonparabolic case can be expressed as

$$w_{0,i} \cong \frac{3kT}{2}(1 + \frac{5}{2}z - 5z^2) \qquad z = \alpha_i kT, \tag{6}$$

where α denotes the nonparabolicity parameter. (6) results from series expansion of the integral representation of the average energy. The parameter γ describes the mobility at low energies i.e. when intervalley scattering does not play any significant role. γ is the energy exponent of the prevalent scattering process in this range. In direct compounds polar-optic phonon scattering yields a value of $\gamma \approx 0.5$. δ can be viewed as the energy at which intervalley transfer starts to be important (at the band edge difference), the exponent β reflects the amount of energy that an electron changes when it moves to a different valley.

Usually the total kinetic energy w is split into a drift and a thermal part, $w = \frac{m^*\langle v \rangle^2}{2} + \frac{3kT_c}{2}$. In parabolic valleys $w = \frac{m^*\langle v^2 \rangle}{2}$, in nonparabolic valleys w is generally higher. In the many-valley case the total energy consists of kinetic and potential energy, $w = w_{kin} + w_{pot} = \sum_i n_i\,(w_i + \Delta E_i)\,/\sum_i n_i$, where n_i is the population of the ith valley. Fig. 3 depicts the energy versus field relations.

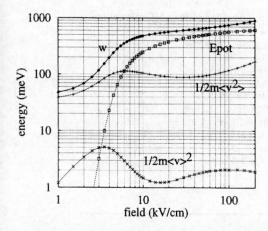

Figure 3. Energy vs. electric field for $Ga_{47}In_{0.53}As$.

4.2.2. The combined mobility. Having defined the valley mobilities, we can calculate the total mobility by weighing the valley contributions by the valley populations, $\mu(w) = \sum_i n_i(w)\,\mu_i(w)\,/\sum_i n_i(w)$. It is known that the distribution function in high fields deviates significantly from the drifted Maxwellian. However, we find it a useful approximation to assume a modified heated Maxwellian leading to a relative population of valley i with respect to the lowest one,

$$n'_i(w) = \frac{n_i}{n_1} = R_i\,\exp(-a\frac{\Delta E_i}{w}), \tag{7}$$

where a is a factor describing the different scaling in energy (the total energy w is used instead of the kinetic energy) and R takes into account valley degeneracy and the carrier masses.

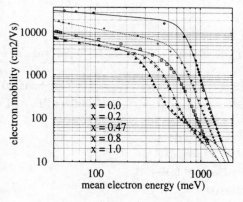

Figure 4. Total mobility for $Ga_xIn_{1-x}As$ at 300K as function of the total energy w with material composition x as parameter.

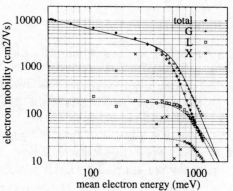

Figure 5. Valley specific mobilities for $Ga_{47}In_{0.53}As$ at 300K as function of the total energy w.

Table 2. Model parameters for $Ga_x In_{1-x} As$ obtained from MC calculation.

$\mu_{0,\Gamma}$ (cm^2/Vs)	$32900 - 101900x + 181820x^2 - 166910x^3 + 62270x^4$
$\mu_{0,L}$ (cm^2/Vs)	$234.8 - 304.4x + 429.7x^2$
$\mu_{0,X}$ (cm^2/Vs)	$27.2 - 91.8x + 204.3x^2$
δ_Γ (meV)	$734.1 - 770.9x + 320.5x^2$
δ_L (meV)	$1108.5 - 1403.5x + 748.7x^2$
δ_X (meV)	$1604.2 - 2528.9x + 1475.8x^2$
β_Γ	$5.82 - 2.85x + 2.13x^2$
β_L	$7.0 - 5.2x + 0.2x^2$
β_X	$5.0 + 1.47x - 4.63x^2$
γ_Γ	$-0.23 - 0.98x + 0.83x^2$
γ_L	0
γ_X	0
a	4.0

5. Results

As an example the new mobility model is demonstrated for $Ga_x In_{1-x} As$, an important channel material used in heterostructure FETs. The parameters β, γ, δ are obtained by a nonlinear least-square fitting algorithm. The result at 300K is given in Table 2. A good correspondence between the new model (lines) and the MC results (symbols) is obtained (Fig. 4 and 5).

Acknowledgments

This work is supported by the laboratories of AMS AG at Unterpremstätten, Austria, Digital Equipment Corporation at Hudson, USA, and SIEMENS AG at Munich, Germany.

References

[1] Jacoboni C and Lugli P 1989 *The Monte Carlo Method for Semiconductor Device Simulation* (Vienna-New York: Springer)

[2] Selberherr S 1984 *Analysis and Simulation of Semiconductor Devices* (Vienna-New York: Springer)

[3] Nag B R 1980 *Electron Transport in Compound Semiconductors* (Berlin-Heidelberg-New York: Springer)

[4] Fischetti M V 1991 *IEEE Trans. on Electr. Dev.* **38** 634–649

[5] Hänsch W and Miura-Mattausch M 1986 *J. Appl. Phys.* **60** 650–656

[6] Chin V W L and Tansley T L 1991 *Solid-State Electr.* **34** 1055–1063

[7] Thobel J L *et al* 1990 *Appl. Phys. Lett.* **56** 346–348

Inst. Phys. Conf. Ser. No 145: Chapter 10
Paper presented at 22nd Int. Symp. Compound Semiconductors, Cheju Island, Korea, 28 August–2 September 1995
© *1996 IOP Publishing Ltd*

On the optimum supply ratio in low temperature epitaxial growth of comound semiconductors

T. Hariu, T. Ohshima* and T. Hamada*

Department of Systems Engineering, Ibaraki University, Hitachi 316, Japan
*Oki Electric Industry Co., Hachioji 193, Japan

Abstract. Crystal quality of epitaxial InSb and InAs grown on GaAs at lower temperature is more critically dependent upon V/III supply ratio. The optimum supply ratio decreases as the growth temperature is reduced, with activation energy 0.4eV for InSb and 0.2eV for InAs. A growth model is proposed here to show that this optimum supply ratio corresponds to the geometrical average of sticking coefficients of group V element at the crystal planes of group III and group V element.

1. Introduction

Lower temperature for epitaxial growth of semiconductors has been pursued to fabricate advanced device structures by reducing the effects of thermal expansion coefficients and chemical reaction between grown layers and substrates. In our efforts devoted to the low temperature epitaxial growth of InSb[1], InAs[2] and their alloys[3], more pronounced importance of V/III supply ratio and the shift of optimum supply ratio to a lower value have been noticed at lower temperature. The purpose of this paper is to show that, together with experimental results on InSb and InAs, the critical dependence of crystal quality of compound semiconductors grown particularly at lower temperature on the supply ratio of constituent elements comes from the temperature dependence of sticking coefficients of more volatile components, i.e. group V elements for III-V compounds on the crystal planes of group III and V elements.

2. Experimental Results

A particular epitaxial growth method in hydrogen plasma, which we call plasma-assisted epitaxy (PAE), was employed here to achieve lower temperature epitaxial growth of InSb and InAs on (100)GaAs down to 270C. One of the important advantages of this growth method is the supply of atomic group V elements in excited states produced by plasma-cracking[4] and then the effective V/III supply ratio is enhanced in addition to leading to simpler growth mechanism.

As the growth temperature is reduced, it has been generally observed that the electronic property as well as the surface morphology depends more critically on V/III supply ratio and then more precise control is required at lower growth temperature. Figs.1 and 2 respectively shows the dependence of electrical properties of InSb and InAs on V/III supply ratio at different growth temperatures. In both cases, the dependence is more critical at lower growth temperature and the optimum supply ratio decreases as the growth temperature is reduced, with activation energy 0.2eV for InAs and 0.4eV for InSb, as shown in Fig.3.

It may be worthwhile to mention here that, although the results of Figs.1 and 2 were obtained with the same RF power applied to excite the plasma, the crystal quality is generally improved by increasing, but not excessively, the RF power. In this case, however, the supply ratio should be further decreased for optimum, corresponding to enhanced plasma-cracking.

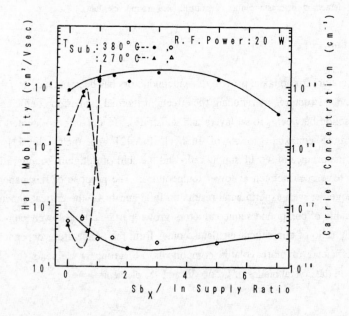

Fig.1 Effect of Sb/In supply ratio on the electrical property of InSb grown on GaAs at different temperature of 270C and 380C.

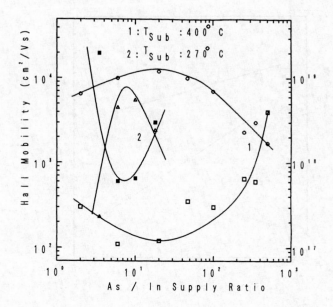

Fig.2 Effect of As/In supply ratio on the electronic property of InAs layers grown on GaAs at different temperature of 280C and 400C.

3. Growth model

In the present epitaxial growth model, different sticking (α) and desorption (ν) coefficients of a group V element, as well as of a group III element, on group III and group V crystal planes are considered, expressed with suffixes V III and V V, respectively. When we have the top surface coverage with group III element (θ_{III}) and with group V element (θ_V), then $\theta_{III} + \theta_V = 1$ but each element can stick to different crystal planes, then $\theta_{III} = \theta_{III III} + \theta_{III V}$ and $\theta_V = \theta_{V V} + \theta_{V III}$. It has been considered that, for example, when a group III element comes to $\theta_{V V}$, an exchange reaction occurs to make a normal stable bond leading to $\theta_{V III}$. Then the time dependent coverage change comes from the unbalance among adsorption, desorption and exchange.

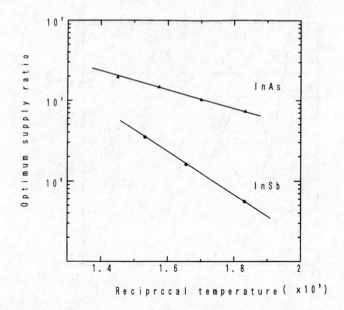

Fig.3 Optimum V/III supply ratio for the growth of InSb and InAs as a function of reciprocal temperature.

$$d\theta_{\mathrm{III\,III}}/dt = \alpha_{\mathrm{III\,III}}\,\mu_{\mathrm{III}}\,\theta_{\mathrm{III}} - \nu_{\mathrm{III\,III}}\,\theta_{\mathrm{III\,III}} - \alpha_{\mathrm{V\,III}}\,\mu_{\mathrm{V}}\,\theta_{\mathrm{III\,III}},$$

$$d\theta_{\mathrm{III\,V}}/dt = \alpha_{\mathrm{III\,V}}\,\mu_{\mathrm{III}}\,\theta_{\mathrm{V\,III}} + \alpha_{\mathrm{V\,III}}\,\mu_{\mathrm{V}}\,\theta_{\mathrm{III\,III}} + \nu_{\mathrm{V\,III}}\,\theta_{\mathrm{V\,III}}$$
$$+ \nu_{\mathrm{III\,III}}\,\theta_{\mathrm{III\,III}} - \nu_{\mathrm{III\,V}}\,\theta_{\mathrm{III\,V}} - \alpha_{\mathrm{V\,III}}\,\mu_{\mathrm{V}}\,\theta_{\mathrm{III\,V}} - \alpha_{\mathrm{III\,III}}\,\mu_{\mathrm{III}}\,\theta_{\mathrm{III\,V}},$$

$$d\theta_{\mathrm{V\,III}}/dt = \alpha_{\mathrm{V\,III}}\,\mu_{\mathrm{V}}\,\theta_{\mathrm{III\,V}} + \alpha_{\mathrm{III\,V}}\,\mu_{\mathrm{III}}\,\theta_{\mathrm{V\,V}} + \nu_{\mathrm{III\,V}}\,\theta_{\mathrm{III\,V}}$$
$$+ \nu_{\mathrm{V\,V}}\,\theta_{\mathrm{V\,III}} - \nu_{\mathrm{V\,III}}\,\theta_{\mathrm{V\,III}} - \alpha_{\mathrm{III\,V}}\,\mu_{\mathrm{III}}\,\theta_{\mathrm{V\,III}} - \alpha_{\mathrm{V\,V}}\,\mu_{\mathrm{V}}\,\theta_{\mathrm{V\,III}},$$

$$d\theta_{\mathrm{V\,V}}/dt = \alpha_{\mathrm{V\,V}}\,\mu_{\mathrm{V}}\,\theta_{\mathrm{V}} - \nu_{\mathrm{V\,V}}\,\theta_{\mathrm{V\,V}} - \alpha_{\mathrm{III\,V}}\,\mu_{\mathrm{III}}\,\theta_{\mathrm{V\,V}},$$

where μ is the supply rate of each element with a different suffix.

In the steady state growth the time average is zero. Then if we choose the V/III supply ratio ($s = \mu_{\mathrm{V}}/\mu_{\mathrm{III}}$) as

$$(\alpha_{\mathrm{III\,V}}/\alpha_{\mathrm{V\,V}} + \nu_{\mathrm{V\,V}}/\alpha_{\mathrm{V\,V}}\,\mu_{\mathrm{III}}) > s > \alpha_{\mathrm{III\,III}}/\alpha_{\mathrm{V\,III}} > 1,$$

then

$$\theta_{\text{III III}} = (\alpha_{\text{III III}} \mu_{\text{III}} / (\nu_{\text{III III}} + \alpha_{\text{V III}} \mu_{\text{V}} - \alpha_{\text{III III}} \mu_{\text{III}})) \theta_{\text{III V}}$$
$$\doteqdot (\alpha_{\text{III III}} / \alpha_{\text{V III}} s) \theta_{\text{III V}},$$

$$\theta_{\text{V V}} = (\alpha_{\text{V V}} \mu_{\text{V}} / (\nu_{\text{V V}} + \alpha_{\text{III V}} \mu_{\text{III}} - \alpha_{\text{V V}} \mu_{\text{V}})) \theta_{\text{V III}}$$
$$\doteqdot (\alpha_{\text{V V}} \mu_{\text{V}} / (\nu_{\text{V V}} + \alpha_{\text{III V}} \mu_{\text{III}})) \theta_{\text{V III}},$$

tend to zero and we get normal layer-by-layer growth ($\theta_{\text{III}} \doteqdot \theta_{\text{III V}}$ and $\theta_{\text{V}} \doteqdot \theta_{\text{V III}}$). Moreover in the case $s \gg (\alpha_{\text{III V}} / \alpha_{\text{V III}})$, we have the growth over the surface stabilized by group V element,

$$\theta_{\text{III}} / \theta_{\text{V}} \doteqdot (\alpha_{\text{III V}} + \nu_{\text{V III}} / \mu_{\text{III}}) / (\alpha_{\text{V III}} s + \nu_{\text{III V}} / \mu_{\text{III}}) \to 0.$$

The ratio R of incorporation rates of group III and group V elements is approximated as follows, with some assumptions ($\nu_{\text{III III}} = 0$, $\nu_{\text{III V}} = 0$, $\theta_{\text{V}} = \theta_{\text{V III}}$, $\theta_{\text{III}} = \theta_{\text{III V}}$),

$$R = (\alpha_{\text{III V}} (\nu_{\text{III V}} / \mu_{\text{III}} + \alpha_{\text{V III}} s + \alpha_{\text{III III}}) + \alpha_{\text{III III}} \nu_{\text{V III}} / \mu_{\text{III}}) / (\alpha_{\text{III V}} \alpha_{\text{V III}} s +$$
$$\alpha_{\text{V V}} s (\alpha_{\text{V III}} s + \nu_{\text{III V}} / \mu_{\text{III}}) - \nu_{\text{III V}} \nu_{\text{V III}} / \mu_{\text{III}}^2)$$

It can be shown that, if the V / III supply ratio is larger than $\alpha_{\text{III V}} / \alpha_{\text{V III}}$ at high temperature, the growth rate of stoichiomeric ($R=1$) compound is determined by the supply rate of group III element, as is usually observed.

At lower growth temperature where the desorption of both elements is negligible ($\nu = 0$), the incorporation ratio is

$$R = (\alpha_{\text{III V}} (\alpha_{\text{V III}} s + \alpha_{\text{III III}})) / (\alpha_{\text{III V}} \alpha_{\text{V III}} s + \alpha_{\text{V V}} \alpha_{\text{V III}} s^2).$$

Then the optimum supply ratio s_m which leads to $R=1$ (stoichiometric) should be

$$s_m = (\alpha_{\text{III V}} \alpha_{\text{III III}} / \alpha_{\text{V III}} \alpha_{\text{V V}})^{1/2} \doteqdot (\alpha_{\text{V III}} \alpha_{\text{V V}})^{-1/2},$$

i.e. geometrical average of sticking coefficients of group V element on group III and group V planes, with a reasonable assumption $\alpha_{\text{III III}} = \alpha_{\text{III V}} = 1$ at low temperature. The above experimentally obtained activation energy is then the average of these values on group III and group V planes.

4. Conclusions

Much more precise control of supply ratio is required at lower growth temperature in order to get better quality epitaxial layers of compound semiconductors. The optimum V / III supply ratio corresponds to geometrical average of sticking coefficients of group V element on group III and group V planes, and it shifts to a lower value due to the increase of sticking coefficients as the temperature is reduced. The present consideration applies to other classes of compounds.

References

[1] Ohshima T, Yamauchi S and Hariu T 1989 *Proc. 16th Int'l. Symp. GaAs and Related Compounds, Inst. Phys. Conf. Ser.* No.106 p241

[2] Hamada T, Hariu T and Ono S 1993 *Proc. 20th Int'l. Symp. GaAs and Related Compounds, Inst. Phys. Conf. Ser.* No.136 p535

[3] Hamada T, Ohshima T, Hariu T and Ono S 1992 *Proc. 19th Int'l. Symp. GaAs and Related Compounds, Inst. Phys. Conf. Ser.* No.112 p163

[4] Hariu T, Yamauchi S, Fang S F, Ohshima T and Hamada T 1994 *J. Cryst. Growth* **136** 157

Inst. Phys. Conf. Ser. No 145: Chapter 10
Paper presented at 22nd Int. Symp. Compound Semiconductors, Cheju Island, Korea, 28 August–2 September 1995
© *1996 IOP Publishing Ltd*

Role of Surfactant for Suppression of Island Formation on Si(001) Surface

Chan Wuk Oh, Young Hee Lee, and Hyung Jae Lee

Department of Physics and Semiconductor Physics Research Center, Jeonbuk National University, Jeonju 560-756, R. O. Korea

Abstract. Using *ab initio* molecular dynamics approach we study the kinetics of Sb/Si(001) system. We first generate a double layer step on Si(001) surface which emulates the {311} facet and show how the diffusion barrier varies with the introduction of Sb surfactant near the step edge. We find that the surfactant significantly enhances the diffusion by reducing the potential barrier near the step edge such that the island formation is severely suppressed and thus the layer-by-layer growth mode is enhanced.

1. Introduction

Fabrication of quantum electronic devices in general requires atomic layer-by-layer growth for the high quality. It was shown recently that the layer-by-layer growth mode is enhanced by surfactant such as As, Sb, and Bi. The surfactant lowers the surface free energy so as to reduce the Ge segregation in Ge/Si growth [1]. The site exchange mechanism has been suggested both experimentally [1] and theoretically [2] to explain the suppression of Ge segregation by a surfactant. On the contrary, it was shown that the island growth rate induced by the strain on Ge/Si(100) system is faster than the suppression rate of islands by a surfactant which is followed by the islands 'smoothing-out' and the island coalescence forming highly twinned structure [3]. It was also shown that the post-growth annealings in Sb flatten drastically the {311} facets of Ge islands [4]. These observations seem to be contradictory to the earlier ones that the surfactant suppresses the diffusion of adatoms on the flat surface. However, this behavior is not clearly understood yet. This phenomenon involves not only the energetics but also the kinetics on the islands. In this paper, we investigate the role of surfactant on the island using *ab initio* molecular dynamics (MD) calculations. The Sb prefers the site at the step edge. This reduces drastically the diffusion barrier near the step edge, enhancing the anisotropic diffusion along the dimer row at the upper terrace. The islands are flattened by the pushing-out mechanism rather than the rolling-over mechanism such that the Sb atoms are always pushed out into the step edge until the islands coalesce to each other.

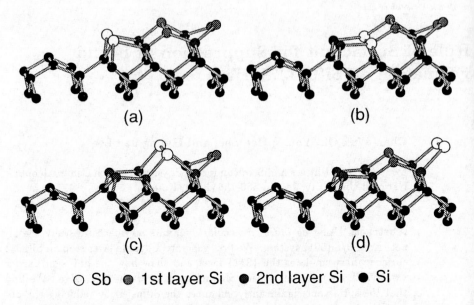

O Sb ⊙ 1st layer Si ● 2nd layer Si ● Si

Figure 1. Various sites for Sb dimer on the D_B step. (a) is taken as a reference energy.

2. Models and calculation method

In this calculation we introduce the rebonded double layer step (D_B) on Si(001) surface to emulate {311} facet of the island which is composed of consecutively rebonded D_B steps [5]. We choose a triclinic cell with lattice vectors $\vec{a}$ ($\sqrt{2}$, 0, 0), $\vec{b}$ ($-\sqrt{2}/4$, $9\sqrt{2}/4$, 1/2) and $\vec{c}$ ($\sqrt{2}/4$, $-\sqrt{2}/4$, 3/2) in units of a lattice constant which include 54 Si and 18 H atoms. The number of layers of the upper and the lower terraces is kept equivalent to prevent any unnecessary effects on both terraces. A detailed description of this structure can be found elsewhere [6]. Our calculation is carried out using of the Car-Parrinello *ab initio* MD approach [7]. We use a plane-wave basis set with a kinetic energy cutoff of 8 Ry and Bloch functions only at Γ point of the supercell surface Brillouin zone. The interaction between ionic cores and valence electrons is described by a fully nonlocal pseudopotential with sp nonlocality. We fix two bottom Si and H layers to prevent any spurious forces. We first search for the electron energy minimization using a steepest decent approach. Ions are moved by the fast relaxation scheme [6]. The energy and forces are converged to 0.5×10^{-5} Ry and 0.005 Ry/Å, respectively.

3. Results

We first search for the stable Sb site at the stepped surface. Figure 1 shows possible sites of Sb dimer. We choose the rebonded site as a reference energy in Fig. 1(a). Sb does not favor the subsurface site as shown in Fig. 1(b), where the subsurface incrase the total

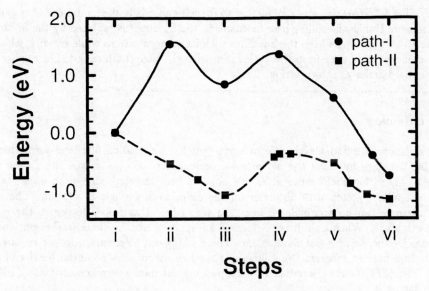

Figure 2. The total energy variations for the Path-I (rolling-over mechanism) and the Path-II (pushing-out mechanism).

energy by 2.48 eV. With the substitution of the Si dimer by the Sb dimer at the step edge significantly lowers the energy by 0.63 eV although the next Si dimer site is also favorable with similar energy gain of 0.57 eV. In order to investigate the kinetics of Si adatoms on the stepped surface we calculate the potential surface with and without the presence of a Sb dimer at the step edge. We use single Si adatom and move it along the top of the dimer row on rebonded D_B step. We fix the y ([110]) direction and relax x and z directions of an adatom. The substrate atoms are fully relaxed at the same time. Once we determine the minimum energy for given y value of an adatom, we move to the next point and repeat the same calculation. The dimerized Si surface without the presence of an Sb dimer has two additional (Schwoebel) barriers with a diffusion barrier of 0.93 eV whereas with the presence of Sb dimer shows a single Schwoebel barrier of 2.1 eV, increasing the barrier height significantly. We consider now the diffusion of a Si dimer in the presence of Sb dimer at the step edge on the upper terace. This seems to be reasonable because the Si dimer can easily be formed on the surface or the Si adatoms can easily be accumulated near the step edge of an upper terrace due to the higher diffusion barrier in the presence of Sb dimer at the step edge. Two pathways may be considered. Both pathways include the same initial step, which is the stable site of a Si dimer on the dimer row, similar to a H site for a single adatomic diffusion [6, 8]. We first consider the rolling-over mechanism that a Si dimer rolls over the Sb dimer at the step edge and reach to the nonrebonded site. The barrier exists with the barrier height of 1.55 eV, which is lower than the value of 2.72 eV for Si dimer to roll over the step edge without the presence of an Sb dimer [6]. This energy difference results from the surface free energy difference between Si and Sb.

We next consider the pushing-out mechanism that Sb dimer is pushed out to the step edge by generating one step further to the left and finally the nonrebonded step edge is formed. Unlike the rolling-over mechanism, the Sb dimer is replaced by the Si dimer. The Sb and Si dimers keep the bondings with its adjacent atoms while moving, which is a significant difference from the Path-I. Surprisingly, these pathways do not require the potential barrier as shown in Fig. 2.

4. Summary

In summery, we find kinetic role of surfactant for suppression of island formation on Si(001) surface by using the *ab inito* MD calculation. Sb favors the site at the step edge. The Schwoebel barrier of 2.1 eV appears near the step edge of an upper terace in rebonded D_B step with Sb dimer for the diffusion of a single Si adatom. Therefore Si adatoms can be accumulated near the step edge of an upper terace at the initial growth stage. When a Si dimer is formed on upper terace, Si dimer pushes out the Sb dimer by the pushing-out mechanism. The pushing-out mechanism do not require the potential barrier, whereas the rolling-over mechanism requires potential barrier of 1.55 eV. The {311} facets are reduced by the pushing-out mechanism and ultimately islands are flattened.

Acknowlegments

This work was supported byt Korea Science and Engineering Foundation (KOSEF) through the Semiconductor Physics Research Center at Jeonbuk National University.

References

[1] R. M. Tromp and M. C. Reuter 1992 *Phys. Rev. Lett.* **68** 954

[2] B. D. Yu and A. Oshiyama 1994 *Phys. Rev. Lett.* **71** 3585 ; 1994 *Phys. Rev. Lett.* **72** 3190 ; 1994 T. Ohno *Phys. Rev. Lett.* **73** 460

[3] H. J. Osten, J. Klatt, G. Lippert, B. Dietrich, and E. Bugiel 1992 *Phys. Rev. Lett.* **69** 450

[4] D. J. Eaglesham, F. C. Unterwald, and D. C. Jacobson 1993 *Phys. Rev. Lett.* **70** 966

[5] A. Oshiyama 1995 *Phys. Rev. Lett.* **74** 130

[6] Y. H. Lee, C. W. Oh, E. Kim, and M. Parrinello *unpublished*

[7] R. Car and M. Parrinello 1985 *Phys. Rev. Lett.* **55** 2471

[8] G. Brocks, P. J. Kelly, and R. Car 1991 *Phys. Rev. Lett.* **66** 1729

Inst. Phys. Conf. Ser. No 145: Chapter 10
Paper presented at 22nd Int. Symp. Compound Semiconductors, Cheju Island, Korea, 28 August–2 September 1995
© *1996 IOP Publishing Ltd*

A Quantum Mechanical Model For Analysing Capacitance- Voltage Profiles In Quantum-Wells.

Dipankar Biswas

Institute of Radio Physics & Electronics, C. U.,

Sudakshina Kundu and Reshmi Datta

Department of Electronic Science, C. U.
92, Acharya Prafulla Chandra Road,
Calcutta - 700 009, INDIA

Abstract: A two - dimensional quantum mechanical model for analysing and interpreting Capacitance - Voltage (C-V) carrier profiles in quantum - wells (Q W) has been developed . During (C-V) profiling when an external field is imposed on the (Q W) the Carrier distribution of the total system changes and depleted carriers from the barrier region move into the well to increase the net carrier concentration in the (Q W) . Our model considers the effects of quantum confinement and the effects of field. The results obtained by iterative solutions of Schrodinger's and Poisson's equations, give a better understanding and explanation of the experiements than the previous three - dimensional models that ignored the effects of quantum confinement.

1. Introduction

Capacitance-voltage (C-V) measurement is carried out on a bulk semiconductor using either a p-n diode or a Schottky-barrier on one side and an ohmic-contact on the other. Usually a reverse-bias is applied to the structure, on which is superposed a small sinusoidal voltage for measurements. For uniformly doped semiconductors the (C-V) curve is parabolic and $1/C^2$ versus V is a straight line. (C-V) measurements are widely used for carrier-profiling of semiconductors and to estimate interface-states (Biswas et al 1989, Kromer 1980).
It has been shown recently that (C-V) measurement with time on (Q.W)s in the form of Deep Level Transient Spectroscopy(DLTS) can be used to determine the conduction and valence- band offsets of (Q.W)s accurately (Debbar et al 1989, Bhattacharya

et al 1990). Workers in the area (Rimmer et al 1991,Tittelbach-Helmrich 1993) have carried out (C-V) measurement on (Q.W)s for determining carrier-distribution inside the well and theoretical models have been developed to fit the experimental results.. The theoretical models mentioned, treat the carriers in' bulk and do not consider the various effects of quantum confinement of carriers in the well exposed to an external field. Thus the models are not very successful in imparting a thorough understanding of (C-V) carrier profiling in a (Q.W) and do not explain all aspects of experimental measurements.

When an external field is imposed on a single (Q.W) with barriers on either side and a Schottky on top, as is necessary for (C-V) profiling, a number of changes take place. The bands bend (Fig.1) and carriers get depleted from the barrier on the Schottky side and collect in the (Q.W) to increase the two-dimensional carrier concentration. This is widely different from the situation in bulk semiconductors where carriers are free to move in all the three directions without any confinement. So, with the impressed field for (C-V) measurements the entire carrier distribution changes(Ahn et al 1986, Bastard et al 1983) and estimation of the carrier profile becomes quite involved. The motion of carriers is limited to two-dimensions only.

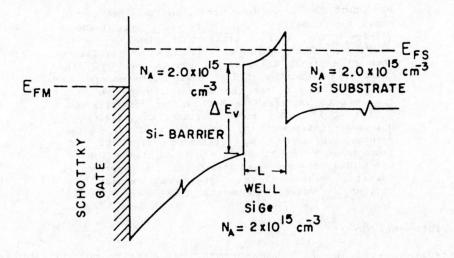

Fig. 1 Valenceband diagram of the Si - SiGe- Si
Structure under reverse bias

Our model accounts for quantum confinement and field dependent phenomena inside the quantum-well (Q.W) along with the band - offset. The carrier distribution and the peak shifts with temperature are computed in this quantum-mechanical model through self-consistent,iterative solutions of the Schrodinger and Poisson's equations.

2. Theory

The cosine wave-function representing the two-dimensional hole

-gas(2DHG) present in the Q.W. is changed in the presence of an external field as given by Bastard. In this paper the wave-function is modified to suit the requirements of a two-dimensional hole-gas (2DHG). From a knowledge of the measured effective-mass of heavy and light holes in the subbands of the Q.W.(Cheng et al 1993, Xu et al 1991) it is evident that for low doping and moderate temprature almost 90% of the holes populate the first heavy hole subband. The wave-function representing the first heavy hole subband is given by

$$\psi_{w+\delta} = N(B)[\cos(\pi z/L+\delta)\exp B(z/xL-1/2x)+\cos(\pi/2+\delta)\exp-q_o(z/L-1/2)] \quad \ldots\ldots 1$$

where $N(B)$ is a normalisation constant and B is a variational parameter obtained by a process of minimisation of the energy term

$$E(B) = E_o[1+B^2/4\pi^2+\emptyset[-1/B-2B/(4\pi^2+B^2)-\coth(-B/2)/2]] \ldots\ldots 2$$

$$\text{Where } \delta = \tan^{-1}(q_o+B/x)/\pi -\pi/2 \ldots\ldots\ldots 3$$

$$q_o^2 = 2m^* L^2(E_o-\Delta E_v)/\hbar^2 \ldots\ldots\ldots\ldots 4$$

L is the total well width. E_o is the ground state energy and ΔE_v is the difference in the valence-band maxima at the (Q.W) and the barrier interface and $\emptyset=qFxL/E_o \ldots\ldots\ldots 5$ where F is the field. Without any field equation 1 reduces to the usual cosine function.

The carrier concentration is obtained by integrating the product of the density of states function (DOS), Fermi function and the probability density. The limits of integration include the quantum well as well as the barrier region over which the carriers are distributed. The 2D-hole concntration is directly related to the amount of carriers depleted from the cap-layer. Maximum 2D-hole concentration is obtained when the entire cap-layer is depleted of holes. Under an applied field the carriers are shifted to a fraction x of the quantum well of width L.

Starting with a trial value of the fractional width xL and a trial hole concentration within the well, the 2D-hole concentration is obtained by iterating.Repeated iterations give a unique set of values for the field F and x.

3. Results

The model was applied to the p-type Si-SiGe-Si quantum well structure of reference (Tittelbach-Helmrich 1993). The calculations were done for different temperatures and band-offsets.

Figure 2 shows the distribution of excess carriers inside the(Q.W.) on the application of reverse electric field. As the field is increased, carriers from the barrier region near the Schottky are pushed into the well. They start piling up inside the quantum well (Q.W.) away from the Schottky -gate due to the existing field across the quantum-well. For a typical value of the valence band offset (30 meV) the peak carrier concentration changes from 5.0×10^{15} cm^{-3} to 7.98×10^{16} cm^{-3} at a field necessary to move the depletion edge into the quantum well during (C-V) measurements.

The changes of peak carrier concentration and peak position with increasing fields for the same band offset are shown in figure 3. At higher fields the shift in the peak position almost stops and the percentage change in the peak carrier concntration decrease3 significantly.

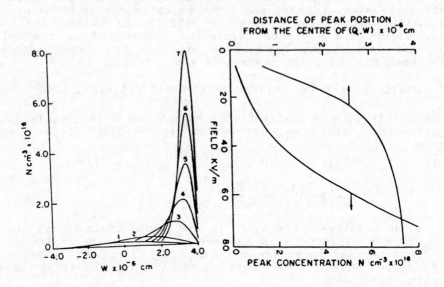

Fig.2 Carrier concentration inside well under reverse bias increaseing in steps of 12 k V/m.

Fig.3 Variation of peak concentration and peak position inside the (Q W) with field.

For a band offset of 30 meV at a field required to move the depletion edge into the quantum well (Q.W), we see from figure 4 that with increasing temperaturs the peak heights decrease and the peak positions shift towards the Schottky end. This is in agreement with the experimental results of Tittelbach-Helmrich.

Figure 5 shows how the peak concentration changes with valence band offset and temperature. It is clear that the peak concentration increases with increase in valence band offset as well as decrease in temperature.

4. Discussions

The results obtained from our model, when compared with the experimental and theoretical results of Tittelbach-Helmrich for the same structure yield interesting findings. The experimental carrier peak heights as obtained from (C-V) measurements $\rightleftharpoons 2 \times 10^{16}$ cm.$^{-3}$ are much lower than our result $\rightleftharpoons 8.0 \times 10^{16}$ cm.$^{-3}$. This is an expected anomaly of (C-V) measurements when there are very high changes in carrier concentrations within a few Debye lengths (Wu et al 1975, Johnson et al 1971) such abrupt high carrier changes are encountered at the (Q.W.). In this case the carrier concentration increases and then decreases about forty times each way within a few Debye lengths.

The theoretical results of Tittelbach-Helmrich fall far short of their experimental results $\rightleftharpoons 5.0 \times 10^{15}$ cm.$^{-3}$. This seems to have arisen from the facts that the extra carriers which are pushed into the quantum well under a reverse bias from the barrier region have not been considered and the fundamental phenomena due to quantum confinement of carrier in the quantum well have been totally ignored.

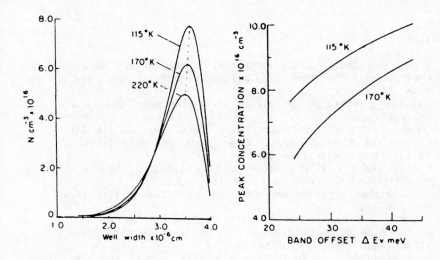

Fig.4 Blown up version of graph 7 of figure 2 at different temperatures.

Fig.5 Variation of peak concentration with band offset for different temperatures.

5. Conclusions

A quantum mechanical model has been developed to explain, analyse and understand(C-V) profiling in a quantum well. The model accounts for the different field dependent phenomena occuring in a quantum well and the effects of the valence band offset (ΔE_v). Self-consistent iterative solutions of the Schrodinger and Poisson's equation yields the actual carrier distribution in a quantum well with an imposed field. It is interesting to note that the carriers pile up inside the quantum well due to imposed fields and the peak concentration of carriers change from $\rightleftharpoons 5 \times 10^{15}$ to $\rightleftharpoons 8 \times 10^{16}$ cm.$^{-3}$ but it is difficult to measure this peak concentration experimentally.

1276

Acknowledgements

The authors express their gratitude to Prof. B.R. Nag, Prof. P.K. Basu and Dr. J.B. Roy of Institute of Radio Physics & Electronics for their useful suggestions. They also thank Sri A.K.Sinha & Sri S. Bhattacharya for their efficient typing. R.D. is indebted to the Council of Scientific and Industrial Research for the financial support.

Reference

Ahn Doyeol and Chuang S L (1986) Appl . Phys . Lett.**49** 1450

Bastard G,Mendez E E,Chang L L and Esaki L(1983) Phys. Rev. B **28** 3241

Bhattacharya P,Debber N and Biswas D(1989)Int.Symp. GaAs and Related Compounds,

Karuizawa , Japan / (1990) Int. Phys. Conf. Ser. No.**106** 351

Biswas D,Berger P R ,Das U,Oh JE and Bhattacharya P(1989)Jr. of Electronic Materials **18** 137

Cheng J P, Kesan V P Grutzmacher D A,Sedwick T O and Ott J A (1993) Appl . Phys. Lett. **62** (13) 1522

Debber N, Biswas Dipankar and Bhattacharya Pallab (1989) Phys. Rev. **40** 1058.

Johnson W C and Panousis P T (1971) IEEE - ED **18** 955

Kromer H, Chien Wu - Yi, Harris J S Jr and Edwell D D (1980) Appl Phys. Lett. **36** 295

Rimmer J S, Nissous M and Parkar A R (1991) Appl. Surf. Sc. **50** 149

Tittelbach - Helmrich K (1993) Semicon. Sc. Technol. **8** 1372

Wu C P, Douglas E D and Mueller C W (1975) IEEE - ED **22** 319

Xu D X, Shen G D, Willander M and Hansson G V (1991) Appl. Phys. Lett. **58** (22) 2500

Inst. Phys. Conf. Ser. No 145: Chapter 10
Paper presented at 22nd Int. Symp. Compound Semiconductors, Cheju Island, Korea, 28 August–2 September 1995
© *1996 IOP Publishing Ltd*

Theoretical Study in the Materials Design of Semiconductor Alloys and Low-Dimensional Quantum System

Bing-Lin Gu

Department of Physics, Tsinghua University, Beijing 100084, China

By means of concentration wave method, a dynamic model of epitaxial growth is presented to determine the ordered structures for bulk ternary III-V alloys. The most stable ordered structures of ternary semiconductors are deduced and compared with the experimental observations. The electronic structures of the ordered phase have been shown. For the long-period superlattice (quantum well), the effect of symmetry and applied electric field on electron-phonon coupling is discussed for understanding elemental properties.

1. Introduction

Recent development of growth techniques, such as molecular-beam epitaxy (MBE), chemical-vapor deposition (CVD) *etc.*, have resulted in the fabrication of many novel structures based on the compound semiconductors, especially multicomponent semi-conductor alloys. These new types of compound materials show some significant novel properties, which can be used to improve the electronic and photonic devices. There-fore, the materials design, or the tailoring band gap, of these new types of semiconductor alloys and devices becomes more and more important in the materials science and has attracted much attention.

Composition, structure and properties are the basic elements in material design, and are related to each other. The structure and properties of the materials can be affected significantly by the composition of alloys. Varying the composition range, these new types of semiconductor systems often undergo structural phase transitions. To understand these alloys, and in particular, their structure and properties, one must first understand their phase transitions. Therefore, the phase stability and phase diagrams of these alloys should be computed and the physical parameters that control growth of these alloys should be identified. The relation between the properties and composition must be considered. The structure, especially the ordered phases appeared in some alloys, can also affect the fundamental properties of the materials, such as the band gap, the lattice phonon modes, and the free carrier mobility.

Some alloy systems : the mixed semiconductor alloys (such as the group (III-V)-IV alloys and II-VI alloys), ternary III-V semiconductor alloys (such as AlGaAs, GaInAs), and low-dimensional superlattices and quantum well (such as GaAs-GaAlAs system), applied to the optoelectronic materials have attracted even more attention. In mixed semiconductor alloys, the metastable alloy (III-V)-IV is known to undergo a order-disorder transition between zincblende and diamond structure,[1] e.g., $(GaAs)_{1-x}Ge_{2x}$, and II-VI compounds, e.g., $CdTe_{1-x}S_x$, show the zincblende-wurtzite transition. In the epitaxial growth of ternary III-V semiconductor alloys, many long-range ordered structures have been observed,[2] which show some novel properties different from the disordered ones. Most interesting is the observed influence of the transition on the band gap as a function of composition x. In both mixed semiconductor alloys and ternary III-V alloys, the bowing of the band gap as a function of x has been noticed and understood well, and the existence of ordered structures often causes the band gap reduction.[3]

In low-dimensional superlattice and quantum well, the presence of heterointerfaces and size effect alter the electronic structure, exciton state and phonon modes, which are important for the improvement of material properties. Among the quantum-well and superlattice systems, symmetric structures have been most extensively studied. More recently, asymmetric quantum-well structures such as GQW (graded-gap quantum-well structure) and LIQW (layer inserted quantum well), *etc* are developed. It has been found that the electron and exciton states in asymmetric structures are quite different from those in symmetric ones, especially when an electric field is applied.[4] It is worthwhile to investigate the dependence of electron relaxation and other properties on applied electric field in various quantum-well structures (symmetric or asymmetric), which would be useful for some device application.

2. Mixed Semiconductor Alloys

Mixed semiconductor alloys are metastable crystals composed to two materials that differ in lattice structures. These alloys undergo order-disorder phase transition as function of composition, e.g., zincblende-diamond transition in (III-V)-IV alloys and zincblende-wurtzite transition in II-VI alloys. Many novel properties are shown in these alloy systems. For example, the bowing of the direct band gap appears commonly, the phase-transition composition x_c depends on both growth conditions and the type of alloy studied, and the band gap bowing is noticeable different grown in different condition, such as MBE and CVD. Here we only focus on the study of (III-V)-IV alloys.

Efforts have been made to understand the structral and electronic properties of (III-V)-IV alloys, such as $(GaAs)_{1-x}Si_{2x}$, $(GaAs)_{1-x}Ge_{2x}$ and $(GaSb)_{1-x}Ge_{2x}$. Early work of Gu *et al.* [1] calculate the thermodynamic quantities of $(GaSb)_{1-x}Ge_{2x}$ alloys using the cluster-variation method (CVM) with the effect of correlations between first neighbors, combined with a correlated virtual-crystal approximation. The calculated bond probabilities are compared with extended x-ray-absorption fine structure (EXAFS) data, and the existence of wrong bonds Ga-Ga and Sb-Sb is demonstrated to exist. Estimating the interaction energies by comparison with EXAFS data, the calculated phase diagram is consistent with the experimental one.

Recently, we propose a random-site model, [5] where in order to eliminate phase separation in $(A^{III}B^{V})_{1-x}C^{IV}_{2x}$ alloy, C^{IV} atoms should be taken to be randomly distributed at all lattice sites. Therefore the phase diagram is calculated using CVM, with the interaction energies of nearest-neighbor atomic pairs computed by a universal-parameter tight-binding method.

Based on the random-site model, both the global features at the absolute zero temperature and the global phase diagrams are determined.[6] We have shown that the model of random site of C^{IV} atoms has four different phase-state configurations in the limit of absolute zero temperature for all combinations of nearest-neighbor interaction energies using pair-correlation aproximation. Furthermore, using the spin-1 Blume-Emery-Griffiths Ising model, the phase diagrams of random-site model are calculated in different typical interaction energy regions, and the crossover behavior near the boundary line between the two interaction energy regions is also investigated. For different (III-V)-IV alloy systems, the interaction energies parameters are different and belong to different energy regions. Thus based on these results we can obtain the global information of phase diagrams of different (III-V)-IV alloys, especially the critical composition x_c.

The contribution of correlations to band-gap bowing have also been studied,[1,5] and the calculations show the V-shaped bowing as a function of composition x of the direct band gap. We compute the energy gap of the alloy in the correlated virtual-crystal approximation, and show that the band gap deepens somewhat and has some curvature in the two branches of the V shape.[1]

3. Ternary III-V Semiconductor Alloys

In epitaxial growth, the ordered phenomena have been observed in nearly all ternary III-V alloy systems: AlGaAs, GaInAs, GaInP etc.[2] Atomic scale ordering in the ternary alloys of type $A^{III}_{1-x}B^{III}_{x}C^{V}$ and $A^{III}C^{V}_{1-x}D^{V}_{x}$ exists on anion (gronp-III) or cation (group-V) fcc sublattice, respectively. The ordering observations have created interest both in the ordering mechanism and in the influence on the properties of the alloy, especially the band gap.[3]

To clarity the origin of ordering and determine the ordered phases, many theoretical researches have been conducted. Initially, first-principles total-energy and phase diagram calculations were conducted for bulk structure.[7] Recently a lot of effort indicates that the atomic ordering takes place at the growth layer surface.[8] Starting from surface thermodynamics, the first-principles total-energy calculations for the surface and subsurface layers and the cluster-variation method were used to study the ground-state ordered structures and finite-temperature thermodynamics.[8]

Some phenomenological models were appiled to study the ordered structures. First we determined the most stable ordered structures for bulk alloys by means of the concentration-wave method.[9] Furthermore, based on a two-dimensional planar model,[10] we deduced the planar most stable ordered structures of a specific growth plane and obtained the possible ordered structures for the alloy through layer-by-layer growth process. We also gave the diagrams of interaction parameters.

In this two-dimensional planar model, only the interactions within a single growth plane are considered. But because of the coupling between different growth layers, it is insufficient to determine the stacking process from 2D planar structures to correct 3D bulk structures by only considering a single plane. Therefore, we consider that the intralayer and interlayer effective interactions of the alloy are the main factors of the appearance of ordered structures, and surface layers and deep layers are different on structure stability in the layer-by-layer growth process of III-V alloys. To consider the interlayer interactions to the third neighborhood, we study three consecutive growth layers which are parallel to the substrate as a whole when they are regarded as three interpenetrating Bravais sublattices. Therefore, we can simplify a three-dimensional problem to a quasi-two-dimensional one. According to this three-interpenetrating-planes method, we can establish a dynamic model of epitaxial growth in ternary III-V alloys and deduced the ordered phases using the concentration-wave method.[11] Furthermore, the interaction relations of the ordered structures are also given and can be applied to determine the existence of the ordered phase in the specific ternary III-V semiconductor alloys.

Table I shows the results of the deduced ordered structures in (001), (110) and (111) substrate growth. The present results can explain some experimental phenomena. In (001) substrate growth, the coexistence of $[001]L1_0$, $[010]L1_0$, and $E1_1$ structures can be explained in Table I. The observed ordered structures $\frac{1}{2}L1_1[\bar{1}11]$, $\frac{1}{2}L1_1[1\bar{1}1]$, and the other two variants of $L1_1$ appear because of the two different deep layer structures,[11] and the energies of two deep layer structures are different and several previous calculations have shown that type $[1\bar{1}1]$ and $[\bar{1}11]$ are selected to meet the minimum energy of the alloy.[8] The case in (110) substrate growth are more complex, and the $[001]L1_0$, $E1_1$, and $\frac{1}{2}[001]$ structures can be predicted. But the others shown in Table I have not been found. We also deduced the ordered structures in (111) substrate growth presented in Table I. The ordered phase in (111) substrate growth has not been reported, and our results about it are just theoretical predictions.

For ternary III-V semiconductor alloys, the most interesting is shown in the technologically attractive possibility of tuning band gaps through ordering. Many experimental and theoretical results have proved that in many ordered phases the band gaps are direct, and the band gap bowing appears commonly. Both the empirical tight-binding model and first-principles calculation are presented to study the effects of ordering on the electronic structure,[3] and give the results of band gap bowing quite close to the experimental ones. The calculations show that band gap bowing is usually bigger in common-cation compounds than in common-anion compounds. The band gap reduction are also well reported in samples showing ordering,[3] which creat interest in the device application.

4. Low-dimensional Quantum System

The ordered patterns described above in the ternary III-V semiconductor alloys can be described as short-period superlattice with the period of one or two atomic layers. In long-period superlattices, especially the low-dimensional superlattices and quantum well, the structure and composition can exert significant influence on the properties.

TABLE I. All of the possible ground state ordered structures of III-V ternary semiconductor alloy according to the present growth model.

substrate	ordered structure types
(001)	$\frac{1}{2}L1_1[\bar{1}11]$, $\frac{1}{2}L1_1[1\bar{1}1]$, $\frac{1}{2}[1\bar{1}0]$ $\frac{1}{2}L1_1[111]$, $\frac{1}{2}L1_1[11\bar{1}]$, $\frac{1}{2}[110]$ $L1_0[100]$, $L1_0[010]$, $\frac{1}{2}E1_1[021]$
(110)	$L1_0[001]$ $\frac{1}{2}L1_1[111]$, $\frac{1}{2}L1_1[11\bar{1}]$, $\frac{1}{2}[001]$ $L1_0[100]$, $L1_0[010]$, $\frac{1}{2}[1\bar{1}0]$ $\frac{1}{2}L1_1[\bar{1}11]$, $\frac{1}{2}L1_1[1\bar{1}1]$, $\frac{1}{2}E1_1[021]$
(111)	$L1_0[001]$, $\frac{1}{2}L1_1[11\bar{1}]$ $L1_0[100]$, $\frac{1}{2}L1_1[\bar{1}11]$ $L1_0[010]$, $\frac{1}{2}L1_1[1\bar{1}1]$

In the present work, effects of well width and electric field on electron-phonon scattering are discussed for symmetric and asymmetric quantum-well structures. One is a symmetrically single quantum-well (SSQW) $Ga_{0.6}Al_{0.4}As/GaAs/Ga_{0.6}Al_{0.4}As$, and the other is an asymmetrically single quantum-well (ASQW) $Ga_{0.6}Al_{0.4}As/GaAs$-$Ga_{0.8}Al_{0.2}$-$As/Ga_{0.6}Al_{0.4}As$.

Effective-mass approximation is employed to calculate the subband wavefunctions of electron states in quantum wells under an electric field F which is applied in the z direction. Assuming that the bound-state approximation is valid[12] and using series expansion method,[13] we can solve the Schrödinger-like equation of electron to obtain energy levels and wave functions.

In quantum-well and superlattice systems, the presence of interface necessarily alters the phonon modes, and then, their interaction with electron is modified. The optical phonon modes can be classified as confined and interface ones. In our work, Huang-Zhu microscopic lattice-dynamic model is used to describe the bulk-like confined phonon modes. Within the framework of the continuum model, the equation of motion and polarization eigenmodes can be obtained for interface phonon modes by use of transfer matrix method.[14,15]

Effects of well width and electric field on electron-phonon scattering in the SSQW and ASQW structures have been for the first time studied. It has been shown that for both of the structures without electric field, the intrasubband and intersubband scattering rates due to interface phonons respectively decrease and increase with increasing well width while those due to confined phonons increase with increasing well width and, then, the relative variation of total intersubband scattering rates are much larger than that of total intrasubband scattering rates in the same quantum-well structures. The dependence of intrasubband and intersubband scattering rates on the applied electric field is quite different for the two structures because of the sensitivity of phonon modes (confined and interface modes) on the structure parameters and that of electron wavefunctions on the structure parameters and applied electric field (not only the absolute value but also the direction). It is found that the applied electric field can largely change the total intersubband scattering rate for the SSQW and ASQW structures with a large well width and that the field-direction effect can be important for the ASQW structure. Therefore, it would be possible to change the electron-phonon coupling in quantum well systems with the use of proper structure parameters.

5. Conclusion

In the matarials design of semiconductor alloys and low-dimensional quantum system, we show the importance of three basic elements: composition, structure and properties. Studying the relations among these basic elements, we can design new types of optoelectronic materials effectively, and improve the quality of devices. Some alloy systems, including mixed semiconductor alloys and ternay III-V semiconductor alloys, as well as the low-dimensional superlattice and quantum well are studied with the aim of material design.

References

[1] Gu B L, Newman K E and Fedders P A 1987 *Phys. Rev.* B **35**, 9135

[2] See, for example, Kuan T S, Kuech T F, Wang W I, and Wilkie E L 1985 *Phys. Rev. Lett.* **54**, 201; Stringfellow G B and Chen G S 1991 *J. Vac. Sci. Technol.* B **9** 2182

[3] Teng D, Shen J, Newman K E and Gu B L 1992 *J. Phys. Chem. Solids* **52**, 1109; Wei S H and Zunger A 1990 *Appl. Phys. Lett.* **56** 662

[4] Duan W, Zhu J L and Gu B L 1994 *Phys. Rev.* B **49** 14403; Zhu J L, Duan W and Gu B L 1994 *Phys. Rev.* B **50** 5473

[5] Gu B L, Ni J and Zhu J L 1992 *Phys. Rev.* B **45** 4071

[6] Ni J and Gu B L 1993 *Phys. Rev.* B **47** 7576; Gu B L, Ni J, Wan J W and Zhu J L 1994 *Phys.* A **206** 454

[7] Scrivastava G P, Martins J L and Zunger A 1985 *Phys. Rev.* B**31** 2561

[8] Osório R, Bernard J E, Froyen S and Zunger A 1992 *Phys. Rev.* B **45** 11173

[9] Gu B L and Ni J 1992 *J. Phys.: Condens. Matter.* **4** 9339

[10] Ni J, Lai X C and Gu B L 1993 *J. Appl. Phys.* **73** 4260

[11] Gu B L, Huang Z F, Ni J, Yu J Z, Ohno K and Kawazoe Y 1995 *Phys. Rev.* B **51** 7104

[12] Bastard G, Mendez E E, Chang L L and Esaki L 1983 *Phys. Rev.* B **28** 3241

[13] Zhu J L, Gu B L and Lou Y M 1989 *Phys. Lett.* **142** 159

[14] Duan W, Zhu J L and Gu B L 1993 *J. Phys. Condensed Matter* **5** 2859

[15] Duan W, Zhu J L, Gu B L and Wang C Y 1995 *Phys. Lett.* A **200** 329

Inst. Phys. Conf. Ser. No 145: Chapter 11
Paper presented at 22nd Int. Symp. Compound Semiconductors, Cheju Island, Korea, 28 August–2 September 1995
© *1996 IOP Publishing Ltd*

Facet terminated growth of quantum functional devices on patterned substrates

R. Tsui, M. Walther, K. Shiralagi and H. Goronkin

Motorola Inc., Phoenix Corporate Research Labs, 2100 E. Elliot Road, Tempe, AZ 85284

Abstract. We report a new approach for forming electrical contacts to quantum devices with very small lateral dimensions. It utilizes epitaxially grown contacting layers on substrates patterned with a double-step mesa structure. By using appropriate step heights, the active region of the device will be located on the middle ledge, with one contact made on top of the mesa and the other contact formed on a layer grown adjacent to the bottom of the mesa. Our concept is demonstrated by forming a functional resonant interband tunneling diode on a 2-μm wide ledge. The growth and characterization of such devices are described.

1. Introduction

There has been considerable interest during recent years in the research of quantum functional devices (QFDs) with sub-100 nm lateral dimensions. A major challenge in such research is the formation of electrical contacts to these small structures. In many cases, the requirements are beyond the capabilities of current lithographic and processing techniques. In this paper, we demonstrate a new approach in the fabrication of QFDs which make use of epitaxially grown contacting layers on non-planar substrates. This technique utilizes the formation of specific growth facets at the pattern edges during epitaxial growth on patterned wafers.

We have previously reported the results of our studies on the faceted growth behavior of InAs/(Al,Ga)Sb heterostructures on patterned wafers [1]. This material system is used to fabricate resonant interband tunneling diode (RITD) structures with high peak-to-valley current ratios (PVCRs) at room temperature [2,3]. We have used these structures to demonstrate a three-terminal QFD by incorporating a RITD with a heterojunction field-effect transistor [4]. The RITD consists of a quantum well (QW) region sandwiched by two n^+ InAs contact layers. The QW region is made up of a GaSb layer 6.5 nm wide confined on each side by an AlSb barrier layer 1.5 to 2.5 nm thick. The InAs-GaSb heterojunction has a Type-II energy band alignment. Resonant tunneling occurs when the occupied conduction band of the InAs aligns with the unoccupied valence band of the GaSb, with electrons tunneling through the valence band states in the GaSb. When the conduction band in the InAs aligns to the bandgap of GaSb, the transmission coefficient becomes a minimum and the current goes to the valley level. Details of the structure and the growth of these RITDs by molecular beam epitaxy (MBE) have been reported elsewhere [3].

2. RITDs on patterned substrates

In the conventional approach to fabricate RITDs, the top ohmic contact layer is first formed using optical lithography and evaporation. It is then used as an etch mask to form a mesa for isolation and to allow the self-aligned evaporation of the bottom ohmic contact layer, resulting in a structure as shown schematically in Fig. 1a. Using the faceted growth behaviors at pattern edges, which are dependent on the adatom species, the crystallographic planes and the deposition parameters, we have already demonstrated functional RITDs that do not need post-growth etching for device fabrication [5]. In these structures, the bottom InAs contact layer grown on top of the pre-patterned mesa is selectively connected with the top InAs contact layer grown in the grooves, as shown schematically in Fig. 1b. This is achieved by chosing an appropriate mesa height, pattern orientation and thickness of the regrown layer structure. PVCRs as high as 16 have been obtained at room temperature, equal to those of RITDs grown and processed conventionally. This approach works well down to lateral dimensions of about 6 μm. For smaller sizes, the facets formed at the pattern edges become dominant. With increasing layer thickness, the (100) planes vanish on top of the diodes and makes the formation of ohmic metals very difficult.

In order to fabricate devices with smaller dimensions, a novel approach that also utilizes the technique of epitaxial regrowth on patterned substrates is employed. On a substrate patterned with a double-step mesa, the faceted growth behavior during a single epitaxial step can be used to form contacting layers that in turn make contact to very small QFD structures, as shown schematically in Fig. 2. By designing the heights of the two steps appropriately, the active region of the device will be located on the middle ledge of the double step such that the top contact is made on the top of the mesa and the bottom contact is formed on a layer grown adjacent to the bottom of the mesa. The area of the top contact is sufficiently large compared to that of the active region that the contact is essentially ohmic.

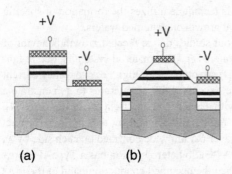

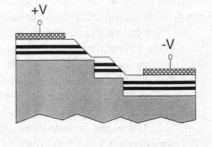

Fig. 1. Schematic cross-sections of RITDs, (a) grown and fabricated in a conventional manner, and (b) grown on a patterned substrate. The two dark stripes represent the AlSb barrier layers on both sides of the GaSb quantum well.

Fig. 2. Schematic drawing of the growth behavior of the RITD on a double-step mesa structure. The contact on the top of the mesa is connected to the upper layer on the middle ledge whereas the bottom layer on the middle ledge is connected to the top layer grown in the groove adjacent to the bottom of the double-step mesa.

3. Experimental procedure

The semi-insulating GaAs substrates were (100) oriented, and were patterned by reactive ion etching through a silicon nitride mask using a $CH_4/H_2/Ar$ mixture. The mask was removed in a SF_6/O_2 plasma after dry etching. Additional details of the patterning process have been described elsewhere [1]. The epitaxial growth was carried out in a Fisons V90H chemical beam epitaxy (CBE) system evacuated by a diffusion pump with a liquid nitrogen cold trap. The InAs layers were grown in CBE mode using arsine and trimethylindium as source materials and elemental silicon as the n-type dopant. For the growth of AlSb and GaSb, elemental Al and Ga were utilized, respectively, in conjunction with elemental Sb. Solid Sb was used due to the lack of a metalorganic or gaseous antimony source with sufficiently high purity and stability. Conventional MBE was used to grow the antimonide layers since that was the technique used in our earlier studies of the growth on patterned substrates [1,5], and MBE is less sensitive to variations in growth conditions such as substrate temperature and V/III flux ratio compared to CBE [6]. In the present study the layers were grown at 495 °C and at the rate of 1.816, 1.524 and 1.78 Å/s, respectively, for InAs, GaSb and AlSb. After growth, ohmic contacts were formed using conventional evaporation and lift-off techniques.

4. Results and discussion

Figure 3 shows a cross-sectional scanning electron micrograph of a RITD structure grown on a substrate with a double etched mesa. The lateral dimension of the middle ledge is about 2 μm, and the height at each of the two steps of the mesa is 0.5 μm. The flat surfaces are (100) planes. The sidewall of the bottom mesa step is less vertical than that for the top step because the patterning process still requires further refinement. The epitaxial thickness on the ledge is higher than those on the neighboring (100) surfaces. This is partly due to the fact that the layer design has not yet been optimized and also because of enhanced migration of In atoms from slower-growing surfaces such as the (101) facet to the (100) planes on the ledge. A similar effect can be observed for growth in the groove adjacent to the bottom step of the mesa.

Fig. 3. Cross-sectional SEM of a RITD structure grown on a substrate patterned with a double etched mesa.

Functional RITDs were obtained on the middle ledges of structures as shown in Fig. 3. The current density–voltage characteristics at room temperature of one such device, measured by connecting to the contacts above and below the 2 μm wide ledge, is shown in Fig. 4a. At a bias voltage of -0.4 V, the peak current density (j_p) is 3.4 kA/cm^2 and the valley current density (j_v) is 1.2 kA/cm^2, resulting in a PVCR of ~ 3. Even though the fabrication process has not been optimized, this result is a confirmation of our concept in the larger, μm-sized regime. Figure 4b shows the corresponding characteristics for the large RITD formed at the top of the mesa. In this case, j_p = 2.4 kA/cm^2 and j_v = 0.16 kA/cm^2, resulting in a PCVR of 15 which is a value typical of those obtained for RITDs grown and processed in a conventional approach. This indicates that the intrinsic qualities of the layers are good, and makes it unlikely that there is degradation by defects or contamination possibly introduced by the patterning process.

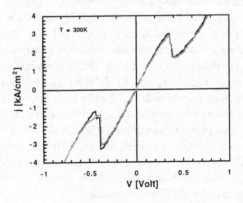

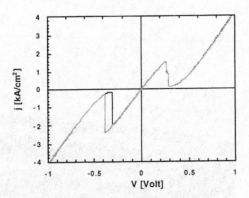

Fig. 4. Current density-versus-voltage characteristics of RITDs grown, (a) on a 2 μm x 50 μm middle ledge of a double etched mesa (top graph), and (b) on the 46 μm x 46 μm mesa top (bottom graph). The measurements were made at room temperature and the two traces in each graph correspond to voltage sweeps in forward and backward directions.

We believe the higher values of j_p and j_v measured for the RITD on the ledge are caused by a leakage path at the upper mesa edge. The existence of such a path is confirmed by the cross-sectional transmission electron micrograph in Fig. 5, in which the lower InAs contact layer on top of the mesa is shown to be partially connected with the bottom contact layer of the diode on the ledge. This is due to the non-optimized layer structure for this particular step height as well as the enhanced migration of In atoms from the (101) facets to the (100) planes on the ledge which resulted in a larger epitaxial thickness in this area. The partial connection of the two layers allowed the flow of a leakage current that caused the current densities for the RITD on the ledge to be up to 1 kA/cm^2 higher than those for the diode on top of the mesa. The increase in j_v is proportionally much higher and resulted in the significant reduction in the PVCR.

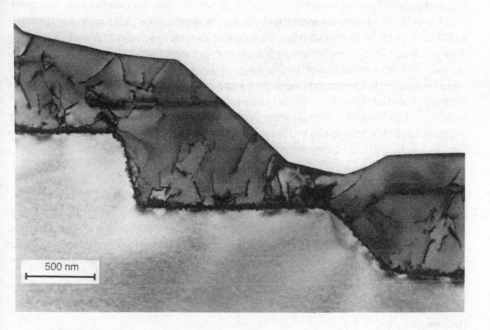

Fig. 5. A cross-sectional transmission electron micrograph of a RITD structure grown on a substrate patterned with a double etched mesa. The lateral dimension of the middle ledge is about 1.5 µm. The thin AlSb/GaSb/AlSb QW region can be seen at about two-thirds of the total epitaxial thickness above the substrate. At the edge of the top step of the mesa, the lower InAs contact layers are connected and resulted in a leakage path for current flow.

The cross-sectional TEM in Fig. 5 also shows a large number of structural defects in the epitaxial layers. The 7% lattice mismatch between the GaAs substrate and the bottom InAs contact layer of the RITD gives rise to a high concentration of misfit dislocations. From our earlier studies of the growth of InAs on GaAs, we have determined the dislocation density to be in the $10^{11}/cm^2$ range at the substrate/epitaxial layer interface, and decreases by about an order of magnitude per µm of epitaxial thickness for the first 2 µm [7]. This means the dislocation density would be in the $10^{10}/cm^2$ range at the surface of our present structures. In fact, dislocations that propagate through the QW region can be seen in Fig. 5. However, we have also found in planar RITD structures that dislocation densities of this magnitude do not seem to degrade the PVCR.

5. Summary

In conclusion, we have demonstrated a new method in the fabrication of QFDs which make use of epitaxially grown contacting layers on non-planar substrates. This should provide a technique to circumvent the difficulty in forming electrical contacts to small device structures.

1288

The approach utilizes the formation of specific growth facets at the pattern edges during epitaxial growth on patterned wafers. Our concept is demonstrated by forming a functional RITD on a 2-μm wide middle ledge on a double-step mesa structure. The PVCR of the device is lower than expected due to formation of a current leakage path during epitaxial growth. A further optimization of the various layer thicknesses of the epitaxial structure should reduce the leakage current significantly and a PVCR comparable to the RITD on top of the mesa should be obtained. To futher reduce the width of the middle ledge, sidewall spacer technology can be used where the ledge width is determined by the wall thickness of the dielectric layer. We believe this will eventually allow one to obtain devices with lateral dimensions in the sub-100 nm regime without the use of fine-line lithography.

Acknowledgment

This work was partially performed under the management of FED (the R&D Association for Future Electron Devices) as a part of the R&D of Basic Technology for Future Industries supported by NEDO (New Energy and Industrial Technology Development Organization, Japan).

References

[1] Walther M, Kramer G, Tsui R, Goronkin H, Adam M, Tehrani S, Rogers S and Cave N 1994 *J. Crystal Growth* **143** 1-6

[2] Söderström J R, Chow D H and McGill T C 1989 *Appl. Phys. Lett.* **55** 1094-1096

[3] Tehrani S, Shen J, Goronkin H, Kramer G, Hoogstra M and Zhu T X 1994 *Proc. 20th International Symp. on GaAs and Related Compounds, Freiburg, 1993* (Bristol: IOP)

[4] Tehrani S, Shen J, Goronkin H, Kramer G, Tsui R and Zhu T X 1995 *Proc. 21st International Symp. on Compound Semiconductors, San Diego, 1994* (Bristol: IOP)

[5] Walther M, Kramer G, Tsui R, Goronkin H, Adam M, Tehrani S and Rogers S 1995 *J. Electronic Materials* **24** 387-390

[6] Shiralagi K, Tsui R, Cronk D, Kramer G and Theodore N D 1995 *Proc. 21st International Symp. on Compound Semiconductors, San Diego, 1994* (Bristol: IOP)

[7] Kramer G D, Adam M S, Tsui R K and Theodore N D 1994 *Proc. 20th International Symp. on GaAs and Related Compounds, Freiburg, 1993* (Bristol: IOP)

Inst. Phys. Conf. Ser. No 145: Chapter 11
Paper presented at 22nd Int. Symp. Compound Semiconductors, Cheju Island, Korea, 28 August–2 September 1995
© *1996 IOP Publishing Ltd*

Three-Terminal Quantum Devices based on Heterojunction Interband Tunneling

J. Shen, S. Tehrani, G. Kramer, H. Goronkin, R. Tsui, S. Allen, and M. Kyler

Motorola, Inc., 2100 E. Elliot Rd. Tempe, AZ 85284, USA

Abstract. We present results on the integration of a Heterojunction Interband Tunneling Diode and Field Effect Transistors in the lattice matched InAlAs/InGaAs/InP system to form quantum multifunctional devices. A three terminal integrated device exhibits the gate-controlled multi-on-off characteristics. The results of a new XNOR operation are also obtained. The speed is limited by that of the FET's and is expected to be at least twice as fast as conventional XNOR circuits because of fewer gate level delays.

1. Introduction

Leakage current through the gate oxide and drain induced barrier lowering will create barriers to scaling in CMOS technology. Around the year 2000, a new technology will have to be introduced into manufacturing labs. Devices that utilize quantum effects are candidates for that future technology.

A number of three-terminal devices have been studied by other groups to achieve negative differential resistance with gain to provide more functionality. Examples are the Resonant Hot Electron Transistor [1], Resonant Tunneling Bipolar Transistor [2,3], a three-terminal device using a double barrier resonant tunnel diode in which the quantum well is modulated by a p+ region from the sides [4], a resonant interband tunneling transistor in which AlGaAs/GaAs layers are grown on the etched sidewall of a GaAs p^+-i-n^+ stack [5], and the integration of Resonant Interband Tunneling Diodes (RITD) with a MESFET [6]. We have also developed a three-terminal Resonant Interband Tunneling Field Effect Transistor (RITFET) [7]. It consists of an RITD and an InGaAs channel FET in which the current is controlled by a Schottky gate. Logic circuits have been demonstrated based on the RITFET [8]. However, the peak-to-valley current (P/V) ratio, the main figure of merit of the device, was limited by valley leakage currents in the InAs/AlSb/GaSb RITD. Here we will report the integration of an n^+ InAlAs/InGaAs/InAlAs/InGaAs/InAlAs p^+ Heterojunction Interband Tunneling Diode (HITD) in the drain region of an GaAs/InAlAs HFET which is lattice matched to InP (Fig 1). This material system is chosen because of the high P/V ratios in the HITD's [9,10] and the excellent FET performances. Ultimately, the P/V ratio determines the speed and power consumption of memory and logic devices based on negative differential resistances.

2. The Material Structures and Processing

Figure 1 shows the cross-section of the device. The structure is grown by MBE on an

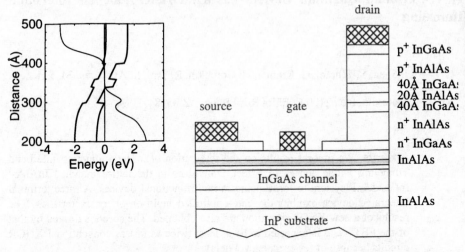

Fig. 1. Cross-sectional diagram illustrating the epitaxial layers of the integrated heterojunction interband tunneling diode and the FET on InP substrate. To form an XNOR another symmetrical FET is added to the right hand side of the diode (see text and Fig. 3.). Also shown on left is the band profiles and carrier distributions in the HITD. Because of the quantum wells at the pn junction, the effective barrier height and depletion length are smaller compared to case in the homojunction Esaki diode.

InP substrate. A 5000Å $In_{0.52}Al_{0.48}As$ buffer layer lattice-matched to InP substrate is first grown. Then a 320Å $In_{0.53}Ga_{0.47}As$ channel layer is grown, followed by a 40Å $In_{0.52}Al_{0.48}As$ spacer layer, a 125Å supply layer (Si: $5\times10^{18}cm^{-3}$), and a 200Å $In_{0.52}Al_{0.48}As$ barrier layer. The diode consists of the n^+ layers, 40Å/20Å/40Å double quantum wells, and the p^+ contact layers.

To fabricate the device, the top metal contact (Ti/Pt/Au) was defined and evaporated. Then the HITD was etched down by using the top contact as a mask. Then bottom ohmic contacts (Ni/Ge/Au) were defined and evaporated and mesa isolation etch was performed. Finally, the gate contact is defined, recess etched, and gate metals evaporated.

Unlike symmetrical resonant tunneling diodes, the HITD has two possible orientations on top of the FET: either p^+ contact on the top or n^+ contact on the top. These different orientations affect both the epitaxial growth and the logic operation of the device. The effect on epitaxial growth lies in the different diffusion lengths of the n (Si) or p (Be) doping species. In the HITD or homojunction Esaki tunnel diode, a sharp step doping profile at the junction is wanted, so a careful choice of growth conditions need to be chosen for the two orientations of the diode. We are currently optimizing these conditions to achieve the best combined performances of the HITD and the FET. When the p^+ contact is on top (Fig. 1), the drain is predetermined to be on that side of the FET. This also establishes the position of the load resistor and the function of the logic operations. In this paper, we present results of one configuration (Fig. 1) and discuss its logic operation.

3. Results and Discussions

The current-voltage characteristics ($I_d \sim V_d$) of the device have been obtained (Fig. 2).

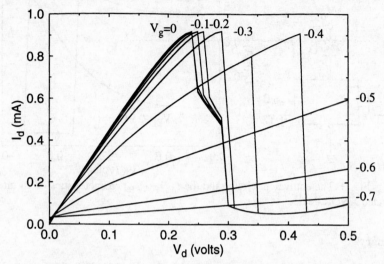

Fig. 2. The I_d-V_d characteristics of the HITFET. Negative differential resistance occurs when the I_d exceeds the peak current of the diode.

At zero gate bias, the FET is on and the drain current I_d increases as the drain bias (V_d) increases. Then when I_d reaches the peak current level of the diode, the current drops to the valley level, giving rise to the negative differential resistance. When a relatively large negative gate voltage is applied (for example, $-V_g > 0.5V$), I_d cannot reach the peak current level of the diode, and the I-V characteristic is monatomic.

The usefulness of the device lies in the gate-controlled multi-on-off characteristics (Fig. 3). When FET2 is turned off (V_B=-0.9V) and V_A increases from -0.9V to -0.5V, FET1 is gradually turned on and the current I_d initially increases and then drops to the valley level when it exceeds the peak current of the diode. On the other hand, when FET2 is on (V_B=-0.3V), the drain current I_d is already high (close to the peak current of the diode) when FET1 is off (V_A=-0.9V). Then when V_A increases to -0.6V, I_d drops to the valley level. If we define V_A=V_B=-0.9V to be logic low, and V_A=-0.6V and V_B=-0.3V to be logic high[1], then we have a mutually exclusive relationship between the inputs and the output: when the inputs are the same (both low or high), the drain current I_d is low; when the inputs are different, the drain current I_d is high. With a proper load and buffering, this is equivalent to the XNOR (or XOR) operation.

The current XNOR device differs from some previous approaches in that it is capacitively coupled to the input signals. The device occupies the space of approximately two FET's, excluding the load resistor. The speed is limited by that of the FET's and is expected to be at least twice as fast as the conventional XNOR circuits because of fewer gate level delays

---

1. In the current device, there is asymmetry between the two FET's. We will need to improve the process control to make the two FET's identical so that the two inputs are identical.

1292

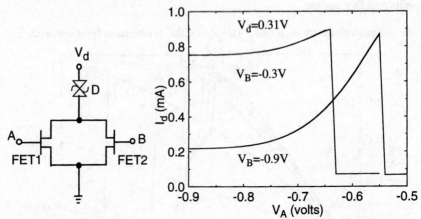

Fig. 3. The equivalent circuit and the I_d-V_A (V_B) characteristics of an integrated XNOR device.

4. Summary

We reported three-terminal device results of the integration of a HITD with InGaAs/InAlAs FET's on InP substrates. The potential of the device lies in the high P/V ratio of the HITD and the good performance of the FET's in this material system.

Acknowledgment: This work was partially performed under the management of FED (the R&D Association for Future Electron Devices) as a part of the R&D of Basic Technology for Future Industries supported by NEDO (New Energy and Industrial Technology Development Organization, Japan).

References

[1] N. Yokoyama, K. Imamura, S. Muto, S. Hiyamizu, H. Nishi, Jpn. J. Appl. Phys. 24, L853 (1985).
[2] F. Capasso, S. Sen, A. C. Gossard, A. L. Hutchingson, J. H. English, IEEE Electron. Dev. Lett. EDL-7, 573 (1986).
[3] A. C. Seabaugh, A. H. Taddiken, E. A. Beam III, J. N. Randall, Y.-C. Kao, B. Newell, IEDM 93-149 (1993).
[4] K. Maezawa, T. Mizutani, Jpn. J. Appl. Phys. 32, 42 (1993).
[5] T. Uemura and T. Baba, Jpn. J. Appl. Phys. 33, L207 (1994).
[6] A. R. Bonnefoi, T. C. McGill, and R. D. Burnham, IEEE Electron Dev. Lett. EDL-6, 636 (1985).
[7] S. Tehrani, J. Shen, H. Goronkin, G. Kramer, R. Tsui, and T. X. Zhu, Proc. 21st Intl. Symp. Compound Semconductors, San Diego, 1994.
[8] J. Shen, S. Tehrani, H. Goronkin, G. Kramer, and R. Tsui, 53rd Device Research Conference Digest, 1995.
[9] D. J. Day, R. Q. Yang, J. Lu, and J. M. Xu, J. Appl. Phys., 73, 1542 (1993).
[10] H. H. Tsai, Y. K. Su, H. H. Lin, R. L. Wang, and T. L. Lee, IEEE Electron Dev. Lett. 15, 357 (1994).

Keyword Index

Author Index